International Workshop on

APPROPRIATE METHODOLOGIES FOR DEVELOPMENT AND MANAGEMENT OF GROUNDWATER RESOURCES IN DEVELOPING COUNTRIES

(February 28 - March 4, 1989)

VOLUME III

Chief Editor
C.P. GUPTA

Editors:
Shakeel Ahmed
V.V.S.G. Rao
M.T. Rajan

NATIONAL GEOPHYSICAL RESEARCH INSTITUTE
(C.S.I.R.)
HYDERABAD - 500 007, INDIA

OXFORD & IBH PUBLISHING CO. PVT. LTD.
New Delhi Bombay Calcutta

ISBN 81-204-0406-8 (Set)
ISBN 81-204-0409-2 (Volume III)

Published by Mohan Primlani for Oxford & IBH Publishing Co. Pvt. Ltd., 66, Janpath, New Delhi 110001 and printed at Rekha Printers, Okhla Industrial Area, Phase II, New Delhi 110020

1-F9-8.5

PREFACE

This volume contains the keynote papers and other significant contributions deliberating on the state-of-the-art techniques in geohydrology and their illustrative applications. Some of them are a critical review/overview of the established techniques, though with a new insight, such as estimation of geohydrological parameters using pumping tests, tracers etc.; interaction between unsaturated and saturated zones; and groundwater management in different hydrogeological situations and so on. Some others have dealt with techniques, though known, but not commonly used in geohydrological studies such as the use of Radon in exploration in fractured zones; 3-D stochastic modeling for fractured media, and use of cross validation in krigings, to name a few.

The papers have been organized in the same sequence of sessions as in Volumes I and II of this Workshop. It is expected that these thought-provoking papers will lead to interactive and purposeful discussions and finally help concretize the operational details, technical guidelines, designs and practices for hazard-free exploration, exploitation, and management of our groundwater resources to meet the ever-increasing needs for water.

C.P. GUPTA

CONTENTS

GEOHYDROLOGIC PARAMETER ESTIMATION: PUMPING TESTS AND TRACER TECHNIQUES

K.R.Rushton, Department of Civil Engineering, University of Birmingham, UK

Abstract

Estimation of the parameters which describe groundwater flow continues to be an important topic. The recommended approach is to identify the important flow mechanisms before attempting to analyse any pumping tests. The limitations of many conventional methods are highlighted and the advantages of numerical methods are discussed. Case studies are presented which illustrate parameter estimation in hard-rock and alluvial aquifers. The use of tracers to estimate recharge and also to identify the flow paths within aquifers is also considered.

INTRODUCTION

Why is parameter estimation so important in groundwater investigations? Why, after more than fifty years of intensive study, is there so much uncertainty about the reliability of the methods? Why is there still so much controversy?

The investigation of groundwater flow has been dominated by the theories of Dupuit, Theim and Theis. Each of these theories assumes that the horizontal components of flow are dominant and that the vertical components are small enough to be neglected. For many practical situations, this assumption is not true. Vertical flow components are of crucial importance in heavily exploited aquifers where the principal flow mechanism is the movement of water from the water table to the pumped wells.

Groundwater investigations are frequently seen as being primarily a determination of certain aquifer parameters with the pumping test used to calculate numerical values of the transmissivity and storage coefficient. The fact that most of the assumptions inherent in pumping test analysis are violated does not seem to deter many workers. Having 'determined' the aquifer parameters and used an approximate method of estimating the recharge, predictions are then made about the resources of the aquifer. However, the one question that is rarely raised is whether the parameters determined in the analysis are the ones which will most strongly influence the future aquifer response.

The identification of the important features of the flow within the aquifer is difficult since these features may change due to the exploitation of the aquifer. Nevertheless, the flow mechanisms must be identified before attempts are made to quantify the aquifer parameters. Methods of determining the magnitude of certain of the aquifer parameters is well established, but for other parameters, such as the effective vertical hydraulic conductivity of low permeability zones, reliable numerical values are difficult to obtain.

A second topic which is considered in this paper is the use of tracers. Tracers in the unsaturated zone are used to obtain estimates of the recharge; the advantages and disadvantage of the various techniques are considered. Reference is also made to the manner in which tracers in the saturated zone can be used to gain additional information about the inflows and outflows.

AQUIFER FLOW MECHANISMS

The identification of the important flow mechanisms in an aquifer system is the first step in any resource investigation. Three examples will be quoted to illustrate the manner in which important features of the flow mechanisms in certain aquifers were not identified initially. Once these features had been identified, the reasons for deteriorating conditions within the aquifers became apparent.

The first example relates to a **hard rock aquifer** close to Hyderabad. Exploitation of the weathered zone by large diameter wells provided only poor yields but when tubewells were drilled into the deeper fractured zone, yields were greatly improved; this zone was used to supply water for grape gardens. Since it was thought that a new supply of water had been found in the fractured zone it was considered that the exploitation of the weathered zone by large diameter wells could continue. However, after a few years, most of the tubewells into the fractured zone failed and also severe falls occurred in the rest water levels for the wells in the weathered zone. Clearly, the flow mechanisms in this hard rock aquifer system were not identified correctly.

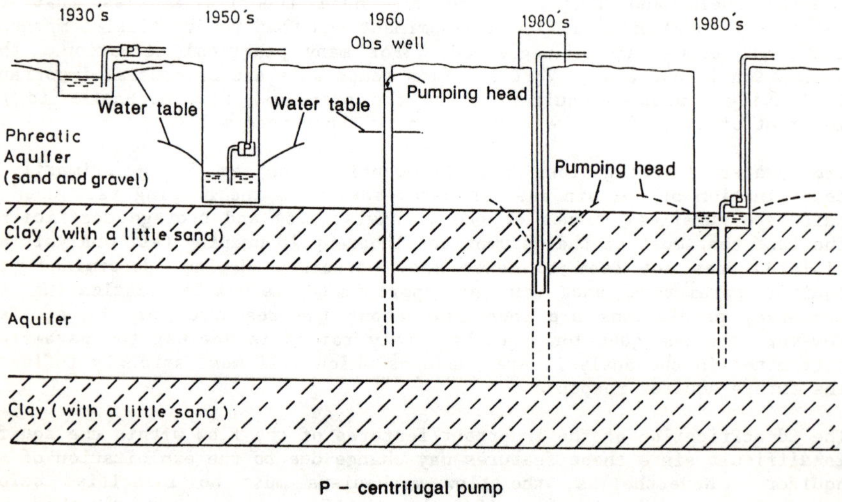

Fig. 1 Continuing exploitation of Mehsana Alluvial Aquifer

In an **alluvial aquifer** in the Mehsana area of Gujarat, the use of large diameter wells in the water table aquifer for double cropping led to rapid falls in the water table. Consequently, exploratory boreholes were drilled into the deeper aquifers, Fig 1. Excellent yields with only small drawdowns were obtained from the exploratory wells and the potential resources of the confined aquifers were quantified assuming that the water moved horizontal from a distant 'common recharge zone'. Again it was assumed that a new source of water had been identified. During the past twenty years, extensive exploitation of the deeper aquifers of the Mehsana area has occurred. The fall in the water table has continued at a rate in excess of one metre each year and the rest levels in the confined aquifers are showing more rapid declines at rates approaching four metres per year. This has required the replacement of many of the earlier wells by deeper wells and has greatly increased the costs of power for lifting the water. Once more, the original concept of the aquifer flow mechanisms appears to be incomplete.

In the **third example**, the reasons for the poor performance of many of the Saline Control And Reclamation Project (SCARP) tubewells in Pakistan are considered. In an effort to control the high water tables in the irrigated alluvial aquifers of Pakistan, an extensive network of tubewells has been provided to lower the water tables. The abstracted water is used for irrigation, usually by mixing with canal water.

Failure of the SCARP tubewells to provide the design yield is a common occurrence, yet deeper replacement tubewells drilled in close proximity to the original wells have provided a good yield, at least for a few years. This suggests that the main cause of failure is a deterioration of conditions close to the well. Deterioration of the well screen does occur but a more serious feature is the design of the gravel pack. The design of the gravel pack depends on the velocity of the water approaching the well; this is calculated as the discharge divided by the area of the slotted portion of the screen. This calculation assumes a uniform approach velocity. However, the approach velocity is proportional to the horizontal hydraulic conductivity; since the hydraulic conductivities typically vary by two orders of magnitude, the actual approach velocity in the more permeable gravel zones will be between 10 and 100 times the average velocity used in the design. Clogging of the gravel pack in the vicinity of these high velocity zones occurs; since these are the zones through which most water approaches the well, far greater pumped drawdowns are required to draw the water into the well through the less permeable zones. Eventually the drawdowns become excessive. Therefore, in understanding the flow patterns in the vicinity of the well, it is not sufficient to use the transmissivity or average hydraulic conductivities; instead the local hydraulic conductivities, as they vary with depth, should be estimated.

These three examples exhibit very different features; therefore it is difficult to make general statements about the relative importance of various parameters. Nevertheless, the following list indicate the parameters and features which may be important in a groundwater investigation:

- any layering within the aquifer unit
- the transmissivity of each layer and the distribution of horizontal hydraulic conductivities with depth

- effective vertical hydraulic conductivities of the low permeability zones
- specific yields and confined storage coefficients for aquifers and aquitards
- recharge actually reaching the true water table including how it varies with time
- any lateral inflows or outflows
- loss of water from springs, to rivers and canals or from overflowing boreholes
- abstraction from the aquifer
- quality of water within the aquifer

The determination of these parameters will be a continuing theme during the Workshop; this paper will consider certain of the methods of estimating aquifer parameters and will also review the use of tracers to estimate recharge and identify aquifer flow mechanisms.

ESTIMATION OF AQUIFER PARAMETERS

The aquifer parameters of hydraulic conductivity, specific storage and specific yield cannot be measured directly; instead some form of *model* is required. In the case of permeameter measurements, the model is that Darcy's Law holds for the one dimensional flow in a cylinder. Therefore, provided that the sample is homogeneous, the hydraulic conductivity in the direction of the axis of the cylinder can be calculated from the flow through and the groundwater head difference across the specimen.

The Theis(1935) equation uses the model of horizontal radial flow through a homogeneous confined aquifer of uniform thickness towards an infinitely small well. Furthermore the aquifer is of infinite extent and before the start of pumping there is no flow within the aquifer. Field results are fitted to this idealised model to obtain estimates of the aquifer transmissivity and storage coefficient; if the field situation is different to the idealised conditions, there are uncertainties about the parameters deduced from the analysis.

Further models have been devised which more nearly represent true field conditions in the vicinity of pumped wells; for some models it is possible to obtain analytical solutions so that type curve techniques can be used (Kruseman and de Ridder 1970, Neumann 1975, Walton 1987); for other models, numerical techniques are required (Rushton and Redshaw 1979). However, before considering in more detail the methods of estimating aquifer parameters, reference is made to the standard values within which the calculated values should lie.

Tabulated Values

From the large number of analyses that have been carried out using a variety of different techniques, the likely ranges of the aquifer parameters for certain typical strata have been identified. These are a good starting point for any study. Table 1 contains a list of parameter values taken from a recent book by Walton (1987).

This table repays careful study since it does provide a summary of recent findings. Two particular issues will be highlighted. The table of

Table 1 Representative Hydraulic Characteristics

I Material	Horizontal Hydraulic Conductivity (m/d)
Gravel	40 - 1200
Basalt	$4x10^{-8}$ - 800
Limestone	$8x10^{-4}$ - 800
Sand	4 - 120
Sand, dune	4 - 20
Peat	0.16 - 4
Sandstone	$4x10^{-3}$ - 2
Loess	$8x10^{-5}$ - 0.8
Clay	$8x10^{-6}$ - $8x10^{-2}$
Till	$2x10^{-5}$ - $4x10^{-2}$
Shale	$4x10^{-7}$ - $4x10^{-3}$
Coal	$4x10^{-2}$ - 40

II Material	Vertical Hydraulic Conductivity (m/d)
Sand, gravel and clay	$4x10^{-3}$ - $4x10^{-2}$
Clay, sand and gravel	$4x10^{-4}$ - $3x10^{-3}$
Clay	$2x10^{-5}$ - $4x10^{-4}$
Shale	$4x10^{-9}$ - $4x10^{-5}$

III Material	Specific Yield
Peat	0.30 - 0.50
Sand	0.20 - 0.40
Gravel	0.10 - 0.35
Loess	0.15 - 0.35
Silt	0.01 - 0.30
Sandstone	0.10 - 0.40
Sandy clay	0.03 - 0.20
Clay	0.01 - 0.20
Igneous, weathered	0.20 - 0.30
Siltstone	0.01 - 0.35
Limestone	0.01 - 0.25

IV Material	Confined Storage Coefficient (m = aquifer thickness)
Clay	$1x10^{-4}$ x m
Sand and gravel	$1x10^{-5}$ x m
Rock, fissured	$1x10^{-6}$ x m

vertical hydraulic conductivities, Section II, indicates that the vertical hydraulic conductivity of a clay containing sand and gravel will have a value of about 1 mm/d. Therefore, with a typical vertical hydraulic gradient in a heavily exploited aquifer of 1:1, the downwards vertical flow can be of the order of 1 mm/d. A second important issue implicit in Section IV of the table is that the confined storage coefficient of an alluvial aquifer 100m thick containing clays and sands will be between 0.001 and 0.01. This is far higher than the confined storage coefficients of consolidated aquifers of about 0.0001.

Estimation of Aquifer Characteristics for Hard Rock Aquifers

Although hard rock aquifers are far from the ideal of a material with a uniform hydraulic conductivity, it is possible to estimate aquifer parameters and then predict the long term yield. The specific example used for illustrative purposes is the weathered-fractured granitic aquifer mentioned in the previous section. The following information is available:

(i) Large diameter wells dug into the weathered zone have a yield of about 50 m^3 each day and take almost two days to refill

(ii) Tube wells drilled through the weathered zone, which is about 10m thick, to penetrate 20 to 30m into the fractured zone have specific capacities within the range 13-130 $m^3/d/m$.

(iii) A pumping test was carried out in one of these tubewells; when pumped at a rate of 288 m^3/d for 7.5 hours, the drawdowns in observation wells in the fractured and weathered zones were as indicated in Fig. 2.

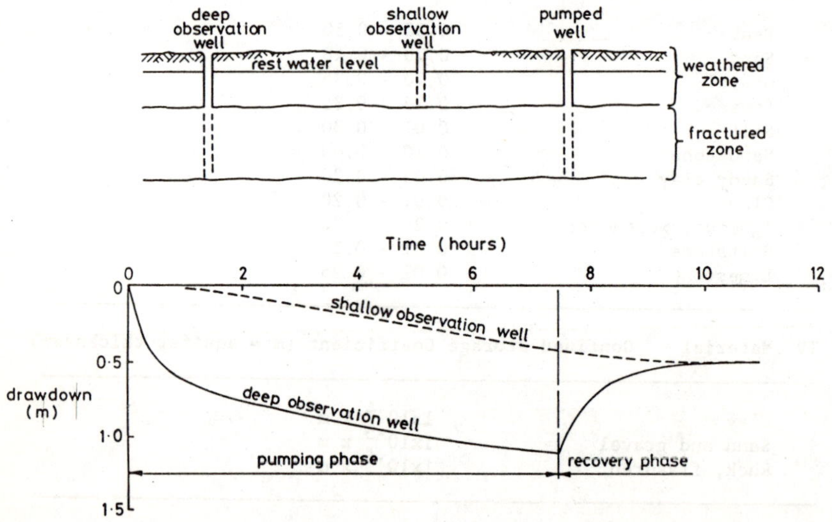

Fig. 2 Field results of pumping test in weathered-fractured aquifer.

As suggested in the Introduction, the first step is to consider what flow mechanisms can occur. The conventional approach is to say that, since the well taps the fractured zone, all the water is drawn from the fractured zone and the tubewell has no influence on the weathered zone. However, the field results of Fig. 2 indicate that this is not the case; after one hour the observation piezometer in the weathered zone starts to fall and the decrease in groundwater head continues after the pump is switched off. The decrease in the shallow piezometer continues until it is the same magnitude as the observation well in the fractured zone; only then does recovery commence. This confirms that water is drawn from the weathered zone.

If the difference in the groundwater head between the fractured and the weathered zone is examined, further important information becomes clear. Figure 2 indicates that the difference between these heads during the pumping phase is almost 1m over a vertical distance of about 10m. With such a high groundwater gradient in the vertical direction, vertical flows are certain to occur even if the vertical hydraulic conductivity is low. Therefore it is important to consider the possibility of horizontal flows in the weathered and fractured zones and vertical flows between these two zones. Consequently the important aquifer parameters when a well is drilled into the fractured zone are as follows:

- horizontal hydraulic conductivity and confined storage coefficient of the *fractured zone*
- effective vertical hydraulic conductivity between the *weathered and fractured zones*
- horizontal hydraulic conductivity and specific yield of the *weathered zone*.

A variety of alternative methods can be used to estimate the above parameters.

(i) Leaky Aquifer Theory If the drawdowns in the observation well are analysed by comparison with leaky aquifer type curves (Jacob 1946), reasonable estimates of the transmissivity and storage coefficients of the fractured zone can be obtained, provided that the results for the earlier times of the pumping test are used. The results for later times are strongly influenced by the fall in the water table.

(ii) Neuman's Unconfined Aquifer Theory Although Neuman's (1975) theory is not meant for layered aquifers (layered aquifers are different to anisotropic aquifers), it can be used to estimate overall horizontal and vertical hydraulic conductivities and confined and unconfined storage coefficients. There will, however, but uncertainties about the reliability of the parameter values.

(iii) Numerical Model A numerical model has been devised which represents horizontal flow in the individual aquifers and vertical flow between the aquifers. Furthermore, a single model simulation can be used to analyse both pumping and recovery phases. Figure 3 indicates the form of the numerical model; full details are given in Rushton and Weller (1985). The model places no restriction on the horizontal or vertical flow components. The actual parameter values are determined using a trial and error procedure whereby each parameter is modified in turn until there is good

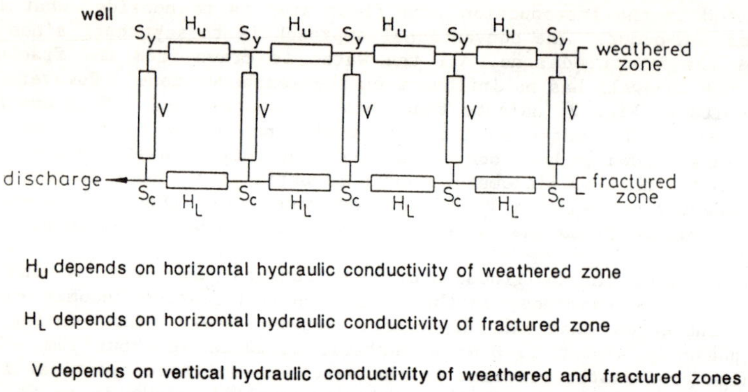

Fig. 3 Schematic diagram of layout for numerical model.

agreement between field and modelled heads for both the pumping and recovery phases.

The values of the parameters which gave the best fit are as follows.

Fractured zone: horizontal hydraulic conductivity = 5.0 m/d
confined storage coefficient = 0.0025
Weathered zone: horizontal hydraulic conductivity = 0.5 m/d
specific yield = 0.01
Effective vertical hydraulic conductivity between fractured and weathered zones = 0.6 m/d.

Another very important conclusion is that the tubewell draws water from the fractured zone but that the fractured zone collects the water from the overlying weathered zone. The tubewell has little influence at distances greater than 250m.

Having devised a numerical model which represents a pumping test, it is possible to *predict future aquifer behaviour*. The numerical model was used to show:

Continuous pumping with submersible pumps from the deep tubewells leads to a dewatering of the weathered zone and within about two years a failure of the tubewells; consequently the submersible pumps are too powerful for the limited resources available.

Large diameter wells in the weathered zone have low yields since they are only able to collect water from about 100m from the well.

If borewells are drilled into the bottom of the dug wells, the yield can be increased since the fractured zone collects water from greater distances. Provided that surface mounted centrifugal pumps are used, there is no danger of too much water being drawn from the system.

Parameters for Alluvial Aquifers

The rapid continuing decline in both the water table and pumping levels in the Mehsana alluvial aquifer has already been introduced and illustrated in Fig. 1. The determination of the horizontal and vertical hydraulic conductivities is essential if successful remedial action is to be implemented. In one particular area of about 1 km^2, more than ten exploratory boreholes have been drilled and a number of pumping and recharge tests carried out. Reference will be made primarily to one of the pumping tests; the implications in terms of artificial recharge are discussed by Rushton and Srivastava (1988).

The difficulties of interpreting the pumping tests became apparent when one pumping test was carried out in a borehole which drew water from two of the permeable zones and drawdowns were measured in the 1st and 2nd confined aquifers, Fig. 4. The response in the lower zone was almost instantaneous but it took more than 15 minutes before drawdowns in the upper zone became apparent. A similar delay occurred during the recovery.

An examination of the groundwater heads in the boreholes under the 'at rest' conditions indicates that the aquifer is not 'at rest' before the start of the test as indicated in the following table.

Borehole	Aquifer	Distance to test well(m)	Head(m)
Observation	1st Confined	164	76.4
Observation	1st Confined	6	72.7
Test	1st & 2nd Confined	-	68.0
Observation	2nd Confined	6	66.8
Observation	2nd Confined	150	66.4

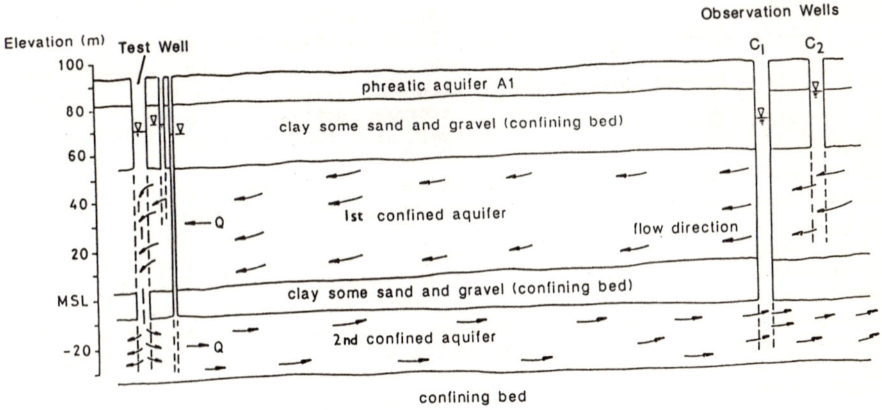

Fig. 4 Initial flow conditions for alluvial aquifer

Groundwater always flows in the direction of falling groundwater heads; the groundwater heads fall in the order that they are listed in the table. Consequently, water flows through the 1st confined aquifer to the test well, through the test well and into the 2nd confined aquifer; this is illustrated in Fig. 4. The occurrence of flows before the commencement of the pumping indicates why there is a delayed response in the upper aquifer following the start of pumping. Before pumping commences, water is flowing through the first confined aquifer towards the pumped well; when the pump starts water continues to flow in the same direction. However, for the 2nd confined aquifer the conditions are reversed changing from flow from the test well to flow towards the test well; hence the rapid response.

Since there is a flow before the test starts, the direct use of the Theis theory is not acceptable. An approximate analysis is possible by estimating the initial flow from 1st confined aquifer into the 2nd confined aquifer to give the 'at rest' head distribution as listed in the table. Then, additional pumping can be superimposed to represent the actual test; a trial and error technique is required to determine how the discharge is distributed between the two aquifers.

A more reliable approach is to use a numerical model. The chosen model is similar to the one used for the weathered-fractured aquifer, Fig. 3, but with the additional feature of leakage from the overlying water table into the 1st confined aquifer. A further important feature is the representation of well losses (reduced transmissivities) in the vicinity of the test well. From the numerical model the following parameters were deduced.

(i) *Initial flow* through test well = 292 m^3/d

(ii) *Overlying aquitard*: effective thickness = 25m
vertical hydraulic cond. = 0.0002 m/d

(iii) *1st confined aquifer*: transmissivity = 50 m^2/d
storage coefficient = 0.001
well loss factor = 2.0

(iv) *Intermediate aquitard*: effective thickness = 20m
vertical hydraulic cond. = 0.0004 m/d

(v) *2nd confined aquifer*: transmissivity = 130 m^2/d
storage coefficient = 0.0045
well loss factor = 6.0

This example illustrates the large amount of information which needs to be learnt about a single borehole tapping two confined aquifers.

The model was subsequently used to represent artificial recharge experiments at the site. Recharge by injection at a rate of 225 m^3/d continued for 250 days but with a gradual build-up in groundwater heads relative to the seasonal fluctuations. This was caused by the well loss factor increasing from 6.0 to 150 during the recharge experiment. The fact that such severe clogging did occur was demonstrated by carrying out a step pumping test at the end of the recharge experiment and after re-development of the well. At the end of the recharge experiment, a discharge of 432 m^3/d caused a drawdown of more than 7m after 8 minutes pumping, whereas after re-development, a test at the same rate led to a pumped drawdown of

1.46m after 2 hours. The results from these tests can be used to demonstrate that the well loss coefficient B in the equation,

$$\text{additional drawdown} = BQ + CQ^2,$$

represents both the formation loss and part of the well loss (Rushton and Rathod 1988).

Deducing Aquifer Parameters from Tests in Large Diameter Wells

Large diameter wells are of great importance in many developing countries and there are a number of techniques for determining the aquifer parameters from pumping tests. However, as has been stressed in all the previous discussions, before selecting the method of analysis it is essential to identify the flow mechanisms; Fig. 5 indicates the important features of a large diameter well.

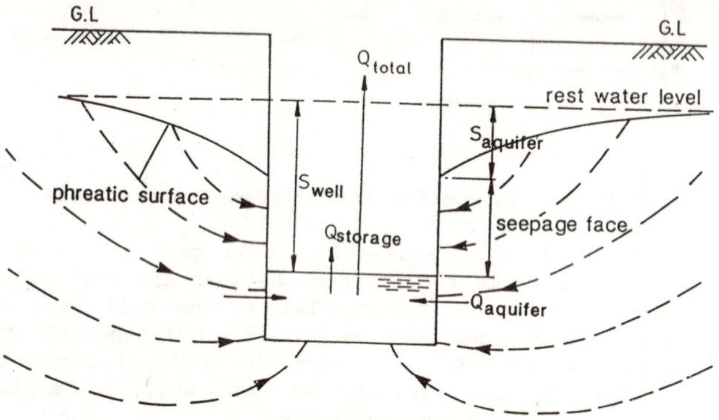

Fig. 5 Flow towards a large diameter well.

The first crucial issue is that the amount of water withdrawn from the aquifer varies during both the pumping and the recovery phase due to the effects of well storage. Indeed, from the change in well water level and the cross-sectional area of the well, it is possible to calculate the change in well storage and therefore deduce the quantity of water actually taken from the aquifer. The solid circles in Fig. 6 show the change in well water level and the calculated *discharge from the aquifer* for a well of 1.88m diameter with a carefully controlled pumped discharge of 136.8 m^3/d for 0.09 day (Sakthivadivel and Rushton 1989). Note that the maximum discharge from the aquifer is less than 90 m^3/d and that, even when the pump is stopped, significant quantities of water continue to flow from the aquifer to refill the well storage.

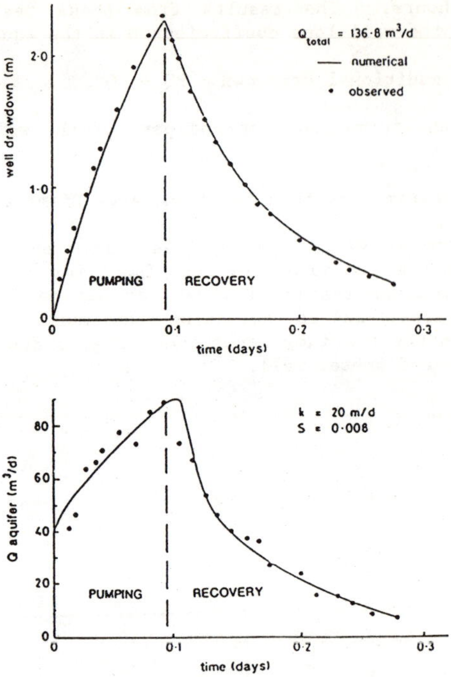

Fig. 6 Well drawdown and flow from aquifer for a large diameter well.

The second important feature is that a seepage face will form on the well face, Fig. 5; this means that there is a difference between the well water level and the water table elevation in the aquifer at the well face (a seepage face may not occur in a dug-cum-bore well). The existence of the seepage face and the sloping water table (or phreatic surface) leads to vertical components of flow in the vicinity of the well. Partial penetration can also lead to vertical flow components.

Any reliable method of examining the response of large diameter wells must be able to include:

(1) the variable withdrawal rate from the aquifer,
(2) both the pumping and the recovery phases,
(3) the presence of the seepage face and
(4) additional drawdowns close to the well due to vertical flows.

The Theis method is clearly not appropriate for the pumping phase nor should the Theis Recovery Method be used since there are substantial flows from the aquifer during recovery. The method of Papadopulos and Cooper (1967) meets the first requirement but fails to meet the others and is most reliable when used for dug-cum-bore wells. A Kernel function technique has been developed by Mishra and Chachadi (1985) to represent both the pumping and recovery phases and their approach has been further developed by

Rushton and Singh (1987) to include the effect of the seepage face. However, the only method which can represent each of the four features is a finite difference numerical model (Sakthivadivel and Rushton 1989); the numerical results from this model are included as full lines in Fig. 6. The parameter values deduced from the analysis are also included on the figure.

Having developed a reliable model of the large diameter well and the associated aquifer, it can then be used to examine the response to pumping for 100 days or more (Rushton and Sathivadivel 1988).

Further Examples

Many further examples can be quoted of recent developments in the determination of aquifer parameters; most of these developments are based on numerical models. Amongst the more important are:

Limestone aquifers containing fissures: Rushton and Rao (1988).

Chalk aquifers with increased permeability in the zone of water table fluctuation: Connorton and Reed (1978).

Sandstone aquifers containing fissures and zones of lower hydraulic conductivity: Brassington and Walthall (1985).

Alluvial aquifers especially the approach velocities to a well: Cookey et al (1987).

Well losses in confined, leaky and unconfined aquifers: Rushton and Rathod (1987).

Decreasing abstraction rates in large diameter well tests: Rushton and Singh (1983).

Unreliability of open observation boreholes: Rushton and Howard (1982).

Interference due to *neighbouring pumping wells* during a pumping test: Rushton (1985).

ESTIMATION OF RECHARGE USING TRACERS

This paper provides a brief resume of the use of tracers in groundwater studies; the primary use of tracers is to estimate recharge to aquifers. A more detailed review is presented by Allison (1988). Consideration will first be given to natural tracers followed by reference to injected tracers; in each case measurements are taken in the unsaturated zone. Subsequently the use of tracers in the saturated zone is considered. Finally, examples are presented of the use of tracers to estimate the groundwater recharge and discharge in an alluvial aquifer.

Natural tracers in the unsaturated zone

Natural tracers in the unsaturated zone are used to estimate recharge. Recharge is defined as the water that reaches the water table by percolation through the unsaturated zone using suitable tracers. If the movement of water through the unsaturated zone can be quantified, it is possible to estimate the amount of water entering from above. During percolation of the water through the unsaturated zone, water in the large pores will move more rapidly than the water in the small pores. However, it is argued by Zimmermann et al (1967) that because of the close proximity of the large and small pores, diffusion will take place leading to equal concentrations in all pores; consequently all molecules of a perfect tracer will move downwards with the same apparent velocity. This *piston flow mechanism* is the basis of recharge estimates using tracers. Yet, due to the variability of most soils and the effect of drying cracks and roots on the hydraulic conductivity distribution, it is likely that there will be significantly different vertical velocities in many situations. Furthermore, the fact that up to half of the water in the soil can be immobile water can lead to unreliable estimates of the recharge (Gvirtzman and Margaritz 1986).

The natural tracers used in recharge studies include ^{3}H, ^{14}C, ^{15}N, ^{18}O, ^{2}H, ^{13}C and chloride. Of these tracers, ^{3}H, ^{2}H and ^{18}O form part of the water molecule and therefore they accurately reflect the movement of the water. For the other tracers, chemical and physical effects may mean that the tracer does not move at the same rate as the water.

Many of the recharge studies are based on *natural tritium (^{3}H)*. Tritium, which has a half life of 12.3 years, is present in precipitation due to nuclear fallout; the tritium content of rainfall differs between the northern and southern hemisphere. Three techniques are commonly used to estimate the recharge from tritium profiles:

(a) From the position of the fallout peak; it is assumed that the water in the profile above the recharge peak represents the recharge since the time of the fallout.

(b) Using the shape of the tritium peak in the soil; a model of the flow process within the soil based on the tritium input function to the unsaturated zone at the base of the root zone can be devised. Comparisons between the calculated and observed tritium profiles are used to estimate the mean recharge.

(c) The total amount of tritium stored in the soil zone and the corresponding volumetric water contents can be used to estimate the mean recharge. This method has advantages over (b) since a very detailed tritium profile with depth is not required and therefore fewer samples are needed.

Chloride (Cl) is another useful natural tracer. The input of chloride to the soil surface occurs both in rainfall and as a dry fallout. Since most plants do not absorb significant quantities of chloride from soil water, the chloride concentration should increase through the soil zone; this can be used to estimate the recharge.

There are, however, certain anomalies in the depth profiles of chloride concentration. Two reasons have been proposed for this response (Peck et al 1981, Johnston 1983, Watson 1982):

-diffusion of the chloride in the soil water,
-extraction of water from roots which extend to depths of 7m or more.

Injected tracers in the unsaturated zone

When injected tracer techniques are used, the tracer is injected at a sufficient depth and the movement of the tracer noted after a fixed period of time. Athavale and Rangarajan (1988) describe recharge estimates made at a number of sites in the hard rock areas of India. Tritium was injected at a depth between 60 and 80cm during June just prior to the onset of the monsoon rainfall. At the end of the monsoon season, depth samples were collected at intervals of 10 or 20cm and the moisture content and tritium activity were determined for each sample. From the movement of the tritium peak and the moisture content distribution between the injection depth and peak tritium value, estimates of the recharge were made.

The calculation assumes that the soil moisture moves down in discrete layers with the additional layer of moisture at the surface pushing down an equal quantity throughout the depth of the soil profile. Although this may be a valid assumption in some instances, there are other situations when the layering of the subsoil and conditions at the bottom of the partially saturated zone mean that the piston flow approach is not valid. Furthermore, the sites were chosen in non-agricultural land distant from any trees and preferably with little vegetation; the results may reflect the conditions in these barren areas but may not provide a reliable indication of the recharge in the cultivated areas.

At this stage of the discussion it is important to differentiate between the *potential* and *actual* recharge. The *potential* recharge is the quantity of water which moves down from the ground surface or soil zone towards the water table. Tracer methods and soil moisture balance techniques can be used to estimate this potential recharge. However, if there is a less permeable zone between the position where the potential recharge is estimated and the true water table, then this low permeability zone can restrict the vertical flow so that the *actual* recharge is *significantly less* than the *potential* recharge, Rushton (1986).

The Mehsana aquifer referred to earlier is an important example. Due to the extension of the canal command to part of the area, it was expected that the water table would begin to rise. This has occurred in the areas with predominatly sandy aquifers. However, over most other areas, the presence clay lenses (which may be thin) means that much of the potential recharge does not reach the true water table, the water table continues to fall with water-logging developing above the clay layer.

Tracers in the saturated zone

Chemical and isotopic studies in the saturated zone can often provide supportive evidence for the presence or absence of recharge to an aquifer; the techniques can rarely provide quantitative information about the recharge. Useful techniques include tritium, ^{14}C, chloride concentrations and also trace elements which can indicate the likely origins of the groundwater.

^{14}C and Tritium are used to determine the age of the water. The fact that the tritium content is low indicates that little water has entered that part of the aquifer within the past 25 years. This does not mean that no recharge has occurred since it is possible that the tritiated water is still moving slowly through less permeable zones. Care also has to be taken when interpreting the ^{14}C ages. The movement of groundwater is so slow (typically 5 to 50m per year horizontally and 0.1 to 3.0m per year vertically) that recent recharge water may take 100's of years before it is identified in pumping wells.

The same is true of chloride concentrations or the presence of trace elements; their main use is in understanding the historical conditions in the aquifer rather than providing information about recent recharge. Edmunds (1981), in a review of hydrogeochemical investigations, lists 21 dissolved species which can be of value in the interpretation of groundwater flow. He suggests four main uses of hydrochemical data:

- Basic geochemical interpretation
- Potability and health
- Groundwater origin
- Agricultural use

Of all the techniques mentioned in this paper, hydrogeochemistry probably has the greatest potential, yet reliable conclusions require careful fieldwork and imaginative but disciplined interpretation.

Field example

The field example concerns the use of various tracers to understand more about the inflows and outflows of the Mehsana alluvial aquifer in Gujarat. Sukhija and Shah (1976) carried out investigations into the change in the *environmental tritium profiles* over an interval of two years. Figure 7 shows the tritium concentration in rainfall; peak concentrations occurred in 1963-64. Tritium profiles for three sites are included in Fig. 8; values are quoted for 1967 and 1969.

Balol is on low ground where there is intense agriculture, the soil is a sandy loam; annual recharge values using methods (a) and (c) are 28mm and 25mm respectively.

Sankeshwar has a sandy clay soil and is in the arid zone in the vicinity of the Rann of Kutch, annual recharge estimates are 17mm and 13mm.

Taranga is on the higher ground where the rainfall is some 30% higher than at Balol; the topsoil is coarse sand. Due to the rapid movement of the tritium profile it is difficult to determine reliable values; the best estimates are 58mm and >27mm.

Injected tritium studies have also been used in the Mehsana area. Gupta and Sharma (1984) report studies in which there was a 0.75 coefficient of correlation between the net moisture transfer and the water table fluctuations at 26 sites. In a more recent study (Sharma and Gupta 1987)

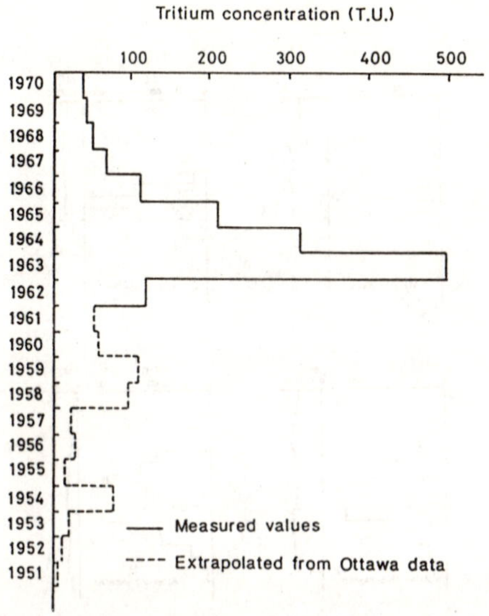

Fig. 7 Variation of environmental tritium concentration with time

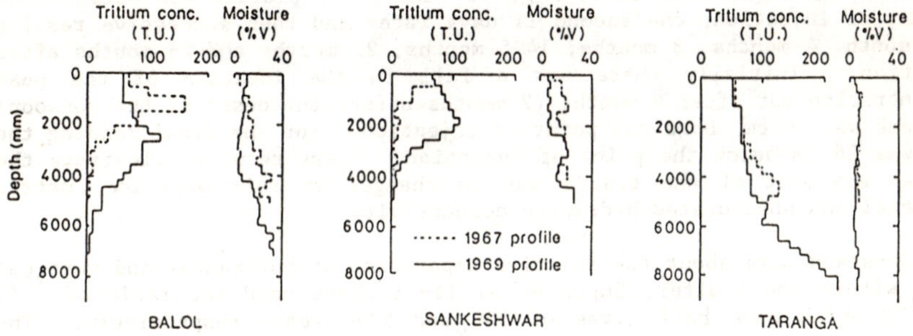

Fig. 8 Tritium concentration and moisture content at three sites.

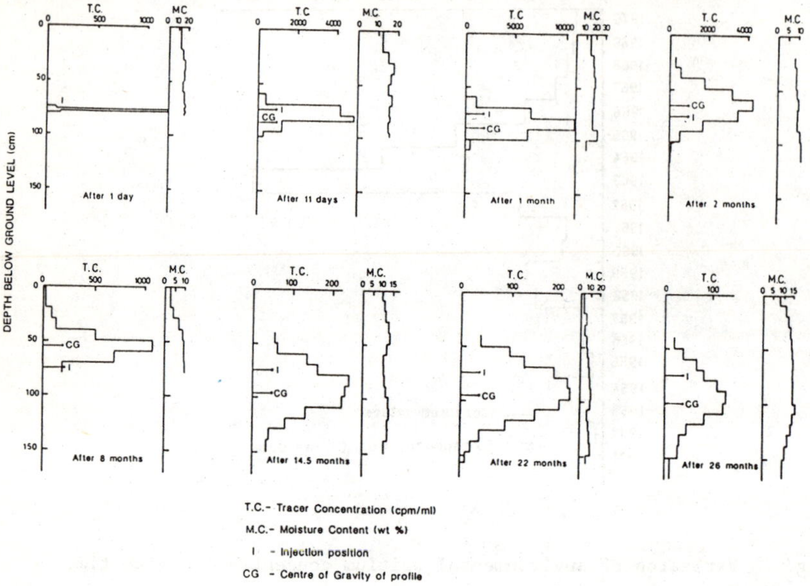

Fig. 9 Tracer concentration and moisture content at different times following injection of tritium.

the possibility of upwards as well as downwards movement of the tracer has been investigated. A set of eight tritium injections were made 1m apart in a non-irrigated field; profiles of tritium concentration and soil moisture were taken on eight occasions, Fig. 9. The first profile was obtained on the day of injection, the second 11 days later and then successive results at 1 month, 4 months, 8 months, 14.5 months, 22 months and 26 months after injection. Initially there was a fall in the location of the peak concentration but after 8 months (2 months before the onset of the monsoon) the peak was 20 cm *above* the point of injection. For the final reading the peak was 26 cm *below* the point of injection. These results illustrate the complex movement of the tracer due to changes in soil moisture, matric potential and unsaturated hydraulic conductivity.

To understand more about the relative importance of horizontal and vertical flows within the aquifer, Gupta et al (1981) have used the *ratio of* ^{32}Si *and* ^{14}C which have half lives of 300 and 5730 years respectively. The ratio of these parameters at different depths within the aquifer indicates that some water in the deeper confined zones originates from horizontal flows from a distant recharge source but other water derives from vertical flow through the less permeable zones.

Uranium isotopes have been used to gain information about the interaction between the Sabamarti River and the phreatic aquifers. Borole et al (1979) used ^{238}U concentrations and the $^{234}U/^{238}U$ activity ratio in boreholes and in the river to estimate the flows from the aquifer to the river. Results indicated that along some stretches flow does occur from the river to the aquifer.

DISCUSSION AND CONCLUSIONS

This review paper has demonstrated that the estimation of aquifer parameters involves far more than the determination of the transmissivity and storage coefficient from pumping tests using standard techniques. The first step is to identify the aquifer flow mechanisms, not only in the present state of the aquifer but for the conditions that are likely to apply due to changes in the exploitation of the aquifer. In certain of the field examples, initial studies did not identify the mechanisms that would apply as exploitation of the aquifers developed; for this reason the potential yields of the aquifers were over-estimated.

Having identified the aquifer flow mechanisms, it is then necessary to carry out controlled tests and select suitable methods of analysis to estimate the magnitude of the aquifer parameters. At present there is too much reliance on pumping tests with the conventional curve fitting techniques. The adoption of numerical methods of pumping test analysis has been very slow; no longer is the non-availability of computers a valid excuse since the programs run successfully on micro-computers.

The most difficult parameter to estimate is the effective vertical hydraulic conductivity. At present the only reliable method is using carefully designed pumping tests with analytical or numerical methods of analysis. However, the accuracy of the estimated values is open to question. The development of a more direct method of determining the effective vertical hydraulic conductivity must have a high priority.

This paper has also referred to the use of tracers to learn more about aquifer features, especially the recharge. The techniques which have been developed are reliable and consistent yet there remains the uncertainty as to whether measurements in the unsaturated zone can give reliable information about the quantity of water actually reaching the water table.

This review has been concerned primarily with individual techniques based on some form of special purpose experiment to estimate certain of the aquifer parameters. Yet there are, in effect, pumping tests and recharge experiments occurring continuously in most aquifers. If records are kept of inflows, well discharges, groundwater hydrographs and river flows, these can be used with some form of regional or local groundwater model to estimate the important groundwater parameters. It is by combining the specific field experiments with the analysis of the overall aquifer response that an adequate understanding of the aquifer response can be developed. The gaining of this understanding can take many years.

References

Allison, G.B. 1988. A review of some physical, chemical and isotopic techniques available for estimating groundwater recharge. *Estimation of Natural Groundwater Recharge, D. Reidel Publishing Company*. 49-72.

Athavale, R.N. and Rangarajan, R. 1988. Natural recharge measurements in the hard rock regions of semi-arid India using tritium injection - a review. *Estimation of Natural Groundwater Recharge, D. Reidel Publishing Company*. 175-194.

Brassington, F.C. and Walthall, S. 1985.Field techniques using borehole packers in hydrogeological investigations. *Quart. Jnl. Eng. Geol*. v. 18, 181-193.

Borole, D.V., Gupta, S.K., Krishnaswami, S., Datta, P.S. and Desai, B.I. 1979. Uranium isotopic investigations and Radiocarbon measurements of river-groundwater systems, Sabarmati Basin, Gujarat, India. *Isotype Hydrology 1978, International Atomic Energy Agency*. v. 1, 181-200.

Connorton, B.J. and Reed, R.N. 1978. A numerical model for the prediction of long term well yield in an unconfined chalk aquifer. *Quart. Jnl Eng. Geol*. v. 11, 127-138.

Cookey, E.S., Rathod, K.S. and Rushton, K.R. 1987. Pumping from an unconfined aquifer containing layers of different hydraulic conductivity. *Hydrol. Sci. Jnl*. v. 32, 43-57.

Edmunds, W.M. 1981. Hydrogeochemical investigations. *Case Studies in Groundwater Resources Evaluation, Clarendon Press Oxford*. 87-112

Gupta, S.K., Lai, D. and Sharma, P. 1981. An approach to determining pathways and residence time of groundwater: dual radiotracer dating. *Jnl. Geophys. Res*. v. 86, 5292-5300.

Gupta, S.K. and Sharma, P. 1984. Soil moisture transport through the unsaturated zone: tritium tagging studies in Sabarmati basin, western India. *Hydrol. Sci. Jnl*. v. 29, 177-189.

Gvirtzman, H. and Margaritz, M. 1986. Investigation of the water movement in the unsaturated zone in an irrigated area using environmental tritium. *Water Resour. Res*. v. 22, 635-642.

Jacob, C.E. 1946. Radial flow in a leaky artesian aquifer. *Trans. Amer. Geophys. Union*. v. 27, 198-208.

Johnston, C.D. 1983. Estimation of groundwater recharge from the distribution of chloride in deeply weathered profiles from south-west Western Australia. *Internat. Conf. on Groundwater and Man, Sydney*. 143-152.

Kruseman, G.P. and deRidder, N.A. 1970. Analysis and evaluaţion of pumping test data. *Intern. Inst. Land Reclam. and Improvement. Wageningen*. Bull. 11, 200pp.

Mishra, G.C. and Chachadi, A.G. 1985. Analysis of flow to a large diameter well by a discrete Kernel approach. *Ground Water*. v. 23, 646-651.

Neuman, S.P. 1975. Analysis of pumping test data from anisotropic unconfined aquifers considering delayed gravity response. *Water Resour. Res*. v. 11, 329-342.

Papadopulos, I.S. and Cooper, H.H. 1967. Drawdown in a well of large diameter. *Water Resour. Res*. v. 3, 241-244.

Peck, A.J., Johnston, C.D. and Williamson, D.R. 1981. Analyses of solute distributions in deeply weathered soils. *Agric Water Manag*. v. 4, 83-102.

Rushton, K.R. 1985. Interference due to neighbouring wells during pumping tests. *Ground Water*. v. 23, 361-366.

Rushton, K.R. 1986. Surface water - groundwater interaction in irrigation schemes. *IAHS Publ.* No 156, 17-27.
Rushton, K.R. and Howard, K.W.F. 1982. The unreliability of open boreholes in unconfined aquifer pumping tests. *Ground Water.* v. 20, 546-550.
Rushton, K.R. and Rathod, K.S. 1988. Causes of non-linear step pumping test responses. *Quart. Jnl. Eng. Geol.* v. 21, 147-158.
Rushton, K.R. and Rao, S.V.R. 1988. Groundwater flow through a Miliolite limestone aquifer. *Hydrol. Sci. Jnl.* v. 33, 449-464.
Rushton, K.R. and Sakthivadivel, R. 1988. Well hydraulics with special reference to hard rock terrain. *Jnl. Applied Hydrol, India.* v. 1, 1-18.
Rushton, K.R. and Singh, V.S. 1983. Drawdowns in large diameter wells due to decreasing abstraction rates. *Ground Water.* v. 21, 670-677.
Rushton, K.R. and Singh, V.S. 1987. Pumping test analyses in large diameter wells with a seepage face by Kernel function techniques. *Ground Water.* v. 25, 81-90.
Rushton, K.R. and Srivastava, N.K. 1988. Interpreting injection well tests in an alluvial aquifer. *Jnl. Hydrol.* v. 99, 49-60.
Rushton, K.R. and Weller, J. 1985. Response to pumping of a weathered -fractured granite aquifer. *Jnl. Hydrol.* v. 80, 299-309.
Sakthivadivel,R. and Rushton, K.R. 1989. Numerical analysis of large diameter wells with a seepage face. *Jnl. Hydrol.* in press.
Sharma, P and Gupta, S.K. 1987. Isotopic investigation of soil water movement: a case study in the Thar desert, western Rajasthan. *Hydrol. Sci. Jnl.* v. 32, 469-483.
Sukhija, B.S. and Shah, C.R. 1976. Conformity of Groundwater Recharge rate by tritium method and mathematical modelling. *Jnl. Hydrol.* v. 30, 167-178.
Theis, C.V. 1935. The relation between the lowering of the piezometric surface and the rate and duration of the discharge of a well using ground storage. *Trans. Am. Geoph. Un.* v. 16, 519-524.
Walton, W.C. 1987. Groundwater pumping tests, design and analysis. *Lewis Publishers, Michegan.* 201pp.
Watson, J.D. 1982. Analysis of salinity profiles in lateritic soils of south-western Australia. *Aust. Jnl. Soil Res.* v. 20, 37-49.
Zimmermann, U., Ehhalt, D. and Munnich, K.O. 1967. Soil water movement and evapotranspiration: changes in the isotope composition of water. *Isotypes in Hydrology, IAEA Vienna.*

PROGRESS AND PROSPECTS IN ENVIRONMENTAL ISOTOPE HYDROLOGY

Mebus A. GEYH

Niedersächsisches Landesamt für Bodenforschung
D-3000 Hannover 51, Federal Republic of Germany

SUMMARY

In recent years various techniques based on environmental isotopes are used widely in applied hydrological studies. Most commonly analyzed are the isotope compositions of the environmental radionuclides tritium and radiocarbon as well as those of the stable nuclides of oxygen and hydrogen. The environmental isotope techniques allow to determine the origin of groundwater, its recharge and discharge, to quantify mixtures of different kinds of groundwater, to estimate the age of groundwater and to follow up its movement. New hydrological chronometers as chlorine 36, argon 39, calcium 41, krypton 81, and krypton 85 as well as the dissolved noble gases have been introduced during the last years. The dating of groundwater via its dissolved organic carbon compounds (DOC) seems to be promissing. In semiarid and arid zones the estimation of very small evaporation rates from the stable isotope composition of pore water in profiles of the unsaturated zone has become important in groundwater balance studies. There is a trend substituting qualitative by quantitative interpretation of environmental isotope data using mass transport calculations. They showed, for instance, that the occurrence of relict ressources of fossil groundwater in semiarid zones complicate the numerical modelling. Last but not least, long-term monitoring of the isotope composition in groundwater abstracted from production wells may yield in recommendations for an optimized groundwater management.

1 INTRODUCTION

Meanwhile the techniques of environmental isotopes are well established in applied hydrogeology (Table 1). Most frequently used are the stable isotopes of oxygen and hydrogen which allow to investigate the origin and fate of the groundwater in the hydrological cycle (GAT and GONFIANTINI 1981; IAEA 1983, 1987). Effects after the recharge as evaporation, interactions with the rock of the aquifer by thermal water, magmatic CO_2, or H_2S can be detected (Fig. 1). The quantitative evaluation two-component mixtures (MAZOR et al. 1973) using isotope and hydrochemical dates is already routine. But the precision is seldom better than + 20 % (GEYH and MICHEL 1981). Long-term monitoring of the stable isotope composition may allow to evaluate even multi-component mixtures quantitatively (MALOSZEWSKI et al. 1983).

The measurement of the environmental radionuclide tritium offers the unique possibility to identify even traces of groundwater recharged during the last 30 years. The main quantity of tritium in the hydrological cycle has its origin from nuclear weapon tests. In 1964 the tritium activity in precipitation approached a maximum and was up to four orders of magnitude above the natural level. Since then it dropped to a low level which is maintained by emissions of the nuclear industry and research (Fig. 2).

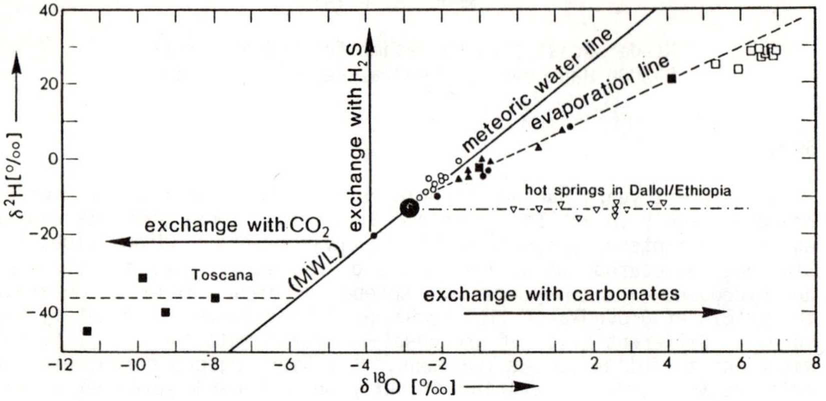

Fig. 1: Plot of δ^2H values above $\delta^{18}O$ values from groundwater of continental origin fit the so-called meteoric water line (MWL), those of evaporated water fall on the evaporation line; thermal water interacting with carbonatic rock from the aquifer have delta data which are on a horizontal straight line similar to groundwater interferred by magmatic CO_2. A vertical line in this plot represents groundwater which contain H_2S.

Radiocarbon analyses are used to estimate the age of the groundwater up to about 40,000 years (MÜNNICH 1957; FONTES 1985) and to recognize hydrochemical interactions in the aquifer.

Only the mentioned isotope techniques are most successful (ANDREWS et al. 1984) and part of common applied hydrological investigations. However, there are others in development (Table 1).

Table 1: Environmental isotopes used or recommended for hydrological studies.

Isotope	$T_{1/2}$	a(spec) dpm/l	sample size (l)	dating range (ka)	sample	detection technique
tritium	12.43 a	250	0.01-2	.001-0.1	GW,MW,BR	PCC,LSC
radiocarbon	5730 a	0.8	2-200	0.1-45	GW	PCC,LSC,AMS
silicon 32	105-172 a	$1.2*10^{-3}$	2000-5000	(0.1-1)	GW	PCC
chlorine 36	301 ka	$20*10^{-5}$	<2	10-3000	GW,(MW)	AMS
argon 39	269 a	$3*10^{-5}$	1000-5000	0.1-2	GW,MW	PCC,(RIS)
calcium 41	103 ka	0.4	0.1	10-1000	GW,(MW)	AMS
krypton 81	213 ka	$1*10^{-8}$	<2	20-2000	GW,MW,BR	RIS
krypton 85	10.76 a	$2.6*10^{-3}$	200-1000	.001-0.1	GW,MW	PCC

Abbreviations: $T_{1/2}$ - half life; a(spec) - maximum specific activity in groundwater in disintegrations per minute and liter; GW - fresh groundwater; MW - mineralized groundwater; BR - brine; PCC - proportional counter counting; LSC - liquid scintillation counting; AMS - accelerator mass spectrometry; RIS - resonant ionization spectrometry

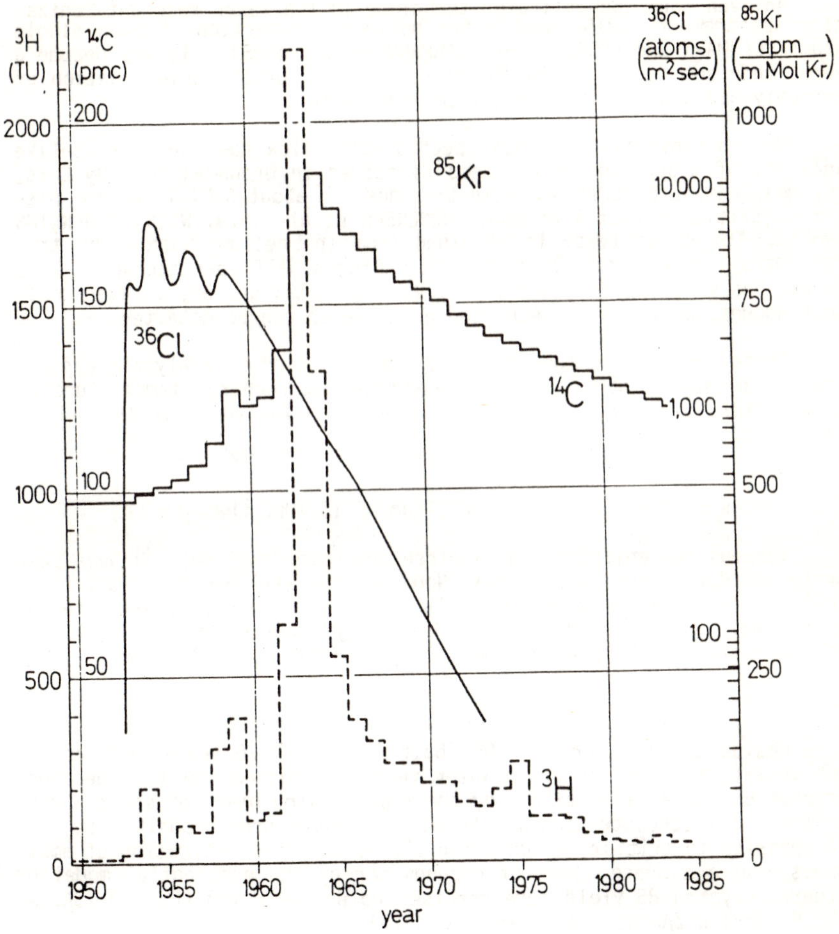

Fig. 2: Changes of the ^{14}C activity (pmc) in atmospheric CO_2 and of the tritium activity (TU) in precipitation in Central Eurcpe due tc nuclear emissions. The krypton-85 activity in the atmosphere stems mainly from spent nuclear fuel. The chlorine-36 fall-out rate was estimated from Dye-3 ice-core data (BENTLEY et al. 1982).

2 PROGRESS IN MEASURING ENVIRONMENTAL ISOTOPES

Most important for isotope hydrology was the introduction of the accelerator mass spectrometry (AMS) during the last years (e.g. GOVE et al. 1987). It allows first time the detection of long-lived environmental isotopes as chlorine 36 at reasonable expenses. By this the dating range of groundwater has become extended to a few million years, very small samples have become detectable and high analysis capacities have become available. About 20 AMS laboratories are now in operation.

An even more sophisticated technique is the laser resonant-ionization spectrometry (RIS) usable for noble gas detection of even single atoms (LEHMANN and LOOSLI 1984; LEHMANN et al. 1985). It may become a very important technique in future especially for the investigation of extremly old and highly mineralized groundwater.

Modern sensitive gas mass spectrometer allow now the quantitative analyses of traces of noble gases dissolved in groundwater. By this, the measurement of tritium activities down to about 0.01 TU became possible (tritium/helium-3 method; TORGERSEN et al. 1979; WEISS & JENKINS 1980). The ^{3}H activity is obtained from the helium 3 grown by tritium decay in a closed water sample during a half year storage. This method has wide applications in the environmental geology as even smallest amounts of recently recharged groundwater can be detected.

These modern mass spectrometers are also used to analyze the spectrum of dissolved noble gases to estimate the recharge temperature of groundwater. This has been application in paleohydrology (RUDOLF et al. 1984).

3 PROGRESS IN EXTENDING THE SPECTRUM OF ISOTOPE CHRONOMETERS

Various new environmental isotope chronometers (Table 1) have been presented during the last years. None of them will become an universal tool as the commonly used ones. Therefore multi-discliplinary studies will remain the basis of isotope hydrology (ANDREWS et al. 1984) and the potential of the new chronometers can be discussed in short.

3.1 Krypton 85

Beside tritium, krypton 85 (half life = 10,76 years) offers the possibility to identify groundwater recharged during the last decades. Krypton 85 is released mainly by the reprocessing plant of spent nuclear fuel. Its concentration in the air increased continuously (Fig. 2). In opposite to tritium, which often allows only the estimation of ambiguous mean residence times of groundwater by the exponential model or others, krypton 85 yield more precise and unique dates (Fig. 3; SALVAMOSER 1986; ROZANSKI and FLORKOWSKI 1979).

The ^{85}Kr technique has an increasing potential for application although the analysis is 10-20 times more expensive than conventional tritium measurements and the extraction of this noble gas from great quantities of groundwater in the field is time-consuming. The reason is the decreasing tritium level which is already very difficult detectable in the southern hemisphere.

3.2 Silicon 32

The silicon-32 method introduced by LAL et al. (1970) has its main field of application in glacier and oceanographic studies today (LAL and SOMAYAJULU 1984) as its application in hydrology is still problematic. The biological degradation of ^{32}Si in the upper soil, adsorption effects and isotope exchange processes are not yet fully understood. The results of more recent studies by FRÖHLICH et al. (1987a) did not improve the situation. Only AMS may do it as extensive analysis programmes can run to understand the fundamental processes of ^{32}Si in the hydrological cycle.

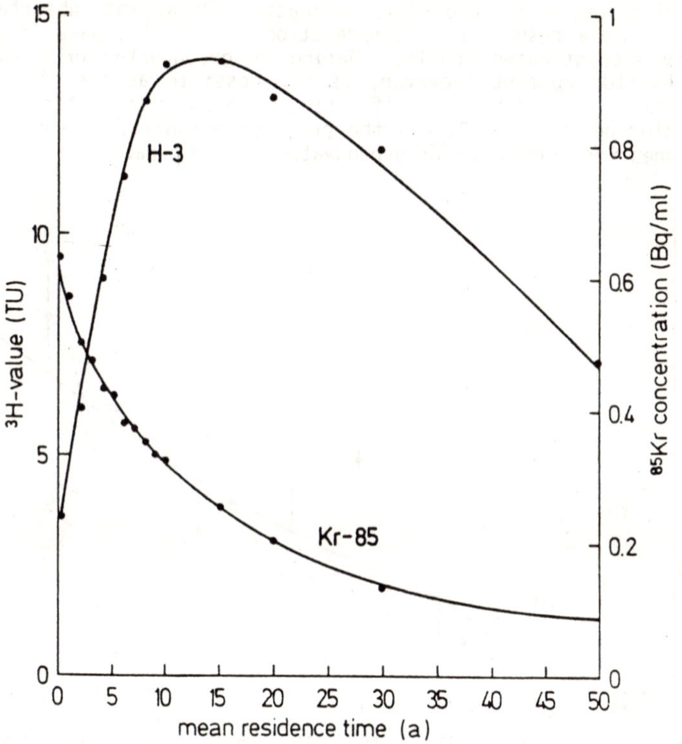

Fig. 3: Relationship between mean residence times (MRT), calculated by the exponential model from the ^3H and ^{85}Kr values of groundwater from southern Germany (SALVAMOSR 1986). Most tritium values yield two MRT while each ^{85}Kr value corresponds to only one and even more precise MRT.

3.3 Chlorine 36

The application of chlorine-36 (half life = 301 ka) in isotope hydrology became possible by the introduction of AMS analysis. A few milligram of cloride is sufficient which is extracted from a few liter of groundwater. Dating of up to several million years and mixing balance studies are possible.

^{36}Cl is produced cosmogenically in the atmosphere, from chlorine 35 in the upper soil and in the deep underground due to neutrons formed mainly by the alpha decay of uranium.

The most instructive case study yielded radiometric piston-flow dates from the fossil groundwater flowing below the Great Artesian Basin in Australia. The hydraulically estimated actual groundwater dates were confirmed (Fig. 4; BENTLEY et al. 1986a).

Other case studies carried out in North America showed that the situation is usually more complicated. Complementary cloride analysis are essential as the chlorine-36 activity changes by more than one process: dissolution of chlorides, radioactive decay and subsurface production. As a result the interpretation of the ^{36}Cl data is possible only by sophisticated models. Dating of groundwater or brines with high chloride content, however, is not possible as the ^{36}Cl content is diluted. It is still open if dating of young groundwater with atom-bomb chlorine-36 (Fig. 2) has the proposed potential in semi-arid and arid zones for estimation of groundwater recharge rates.

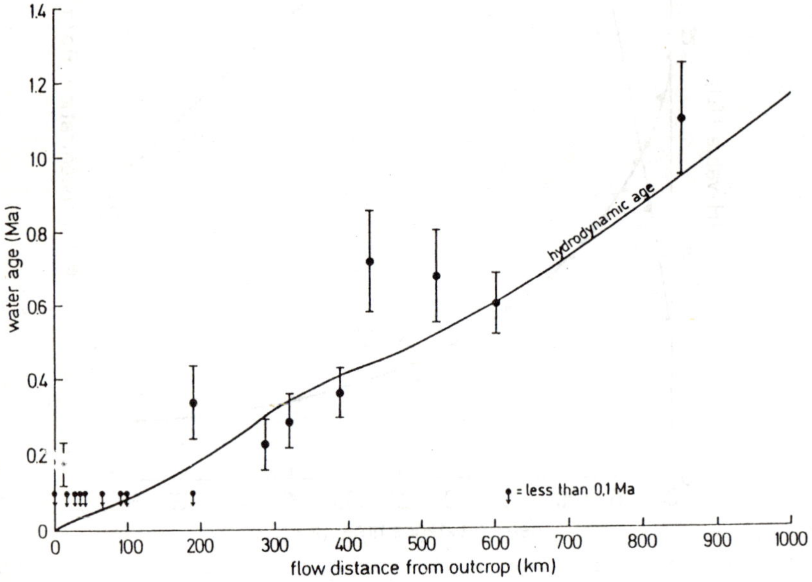

Fig. 4: Concordant ^{36}Cl and hydrodynamic groundwater ages along a southern flow line in the Great Artesian Basin in Australia (BENTLEY et al. 1986b).

3.4 Argon-39 method

The argon-39 method (LEHMANN and LOOSLI 1984) has its potential in groundwater dating up to 2000 years (Loosli and Oeschger 1979; Loosli 1983). This cosmogenic noble gas does not react hydrochemically as e.g. radiocarbon. However, subsurface production from potassium 39 by neutron/gamma-reactions is often complicating the interpretation of the isotope data (ANDREWS et al. 1984; FORSTER et al. 1984; GEYH et al. 1984). In addition, the vacuum extraction of the dissolved trace gases in the field from several 1000 liter of water is costly as the preparative work in the laboratory and the long measurement time. RIS may once improve the situation (LEHMANN and LOOSLI 1984).

3.5 Calcium-41 method

Calcium 41 (half life = 103 ka) occurs as cation in groundwater and may cover a dating range of one million years. It is now measurable by AMS (HENING et al. 1987). But basis research is needed as the natural calcium isotope cycle in groundwater is not yet understood.

3.6 Krypton 81

The cosmogenic krypton 81 (half life = 213 ka; LEHMANN et al. 1985) is present in groundwater in only hundreds of atoms which can be detected, however, by the high sensitive but very expensive laser resonant-ionization-spectrometry (RIS). Dating of groundwater should be possible up to about two million years. More important is that also mineralized water and brines may become datable as krypton neither reacts hydrochemically nor is produced in the subsurface.

3.7 Uranium isotopes

There have been many attempts to date groundwater by means of uranium isotopes (IAEA 1983, 1987). As the $^{234}U/^{238}U$ activity ratio is higher in young aquifers than in old ones because ^{234}U is more soluble than the parent isotope ^{238}U an estimate of the relative geological age of the aquifer can be obtained (WAKSHAL and YARON 1974) as well as an effect of earthquakes (FINKEL 1981).

The main difficulty in interpretation of uranium data arises from the complex hydrochemistry of the dissolved uranium including secondary dissolution, adsorption and retardation processes. In aquifers three zones of different uranium chemistry are distinguished (Fig. 5): in the oxidation zone near the catchment area the $^{234}U/^{238}U$ activity ratio increases rather linearly with the residence time of the groundwater. Within the relative narrow transition zone the situation is intricated. In the reduction zone the uranium content is low but constant. The uranium 234 activity decreases with the water age. Insofar, a potential for groundwater dating does exist in both the oxidation and the reduction zones (e.g. FRÖHLICH et al. 1984).

The processes of the complex uranium chemistry in aquifers are consicely discussed by OSMOND and COWART (1976), OSMOND et al. (1983) and CHALOV (1983). Case studies often yielded ambiguous results as for instance in STRIPA in Schweden where costly multi-disciplinary studies were undertaken for the set-up of a radioactive waste disposal (ANDREWS et al. 1982) or in a sandstone Aquifer in UK (ANDREWS et al. 1984). Uranium isotope analysis, however, can be applied successfully for the genetic classification of groundwater or the mass balance evaluations of mixtures of different kinds of groundwater (OSMOND et al. 1974).

3.8 ^{14}C groundwater dating by the dissolved organic compounds (DOC)

The problems envolved in ^{14}C groundwater dating via the dissolved inorganic carbon compounds (TDIC) is a topic of continuous discussion (e.g. FONTES 1985). The reliability of ^{14}C groundwater dates depends upon the reliability of the estimation how much fossil carbon from the top soil carbonate participated in the groundwater chemistry. Hydrochemical or stable carbon isotope analysis may help but does not solve the problems.

Attempts to measure the ^{14}C activity of the dissolved organic compounds entering during the groundwater recharge seem to be promising (WASSENAAR et al. 1988). As an example, we found the following results from samples collected in an northern German aquifer:

well	^{14}C age(TDIC) BP	^{14}C age (DOC) BP	difference year
1	11855 ± 125	6740 ± 100	5115 ± 160
2	15510 ∓ 135	9945 ∓ 430	5565 ∓ 450

The ^{14}C age difference between these two fractions is rather constant proving that hydrochemical changes of the ^{14}C activity did not occur in the aquifer. The DOC fraction seem to reflect the actual groundwater age. However, we also obtained results reflected a more complicated picture.

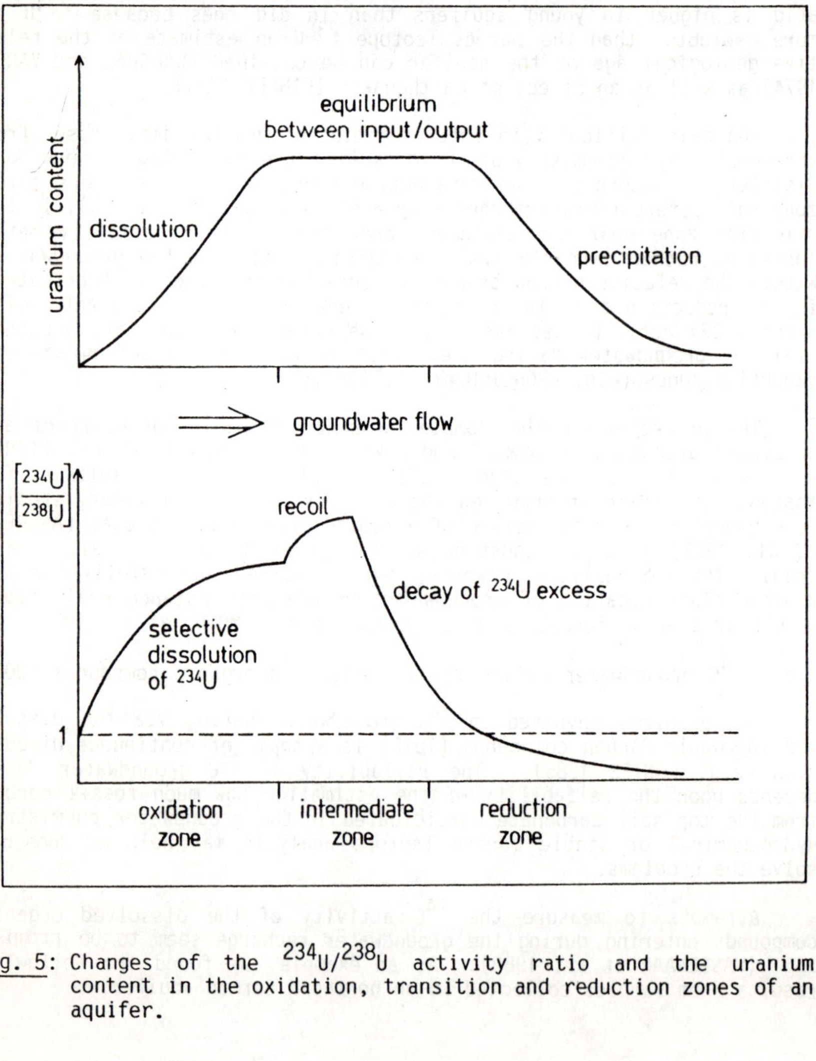

Fig. 5: Changes of the $^{234}U/^{238}U$ activity ratio and the uranium content in the oxidation, transition and reduction zones of an aquifer.

4 PROGRESS IN MODEL INTERPRETATION

4.1 RECHARGE AND EVAPORATION RATES OF THE UNSATURATED ZONE

In arid and semiarid zones the rate of groundwater recharge is determined by the tritium tagging method widely and successfully applied in India (e.g. ATHAVALE et al. 1983). The other important parameter of the water budget, the evaporation rate of groundwater from the deep aquifer down to mm/a can now be estimated from the stable isotope composition of oxygen und hydrogen in the pore water of the unsaturated zone and its pore water content (BARNES and ALLISON 1988). This is due to isotope fractionation occurring during the steady-state moisture transfer from the groundwater surface to the atmosphere by capillary rise up to the evaporation front followed by the equivalent water vapor flow through the dry soil layer above it (CHRISTMANN and SONNTAG 1987).

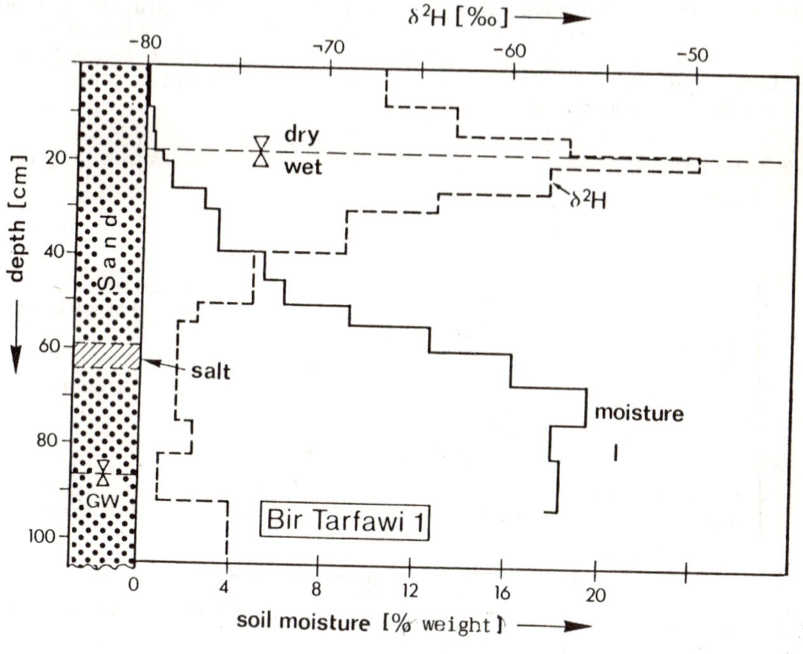

Fig. 6: Change of the isotope composition of oxygen and hydrogen in pore water and of the pore water content with depth as the result of the steady state moisture transfer from the groundwater to the atmosphere (CHRISTMANN and SONNTAG 1987).

4.2 QUANTITATIVE ISOTOPE DATA INTERPRETATION

4.2.1 RELICT RESOURCES OF FOSSIL GROUNDWATER

There are increasingly attempts to substitute the qualitative interpretation of isotope data by quantitative ones. This is already routine in mixing studies by stable isotopes (e.g. MALOSZEWSKI et al.

1983; STICHLER et al. 1986). However, the quantitative interpretation of tritium and ^{14}C dates is a new field. Mass transport calculations were carried out to determine the hydraulic conductivity of an aquitard in a multi-storage aquifer system and to estimate the groundwater flux between adjacent aquifers (GEYH et al. 1984).

A more recent case study was undertaken in the semiarid area at Karthoum, Sudan (FRÖHLICH et al. 1987b). The decreasing tritium activity in shallow groundwater with increasing distance from the Nile river was first interpreted as a support for the hydrodynamic concept of a numerical model. It proposes that the shape of the groundwater table is determined by a steady-state groundwater recharge by seeping Nile water. A rather extensive isotope and hydrochemical survey including ^{14}C analysis questioned this concept. The mass transport evaluation of these data showed that only 10 % of the proposed groundwater seepage may occur. According to the isotope data the shape of the present groundwater table is mainly due to the occurance of a relict resource of fossil groundwater recharged during a former pluvial and only as a part due to present Nile water seepage (Fig. 7; FRÖHLICH et al. 1987b). The dispute on the validity of these different concepts will be terminated by studying the reaction of the groundwater table in relation to the groundwater abstraction.

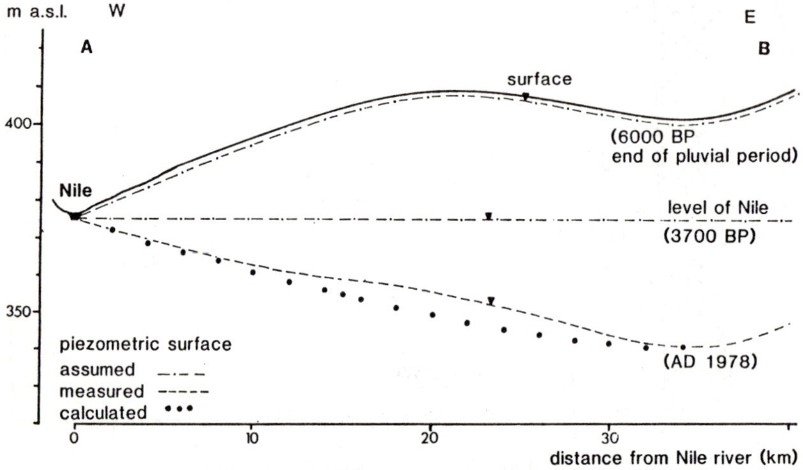

Fig. 7: The measured groundwater table in the Karthoum area and the simulated one (dotted line) applying mass transport calculations to the ^{14}C dates (FRÖHLICH et al. 1987b).

Another case study was undertaken in the semiarid Arzraq-spring area, Jordan where similar hydrodynamic conditions exist. The groundwater recharge rate estimated by a numerical model is the basis for the recommended maximum rate of groundwater abstraction. The latter is by about one order of magnitude larger than that derived from the mass transport calculation based on isotope data.

Based on this experience it seems to be necessary especially for semiarid and arid zones to take into account the occurrance of relict resources of fossil groundwater in numerical models. The assumption of hydraulic steady-state conditions in such regions may be incorrect. The near future will show if this statement must be followed.

4.2.2 TEMPORAL VARIABILITY OF ^{14}C DATA

It is already an often made observation that ^{14}C and other isotope data of heavily pumped wells vary with time. This was called "mining effect" of fossil groundwater (GEYH and BACKHAUS 1979). Inspite of the serious consequences for the interpretation of isotope data the majority of environmental isotope studies do not consider this aspect even for heavily exploited aquifer systems.

The constancy of ^{14}C groundwater ages over years may be interpreted as reflection of rather undisturbed hydraulic conditions. Then, the groundwater withdrawal is only small compared to the naturally available resource. Increasing ^{14}C groundwater ages may forecast the approach of fossil groundwater which is often highly mineralized. In opposite, a decrease of groundwater ^{14}C ages may indicate an increasing inflow rate of groundwater from the shallow aquifer which is often polluted.

A first quantitative survey was undertaken in the Azraq area, Jordan (FRÖHLICH et al. 1987b). The long-term trend of the ^{14}C data reflecting a decrease of water ages since 1979 (Fig. 8) has been interpreted as an increasing flux of young basalt groundwater substituting the fossil fresh water flux from the B4 formation. This would have consequences for the quality of the groundwater and the recommendations for optimum groundwater management. A fall of the groundwater table below an economically acceptable depth may occur in near future.

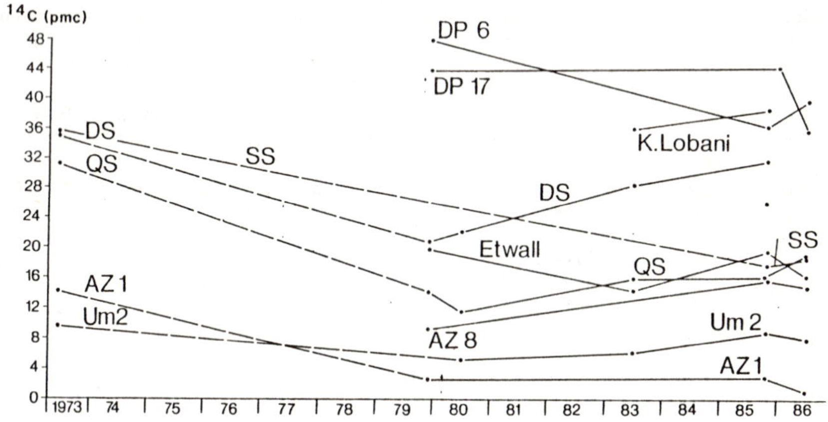

Fig. 8: Change of the ^{14}C data of various production wells in the Azraq area/Jordan due to the change in the composition of groundwater from the basalt and the B4 formation (FRÖHLICH et al. 1987b).

5 CONCLUSION

The techniques based on environmental isotopes have profitted from new analysis techniques as AMS, RIS, and modern mass spectrometry. Additional chronometers became available by the measurement of long-lived cosmogenic isotopes. In addition, the recording of trace constituents in groundwater allows to detect even very small admixtures of young groundwater, to estimate recharge temperatures as well as to date via DOC.

Progress is also visible in the interpretation of isotope data. Reliable estimates of evaporation rates are obtained from the stable isotope composition of the pore water in the top soil. Mass transport calculation with ^{14}C data yield new insight in the groundwater dynamics in semiarid and arid zones. Relict resources of fossil groundwater may modify the shape of the groundwater table and may exclude the assumption of steady-state conditions in numerical modelling.

6 LITERATURE

BARNES, C.J., and Allison, G.B., 1988. Tracing of water movement in the unsaturated zone using stable isotopes of hydrogen and oxygen. J. Hydrol., v. 100, p. 143-176.

Andrews, J.N., Giles, I.S., Kay, R.L.F., Lee, D.J., Osmond, J.K., Cowart, J.B., Fritz, P., Barker, J.F., and Gale, J., 1982. Radioelements, radiogenic helium and age relationships of groundwaters from the granites at Stripa, Sweden. Geochim. Cosmochim. Acta, v. 46, p. 1533-1543.

Andrews, J.N., Balderer, W., Bath, A.H., Clausen, H.B., Evans, G.V., Florkowski, T., Goldbrunner, J.E., Ivanovich, M., Loosli, H., and Zojer, H., 1984. Environmental isotope studies in two aquifer systems: a comparison of groundwater dating methods. Isotope Hydrology 1983, p. 535-576; IAEA, Vienna.

Athavale, R.N., 1983. Nuclear tracer techniques for measurement of natural recharge in hard rock terrains. In: Int. Workshop on Rural Hydrogeology and Hydraulics in Fissured Basement Zone, p. 71-80; Universty of Roorkee.

Bentley, H.W., Phillips, F.M., Davis, S.N., Gifford, S., Elmore, D., Tubbs, L.E., and Gove, H.E., 1982. Thermonuclear ^{36}Cl pulse in natural water. Nature, v. 300, p. 737-740.

Bentley, H.W., Phillips, F.M., and Davis, S.N., 1986. Chlorine-36 in the terrestrial environment. In: Fritz, P. and Fontes, J.-Ch., edts. Handbook of Environmental Isotope Geochemistry, vol. 2, The Terrestrial Environment, p. 427-480; Elsevier, Amsterdam, London.

Bentley, H.W., Phillips, F.M., Davis, S.N., Habermehl, M.A., Airey, P.L., Calf, G.E., Elmore, D., Gove, H.E., and Torgersen, T.L., 1986a. Chlorine-36 dating of very old groundwater 1. The Great Artesian Basin, Australia. Water Res. Res., v. 22 (13), p. 1991-2001.

Chalov, P.I., 1983. Uranium disequilibrium as an indicator of processes in the hydrosphere. Water Res., v. 9, p. 466-479.

Christmann, D., and Sonntag, C., 1987. Groundwater evaporation from east-Saharian depressions by means of deuterium and oxygen-18 in soil moisture. Isotope Techniques in Water Resources Development, p. 189-204; IAEA, Vienna.

Finkel, R.C., 1981. Uranium concentrations and $^{234}U/^{238}U$ activity ratios in fault-associated groundwater as possible earth-quake precursors. Geophys. Res. Letters, v. 8, p. 453-456.

Fontes, J.Ch., 1985. Some considerations on groundwater dating using environmental isotopes. In: Hydrogeology in the Service of Man, p. 118-154; Mem. IAH, Cambridge.

Forster, M., Moser, H., and Loosli, H., 1984. Isotope hydrological study with carbon-14 and argon-39 in the Bunter sandstones of the Saar region. Isotope Hydrology 1983, p. 515-533; IAEA, Vienna.

Fröhlich, K., Gellermann, R., and Hebert, D., 1984. Uranium isotopes in a sandstone aquifer: interpretation of data and implications for groundwater dating. Isotope Hydrology 1983, p. 447-466; IAEA, Vienna.

Fröhlich, K., Franke, T., Gellermann, R., Hebert, D., and Jordan, H., 1987a. Silicon-32 in different aquifer types and implications for groundwater dating. Isotope Techniques in Water Resources Development, p. 149-163; IAEA, Vienna.

Fröhlich, K., Geyh, M.A., Verhagen, B.Th., and Wirth, K., 1987b. Isotopenhydrologische Methoden zur Begutachtung von Grundwasser in Trockengebieten. pp. 179; Weltforum, Cologne.

Gat, J., and Gonfiantini, R., edts., 1981. Stable Isotope Hydrology. Deuterium and Oxygen-18 in the Water Cycle. Technical Reports Series, No. 210, pp. 337; IAEA, Vienna.

Geyh, M.A., and Backhaus, G., 1979. Hydrodynamic aspects of carbon-14 groundwater dating. Isotope Hydrology 1978, p. 631-643; IAEA, Vienna.

Geyh, M.A., and Michel, G., 1981. Isotopen- und hydrochemische Betrachtungen über die Süßwasser/Salzwasser-Zone am Nordostrand des Münsterländer Beckens. Z. dtsch. geol. Ges., v. 132, p. 597-611.

Geyh, M.A., Backhaus, G., Andres, G., Rudolph, J., and Rath, H.K., 1984. Isotope study on the Keuper sandstone aquifer with a leaky cover layer. Isotope Hydrology 1983, p. 499-513; IAEA, Vienna.

Gove, H.E., Litherland, A.E., and Elmore, D., 1987. Accelerator Mass Spectrometry. Nucl. Instr. Meth. Phys. Res., v. B 29 (No. 1,2), pp. 455.

Henning, W., Bell, W.A., Billquist, P.J., Glagola, B.G., Kutschera, W., Liu, Z., Lucas, H.F., Paul, M., Rehm, K.E., Yutema, J.L., 1987. Calcium-41 concentration in terrestrial materials: prospects for dating pleistocene samples. Science, v. 26, p.2-727.

IAEA, 1984. Isotope Hydrology 1983, pp. 873; IAEA, Vienna.

IAEA, 1987. Isotope Techniques in Water Resources Development, pp. 813; IAEA, Vienna.

Lal, D., Nijampurkar, V.N., and Rama, S., 1970. Silicon-32 hydrology. Isotope Hydrology 1970: 847-863; IAEA, Vienna.

Lal, D., and Somayajulu, B.L.K., 1984. Some aspects of the geochemistry of silicon isotopes. Tectonophysics, v. 105, p. 383-397.

Lehmann, B.E., and Loosli, H.H., 1984. Use of noble gas radioisotopes for environmental research. Int. Phys. Conf. Ser. No., vol. 71, p. 219-226.

Lehmann, B.E., Oeschger, H., Loosli, H.H., Hurst, G.S., Allman, S.L., Chen, C.H., Kramer, S.D., Payne, M.G., Phillips, R.C., Willis, R.D., and Thonnard, N., 1985. Counting ^{81}Kr atoms for analysis of groundwater. J. Geophys. Res., v. 90 (B13), p. 11557-11551.

Loosli, H.H., 1983. A dating method with ^{39}Ar. Earth Planet. Sci. Letters, v. 63, p. 51-62.

Loosli, H.H., and Oeschger, H., 1979. Argon-39, carbon-14, and krypton-85 measurements in groundwater samples. Isotope Hydrology 1978, v. 2, p. 931-945; IAEA, Vienna.

Maloszewski, P., Rauert, W., Stichler, W., and Herrmann, A., 1983. Application of flow models in an alpine catchment area using tritium and deuterium data. J. Hydrol., v. 66, p. 319-330.

Mazor, E., Kaufman, A., and Carmi, I., 1973. Hammat Geder (Israel): Geochemistry of a mixed thermal spring complex. J. Hydrol., v. 18, p. 289-303.

Münnich, K.O., 1957. Messung des ^{14}C-Gehaltes von hartem Grundwasser. Naturwissenschaften, v. 34, p. 32-33.

Osmond, J.K., and Cowart, J.B., 1976. The theory and uses of natural uranium isotopic variations in hydrology. Atomic Energy Rev., v. 144, p.

621-679.
Osmond, J.K., Kaufman, M.I., and Cowart, J.B., 1974. Mixing volume, calculations, sources and aging trends for Floridan aquifer water by uranium isotopic methods. Geochim. Cosmochim. Acta, v. 38, p. 1083-1100.
Osmond, J.K., Cowart, J.B., and Ivanovich, M., 1983. Uranium isotopic disequilibrium in ground water as an indicator of anomalies. Int. J. Appl. Radiat. Isot., v. 34, p. 283-308.
Rozanski, K., and Florkowski, T., 1979. Krypton-85 dating of groundwater. Isotope Hydrology 1978, v. 2, p. 949-959; IAEA, Vienna.
Rudolf, J., Rath, H.K., and Sonntag, C., 1984. Noble gases and stable isotopes in ^{14}C dated paleowaters from Central Europe and the Sahara. Isotope Hydrology 1983, p. 467-477; IAEA, Vienna.
Salvamoser, J., 1986. Quantitative separation of admixed young groundwater and surface water with the ^{3}H-^{85}Kr method. In: Conjunctive Water Use. IAH Publ., v. 156, p. 355-363.
Stichler, W., Maloszewski, P., and Moser, H., 1986. Modelling of river water infiltration using oxygen-18 data. J. Hydrol., v. 83, p. 355-365.
Torgersen, T., Clarke, W.B., and Jenkins, W.J., 1979. The tritium/helium-3 method in hydrology. Isotope Hydrology 1978, vol. 2, p. 917-929; IAEA, Vienna.
Wakshal, E., and Yaron, F., 1974. $^{234}U/^{238}U$ disequilibrium in waters of the Judea group (Cenomanian-Turonian) aquifer in Galilee northern Israel. Isotope Techniques in Groundwater Hydrology 1974, vol. 2, p. 151-176; IAEA, Vienna.
Wassemaar, L.I., Aravena, R., and Fritz, P., 1988. Dissolved organic carbon and its relation to geochemical processes in an unconfined sand aquifer, southwestern Ontario. In: 13th. Int. Radiocarbon Conf., Dubrovnik.
Weiss, W., and Jenkins, W.J., 1980. Tritium-helium-3 dating of natural waters. ZFI Mitt., v. 29, p. 292-301.

to be presented on the

International Groundwater Workshop 1989 "Appropriate Methodologies for Development and Management of Groundwater Resources in Developing Countries", NGRI, Hyderabad, India from February 28 to March 4, 1989.

EXPLORATION OF FRACTURED ZONES BY RADON

DETERMINATION IN THE SOIL

TH. POINTET - BRGM (FRANCE)

ABSTRACT

The statistical study of yields obtained from several tens of thousands of boreholes in the African crystalline rocks shows that 15 % of them give more than 5 m3 per hour. The usual geological analysis however does not discover reasons why one borehole has a small or a high yield. By these methods alone, it was impossible to propose siting rules which would significantly increase the possibility of obtaining large yields. The water demand is increasing at the present time. The predictable development of irrigated farming and potable water supply of small towns from groundwater supposes that one will be able to find high individual yields in the basement zones with high enough success rates for the cost of research not to be prohibitive. On the other hand, 65 % of the population of West Africa is concentrated in the basement regions, and it is therefore this basement which is the hydrogeologist's main aim. By turning towards less traditional techniques inspired by structural geology and the study of fracture networks, it has been observed that the Radon contained in the gases of the soil acted as an efficient natural tracer : it underlines the presence of permeable zones in a fractured massif. Systematic prospecting by RADON dosing in the soil gases brings out locally-concentrated anomalies on a usually marked groundnoise. The boreholes drilled immediately below these anomalies are usually very productive. Applied to prospecting, the method was tested in Burkina-Faso along with the electric and electromagnetic geophysical methods, and with siting techniques bases on the morpho-structural analysis. In environments of crystalline and metamorphic rocks, the water yields obtained from boreholes sited on Radon anomalies, are distinctly higher than those from boreholes sited by other methods.

I - INTTRODUCTION

The exploration of aquifer environments in basement rocks, like most aspects of the hydrogeology of fractured environments, is marked by their discontinuous character. To a certain extent, it is an advantage if the favourable fractured zone displays a sharp contrast with a virtually impermeable environment.

The detection of 'fracturd compartment/whole rock' contrasts is a widely used prospecting principle. This is the domain of structural geology and the compilation of indirect data. Geophysical prospecting, for example, looks for subsurface electric conduction anomalies, and anomalies in electromagnetic fields, along bedrock discontinuities.

Based on the fact that the granitic plutons, and other crystalline or non crystalline rocks as well, contain very low concentrations of uranium minerals,uniformely distributed, which release radon by their decay, that this radon easily migrates and tends to reach the surface, and that, finally, it is the only radioactive gas ( ); it present in nature, it is likely that the paths of this migration are the permeable zones of the basement. Moreover, radon is relatively easy to detect.

1.1 - Radon Radioactivity

Radon, with atomic number 86, is an inert, short-lived radioactive gas. The radon group is formed of three isotopes, radon sensu stricto 222/86 Rn, thoron 220/86 Rn, actinon 219/86 Rn. Among them, only isotope 220 and especially 222 are of geological importance. The three isotopes of radon fit into the relationship of natural radioactive families, between an isotope of radium and an isotope of polonium :

$$_{88}Ra \rightarrow {}_{86}Rn \rightarrow {}_{84}Po$$

1.2 - Radon as a the marker of water bearing fractures

Radon atoms replace, in the cyrstalline network, the radium atoms from which they are derived. When being formed, the radon atom deeps part of the energy produced by disintegration. This residual energy brings about a displacement of the radon atom varying from 20 to 70 nanometers. This displacement is enough to allow the liberation of the atoms located on the crystal surface, into the inter-granunar space. Surfaces on which radium atoms seem to concentrate.

Emanation power is the name given to yhe relation between the quantity of radon liberated from the host rock to the quantity of radon which remains trapped.

Factors influencing the emanation power.

The emanation power is all the greater as the contact surface between the rock crystals and the outside is large. In other words, the more a rock is divided up, the more possibilities the radom will have to escape from it.

For instance, the emanation power from an NaCl cyrstal is 1.7 %, whereas it can reach 99 % for a finely crashed uraniferous rock.

1.3 - Migration of radon in the rock

Radon is soluble in water, and can therefore be trans ported by groundwaters. The solubility of radon in water decreases with temperature.

The emanation power increases with the permeability of the host rock : but it decreases with the water content of the unsaturated zone.

The radon liberated from the crystals by its power of emanation migrates into the rock by means of the intergranular spaces and discontinuities (fractures).

This migration will be all the easier, the more discontinuities exist. Moreover, if the rock is water-bearing, the migration process will be intensified by the radon being carried mechanically by moving water. The more permeable the medium, the more effective this transport will be ; consequently more radon must be given off at an aquiferous fracture than from sound rock for :

. the dividing up of the fractured rock and the presence of water considerably increase the radon's emanation power,

. the discontinuities and the water movement favour the migration of the radon through the rock mass. It can be supposed that the more interconnected fractures there are, the greater will be the escape from the ground.

Therefore, by taking gas samples from the ground, it should be possible to discover positive radon concentration anomalies which should be located vertically above the basement fractures. The extent and size of these anomalies should be linked with the fracturing intensity.

2 - FIRST RESULTS

2.1 - Prospecting carried out in France

Prospections of an experimental nature were carried out by BRGM in Brittany and in Corsica on sites, in crystalline zones, which were well known as regards fracturing.

Profiles were set up so as to intersect water-bearing fractures recognized by photo-interpretation or by electrical and electromagnetic geophysical methods.

These first experiments gave the following conclusions :

- the radon anomalies lie, to within a few metres, over the water-bearing anomalies positioned by photos or geophysical methods and confirmed by drilling. They are narrow (5 to 20 m) and are 5 to 10 times larger than the average concentration level,

- the more permeable the fracture, the stronger is the anomaly,

- the gas emanations are discontinuous on one and the same fracture. The anomalies are probably restricted to the segments of accidents for which there is the most open fracturing, i.e. to the most favourable zone for obtaining high yields.

The method has been exploited from then on and twenty or so sites in France were prospected in the last year and a half. The influence of multiple morphologic, geologic and climatic parameters was observed.

2.2 - Prospecting carried out in Burkina Faso

The project which is in progress - "study of the water resources of the crystalline basement", financed ty the European Community and French and Italian Aid, foresees radon prospecting on 3 experimental fields.

Under such semi-arid conditions, the results obtained and the first conclusions on the effectiveness of the radon prospecting can be resumed as follows :

Sanon Site

This first experimental site, located 15 km north-west of Ouagadougou, consists of an ancient granitic bedrock covered by a thick saturated weathered·layer (30-45 m).

The radon prospecting carried out on this site (700 gas samples taken from the ground) discovered ten or so anomalies 3 to 4 times greater than the regional average concentration (cf. figures 1).

The results of 18 boreholes which discovered and tested the crystalline basement on the Sanon site are given in the table below, distinction being made between the boreholes sited on radon anomalies and those sited by means of classic techniques (photo-interpretation, geophysics).

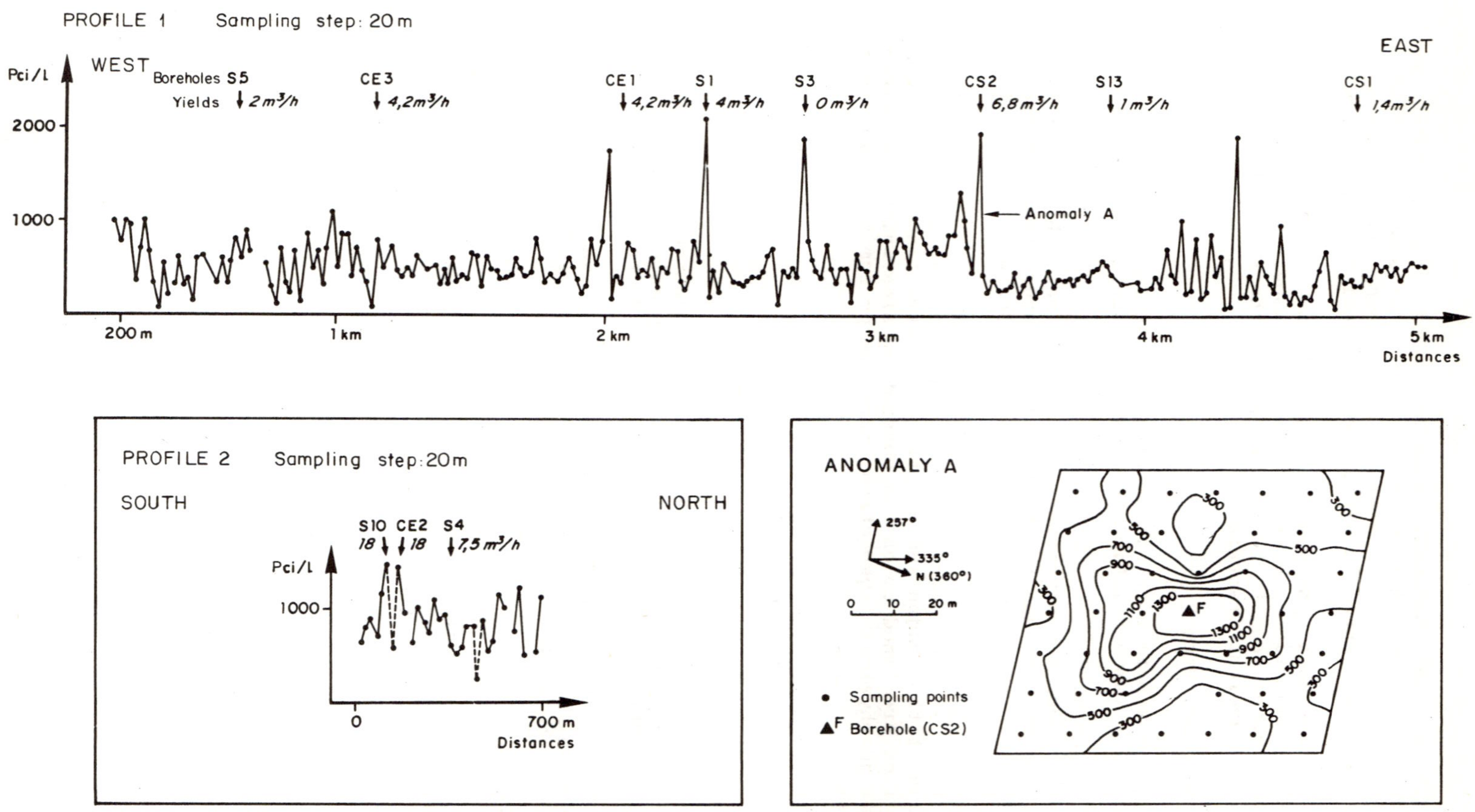

Fig:1 — RADON PROSPECTION IN SANON SITE

It is noted that only those boreholes sited according to the radon prospecting results are capable of supplying 5 m3/h or more, under conditions which are economically acceptable for irrigation (specific yield over or equal to 0.25 m3/h/m).

In that case the weatered zone being from 30 to 40 m thick the emanation process seems to be something different. Radium sources seem to be located within the regolith, on the traces of ancient fractures.

Katchari Site

The Katchari site is located in the Sahelian zone, the bedrock is amphibolitic here with a thin dry-weathered covering.

The radon concentrations are seen to be much smaller than at Sanon. This is due to the lithologic nature of the bedrock, the green rocks being less rich in uraniferous minerals than granite.

There is also seen to be an excellent agreement of the results of the radon and electrical prospecting ; the radon anomalies are lined up on the conducting axes which, in electrical geophysical prospecting, mark the presumably favourable zones.

The 2 boreholes sited on the radon anomalies (K1, K2) gave a good yield : 6 and 11 m3/h

Barogo Site

In the Barogo fractured schists and greenshists, some 300 gaz samples were analysed. A single sharp anomaly was discovered, constrasting on a ground noise of 100 to 200 pci/l.

The borehole sited on this anomaly gave an instaneous production of 25 m3/h, which is the best productivity on a well for the whole region (figure 2).

CONCLUSIONS

From the results on the sites, it appears that the boreholes sited by the radon method are more productive than those sited by classic methods, as the summary table below shows :

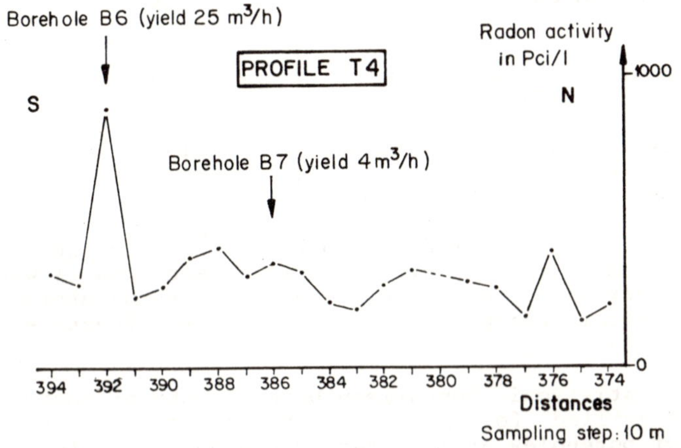

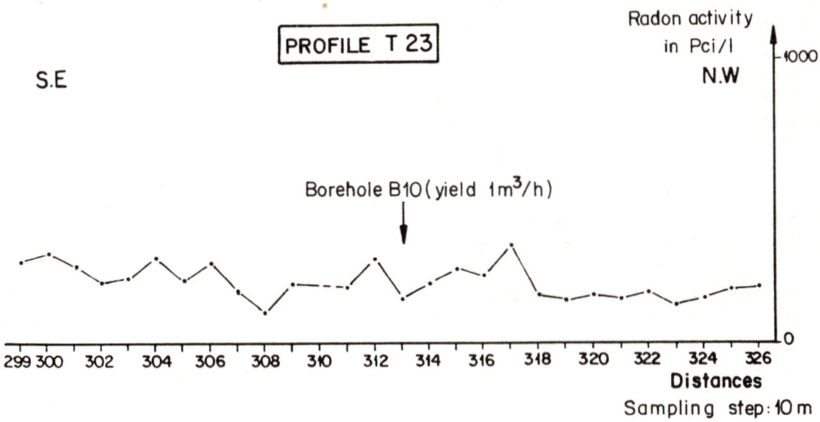

Fig: 2 — RADON PROSPECTION IN BAROGO SITE

	Radon siting	Photo-geophysical siting	Total
Number of boreholes drilled	9	20	29
Number of boreholes with Qsp > 0.25m3/h/m	5	3	8
Success rate for Q > 5 m3/h	56 %	15 %	28 %

The sucess rate for an exploitable target yield higher or equal to 5 m3/h (Qsp > 0.25 m3/h/m) reaches 56 % when the boreholes are sited on radon anomalies. Photo-interpretation and geophysical prospecting are far less effective (15 % success rate).

The statistical comparison between the productivity of boreholes drilled for this project and the productivity of hundred of boreholes drilled during the near past for village water supply in same geological features underlines the effectiveness of the radon technique (fig 3).

The lack of relief and the regular thickness of the regolith, controlled through the geophysics, were favourable and simplified features for these tests.

Prospections conducted in hilly regions in temperate zone show the influence of more parameters and underlines the necessity to understand the geologic structure of massives for an efficient interpretation of the radon distribution.

The effectiveness of radon prospecting will only be confirmed when the number of boreholes sited by this method will be large enough to be significant.

Finally, there will have to be progress in the field of understanding the phonomena studied : identification and location of the radon-emanating sources, gaz migration through the cover formation, definition of the criteria for detecting anomalies which are not connected with fracturing. These various lines of research will be developed in the near future.

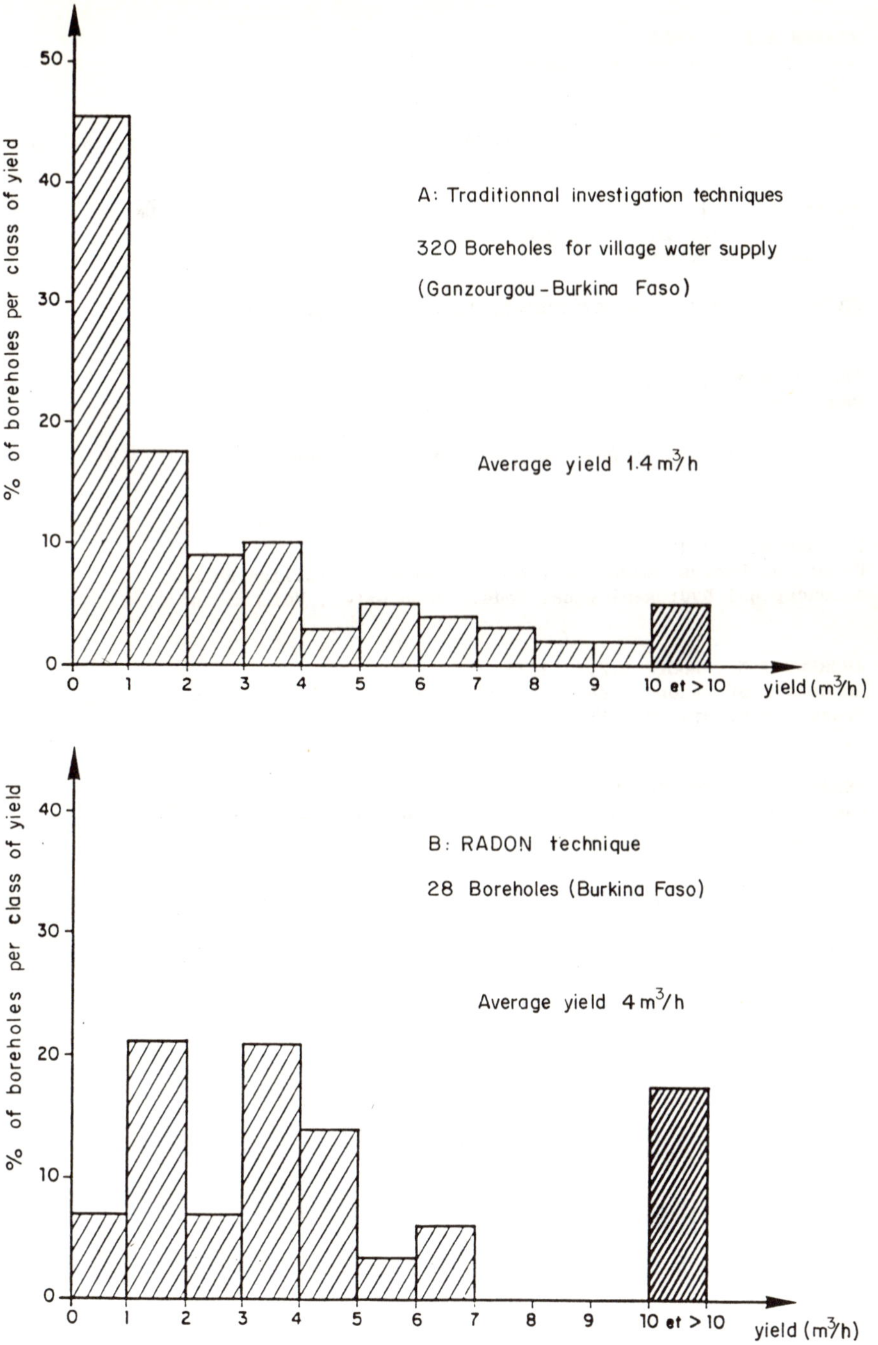

A: Productions 7 boreholes sited with traditionnal techniques for village water supply
B: Productions 7 boreholes sited with the Radon techniques

Fig: 3 — EFFICIENCY OF PROSPECTING TECHNIQUES

- **TANNER A.B. : 1964**
  Radon Migration in the Ground
  A Review

  In The Natural Radiation Environment. Ed. by Adams (T.A.S.) - p. 161-190.

- **TANNER A.B : 1980**
  Radon Migration in the Ground
  A Supplementary Review

  In Natural Radiation Environment III. Vol. 1, pp. 5-56.

- **TELFORD W.M. : 1983**
  Radon mapping in the search for Uranium

  In Development in Geophysical Exploration Methods
  Ed. by A.A. FITCH

- **LEGRAND M.E. : 1987**
  Radon and Radium Emanation from fractured Crystalline Rocks
  A conceptual hydrogeological Model Groundwater. Vol. 25, n° 1.

- **ANDREWS J.N., WOODS D.F. : 1972**
  Mechanism of radon release in rock matrices and entry into groundwaters.
  Trans. Inst. Min. Metall. Sec. B81, 198-209.

- **ISRAEL M., BJORNSSON S.: 1967**
  Radon (Rn 222) and Thoron (Rn 220) in soil air over faults.
  J. Geophys., 33 (1) : 48-64.

NATURAL RECHARGE OF BASEMENT AQUIFERS IN A SAHEL AND SUDAN CLIMATE: Experimental study in Burkina faso

F. MILVILLE, M. NORMAND, T. POINTET
Bureau de Recherches Geologiques et Minieres - FRANCE

C. MERLO, F. URANI
AQUATER, Italian consultant, ITALY

ABSTRACT

Two projects, have been developed in Burkina Faso, in order, first, to evaluate by multidisciplinary methods the recharge of basement aquifers in sahelian and sudanese climates, and second, to perform life-size tests of exploitation of large yields destined for the irrigation of small areas. The recharge study was carried out on ten pilot stations, equipped within the scope of the first project. Monitoring over a two year period of data recordings provided the numerical informations required for the constitution and the calibration of lumped-up rain/water level models which allowed the quantitative analysis of the water transit from the surface to the water saturated underground. Roughly 300 samples were collected for isotope analysis, to attempt at dating the water and to study the movement of water masses within the fissured medium. A striking phenomenon of recharge was thus observed and characterized, in both climates concerned. It supposes the infiltration of amounts of water comprised between, 50 and 150 millimeters for the different latitudes in which the tests were conducted. We also observed, in particular by water isotopes, a partitioning effect of the drainage basins into sub-units which is more and more marked towards the dry zones. This means that the aquifers to be exploited in the sahel zones will be even more fragile.

RECHARGE, ITS VALUE AND DEFINITIONS

In the past ten years, the production of groundwater tapped in basement rocks in Sudan/Sahel zones has expanded considerably. Thousands of boreholes have been drilled and tapped in areas considered poor in water. The experience gained in village water supply projects by BRGM and other organizations helps to site the boreholes more accurately and to guarantee a statistical success ratio during implementation. However, predicting the duration of such production, and the flow associated with it requires an assessment of the renewable resource, namely recharge. This knowledge is indispensable to determine a production limit, and to guarantee its permanence, beyond the dry season, or even better, over a series of low rainfall years. It is evidently possible to tap more than the average recharge by overproduction, but this involves a mining type of operation because non-renewable reserves are affected.

1 - FRAMEWORK OF THE STUDY

1.1 - Study method

The method selected is multidisciplinary. Ten experimental sites were equipped and tested. They were selected in three types of climate, Sudan, humid Sahel and Sahel. Equipped with a climatic station and three boreholes on the average, with each having been covered by tests and follow-up by the CIEH (Comite Interafricain d'Etudes Hydrauliques), the sites were analyzed by means of global hydrological models in order to determine, during the hydrological year 1985, the order of magnitude of the terms of the water balance which define exchanges between the subsurface environment and the surface environment.

Throughout the country, ninety pumping stations on tapped boreholes were covered by three sampling campaigns before and after the rainy season in the years 1984 and 1985. The 300 samples collected were analyzed to determine the tritium concentrations, which indicate the age of the groundwaters, and the deuterium and oxygen 18 concentrations, which are influenced by evaporation between the stage when the rain reaches the ground and the stage when this water has recharged the aquifer.

Continuous monitoring was carried out over two recharge periods. The quantitative results were compared with the qualitative results of isotopic analysis, in order to determine the mode of renewal of the resource, its permanence, and to identify the requirements for secure groundwater tapping.

The installation of a piezometric and climatic monitoring network, which is now continually supervised, is an unprecedented achievement in West Africa.

1.2 - Siting

The project was conducted in Burkina Faso (figure 1) for three main reasons :

- this state offers a typical climatic variety from the Sudan zone (1500 mm) to the Sahel zone (400 mm),
- the geological and geomorphological contexts are varied,
- the country has already been covered by many village water supply programmes and offers a large network of water points equipped with pumps.

In addition, the CIEH has its head offices at Ouagadougou and could therefore be associated directly with the project.

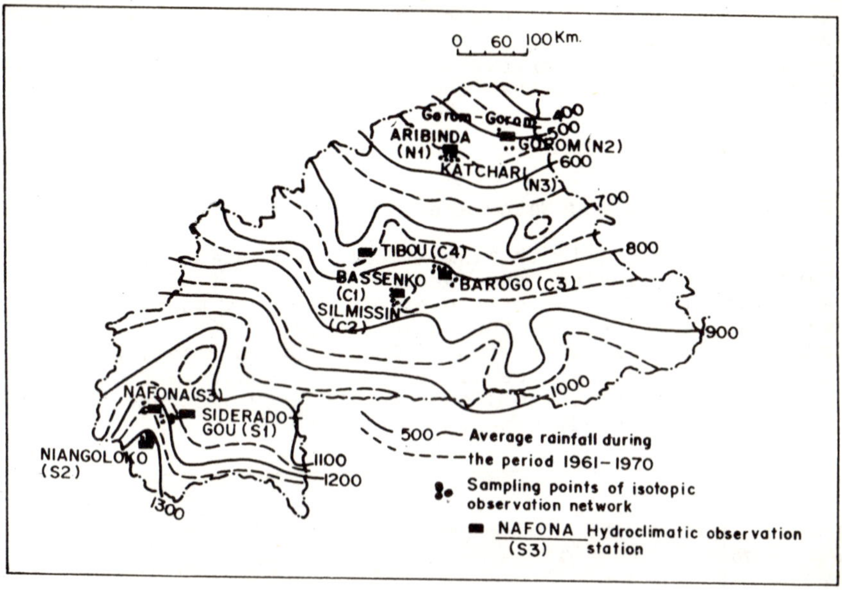

Figure 1: LOCATION OF OPERATION

2 - RECHARGE ANALYSIS BY DETERMINATION OF THE ISOTOPES OF WATER

2.1 - Methodology

2.1.1 - Tritium

This radioactive isotope of hydrogen is a very low energy emitter (Emax = 18.6 keV) which was injected massively into the atmosphere from 1952 to 1963 by aerial thermonuclear explosions. The average annual activities of tritium in rainfall grew considerably during this period due to the more than 200 kg of tritium released into the atmosphere by these tests. Since the moratorium on atmospheric tests, the activities have been declining steadily. Through its presence in the water molecule, tritium marks the natural cycle of the water masses, precipitation, seepage, atmospheric or subsurface circulation, once the input function of the system has been defined, namely the isotopic composition of the precipitation.

2.1.2 - Oxygen 18 and deuterium

Based on a constant composition observed in sea-water, the concentrations of these two stable isotopes in natural waters very in accordance with two physical mechanisms, the change in temperature during rainfall, and the evaporation effect undergone by water on the ground during infiltration, and in the ground at shallow depth due to recovery by the atmosphere. Evaporation causes an increase in the concentrations of the two heavy isotopes. Moreover, this increase is comparatively higher for oxygen 18 than for deuterium.

2.2 - Results

2.2.1 - Tritium and Water age

The atmospheric tritium signal was seasonally variable, particularly during the period of high activity. The annual amplitude in 1966 was of the same order of magnitude as the average decrease in the signal from 1966 to 1980. This consequently excludes the possibility of dating a water accurately, and best recharge episodes could be related to the present period (1980 to 1986), to the 1950-1980 period, or to the previous period, namely before 1950.

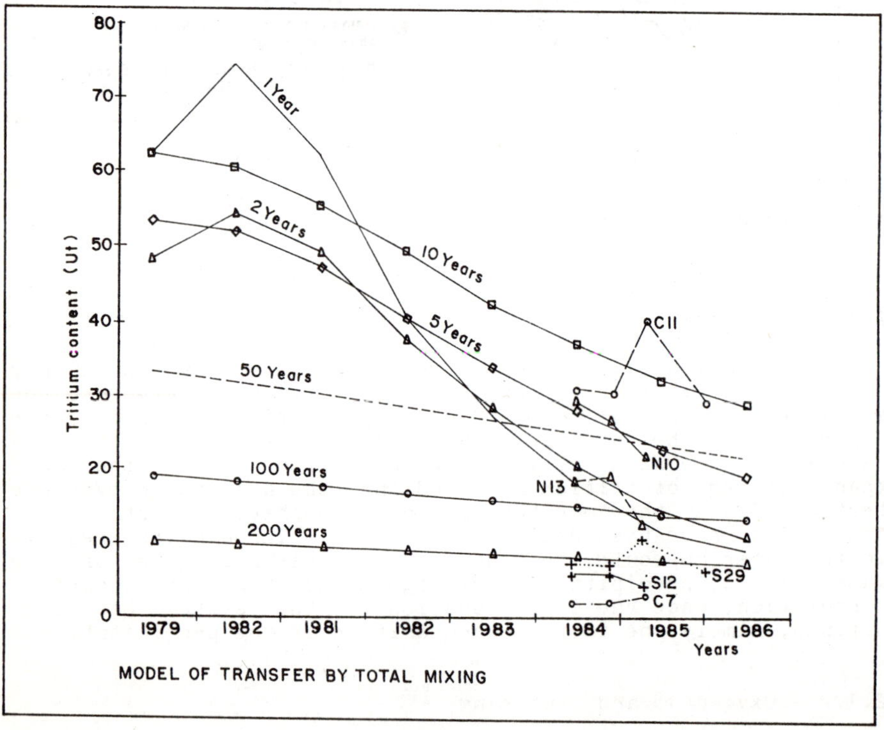

Figure 2: VARIATION IN CALCULATED TRITIUM CONTENTS OF GROUNDWATERS AS A FUNCTION OF TIME FOR DIFFERENT VALUES OF TRANSIT TIME (in years) USING A TOTAL MIXING MODEL - EXPERIMENTAL VALUES OF TRITIUM CONTENT IN SAMPLED WATERS ARE ILLUSTRATED ON THE RIGHT SIDE OF THE FIGURE SUR-IMPOSED WITH THE SIMULATED CURVES (N=North st., C=Centre st., S=South st.)

The analysis of tritium in groundwaters revealed the existence of higher contents, with a distribution histogram spread out more in the north than in the south of Burkina Faso. More recent waters are therefore roughly present in the boreholes in the north than in the south, where 72% of the samples display contents lower than 5 Ut, corresponding to pre-1950 waters. The tritium contents are generally more dispersed and lower in granites than in the other lithological facies, reflecting lower circular velocities, and hence greater mixing and residence time (figure 2).

2.2.2 - Oxygen 18 and Deuterium

The detailed analysis of the characteristics of stable isotopes in the rainfall in Bamako and N'Djamena showed that these parameters undergo an average seasonal cycle, but with very wide interannual fluctuations, which prevents any reconstruction of the values since 1979, the date of the latest published measurements. Added to the seasonal variation is another factor : an initial isotopic fractionation occurs in evaporation. In the tropical forests (Guinea, Ivory Coast, Ghana), the isotopic enrichment is not the same as in a water of oceanic origin.

As for tritium, we reconstructed the weighted annual contents of oxygen 18 and deuterium in the recharge. The analysis of the results of these calculations reveal a wide interannual variability of the values, and random differences between the values of the north and south zones. The surface water samples in this survey and during previous surveys clearly show an isotopic enrichment increasing after the rainy season.

The analysis of the relationship between deuterium and oxygen 18 in groundwaters shows that at the confidence level of 80%, no isotopic enrichment by evaporation can be identified in comparison with the rain water contents measured at N'Djamena, and especially at Bamako where the relationship is weaker. These waters therefore infiltrated very rapidly without being subjected to evaporation. The dispersion of the experimental points is wider in the northern zone than in the southern zone, where it is greater in granites than in schists (figure 3).

The spatial distribution of stable isotope contents reveals considerable heterogeneity. At each point, however, a certain stability in time reflects the fact that they result from a mixture. Between June 1984 and April 1985, however, the north zone shows a significant increase in deuterium contents and a significant drop in oxygen 18 contents.

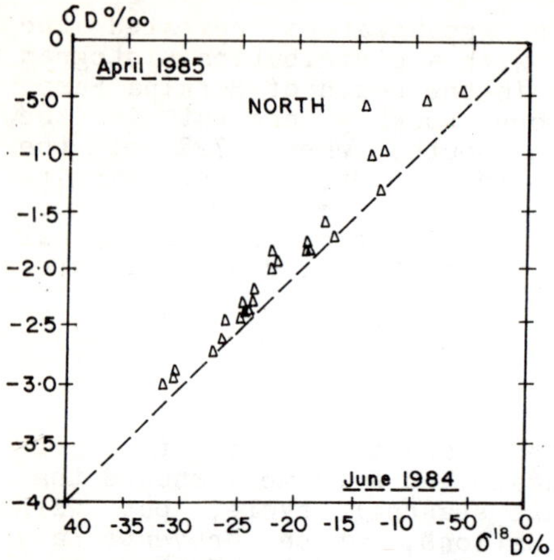

RELATIONSHIP BETWEEN FIRST THIRD SURVEYS NORTH ZONE

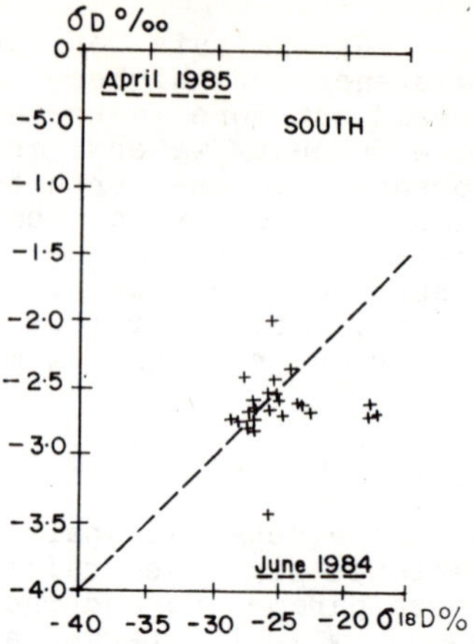

RELATIONSHIP BETWEEN FIRST AND THIRD SURVEYS, SOUTH ZONE

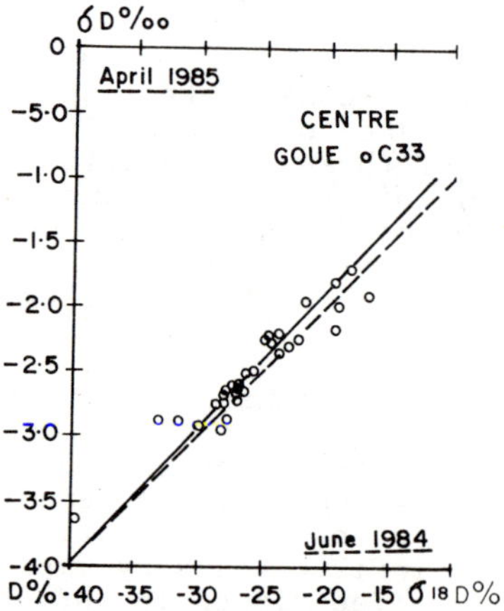

RELATIONSHIP BETWEEN FIRST AND THIRD SURVEYS, CENTRE ZONE

Figure 3: RELATIONSHIP NETWEEN DEUTERIUM CONTENTS MEASURED IN JUNE 1984 AND APRIL 1985 IN GROUNDWATERS FOR THE SOUTH, CENTRAL AND NORTH ZONES

The homogenization of the stable isotope contents observed in the south is probably due to the action of a porous filter in the circulation of the waters, in which the weathered and decayed materials play a predominant role. The differences in the dispersion of the values observed between north and south can be explained by the irregularity of the weathered facies. The weathered materials in the north show variable thickness and irregular extension. The piezometric level is located in the weathered materials or below. In the south, the weathered layer is thicker and has a more regular extension than in the north, and the piezometric level is generally located in the weathered layer.

It must be acknowledged that, whilst the isotopic analysis failed to allow a quantitative assessment of the recharge, it nevertheless offered a better understanding of the hydraulic operation of the fractured formation, and revealed the extremely important role played by weathered materials in water transfer mechanisms in aquifers and during recharge.

3 - SIMULATIONS OF RECHARGE OF EXPERIMENTAL STATIONS

3.1 - Choice of the model : the gardenia software

This is a lumped-up model that establishes the relationship between the rains and the aquifer levels, and sets the different terms of the water balance in unsteady state conditions, whilst making sure that this balance is in dynamic equilibrium.

The model consists of three superimposed reservoirs. The upper reservoir is characterized by its maximum retention capacity. It represents the retention effect of the first few metres of ground. This reservoir is supplied by rainfall and provides the water for evapotranspiration. No runoff or infiltration can be observed before it is saturated. It accounts for the effect of interception in the soil depressions and illustrates the 'valve effect' of the unsaturated soil according to its moisture content. The second reservoir allows transfer to the aquifer through the unsaturated zone, and the distribution between runoff and seepage. The higher the level in this reservoir, due to high rainfall, for example, the higher the proportion of runoff. Drainage by percolation obeys an exponential law related to the water head in this reservoir. The third reservoir, supplied with water by the intermediate reservoir, is drained by gravity in the form of a subsurface flow, obeying an exponential law.

Day by day, the total flows calculated represented the terms of the water balance : evapotranspiration, runoff, seepage and aquifer recharge. The optimal values of the parameters, which cannot be measured in the field, are determined by iterative process from the initial values, relying on the search for the best adjustment between the measured piezometric levels and the calculated piezometric levels.

The daily time step was selected for all the model calibration simulations.

3.2 - Use of the models on nine stations

The 1985 piezometric series on nine sites were simulated successfully by assigning numerical values to the storage coefficients that were close to the regional averages determined by other means, and by assigning regional values of 10 to 30 mm to the water retention capacities of the soil (figure 4).

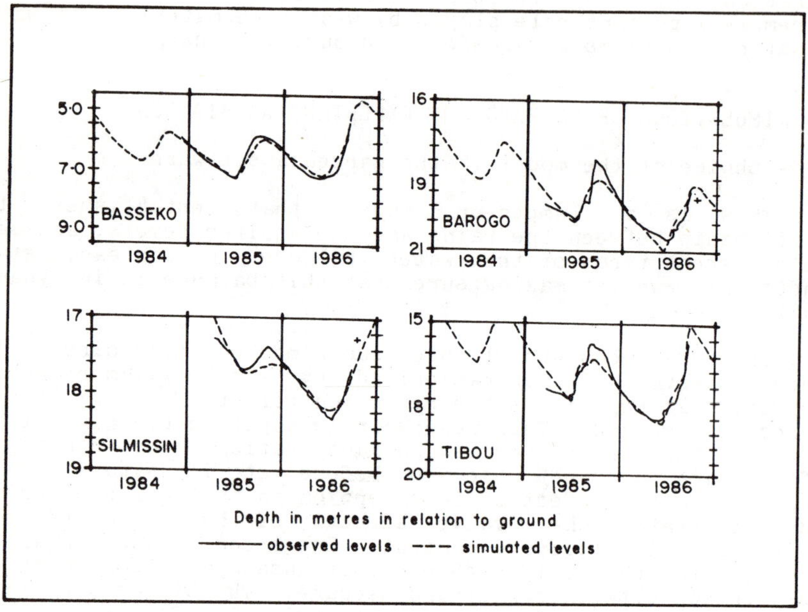

Figure 4: SIMULATION OF PIEZOMETRIC LEVELS OF STATIONS IN THE CENTRE REGION

The 1985 recharges, range between 47 and 267 mm. From one station to another, and within the same region, these water amounts vary widely. however, a spread is observed in the values, which are increasingly low from the south to the north. The variations in recharge between dry and wet years calculated by extension to the 1954/1986 period are of the same order of magnitude. By contrast, the fraction of rains accounting for recharge grows steadily from the south to the north; varying on the average from 25 to 36% in this direction.

The average runoff, which is certainly inaccurate due to the lack of comparative data on the scale of the aquifer unit, and calculated on the basis of simulations extended to the 1954/1986 period, varies from 134 mm in the south to 111 and 137 mm for two stations in the centre, and 69 mm for one station in the north. These values, which are much higher than the total runoff expressed over large basins, imply that these waters could partly have been lost by evaporation, and that only part of them reaches the large waterways (figure 5).

4 - QUANTITATIVE ESTIMATE OF RECHARGE

The simulations led to the selection of retention capacities of 10 mm for the north, 30 mm for the south, and storage coefficients spread between 1% and 5%.

It should be recalled that the recharge was determined from observations made on boreholes drilled in fractured zones, which represent permeability anomalies in a rocky environment which is of very low permeability on the whole. This siting method, which is the same in the north and the south, introduces a levelling of the calculated values of annual recharge. Nevertheless, some gradation occurs across the different latitudes, reflecting a mean of 94 mm in the north and 198 mm in the south.

The reconstruction of the level/time data obtained by extending the use of the models to the years 1954 to 1984 for four stations reveals a generalized drop in the groundwaters from 1980, especially pronounced in Barogo and Silmissin, in the central region.

The recharge evaluated for the year 1985 ranges between 47 and 125 mm for the sites in the north. The extension of the simulations to the Katchari station shows that this recharge is never zero, and the calculated minimum is 99 mm.

In the central region, this recharge ranges between 125 and 177 mm, with a simulated minimum of 130 mm for Barogo and 175 mm for Silmissin. In the south region, the values spread between 12 and 267 mm, with a simulated minimum at Sideradougou of 160 mm in 1983/84. On these four stations, the ratio recharges/total rain vary about an interannual mean as follows :

- 27% for Katchari (N),
- 31% for Silmissin (C),
- 27% for Barogo (C),
- 25% for Sideradougou (S).

The numerical results are compiled in Table 1.

5 - SUMMARY OF RESULTS AND EXPRESSION OF THE PRODUCIBLE RESOURCE

5.1 - The weathered materials are saturated at their base

When the recharge is reflected by a fluctuation in the water level situated in the weathered layers, a roughly uniform distribution of the water depth can be assumed. The annual

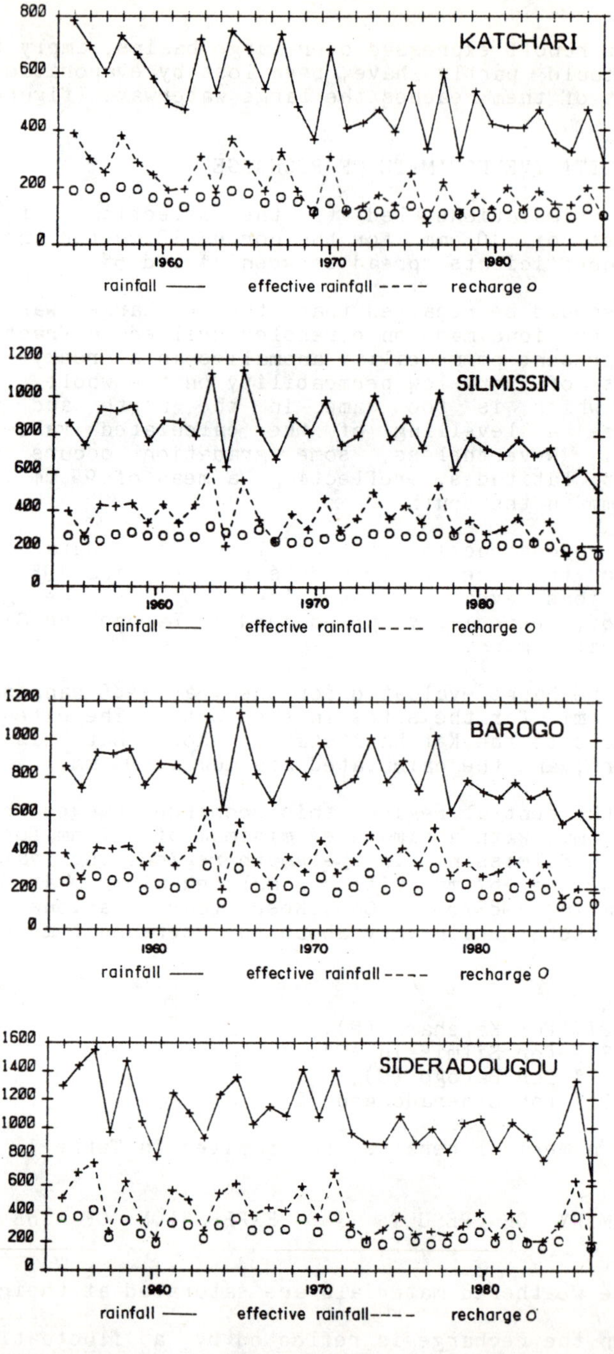

Fig: 5 — RECONSTRUCTION OF THE VARIATION IN ANNUAL RECHARGE BY EXTENSION OF DATA FOR FOUR PROJECT SITES: KATCHARI (NORTH), SILMISSIN AND BAROGO (CENTRE), AND SIDERADOUGOU (SOUTH) 1954/1986 PERIOD, ANNUAL VALUES IN mm

		SOUTH ZONE	CENTRE ZONE	NORTH ZONE
	CLIMATE	SUDAN	WET SAHEL	SAHEL
PARAMETERS	RAINFALL (53-85) (mm)	Bobo-Dioulasso m* = 1 108 cv* 19 %	Ouagadougou m = 831 cv 17 %	Dori m = 535 cv 26 %
	MEAN TEMPERATURE (59-83)	27,0°C	28,2°C	27,0°C
	POTENTIAL EVAPOTRANSPIRATION (TURC) (59-83)	m = 1 874 cv 4 %	m = 1 999 cv 4 %	m = 2 173 cv 3 %
	TOTAL FLOW INDEX (mm) (EFFECTIVE RAINFALL RFU MAX = 50 mm)	m = 395 cv 42 %	m = 214 cv 48 %	m = 83 cv 80 %
AND HYDROGEOLOGY	DEPTH OF PIEZOMETRIC LEVEL IN RELATION TO GROUND (m)	6 to 29	5 to 32	3 to 55
	SITUATION OF STATIC LEVEL IN :			
	. WEATHERED MATERIALS	91 %	88 %	53 %
	. FRACTURED BASEMENT	9 %	12 % { 83 % shales; 7 % granite	47 % { 81 % shales; 9 % granite
	WEATHERED THICKNESS (m)	15 to 50	10 to 50	8 to 50
	ELECTRIC CONDUCTIVITY OF GROUNDWATER (µmHg/cm)	178 cv 37 %	223 cv 39 %	387 cv 50 %
		increase in groundwater salinity towards the north reflecting the extent of recovery by evaporation which is higher towards the north		
	CONTENT IN 1984/1985 RAINS	estimated at 5 to 8 UT		
	CONTENTS IN GROUNDWATERS	6,2 UT cv 186 % (background at 98)	10,4 UT cv 98 % (background at 40)	25,8 UT cv 61 % (background at 81)
	. DISTRIBUTION OF CONTENTS	27 % < sensitivity threshold I UT; 72 % < 5 UT	19%< sensitivity threshold I UT; 45 % < 5 UT	2%< sensitivity threshold I UT; 9 % < 5 UT
	. 'AGE'	most waters more than 30 years old	water equally divided between more and less than 30 years	minority of waters more than 30 years old
	. GROUNDWATER CIRCULATION MODES	two types of circulation: rapid in drains (fractures), slow in auxiliary systems (weathered materials and finely-fractured basement zones)		
	CONTENT IN RAINS	Highly variable d D = 6,31 d ^{18}O + 4,19 ± 10 (IC** à 80 %) A N'Djamena : d ^{18}O varie de - 11 à + 18 et d D de - 78 à + 81 Average annual rainfall d D - 19,9 et d ^{18}O - 3,7 Effective rainfall d D - 23,2 à - 28,4 et d ^{18}O - 3,9 à - 5,0		
	CONTENTS IN GROUNDWATERS			
	. STATISTICAL AVERAGE OF CONTENTS { Deuterium	m = - 24,8 cv 16 % (- 10,8 à - 28,6)	m = - 25,5 cv 19 % (- 13,8 à - 39,5)	m = - 21,1 cv 32 % (- 5,9 à - 32)
	. STATISTICAL AVERAGE OF CONTENTS { Oxygen 18	m = - 4,5 cv 15 % (- 2,1 à - 5,4)	m = - 4,4 cv 15 % (- 2,1 à - 5,8)	m = - 3,7 cv 33 % (- 0,6 à - 5,2)
		. higher D and ^{18}O contents in the north than in the centre and south, greater variability		
	. Relationship D = f (^{18}O)	at the 80% confidence level, the groundwater contents do not reveal any isotopic enrichment by evaporation in comparison with rainwater		
	ANNUAL RECHARGE CALCULATED IN 1985	Niangoloko : 129 mm Sideradougou : 267 mm	Basseko : 177 mm Silmissin : 133 mm Barogo : 134 mm Tibou : 125 mm	Gorom-Gorom : 47 mm Aribinda : 111 mm Katchari : 125 mm
	MEAN REGIONAL RECHARGE CALCULATED IN 1985	198 mm	142 mm	94 mm

= mean value
= coefficient of variation (ratio of standard deviation to mean)
C = confidence interval

Table 1

Comparative numerical data on the three survey zones

recharges can be calculated for a given environment, i.e. constant, according to the following criteria :

- geological type,
- weathered thickness,
- morphology and drainage pattern,
- climate.

To pass from the recharge to the producible resource, a coefficient A lower than 1 must be introduced, reflecting the fact that it is impossible to completely tap the renewable resource. In the following discussion the coefficient A is assigned the value 0.5.

The recharges calculated for the stations at which the water levels reach the weathered materials are as follows (Table 2) accompanied by the annual productible yields.

TABLE 2 (A = 0,5)

Station	Type of weathered material	Calculated recharge 1954/1986 mean (mm)	minimum (mm)	1985 (mm)	Minimum producible volume per bectare* (m3/hectare per year)
NORTHERN ZONE					
- Aribinda	amphibolites and weathered shales			111	555
- Katchari	weathered tonalite	136	99	125	495
CENTRAL ZONE					
- Basseko	clays			177	885
- Silmissin	clays	251	(175)	133	665
- Barogo	clays	221	130	134	650
- Tibou	clays			125	625
SOUTHERN ZONE					
- Sideradougou	clays	266	160	267	800

* One hectare is the theoretical area influenced by the effects of pumping.

5.2 - The static level does not reach the weathered layer

This case was found at the sites of Gorom-Gorom (N) and Niangoloko (S), and generally in 47% of the boreholes in the north zone, 12% of the boreholes in the central zone, and 9% of the boreholes in the south zone.

The recording uncertainty assumed by the simulations is higher than for the foregoing cases. The formations are totally discontinuous. The borehole taps the water through the interconnected fracture network, and the reservoir for this water is exclusively the fractured and microfractured basement.

The recharges calculated for 1985 are 47 mm for Gorom-Gorom and 129 mm for Niangoloko. The renewable resource is 470 m3/hectare and 1290 m3/hectare per year respectively.

The producible resource can be deducted from the renewable resource by using the same coefficient A. However, it is highly improbable that a constant value can be assigned to this coefficient, because the effects of the discontinuity of the fractured basement are stronger. The specific features of each site are consequently increasingly important.

6 - RUNOFF

Calculated during previous surveys by the monitoring of the gauging stations placed at the outlets of the major basins, runoff is expressed in mm by the following figures in Table 3.

TABLE 3

Basin	Station	Area monitored (Km2)	Runoff related to basin area (mm)	Period
Gorouol	Dolbel	7500	18.7 to 54.7 (average 34.4)	1961/1973
Jalafanka	Belongs to the endoreic basin	0.8	138	1977/1981
Polaka	of Mare d'Oursi	9.2	52	1976/1980
Massili	Lumbila	2100	3.7 to 47.7 (average 19.3)	1956/1970

Apart from their wide variability from one year to another, the total runoff water depths can be observed to be lower if the catchment basin is smaller. The values calculated per plot are hence logically among the highest. The order of magnitude of the runoff per plot determined using the simulations, although inaccurate due to the lack of field data, is compatible with these figures.

Region	Station	Runoff per plot (average annual) (mm)	Period
north	Katchari	69	1954/1986
centre	Silmissin	111	1954/1986
centre	Barogo	137	1954/1986
south	Sideradougou	134	1954/1986

The experiments on ninety sampling locations and ten equipped stations revealed the advantages and drawbacks of the methods that had been selected to conduct this investigation, the first of its type, with this kind of funding. The need emerged to identify, observe and analyze the slow underground flows on the scale of the hydrographic basins, which, despite the low gradients, involve yields that assume a significant share in the water balances.

We now have major series of observations on the scale of the small aquifer unit, which opens up the subject of the atitude of basement aquifer formations to receive and satisfy the different groundwater tapping methods.

ACKNOWLEDGEMENT

The project was funded by the European Economic Community, the French Ministry for Cooperation and Development, the French Ministry of the Environment (through the fund for action on the quality of life), and the Italian Ministry of foreign Affairs.

The operation was entrusted to the BRGM/AQUATER Group with the participation of the Directorate of Water and Rural Facilities of Burkina Faso and the Inter-State Committee for Water Studies (CIEH). Under a contract with the French Ministry of Cooperation and Development, BRGM was charged with selecting the wells to be sampled and the sites of the pilot stations.

Under contract with the Commission of European Communities, the BRGM/ AQUATER Group, under the direction of BRGM, was charged with geophysical studies, execution and follow-up of the boreholes, supplies, installation and periodic readings of meteorological instruments, the interpretation of the overall data, and the preparation of the final report.

REFERENCES

International Atomic Energy Agency - Environmental Isotope Data World Survey of Isotope Concentration in Precipitations (1952-1983), Technical Reports Series No. 264, IAEA, Vienna (1969-1986).

Chevallier, P. (1985) - analyse et Modelisation du Ruissellement en Region Sahelienne (mare d'Oursi, Burkina Faso) - La Houille Blanche n 6/7 - pp. 574 - 580.

C.I.E.H. - Etude de la Recharge par les Pluies des Aquiferes de Milieu Fissure - Compte Rendu des Observations Recueillies sur les Stations Pilotes :

- (mai 1985) - Rapport n 1
- (novembre 1985) - Rapport n 2
- (mars 1986) - Rapport n 3

Stewart, M.K. (1975) - Stable Isotopic Fractionation due to Evaporation and Isotope Exchange of Falling Water Drops. Applications to Atmospheric Processes and Evaporation of Lakes. J. Geophys. Res. 80, pp. 1133 - 1146.

HYDROGEOLOGICAL PARAMETER ESTIMATION TECHNIQUES: LOCAL TEST, MODEL FITTING, STOCHASTIC METHODS, AND THE USE OF TRACERS

G. de MARSILY
Professor at the University Paris VI and at the Paris School of Mines, Laboratoire de Géologie Appliquée, 4, pl. Jussieu, 75252 PARIS CEDEX 05, France.
P.LACHASSAGNE
Paris School of Mines, Informatique Géologique, 35, rue St Honoré, 77305 FONTAINEBLEAU, France

ABSTRACT

For most hydrogeological problems, the permeability or transmissivity of the aquifers is the most important parameter to be determined. To do so, the hydrologist can measure a variety of magnitudes in the field, but he can never directly measure a permeability: he has to rely on a "model" to interpret the measurements and calculate a permeability or transmissivity. A very large number of "models" is available, with varying underlying assumptions, for instance : (i) rock sampling and laboratory measurements; (ii) local flow tests (slug test, injection or pump test); (iii) analysis of natural variations of water levels; (iv) inverse modelling; (v) correlation with another parameter (e.g. electrical transverse resistance, or thickness, or grain size); (vi) use of environmental or artificial tracers. Each of these methods determines a parameter value which is representative of a certain volume of rock. It is useful to first define a local parameter value (using the concept of Representative Elementary Volume, REV, or better the concept of Ensemble Averages of stochastic processes; it is clear then that such local values are frequently highly variable in natural media. One can then try to determine which type of average of these local values is estimated by each method. For instance, does a pump test estimate the arithmetic or the harmonic mean of the local transmissivity values ? What is the case for a tracer test ? Then the next question is what type of average is needed for solving the problem at hand; do we need to determine an average from several estimations, representative of a larger rock volume? Do we on the contrary need to understand the variability in space of the parameter of interest? This paper reviews the theoretical framework in which these questions are best addressed, and the methods which can be used to determine averages of local values. Geostatistical concepts are introduced and references given.

AVERAGING PERMEABILITIES

One may first recall that the averaging of permeabilities is not necessarily a straightforward operation : it is well known for instance that when different parallel layers of a porous medium with a permeability K_i and a thickness e_i are assembled, the average permeabilty K of the medium is the harmonic mean ($\Sigma e_i / K = \Sigma (e_i / K_i)$) if the flow is orthogonal to the layers, or the arithmetic mean $((\Sigma e_i)K = \Sigma(e_i K_i))$ if the flow is parallel to the layers. In a more general way, if N measurements of permeability K_i are available in an aquifer, we will call the harmonic mean $K_h = N / \Sigma (1/K_i)$, the arithmetic mean $K_a = (1/N) \Sigma K_i$, and the geometric mean $K_g = Exp(1/N \; \Sigma \; Ln \; K_i)$, i.e. the arithmetic average taken in the log space. One always has the following inequality between the three : $K_h < K_g < K_a$. It is also possible to define a unique averaging formula, as e.g. in JOURNEL et al, 1986:

$$K^m = (1/N) \Sigma K_i^m$$

with m=-1 for the harmonic average, m=+1 for the arithmetic average, and m 0 for the geometric average (at the limit), but all other possible values of m will also produce an average. The issue we would like to discuss is the type of average which is obtained from different tests integrating the permeability over a large volume of rock, and also the type of average which is needed to predict flow or transport.

It is, however, necessary to first discuss the type of distribution that is generally observed in the field. Several authors, e.g. KRUMBEIN (1936), LAW (1944), WALTON & NEIL (1963), FARENGOLTS & KOLYADA (1969), ILYIN et al (1971), JETEL (1974), FREEZE (1975), ROUSSELOT (1976), NEUMAN (1982), have observed that permeabilities are generally log-normally distributed in a given formation; this means that if the logarithm of the permeability is taken, then this magnitude has in general a Gaussian frequency, if a large number of measurements is available. If we now consider that the permeability in space is a random function, i.e. that in two different locations, the permeabilities will in general be different, and that they are independent of each other, then some theoretical results can help us to determine the type of average we need. MATHERON (1967) for instance has shown that (i) if the flow is uniform (parallel flow lines on the average), whatever the number of dimensions of the space, the distribution of the permeability and its spatial correlation, the average permeability always ranges between the harmonic mean and the arithmetic mean of the local permeabilities; (ii) if the probability distribution function of the permeability is log-normal and if the flow is uniform and in two dimensions, the average permeability is exactly equal to the geometric mean; (iii) it is not possible to define an average permeability in steady state for radial flow. Similarly, GELHAR (1976), BAKR et al (1978), GUTJAHR et al (1978) have given linearized approximations of the average permeability in uniform flow, for a log-normal distribution function of the permeability :

- in 1 dimension : $K = K_g(1 - \sigma^2/2)$
- in 2 dimensions: $K = K_g$
- in 3 dimensions: $K = K_g(1 + \sigma^2/6)$

where K_g is the geometric mean and σ^2 the variance of Ln K .

A last result concerns the dependence or independence of the different permeability measurements in space. If a large number of permeability measurements is available from different locations in space in the same formation, it is easy to observe that the permeability measurements are not independent : close to a location where a high permeability has been measured, the probability is greater that one will measure a high value again, and vice versa. However, this dependence is a function of the distance to the first measurement point, and, after a certain distance, the measurements appear in general to become independent again. MATHERON (1965, 1970,1971) has given the name of "regionalized variables" to magnitudes which display this type of behaviour in space, e.g. the concentration of metal in an ore body. They are variables typical of a phenomenon developing in space (and/or time) and possessing a certain structure. Here, the term "structure" refers to this spatial correlation which, of course, is different from one magnitude to the other and from one formation to the next. The theory of geostatistics, also developed by MATHERON, is a means of estimating regionalized variables from local measurements. It is beyond the scope of this paper to present this theory, the reader is referred to e.g. MARSILY (1986) for the application to hydrology, or JOURNEL & HUIJBREGTS (1978) for mining, or to MARSILY & AHMED (1987) for a general presentation for groundwater hydrology. Let us remember, however, that the "variogram", which can be calculated from the data, is the special tool developed to determine the spatial structure, or correlation, of the magnitude of interest.

It can be shown that the above results on the averaging of permeabilities still apply even if these are spatially correlated : the geometric mean is the correct average for two-dimensional uniform flow, and has to be weighted as a function of σ^2 for other dimensions, in the first order.

AVAILABLE METHODS FOR PERMEABILITY MEASUREMENTS

We will first briefly review here the most common methods currently used to estimate the permeability or transmissivity of a formation in the field; we will also discuss the size of the domain investigated by the tests, and the type of average obtained by the test, if the tested domain is not homogeneous.

Rock sampling and laboratory measurements.

If "undisturbed" rock samples or cores are taken from a formation, it is usually assumed that the size of the sample is on the order of the Representative Elementary Volume (BEAR, 1972), and that the permeability measured in the laboratory on the sample is representative of a "local" permeability in situ. If the sample is layered, its anisotropy can be determined by performing the test in two orthogonal directions, in order to obtain the horizontal and vertical permeabilities. Now suppose that N samples have been taken from a single well over the vertical in the formation; can we predict the average permeability (or estimate the transmissivity) of this formation ? Let us first assume that we have a continuous recording of the permeability over the well (a fully cored reservoir where all sections of the core have been tested in the laboratory, for instance). Then the local average horizontal permeability (or transmissivity) around the well will be given by the arithmetic mean of the local horizontal measurements. This is because in the immediate vicinity of the well, the flow is indeed orthogonal to the direction of the well, and the various layers act strictly in parallel. Similarly the average vertical local permeability (e.g. for the estimation of upconing) will be given by the harmonic mean of the vertical permeabilities. But we may want a more general estimate of the average permeability within the whole formation, and not just at the well bore. If the formation is strictly layered with uniform and continuous layering, the arithmetic and harmonic means will indeed be respectively the correct averages; but if this is not the case, and the formation is made of lenses of rocks of different permeabilities, with a complex geometry and interweaving, the arithmetic mean will overestimate the average horizontal permeability, since the lenses are not strictly in parallel, but sometimes the flow has to go from one lense to the next, in series. Similarly, the harmonic mean will underestimate the average vertical permeability. For uniform flow, we may then apply the theoretical results and estimate the average permeability using the geometric mean, weighted by the appropriate function of the variance of LnK, depending on the dimensionality of the flow of interest (in general, the flow is three-dimensional). This should be done separately both for the horizontal and the vertical permeabilties.

Now if the sampling is not continuous, an estimation technique taking into account the spatial correlation of the permeability will be needed. Indeed, a sample must be considered representative of an area and, in some way, the averaging must be weighted by this area. This is particularly true if the sampling intervals are not uniform. One simple way to make such an estimation is to use kriging for the mean, i.e. to look for an optimal set of weights λ_i which will give the unbiased and optimal estimate of the mean $\overline{Z}$ of the magnitude Z by $\overline{Z} = \Sigma \lambda_i Z_i$, Z_i being the measurements. The variogram of Z must first be obtained, from these measurements. Now to estimate respectively the average horizontal and vertical permeabilities at the well bore, Z will be taken as K or 1/K; kriging will thus give us directly the arithmetic and harmonic means. To estimate the permeabilities within the formation, we will take Z = LnK, and determine directly by kriging the geometric mean. We can then apply the weighting functions depending on the variance of LnK and the dimensionality of the flow, as before (note that the sill of the variogram is precisely the variance of Z). It is thus clear that the variogram of the permeability must be calculated not only for K, but also for 1/K and LnK. Note that kriging, using the appropriate K, 1/K or LnK, can also produce the estimate of K along the well bore and not only the mean.

Another thing to remember about discontinuous sampling is that in highly pervious media, where the low permeability layers are thin strata distributed in the medium and difficult to sample, the vertical permeability will generally be overestimated, whereas for impervious media, where the high permeability layers are thin strata (or fractures), the horizontal permeability will generally be underestimated, unless geophysical techniques are used to locate these anomalies.

Local flow test.

Slug tests or injection tests over a short packed section of an aquifer can be considered as giving a local value of the permeability over the section; if the portion of the well, that is tested, is long compared to its diameter, the test already gives an arithmetic average over that section. However, only the horizontal permeability will be obtained, and not the vertical one (tests to build a special triple packer system to estimate the vertical permeability in a borehole by forcing fluid vertically from one chamber to the next inside the formation have so far proved unsuccessful, BGS, 1988). The average permeability is indeed the arithmetic one since the flow is forced to be orthogonal to the well bore. If the whole aquifer thickness is tested by successive slug or injection tests in successive packed sections, one can estimate the local average permeability by the arithmetic mean of the measurements; if incomplete testing has been done, kriging in the mean using K for Z can also be applied. If the average permeability of the formation is required, and not just the one at the well bore, kriging should be done with LnK, and posssibly weighted; however, it is preferable in that case to have very short test sections, since a long test section produces an arithmetic average over that section which cannot be transformed into a series of local values to be used for estimating the geometric mean.

A very interesting case is that of the classical pump test. Let us consider two cases: steady state and transient. In steady state, the interpretation of a pump test is made with the DUPUIT (or THIEM) formula using the drawdown at two different distances from the axis, preferably two piezometers, or possibly the well and one piezometer. LACHASSAGNE (1989) has shown that the average transmissivity which is determined by such a test is close to the geometric mean of the local transmissivities of the area between the two piezometers; in particular, if the drawdown at the well is considered, the transmissivity obtained is only representative of the value in the local area surrounding the well. It is no wonder then that different values of transmissivities are obtained if different piezometers are used, since little spatial integration is done by the radial flow. Estimates of average transmissivities over larger areas should then be made using the geometric mean (without weighting factor since transmissivities are defined in 2 D) of the local values. A better estimate will be obtained if kriging for the mean is made with LnT. A word of caution must be said about the correct definition of a transmissivity obtained by a pump test. Classically, the transmissivity is defined as the average K times the thickness of the aquifer, or preferably the integral of K over this thickness. In a very short test (slug test or injection test over the whole section of the aquifer), we have seen that this is indeed what is determined by the test, since in the immediate vicinity of the well the velocity is orthogonal to the well bore. But for a longer test, where the domain tested extends radially to a piezometer, the flow is no longer strictly horizontal, unless the formation is perfectly layered : the velocity varies in magnitude and direction according to the spatial distribution of the local permeabilities; we then know from theory that the average permeability is close to the geometric mean multiplied by $(1+\sigma^2/6)$ in three dimensions. The transmissivity is then the product of this average permeability by the aquifer thickness.

However, in transient state, the picture is different. LACHASSAGNE (1989) has also shown that, whatever the position of the piezometer used for the interpretation, the value of the transmissivity that is determined will, in general, vary with time during the test, in a heterogeneous aquifer. At early times, the interpretation of the test will give an average transmissivity which is relatively close (although not identical) to the geometric mean in the area of the well; but at very long times, the transmissivity obtained from the interpretation will be very close to the actual geometric mean of the entire aquifer; it is therefore not surprising to observe, for instance in a semi-log plot, several straight lines with different slopes, each of them representative of different portions of the aquifer, where the integration over a very large area is obtained for the later lines. It may however be difficult to distinguish them from boundary effects, if the aquifer is not of very large extent, or even to select a straight line in a plot where the drawdown versus log-time is a curve with a changing slope; it must be realized however that the earliest straight line that can be distinguished in such a plot is not representative of the true transmissivity of the aquifer. Note that TOTH (1987) mentioned that in Alberta, very long pumping tests performed in the same formation ended up giving very similar values of the transmissivity, if the interpretation was made using the final portion of the plot and not the earlier one, which gave different values.

It is very important to realize that the transmissivity which is thus obtained from a long pump test is representative of an average over an area which is a function of the duration of the test, or rather of the portion of the test which is used for the interpretation. Now when such transmissivity values are later used for kriging to estimate transmissivity maps (or even an average transmissivity, when kriging for the mean), it is not acceptable to use directly data which are not representative of areas of the same size simultaneously. One must then prescribe that pumping tests in an aquifer must not only have the same duration, but also be interpreted in the same time frame. More precisely, the area investigated by a test is a function of the dimensionless time Tt/Sr^2; for a given time t, this area is thus a function of T/S. Suppose that two pump tests of the same duration are performed in an aquifer, and interpreted in the same time range (t′,t"); if T1/S1 and T2/S2 are the two values obtained from each well, and if they are significantly different, then the areas to which these values should be assigned around each well are significantly different, and their transmissivities cannot be used directly in kriging. It is thus better to re-interpret one of the tests at a different time range (t",t*) in order that the values of the ratios T1(t'+t")/S1 and T2'(t"+t*)/S2' be similar; T2' and S2' will in general be different from T2 and S2. If this is not possible, one should try to group all the tests that have been made and interpreted over a similar area, calculate a variogram for each series of tests, and convolute or deconvolute the variograms to obtain compatible variograms at each scale (see e.g. MARSILY, 1986, p. 298-299, or JOURNEL & HUILBREGTS, 1978, this problem is called "change of support" in geostatistics); an alternative is to use cokriging.

Note that as a first approximation, it is probably acceptable to consider that the radius r of the circular area around a well, which is investigated by a pump test interpreted at the time range (t',t"), is given by $r^2 = T(t'+t")/S$, based on the effect of a doublet of well in an infinite aquifer.

Analysis of natural variations of water levels.

Information on aquifer parameters can be obtained from the interpretation of natural variations of the piezometric head due to e.g. climatic variations. The most common approaches are response to barometric variations in confined aquifers (e.g. ARDITY,1978), response to a variation of the water level in a river in a water table aquifer, or response to the annual recharge. Examples can be found in DEGALLIER (1975), and are sumarized in MARSILY (1986). For barometric variations, the response observed in a well will give information on the compressibility and the local transmissivity at the well (like a slug test). With several wells, the average transmissivity can be obtained with the geometric mean, or by kriging with LnT. For water level variations due to recharge or river fluctuations, the anaytical solution used to interpret the test is in general the one-dimensional error function, which depends on the ratio T/S. The spatial integration is thus done between the piezometer and the river, or on the whole one-dimensional flow line between the watershed boundary and the river in the case of recharge. In both cases, the flow can be taken as uniform. If we can assume that the storage coefficient (or specific yield) S is more or less constant, then the averaging over the transmisssivity is really two-dimensional, and will thus give us directly the geometric mean of the transmissivity of the investigated area, and not the local value at the observation well. If we want to use kriging of these values, it is not possible to consider them as point values at each piezometer, but as average values over a large area surrounding each piezometer, in relation to the position of the boundaries. It is thus not possible to use them directly with data coming from a pump test, since their variogram will be different. This problem was already discussed in the previous section.

Inverse modelling

The inverse problem in groundwater hydrology is defined as the calibration of an aquifer model based on the available piezometric head information, together with measured transmissivities and flow rates. The reader is referred e.g. to the work of EMSELLEM & MARSILY, 1971, COOLEY, 1977, 1979, 1982, NEUMAN, 1973, 1980, NEUMAN & YAKOWITZ, 1979, NEUMAN et al, 1980, YEH et al, 1983, 1988, MARSILY et al, 1984, CARRERA, 1984, TOWNLEY & WILSON, 1985. By definition, the unknowns of the inverse problem are the values of the transmissivities of the aquifer averaged either over an element of the model, or even a zone including several elements. The size of this element or zone can be larger or smaller than the size of the area investigated by a pump test, if such pump test values are used jointly in the estimation. For the model, the average value which is needed in a zone is definitely the geometric mean of the local transmissivity, since in general the flow in a model is more or less locally uniform and two-dimensional. The following precautions should thus be taken :(i) local values of transmissivities determined from a pump test, should never be directly used in a model unless the area representative of the test is of the same order of magnitude as the area of the zone to which it is assigned; (ii) if kriging is used as a means to interpolate or fit the transmissivity (as in e.g. MARSILY et al, 1984, NEUMAN et al, 1980), one should use Z=LnT as the unknown, in order to produce directly geometric averages over the meshes or zones; kriging should be done estimating directly averages over the meshes or zones; when returning to the T values, it is neither necessary nor appropriate to modify the transformation T=Exp(Z) to account for the bias between a normal and lognormal distribution, e.g. T=Exp(Z+Var(Z)/2); doing so would deviate T from the geometric mean which is precisely the type of average we want; (iii) the variograms to be used for kriging must be different for each of the areas to which they apply : starting from the "local" transmissivity values related to slug tests (considered punctual), they define the variogram of the point values; at any other scale, the corresponding variogram is then the double integral over the areas of that scale (see e.g. MARSILY, 1986, p. 299; one can also build the variogram to be used for estimations based on measurements and estimates at different scales).

Correlation with another parameter

As measurements of transmissivities are in general not very numerous in aquifers, due to their costs, it is of great interest to try to relate the transmissivity to other parameters which can be more easily obtained in the field. Among these are electrical transverse resistance of the formation, thickness, grain size of the sediments, specific drawdown in wells, head measurements in piezometers, etc. It has been shown that the best framework in which to use jointly such different sources of information is that of geostatistics, more precisely co-kriging. The reader is referred to the work of AHMED (1987), AHMED & MARSILY (1987,1988), AHMED et al (1988). Part of this work will also be presented at this conference by AHMED.

A question which still needs to be addressed is that of the area over which such related measurements have to be taken, and the type of average which is obtained through them. For instance, if the transverse electrical resistance is measured with a Schlumberger quadripole electric sounding, what is the size of the area investigated, as a function of the size of the quadripole, and what type of average is obtained if the medium is heterogeneous ? Is the correlation between this transverse resistance and the transmissivity size dependent ? Clearly, such questions need to be addressed.

Use of tracers

Environmental or artificial tracers have often been used to determine the age of water and thus its velocity; Tritium, Radiocarbon, 36Cl, Helium, among others, have been frequently used for that purpose, e.g. HANSHAW & BACK (1974), FONTES (1982), GARNIER (1988), MICHELOT (1988), MARTY et al (1988). It is of interest to try to evaluate the type of average which is obtained when such a velocity over a long flow line is transformed into an average permeabilty by dividing it by the average hydraulic gradient; most of the time, is very difficult to reconcile such estimates of permeability with those obtained from pump test or inverse problem modelling. It is easy to show (see e.g. MATHERON & MARSILY, 1980) that if the medium is perfectly stratified and the porosity uniform, the average tracer velocity will give the arithmetic mean of the permeability if the flow is parallel to the layering, and the harmonic mean if the velocity is orthogonal to the layering. For random media, DIEULIN et al (1981) have shown that the Eulerian and Lagrangian velocities have the same probability distribution functions if the porosity is uniform (based on an unpublished work of G. MATHERON), therefore they have the same mean. Since average Eulerian velocities in a random medium represent a geometric mean of the average permeability (weighted by the dimensionality factor), an average Lagrangian velocity should give an acceptable estimate of the permeability. Where the problem really lies is in the sampling of the tracer. Measuring an average Lagrangian velocity in a well would mean dividing the well into small sections representing a flowtube, sampling the water in each section to determine the age of that water (thus the velocity along this flowpath since the outcrop), and then taking the geometric mean of these velocities to obtain the average velocity to be divided by the gradient to obtain the average permeability. But this is very seldom what is done in practice : in general, the well is pumped, and an average sample is taken, the age of which is determined. It is clear that when a well is pumped, the flow-tubes that locally have a higher velocity will produce more than the others, and the water sample will be highly biased towards the layers with a higher permeability; furthermore, at the well, we have seen that the average permeability is an arithmetic one, not a geometric one : these two circumstances show that using tracers to determine average permeabilities in a heterogeneous aquifer is not a reliable method.

CONCLUSION

In this paper, we have tried to show that the determination of aquifer parameters in the field should be made with a clear understanding of the type of average that the instruments, method or device generate automatically and almost secretely; this is of course relevant for aquifer systems which are heterogeneous, while homogeneous systems can be studied classically. Making predictions on the behaviour of the aquifers often requires another type of average than the one which is produced by the method or instrument. For that purpose, geostatistics can be of a great help to estimate both these averages as well as local values at any scale of interest (e.g. the scale of the mesh of a model). We have seen that it is important to apply geostatistics to the correct variable, e.g. K, 1/K, or LnK.

Among the methods which we have reviewed, it seems that the slug test or injection test over a small section of a well or piezometer should be considered as the basic method to determine aquifer properties in the field; such tests should then be performed at numerous locations, to determine both the type of distribution and the variogram of the local permeability values. In low permeability media, however, such tests should be continuous in order not to underestimate the permeability, which could result from a few highly conductive layers that could be missed from an incomplete samplig. These results can then be used to estimate (with the appropriate average) the local transmissivity at the well, or the regional transmissivity in the area. Long term pump test should be considered as a means to validate the averages which could be obtained from the local measurements; they are indeed very costly, and do not provide sufficient information on the variability of the system, unless a large number of piezometers are available.

The use of environmental tracers may be helpful as analogues for contaminant transport, but do not seem at present to be able to provide indirect estimates of aquifer permeability, unless the sampling techniques are completely modified.

Further work is, however, needed to validate these ideas in the field.

REFERENCES

AHMED, S. (1987) Doctoral thesis, Paris School of Mines, Fontainebleau.

AHMED, S., MARSILY, G. de (1987) Comparison of geostatistical methods for estimating transmissivity using data on transmissivity and specific capacity. Water Reour. Res., 23(9), 1717-1737.

AHMED, S., MARSILY, G. de, TALBOT, A.(1988) Combined use of hydraulic and electrical properties of an aquifer in a geostatistical estimation of transmissivity, Groundwater, 26(1).
AHMED, S., MARSILY, G. de (1988) Geostatistical estimation of aquifer transmissivity using jointly water level data. Int. Symp. Geostatistics, Avignon, France, to appear.
ARDITY, P.C. (1978) The earth tide effect on petroleum reservoirs. Ph.D. thesis, Stanford University, California.
BAKR, A.A., GELHAR, L.W., GUTJAHR, A.L., McMILLAN, J.R. (1978) Stochastic analysis of spatial variability in subsurface flow. Part 1: Comparison of one and three dimensional flows. Water Resour. Res., 14(2), 263-271.
BGS (1988) Personnal communication, British Geological survey, Nottingham, UK.
CARRERA, J. (1984) Estimation of groundwater parameters under transient and steady state conditions. Ph.D. Thesis, Univ. of Arizona, Tucson.
COOLEY, R.L. (1977) A method of estimating parameters and assessing reliability for models of steady state groundwater flow, 1, Theory and numerical property. Water Resour. Res. 13(2), 318-324.
COOLEY, R.L. (1979) A method of estimating parameters and assessing reliability for models of steady state groundwater flow, 2, Application of statistical analysis. Water Resour. Res., 15(3), 603-617.
COOLEY, R.L. (1982) Incorporation of prior information into non-linear regression groundwater flow models, Theory. Water Reour. Res., 18(4), 965-976.
DEGALLIER, R. (1975) Doctoral Dissertation (Thèse d'Etat), Université de Montpellier, France.
DIEULIN, A., MATHERON, G., MARSILY, G. de, BEAUDOIN, B. (1981) Time dependence of an equivalent dispersion coefficient for transport in porous media. Proc. Euromech 143, Delft, 1981 (Veruijt, A., Barends, F.B.J., Eds., 199-202, Balkema, Rotterdam, The Nederlands.
EMSELLEM, Y., MARSILY, G. de, (1971) An automatic solution for the inverse problem. Water Resour. Res., 7(5), 1264-1283
FARENGOLTS, Z.D., KOLYADA, M.N., (1969) Opyt primeneniyae metodo mathematcheskoy statitiki slya izutcheniya zakonov raspredeleniya gidrogeologichestikh parametrov. Trudy Vsegingeo 17,76-112.
FONTES, J.C. (1982) In J. of Hydrology.
FREEZE, A.L. (1975) A stochastic conceptual analysis of one-dimensional groudwater flow in non-uniform, homogeneous media. Water Resour. Res., 11(5), 725-741.
GARNIER, J. M. (1988) Doctoral thesis (thèses d'état), University Paris-Sud.
GELHAR, L.W. (1976) Effects of hydraulic conductivity variations on groundwater flows. Proc. Int. Symp. Stochastic Hydraulics 2nd Int. Assoc. Hydralic Res., Lund, Sweden.
GUTJAHR, A.L., GELHAR, L.W., BAKR, A.A. McMILLAN, J.R. (1978) Stochastic analysis of spatial variability in subsurface flow. Part 2: Evaluation and application. Water Resour. Res., 14(5),953-960.
HANSHAW, B., BACK, W. (1974) Determination of regional hydraulic conductivity through use of 14C dating of groundwater. Mem. Int. Assoc. Hydrogeol., Montpellier, 10(1),195-198

ILYIN,N.L., CHERNYCHEV, S.N., DZEKSTER, E.S., ZILBERG, V.S. (1971) Ostenka tochnosti opredeleniya vodopronitsayemosti gornykh porod. Nedra, Moscow.

JETEL, J. (1974) Complément régional de l'information sur les paramètres pétrophysiques en vue de l'élaboration des modèles de systèmes aquifères. Proc. Meet. Int. Assoc. Hydrogeol., Mem. 10(1), 199-203.

JOURNEL, A., HUIJBREGTS, C. (1978) Mining Geostatistics . Academic Press, New York.

JOURNEL, A.G., DEUTSCH, C.V., DESBARAT, A. (1986) Power averaging for block effective permeability. Soc. Pet. Eng. SPE 15128, 56th California regional meeting.

KRUMBEIN, W.C. (1936) Application of the logarithmic moments to size frequency distributions of sediments. J. Sedim. Petrol. 6(1), 35-47.

LACHASSAGNE, P. (1989) Thesis in progress, Paris School of Mines, Fontainebleau, to appear.

LAW, J. (1944) A statistical approach to the intersticial heterogeneity of sand reservoirs. Trans. Am. Inst. Min. Metall. Pet. Eng. 155, 202-222.

MARSILY, G. de, LAVEDAN, G., BOUCHER, M., FASANINO, G. (1984) Interpretation of interference tests in a well field using geostatistical techniques to fit the permeability distribution in a reservoir model. In: Geostatistics for Natural Resources Characterization, Part 2, (Verly et al, Eds.), Proc. NATO-ASI Ser. C, 182.

MARSILY, G. de (1986) Quantitative Hydrogeology. Groundwater Hydrology for Engineers. Academic Press, New York.

MARSILY, G. de, AHMED, S. (1987) Application of kriging techniques in groundwater hydrology. Geol. Soc. of India, 29(1), 57-82.

MARTY, B., CRIAUD, A., FOUILLAC, C. (1988) Low enthalpy geothermal fluids from the Paris sedimentary basin. 1. Characteristics and origin of gases. Geothermics, 17(4), 619-633.

MATHERON, G. (1965) Les variables régionalisées et leur estimation. Masson, Paris.

MATHERON, G. (1967) Eléments pour une théorie des milieux poreux. Masson, Paris.

MATHERON, G. (1970) La théorie des variables régionalisées et ses applications. Cahiers Cent. Morpho. Math., Paris School of Mines, Fontainebleau.

MATHERON, G. (1971) The theory of regionalized variables and its applications. Cahiers Cent. Morpho. Math., Paris School of Mines, Fontainebleau.

MATHERON, G., MARSILY, G. de (1980) Is transport in porous media always diffusive ? A counter example. Water Resour. Res., 16(5), 901-917.

MICHELOT, J.L. (1988) Doctoral thesis (tèse d'Etat), Université Paris-Sud.

NEUMAN, S.P. (1973) Calibration of distributed parameter groundwater flow models viewed as a multiple objective decision process under uncertainty. Water Resour. Res., 9(4), 1006-1021.

NEUMAN, S.P. (1980) A statistical approach to the inverse problem of aquifer hydrology. 3. Improved solution method and added perspective. Water Resour. Res., 16(2), 331-346.

NEUMAN, S.P. (1982) Statistical characterization of aquifer heterogeneities: an overview. In "Recent trends in Hydrogeology", Narasimhan, T.N. Ed., Spec. Pap. Geol. Soc. Amer. 189, 81-102.
NEUMAN, S.P., YAKOWITZ, S. (1979) A statistical approach to the inverse problem of aquifer hydrology. 1. Theory. Water Resour. Res., 15(4), 845-860.
NEUMAN, S.P., FOGG, G.E., JACOBSON, E.A. (1980) A statistical approach to the inverse problem of aquifer hydrology. 2. Case study. Water Resour. Res., 16(1), 33-58.
ROUSSELOT, D. (1976) Proposition pour une loi de distribution des perméabilités ou transmissivités. Rep. Bur. Rech. Géol. Min., Serv. Géol. Jura-Alpes, Lyon.
TOWNLEY, L.R., WILSON, J.L. (1985) Computationally efficient algorithms for parameter estimation and uncertainty propagation in numerical models of groundwater flow. Water Ressour. Res., 21.
TOTH, J.(1987) Personnal communication, Vancouver, B.C.
WALTON, W.C., NEIL, I.C. (1963) Statistical analysis of specific capacity data for a dolomite aquifer. J. Geoph. Res. 68(8), 2251-2262.
YEH, W.W., YOON, Y.S., LEE, K.S. (1983) Aquifer parameter identification with kriging and optimum parametrization. Water Resour. Res., 19(1), 225-233.
YEH, W.W., et al (1988) Water Resour. Res.

STOCHASTIC SPATIAL PREDICTION OF HYDROGEOLOGIC PARAMETERS: ROLE OF CROSS-VALIDATION IN KRIGINGS

S. AHMED AND C.P. GUPTA

National Geophysical Research Institute
Hyderabad-500007, INDIA

ABSTRACT

Stochastic treatment of spatial processes has been realized mainly with the theory of regionalized variables and it is now being used quite extensively in many fields of earth science as well as other sciences.

The main handicap in applying kriging to hydrogeological problems is the scarcity and sparsity of observed data. However, a few modifications and additions to the existing techniques permit to utilize hydrogeological data successfully in spatial prediction of aquifer parameters. One such modification in kriging techniques is called "kriging with an external drift". This technique has been found quite useful in arriving at the estimation of hydrogeologic parameters. The use of several properties of a system together in estimation has been proved to be useful both physically and mathematically.

Cross-validation, though very cumbersome and not very useful when the data are numerous as in case of mining, it is much more useful and almost necessary when there are less data as in case of hydrogeologic parameters. It is found that cross-validation checks ambiguous structural models, decides the appropriate neighbourhood, and selects the suitable combination of the additional information. The process of cross-validation and estimation have been discussed when measured values contain errors and uncertainities. It has been found possible to employ cross-validation in deciding the best external drift function when there are more than two types of additional informations. Even in the method of kriging with an external drift, a conditional variogram which is a combination of several variograms and cross-variograms should be used but practical examples show that a modified variogram of the main variable through cross-validation can be easily utilized for estimation. An important role of cross-validation is to sort the unreliable data out which are called outliers in geostatistical language.

INTRODUCTION

Knowledge of system characteristics and its stresses is essential to perform any quantitative and qualitative study. Collection of the relevant information through field or laboratory experiments is constrained by a number of factors like feasibility and cost of carrying out the experiment. Therefore spatial prediction of the parameters at the unmeasured locations is unavoidable. In geohydrology, though for aquifer modeling a spatial distribution of transmissivity close to reality is helpful, it is also used to locate the production wells. Stochastic spatial prediction can be defined as the estimation of a parameter at a point using its spatial structure (or correlation) and the measured values in neighbourhood of the unmeasured point. Sometimes, values of many parameters are used simultaneously in estimating a parameter.

KRIGING AS A STOCHASTIC SPATIAL PREDICTOR

A kriging estimate of any variable in space, in its simplest form, is the weighted average of the measured values to its surrounding which can be written as:

$$z^*(x_o) = \sum_{i=1}^{N} a_i \, z(x_i) \qquad \ldots(1)$$

where $z^*(x_o)$ is the estimated value of the variable $Z(x)$ at the point x_o (x_o being spatial coordinate in one, two or three dimension), $z(x_i)$ are the measured values at $x_i (i=1,\ldots N)$, N is the total number of measured values and a are the associated weights.

In order to calculate the weights a_i, some statistical analysis is carried out using spatial variability of the variable. As kriging estimator is linear, unbiased and optimal estimator, it should satisfy the following two conditions besides the condition of linearity (cf equation 1):

(i) Unbiasedness $$E\,[z(x_o)-z^*(x_o)] = 0 , \qquad \ldots(2)$$

which gives $$\sum_{i=1}^{N} a_i = 1 , \text{ and} \qquad \ldots(3)$$

(ii) optimality $$\text{var}\,[z(x_o)-z^*(x_o)] \ \text{min} \qquad \ldots(4)$$

developing above equation with the help of equation (1) we get a matrix relation with the kriging weights and the covariance function as follows

$$[a_i] = [Cij]^{-1} [Cio] \qquad ...(5)$$

where [Cij] is a square matrix of covariance functions between any two points x_i and x_j ; and [Cio] is a column matrix of covariance between estimation point x_o and observation points x_i. Thus the kriging estimates can be written as:

$$z^*(x_o) = [\,[Cij]^{-1}[Cio]]^T [z_i] \qquad ...(6)$$

Another information associated with the estimation is the confidence interval:

$$\sigma^2(x_o) = Coo - [[Cij]^{-1}[Cio]]^T [Cio] \qquad ...(7)$$

Matrices [Cij] and [Cio] can be written as follows

$$[Cij] = \begin{bmatrix} C11 & C12 & C13 & \cdots\cdots & C1N & 1 \\ C21 & C22 & C23 & \cdots\cdots & C2N & 1 \\ \vdots & \vdots & \vdots & \cdots\cdots & \vdots & \vdots \\ \vdots & \vdots & \vdots & \cdots\cdots & \vdots & \vdots \\ \vdots & \vdots & \vdots & \cdots\cdots & \vdots & \vdots \\ CN1 & CN2 & CN3 & \cdots\cdots & CNN & 1 \\ 1 & 1 & 1 & \cdots\cdots & 1 & 0 \end{bmatrix} \qquad ...(8),$$

$$[Cio] = \begin{bmatrix} C1o \\ C2o \\ \vdots \\ \vdots \\ \vdots \\ CNo \\ 1 \end{bmatrix} \qquad ...(9) \qquad \text{and} \quad [ai] = \begin{bmatrix} a1 \\ a2 \\ \vdots \\ \vdots \\ \vdots \\ aN \\ v \end{bmatrix} \qquad ...(10)$$

where v is the Lagrange multiplier to introduce the unbiased condition.

Delhomme(1974,76,78,79), Marsily(1986), Marsily and Ahmed (1987a), etc. have described in detail the theory and applicability of kriging methods. Herein, main emphasis will be given to the multivariate kriging and the cross- validation.

It may however, be desirable to briefly describe structural models. Following functions are usually defined as the structural model of a variable :

(a) For 2nd order stationary function, a covariance function is given by:

$$C_{ij} = C(x_i, x_j) = E\,[(z(x+h)-m)(z(x)-m)] \quad \ldots(11)$$

Another function known as centralized covariance is given by:

$$C_c(x_i, x_j) = E\,[z(x+h)\, z(x)] \quad \ldots(12)$$

(b) In case of Intrinsic random functions of zeroth order, the structural models are defined in the form of variograms as:

$$g(x_i, x_j) = (1/2)\, E\,[(z(x+h)-z(x))^2]. \quad \ldots(13)$$

(c) In case of Intrinsic Random Funcion of higher order (IRF-k), the same is defined in the form of a generalised covariance as:

$$K(h) = A_o\, \delta(h) + A_1 h + A_2 h^2 \log h + A_3 h^3 + A_4 h^5 \quad \ldots(14)$$

$$\text{where } h = |x_i - x_j|\,. \quad \ldots(15)$$

The variograms and covariances are approximately fitted with simple and smooth mathematical functions known as theoretical structural models.

To calculate the structural model, in practice, the expectation is replaced by an arithematic mean using tolerance limits as follows (cf equation 13):

$$g(h, \theta) = (1/2N(h)) \sum_{i=1}^{N(h)} (z(x_i + \tilde{h}, \tilde{\theta}) - z(x_i, \tilde{\theta})) \quad \ldots(16)$$

where $\tilde{h} = h \pm \Delta h$

$\tilde{\theta} = \theta \pm \Delta\theta$

N(h) is the number of pairs for lags ranging from $h+\Delta h$ to $h-\Delta h$. The variability is different in different directions due to anisotropic behaviour of the parameters and the structural model should be calculated in different directions separately. The pairs are taken in a particular direction in which the variability is calculated. Similarly, one could consider a range of angles from $\theta+\Delta\theta$ to $\theta-\Delta\theta$.

KRIGING WITH AN EXTERNAL DRIFT

Drift in a general sense, is a deterministic (linear quadratic or even higher order polynomial) function of spatial coordinates. These types of drift functions are called internal drift or simply drift. However, according to Matheron (1971) a drift could be a random function also and could be a function of a different variable. This type of drift is called an external drift. Sometimes, two variables are very closely related and hence one could represent the drift of the other. The expected value of a variable, knowing other variables simultaneously can be written as the conditional expection in the following manner:

$$E\,[z(x)/\ y(x)] = A + B\ y(x) \qquad \forall\ x \qquad \ldots(17)$$

Similarly if data on many variables are available a, combination of these values can be used to define the external drift of the main variable as:

$$E\,[z(x_i)/\ y_j\,(x_i)] = A_o + \sum_{j=1}^{M} A_j\, y_j\,(x_i) \quad \forall\ i\ , \qquad \ldots(18)$$

where M is number of other variables.

An estimator in this case is written as usual (equation 1) but the weights associated to the measured values will be different as the conditions of unbiasedness in this case give the following equations:

$$\sum_{i=1}^{N} a_i = 1 \qquad \ldots(19)$$

$$\sum_{i=1}^{N} a_i\, y_j\,(x_i) = y_j\,(x_o) \quad \forall\ j \qquad \ldots(20)$$

The system of kriging equations is modified as:

$$\begin{bmatrix} C11 & C12 & \ldots & C1N & 1 & y1(x1) & \ldots & yM(x1) \\ C21 & C22 & \ldots & C2N & 1 & y1(x2) & \ldots & yM(x2) \\ \vdots & \vdots & \ldots & \vdots & \vdots & \vdots & \ldots & \vdots \\ \vdots & \vdots & \ldots & \vdots & \vdots & \vdots & \ldots & \vdots \\ CN1 & CN2 & \ldots & CNN & 1 & y1(xN) & \ldots & yM(xN) \\ 1 & 1 & \ldots & 1 & 0 & 0 & \ldots & 0 \\ y1(x1) & \ldots & \ldots & y1(xN) & 0 & 0 & \ldots & 0 \\ y2(x1) & \ldots & \ldots & y2(xN) & 0 & 0 & \ldots & 0 \\ \vdots & \ldots & \ldots & \vdots & \vdots & \vdots & \ldots & \vdots \\ yM(x1) & \ldots & \ldots & yM(xN) & 0 & 0 & \ldots & 0 \end{bmatrix} \begin{bmatrix} a1 \\ a2 \\ \vdots \\ \vdots \\ aN \\ vo \\ v1 \\ v2 \\ \vdots \\ vM \end{bmatrix} = \begin{bmatrix} C1o \\ C2o \\ \vdots \\ \vdots \\ CNo \\ 1 \\ y1(xo) \\ y2(xo) \\ \vdots \\ yM(xo) \end{bmatrix} \qquad (21)$$

The expression for the variance of the estimation error can be written as:

$$\sigma^2(x_o) = Coo - \sum_{i=1}^{N} a_i C_{io} - \sum_{j=1}^{M} v_j y_j (x_o) - v_o \quad \ldots(22)$$

It can, however, be seen that the values of all the additional variables are required not only at all the estimation points but also at those locations where values of the main variable are known. In practice, some of these values may not be available and we may have to use their estimated values, though it would affect the accuracy of estimation. The estimation variance in this method is generally higher than that in Cokriging or any other multivariate kriging methods (Ahmed and Marsily, 1987a). This is due to the fact that the unbiased conditions are more (equation 20) and the data to minimize the variance are less.

Another important point to be borne in mind here is the calculation of the conditional variogram. A conditional variogram of a variable having other variables as external drift, is a function of all the individual variograms as well as the cross-variograms (Ahmed, 1987). However, most of the authors e.g., Galli and Meunier (1987), Ahmed and Marsily (1987b) etc. have used the variograms of the main variable after a slight modification through cross-validation.

CROSS-VALIDATION: A PRACTICAL TOOL

Cross-validation provide a check on reliability of inputs and the method used in estimation. Of course, one should not get mixed up with the reliability of the estimated values which also is linked with cross-validation. Cross-validation proves that the method and other inputs used are capable of reproducing the measured values. The estimated values are thus compared with corresponding measured values.

As the main purpose of cross-validation is to reproduce the observed values by estimation. A measured value is removed from the data set one by one and the same is estimated at that point using the remaining values and the estimation inputs. This is repeated for all the measurement points and following values are calculated on an average basis:

$$(1/N) \sum_{i=1}^{N} (z_i - z_i^*) \approx 0 \quad \ldots(23)$$

$$(1/N) \sum_{i=1}^{N} (z_i - Z_i^*)^2 \simeq \min \quad \ldots(24)$$

$$(1/N) \sum_{i=1}^{N} (z_i - z_i^*)^2 / \sigma_i^2 \simeq 1.0 \quad \ldots(25)$$

$$|z_i - z_i^*| / \sigma_i \leq 2.0 \ \forall \ i \quad ...(26)$$

If the estimated values are not comparable with the measured/observed values, various steps/inputs involved are progressively modified until satisfactory match is achieved.

Equation (24) gives the variance of the error (difference between the observed and the estimated values) which when divided by the kriging variance should be nearly equal to unity (equation 25). We can also take advantage of the properties of the error as they usually follow the normal distribution. Thus an estimated value to be in the 95% confidence interval should lie within the range of:

$$z = z^* \pm 2\sigma \quad ...(27)$$

$$\text{or } |z - z^*| / \sigma \leq 2 \quad ...(28)$$

Thus, one can analyze at each measured points whether the estimated values satisfy the equation (26). A correlation cloud is also drawn sometimes between estimated and the measured values. A satisfactory cross-validation test also shows that the measured values (z) and the corresponding estimated values (z^*) should be well correlated. Also, the error (z^* -z) should be uncorrelated with the values of z (Myers, 1984).

The following iterative steps may result into a fast convergence:

(i) The structural models may be calculated in different directions to infer the presence of an anisotropy. Then an experimental variogram or covariance is fitted with a theoretical one. A least-squares fitting through the points of experimental model can be done giving more weights to the point associated with large number of pairs and to one which are nearer to the origin. The value of a structural model is more important near the origin. There are certain number of mathematical function viz., spherical, exponential, gaussian etc which are authorised models to be fitted. A model type and other parameters viz., sill, nugget and range of the structural model can be approximately worked out.

(ii) Range is an important parameter of the structural model and therfore, a range is varied by ± 100%. When a range is gradually varied from a low value to higher, the value of left hand side of equation (24) decreases and after acquiring a minimum value, it again increases. The corresponding range to this minimum value gives the range which can produce the best estimation. Figure 1 shows a plot of range versus the value of equation (24) (In this case an example of log of electrical transverse resistance values in an alluvial aquifer in Tunisia have been taken for different illustrations [Ahmed et al., 1988]). The minimum value of equation (24) was obtained when range was fixed at 6 km. Thus, a range is more or less decided.

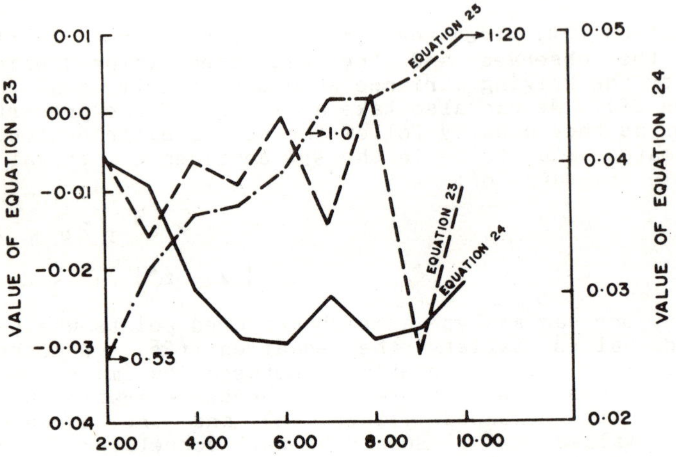

Fig. 1 Values of equations 23,24 & 25 with defferent ranges

(iii) As a matter of fact, the change in the value of sill does not effect the estimated values but changes the kriging variance. One can see that equation (24) is invariant until we change the sill by introducing or removing a nugget effect (figure 2).

In general, we can arrive at a better structural model after certain iterations which produce satisfactory values of cross-validation tests.

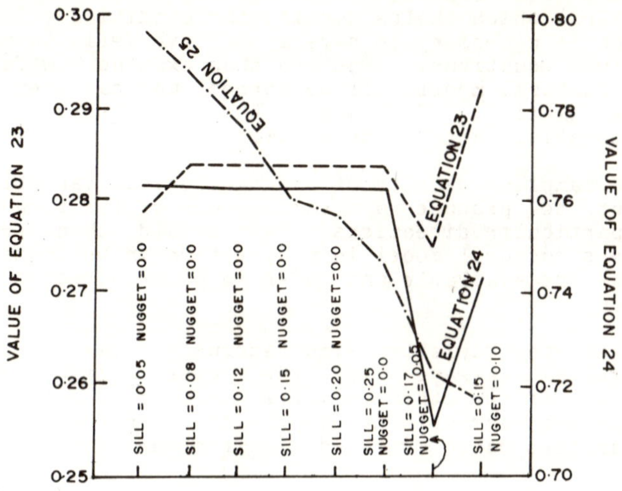

Fig. 2 Values of equations 23,24 & 25 with defferent sills

Infering a Structural Model and Checking its Ambiguity:

It may be seen that a cross-validation test checks the validity of several approximations used at different stages. Let us start with the structural model wherein at least the following three basic approximations are involved during the structural analysis of a variable:

(i) Using ergodic hypothesis, the expectation is approximated by an arithematic mean.

(ii) Instead of using an exact lag (h), a range of it with a tolerance of 25% to 50% are used. Similar is the case with the angle when an anisotropic models are calculated.

(iii) The experimental structural models viz., variogram etc., thus computed are approximated by a smooth and simple mathematical function which are fitted through the points of the experimental variogram manual or by least-squares method.

In fact the (ii) condition is necessary to have more number of pairs which helps justify the condition (i). The (iii) condition is necessary to have a positive definite matrix of the kriging equations. One can thus examine how far the theoretical structural model will be correct and can then make a reliable estimation. Such a test through cross-validation is, thus an essential part of the estimation.

Often the sampling density of hydrogeological variables is not sufficient to produce a clear picture of the structural model in any particular directions. We should then try to calculate the structural model in a direction in which maximum number of pairs are available and infer in other directions by cross-validaion.

While using many variables simultaneously, some variables are undersampled (always is the case with aquifer transmissivity). If the other variable (say Y) is linearly related to the main variable Z (say Z=a+bY) then the variogram of the main variable can be related with the variogram Y as:

$$g_z = b^2 g_y \qquad \ldots(29)$$

In this case the range of two variograms may be identical but the sill values will be different. If the variogram of Y is tested to reproduce the corresponding observed data, the variogram of Z (with same range) can be infered through cross-validation (Ahmed and Marsily, 1987b).

While using cokriging (with two variable for example), the individual variograms are first tested separately using ordinary kriging and then a cokriging cross-validation is performed to infer the cross-variogram without further modifying the other variograms (Ahmed and Marsily, 1988). Ambiguity in using nested structures in variogram models may also be tested by cross-validation.

Deciding the Optimum Neighbourhood

The observed data can be utilized in two different ways while performing kriging estimation i.e., (i) by taking all the measured values or (ii) a certain number of measured values in the nearest surrounding. The former is called the use of unique neighbourhood and the latter, the moving neighbourhood. It is not necessary to utilize all the measured values in the estimation since distant values may not influence the estimation. However, if all the measured values are utilized in estimation, the kriging matrix (equation 6) has to be inverted only once. On the contrary, in case of moving neighbourhood, the kriging matrix has to be inverted each time (equations 6 and 7 are to be rewritten). Of course, a large matrix will take a very long time for inversion. Still the uncertainty remains that how many points should be considered to have an appropriate neighbourhood? Once we decide the structural model, we can perform cross-validation using different neighbourhoods. Figure 3 shows the value of equations 23, 24 and 25 remain almost unchanged after a certain neighbourhood value. Thus an optimum neighbourhood can also be decided through cross-validation.

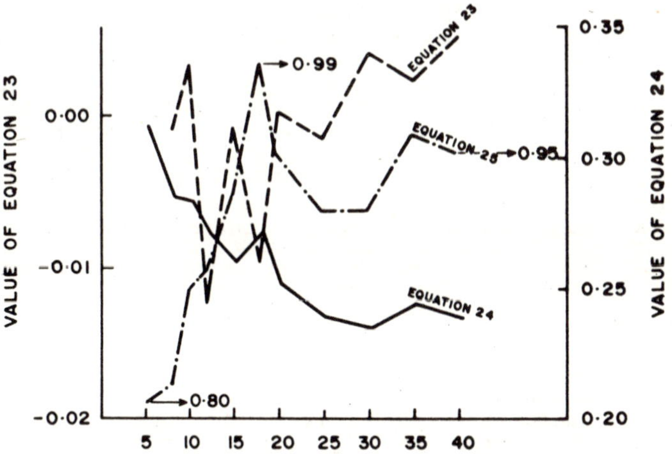

Fig. 3 Values of equation 23,24 & 25 with defferent neighbourhood

Selecting Suitable Combination of Additional Informations/ External Drift Function

Even if we assume that the structural models have been tested by reproducing the measured values in the limits given by equations 23, 24 and 25, different combinations of additional variables may also be tried and the results may be compared to select the most suitable combination of additional variables while using cokriging or a suitable external drift function while using the method of external drift. Ahmed and Marsily (1987b) have used the method of external drift in estimating aquifer transmissivity in an aquifer in the Upper Rhine valley in France. They found that even though the data on electrical transverse resistance was much more as compared to the data on specific capacity, the use of the latter as external drift yielded better results than from electrical transverse resistance.

Sorting out the Unreliable data

Instead of analyzing average statistics (given by equations 23,24 and 25), if we analyze individually (ref: equation 26), we find that at certain points the constraint set by equation (26) is not satisfied i.e., the value of equation 26 is much higher than 2 which appreciably affects the other statistics also. If we ignore them from the data set or do not take them into account while calculating the averages, the modified results satisfy all the constraints. These values are called the outliers. In the begining there may be many points which do not satisfy equation 26 but with the improvement in the structural model this number reduces. But somtimes, even after all possible trials, a few values give results beyond the prescribed limits. We may have to delete them from the data set assuming the possibility that they are the unreliable or erroneous data either wrongly measured/interpreted or wrongly located.

Use of Erroneous data

Besides sparseness of data, quite often the quality of hydrogelogical data is also poor. In such situations, it becomes necessary to appropriately account for the errors while performing any analysis.

If a measured value is erroneous, the error affects all the steps involved in the estimation. For example, the variogram calculated using this data will not give the true variogram. Let us assume that the measured value of a variable Z consists of two parts viz., true part (S) and the error (e) as:

$$Z = S + e \qquad \ldots(30)$$

the computed variogram g_z can be written as:

$$g_z = g_s + \text{var}(e) \qquad \ldots(31)$$

where g_s is the true variogram. Here we assume that the errors are uncorrelated among themselves as well as with the variable. Of course, 2nd term at the right hand side of the equation (31) is an additive term, but it is not possible to remove it as we don't know its value. This is one of the reasons of the presence of the nugget effect in the structural model. Since a nugget effect is also the result of the high variability of the variable near the origin, we cannot neglect it as suggested by De Smedt et al. (1985). However, the true variogram could be inferred by reducing the nugget effect through the process of cross-validation.

CONCLUSIONS

Estimation of geohydrological parameters in an area from a discrete set of data observed at a few points is inevitable in view of the high cost of field investigations required for data collection. The method of cross-validation can be a very potential aid to kriging geohydrological data and thereby estimate value of a parameter using data on other better known variables.

In fact cross validation has been a known technique in ascertaining correctness of the estimates of a particular variable making use of its functional relationship with some other parameters in the same domain. However its application in such cases as field of mining has been rather limited since the volume of data is huge and accordingly the computation costs are prohibitive. However, in the field of Geohydrology where the data are relatively much sparse, this technique offers a good promise. Cross-validation may also be used in checking the reliability of measured data and thus helps in identifying spurous data which could perhaps be eliminated in the kriging process.

The process of cross-validation can be especially useful in estimating certain derived parameters such as transmissivity from such data as may be directly observed for example the saturated thickness in an aquifer or data on electrical resistivity which may have even a remote symbiotic relationship.

ACKNOWLEDGEMENTS

Authors are grateful to the Director NGRI for his kind permission to present this paper. Various helps from Messers P.T. Varghese, Sunder Rao and J. Gabrial in preparing the manuscript are thankfully acknowledged.

REFERENCES

Ahmed, S., Estimation des Transmissivites des Aquiferes par Methodes Geostatistiques Multivariables et Resolution Indirect du Probleme Inverse, Doctoral thesis, Ecole des Mines de Paris, 1987.

Ahmed, S. and G. de Marsily, Comparison of Geostatistical Methods for Estimating Transmissivity using Data on Transmissivity and Specific Capacity, Water Resour. Res., 23(9): 1717-1737, 1987a.

Ahmed, S. and G. de Marsily, Multivariate Geostatistical Approach in Estimating Aquifer Parameters, In proc. Int. Symp. on "Groundwater and the Environment", S. Awadalla and I. M. Noor(eds.), Malaysia, June 22-26, 1987b.

Ahmed, S. and G. de Marsily, Some Applications of Multivariate kriging in Groundwater Hydrology, Science de la Terre, Serie Informatique, Vol 28: 1-25, 1988.

Ahmed, S., G. de Marsily and A. Talbot, Combined use of Hydraulic and Electrical Properties of an Aquifer in Geostatistical Estimation of Transmissivity, Ground Water 26(1): 78-86, 1988.

Delhomme, J.P., La Cartographie d'une grandeur physique a partir de donnees de differentes qualites, In proc. of Montpellier, France meeting, Vol 10: 185-194, 1974.

Delhomme, J.P., Application de la Theorie des Variables Regionalisees dans les Science de l'Eau, Doctoral Thesis, University of Paris VI, France, 1976.

Delhomme, J.P., Kriging in Hydrosciences, Adv. Water Resour., 1(5): 251-266, 1978.

Delhomme, J.P., Spatial Variability and Uncertainty in Groundwater flow parameters: A Geostatistical Approach, Water Resour. Res. 15(2): 269-280, 1979.

De Smedt, F., M. Belhadi and I.K. Nurul, Study of Spatial Variability of the hydraulic Conductivity with different Measurement techniques, In proc. of Internat. Symp. on "The Stochastic Approach to Subsurface Flow", edited by G. de Marsily, Montvillargenne, France, June 3-6, 1985.

Galli, A. and G. Meunier, Study of a Gas Reservoir using the External Drift Method, In Geostatistics case studies, edited by G. Matheron and M. Armstrong, D. Reidel Pub. Co., 105-120, 1987.

Marsily, G. de, Quantitative Hydrogeology: Groundwater Hydrology for Engineers, Academic Press, 1986.

Matheron, G., The Theory of Regionalised variable and its Applications, CGGM, no. 5, Ecole des Mines Publication, 1971.

Myers, D.E., Personal Communication, 1984.

GROUNDWATER FLOW IN UNSATURATED/SATURATED ZONES

H. J. Morel-Seytoux
Colorado State University
Fort Collins, Colorado 80523, USA

and Maître de Recherches
Ecole Nationale Superieure des Mines de Paris

and

C. Miracapillo
College of Engineering
University of Rome, La Sapienza, Italy

ABSTRACT

The unsaturated zone play a crucial role in the hydrologic cycle as the link between surface and ground waters. Though there are many aspects to this interaction, a discussion limited to a single aspect, that of aquifer recharge, will illustrate nevertheless most essential elements in the linkage. Typically intermittent streams (wadis) and (artificial) infiltration basins are not in permanent hydraulic connection with the underlying water-table aquifer. As water becomes available in the wadi or in the basin, infiltration proceeds. Due to the frequent presence of a clogged layer the water flow below the wadi (or basin) bed is unsaturated. Recharge occurs only after the unsaturated wetting front reaches the water-table. On the other hand, long after infiltration has ceased at the soil surface, recharge continues from the unsaturated storage in excess of field capacity, which has accumulated above the water table during the infiltration phase. The three-dimensional character of the flow is rendered by a technique that consists of matching a flow solution in a vertical cross-section with a flow pattern in a horizontal plane. Practical examples illustrate the influence of the parameters on the water table level below the infiltrating area and on the lateral recharge rate into the part of the aquifer which is not overlain by the river or basin bed. Comparison with field observations shows the technique to be accurate, inexpensive and easy to apply.

INTRODUCTION

It is rather clear that different approaches to the interaction between surface and ground waters are needed depending upon the spatial and temporal scales involved. At the extremely large scales in time and space, mass balance is the key instrument for analysis. In this paper the considered time scale is the operational one, which ranges from a few hours to a few days. The considered spatial scale is roughly that of a few typical wadi widths or aquifer thicknesses. It is a macroscopic scale much larger than the Darcy scale (say 20-50 cm) but much smaller than the regional scale (say 1 km).

There is clearly a great need for techniques of observations and for development of simple reliable measurement instruments in the unsaturated and saturated zones. Though the emphasis in this paper is on methods of analysis, that emphasis should not be misconstrued as a priority need. As much as possible theoretical analysis and field observations should be undertaken cooperatively. In fact the motivation for the reported work was the practical need (1) to estimate natural aquifer recharge from wadis occurring during seasonal floods in Saudi Arabia for the ultimate purpose of designed maximization of the recharge and its conservation and (2) to design and operate optimally artificial infiltration basins in the mediterranean area. However in this paper the discussion shall not be geared to these special cases but to general applicability. The present paper is essentially a summary of a much more comprehensive report (Morel-Seytoux and Miracapillo, 1988). For details the reader should consult the report.

The paper discusses the phenomenon of infiltration as pertinent to recharge, the infiltration-induced recharge and the drainage-induced recharge, the rise of the mound (i.e. the water table directly below the infiltrating area) and the lateral recharge, i.e. the recharge in the aquifer away from the infiltrating area. The sensitivity of the response of the system to some parameters is shown and finally a comparison of the theory with field experiments is displayed graphically. It is clear that the proposed simplified description is more than adequate and very practical.

Often at the basin or wadi bottom there is a relatively clogged thin layer of lower conductivity than the underlying soil. At time zero, when supply is initiated or the flood arrives, the ponded depth is zero. Because the crust layer is quite thin, it is assumed that it saturates in an amount of time which is negligible. For simplicity it is assumed that the shape of the water content profile below the crust (i.e. in the soil) is practically rectangular. More rigorous studies (Morel-Seytoux and Billica, 1984) have shown that this is a reasonable assumption. As time proceeds and infiltration continues the rectangular profile extends downward and usually widens.

MATHEMATICAL FORMULATION FOR PREDICTION OF INFILTRATION

The ponded depth y(t) over the wetted plane area A (see Figure 1) depends on the net supply rate to the pond or wadi reach, the infiltration rate etc. From mass conservation, change in pond storage and therefore depth is easily assessed, leading after integration to the formula:

$$y(t) = \frac{1}{A}\int_{o}^{t} Q_S(\tau)d\tau - \int_{o}^{t} e(\tau)d\tau - \int_{o}^{t} i(\tau)d\tau + y(o) \tag{1}$$

where $Q_S(t)$ is the supply discharge to the basin, e(t) is the evaporation rate and i(t) is the infiltration rate into the soil. The supply discharge $Q_S(t)$ and the evaporation rate e(t) are provided inputs. The infiltration rate depends upon the crust and soil hydraulic properties and upon the ponded depth.

Application of Morel-Seytoux's v-integral equation to a heterogeneous medium under a ponded surface condition (Morel-Seytoux, 1973, 1983) under the assumption that clogged changes only the saturated hydraulic conductivity and not the water retention curve and given that the crust is saturated, yields the expression for the infiltration rate:

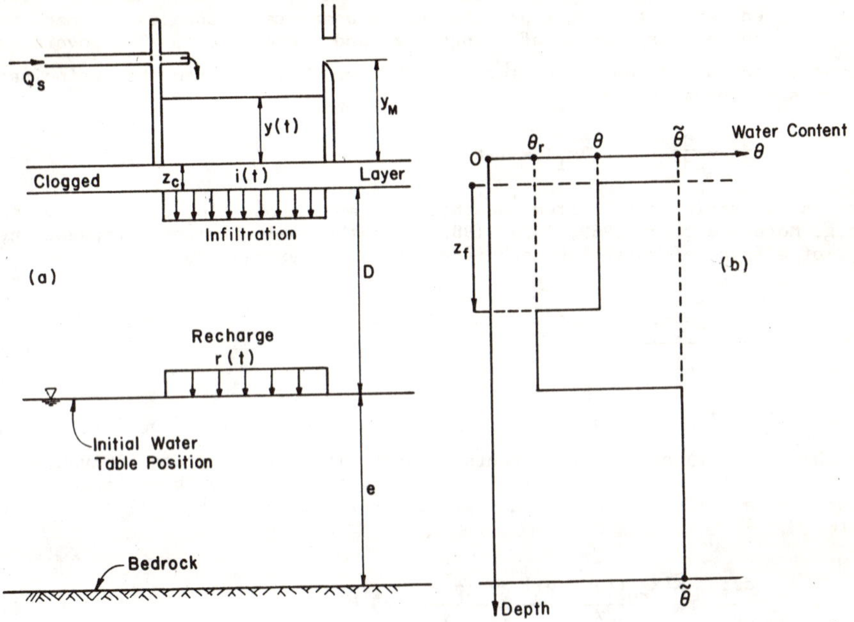

Figure 1. Geometry and water content profile for recharge system under clogged conditions.

$$i(t) = \left(\frac{\tilde{K}}{\beta}\right) \frac{\left[(\tilde{\theta}-\theta_i)\ [H_c + z_c + y] + I(t)\right]}{(\tilde{\theta}-\theta_i)\ \frac{\tilde{K}\ z_c}{\beta\ K_c} + I(t)} \tag{2}$$

where $\tilde{K}$ is hydraulic conductivity of the soil (below the crust), β is the total viscous and shape correction factor (Morel-Seytoux, 1987), $\tilde{\theta}$ is water content at natural saturation, θ_i is initial water content, H_c is the usual effective capillary drive i.e.

$$H_c = \int_0^{h_{ci}} f_w\ dh_c \tag{3}$$

with the usual two-phase flow notations (e.g. Morel-Seytoux, 1987), that is h_c is capillary pressure head, f_w is the "little-f" function of water content, h_{ci} is capillary pressure head under the initial condition of water content at some depth below the wetting front, z_c is the crust thickness, D is the depth of soil below the crust to the water table, $\tilde{K}_c$ is the saturated hydraulic conductivity of the clogged layer, and where I(t) is the cumulative infiltration depth up to time t. During the infiltration phase until the wetting front reaches the water table at a depth D below the clogged layer, the water profile in the unsaturated zone is assumed to be of rectangular shape of length z_f and width $(\theta-\theta_i)$. The governing equation for water content evolution with time, derived from the principal of mass conservation is:

$$\frac{dI}{dt} = \frac{d}{dt}[(\theta-\theta_r)z_f] = i(t) \tag{4}$$

It is a known result from the theory of multiphase flow in porous media (e.g. Morel-Seytoux, 1969, 1973, 1987) that the velocity of propagation, V_f, of a front of water contents is given by the expression:

$$V_f = \frac{v_w^+ - v_w^-}{\theta^+ - \theta^-} \tag{5}$$

where v_w^+, θ^+ and v_w^-, θ^- are the respective values of water velocities (in the Darcy sense) and water contents on both sides of the front. Given the fact that the capillary influence on the water velocity is negligible (or since for a uniform water content profile $\partial\theta/\partial z$ is zero) Eq. (5) takes the form:

$$\frac{dz_f}{dt} = \frac{\tilde{K}[k_{rw}(\theta) - o)]}{\theta - \theta_r} = \frac{\tilde{K}}{(\theta-\theta_r)}(\theta^*)^p \tag{6}$$

Equation (6) can be used to eliminate dz_f/dt from Eq. (4) and z_f can also be eliminated since $I = (\theta-\theta_r)z_f$. The resulting equation for k_{rw}, the relative permeability to water, usually expressed as a power function (of exponent p) of normalized water content $\theta^* = (\theta-\theta_r)/(\tilde{\theta}-\theta_r)$ is:

$$\frac{dk_{rw}}{(\theta^*)^{p-1}\,\theta^*(i^* - k_{rw})} = \frac{dk_{rw}}{k_{rw}(i^* - k_{rw})} = \frac{p\,\tilde{K}\,dt}{I(t)} \tag{7}$$

Integration of Eq. (7) between times t^o and t when relative permeabilities are respectively k_{rw}^o and k_{rw} and approximation of i^* and I during the time interval by a mean value over the interval, yield:

$$k_{rw} = \frac{C(t)}{C(t) - 1}\, i^* \tag{8}$$

with the function of time C(t) defined as:

$$C(t) = \frac{k_{rw}^o}{k_{rw}^o - i^*}\, e^{\frac{pi(t-t^o)}{\overline{I}}} \tag{9}$$

where $\bar{I} = I^o + 0.5\ i(t - t^o)$ and $i^* = i/\tilde{K}$ is the normalized infiltration rate. Once k_{rw} determined at time t, θ^*, and thus θ, and z_f are known.

MATHEMATICAL FORMULATION FOR PREDICTION OF RECHARGE DURING INFILTRATION PHASE

Once connection established with the water table, for times t greater than t_R, recharge occurs. Due to its unsaturated nature after the front has hit the water table, the recharge flux r splits into two parts, one reflected (filling upward the residual pore space still available after the passage of the descending unsaturated front) and one transmitted (to the aquifer not overlain by the spreading basin). In Figure 2b the true (unknown) position of the water table at time t is indicated. That profile

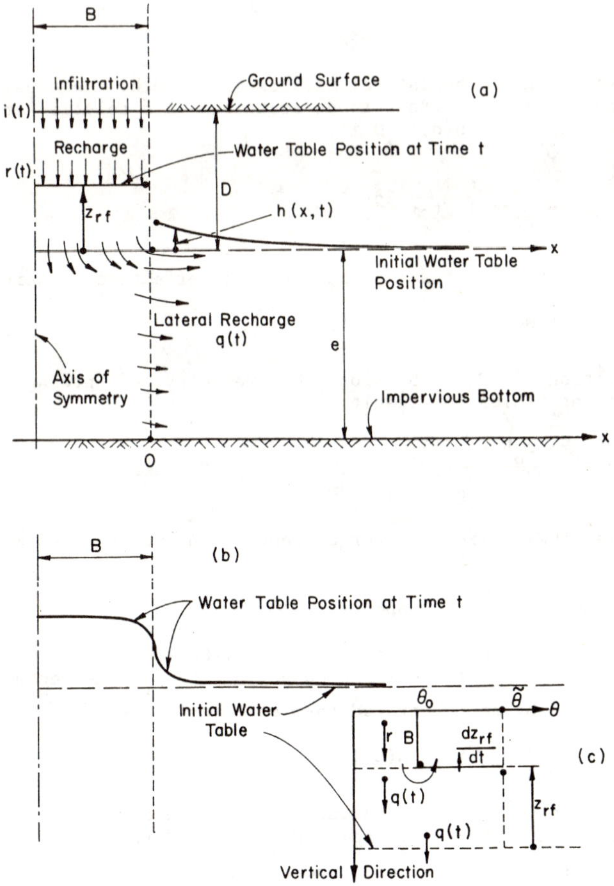

Figure 2. Schematic view of the geometry of the problem.

is approximated below the basin (Figure 2a) by a plateau at a distance z_{rf} above the datum for water table elevations and by the profile h(x,t) in the region not below the basin. The origin for the x axis is at the limit of the basin of width 2B. For reason of symmetry only one half of the system is considered and displayed. The rapid change in water table elevation in the region of sharp curvature of the flow is represented by a finite drop at x = 0 from the value $z_{rf}(t)$ to h(o,t). The problem is decomposed into two parts: (1) the determination of the evolution of lateral rise h(x,t) due an imposed lateral flux q(t) at x = 0, and (2) the evolution of the rising front $z_{rf}(t)$ given a prescribed recharge velocity r and specific discharge q(t). Note that r has dimension of a velocity whereas q(t) which is the lateral discharge per unit length in the longitudinal direction has dimension of a transmissivity. Then the two solutions which both depend on q(t) are coupled through a third equation, the matching equation, that relates the variables z_{rf}, h(o,t) and q(t).

Determination of Lateral Rise, h(x,t)

The evolution of the lateral rise is governed by the usual linear form of the one-dimensional Boussinesq equation. Its solution at x = 0 since q(o) = 0, and $z_{rf}(o)$ = h(o) = 0 is:

$$h(t) = \frac{2}{\sqrt{\phi T \pi}} \int_o^t \sqrt{t-\tau}\, \frac{\partial q(\tau)}{\partial \tau}\, d\tau \tag{10}$$

where T is horizontal transmissivity and ϕ is effective porosity.

Determination of Mound Rise, $z_{rf}(t)$

Application of Eq. (5) for the velocity of propagation of the reflected front yields the result:

$$\frac{dz_{rf}}{dt} = \frac{r_o B - q(t)}{(\bar{\theta} - \theta_o)B} \tag{11}$$

where θ is the uniform water content in the unsaturated zone above the rising mound.

Coupling of the Various Components

Using the flow net approach (e.g. Morel-Seytoux, 1985) and under the conditions that B and e are large compared with z_{rf} the recharge rate q(t) can be expressed by an integrated form of Darcy's law as:

$$q(t) = K[z_{rf}(t) - h(t)] \tag{12}$$

where K is an effective conductivity related to the hydraulic conductivities in the horizontal and vertical directions K_H and K_V and to the geometrical characteristics B and e, namely:

$$K = \frac{K_H}{1 + \frac{K_H e}{K_V B}} + \frac{K_V}{1 + \frac{K_V B}{K_H e}} = \frac{(B+e)\, K_H K_V}{BK_V + eK_H} \tag{13}$$

Substitution of the expressions for z_{rf} and h from Eqs. (11) and (10) into Eq. (12) leads to an integro-differential equation for the unknown q(t), specifically for the circular case:

$$q(t) + \frac{2K}{R} \int_{t_R}^{t} \frac{q(\tau)d\tau}{[(\bar{\theta}-\theta(\tau)]} + K \int_{t_R}^{t} K_h(R,t-\tau) \frac{\partial q(\tau)}{\partial \tau} d\tau = K \int_{t_R}^{t} \frac{r(\tau)\, d\tau}{\bar{\theta}-\theta(\tau)} \tag{14}$$

where R is the radius of the circular infiltrating area and $K_h(r,t)$ is the unit step (kernel) response of lateral rise at distance r and time t due to a lateral flux. At time t_R, $q(t_R)$ is equal to zero and at that time the recharge rate, $r(t_R) = \tilde{K}\, k_{rw}(\theta_R)$.

Water Content Equation

Let S denote the amount of water in storage in the unsaturated zone, expressed as a water depth. If z_f were constant (i.e. water table is steady) the expression of conservation of mass for the unsaturated zone would be written as:

$$\frac{dS}{dt} = \frac{d}{dt} [(\theta-\theta_r)\, z_f] = z_f \frac{d(\theta-\theta_r)}{dt} = i - \tilde{K}\, k_{rw}(\theta) \tag{15}$$

This equation can be rewritten in terms of k_{rw} as the dependent variable with the result:

$$\frac{dk_{rw}}{k_{rw}(i^* - k_{rw})} = \frac{p\,\tilde{K}}{S(t)}\, dt \tag{16}$$

Equation (16) is the same as Eq. (7) except that S(t) appears in the place of I(t). Thus a similar solution obtains.

Numerical Procedures

The numerical discrete form of Eq. (1) is:

$$h(n) = \frac{1}{A} \sum_{\nu=1}^{n} \bar{Q}_s(\nu) - \sum_{\nu=1}^{n} \bar{e}(\nu) - I(n-1) - i(n-1) \tag{17}$$

where y(n) is ponded depth at time n periods. Given the state of the system at time (n-1) and the inputs during the nth period, the right-hand side of Eq. (17) is known and y(n) can be calculated. The value of i(n) is obtained next from Eq. (2) as:

$$i(n) = \frac{\boxed{\frac{\tilde{K}}{\beta}} \left[(\tilde{\theta}-\theta_i)\ [H_c + z_c + y(n)] + I(n-1) + i(n-1) \right]}{I(n-1) + i(n-1) + \dfrac{(\tilde{\theta}-\theta_i)\ \tilde{K}\ z_c}{\beta\ K_c}} \tag{18}$$

Naturally the cumulative infiltration depth up to time n is given by the formula:

$$I(n) = (1-\lambda)\ i(o) + \sum_{\nu=1}^{n-1} i(\nu) + \lambda\ i(n) \tag{19}$$

The value $\lambda = 1$ corresponds to a fully implicit procedure, $\lambda = 0.5$ to a Crank-Nicolson scheme, etc. Next the value of k_{rw} at time n is evaluated from the modified form of Eq. (9), namely:

$$C(n) = \frac{k^o_{rw}}{k^o_{rw} - \bar{i}^*(n)}\ e^{\frac{p\ \bar{i}(n)}{\bar{S}(n)}} \tag{20}$$

where $k^o_{rw} = k_{rw}(n-1)$, $r(n-1) = \tilde{K}\ k^o_{rw}$, $\bar{i}(n) = \dfrac{i(n-1)+i(n)}{2}$, $\bar{i}^*(n) = \bar{i}(n)/\tilde{K}$, $\bar{S}(n) = S(n-1) + 1/2\ [\bar{i}(n) - r(n-1)]$. The value of k_{rw} at time n, $k_{rw}(n)$ is obtained from the modified form of Eq. (8), namely in discrete form:

$$k_{rw}(n) = \frac{C(n)}{C(n) - 1}\ \bar{i}^*(n) \tag{21}$$

Once $k_{rw}(n)$ evaluated, $\theta^*(n)$ and $\theta(n)$ are known and the recharge rate at time n periods is $r(n) = \tilde{K}\ k_{rw}(n)$.

The discrete form of Eq. (14) is:

$$q(n) + \frac{2K}{R(\tilde{\theta}-\theta_r)} \sum_{\nu=t_R+1}^{n-1} \frac{q(\nu)}{[1-\theta^*(\nu)]} + \frac{2\lambda K}{R(\tilde{\theta}-\theta_r)} \frac{q(n)}{[1-\theta^*(n)]}$$

$$+ K \sum_{\nu=t_R+1}^{n} \Delta(n-\nu+1)[q(\nu) - q(\nu-1)] = \frac{(1-\lambda)\ K\ r(t_R)}{(\tilde{\theta}-\theta_r)[1-\theta^*_R]}$$

$$+ \frac{K}{(\tilde{\theta}-\theta_r)} \sum_{\nu=t_R+1}^{n-1} \frac{r(\nu)}{[1-\theta^*(\nu)]} + \frac{\lambda K\ r(n)}{(\tilde{\theta}-\theta_r)\ [1-\theta^*(n)]} \tag{22}$$

where $\Delta(\cdot)$ is an "integrated discrete kernel" (Morel-Seytoux et al., 1988; Morel-Seytoux and Miracapillo, 1988). Again $\lambda = 1$ corresponds to a fully implicit procedure, $\lambda = 0.5$ to a Crank-Nicolson scheme, etc. The position of the reflected front at time n is given by integration of the velocity of propagation equation, namely:

$$z_{rf}(n) = z_{rf}(n-1) + (1-\lambda) \left[\frac{i(n-1) - r(n-1)}{\tilde{\theta} - \theta(n-1)}\right] + \lambda \left[\frac{i(n) - r(n)}{\tilde{\theta} - \theta(n)}\right] \quad (23)$$

Due to the various approximations involved in the procedures it is good to reevaluate storage at time n from a mass balance consideration rather than from the value of $\theta(n)$. Indeed let L(n) be the cumulative depth of lateral recharge expressed as a depth using the wetted recharge area, A, as the area to convert volume to depth. Let R(n) be the cumulative recharge depth at time n periods. Mass balance yields for S(n) the expression:

$$S(n) = I(n) - L(n) - (\tilde{\theta}-\theta_r)\, z_{rf}(n) \quad (24)$$

and $\theta(n)$ is reestimated as:

$$\theta(n) = \theta_r + S(n)/[D - z_{rf}(n)] \quad (25)$$

The complete state of the system at time n has been obtained since knowing the state of the system at time (n-1), y(n) is calculated from Eq. (17), i(n) is calculated from Eq. (18), I(n) is calculated from Eq. (19), C(n) is calculated from Eq. (20), $k_{rw}(n)$ from Eq. (21), and $\theta(n)$ is thus known and $r(n) = \tilde{K}\, k_{rw}(n)$. Naturally $R(n) = R(n-1)+(1-\lambda)\, r(n-1) + \lambda\, r(n)$. Equation (22) provides q(n). Once q(n) is known, L(n) can be calculated. In the circular geometry case since the infiltration area is πR^2 but the lateral recharge area is $2\pi Re$, then:

$$L(n) = \frac{2\pi R}{\pi R^2} \sum_{\nu=1}^{n} q(\nu) = \frac{2}{R} \sum_{\nu=1}^{n} q(\nu) \quad (26)$$

Once L(n) determined, and $z_{rf}(n)$ calculated from Eq. (23), S(n) can be calculated from Eq. (24). If desired h(n) can be calculated since:

$$h(n) = z_{rf}(n) - q(n)/K \quad (27)$$

and the final value of $\theta(n)$ is given by Eq. (25). The complete description of the system at time n has been obtained.

MATHEMATICAL FORMULATION FOR PREDICTION OF RECHARGE DURING THE DRAINAGE PHASE

In this case the previous procedures for the infiltration phase apply except that now y(t) and i(t) are identically zero for $t \geq t_D$, where t_D is the time when recharge is fed strictly from drainage of the unsaturated zone. Equations (17) and (18) are no longer used, Eq. (19) for I(n) is still valid but simply states that $I(n) = I(t_D)$ for $n \geq t_D$. C(n) is determined from Eq. (20) in which all i are zero for arguments $n \geq t_D$. Equation (21) provides $k_{rw}(n)$ and thus $\theta(n)$ and $r(n) = \tilde{K}\, k_{rw}(n)$, Eq. (22) provides q(n), Eq. (23) provides $z_{rf}(n)$, Eq. (26) provides L(n) for the circular case, Eq. (24) in which $I(n) = I(t_D)$ for $n \geq t_D$ provides S(n) and a complete cycle of calculations has been completed.

RESULTS

Influence of Anisotrophy

The results displayed on Figures 3 to 6 apply to a condition of a steady infiltration rate for 12 hours after the descending unsaturated wetting front has hit the water table. They apply for the case of a linear geometry. It is representative of what would happen in a wadi during a significant flood.

The ratio, $\rho_{HV} = K_H/K_V$, is a measure of the anisotropy of the aquifer. Typically K_H is greater than K_V. Figures 3 and 4 show, respectively for $\rho_{HV} = 1$ and $\rho_{HV} = 10$, the evolution of mound elevation below the infiltration zone, $z_{rf}(t)$, for three different cases. In one case (labeled a on the figures) the recharge rate is constant and equal to the infiltration rate over a total simulation period of 48 hours. Time is measured in all cases from the moment the descending wetting front has reached the initial water table. In the second and third cases, infiltration is discontinued at the surface after 12 hours. In the second

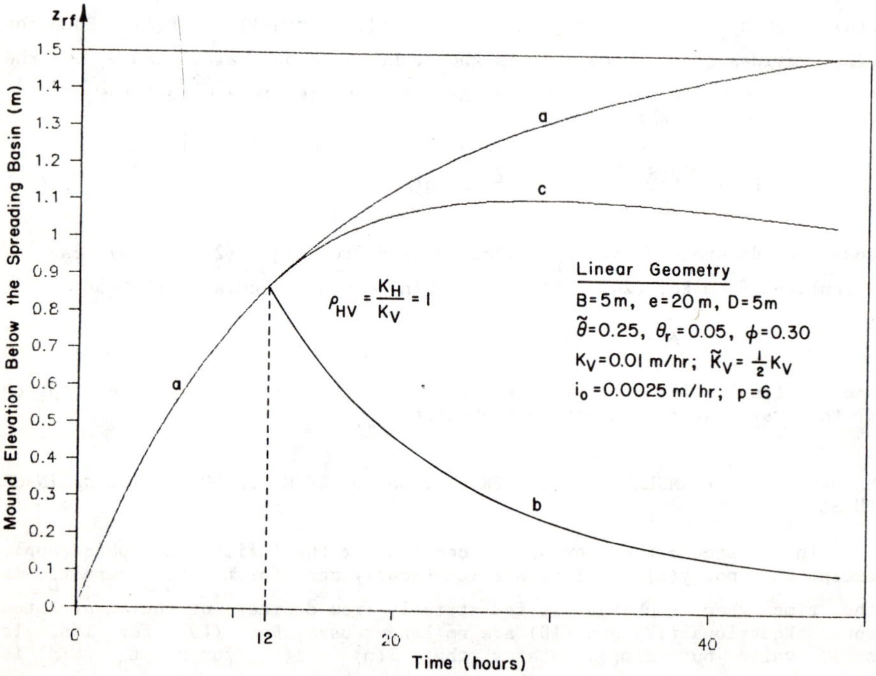

Figure 3. Evolution of the mound elevation below the basin: a - continuous steady infiltration rate, b - recession after 12 hours ignoring redistribution of moisture, c - allowing for redistribution of moisture.

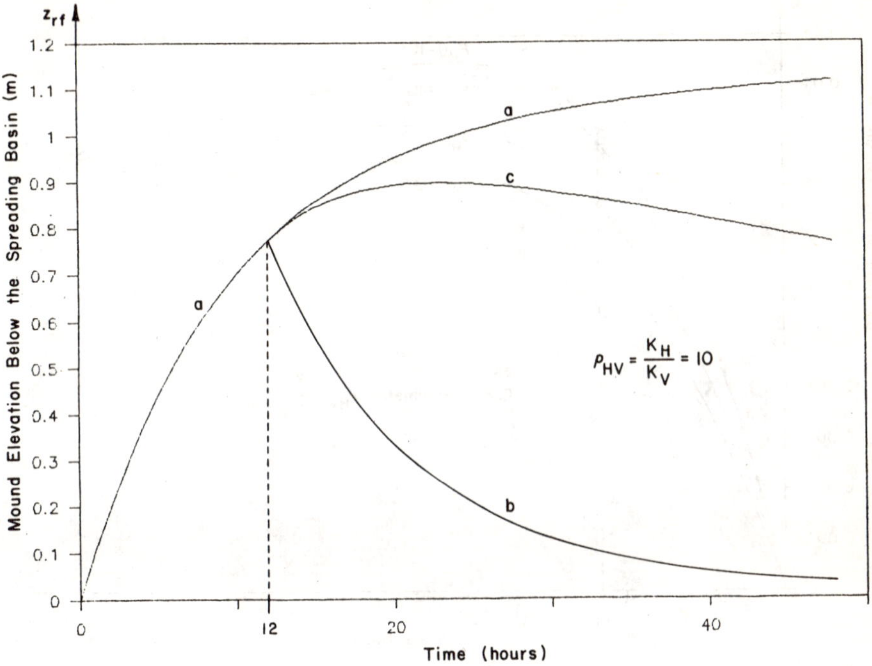

Figure 4. Evolution of the mound elevation below the basin: a - continuous steady infiltration rate, b - recession after 12 hours ignoring redistribution of moisture, c - recession after 12 hours taking into account redistribution of moisture.

case (labeled curve b in the Figures) it is assumed that water content retains its former value above the water table (i.e. θ_o = 0.2282) as if that water could not drain. It turns out that such assumption is clearly unrealistic. In the more realistic third case (labeled c in the Figures) infiltration is discontinued at the surface after 12 hours but the unsaturated zone is allowed to drain and its uniform value of water content decreases with time. It is clear from Figures 3 and 4 that curve b is a poor approximation for curve c regardless of the value of the anisotropy ratio. For this reason only results for the realistic situation of drainage from unsaturated storage are displayed in the following figures. Figures 5, and 6 show respectively plots of the lateral recharge rate q(t), and of the overall recharge rate, r(t).

Comparison with Field Experiments

The data of Bianchi and Haskell (1966, 1968) were used to test the current theory. Table 1 provides a summary of the information necessary for the theoretical calculations and for the interpretation of the field observations. Previous investigators have tried to reproduce the rising phase of the hydrograph on Figure 7 with mediocre success and essentially none attempted to model the recession. The anisotrophy ratio was used as a calibration parameter. A value of ρ_{HV} equal to 30 gave the best match for

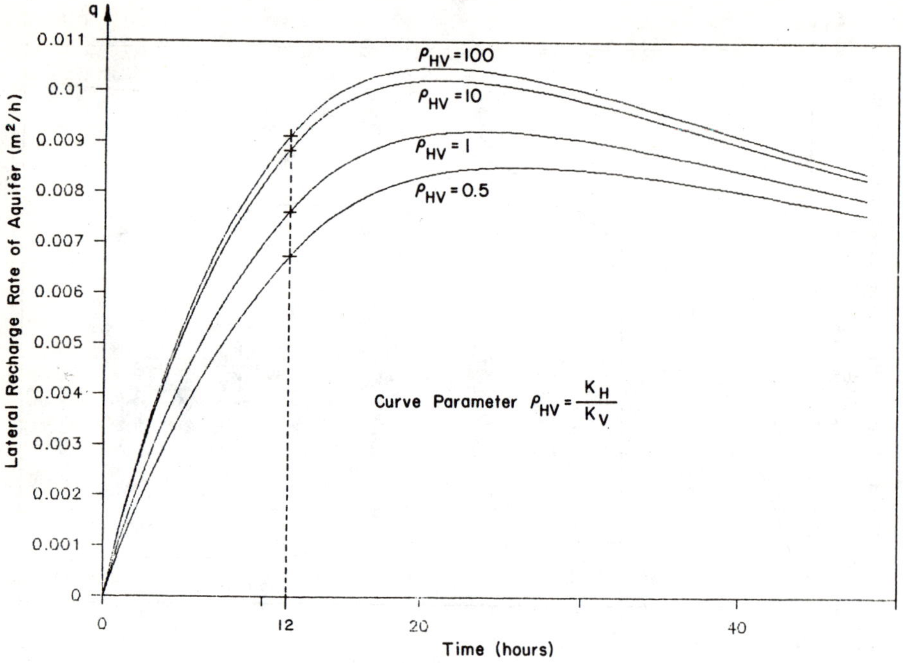

Figure 5. Evolution of the lateral recharge rate to the aquifer not below the spreading basin or wadi bed (infiltration period = 12 h).

the rising part of the hydrograph. Parameter values were not changed for the prediction of the recession which can thus be viewed as a verification run. The agreement is very good. Note that the recession prediction by Bianchi and Haskell using the Dupuit-Forcheimer assumption did not start where the prediction on the rising stage ended but at the observed value. This is why the solid line displays a jump around 32 hours. Otherwise the recession prediction would have been quite poor.

CONCLUSIONS

The method presented in this paper to predict aquifer recharge has several advantages. First of all it is simple and clearly physically based. The boundary condition for the process is the realistic one of a prescribed supply of water in the pond from which infiltration is derived. The frequent presence of a clogged layer and its effect on the infiltration rate are realistically described in the method. Second it is applicable for a complete cycle of recharge operations, that is a period of infiltration without recharge yet, a period of simultaneous infiltration and recharge and the continuation of recharge from unsaturated storage after infiltration stops. The method quite simply permits estimation of the relative amounts of water in saturated storage below the infiltrating area and away from it and in unsaturated storage between the ground surface and the transient mound elevation. This knowledge is very useful in planning the design of the recharge facility and its operations in time.

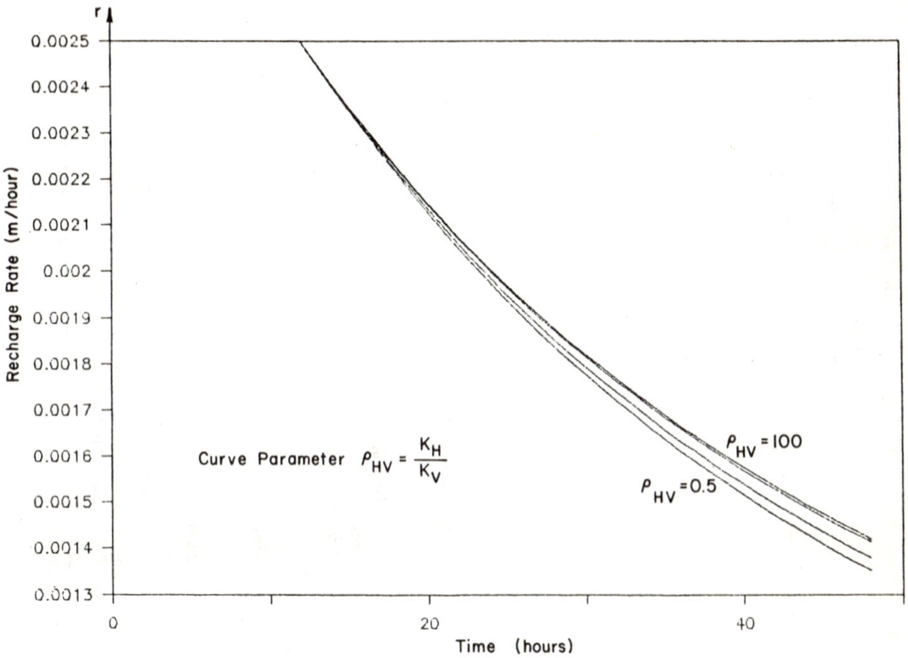

Figure 6. Evolution of recharge rate (infiltration period = 12 h).

The method accounts for the fact that below the infiltrating area the fillable pore space available to a rising water table for storage is different than that away from the infiltrating area. In fact that fillable pore space is not only different from the classical "specific yield" but varies with time as drainage occurs from the unsaturated storage above the rising or dropping water table. The method accounts very simply for anisotropy between the horizontal and vertical directions. This is a useful feature since significant anisotropy is the rule rather than the exception.

Comparison of the method with field observations is very good. In fact none of the previous approaches could match the observations well during the drainage phase and most failed also for the prediction of the mound elevation during its rise. The numerical examples show that anisotropy of the medium influences the mound elevation below the irrigated zone but much less the discharge. With a decrease in horizontal conductivity, the hydraulic head increases, as the mound height rises more rapidly. The numerical example, in fact, confirms that the effect of a lower horizontal conductivity is almost compensated by a higher hydraulic head, due to a more rapid rise of water table below the spreading basin. From the numerical examples it is clear that the role of the unsaturated zone cannot be neglected in the drainage phase as recharge of the aquifer occurs from redistribution of moisture in the unsaturated zone until a new equilibrium is established.

Table I. Physical Constants, Methods, Values and Notation

Constant	Description of the Method	Evaluation Pond #1	Pond #2	Notation*
Plot width (ft)		296	296	
Equivalent radius (ft)	From plot area	166.4	166.4	R
Saturated thickness (ft)	Well logs, electric logs, core sampling	(15)	80	e (pond 2)
Thickness of saturated zone above impeding layer (ft)	Well test	16	--	e (pond 1)
Depth from ground surface to initial water table location (ft)	Piezometers	25	21	D
Recharge rate (ft/day)	Average infiltration rate over flooding period	0.32	0.35	$\bar{i}$
Hydraulic conductivity (ft/day)	Well test data; Theis nonequilibrium method average from four lines of partially penetrating observation wells	104	26	K_H
Fillable void space	(1) Pore space minus water content before passage of wetting front; gamma ray and neutron probe techniques	0.26	0.17	ϕ
	(2) Pore space minus water content after passage of wet front; gamma ray and neutron probe techniques	0.12	0.052	$(\tilde{\theta}-\theta_o)$
	(3) Pumping test data; Theis nonequilibrium method average from four lines of partially penetrating observation wells	0.089	0.022	

*Notation refers to the method described in this paper.

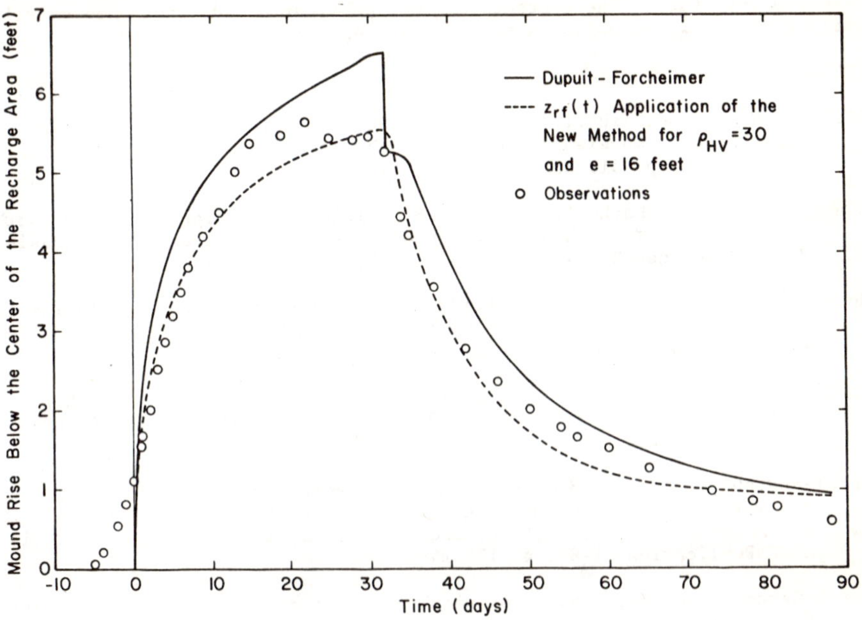

Figure 7. Pond No. 1: Application of the method and comparison with D-F theory and data (Bianchi and Haskell, 1968).

As mentioned in the introduction, only one aspect of the interaction between surface and ground waters was discussed. There was no discussion of the stream-aquifer interaction when hydraulic connection is permanent. Actually a valuable description of this situation previously published in the literature (e.g. Morel-Seytoux and Daly, 1975; Morel-Seytoux, 1985) is very similar to the one discussed in this paper. In particular the use of the "discrete kernel" approach is very useful for simulation and optimal conjunctive management of surface and ground water.

ACKNOWLEDGMENTS

The research on which this paper is based was made possible by a visit of Professor Morel-Seytoux at King Abdulaziz University in April 1987, by his year as a Professor Invité at the Ecole Polytechnique Federale of Lausanne, Institut de Genie Rural, Hydrologie et Amenagement (IGRHAM) in 1987 and by a visit of Ms. Miracapillo at Colorado State University in 1988, thanks to a doctoral scholarship from the Italian government. The authors would like to acknowledge useful discussions with Dr. Abdulrazzak and Dr. Sorman of King Abdulaziz University, Dr. Benedini of Water Research Institute (IRSA), Italian National Research Center, Professor Margaritora of College of Engineering, University of Rome, La Sapienza, and IGRHAM colleagues Prof. Musy and Dr. Mermoud. Partial financial support was also provided by the Colorado Experiment Station and by the U.S. Department of Interior, Geological Survey, through the Colorado State Water Resources

Research Institute. The contents of this publication do not necessarily reflect the view and policies of the U.S. Department of Interior.

REFERENCES

Bianchi, W. C. and Haskell, Jr., E. E. 1966. Air in the vadoze zone as it affects water movements beneath a recharge basin. Water Resour. Res. v. 2, no. 2, p. 315-322.

Bianchi, W. C. and Haskell, E. E. 1968. Field observations compared with Dupuit-Forcheimer theory for mound heights under a recharge basin. Water Resources Research. v. 4, no. 5, p. 1049-1057.

Morel-Seytoux, H. J., 1969. Introduction to flow of immiscible liquids in porous media. Chapter in "Flow Through Porous Media" R. J. M. de Wiest, editor, Academic Press, 1969, p. 455-516.

Morel-Seytoux, H. J., 1973. Two-phase flows in porous media. Chapter in "Advance in Hydroscience", v. 9, V. T. Chow, editor, Academic Press, 1973, p. 119-202.

Morel-Seytoux, H. J., 1983. Infiltration affected by air, seal, crust, ice and various sources of heterogeneity. Proc. ASAE National Conference on Advances in Infiltration, December 12-13, 1983, Chicago, Illinois, ASAE Publication 11-83, p. 132-146.

Morel-Seytoux, H. J., 1985. Conjunctive use of surface and ground waters. Chapter in "Artificial Recharge of Groundwater", T. Asano, editor, Butterworth Publishers, Boston, 1985, p. 35-68.

Morel-Seytoux, H. J., 1987. Multi-phase flows in porous media. Chapter in "Developments in Hydraulic Engineering", P. Novak, editor, Elsevier Applied Science, p. 103-174.

Morel-Seytoux, H. J. and Daly, C. J. 1975. A discrete kernel generator for stream-aquifer studies. Water Resour. Res. Jour. v. 11, no. 2, April 1975, p. 253-260.

Morel-Seytoux, H. J. and Billica, J. A. 1984. A two-phase numerical model to predict infiltration rate and water content profiles. HYDROWAR Reports Division, HYDROLOGY DAYS Publications, 1005 Country Club Road, Fort Collins, CO 80524, 109 pages.

Morel-Seytoux, H. J. and Miracapillo, C. 1988. Prediction of infiltration, mound development and aquifer recharge from a spreading basin or an intermittent stream. HYDROWAR Reports Division, Hydrology Days Publications, 1005 Country Club Road, Fort Collins, Colorado 80524, September 1988, 96 pages.

Morel-Seytoux, H. J., Miracapillo, C. and Abdulrazzak, M. J. 1988. A reductionist physical approach to unsaturated aquifer recharge. Proc. Inter. Symposium on Interaction between Ground Water and Surface Water, Ystad, Sweden, May 30-June 3, 1988. p. 181-187.

ESTIMATING GROUNDWATER RECHARGE USING SURFACE AND SOIL WATER BALANCES

J.A. Hudson, Institute of Hydrology (Plynlimon), Staylittle, Llanbrynmair, Powys, SY19 7DB, Wales, U.K..

ABSTRACT

Groundwater resources in Britain are divided between those in the chalk, limestone and permo-triassic sandstone aquifers of the south and east and the smaller, less continuous, yet locally important, drift and gravel aquifers of the upland north and west. Assessment of the worth and sustainable yields of these sources has depended on the development of techniques for estimating the rate of renewal of groundwater by recharge; these methods rely heavily on the use of surface and soil water balances. The framework, if not necessarily the detail, of these methods is of equal relevance to problems of groundwater exploitation in the developing countries. The paper attempts to assess the relative merits of the available physical, mathematical and chemical techniques, and suggests scenarios for their deployment. The main objective should be the provision of comprehensive data collection networks for monitoring precipitation, evaporation, soil moisture content and tension, transpiration, soil moisture drainage and streamflow, in order to construct spatially and temporally relevant models of groundwater recharge.

INTRODUCTION

Although groundwater supplies only 30% of the total water requirements in Britain the geological split between the hard rocks of the upland north and west of the country and the soft aquifers of the lowland south and east means that many major industrial and agricultural areas rely entirely on subsurface supplies (see figure 1). Of the remaining 70%, much is surface derived, with dry weather supplies depending on storage of winter runoff in man-made reservoirs. However, in severe conditions of drought, baseflow in streams, and hence inflows to upland reservoirs, are dependent on localised perched aquifers within topographical catchments. Many smaller supplies depend on wells sunk into these stores or, alternatively, water pumped from floodplain gravels recharged from river baseflow (see also Davey, 1982).

The importance of subsurface supplies in Britain has been underpinned by their traditional high quality. Filtration, ion exchange and microbiological action within the soil and unsaturated zones act as a natural water treatment plant, while the calcareous nature of many aquifers acts as a natural buffer to the acid rainfall falling over much of Britain. Moreover, abstraction from many of these groundwater sources has been a cheap source of supply because of the vast resources of cleanwater residing only short distances below the ground surface and the semi-artesian nature of many of the more important sources, for instance those underlying the London Basin (Marsh & Davies, 1983). More recently, however, these supplies have come under pressure. Increasing demand for water has led to greater exploitation of near surface supplies, while increased demand for arable produce has led to larger amounts of inorganic fertilizer and pesticides being applied to soil, and a corresponding rise in concentrations of these materials being found in groundwater supplies.

Figure 1. Distribution of the major aquifers in Britain

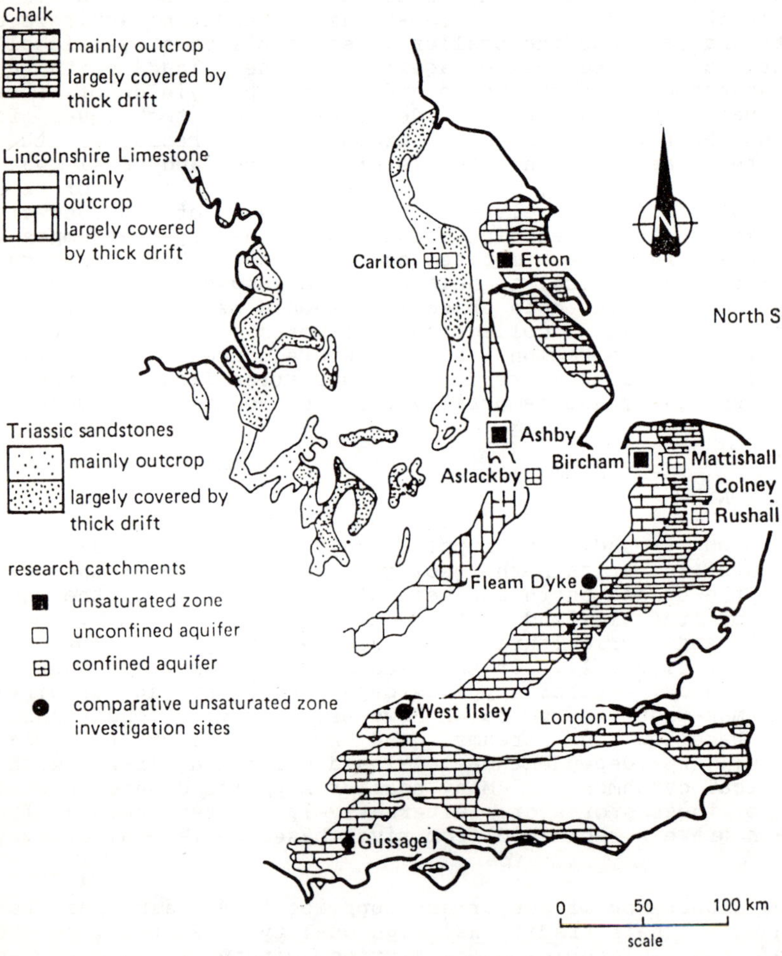

The twin threat of pollution and over-exploitation has led to the need for opening up new areas of groundwater and deeper drilling in existing areas of supply. In addition, patterns of land use are changing as a response to demographic and economic factors such as industrial and domestic development on agricultural land. Set-aside of agricultural land, because of overproduction of food stuffs within the European Economic Community, is becoming a real possibility. The future of this set-aside land is unclear at present but may well involve increased planting of woodland or reversion to scrub. This may be beneficial in terms of water quality but may also lead to a reduction in available resources because of changes in evaporation rates from the new vegetation.

Traditionally, the exploitation of water resources, particularly groundwater, has been haphazard and demand led, rather than relying on sound scientific principles backed up by comprehensive monitoring of existing supplies and hydrological surveying of potential future supplies. Over the last twenty years the picture in Britain has changed radically, firstly with the setting up of the surface and groundwater hydrology units of the Natural Environment Research Council, and subsequently with the formation of the Water Authorities under the 1973 Water Act. For the first time, a co-ordinated programme of resources research and exploitation became possible, in particular inter-basin response to variable demand and problem solving. Implementation of these schemes, depended to a large extent on the collection and interpretation of relevant data on recharge to groundwater stores, forecasts of demand and the effects of changes in basin land use on quantity and quality of supply.

With this need as the catalyst, the last twenty years has seen the development of numerous new research techniques designed to fill this information gap. This paper takes a critical view of the relevant advantages and disadvantages of the various techniques, concentrating in particular on the transportability of the technology and expertise to developing countries, where basic hydrological information such as precipitation, river flow, soil moisture, water table level and existing rates of exploitation may not be known to any degree of accuracy. On account of remoteness, the impracticality of installing the dense networks of hydrometric equipment common in industrialised countries, and the lack of access to adequate computing facilities, many of the more sophisticated groundwater recharge estimation techniques, particularly mathematical models, have to be simplified to cope with sparse data (see McCulloch, 1969; Piper et al., 1979).

HYDROMETRIC REQUIREMENTS FOR RECHARGE ESTIMATION

Underlying all hydrological techniques is the basic concept of conservation of water in its gaseous, liquid and solid phases. The water balance technique appears in many guises, either implicitly in physical models of hydrological processes or explicitly, in that independent water balance data from selected basins or geological units are used to calibrate and optimise models prior to their application elsewhere. Inevitably, when models are transported, they still require input of measured data from one or more of the fluxes that go to make up the water balance.

In terms of hydrological research in Britain, the water balance method has most often been used to estimate evaporation rates from watertight upland catchments (IH, 1976; Hudson, 1988), although not without difficulties and pitfalls (Newson,1979). Clay-covered catchments in lowland southern England have also received the same treatment successfully (Edwards & Rodda, 1970). Estimates of evaporation are gained from use of the equation:

$$AE = P - Q - dS$$

where AE is actual evaporation, P is precipitation input, Q is surface flow from the catchment, and dS is the change in storage within the catchment. In geologically watertight catchments, when the water balance is accumulated over long periods, the change in storage term (dS) tends to insignificance, which greatly simplifies the calculation. For this reason, and also to take into account long term variations in the climate, water balances in research catchments tend to be more accurate, and to have greater value, when run for decades rather than years.

When specific water resource problems arise it is usually too late to install and run, for a sufficiently long period, the hydrometric networks required to answer the questions posed. Invariably, it will be models developed over the long term in research catchments that have to be used to predict the likely effects of the courses of action available. This highlights the need for the setting up of densely instrumented research catchments, representative of all major geological, vegetational and climatic units. This is one of the major objectives that have occupied the activities of research organisations in the U.K. and elsewhere for the last two decades, as part of the International Hydrological Decade, 1965-1974 and, more recently, its successor, the International Hydrological Programme (see Rodda, 1988).

The quantification of the main characteristics of basin hydrology can only be achieved by long term research, yet the exploitation of resources depends on a day to day knowledge of the state of surface or ground water reservoirs. Groundwater recharge models based on the water balance must therefore be able to cope with shorter time scales than the decades preferred for research studies, preferably down to sub-daily time increments. In such circumstances changes in catchment storage cannot be ignored. In hard rock, watertight catchments, two storage terms need to be considered, dS and dG, which refer to the change in storage in the soil profile and underlying aquifers respectively (a positive value indicates recharge to the store and a negative value discharge from the store). The water balance equation then becomes:

$$AE = P - Q - dS - dG$$

To use this equation to estimate evaporation rates, precipitation and discharge in rivers must be estimated to a high degree of accuracy. Soil moisture and groundwater storage change only become important for within-year periods although, following extreme conditions of drought, soil moisture deficits can persist across annual boundaries. Groundwater and soil moisture may also exhibit long term trends, particularly if groundwater is over-exploited or soil structure is altered by, for example, ploughing or changes in cover crop. The above equation therefore assumes a certain continuity of land use and persistence of climatic pattern within the catchment when it is used as a basis for research models of evaporation.

INPUT ESTIMATION

When using surface and soil water balances to estimate groundwater recharge, probably the biggest source of error is introduced during estimation of the input of rain and snow to the soil and canopy store. There is a quantum leap from the simple expedient of placing a collector of known orifice size on a site and measuring the volume caught, to designing and siting a rain gauge to catch the precipitation, snow as well as rain, that would have fallen at the point had the gauge not been there. The problem is compounded by having to install networks of gauges, sometimes of more than one type to best collect different forms of precipitation, in order to sample adequately the spatial variation in rainfall to a prescribed accuracy. For water balance residuals, such as evaporation or groundwater recharge, to mean anything in accuracy terms, all the other terms in the equation have to be estimated to within 5% of their "true" value.

Attempts to attain this sort of accuracy in research catchments in upland Britain, which are characterised by uneven topography, high wind speeds and mixed liquid and solid precipitation, has led to pioneering work with the ground level raingauge. Use of this gauge was made necessary by the known tendency for U.K. Meteorological Office standard gauges, (or any gauge set above ground level), to undercatch relative to true precipitation, proportionally to wind speed at the gauge orifice (Rodda, 1967). However, in times of snowfall, the ground level gauge overcatches relative to "true" precipitation (Clarke et al., 1973), and parallel networks of sheltered standard gauges or shielded snow gauges are recommended for year round operation (see Blackie et al., 1986). Precipitation is known to vary dramatically over short distances in areas of uneven topography, and point rainfall is a function of: altitude, distance in the direction of the prevailing wind (a vector combination of latitude and longitude), whether on a windward or leeward site (aspect) and, to a lesser extent, ground slope. To cover this spatial variation, domains sampling all areally significant combinations of each of these characteristics have to be gauged, and this has given rise to precipitation gauge networks of density as high as 1 gauge per 48ha, (Hudson, 1988; Clarke et al., 1973; Blackie et al., 1986).

Clearly this density of instrumentation is only practical in research catchments, and would certainly not be feasible in remote areas of developing countries. Fortunately, experience with these dense networks of gauges has shown that, under most circumstances, less dense networks will give accuracy which is still probably better than can be achieved with other variables in the water balance equation (Richards, 1975; IH, 1976). This is particularly relevant in areas of the flat or gently sloping ground with which some aquifers are associated, but caution must be taken with topographically-rough outcrop aquifers where rainfall can be very variable.

SURFACE FLOW MEASUREMENT FOR RECHARGE STUDIES

Clearly, not all the rainfall that reaches the ground surface goes into the soil moisture store, and not all the drainage water that percolates beyond the rooting zone runs directly to aquifers. In areas of high permeability such as that encountered in the volcanic series of the Chyulu aquifer (Gunston, 1988) in Kenya, lateral drainage in the unsaturated zone is non existent and this simplifies the calculation of recharge to a simple soil model. However in most situations, particularly where soils are lying on steep slopes in topographic watersheds, lateral flow predominates over vertical flow, so some water will run to channels and exit the surface catchment without ever becoming part of the ground water system. To complicate matters, a variable proportion of streamflow will be made up of piezometric drainage from groundwater, so the streamflow estimate cannot be used as measured as a flux in a soil balance model, or else some water will be counted twice.

The first priority in establishing catchment balances of this type is in estimating streamflow accurately; the second is in devising techniques of splitting the hydrograph volumetrically into rapid surface runoff, throughflow (lateral drainage) and groundwater flow (baseflow). Continuous and accurate measurements of streamflow depend to a large extent on the resources applied and the effort, care and accuracy put into the stream gauging technique. Permanent structures of concrete and/or metal are preferred for the long term flow gauging required for water balance studies. Structure design depends on the channel slope and sinuosity, its flow range, the calibre of bedload in the stream and on the practical considerations of cost, access and availability of skilled labour for construction and maintenance. A review of streamflow structures suitable for upland areas is given by Reynolds, Hudson & Leeks (1989), detailed design criteria are available from various British Standard Institution publications, (summarised in Ackers et al.,

1978), and a comprehensive guide to design, maintenance and calibration of one particular set of instruments is given by Smart (1977).

Methods of baseflow separation are numerous and many are highly subjective. Most methods are stochastic and therefore independent of any foreknowledge of the catchment behaviour, although the best probably have a conceptual input of the point on the recession where true baseflow takes over completely from surface runoff and throughflow. If baseflow can be assessed with confidence, then the relationship between flow and available groundwater storage will be unique for a particular groundwater system, given no change over time in soil or aquifer structure. This unique relationship between streamflow and storage has to be descibed mathematically before it can form part of a water balance estimate of recharge. Conventional wisdom suggests that, in a simple system, the storage is linearly proportional to the logarithm of flow. Most systems, however, are not composed of single stores but of many stores which dry up successively over the hydrograph recession. A situation of this kind was evident from a study of recession curves in the Wye, a grassland catchment being studied as part of the Plynlimon experiment, mid-Wales, U.K. (see figure 2).

Figure 2. Baseflow recessions for the Wye catchment, Plynlimon, showing recession and aquifer storage coefficients.

K VALUES (Recession Slope) $K = \frac{\log_e Q_b - \log_e Q_0}{t}$ Sec^{-1}

BASE FLOW RANGE (cumecs)	Recession 1 1/3/76 → 6/3/76	Recession 2 15/4/76 → 30/4/76	Recession 3 21/6/76 → 3/7/76	Recession 4 21/7/76 → 28/8/76	Recession 5 30/8/76 → 7/9/76	Recession 6 18/11/76 → 25/11/76
> 0·1	$\cdot 8627 \times 10^{-6}$	$1{\cdot}3962 \times 10^{-6}$				$1{\cdot}4125 \times 10^{-6}$
0·06 → 0·1		$\cdot 6733 \times 10^{-6}$	$1{\cdot}5327 \times 10^{-6}$			
0·035 → 0·06			$\cdot 6572 \times 10^{-6}$	$\cdot 9355 \times 10^{-6}$		
< 0·035				$\cdot 1623 \times 10^{-6}$	$\cdot 2809 \times 10^{-6}$	

O Rainfall During Period

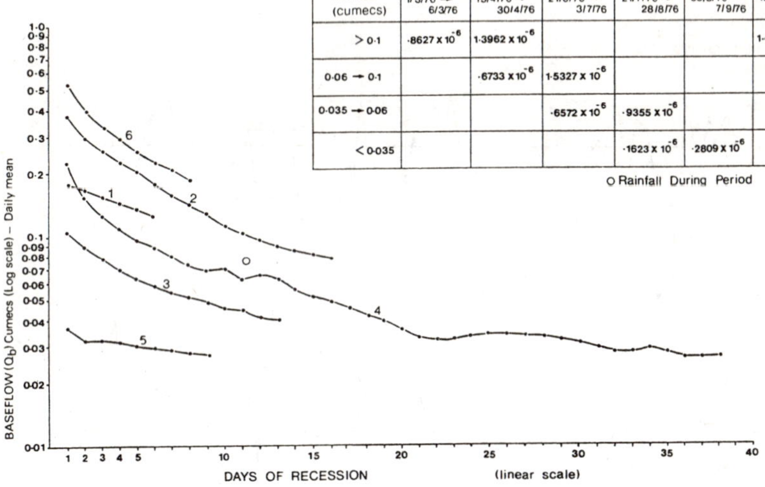

At the outset of the Plynlimon experiment, in the mid 1960s, it was thought that storage in upland catchments was minimal and probably confined to soil moisture and perched groundwater lenses in peat bogs which could be measured by soil moisture techniques. Attempts to distribute known annual evaporation rates over the year, by introducing a soil moisture change term, left large errors, resulting in evaporation being overestimated in winter (wet season) and underestimated in summer (dry season), yet annual estimates of evaporation by water balance and meteorological methods were in reasonable agreement (Hudson, 1988). The discrepancies could not be explained in terms of energy balance errors, as had been suggested by Edwards and Rodda (1970) for a clay catchment in eastern England. The conclusion was that storage in the Wye catchment was roughly twice that measured in soil moisture, and that the discrepancy represented recharge to and discharge from groundwater stores (see figure 3). Futhermore, this finding highlighted the importance of groundwater storage for maintaining baseflow in areas hitherto considered predominantly surface water catchments. This has implications, not only for many small borehole sources of water in the British uplands, but also for recharge of surface water reservoirs in drought periods and for recharge of river gravels used extensively for water supply. More recently this finding has also indicated the importance of groundwater chemistry as a control of streamflow quality, particularly with regard to the increased base saturation found at depth, which helps to ameliorate the serious acidification problems evident in much of upland Britain (Reynolds et al., 1986).

Figure 3. Within-year evaporation rates for the Wye catchment, showing recharge and discharge to shallow groundwater.

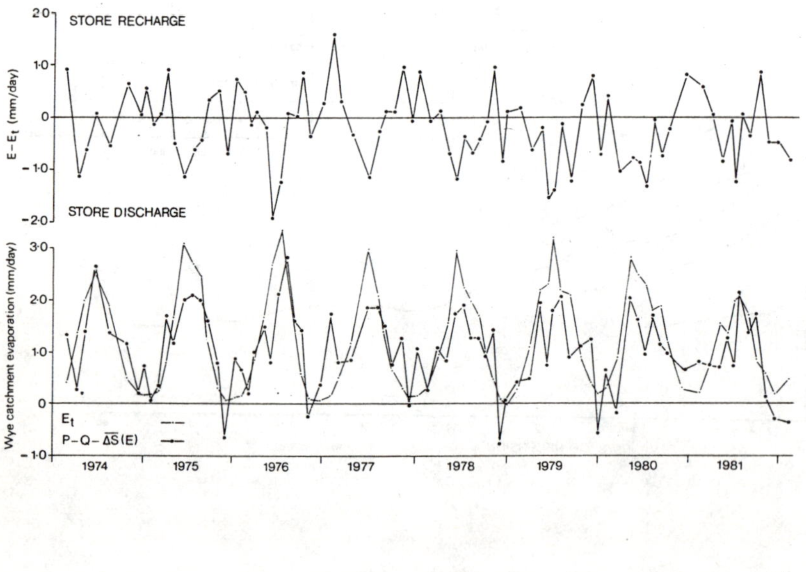

EVAPORATION - MEASUREMENT OR ESTIMATION?

The water balance equation may be rearranged and the same evaporation model later utilised elsewhere to derive deep percolation (recharge) estimates to groundwater stores. Fortunately, evaporation is a relatively conservative process, which is largely independent of groundwater storage, as rooting systems of crops rarely reach far into the unsaturated zone. Consequently, evaporation models, derived as collaborative ventures between water balance experiments in watertight catchments, and process studies considering just the soil and vegetation compartments, can equally be extended to cover the same processes in catchments with considerable deep percolation, even those including large scale aquifers. This depends on links being established between climate, direct evaporation from wet vegetation, transpiration and soil moisture availability.

In terms of groundwater recharge, evaporation can be considered as a loss of resource. In all parts of the world some transpiration loss is inevitable, and indeed desirable because it indicates the presence of a growing crop, with all the benefits for agricultural productivity and soil conservation that this implies. However, direct evaporation of rainwater lying on vegetation is a loss that has no obvious benefit, except perhaps for flood control or for inducing increased precipitation downwind, and may have disbenefits to the crop because of the suppression of transpiration required for growth (Hudson, 1988). Interception of water cannot be avoided if that crop is a subsistential or commercial necessity. In many parts of the world, however, the loss of recharge or riverflow water from direct evaporation may be less acceptable than the loss of a productive crop. Economic and social decisions of this kind can only be made on the basis of sound quantitive evidence of the extent of the problem and the likely side effects of change. It may be possible, theoretically if not practically, to control the loss of water by careful choice of land use, perhaps by utilizing different planting patterns in particularly susceptible areas, or by deliberately planting short rather than tall crops in areas of low precipitation.

Most of the evaporation models based on meteorological controls are derived from Penman's famous equation, stated in 1948 and developed in 1949, which in simple terms used mean daily estimates of net radiation, wind speed and wet bulb depression to give daily estimates of evaporation from short, green crops in northern latitudes. The model is partly physical and partly empirical, the physical part utilising the Bowen ratio to partition net radiation into sensible heat and latent heat fluxes from the vegetation surface, (see also Gash & Stewart, 1975), and the empirical part relating the diffusion of vapour from the surface as a function of humidity gradient and wind speed. Later versions of the model also embody stomatal and aerodynamic controls on vapour diffusion (Monteith, 1965).

Many modifications have been made to the original Penman model to cope with situations outside the scope of the original research. One is the introduction of a routine to reduce the estimate of transpiration when soil moisture declines to the point that the crop can no longer overcome soil tension in trying to draw water through the roots. This is the concept of the root constant (Penman, 1949). At this point the stomata may start to close to conserve water held within the plant. Also, in many upland areas of Western Europe, Penman's equation seems to predict higher evaporation rates than indicated by water balances in the same area. Circumstantial evidence points to low temperatures in spring and autumn restricting growth and hence water uptake, even when average daily meteorological conditions suggest that transpiration is likely. Work is in progress using a system of weighing lysimeters, weather stations and soil moisture measurements to quantify this process and find causal explanations for the phenomenon.

Perhaps the main restriction of Penman's model was the need to develop an equation that only required the simple daily estimates of climatic variables that were taken regularly, at 2m height, at U.K. Meteorological Office sites, and would be likely to be available in developing countries (Thom & Oliver, 1977). Recent advances in electronics, allowing the digitising of analogue electrical signals from environmental sensors and the logging of digital data, now enable researchers to collect meteorological data from automatic weather stations down to increments of seconds rather than days, (Strangeways, 1972). The same detail can be obtained for the vegetation canopy processes of input and drainage, while for short and intermediate crops such as grass or heather, weighing lysimetery is available to estimate canopy storage, (Hall, 1985; Calder et al., 1984). On a larger scale, whole trees have been cut and weighed to give short timescale estimates of canopy storage of both water and snow (Roberts, 1978), while gamma ray attenuation has also been used for storage profiling within tall canopies, (Calder & Wright, 1986).

Using these methods, research models of evaporation have been built that can now predict short and long term losses by interception or biological transpiration from a whole variety of crops, including tall crops such as forests. It was probably the debate over evaporation losses from forests in the U.K. in the mid 1950s, (Law, 1956), which sparked off the quest for better research models, and these models can now be used worldwide, providing suitable meteorological data is available at the site under scrutiny.

One major breakthrough in evaporation measurement has been the development of the Institute of Hydrology "Hydra", an instrument that can estimate the flux of water vapour from a vegetation surface directly (Shuttleworth et. al., 1988; Lloyd et al., 1984). The technique uses state of the art sensors - sonic anemometers, infra-red hygrometers etc., to estimate vertical components of sensible heat transfer, wind speed and humidity fluctuations and calculates evaporation by the eddy correlation technique. For operation in developing countries in particular it has great advantages: the ability to work at remote sites under battery power, the ability to estimate evaporation over crops under water stress and over tall crops where existing micro-meteorological methods are notoriously suspect. So far it has been used successfully in the Amazon Basin, Niger, South West France and the Sahel (e.g. Shuttleworth, 1988; Shuttleworth et al., 1984). It is particularly useful in arid climates where its major disadvantage, its inability to operate during and just after rain when the sensors are wet, is not too important. This restricts its use in wet climates for continuous measurements of evaporation, but does not detract from the benefits of its operation for short periods when transpiration is the major flux from the soil. The technique has not yet proved itself in areas of high relief, such as the British uplands, where the required, even fetches of 200-400 metres are not always available, and where it is difficult to fulfill the major requirement of siting the instrument perpendicular to the prevailing air stream.

PRACTICAL EXAMPLES OF RECHARGE ESTIMATION

The impact of errors in assessment of the water balance components is shown by a study recently tackled by the Institute of Hydrology and the British Geological Survey in the Chyulu Hills in Kenya, an aquifer outcrop of volcanics rising above a low lying plain. Water supply for the area has traditionally been derived from perennial springs discharging from the volcanics over the plain deposits. Development of the aquifer depends on an assessment of the reliability of a supply that is recharged only intermittently, when rainfall exceeds the evaporation rate for long enough for drainage to occur below the rooting zone. Once this process has started, the hydrogeological evidence suggests that vertical drainage to the aquifer is rapid because of the high transmissivity of these volcanics, (Makin & Pratt, 1984; Wright, 1988). Estimation of recharge rates has depended on two methods: an overall aquifer water balance and development of a soil recharge model (Eeles & Blackie, 1988).

The difficulties involved in establishing an aquifer water balance for this site amply illustrate the hydrometric problems facing exploiters of groundwater in many developing countries. The available raingauge network was sparse (see also McCulloch, 1969; Edwards, 1972; Piper et al., 1979), with long term data available from only 11 gauges (1 gauge per 198 square kms). Over the period of study 12 extra gauges were installed. Long term rainfall records were synthesized, initially by developing a multiple regression model of rainfall on altitude and location (a quadratic surface was also fitted, a method which showed some promise - Eeles et al., 1988), and then by back-extrapolating the rainfall record for each site using the relationship of these gauges with the long term station at Dwa (records available since 1917). The random, localised and unpredictable pattern of convectional storms in this area made any assumption of causality a tenuous one, but this was the best estimate that could be made with the data available (see also Richards, 1971). A similar problem arose with spring discharge. Only 6 springs had a historical discharge record and 7 new sites were installed for the period of the study, (Gunston,1988). A spring discharge record for each of the new sites was synthesised by a regression method.

Evaporation from the area was a major term in the water balance equation and estimates needed to be of the highest accuracy in order to prevent the passing of errors into the recharge term in the equation. Only two automatic weather stations were available, situated at either end of the altitude range of the Chyulu Hills (995m to 1800m). Evaporation rates were estimated by the Penman equation using the root constant concept of limiting evaporation in times of stress. Areal evaporation was estimated by linear altitudinal interpolation, which involves an even more tenuous assumption of causality than for rainfall; this, at least, is known to vary with altitude because of orographic cooling effects. Evaporation is dependent on the variation in its own controlling variables: windspeed, net radiation and humidity deficit, which are not necessarily altitudinally dependent. In fact, evidence for an inversion of Penman estimates, with values increasing with altitude, is emerging from the I.H. Balquhidder study in highland Scotland (Blackie et al., 1986). Net radiation in particular depends on the nature of the vegetation and the angle of insolation, while windspeed is variable and dependent on topography, though there will be a general increase in windspeed with altitude because of the decreasing effect of boundary layer friction (ie. longer fetch). The Chyulu study recognised these limitations and a call was made for more intensive networks to be used in future studies.

The Chyulu Hills outcrop, like many upland areas of the world supports a considerable amount of tall vegetation, especially forests developing in the high altitude humid to sub-humid climate. Estimation of evaporation from forests has proved notoriously difficult and has tested the ingenuity of researchers around the world for the past twenty or thirty years. The Penman equation was designed for short green crops like grazed grassland, as it describes transpiration under stomatal control rather than the direct drying of a wet forest canopy. Various models of interception, both complex research and simpler operational types, have been derived. In Britain, physical models such as those of Rutter et al. (1975), Gash (1979), Calder (1977) and empirical, operational models such as Calder & Newson (1979), have been developed to describe a process which consistently allows twice the evaporation rate from plantation coniferous foresty as from grassland, or as predicted from solar derived available energy, at least in the wet and windy upland areas of western and northern Britain (Hudson, 1988).

Most of the problems associated with estimating evaporation from tall crops, and hence recharge to underlying aquifers or discharge to streams, is that models have been derived which try to treat the process as a single meteorologically-dependent process rather than two separate but interacting processes. The physical problem of scale in short vegetation makes it difficult to split the two processes but, in tall vegetation, instrumentation can be placed beneath the canopy to estimate net rainfall, and a canopy balance can be calculated if gross rainfall is known. This is problematical in itself, because of the difficulty associated with placing rain gauges above trees in a very turbulent air stream and expecting them to measure both rainfall and snow with equal accuracy (see Newson & Clarke, 1976). The instrumentation for measuring net rainfall varies according to the purpose of the study. Statistical sampling of point throughfall using raingauges or troughs, and selection of trees for stemflow measurements, gives a better idea of the variation in net rainfall patterns. This can be critical if soil properties also vary, in which case estimation of infiltration relies on knowing how much water is falling on a particular soil type or structure. Under vegetation such as trees, soil structure can vary markedly with distance away from the trunks and depends to a large extent on root density. For water quality studies also, stemflow and throughfall have to be split because of their disparate chemistries. For most hydrometeorological studies, however, it is sufficient to know the total volume of water hitting unit area of ground and this has led to the design of plastic "catch-all" gauges strung between and sealed to all the trees in a known area (Calder & Rosier, 1976).

ESTIMATING RECHARGE FROM SOIL WATER BALANCES

In areas of low topography, where lateral movement within the soil is minimal and vertical drainage predominates, the soil water balance alone can be used to estimate point recharge to aquifers. To assess the combined impact of these estimated point recharges on an aquifer, requires some idea of the spatial variation in soil physical properties, particularly hydraulic conductivity and its variation with depth (see Wikramaratna & Reeve, 1984; Wellings & Cooper, 1983). Many soils with the potential to take high infiltration rates of water at the surface, may also have impermeable layers, for instance iron pans, at some depth below the surface, which will control the recharge rate to the underlying aquifer. For this reason surface infiltration measurements using ring infiltrometers, (see Hills, 1970 for a review of instrumentation available), are of little use in isolation but may form an important part of an overall soil drainage model. They could be particularly useful for assessing initial soil wetting before the layer of low transmissivity takes over control.

As an example of this approach, Wheater et al. (1988) estimated recharge to shallow flood plain aquifers in Saudi Arabia, as a transmission loss from ephemeral wadi flows. A single ring infiltrometer surrounding a neutron probe access tube was used to assess surface infiltration rates, and profiles of soil moisture content were taken over time, following application of water to the ring and allowing a head of water to develop. Initial rates of infiltration through the sandy wadi bed were rapid but were then held up by silt lenses at a depth of about 1m. Regional estimates of infiltration from phreatic surface survey across the flood plain indicated that recharge was overestimated by this method, probably because, in the natural recharge situation of flash flood channel flow, siltation on the wadi bed would inhibit infiltration, even though natural heads would exceed those used in the ring infiltrometer test. The principle of wadi transmission infiltration was further investigated by I.H. staff in Oman, with a view to exploiting the recharge process by ponding of wadi flows. It was hoped that, by increasing heads behind the artificial barrages and encouraging precipitation of fine sediment at the entrance to the artificial lake, this would ensure enhanced infiltration. A thorough hydrological survey had to be made a priority in order to quantify the likely impact on downstream irrigation practices, (IH, 1988).

Similar problems were encountered by Rushton et al. (1988), when trying to construct a recharge/abstraction balance for the Liverpool sandstone aquifer. Potential recharge rates were calculated by a soil moisture balance method, (Rushton and Ward, 1979), yet actual recharge to the aquifer had to be reduced by a factor which was dependent on the thickness and clay content of drift deposits lying between the soil and the aquifer. This was complicated further at grid points where the water table lay within the drift profile. This highlights the need for comprehensive geological survey as well as water flux measurement to get a true picture of recharge to an aquifer.

The use of surface infiltration as a measure of recharge is fraught with difficulties and uncertainties, partly because of the artificial effects of ponding, which is necessary for volumetric work and for ensuring potential rates of infiltration. Also, surface infiltration is generally more variable than transmissivity lower down the profile, which results from the nature of the surface, controlled as it is by the cumulative effect of raindrop and human compaction of the natural soil surface structure, and by the vegetation cover.

A better approach involves estimation of soil water fluxes, allowing recharge to be estimated by the water balance equation. To do this successfully requires accurate estimates of rainfall input, evaporation output and soil moisture change. Rainfall and evaporation estimates have already been dealt with, yet soil moisture has so far only been mentioned in connection with catchment water balances. The techniques used for soil moisture determination can be destructive or non destructive. Carefully arranged gravimetric sampling of soils, when their water contents are lower than the limit of gravitational drainage (less than field capacity), is potentially the most accurate and precise method of soil moisture determination. However, this method neither gives an accurate picture of water contents between saturation and field capacity, because of the tendency of samples to drip after extraction from the ground, nor does the total extraction of moisture by oven drying give any clue to the water in the sample that would be available for capilliary suction by vegetation. An additional problem is that long term monitoring of soil moisture by this method is impossible, as the original sample is destroyed, and subsequent samples may have sufficiently different structural properties to mask "real" changes over time. The solution seems to involve a combined approach of destructive and non destructive sampling, whereby indirect techniques of soil moisture estimation - gamma radiation attenuation, neutron energy modification (Bell & Eeles, 1967; Bell & McCulloch, 1966) or a capacitance technique (Bell et al., 1987), are used to give variations of soil water content over time and with depth, while gravimetric estimates are used to calibrate the indirect methods. Gravimetric methods often entail sampling of soil in different horizons and reconstituting this in a large enough volume of homogenous material to completely encompass the sphere of influence of the indirect method being used (Bell, 1976; Vachaud et al., 1977). This is particularly important with the neutron scattering method, where changes in soil type between horizons can result in the use of the wrong soil type calibration, which can introduce errors into the estimation of water content.

The main advantage of the neutron scattering technique lies not in its estimation of total soil water content but in its precise and accurate estimation of moisture content relative to an arbitrary field capacity (soil moisture deficit). Of particular relevance to soil and catchment water balances, is the accurate estimation of soil moisture change over time. This accuracy is dependent on the fact that, while absolute estimates of hydrogen concentration may include organics and material other than water, changes in hydrogen concentration will be almost entirely due to moisture changes. Where soil moisture is measured, therefore, and evaporation can be estimated by the Penman equation, or some derivative which takes account of differences in vegetation, the root constant concept can be used to quantify the decline in evaporation to below potential rates when soil water is no longer easily extractable by the plant (under stress). Stress is defined as starting when the capilliary action of transpiration sets up tensions in the soil which limit uptake because tension has to be overcome by the available atmospheric energy. From this point, it is usually assumed that there is a linear decline in actual evaporation until maximum soil moisture deficit is reached.

Because of the circular argument involved in its definition, the root constant is a difficult value to fix by reference to soil moisture content alone. Soil water tension profiling, therefore, assumes prime importance, if only to enable estimation of the soil moisture value (or range of values) below which tensions start to rise. In dry periods, which are of major importance to water resource assessment, the classic zero flux plane may be encountered (see figure 4). This is defined by a reversal of tension gradient with depth and indicates the plane, above which upwards capillary movement occurs by transpiration, and below which drainage occurs to underlying aquifers, (Richards et al., 1956). Clearly the method cannot easily be used in wet periods when soil moisture content is at field capacity or above but it is useful for recharge estimation below field capacity. Thus, in Britain, it can be used to give good estimates of transpiration during summer months when deficits are in evidence, which is the time when transpiration is a major flux in the water balance.

Figure 4. Definition of the Zero Flux Plane.

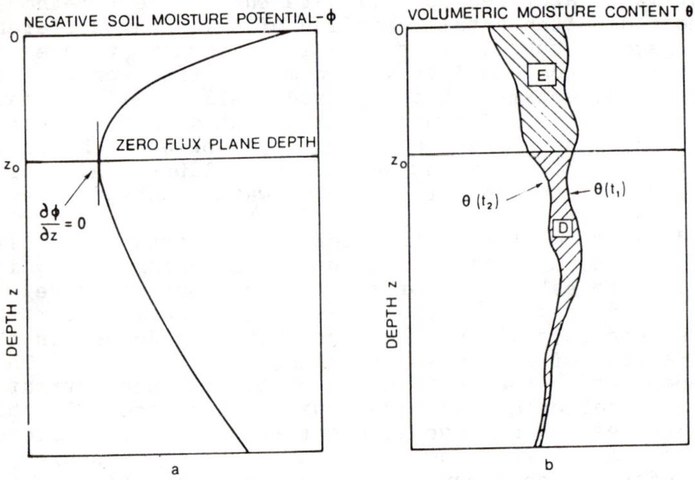

Potential and moisture content profiles under summer conditions illustrating the ZFP technique. Areas marked E and D are the amounts of evaporation and drainage between times t_1 and t_2.

Cooper (1980) cites as the main advantage of the ZFP method, the fact that in Darcy's equation:

$$V = -K\, d\phi/dz$$

where V is the volume of water passing through unit area of soil in unit time, z is the depth and ϕ is the moisture potential, the unsaturated hydraulic conductivity coefficient (K) is not required. Having precluded the need to estimate K, which can vary by a factor of 100,000 over the normal moisture content range, and spatially over a few metres, drainage becomes simply a function of moisture content gradient. However, soil tension measurements are still required to define the ZFP depth (z0). Cooper (1980) also gives practical guidance on the construction of tensiometers, which were first suggested by Richards (1949), including the interfacing of these with pressure transducers for easy data logging. A more recent and cost effective technique is the use of gypsum resistance blocks, described by Wellings et al., (1985) and Strangeways, (1983).

Soil moisture content and tension metering are both difficult to achieve in practice, particularly the latter, as they are both expensive and time consuming techniques that are, consequently, only in general use for research. Besides this, in Britain, large variations in soil type over short distances make it difficult to achieve adequate spatial sampling of soil moisture change (Hudson, 1988). If cost and difficulty can preclude their widespread use as a spatial technique in a small country like Britain, these drawbacks would be even more restricting in many developing countries. Potentially, the use of satellite data shows some promise for spatial extrapolation of information from intensively studied sites (Gurney, 1978), although widespread ground truth measurements will always be required for calibration purposes.

The routine monitoring of soil moisture content and tension has been introduced on a limited basis by the Severn Trent Water Authority, as part of the Shropshire conjunctive groundwater scheme. Water pumped from an extensive permo-triassic sandstone aquifer is being used to top up the low flows in the River Severn, to meet statutory flow requirements in drought periods (see Birtles & Morel, 1980 for a practical insight into river regulation by groundwater extraction). Even though the resource is infrequently used , concern was expressed over the effects on soil moisture of a drop in the water table and ultimately the productivity of crops overlying the aquifer. Provision of comprehensive soil moisture and soil tension networks has allayed fears in the local community, showing that the soil moisture estimation methods described can be useful operational as well as research techniques.

If measurement is not possible, soil moisture can be estimated as a product of a soil water balance. The U.K. Meteorological Office runs a model to predict soil moisture deficit (MORECS - Met. Office Rainfall and Evaporation

Calculation System) on a routine basis (Grindley, 1969; Thompson et al., 1981), mainly to help farmers plan their irrigation requirements or fertiliser applications. It requires rainfall and evaporation estimates on a regional basis (40Km x 40km grids) and estimates soil moisture deficit from a soil balance accounting routine. It assumes that precipitation in excess of evaporation wets the soil up to saturation, it applies a drainage function following rain until field capacity is reached and then assumes that evaporation in excess of rainfall will cause a deficit. Errors are therefore cumulative; however, each time the soil wets to field capacity, this effectively starts a new accounting period and errors are zeroed. The technique in its operational form does not give good spatial resolution, especially in Britain where soils, rainfall, crops and meteorological conditions can vary markedly over short distances. The same technique used in those developing countries where spatial environmental changes are on a larger scale may give adequate resolution, however it is unlikely that estimates of rainfall and evaporation will be undertaken in these areas on as fine a scale as they are at present in Britain.

When using MORECS or similar techniques, the main difficulty lies in the calculation of evaporation, (see Calder et al., 1983). Where data from automatic or manual weather stations are available, a suitable form of the Penman equation can be used. Alternatively, if some knowledge of annual evaporation is known, perhaps from water balance studies nearby, then short timescale evaporation can be approximated as a sinusoidal function of day number around a regional mean annual evaporation. This is a fairly crude technique which does not take into account year to year fluctuations in climatic conditions, but one which may have some merit in areas of stable climate. Estimates of SMD from MORECS have been checked against soil measurements by neutron probe at various sites in Britain (Gardner & Bell, 1980; Gardner, 1983) and the relationship was found to be reasonable, although the greater the SMD the greater the difference between the two methods. This might be expected considering the uncertainty surrounding the applicability of the evaporation estimate, especially when tensions are increasing in the soil.

RECHARGE ESTIMATION USING LYSIMETERS

The problems inherent in the soil water balance technique of recharge estimation, have led to efforts to reduce the spatial uncertainty associated with measurement of the fluxes involved. By enclosing the area under study with artificial barriers (i.e. a lysimeter), it is possible to estimate rainfall, vertical drainage, soil moisture change and soil tension change to high degrees of accuracy and precision and to relate these to a known volume of soil. The problem lies in deciding whether the fluxes then measured are the true fluxes that would have occurred in the absence of the lysimeter. The design of a lysimeter is site dependent; the size, shape, orientation, depth and constructional materials depend on the soil type, underlying material, vegetation spacing, slope and both natural and artificial drainage patterns, so it is difficult to discuss the choice of lysimeter type except in terms of broad principles.

Lysimeters can vary from flower pot size to hundreds of square metres in area, and from a few centimetres to 5 metres deep, (Kohnke et al., 1940). Some principle design requirements must be met (see Slatyer & McIlroy, 1961; Calder, 1976).

(1) The dimensions should be large enough to ensure a fully representative sample of the plant community and minimize boundary effects.
(2) The container walls should be designed to minimise the gap between the internal and external crop.
(3) Water table differences between the lysimeter and external soil should be minimized.
(4) Soil structure within the lysimeter must resemble that outside.
(5) The site should be representative of the area outside or the lysimeter should be replicated in order to sample the variations in crop and soil.

When using lysimeters to estimate recharge the most important considerations are, firstly, whether lateral flow is a significant process in the natural profile, in which case a lysimeter will probably give results unrepresentative of the natural situation. Recharge will probably be overestimated, as water that would have moved laterally, possibly to a surface channel further downslope, will be forced downwards. However lateral flow will also be prevented from entering the profile from an upslope direction, which will balance the situation to some extent. Secondly, a decision has to be made whether to provide a seal at the base of the column in order to collect drainage water, or whether to leave the base open to allow natural recharge. The sealed type of lysimeter is more useful for water balance studies, but is only relevant for recharge estimation if, for instance, it is necessary to study the drainage through a layer of low permeability underlain by layers of higher permeability. It is important that the restriction of flow through the drainage offtake does not create a smaller flux than that through the strata under consideration.

Bottomless lysimeters - "natural lysimeters" - have been used for hydrological and hydrochemical studies by the Institute of Hydrology in mid-Wales (see Calder, 1976; Roberts et al., 1986). Their primary function was to use a supposedly low permeability clay layer as a natural base membrane to enable use of a soil water balance to give estimates of forest evaporation (Calder, 1976) and drainage losses of nutrients (Roberts et al.,1986). In the first example the technique worked well, with losses of water through the clay found to be insignificantly small (using laboratory permeability tests) compared to the magnitude of the other measured fluxes. In the second example it proved impossible to produce a soil water balance. However, the fact that flow tapped from the peat/clay interface appeared to be greater than rainfall input, highlighted the need for further hydrogeological studies to be undertaken of the stoney blue clay (drift) on which the lysimeter had been installed. Arrays of piezometers were installed and read (Allen & Bird, 1981) which indicated the existence of artesian heads in the drift and some areas of high transmissivity within the lysimeter. These were allowing rapid upward movement of water into the lysimeter, probably along widespread fissure lines within the clay. This finding has led to a complete reappraisal of the impact of these remnant drift deposits in upland Welsh catchments on flow generation, control of stream chemical quality (Reynolds et al., 1986) and maintenance of low flow in drought periods, (Hudson, 1988), while at the same time showing the potential benefits of this type of lysimeter as a research tool.

With most bottomless lysimeters there is a danger that recharge may be overestimated, because of the divergence of flow once the depth is reached where the constraining influence of the lysimeter walls no longer operates. The same criticism has already been applied to the ring infiltrometer, which is essentially a near-surface, open-based lysimeter. A solution widely adopted involves use of the concentric ring infiltrometer, which confines volumetric measurement to the central ring and divergence to the outer ring. With large lysimeters such a modification is impractical. One solution adopted for recharge estimation in both chalk and sandstone aquifers has been to install lysimeters around an undisturbed monolith of soil and rock, the walls of which reach at least to the water table, and preferably beyond. By this method recharge is estimated directly, negating many of the uncertainties associated with shallower lysimeters. It is also possible to control the water table within the lysimeter using a system of pumps, the flow through which, if measured, can be used to give estimates of fluxes within the lysimeter, particularly evaporation (Gilman, in I.H., 1988) or recharge (Kitching et al., 1977). It should be remembered that lysimeters are expensive to construct and maintain and should probably be used only as a last resort when less manipulative techniques have been tried.

ESTIMATING RECHARGE USING MATHEMATICAL MODELS

One of the main criticisms of mathematical modelling as a means of describing quantitatively the behaviour of hydrological systems is that, the more sophisticated they become, the more site specific they also tend to be. This means that the model which gives the best fit with data from a particular catchment is likely to be the one which requires the most data, effort, time and computing power when used for predicting the behaviour of adjacent ungauged catchments. In other words its transportability is compromised by its sophistication. For this reason, the most elegant, distributed, three dimensional models are of little practical use for application in undeveloped areas unless a hydrometric, cartographic, computational and analytical infrastructure has been put in place. This usually tends to happen only after exploitation of the water resources of the area has started and problems have already arisen.

Nevertheless, models of varying complexity can be used to estimate recharge. Both the catchment and soil water balance methods of recharge estimation, already discussed, are models in the strict sense of the word, as are many of the techniques of estimating areal rainfall, soil water drainage, evaporation and other components of the hydrological cycle. In this section of the paper the description "model" is restricted to those combinations of mathematical equations that are used to describe the whole of an hydrological system, rather than just individual processes. Beginning with a simple black box model, the catchment or aquifer water balance allows the prediction of cumulative recharge for annual periods or longer, given some prior knowledge of the proportions of rainfall that are either lost by evaporation or become streamflow. More sophisticated models are also based on this water conservation principle, but are capable of using better temporal and spatial resolution of processes to predict a temporally (and sometimes spatially) distributed output. A simple within-year water balance model was used by Hudson (1988), to estimate the temporal distribution of recharge to, and discharge from, glacial drift aquifers in the Plynlimon catchments, mid-Wales.

A slightly more sophisticated model is exemplified by the I.H. lumped conceptual model, developed by Dickinson and Douglas (1972), among others, using data from the Cam catchment in Cambridge shire, U.K., and further refined by Blackie and Eeles (1988) for use on many other U.K. and overseas catchments. The model is one-dimensional and consists of discrete tanks or stores, as shown in figure 5 (from Douglas, 1974), superimposed vertically as a cascade, with by-pass mechanisms included for when the stores become full. Fluxes to and from these stores are expressed mathematically as linear or non-linear functions of storage, depending on the routine chosen for the canopy, surface, soil or groundwater compartments. The operation of individual stores is described by a physical process but, if this is not possible, the store parameters are optimised iteratively by a least squares technique, and a sensitivity analysis performed to gain an insight into the relative importance of individual parameters. Groundwater recharge is given as a transitional flux in the model and is assumed to be the vertical output of a single soil moisture tank. This is obviously an oversimplification of the behaviour of a real catchment, which would tend to operate as a parallel series of stores discharging to groundwater at different times. The main criticism of this type of model, therefore, is that it bears mainly mathematical and little physical relation to the real world. In most catchments, there are horizontal (downslope or converging) cascades as well as vertical cascades. Nonetheless, simple models of this type can be optimised to fit catchment behaviour successfully, and they are particularly useful for infilling gaps in historical streamflow or recharge records (see Eeles, 1978).

Physically realistic, three-dimensional, distributed models are also being developed, to solve the sort of recharge estimation problems found in areas with spatially-variable surface, soil and groundwater conditions. These incorporate routines which allow the mathematical synthesis of precipitation, interception, transpiration, soil moisture change, groundwater etc, which are in reality a series of continuous surfaces, into finite elements or cells, so that mechanisms operating at various points in the catchment can be described. A vertical tank model then operates on each element, with an input from, and an output to, adjacent elements; these depend on the general directions of flow at the various vertical levels (flow routing). The literature concerning these models is extensive. Within the U.K., the two most widely researched and best developed are the I.H. Distributed Model (see Beven & Kirkby, 1979; Beven, 1978; Beven & O'Connell, 1982) and the Systeme Hydrologique Europeen Model (S.H.E.) (see Beven et al., 1980; Bathurst, 1980).

Figure 5. A conceptual, lumped, one-dimensional catchment model.

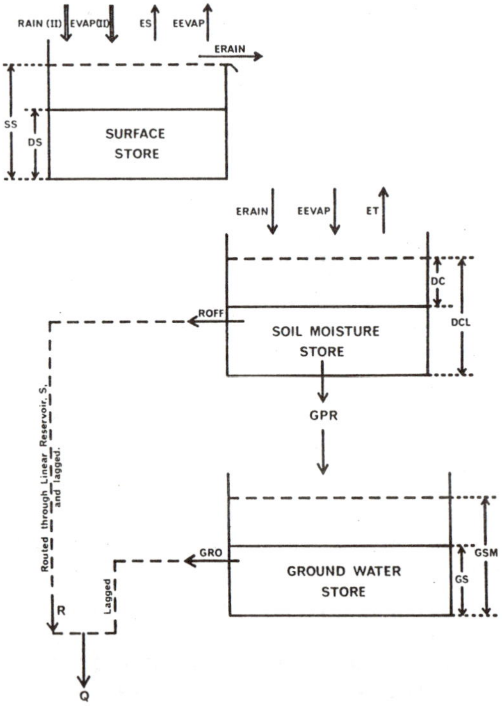

A novel approach to the modelling of recharge, that does not rely on surface and unsaturated zone estimates of downward percolation, has been developed by Smith and Wikramaratna (1981), who used groundwater observations to produce a model which quantified recharge inputs to the aquifer as a function of change in storage and flow from the aquifer. The problem with aquifers already in use as sources of water, lies in quantifying accurately a resource utilisation rate that can be put into the aquifer balance equation. Given the present state of data collection at many water supply boreholes in Britain, crucial errors could result using this method, a problem that has recently been addressed for the Severn Trent borehole network by Skinner (1988), with promising results. Similar problems will occur in developing countries where the existing rate of extraction will probably not be known to any degree of accuracy, although as-yet undeveloped sources of supply can be assessed using an aquifer balance model of the type just mentioned.

RECHARGE ESTIMATION USING CHEMICAL TRACERS

Although the pollution of groundwater is the single most serious problem facing water engineers and scientists in developing countries, a useful side effect has been the increased ability to study water movements through the unsaturated zone using pollutants (and natural non-polluting chemicals) as tracers. The drilling of test boreholes has involved the taking of cores which can be centrifuged, and the extracted water subjected to chemical analysis. While cations tend to remain in and react with the soil, anion concentrations can be useful when plotted as concentration profiles. In the case of seasonal pollutants, exemplified by the wet season peak in nitrate concentrations under arable crops, an assessment of the speed of descent of the anion wave can give the times of travel of both water and pollutants through the unsaturated zone.

With biologically reactive anions like nitrate, and geologically reactive chemicals such as sulphate, quantitative estimation of pollutant and water movement is impossible (except where it can be shown that modification is minimal). However chloride is present in abundance, particularly in areas with maritime climates, where it tends to be of meteoric origin. It is basically unreactive, unless chloride bearing rocks are present, or KCl fertiliser has been added, and can be used in a chemical balance model to quantify drainage through the unsaturated zone, (see Reynolds & Pomeroy, 1988; Thomson & Foster, 1986; Wright, 1988, for use of the method on Plynlimon, Mid Wales, on Bermuda and on the Chyulu aquifer, Kenya, respectively). Reynolds & Pomeroy cite the underestimation of Cl input as aerosols or occult precipitation (cloud condensation) as the main drawback of the method, and also the existence of long term oscillations in surface flow concentrations, which may indicate some cycling of Cl by vegetation. Long term interactions of groundwater and surface flow also complicate the Cl picture making mass balances of limited use in the short term. Periods of 5 years and over give a better picture of average conditions. Cl tracing has advantages, because it is an abundant, universal and consistent input, with a large concentration range that is reflected in soil and unsaturated zone profiles of Cl concentration.

Isotope techniques are also useful for water tracing. Smith et al., (1972) suggest the use of tritium as a non reactive tracer. This can be a particularly useful technique for constructing historical records of water movement in the unsaturated zone in areas affected by fallout from bomb tests carried out in the 1950's. Its continued use in the long term is dependent on future above-ground nuclear tests, which would not be advocated by environmental researchers, even though advantage has been taken of previous tests. Deuterium is another stable isotope of hydrogen that has been used for estimating recharge (Wellings, 1982) while, in the broader groundwater research sphere, the ratio of deuterium to oxygen-18 has been used for classifying ancient groundwaters according to their likely origins. For near-surface tracing, it should be possible theoretically to use the deuterium:oxygen-18 ratio to split recharge into water derived from rainfall, snowfall or irrigation. Having assessed the relative importance of these sources, a use for the technique can be envisaged in developing countries where irrigation is widely practised, not primarily as a means of artificial recharge, but where recharge comes about as a fortunate bi-product (provided pollution is not the end result).

TOWARDS A METHODOLOGY FOR RECHARGE ESTIMATION IN DEVELOPING COUNTRIES.

In spite of the claims made for the transportability of sophisticated catchment models, which incorporate both surface runoff and groundwater recharge components of loss from the surface catchment, they remain of little use in areas where data is not yet available on which the models can be calibrated and run. Appropriate data is hard enough to find in developed countries, which generally have a dense hydrometric infrastructure in place, let alone in developing countries where large scale exploitation of resources is only just beginning. Even where reliable, accurate, precise, long term records of rainfall, snowfall, occult deposition and streamflow are available, the ability to construct physical models is constrained by the lack of information on the spatial variability of vegetation, soils and groundwater stores. Remote sensing from aeroplanes and satellites might eventually solve some of the spatial problems but the techniques are as yet in their infancy. Low resolution is the main restriction on their use for detailed hydrological purposes, closely followed by a lack of personnel with the interpretive skills for assessing hydrological characteristics of catchments from the available multi-spectral images.

For many applications, simple, one-dimensional, tank models are all that can be contemplated with the spatial resolution of data available, yet even these require some knowledge of the vertical variation in water fluxes. Interception needs to be split from net rainfall, surface runoff from surface infiltration and transpiration from soil drainage. Approximate quantities and timing can often be extrapolated from studies undertaken in other areas, but there is always the danger that the controls on these fluxes are not the same as those for the area in which the original model was developed. This is especially true if the original model was partly empirical and not wholly physically based.

Clearly, catchment modelling has a role in recharge estimation in developing countries, but not unless regional variations can be generalised by physical rather than empirical factors. It is imperative, therefore, that models derived in the industrialised countries should be tested in research catchments set up in developing countries. To take an extreme example, evaporation models developed in temperate humid climates can and should be calibrated to work in tropical arid climates. At present there is a long way to go and it is obvious that until comprehensive hydrometric networks are available world-wide, initially as densely instrumented experimental watersheds, and later as water resource operational networks, little progress can be made with catchment modelling.

Modelling of compartment processes such as canopy evaporation, transpiration and soil drainage is easier, however, and the appropriate techniques have been transferred successfully from developed to developing countries whenever specific water resource problems have arisen. To extrapolate these compartment models requires networks of hydrometric instrumentation, such as raingauges, streamflow gauges, weather stations, neutron probe access tubes and soil tensiometers , an objective which has not yet even been fully achieved on a regional basis in Britain. To give an indication of what is feasible, however, models of evaporation loss from coniferous forests, originally derived from a combination of catchment water balances, natural lysimetry and canopy process studies, have now been extrapolated to those parts of upland Britain likely to be affected by increased forest planting over the next 25 years. Because the original research was so comprehensive, estimation of evaporative loss is now possible using only automatic weather stations and plastic net rainfall gauges underneath the canopy. These estimates have been of great benefit to the water industry in Britain, enabling better use to be made of existing resources. It is envisaged that in many developing countries a similar pattern will emerge. A period of intensive experimental work on densely instrumented areas, will be followed by a period of regional work, utilizing those data found to be essential for quantifying the processes involved. It is hoped that, by passing on the research lessons learnt painfully over many years in the industrialised countries, the development process will be made that much easier.

Whatever catchment or compartment models are used, they will almost certainly be based on a form of the equation of water conservation or water balance. The setting up of a network of representative basins has been a priority within Europe for some time, for example the FREND (Flow Regimes from Experimental and

Network Data) project (see Gustard, in I.H., 1988; Rodda, 1988), as much can be learnt by assessing the similarities and differences between catchments in different geological and climatic regions. This should be expanded to include a wide range of catchments throughout the rest of the world, it being particularly important to include a selection of catchments in both the humid and arid tropics, where many of the worst environmental problems are manifest. The main water balance fluxes: rainfall, evaporation, surface streamflow, groundwater recharge and baseflow, need to be quantified over the long term, and the sites made available for process work on evaporation, soil moisture movement and groundwater behaviour. Whatever physical techniques are developed it is also likely that chemical tracer techniques will provide important support for future studies.

ACKNOWLEDGEMENTS

The author wishes to express his thanks to his colleagues, past and present, at the Institute of Hydrology for the vast amount of work achieved over the last twenty years, much of which has some relevance to recharge estimation and has therefore been described or at least mentioned in this paper. For researchers setting out to instrument sites or catchments in developing countries the reference list appended should provide a wealth of information. The author makes the point that, although the list contains work which is nearly all derived from the Institute of Hydrology, this is not to detract from the crucial contribution of researchers from elsewhere around the world. It is simply a means of limiting a bibliography what would have become unmanageable. Full recognition of the remaining research will have been given within the quoted references if further details are required.

REFERENCES

Ackers, P., White, W.R., Perkins, J.A. and Harrison, A.J.M. (1978) Weirs and flumes for flow measurement. John Wiley and Son Ltd..

Allen, D.J. and Bird, M.J. (1981) A report on the permeability investigations at the Institute of Hydrology lysimeter, Plynlimon, Wales. IGS Rept. No. WD/ST/81/16, Wallingford, Oxon., U.K..

Bathurst, J.C. (1980) Modelling surface flow / subsurface flow interactions with the Systeme Hydrologique Europeen. Paper to Symposium AS, 7th Ann. Meeting EGS, Budapest.

Bell, J.P. (1976) Neutron probe practice. Inst. Hydrol. Rept. No. 19, Wallingford, Oxon., U.K..

Bell, J.P., Dean, T. and Hodnett, M. (1987) Soil moisture measurement by an improved capacitance technique. Part II: Field techniques, evaluation and calibration. J. Hydrol., 93, 79-90.

Bell, J.P. and Eeles, C.W.O. (1967) Neutron random counting error in terms of soil moisture for non-linear calibration curves.

Soil Sci., 103(1), 1-3.
Bell, J.P. and McCulloch, J.S.G. (1966) Soil moisture estimation by the neutron scattering method in Britain. J. Hydrol., 4, 254-263.
Beven, K. (1978) The hydrological response of headwater and sideslope areas. Hydrol. Sci. Bull., 23(4), 419-437.
Beven, K. and Kirkby, M. (1979) A physically based, variable contributing area model of basin hydrology. Hydro. Sci. Bull., 24(1), 43-69.
Beven, K.J. and O'Connell, P.E. (1982) On the role of physically based distributed modelling in Hydrology. Inst. Hydrol. Rept. No. 81, Wallingford, Oxon., U.K..
Beven, K., Warren, R. and Zaoni, J. (1980) S.H.E.: towards a methodology for physically based distributed forecasting in hydrology. Proc. Int. Symp. on Hydrological Forecasting, Oxford, IAHS Publ. No. 129, 133-137.
Birtles, A.B. and Morel, E.H. (1980) Factors affecting cost optimisation of groundwater development schemes for river regulation. J. Hydrol., 48, 197-219.
Blackie, J.R. and Eeles, C.W.O. (1985) Lumped catchment models. In: Hydrological forecasting, (eds. Anderson and Burt), Wiley, Chichester, 311-345.
Blackie, J.R., Hudson, J.A. and Johnson, R.C. (1986) Preliminary results of phase 1 of the Balquhidder experiment. Inst. Hydrol. Progress Report, Wallingford, Oxon., U.K..
Calder, I.R. (1976) The measurement of water losses from a forested area using a "natural" lysimeter. J. Hydrol., 30, 311-325.
Calder, I.R. (1977) A model of transpiration and interception loss from a spruce forest in Plynlimon, central Wales. J. Hydrol., 33, 247-265.
Calder, I.R., Hall, R.L., Harding, R.J. and Wright, I.R. (1984) The use of a wet surface weighing lysimeter in rainfall interception studies of heather (Calluna Vulgaris). J. Climate and Applied Meteorology, Vol. 23, 461-474.
Calder, I.R., Harding, R.J. and Rosier, P.T.W. (1983) An objective assessment of soil moisture deficit models. J. Hydrol., 60, 329-355.
Calder, I.R. and Newson, M.D. (1979) Land-use and upland water resources in Britain - a strategic look. Wat. Resources Bull., 15(6), 1628-1639.
Calder, I.R. and Rosier, P.T.W. (1976) The design of large plastic sheet net rainfall gauges. J. Hydrol., 33, 403-405.
Calder, I.R. and Wright, I.R. (1986) Gamma ray attenuation studies of interception from Sitka Spruce: some evidence for an additional mechanism. Wat. Resources Res., 22(3), 409-417.
Clarke, R.T., Leese, M.N. and Newson, A.J. (1973) Analysis of data from the Plynlimon raingauge networks, April 1971-March 1973. Inst. Hydrol. Rept. No. 27, Wallingford, Oxon., U.K..
Cooper, J.D. (1980) Measurement of moisture fluxes in unsaturated soil in Thetford Forest. Inst. Hydrol. Rept. No. 66, Wallingford, Oxon., U.K..
Davey, J.C. (1982) Hydraulic properties of a major alluvial

aquifer: an isotropic, inhomogeneous system - comments. J. Hydrol., 57, 203-204.

Dickinson, W.T. and Douglas, J.R. (1972) A conceptual runoff model for the Cam catchment. Inst. Hydrol. Rept. No. 17, Wallingford, Oxon., U.K..

Douglas, J.R. (1974) Conceptual modelling in Hydrology. Inst. Hydrol. Rept. No. 24, Wallingford, Oxon., U.K..

Edwards, K.A. (1972) Estimating areal rainfall by fitting surfaces to irregularly spaced data. Proc. Symp. "Distribution of Precipitation in Mountainous Areas", WMO Publ. No. 326.

Edwards, K.A. and Rodda, J.C. (1970) A preliminary study of the water balance of a small clay catchment. J. Hydrol. (NZ), 9(2), 202-218.

Eeles, C.W.O. (1978) A conceptual model for the estimation of historic flows. Inst. Hydrol. Rept. No. 55, Wallingford, Oxon., U.K..

Eeles, C.W.O. and Blackie, J.R. (1988) The development of a recharge model. In: The Chyulu Hills Aquifer Study Final Report, British Geological Survey, Wallingford, Oxon., U.K..

Eeles, C.W.O., Munir, E. and Blackie, J.R. (1988) The development of a rainfall model. In: The Chyulu Hills Aquifer Study Final Report, British Geological Survey, Wallingford, Oxon., U.K..

Gardner, C.M.K. (1983) A comparison of measured soil moisture deficits with those estimated by the Meteorological Office system, MORECS: a brief report. Agric. Wat. Management, 6, 307-316.

Gardner, C.M.K. and Bell, J.P. (1980) Comparison of measured soil moisture deficits with estimates by MORECS. Proc. Symp. "The Influence of Man on the Hydrological regime with Special Reference to Representative and Experimental Basins", IAHS Publ. No. 130, 337-341.

Gash, J.H.C. (1979) An analytical model of rainfall interception by forests. Q. J. Roy. Met. Soc., 105, 43-55.

Gash, J.H.C. and Stewart, J.B. (1975) The average surface resistance of a pine forest derived from Bowen ratio measurements. Boundary Layer Meteorology, 8, 453-464.

Grindley, J. (1969) The calculation of actual evaporation and soil moisture deficit over specified catchment areas. Hydrol. Mem. No. 8, Met.Office, G.B..

Gunston, H. (1988) Water abstraction from Chyulu springs and rivers. In: The Chyulu Hills Aquifer Study Final Report, British Geological Survey, Wallingford, Oxon., U.K..

Gurney, R.J. (1978) The estimation of soil moisture content and actual evapotranspiration using thermal infra-red remote sensing. In: Remote sensing and national mapping (eds: Allan and Harris), Proc. 5th annual conf., Remote Sensing Soc., Durham, 101-109.

Hall, R.L. (1985) Further interception studies of heather using a wet surface weighing lysimeter system. J. Hydrol., 81, 193-210.

Hills, R.C. (1970) The determination of the infiltration capacity of field soils using the cylinder infiltrometer. B.G.R.G. Tech. Bull., No. 3, 24pp.

Hudson, J.A. (1988) The contribution of soil moisture storage to the water balances of upland forested and grassland catchments. Hydrol. Sci. J., 33, 3, 289-309.

Institute of Hydrology (1976) Water balance of the headwater catchments of the Wye and Severn, 1970-75. Inst. Hydrol. Rept. No. 33, Wallingford, Oxon., U.K..

Institute of Hydrology (1988) Research Report 1984-87. Wallingford, Oxon., U.K..

Kitching, R., Shearer, T.R. and Shedlock, S.L. (1977) Recharge to Bunter sandstone determined from lysimeters. J. Hydrol., 33, 217-232.

Kohnke, H., Dreibelbis, F.R. and Davison, J.M. (1940) A survey and discussion of lysimeters and a bibliography on their construction and performance. U.S. Dept. Agr. Misc. Publ. No. 372, 1-68.

Law, F. (1956) The effect of afforestation upon the yield of water catchment areas. J. British Waterworks Ass., 38, 484-494.

Lloyd, C.R., Shuttleworth, W.J., Gash, J.H.C. and Turner, M. (1984) A microprocessor system for eddy-correlation. Agric. and Forest Met., 33, 67-80.

Makin, M.J. and Pratt, D.J. (1984) Land use and development in the Chyulu area of Kenya. Project Report, No. 135, Land Resources Development Centre, Tolworth, London, 77p.

Marsh, T.J. and Davies, P.A. (1983) The decline and partial recovery of groundwater levels below London. Proc. Instn. Civil Engs., Part 1, 74, 263-276.

McCulloch, J.S.G. (1969) The estimation of factors controlling streamflow in areas with limited data. "Water Resource Evaluation from Scarce Data". 2nd seminar on Hydrology and Water Resources Development, Central Treaty Organisation.

Monteith, J.L. (1965) Evaporation and the environment. Symp. Soc. of Experimental Biology, vol. 19, 205-234.

Newson, A.J. and Clarke, R.T. (1976) Comparison of catch of ground level and canopy level raingauges in the Upper Severn experimental catchment. Met. Mag., 105, 2-7.

Newson, M.D. (1979) The results of ten year's experimental study on Plynlimon, mid-Wales, and their importance for the water industry. J. Inst. Wat. Engrs. and Sci., 33, 321-333.

Penman, H.L. (1948) Natural evaporation from open water, bare soil and grass. Proc. Roy. Soc. London, A193, 120-146.

Penman, H.L. (1949) The dependence of transpiration on weather and soil conditions. J. Soil Sci., 1, 74-89.

Piper, B.S., Plinston, D.J. and Sutcliffe, J.V. (1979) Assessment of water resources from scarce hydrological data. I.C.E. Conference on Water Resources - A Changing Strategy, London.

Reynolds, B., Hudson, J.A. and Leeks, G.J.L. (1989) Estimating solute and sediment losses in small upland streams. In: Field Methods in Terrestrial Ecosystem Nutrient Cycling, (eds Harrison and Ineson), Elsevier, Amsterdam, (in press).

Reynolds, B., Neal, C., Hornung, M. and Stevens, P.A. (1986) Baseflow buffering of streamwater acidity in five mid-Wales catchments. J. Hydrol., 87, 167-185.

Reynolds, B. and Pomeroy, A.B. (1988) Hydrogeochemistry of

chloride in an upland catchment in mid-Wales. J. Hydrol., 99, 19-32.
Richards, D. (1971) A computer program to use orthogonal polynomials in two variables for surface fitting. Inst. Hydrol. Rept. No. 13, Wallingford, Oxon., U.K..
Richards, D. (1975) The effects of reducing raingauge network density on goodness of conceptual model fit and prediction. Inst. Hydrol. Rept. No. 28, Wallingford, Oxon., U.K..
Richards, L.A. (1949) Methods of measuring soil moisture tension. Soil Sci., 68, 95-112.
Richards, L.A., Gardner, W.R. and Ogata, G. (1956) Physical processes determining water loss from soil. Soil Sci. Soc. Amer. Proc., 20, 310-314.
Roberts, G., Hudson, J.A. and Blackie, J.R. (1986) The effect of grassland improvement on the quality of upland streamflow. Inst. Hydrol. Rept. No. 96, Wallingford, Oxon., U.K..
Roberts, J.M. (1978) the use of the "tree cutting" technique in the study of the water relations of Norway spruce, Picea abies (L) Karst. J. Exp. Bot., 29(109), 465-471.
Rodda, J.C. (1967) The systematic error in rainfall measurement. J. Instn. Wat. Engs., 21, 173-177.
Rodda, J.C. (1988) Developments in international hydrology. Paper presented to the British Hydrological Society, Institute of Civil Engineers, London, 5th December, 1988.
Rushton, K.R., Kawecki, M.W. and Brassington, F.C. (1988) Groundwater model of conditions in Liverpool sandstone aquifer. J. Inst. Wat. Env. Man., Vol. 2, No. 1, 67-84.
Rushton, K.R. and Ward, C. (1979) The estimation of groundwater recharge. J. Hydrol., 41, 345-361.
Rutter, A.J., Morton, A.J. and Robins, P.C. (1975) A predictive model of rainfall interception in forests. 2: generalisation of the model and comparison with observations in some coniferous and hardwood stands. J. Appl. Ecol., 12, 367-380.
Shuttleworth, W.J. (1988) Evaporation from Amazonian rain forest. Proc. Roy. Soc. London., B233, 321-346.
Shuttleworth, W.J., Gash, J.H.C., Lloyd, C.R., McNeil, D.D., Moore, C.J. and Wallace, J.S. (1988) An integrated micro meteorological system for evaporation measurement. Agric. and Forest Meteorology, 43, 295-317.
Shuttleworth, W.J., Gash, J.H.C., Lloyd, C.R., Moore, C.J., Roberts, J.M. et al. (1984) Eddy correlation measurements of energy partition for Amazonian forest. Q. J. Roy. Met. Soc., 110, 1143-1162.
Skinner, A.C. (1988) Practical experience of borehole performance evaluation. J. Inst. Water Env. Management, Vol. 2, No. 3, 332-340.
Slatyer, R.D. and McIlroy, I.C. (1961) Practical micro climatology, UNESCO, Paris.
Smart, J.D.G. (1977) The design, operation and calibration of the permanent flow measurement structures in the Plynlimon experimental catchments. Inst. Hydrol. Rept. No. 42, Wallingford, Oxon., U.K..
Smith, D.B., Wearn, P.L., Richards, H.J. and Rowe, P.C. (1970) Water movement in the unsaturated zone of high and low

permeability strata by measuring natural tritium. "Isotope Hydrology, 1970", IAEA, Vienna, 73-87.

Smith, P.J. and Wikramaratna, R.S. (1981) A method for estimating recharge and boundary flux from groundwater level observations. Hydrol. Sci. Bull., 26, 113-136.

Strangeways, I.C. (1972) Automatic weather stations for network operation. Weather, (10), 403-408.

Strangeways, I.C. (1983) Interfacing soil moisture gypsum blocks with a modern data logging system using a simple, low cost, DC method. Soil Sci., 136(5), 322-324.

Thom, A.S. and Oliver, H.R. (1977) On Penman's equation for estimating regional evaporation. Q. J. Roy. Met. Soc., 103(436), 345-357.

Thompson, N., Barrie, I.A. and Ayles, M. (1981) The Meteorological Office Rainfall and Evaporation Calculation System : MORECS. Met. Office Hydrol. Memo. No. 45, 72pp.

Thomson, J.A.M. and Foster, S.D.D. (1986) Effect of urbanisation on groundwater of limestone islands: an analysis of the Bermuda case. J. Inst. Wat. Engs. and Scientists, Vol. 40, No. 6, 527-540.

Vachaud, G., Royer, J.M. and Cooper, J.D. (1977) Comparison of methods of calibration of a neutron probe by gravimetry or neutron capture model. J. Hydrol., 34, 343-356.

Wellings, S.R. (1982) The use of deuterium oxide as a water tracer in a study of the hydrology of the unsaturated zone of the English chalk. In: Stable Isotopes (eds: Schmidt, Fortel and Heinzinger), Elsevier, Amsterdam, 167-171.

Wellings, S.R., Bell, J.P. and Raynor, R.J. (1985) the use of gypsum resistance blocks for measuring soil water potential in the field. Inst. Hydrol. Rept. No. 92, Wallingford, Oxon., U.K..

Wellings, S.R. and Cooper, J.D. (1983) The variability of recharge of the English chalk aquifer. Agric. Wat. Management., 6, 243-253.

Wheater, H.S., Parissopoulos, G.A. and Hamilton, G.S. (1988) Groundwater recharge from ephemeral flows; observations from south west Saudi Arabia. Paper presented to the British Hydrological Society, Institution of Civil Engineers, London, 5th December 1988.

Wikramaratna, R.S. and Reeve, C.E. (1984) A modelling approach to estimating aquifer recharge on a regional scale. Hydrol. Sci. J., 29(3), 327-337.

Wright, E.P. (1988) Conceptual hydrogeology and groundwater recharge. In: The Chyulu Hills Aquifer Study Final Report, British Geological Survey, Wallingford, Oxon., U.K..

THE INTERACTION BETWEEN THE GROUNDWATER AND UNSATURATED SOIL-WATER FLOW REGIONS

E.G. Youngs*

9 Roundwood Park, Harpenden, Herts. AL5 3AB, England.

Abstract: The physics of water movement in saturated and unsaturated soils is outlined and used as the basis of a rational approach to problems involving the interaction between the groundwater and unsaturated soil-water flow regions. Two examples of applications are given, using physical theory: first, the effect of evaporation on water-table heights in unconfined aquifers under an artesian head is analysed; and second, the movement of water-tables in low-lying wet lands as a response to meteorological input is computed. The limitations of classical theory due to complications of soil heterogeneity, of soil swelling and shrinking, and of soil aggregation are discussed.

INTRODUCTION

The application of the physics of water movement through soils provides a rational basis for the management and development of water resources. In unconfined aquifers there is water movement both in the saturated groundwater region below the water table and in the unsaturated soil-water region above, with soil-water flow continuous across the water-table boundary. Thus the flow region that should properly be considered in problems concerned with water resources in unconfined aquifers is one embracing both the groundwater and the unsaturated soil-water regions, especially since the position of the water table dividing the two, and hence the extent of each region, depends on the conditions imposed on the boundaries of the combination. However, both the analysis of groundwater flow and that of flow in unsaturated soils have advanced separately without detailed consideration of the other, since such detailed analysis usually introduces complications that hinder the easy solution of problems. Nevertheless, increasingly it is being recognized that as far as the water flow is concerned the water-table boundary is an arbitrary dividing boundary and that more attention has to be paid to the interaction between the flows occurring in these two often seemingly distinct flow regions.

The application of the physics of water movement in porous media to groundwater problems is well established. For over a

---

*Visiting Fellow at Silsoe College, Cranfield Institute of Technology, Silsoe, Bedford, MK45 4DT, England; formerly at the Agricultural Research Council Unit of Soil Physics, Cambridge, and at Rothamsted Experimental Station, Harpenden, Herts.

century the fundamental law of Darcy (1856) has been used to calculate groundwater flows. The work of Dupuit (1863) and Forchheimer (1886), based on this law, produced the approximate theory of groundwater flow in shallow unconfined aquifers that simply gives estimates of the horizontal seepage and of the height of the water table. Thus Colding (1872) was able to give a formula for the maximum water-table height in lands drained by rows of parallel sinks, and even today the theory is still widely used. Slichter (1899) placed the analysis of groundwater flow problems on a sound theoretical basis when he showed that groundwater flow could be regarded as a process that was governed by potential theory, allowing the substantial body of theory associated with this branch of physics to be applied to groundwater flow calculations. Since then groundwater physics itself has been a major contributor to potential theory (Polubarinova-Kochina, 1962; Aravin and Numerov, 1965; Harr, 1962), especially because of the peculiar boundary condition of the phreatic surface in unconfined aquifers that is not met with in other branches of physics concerned with potential. The work of Charnyi (1951) showing that seepage rates through earth banks were correctly, although fortuitously, calculated from the Dupuit formula, led to the seepage analysis (Zaoui, 1964; Youngs, 1965; 1980; 1986) that allows the horizotal seepage to be exactly calculated in unconfined aquifers in which the hydraulic conductivity varies spatially. In addition to this analytical work, electric analogues and Hele-Shaw models have been used extensively in groundwater studies, and today computer-based numerical solutions of Laplace's equation for groundwater flow problems are commonplace. However, in most of this work concerned exclusively with the groundwater region, the flux contribution from the unsaturated soil-water region across the water table and the effect of this zone on changes in the extent of the groundwater aquifer are treated in an approximate way with little consideration to the advances that have been made in the physics of water movement in unsaturated soils.

The physics of water movement in unsaturated soils, in contrast to that in saturated soils, has developed more slowly and is not so generally applied in computations concerned with practical problems. The physical undertanding of water flow in unsaturated soils is based on Buckingham's (1907) concept of capillary potential of the soil water held in lenses between particles in unsaturated soil, that culminated in Richards' (1931) formulation of an equation to describe water flow in unsaturated soils. While Darcy's law was founded from experiment, Richards' equation was developed from theoretical concepts and so required experimental validation before becoming established. The difficulty of performing accurately experiments on flow in unsaturated soils made this validation a painstaking task. Engineeers concerned with groundwater were only concerned with the contribution unsaturated soil made to the groundwater aquifers and this was estimated sufficiently for their purposes from empirical concepts. The advances in the study of unsaturated soil behaviour came from agricultural science where the interest was more in the state of the soil water as it influences crop performances rather than in water transport. Nevertheless, the work on water movement in unsaturated soils was fundamental to irrigation and drainage practices used in agricultural management, so that agricultural scientists became involved in work to establish Richards' equation as the basis for describing this movement (see, for example, Childs, 1969).

Recently there has been a wider realisation that the unsaturated soil-water zone plays an important role in the hydrological cycle (Youngs, 1988). It not only contains vast amounts of stored water that is released over time by drainage and evaporation and plant use, it also acts as a buffer between the atmosphere and groundwater, influencing runoff and regulatimg the input to the groundwater zone. Much of the work done by soil physicists in agricultural science was hydrological, and now the interest in soil-water physics is more in hydrology than in agricultural science. Soil-water movement is becoming much more the concern of environmental science.

With this interest in soil physics by hydrologists and environmentalists, it becomes more important to consider the interaction between the groundwater and unsaturated soil-water zones. The work done in soil physics for agriculture was inevitably slow when it was not considered to be central to the major objectives of agricultural science. However, the studies on the physics concerned with agricultural water management, although directed towards narrow ends, have established fundamental principles on which a comprehensive understanding of soil-water movement can proceed, that combines our knowledge of of the physics of water movement in the unsaturated soil-water zone with the large body of knowledge concerned with groundwater. This paper reviews the present state of knowledge concerning the interaction between the groundwater and unsaturated soil-water regions and discusses examples of its use in solving practical problems. Complicating factors that presently limit the application of theory are also discussed.

THE PHYSICAL PRINCIPLES OF SOIL-WATER MOVEMENT

Water movement in both saturated and unsaturated soils is caused by gradients of hydraulic head, moving from positions of high hydraulic head to positions of low. The hydraulic head is given by the sum of two components:

$$h = z + p \tag{1}$$

where z is the elevation of the point in question above a datum level and p is the soil-water pressure head, measured by the pressure of bulk water in equilibrium with the soil water. In the groundwater region p, as measured by the pressure of the bulk water in piezometers, is positive. In unsaturated soils, where the water is held in the form of lenses in the interstices between soil particles by surface tension forces giving rise to a pressure drop across the curved menisci, p is negative, as measured by the pressure of bulk water inside tensiometer cups in which the water is in communication with the water in the soil through their ceramic walls. The rate of flow of water v is given by the law Darcy (1856) derived from experiments on flow through saturated columns of sand:

$$v = -K\nabla h \qquad (2)$$

where K is the hydraulic conductivity of the soil. While this law was formulated for saturated soils, Richards (1931) argued that it should also hold for unsaturated soils with a lower value of K, dependent on the water content because of the fewer and smaller channels conducting water in unsaturated soils. This was confirmed in the laboratory experiments of Childs and Collis-George (1950) on flow down columns of unsaturated porous materials.

The gradient of hydraulic head that produces the soil-water movement is usually accompanied by a gradient of soil-water pressure. In saturated soils the hydraulic conductivity is constant and does not change with the soil-water pressure. However, in unsaturated soils the soil-water content is dependent on the soil-water pressure, so that a change in the latter implies a change in soil-water content and hence a change in the hydraulic conductivity. For an inert soil with water content θ at (x,y,z) at time t:

$$\frac{\partial\theta}{\partial t} = -\nabla.v \qquad (3)$$

so that equation (2) becomes:

$$\frac{\partial\theta}{\partial t} = \nabla.K\nabla h = \nabla.K\nabla p + \frac{\partial K}{\partial z} \qquad (4)$$

which is a form of Richards' equation. In equation (4) θ, K and p are inter-related hysteretically, with K decreasing markedly with the water content of the soil. Equation (4) is the basis on which soil-water behaviour is calculated from soil-water properties.

Equation (4) reduces to Laplace's equation:

$$\nabla^2 h = 0 \qquad (5)$$

for a saturated soil, which is the equation used for solving groundwater flow problems. Equation (4) is thus the general equation describing flow in both saturated and unsaturated soils and so can be used to compute the soil-water flow in the complete flow region comprising both the groundwater and unsaturated soil-water regions associated with an unconfined aquifer. This has become possible only recently with advances in numerical methods with the present large computers now available, for example in the hydrological model described by Abbott *et al.* (1986). However, difficulties still abound in dealing with the large-scale three-dimensional flow problems often encountered because of problems in obtaining the necessary field data of physical measurements and also because of problems associated with using numerical procedures on large flow regions with spatially varying properties. Thus approximate methods of dealing with the interaction between the groundwater and unsaturated soil-water regions will continue to be used and developed. These require examination and assessment in the light of our knowledge of soil physics.

Particularly difficult is the solution of Richards' equation in two or three dimensions with the soil-water properties varying over orders of magnitude. However, it is found that the water flow from the soil surface through the unsaturated soil to the water table is practically vertical (Childs, 1945), so that a one-dimensional form of Richards' equation can be used that makes it easier to accommodate the functional relationships between K, p and θ. Below the water table the flow is usually far from one-dimensional, but the simpler Laplace equation that describes flow in saturated soils is more easily solved for multi-dimensional flows, particularly for two-dimensional problems that can be solved using conformal transformation methods making use of the hodograph to map the mixed boundary conditions at the water table. Thus it can be convenient to separate the flow regions for analytical purposes although physically the flow region comprises the two. Part of the solution is then the position of the water-table boundary where p = 0.

STEADY-STATE FLOW

Steady-state flow in unconfined aquifers arises when there are sinks and sources of constant strength, such as drainage channels, wells, rivers and canals, in the groundwater region, and when there is a steady distributed flux in the form of precipitation or evaporation through the soil surface that moves through the unsaturated soil to the water table. With the physically reasonable assumption that water moves vertically in the unsaturated soil-water zone to and from the water table, the steady flux across the water table into the groundwater zone is the same as the steady flux across the soil surface. Childs (1945) used physical arguments to show that the flow was nearly vertical, whilst Gureghian and Youngs (1976) confirmed, using finite element calculations in the complete flow region, that little error resulted in using the approximation. With the boundaries other than the water-table boundary known, the potential problem can be solved in the groundwater zone, with the position of the water table emerging as part of the solution. This can be done either by using analytical methods or by using electric analogues or the Hele-Shaw viscous flow analogue.

The approximate Dupuit-Forchheimer analysis provides a simpler means of estimating the phreatic surface. However, its assumption of horizontal flow everywhere and its neglect of surfaces of seepage make the calculated water-table heights inaccurate, particularly in the region near surfaces of seepage in the groundwater zone. The seepage analysis (Zaoui, 1964; Youngs, 1965; 1980; 1986) takes into account the vertical component of flow in the groundwater region. With the seepage potential E defined by:

$$E = \int_0^H Kh \, dz \qquad (6)$$

the components of seepage Q_x and Q_y in the x- and y- directions are given by:

$$Q_x = -\frac{\partial E}{\partial x} \quad , \quad Q_y = -\frac{\partial E}{\partial y} \tag{7}$$

E is described by Poisson's equation:

$$\frac{\partial^2 E}{\partial x^2} + \frac{\partial^2 E}{\partial y^2} = V(x,y) \tag{8}$$

where $V(x,y)$ is the flux, measured positive upwards, through the water table at the position (x,y). The two-dimensional form of equation (8), subject to boundary conditions in E, replaces the three-dimensional Laplace equation for groundwater flow that describes h. Values of E are thus more easily found than values of h, so that horizontal seepage rates are simply found through equation (7). However, the water-table height cannot be found exactly, although close bounds can often be deduced using physical arguments.

When the water-table position is known, the soil-water pressure and content in the unsaturated soil-water zone above it can be found from Richards' equation. The height z above the water table where the soil-water pressure is p, is given by:

$$z = -\int_0^p \frac{dp}{1 + V/K(p)} \tag{9}$$

where V is the steady flux moving vertically through the soil profile, measured positive upwards. When there is precipitation so that V is negative, the water content and soil-water pressure approach constant values as z becomes large (Childs, 1945; Youngs, 1957). When there is evaporation so that V is positive, the solution of equation (9) shows that the soil limits the evaporation for a given depth of the water table below the soil surface (Gardner, 1958; Gardner and Fireman, 1958). For example, if $K(p)$ is given by:

$$K = K_o \exp(cp) \tag{10}$$

equation (9) can be integrated and it is found that the limiting evaporation e is related to the depth Z of the water table by:

$$e = K_o/[\exp(cZ) - 1] \tag{11}$$

NON-STEADY STATE FLOW: THE MOVING WATER TABLE

The exact calculation of the soil-water movement from equation (4) that would give the position of the phreatic surface as well as the soil-water profile development in the unsaturated soil-water zone, is not simple. By making the assumption of one-dimensional vertical flow in the unsaturated zone, some progress has been made in including Richards' equation in general groundwater hydrology models (for example, Abbott *et al.*, 1986). Generally, however, recourse is made to more approximate analyses.

In non-steady state conditions, there is a contribution to the flux through the water table due to the change in the amount of water stored in the unsaturated soil. Thus the flux varies with the time-dependent soil-water distribution above the water table, producing a non-stationary water table because of this varying boundary condition at the upper boundary of the groundwater region. Boussinesq (1904) in effect supposed that the soil water in the unsaturated soil-water zone was in equilibrium with the moving water table at all times, so that with a deep water table a unit fall of the water table was accompanied by a constant amount of water draining from the soil profile, known as the *drainable porosity* or the *specific yield* S. The flux through the water table is thus $S\ \partial H/\partial t$, which, when used in the Dupuit-Forchheimer analysis of groundwater seepage in shallow aquifers, leads to Boussinesq's equation to describe the changing water-table height H in unconfined aquifers in non-steady state conditions:

$$K[\frac{\partial}{\partial x}(H\frac{\partial H}{\partial x}) + \frac{\partial}{\partial y}(H\frac{\partial H}{\partial y})] = S\frac{\partial H}{\partial t} + V_s \qquad (12)$$

where V_s is added to take into consideration the contribution of the flux through the soil surface to the flux through the water table. This is the equation that has been mostly used to calculate the drawdown of the water table to wells and drainage ditches (see, for example, Polubarinova-Kochina, 1962; Aravin and Numerov, 1965). Usually Boussinesq's equation is linearised in order to make analytical solutions possible.

A unit change in the water-table height does not in general produce a constant change in the amount of water stored in the soil profile. Quite generally the specific yield can be defined by:

$$S = \frac{\partial}{\partial H}[\int_H^Z \theta\ dz] \qquad (13)$$

where H is the water-table height above the given datum level and Z is the height of the soil surface. Above a stationary water table, the equilibrium soil-water pressure head is balanced by the gravitational head, given by the height of the water table, so that the soil-water profile is given by the relationship between the soil-water content and the soil-water pressure head, the so-called *soil moisture characteristic* (Childs, 1969). Thus, when the water table is close to the soil surface, S can never be regarded as a constant (Childs, 1960); it becomes equal to the difference between the saturated soil-water content and that at the soil-water pressure head equal to H − Z, when the water table moves very slowly so that the profile is in equilibrium with the water table (dos Santos Junior and Youngs, 1969). Thus Boussinesq's assumption of a constant specific yield implies not only a slowly moving water table to produce an equilibrium soil-water profile above, but also that the water table is deep enough so as to produce complete gravity drainage of the soil profile.

The solution of Boussinesq's equation gives the phreatic surface at a given time t in the form:

$$H = H(x,y,t) \tag{14}$$

with a flux V through the water table:

$$V = S(H)\frac{\partial H}{\partial t} + V_e \tag{15}$$

varying over the area, with V_e given by the rainfall or evaporation rate. The flow in the groundwater zone is assumed to be completely horizontal. If this condition is not approached, solutions are of doubtful value. At any time the groundwater flow is given properly by solutions to Laplace's equation with the flux through the water table being the flux derived both from the surface flux and the change in water content of the soil profile, so that the non-steady state water table regime may be considered to be a continuous succession of steady states (Muskat, 1937). Analytical solutions with the variable flux given by equation (15) are not available, although it is possible to include this difficult condition in analogue and numerical studies. However, it is often the case that the phreatic surface changes at approximately the same rate over the whole area, so that if the specific yield can be considered uniform the flux V becomes momentarily a constant and analytical solutions to many problems become available for the non-steady state.

The use of the concept of the non-steady state groundwater situation being considered as a continuous succession of steady states is made particular use of in studies concerning water-table movements when there are sinks and sources in the groundwater zone with precipitation or evaporation through the soil surface. In such aquifers the water table has a maximum or minimum height H_m that is of particular interest. For a steady state with a uniform flux V through the water table:

$$V = f(H_m) \tag{16}$$

where $f(H_m)$ is a function of H_m for the aquifer involving the soil's hydraulic conductivity and the geometry of the flow region. The drainage equations used by drainage engineers are examples of such a functional relationship. If we now assume that the rate of change of the water-table height is approximately the same over the whole area and that $S = S(H_m)$, the value at H_m, equation (16) becomes:

$$V = S(H_m)\frac{\partial H_m}{\partial t} + V_e \tag{17}$$

or

$$\Delta t = \int_{H_1}^{H_2} \frac{S(H_m)\,dH_m}{f(H_m) - V_e} \tag{18}$$

where the maximum water-table height changes from H_1 to H_2 during the time interval Δt. This offers a simple means of modelling the water-table response in unconfined aquifers.

APPLICATION EXAMPLES

1. The effect of evaporation rates on water-table heights.

Evaporation from soil surfaces, fed by capillary rise from a water table, has a controlling influence on the height of the water table in unconfined aquifers under an artesian head. It is an important factor that must be taken into account in changing the use of land; for example, in Saudi Arabia it is found that the water table rises considerably when evaporation is prevented, and this rise is accompanied by land subsidence, making construction work problematic (James and Little, 1989). The water-table rise in such situations can be calculated by considering the vertical movement of water through saturated soil from the artesian supply at depth to the water table, and thence through the unsaturated soil by capillary rise to the soil surface where it leaves by evaporation.

Suppose there is an artesian head under soil of depth L, rising to a level h_L below the soil surface. When a steady state ensues with an evaporation rate e through the soil surface, the water moving upwards causes the level of the water table to be depressed to a depth Z below the surface. Then, by Darcy's law:

$$e = K_o(Z - h_L)/(L - Z) \tag{19}$$

or

$$Z = \frac{L(h_L/L + e/K_o)}{1 + e/K_o} \tag{20}$$

where K_o is the hydraulic conductivity of the saturated soil, so that the water-table depression ΔZ due to the evaporation is:

$$\Delta Z = Z - h_L = \frac{e/K_o(L - h_L)}{1 + e/K_o} \tag{21}$$

The water-table depression ΔZ is shown as a function of the evaporation rate in Figure 1 for various artesian heads.

The evaporation may be controlled by meteorological conditions, in which case e may be computed from the weather records, for example, by using Penman's equation. However, when the water table is deep, the evaporation is limited by the soil, and in this case e can be calculated from Richards' equation in the form of equation (9), knowing the relationship K(p) between the hydraulic conductivity and the soil-water pressure. For the exponential relationship of equation (10), e is given by equation (11) for the water-table depth Z. Putting this value in equation (20), the depth of the water table in the artesian situation with evaporation from the soil surface can be found by trial and error from:

$$Z = L[(h_L/L - 1)\exp(-cZ) - 1] \quad (22)$$

If evaporation is prevented from lands over artesian aquifers that under normal conditions lose water to the atmosphere, for example by paving or building construction, the water table will rise towards an equilibrium level h_L below the soil surface. James and Little (1989) show how the rate of rise of the water table in this situation can be obtained using the approximate Green and Ampt (1911) theory of soil-water movement.

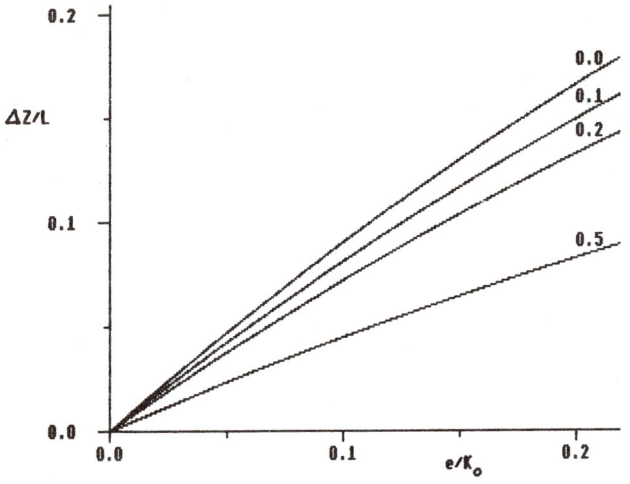

Figure 1. The relationship between the water-table depression ΔZ, expressed as a fraction of the soil depth L, and the steady-state evaporation rate e, expressed as a fraction of the hydraulic conductivity K_o of the saturated soil, for different values of the artesian head h_L, shown as a fraction of the soil depth L by the numbers by the graphs.

2. Water-table movement in flat low-lying wet lands.

Water tables in flat low-lying wet lands are regulated by networks of water-filled ditches that are a feature of the management of such areas. Their level is determined not only by the ditch-water level but also by the rainfall and evaporation through the soil surface that give rise to seepage in the groundwater zone between the water table and the ditches, and by the hydraulic properties of the soil. In the case of rainfall, the water percolates vertically downwards to the water table through the unsaturated soil; in the case of evaporation, the water moves upwards by capillary rise. For a particular field and ditch-water level, there will be a relationship in the form of equation (16) that can be used in the expression under the integral sign in equation (18). With V_B known from weather data if it can be assumed that the flux through the surface communicates rapidly with the water table with no lateral component of water flow, and with the specific yield obtained from experiment, equation (18) can be integrated numerically to find the water-table response to the meteorological input.

Drainage equations giving the form of $f(H_m)$ in equation (18) form the basis of land-drainage theory. Many such equations have been proposed. Ten of the most commonly used were compared by Lovell and Youngs (1984), and the physical assumptions and mathematical approximations critically examined. A convenient power-law empirical equation that represents well the steady-state drainage situation for drains large enough not to impede flow so as to cause head loss that would increase the water-table height (Youngs, 1985), has been modified by Youngs *et al.* (1989) to describe the steady-state water-table height, maintained by rainfall or evaporation, in a rectangular area of wet land surrounded by water-filled ditches. Thus, with a ditch water level H_o, the water-table height H_m in the middle of the rectangular area is given by:

$$\frac{H_m}{D} = i\{i[-A_m\frac{V_B}{K_o} + (\frac{H_o}{D})^{\alpha}]\}^{1/\alpha} \qquad (23)$$

where 2D is the length of the shorter side of the rectangular area, A_m a geometrical factor dependent on the ratio of the lengths of the sides of the rectangle (Childs and Youngs, 1961; Youngs, 1980), K_o the hydraulic conductivity of the saturated soil, and α is a factor dependent on the depth d below the base of the ditch to an imperemable barrier given by:

$$\alpha = 2(d/D)^{d/D} \quad , \quad 0 < d/D < 0.35$$
$$\alpha = 1.36 \quad , \quad 0.35 < d/D < \infty \qquad (24)$$

In equation (23) $i = 1$ if $(H_o/D)^{\alpha} > V_B/K_o$ and $i = -1$ if $(H_o/D)^{\alpha} < V_B/K_o$. The factor A_m has a value of 0.586 for a square area and a value of 0.911 for a rectangular area with the longer side double the length of the shorter.

With H_m given by equation (23), the water-table movement in a field on the Somerset Levels in England was computed using equation (18) with measured hydraulic soil properties and available rainfall and evaporation data, assuming the non-steady state water tables were a continuous succession of steady states. The soil was a very permeable peat, and the water table was regulated by the level of water in ditches surrounding the rectangular field that measured 115m by 230m. Figure 2 shows the computed water-table response near the centre of the field obtained by inputting daily meteorological data that gave the flux V_B, the net upward flux through the soil surface given by the difference between the evaporation and the rainfall. The evaporation was assumed to be the potential evaporation, calculated by Penman's equation, unless the soil was limiting the evaporation. Then it was calculated for the water-table depth below the soil surface from equation (9) with the known relationship between the hydraulic conductivity and the soil-water pressure for the unsaturated soil.

Also shown in Figure 2 are observed water-table heights obtained from dipwell observations. It is seen that good agree-agreement was obtained between the simulated and observed values.

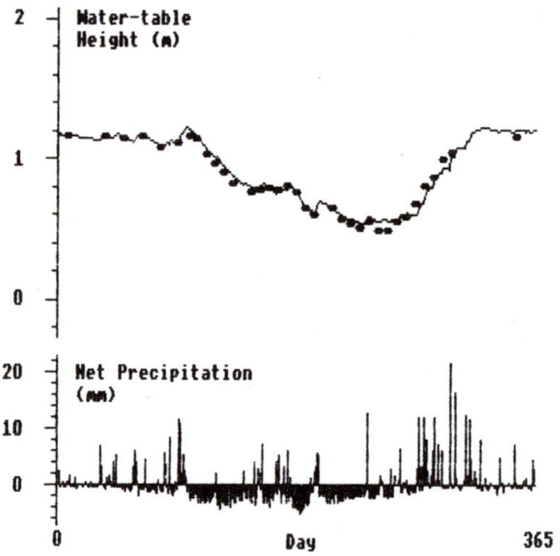

Figure 2. The water-table height in the centre of a low-lying field surrounded by water-filled ditches on the Somerset Levels in England during 1987. The full line is calculated from equation (22) using measured hydraulic soil properties and weather records; the net precipitation (= $-V_s$) is given by the difference between the measured rainfall and estimated evaporation. Measured dip-well results are shown by the filled circles.

DISCUSSION

The two examples given above indicate how the theory of water movement in soils can be applied in problems involving the interaction between the groundwater and the unsaturated soil-water zones. The theory is based on continuum physics embodied in Darcy's law and Richards' equation for flow through inert homogeneous porous materials. The soils involved in the examples were unstructured uniform soils, so that the soil-water movement in them would be expected to behave in a similar fashion to that in the uniform porous materials for which the theory was developed. However, many workers have drawn attention to the inadequacy of the classical theory in many real situations when there is soil heterogeneity, when there is soil swelling and shrinking, and when there is soil aggregation; for example, a recent review of soil physics and hydrology (Youngs, 1988) has high-lighted the difficulties in applying classical theory to field problems concerned with water movement in unsaturated soils.

Porous materials are in general by nature inherently heterogeneous at the scale of the size of the soil particles with the pore space made up of a network of interconnecting pores of different shapes and sizes. Darcy's law implies that a porous body is made up of a large number of groups of pores that are repeated throughout the flow region, and measurements of soil properties and results of calculations relate to volumes greater than some elementary volume that is representative of the pore space (Youngs, 1983). Soil regions smaller than this representive elementary volume do not give consistent soil properties and calculations of soil-water flow on such small flow regions cannot be expected to give accurate predictions.

Besides this micro-structure heterogeneity, heterogeneity in soils manifests itself at many larger scales. Many soils have a macrostructure due to aggregate formation as a result of natural and man-made disturbances, giving a much larger representative elementary volume which contains a representative sample of the macropore space. Hydraulic soil properties, such as hydraulic conductivity, must be made on this much larger volume. Since Richards' equation has been established by considering the soil to behave as a continuum, the macropores have to be viewed as boundaries when considering unsaturated soil-water flow. At a larger scale still, there is heterogeneity at the scale of the field and of the region, and this needs to be considered in large-scale studies of groundwater aquifers.

While stochastic heterogeneity on a field or regional scale is not easily included in theoretical analyses of soil-water movement in the groundwater and unsaturated soil-water regions, deterministic heterogeneity, such as that shown by layered soils, are more easily included when the spatial dependence of the hydraulic soil properties can be measured. For example, there is no difficulty in including the variation of hydraulic soil properties with depth in one-dimensional steady-state calculations of soil-water movement, and the seepage analysis of groundwater flow in unconfined aquifers (Zaoui, 1964; Youngs, 1965; 1980; 1986) particularly includes the spatial variation of hydraulic conductivity.

The variation of hydraulic soil properties has a marked effect on water movement in unsaturated soils, since hydraulic conductivities of coarse-textured soils, although much larger than those of fine-textured soils when saturated, become much less as they desaturate at a relatively small negative soil-water pressure when the finer textured soils are still saturated. Because of the continuity of soil-water pressure across the boundary between two soils, regions of coarse-textured soils can remain dry while in contact with wet finer textured soil. Thus a substratum of coarse-textured soil beneath a finer one does not allow downward drainage to continue once the soil-water pressure of the top soil has decreased to a soil-water pressure at which the coarser soil becomes unsaturated. Also, with a coarse-textured soil beneath a fine one, unstable flow can result with fingers of saturated soil appearing in the coarse soil (see, for example, Miller and Gardner, 1962) 1962).

With surface-active fine-textured soils, swelling and shrinking occur that require the overburden pressure to be considered as part of the driving force for soil-water flow. The theory of flow in swelling soils is not well developed. A one-dimensional theory (Smiles and Rosenthal, 1968) assumed such soils to behave as slurries and that Darcy's law describes the flow relative to the soil particles, resulting in the description of the soil-water movement in the form of a non-linear diffusion equation with a material coordinate system. However, this theory cannot deal with the three-dimensional situation that arises when swelling and shrinking cause cracks to appear. These cracks have a profound effect on the transport of water from the soil surface, with high precipitation rates producing preferential flow down them, wetting up the soil at depth as well as from the soil surface (Beven and Germann, 1982). Vertical fissures produced by burrowing insects and the roots of growing plants have a similar effect. The water movement in such fissured soils has to be modelled considering the relatively unimpeded flow down the channels with water being absorbed from their surfaces more slowly into the soil matrix (Jarvis and Leeds-Harrison, 1987).

To apply the classical theory of soil-water movement, using Darcy's law and Richards' equation, the soil needs to conform as nearly as possible to the inert uniform porous material that is assumed in the establishnent of this theoretical approach. Our present theory needs to be extended to take into account more thoroughly the complicating factors of soil heterogeneity, of soil swelling and shrinking, and of soil aggregation. These complications need to be considered when making measurements of soil properties for predicting water flow in soils. In particular, measurements of hydraulic soil properties have to sample volumes larger than the soil's representative elementary volume, and the latter can be found only by repeated measurment on different sized samples. Indeed, it would seem that the collection of good field data could remain a major problem in the use of our physical knowledge of soil-water behaviour. Nevertheless, the progress that has been made over the last century in the physics of water movement in both saturated and unsaturated soils provide us with a basic physical understanding of the interaction between the groundwater and unsaturated soil-water flow regions, and allows us to examine our traditional procedures for dealing with the contribution of the flow from the unsaturated soil-water zone to groundwater resources.

REFERENCES

Abbott, M.B., Bathurst, J.C., Cunge, J.A., O'Connell, P.E. and Rasmussen, J., 1986. An introduction to the European Hydrological System - Système Hydrologique Européen, "SHE". J. Hydrol., 87; 45-77.

Aravin, V.I. and Numerov, S.N., 1965. Theory of Fluid Flow in Undeformable Porous Media (translated from Russian by A. Moscana). Israel Program for Scientific Translation, Jerusalem.

Beven, K. and Germann, P., 1982. Macropores and water flow in soils. Water Resour. Res., 18; 1311-1325.
Boussinesq, J., 1904. Recherches théoretique sur l'écoulement des nappes d'eau infiltrées dans le sol et sur débit de sources. J. Math. Pure et Appl., 10; 1-16, 363-394.
Buckingham, E., 1907. Studies on the movement of soil moisture. U.S. Dept. Agr. Bur. Soils Bull., no. 38.
Charnyi, L.A., 1951. A rigorous proof of the Dupuit formula for unconfined seepage with a seepage face, (in Russian). Dokl. Akad. Nauk. SSSR, 79, 937-940.
Childs, E.C., 1945. The water table, equipotentials and streamlines in drained land: III. Soil Sci., 59; 405-415.
Childs, E.C., 1960. The non-steady state of the water table in drained land. J. Geophys. Res., 65; 780-782.
Childs, E.C., 1969. An Introduction to the Physical Basis of Soil Water Phenomena. John Wiley, London.
Childs, E.C., and Collis-George, N., 1950. The permeability of porous materials. Proc. Roy Soc., 201A; 392-405.
Childs, E.C., and Youngs, E.G., 1961. A study of some three-dimensional field-drainage problems. Soil Sci., 92; 15-24.
Colding, A., 1872. Om lovene for vandets bevaegelse i jorden. Kong. Danske Vidensk. Selkabs. Skr. 5 Raekke, 9; 563-621.
Darcy, H., 1856. Les fontaines publiques de la ville de Dijon. Dalmont, Paris.
Dupuit, J., 1863. Etudes Théoriques et Pratiques sur le Mouvement des Eaux dans les Canaux Découverts et a Travers les Terrains Perméales (2nd. edn.). Dunod, Paris.
Forchheimer, Ph., 1886. Ueber die Ergiebigheit von Brummen-Anlagen und Sickerschlitzen. Z. Archit. Ing.-Verein Konigreich Hannover, 32; 539-563.
Gardner, W.R., 1958. Some steady state solutions of the unsaturated moisture flow equation with application to evaporation from a water table. Soil Sci., 85; 228-232.
Gardner, W.R., and Fireman, M., 1958. Laboratory studies of evaporation from soil columns in the presence of a water table. Soil Sci., 85; 244-249.
Green, W.H., and Ampt, G.A., 1911. Studies in soil physics. I. The flow of air and water through soils. J. Agr. Sci., 4; 1-24.
Gureghian, A.B., and Youngs, E.G., 1976. The use of the finite element method in obtaining soil-water pressure and soil-water content distributions in drained land with a uniform rainfall incident on the surface. Proc. I.M.A. Conf. Mathematics of Finite Elements and Applications, Brunel University, 289-297.
Harr, M.E., 1962. Groundwater and Seepage. McGraw-Hill, New York.
James, A.N., and Little, A.L., 1989. Sabkha at Jubail, Saudi Arabia: geotechnical aspects. Part I: Properties affecting foundations. Part II: Design. Q. J. Eng. Geol. (in press).
Jarvis, N.J., and Leeds-Harrison, P.B., 1987. Modelling water movement in drained clay soils. J. Soil Sci., 38; 487-509.
Lovell, C.J., and Youngs, E.G., 1984. A comparison of steady-state land-drainage equations. Agric. Water Manage., 9; 1-21.
Miller, D.E., and Gardner, W.H., 1962. Water infiltration into

stratified soil. Soil Sci. Soc. Am. Proc., 26; 115-119.

Muskat, M., 1937. The Flow of Homogeneous Fluids through Porous Media. McGraw-Hill, New York.

Polubarinova-Kochina, P.Ya., 1962. Theory of Ground Water Movement (translated from Russian by J.M.R. de Weist). Princeton University Press, Princeton, N.J.

Richards, L.A., 1931. Capillary conduction of liquids through porous mediums. Physics, 1; 318-333.

dos Santos Junior, A.G., and Youngs, E.G., 1969. A study of the specific yield in land-drainage situations. J. Hydrol., 8, 59-81.

Slichter, C.S., 1899. Theoretical investigations of the motion of ground waters. U.S. Geol. Surv. 19th Ann. Rep. Pt. II; 295-384.

Smiles, D.E., and Rosenthal, M.J., 1968. The movement of water in swelling materials. Austr. J. Soil Res., 6; 237-248.

Youngs, E.G., 1957. Moisture profiles during vertical infiltration. Soil Sci., 84; 283-290.

Youngs, E.G., 1965. Horizontal seepage through unconfined aquifers with hydraulic conductivity varying with depth. J. Hydrol., 3; 283-296.

Youngs, E.G., 1980. The analysis of groundwater seepage in heterogeneous aquifers. Hydrol. Sci. Bull., 25; 155-165.

Youngs, E.G., 1983. Soil physical theory and heterogeneity. Agric. Water Manage., 6; 145-159.

Youngs, E.G., 1985. A simple drainage equation for predicting water-table drawdowns. J. Agric. Engng. Res., 31; 321-328.

Youngs, E.G., 1986. The analysis of groundwater flows in unconfined aquifers with non-uniform hydraulic conductivity. Trans. Por. Med., 1; 399-417.

Youngs, E.G., 1988. Soil physics and hydrology. J. Hydrol., 100; 411-431.

Youngs, E.G., Leeds-Harrison, P.B., and Chapman, J.M., 1989. Modelling water-table movement in flat low-lying wet lands. Hydrol. Process. (submitted for publication).

Zaoui, J., 1964. Les écoulements en milieu poreaux et l'hyothèse de Dupuit. La Houille Blanche, 3, 385-388.

GEOHYDROLOGICAL DATA ACQUISITION: SOME IMPERATIVES

C.P. Gupta

National Geophysical Research Institute
Hyderabad -500 007, INDIA

ABSTRACT

The objective of a Geohydrological Data Acquisition System (GDAS) is the generation of an appropriate data base which could be transformed into a meaningful information system for a hazard free development of groundwater resources. The design of a GDAS should take into account (i) essential features of the geohydrological regime, (ii) specific needs of the users, and (iii) the available financial resources for the establishment, operation, and management of the system. Standardization of the methodology and unrestricted accessibility to data base are other pre-requisites for communication and meaningful utilization of the geophydrological data.

INTRODUCTION

There has been a growing realization of the need for a reliable and credible data base for an objective and rational management of groundwater resources. The need is felt all the more strongly in developing countries where the demand on groundwater resources has been incessantly mounting up due to a fast growth of population and increased agricultural activities. The growth of industrialization and other man-made activities have been giving rise to problems of pollution also. The hazardous effects accruing from indiscriminate exploitation of groundwater resources have been manifesting in the form of decreasing yields of wells, deteriorating water quality, sea water intrusion in coastal aquifers, land subsidence, reduction of base flow of rivers and streams etc.

In such situations, it is extremely important to have a good quality geohydrological data base which could help evolve a sound strategies for groundwater management through

i. defining aquifer geometry,

ii. determining hydrogeological characteristics of an aquifer,

iii. estimating groundwater inflows and outflows,

iv. assessing groundwater abstraction,

v. quantifying hydrochemical constraints,

vi. designing for the augmentation of recharge to an aquifer, and finally

vii. optimizing the use of groundwater.

A list of main parameters/features which should be monitored through GDAS is given in Table 1.

TABLE 1

MAIN PARAMETERS TO BE MONITORED THROUGH A GEOHYDROLOGICAL DATA ACQUISITION SYSTEM (Spatio-Temporal Variations at Appropriate Scales)

CATEGORY I	CATEGORY II
CLIMATOLOGICAL	GEOPHYSICAL (SITE SPECIFIC)
PRECIPITATION (RAINFALL AND SNOW) EVAPORATION, WIND SPEED, TEMPERATURE, HUMIDITY, RADIATION	SURFACE GEOPHYSICAL SURVEYS (ELECTRICAL, ELECTROMAGNETIC SEISMIC, GRAVITY AND MAGNETIC)
SOIL	BORE HOLE GEOPHYSICAL DATA (SP, ELECTRICAL LOGS, NEUTRON, NATURAL GAMMA, GAMMA-GAMMA, CALIPER LOG, TEMPERATURE ETC.)
TEXTURE, POROSITY, SOIL MOISTURE pH, EC, SOIL WATER CAPACITY, INFILTRATION CAPACITY	
	HYDROGEOLOGICAL
GEOLOGICAL	AQUIFER CHARACTERISTICS: (HORIZONTAL AND VERTICAL HYDRAULIC CONDUCTIVITIES, AND SPECIFIC YIELD STORATIVITY, TRANSMISSIVITY, ETC.)
FORMATIONS, LITHOLOGY, STRUCTURES, BASEMENT DEPTH AND CHARACTERISTICS	
HYDROLOGY	WATER LEVELS IN DIFFERENT AQUIFER ZONES
STREAM DISCHARGE, WATER LEVELS, STORAGE IN LAKES, PONDS AND RESERVOIRS, SPRING DISCHARGE	SUBSURFACE FLOW RATES (HORIZONTAL AND VERTICAL)
	RECHARGE RATES, EXCHANGE WITH SURFACE WATER BODIES
VEGETATION	
CROPPING PATTERN, FORESTS ETC.	QUALITY
SOCIO-ECONOMICAL	CHEMICAL QUALITY (pH, EC, MAJOR IONS)
POPULATION, LAND-USE, WATER DEMAND, BENEFITS AND COST ESTIMATES	BIOLOGICAL QUALITY (MNP OF COLIFORM, BOD, COD AND OTHER MICRO-ORGANISMS)

DESIGN CONSIDERATIONS FOR GDAS

An appropriate Geohydrological Data Acquisition System (GDAS) should, therefore, be designed for generation of the required data base. The approaches and strategies for geohydrological data acquisition depend upon a number of factors, the most important of them are the following:

i. The geohydrological data must meet the needs of users and help them make right decisions on issues which may range from determining the number and location of irrigation wells in a particular area, fixing of rates and duration of pumping, designing of schemes for drinking water supply, protection of an aquifer from pollution, to implementation of geohydrological research projects. Quite often, these decisions are taken arbitrarily without the help of a quantitative data base.

ii. The design of GDAS and scale of operations should conform to the physical framework of the geohydrological set-up taking into account the prevalent meteorological and hydrogeological conditions. The dynamic nature of the groundwater regime should also be kept in view. As is well known, the input and output stresses on an aquifer widely vary in time and thereby modify the geometry, boundaries and storage of an aquifer. The conceptualization and quantification of an aquifer system from a discrete set of data collected from a few observation points is, thus, rendered more difficult and complex. The GDAS should aim at better measurement accuracies and wider base with the progressive development of these resources.

iii. In the context of many tropical countries particularly for India, the vagaries of monsoon and recurrence of droughts are some other additional features which should be reckoned with while monitoring the geohydrological data.

iv. The most important consideration is the cost-effectivity of GDAS. In developing countries where financial resources are scarce, special caution need to be exercised in sophistication of the hardware and software for acquisition, storage, retrieval, and communication of data. The density of data network and frequency of observations should also have an important bearing on the cost-benefit analysis.

It is against this backdrop, that a situation-specific Geohydrological Data Acquisition System should be planned and implemented taking into account (i) the hydrogeological conditions, (ii) an envisaged time-frame, and most important of all, (iii) the available financial resources. This will, of course, be a dynamic process and will require modification/updating of the system depending upon the level of development of groundwater resources.

Details of geohydrological data acquisition are well documented in the National Handbook of Recommended Methods for Water Data Acquisition brought out by the U.S. Geological Survey (1977 and subsequent updates). Baker, Jr. (1977) has given techniques and operational details of computer storage and dissemination of hydrogeologic data. These aspects are not discussed here.

DENSITY OF DATA NETWORK

As the observations are made at a discrete set of observation points, it is inescapable to resort to interpolation of observed/estimated values of characteristic parameters to obtain value of these parameters anywhere else in the area. The density of the data network should be such that the interpolation should give reasonably accurate parametric values and thereby help quantify the invisible groundwater regime. The scale of geohydrological data monitoring network should, normally, be comparable or preferably smaller than the scale of inhomogeneities. An optimal network may be defined as the system which gives the "best" estimate of the true value of a parameter at any location. The "best", in turn, has to be defined as an appropriate function of accuracy and cost in a particular situation.

A number of deterministic as well as stochastic models have been developed to optimize the data network. Still, as a first order criterion, the data network should conform to the physical features of the groundwater regime. In a country like India, the monitoring of data for recharge estimation should have a closely spaced network during monsoon season than in the remaining part of the year. In general, the data network should be denser near the inhomogeneities as compared to regions where the hydrogeological conditions are relatively uniform. For example, closely spaced monitoring is warranted near a contaminant plume.

In cases where there is a number of aquifer storeys, water levels, hydrogeological parameters, groundwater abstraction etc., should be separately monitored for each individual aquifer zone. Usually in such situations, determination of both vertical and horizontal hydraulic conductivities is also important.

The density of data network may be classified, according to the scale of operations, under three groups (Brouwer, 1981) viz.,

i) Global or Base Line Network (Level I)
ii) Regional Network (Level II)
iii) Local Water Management Network (Level III)

The approaches, instrumentation, and strategies for geohydrological data acquisition for each scale should be such that one set of data could telescope into the other.

The spatial density of observation points as well as the frequency of measurement of time variant parameters should progressively increase from a global class of network to that of a local class. It has been found that the method of kriging could help select locations for additional data collection points in an objective fashion (Pucci Jr. and Murashige, 1987).

For each distinct meteorological, topographical and geological unit an individual Base-line Network should be established and operated on a long term basis to serve as the reference network for smaller scale networks. A Regional Network (Level II) should aim at the preparation of regional water surface map for different aquifer zones and an overall estimate of groundwater potential in an area. The Local Network (Level III) provides a data base for prognosing the response of an aquifer to various input and output stresses, particularly the artificial ones. A brief description of information to the extracted from the data at three levels is given in Table 2.

TABLE 2

SALIENT INFORMATION TO BE EXTRACTED FROM GEOHYDROLOGICAL DATA ACQUISITION SYSTEM

(After Peek, 1980)

I LEVEL [BASELINE NETWORK] (Scale = 1:250,000)	II LEVEL [REGIONAL NETWORK] (Scale = 1:100,000)	III LEVEL [LOCAL NETWORK] (Scale = 1:50,000 or 1:20,000)
GENERAL IDENTIFICATION OF	*QUANTITATIVE DETAILS OF*	*DATA WITH ERROR ESTIMATES FOR*
i) Climate	i) Hydrogeologic systems and significant sub-units giving hydrology, water quality and hydraulic characteristics	i) Conceptual and mathematical modeling of the aquifer unit
ii) Topography, geological settings and geomorphology	ii) Regional water table/ piezometric water level configuration	ii) Water development and management alternatives
iii) Principal hydrogeologic units and their geometry	iii) Potential yields of principal aquifer units	iii) Response of the aquifer to (a) artifical stresses and (b) management alternatives
iv) Range of well depths and yields	iv) Effects of man-induced stresses on a) hydrologic conditions b) water quality	iv) Effect of man-made activities on water quality
v) Range of water level fluctuations	v) Data requirements for groundwater management	v) Monitoring strategies for management of quantity and quality
vi) Water use, type and number of users		
vii) Range of water quality for main units		

The geohydrological data, as briefly listed in Table 1, could be classified under two categories. The one which may be obtained from field investigations carried out under allied disciplines. For example, the satellite imageries or aerial photographs generated for other purposes (including resource surveys) could be used, with advantage, for the preparation of base line information for geohydrological investigations. Similarly, the investigations by soil scientists and agronomists provide extremely useful data/information about movement of water in unsaturated zone which has an important bearing on the rate at which water can percolate from the ground surface to the aquifer. The data on soil moisture movement and chemical quality are useful in monitoring the pollutant movement as well as evolution of remedial measures.

The second category of data needs special kind of directed investigations and measurements. These mainly include site-specific geophysical exploration, hydrogeological investigations, and water quality monitoring. The primary source for this category of data are wells tapping different aquifer zones. While some of the existing wells could be used for data collection, it is desirable to construct some observation wells which could be specially designed to monitor various parameters from different aquifer zones. It should be possible to (i) monitor water levels in different zones, (ii) sampling of water from different depths for quality monitoring, (iii) pump testing of different aquifer zones, etc. Location of some of the observation wells may be in the vicinity of sources and sinks viz., pumped wells, streams and other surface water bodies, for short range monitoring of the effects of these stresses.

Peek (1980) has given an empirical relation, as illustrated in Figure 1, for spacing of wells for monitoring of data relevant to different levels of networking.

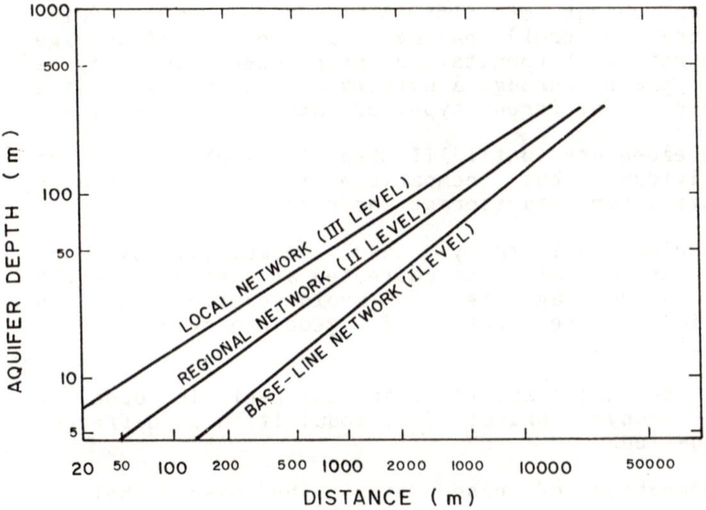

Fig. 1. Average Well Spacing for Data Monitoring

CONCLUDING REMARKS

It is surprising that in spite of a good awareness about the need for geohydrological data acquisition, the data base particularly at levels II and III have been rather scanty in most of the developing countries. If we have to rationally manage our groundwater resources, it is imperative that the following should be accomplished within a reasonable time frame:

I. Standardization of

  i. formats for data collection and compilation,
  ii. instruments and methods of measurements, and
  iii. formats for data presentation.

  relevant to various geohydrological environs.

II. Compatibility of hardware and software for data storage at different centres enabling easy communication and exchange of data.

III. Establishment of three levels of data acquisition networks and their management in a coordinated manner.

IV. Establishment of distributed data banks (with central control) having compatibility of storage and retrieval formats and their interfacing (may be in phases) through a network for an easy and quick access to different types of data by the users.

V. Increased use (until III step is implemented) of individual but compatible micro/mini computer systems for data storage and retrieval.

VI. Ensuring uniformity and compatibility of geohydrological data collected at different levels by various agencies to avoid duplication and minimize the cost of geohydrological data acquisition.

VII. A cost-benefit analysis, and its periodic updating for geohydrological data acquisition in different situations.

VIII. Preparation of base maps (hydrogeomorphology, hydrogeology, water quality) on a scale of 1:250,000.

ACKNOWLEDGEMENT

The author is grateful to Director, NGRI for permission to publish the paper. Sincere thanks are due to Mr. V.V.S. Gurunadha Rao, Dr. V.S. Singh and Dr. M. Thangarajan for thought provoking discussions and to Mr. G.R. Babu and Mr. Joseph Gabriel for help in preparation of the manuscript.

REFERENCES

Brouwer, G.K., 1981. Geohydrological Information Systems, Groundwater Survey TNO Report, PN 81-01, pp. 105.

Peek, H.M., 1980. Classification of Groundwater Studies - A Critical Need for Research and Development in the Eighties, Groundwater, v. 18, no. 4, pp. 326-330.

Pucci Jr., A.A. and Murashige, Jo Ann. E., 1987. Applications of Universal kriging to an aquifer study in New Jersey, Groundwater, v. 25, no. 6, pp. 672-678.

Baker, Jr., C.H., 1977. Computer Storage and Dissemination of Hydrogeologic Data by the United States Geological Survey, IAHS Memoires v. XIII, Part 1 (Birmingham Congress, U.K.), pp. B14-B21.

GEOHYDROLOGICAL DATA BASE FOR WATER MANAGEMENT

Prof. Dr. B. H. Briz-Kishore
Director of Water Resources
Jawaharlal Nehru Technological University
Hyderabad - 500 028 (India)

1. ABSTRACT

Effective Water Management acts as a catalyst for extensive socioeconomic development and administrative policy modernisation. The extent of modernisation depends upon the stimulating impact of the data processing itself, and it is in this sense the geohydrological data base development as part of the water management process becomes one of the most crucial building blocks of public administration. It is well known that the data required for different applications in water management is copied from original field records several times. This kind of copying the same data for different applications at different times results in the redundant storage of data under the same processing environment. Conventional methods involving different physical file structures pose severe limitations such as rigidity for carrying out any extensions and necessity for frequent reorganization. Hence it is essential to formulate an efficient data storing and interacting mechanism removing all the prevailing lacunae and limitations of conventional methods of storing through recently emerging data base systems.

2. INTRODUCTION

System data collection is generally directed to a specific purpose and has utility onlyto meet that particular need. But sometimes the data is gathered and stored to build up an information base for future unspecified uses. The planning, design and operation of water resources systems depend heavily on an inventory of accurate and relevant data. This is very often time dependent and hence historical records are often essential for any project. In many cases, the value of information gained from the data is a function of the length of logical record. Hence it is apparent that the data collection procedures may be ensured with long continuous records having broad areal coverage, with the idea that the futuristic needs, which are unrecognized today, may derive information subsequently from these records fairly well.

The collection of data should be undertaken with a good understanding of the requirement and completeness, consistently maintaining high degree of precision and accuracy. However the quality of data is impaired if there are gaps in the records. Records even with high quality may still become useless if they are not continuous. It is therefore important to place on record extra infromation also in addition to the actual data to facilitate the future users in evaluation of the extent of changes in environmental factors.

3. GEOHYDROLOGICAL MANAGEMENT

A Geohydrological management system is classified into two major areas involving water resources information and socioeconomic development, whose effective combination is usually referred to as water management system. These two broad areas are further classified into three subdivisions each, depending on the requirements of data. The entire conceptualisation is diagrammatically presented in Fig.1 and the data elements in each of the subdivision are indicated in Table 1.

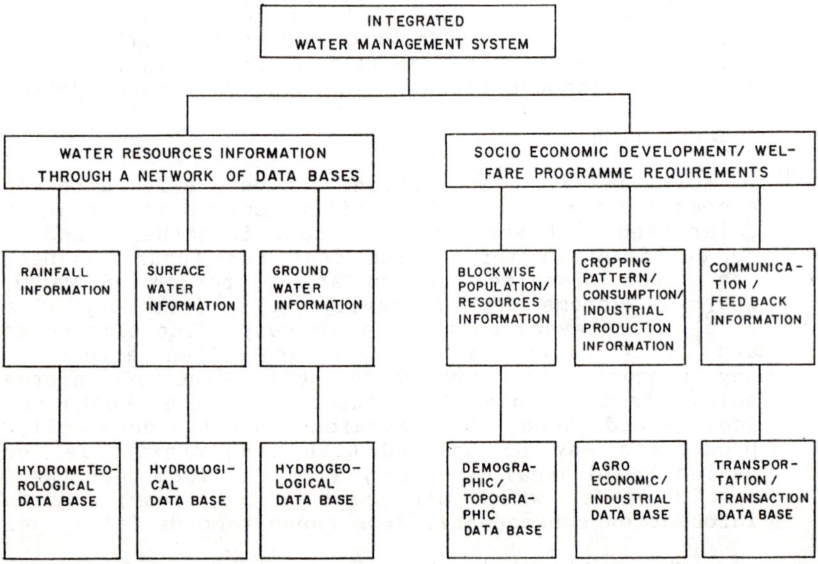

Fig. 1: A Data Base Approach for Itegrated Water Management System

HYDROMETEOROLOGICAL DATA BASE	HYDROLOGICAL DATA BASE	HYDROGEOLOGICAL DATA BASE
AVERAGE TEMPERATURE	COASTAL FEATURES	WELL INVENTORY
HUMIDITY %	LAKE/RIVER/TRIBU TORIES	WELL TYPE
RAINFALL	DRAINAGE DETAILS	ENERGIZATION TYPE
WIND VELOCITY	RESERVOIRS	AMOUNT OF WATER LIFTED
PRESSUURE(ATMOSPHERE)	FLOODS	TRANSMISSIVITY
EVAPOTRANSPIRATION	PERCOLATION COEFFICIENT etc.	STORAGE COEFFICIENT
		WATER LEVELS
		MASS BALANCE etc.

DEMOGRAPHIC/TOPO-GRAPHICAL DATA BASE	AGRO ECONOMIC/ INDUSTRIAL DATA BASE	TRANSPORTATION DATA BASE
GEOGRAPHICAL AREA	CROP PARTICULARS	INFRA STRUCTURE
FOREST AREA	LIVE STOCK POPULATION	COMMUNICATION SYSTEMS
AREA REQUIRING SOIL CONSERVATION	MAJOR/MINOR IRRI-GATION SCHEMES	ROADS & BUILDINGS SPECIAL DATA
MINERAL RESOURCES	IRRIGATED AREA	MARKETING FACILITES
MALE/FEMALE/CHILD POPULATION	CROPWISE WATER REQUIREMENTS	NAVIGATIONAL FACILI-TIES/REQUIREMENTS
FUNCTIONAL CATEGORI-ZATIONAL OF VILLAGES	TYPES OF URBAN/ RURAL INDUSTRIES	
PHYSICAL PROPERTIES OF ROCKS	ECONOMIC STANDARDS OR WORKING PERSONAL	
CHEMICAL AND MECHA-NICAL PROPERTIES OF ROCKS etc.	CLASSIFICATION OF FARMS ACCORDING TO SIZE OF HOLDINGS	
	EDUCATIONAL FACILITIES	
	PUBLIC HEALTH PARTICULARS	
	ENERGY REQUIREMENTS	
	POLLUTION DETAILS etc.	

Table 1: Data elements in various Data Bases of Water Management System

The data of these elements have to be effectively stored and integrated for purposeful implementation in any water resources development programme. A model of such an ideal data storage and retrieval system through computerised data base organisation is described in the following sections.

4. CONSTITUENTS OF COMPUTERISED DATA BASE

The data of individual parameters is generally collected at different coordinates of space and time under the custodianship of various organisations. These organisations have well identified data collecting units specifically designated for this purpose. The Units collect the data through different processes of observation, measurement and calculations, and record the same in data manuscripts as per the proforma evolved. Normally each manuscript gives provision for one or more key parameters with reference to which the remaining data is organised and presented. For example, the rainfall data containing the intensity of rainfall is presented with respect to the rain gauge station and time which act as keys in this case.

Because of severe limitations of manual methods in the storage and retrieval of data, the computerisation is resorted to (Briz-Kishore,1983, 1984, 1987 a) which however involves several stages, as described below:

4.1 Creation of Data Files

The data that is initially captured into the manuscripts is transferred to magnetic media like floppy discs and magnetic tapes with the help of data entry machines and/or personal computers. The recorded data is subjected to rigorous validation checks(Fig.2)for possible elimination of human errors and also for filtering the wrong data that has crept into the process through various means. The validated data is arranged in a systematic way by sorting the same with respect to the key fields and merged at appropriate locations to form computer data file as shown under data file creation in the same Fig.2. Each line of data so arranged is called a data record and consists of values for several related parameters in different fields. The contents of master file can be opened for retrieval as and when necessary for different applications.

4.2 Updation of Data Files

Most of the systems generate additional data under different conditions which could not be envisaged at the first instance, while some systems originate data at regular intervals. In order to define and understand the water system thoroughly, it is essential to incorporate the feedback/additional data also in the earlier created data files for maintaining upto date information. For this purpose a transaction file is created with the additional/feedback data, following the necessary precautionary measures of data validation checks, as indicated in Fig. 2.

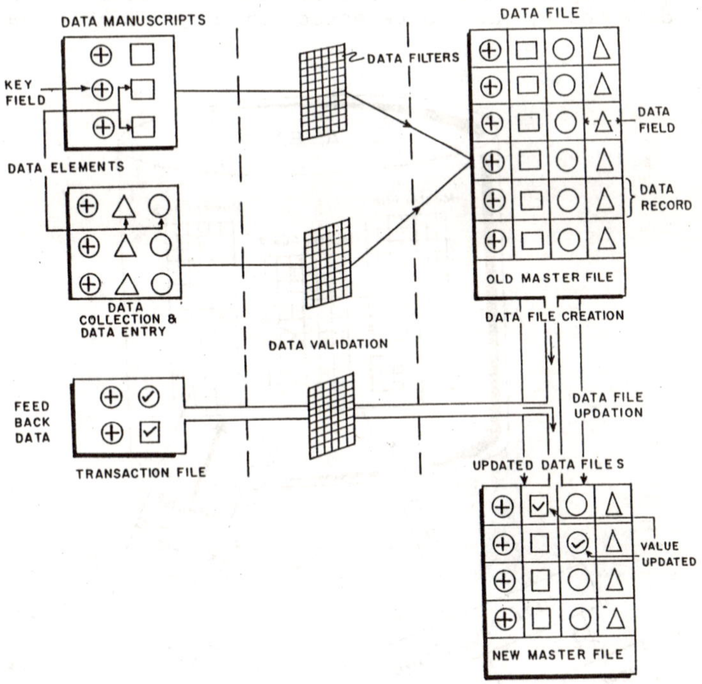

Fig. 2: Data Collection, Validation, File Creation and Updation Procedures

The transaction file contains, in addition to new data records, provision for special commands like insertion, deletion and replacement for taking appropriate action with respect to the earlier created data file (which is hereafter termed as old master file). While the insertion command indicates that the new data record is to be appended at the specified address in the same form without any change, the replacement command indicates that part of the old file has to be modified with new data. The deletion command demands elimination of a group of records from the old master file which have lost their significance hereafter. The changes in the old master file as indicated by the records of transaction file are affected through a special procedure called "updation" as indicated in the same Fig. 2.

4.3 Operation and Updation of Data Base

While the above process creates one master file, creating several number of such master files with a simplified updating and retrieving mechanism constitutes a data base, as represented in Fig. 3. The data base

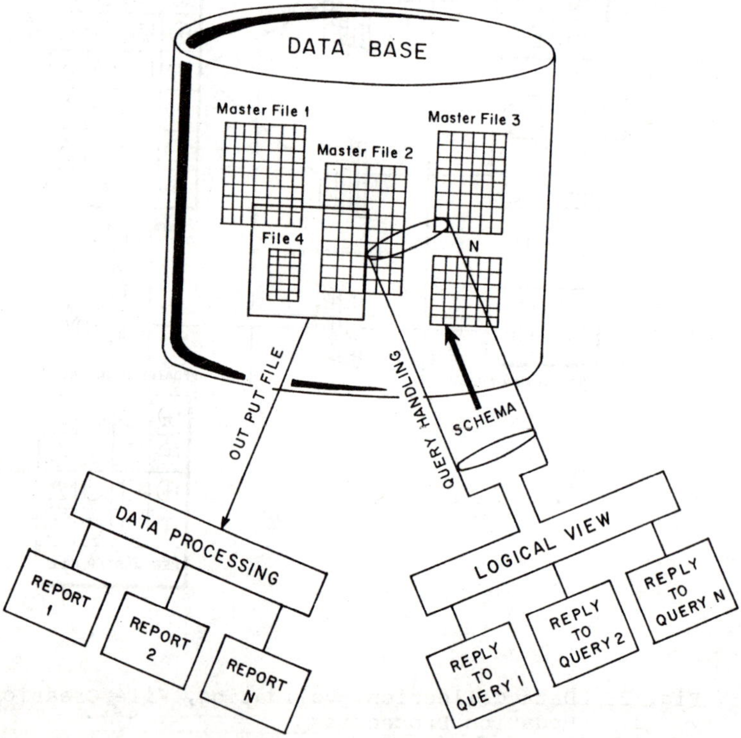

Fig. 3: Data Base Management With Normal Data Processing & Query Handling System

files convey a special meaning through several logical views of the same data in different combinations for understanding the system in different aspects. In addition to normal data processing for report generation, the system facilitates a powerful query handling by telescopically scanning different master files simultaneously. Further the data base concept relieves the users from writing exhaustive set of commands for sorting and merging which occupy a major portion of work load of programming, debugging and execution.

5. DATA BASE MODELS

For efficient interaction and retrieval among several files, the data base can be designed and developed in three different ways, namely, hierarchical, network and relational models. (Briz-Kishore, 1986)

The hierarchical model can be explained very easily from the term "hierarchy". The hierarchy in the model lies with the overall control of the functions through a single defined element and its relation may be explained as "one to many correspondence". As an example, the regional hydrological data is dependent on basin data which in turn depends on wellwise data, basin and rain gauge station characteristics. Any required information pertaining to these elements has to be retrieved through regional data only, thus maintaining the hierarchical status of the model.

When different elements in the existing system are not accepting any hierarchy and retain their own identity and at the same time maintain coexistence with other related elements, the combinations result in the network model. In any basin the data pertaining to water levels, recharge and discharge are interrelated and the pertinent information regarding any one of the elements and their relations among them can be elucidated as classical example of network systems.

The relationships in hierarchical or network models can be represented in the form of charts and tables which is known as relational model. The well number versus water level observations can be well explained through relational model. (Briz-Kishore, 1987 b)

5.1 Development of Geohydrological data base

The geohydrological system dealing with integrated water resources involves three such units of data base to handle its three main constituents, viz, Hydrometeorological, Hydrological and Hydrogeological subsystems, as shown in Fig. 4. As can be seen from the figure, these data bases individually comprise of master data files belonging to the concerned set of parameters of these sub-systems as presented in Fig. 3. At the same time, the data base units have a provision to exchange the data among themselves for the parameters which are commonly required for their precise assessment and evaluation.

For example, the mass balance of the hydrological data base has its inputs in the rainfall data of hydrometeorological system, which in turn has its inputs in evaporation and evapotranspiration aspects of the surface and ground water systems. Further, the interaction between

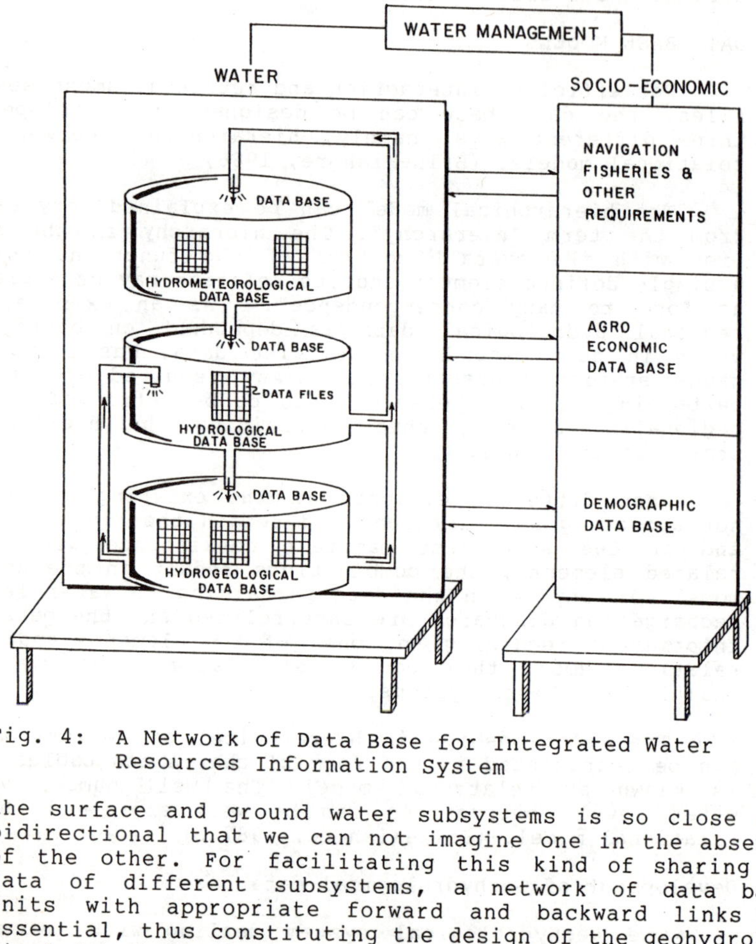

Fig. 4: A Network of Data Base for Integrated Water Resources Information System

the surface and ground water subsystems is so close and bidirectional that we can not imagine one in the absence of the other. For facilitating this kind of sharing of data of different subsystems, a network of data base units with appropriate forward and backward links is essential, thus constituting the design of the geohydrological data base.

5.2 Water Management through geohydrological data base

The geohydrological data base, capable of providing integrated information on water resources, constitutes only one part of the water management. The other equally important part is constituted by the socioeconomic information system of the society, which may be broadly subdivided into demographic, Agroeconomic and Navigationfisheries data base units as shown in the same Fig. 4. While it is an established fact that the entire civilisation of human societty depends mainly on the water resources, it is equally true that the planning, allocation and utilisation of water resources depend on the socio-economic developmental schemes of the society as well, thus establishing a strong mutual interaction.

Hence the water management will be effective and efficient only when its planning, execution and monitoring are based on the sound information generated by the geohydrological data base in consonance with the socio-economic programme data base.

COMPUTER CONFIGURATION

The design and development of such a system can be implemented(Briz-Kishore,1988)on any centrally located supermini computer of recent generation, with six PC/XT terminals operated for handling the six units of the geohydrological data base. Further each data file will have more than 1000 data records, with average record length of 120 bytes for handling different combinations of information, thus requiring a memory space of about 1 MB for buffers and the same amount of memory requirement (1 MB) for operating system and language processors. Thus the main memory requirement amounts to about 2 MB.

The system configuration requires two winchester discs of about 140 MB capacity to continuously update and retrieve the data and also for sort/merge and query handling. In view of enormous amount of data to be handled, it is always advisable to go for high speed printers, prefereably 600 lpm. Thus a typical computer configuration as arrived is shown in Table 2.

Description	Capacity
Central Processing Unit	32 bit super mini computer
Main Memory	2 MB 4 MB
Winchester Disc	2 Nos. (160 MB each)
Magnetic tape	1 No. (Streamer type)
Terminals	6 PC/XT terminals with communication multiplexer
Line Printer	600 lines per minute

Table 2: Typical Computer Configuration

7. CONCLUSIONS

The Computerised data base will provide integrated information on mass balance at any point of time for pragmatic planning, efficient execution and serious monitoring of several socio-economic schemes for the benefit of the society. This approach also enables to overcome the problems in the manual methods of data handling and removes all redundancies, inconsistencies and inaccuracies. Further, the system helps in fast review and evaluation of various resources through instantaneous information for more effective reorganisation and reorientation. The system, can be operated even by non-technical persons and is capable of responding to several queries simultaneously to stimulate the information through graphic outputs, when well defined documentation is available.

References

Briz-Kishore B.H.(1983);An approach for usage of micro computers in ground water management studies; Journal of Association of Exploration Geophysicists, India, Vol. 3 No.4.

Briz-Kishore B.H.(1984);An automated information system for hydrogeological & Basin Characteristics; Geophysical Research Bulletin, Vol. 22 No. 2 & 3.

Briz-Kishore, B.H. (1986) Hydrological Data Base A Modern Tool for Effective Information, Institution of Engineers (India), Vol. 67.

Briz-Kishore, B.H.(1987-a) A New Design on Personal Computer for storing and retrieving hydrometeorological data; Hydrology Journal of IAH, Vol. x No 1.

Briz-Kishore, B.H.(1987-b) Network for Ground Water Monitoring; Jal Vigyan Sameeksha, Vol. II No.2.

Briz-Kishore, B.H.(1988); A Compact SOCRATE Program for an Efficient Water Management Information System; Institution of Engineers (India),Vol. 68.

A 3-D STOCHASTIC DISCRETE FRACTURE NETWORK MODEL FOR SIMULATING FLOW AND TRANSPORT THROUGH FRACTURED MEDIA

M.C. CACAS, E. LEDOUX, G.de MARSILY* and B. TILLIE

CENTRE D'INFORMATIQUE GEOLOGIQUE
ECOLE DES MINES DE PARIS
FONTAINEBLEAU, FRANCE

* also at Universite Pierre et Marie Curie, Paris, France

ABSTRACT

A large-scale investigation of fracture flow was recenlty conducted in a granite uranium mine at Fanay-Augeres, France. Its aim was to develop a methodology for the investigation of possible nuclear waste repository sites in crystalline environments, and thus to determine what measurements to make, and what models to use in order to predict the flow rate in the medium, i.e., its average value and its variability at different scales, as well as the transport. Four types of data were collected: (i) geometry of the fracture network; (ii) local hydraulic properties measured by injection tests in boreholes; (iii) global hydraulic behaviour from flow rate and piezometric head distribution at a 10^6 m^3 scale; (iv) tracer tests performed at a scale of up to 40 meters. A stochastic fracture network model was developed. The fractures are represented by discs the geometry of which is inferred from information (i). The flow through the disc network is simulated by assuming that it occurs through small "channels". The model calibration consisted in adjusting the channel hydraulic conductivities on information (ii). The model was then used to predict information (iii). This attempt to validate the parameters and the model structure appeared to be positive. A particle transport simulator is coupled to the flow model, based on a particle-following method using the advection-biased random-walk technique. Microscopic dispersion in the fractures and slowing-down effects due to unevenness of the flow paths are taken into account. The transport model was calibrated on tracer tests (iv). The model appears to be able to account for the large-scale channelling phenomenon observed in situ. Moreover, its dispersive properties controled by the network geometry and the hydraulic conductivities assigned to its flow elements are reasonably comparable to those of the real site. This feature tends to confirm the validity of the preliminary hydraulic calibration of the model, and thus to justify the approch used to simulate hydraulic and transport phenomena in fractured media which is based on the assumption of a strong channelling effect in the fractures.

THE FANAY-AUGERES EXPERIMENT:
A LARGE SCALE FLOW AND TRACER TEST IN FRACTURED ROCKS

We will first present the Fanay-Augeres data, then the model, its calibration, and how it was partially validated. Fanay-Augeres is an active uranium mine in central France, near Limoges. The ore is contained in dykes in highly fractured granite. The experimental site was selected in an ore-free horizontal drift connecting two areas of the mine, 150 m below ground. The drift had been opened by conventional drilling and blasting, and the test section did not intersect any major fractured zone. A major fracture plane was, however, known to exist above the drift, but it was found to be clay-filled and impervious. The design of the experiment was quite similar to the "ventilation test" experiment performed at the Stripa mine in Sweden (WITHER-SPOON et al, 1981).

1.1 Data Collecting

The measurements were performed in a 100 m-long test section of the drift. Ten 50m-long radial boreholes were fully cored in three planes perpendicular to the drift (Figure 1). The following data were collected:

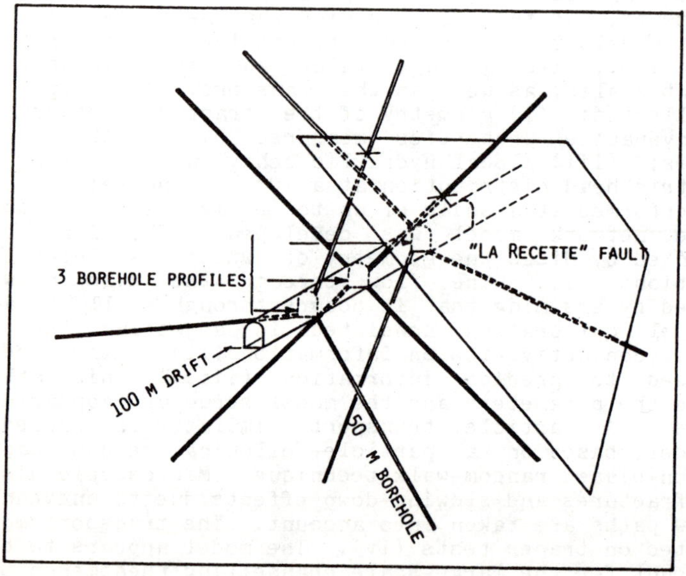

Figure 1: THE FANAY-AUGERES EXPERIMENT POSITION OF DRIFT AND BOREHOLES (figure from the BRGM report 87 SGN 502 STO [1987])

- i. A systematic survey was made of the eastern wall of the drift of the test section, where each fracture was numbered, and the following properties determined: type of element, aperture, fracture filling, presence of water, decompression, rugosity, length, position and orientation.

ii. The structural characteristics of all the discontinuities encountered by the exploration hole were studied on core samples, and their spatial orientation was determined partly by a core-orientator, partly by taking prints of the boreholes. The data gathered in these different ways (orientation, distance between fractures,....) were added to the data collected on the wall of the drift.

iii. The permeability in the ten boreholes was determined by injecting water at several prescribed pressures either behind a single packer or between double packers. Three lengths were chosen for the chambers: ten tests were performed in 50m-long chambers, 50 in 10m-long chambers, and 180 in 2.5m long chambers.

iv. The inflow coming to the drift over the 100m section was also recorded continuously. Since this inflow is rather high (3 to 6 l/mn), the flux of water evaporated in the drift ventilation flux can be neglected, contrary to what happened in Stripa.

v. The piezometric head around the drift was measured in the 10 boreholes, in 68 5m-long sections isolated with double packers (7 per borehole, ranging from 5.5 to 47.5m from the drift wall).

vi. Tracer tests were carried out by injecting various tracers into ten injection chambers located in two of the ten drilled boreholes. The sampling was made in the water flowing into the gallery. The tracer concentration was registered versus time, and three sampling points were used.

Data (i), (ii) and (iii) can be considered as small scale measurements. Data (iv) and (v) are large-scale measurements, and data (vi) are roughly intermediate.

1.2 Statistical Data Analysis

The geometric data analysis was initiated by LASSAGNE (1983). He defined fifteen directional fracture sets from the 1000 orientation measurements performed on the wall of the drift. We combined these fifteen sets into five larger sets, including the orientation measurements performed both in the drift and along the cores. A residual set was also defined; it includes the fractures which do not belong to any of the five main sets.

A directional fracture set can be represented by the ensemble of the directions of all the perpendiculars to the fractures it includes. A main direction (a directional pole) can be assigned to this ensemble, and we can study how the perpendiculars are distributed around this main orientation. As in DERSHOWITZ (1985), we used for four sets a Fisher-von-Mises distribution of the perpendiculars around the pole. The Fisher-von-Mises probability density function is defined as follows (MARDIA, 1972):

$$f(a) = \frac{k}{2 \cdot \sinh(k)} \exp(k \cdot \cos a) \cdot \sin a$$

k: parameter of the pdf
a: angle between the mean direction and the fracture pole

The fractures of the other two sets are nearly uniformly oriented over the orientation range. The parameters of the fitted probability density function are presented in Table 1.

TABLE I

Set	Density in m	Mean of the ln (length in m)	Standard deviation of the ln (length in m)	Fisher distribution parameter
1	1.2	-0.1	0.75	5.
2	0.2	0.15	1.	3.4
3	1.5	-0.15	0.75	uniform
4	1.95	-0.3	0.85	6.3
5	1.49	0.	0.71	27.0
residual	0.8	-0.02	0.85	uniform

The trace-length distribution is then considered set by set and inferred from the trace-length measurements performed on the drift wall. The fracture traces are one meter long on average. The empirical trace-length distribution can reasonably be approximated by a log-normal distribution whose parameters are presented in Table 1.

The fracture density was evaluated only from the fracture frequency observed along the cores (8 fractures per m on average), because the fracture trace density observed on the drift was is biased: the smallest fractures (<20 cm) were not recorded, and the fractures with a very thin aperture were not visible. The fracture density in terms of the number of fracture centers per m3 is a function of the fracture frequency along the cores. There are two cases in which the fracture density can be deduced exactly from this fracture frequency: when the fractures are randomly orientated and when they are parallel (ROULEAU, 1984). We made the assumption that all the fractures in a set can be considered as parallel and we calculated the fracture frequency of each directional set along each core. These frequencies were corrected according to the bias induced by the angle between the core and the average orientation of the fracture set. The fracture frequencies calculated in this way are given in Table 1. It appears that the medium is widely fractured (7 fractures per m3). As the fractures are also quite extensive, we can conclude that the fracture network is well connected.

The 180 injection flow-rates given by the injection tests in the 2.5m long chambers can also be treated statistically. Figure 2 shows the histogram of the logarithm of the flow-rates. The empirical distribution of the flow-rates can be roughly approximated by a log-normal distribution. This result takes into account that one third of the injection tests provided so small an injection flow rate that it could not be measured.

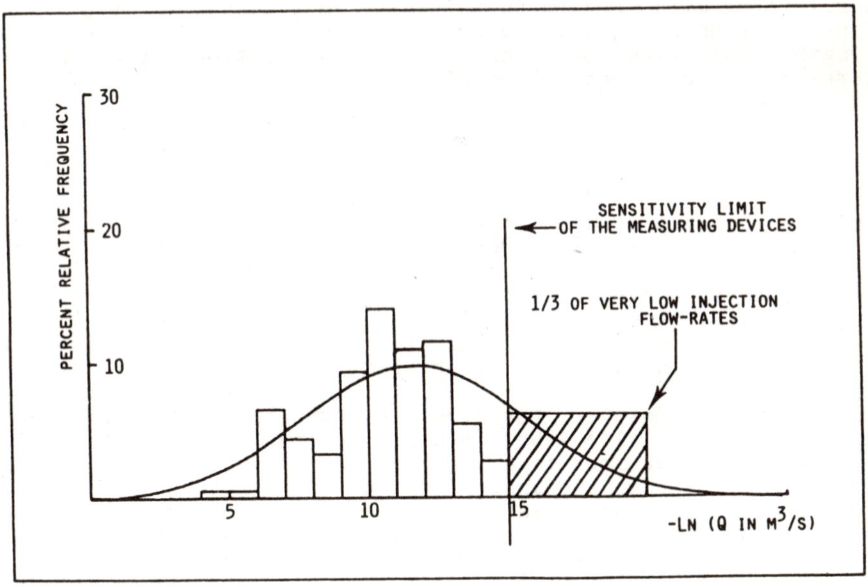

Figure 2: HISTOGRAM OF THE MEASURED INJECTION FLOW RATES

FRACTURE NETWORK MODEL

A key assumption in this study is that the global behaviour of the fractured medium does not depend on the exact geometry of the fracture network but on its statistical properties. Therefore, our goal is only to generate fracture networks whose geometry is statistically similar to the observed one.

The fractures are represented by disks located randomly in space, as in the Beacher model (BEACHER et al, 1977). Each directional family is represented by a set of disks with a distribution of radii and directions inferred from the statistical analysis of in-situ data.

The fracture centers are located in space by a Poisson process. The imposed fracture density is 5 fractures per m3. This value is lower than the density observed in the field (7 fractures per m3): we assumed that a third of the fractures were filled or had very low conductivity (as half of the observed fracture traces were dry on the drift wall) in order to simplify the computation. We will see later on that we could reduce that number even further to 2.5 fractures per m3.

The distribution of radii is taken as the same distribution of trace-lengths as observed on the wall of the drift, which is log-normal in the case of Fanay-Augeres. In fact, a log-normal distribution of radii does not lead to the same distribution of trace-lengths, but the error seems to be negligible with regard to the measurements uncertainty. This approximation was justified by testing with the model the trace length distribution on a wall cutting through a three-dimensional network of disks.

2.1 Flow Modelling in the Network

Both the two and the three-dimensional discrete fracture network models presented in the past few years e.g. LONG et al (1982), ROULEAU (1984), ROBINSON (1984), ANDERSSON (1987), DERSHOWITZ (1984) are based on the assumption that the flow in the fractures occurs between two smooth parallel plates. It takes place over the entire area of each disk and the boundaries are defined by the contours of the fractures and by the intersections with the other elements of the network.

There is, however, the recent evidence that "channelling" takes place inside the fracture planes. GENTIER (1986) analyzed the flow area in a fracture in a granite core sample by observing the two inner faces of the fracture. She found that many flow-paths appear, and that a wide area of the fracture is not water conductive, especially when the fracture is loaded. Such a phenomenon can also be seen in underground drifts where the out-flow points are made visible by the red ferrous iron compounds which precipitate (NERETNIEKS, 1987). TSANG (1987) also suggested that flow within a fracture plane should be considered as a series of parallel one-dimensional "flow-tubes". BOURKE et al (1987) observed such flow tubes in an in-situ experiment on a fracture plane. In Fanay-Augeres, these out-flows are punctual and show that the water flow is more linear than two dimensional.

The new technique which we developed for modelling the flow in the fractures is inspired by these observations. The nodes of the network are no longer the intersections between the fractures (as in LONG, 1983) but their centers. We assume that the flow occurs through bonds joining the centre of each disk to the centre of the adjacent disks, provided that the fractures are connected (Figure 3). The bonds are made up of two parts, one for each fracture, meeting at the intersection between the two disks. Each part is assumed to be equivalent to the set of channels inside the fracture which ensure the hydraulic linkage with the connected fracture. Each bond is assigned a hydraulic conductivity which is a given property of the fracture to which it belongs. The resulting hydraulic resistance of the entire bond is thus a function of the conductivity values in each adjacent fracture and of the distances from the two fracture centers to the middle of the intersection line.

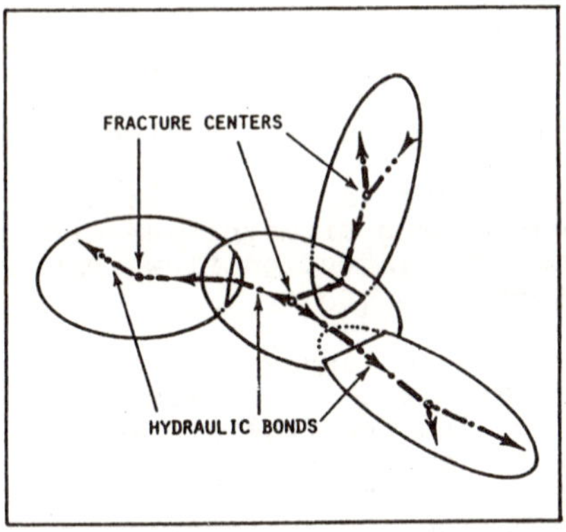

Figure 3: FLUID FLOW SIMULATION

The head is computed by writing a mass-balance equation at each node, i.e each fracture center, where the flow bonds converge. This leads to the solving of a system of linear equtions with as many unknowns as the number of fractures connected to the network. In a well-connected network as in Fanay-Augeres, each fracture is connected with, on average, five other fractures. This means that flow computation is five times faster than in the classical methods where the intersections between fractures are taken as the nodes.

This model may appear unrealistic for a fracture including a great number of intersection lines, or when the fracture is too large: the flow bonds may cross intersection lines or be much longer than in the real situation because we impose on them to connect at the fracture center. However, as will be explained in the next section, the bond conductivities are not considered as external data independent of the model structure, but are calibrated on the hydraulic injection test data. Therefore the fitted conductivity values implicitly include the bias induced by the possible excessive length of the bonds.

It must also be remembered that the assumption of circular-shaped fractures is not of great consequence in our modelling technique. It only plays a role in defining the connections between the fractures, but the flow conputation does not depend on the contours of the fractures.

2.2 Parameters of the Model

In classical fracture network models, the fracture conductivities are introduced as data. In our case, the hydraulic conductivity of the bonds cannot be directly deduced from in-situ observations, knowing that this equivalent conductivity includes the aperture irregularities and tortuosity of the fractures and the channeling phenomenon. That is why we introduced the bond hydraulic conductivities as a parameter of the model. In Fanay-Augeres, the presence of moisture was recorded for each fracture trace, and one realizes that all the five sets include conductive fractures. Therefore, no fracture set can be neglected in the fracture network. One also notices that despite the very important connectivity of the medium (around seven metric fractures per m3), there is a four orders of magnitude difference between the largest and the smallest injection flow rate measurements in the 2.5m long chambers!

We finally decided to assign a single distribution of conductivities to all the fractures. The fitting of one conductivity distribution per fracture set would have led to the adjustment of too many parameters in view of the number of tests available for calibration (180). In a first step, a log-normal distribution of fracture conductivities was prescribed. Then, the two first moments of the distribution had to be fitted: the mean and the standard deviation of the logarithm of the conductivities. The log-normal distribution was chosen because of the log-normal distribution of fracture apertures generally observed in the field.

2.3 The Calibration

The fractured medium model should be able to account for the large-scale spatial variability although it is fitted on small-scale measurements. The small-scale permeability fluctuations are given by the 180 injection tests performed in 2.5m long chambers along the boreholes. As we saw earlier, these are nearly log-normally distributed. Our idea was then to simulate these injection tests and try to adjust the bond conductivity distribution in order to get the same distribution of computed injection flow-rates as the one observed. This can be seen as using the ergodic assumption.

200 independant three-dimensional fracture networks were generated with different random generator seeds, inside which an injection test was simulated with the boundary conditions described in Figure 4. Then, we established that the computed injection flow-rates were log-normally distributed, just like

the measured ones. At this stage, it is very easy to make the parameters fit in order to get the same mean and standard deviation of the computed flow-rates as the measured ones. This is done by first adjusting the standard deviation of the conductivity logarithms and then the mean, which does not interfere with the fitting of the standard deviation. Since the fitting of a log-normal distribution of conductivities turned out to be successful, we kept this distribution all through this work.

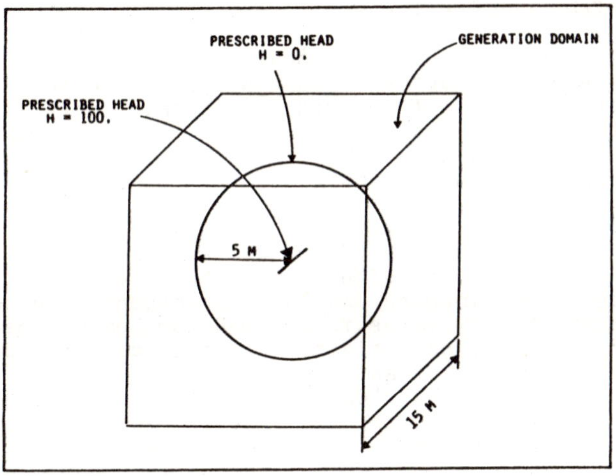

Figure 4: AN INJECTION TEST SIMULATION

The mean m and the standard deviation σ of the conductivity logarithm are: m=-10.4m/s and σ =7.4m/s. The fact that the value of σ is large means that the fracture conductivities are broadly distributed. There is an eight orders of magnitude difference between the first 10% and the last 10% of the conductivities! In the case of such a broad distribution of conductivities and a strongly connected medium the less conductive fractures are negligible. CHARLAIX et al. (1987) showed that the flow takes place in the more conductive fractures and that the equivalent permeability of the medium is controlled by the conductivity of the less conductive among the more conductive fractures which are necessary to reach the percolation threshold.

Therefore, a large number of the less conductive generated fractures can be neglected in the flow computation in order to decrease the computing effort. After several tests, it was shown that the fracture density which must be used is 2.5 fractures per m3.

GOING FROM SMALL TO LOCAL AND GLOBAL SCALE

Because of computing time and memory limitations, the global-scale hydraulic properties of the medium cannot be studied directly by simulating the flow in a very large network. We are limited to using the model for studying the local hydraulic properties of the medium, such as anisotropy or spatial conductivity variability. Then, with the help of the ergodic assumption, the global-scale hydraulic properties can be derived from the local ones.

We first defined the local scale as the size of a domain containing enough fractures to constitute a network. As the Fanay granite is highly fractured, the local scale is around ten times the average fracture length, i.e. 10m in our case.

The anisotropy was investigated by simulating seventeen realisations of a 15*15*15m cube into which a parallel flow from different directions was simulated by rotating the flow boundaries. The parallel flow is simulated in a 10*10*10m "sub-cube" with two opposite prescribed head faces and a no-flow condition imposed on the four others. An equivalent hydraulic conductivity value is derived for each position of the limits and a conductivity ellipse can be drawn. No preferential direction of anisotropy seems to emerge from the 17 permeability ellipses. One can therefore conclude that the medium appears isotropic at the global scale, even if an anisotropic behaviour can be observed at the local scale.

The permeability values of the 17 simulations introduced previously are now used to estimate the global-scale permeability of the medium. We assume the permeability field to be one realization of an ergodic random process where each local permeability is randomly sampled in the spatial distribution function of the local permeabilities.

According to theoretical results established e.g. by MATHERON (1967), GUTJAHR et al (1978), DAGAN (1982), for parallel flow conditions in a three- dimensional porous medium, the global permeability value always lies between the geometric mean Kg and the arithmetic mean Ka of the local permeabilities and is given in the first order by:

$$K = K_g \, (1 + \sigma^2/6)$$

where σ^2 is the variance of the logarithm of the permeability. In our case, the fractured medium is three-dimensional and isotropic. Moreover, under the hypothesis of ergodicity, we can consider that the 17 local-scale network realizations generated earlier, are representative of the local variability of the medium. Then, we can apply the above rules to the 17 permeability values. We get:

- geometric mean of the permeability values = 1.5 10(-8) m/s
- arithmetic mean of the permeability values = 2.3 10(-8) m/s

Then, in the case of parallel flow, the global permeability value of the medium belongs to the following range:

1.5 10(-8) m/s < global permeability < 2.3 10(-8) m/s

and to the first order:

K = 1,6.10(-8) m/s

VALIDATION BASED ON LARGE-SCALE MEASUREMENTS

The fracture network model which was developed and calibrated has, by definition, the correct variability of the hydraulic conductivity at the scale of the injection tests, since this is the information used for calibration. How can we validate the model, i.e. both its structure and its parameter values ?

i. By a change of scale; in practice, we can only determine a "global" hydraulic conductivity in situ by using the measurements of the flow coming to the drift and of the obSERvED piezometric field. We will compare this global conductivity to the global conductivities determined with the model in the previous section.

ii. By using the results of tracer tests.

To determine the global hydraulic conductivity of the real medium, an equivalent porous medium model of the regional flow around the drift was used to compute the hydraulic conductivity required to reproduce simultaneously both the measured drainage flow rate and the measured piezometric field. As this piezometric field is approximately the same over the three sections of the investigated domain, the global flow around the drift can be simulated in 2-D in a plane perpendicular to the drift axis.

The shape of the simulated domain is a ring whose internal boundary is the drift contour where zero pressure is prescribed. The external boundary is a circle centered on the drift axis, where the measured heads are prescribed (Figure 5). The conductivity is fitted so that an ordinary finite element model can reproduce the measured drainage flow rate in the drift. This test was made for seven different values of the external boundary radius, since the head measurements were made in 68 chambers located at seven distances from the drift axis: 5.5m, 12.5m, 19.5m, 26.5m, 33.5m, 40.5m and 47.5m. From this, we obtain seven conductivity values corresponding to seven radii of the simulated domain. The conductivity variations versus the radius of the domain are shown in Figure 6.

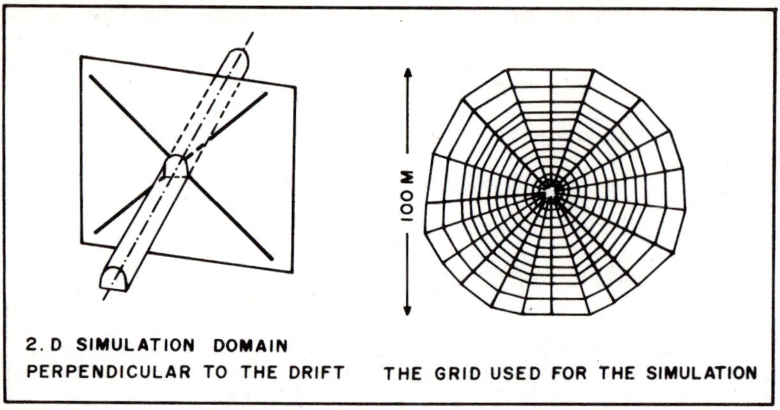

Figure 5: REGIONAL FLOW SIMULATION

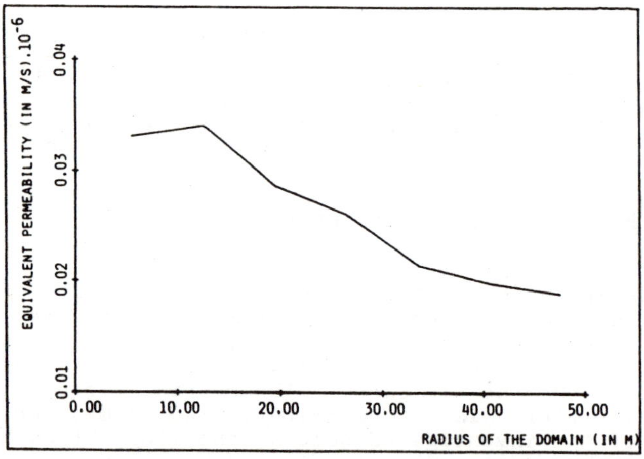

Figure 6: EQUIVALENT PERMEABILITY AS A FUNCTION OF SCALE

It appears that the equivalent hydraulic conductivity of the medium is a function of scale. Furthermore, this permeability seems to reach a constant value when the observation scale arrives at its largest value. The following conclusions can be drawn :

- The scale effect can be observed in-situ; it is revealed by the modification of the equivalent conductivity when the scale of observation is changed.

- It appears that there is a slight decrease in conductivity at a short distance from the drift (below 12.5 m). This could be due to the closing of the joints owing to the increase in tangential stress at the drift wall.

- The size of the representative elementary volume seems to be reached at the hectometric scale. A single hydraulic conductivity value can be assigned to the medium. According to the data collected in the investigated domain, this conductivity value is:

$$K = 1.8*10(-8) \text{ m/s}$$

This value falls exactly within the range of the global conductivity obtained with the model (1.5 to 2.3*10(-8)m/s).

THE TRACER TESTS

6 different chemicals (Amino G, Zinc E.D.T.A., Rhodamine Wt, Iodide INa, Fluoresceine and Nickel E.D.T.A.) were injected in 10 5m-long chambers located along two 50m-long boreholes perpendicular to the drift (Figure 7). These tracers are assumed to be chemically stable and not to react with the medium. The tracers were deposited without any significant pressure increase in the chambers. Then, the tracer concentration of the water seeping through the drift was monitored for 30 to 150 days. The water samples were taken at 3 locations (Figure 7):

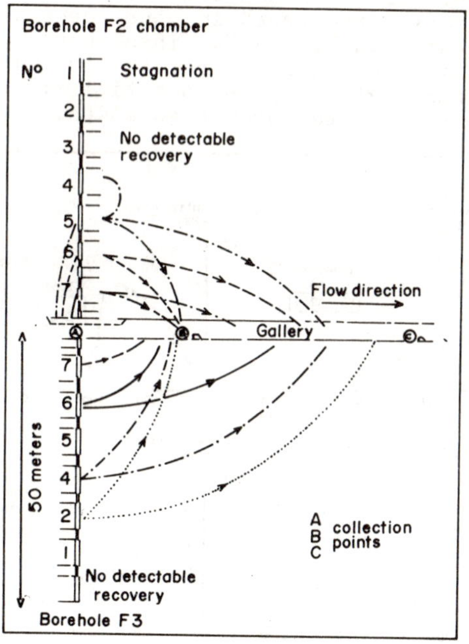

Figure 7: THE TRACER TESTS PERFORMED IN THE FANAY-AUGERES GRANITE

- at the roof of the gallery, just at the level of the injection section;

- in the drainage ditch collecting the water flowing to the gallery, 25 meters downstream from the injection section;

- in the drainage ditch, 50 meters downstream from the injection section.

The 10 injection experiments yielded erratic results:

- the tracer stagnated in chamber 1-F2 (see notations on Figure 7);

- the tracers injected in chambers 3-F2 and 1-F1 were never recovered in the drift;

- injection in chamber 7-F3 was disturbed by a technical failure;

- the breakthrough of the tracer injected in chamber 5-F2 was multimodal and rather difficult to interpret.

Only five out of the ten injections led to unquestionable breakthrough buts with various time-scales. The breakthroughs curves of the tracers recovered in the central part of the drift are presented in Figure 8. These curves are peak-shaped with a long tail but appear relatively smooth. We will characterize them later by two quantities: the time of peak arrival and the total breakthrough duration.

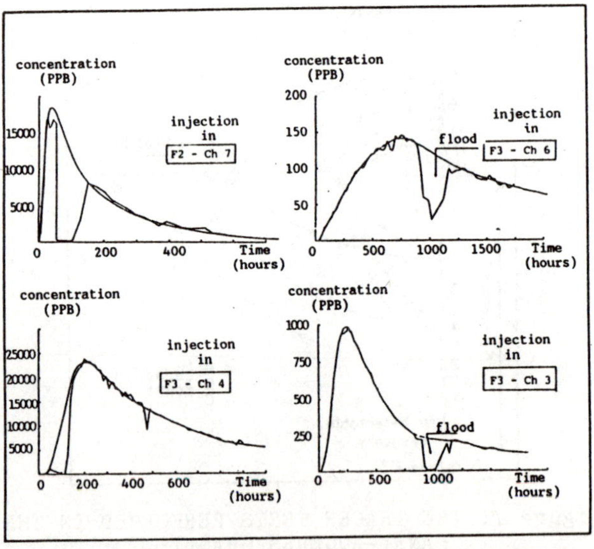

Figure 8: BREAKTHROUGH CURVES OBSERVED AT THE COLLECTING POINT B

The tracer breakthroughs are also characterized by the very low percentage of tracer that returned to the drift. Even for the tracer injected in chamber 7-F2, 10 m from the drift, only 45% of the injected tracer was collected in the drain. Other tests vary between 0% and 23% recovery.

5.1 The Transport Model

In systems like the fracture network of Fanay-Augeres, tracer motion is mostly governed by the flow velocity field. The transport model is thus coupled to the flow simulation.

Three approaches at least can be considered when simulating transport through such a network of "pipes". The first one, presented e.g., by ROUX et al (1986) consists in solving the convection-diffusion equation in each elementary bond of the system and then using a transfer matrix algorithm to convolute the bond responses to the tracer injection. The second method consists in simulating a particle advection-biased random-walk through the system of pipes. The last method is much easier to apply. It was used e.g. by ROBINSON (1984), SCHWARTZ et al (1983), SMITH and SCHWARTZ (1984) and TSANG and TSANG (1987). A third technique called "particle-following" method was proposed e.g. by ROBINSON (1984). In its original form, it consists in following discrete particles added to the flow, whose trajectories are drawn at random according to the flow pattern given by the hydraulic simulations. The main assumptions are: (i) the particles are moved by advection only; longitudinal diffusion is neglected; (ii) chemical reactions and sorption are not taken into account; (iii) diffusion into the rock matrix is neglected.

The particle trajectories are drawn at random according to the following rules:

- The particle trajectory is divided into elementary steps corresponding to the displacement between two adjacent nodes (fracture centers). An average travel time is computed at each step. It is inferred from the flow-rate and the volume of the bond through which the particle travels. The total residence time of a particle in one domain is given by the summation of the elementary travel times computed at each step, from the entrance to the exit point.

- When the particle arrives at a new fracture center, the next bond through which it is going to run is drawn at random. We assume that there is a perfect mixing of the different fluxes entering into a fracture. Therefore, the probability that the particle follows a given bond is proportional to the ratio of its flow rate to the total flux that goes through the fracture. This assumption of perfect mixing at each fracture node seems to be supported by the work of HULL et al (1987), and NOVAKOWSKI et al (1985).

5.2 The Average Residence Time

The difficulty in assigning an average residence time to a particle at each step is to estimate the volume of the channel that links the two fracture centers. The hydraulic conductivity of this channel is determined by randomly sampling in a calibrated distribution function, but an infinity of different shapes and volumes of the channel can lead to the same hydraulic conductivity value. The lower limit of these volumes is obtained for a straightforward smooth pipe. Its length is inferred from the fracture centers and the intersection segment coordinates. Then, its volume can be deduced from the total length and the hydraulic conductivity of the pipe, transformed into an equivalent radius assuming Poiseuille flow.

Of course, in a real rock mass, the channels that insure the hydraulic linkage between two interconnected fractures are not smooth and linear pipes. For the same hydraulic conductivity, the volume of the real hydraulic path is much larger than the volume of the equivalent linear pipe. This feature led us to assign to the bonds the volume of the corresponding linear pipe multiplied by a uniform coefficient Cr. We called this coefficient the "slowing-down" coeffecient because the average particle velocity decreases when the coefficient is increased; it is a very sensitive parameter of our migration model.

5.3 Hydrodynamic Dispersion Inside the Fracture Planes

The transport of solute in the fracture planes involves hydrodynamic dispersion along the pathways from one fracture to the next. The method used to estimate the average residence time in the bonds is presented above. Longitudinal microscopic dispersion can be incorporated approximatively into the model by randomly modifying this average residence time in each bond. The final residence time is obtained by randomly drawing the velocity deviation of the particles in a Gaussian distribution function with a mean of 0. and a standard deviation σ given by:

$$\sigma = 1/T*\mathrm{sqrt}(2*D*T)$$

where
- T is the average residence time
- D is the dispersion coefficient in the fracture plane

The dispersion created in this way is an approximation of the monodimensional diffusional model of dispersion. D will be calibrated on the global breakthrough curves. It is a parameter of the particle transport model. Sensitivity tests showed that it only smoothes the residence time curves, and does not modify their shape.

5.4 Model Calibration

As the transport model was shown not to be very sensitive to the microscopic dispersion, we did not try to finely adjust it on in-situ observations. We fixed it to a value of D=0.8m/s, which leads to smooth simulated breakthroughs like those observed in situ.

To fit the slowing-down coefficient, the only experimental data we have are the tracer breakthrough curves in the gallery. These tests pointed out the heterogeneity of the migration time-scale at the investigated scale (around 10 meters). Unfortunately, we cannot quantify this heterogeneity in statistical terms, because only five of the tracer tests led to unquestionable results. Moreover, they are not comparable since they were performed with different experimental conditions. Therefore, the approach used above for calibrating the hydraulic model can no longer be followed.

Our calibration approach is then to perform a large number of realizations of the five in-situ tracer tests, taking into account the different experimental conditions. Then, we adjust a single slowing-down coefficient for all the simulations, until we obtain, for all the five tests simultaneously, the best possible adequation between the observed in-situ breakthrough times and the distribution of the breakthrough times given by the realizations. Unfortunately, one of these tests involves such a large fracture network that the simulations could not be performed because of the excessive computing efforts required. Only 4 tests could therefore be simulated and for each one the distribution of the transport model time-scales could be estimated. Around 20 realizations of each of the tracer injections in chambers 7-F2, 6-F2, 6-F3 and 4-F3 were performed with a slowing-down coefficient Cr=1. These simulations gave the 4 sets of 20 dates for the peak of arrival.

The tracer tests were simulated by a pulse injection of particles in a segment that represents the injection chamber. The experimental conditions of the 4 tracer tests were precisely represented on the model, both as to boundary conditions and geometry. We emphasize that the distance between the injection chamber and the gallery was exactly the same in the model and in the real medium. Figure 9 shows an example of such simulations. The experimental breakthrough times are compared to the histograms of the 20 simulated ones.

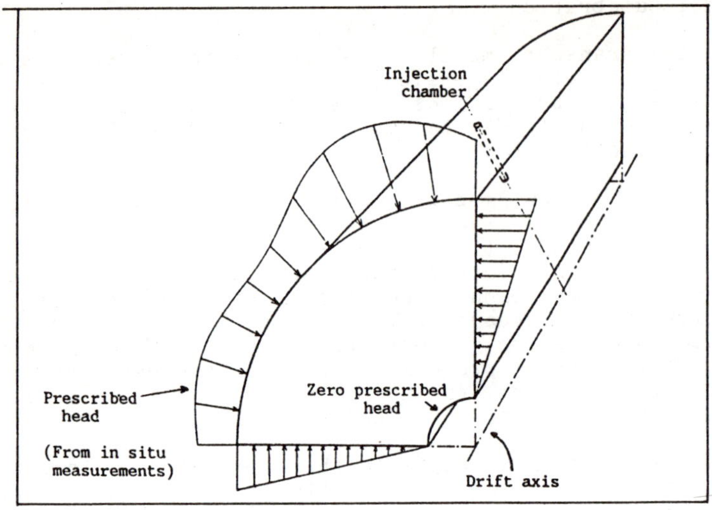

Figure 9: BOUNDARY CONDITIONS USED FOR SIMULATING THE TRACER INJECTIONS

Since Cr is only a time-scale multiplying factor, no additional computation is required for the calibration. The Cr value is chosen so that the central part of the 4 histograms using the same Cr value for all 4 calculations, are as close as possible to the 4 in-situ values of the peak time of arrival. This objective seems to be reached for a slowing-down coefficient of 32. (see Figure 10). Our underlying assumption here is that there is no reason why the 4 experiments should have simultaneously a systematic bias, and all lie on the same side of their respective distributions.

In the calibrated flow model, the average radius of the pipes is 3.10(-5)m. The average radius of the straight pipe in terms of volume is thus 17.10(-5)m, which is still very small. However, a set of 6 tortuous smooth pipes twice as long as the initial equivalent straight pipe, but with globally the same equivalent hydraulic conductivity have a total volume 7 times larger (their diameter is 1/1.7 of that of the initial pipe). Then, if we add dead ends or stagnant void regions that multiply the total volume by a factor 5., the residence times are multiplied by 35. Therefore, a slowing-down coefficient of 32 appears to be realistic.

With a ribbon-path having the same behaviour, the results could be explained with the following dimensions:

- width   l = 0.015 meters = 1.5 cm
- aperture e = 6.0*10 (-6)m = 6μm

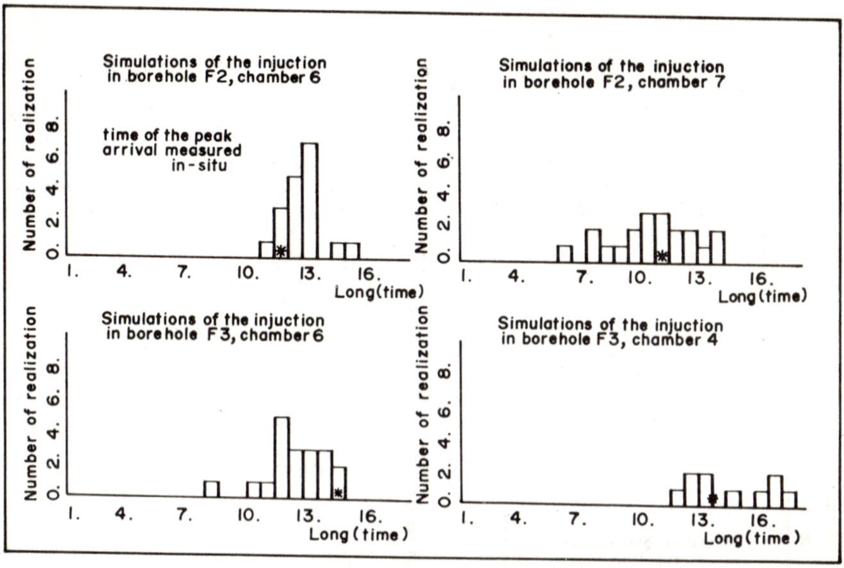

Figure 10: HISTOGRAMS OF THE TIME OF THE PEAK ARRIVAL ASSOCIATED TO THE 4 SETS OF SIMULATION OF THE TRACER INJECTION IN F2-Ch6, F2-Ch7, F3-Ch6, F3-Ch4 - Cr=32

These values appear to be very small. If we take a tentative value of the ribbon-path width of l=0.1m with the prescribed average hydraulic conductivity, the slowing-down coefficient would be Cr=120, which is 4 times too large. This result is very important since it seems to show that a two-dimensional flow inside the whole fracture plane leads to particle residence times that do not fit the observed average migration velocities in Fanay-Augeres. It appears to us as a very significant step towards the validation of our preliminary assumption of channel flow inside the fracture planes.

5.5 Validity of the Coupled Hydraulic-Transport Model

As the dates of the peaks of arrival given by the model are now assumed to be calibrated, we must question whether or not the breakthrough durations given by the model fit the durations observed in the field. We consider whether the breakthrough durations of the tracer tests are simultaneously acceptably close to the distribution of the breakthrough durations of the ensemble of realizations and simulations of these tests. For this purpose, the breakthrough durations of the simulations corresponding to each of the 4 tracer tests were arranged in 4 histograms presented in Figure 11. Then, the observed breakthrough durations were compared to these histograms. It appears that the 4 experimental breakthrough durations are all included in the breakthrough duration range estimated from the simulations and reasonably close to the central value. It means that our model is able to give simultaneously satisfactory dates of the peak of arrival and of breakthrough durations without any extra calibration of the hydraulic parameter σ . This shows that the initial calibration of the hydraulic model is still appropriate for simulating the particle transport. This result appears to us as a second validation of the hydraulic model. It also shows that the coupling of hydraulics and transport seems to be justified.

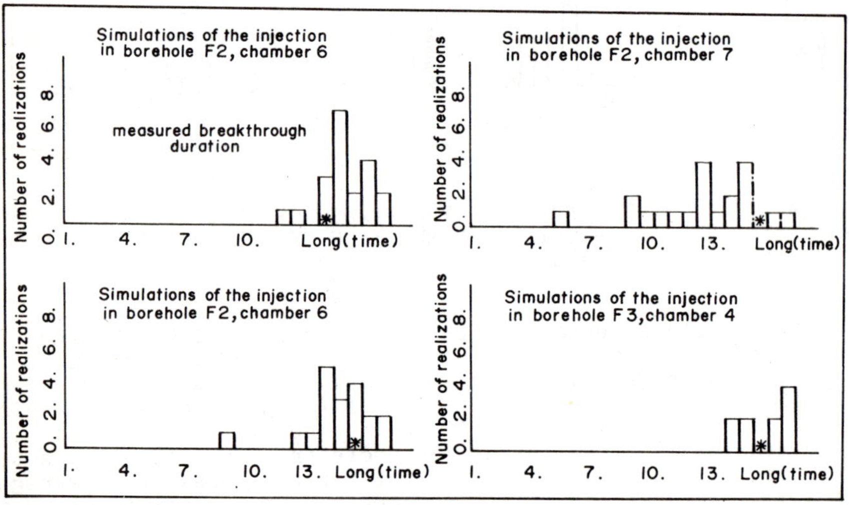

Figure 11: HISTOGRAMS OF THE BREAKTHROUGH DURATIONS ASSOCIATED TO THE 4 SETS OF SIMULATION OF THE TRACER INJECTIONS IN F2-Ch6, F2-Ch7, F3-Ch6, F3-Ch4 - Cr=32

CONCLUSION

The stochastic discrete fracture network model that we have developed incorporates information on fracture geometry (inferred from fracture mapping), fracture hydraulic properties (inferred from small-scale injection tests), and fracture transport properties (inferred from tracer tests). Although the model is stochastic, its parameters could be calibrated from observations and experiments using ergodic assumptions with a large number of independent measurements (injection tests) or with a limited number of measurements (tracer tests).

The model is capable of predicting both the large-scale properties of the medium and its variability at any scale starting from that of a single fracture. The model structure and the calibration methodology were validated with two independent sources of information:

- large-scale hydraulic behavior, which validates the mean model properties.

- intermediate-scale tracer tests which validate the variability of the model properties.

Futher validation would certainly be very desirable but difficult to conceive.

One of the essential features of this model is the assumption of strong channel flow within the fracture planes. The existence of such channelling in crystalline rocks has been suggested recently (e.g. by GENTIER (1986), BOURKE (1987), NERETNIEKS (1987), RASMUSON and NERET-NIEKS (1986), ABELIN et al (1985), but our results also clearly show that it would not be realistic to interpret the tracer experiments whithout assuming the existence of such channels. In Fanay-Augeres, their dimension seems to be on the order of a few cm in width, i.e. a few percent of each fracture area. Their shape and geometry are, however, unknown.

ACKNOWLEDGMENTS

This work was jointly sponsored by the Commission of the European Communities and the French Atomic Energy Commission in the framework of the research program MIRAGE for the disposial of radioactive waste in crystalline formations.

REFERENCES

ABELIN, H., I. NERETNIEKS, S. TUNBRANT, and L. MORENO, Migration in a Single Fracture: Experimental Results and Evaluation, final report, Stripa Project, Stockholm, Sweden, May 1985.

ANDERSSON, J., and B. DVERSTORP, Conditional Simulations of Fluid Flow in Three-dimensional Networks of Discrete Fractures, Water Resour. Res., 23(10), 1876-1886, 1987.

BAECHER, G.B., N.A. LANNEY, and H.H. EINSTEIN, Statistical Description of Rock Properties and Sampling, Proceedings of the 18th U.S. Symposium on Rock Mechanics, American Institute of Mining Engineers, Ref 5C1-8, 1977.

BOURKE, P.J., Channelling of Flow through Fractures in Rock, Report from the Chemistry Division, Harwell Laboratory, Ref: HL87/1124 (C22), May 1987.

CHARLAIX, E., E. GUYON, and S. ROUX, Permeability of a Random Array of Fractures of Widely Varying Apertures, Transport in Porous Media, 2(1), 31-43, 1987.

DAGAN, G., Analysis of Flow through Heterogeneous Random Aquifers. 2 Unsteady Flow in Confined Formations, Water Resour. Res., 18(5), 1571-1585, 1982.

DERSHOWITZ, W.S., Rock joint Systems, Ph.D. Thesis, Department of Civil Engineering, Massachussets Institute of Technology, Jan. 1985.

GENTIER, S., Morphologie et Comportement Hydromecanique d'une Fracture Naturelle dans un Granite sous Contrainte Normale. Etude Experimentale et Theorique, These de Doctorat a I' Universite d'Orleans, Mar. 1986.

GUTJAHR, A.L., L.W. GELHAR, A.A. BAKR, and J.R. McMILLAN, Stochastic Analysis of Spatial Variability in Subsurface Flow, Part 2: Evaluation and Application, Water Resour. Res., 14(5), 953-960, 1978.

HULL, L.C., J.D. MILLER, and T.M.CLEMO, Laboratory and Simulation Studies of Solute Transport in Fracture Networks, Water Resour. Res., 23(8), 1505-1513, 1987.

LASSAGNE, D., Essai de Caraterisation du Milieu Fracture en Massif Granitique, Site de Fanay-Augeres (Haute VIenne). Report from the B.R.G.M. Ref 83SGN576 GEG, 1983.

LONG, J.C.S., Investigation of Equivalent Porous Medium Permeability in Networks of Discontinous Fractures, Ph.D. Thesis, Lawrence Berkeley Laboratory, Earth Sciences Division, University of California, Ref LBL.16259, Apr. 1983.

LONG, J.C.S., J.S. REMER, C.R. WILSON, and P.A. WITHERSPOON, Porous Media Equivalents for Networks of Discontinuous Fractures, Water Resour. Res., 18(3), 645-658, 1982.

MARDIA, K.V., Statistics of directional Data, Academic Press, New-York, 1972

MATHERON, G., Elements pour une Theorie des Milieux Poreux, Masson Ed., Paris, 1967.

NERETNIEKS, I., Channeling in crystalline rocks. Its possible impact on transport of radionuclides from a repository, Colloque international "Impact de la physico-chimie sur l'etude, la conception et l'optimisation des procedes en milieu poreux naturel", Nancy, 10, 11 et 12 Juin 1987.

NOVAKOWSKI, K.S., V.E. GORDON, A.L. DAVID, and G.R. KENNETH, A., Field Example of Measuring Hydrodynamic Dispersion in a Single Fracture, Water Resour. Res., 21(8), 1165-1174, 1985.

RASMUSON, A. and NERETNIEKS, I., Radionuclide transport in Fast Channels in Crystalline Rock, Water Resour. Res., 22(8), 1247-1256, 1986

ROBINSON, P.C., Connectivity, Flow and Transport in Network Models of Fractured Media, Ph.D. Thesis, St. Catherine's College, Oxford University, Ref: TP 1072, May 1984.

ROULEAU, A., Statistical Characterization and Numerical Simulation of a Fracture System, Application to Groundwater Flow in the Stripa Granite, PHD Thesis, Waterloo, Ontario, 1984.

ROUX, S., MITESCU C., CHARLAIX E. and BAUDET C., Letter to the editor: Transfer matrix algorithm for convection-biased diffusion, J.Phys. A: Math. Gen. 19, L687-L692, 1986.

SCHWARTZ, F.W., L. SMITH, and A.S. CROWE, A stochastic Analysis of Macroscopic Dispersion in Fractured Media, Water Resour. Res., 19(5), 1253-1265, 1983

SMITH, L., and F.W. SCHWARTZ, An Analysis of the Influence of Fracture Geometry on Mass Transport in Fractured Media, Water Resour. Res.,20(9), 1241-1252, 1984.

TSANG, Y.W., and C.F. TSANG, Channel Model of Flow through Fractured Media, Water Resour. Res., 23(3), 467-479, 1987.

WITHERSPOON, P.A., N.G.W. COOK, and J.E. GALE, Geologic Storage of Radioactive Waste; Field Studies in Seden, Science, 211(4485), 894-900, 1981.

A NEW APPROACH TO TRACER TRANSPORT ANALYSIS: FROM FRACTURE SYSTEMS TO STRONGLY HETEROGENEOUS POROUS MEDIA

Chin-Fu Tsang

Earth Sciences Division
University of California
Lawrence Berkeley Laboratory
1 Cyclotron Road
Berkeley, CA 94720

ABSTRACT

Many current development and utilization of groundwater resources include a study of their flow and transport properties. These properties are needed in evaluating possible changes in groundwater quality and potential transport of hazardous solutes through the groundwater system. Investigation of transport properties of fractured rocks is an active area of research. Most of the current approaches to the study of flow and transport in fractured rocks cannot be easily used for analysis of tracer transport field data. A new approach is proposed based on a detailed study of transport through a fracture of variable aperture. This is a two-dimensional strongly heterogeneous permeable system. It is suggested that tracer breakthrough curves can be analyzed based on an aperture or permeability probability distribution function that characterizes the tracer flow through the fracture. The results are extended to a multi-fracture system and can be equally applied to a strongly heterogeneous porous medium. Finally, the need for multi-point or line and areal tracer injection and observation tests is indicated as a way to avoid the sensitive dependence of point measurements on local permeability variability.

INTRODUCTION

In the utilization and management of groundwater resources, a problem that has come into increasing attention in both developed and developing countries is the need of monitoring and maintaining groundwater quality. Thus for the long term utilization of groundwater, some of the properties of the groundwater system that should be obtained are those associated with solute (or contaminant) transport, such as effective porosity and dispersivity. These are normally obtained by tracer tests in which solutes or tracers are deposited at one location in the aquifer and the concentration of their arrival measured at another location. Methods have been developed over the years to analyze such tracer transport measurements for a homogeneous porous aquifer of constant thickness, in terms of the so-called advective-dispersive equation.

The methodology becomes rather complicated if one has to deal with a heterogenous porous medium. Stochastic methods (e.g., Gelhar, 1987 and Dagan, 1987) are developed for this if the heterogeneity is not too strong (i.e., the variance of permeability distribution in the aquifer is not large) and a representative elementary volume (REV) can be defined for the heterogeneous system. However, no methods have been fully established for a strongly heterogeneous system for which no REV can be defined. A particular type of strongly heterogeneous systems is the fractured porous medium, in which the porous medium is intersected by a system of fractures of various orientations, spacings, and sizes. This presents a difficult problem in tracer transport analysis that is being actively investigated by a number of research groups worldwide.

To put the problem in perspective, it should be pointed out that tracer transport is much more sensitive to the presence of fractures, or in general strong permeability heterogeneity, than heat transfer and pressure transmission through the system. Heat transfer is a diffusive process that often averages out formation heterogeneity. Pressure is not as effective a diffusive process, but is still less sensitive than tracer transport. For example, a series of connected fractures can easily form a fast path for solute transport.

The present paper attempts to introduce a new but simple model for the analysis of tracer transport for fracture systems or strongly heterogeneous porous media. The work reported is a continuing co-operation with Yvonne Tsang, Frank Hale at Berkeley and with Ivan Neretnicks and Luis Moreno at Stockholm (Tsang and Tsang, 1987; Tsang, et al., 1988a; Moreno, et al., 1988; Tsang, et al. 1988b; Tsang and Tsang, 1989; Tsang, 1989 and Tsang and Tsang, 1988). In the next section current approaches to modeling flow and transport in fractured porous media will be briefly reviewed. References are given for further study. Then results on flow and transport through single fractures will be presented. Single fractures have variable apertures and hence variable permeabilities. Thus flow through single fractures with a distribution of variable permeability is similar to the problem of flow through a two-dimensional heterogeneous porous medium. Extension of the results to multiple fractured systems will then be discussed. Finally the need of multiple-point or line/areal measurements of tracer transport is demonstrated.

CURRENT APPROACHES TO THE ANALYSIS OF FLOW AND TRANSPORT IN FRACTURED POROUS MEDIA

Three major approaches have been used to address the problem of flow and transport through fractured porous media. The first is the equivalent porous medium approach. This approach assumes that the fractures are sufficiently dense and random in the porous medium that the total system can be represented by a new porous medium with an effective permeability and porosity. This approach may be useful in giving average transport behavior. Recently two interesting developments have been made in this area. Pruess and co-workers (1985) considered flow of two-phase fluid in fractured porous media. Taking advantage of the fact that at low liquid saturation, the gas phase fills the fractures and the effective liquid flow permeability is that of the porous matrix and at high saturation liquid fills the fracture and thus flows with fracture permeability, they are able to design an equivalent porous medium with a saturation-dependent permeability function. The other development is by Neuman (1987) who considered single phase flow in a fractured porous medium and developed a stochastic approach with permeability as a stochastic variable. The approach is applicable in the sub REV scale, i.e., applicable even if the scale of heterogeneity is such that a REV is larger than the region of study.

The second approach to modeling flow in fractured porous media is the double porosity method (Barenblatt, et al., 1960; Warren and Root, 1963; Duguid and Lee, 1977). Here the fractures and porous matrix are considered to be two overlapping continua, each with its own flow equation. The exchange between the two continua is given by a source-sink term in the two flow equations, whose magnitude is proportional to the local pressure difference between the two continua. Pruess and Narasimhan (1985) extended this into a multiple interacting continua or MINC model and successfully applied it not only to flow but also to heat transport problems. The third approach in the fracture network model (Irmay, 1964; Romm, 1966; Snow, 1965), which ignores the role of porous matrix and considers the fractures to form a network through which flow can occur. Recent work considers statistical distributions of these networks with variable fracture density, length, orientation and aperture (Long et al., 1982, 1985; Smith et al., 1987, Endo et al., 1984; Cacas et al., 1988). All the above approaches have their strengths and weaknesses. In most cases, it would be quite difficult to apply them to the analysis of tracer transport field data. A new approach is introduced in the next section.

CHARACTERIZATION OF TRACER TRANSPORT IN 2-D STRONGLY HETEROGENEOUS PERMEABLE MEDIA

The new approach is based on a detailed study of flow and transport in a two-dimensional (2-D) plane with a distribution of strongly varying permeability. This can be interpreted as an heterogeneous porous aquifer of uniform thickness or as a single fracture with variable apertures.

Most theoretical studies of fluid flow in the fractured medium make the assumption that each fracture may be idealized as a pair of parallel plates separated by a constant distance which represent the aperature of the fracture. In fact, the aperture within a rock fracture is far from being constant, and the wide range of aperture values in a single fracture gives rise to a very heterogeneous system. Hence one would expect that the fluid flow would concentrate in a few preferred paths of least fluid resistance. Evidence that flow in fractures tends to coalesce in preferred paths has been found in the field (e.g. Abelin et al., 1985; Neretnicks, 1987; and Bourke, 1987). Motivated by these data, Tsang and Tsang (1987) proposed the variable-aperture channel model for transport through fractured media. The properties of the channel model were further developed by Tsang et al. (1988b) and the channeling characteristics of flow through a two dimensional single fracture were analyzed in detail by Moreno et al. (1988).

We use an aperture density distribution to characterize the range of aperture values in a single fracture. Based on measurement of aperture values in laboratory core samples of fracture (Gentier, 1986; Gale, 1987), we assume that the aperture values obey a log normal distribution, which is characterized by two parameters: the mean log aperture log b_o and the standard deviation (in log b) σ_b. In addition to the aperture density distribution parameters, the spatial correlation length λ and its anisotropy ratio λ_x/λ_y (where $\lambda^2 = \lambda_x^2 + \lambda_y^2$) for the variable apertures are needed to define the fracture. Given the aperture distribution, the single fracture with variable apertures may be generated by means of standard geostatistical methods similar to a procedure described in Moreno et al. (1988), the only difference being that in that earlier work the correlation length was assumed to be isotropic.

Figure 1 show three schematic representations of the fractures so generated. Each of the three square region in Figure 1 represents a single fracture which is discretized into square meshes to which different aperture values are assigned. The five shadings used in the figure illustrate the varying aperture values in five steps from $b < 29$ μm to $b > 116$ μm, with the darkest shading representing the smallest apertures. The three realizations in Figure 1 are based on the same input log normal distribution with $\log b_o = 1.7$, where the aperture b_o is in μm and $\sigma = 0.43$. They have the same correlation length $\lambda = 0.2L$, where L is the dimension of the single fracture, but different correlation anisotropy ratios of $\lambda_x/\lambda_y = 1, 3,$ and 5 respectively in Figures 1a, 1b, 1c. Variograms have been computed for the apertures of all the fractures to confirm that the aperture variation obeys the required correlation length and anisotropy ratios. The increase of the correlation anisotropy ratio from 1 to 5 in the Figures 1a, 1b and 1c is also visually evident.

For the flow calculation, Darcy's law is assumed valid locally for each grid block where the permeability is assumed to be proportional to the square of the local aperture. Then the fluid pressure at each node may be computed with a constant pressure difference maintained across the single fracture on two opposite fracture boundaries, and no flow conditions imposed on the two remaining boundaries. From the calculated fluid potential at each node, the fluid flow rate between each pair of nodes within the fracture may be computed. As an example, the flow rates so calculated for the fracture as shown in Figure 1b are displayed in Figure 2. Figure 2a shows the fluid flow when the pressure difference is applied on the left and right boundaries of the fracture and no flow boundary conditions are imposed on the top and bottom boundaries; Figure 2b shows the fluid flow when the pressure difference is applied on the top and bottom boundaries. The calculated flow rates between nodes cover a range of four orders of magnitude as a result of the log normal aperture distribution of the fracture. Five levels of shadings: space, dash line, solid line, thicker solid line and thickest solid line are used to represent the ascending values of the flow rates.

It is apparent from Figure 2 that although flow occurs in the entire plane of the fracture, the majority of flow takes place in a few selected pathways. With both geometrical configurations of boundary conditions (pressure difference from left to right and pressure difference from top to bottom), preferred paths of large flow rates are present, though the flow pattern is quite different for the two flow directions. These preferred flow paths are what we call channels, along which the apertures are variable. Thus these channels are not large physical flow-tubes embedded in a fracture plane, but rather preferred paths, in the sense of stream-tubes in the hydraulic potential theory, due to the strong variation of the aperture and hence permeability value. The channeling effect described here occurs in addition to the physical or structural large-permeability

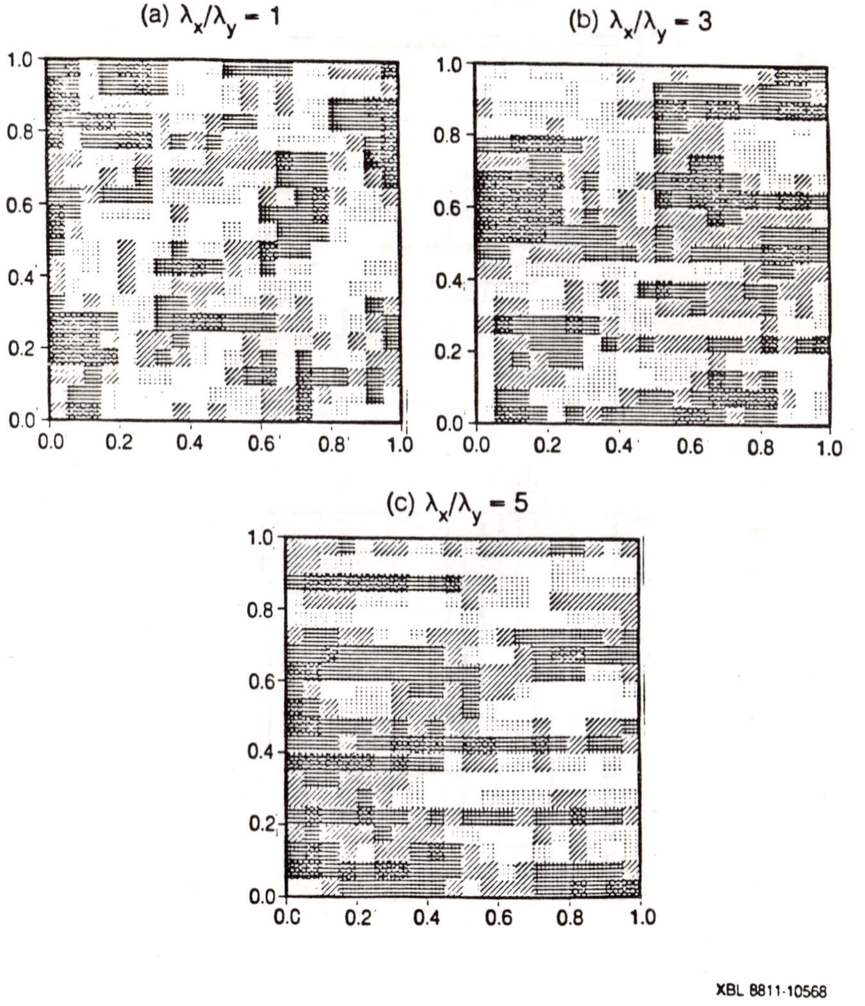

Figure 1. Statistically generated apertures with spatial correlation length of 0.2L in the plane of a single fracture of linear dimension L. The aperture values are partitioned into five classes with the smallest apertures corresponding to the darkest shading. The three realizations have different anisotropy ratio for the spatial correlation length: (a) $\lambda_x/\lambda_y = 1$, (b) $\lambda_x/\lambda_y = 3$, and (c) $\lambda_x/\lambda_y = 5$.

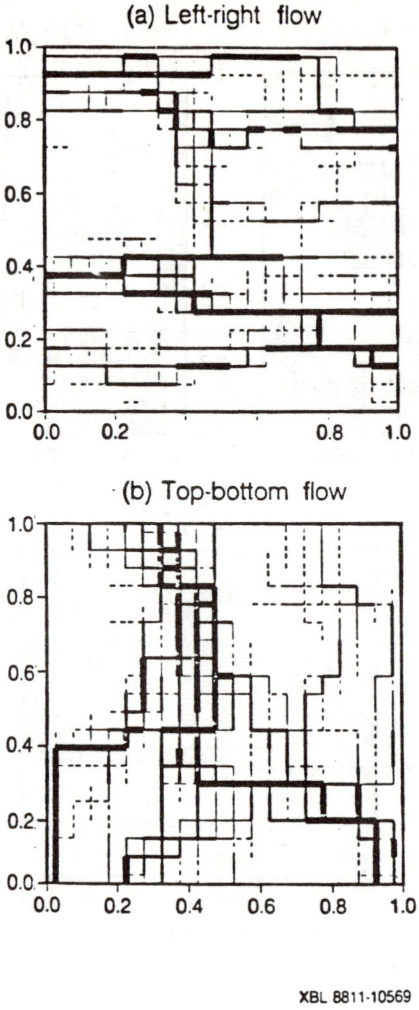

Figure 2. Fluid flow rates for the fracture with aperture variations as shown in Figure 1b. The thickness of the lines is proportional to the flow rates. Pressure difference is applied from (a) left to right, and (b) top to bottom.

channels in the fractured medium, such as those observed near the intersection of orthogonal fractures (Abelin et al., 1988).

We can see that the pattern of large flow rate channels (Figure 2a) observed for the boundary conditions with the pressure difference applied to the left and right boundaries is quite different from that (Figure 2b) observed for the boundary conditions of pressure difference applied to the top and bottom boundaries of the same fracture. Similar results are obtained for the fractures in Figures 1a and 1c. If the pattern of flow channeling appears to be quite different for overall flow in orthogonal directions for the same fracture, can we find parameters which are common to flow in both directions and which characterize the fluid and transport in the fracture? To study this, we calculate tracer transport through the fracture using the particle tracking method, which was first applied to study the dispersion of a heterogeneous porous medium by Smith and Schwartz (1980), and recently applied specifically to a fracture with variable apertures by Moreno et al. (1988). A sufficiently large number of particles, in our case four thousand, are introduced on the high pressure side of the fracture and are allowed to travel through the system with the probability of entering a particular flow section in the fracture proportional to the local flow rate. When a particle comes to an intersection with two or three outgoing flow sections, a Monte Carlo method is used to assign the particle to one of the flow directions with probability weighted by their respective flow rates. As each of the particle travels through the fracture, the aperture values along its flow path are recorded. These build up the statistics of the aperture values of the flow channels. Aperture density distribution functions fitted to these aperture values are shown in Figure 3. The solid line gives the probability density distribution of apertures over the entire 2D plane of the fracture. The dot-dash line and the short dash line give the aperture density distributions which describe the apertures along the actual flow paths of the particles for the boundary conditions of overall left-right flow and top-bottom flow respectively. All three curves are normalized to unit total integrated probability. Figures 3a, 3b, 3c correspond to the results for the fractures with aperture variation as shown in Figures 1a, 1b, 1c respectively.

It is clear from the distributions shown in Figure 3 that the tracer particles favor large apertures and avoid small apertures of the 2D fracture. However, they cannot avoid the small apertures altogether and thus there is a finite probability of flow through apertures one or two orders of magnitudes smaller than the large apertures most particles travel through. The result is that solute transport is described by a distribution function that is different from that of the 2D fracture: the former has a larger mean k value and a smaller standard deviation (see Figure 3).

Now Figure 3 also shows that the aperture distribution functions along the channels are almost identical for the two orthogonal flow geometries (left-right and top-bottom flows). This is not obvious a priori, since the channels in the two cases represent rather different flow paths (Figure 2a and 2b). The result is even less obvious when one considers that the anisotropy ratio of the correlation length is a factor of 5 for the case in Figures 1c and 3c. We believe the insensitivity of the aperture values of the flow paths in the two orthogonal directions of overall flow arises from the fact that the correlation length chosen here is 0.2L. That is, the scale of heterogeneity is on the order of one fifth of the flow region, or measurement. Hence the correlation length is small enough for the particles flowing under different boundary conditions to sample statistically similar values of aperture along the flow paths. In the limit that the scale of heterogeneity is increased to the size of the flow region, then the large anisotropy ratio in the aperture

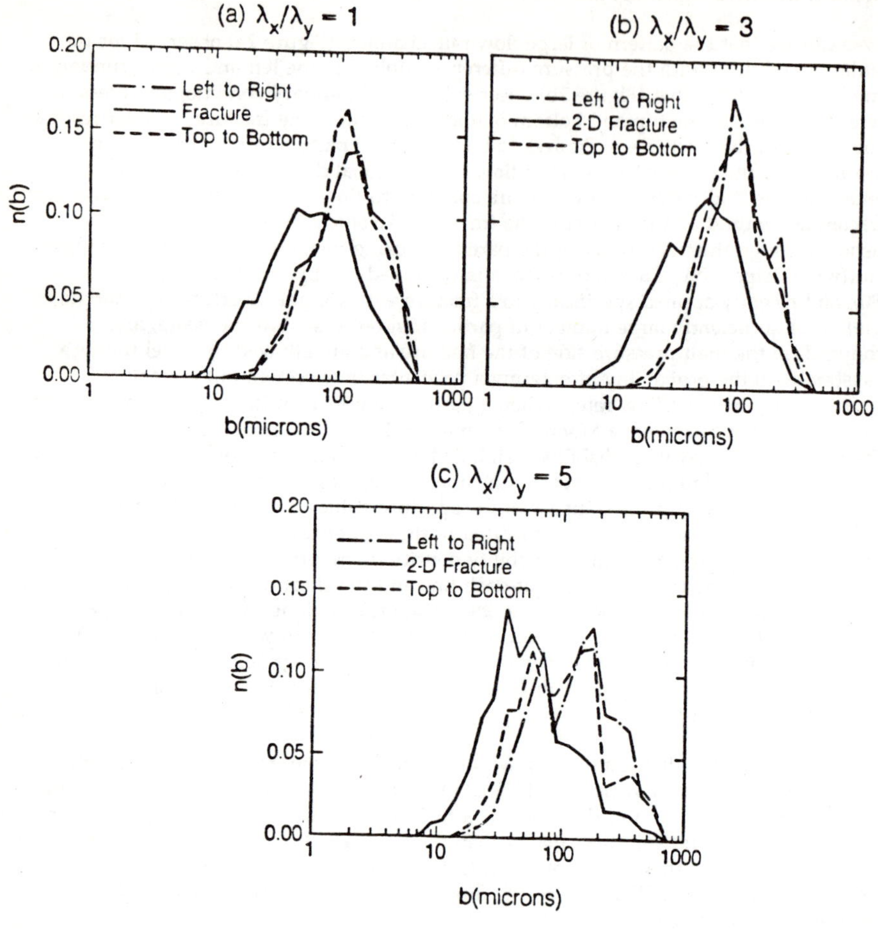

XBL 8811-10570

Figure 3. The aperture probability density distributions for apertures over the 2D fracture, along flow paths with left-right pressure difference, and along flow paths with top-bottom pressure difference for the fractures shown in (a) Figure 1a, (b) Figure 1b, and 9c) Figure 1c.

correlation length would give rise to a highly stratified medium through which the flow should consist of major conduits of very large apertures in the direction parallel to the large correlation length, but many smaller apertures are sampled in the orthogonal direction. When this happens, the aperture density distribution of the particle flow paths will be very sensitive to the geometric configuration of the boundary conditions. But as long as the spatial correlation length is reasonably small (less than 1/5 of the measurement), we believe that even over a range of anistropy ratios of the correlation, flow characteristics in terms of aperture distributions (Figure 3) are the same in the two directions. Thus we arrive at the interesting result that channels of tracer transport in the fracture system is characterized by an aperture density distribution function, independent of specific deterministic flow paths, or the anisotropy ratio, as long as the correlation length of the heterogeneity is not too large.

The above discussions were presented in terms of flow and transport through a single fracture with variable aperture. The discussion could also easily be expressed for a strongly heterogeneous porous medium defined by a permeability probability distribution where the scale of heterogeneity is expressed in terms of a correlation length in the spatial variation of k. In our single fracture calculations, the variable is the fracture aperture b which relates to the local permeability by the relation $k = b^2/12$. In the porous medium, the fluid volumetric flow rate is proportional to kA, where A is the cross sectional area to the flow. Thus, the equivalent kA varies as $b^3/12$. A wide range of values in kA would give rise to the phenomenon of channeling. Therefore all the above observations for the fluid flow behavior in a single fracture with variable aperture is directly applicable to a highly heterogeneous porous medium (one with a strong spatial variation in permeability). One may then make the following remarks. The heterogeneous medium would give rise to flow channeling, and provided that the largest scale of heterogeneity is less than a few tenths of scale of measurement, the least resistive channels for tracer transport are characterized by a permeability distribution, which differs from the permeability distribution of the entire medium. The distribution along the channel typically takes on a larger mean and a smaller spread. Therefore, the dispersion of the system is not related to the spread of the permeability distribution over the entire region, but rather to the spread of the subset of permeability values along the preferred flow channels. (See Figure 3.)

EXTENSION TO MULTI-FRACTURE SYSTEMS

A multi-fracture system is composed of a network of intersecting single fractures. A simple example is shown in Figure 4. Since within each single fracture, the flow channeling occurs as demonstrated in the last section, transport through the multiple fracture system is also through channels as is apparent in this figure. If all the fractures in the multi-fracture system have the same basic properties geologically, i.e., they can be related to the same aperture density distribution, the channelled transport through the multi-fracture system is characterized by the same particle-path aperture density distribution with its mean log b_0 and spread σ_b. If the fractures in the system are composed of two, three or more sets of fractures, each related to its own aperture density distribution, then the aperture density distribution characterizing the tracer transport through the system will be a composite, or union, of all the distributions for the different sets. An example of channeling flow in a series of 2 fractures is shown in Figure 5. It is easy to see that such calculations can be done to many connected fractures. We have obtained results for up to 100 fractures in series.

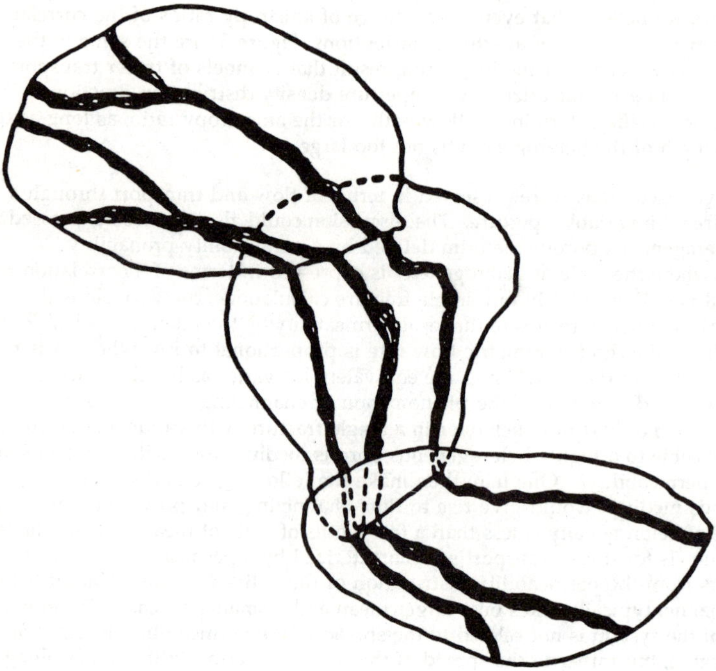

Figure 4. Schematic picture of three intersecting fractures with flow channeling

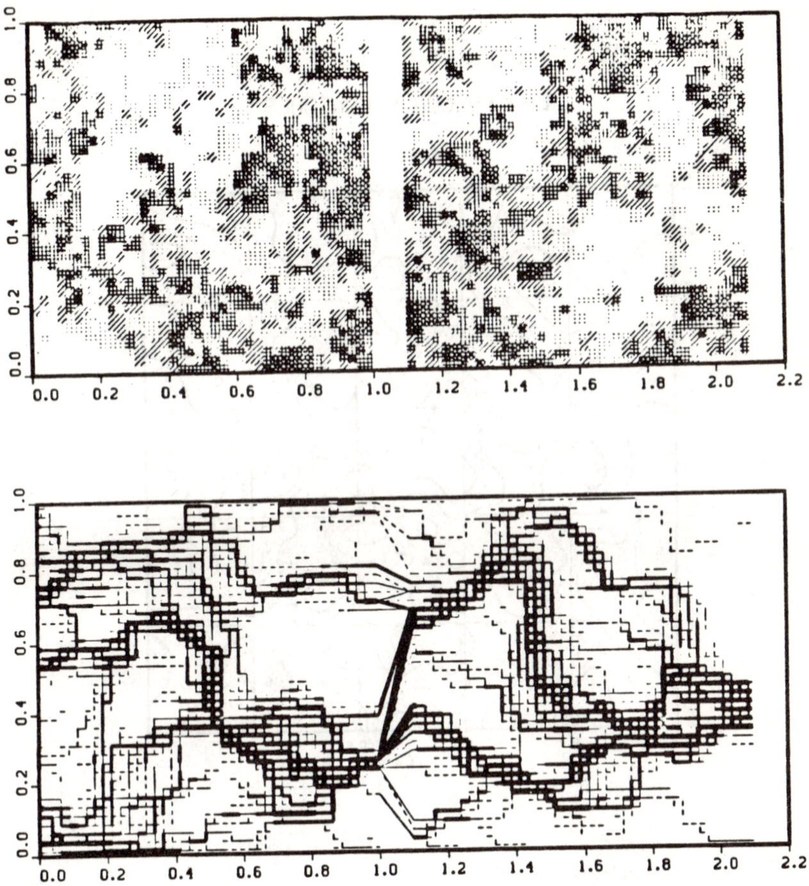

Figure 5. Flow channeling in two intersecting fractures. Aperture distribution is shown on top and flow lines are shown in the bottom.

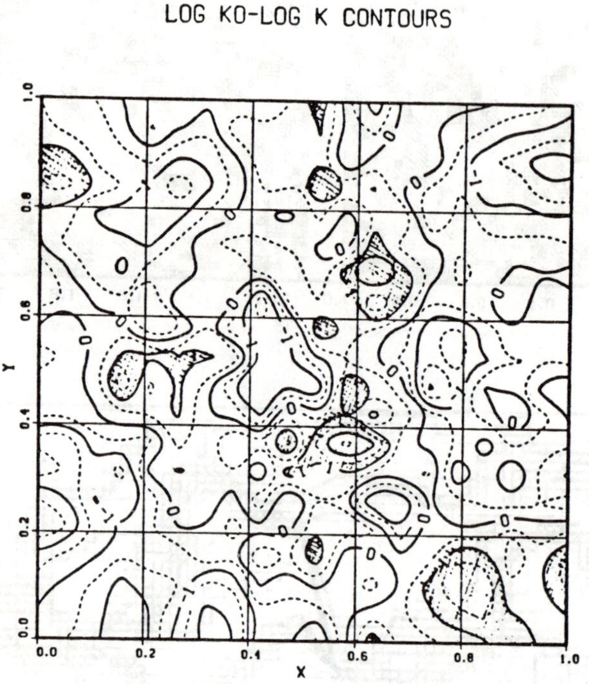

Figure 6. Highly heterogeneous 2-D porous medium with contours in log k.

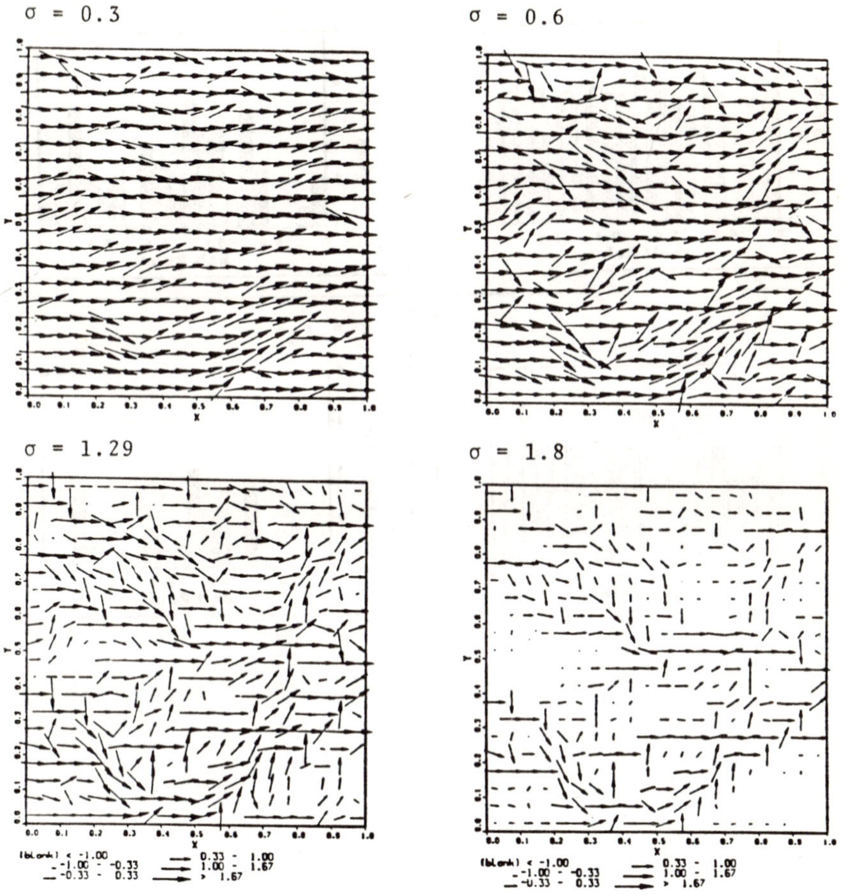

Figure 7. Flow through the heterogeneous system shown in Figure 6, with rates proportional to arrow sizes for four different values of permeability standard deviation σ_k = 0.3, 0.6, 1.29, 1.8.

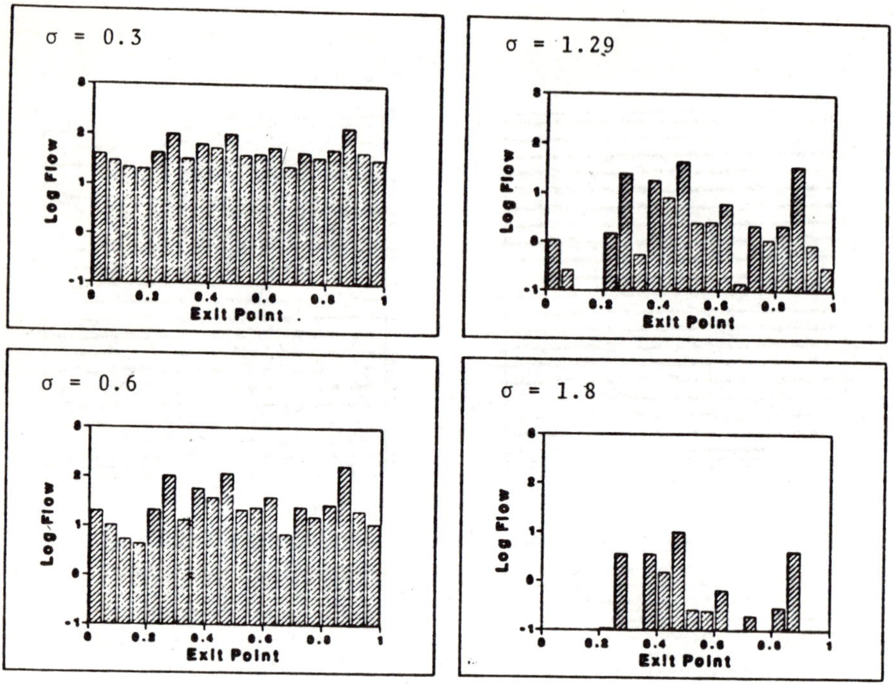

Figure 8. Exit flow rates at different points along the exit line corresponding to Figure 7, for $\sigma_k = 0.3, 0.6, 1.29, 1.8$.

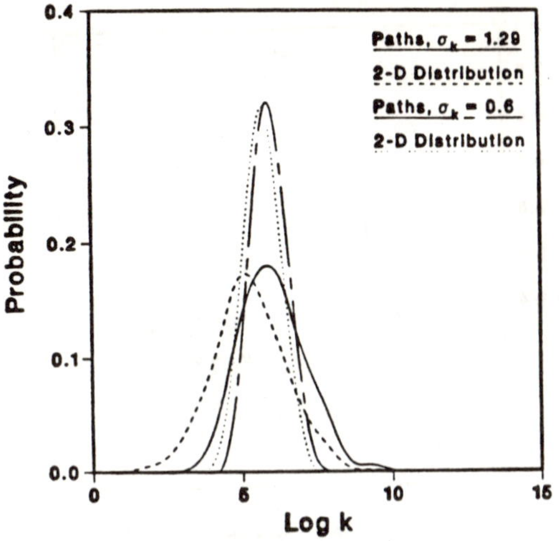

Figure 9. Comparison of permeability probability distribution function in 2D medium with that along particle flow paths, for $\sigma_k = 0.6$ and 1.29.

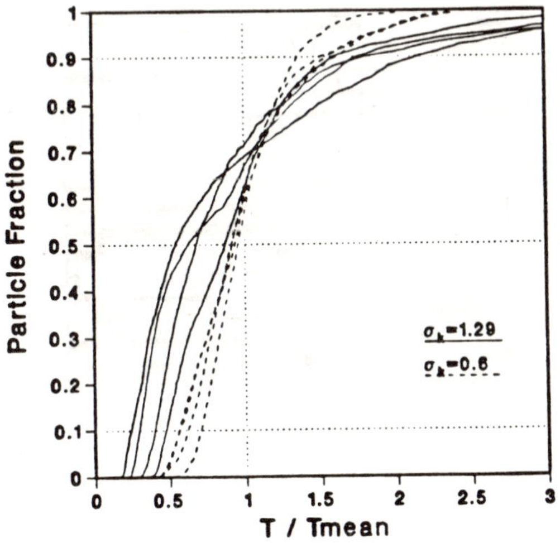

Figure 10. Tracer breakthrough curves for $\sigma_k = 0.6$ and $\sigma_k = 1.29$. Four realizations are shown for each σ_k value. Figure 6 is the permeability plot for one of the four realizations with $\sigma_k = 1.29$.

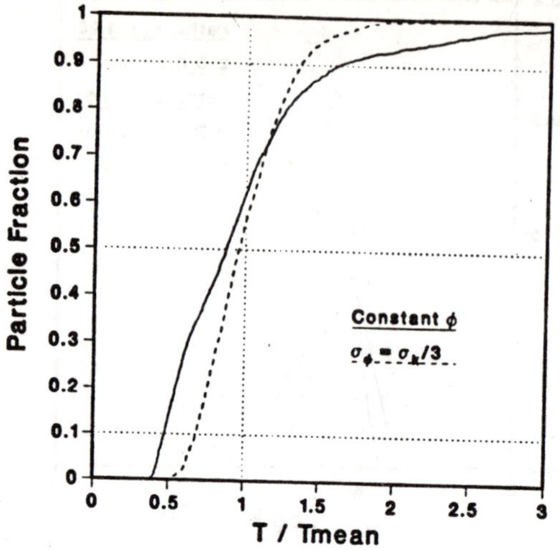

Figure 11. Tracer breakthrough curves for a heterogeneous medium shown in Figure 6, using two assumptions of porosity variability: $\sigma_\phi = 0$ and $\sigma_\phi = \sigma_k/3$.

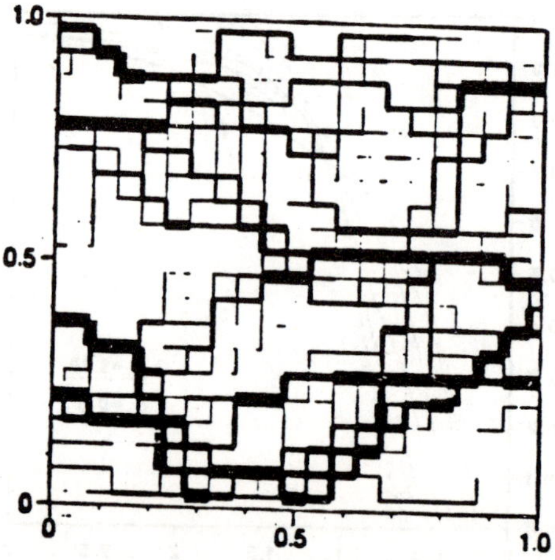

Figure 12. Fluid flow rates for the fractures with aperture variations as shown in Figure 6. The thickness of the lines is proportional to the square root of the flowrates. Pressure difference is applied from left to right.

DEPENDENCE OF TRACER TRANSPORT AND CONCENTRATION BREAKTHROUGH CURVES ON PERMEABILITY VARIANCE

In the last section we demonstrated that tracer transport through a fracture system is characterized by a mean aperture, log b_o and the spread σ_b in log b. This can be directly translated to permeability distribution by the equation kA ~ b^3, with the associated mean and permeability spread σ_k in log k. In this section the dependence of flow and transport through a heterogeneous system as a function of σ_k is presented. Given the permeability distribution function and a spatial correlation length we can generate as before a realization of permeability spatial distribution as shown in Figure 6. This figure corresponds to the case of $\sigma_k = 1.29$ and $\lambda = 0.1$, and is shown in a contour plot form. A pressure difference is then applied to the left and right boundaries, while the top and bottom boundaries are closed. The flow lines are calculated and represented in Figure 7a-d by arrows at each location with the sizes of the arrows proportional to the logarithm of local flow rates. Figure 7c is the result of flow field for the heterogeneous system in Figure 6. If the value of σ_k is changed, the flow is strongly affected. Figure 7a,b,c and d shows the results for σ_k = 0.3, 0.6, 1.29 and 1.8 respectively. The change of flow field from the almost uniform flow to channeled flow corresponds to low σ_k (almost constant k) case to large σ_k as the strongly heterogeneity case. This is even more apparent when we plot the flow rates along the exit line for these 4 cases (Figure 8).

Now a tracer transport calculation is done by particle tracking method. The permeability probability distribution of the 2D distribution and that along the particle paths are shown in Figure 9 for two different values of σ_k = 0.6 and 1.29. The curves for $\sigma_k = 0.6$ is narrower than that for $\sigma_k = 1.29$, and the shift between particle path and the 2D distribution is also smaller for the smaller σ_k value. According to the last section it is these particle-path permeability probability distribution functions that characterize the tracer transport.

The particle tracking method affords us also a way to calculate the tracer concentration breakthrough curves along the exit line on the right side of Figure 6. A step function of concentration is applied by introducing tracer particles on the left side and collected along the right boundary. This is done for a number of realizations of the permeability distribution. The results are shown in Figure 10 for two values of $\sigma_k = 0.6$ and $\sigma_k = 1.29$, with four realizations in each case. Thus more dispersion (a larger initial slope) is noticed for larger input σ_k. The band covering the different realizations for each σ_k value represents the intrinsic uncertainties of these curves due to statistical variability. The above calculation was made assuming the porosity, ϕ, of the heterogeneous porous medium to be constant. For a single fracture of variable aperture, the spread in porosity or aperture is related to that of permeability by $\sigma_\phi = \sigma_k/3$. The results of these two assumptions on porosity are compared in Figure 11. It is interesting to note that the constant ϕ case has more dispersion for the concentration breakthrough curves.

Given either a constant porosity or a variable porosity with a finite value of σ_ϕ, the tracer breakthrough curves such as those shown in Figure 10 can be correlated with σ_k, a parameter characterizing the heterogeneous medium. Such an approach has been applied with some success to a few sets of field data. (See: Tsang et al., 1988b and Tsang and Tsang, 1988.)

THE NEED FOR MULTIPLE-POINT OR SPATIALLY AVERAGED RESULTS FOR TRACER TRANSPORT

So far we have considered tracer transport through a heterogeneous system with tracer injection along the left boundary line and tracer observation along the right boundary line of the system. However often point tracer injection and observation are made. For heterogeneous systems, this kind of measurement may not be adequate since tracer transport through the medium is strongly dependent on local variations of permeability and porosity. In other words, if one makes point measurements in which the measurement (or sample) size is less than or comparable to the spatial correlation length of the variability, the result is so sensitive to statistical variations that it is cannot be analyzed.

To illustrate this, consider tracer transport between two sides of a single fracture with variable aperture shown in Figure 6. the spatial correlation length, λ, in this particular case is 0.1 L, where L is the linear dimension of the fracture. Assuming a pressure difference is applied on the two sides of the fracture, flow is established with paths shown in Figure 12, where the thickness of the lines are drawn as proportional to the square root of the local flow rates. This Figure is just a different way to present Figure 7c.

Now, let tracer particles be released only within a section length $s < L$ at a few high-flow locations along the high pressure side and collected at a few high-flow positions on the low pressure side (see Figure 13). The collection length is also assumed to be s. Calculations were made for $s/L = 0.1, 0.2, 0.3,...,0.8$. Figure 14 shows tracer breakthrough curves for $s/L = 0.1$ for one of the realizations with $\lambda/L = 0.1$. Here tracers are released at two alternative locations, A and b, and collected at three alternative locations, D, E and F (Figure 13). These locations are chosen because of their relatively large local flow rates. At each section, concentration, C, was calculated by dividing the accumulative exit particle number by the local flow rate. The maximum input concentration, C_o, is defined as the total number of tracer particles used in the calculation divided by the flow rate in the input section. The plot in Figure 14 is C/C_o versus time of arrival t, at the exit sections. Note that C/C_o does not reach a value of 1 because of mixing of water from the tracer-injection section with the remainder of water inflow from the high pressure side. Also in this figure, only five curves are shown, though there are six possible flow connections. The reason is that the flow from point B to point D is so slow that its first arrival time is out of the range of this graph.

Several comments on the calculations may be made. First, the exit flow rates at the three exit sections are quite different. They are given by Q = 26, 53, 37 units, respectively. Second, the initial arrival times of the six possible tracer paths do not correlate with the linear distances between the tracer-injection section and the collection section. For example, the earliest arrival time corresponds to the case with linear transport of 1.04 L (from B to E), and this is about half of the arrival time for a case with a slightly shorter linear transport distance of 1.005 L (from B to F). Such non-correlation between initial arrival times and linear transport distances has often been observed in the field. A recent example may be found in data from the Fanay-Augeres field tracer experiment (Cacas et al., 1988).

One may define a measure for the dispersion displayed in Figure 14 as given by $(t_{0.9} - t_{0.1})/t_{0.5}$, were t_x is the time of arrival when the concentration reaches a fraction x of the maximum concentration value at large times. This empirical definition of dispersion has the advantage that it is independent of a model for the system. Taking the cases

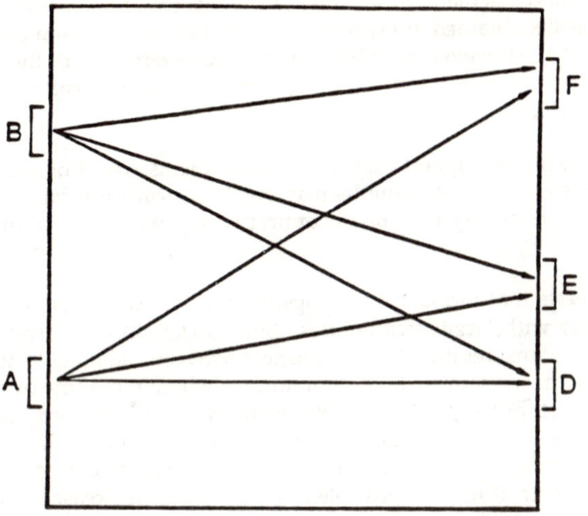

Figure 13. Schematic diagram illustrating "point" measurements of tracer transport. Based on one of the realizations of a fracture with variable apertures with an applied left-right pressure difference, tracers are deposited at A and B and collected at D, E, and F. Points A, B, D, E and F are chosen at locations of large entrance or exit flow rates and each is associated with a section length of 0.1L.

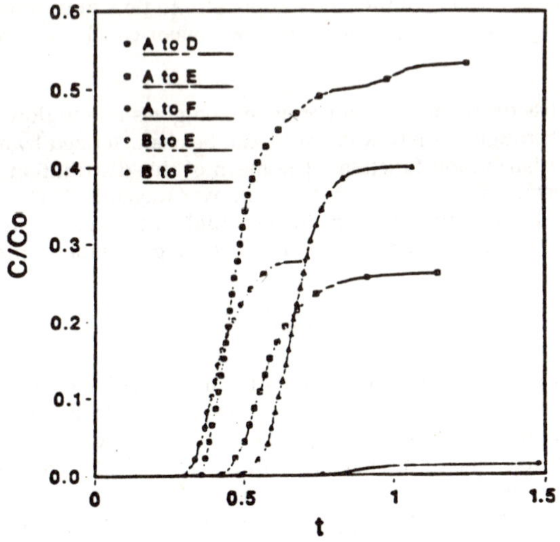

Figure 14. Tracer breakthrough curves for cases shown in Figure 13, with the section length for tracer entrance or exit equal to 0.1L.

shown in Figure 13, and calculating the value of the dispersion measure as defined above for each case, the results obtained are spread out over a factor of 3 from each other (see Figure 15 for s = 0.1 L). However, carrying out the same exercise with the tracer collection section length increased from s/L = 0.1 to 0.8 in steps of 0.1, the spread of the dispersion measure is greatly reduced. This is shown in Figure 15.

The results in Figure 15 suggest that line measurements with line section lengths corresponding to a few correlation lengths may yield less sensitive values for the dispersion measure, which may then be the appropriate predictive quantity for the model.

This study suggests that formation heterogeneity can result in calculated tracer transport dispersivity with large uncertainties. Such uncertainties, however, can be greatly reduced if the consideration is not transport between points but rather those between line or areal tracer sources and observations, where these line or areas cover two or more spatial correlation lengths of the heterogeneity. In other words, such averaging or summation reduces significantly the statistical range of predicted dispersivity values. Thus, it is perhaps impossible to predict the tracer concentration at a point in space at an instant of time. However, it may be possible to predict, within a reasonable range of uncertainty, tracer transport behavior between a region of tracer sources and an observation region. In many problems of current interest, such as the performance assessment of a nuclear waste repository, perhaps this is what is required.

CONCLUSIONS AND SUMMARY

In this paper we reviewed briefly the current approaches in the modeling of flow and transport in fractured rocks. A detailed study of the permeability heterogeneity of a single fracture shows a channeling behavior of tracer transport which can be extended to a multi-fracture rock system. Results of the study apply also to flow and transport through a strongly heterogeneous porous medium. Two major conclusions may be given as follows:

(1) Heterogeneity in fractures and also in porous media gives rise to flow channeling effects. Tracer transport through these flow channels can be characterized by a permeability probability distribution function. The mean of this distribution function is larger than that of the permeability probability function over the entire 2-D heterogeneous medium. Tracer concentration breakthrough curves can then the analyzed to obtain the parameters (mean and variance) of the permeability probability distribution function.

(2) Since this is a statistical approach to interpret tracer transport data, it is shown that it would be impossible to make accurate predictions of concentration of tracer arrival at a point in time and space, due to local permeability variations. It is demonstrated that multiple-point or line/areal measurements of tracer transport should be made over a few spatial correlation lengths, that would average the variability and may yield dispersion parameter characteristic of the medium. In other words, this new approach may not be able to make point predictions of tracer transport, but may be able to predict tracer concentration over a region of space. In many practical applications, perhaps this capability is all that is required.

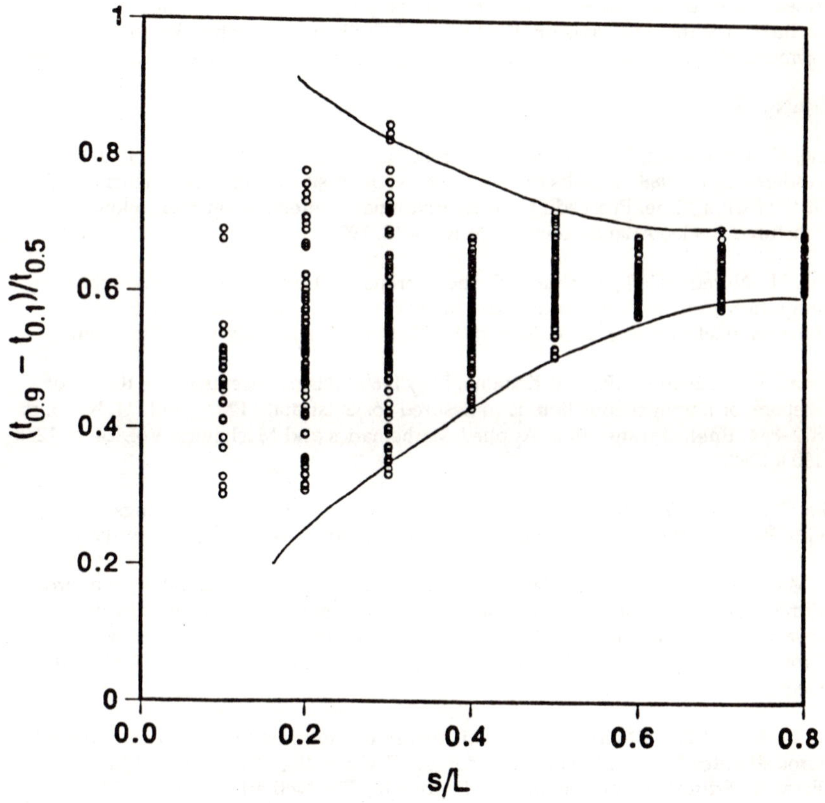

Figure 15. The tracer dispersion measure $(t_{0.9} - t_{0.1}),/t_{0.5}$, as a function of the section length for tracer entrance or exit. At each S/L value the points represent results from different alternative injection-collection pairs.

ACKNOWLEDGEMENTS

Continued co-operation and discussions with F.V. Hale, Y.W. Tsang, L. Moreno and I. Neretnicks are much appreciated. Most of the work on which this report is based was performed under the auspices of the Office of Facilities Siting and Development, Siting and Facilities Technology Division, the Office of Civilian Radioactive Waste Management of the Department of Energy through contract number DE-AC03-76SF00098.

REFERENCES

Abeline, H., Bergersson, L., Gidlund, J., Moreno, L., Neretnieks, I., Widen, H., and Andersson, J., 1988. Results of some large scale in situ tracer experiments in a drift at the Stripa Mine, Proceedings of International Conference on Fluid Flow in Fractured Rocks, Atlanta, Georgia, May 16–18, 1988.

Abeline, H., Neretnieks, I., Tunbrant, S. and Moreno, I., 1985. Final report on the migration in a single fracture, Experimental results and evaluations, Sven. Kärnbränslefösorjning Tech. Report. 85-03, Nucl. Fuel Safety proj., Stockolm.

Barenblatt, G.I., Zheltov, I.P., and Kochina, I.N., 1960. Basic concepts in the theory of seepage of homogeneous liquids in fissured rocks (strata): PMM, Vol. 24, No. 5, p. 852–864; English translation, Applied Mathematics and Mechanics, Vol. 24, p. 1286–1303, 1960.

Bourke, P.J., 1987. Channeling of flow through fractures in rock. Proceedings of GEOVAL-87, International Symposium, Stockholm, Sweden, April 7–9, 1987.

Cacas, M.C., Ledoux, E., and de Marsily, G., 1988, Calibration and validation of a three-dimensional stochastic network model on a large scale experiment with flow measurements and tracer tests performed in the uranium mine of Fanay-Augeres, Proceedings of International Conference on Fluid Flow in Fractured Rocks, Atlanta, Georgia, May 16–18, 1988.

Dagan, Gedeon, 1987. Review of stochastic theory of transport in groundwater flow, in Groundwater Flow and Quality Modelling, E. Custodio, A. Gurgui and J.P. Lobo-Ferreira, Editors, Reidel Publishers, Dordrecht, The Netherlands, pp. 1–32.

Duguid, J.O., and Lee, P.C.Y., 1977. Flow in fractured porous media. Water Resources Research, Vol. 13, No. 3, p. 558–566.

Endo, H.K., Long, J.C.S., Wilson, C.R. and P.A. Witherspoon, 1984. A model for investigating mechanical transport in fractured networks, Water Resources Research, Vol. 200, No. 10, pp. 1390-1440.

Gale, J.E., 1987. Comparison of coupled fracture deformation and fluid flow models with direct measurements of fracture pore structure and stress-flow properties, Proceedings of 28th U.S. Symposium of Rock Mechanics, Tuscon, Arisona, 19 June-1 July, p. 1213–1222.

Gelhar, L.W., 1987, Stochastic analysis of solute transport in saturated and unsaturated porous media, in Advances in Transport Phenomena in Porous Media, Jacob Bear and M. Yavnz Corapcioglu, Editors, Martinus Nijhoff Publishers, Dordrecht, The Netherlands, pp. 657–700.

Gentier, S., 1986. Morphologie et comportment hydromecanique d'une fracture naturelle dans un granite sous contrainte normale, Doctoral Thesis, U. d'Orleans, France.

Irmay, 1964. Theoretical models of flow through porous media. RILEM Symp. on the Transfer of Water in Porous Media, Paris, April.

Long, J.C.S., Gilmour, P., and Witherspoon, P.A., 1985. A Model for steady fluid flow in random three-dimensional networks of disc-shaped fractures, Water Resources Research, Vol. 21, No. 8, pp. 1105–1115.

Long, J.C.S., Remer, J.S., Wilson, C.R., and Witherspoon, P.A., 1982. Porous media equivalents for networks of discontinuous fractures, Water Resources Research, Vol. 18, No. 3, pp. 645–658.

Moreno, L., Tsang, Y.W., Tsang, C.F., Hale, F.V., and Neretnieks, I., 1988. Flow and tracer transport in a single fracture: A stochastic model and its relation to some field observation, Water Resources Research, Vol. 24, No. 12, pp. 2033–3048.

Neretnieks, I., 1987. Channeling effects in flow and transport in fractured rocks—some recent observations and models, Proceedings of GEOVAL-87, International Symposium, Stockholm, Sweden, April 7–9, 1987.

Neuman, Shlomo, P., 1987. Stochastic continuum representation of fractured rock permeability as an alternative to the REV and fracture network concepts, in Groundwater Flow and Quality Modelling, E. Custodio, A. Gurgui and J.P. Lobo-Ferreira, Editors, Reidel Publishers, Dordrecht, The Netherlands, pp. 331–362.

Pruess, K. and Narasimhan, T.N., 1985. A practical method for modeling fluid and heat flow in fractured porous media, Society of Petroleum Engineering Journal, Vol. 25, No. 1, pp. 14–26.

Pruess, K., Tsang, Y.W., and Wang, J.S.Y., 1985. Modeling of strongly heat-driven flow in partially saturated fractured porous media, in Proceedings of Conference on Hydrogeology of Rocks of Low Permeability, Tucson, Arizona, pp. 486–497.

Romm, E.S., 1966. Fluid flow in fractured rocks, Nedra, Moscow.

Smith, L., Schwartz, F., and Mase, C., 1987. Applications of stochastic methods for the simulation of solute transport in discrete and continuum models of fractured rock systems, Proceedings of Conference on Geostatistical, Sensitivity and Uncertainty Methods for Ground-Water Flow and Radionuclide Transport Modeling, San Francisco, California, September 15–17, 1987, pp. 425–440.

Smith, L. and Schwartz, F.W, 1980. Mass transport. 1, A stochastic analysis of macroscopic dispersion, Water Resources Research, Vol. 16, No. 2, pp. 303–313.

Snow, D.T., 1965. A parallel model of fractured permeable media. Ph.D. Thesis, University of California, Berkeley, 330 p.

Tsang, C.F., 1989. Tracer travel time and model validation, International Journal of Radioactive Waste Management and the Nuclear Fuel Cycle, in press.

Tsang, C.F., Tsang, Y.W., and Hale, F.V., 1988a. Tracer transport in fractured rocks, Invited Paper, Proceedings of the International Conference in Flow in Fractured Rocks, Atlanta, Georgia, May 16–18, 1988.

Tsang, Y.W. and Tsang, C.F., 1987. Channel model of flow through fractured media, Water Resources Research 23(3), 467–479, 1987.

Tsang, Y.W. and Tsang, C.F., 1988. Interpretation of large scale tracer transport by a multi-fracture channel model, EOS. Transactions of the American Geophysical Union, Vol. 69, No. 44, p. 1176.

Tsang, Y.W. and Tsang, C.F., 1989. Flow Channeling through strongly heterogeneous media. Submitted to Water Resources Research, January, 1989.

Tsang, Y.W., Tsang, C.F., Neretnieks, I., and Moreno, L., 1988b. Flow and tracer transport in fractured media—A variable aperture channel model and its properties, Water Resources Research, Vol. 24, no. 12, pp. 2049–206.

Warren, J.E. and Root, P.J., 1963. The behaviour of naturally fractured reservoirs, Soc. Petrol. Eng. J., Vol. 3, p. 245–255.

NUMERICAL GROUNDWATER QUALITY MODELLING: THEORETICAL BASIS AND PRACTICAL APPLICATIONS

W. K. H. Kinzelbach, Kassel University, FRG

1. TYPES OF PROBLEMS REQUIRING MODELS

The major tasks in which a model of the transport of pollutants and their transformation in the saturated zone can be applied are:

- interpretation of measured concentration data
- understanding of mechanisms of transport
- understanding of geochemical and microbiological reactions
- estimation of pollutant degradation rates
- identification of major pollutant sources
- predictions of future concentration distributions
- establishment of pollutant balances
- determination of management schemes
- planning of remedial action
- planning of preventive action
- determination of standards and policies
- planning of monitoring programs.

2. AVAILABLE TYPES OF MODELS AND EXISTING MODELLING EFFORTS

The basis for any model is the mass balance equation in its differential form. It requires three ingredients:

- a spatially averaged flow field as a basis for advective transport,
- a model of mixing processes due to diffusion and dispersion, and
- a chemical model accounting for adsorption and reactions.

Advection is defined as the bulk motion with the average saturated water flow, while dispersion summarily describes all deviations from average motion on a scale smaller than the scale resolved by the advection term. Dispersion and diffusion manifest themselves as processes which lead to transverse and longitudinal spreading of an initial concentration distribution. The simplest and widely used dispersion model consists of a Fickian diffusion term which in contrast to molecular diffusion is anisotropic, longitudinal dispersion being always stronger than transverse dispersion. Adsorption exerts a delaying influence. In the simplest model it is described by a retardation factor. Chemical transformation leads to attenuation of the original concentration distribution. An often used model of decay is the first order decay reaction. The transport equation in its most simple form reads:

The equation is a partial differential equation of second order. It requires initial and boundary conditions. If the advection term is disregarded the equation is a good-natured parabolic diffusion equation, describing the propagation of a disturbation in all spatial directions until a boundary is encountered. If the diffusion term is disregarded the equation is of hyperbolic nature, describing the propagation of disturbations in space only along characteristic lines only. The solution of this problem is much more difficult than that of the parabolic problem. Which type of behaviour is exhibited by the complete transport equation depends on the ratio of dispersive and convective fluxes, i. e. the Peclet number. Hyperbolic properties prevail for large Peclet numbers while parabolic properties dominate at small Peclet numbers.

3. METHODS OF SOLUTION

Three levels of solving the transport equation can be distinguished:

- analytical solutions
- pathline solutions
- full numerical solutions

Analytical solutions are used in the interpretation of tracer tests and column experiments as well as in the rough estimation of transport parameters in the field. Their use is quite limited due to the departure of nature from the idealized conditions required for analytical solutions.

All transport models require as a basis a model of saturated water flow or a discretized regional water balance. From a mere flow model pathline solutions describing the purely advective transport tangential to the velocity field can be derived as well as directions of average pollutant movement, average arrival times, isochrones and catchment boundaries. An example is shown in figure 1, where a horizontally two-dimensional flow model is used to determine catchment areas of drinking water wells and decontamination wells. The model already suffices for the comparison of alternative remediation schemes, decisions on the design of purging wells, well operation in a partially polluted aquifer, the determination of protection zones, and related tasks. The flow model essentially gives the average transport if mixing by diffusion-dispersion is neglected. Homogeneous chemical reactions can be incorporated as functions of travel time along pathlines.

A number of numerical methods are available for the solution of the full transport equation. Four types of numerical solution methods have been used in a number of numerical codes. These are

- standard finite differences
- standard finite elements
- method of characteristics
- random walk method.

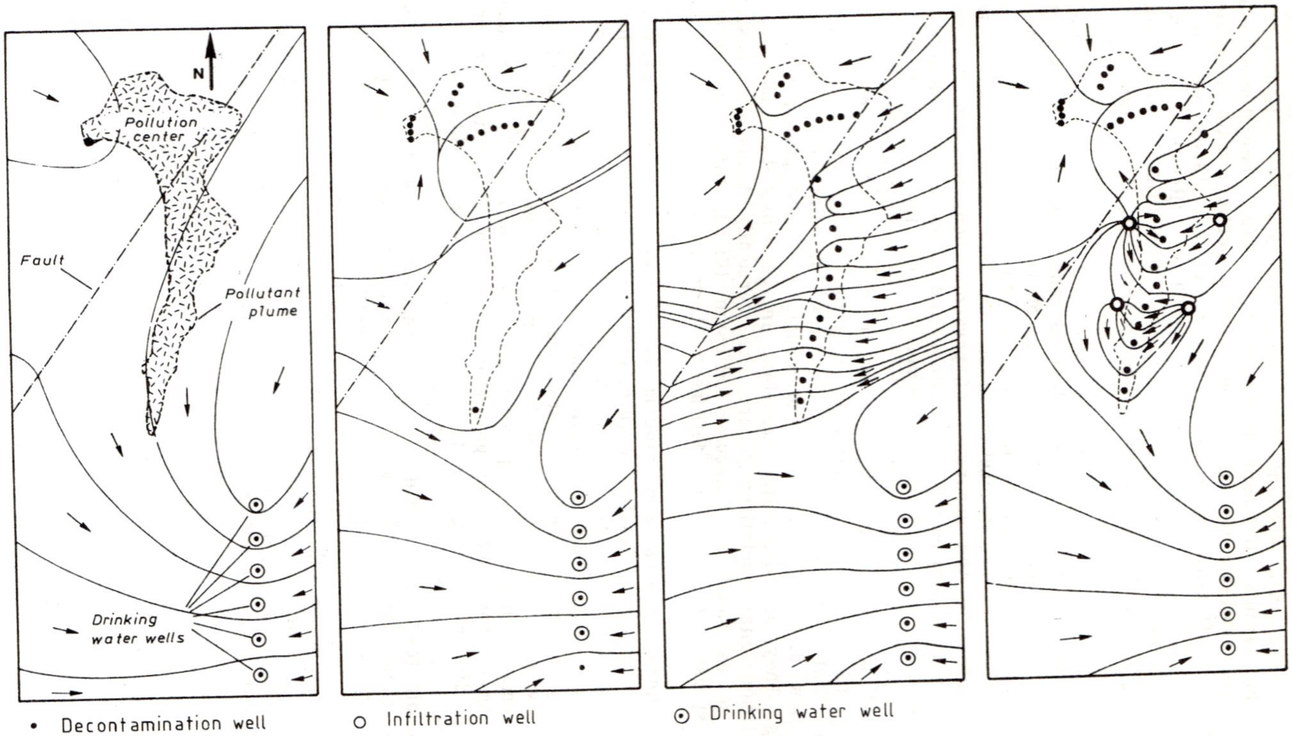

Figure 1: Comparison of alternative remediation schemes for an aquifer polluted by chlorinated hydrocarbons on the basis of computed catchment areas of wells

All methods yield satisfactory results provided the limitations of each method are taken into account. For equal accuracy the computational effort may be largely different (Kinzelbach, 1987a). Two-dimensional model codes are available for all methods in the public domain. Examples are the finite difference code SWIFT (Dillon, et al., 1978), the finite element codes SUTRA (Voss, 1984) and SEFTRAN (Huyakorn, et al., 1984), the method of characteristics code USGS-MOC (Konikow and Bredehoeft, 1978), and the random walk code by (Prickett, et al., 1981).

Numerical problems encountered in standard finite difference and finite element methods are numerical oscillations and numerical dispersion. Their application requires a careful planning of the discretization grid to guarantee accurate results. Generally the Courant-number should be less than 1 and the grid-Peclet-number should be less than 2, the latter being a guarantee for the dominance of parabolic behaviour on the grid cell level. For large ratios of longitudinal and transverse dispersivities there may still be important lateral numerical diffusion, which in standard models can be eliminated only at the cost of a spatial discretization on the order of the lateral dispersivity in regions with steep lateral gradients. A more economic suppression of lateral numerical diffusion is possible in finite element models which allow an alignment of element sides and flow direction. This class of models is called principal direction models. They are especially important in vertically two-dimensional or three-dimensional modelling where the vertically transverse mixing effect can be as small as molecular diffusion (Frind, 1981) and where numerical dispersion is particularly detrimental to model results.

Method of characteristics and random-walk show their advantages in cases where the discretization requirements of the standard methods cannot be met. While the method of characteristics divides a time step into an advective part which is usually solved by particle tracking and a dispersive step which is treated by standard finite difference techniques, the random walk method represents both advective and dispersive step by particle movements. The dispersive step is simulated by a random particle movement the stochastic properties of which are related to the size of the dispersion coefficients. Concentrations are obtained from distributions of a large number of particles. The random walk method virtually eliminates both longitudinal and transverse numerical diffusion. The method shows its greatest strength in situations where the ratio of longitudinal and transverse dispersivities is very large. Figure 2 compares the simulation of a plume by an upwind-weighted finite difference method and the random walk method. Although the grid Peclet number is smaller than 2 the finite difference scheme introduces considerable lateral dispersion. In horizontally two-dimensional agricultural applications the random walk method is not efficient. Due to the ubiquitous occurrence of agricultural pollutants, it requires particles everywhere in densities which allow a significant evaluation of concentrations. Further, the method is less amenable to the introduction of nontrivial non-linear chemical reactions. A more detailed comparison of the four basic model types and their properties is presented in (Kinzelbach, 1987b).

4. DIMENSIONALITY

Any real world problem is spatially three-dimensional and time-dependent. Yet, approximations in lower spatial dimensionality are often

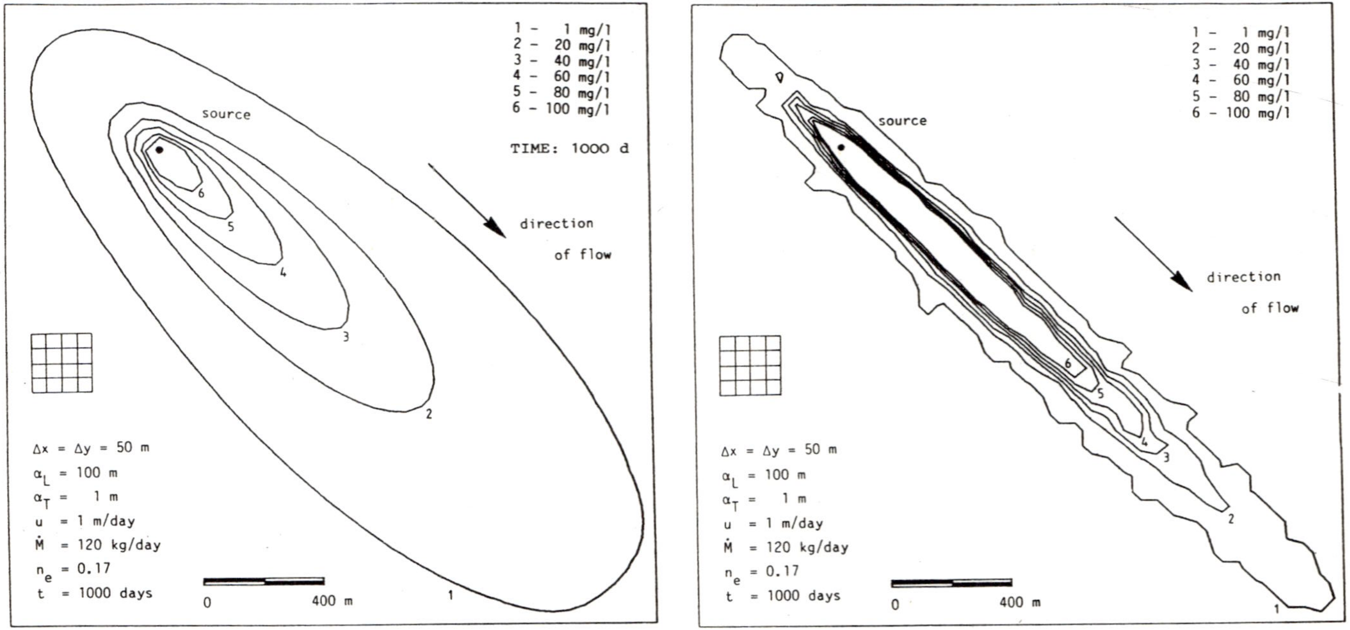

Finite Difference Method (upwind) | Random–Walk–Method

Figure 2: Comparison of two numerical methods: Upstream-weighted finite differences and random walk method feasible.

4.1 Zero-dimensional models

The most basic model of a catchment of a well or a spring is the zero-dimensional model (e.g. Mercado, 1976, Thiery, 1984, Strebel, et al., 1985). It approximates the catchment or a well-defined part of the catchment as one single or several well-stirred vessels. It is able to describe temporal changes of concentrations at the well as a function of gradually varying inputs to the catchment which are fairly homogeneously distributed in space. It is suited for the treatment of non-point pollutants from agriculture and should in these cases be the first step in a pollutant balance calculation. The size and shape of the catchment or the fluxes into the balance element chosen must be determined by means of a flow model.

On the basis of a known catchment a zero-dimensional model calculates a balance of inputs and outputs (Figure 3) or concentration histories at a well taking the whole catchment of the well as one well-stirred reactor.

A generalization of the zero-dimensional model are the multi-box or mixing-cell models. Increasing the number of boxes eventually leads to a full-fledged finite difference or finite element model.

4.2 One-dimensional models

Models in one spatial dimension are used in the interpretation of column experiments in the laboratory. Applications in the field are feasible if concentration gradients transverse to the direction of flow are very small. A streamtube can under these circumstances be treated as a long column.

4.3 Horizontally two-dimensional models

A horizontally two-dimensional model can handle the superposition of spatially and temporarily varying inputs in their impact on the water quality in pumping wells. It allows the calculation of concentration distributions averaged over the saturated aquifer depth as well as concentration histories at pumping wells or observation points. It also allows evaluation of the influence of any input into the flow field on the concentrations in particular wells. It is a tool for regional remediation and management.

Figure 4 shows a pollution plume and concnetration histroies at wells simulated with a horizontally two-dimensional transport model (Kinzelbach and Kauffmann, 1987). The calibrated model was used to determine optimal strategies of managing the partly polluted aquifer.

4.4 Vertically two-dimensional models

Vertically two-dimensional models can be applied in agriculture related transport problems as the horizontal gradients of nitrate or pesticides are usually small. The main objective of vertically two-dimensional transport models is the evaluation of vertical transport and vertical

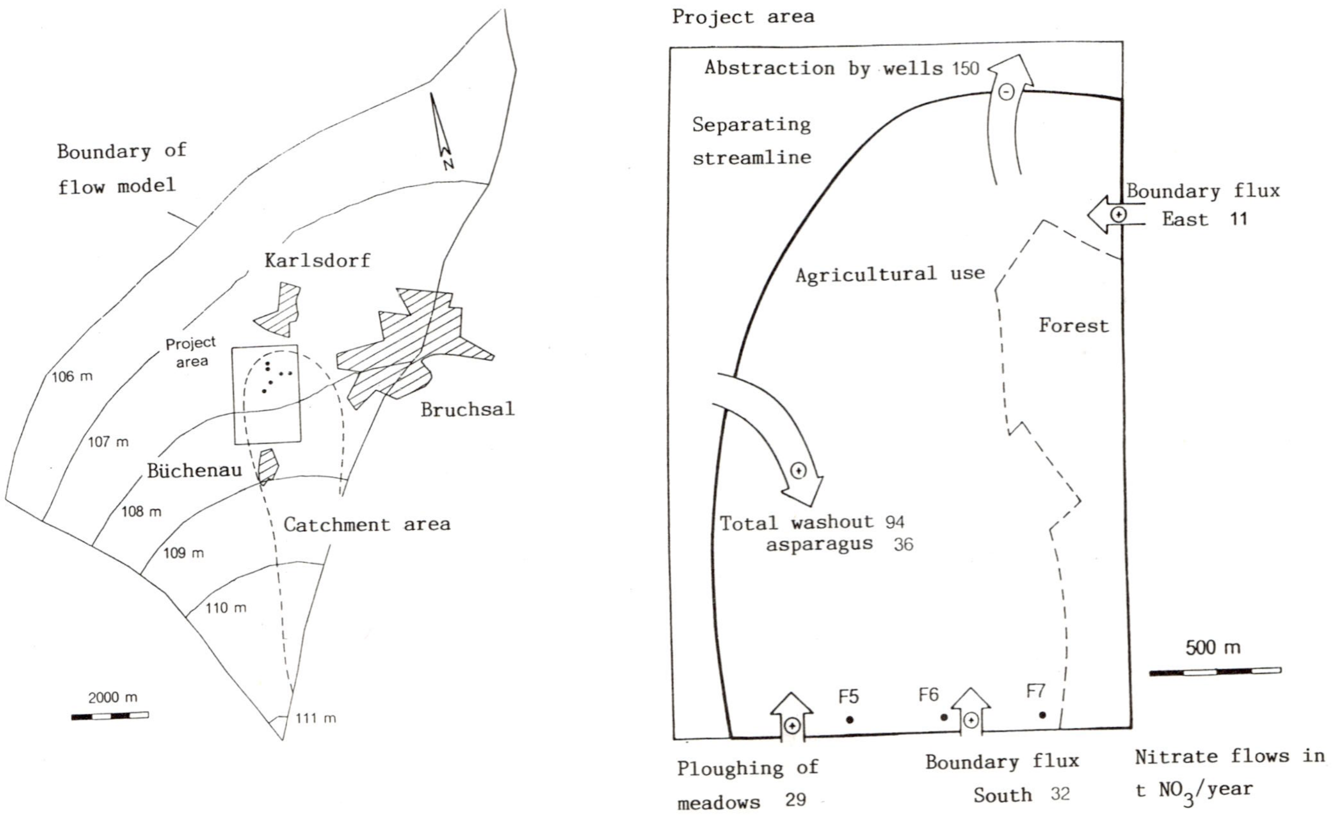

Figure 3: Nitrate balance in an intensively farmed region

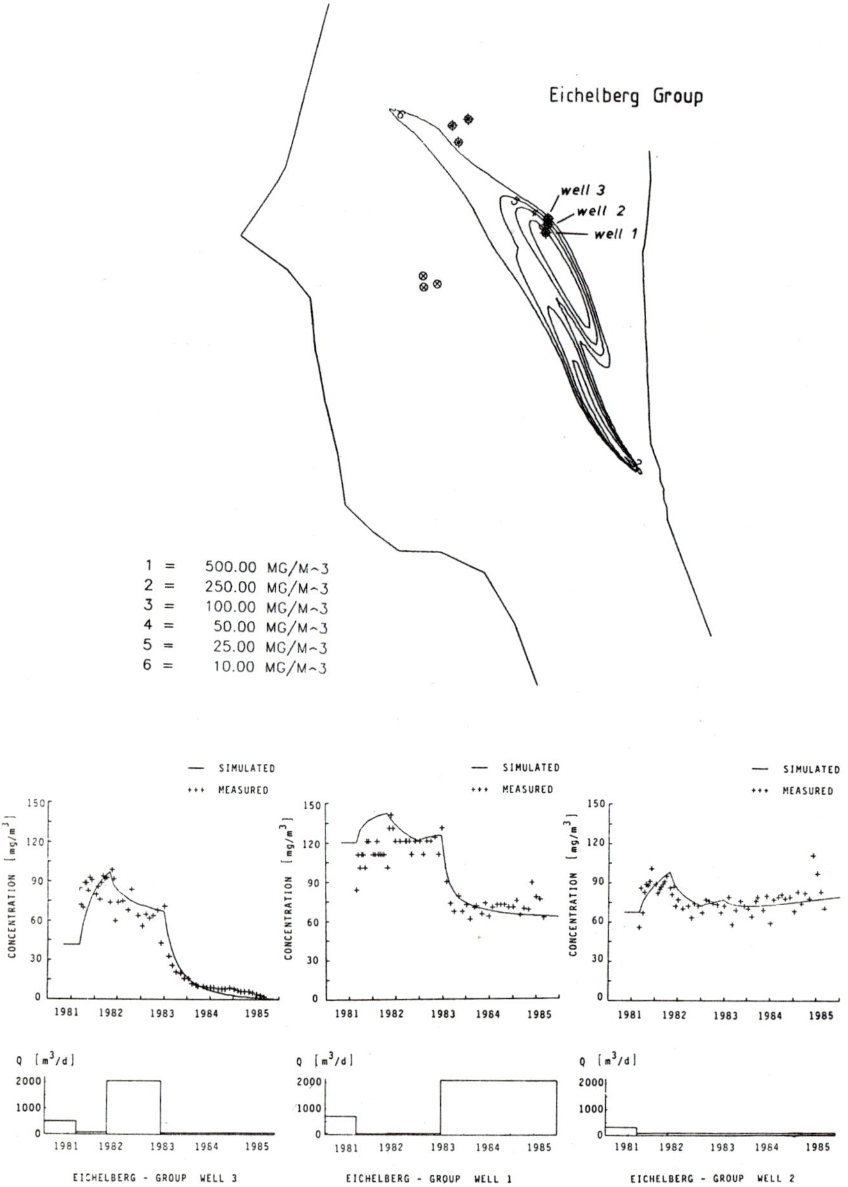

Figure 4: Application of a horizontally two-dimensional model to the simulation and management of a partially polluted aquifer in the Upper Rhine valley: Computed plume and concentration histories at wells

mixing, the interpretation of vertical concentration profiles and the chemical development of a pollutant input in all cases where vertical averaging leads to erroneous results.

In several studies the transport of nitrate from agriculture was modelled in vertical cross-sections along streamtubes. Figure 5 (Obermann, 1981) shows the vertically 2-D flow field represented by pathlines and isochrones. The latter could be compared with results from tritium measurements. In a nitrate study of the Fuhrberger Feld near Hannover Böttcher et al. (1985) carried out a comparable study which used more dense information on the vertical distributions of nitrate, sulfate, and tritium. In Kinzelbach, et al. (1989) a full transport model including bacterial denitrification processes was applied to a vertical cross-section.

4.5 Three-dimensional models

Three-dimensional transport models are required in local problems where the three-dimensional nature of the flow field cannot be neglected, such as in the design of remedial in-situ measures. Also in large-scale field problems where the flow field could be viewed as two-dimensional, chemical reactions occurring at varying depths of the aquifer (depending on oxygen concentrations) may necessitate a resolution of the vertical aquifer dimension (Figure 6) as non-linear reaction kinetics require local concentrations rather than vertical averages.

A three-dimensional model imposes serious demands on the computer which are not easily fulfilled even with present computers. Still, the problem of computation is more easily solved than problems posed by the increased requirements in terms of input data, interpretations of output and calibration of parameters.

For large-scale problems an interesting and pragmatic model combination consists of a 3-D flow model with a one- or two-dimensional chemical transport model along single stream-tubes or bundles of stream-tubes, the dimensionality of the transport model depending on whether the pollutant source is distributed or point-like.

5. MACRODISPERSION

The modelling of macrodispersion in the non-asymptotic stage is as yet an unsolved and maybe unsolvable problem. The use of Fickian dispersion models with constant dispersivities is a crude tool only in which the dispersivities are mere fitting parameters.

While the Fickian model proved adequate in experiments on laboratory scale, field tests show that it is inadequate for the description of macrodispersion. It turns out that the apparent dispersivity is not a constant but rather increases with displacement distance. Stated differently, observations show that the variance of a tracer cloud grows faster than linearly in time. This observation is related to the geological heterogeneity scale, which increases with the scale of the problem.

To interpret these findings stochastic transport and dispersion theories were formulated (Gelhar et al., 1979, Mathéron and de Marsily, 1980, Gelhar and Axness, 1983, Dagan 1984). In principle, if the three-dimen-

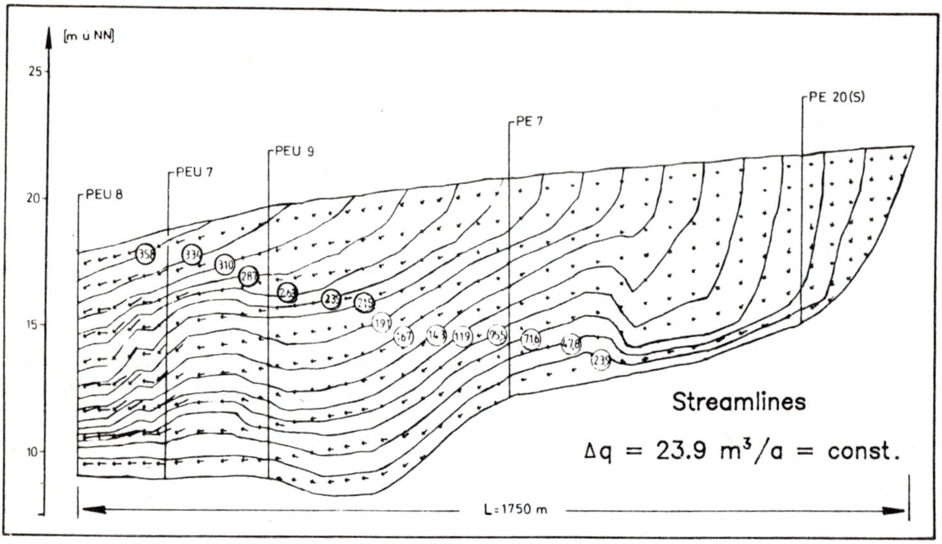

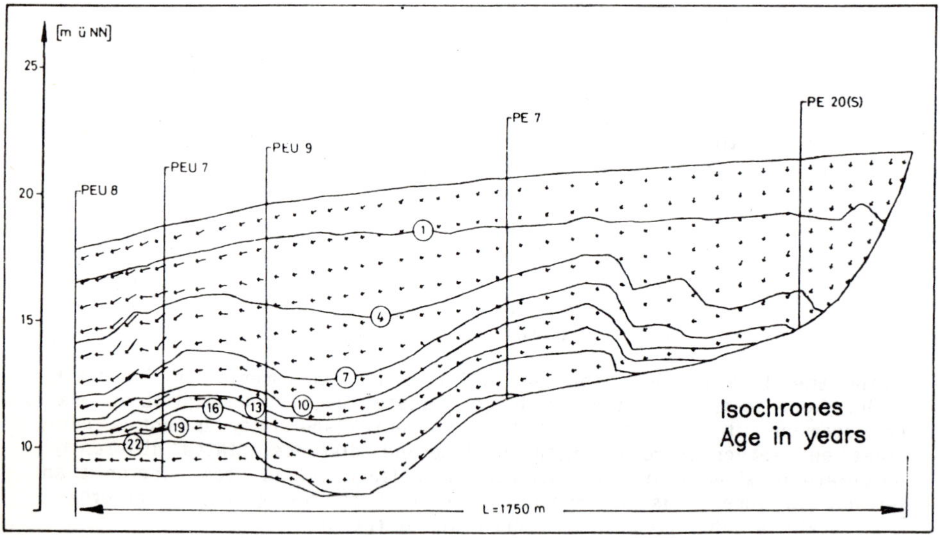

Figure 5: Isochrones and pathlines in a vertical cross-section through the catchment of Mussum water works (Obermann, 1981)

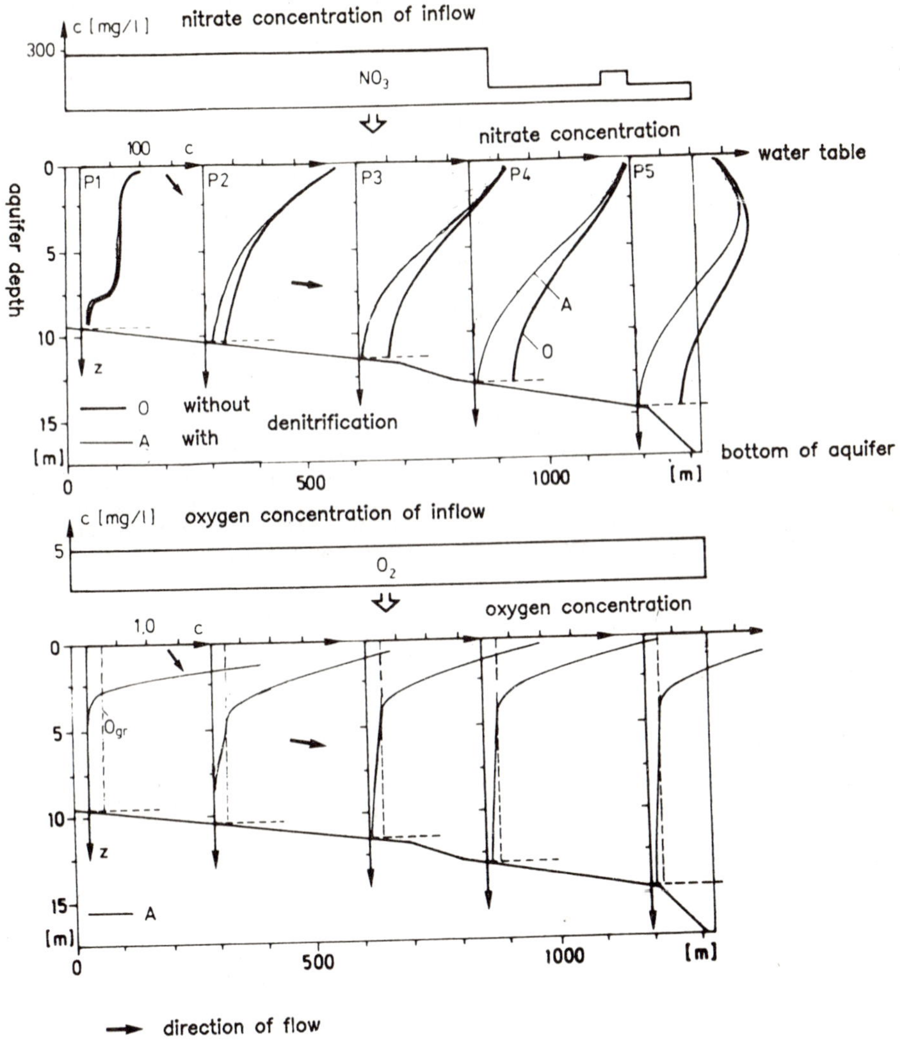

Figure 6: Simulated chemical profiles in a vertical cross-section of an aquifer (Herzer, 1988) (O_{gr} limiting oxygen concentration for denitrification)

sional aquifer heterogeneity structure was known in all details, the solute transport could be predicted accurately because the dispersion component would only have to account for the well understood micro-scale spreading effect. Such knowledge will never be available in practical field work. Aquifer properties will only be known at a limited number of points and the variation in between is subject to uncertainty due to the unknown irregular spatial variations. This unknown variation in aquifer properties is responsible for dispersion. The uncertainty of the actual heterogeneity introduces uncertainty into the model predictions.

The stochastic treatment of the transport equation has been approached by spectral methods (Gelhar and Axness, 1983) and particle-tracking Lagrangian methods (Dagan, 1984). These methods give insight rather than they allow treatment of specific field cases. An approach which is more amenable to practical application employs Monte-Carlo-techniques. Such methods are less restrictive in terms of underlying assumptions. They do not require infinite extension of the aquifer nor do they require smallness of the variations of hydraulic conductivity compared to the mean value. But the results are more site specific and the large number of replicates involved requires significant computational resources. Smith and Schwarz (1980) have used Monte Carlo techniques to demonstrate the dependence of the dispersivity on displacement distance (or alternatively time). The type of work which comes closest to practical needs is the Monte-Carlo modelling as presented by Gorelik (1987). A pragmatic device of introducing scaling dispersivities into random walk simulations of transport is given by Kinzelbach (1987c).

The stochastic techniques are still in the developing phase and more field validations are called for before they can be used on a routine basis in practical aquifer contamination problems. Also, it must be remembered that the stochastic approach requires the knowledge of the characteristic properties of the distributions of uncertain input parameters. These can only be obtained by extensive field work which oftne times cannot be justified. If they cannot be provided, the whole approach becomes doubtful as the quantification of uncertainty in the results rests upon an assumed uncertainty of input parameters. Further, the odd unknown highly pervious or impervious lens makes predictions in the near field very much uncertain.

6. MODELLING OF CHEMICAL PROCESSES

Although the concepts and techniques of coupling transport models and chemical models are well known (Rubin, 1983) the modelling of chemical processes in aquifer materials remains inherently a complicated task, since the possible number of reaction processes and participating constituents are large. Many reactions have not yet been identified or formulated in quantitative chemical terms. Moreover, if a quantitative theory has been developed, the formulations may turn out to be very complicated and hardly operational, because single reactions cannot be isolated due to the interactions with other processes. In general, any chemical modelling effort of aquifer conditions has a data problem, which is much more serious than that of tracer transport modelling. The available field data will usually not allow for even an approximate assessment of the parameters, and since the field modelling experience is limited, a particular application cannot take advantage of other applications.

Compared to the data problem the task of numerical modelling of chemical processes coupled to transport is relatively easy to solve. Usually the split-operator method is applied, which divides a time step into a substep of advective-dispersive transport and a substep of chemical reactions. Figure 7 shows a computed concentration distribution of adsorptive cation transport. After the injection of a potassium ion solution into an equilibrium background the potassium ions are strongly adsorbed while a cloud of ions displaced from the matrix moves downstream at the velocity of flow (Schulz, 1981). This type of behaviour cannot be reproduced with simple retardation models.

Only a few well-documented field applications of geochemical models have been reported, and no conclusive evidence of the predictive capabilities can be provided yet. The major part of the theoretical framework on geochemical modelling developed so far has been related to small scale homogeneous soil columns in the laboratory. Transferring these model developments to field conditions is not straightforward because of the natural geological heterogeneity. One may fear that the parameters and relations pertinent to the process are an obvious parallel. It appears that more research should be directed towards developing a theoretical framework that relates results of laboratory experiments to field conditions instead of adding still more refinements to the small-scale description of equilibrium controlled reactions. It is intuitively clear that the behaviour of chemicals in groundwater flow will be very different depending on whether correlations between the chemical inhomogeneity structure and the permeability distribution exist or not. Garabedian (1987) has applied a stochastic approach to large-scale reactive transport, and more developments along these lines and toward the concept of effective parameters, Gelhar (1986), are required.

Chemical transport modelling spans a spectrum from zero-order reaction and first-order reaction to transport of multiple interacting components. The simple models suffer from the fact that they are not predictive. The reaction rates must in every application be estimated from observation data. The complicated models suffer from the fact that parameters are too numerous and are usually not known in the particular field application. Complicated chemical models will for the forseeable future be more valuable for conceptualization of the processes in the underground, parameter sensitivity studies and answering fundamental questions than for actual simulation of field problems.

7. DATA REQUIREMENTS AND PARAMETER IDENTIFICATION

The modelling of flow and transport requires a large amount of data from groundwater recharge, transmissivities or permeabilities to solute input rates and degradation rates. Usually only few of these are readily available. Some have to be estimated by means of the model and sufficient observation data on heads and concentrations. The usual flow model calibration which may be satisfactory in modelling heads is generally not accurate enough for transport modelling. Figure 8 shows a case where optimal fitting of observed heads leads to a wrong plume (Kauffmann and Kinzelbach, 1989). Therefore the calibration of the flow model must be improved by combining it with the calibration of the

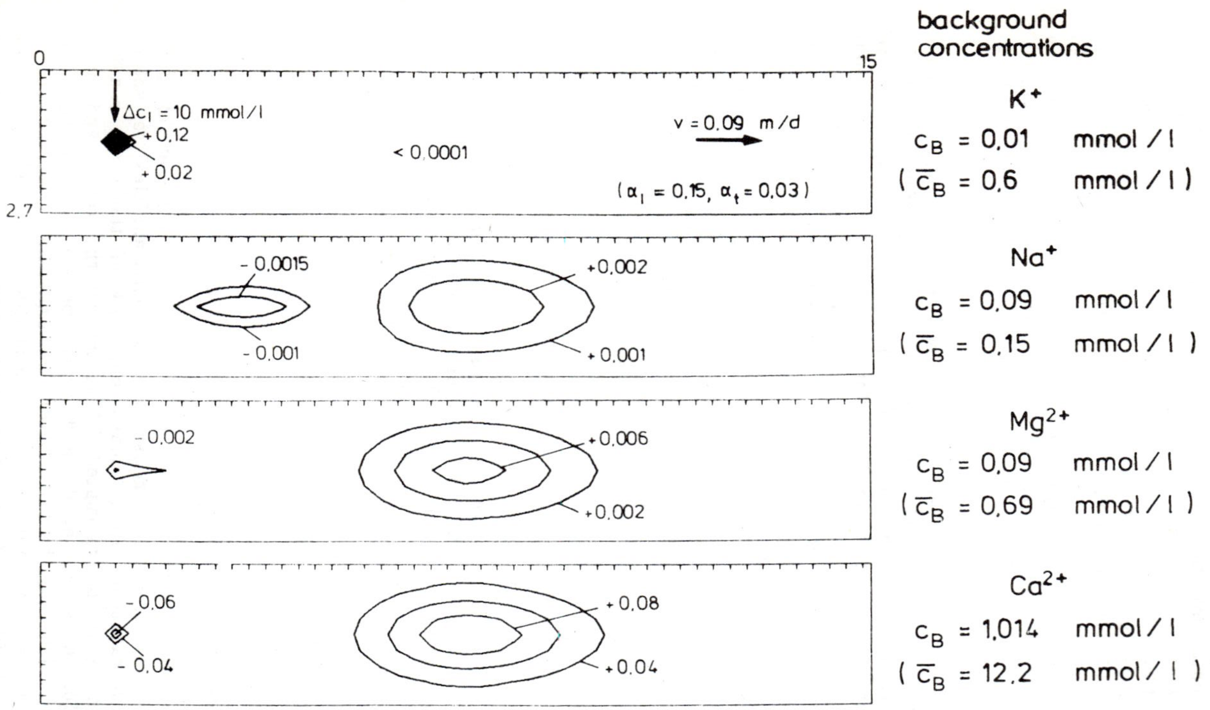

Plan view of the major cation concentrations (relative to background) 70 days after injection of Potassium (c_B...pore water concentration, $\bar{c}_B$...adsorbed concentration)

Figure 7: Coupling of cation exchange to transport

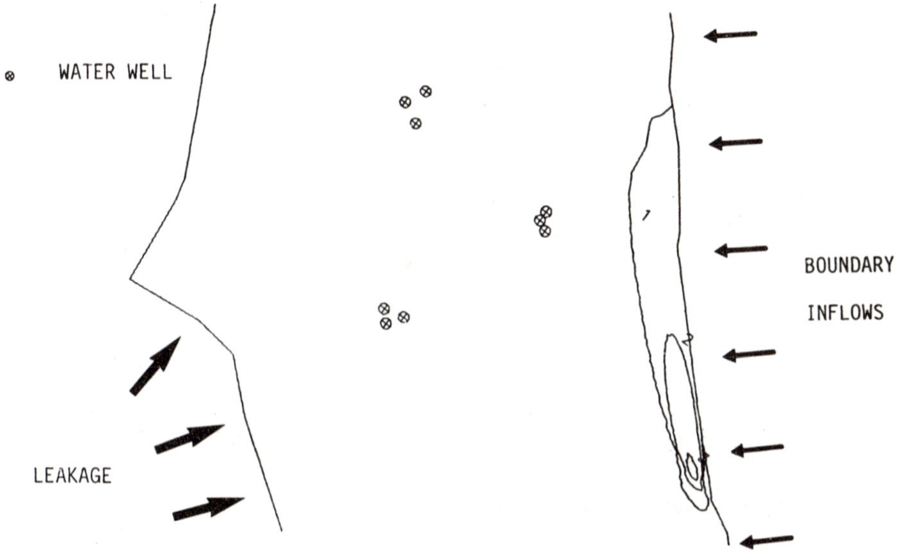

Flow model optimally calibrated from heads

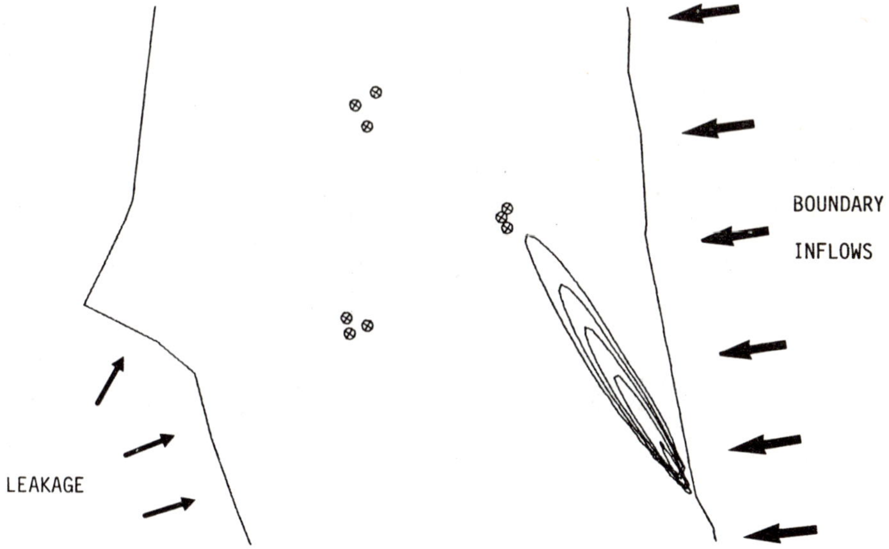

Flow model corrected to give true plume direction

Figure 8: Comparison of simulated plumes from "optimally" calibrated flow model and corrected flow model transport model. In a first step, the calibration of the flow field could take into account information on average flow directions as given by the axis of a known plume. Even better conditions for model calibration exist if strong controllable hydraulic stresses are exerted upon the aquifer which modulate concentration responses and thus allow a unique separation and identification of flow and transport parameters.

The most severe data problem in agricultural applications is the usually insufficient knowledge of source terms, especially of source terms in the past. Estimation of source terms from observed concentration data is usually not unique and introduces a large amount of uncertainty into the model. Similarly, it is difficult to estimate source terms in accidental spills of industrial pollutants as the pollution is often times discovered only years later.

Finally, it must be stressed that uniqueness of model parameters can usually not be assured through calibration alone. Therefore all sources of prior knowledge outside of the actual model must be tapped in an attempt at least to rule out certain ranges of parameter values or to reduce ambiguity by constraining parameters to physically reasonable intervals during calibration.

8. OBSERVATIONS ON LIMITATIONS OF NUMERICAL MODELLING

Transport models are crude images of reality. Our knowledge of source and reaction terms, boundary and initial conditions, and the distributed parameters for flow and transport, will never be complete while data acquisition bears a cost. The predictive capability of models is therefore quite limited. Still, there are many applications where modelling is of use. First, a model allows a reproducible interpretation and interpolation of data. Second, even a model of limited accuracy can be used in decision making by worst case techniques and comparison of alternative courses of action where different tendencies might be reflected more accurately than absolute concentration values.

Sensitivity analysis enables the modeler to determine the effect of changing inputs and parameters on model outputs. Stochastic modelling provides a logical extension of sensitivity analysis for uncertain inputs, where their assumed probability distributions are used to generate series of random values used in a Monte-Carlo framework. This provides a suite of model predictions, giving a distribution for the likelihood of particular consequences arising rather than one definite prediction.

The difficulties in establishing model uniqueness may often be alleviated as for some parameters there are wide parameter ranges which lead to model predictions having the same consequence for decision making. The decision maker can confidently use an inaccurate model if the recommended course of action is stable towards rather large parameter variations.

Finally, it should be stressed that the art of successful modelling consists to a large degree of identifying interesting questions which under given data constraints can be answered by available modelling techniques with sufficient certainty. There may well be cases where modelling is not required - because the answer is evident - or not useful - because the uncertainty of results is too large.

LITERATURE

Böttcher, J., Strebel, O., Duynisveld, W. H. M. (1985). Vertikale Stoffkonzentrationsprofile im Grundwasser eines Lockergesteinsaquifers und deren Interpretation (Beispiel Fuhrberger Feld). Z. dt. geol. Ges., 136, pp. 543-552.

Dagan. G. (1984). Solute transport in heterogeneous porous formations. J. Fluid Mech., 145, pp. 151-177.

Dillon, R. T., Lantz, R. B. and Pahwa, S. B. (1978). Risk methodology for geological disposal of radioactive waste: The Sandia waste isolation flow and transprot (SWIFT) model. Report SAND 78-1267, Sandia National Laboratory.

Frind, E. O. (1982). The principal direction technique: A new approach to groundwater contaminant transport modelling. In: finite Elements in Water Resources, Proceedings of the 4th International Conference, Hannover. Springer Verlag, Berlin, pp. 13/25-13/42.

Garabedian, S. P. (1987). Large-scale dispersive transport in aquifers: Field experiments and reactive transport theory. Doctoral thesis. Massachusetts Institute of Technology.

Gelhar, L. W., A. L. Gutjahr and R. L. Naff (1979). Stochastic analysis of macrodispersion in a stratified aquifer. Water Resources Research 15(6), pp. 1387-1397.

Gelhar, L. W. and C. L. Axness (1983). Three-dimensional stochastic analysis of macrodispersion in aquifers. Water Resources Research, 15(6), pp. 161-180.

Gelhar, L. W. (1986). Stochastic analysis of solute transport in saturated and unsaturated porous media. In: Fundamentals of transport phenomena in porous media. NATO Advanced Study Institute, eds. J. Bear and M. Y. Corapcioglu, Reidel Publishers, The Netherlands.

Gorelik, S. (1987). Sensitivity analysis of optimal ground water contaminant capture curves: Spatial variability and robust solutions. In: Solving Groundwater Problems with Models, Proceedings of the NWWA conference, Denver.

Herzer, J., 1989. Numerische Modellierung von Stofftransport und Reaktion im Grundwasser. Dissertation, Universität Stuttgart (to appear).

Huyakorn, P. S., Broome, R. W., Lester, B. H., Kretschek, A. G., and Mercer, J. N. (1984). SEFTRAN. Geotrans Inc., Herndon VA.

Kauffmann, C. and Kinzelbach, W. (1989). Parameter esimation in contaminant transport modelling. Contaminant Transport in Groundwater,

Proceedings of a Symposium, Balkema Publishers (in print).

Kinzelbach, W. (1987a). Methods for the simulation of pollutant transport in groundwater - A model comparison. In: Solving Groundwater Problems with Models, Proceedings of the NWWA Denver Conference.

Kinzelbach, W. (1987b). Numerische Methoden zur Modellierung des Transports von Schadstoffen im Grundwasser. Oldenbourg Verlag, München.

Kinzelbach, W. (1987c). The random walk method in pollutant transport simulation. In: Advances in analytical and numerical groundwater flow and quality modelling. NATO ASI Series, Reidel Publishers, Holland.

Kinzelbach, W. and Kauffmann, C. (1987). Application of a flow and transport model to the exploration and management of partially polluted aquifers. In: Groundwater Contamination, Proceedings of an International Conference, Reidel Publishers, Holland.

Kinzelbach, W., Schäfer, W., and Herzer, J. (1989). Numerical modelling of nitrate transport in a natural aquifer. Contaminant Transport in Groundwater. Proceedings of a Symposium, Balkema Publishers (in print).

Konikow, L. F. and Bredehoeft, J. D. (1978). Computer model of two-dimensional solute transport and dispersion in groundwater. Techniques of Water Resources Investigations of the United States Geological Survey, Book 7, Chapter C2. United States Government Printing Office, Washington, 90 p.

Mathéron, G. and G. de Marsily (1980). Is transport in porous media always diffusive? A counter example. Water Resources Research 16(5), pp. 901-917.

Mercado, A. (1976). Nitrate and chloride pollution of aquifers, regional study with the aid of a single cell model. Water Resources Research 15(12) pp. 731-747.

Obermann, P. (1981). Hydrochemische/hydromechanische Untersuchungen zum Stoffgehalt von Grundwasser bei landwirtschaftlicher Nutzung. Besond. Mitt. Dt. Gewässerkundl. Jahrb., 42.

Prickett, T. A., Naymik, T. G. and Lonnquist, C. G. (1981). A "random-walk"solute transport model for selected groundwater quality evaluations. Illinois State Water Survey, Bulletin 65, 103 p.

Rubin, J. (1983). Transport of reacting solutes in porous media: relation between mathematical nature of problem formulation and chemical nature of reactions. Water Resources Research, 19(5), pp. 1231-1252.

Schulz, H. D. (1981). Zweidimensionales Transport-Reaktionsmodell für Ionen im Grundwasser. Z. dt. geol. Ges., 132, pp. 293-502.

Smith, L. and F. W. Schwartz (1980). Mass transport, I. Stochastic analysis of macrodispersion. Water Resources Research 16(2), pp. 303-313.

Strebel, O., Böttcher, J., and Kölle, W. (1985). Stoffbilanzen im Grundwasser als Hilfsmittel bei Klärung und Prognose von Grundwasserqualitätsproblemen (Beispiel Fuhrberger Feld). Z. dt. geol. Ges., 136(2).

Thiery, D. (1984). Modelisation des transferts de nitrates dans cinq bassins versants a Rembercout-Aux-Pots (Meuse). Technical Note Nr. 84/01, Bureau de Recherches Géologiques et Minières, Service Géologique National, Orléans, 48 p.

Voss, C. I. (1984). A finite element simulation model for saturated-unsaturated, fluid density-dependent groundwater flow with energy transport or chemically-reactive single-species solute transport. U.S.G.S., Reston, Virginia, Open file report, Scientific Publications Co. Washington, D.C.

Prof. Dr. W. Kinzelbach, GhK-Universität,

FB 14, Moritzstr. 21, D-3500 Kassel, FRG

GROUNDWATER POLLUTION PROTECTION : A MULTIDISCIPLINARY ACTION TO BE LED BY A PRACTISING HYDROGEOLOGIST AND BACKED UP BY AN ENFORCEMENT OF LAW

N. CRAMPON
Laboratoire d' Hydrogéologie, SN5, Université de Lille 1
59655 Villeneuve d'Ascq - FRANCE

ABSTRACT - Fighting against groundwater pollution already needs a proper knowledge of polluting activities, of their localization and of contaminant substances behaviour with exchange and transport processes in unsaturated then in satured undergrounds. But, first at all, a pollution is a water quality alteration propagated into a special context. This has to be studied on geological, hydrogeological and hydrogeochemical levels. Monitoring networks and mapping allow to specify the source, the displacement and the distorsion of the polluting cloud. To predict the concentration variation over distance and time, models fitted on field data (especially tracing tests) need to be used. Finally, preventing pollution is getting adequate provisions made and applied. Pollution removal is always more difficult and more expensive than prevention, in spite of a few rehabilitation processes that are now being developed. All these actions can only be co-ordinated by an experienced hydrogeologist.

1. INTRODUCTION

A lot of developed countries are hit by industrial pollution and their rural areas are more and more contaminated by nitrate. Drinking water is getting scarcer and scarcer, hence more and more expensive. In 1950, eleven among the fifteen largest towns in the world were situated in industrialized countries. In the year 2000 there will be 42 towns inhabited by more than 5 million people in developing countries versus 16 towns in developed countries. Such a growth is incorrectly followed by water-supply and sanitation equipments. In the rural areas of developing countries, salubrity is not much better and the shallow wells are insufficiently protected.

Two billion people cannot use potable water. In developing countries 80 per cent disease outbreaks occur due to incorrect water. Finding a remedy is the aim of U.N. International Drinking Water-Supply and Sanitation Decade (1981 - 1990).

Some industries, at first mining ones then processing ones, are sometimes these developing countries' only wealth. The economical priority often prevails over the groundwater protection. Many liquid effluents are spilt and solid waste dumps are discharged without any treatment on ground surface.

However, the often thin and sometimes non-existent soils allow little filtration, especially in case of exceptional high rainfall as it happens in tropical countries.

2. PRINCIPAL ACTIVITIES CAUSING POLLUTION RISK

Foster (1985) set up a clear classification of these activities, pointing out those that seemed more significant to him in developing countries.

2.1. Sanitation

Sewers are scarce, but where effluents are discharged directly on ground surface or in watercourses without any treatment, risks are not negligible. Unsewered (on-site) sanitation can be adequate for excreta disposal in rural areas and small towns. But in many countries, especially in Asia, a lot of people are currently without any form of sanitation. In very densely populated settlements, spacing between excreta disposal unit and drinking groundwater source is too short. The lack of sanitation is a proven cause of pathogen transmission in numerous disease outbreaks. Major conurbations also lay a serious problem. Thus, in Mexico, a more than 18 million people town whose population increases by 2000 every day, 42 000 tons of waste and 2 000 tons of excreta are dumped in the open air every week (Girardon, 1988).

As far as nitrogen compounds are concerned, one must remember one person throws off 14 to 15 g N/day (5 kg/year) which, if oxidised and leached, provide 22 to 24 kg $NO3$/year. Indeed, dilution and reduction may reduce that amount. But, in relatively humid climates, it is often difficult to make a distinction between a domestic nitrate pollution and an agricultural one. Nevertheless, either the examination of nitrate distribution maps (Crampon, 1983-b & 1988) or 14 and 15 Nitrogen isotopic ratios (Létolle & Mariotti, 1975) can help in such a distinction.

2.2. Agricultural land-use

The use of organic fertilizers (dung, liquid manure) and the irrigation with liquid wastes are, there again, a threat of microbial contamination. They also contribute to the increase of nitrate in groundwater. Those traditional practises are decreasing in Europe (except intensive breeding), but remain current in developing countries.

The need for practising profit - earning crops, the mechanization in equipment, are modifying soil treatments. Deeper ploughing, with the slope ploughing, heavy engine use, non-hiding straw, bare soils at certain times, all these practises tend to reduce the organic content of the soils, to break up its structure, therefore to exhaust it and to make its erosion easier. The infiltration of water more and more polluted by chemical fertilizers and pesticides, will be all the more facilitated.

In France, the rate of yield per hectare has roughly decu-pled from 1900 to 1980, whereas, at the same time, the chemical fertilizer consumption has been multiplied by 22.4 (Crampon, 1983-b). Agricultural pollution seriously begins to worry some governments, but the vertical transfer to the water-table may be very slowed down, for example, by about 0.5 to 1m/year in the France chalk and in the U.K. one according to Cole & Wilkinson (1976), Arnoult (1981) and Roux (1988). The actual agricultural contamination will then be wholly established in several years. May those mistakes not be repeated in developing countries. Whatever the source, an organic or a chemical one, nitrate leads to toxic compounds in the system : carcinogen nitrosamines, asphyxiating methemoglobin in young infants and ruminants (Fritsch & de St Blanquat, 1985).

The use of increasing numbers and quantities of pesticides and herbicides in agriculture and forestry for the control of detrimental organisms is another cause for groundwater contamination, fortunately it is not major or widespread yet. Matthess et al. (1985) present a rather comprehensive list of these synthetic organic compounds, with their solubility and main drinking water standards (WHO and EEC). Organophosphates are more rapidly biodegradable and go down through soils hardly more than 0.30m. On the other hand, chlorinated hydrocarbons may occasionaly be found in 40 m depth (Van der Molen & Van Ommen, 1988) and in groundwater (Matthess et al., 1985). The use of weakly-adsorbed compounds of significant solubility can give rise to serious contamination, especially where irrigation is practised, with coarse-textured soils of low organic content and shallow water-table.

2.3. Urban and industrial activities

Among pollution risks can be quoted : industrial sewages, solid wastes, raw material, storage, hydrocarbon and chemical tanks and pipelines, traffic with hazardous risks of toxic products spilling, road contamination (salt for snow, tetraethyl lead in gases from cars), radioactive products and wastes, mining and quarrying with their wastewater and residues. Nowadays, to this should be added acid rain from the atmospheric pollution (Van der Molen & Van Ommen, 1988).

Plants, thermal and nuclear power-stations, refineries, pipelines, are built more and more suitably with respect to environmental laws, groundwater included. It is not the same, far from it, with industrial wastes, sometimes illicit and often deposited without any care with regard to groundwater. Thus, we know gravel-pits partially filled with siderurgy solid wastes where the water pH has reached the 12 value. We also know a former limestone quarry filled up with moter-oil reuse residues containing phenol, heavy metals and sulphonates (Crampon & Lempereur, 1983).

Heavy metals (with a density over 5 g/cm3) may be found in many raw materials, industrial products, hydrocarbons, pesticides, mining residues. The most important ones in groundwater are Cd, Hg, Pb, Cr and Zn, where the acid pH is less than 6.7 (Roux, 1988).

Mining or quarrying dumps may also generate contamination by leaching : sulfates by oxidation of pyrite included in shales (Northern France), saline pollution in alluvial aquifer from KCl and Na Cl residues piles (Eastern France).

2.4. Conclusions

With regard to all these threats, the right thing to do is to get a better knowledge of the chemical characteristics of the various contaminants : solubility, pH and Eh equilibrium, interaction in water and in relation to soil or rock minerals, complexation, miscibility with water, biodegrability, bacteria and viruses persistence (see Matthess et al., 1985).

3. QUANTITATIVE AND QUALITATIVE GEOHYDROLOGICAL ENVIRONMENT

"Contaminant hydrology is the science which studies the transport and behaviour of contaminants in the underground" (Van Waegeningh, 1985). Previous to the modelling of these transport and behaviour, a detailed geological and hydrogeological reconnaissance survey is needed. Many results of modelling remain disappointing due to the lack of such field studies.

3.1. Geology

The geolocal structure must be studied on the complete aquifer scale, then more accurately on a local scale where a pollution happens i.e. : the lithology of layers or massif, the lower, upper and lateral boundaries of the aquifer, facies changes, dips and folds, schistosity, comprehensive and local fracturation.

As for porous aquifer, one should properly know, not only the grain size distribution and mineralogy, but also the thickness and the composition variability of strata, the occurence of paleochannels and of any different causes of heterogeneity. The determination of open and effective porosity will be particularly useful at all scales.

The rock matrix of fissured aquifers will retain the same attention, mainly matrix porosity and adsorbing minerals. The main point of the survey will relate to fissuration (extents, direction, dip, aperture, filling and frequency for each family), its distribution in aquifer and its relationship with regional structure. It is well known that the spatial fissure distribution has a strong influence on the hydraulic characteristics of the aquifer (Carlier, 1983 ; Skinner, 1985).

Theoretically, it is even possible to calculate the hydraulic conductivity tensor from elementary equations in fissure (Louis, in Mangin, 1975 ; Carlier, 1983). The differences in fissure type, geometry and scale pose considerable problems for the design of investigation programs for fissured groundwater systems and the interpretation of data derived from them. In some cases, such as the chalk in the U.K. or in Northern France, the equivalent (anisotropic) medium concept can be applied, whereas, in other cases, the scale of discontinuities will be larger than the investigation one and the medium will remain very heterogeneous.

In karstic systems, the carbonate dissolution has developed cavities which affect groundwater storage and flow much more than the other voids (pores and fissures) do. On the investigation scale the heterogeneity is very high. The study of these cavities is particularly relevant to speleology, underground hydrology (flowrates, water chemistry, tracer experiments) and structure analysis.

Where an unsaturated zone is present, in the same rocks above the water-table (unconfined aquifer), the survey of this zone is at least as important as the one of the saturated ground, especially concerning all that can make the infiltration faster (opened fractures, sinkholes, shafts, etc...) or slow it (overlying formations).

The boundary grounds will also be examinated with regard to lithology, geometry, permeability and porosity in order to estimate the exchange possibilities (diffusion, adsorption - desorption, leakage).

3.2. Groundwater recharge, storage and flow

Even if it is very approximative a global approach of the resource is necessary. It is based on an accurate mapping of the outcrops, of the slopes and of the vegetation (recharge area, runoff coefficient). The variations in time and space of the rainfall, of the temperatures and even of the insolation allow assessing the evapotranspiration, the soil storage and the water-surplus. Obviously this water balance assessment of recharge has to be completed with the inventory of outputs (pumping wells, springs) and, if necessary, with the estimated exchanges between surface water and aquifer.

The piezometric mapping is a useful tool where the point data are numerous, equally and accurately located. It allows to determine hydraulic gradients, flow directions, recharge and abstraction areas. Nevertheless, in a strongly anisotropic aquifer, flow directions are no longer orthogonal with equipotential lines (Carlier, 1983 ; Skinner, 1985). The natural fluctuation of the water-table in response to rainfall is a useful source of information : effective porosity estimate, if rainfall excess is known, vertical pressure transfer time through unsaturated zone, storage in the same zone (Crampon et al. , 1987 ; Chevalier, 1988).

An unceasing monitoring of piezometric responses in function of the intensity and duration of the rains can also put forward different conditions of recharge, especially in fissured media (Skinner, 1985). It is the same for the flowrate response of the springs, particularly in karstic system (Chevalier, 1988) or simply fissured (In New Zealand basalts : Bert, 1984).

The hydrodynamic (or hydraulic) characteristics (transmissivity and storage coefficient) are macroscopic paramaters valid for local scale, the one where the medium can be considered as homogeneous (isotropic or not), the one in the Darcy law field of application. Besides, transmissivity integrates on all the aquifer thickness the different hydraulic conductivity values. It may happen that owing to the heterogeneity of the aquifer, these values vary according to the scale of investigation (there is no Representative Elementary Volume or the latter, if it should exist, would be larger than the aquifer size). The study of that random spatial variability is now tackled with a stochastic approach (Gelhar, 1979). In a medium in which the heterogeneity displays the property of self-similarity (the irregularity is statistically the same at any scale that the object is viewed at), a fractal method may be used (Mandelbrot, 1985). Such models begin to be applied in fracturation and flow studies (Thomas, 1987 ; Thomas & Blin-Lacroix, 1987). More currently, deterministic models (finite-difference or element) are still being used. But it often happens that the transmissivity and storage coefficient values used on the mesh scale to obtain the model fitting are rather different from the ones experimentally obtained by pumping test. And yet these are the models that more easily enable to quantify the vertical flow and hence precise the aquifer vulnerability. It is also possible to set up vertical flow models through unsaturated zone, but that requires knowing at all times and everywhere the moisture from which then depends the hydraulic conductivity and the potential. Such measurements are only conceivable in an experimental station.

3.3. Groundwater quality monitoring

The water-supply monitoring analysises are often to short as for the number of parameters, to allow an actual survey of the waterground quality variability (Roux, 1988) ; what's more they are often too irregularly distributed to enable a global assessment of the groundwater quality. The ideal solution would be to have at hand, at the same time, a great number of very well geographically scattered samples and analysed as thoroughly as possible (i.e. major ions, saturation indexes, pH, Eh, etc. and even some heavy metals). Though such an inventory is expensive it is the necessary starting point for an efficient monitoring of the groundwater quality. It hence becomes possible to draw quality maps, according to the types of parameters, or maps of synthesis obtained by quality criteria defined by several parameters in function of prescribed standards (for example by the WHO or the EEC).

Hence the values can be kriged (Droz, 1985) and the maps automatically drawn if all the analysis results are entered into computerized files that then become real data banks.

These hydrogeochemical maps allow detecting anomalies whose origin may be caused by pollution or, on the contrary, by a natural change in water mineralisation linked to the geological setting. Monitoring networks must be designed according to the quality space distribution and the quality time variability, as well as to the water-catchment location (Roux, 1988) :

- Regional quality monitoring networks with a limited number of sampling points, whose objective is to follow the whole evolution of the groundwater quality and, if it occurs, of diffuse pollutions.
- Localized pollution monitoring networks aiming at warning and, if necessary, at rehabilitation, placed between the point sources of pollution and the water-catchment.

The sampling frequency depends on the local or regional evolution conditions of quality and therefore on those of the flow. This frequency should be high in case of a temporary and punctual pollution due to an accidental spillage (a traffic accident for instance). The analysis type will obviously be adapted to the contaminant nature.

It is also to be noticed that the pollution monitoring in terms of time and space may be a means of determining the hydrodispersive characteristics on the aquifer scale, instead of tracing test, and to facilitate the transport modelling aiming at prevention.

4. BEHAVIOUR OF CONTAMINANTS IN GROUNDWATER AND MODELILING

4.1. Behaviour and transport

Among the mechanical, physico-chemical, radioactive and biological **processes** that may occur either in movable or unmoved liquid phase, or in solid phase, there exist phenomena either non or incompletely conservative such as : filtration (pores too narrow with regard to particles, diffusion into unmoved water zone, adsorption), radioactive decay, adsorption-desorption, ionic exchange, chemical and biochemical reactions, molecular diffusion in unmoved phase.

The filtration efficiency could be described by a decrease of the initial concentration according to a law similar to the radioactive decay one (Matthess et al., 1985). The equilibrium between the concentrations of dissolved and adsorbed species is described by a linear or non-linear isotherm whose constants are specific to the investigated adsorbing substances. Besides, Langmuir's isotherm uses a representative constant of the adsorption maximum on an adsorbent. Exchanges may be wholly reversible or partly reversible (desorption and adsorption isotherms are different). This process causes a retardation of contaminant transport with respect to water transport and may be described by a retardation factor in the simplest cases. The molecular diffusion in porous matrix may be propagated very deeply and cause such a late retrodiffusion that the phenomenon may be considered as nearly irreversible.

The molecular diffusion also intervenes during the water transport and its effect is added to the one of the kinemactic (or hydromechanic) dispersion due to the heterogeneity of microscopic velocities (de Marsily, 1981) and travels (Wang, 1987). A comprehensive coefficient of apparent dispersion (Wang, 1987) or hydrodynamic dispersion (Zuber, 1986-b) describes the whole. This coefficient (or tensor) is linked with dispersivity and water velocity (quotient of Darcy velocity to kinematic or effective porosity), the latter describing the convection process (piston flow assumption). There still remains a major problem namely the one of the physical signification of the dispersivity and of its connection with the structure and the investigation scale, especially in heterogeneous media (Gelhar et al., 1979 ; Matheron & de Marsily, 1980 ; Güven et al., 1984 ; Dagan, 1984 ; Wang, 1987 ; Crampon et al., 1987-b ; Güven & Molz, 1988).

The immiscible fluid convective transport can be very different from the water one since it depends on its own hydraulic conductivity linked to the density and the dynamic viscosity specific to this fluid. Matthess et al. (1985) show that the fluid/water conductivity rates can vary to 0.15 to 3, with the hydrocarbons.

When they do not allow a complete stoppage of the contaminant transport by fixation or degradation, the non conservative processes have the following **effects** : i) a transport duration that may be higher than the persistence of some micro-organisms and of biodegradable or radioactive substances whose concentrations can then become innocuous, ii) a dilution owing to a dispersion in time and space such as the concentration has become, at the catchment point, inferior to the threshold concentration defined by the standard in use. But the effect of these processes will be very different according to the infiltrated grounds and to the flow possibilities. In particular, the slower the velocity the more developed the exchanges and the diffusion can be.

Soils, especially fine-textured ones, with humic substances, clay minerals, hydroxides, plant-roots and micro-organisms, are privileged media for filtration, adsorption, reduction-oxidation, complexation, precipitation or dissolution processes. As most of the overlying formations are stratified, as the equivalent vertical hydraulic conductivity is always very inferior to the horizontal one, and as this vertical conductivity is itself inferior in unsaturated media than in staturated ones, it hence happens that the vertical transport in unsaturated zone is normally far slower than the horizontal transport in aquifer.

Nevertheless, one must pay attention to the possibilities of fast preferential pathways through cracks or preferred channels (Van der Molen & Van Ommen, 1988). Skinner (1985) emphasizes the point that in fissured rocks the majority of the recharge at low infiltration rates will flow through the rock matrix, but at higher infiltration rates (high rainfall periods) a significant proportion of the flow will be through any fissures present. The vertical flow velocities may be estimated from flux densities (rainfall excess) divided by the fraction of water-filled pores (Van der Molen & Van Ommen, 1988). On the contrary we have not yet found in literature any values or estimates on the dispersivities associated to such a transient flow in unsaturated media. Some examples of transit time and dispersivity will also be welcome in cases of leakage between aquifers whose hydrogeochemical consequences we often only know (Droz, 1985).

In saturated media also, contaminant behaviour will depend a lot on water transit velocity, on the presence of adsorbing substances (particularly clay minerals) and of micro-organisms, and on the unmoved water volume : bacterial nitrate reduction to nitrite where the chalk aquifer becomes confined (Roux, 1988), variability of macrodispersivity and restitution ratio according to flowrate in Northern France chalk too (Wang, 1987 ; Porel, 1988). Eventually, it is to be remembered with Zuber (1986-b) that contaminant (or tracer) transport velocity may be much lower than the water one, especially in fissured porous media (see explanation of retardation factor).

4.2. Transport modelling

This area being very wide, the reader could refer to the critical review presented in the proceedings of the "meeting on mathematical models for interpretation of tracer data in groundwater hydrology" (Vienna, 1984, IAEA - TECDOC-381, 1986).

Compartmental models, with successive compartments, are normally used for a long time to simulate relationship between rainfall and discharge in outlet, in a drainage basin, taking into account the provisional storage in unsaturated media, for instance. In a complementary way, Thiery & Seguin (1985) have conceived a transfer model of nitrate that starts from the modes of fertilizer spreading, from the type of crops, from mineralisation processes and from the release of nitrate by the cropping residues. The advantage of such a model is calculating concentrations from input conditions, exchanges and unsaturated zone. But concentration can be computed in a single point (outlet or piezometer).

The system analysis theory provides a method that can be used to formalize input/output relationship for solute transport. This type of analysis is not bound to the assumption of physical mechanisms governing the solute behaviour in the groundwater flow system. In fact, the latter is only considered as a "black-box" changing an input signal into an output signal by means of a transfer (or weighting) function. Margrita et al. (1984) and Zuber (1986-a) suggest extending these methods to variable flow systems by using a flux weighting function that could be very useful in karst, except that this function should not be modified with hydraulic regime. This lumped-parameter approach makes it possible no parameter distribution. However, in the simplest cases, the method of moments allows to link the distribution of transit time and the hydrodispersive characteristics (Zuber, 1974 ; Wang, 1987 ; Sposito & Jury, 1988).

In deterministic models, each specific process is quantified in the differential equation describing the mass balance. In the simplest cases (homogeneous media, steady flow, instantaneous or continuous injection), analytic solution have been set up according to specific boundary and initial conditions. Wang (1987) presents a very useful overview of these analytic solutions. Only one equation thus allows either to predict the concentration evolution in time, on a given distance from a point-source of pollution, or reciprocally to deduce macrodispersivity and kinematic porosity values from a tracing test. Dimensionless solutions (with or without type curve) allow an easier and faster use of analytic solutions in field (Sauty, 1977, 1978 ; Crampon, 1983-a, 1984 ; Wang et al., 1987 ; Porel, 1988).

Obviously, the deterministic numerical models allow a better adjustment to actual cases of contaminant transport and behaviour whatever the conditions of input (pollution), flow and, theoretically, medium heterogeneity. The dispersion and exchange equations are translated either in finite-differences of finite-elements, with as many meshes or nodes as necessary to present spatial evolution of phenomena (in 2D, sometimes 3D). However it is often difficult to collect sufficient significant parameter values to enter in each mesh or node. Furthermore, when simulated phenomena are too numerous, the model, even when fitted, may no longer lead to any physical signification, if it is not followed by a validation stage. At present, it is still difficult to model stratified porous media and fissured media owing to the increase in macrodispervity with the distance (Nieretnieks, 1986). Besides, in porous fractured rocks, it does not seem appropriate to try including the effects of matrix diffusion in the mechanism of hydrodynamic dispersion. Thus, the fissured chalk model designed by Wang (1987) separately computes both phenomena. A few models of transport in fissured media are found on 1D or 2D fissure networks (some 3D models are being developed), but all these models require either a full knowledge of the fissures and of their properties or information of the statistical distribution of the fissures properties. Their use is generally limited by the lack of data.

For the last few years, the stockastic approach has been proposed by several authors to allow a more general formulation of mass transport. Stochastic (or random or probabilistic) processes can only describe the probability of an outcome of such a process at any time as one of a range of possibilities (Nir, 1986). The stochastic methods use Monte Carlo simulation associated with particle tracking methods in velocity fields of a known statistical structure. They allow the simulation of solute movement in variable velocity fields without a priori reference to the convection-dispersion model on the regional scale, provided that the measurements necessary to compute the covariances are available. Finally, stochastic approaches give the values of concentration with a confidence interval related to the variability of the medium. In heterogeneous media with self-similarity, a fractal concept of dispersion may be used and may offer one possible explanation for the scale-dependence of field-measured dispersivities (Wheatcraft & Tyler, 1988).

In spite of the design diversity of transport models, some progress is still be made, especially to have a better assessment as for diffusion in unmoved water (porous matrix or closed fissures), exchanges by sorption-desorption (Gupta & Greenkorn, 1974) and above all the vertical transport in unsaturated media that should be connected to horizontal transport in saturated media (Van Waegeningh, 1985).

5. REMEDIAL MEASURES

It mainly deals with preventive measures since it is still easier to avoid that a pollution should occur, or to forecast the contaminant dispersion than to decontaminate an aquifer.

5.1. Preventive measures

The background of the **technical preventive measures** is in fact what we have reminded in the preceding chapters : namely to have a thorough knowledge of the main activities and of the pollution sources (§ 1), of the possible behaviour of contaminants in function of the various possible exchange, degradation and transport processes (§ 4-1), to apply the better adapted modelling methods (§ 4-2) and obviously to get as much information as possible on the actual media thanks to a deep geological and hydrogeological survey (§ 3). Four different approaches then seem necessary to us to control groundwater quality :

a) the periodical following of water quality and flow conditions by means of quality monitoring networks and if necessary pollution monitoring networks (§ 3-3),

b) the tracing test performance from various upstream points to main water catchments and in various hydroclimatic states, in order either to check the reality of the assessed flowpaths (notably in fissured or karstic media), or to precise the recharge area, or to obtain the transit time distribution. Above all, the tracing tests enable to try to have an idea on the values and on the variability of the hydrodispersive characteristics either aiming at a determinisitic modelling or determining the isochronous lines around a water catchment and settling protection zones there.Tracing tests are also advisable in hydrodispersion as well as pumping tests in hydrodynamics. Finally one can dream of more quantitative vulnerability maps including indications about effective water velocities and hydrodispersive characteristics (just like in the transmissivity or storage coefficient maps).

c) flow and transport modelling referring either to the whole aquifer or to the most vulnerable sectors. When an accident happens one would only have to enter the particular conditions of the spillage (mass or flowrate input, natural hydraulic gradient, pumping flowrate and eventually the special distribution coefficient of the contaminant),

d) the realization of specific studies about the major risk of a sector (contaminant and medium), in order to precise the catchment vulnerability versus this contaminant ; for instance : binding heavy metals in the banks of a river or solubility of these same metals in function of groundwater pH (Roux, 1988), vertical evolution of pesticide or nitrate concentrations, etc...

By his knowledge of the underground medium and of the phenomena, the hydrogeologist must also contribute to prevention by helping the Administration prepare or apply more adequate **rules**. But the technical constraints he puts forward must remain totally independent from the economic, social or political ones among which only executives are to choose from in priority.

In many countries there exists a legislation corresponding to each specific activity with a clause that allows avoiding a groundwater pollution (retention basin under storage tanks, forbidding discharge of severage or industrial effluents directly in aquifer, waste dumps forbiden in quarries, etc.). Rather than severe restrictions (quantity, flow, distance), the rules should be adapted to the vulnerability conditions of the aquifer, to its quality level and to the long termed fore cast objectives for its use. The vulnerability and hydrogeochimical maps are useful in that respect. But more and more, and especially in industrial countries, an inquiry previous to any permit is asked an hydrogeologist expert who will have to specify the measures to be taken corresponding to the local hydrogeological context. In the most important cases it deals with a real impact study (initial environmental state, its foreseen degradation and the suggested protection and rehabilitation measures).

More and more drinking water-supply protection zones are defined. By taking catchment flowrates into account, three pieces of information have to be obtained for demarcating these protection zones :

- the drawing of streamlines to know the water withdrawal area, by means of piezometric map, hydrodynamic model or tracing tests,

- the drawing of isochronous lines thanks to tracing tests or calculation of influence superposition, taking into account the aquifer characteristics, such as thickness, effective porosity, natural gradient, transmissivity and discharge (Bear, 1972 ; Skinner, 1985 ; Sauty, 1987),

- the choice of a delay time according to the objectives : either removal of pathogens and rapidly degradable chemicals, or disposal of delay time for warning and implementing of rehabilitation measures, in case of severe pollution of a persistent compound within the recharge area.

Although two remarks should be made :

. The isochronous line calculation should take into account not only the flow in saturated media but also the vertical transport duration in unsaturated zone, in so far as there is no risk of reduction in time of the vertical transfer owing to local channeling or works.
. The choice of a delay time should depend on pathogen persistence conditions, themselves dependent on the climate and on the organic matter content of soil or of water (in Europe, this delay time is usually 50 to 60 days). As for warning delay time, it varies a lot according to the countries : for instance, in the Netherlands, there are two protection areas with 10 and 20 years delay time (Van Waegeningh, 1985). Of course, one must then define the banning or the restrictions of the various activities inside these zones, have them locally imposed and then have their right application controlled.

4.2. Curative measures

The rehabilitation measures of contaminated groundwater include the cleaning up of the aquifer as well as applying new purification processes of pumped water. These methods are improving day by day and we cannot make an exhaustive overview here.

In regard to that, Schwartz (1988) reminds the pumping techniques and the equipments designed to minimize the mixing between free product and water, in case of nonaqueous phase contaminants. Soil venting is also a technique that is used to evaporate volatile organic compounds (e.g. volatile fuels) from the unsaturated zone. Another approach to clean up compounds that are biodegradable under aerobic conditions is enhanced bioreclamation : to speed up aerobic biodegradation reactions by bacterial growth by providing additional oxygen with the necessary nutrients (hydrogen peroxide, nitrate and phosphorus).

Among the cheapest treatments realizable on pumping water contaminated by nitrate, Boussaïd et al.(1988) propose a deni-trification process which is a heterotrophic-type one using finely chopped straw as the organic base : oxygen of nitrate takes the place of dioxygen for oxidation of organic matter. Finally, the water so treated is seeped again into the unsaturated zone in which it is slowly sweetened before rejoigning groundwater.

5. CONCLUSIONS

A certain number of lacks or weaknesses appear in the means and methods to fight against groundwater pollution.

1°- On the level of phenomena themselves, the parameters necessary to the characterization of transport and of the exchanges or of the contaminant degradation need studying at successive different scales : on samples in laboratory, on experimental site and above all on the aquifer scale. A catalogue of these data would be useful to all of us and would allow a better adaptation of the transport models (especially with regard to the scale-dependent properties). Such fundamental and applied researches need a close multidisciplinary collaboration (chemists, geochemists, fluid mechanics, soil physicists, pedologists, agronomists, biologists and obviously hydrogeologists.

2°- On the level of the methods, a great effort of adaptation of the models to reality remains to be made (see Schwartz, 1988, among others). It would notably be suitable to conceive nested models (with multiple subroutines) allowing a better adaptation to the specific problems of field (taking the matrix diffusion into account, for instance) and above all incorporating vertically downward transport through unsaturated zone. The definition and the mapping of vulnerability should more and more go through a quantitative analysis : vertical transit times down to the water-table, flow directions in aquifer, horizontal effective velocities and, why not, macrodispersivities.

3°- As for the choice of the rules and of the activities to be settled in a region, only the intervention of a trained hydrogeologist can promote a correct analysis of the measures to be taken for the safeguard of groundwater quality and the procedure to be observed in case of pollution. Although it is necessary that the control, intervention and regulation means should be really applied by the competent public services.

In conclusion, the hydrogeologist can only be the keystone of the battle against groundwater pollution by representting the link between the scientists in the various fields that must be applied to hydrogeological conditions and by suggesting adapted means of study, control and intervention.

References

Arnoult, Ph., 1981. Gestion quantitative des eaux souterraines en zone agricole. Application à la nappe de la craie de l'Artois et du Cambrésis. Thèse 3e cycle, Lille, 182 p.

Bear, J., 1972. Dynamics of fluid in porous media. Amer. Elsevier Env. Sci. Serv., New York, 764 p.

Bert, J.L., 1984. Etude et modélisation d'un aquifère basaltique fissuré : le plateau de Pukekohe, Nouv. Zelande. Thèse 3e cycle, Lille, 253 p.

Boussaïd, F., Morvan, J., Martin, G., and Landreau, A., 1988. Expérimentations sur un procédé de dénitrification des eaux souterraines en vue de la production d'eau potable à l'usage de petites collectivités. Hydrogéol., 1-1988, p.45-54.

Carlier, E., 1983. Influence de la fracturation sur le comportement hydrodynamique et hydrodispersif des calcaires paléozoïques de l'Avesnois (Nord). Thèse 3e cycle, Lille, 224p.

Chevalier, J., 1988. Hydrodynamique de la zone non saturée d'un aquifère karstique : étude expérimentale (site du Lamalou - Languedoc). Thèse 3e cycle, Montpellier, 212 p.

Cole, J.A., and Wilkinson, W.B., 1976. Un aperçu des recherches récentes sur la pollution des eaux souterraines au Royaume-Uni. XIVe Journées de l'Hydraul., Paris, Q IV, Rapport 8, 8p.

Crampon, N., 1983-a. Abaque de détermination rapide des caractéristiques hydrodispersives lors de traçages par injection brève. Hydrogéol., Géol. Ing., 4-1983, p.287-292.

Crampon, N., 1983-b. Qualité de la nappe de la craie dans les zones rurales de l'Artois et du Cambrésis. Essai de typologie de la pollution azotée. Rapport Ch. Rég. Agr.Nord, 68p.

Crampon, N., 1984-a. Prévision rapide de l'extension d'une pollution dans quelques cas d'écoulement et de mode d'injection. J. Hydrol., v.70, p.149-165.

Crampon, N., 1984-b. Prévision du temps d'arrivée et éventuellement de passage d'une pollution en eau souterraine. Hydrogéol., Géol.Ing., 3-1984, p. 233-244.

Crampon, N., 1988. Eaux souterraines : un problème d'échelle - Hydrogéologie : une science d'interface. Ann. Soc. géol. Nord, CVI, p.309-321.

Crampon, N., and Lempereur, R., 1983. Etude de la vulnérabilité des eaux souterraines susceptibles d'être captées sur le site de Moranrieux (Avesnois). Rapport SIDEN, 118 p.

Crampon, N., Carlier, E., Colbeaux, J.P., Droz, B., and Lempereur, R., 1987-a. Guides structuraux de l'écoulement souterrain dans les formations paléozoïques carbonatées de l'Avesnois (Nord de la France). Bull. Soc. géol. de France, (8), t.III, n°4, p. 783-796.

Crampon, N., Wang, H.Q., Porel, G., and Carlier, E., 1987-b. Variabilité des paramètres hydrodispersifs dans la craie fissurée (site expérimental de Béthune, France). IUGG, 19th Gen. Assembly, Vancouver, Abstracts V3, HW 2-23, p.960.

Dagan, G., 1984, Solute transport in heterogeneous porous formations J. FLuid Mech., v. 145, p. 151-177.

Droz, B., 1985. Influence de la structure et de la nature des terrains du Valenciennnois sur la qualité de la nappe de la craie (Nord de la France). Apport du krigeage à l'hydrochimie régionale - Gestion qualitative des eaux souterraines. Thèse 3e cycle, Lille, 205 p.

Foster, S.S.D., 1985. Groundwater pollution protection in developing countries. Intern. Contrib. to hydrogeol., UNESCO-IAH, v.6, 9. 167-199.

Fritsch, P., and de St Blanquat, G., 1985. La pollution par les nitrates. La Recherche, n° 169, v.16, p.1106-1115.

Gelhar, L.W., 1979. Stochastic subsurface hydrology : from theory to applications. Water Resour. Res., V.22(9), p. 1355-1455.

Gelhar, L.W., Gutjahr, A.L., and Naff, R.L., 1979; Stochastic analysis of macrodispersion in stratified aquifer. Water Resour. Res., v.15(6), p.1387-1397.

Girardon, J., 1988. La mort lente : Mexico, le ciel se couvre. L'Express Aujourd'hui, n° M 1488-14, p. 20-22

Gupta, S.P., and Greenkorn, R.A., 1974. Determination of dispersion and nonlinear adsorption parameters for flow in porous media. Water Resour. Res., v.10(4), p.839-846.

Güven, O., Molz, F.J., and Melville, J.G., 1984. An analysis of dispersion in stratified aquifer. Water Resour. Res., v.20(10), p.1337-1354.

Güven, O., and Molz, F.J., 1988. A filed study of scale-dependent dispersion in a sandy aquifer. Comment. J. Hydrol., v.101, p. 359-365.

Létolle, R., and Mariotti, A., 1974. Utilisation des variations naturelles d'abondance de l'azote 15 comme traceur en hydrogéologie. IAEA, SM 182/40.

Mandelbrot, B., 1985. Les objets fractals : forme, hasard et dimension. Flammarion éd., Paris, 2e éd., 204 p.

Mangin, A., 1975. Contribution à l'étude hydrodynamique des aquifères karstiques. Thèse Sc., Dijon, 261 p.

Margrita, R., Guizerix, J., Corompt, P., Gaillard, B., Calmels, P., Mangin, A., and Bakalowicz, M., 1984. Réflexions sur la théorie des traceurs : applications en hydrologie karstique. Symp. Isotope Hydrology, 1983, Vienna, IAEA, SM-270/84, p.653-678.

Marsily, G. de, 1981. Hydrogéologie quantitative. Masson éd., Paris, 215 p.

Matheron, G., and Marsily, G. de, 1980. Is transport in porous media always diffusive ? A counterexample. Water Resour. Res. v. 16(5), p. 911-917.

Matthess, G., Pekdeger, A., and Schröter, J., 1985. Behaviour of contaminants in groundwater. Intern. Contrib. to Hydrogeol. , UNESCO-IAH, v.6, p.1-86.

Neretnieks, I., 1986. Solute transport in fractured rocks. IAEA, TECDOC-381, p. 139-164.

Nir, A., 1986. Role of tracer methods in hydrology as a source of physical information - Basic concepts and definitions - Time relationship in dynamic systems. IAEA, TECDOC-381, p.7-44.

Porel, G., 1988. Transfert de soluté en aquifère crayeux. Causes de modifications des résultats de traçages. Thèse, Lille, 327 p.

Roux, J.C., 1988. Contrôle et protection de la qualité des eaux souterraines. Opérations récentes de Service Public réalisées en France par le BRGM. Hydrogéol., 1-1988, p.5-33.

Sauty, J.P., 1977. Contribution à l'identification des paramètres de dispersion dans les aquifères par l'interprétation des expériences de traçages. Thèse Dr. Ing., Grenoble, 157 p.

Sauty, J.P., 1978. Identification des paramètres hydrodispersifs dans les aquifères par interprétation de traçages en écoulement cylindrique convergent ou divergent. J. Hydrol., v. 39, p. 96-103.

Sauty, J.P., 1987. Utilisation de traceurs pour définir les périmètres de protection. Rapport BRGM, 87 SGN 287 EAU, 109 p.

Schwartz, F.W., 1988. Contaminant hydrology - Dollars and senses. J. Hydrol., v.100, p. 453-470.

Skinner, A. Ch., 1985. Groundwater protection in fissured rocks. Intern. Contrib. to Hydrogeol., UNESCO-IAH, v.6, p.123-143.

Sposito, G., and Jury, W.A., 1988. The lifetime probability density function for solute movement in the subsurface zone. J. Hydrol. , v. 102, p.503-518.

Thiery, D., and Seguin, J.J., 1985. Modélisation globale des transferts de nitrates dans un bassin hydrogéologique pour prévoir l'évolution des concentrations dans les eaux souterraines. Rapport BRGM, 85 SGN 663 EAU, 182 p.

Thomas, A., 1987. Structure fractale de l'architecture de champs de fractures en milieu rocheux. C.R. Acad. Sc. Paris, t.304, sér.II, n°4, p. 181-186.

Thomas, A., and Blin-Lacroix, J.L., 1987. Stochastic fractals models of fractures and fracture fields. Application to hydraulic models of fractured rock masses. "Workshop on forced fluid through strong fractured rock masses", Garchy.

Van der Molen, W.H., and Van Ommen, H.C., 1988. Transports of solutes in soils and aquifers. J. Hydrol., v.100, p.433-451.

Van Waegeningh, A.G., 1985. Protection of the groundwater quality in porous permeable rocks. Intern. Contrib. to Hydrogeol., UNESCO-IAH, v.6, p.111-121.

Wang, H.Q., 1987. Modélisation des transferts de masse en milieu saturé à double porosité. Application aux écoulements convergents en craie fissurée semi-confinée et multicouche. Thèse, Paris-Sud, 271 p.

Wang, H.Q., Crampon, N., Huberson, S.,and Garnier, J.M., 1987. A linear graphical method for determining hydrodispersive characteristics in tracer experiments with instantaneous injection. J. Hydrol., v.95, p.143-154.

Wheatcraft, S.W., and Tyler, S.W., 1988. An explanation of scale-dependent dispersitivity in heterogeneous aquifer using concepts of fractal geometry. Water Resour. Res., v.24(4), p.566-578.

Zuber, A., 1974. Theoretical possibilities of two-well pulse methods. Isotop. Techn. in Groundwater Hydrol., v.2, IAEA, Vienna, p. 227-294.

Zuber, A., 1986-a. On the interpretation of tracer data in variable flow systems. J. Hydrol., v.86, p. 45-57.

Zuber, A., 1986-b. Review of existing mathematical models for interpretation of tracer data in hydrology. IAEA, TECDOC-381, p.69-116.

GROUNDWATER CONTAMINATION: PROCESSES, CHARACTERIZATION, ANALYSIS, AND REMEDIATION

Prepared by:

J.W. Mercer, D.A. Giffin, Jr., J.C. Herweijer, and P. Srinivasan

GeoTrans, Inc.
250 Exchange Place
Suite A
Herndon, Virginia 22070
USA

Prepared for:

International Workshop on
Appropriate Methodologies for Development
and Management of Groundwater Resources
in Developing Countries

ABSTRACT

The ultimate goal at a site involving groundwater contamination is remediation. To properly remediate a site, however, an understanding of the controlling processes is necessary. This requires site characterization and analysis of the resulting data. At many sites, the primary mode of transport is advection. Therefore, characterizing the groundwater flow system (i.e., hydraulic conductivity and hydraulic gradient) is critical to understanding the contamination distribution. Other important processes include sorption, chemical and biological reactions, and hydrodynamic dispersion. Sorption and reactions affect remediation. Desorption increases the time required to remediate, whereas chemical and biological reactions can aid remediation (e.g., bioreclamation). For problems involving groundwater contamination, the current practice of remediation involves some form of pump-and-treat technology. Depending on the controlling processes, this technique generally requires long time frames (10-100 years). Therefore, current research examines innovative technologies such as bioreclamation, fixation, soil venting, and in situ heating.

INTRODUCTION

Characterization of a waste site involves data gathering and analysis to determine the processes controlling the transport of wastes from the site. It provides the understanding to predict future site behavior based on past site behavior. This can encompass the characterization of the waste itself as well as various transport pathways such as air, surface water, biota, and groundwater. Groundwater, the area on which this discussion focuses, is often the most significant and least apparent transport pathway.

Ideally, site characterization follows the scientific method and is performed in phases as shown in Figure 1 (Bouwer et al., 1988). First, a hypothesis or conceptualization is made concerning site or system behavior. Based on this hypothesis, a data collection program is designed, data are collected, and an analysis or assessment is made. Using the results of the analysis, the hypothesis is refined and additional data may be collected. As the knowledge of the site increases in complexity, the working hypothesis may take the form either of a numerical or analytical model. Data collection continues until the hypothesis is proven sufficiently to form the basis for decision making.

Because the ultimate goal of site characterization is to make informed decisions, the first step is to define study objectives. A partial list of objectives, provided by Cartwright and Shafer (1987), includes the following: (1) to assess the background or "ambient" water quality (how was the water before contamination?), (2) to establish the impacts of certain facilities, practices, or natural phenomena on water quality (what is the extent of contamination?), and (3) to predict future groundwater quality trends under a variety of conditions (what would be the impact of various remedial actions?).

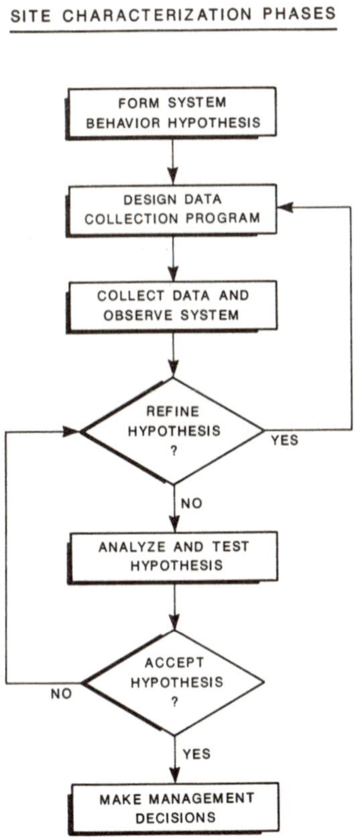

Figure 1. Site characterization phases (from Bouwer et al., 1988).

Whatever the objective, groundwater site characterization has two major components: assessment of the groundwater flow system and assessment of the contamination in the groundwater. All too often, emphasis is placed on the latter component, which is groundwater quality monitoring. Everett (1980) defines monitoring as a scientifically designed surveillance system of continuing measurement and observations. At many waste sites, groundwater quality data are abundant; however, water-level data used to determine advective transport are limited. This is unfortunate because water-level data are equally important. In addition, they are easier and less expensive to collect than water quality data. Equally important for contaminant movement, but more difficult and expensive to assess, is a detailed description of heterogeneous hydraulic conductivity field and resulting tortuous flow paths.

When groundwater is a concern at waste sites, characterization and remediation of the site are generally more time-consuming and expensive. This is because wells must be drilled to reduce uncertainties associated with the subsurface. The time required to clean up groundwater is lengthy, primarily due to slow groundwater velocities. Because of the expense and long time frames, and the variety of available remediation and treatment technologies (OTA, 1984), selecting the appropriate remedial action is both difficult and important. Mercer et al. (1985) discuss the use of groundwater models to aid in this selection process. They point out that only long-term monitoring will verify whether engineering design criteria are met and if system behavior is sufficiently understood.

Flow and transport processes are discussed throughout this paper. The first portion emphasizes flow system and contamination characterization. The next portion presents ways to analyze the data resulting from the characterization process. Innovative ways to analyze water-level data and collect hydraulic conductivity data are presented in two examples. The final portion of this paper concerns remediation, with emphasis on innovative technologies.

FLOW SYSTEM CHARACTERIZATION

Flow system characterization begins with an understanding of controlling processes and of the data required to define those processes. Groundwater flows from areas of recharge to areas of discharge. Natural recharge occurs from precipitation and surface water bodies; artificial recharge results from manmade actions such as irrigation and well injection. Groundwater discharges naturally to springs and other surface water bodies, e.g., rivers, lakes, and the oceans. Under natural conditions, groundwater moves very slowly, its flow velocity ranging from a fraction of a foot per year to several feet per day.

The average flow path is reflected in the potentiometric surface. When considering the actual shape of individual paths, however, tortuous shapes appear caused by heterogeneities at the pore scale to large-scale geological boundaries. Characterization of this heterogeneity means determination of the scale dependency or nested scales (Burrough, 1983). A general understanding of the internal geological/sedimentological assemblage of an aquifer helps determine which scale heterogeneity, with anomalous hydraulic conductivities, will impact flow. A well network should be designed to measure the relevant geostatistical properties of the heterogeneous hydraulic conductivity field. Methods of quantifying well networks are not widely used but do exist. For example, van Geer (1987) shows how Kalman filters can be used and Olea (1982) evaluates which networks are optimal for kriging.

Shallow aquifers are usually important sources of groundwater for water supply. These upper aquifers are also the most susceptible to contamination. Contaminants may enter such an aquifer in one of the following ways: (1) artificial recharge or leakage through wells, (2) infiltration from precipitation or irrigation return flow through the unsaturated zone above the water table, (3) induced recharge from influent streams and lakes or other surface water bodies, (4) inflow through aquifer boundaries and leakage from overlying or underlying formations, and (5) leakage or seepage from impoundments, landfills, or miscellaneous spills.

Water and contaminants carried with it may leave an aquifer in the following ways: (1) groundwater leakage from the aquifer into adjacent strata, (2) groundwater withdrawal by pumping and drainage, (3) seepage into effluent streams and lakes, (4) spring discharge, and (5) evapotranspiration. Data required to assess these processes are shown in Table 1. In general, these data requirements include a geometric description of the site (layering and hydraulic boundaries), storage and transmissive properties, and source/sink information, such as wells. More specific lists of data with ranges of values are provided in Mercer et al. (1982). For any particular site, it is rare to have all information available. Data gaps can be addressed by a field collection program, but to some extent must be filled based on experience.

The first step in designing a field program is to review existing data for the site and nearby locations. In the United States, sources of information include the U.S. Geological Survey (Mercer and Morgan, 1981), state geologic, environmental and water agencies, local water districts, and city, county, and state health departments. Other federal agencies that may provide data include the U.S. Environmental Protection Agency (EPA) (e.g., the STORET computerized information storage system), U.S. Bureau of Reclamation, U.S. Army Corps of Engineers, and U.S. Soil Conservation Service.

A summary of hydrogeologic investigation techniques is given in U.S. EPA (1986) and Sisk (1981). Procedures for well installation and aquifer testing are described in Ford, Turina, and Seely (1983). Methods of analysis of aquifer and slug tests are provided in Kruseman and de Ridder (1976), Lawrence Berkeley Laboratory (1977), and Lawrence Berkeley Laboratory (1978). In general, as the scale of the observation increases, the range of measured properties such as hydraulic conductivity tends to increase because of the heterogeneous nature of geologic materials. Particularly, groundwater flow rates estimated from measurements on cores may underestimate groundwater flow rates in the area if flow is in fractures or in other more permeable layers. Because of seasonal changes in groundwater, a minimum of one year should be devoted to characterization. As the site complexity increases, this period will increase proportionately. The number of boreholes required is influenced by several factors, the most important being heterogeneities in the aquifer materials.

I. Physical Framework
   1. Hydrogeologic map showing areal extent and boundaries of aquifer.
   2. Topographic map showing surface-water bodies.
   3. Water-table, bedrock-configuration, and saturated-thickness maps.
   4. Hydraulic conductivity map showing aquifer and boundaries.
   5. Hydraulic conductivity and specific storage map of confining bed.
   6. Map showing variation in storage coefficient of aquifer.
   7. Relation of stream and aquifer (hydraulic connection).

II. Stresses on System
   1. Type and extent of recharge areas (irrigated areas, recharge basins, recharge wells, impoundments, spills, tank leaks etc.).
   2. Surface-water diversions.
   3. Groundwater pumpage (distributed in time and space).
   4. Stream flow (distributed in time and space).
   5. Precipitation and evapotranspiration.

III. Observable Responses
   1. Water levels as a function of time and position.

IV. Other Factors
   1. Economic information about water supply.
   2. Legal and administrative rules.
   3. Environmental factors.
   4. Planned changes in water and land use.

Table 1. Data pertinent to the prediction of groundwater flow (after Moore, 1979 and Bouwer et al., 1988).

CONTAMINATION CHARACTERIZATION

As with flow-system characterization, contamination characterization begins with understanding the controlling processes and the data required to define those processes. Nonreactive (conservative) dissolved contaminants in saturated porous media are controlled by the following factors:

1. Advection (or convection): This mechanism causes contaminants to be transferred by the bulk motion of the groundwater.

2. Mechanical (or kinematic) dispersion: This process involves mechanical mixing caused by pore-scale heterogeneity.

3. Molecular diffusion: This mechanism causes the contaminant molecules or ions to move opposite the direction of concentration gradient. This movement is caused by the random kinetic motion of the ions or molecules.

The combined effect of mechanical dispersion and molecular diffusion is known as hydrodynamic dispersion. Dispersion is the overall process that causes the zone of contaminated groundwater to occupy a greater volume than if the contaminant distribution were influenced only by advection. If a slug of contaminant enters the groundwater system, advection causes the slug to move in the direction of groundwater flow. Hydrodynamic dispersion causes the volume of the contaminated zone to increase and the maximum concentration in the slug to decrease.

For reactive contaminants, there are additional processes which affect transport. They include adsorption, desorption, and chemical or biological reactions.

1. Adsorption or desorption: These processes involve mass transfer of contaminants. Adsorption is the transfer of contaminants from the groundwater to the soil. Desorption is transfer of contaminants from the soil to the groundwater.

2. Chemical reactions: These processes involve mass transfer of contaminants caused by various chemical reactions (e.g., precipitation and dissolution, oxidation and reduction, etc.). For some contaminants, degradation is also an important process that may need to be characterized.

The processes of adsorption-desorption and chemical reactions play important roles in controlling the migration rate as well as concentration distributions. These processes tend to retard the rate of contaminant migration and act as mechanisms for concentration attenuation. Because of their effects, the plume of a reactive contaminant expands more slowly and the concentration changes slower than that of an equivalent nonreactive contaminant.

The data requirements for contamination characterization are listed in Table 2. These requirements are in addition to the requirements in Table 1. The flow system must first be understood in order to characterize advective transport. More specific lists of data with ranges of values are provided in Mercer et al. (1982). These data requirements provide a broad view of the factors affecting contaminant transport from a site.

I. Physical Framework
   1. Estimates of the parameters that comprise hydrodynamic dispersion.
   2. Effective porosity distribution.
   3. Information on natural (background) concentration distribution (water quality) in aquifer.
   4. Estimates of fluid density variations and relationship of density to concentration.

II. Stresses on System
   1. Sources and strengths of pollutants.

III. Chemical/Biological Framework
   1. Mineralogy of media matrix.
   2. Organic content of media matrix.
   3. Groundwater temperature.
   4. Solute properties.
   5. Major ion chemistry.
   6. Minor ion chemistry.
   7. Eh-pH environment.

IV. Observable Responses
   1. Areal and temporal distribution of water quality in aquifer.
   2. Stream flow quality (distribution in time and space).

Table 2. Data pertinent to prediction of the pollutants in groundwater (in addition to those in Table 1) (after Bouwer et al., 1988).

ANALYSIS OF DATA

The assessment of groundwater quality at any scale involves estimating chemical variables distributed in three-dimensional space. A key consideration in establishing an effective and efficient groundwater quality monitoring program is the spatial distribution of sampling locations. Care must be taken in designing monitoring well networks to avoid biasing any inferences made from the resulting data.

Knowledge of the hydrodynamics of the groundwater system(s) being monitored is also critically important for the design of monitoring networks. For certain groundwater monitoring program objectives, an optimum monitoring network for a relatively homogeneous porous flow environment is different from that for a discretely fractured porous medium. However, for other monitoring objectives, the fundamental differences between flow regimes may have very little impact on the design of an optimum sampling network.

Proper groundwater sampling and analysis are equally important to assure effective groundwater monitoring. A quality assurance program composed of well-conceived and effectively implemented quality control procedures should be followed (Cartwright and Shafer, 1987). This minimizes both systematic and random errors. Groundwater samples collected according to these procedures will produce reliable analytical determinations. As with monitoring network design, understanding the overall objectives of the monitoring program is a key factor in determining sampling and analysis requirements.

The results of laboratory analyses are only as reliable as the samples, field standards, and blanks received (Cartwright and Shafer, 1987). To assure that representative samples are provided to the laboratory, careful thought and practice must be part of any sampling program. A representative sample accurately reflects in-situ conditions in or near the sample point at the time the sample is collected. The sampling program should include well purging, sample collection, and sample preservation. Barcelona et al. (1985) have prepared an extensive guide to the practical aspects of groundwater sampling.

Parameter selection is an important aspect of the design of a sampling program. The types of hydrochemical measurements made affect the sampling equipment and the sampling methodology chosen. Barcelona et al. (1985) recommend slightly more chemical and hydrologic data than immediately required to aid subsequent interpretation.

The frequency of sample collection is important when designing an optimum groundwater quality monitoring program (Cartwright and Shafer, 1987). Sampling frequency affects the cost of the monitoring program and the appropriateness of any inference(s) made from the resulting data. Sample collection and analysis should not occur so often as to result in redundant information which would increase costs with no marginal gain in useful information. Conversely, sample collection should not be so infrequent as to negate the ability to forecast trends in groundwater quality. Groundwater sampling frequency should be based on the objectives of the monitoring program and the dynamics of the system being monitored. As discussed, groundwater movement is relatively slow, and there is little need to sample every few meters of a flow path. Travel time should be taken into account for sampling frequency.

During the past decade, applying geostatistical principles (i.e., structural analysis, kriging, and conditional simulation) to interpret groundwater data has increased. Geostatistical techniques are used to evaluate the spatial variability of groundwater flow parameters, particularly hydraulic head and transmissivity. However, less work has been conducted on applying geostatistics to interpret hydrochemical data and groundwater quality monitoring network design. Samper and Neuman (1985) performed a geostatistical analysis of selected chemical variables. Their results show that geostatistical approaches may be valid to evaluate groundwater chemical data, particularly on a regional scale (Cartwright and Shafer, 1987).

The principles of geostatistics may be appropriate for interpolation of point data to estimate the spatial distribution of certain aspects of groundwater quality. Kriging provides a measure of the error of estimation, which can be mapped and used to select locations for additional sampling points. The error maps show where the interpolated values deviated from the expected statistical structure, thus indicating the best locations to place additional wells (Virdee and Kottegoda, 1984). However, this information can only serve as a guide because of other constraints on well location such as environmental concerns, political issues, and economic limitations (see Table 1). Nevertheless, a near-optimal monitoring network can be developed for a predetermined level of reliability.

Using geostatistics to design monitoring networks and interpolate data has limitations. Application of kriging to groundwater investigations suffers from the lack of sufficient data to perform the structural analysis. Hughes and Lettenmaier (1981) suggest that a minimum sample size of 50 is required before kriging is superior to more traditional interpolation schemes (e.g., the least squares method). Even with sufficient data and suitable statistical support, structural analysis is highly subjective. Further, the theoretical basis for applying geostatistics is using a regionalized variable, which is defined as a spatially correlated random variable. To date, there have not been any definitive studies that validate the assumption that hydrochemical properties of groundwater behave as regionalized phenomena (Cartwright and Shafer, 1987).

EXAMPLES OF ANALYSIS

Water-Level Data Analysis

There are several ways to analyze water-level data, with groundwater modeling being the most common. This usually provides only a steady-state snapshot of flow directions with time. An alternative approach is presented here that distinguishes overall changes of direction during a certain time frame.

This approach can be very useful in efficiently locating wells and sampling points, especially in areas of highly horizontal flow where primary flow directions are difficult to discern.

Water levels were collected from Stevens Water-Level Recorder setup at three stations that most represented the entire site. An additional station was maintained to verify that there was no significant vertical gradient. Three-point problem calculations of the horizontal gradients were made of this data at the points representing the various hydrological cycles affecting the site (pumping, tides, canal stages, etc.). A summary of these calculations is best depicted using a rose diagram (Figure 2) showing the magnitude and directions of the apparent horizontal groundwater flow during the period of measurement. The diagram determines monitoring locations in areas where gradients are low and flow directions are not easily determined. This can mean substantial reductions in effort and cost because a comprehensive radial pattern of wells would no longer be required. As evidenced in Figure 2, the flow was primarily north-northeast and north-northwest for the period of measurement.

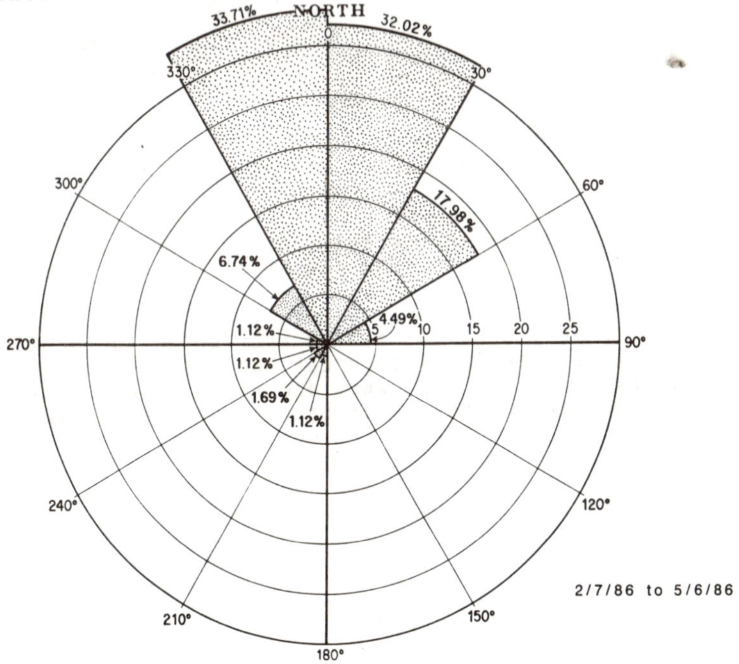

Figure 2. Rose diagram based on three-point gradient calculations depicting apparent horizontal groundwater flow directions (from Giffin, 1987).

Although the rose diagram is effective in displaying flow distribution direction over a period of time, it does not show the sequence in which flow direction changes with time. To show this effectively, plot flow direction and gradient in a Cartesian coordinate system vs. time as shown in Figure 3. It is apparent from Figure 3 that the fluctuations in the flow direction exhibit a certain trend and seasonality.

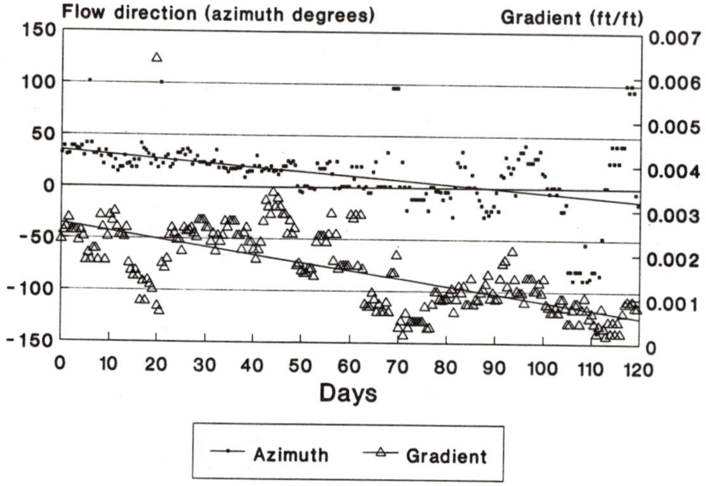

Figure 3. Flow direction fluctuations shown as gradient and azimuth (from Giffin and Srinivasan, 1988).

The periodical behavior may be attributed to the water-table variation due to external influences such as precipitation, pumping, tidal effects, and canal and river stages. By having a clear understanding of these influences, one may be able to separate (or filter) their contributions from the fluctuations shown in Figure 3. This filtering should result in a residue which cannot be attributed to any known external influence. Therefore, it may be considered as noise.

From this analysis, synthesized data may be generated using the trend, various frequencies, noise, and boundary conditions. This synthesized data should accurately represent the real world and can be used to predict solute plume generation. Futhermore, there may be a relationship between particle transport simulation using the above-mentioned synthesized flow domain, and that simulated using macrodispersion theory (see e.g., Gelhar and Axness, 1983).

Network Design for Hydraulic Conductivity Data

For this example, an evaluation is conducted to determine how 30 wells (or less) may be best located on a one-acre test site to provide maps and variograms of hydraulic conductivity. The hydraulic conductivity distribution is required to design a bioreclamation system. Data are collected using the borehole flowmeter technique for measuring hydraulic conductivity variability (see e.g., Hufschmied, 1983). Three different cases were considered for siting gridded wells (rectangular 4x4 and 3x3, hexagonal 4x4) each combined with 3 different methods for randomly siting the remaining wells. Figure 4 shows a map and table depicting all cases investigated.

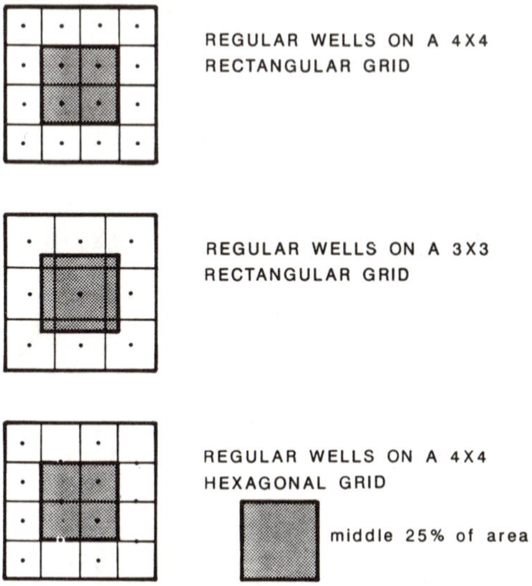

REGULAR WELLS			RANDOM WELLS: OVER WHOLE AREA	RANDOM WELLS: STRATIFIED	RANDOM WELLS: ONLY IN MIDDLE 25% OF AREA
	4X4 RECTANGULAR	16	R41	R42	R43
	3X3 RECTANGULAR	9	R31	R32	R33
	4X4 HEXAGONAL	14	H41	H42	H43

Figure 4. Options for network design for one-acre test site.

Figure 5 shows an example for the general case including a map of the location of the wells (5A), graphs of the frequency distribution of the lags (5B), the deviation of the average lag in an interval from the desired lag (s2) (5C), and the standard deviation of the lags in an interval (s3) (5D). The optimal case was found by evaluating the previously discussed combinations of different grids and modes of randomly siting wells. As any calculation involves different sequences of random numbers, comparisons have to be based on more than one realization. By calculating two realizations, the trend was assessed leading toward an optimal case. For this optimal case (a rectangular grid: 9 gridded wells, with randomly sited wells in the middle 25% of the area), a sensitivity analysis was run to minimize the amount of randomly-sited wells (see Table 3). For 12 randomly-sited wells, which appear to be the threshold, a set of realizations was generated to determine the optimal case. Although this method is based on trial and error without objective convergence, it is a tool to maximize information economically.

GROUNDWATER REMEDIATION

According to Hall (1988), to restore a contaminated aquifer today, pumping contaminated water to the surface for treatment uses state-of-the-art technology. An example analysis of pump-and-treat technology is presented in Ward et al. (1987) for the Chem-Dyne site. The hydrology and contaminant characteristics at the site must be understood so the perceived success of pump and treat is not misunderstood. Failure to understand the processes controlling contaminant transport can result in extremely long pumping periods and, consequently, costly and inefficient remediation.

Because of the problems associated with pump-and-treat remediation, there is an effort to develop alternative or innovative treatment technology (U.S. EPA, 1988a). Several of these technologies are presented in U.S. EPA (1988b). Promising technologies include soil venting, in situ heating, bioreclamation, and fixation which are discussed below.

Soil Venting

While losses of petroleum products to the subsurface can have significant consequences to the quality of local groundwater supplies, it is often the migration of hydrocarbon vapors that have the most immediate environmental impact in urban areas (O'Connor et al., 1984). Soil venting or stripping involves the forced movement of air through soils contaminated with volatile organic compounds. The resulting vapors rise through the soil and are collected on the surface. Simple techniques have been developed to control subsurface hydrocarbon vapors and are

A

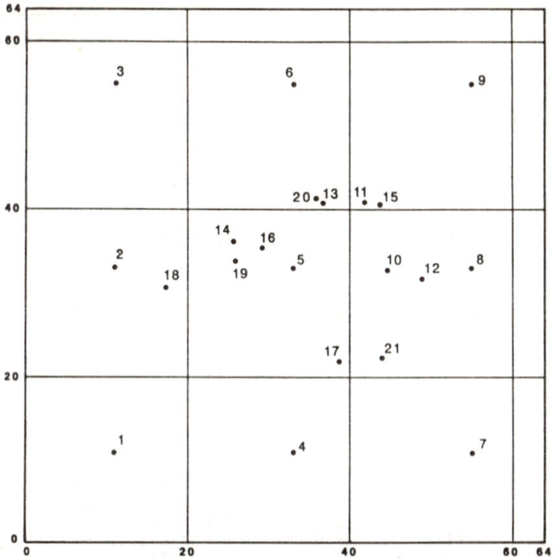

	x_i	y_i
1	11	11
	11	33
3	11	55
	33	11
5	33	33
	33	55
7	55	11
	55	33
9	55	55
	44.5	32.5
11	41.9	40.7
	48.8	31.6
13	36.5	40.5
	25.5	36.1
15	43.8	40.5
	29.0	35.3
17	38.6	21.8
	17.2	30.6
19	25.8	33.5
	35.8	41.0
21	43.8	22.4

B

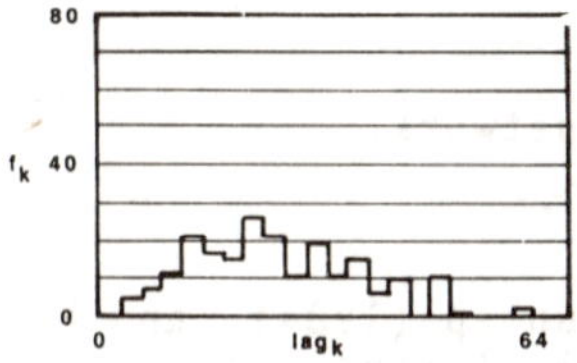

C

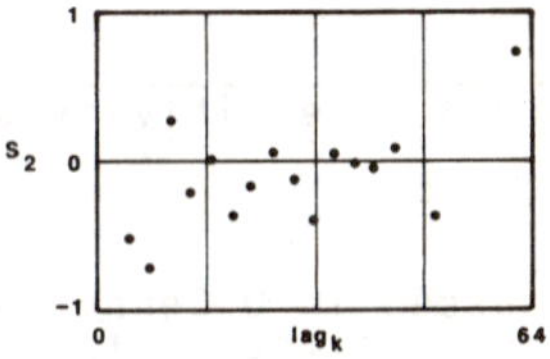

Final results of well planning on 1-acre test site.

A: Map showing optimal sites at 21 wells.

B: Histogram showing no. of observations for 3 m lag intervals (lag_k).

C: Difference between average distance between well pairs and required lag for all intervals.

D: Standard deviation of distance between well pairs.

D

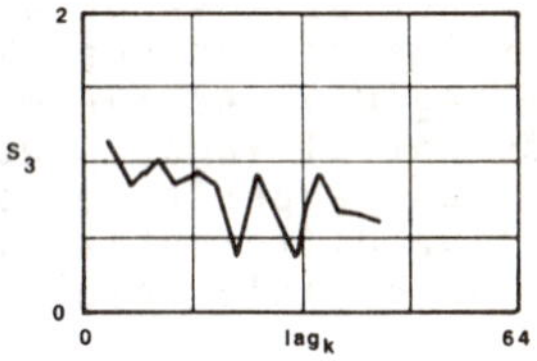

Figure 5. Well network for one-acre test site.

Type of Network	Distance Index	s0	s1	s2	s3	case#
3*3 grid, 9 regular and X random well locations in middle 25% area (case# R33)						
X=21	1.1	0	0	1.8	7.2	R33/21
	1.9	0	0	1.4	7.0	
X=20		0	0	1.1	6.8	R33/20
		0	0	1.9	7.4	
X=18		0	0	1.1	6.8	R33/18
		0	0	1.1	7.4	
X=16		0	0	1.2	7.5	R33/16
		1	3	1.7	7.3	
X=14		1	3	1.4	7.9	R33/14
		1	3	1.7	8.2	
X=12	1.1	2	8	2.8	7.9	R33/12.1
	1.1	1	4	1.9	7.2	R33/12.2
	1.1	4	14	2.2	6.4	R33/12.3
	1.2	1	6	2.1	7.7	R33/12.4
		2	11	3.5	5.9	R33/12.5
		3	10	2.0	7.3	R33/12.6
		4	16	1.9	7.7	R33/12.7
		3	13	1.6	7.5	R33/12.8
		2	9	2.4	7.0	R33/12.9
		3	10	1.7	7.8	R33/12.10

Table 3. Sensitivity analysis on well network.

discussed in O'Connor et al., (1984), Dunlap (1984), and Marley and Hoag (1984). In general, two principal types of vapor management systems are available: positive differential pressure systems, which induce vapor flow away from the control points, and negative differential pressure systems, which induce vapor flow toward the control points. The vapor management methods may be either passive or active. Passive methods use naturally occurring differences in vapor pressures to induce the required flow regime. Active methods require the artificial generation of differential vapor pressures to accomplish the same flow pattern. Practical experience has demonstrated that active generation of negative differential vapor pressures typically provides the most favorable field results.

In Situ Heating

An important problem in soil and groundwater contamination is immobile separate phase fluid at residual saturation. In situ heating, as a means to enhance volatilization in soils, may be explored. Recent in situ tests using radio-frequency (RF), electromagnetic energy to vaporize hydrocarbons in soil have successfully removed up to 99% of the hazardous chemicals (Dev and Downey, 1988). The method involves using a system of electrodes implanted in vertical or horizontal holes drilled into the soil. Electromagnetic energy directed from the electrodes into the ground heats the soil to a predetermined temperature that volatilizes trapped hydrocarbons. Vaporized contaminants and soil moisture are vented to a surface vapor barrier, then condensed for disposal or incineration. The frequency of operation is based upon the dielectric properties of the soil matrix and the size of the area requiring treatment.

According to Dev and Downey (1988), a pilot test program at Volk Field Air National Guard Base, Camp Douglas, Wisconsin involved heating more than 500 cu. ft of contaminated soil to 150°C in an abandoned fire training pit. Borings indicated that average petroleum hydrocarbon concentrations of 4,000 mg/kg extended down 12 ft to groundwater. A more specific gas chromatography analysis showed an accumulation of heavy oils in the upper 12 in. and an increase in lighter fuel components in deeper samples.

The field test lasted 12 days. Within the first eight days, the heating system raised the soil temperature in the pit to between 150°C and 160°C. This temperature was maintained over the remaining four days of the test. Final analysis indicated that 94% of semivolatile aliphatic and 99% aromatic fraction present in the originally contaminated soil had been removed.

Bioreclamation

An overview of bioreclamation as a groundwater remediation technique is presented by Wilson et al. (1986). In the saturated zone, oxygen needed for aerobic biodegradation is available in a dissolved state at relatively low concentrations. Many of the bioreclamation systems used today are derived from a patent process at Suntech (Raymond, 1974). The process consists of injecting oxygen into the contaminated zone to enhance natural biodegradation. A control of groundwater flow in the saturated zone is critical for transporting oxygen and nutrients to the contaminated zone and enhancing the degradation process. Therefore, the limiting process in bioreclamation is convective transport of oxygen and nutrients, not rate limitations due to degradation. Hence, site characterization providing a detailed definition of hydraulic conductivity is required for a bioreclamation design to be successful.

Fixation

Fixation technology is discussed through a case study of the 30-acre Pepper's Steel clean-up site, located near Miami and

Medley, Florida, where the Miami Canal borders the site. Groundwater in the Biscayne aquifer is about 5-6 ft below land surface. Soils above the aquifer were contaminated as a result of prior business operations at the site. Polychlorinated biphenyls (PCBs) and heavy metals (lead, arsenic) were found in significant concentrations to warrant action.

The two primary goals of site clean up were as follows:

- Collect and dispose of oils containing PCBs that are uncovered during site excavations, and

- Treat or dispose of soils that are contaminated with PCBs and heavy metals.

After reviewing several remedial options, solidification/ stabilization was selected. In accordance with regulations, PCB-contaminated oils were removed and disposed of at an approved facility away from the site. All of the contaminated soils were solidified on site with a proportioned mixture of fly ash and cement. Solidification changes the physical characteristics of the waste and decreases surface area of pollutants available for leaching. Through stabilization, the wastes become less water soluble and less toxic. The PCBs are trapped in the cement mixture and the heavy metals (arsenic and lead) become insoluble metal silicates.

Details of the clean up include (USEPA, 1987):

- Approximately 60,000 cubic yards of contaminated soils were excavated.

- All free oil uncovered during excavation were collected and sent for treatment or disposal.

- Soils contaminated with PCBs and heavy metals were stabilized/solidified with a cement mixture.

- Solidified materials were placed back on the Pepper's Steel site and covered with 12 inches of crushed limestone.

- Surface water was controlled by grading the site and placing drains around the solidified material.

- Groundwater is monitored on an annual basis.

SUMMARY

This paper presents an overview of groundwater contamination, emphasizing characterization and analysis. Sites involving groundwater contamination generally require more time and effort to characterize and remediate than sites not involving groundwater contamination. Site characterization is critical to selection and implementation of remediation. Both site characterization and remediation should be performed in phases. This will insure that later phases are modified based on knowledge gained from earlier phases. Because of the long time frames required for pump-and-treat technologies, research on groundwater remediation focuses on new innovative technologies.

REFERENCES

Barcelona, M.J., J.P. Gibb, J.A. Helfrich, and E.E. Garsek, 1985. Practical guide for ground water sampling, ISWS Contract Report 374, Illinois State Water Survey, Champaign, IL.

Bouwer, E., J.W. Mercer, M. Kavanaugh, and F. DiGiano, 1988. Coping with groundwater contamination, Journal Water Pollution Control Federation, 60(8), p. 1414-1428.

Burrough, P.A., 1983. Multiscale sources of spatial variation in soil. I. The application of fractal concepts to nested levels of soil variations, Jour. Soil Sci., 34, p. 577-597.

Cartwright, K. and J.M. Shafer, 1987. Selected technical considerations for data collection and interpretation-ground water, National Water Quality Monitoring and Assessment, National Academy Press, Washington, D.C., p. 33-56.

Dev, H., and D. Downey, 1988. Zapping hazardous wastes, Civil Engineering, (August), p. 43-45.

Dunlap, L.E., 1984. Abatement of Hydrocarbon Vapors in Buildings, <u>in</u> Petroleum Hydrocarbons and Organic Chemicals in Ground Water, National Water Well Association, Worthington, OH, p. 504-518.

Everett, L.G., 1980. Ground water monitoring, Schenectady, N.Y. General Electric Co. Technology Marketing Operation.

Ford, P.J., P.J. Turina, and D.E. Seely, 1983. Characterization of hazardous waste sites - a methods manual, II, Available sampling methods, EPA Report No. 600/4-83-040, NTIS No. PB84-126920, U.S. EPA, Las Vegas, NV.

Gelhar., L.W., and C.L. Axness, 1983. Three-dimensional stochastic analysis of macrodispersion in aquifers, Water Resources Research, 19(1), p. 161-180.

Giffin, D., 1987. Groundwater elevations and flow directions at the Peppers Steel and Alloy Site in Medley, Florida, Report to U.S. Environmental Protection Agency.

Giffin, D., and P. Srinivasan, 1988. Quarterly flow analysis at Peppers Steel and Alloy Site, Report to U.S. Environmental Protection Agency.

Hall, C.W., 1988. Practical limits to pump and treat technology for aquifer remediation, Proceedings Water Pollution Control Federation Groundwater Quality Protection, Pre-Conference Workshop, Dallas, TX, p. 8-12.

Hufschmied, P., 1983. Die Ermittlung der Durchlassigkeit von Lockergesteins-Grundwasserleitern, eine Vergleichende Unterscuchung Verschiederner Felmethoden, Doctoral Dissertation No. 7397, ETH Zurich, Switzerland.

Hughes, J.P. and D.P. Lettenmaier, 1981. Data requirements for kriging: Estimation and network design, Water Resources Research, 17(6), p. 1641-1650.

Kruseman, G.P. and N.A. de Ridder, 1976. Analysis and evaluation of pumping test data, International Institute for Land Reclamation and Improvement, Bulletin 11, Wageningen, The Netherlands.

Lawrence Berkeley Laboratory, 1977. Invitational Well-Testing Symposium Proceedings, LBL-7027.

Lawrence Berkeley Laboratory, 1978. Second Invitational Well-Testing Symposium Proceedings, LBL-8883.

Marley, M.C., and G.E. Hoag, 1984. Induced Soil Venting for Recovery/Restoration of Gasoline Hydrocarbons in the Vadose Zone, in Petroleum Hydrocarbons and Organic Chemicals in Ground Water, National Water Well Assoc., Worthington, OH. p. 473-503.

Mercer, M.W. and C.O. Morgan, 1981. Storage and retrieval of ground-water data at the U.S. Geological Survey, Ground Water, 19(5), p. 543-551.

Mercer, J.W., S.D. Thomas, and B. Ross, 1982. Parameters and variables appearing in repository siting models, U.S. Nuclear Regulatory Commission NUREG/CR-3066.

Mercer, J.W., C.R. Faust, R.M. Cohen, P.F. Andersen, and P.S. Huyakorn, 1985. Remedial action assessment for hazardous wastes sites via numerical simulation, Waste Management & Research, (3), p. 377-387.

Moore, J.E., 1979. Contribution of ground-water modeling to planning, Journal of Hydrology, (43), p. 121-128.

O'Connor, M.J., J.G. Agar, and R.D. King, 1984. Practical Experience in the Management of Hydrocarbon Vapors in the Subsurface, in Petroleum Hydrocarbons and Organic Chemicals in Ground Water, National Water Well Association, Worthington, OH, p. 519-533.

Olea, R.A., 1982. Optimization of the High Plains Aquifer observation network, Kansas, Groundwater Series 7, Kansas Geological Survey.

OTA, 1984. Protecting the nation's groundwater from contamination, U.S. Office of Technology Assessment, OTA-O-233, Washington, DC.

Raymond, R.L., 1974. Reclamation of hydrocarbon contaminated groundwaters. U.S. Patent Office, 3,846,290, patented November 5, 1974.

Samper, F.J. and S.P. Neuman, 1985. Geostatistical analysis of hydrochemical data from the Madrid Basin, Spain (abstract), Eos, Transactions AGU, 66(46).

Sisk, S.W., 1981. NEIC manual for groundwater/subsurface investigations at hazardous waste sites, U.S. Environmental Protection Agency Report EPA-330/9-81-002, National Enforcement Investigations Center, Denver, CO.

U.S. Environmental Protection Agency, 1988a. The Superfund innovative technology evaluation program: Technology profiles, EPA/540/5-88/003, Nov., 83 p.

U.S. Environmental Protection Agency, 1988b. Assessment of international technologies for Superfund applications, EPA/540/2-88/003, 50 p.

U.S. Environmental Protection Agency, 1986. RCRA ground-water monitoring technical enforcement guidance document (TEGD), U.S. Environmental Protection Agency OSWER-9950.1.

U.S. Environmental Protection Agency, 1987. Protecting the Biscayne Aquifers: Actions to be taken at the Pepper's Steel and Alloy Site, prepared by CH_2M Hill.

van Geer, F.C., 1987. Applications of Kalman filtering in the analysis and design of groundwater monitoring networks, TNO Institute of Applied Geoscience, Delft, The Netherlands.

Virdee, T.S. and N.T. Kottegoda, 1984. A brief review of kriging and its application to optimal interpolation and observation well selection, Journal des Sciences Hydrologiques, 29(4), p. 367-387.

Ward, D.S., D.R. Buss, J.W. Mercer, and S.S. Hughes, 1987. Evaluation of a groundwater corrective action at the Chem Dyne hazardous waste site using a telescopic mesh refinement modeling approach, Water Resources Research, 23(4), p. 603-617.

Wilson, J.R., L.E. Leach, M. Henson, and J.N. Jones, 1986. In situ biorestoration as a groundwater remediation technique. Ground Water Monitoring Review, Fall 1986, p. 56-64.

COUPLED MODELLING OF WATER CYCLE AND NITRATE TRANSPORT IN A HYDROLOGICAL SYSTEM

Q.Z. GENG, G. GIRARD, E. LEDOUX

ABSTRACT

Nitrogen input due to agriculture is the major cause of groundwater pollution by nitrates. Nitrogen cycle which is responsible of the transfer of pollutant from the soil to the aquifers is governed by multiple and complex phenomena which have to be simplified before attempting its modelling. This paper proposes a nitrate leaching conceptual model into the soil taking into account nitrogen input and agricultural techniques effects. This model is compatible with a spatially discretized representation of both surface and ground flow model already developed by the authors. This model has been validated to a lysimetrical case scale as well as to the scale of a small french cultivated watershed.

Results are meaningful and allow future utilization in studies for prevision of the ground water quality related to agriculture.

INTRODUCTION

Nitrogen, particularly under its nitric species (NO_3-) is one of the principal elements used in the nutrition of beeings. However, this need is limited and nitrogen may become a pollutant when its quantity overcomes a certain amount. In potable water for man, for instance, this amount is limited to 50 mg/l of nitrates in C.E.E. countries as defined wy WHO.

At the scale of the earth, the main input of nitrogen in ground comes from natural fixation from the one contained in the air; however, it is the production of fertilizers which predominates in populated areas. Nowadays, it seems clear that the major nitrogen pollution in the water is caused by the intensive agricultural activities.

Even though the situation has become alarming, there is no evidence of a solution in the near future. Due to the fact that the nitrates before becoming pollutants are nutrients, a balance between the soil productivity and the quality of the environment is not easy to measure and to preserve.

This problem is therefore the one of an optimization in the use of nitrates in the soil-water-plant system where a lot of important natural and human factors are included which can be resumed by a competition between the nutrient intake by the plants and the one dragging by water. Modelling may contribute to that objectif.Our objectif is to present a model that could, first, simulate the behaviour of an hydrogeological and agricultural system and, secondly, provide the system managements with the information on the future of the potential pollution of that system under different scenarios of human activities. It is rather difficult for the manager to obtain precise and sufficient information whenever they want to use empirical models. That is why one should use the phenomenological models where the variables and parameters have physical meaning and are measurable.

But to this scale, certain more rigorous phenomenological approaches are not applicable in practice, either because of lack of data, or because of a large set of parameters to be introduced in the use of the model. This leads to a compromise of suiting the phenomenological and parametric approaches according to the need and a pragmatic oncoming.

1. GENERAL ORGANIZATION IN THE MODELLING OF NITRATES TRANSFER

The modelling considers the representation of different phenomena (see fig.1):

- hydric balance and water runoff;
- nitrogen cycle and balance as well as leachate of nitrates into the soil;
- nitrates migration in the aquifer system.

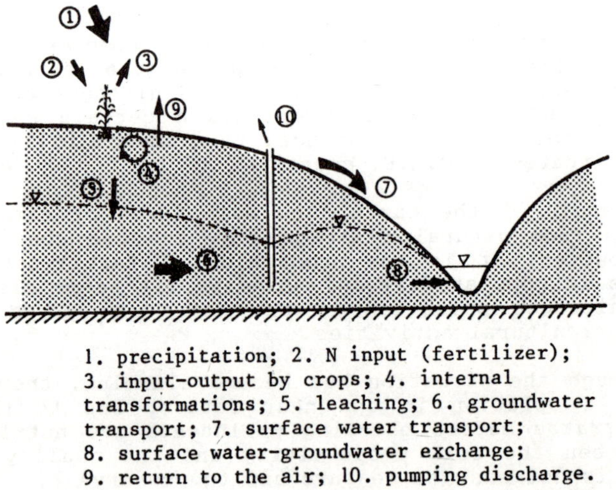

Figure 1. Nitrates transfer scheme in a hydrological and agricultural system at a regional scale

The modelling of the water flow and the migration of a soluble conservative substance in an aquifer has been performed in several studies. We have begun our study from the spatially discretized hydrological models already developed and known as "the coupled model" MC (Ledoux, 1980) and NEWSAM(Marsily et al., 1976) model. We then developed a model representing the nitrogen balance and nitrates leachate called MORELN which uses the same spatial discretization as the before mentioned models and is totally compatible with them.

The basic idea is to provide the MC model with a function simulating the transport of a pollutant of agriculture origin, like nitrates.

In a complete hydrological system like the one represented by the MC model, one can distinguish, according to time and space residence of a pollutant, four different categories in the transport:

- transport in the soil, essentially the nitrogen mass balance determined by agricultural applications, water balance and the climatic factors;
- transport in the unsaturated zone, which traduces the lag-time in leaving the soil and entering the aquifer;
- transport in the aquifer system, governing the spatial distribution and the evolution of pollutant concentration in time;
- transport in the surface, more particularly in the drainage system interacting with the groundwater aquifer.

The key point is the first type of transport where the nitrates production is regulated defining the input function of the pollutant into the system. The elaboration and validation of the model representing this function constitutes the essential part of the research.

2. THE NITROGEN CYCLE IN THE SOIL

The nitrogen cycle in the soil is a complex phenomena which includes the interaction of several processes which we classified in three categories (see fig.2).

- Processes governing the nitrogen input:
  - . biological fixing of atmospheric nitrogen by plants;
  - . input of mineral nitrogen by precipitations;
  - . artificial contributions (agricultural practices, spoils);

- Processes governing the nitrogen output:
  - . bacterial denitrification;
  - . volatilization of gaseous nitrogen;
  - . leaching of nitrates by infiltration.

Processes governing internal transformation:

- ammonification from organic nitrogen to mineral nitrogen by bacteria;
- nitrification from ammoniacal nitrogen to nitric nitrogen by bacteria;
- reorganization, assimilation by micro-organisms;
- adsorption-desorption of ammoniac by the soil.

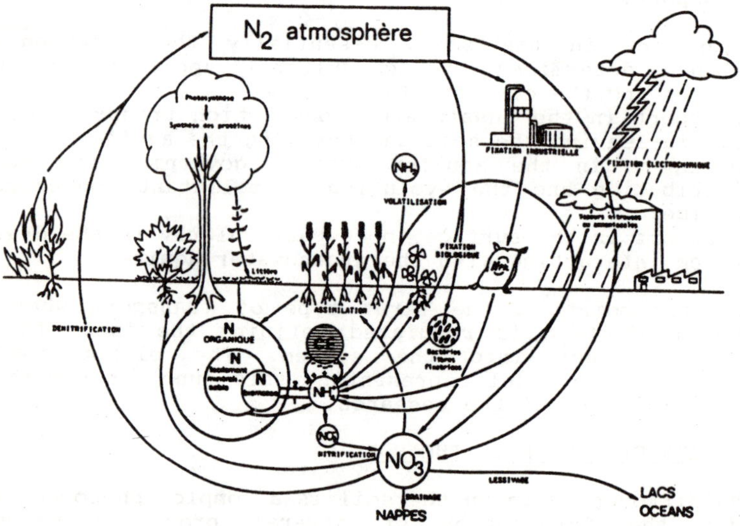

Figure 2. General representation of nitrogen cycle (from Mariotti, 1982)

3. MODELLING OF THE NITROGEN CYCLE IN THE SOIL

The nitrogen cycle in the soil is extremely complex. Because of the lack of knowledge of certain phenomena and the lack of data to be used in a very detailed modelization, we will adopt some simplifying assumptions in order to make possible the modelling exercices.

Generally, there exists no data on the composition of organic nitrogen in the soil. So we will consider it as homogeneous and under a unique forme where the mean composition will reflect the mineralizable part at the time scale considered by the simulation.

The state of the knowledge and data availability does not allow us to take into account the soil-atmosphere interchange phenomena in the modelling. For this reason, we will suppose that the input (precipitation + biological fixing) and the output (denitrification + volatilization) terms globally compensate;

Because of the rapidity in the nitritation according to the nitratation, we will consider that nitrification is done in a single stage governed by the slowest speed (nitratation). In other words, we will neglect the (NO_3-) species in the representation of the nitrogen cycle.

Finally, we assume that the processes of the uptake by plants and leaching by water only concerns the nitrates.

The simplified nitrogen cycle may be represented as follows:

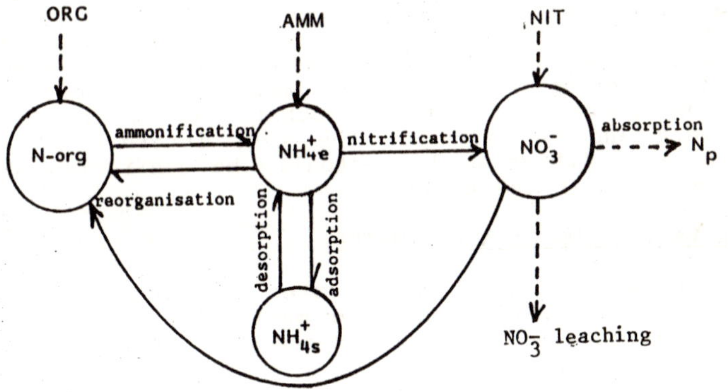

Figure 3. Diagram of the simplified nitrogen cycle for the modelling

Each part of the process described above is represented mathematically, and implemented in the model in space and time, allowing the estimation of a nitrates leaching term which is then introduced into the ground transport model (Geng, 1988).

There are two categories of data introduced in the MORELN model:

. the data from the hydrological model MC, including precipitation, runoff, infiltration and moisture storage;
. the agricultural and pedological type of data due to the model MORELN. These are the contributions from fertilizers, irrigation practices, thickness, density and water capacity of the superficial soil.

The parameters are divided into three categories:

- the nitrogen potential consumption used by the field crops and the absorption coefficient. The former can be evaluated from the field and the latter can be determined by calibration of the model;

- the kinetic coefficients of the four internal biochemical transformations. They play the indirect role in the calcul of the leaching flux and the uptake of nitrates. Their estimation by calibration of the model is less rigorous due to the fact that these four processes are not precisely known in general;

- the correction coefficients of the preceeding coefficients as function of the soil temperature. In this case, reported values will be taken.

The results in the output of the model are principally the leaching flux of nitrates and the consumption in nitric nitrogen by the crops. These are the two most important results which characterize the behaviour of the soil-water-plant system and the easiest to check. Additionally, the model also estimates the reserve of nitrogen substances of the different compartments and the flow of the four internal biochemical transformations, which will eventually be confronted with the the observations in the field.

4. MODEL VALIDATION ON A LYSIMETER

A first validation of the nitrates leaching model was performed using the data of an experiment in a lysimeter by the National Institute of Agronomic Research (INRA) in Versailles (France).

The box used is a parallelepiped of 2 m² in the surface and over 1.5 m depth. The fill material is homogeneous over the depth and is constituted mainly by silty soils.

The box has a free drainage open device and the water is capted at the bottom. Measurement of the drained water was done daily, while the chemical analysis of water was done rather irregularly, varying from some days to two weeks.

The crop rotation in this box is wheat-corn. The occupation period of wheat goes from November to August between the dates of sowing and harvest. For corn, this period goes from end of April to October.

The box receives the fertilizers essentially in spring, representing a nitrogen input equivalent to a 130 and 160 kg N/ha dose by crop cycle. There is an additional and frequent input of nitrogen in autumn by reincorporating the residual after the harvest accompanied sometimes by a complementary input of ammoniacal nitrogen evaluated to some tenths of kg N/ha.

Results over 10 years of simulation are shown in Table I:

Year	73-73	74-74	75-75	76-76	77-77	78-78	79-79	80-80	81-81	82-82
Type of culture	C	W	C	W	C	W	C	W	C	W
Observed drainage (mm)	91	101	37	127	224	94	168	71	187	122
Computed drainage (mm)	86	103	31	126	222	101	115	62	180	107
Observed N leaching (kg N/ha)	7.0	8.4	3.1	28.	67.	2.8	15.	1.2	21.	5.7
Computed leaching (kg N/ha)	9.1	7.8	3.0	6.6	30.	3.7	11.2.3		20.5.6	

C = corn   W = wheat

Table I. Comparison between annual observed and estimated leaching.

Due to the fact that leaching of NO_3^- is strongly influenced by water drainage, the trend of the leaching curve follows very closely the one of the drainage.

In the two years (76-77 and 77-78) following a great drought period (75-76), the difference between the estimations and observations is important, estimated leaching is greatly inferior to observed leaching (fig.4). This can be explained by an accretion of mineralization of organic nitrogen as shown by some experiences in the laboratory (Mariotti, 1982), where the speed of the mineralization of samples pre-dried in the air is accelerated compared to normal samples. This phenomenon, known as "flush effect" is explained by the destruction of soil biomass (Mariotti, 1982). We think that

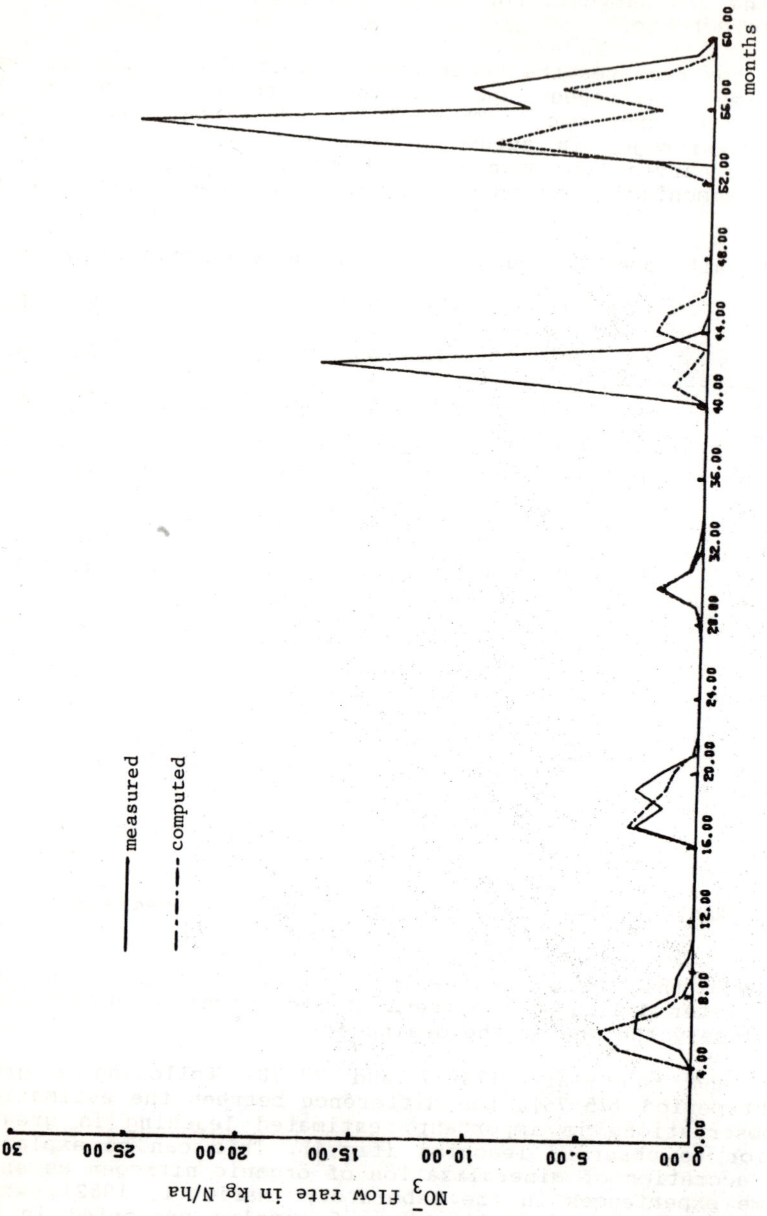

FIGURE 4. NITRATE LEACHING FROM A LYSIMETER

such a phenomenon has indeed happened in the lysimetric box following the great drought of 1976. This is not taken into account in the model, which considers that the kinetic of mineralization is unchangeable.

Agricultural data	Type of culture	Observed consumption	Estimated consumption
08/05/74-30/10/74	Corn	157.1	124.6
05/12/74-29/07/75	Wheat	158.6	163.8
23/04/76-16/09/76	Corn	107.3	158.9
26/10/76-16/08/77	Wheat	214.7	147.9
12/05/78-31/10/78	Corn	243.2	186.4
22/11/78-14/08/79	Wheat	129.7	133.6
29/04/80-20/10/80	Corn	209.2	157.4
29/10/80-12/08/81	Wheat	136.3	152.1
21/04/82-18/10/82	Corn	201.1	202.6
23/11/82-03/08/83	Wheat	138.8	155.7

Table II. Comparison between estimated and observed consumption.

Besides leaching, another important result in the model is the estimation of nitrogen consumption by the crops. Table II shows the comparison between observed and estimated consumption. We can see that estimated values are in general very close to the observed ones. Like for leaching, the model gives smaller values compared to the observations for the two years after the drought period.

We can say that in general the model has correctly simulated the water and the NO_3- flow through the soil as well as the nitrogen consumption by the crops. Values of parameters concerning the kinetics of nitrogen biochemical transformation in the soil are quite consistent with those cited in the literature. A conclusion in our approach is the proposal of the values for the parameters which characterize the nitrogen absorption by the roots which should be critisized in other applications.

Extreme climatic events may considerably disturb the nitrogen cycle in the soil. The great drought of the year 1976 has greatly amplified the mineralization and nitrification in the soil during the two following years. The model should be modified and adapted in order to take those disturbances into account in which the mechanisms are not yet fully quantified.

To a lysimetric box scale, the soil behaviour with respect to drainage and leaching is very sensitive to changes in soil occupation (crops rotation) and to tilling.

5. MODELLING NITRATES TRANSPORT IN A SMALL EXPERIMENTAL WATERSHED

The example chosen is the watershed of the Noë-Sèche, located in Bretagne, in the west of France. This basin, situated in a granitic region, has been studied in detail, investigating the agricultural activities and the nitrogen balance during the year of 1983.

The site is a small drainage basin in which the surface area is only of 576 ha. Even though there seems to be a groundwater aquifer relatively developed compared to other basins in the region, its importance is limited according to nitrates transport as compared to other processes of rapid transport manifesting during the floods. This hypothesis seems conforted by measurements of nitrates flux in the river in which high values coincide with high rates of flow. Besides, we do not dispose of any data to characterize that aquifer. So we will admit that the dominant phenomenon to the basin scale is production and not displacement; in such a way, we consider, in the modelling, the basin to be a great box lysimetric of polyculture.

AVAILABLE DATA

- Flow rates and nitrates content:

Data are available on a daily basis for the period March 1983-February 1984, flow rates are measured in a gauging station at the outlet of the Noë-Sèche basin.

Water chemical analyses give concentration in NO_3-. This analysis is done every three days in average. In the simulation, the values of the preceding days will be given to those days without data.

With these two series of data, we will estimate a flow of NO_3- that will be used in the calibration process.

- Meteorological data:

A rain gauge in the site has registered, for that period, a precipitation of 967 mm.

Decadal data of temperature in the soil and potential evapotranspiration (Penman) are available in a nearby meteorological station.

- Agricultural data: Nitrogen inputs into the basin

In the first place, we have a daily schedule of fertilizers application in the form of mineral and organic fertilizers in sites not spatially localized.

Because of the three types of chemical forms of nitrogen used in the model (N organic, N ammoniacal and N nitric), the inputs should be converted into those three forms.

In the second place, besides those inputs valorized to 100% for the crops, there also exists other quantities of nitrogen coming from different parts (dejections in the highways, different leakage of effluents, etc...). Those nitrogen quantities are considered as "direct losts" for the crops, but they should be considered as inputs in our modelled hydrogeological system, it is convenient to take them into account in the total balance.

Those quantities (N) have only been estimated annually. In order to distribute them in the time (on a daily basis), we have supposed that the contributed quantity is proportional to the number of grazed animals.

Finally, we have also taken into account the nitrogen contained in the crops roots, reincorporated into the soil under the form of organic N after the harvest.

- Data concerning field crops practices:
These data are used to estimate the potential consumption of nitrogen by the crops. They are available in the basin for the cereal crops.

Concerning the meadows, things become more difficult: we do not know the rythm and the intensity of the grazing, by the effect of mowing which perturbs the growing of meadows and so their nitrogen consumption.

Therefore, we have decided to use the results of a "grazing simulation".

MODEL CALIBRATION AND ANALYSIS OF RESULTS

- A) Calibration of the flow rate:
Figure 5 shows the flow rate calibration results with a daily time-step. The model reproduces correctly, in the whole, the evolution of the measured flow rate except for two periods:

- during the month of May, marked by an underestimation of the estimated flow rates;
- at the starting of autumn, where the estimated flow rate curve rises more slowly than the observed one.

- B) Calibration of NO_3- transport:
In this part of the calibration process, we will rather insist on the NO_3- flow transported by the Noë-Sèche basin than on the concentrations for the following reasons:

- the evolution of the behaviour in the basin lies, we believe, on the regular seasonal variations of the NO_3- flow rather than on the daily irregular fluctuations in concentration. The global model of the basin does not seem to have a good performance in simulating this last type of variability.

FIGURE 5. FLOW AT THE OUTLET OF THE WATERSHED

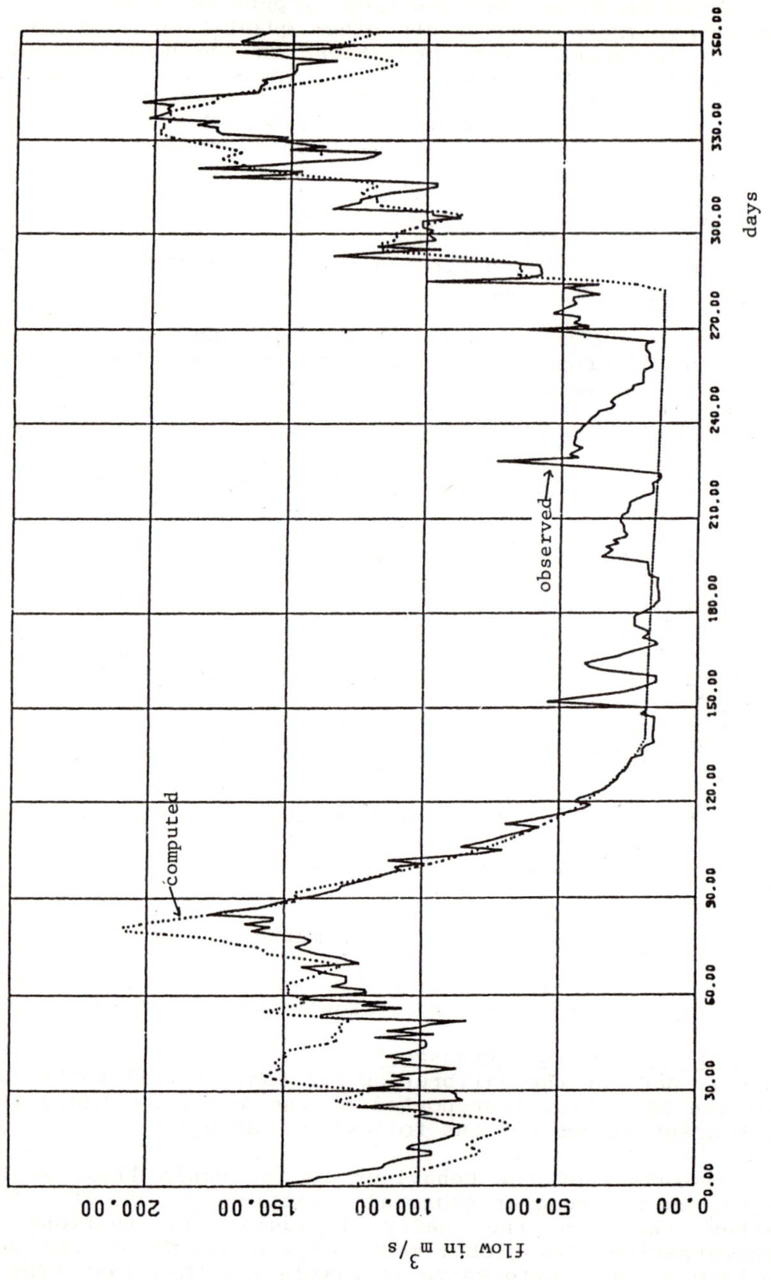

It is the impact of agricultural activities of the basin on the upstream area, e.g. the quantity (flow) of NO_3^- going out of the basin which is most worrying. Figure 6 shows the comparison between the observed and estimated NO_3^- flow on a daily time-step basis. It is clear, from the figure, that the leached NO_3^- flow curve is quite well reproduced, except for two periods:

- April-May, where the model estimates a higher NO_3^- flow as compared with the observed one. This may probably be a consequence, in the estimation, of the imprecision of the potential consumption of NO_3^- by plants during that period.
- As in the flow rate, the estimated NO_3^- flow curve rises more slowly than the one observed in the period starting in autumn. It then shows clearly the influence of hydrogeological factors affecting the evolution of pollution. Due to the fact that the concentration is rather constant, it is indeed essentially the flow rate which determines the flow of NO_3^-.

As a conclusion of the modelling exercices, we believe that:

1) The annual nitrogen balance over the basin is very lean:

	Observed (kgN/ha)	Computed
Input in organic N	52	
Input in ammoniacal N	83	
Input in nitric N	43	
Total input	178	
Output from the Noë-Sèche	50	49
Output from crops	198	201
Total output	248	250
Total balance:	-70	-72

2) The NO_3^- exportation by the Noë-Sèche basin esti

mated in monthly and yearly terms agrees well with the measurements. The underestimation by the model for the months of October and November is quite understandable due to the fact that the flow rate, for the same period, has also been underestimated.

3) The NO_3^- consumption by the crops is well simulated. The simultaneous calibration of both the consumption by plants and the leached nitrates flow allows us to have a better control of the simulation of the whole nitrogen balance.

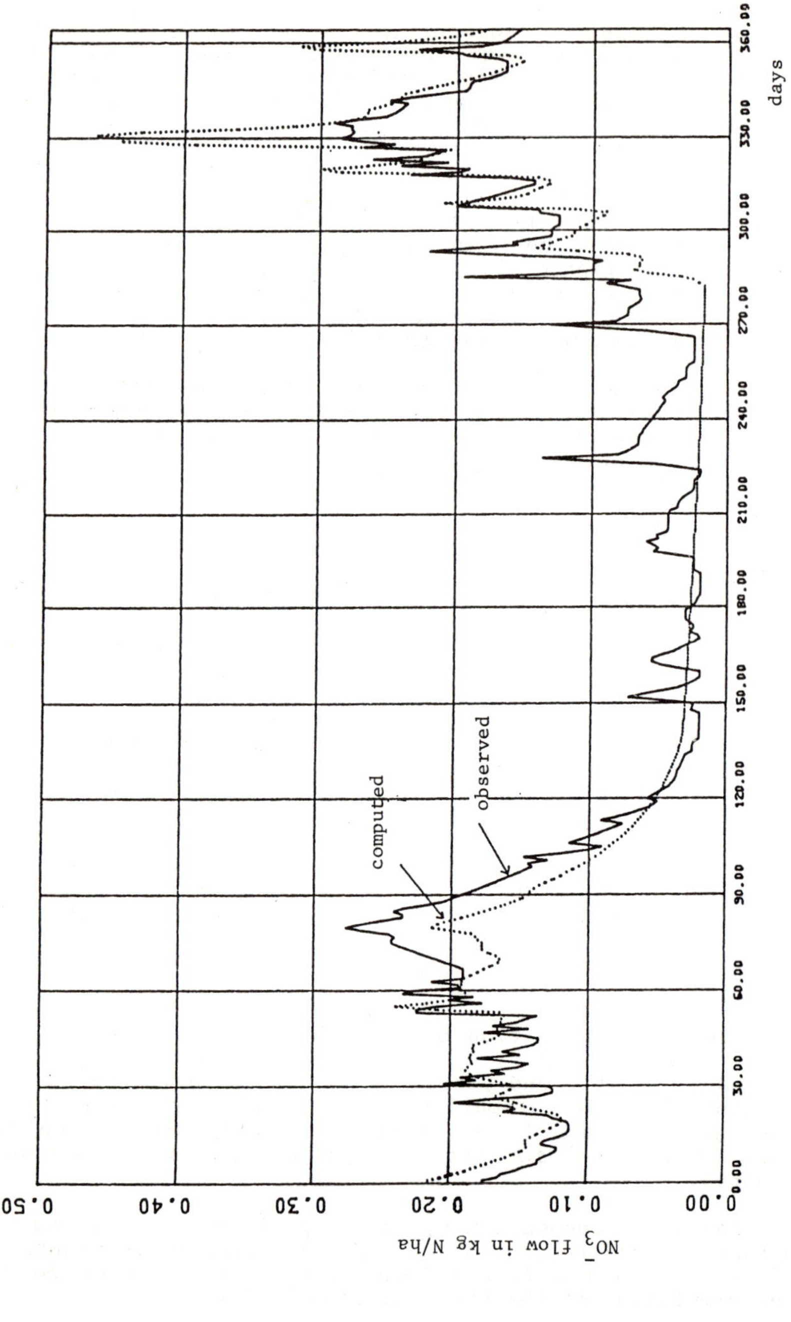

FIGURE 6. NITRATE LEACHING AT THE OUTLET OF THE WATERSHED

CONCLUSIONS

The proposed model for the transport of nitrates on a watershed lies essentially in a conceptualization of the nitrogen cycle in the soil, compatible with a spatially discretized hydrological model of a joint simulation of both surface and ground flow already developed by the authors.

The model has been validated to a parcell scale on the drainage and nitrates flow data obtained over several years on lysimeters. The extrapolation of the approach to a small drainage basin in polyculture has given good results showing a certain robustness of the nitrates leaching model concerning the parameter values considering the space scale in the study.

A foreseen utilization of such a model is the study of the influence of agricultural applications on the long range evolution of groundwater quality.

BIBLIOGRAPHIE

Marsily, G. de, Ledoux, E., Levassor, A., Poitrinal, D., Salem, A., 1978. Modelling of large multilayered aquifer systems: theory and applications. J. Hydro., v. 36, p. 1-34.

Ledoux, E., 1980. Modélisation intégrée des écoulements de surface et des écoulements souterrains sur un bassin hydrologique. Thèse de Docteur-Ingénieur, Ecole des Mines de Paris.

Mariotti, A., 1982. Apports de la géochimie isotopique à la connaissance du cycle de l'azote.

Geng, Q.Z., 1988. Modélisation conjointe du cycle de l'eau et du transfert des nitrates dans un système hydrologique. Thèse de Doctorat, Ecole des Mines de Paris.

ARTIFICIAL RECHARGE OF GROUND WATER
STATE OF ART AND LIMITATIONS

By

P.N. PHADTARE
CENTRAL GROUND WATER BOARD.
AHMEDABAD

A B S T R A C T

Artificial recharge of ground water is the process of enhancing movement of water into the under ground formation by constructing infiltration facilities (infiltration ponds, channels and pits or injection wells, etc.), or by inducing recharge from surface water bodies (by lowering ground water levels or improving the infiltration capacity of the soil. Artificial recharge is done to replenish over exploited aquifers and to store surplus surface water from wet season to dry season or from wet years to dry years. In this respect artificial recharge is an important element in water resources management. An other reason for artificial recharge is improving water quality of ground water or reclaimed waste water or checking ingress of saline water. Artificial recharge is carried out on a large scale only in a few countries around the world. In India very little has been done thus far in artificial recharge, but the scope for this technology is very large.

This paper describes various methods of artificial recharge as practiced in India with actual case histories. The pre-requisites for the different types of artificial recharge techniques and the criteria for site selection have been discussed in the Indian context. The periodic maintenance aspect and cost economics of various recharge structures are given with a view to assist the planners in taking policy decisions.

INTRODUCTION

Ground water, the elixir of organic life, unlike other minerals found in the crust of the earth, is a replenishable resource. Rainfall is the source of annual ground water replenishment and the movement of rainwater down to the ground water reservoir takes place by several ways, such as, direct infiltration through surface soils and rocks, vertical percolation from surface water bodies and sub-surface movement through the permeable media. In the tropical countries like India, the recharge through rainfall takes place only during the short period of monsoon but due to slow sub-surface movement of ground water, it is available for use throughout the year, in almost all the areas except the highly permeable cavernous limestones, bouldery formations, areas of very high topographic gradients and areas of excessive ground water draft. Excessive draft is observed in three types of areas i) Thickly populated and industrial areas ii) Highly progressive agricultural areas and iii) Thickly forested areas which discharge ground water due to evapotranspiration.

Additional recharge by artificial methods is necessary where the ground water resources are inadequate, that is, where the ground water draft, either due to extraction by wells or by sub-surface run-off, exceeds annual replenishment. Overexploitation of ground water has become a common phenomena in most parts of the country on account increasing population and heavy demands for irrigation and industrial uses. Artificial recharge is an important aspect of ground water management as it provides storage space free of cost, avoids evaporation lossess and allows use of stored water during dry period. Unlike surface reservoirs, the underground reservoirs can be used for indefinite period, if properly operated. They are also free from major adverse effects of natural catastrophes. Artificial recharge is also adopted where there is need for improving the quality of ground water or for arresting the ingression of inferior quality water from adjoining areas.

ARTIFICIAL RECHARGE METHODS

Artificial recharge is achieved by three methods, namely, by speading, by induced recharge and by injection. The suitablility of a particular method is based on the hydrogeological conditions, quality of source water and proposed use of recharged water as indicated below.

a) Highly permeable surface formations	Spreading method
b) Shallow aquifers	"
c) Source water of inferior quality	"
d) Deep aquifers with permeable or semi-permeable overburden	Spreading method
e) Aquifers with limited storage capacity	Induced recharge method
f) Deep aquifers with impervious overburden	Injection method
g) Source water of very good quality	Injection method and/or spreading method

Pre-requisites for Artificial Recharge:

Artificial recharge of ground water becomes essential where shortage of water, either for irrigation or domestic or for industrial purposes, is felt. The excessive demands in such areas are normally metwith from ground water reservoirs resulting in over exploitation. The over exploitation creates space for storage of artificially recharged water. Unless excess surface water and aquifers of suitable nature are available, artificial recharge schemes cannot be implemented. Over exploitation of ground water causes salinity ingress in the coastal areas. In certain areas having marine sedimentary deposits, the ground water has inherent salinity. In areas having semi-arid to arid climate ground water salinity is a major problem due to excessive evaporation which causes precipitation of salts from surface waters. The areas with interdunal drainage also have ground water salinity as there is no possibility of flushing of the salts. Artificial recharge together with other water management practices is essential in such areas for improving quality of ground water. Compatibility of source water and ground water of the aquifer to be recharged is another important pre-requisite as the chemical precipitation and bacterial growth, caused by the intereaction of waters of two different types, causes clogging and reduces rate of recharge.

The aquifer suitability depends mainly upon the storage coefficient, availability of storage space and permeability. Very high permeability results in loss of recharged water due to sub-surface run-off whereas low permeability reduces recharge rate. In order to have good recharge rate and to retain the water recharged for a sufficient period to allow its usage during lean period, moderate permeabilities are needed. Older alluvium, burried channels, alluvial fans, dune sands, glacial outwash, are some of the unconsolidated formations which are favourable for artificial recharge practices. As regards hard rocks, fractured, weathered and cavernous rocks are capable of allowing higher intake of water.

Technical feasibility of artificial recharge scheme depends upon storage capability, intake capacity, retention capacity of the aquifers and quality of source water, whereas economic viability depends upon the cost of artificial recharge structures, the cost of transport of source water and the cost of maintenance and operation. However the limits of costs vary with the proposed use of the recharged water, such as irrigation, domestic or industrial use. In artificial recharge schemes, the cost escalation is attributed mainly to source water treatment and clogging of structures and subsequent cleaning.

Artificial Recharge Methods:

(i) Spreading method : Artificial recharge by spreading method comprises increase in the surface area of infiltration, thereby allowing more surface for infiltrating waters. It is effected by channel spreading, spreading channels and recharge basins/ponds/pits. Channel spreading comprises bunding of existing stream or river channels either by bunds (check dams) across the channels or by constructing small "L" shaped levees within the channels. Spreading channels comprise unlined channels of gentle gradient which either hold the water or allow the movement of water at very gentle velocity in order to effect maximum infiltration. Recharge basins/ponds/pits help in collecting the surface run off from storm drainages and allowing their collection at places where it can effectively infiltrate into the the sub-surface ground water reservoirs. The levees, bunds, ponds, recharge basins, etc. which accumulate and keep the water stagnant for a longer period, allow more time for infiltration of water, thereby preventing losses due to run-off.

Artificial recharge by spreading has the disadvatages of loss of surface waters by evaporation and decrease in infiltrating capacity of spreading structures by deposition of silt from the source water, deposition of dust from the atmosphere and growth of vegetation on the infiltration surfaces. However, it has following important advantages over the injection method: The infiltration is almost like natural rainfall infiltration and extreme purity of infiltrating water is not needed. Normal storm run-off or primarily treated drainage waters (removal of harmful chemical constituents and suspended materials) can be utilised for artificial recharge. Cleaning of the infiltration surface once in a while is an easy process as it involves only scrapping of the surface.

Spreading methods are useful only for recharge to unconfined aquifers. The meagre downward leakage through the aquitards and slow sub-surface lateral movement of the ground water, make it uneconomical to artificially recharge the confined or semi-confined aquifers either by downward infiltration or lateral infiltration through the recharge zone. Spreading method which involves stagnation of water, sometimes creates environmental problems. It is, therefore, necessary to adjust supply of water to these structures in such way that large pools of stagnated water are not produced. This is however difficult as most of the artificial recharge experiments are based on collection of storm run-off. The recharge rates also need to be controlled as higher ground water mounds reduce the infiltration rate. The best way is, therefore, to store water underground with the help of sub-surface dykes, so that not only the excess water flows out but it flushes the accumulated silt also. The sub-surface strutures are more immune to natural catastrophes. Evaporation losses are also minimum as no water is stored above the surface.

(ii) Induced Recharge Method: The natural conditions like hydraulic gradient and vertical head are changed to allow rapid movement of infiltrated waters and to allow more space for storage, respectively. This situation has not arisen in our country so far but it was under active consideration whether such induced recharge can be experimented along banks of Ganges to direct flood waters underground, thereby effecting flood control measures. However conditions similar to these are getting created in more areas all over India on account of overexploitation of ground water. In certain heavy rainfall areas where excess rainfall results only in excess surface run-off on account of limited storage capcity of aquifers (especially hard rock areas), the overexploitation has helped in higher recharge, that is, though the pre-monsoon levels show progressive declines, the post monsoon levels are always attained.This suggests that overexploitation should be permitted in such areas.

(iii) Injection method: Water is fed directly into the depleted aquifers by providing a conduit access, such as tubewell or shaft or connector wells. Recharge by injection is the only method for artificial recharge of confined aquifers or deep seated aquifers with poorly permeable overburden. The recharge is instantaneous and there are no transit lossess and evaporation lossess. Injection method is also very effective in case of highly fractured hard rocks and karstic limestones but very high permeabilities are harmful as they do not allow the water to be retained for a longer period to make its use during dry season. However, requirement of extreme purity of the source water as well as compatibility of source water and water from the aquifer to be injected, are the precautions necessary, otherwise frequent clogging of injection structures, by bacterial growth, chemical precipitation or deposition of silt, results in heavy expenditure on well cleaning. As pumping wells are self cleaning, dual purpose injection wells that is injection cum pumping wells are more efficient. Connector injection well where saturated shallow aquifer and overexploited confined aquifers are tapped in a single well, allows free fall of water from shallow aquifer into the deeper aquifer, thereby reducing cost of injection. Injection method is also used as a "Pressure Barrier Technique" to arrest or reverse saline water ingression.

CASE HISTORIES

Technological developments in the pumping method and well construction have resulted in large scale exploitation of ground water. In many countries, which have to face the vagaries of monsoon, dependability on ground water has increased tremendously. In arid or semi-arid regions, availability of surface water is either lacking or is inadequate and storage of surface water is also uneconomical on account of high evaporation losses. Proper storage and management of available ground water resources therefore is very essential in such sources of replenishment of the ground water reservoirs in the arid and the semi-arid regions because the intensity of normal rainfall is grossely inadequate to produce any moisture surplus under normal infiltration conditions. Although artificial ground water recharge methods have been familiar in the developed nations over the past several decades, their importance was felt in the developing nations like India, only during past two or three decades. Methods like nala bunding, percolation tanks, trenching along hill slopes and around the hills, etc., are being practised in many parts of the country but full advantage of such projects could not be obtained for want of scientific orientation of the programmes. A thorough and detailed knowledge of the geological, hydrological and morphological features of the area are necessary for selecting sites for recharge structures.

The Central Ground Water Board (CGWB) tried to develop methodologies by including artificial recharge experiments as a component in each of its several ground water resources evaluation and exploration projects during past two decades. During 1980-85, the CGWB implemented a full scale Artificial Recharge Project with the assistance of the United Nations Developments Programme and established technical feasibility and economic viability of various artificial recharge methods in the semi-arid and drought affected areas. The findings of these various experiments are summarised below.

i) Studies on artificial recharge through percolation tanks in the Sina and the Man river basins, Maharashtra.

Both the Sina and the Man basins comprise mainly basaltic flows with narrow and shallow alluvial patches. Studies were carried out on seven percolation tanks, to find out their effectiveness and suggest method for improving their utility. The average recharge was 50% of the capacity of the tank, provided the tank bottom was renovated by scrapping the surface, every year before monsoon and the location was selected in such a way that the reservoir area covered vesicular or fractured basalts. Nala (stream) bunding, where the entire storage of surface water was restricted to the course of the Nala, was found to be more effective and economical as the surface exposed to evaporation was on an average 10% of that of normal percolation tank. The rate of infiltration varied from 50 to 70% of the capacity.

A connector well tapping phreatic alluvial aquifer of 6 metres thickness and deeper confined basaltic aquifer (63 metres depth), allowed free flow under gravity from phreatic aquifer to the confined aquifer, @ 0.19 MCM per year. The water level in the phreatic aquifer, which was saturated on account of nearness to a small surface reservoir, was 3 metres below ground level (bgl) whereas the piezometric level in confined aquifer was 30 m. bgl. It was found out that, the above recharge effected at a cost of Rs. 30,000/- (cost of bore hole) was very less compared to the cost of percolation tank needed for similar recharge (Rs. 4,00,000).

ii) Studies on artificial recharge through percolation tanks and sub-surface dykes in the Noyil Ponani and Vattamalai river basins, Tamil Nadu & Kerala.

Nine percolation tanks were selected in the semi-arid regions of above rivers basins and data on evaporation, change in storage surface run-off and water spread were collected. It was observed that the rate of percolation was as high as 163 mm/day at the beginning of the rainy season but reduced to hardly 15 mm/day after few months. The reduction in infiltration rate was mainly due to accumulation of silt at the bottom. Periodic desilting was therefore found to be essential.

Sub-surface dykes of 1 to 4 metres height were found effective in augmenting the ground water resources, particularly in the hard rock areas underlain by fractured aquifers.

iii) Artificial recharge studies in the Ghaggar river basins, Punjab.

Artificial recharge experiments through injection well were carried out, using canal waters. The injection rate was 43.8 litres per second (lps) at injection pressure of 1 atmosphere (atm). The pressure increased to 2 atm after 5 hours but remained constant thereafter. However, recharge under gravity was found to be 5.1 lps. The recharge rate also showed gradual reduction to 3.5 lps after few days.

On the basis of repeat tests, it was concluded that the recharge rate obtainable with injection under pressure is about 10 times greater than the rate obtainable by gravity flow. The increase in pressure during injection was found to be due to clogging. In order to avoid clogging of well, control on quality of source water was recommended. Periodic cleaning of well was suggested whenever the pressure increased beyond 6 atm or showed tendency towards sudden rise.

Studies were also conducted to find out possibilities of induced recharge from the Ghaggar river. Based on the above studies, construction of well-field, not less than 100 m from the river, with spacing of 200 m was suggested. In order to increase the effectiveness of well field so proposed, periodic scrapping off the clay film and clogged upper part of the flood plain aquifer was suggested.

iv) Pilot Project for Artificial Recharge, Gujarat.

Two areas were selected in Gujarat to study, experiment and conduct artificial recharge schemes. In the Central Mehsana area of North Gujarat, where large scale overexploitation of ground water has resulted in heavey declines during past three decades, artificial recharge was carried out by injection wells, connector wells, and by infiltration channels and ponds. Surplus ground water from the flood plain aquifers of the major rivers in Mehsana area and the tail-end releases from the Dharoi canal system were utilised. In the coastal areas of Saurashtra, Gujarat, where overexploitation of ground water has resulted in salinity ingress, artificial recharge experiments were carried out with the help of injection wells and recharge basins. Storm run-off and tail-end releases from the canal system of Hiran Irrigation Project were used for these experiments in the coastal area. In addition to this, in the coastal area, detailed observations were made at the existing Tidal Regulators and Check Dams to study their effectiveness. Injection of water from the phreatic aquifers into the deeper overexploited aquifers was found feasible in the entire Central Mehsana area. The spreading methods like spreading channels, recharge pits and ponds were found to be economically more viable than the injection method. As regards injection method, connector wells, when used as dual purpose wells were found to be more economical as recharge under gravity reduced the cost of recharging and periodic pumping avoided clogging of wells. In the coastal Saurashtra area, where the aquifers are highly cavernous, rapid underflow of recharged water to the sea does not make the artificial recharge as a viable proposal. However, the tidal regulators were found to be very useful in creating barriers of freshwater along the creeks and coastal depressions. These barriers prevent the sea water ingression very effectively.

The cost of recharge scheme depends mainly on treatment of source water, transportation of source water and stability of recharge structure. In the above experiments, all these factors were critically studied and simple methodologies were tried to overcome these problems, such as (1) Providing higher slopes for sides of spreading structures to avoid silt accumulation (2) Providing gravel pack in recharge pits to avoid caving of sides and to prevent deposition of silt during dust storm (3) Passage of source water through gravel-sand channel for desilting (4) Provision of desilting pits (5) Proper gravel pack and thorough well development in the phreatic zone of injection well to allow silt free water to enter deeper aquifers.

v) During the course of Technology Mission work in Jamnagar Distirct of Gujarat it was found that basaltic dykes cutting across streams create good sub-surface storage.

ECONOMICS OF ARTIFICIAL RECHARGE SCHEMES

The cost of construction and cost of operation of the recharge structures, except in case of injection well in alluvial area and tidal regulators are reasonable compared to costs of such experiments in other countries. However, considering the requirement of water for irrigation purposes, which is on an average, 5000 m3/hectare, the additional cost of cultivation on account of irrigation by recharged water works out from Rs.125/- to Rs.550/- per crop per hectare, which is quite high. In case of domestic water supply, however, the cost works out to Rs.0.75 to Rs.5/- per head per year which is reasonable, especially in areas where there is shortage of water. The cost and initial investment is many times less than that required for supply through tankers which is being adopted in case of shortage of water.

The table No.1 shows the salient features of the economics of various artificial recharge methods.

Artificial Recharge Structure	Initial Cost of structure per 1000 Lt/year recharge	Operational cost per 1000 Lit. of recharged water
	Rs.	Rs.
i) Injection well (Alluvial area)	2400/-	0.700
ii) Injection well (Hard rock)	150/-	0.150
iii) Spreading channel (Alluvial area)	220/-	0.065
iv) Recharge pit (Alluvial area)	225/-	0.070
v) Recharge pond or percolation pond (Alluvial area)	45/-	0.045
vi) Percolation tank (Hard rock area)	37/-	0.036
vii) Vasant Bandhava or check dam	9/-	0.022
viii) Tidal regulator	300/-	0.500

Table 1 : Economics of Artificial Recharge Structures (costs, as on 1/1988)

However, these works when taken up under scarcity or relief works, where expenditure on labour component is a liability of the State Government, the real investment which is required is for the material component. The structure thus becomes economically viable. Many such schemes can be taken up under scarcity work, if planned in advance, because drought relief works are needed almost every 5 to 7 years, which is the normal drought frequency in the country.

SELECTION OF SITES AND DESIGNS FOR ARTIFICIAL RECHARGE STRUCTURES

i) Injection or Connector Wells: The selection of sites for these structures depends upon the configuration of the confined aquifers, hydraulic gradient and location of source of excess surface water. It is always better to construct it closer to source to save cost of transportation of water to be used for recharge. However, the slow sub-surface movement of ground water does not allow the recharge to be carried out at a distance far off from the source of exploitation. The designs of injection wells and connector wells are not much different from those of thc normal tubewells and depending upon the aquifer characteristics, slot sizes, casing sizes and gravel packings are to be selected. These wells are required only for recharging deep confined aquifers. There is no need for recharging shallow semi-confined aquifers between depth range of 40 to 100 metres by injection method as the recharge to shallow un-confined aquifers by spreading method takes care of recharge to such aquifers by vertical leakage through leaky aquitards.

ii) Artificial recharge by spreading methods however needs elaborate field studies for selection of site and selection of the type of structure:

a) Channel spreading: This comprises changing of pattern of the surface flow in the river channel with the help of "L" shapped levees (sand bunds) so that instead of flowing rapidly downstream, the river water becomes sluggish, adopts longer route and thus provides more time for its infiltration. In our country where the rivers are ephemeral and prone to flash floods due to restricted period of monsoon, this method is not helpful.

b) Spreading channels: They comprise passage of water through unlined canals which in turn recharge the ground water reservoir. The spreading channels are similar to unlined canals with side slopes of 1 in 1 and very gentle floor gradient. The gentle floor gradient allows very gentle movement of water thereby allowing more time for infiltration and also reducing erosional action of water. Moderately high slopes of sides make these structures more stable and prevent wind blown silt from depositing on sides. The channel however needs periodic scrapping of floor to remove accumulated silt and weeds at the bottom. These structures are costly and need very high order of maintenance and supervision. They are not suitable for augmenting drinking water supply where requirement of recharge is limited and high costs are prohibitive. These channels are very effective as irrigation cum spreading channels and are very gainfully used in certain river basins - e.g. the flood water spreading channels of the Mahanadi basin in Chattisgarh region of Madhya Pradesh, the Ganga canals, etc. Instead of this, contour trenching which is a spreading channel system of limited extent, is more useful as it not only reduces surface run-off but helps in preventing soil erosion. These are ideally suited for valley slopes in hilly areas where surface run-off is very high. Planting of trees, along contour bunds or trenches, further helps in reducing surface run-off and soil erosion.

c) Recharge pond: It has been observed that the recharge ponds, percolation tanks, check dams or sub-surface dykes, which come under this category, create a ground water mound down to the full thickness of the aquifer. This ground water mound spreads around as a dome upto 100 to 1000 metres from the recharge structure, depending upon availability of water for recharge. These are the cheapest modes of artificial recharge but selection of a proper structure needs careful consideration. As already stated, artificial recharge for irrigation purposes is undesirable and as the requirement for domestic purposes varies according to population, it is better to select the structure according to requirement. For domestic wells of villages with population of less than 500, a normal village pond where not much elaborate arrangements for overflow are needed, is more than sufficient. However, it should be ensured that sufficient storm run-off is available for filling up the pond. If not, water from surrounding areas should be diverted to the pond with the help of minor trenches. Check dams are desirable only where the recharge requirement is very high. Out of the above 4 structures, that is, pond, check dam, percolation tank and sub-surface dyke, the sub-surface dyke is the most suitable structure as it is safe from flood havocks, does not need elaborate overflow arrangements and periodic disilting. Even the problem of forest sub-mergence, which is becoming main hurdle in getting clearance for artificial recharge structures, does not arise in case of sub-surface dykes. The silt from the sub-surface near the sub-surface dyke is flushed away during flash floods whereas the entire storage of water being underground, evaporation losses are also insignificant. The construction needs a 30 to 60 cm wide concrete or cement brick or cement masonary dyke extending down to impermeable basement or compact foundation. Two such sub-surface dykes of 100 m length each, within 300 m from the water supply well are enough to store water required for a village with population of 500. As the domestic wells are located close to the village, there is need for constructing these structures close to the village only. One of the structure should be upstream whereas other should be downstream. If only one structure is to be constructed, it should be downstream of the well. This is because the ground water mound created acts as a sub-surface barrier and reduces sub-surface outflow of ground water from around the well. During dry season, when the pumping water level in the well is low, the hydraulic gradient is reversed and the water is drawn from the ground water mound downstream.

The sites selected for construction of ponds, tanks, etc. during scarcity works are normally those where mannual excavation is possible. Such sites are nothing but sites with intense weathering. As intensive weathering is manifestation of higher fracture porosity, the suitability of site is automatically justified. In alluvial area, hydrogeological investigations for site selection are not necessary on account of homogeneous nature of the formations.

iii) Basinwise Development for Artificial Recharge of Ground Water:

In areas where rainfall is scanty and drought frequency is very high, conservation of every drop of rain water is essential. Out of the scheduled drought prone areas, worst affected areas can be selected for such schemes. The scheme for a particular basin comprises gully plugging in minor streams, sub-surface dykes or percolation tanks along tributaries streams, contour bunding or trenching along hill slopes, farm ponds in the foot hill zones and wherever possible, check dam-cum-minor irrigation dams on the main stream. Land levelling (terracing) and afforestation along hill slopes also form part of basin development plan because they help in higher rainfall infiltration.

The most important factor in the design of artificial recharge structure is the consideration for its stability against probable higher storm run-off inflow during years of excess rainfall, occasional flash floods and safeguard against accumulation of silt and organic matters at the bottom. It was observed that the infiltration capacity of ponds reduces by almost 25%, each year, due to silting and by the end of 5th year it becomes almost 10% of the total storage. The 90% of stored water is thus lost on account of evaporation only. Providing side slopes of 1:1 for all recharge structures helps in maintaining constant infiltration through sides, because the steep slope does not allow accumulation of silt and also makes the sides quite stable. Dry stone pitching is another safety measure against degradation of the structures.

For basinwise development, it is better to take up work in small basins where it would be easy to group the people in one society and to educate them. These schemes depend mainly on the co-operation of people as multiple activities like land development, water harvesting, afforestation, etc. are involved. Moreover the achievements depend upon peoples participation and active contribution. The problem of shortage of funds if any can also be overcome, because close co-operation helps in people willing to come forward and take over management and offer "shramadan". As the normal area of a village averages to 1000 to 1500 hectares, a basin with 1250 hectares area will be ideal.

iv) Water Harvesting Structures: Storage of rain waters in underground tanks or surface tanks is being practiced in the desert areas of western Rajasthan and also in the high rainfall areas of north-eastern region. The Technology Mission has designed rain water collection system for tiled roofs, thached roofs etc. In fact this system can be adopted in both rural and urban areas for houses with slanting or flat topped roofs. However, storage of the water in the tanks for use during the long period of dry season which normally extends for 7 to 8 months in major parts of the country, makes it necessary to take great care for keeping the water fresh and usable. Instead, structures should be designed for storing this water in underground rock formations and their utilisation by hand pumps or pumps. In hard rock areas, this water can be stored in small diameter (1 to 2m) shallow open wells (5 to 10m deep) from where it will pass into the underground formations. With normal minimum intake capacity of 1 LPS for wells in hard rock, water can be stored @ 80 m3/day (80,000 lit/day) whereas in alluvial area, well of same dimension can recharge water @ upto 200 m3/day. With slow sub-surface movement of ground water, the water thus stored remains within the vicinity of recharge structure for a considerable period and is available for use during dry period. For wells in alluvial areas, concrete rings with weepholes for infiltration of water will have to be provided to prevent the well from caving. The water can be directly pumped from these wells also but installing hand pump on a shallow well a little distance away from this recharge shaft will provide naturally filtered water.

In urban areas the multi-storied housing societies can have one recharge-cum-pumping well each where fresh rain water from the roof tops can be diverted. During rainy reason, such medium deep tubewells can recharge the ground water aquifer to the extent of 80 to 240 thousand litres per day. If the water is properly filtered before diverting underground through tubewells, the water can be used for domestic purposes also - if not for drinking, atleast for washing purposes. This will be the supply in addition to the normal supply by Municipal Corporation for drinking purposes.

In short, the suitability of an artificial recharge structure can be tabulated as follows:

Lithology	Topography	Type of structure feasible
Alluvial or hard rock to 40m depth.	Plain area or gently undulating area	Spreading pond, sub-surface dyke, minor irrigation tank, check dam, percolation tank, or unlined canal system.
Hard rock down to 40 m depth.	Valley slopes	Contour bunding or trenching.
Hard rocks	Plateau regions	Recharge ponds
Alluvial or Hard rock with confined aquif. (40m depth).	Plain area or gently undulating area.	Injection wells or connector wells.
-do-	Flood plain deposits	-do-
Hard rock	Foot hill zones.	Farm ponds or recharge trenches.
Hard rocks or alluvium.	Forested area.	Sub-surface dykes.

PERIODIC MAINTENANCE

Periodic maintenance of artificial recharge structures is essential because the infiltration capacity reduces rapidly due to silting, chemical precipitation and accumulation of organic matter. In case of spreading structure, the annual maintenance comprises scrapping of the infiltration surfaces and removing the silt and organic matter whereas in case of injection wells and connector wells, periodic pumping and/or mild acid treatment for removal of incrustations and bacterial growths on slots is essential. By converting the injection or connector well into a dual purpose well that is pumping-cum-injection well, the interval of periodic cleaning can be prolonged but in case of spreading structures, except for sub-surface dykes, annual desilting is a must. As the artificial recharge schemes are carried out as relief works, no budgetory provision is kept for its periodic maintenance and it is observed that normally these structures are neglected in later phases. They only get the face lift during the subsequent drought relief work and fortunately, the droughts being frequent (5 to 7 year frequency), some maintenance does take place.

However, if it is decided to construct these structures only for augmenting the supply of domestic wells, the structures should form part of domestic water supply scheme and provision for its periodic maintenance be carried out by Public Health Engineering Departments or Gram Panchayats, whosoever maintains the well.

The basinal development where the recharge is mainly for irrigation purposes, being a multidisciplinary scheme, the maintenance of minor irrigation tanks is normally carried out by the irrigation department, that of the contour bunding and trenching (along with afforestation) by the forest department and that of the farm pond and land levelling by the cultivators themselves.

CONCLUSIONS

i) For artificial recharge of ground water, rock formations with moderate permeabilities are most desirable. Low permeabilities reduce intake rate whereas high permeabilities do not allow retention of recharged water for a longer time to make use of it during dry season.

ii) Artificial recharge should be carried out only for augmenting domestic water supply as it is not economically viable for irrigation purposes.

iii) Artificial recharge for irrigation purposes should never be the main objective of any scheme. It should always be a byproduct, e.g.

a) Minor irrigation tank which along with irrigation, increases ground water recharge by way of reservoir and canal seepages, recycled irrigation water and added rainfall infiltration due to levelling of fields for irrigation purposes.

b) Afforestation along hill slopes by contour trenching and contour bunding which is normally carried out along with such plantations, helps in reducing the surface run-off.

c) Basinal development schemes (construction of farm ponds, afforestation, contour trenching ,contour land levelling and construction of check dams cum minor irrigation dams) where proper maintenance of structures is guaranteed due to involvement of Government agencies and cultivators.

iv) Injection wells or connector wells are costly schemes requiring high order of quality control and as such are not desirable. However, where plenty of surplus ground water is available in phreatic aquifers and there is scope for accommodating this into the deeper aquifers, dual purpose that is pumping (for irrigation or domestic supply) cum recharge wells may be constructed.

v) Among spreading methods, sub-surface dykes are most desirable as they need very little maintenance. They are safe from natural catastrophes. The evaporation losses are minimum and environmental problems arising out of water stagnation are avoided. There is no submergence of land or forest also.

vi) If proper maintenance of existing recharge structures (ponds, trenches, percolation tanks) is carried out, there is no need for any further work in this direction as the present ground water development in the country is hardly 50% of the resources available. At the most, schemes like basinal development, farm ponds, etc. may be planned in advance and executed during drought relief works.

vii) Every domestic well should be supported by artificial recharge structure and its initial expenditure on construction and later expenditure on maintenance should be included in the domestic water supply scheme. This will take care of proper periodic maintenance of artificial recharge structure.

viii) For Urban Water Supply, where water is normally pumped from surface water reservoirs or from intake well (radial well or infiltration gallery) in the river bed, it would be advisable to construct the well in such a way that it would pump water from the ground water mound formed due to seapages and not from the storage in the river bed or reservoir. This will not only increase scope for induced recharge but also reduces the evaporation losses due to constant diversion of surface storage into underground.

ix) The drought relief works should be planned in advance (due to high drought frequency in the country these works are almost necessary, once in 5 to 7 years) and should include contruction of new artificial recharge structures, renovation of old ones and basinal development programmes.

x) Availability of rainfall water from the roof tops is so great, that if properly used for artificial recharge will not only reduce the pressure of demand on domestic water supplies but will also reduce the problem of disposal of storm run-off from city areas.

REFERENCES

Brown, R.F., Signor,D.C. and Wood,W.W. Artificial Ground Water Recharge as a water management technique. Report 220, Texas Department of Water Resources, U.S. Geol. Survey, 1978.

Dutt,D.K. and Charlu,T.G.K. Ground Water Development in India, Rural Elec. Corporation, New Delhi, 1982.

Pathak,B.D. General Review of Artificial Recharge Works in India, Seminar Volume,International Seminar on Artificial Recharge of Ground Water, Ahmedabad, India, 1985.

Phadtare,P.N. Artificial Recharge Studies in the Mehsana and coastal areas of Gujarat State, India.

i) Seminar Volume, Int. Sem. on Art. Rec. of G.W., Ahmedabad 1985.
ii) Terminal Report-Project findings and recommendations, UNDP, CGWB, 1986.

Raju, K.C.B. Artificial Recharge through percolation tanks and sub-surface dykes, Seminar Volume, International Seminar on Artificial Recharge of Ground Water, Ahmedabad, 1985.

Sinha, B.P.C. Artificial Recharge Studies in the Ghaggar River Basin, India, Seminar Volume, International Seminar on Artificial Recharge of Ground Water, Ahmedabad, 1985.

Todd, D.K. Ground Water Hydrogeology, 2nd Ed., John Wiley, New York, 1980.

Todd, D.K. Annotated Bibliography on Artificial Recharge of Ground Water U.S.G.S. Water Supply Paper, 1477, 1959.

Todd, D.K. Artificial Recharge and Ground Water Management, Seminar Volume, International Seminar on Art. Rec. of Ground Water, Ahmedabad, 1985.

SYSTEMS FOR ARTIFICIAL RECHARGE OF GROUND WATER

HERMAN BOUWER

Director, U.S. Water Conservation Laboratory
4331 East Broadway Road, Phoenix, AZ 85040

INTRODUCTION

With artificial recharge of ground water, surface water is infiltrated into the ground for storage in aquifers and eventual recovery from wells. Infiltration and flow to aquifers can be achieved with infiltration facilities on the surface (Fig. 1). Such systems require permeable surface soils to obtain adequate infiltration rates, vadose zones without clay or other flow-restricting layers that would inhibit the flow to the aquifer, and transmissive aquifers that are unconfined. Where these conditions do not exist, or where suitable land is too expensive, artificial recharge of ground water can be achieved with wells.

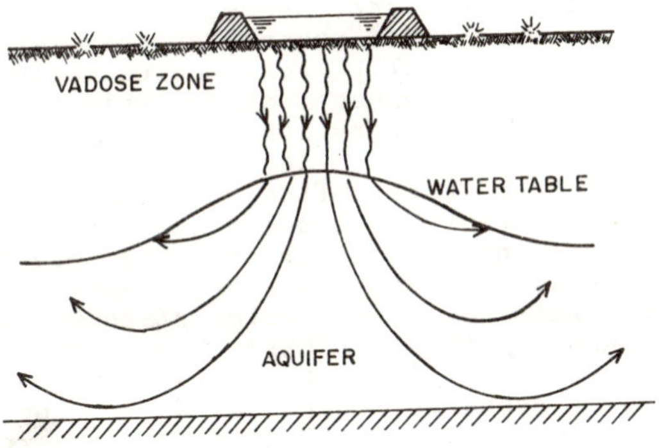

Fig. 1. Schematic of groundwater recharge system showing infiltration basin, vadose zone with wetted zone (wiggly lines), groundwater mound, and flow lines in aquifer.

RECHARGE WELLS

Recharge or "injection" wells are similar in construction to pumped wells, using screened section(s), gravel packs (in unconsolidated aquifers), and grouting. Before injection, the water needs to be carefully treated to remove essentially all suspended materials. Even then, injection wells in unconsolidated aquifers eventually clog up at the interface between the well and aquifer. This requires periodic pumping and/or redevelopment (surging, jetting) of the well. Because of clogging, the specific capacity of wells for injection into unconsolidated aquifers is only about half the specific capacity for pumping. Injection wells in fractured-rock aquifers or in limestone with solution channels or other well-developed secondary porosity have injection rates that are closer to pumping rates. Water for injection wells should be applied through a relatively small pipe in the well that ends below the water level. This is to avoid free fall of the water in the well and resulting entrainment of air in the water. Dissolved air in the recharge water could cause problems of "air binding" in the aquifer as air goes out of solution and forms entrapped air in the aquifer if the recharge water is colder than the ground water. The entrapped air can significantly reduce the hydraulic conductivity of the aquifer around the well and, hence, the injection rate. Ground water recharge with injection well usually is much more expensive than recharge with spreading or infiltration basins, often by an order of magnitude. Therefore, the rest of this paper will be devoted to design and management of infiltration systems for ground water recharge.

INFILTRATION SYSTEMS

Infiltration facilities for ground water recharge can be divided into in-channel and off-channel systems. In-channel systems consist of weirs, dams, and levees (T-dikes or L-dikes) to increase the wetted area and, hence, the infiltration in the stream bed or flood plain. Off-channel systems are basins in old gravel pits or specially constructed basins in areas of permeable soil. Design and management criteria to maximize the hydraulic capacity of infiltration basins depend on water quality, climate, and soil. Thus, these criteria are site-specific and they must often be evaluated by on-site experimentation. Factors to be studied are optimum schedules of flooding, drying, and cleaning of the basins; optimum pretreatment of the water; optimum water depth; and optimum velocity of the water (basins with stagnant water versus channels with flowing water). There are also environmental factors to be considered (insects, algae, odors). Sources of water for artificial recharge of ground water include surplus water in streams and rivers (including possibly increased flow due to cloud seeding or water harvesting techniques), storm water runoff, surplus water in aqueducts or water transfer projects (California Aqueduct, Central Arizona Project), and sewage effluent or other waste-water. Some water sources are continuous and permit year-around operation of the infiltration basins. Others are seasonal or hapazard.

IN-CHANNEL SYSTEM MANAGEMENT

If the streams in in-channel systems have sufficient flow and the system of spreading the water in the channel is such that water depths are small and velocities are relatively high, the system can be essentially self cleaning. Fine suspended solids are then kept mostly in suspension, and scour helps prevent formation of permanent clogging layers. If the channels are subject to high-runoff events, the spreading facilities (weirs, dams, T-dikes, etc.) may be damaged and must be reconstructed. However, the floods also wash accumulated sediment down-stream and leave a clean, rejuvenated streambed for more infiltration and ground water recharge.

BASIN MANAGEMENT

Infiltration rates in flooded basins decrease with time due to accumulation on the bottom of sediment that was suspended in the water. Biological activity in the water (growth of algal cells that form a filter cake on the bottom upon infiltration), and on the bottom (bacterial and algal activity) can also reduce infiltration through formation of organic clogging layers. Thus, the basins must be regularly dried and cleaned to restore infiltration rates. If the clogging material consists primarily of silt, clay, or other inorganic matter, it must be removed by scraping, raking, or other procedure that removes only the clogging material. Disking the clogging material into the subsoil gives temporary improvement, but ultimately the entire soil layer to the depth of disking may have to be removed because of accumulation of fine particles. If the clogging material is primarily organic (sludge, bacteria, algae), drying alone can give considerable recovery of infiltration rates due to the decomposition, shrinking, cracking, and curling-up of the material. Cleaning the basin bottoms may then not be necessary for every drying period but may be done only occasionally like once or twice a year. The best combination of drying and cleaning schedules must be determined on-site, especially for projects in new areas where there is no local experience with management of infiltration basins.

Sometimes, flooding and drying cycles are controlled by life-cycles of insects. To avoid nuisance problems, flooding periods may have to be only a few days to prevent hatching of insect eggs and emergence of adult insects (for example, the midge flies in California).

Where the water for recharge basins contains considerable suspended material, it can be more economical to remove this material in presedimentation or desilting basins with possible use of coagulants to enhance settling of the solids. However, this costs money. On the other hand, not removing suspended solids first and letting them all accumulate on the bottom of the infiltration basin costs money, too, in the form of frequent drying and cleaning of the basins and loss in infiltration capacity. Thus, there is an optimum combination of pretreatment and drying and cleaning of the basins and loss in infiltration capacity. This economic optimum must be determined for each individual system where presedimentation appears desirable.

If essentially all suspended solids are removed in a presedimentation facility, the water will be very clear. After the water enters the infiltration basin, sunlight can then penetrate to greater depths. This could cause rapid growth of algae and stimulate other biological activity that could lead to serious clogging. Thus, effective desilting could accelerate clogging of the basin by organic material. To minimize this type of clogging, the water depth in the basins should be small (possibly 10 to 20 cm), so that there is a high rate of turnover of the water in the basins and suspended algae have "no time" to grow.

Where surface water is available for artificial recharge of ground water during most of the year or the entire year, there may be an interest in using the infiltration basins also for recreational purposes. Such use places constraints on the management of the basins for maximum hydraulic loading. Regular drying and cleaning may then be more difficult. Presedimentation may be desirable to minimize sediment accumulation on the bottoms of such basins.

A choice can be made between infiltration basins that have essentially stagnant water where even the finest suspended material can settle out, and infiltration channels where the water is kept moving to create enough turbulence to keep the fine material in suspension. On-site testing needs to be done to see which system is better and gives the highest infiltration rates. If the channel system with moving water is used, a few infiltration basins may have to be constructed at the end of the channels to catch any residual flow.

EFFECT OF WATER DEPTH

Intuitively, one would think that a large water depth in infiltration basins gives higher infiltration rates than a small water depth. This may not always be so, however (Bouwer in press, a). If the ground water table is above the bottom of the basin, as can happen if the basins are in old gravel pits or where ground water tables are high, then an increase in water depth could produce a significant increase in infiltration rate. If the ground water table is a considerable distance below the bottom of the basin, an increase in water depth will produce only a small increase in infiltration rate if the basin bottom and banks are clean (not covered by sediment or other clogging material). This can be demonstrated by applying Darcy's equation to the flow from the basin to the ground water. If, however, the wetted perimeter of the basin is covered by a well-developed clogging layer (organic or inorganic), the entire head due to water depth in the basin is dissipated across the clogging layer, and the infiltrated water moves as unsaturated flow to the underlying ground water. Applying Darcy's equation to the flow through the clogging layer (Bouwer, 1982) then shows that for this case there is an almost linear relation between water depth in the basin and infiltration rates. In that case, for example, doubling the water depth would essentially double the infiltration rate. However, there are other effects that can negate this linear relation.

The first effect is compaction of the clogging layer due to an increase in the seepage force across this layer as the water depth and, consequently, the head loss across the clogging layer, are increased (Fig. 2). The compaction of the clogging layer produces a significant decrease in the hydraulic conductivity of this layer, causing a lower increase in infiltration rate than expected from the hydraulic head (water depth) increase alone, and perhaps even a decrease, depending on the type of clogging material (Bouwer in press b).

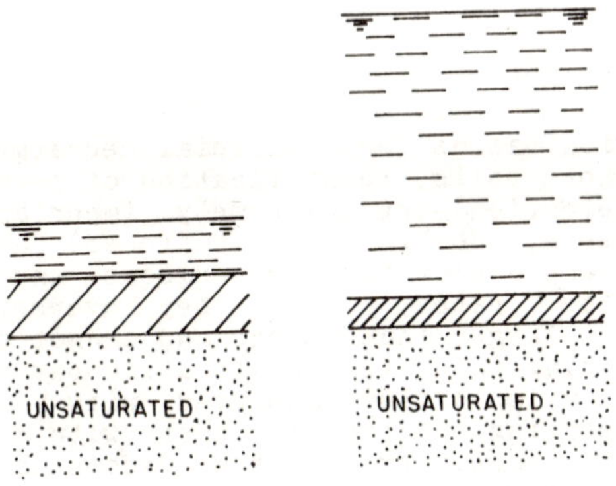

Fig. 2. Compaction of clogging layer (hatched) by increasing water depth from small (left) to large (right).

The second effect is that if increasing the water depth does not produce a proportional increase in infiltration rate, the turnover rate of the water in the basin decreases, which could promote the growth of suspended, unicellular algae in the water due to longer exposure to sunlight. The algae will then be filtered out on the bottom as water infiltrates and form a filter cake on the clogging layer, which further reduces the hydraulic conductivity of this layer and, hence, the infiltration rate. This, in turn, reduces the turnover rate in the basin even more, increasing the exposure of suspended algae to sun-light which increases the growth of algae and further clogs the bottom layer with the algal filter cake, etc., etc.

The third effect is that at high algae concentrations, uptake of carbon dioxide from the water for photosynthesis becomes significant, causing the pH of the water in the basins to increase to values as high as 9 or 10. At these pH values, calcium carbonate will precipitate out and accumulate on the bottom, further aggravating the clogging process and reducing infiltration rates even more.

In view of these processes, shallow basins may actually produce higher infiltration rates than deep basins where the wetted perimeter of the basin is covered with a clogging layer, and such layers are the rule rather than the exception. Because a number of factors govern the relation between water depth and infiltration rate, the water depth giving maximum infiltration rates must be evaluated by on-site experimentation. If deep basins are considered or changeovers from shallow basins to deep basins are contemplated, the local conditions should be thoroughly investigated, and studies with test basins should be made to make sure that the deep basins will produce the desired results.

SITE SELECTION AND HYDRAULIC LOADING RATES

Since infiltration basins for artificial recharge of ground water require permeable soils, identification of permeable soil profiles and site selection are extremely important. Small cylinder infiltrometers, double-ring (buffered) as well as single-ring systems, are useful for comparisons and measuring relative infiltration rates, but they overestimate the infiltration rates for larger inundations and cannot be used to predict hydraulic loading rates for infiltration basins (Bouwer, 1986). Such prediction is better achieved with larger test basins and by supplementing infiltration tests with measurements of soil hydraulic conductivity in the vadose zone. For unclogged basins and deep ground water tables, basin infiltration rates are approximately equal to the average resaturated hydraulic conductivity (harmonic mean, see Bouwer, 1978, pp. 56-60 and pp. 253-254) of the vadose zone or upper portion thereof. Thus, hydraulic conductivity measurements can give a good estimate of maximum hydraulic loading rates attainable with the basins.

Methods available for <u>in-situ</u> measurement of resaturated hydraulic conductivity in the vadose zone include the air-entry permeameter, double-tube method, infiltration gradient technique, and reverse auger hole or well pump-in method (Bouwer, 1978, pp. 123-130). Hydraulic conductivity of stony or gravelly materials can be estimated from the hydraulic conductivity of the soil between the gravel or boulders and the volume fraction of the rock or void ratios of the soil and rock matrix (Bouwer and Rice, 1984).

Permeable soils typically have hydraulic conductivities in the range of 1 m/day or 3 ft/day (fine loamy sands) to 10 m/day or 30 ft/day and sometimes even higher (sands, and sand and gravel mixes). Because of clogging, infiltration rates of recharge basins tend to be less than the resaturated hydraulic conductivity of the underlying soil materials. Actual infiltration rates during flooding thus generally vary from about 0.3 m/day (1 ft/day) to 3 m/day (10 ft/day). For year-round operations and including time for drying and cleaning the infiltration basins, hydraulic loading rates or accumulated infiltrations typically range from 30 to 300 m/yr (100 to 1000 ft/yr).

GROUND WATER MOUNDS

When the infiltrated water joins the underlying unconfined aquifer, a ground water mound is formed (Fig. 1), and the recharge water moves mostly laterally through the aquifer to produce smaller ground water table rises further away. The rise of ground water mounds during infiltration and the fall of the ground water mound during drying can be predicted with Hantush's equation (Bouwer, 1978, pp.283). This equation can also be used to calculate the effect of the infiltration system on ground water levels at various distances from the infiltration basins. If there are complicating factors in the aquifer system, such as a natural ground water table slope and other recharge or discharge mechanisms (losing streams, wells, springs, uptake of ground water by vegetation, etc.), the effect of artificial recharge on ground water levels can be estimated by modeling the aquifer system, using finite difference or finite element analysis techniques.

Aquifers should be sufficiently transmissive to keep ground water mounds below the bottom of infiltration basins if reductions in infiltration rates are to be avoided (Bouwer, 1988). A long, narrow infiltration basin or system of basins produces lower ground water mounds than square or round systems with the same area and hydraulic loading.

BASIN LAYOUT AND DESIGN

Where hydraulic loading rates need to be maximized, infiltration systems for artificial recharge of ground water should consist of a number of basins. Depending on topography, such basins can have a surface area of 0.1 ha (0.25 acre) or less to 10 ha (25 acres) or more. Each basin must have its own water supply and drainage so that each basin can be flooded, dried, and cleaned according to its best schedule. Basins should never be in series so that the outflow from one basin is the inflow for the lower basins because in such systems, basins cannot be dried and cleaned individually. Often, the first few basins are "sacrificed" as presedimentation facilities. Deep basins often are not dried and cleaned frequently enough to maintain high infiltration rates. When such basins are finally cleaned, infiltration rates are very low, and letting the water disappear by infiltration alone will take much too long. Thus, such basins should have their own drainage systems so each basin can be quickly dewatered for drying and cleaning. Shallow basins with high infiltration rates do not need special drainage systems because the water usually disappears by infiltration a few days after the inflow is stopped. Basins should be constructed with horizontal or well-graded bottoms to prevent low places where water can stand for long periods and interfere with infiltration recovery and cleaning operations.

Where ground water mounds (including perched mounds) could rise above the bottom of the infiltration facilities and reduce infiltration rates, the basins should be laid out so that there is no large near-square or near-round block of inundated area. Thus, basins should be small and/or long and narrow, and adjacent basins should not be flooded at the same time.

WATER QUALITY

As the infiltrated water moves through the vadose zone and aquifer, some quality parameters may be improved, and some may be adversely affected. Constitutents that are partly or almost completely removed from the water as it moves through the vadose zone and aquifer include suspended solids, bacteria, viruses, other microorganisms, biodegradable material (BOD), nitrate phosphate, and some synthetic organic compounds, particularly the nonhalogenated hydrocarbons (see paper, "Ground water recharge with sewage effluent" by Herman Bouwer). Because the soils in ground water recharge systems normally are quite coarse and permeable, there is little or no clay, and ion exchange will be insignificant. Hence, the ionic composition of the water after it has moved through the vadose zone and aquifer generally will be about the same as that of the water entering the infiltration basins.

Adverse effects include mobilization of iron and manganese from the vadose zone and aquifer as oxygen levels are reduced, and leaching of trace elements (including selenium, arsenic, boron, cadmium, molybdenum, and mercury) from the vadose zone. Leaching of trace elements may be significant where soils are relatively fine and almost marginal for artificial recharge of ground water, and have not had a long history of infiltration. Such soils include basin and valley soils and marine deposits, i.e., the San Joaquin Valley in California where selenium is leached from the soil and appears in the drainage water from the irrigated fields. Alluvial fans, stream channels, and floodplains generally would not be expected to have problems of leaching of trace elements, but potential sites should be checked regardless to avoid unpleasant surprises later on.

There may also be concern for humic and fulvic acids and algal compounds and metabolites that may already be in the water before infiltration. These organics may not be completely removed in the vadose zone and aquifer. Since they react with chlorine to form trihalomethanes, special consideration may be required when the water is pumped from the aquifer and needs to be disinfected for drinking (Fam and Shenstrom 1987). However, unpolluted, pristine ground water also contains organic carbon, mostly as fulvic or humic acids and typically at concentrations of about 0.2 to 0.7 mg/l (Thurman 1979). Such water is commonly disinfected with chlorine when used for public water supplies, without giving much thought to the possibility of forming THM's.

From an operational standpoint, the most important quality parameters of the water going into infiltration basins are the total dissolved solids content (TDS) and the sodium adsorption ratio (SAR), calculated as Na/[(Ca+Mg)/2] with the concentrations expressed in meq/l. TDS and SAR control whether clay in the soil is flocculated or dispersed. A flocculated state is preferred because a soil with such a clay is much more permeable than a soil with dispersed clay. A low SAR and a high TDS favor flocculation, where a high SAR and a low TDS favor dispersion of clay (see Mc Neal's graph in Bouwer, 1978, p. 44). Soils below infiltration basins generally are sandy or gravelly and contain little or no clay. The same is true for aquifers. Thus, SAR and TDS will have little or no effect on the hydraulic conductivity of vadose zones and aquifers, but they will have an effect on the hydraulic conductivity of the sediment layers that form on the basin bottoms. Such layers consist of fine materials and often contain clay that was suspended in the water. Clogging due to inorganic sediment accumulations on the basin wetted perimeter thus tends to be more severe where SAR and TDS cause the clay to be dispersed than where the clay is flocculated, necessitating more frequent cleaning operations for the former. Sometimes, dispersing SAR and TDS values can mobilize clay particles in the aquifer system. These particles can then migrate through the aquifer and move to wells where they increase the turbidity of the pumped water.

PILOT PROJECTS AND FLEXIBILITY

There are hundreds of successful artificial ground water-recharge projects in the U.S. alone, and many more in the rest of the world. Recharge systems are site-specific and what works well in one place may not be the best in another. Thus, when artificial recharge of ground water is considered in areas where there is no previous experience with such systems, it is always desirable to start with a small project to obtain local experience with artifiicial recharge of ground water and to develop design and management criteria for the full-scale project. This prevents costly mistakes and can save large amounts of money later on. Even then, prototype or full scale infiltration systems for artificial recharge of ground water should be designed and constructed with enough flexibility to permit changes in water depth, flooding-drying-cleaning schedules, configuration of flooded and drying basins, and other operational aspects when adjustments need to be made.

APPENDIX I. -- REFERENCES

Bouwer, H., 1978. Ground Water Hydrology. McGraw-Hill Book Co., New York, N.Y.

Bouwer, H., 1982. Design considerations for earth linings for seepage control. Ground Water, 20(5), 531-537.

Bouwer, H., 1986. Intake Rate: Cylinder Infiltrometers. Chapter in Amer. Soc. of Agronomy Monograph on Methods of Soil Analysis, no.9, Part 1, Physical and Mineralogical Methods, 2nd ed., 825-844.

Bouwer, H. (In press, a). Surface water - ground water relations for open channels. Proc. Irrig. and Drainage Specialty Conf. ASCE, Lincoln, Nebr., 19-21 July 1988.

Bouwer, H. (In press, b). Effect of water depth in ground water recharge basins on infiltration rate. J. Irrig. and Drain. Engg., ASCE.

Bouwer, H., and Rice, R.C., 1984. Hydraulic properties of stony vadose zones. Ground Water, 22(6), 696-705.

Fam, S., and Shenstrom, M.K., 1987. Precursors of nonvolatile chlorination by-products. J. Water Poll. Contr. Fed., 59, 969-978.

Thurman, E. M., 1979. Isolation, characterization, and geochemical significance of humic substances from groundwater, Thesis presented to the Univ. of Colo., Boulder, Colorado, in partial fulfillment of the requirements for the degree of Doctor of Philosophy.

AUTHOR INDEX

International Workshop on

APPROPRIATE METHODOLOGIES FOR DEVELOPMENT AND MANAGEMENT OF GROUNDWATER RESOURCES IN DEVELOPING COUNTRIES

(February 28 - March 4, 1989)

VOLUME II

Chief Editor
C.P. GUPTA

Editors:
Shakeel Ahmed
V.V.S.G. Rao
M.T. Rajan

NATIONAL GEOPHYSICAL RESEARCH INSTITUTE
(C.S.I.R.)
HYDERABAD - 500 007, INDIA

OXFORD & IBH PUBLISHING CO. PVT. LTD.
New Delhi Bombay Calcutta

ISBN 81-204-0406-8 (For Set)
ISBN 81-204-0407-6 (Vol. I)
ISBN 81-204-0408-4 (Vol. II)

Published by Mohan Primlani for Oxford & IBH Publishing Co. Pvt. Ltd., 66, Janpath, New Delhi 110001 and printed at Rekha Printers, Okhla Industrial Area, Phase II, New Delhi 110020

1-F9-8.5

FOREWORD

Groundwater is the great hidden component of the hydrological cycle comprising the most important natural resource of our planet. The history of groundwater utilization goes as far back as the history of mankind itself. But the stupendous growth of population, agriculture, and industry has compelled an indiscriminate exploitation of this resource. As a result, a variety of environmental hazards such as declining water levels in the aquifers, reduction of groundwater effluence to rivers and streams, deterioration of water quality, land subsidence etc., have set in despite vast strides that the multidisciplinary science of geohydrology has taken over the last few decades.

Utmost care has, therefore, to be exercised in the exploitation and management of this precious resource. In order to evolve a pragmatic and scientific plan for the management of groundwater resources, one needs to quantify the characteristic hydrogeological, hydrometeorological, hydrological, hydrogeochemical and other relevant parameters; establish an optimal network for data acquisition, storage and retrieval; and adopt appropriate methodology for data interpretation and analysis. Already, a lot of work has been done to evolve and apply appropriate techniques related to these aspects, particularly in the developed countries.

It is heartening that an International Workshop is being organized to share the experiences and knowledge of experts from developed as well as developing countries. I hope that it would culminate into making concrete suggestions on such techniques and strategies as could be easily replicated in the developing countries with their limited resources of man, material, money, for optimal harnessing and proper management of their groundwater resources.

Hyderabad
January 6, 1989

Hari Narain

PREFACE

Water is essential for existence. An assured supply of wholesome water has, therefore, been the prime concern of all societies. In our own time the increasing stress on the supportive and assimilative capacities of water systems has led to a growing realization that it is only through a scientific approach to their management, that we can hope to sustain the quality and regular availability of this basic human requirement. The need for appropriate integrated studies in developing countries is all the more important in view of a burgeoning population, widespread aridity, scarcity of fiscal resources and vagaries of hydrometeorological conditions.

Since groundwater, surface water and land comprise a complex interlocked system, they must be considered and managed in their totality, towards making the most optimal use of this precious resource. With the advent of remote sensing and sophisticated data processing techniques, the speed and scale of investigations of visible resources such as land and surface water have greatly improved. An understanding of the invisible groundwater regimes and quantification of their characteristic parameters, however, continue to remain mostly inferential and indirect. Groundwater resources are now being increasingly tapped to meet societal needs on account of their possible availability in the vicinity of the consumer. However, it is important to recognize that although groundwater has ubiquitous occurrence and vast magnitude, its potential as a perennial source is subject to wide spatio-temporal variations. It is renewable but not inexhaustible. It is underground but not entirely protected from pollution.

Some of the related issues that the ensuing Workshop IGW-89 would address, include identification of the state-of-the-art techniques for geohydrological parameter estimation, acquisition of data, analysis of flow in the unsaturated zone, studies of contaminant migration and augmentation of groundwater. All these are essential pre-requisites for reliable estimation and rational management of this vital resource. Cost-effectiveness of these techniques is especially important for developing countries.

I sincerely hope that deliberations of this Workshop will go a long way in enhancing our collective technical competence, facilities and infrastructure on the one hand and socio-political awareness of these problems, on the other. For thus alone, can we successfully face the formidable challenge of meeting the ever-increasing demand for water.

Hyderabad
January 19, 1989

Vinod K. Gaur

SPONSORS FOR IGW-89

EMBASSY OF FRANCE
2, AURANGZEB ROAD
NEW DELHI-110 001, INDIA

INTERNATIONAL CENTRE FOR TRAINING AND EXCHANGES IN GEOSCINCES
103 RUE DE LILLE
75007 PARIS, FRANCE

AUSTRALIAN INTERNATIONAL DEVELOPMENT ASSISTANCE BUREAU
C/O THE AUSTRALIAN HIGH COMMISSION
1/50 G, SHANTI PATH
CHANAKYAPURI
NEW DELHI - 110 021, INDIA

COSTED
24, GANDHI MANDAPAM ROAD
MADRAS - 600 025, INDIA

ASSOCIATION OF GEOSCIENTISTS FOR INTERNATIONAL DEVELOPMENT
NATURAL RESOURCES DIV.
U.N. BUILDING, RAJADAMNOEN AVE.
BANGKOK 10200, THAILAND

DEPARTMENT OF RURAL DEVELOPMENT
MINISTRY OF AGRICULTURE
KRISHI BHAVAN
NEW DELHI - 110 001, INDIA

NEYVELI LIGNITE CORPORATION
NEYVELI - 607 801, INDIA

RAIL INDIA TECHNICAL AND ECONOMICS SERVICES
646, ANNA SALAI
MADRAS - 600 006, INDIA

CMC LIMITED
POSNETT BHAVAN, TILAK ROAD
HYDERABAD-500 001, INDIA

UNIVERSITY GRANTS COMMISSION
BAHADUR SHAH ZAFAR MARG
NEW DELHI - 110 002, INDIA

NATIONAL REMOTE SENSING AGENCY
BALANAGAR
HYDERABAD - 500 037, INDIA

TAMILNADU WATER AND DRAINAGE BOARD
CHEPAUK
MADRAS - 600 005, INDIA

UPTRON INDIA LTD
10, ASHOK MARG
LUCKNOW (U.P.), INDIA

PANCHAYAT RAJ ENGINEERING DEPARTMENT
SADHANA BUILDING, HYDERGUDA
HYDERABAD - 500 029, INDIA

NATIONAL BANK FOR AGRICULTURE AND RURAL DEVELOPMENT
P.B.NO.63520, WORLI
BOMBAY - 400 018, INDIA

AGRICULTURAL ENGINEERING DEPARTMENT
GOVT. OF TAMILNADU
NANDANAM
MADRAS-60035, INDIA

INTEGRATED GROUNDWATER SERVICES PVT. LTD.
71, HUDA COMPLEX, TARNAKA,
HYDERABAD - 500 007, INDIA

TERRA SCIENCE,
16-47, PRASANTHI NAGAR,
HYDERABAD -500 039, INDIA

STEERING COMMITTEE

PROF. V.K. GAUR DIRECTOR, N.G.R.I. HYDERABAD-500 007, INDIA	CHAIRMAN
DR. S. PLATTARD SCIENTIFIC ATTACHE EMBASSY OF FRANCE NEW DELHI-110 001, INDIA	MEMBER
DR. M. LELEU DIRECTOR CIFEG 103, RUE DE LILLE 75007 PARIS, FRANCE	MEMBER
MR. S.R. SENGUPTA GENERAL MANAGER (MINES) NEYVELI LIGNITE CORPORATION NEYVELI-607 803, INDIA	MEMBER
MR. G. GHOSH DIRECTOR, TECHNOLOGY MISSION DEPARTMENT OF RURAL DEVELOPMENT KRISHI BHAVAN NEW DELHI-110 001, INDIA	MEMBER
MR. K.N. JOHRY ADVISOR AND HEAD, I.S.C., CSIR NEW DELHI-110 001, INDIA	MEMBER
PROF. S. RADHAKRISHNA DEPARTMENT OF PHYSICS INDIAN INSTITUTE OF TECHNOLOGY MADRAS-600 036, INDIA	MEMBER
DR. SHAKEEL AHMED SCIENTIST, N.G.R.I. HYDERABAD-500 007, INDIA	JOINT SECRETARY
DR. C.P. GUPTA SCIENTIST, N.G.R.I. HYDERABAD-500 007, INDIA	SECRETARY

ORGANIZING COMMITTEE

PROF. V.K. GAUR DIRECTOR, N.G.R.I. HYDERABAD-500 007, INDIA	CHAIRMAN
MR. MOHD. INAMUL HAQ ADVISOR TECH. MISSION ON DRINKING WATER D.R.D., KRISHI BHAVAN NEW DELHI-110 001, INDIA	MEMBER
MR. K.M. SUBRAMANYAM DIRECTOR, A.P.STATE GROUNDWATER DEPT. SOMAJIGUDA HYDERABAD-500 463, INDIA	MEMBER
MR. T. HANUMANTHA RAO DIRECTOR GENERAL WALMATARI DIRECTORATE OFFICE SECRETARIAT BUILDINGS HYDERABAD-500 022, INDIA	MEMBER
MR. N.K. BALASUNDER DY. GENERAL MANAGER (HYDROLOGY) NEYVELI LIGNITE CORPORATION NEYVELI-607 803, INDIA	MEMBER
MR. K.V.J.R. KRUPANIDHI DIRECTOR, C.G.W.B. 3-6-291, HYDERGUDA HYDERABAD-500 029, INDIA	MEMBER
PROF. B.H. BRIZ KISHORE DIRECTOR (PLANNING AND CO-ORDINATION) J.N.T.U. HYDERABAD-500 028, INDIA	MEMBER
DR. P.V.L.N. RAO MANAGING DIRECTOR,NRDCS LIMITED P.B. NO. 1067, HIMAYATNAGAR HYDERABAD-500 029, INDIA	MEMBER
DR. A. APPA RAO SCIENTIST, N.G.R.I. HYDERABAD-500 007, INDIA	MEMBER
DR. M. THANGARAJAN SCIENTIST, N.G.R.I. HYDERABAD-500 007, INDIA	JOINT SECRETARY
DR. C.P. GUPTA SCIENTIST, N.G.R.I. HYDERABAD-500 007, INDIA	SECRETARY

SCIENTIFIC ADVISORY COMMITTEE

PROF. G. DE MARSILY — CHAIRMAN
UNIVERSITY PARIS VI AND
PARIS SCHOOL OF MINES
PARIS, FRANCE

PROF. K.R. RUSHTON — MEMBER
DEPARTMENT OF CIVIL ENGINEERING
UNIVERSITY OF BIRMINGHAM
BIRMINGHAM B15 2TT, U.K.

DR. T. POINTET — MEMBER
BRGM, B.P. 6009
45060, ORLEANS, FRANCE

DR. K. TRIPPLER — MEMBER
BGR, STILLEWEG 2
3000 HANNOVER 51
FEDERAL REPUBLIC OF GERMANY

PROF. B.B.S. SINGHAL — MEMBER
DEPARTMENT OF EARTH SCIENCES
UNIVERSITY OF ROORKEE
ROORKEE-247 667, INDIA

PROF. S. RAMASESHAN — MEMBER
DEPARTMENT OF CIVIL ENGINEERING
INDIAN INSTITUTE OF TECHNOLOGY
KANPUR-208 016, INDIA

DR. A.S. CHAWLA — MEMBER
DIRECTOR, W.R.D.T.C.
UNIVERSITY OF ROORKEE
ROORKEE-247 667, INDIA

PROF. P.G. SASTRY — MEMBER
DEPARTMENT OF CIVIL ENGINEERING
REGIONAL ENGINEERING COLLEGE
WARANGAL, INDIA

PROF. R. SAKTHIVADIVEL — MEMBER
DIRECTOR, CENTRE FOR WATER RESOURCES
GUINDY ENGINEERING COLLEGE
ANNA UNIVERSITY
MADRAS-600 025, INDIA

PROF. V. LAXMINARAYANA — MEMBER
DEPARTMENT OF CIVIL ENGINEERING
INDIAN INSTITUTE OF TECHNOLOGY
KANPUR- 208 016, INDIA

MR. S. GOWRISANKARAN DEPUTY GENERAL MANAGER NEYVELI LIGNITE CORPORATION NEYVELI-607 803, INDIA	MEMBER
DR. DEEPAK KASHYAP READER, INSTITUTE OF HYDROLOGY UNIVERSITY OF ROORKEE ROORKEE-247 667, INDIA	MEMBER
DR. SHAKEEL AHMED SCIENTIST, N.G.R.I. HYDERABAD-500 007, INDIA	MEMBER
MR. V.V.S. GURUNADHARAO SCIENTIST, N.G.R.I. HYDERABAD-500007	JOINT SECRETARY
DR. C.P. GUPTA SCIENTIST, N.G.R.I. HYDERABAD-500007, INDIA	SECRETARY

CONTENTS

VOLUME I

S-1A : GEOHYDROLOGIC PARAMETER ESTIMATION : RECENT ADVANCES IN CONVENTIONAL TECHNIQUES; USE OF TRACER TECHNIQUES

VOLUME II

S-4: MANAGEMENT OF GROUNDWATER RESOURCES PARTICULARLY IN HARD ROCKS: PARAMETERIZATION OF GROUNDWATER FLOW REGIME; MODELING OF COMPLEX AQUIFERS; SIMULATION OF FLOW IN FRACTURED ROCKS; ASSESSMENT OF DYNAMIC POTENTIAL OF AQUIFERS

SESSION S-4

MANAGEMENT OF GROUNDWATER RESOURCES PARTICULARLY IN HARD ROCKS : PARAMETERIZATION OF GROUNDWATER FLOW REGIME ; MODELING OF COMPLEX AQUIFERS ; SIMULATION OF FLOW IN FRACTURED ROCKS ; ASSESSMENT OF DYNAMIC POTENTIAL OF AQUIFERS

AN OVERVIEW OF FLOW SIMULATION MODELS IN FRACTURED AQUIFERS

K.Sridharan, M.S.Mohan Kumar and M.Sekhar
Indian Institute of Science, Bangalore 560 012

ABSTRACT

An overview of ground water flow simulation models in fractured rock aquifers is presented under the following four classifications - fracture based models, equivalent porous medium models, double porosity models and leaky aquifer models. Fracture based models use the concept that the entire flow occurs through the network of fractures and are based on the analogy with flow between parallel plates. Double porosity models are considered in detail with particular reference to the alternate approaches used in treating the flow transfer term between the blocks and fractures. The three alternatives available are the quasi-steady, unsteady parallel fracture and unsteady spherical block fracture models. The aquifer-water table aquitard system where the aquitard corresponds to the weathered zone and the aquifer to the fractured zone has considerable conceptual similarity with the double porosity model. As an extension of this approach, a general two domain model is proposed, allowing lateral and vertical flow in the upper domain, lateral flow in the lower domain, flow transfer between the domains, anisotropy, well storage and partial penetration of the well.

INTRODUCTION

The study of ground water occurance and movement in hard rock regions has started receiving considerable attension in recent years. These studies are of particular importance to India as nearly 2/3rd of the land area is covered by hard rocks. Mathematical modelling has become an important tool in the assessment and development of ground water resources, due to the tremendous progress made in analytical modelling in the 1950's and 1960's and numerical modeling since the late 1960's to the present time. While these models have been successfully applied for a number of field situations, there has been a constant and inconclusive debate on the applicability of porous media flow concepts for hard rock aquifers. The obvious differences between hard rock formations and the simple grains-pores picture, and the presence of fractures which may act as planes of mechanical discontinuity have raised basic questions on the validity of continuum approach based on concepts of flow through porous media.

This paper presents an overview of ground water flow simulation models in fractured rock aquifers, under the following four classifications: 1. Fracture based models, 2. Equivalent porous medium models, 3. Double porosity models and 4. Leaky aquifer models. While flow through fractured systems is of importance in geotechnical engineering, petroleum engineering and ground water hydrology, the stress in the present paper is with respect to ground water applications. A general two domain model is proposed as a possible representation which can be used both for dug wells and bore wells and can be also adapted conveniently for regional studies.

FRACTURE BASED MODELS

Fracture based models are based on a concept that the entire flow occurs through the network of fractures and the blocks surrounding the fractures may be treated as impervious. Studies based on this concept focus attention on individual fracture properties and fracture geometry and their relation to the flow conductivity and storage properties. Many of these studies consider the effect of single fracture of specified extent and orientation. Network models which attempt to duplicate the actual geometry of the fracture system by specifying individual fracture locations, orientations and aperture variations do not have much relevance in ground water hydrology. However, such models may have to be considered in sensitive studies involving underground disposal of nuclear waste.

The ability of a fracture to transmit water depends primarily on the size of the opening or fracture aperture. Many investigations have used flow between parallel plates as a model for flow in fractures (Neuzil and Tracey, 1981). Based on this analogy, the flow through a fracture, Q_f is obtained as,

$$Q_f = \frac{\gamma \ell b^3 J}{12 \mu} \qquad (1)$$

where b is the fracture aperture, ℓ is the length of the fracture normal to the flow, J is the gradient of piezometric head, γ is the specific weight of the fluid and μ is the dynamic viscosity. Eqn.(1) is commonly referred to as the 'cubic law' as the flux is proportional to the cube of the aperture size. Based on eqn.(1), the fracture hydraulic conductivity, K_f is obtained as,

$$K_f = \frac{b^2 \gamma}{12 \mu} \qquad (2)$$

Eqn.(2) may be used to obtain an idea of hydraulic conductivity of a fracture. For fractures of width 4 microns, 0.1mm, and 1 mm, the value of K_f of the individual fractures works out to 1.3, 10^3 and 10^5 m/day respectively. Thus it might be seen that the hydraulic conductivity of fractures may vary over a wide range.

It is necessary to distinguish between the hydraulic conductivity of the individual fracture, K_f and the conductivity of the fractured rock as a whole, K, which is the average value over the bulk volume, including the fractures and blocks. If the flow is one dimensional, the equivalent hydraulic conductivity of the fractured rock is given by

$$K = \frac{b}{b+L} K_f \qquad (3)$$

where L is the average linear dimension of the impervious block surrounding the fracture of width b. If L/b is of the order of 1000, the hydraulic conductivity of the rock as a whole is only 1/1000 th of the fracture conductivity.

The porosity of the rock containing fractures surrounded by impervious blocks, is given by,

$$n = \frac{(L+b)^3 - L^3}{(L+b)^3} \qquad \frac{3b}{L} \qquad (4)$$

If L/b is of the order of 1000, n is about 0.3%. Hence, if the blocks are impervious, the storage capacity of the fractured rock is very small and will be without practical interest in ground water exploration, if there is not a continuous source of supply. However, this might not be true for oil exploration.

Fractures in rocks have a variable aperture which requires that the cubic law (eqn.1) has to be modified to account for these variations. Neuzil and Tracy (1981) proposed an idealised fracture model based on a fracture opening which varies continuously normal to the flow, but with constant aperture in the flow direction. By using this model, they obtained the following equation for flow through the fracture.

$$Q_f = \frac{\gamma}{12\mu} \ell J \int_0^\infty b^3 f(b)\, db \qquad (5)$$

where ℓ is the total length of opening normal to the flow, b is the varying aperture size and f(b) is the normalised aperture frequency distribution. Based on this model, the average hydraulic conductivity is given by,

$$\overline{K}_f = \frac{\gamma}{12} \frac{\int_0^\infty b^3 f(b)\, db}{\int_0^\infty b\, f(b)\, db} \qquad (6)$$

The exact form of aperture frequency distribution, f(b) is generally unknown. It depends on fracture orientation relative to the rock fabric, the mode of fracturing, the stress history of the fracture and the size scale being considered. Also, the aperture frequency distribution may change spatially as a result of spatial changes in stress. Snow (1970) notes that many arguments point to the suitability of the log normal distribution for describing the frequency of fracture apertures in rocks. As the log normal distribution has two parameters, the flow rate through the fracture is determined by two parameters, rather than a single parameter representing aperture size.

Wilson and Witherspoon (1970) have given the following expression for the effective aperture size for a fracture, wherein the aperture varies continuously along the length of the fracture. This effective aperture size is given by,

$$b_{eff} = [\int_0^{\ell_0} d\ell \; (\int_0^{\ell_0} d\ell \; / \; b(\ell)^3 \;)^{-1} \;]^{1/3} \qquad (7)$$

where b is expressed as a function of length ℓ along the fracture and ℓ_o is the total length. Sagar and Runchal (1982) have derived equations for equivalent permeability of a fracture system that consists of nonextensive fractures having arbitrary orientation with respect to the rock block considered. By specifying the field data in terms of mean and variance and by using second order statistical analysis, the mean and variance of equivalent permeability are calculated.

Witherspoon et al (1980) have experimentally studied the validity of the cubic law for flow in a fracture where the surfaces are in contact and the aperture is being decreased under stress. The cubic law was found to hold good whether the fractures were open or closed and the results are not dependent on rock type. The cubic law seems to hold good regardless of the loading path and how often the loading process is repeated. The effects of deviations from the ideal parallel plate concept only cause an apparent reduction in flow and are taken care of by the factor f, defined by,

$$Q_f = \frac{\gamma}{12\mu} \; \ell \; b^3 \; J \; f \qquad (8)$$

where $f > 1.0$.

Models of fractured media, consisting of vertical fractures, have been developed and applied to analyse a fractured reservoir performance by several investigators. Gringarten (1982) has presented solutions for drawdown for a well intersecting a single plane vertical fracture (Fig.1a) and a well located along the axis of a plane horizontal fracture (Fig.1b). The solutions consider anisotropy, with the conductivity K_f representing the increased conductivity due to the presence of the fracture and the conductivity K_m representing the matrix or block conductivity. In the vertical fracture model, far enough from the pumping well, the behaviour of the system becomes identical to that of the line source well in an anisotropic porous medium. In the horizontal fracture model, beyond a sufficient distance from the pumping well, the drawdown is the same as that due to a line source well in a homogeneous reservoir with isotropic radial conductivity, K_f. Generally, these studies on the hydraulics of a single fracture are more relevant to application in oil wells.

Long and Billaux (1987) have developed a technique for processing field data into a fracture network model which

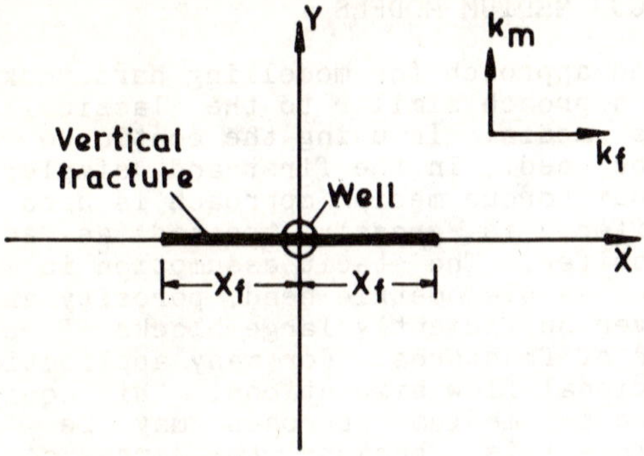

Fig.1a. Vertical fracture

accounts for observed spatial variability. In order to generate fractures in a hetrogeneous region, the region is devided into homogeneous subregions. In each subregion and for each fracture set, the fracture density (number of fractures per unit area), orientation, length and aperture distributions are specified. The properties of each subregion are predicted through geostatistics based on observed field data. The purpose of geostatistics is to estimate the value of a given regionalised variable, such as the average fracture length or fracture density, at all points in a specified field, when the value of that variable has been measured at a finite number of points in the field. The two dimensional network model is applied based on data from a uranium mine in France, which is used for nuclear waste storage research. In general, with regard to application to ground water, the scope of models considering fracture flow alone, appears to be of restricted nature.

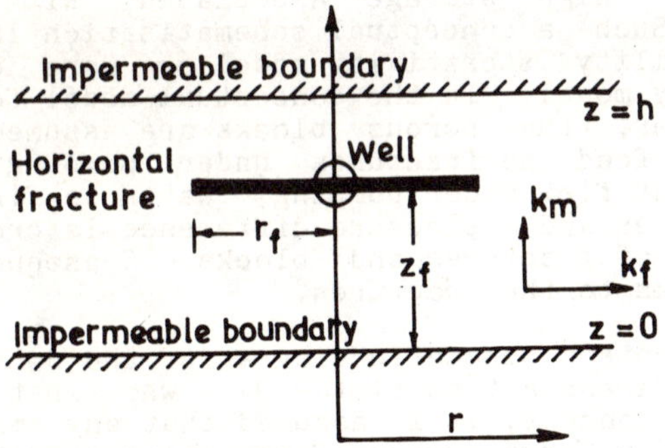

Fig.1b. Horizontal fracture

EQUIVALENT POROUS MEDIUM MODELS

The second approach for modelling hard rock aquifers is the continuum approach similar to the classical study of flow through porous media. In using the continuum approach, two methods may be used. In the first and simpler method, the single continuum porous medium approach is used in which the fracture aquifer is grossly treated as an equivalent, anisotropic aquifer. The tacit assumption is that physical quantities, such as piezometric head, porosity and permeability are averaged over sufficiently large blocks of rock containing a large number of fractures. For many applications involving large scale regional flow simulations, this equivalent, single continuum porous medium approach may be justified as reasonable. This is because the area of interest is sufficiently large to smooth out hetrogeneities resulting from the presence of fractures. An unconfined, two dimensional, anisotropic model can be quite successfully used for regional ground water studies in hard rock areas. The direction of major fracture orientation may be used as the maximum transmissivity direction, while the direction normal to it, may be used as the minimum transmissivity direction. Another application in which the equivalent porous medium approach may be useful is in the study of dug well hydraulics, where the large diameter dug well taps the shallow, weathered zone, which is seperated from the deeper fractures by massive rock.

DOUBLE POROSITY MODELS

In the second method based on the continuum approach, the fractured rock mass is assumed to consist of two interacting, overlapping continua, a continuum of low permeability, primary porosity blocks and a continuum of high permeability, secondary porosity fissures. The secondary permeability associated with the fissures is usually high, whereas the secondary porosity is often very low. On the other hand, the blocks are characterised by relatively high storage associated with very low permeability. Such a conceptual schematisation is referred as double permeability storativity model or more commonly, as double porosity model. In the conceptual model for a double porosity aquifer, the porous blocks are assumed to act as sources which feed the fractures under pumping conditions. During transient flow under pumping, water is first released from the fractures and a pressure difference is created between the two media of fractures and blocks. Consequently, water moves from blocks to the fractures.

Fig.2 presents a schematic diagram of the double porosity model. As per Barenblatt et al (1960), who first proposed the double porosity concept, it is assumed that any small volume of rock consists of a large number of randomly distributed, sized and oriented fractures. With this approach, the drawdown s_1 in the fracture at a distance r from the pumping well at a time t, and the drawdown s_2 in the porous block are regarded as two values of drawdown at the location r in the double porosity medium. Thus, at each point, we have to consider two representative elementary volumes, one for the fracture system and one for the block system (Fig.2). The mathematical representation of the system requires two mass conservation equations, one for the fracture regime and another for the porous block regime. These to equations are coupled by a fluid transfer term which depends on the pressure difference between the block regime and fracture regime at the location.

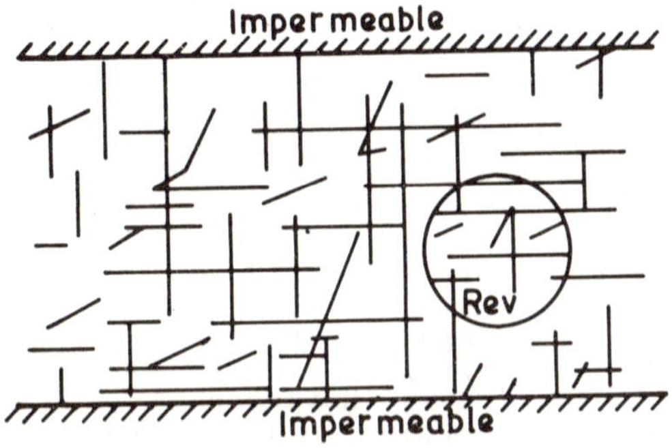

Fig.2. Confined double porosity aquifer with representative elementary volume (Rev)

If the fluid transfer term is taken to be dirctly proportional to the potential difference between the blocks and fracture domains, the following equations are obtained, with subscript 1 referring to the fracture domain and subscript 2 referring to the porous block domain.

$$K_1 \nabla^2 s_1 = S_{s1} \frac{\partial s_1}{\partial t} + \alpha (s_1 - s_2) \tag{9}$$

$$K_2 \nabla^2 s_2 = S_{s2} \frac{\partial s_2}{\partial t} - \alpha (s_1 - s_2) \tag{10}$$

where K is the hydraulic conductivity, S_s is the specific storage, α is a dimensional constant proportional to the conductivity of the block and the specific surface of the fracture - block contact surface, s is the drawdown, and ∇^2 is the Laplacian operator in (x,y) or polar coordinates. Barenblatt et al (1960) simplified eqns.(9) and (10) by neglecting the storativity of the fracture domain (1st term on right side of eqn.9) and the flow in the block domain (left side of eqn.10).

Warren and Root (1963) and Kazemi et al (1969) solved the single well pumping problem, considering the storativity effect of the fracture domain, but again neglecting the flow in the block domain. These solutions show that after a sufficiently large duration of pumping, the heterogeneous fractured aquifer behaves as a homogeneous one with equivalent transmissivity $T = T_1$ (that is, fracture transmissivity) and $S = S_1 + S_2$, where S is the storage coefficient. However, during the early stages of pumping, two additional parameters λ and ω influence the drawdown pattern and these are defined by,

$$\lambda = \alpha' \frac{T_2}{T_1} \tag{11}$$

$$\omega = \frac{S_1}{S_1 + S_2} \tag{12}$$

where α' which is directly proportional to the dimensional constant α is a characteristic coefficient of the fractured rock.

Double porosity models differ in the treatment of the flow transfer term from the blocks to the fractures. The three alternative models proposed in the literature are referred as quasi-steady, unsteady parallel fracture and unsteady spherical block fracture flow models (Huyakorn et al 1983). For an isotropic aquifer, the vertically averaged equation of flow in the fracture domain may be written as,

$$\frac{\partial}{\partial x}\left(T_1 \frac{\partial h_1}{\partial x} \right) + \frac{\partial}{\partial y}\left(T_1 \frac{\partial h_1}{\partial y} \right) = S_1 \frac{\partial h_1}{\partial t} - \Gamma + q \tag{13}$$

Where h_1 is the piezometric head in the fracture, T_1 and S_1 are the transmissivity and storage coefficient of the fracture domain, q is the net withdrawal rate per unit area and Γ is the flow rate from the blocks to the fractures per unit area.

In the quasi - steady state model, the leakage term from the blocks to the fractures is taken as per the assumptions of Barenblatt et al (1960), wherein the leakage is taken to be linearly dependent on the head difference between the blocks and fracture. The leakage term for this model can be shown to be,

$$\Gamma = - C_L \int_0^t \frac{\partial h_1}{\partial \tau} e^{-C_L (t - \tau)/S_2} d\tau \tag{14}$$

where C_L, the leakage parameter, depends on the block conductivity and the specific surface of the fracture - block surface contact.

The unsteady parallel fracture flow model is based on the parallel fracture representation of the system. In this model, proposed by Kazemi et al (1969), the rock is idealised as alternating layers of blocks and fractures as shown in Fig.3. The model considers vertical flow in the block by solving the one dimensional flow equation for the block. The resulting solution for the leakage term is given by,

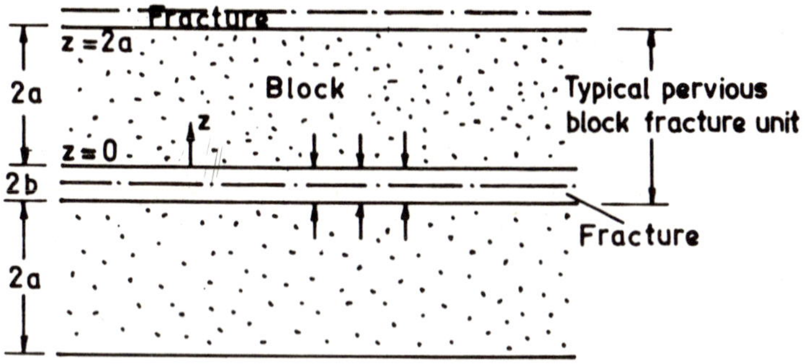

Fig.3. Double porosity parallel fracture model

$$\Gamma = - \frac{2 H K_2}{a(a+b)} \left[\sum_{n=0}^{\infty} \int_0^t \frac{\partial h_1}{\partial \tau} e^{-\alpha_n (t-\tau)} d\tau \right] \qquad (15)$$

where

$$\alpha_n = \frac{\pi^2 (2n+1)^2 K_2}{4 S_{s2} a^2} \qquad (16)$$

H is the aquifer thickness, a is half thickness of the block and b is half of the fracture aperture (Fig.3).

In the unsteady spherical block fracture flow model, the porous blocks are idealised as a series of equal sized spheres (Fig.4), the number of these spheres being such that the secondary fracture porosity is the same for the actual fractured medium. The leakage term for this model is derived in a similar manner to that for the unsteady parallel fracture flow model and is given by,

$$\Gamma = - \frac{6 H K_2}{a(a+b)} \left[\sum_{n=1}^{\infty} \int_0^t \frac{\partial h_1}{\partial \tau} e^{-\alpha_n (t-\tau)} d\tau \right] \qquad (17)$$

where a is now the radius of the spherical block, and

$$\alpha_n = \frac{\pi^2 n^2 K_2}{S_{s2} a^2} \qquad (18)$$

It is seen from eqns.(14), (15) and (17), that for all the three models, the flow transfer term between the block and the fracture, contains a convolution integral, similar to the case of "delayed yield" in unconfined aquifer. Boulton and Streltsova (1977,1978) have used a concept similar to that of

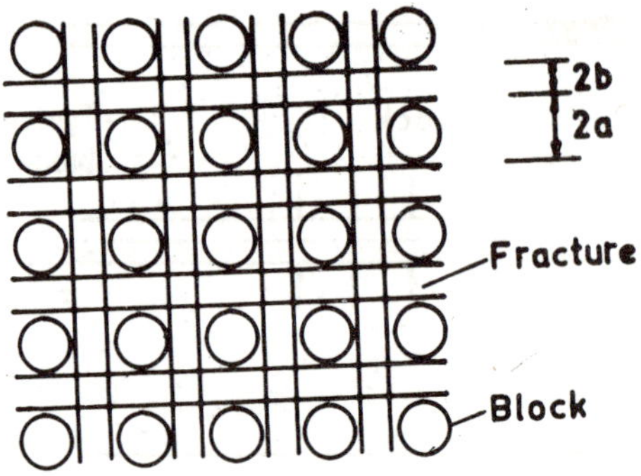

Fig.4. Double porosity spherical block fracture model

delayed yield to represent the flow transfer term from the blocks to the fractures. The effect of the difference in modelling the block to fracture flow transfer term by quasi - steady state or unsteady state models is shown schematically in Fig.5. In the absence of well bore storage effects, at early times, water is primarily derived from the fractures and the drawdown follows the familiar Theis type curve. Due to transfer of water from the blocks to the fractures, the drawdown curve deviates from the Theis curve. In the unsteady state model, there is a gradual transition resulting in early deviation from the Theîs curve (Fig.5).

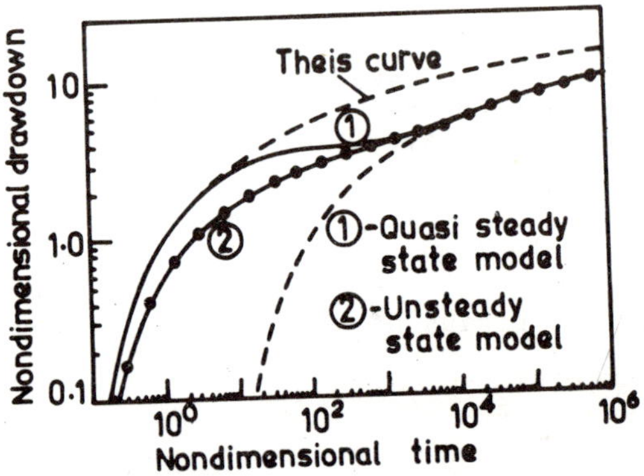

Fig.5. Schematic diagram for type curves for double porosity system

Moench (1984) has attempted to unify the prevailing theories of flow to a well in a double porosity reservoir by accounting for fracture skin. Fracture skin, which is a thin skin of low permeability material deposited on the surface of the blocks, which impedes the free exchange of fluid between the blocks and the fractures. Depending on the relative value of the hydraulic conductivity of the fracture skin, the quasi-steady or unsteady state model becomes applicable. Studies are reported in literature to consider the effects of partial penetrating well and well storage in a double porosity medium.

Narasimhan (1982) has proposed a unified approach within a single conceptual framework, for modelling flow in a variety of fractured rock systems. The framework is the integral form of the mass conservation equations, which are directly replaced by discretised equations based on an integral finite difference method, termed IFDM. Using the method, the three commonly used conceptualisations, namely, equivalent porous medium approach, discrete fracture approach and the double porosity approach, are analysed.

LEAKY AQUIFER MODELS

There is considerable conceptual similarity between the double porosity model and leaky aquifer model. The aquifer-aquitard system conceptually corresponds to the fracture-block system. There is a widespread feeling among hydrogeologists in India that ground water occurrence in hard rock areas is characterised by the presence of a water table in the shallow zone. It is also acknowledged that the fractures occur a little deeper, below the weathered rock zone. Besides, in hard rock areas, the fractures close below 60 to 100 m depth, yielding little water further deeper. These observations can be simulated by a leaky aquifer model with a water table aquitard, wherein the water table level varies, influenced by recharge from rainfall or other sources and leakage to the fracture zone. The upper weathered zone can be treated as a water table aquitard and the lower fractured zone as the aquifer. Such a model, which may be well suited for ground water simulation, may not be applicable for oil reservoirs, thus lacking general applicability as in the case of a double porosity model.

Cooley and Case (1973) have shown that Boulton's convolution integral represents the velocity at the bottom of an incompressible unconfined aquitard overlying a pumped aquifer. Thus the Boulton's convolution integral can be used as a source term in the aquifer differential equation to represent the influence of water table aquitard on the aquifer. It was already noted that the leakage or flow transfer term between the block and the fractures is represented by a convolution integral in the double porosity model. Even the concepts of quasi-steady and unsteady state block to fracture flow have evolved from leaky aquifer concepts.. There are leaky aquifer models which correspond to quasi-steady and unsteady parallel fracture models.

The leaky aquifer model has some clear advantages. It can effectively consider the presence of a varying water table. As the leaky aquifer concept is classical in ground water hydrology, the physical meaning of the parameters associated with the model is well understood and hence it may be possible to make a reasonable estimate of the parameter values even with limited data. The third advantage is the ease with which the same model can be applied to nonhomogeneous situations involving regional applications. Sridharan et al (1986) have applied the leaky aquifer model with a varying water table, for regional ground water simulation in the Vedavati river basin in Karnataka and Andhra Pradesh. A schematisation of the model is shown in Fig.6.

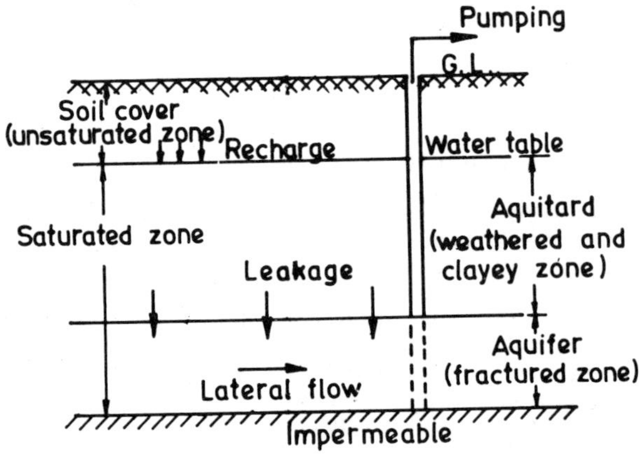

Fig.6. Schematic diagram of aquifer-water table aquitard model

CONCLUSIONS

An overview of ground water flow simulation models in fractured rock aquifers is presented under four classifications. The fracture based models require detailed field data which can only be obtained on a local scale. Hence these models may be more suited to oil well applications and for study of hazardous contaminant transport, such as nuclear waste. The double porosity models may be more relevant for ground water applications, particularly with respesct to well hydraulics. The leaky aquifer model with a aquifer-water table aquitard system has conceptual similarity with double porosity model. A general two domain model, allowing lateral and vertical flow in the water table domain, flow transfer between the domains, and which considers the effects of anisotropy, well storage and partial penetration may represent ground water flow in a fractured rock system. Such a model has the advantage that it can be applied for dug wells or bore wells and with some simplifications, can be used for regional applications. Numerical solutions have been recently obtained at the Indian Institute of Science for models incorporating some additional simplifications to this general two layer system.

REFERENCES

Barenblatt, G.E., Zheltov, I.P., and Kochina, I.N., 1960. Basic concepts in the theory of seepage of homogeneous liquids in fissured rocks. Journal of Applied Mathematical Methods (USSR), v.24, p.1286-1303.

Boulton, N.S., and Streltsova, T.D., 1977. Unsteady flow to a pumped well in fissured water bearing formation. Journal of Hydrology, v.35, p.257-269.

Boulton, N.S., and Streltsova, T.D., 1978. Unsteady flow to a pumped well in an unconfined fissured aquifer. Journal of Hydrology, v.37, p.349-363.

Cooley, R.M., and Case, C.M., 1973. Effects of water table aquitard on drawdown in an underlying pumped aquifer. Water Resources Research, v.9, p.434-447.

Gringarten, A.C., 1982. Flow-test evaluation of fractured reservoirs. In: Narasimhan, T.N., Recent trends in hydrogeology, special paper, Geological Society of America, p.237-263.

Huyakorn, P.S., Lester, H., and Faust, C.R., 1983. Finite element techniques for modelling ground water flow in fractured aquifers. Water Resources Research, v.19, p.1019-1035.

Kazemi, H., Leith, M.S., Thomas, G.W., 1969. The interpretation of interference tests in naturally fractured reservoirs with uniform fracture distribution. Society of Petroleum Engineers Journal, v.12, p.463-472.

Long, J.C.S., and Billaux, D.M., 1987. From field data to fracture Network modeling: An example incorporating spatial structure. Water Resources Research, v.23, p.1201-1216.

Moench, A.F., 1984. Double-porosity models for a fissured ground water reservoir with fracture skin. Water Resources Research, v.20, p.831-846.

Narasimhan, T.N., 1982. Multidimensional numerical simulation of fluid flow in fractured porous media. Water Resources Research, v.18, p.1235-1247.

Neuzil, C.E., and Tracy, J.V., 1981. Flow through fractures. Water Resources Research, v.17, p. 191-199.

Sagar, B., and Runchal, A., 1982. Permeability of fractured rock: Effect of fracture size and data uncertainties. Water Resources Research, v.18, p.266-274.

Snow, D.T., 1970. The frequency and aperture of fractures in rock. International Journal of Rock Mechanics and Mineral Science, v.7, p.23-40.

Sridharan, K., Lakshmana Rao, N.S., Mohan Kumar, M.S., and Ramesam, V., 1986. Computer model for vedavati ground water basin. Part 2. Regional model. Sadhana, v.9, p.43-55.

Warren, J.E., and Root, P.J., 1963. The behaviour of naturally fractured reservoirs. Society of Petroleum Engineers Journal, v.9, p.245-255.

Wilson, C.R., and Witherspoon, P.A., 1970. An investigation of laminar flow in fractured porous rocks. Publ.70-6, Department of Civil Engineering, University of California, Berkeley, p.1-178.

Witherspoon, P.A., Wang, W.S.Y., Iwai, K., and Gale, E., 1980. Validity of cubic law for fractured flow in a deformable, rock fracture. Water Resources Research, v.16, p.1016-1024.

NONDEFEATIST DERIVATION OF FINITE DIFFERENCE SCHEMES

P.S. MOHARIR

Department of Earth Sciences
University of Roorkee, Roorkee, India-247 667

Abstract: Groundwater modelling requires solving a partial differential equation subject to initial and boundary conditions. In finite difference modelling, first the derivatives are replaced by suitable finite difference approximations, and then the resultant numerical scheme is tested for stability. It may look natural to do one's best in the first stage to improve the approximation. However, it is a surprising, unfortunate but a well-established fact that a more accurate finite difference approximation for the derivatives does not automatically imply a stabler finite difference scheme. Therefore, the two-stage procedure above must be regarded as difeatist, as under it one does his best in the first stage and when the verdict at the second stage is unfavourable, simply abandons the scheme and tries another. It is recommended here that the twin objectives of stability of the finite difference scheme and the accurate approximation of the net differential operator must be dealt with simultaneously.

INTRODUCTION

Flow of water through porous media can be represented by a partial differential equation. Therefore, groundwater modelling for prognostics consists of solving that partial differential equation subject to initial and boundary conditions. Of course, abstraction and /or recharge are also to be represented by a suitable source term. But elucidation of the concepts intended to be introduced does not require dealing with the whole situation. Therefore, they would be explained in the context of the partial differential equation

$$\partial^2 h / \partial x^2 = \alpha \, \partial h / \partial t \tag{1}$$

which may be taken to constitute the 'toy domain' (Hofstadter 1986) for the groundwater simulation.

In deriving the finite difference scheme, the conventional procedure is to replace the derivatives by their finite difference approximations of desired accuracy. For example, using three-point central approximation for the second derivative and the two-point forward approximation for the first derivative in Eq. (1) gives

$$[h(m-1,p) - 2h(m,p) + h(m+1,p)]/a^2$$
$$= \alpha \, [h(m,p+1) - h(m,p)]/b \tag{2}$$

or

$$h(m,p+1) = \beta\ [h(m-1,p) + h(m+1,p)] + (1-2\beta)\ h(m,p) \tag{3}$$

as the corresponding explicit finite difference scheme. Here h(m,p)stands for h (x = ma,t = pb), a and b are uniform discretization intervals along x and t axes respectively, m and p are integers and

$$\beta = b/\alpha\ a^2 \tag{4}$$

Alternatively, if two-point backward finite difference approximation is used for the first derivative with respect to time, the corresponding implicit finite difference scheme is

$$h(m,p-1) = -\beta\ [h(m-1,p) + h(m+1,p)] + (2\beta-1)\ h(m,p) \tag{5}$$

Derivation of such finite difference schemes is only the first stage. The second stage is to derive the stability properties of these schemes. The explicit scheme of Eq. (3) is known to be conditionally stable, whereas the implicit scheme of Eq.(5) is unconditionally stable. It is well-known that a central finite difference approximation for the first derivative is more accurate than either the forward or the backward approximations used above. It is also known that the finite difference scheme that results by its use is unconditionally unstable (Noye 1978). This result is anti-intuitionistic, because attempt to do better in the first stage results into worse results at the second stage. Such an experience is not very uncommon to the observant, and has led to formulation of the principle of suboptimization (Hitch 1953, Machol and Miles Jr. 1973) according to which subsystems may have to perform suboptimally so that the system can have an optimal performance or alternatively, optimality in a subsystem may inflict not merely nonoptimality but even malignance in the system as a whole (Moharir 1987). This implies that the total task of devising a satisfactory finite difference scheme should not be serially organized, so that good approximations to derivatives are chosen first and the stability analysis conducted later to pass an a posteriori judgment. The next section offers an alternative approach.

NONDEFEATIST APPROACH

Having agreed that the stability of the scheme and satisfactory approximation of the derivative operator must be dealt with simultaneously, a procedure which will allow this has to be developed. Such a procedure is illustrated with two examples in this section.

An Explicit Scheme

Consider a finite difference scheme

$$h(m,p+1) = \mu\,[h(m-1,p) + h(m+1,p)] + \nu\, h(m,p) \tag{6}$$

to solve Eq.(1) where μ and ν are to be selected to satisfy the twin objectives above. Applying von Neumann's method of stability analysis (Noye 1978), the condition of stability of the scheme of Eq.(6) can be shown to be

$$-1 \le 2\mu \cos\theta + \nu \le 1 \text{ for all } \theta \tag{7}$$

Consider $\mu - \nu$ plane with μ as an abscissa and ν as an ordinate. Use the elementary fact that $-1 \le \cos\theta \le 1$. It then follows that stability is guaranteed only in the region below straight lines passing through (0,1) and having slopes ±2, and above the straight lines passing through (0,-1) and having slopes ±2. That is, the region in the $\mu - \nu$ plane which guarantees stability is a diamond ABCD where the points A,B,C,D are (0,1),(1/2,0),(0,-1) and (- 1/2,0) respectively.

Next ,using Taylor's series, is follows that Eq.(6) simulates

$$b\,\partial h/\partial t + (b^2/2)\,\partial^2 h/\partial t^2 = (2\mu+\nu-1)h + \mu a^2\,\partial^2 h/\partial x^2 + (\mu a^4/12)\,\partial^4 h/\partial x^4 + \dots \tag{8}$$

i.e.

$$\partial^2 h/\partial x^2 = -(2\mu+\nu-1)\,h/\mu a^2 + (b/\mu a^2)\,\partial h/\partial t + (b^2/2\mu a^2 - \alpha^2 a^2/12)\,\partial^2 h/\partial t^2 + \dots \tag{9}$$

where Eq.(1) has been used to convert the fourth spatial derivative into the second temporal derivative. If Eq. (6) is to simulate Eq.(1), we need

$$2\mu + \nu - 1 = 0 \tag{10}$$

$$\mu = b / a^2 \alpha \tag{11}$$

and

$$\mu = 6\, b^2 / a^4 \alpha^2 \tag{12}$$

which may be called conditions for unbiasedness, proper diffusivity and good convergence, respectively. Condition for unbiasedness is satisfied on the entire straight-line segment AB of the stability region. From the condition of proper diffusivity seen in conjunction with the stable region in the $\mu - \nu$ plane, it follows that the scheme of Eq. (6) can only be conditionally stable for

$$\mu = b/a^2 \alpha \le 1/2 \tag{13}$$

This ia a well-known (Noye 1978) result. Using the conditions for proper diffusivity and good convergence together, it follows that we should choose

$$\mu = b/a^2 \alpha = 1/6 \tag{14}$$

The condition for unbiasedness then yields

$$\nu = 2/3 \tag{15}$$

Therefore, the best point to operate the scheme of Eq. (6) in the $\mu - \nu$ plane is (1/6,2/3) which is on the boundary of the stability diamond.

An Implicit Scheme

Consider a finite difference scheme

$$h(m,p-1) = \mu[h(m-1,p) + h(m+1,p)] + \nu h(m,p) \tag{16}$$

to solve Eq. (1). The stability condition is

$$-1 \leq (1-\nu)/2\mu \cos\theta \leq 1 \quad \text{for all } \theta \tag{17}$$

With little mundane exercize, keeping in mand the sign of $\mu \cos\theta$, it can be seen that the stability conditions are satisfied only if in the half plane $\mu \geq 0$ the points are on the right of straight lines with slopes ±2 passing through (0,1) or in the half plane $\mu \leq 0$, they are on the left of straight lines with slopes ±2 passing through (0,1).

Next, Eq. (16) simulates

$$-b\partial h/\partial t + (b^2/2)\partial^2 h/\partial t^2 = (2\mu + \nu - 1)h$$
$$+ \mu a^2 \partial^2 h/\partial x^2 + (\mu a^4/12)\partial^4 h/\partial x^4 + \ldots \tag{18}$$

i.e.

$$\partial^2 h/\partial x^2 = -(2\mu + \nu - 1)h/\mu a^2 - (b/\mu a^2)\partial h/\partial t$$
$$+ (b^2/2\mu a^2 - a^2\alpha^2/12)\partial^2 h/\partial t^2 \tag{19}$$

If Eq. (16) is to simulate Eq. (1), the conditions for unbiasedness, proper diffusivity and good convergence, respectively are

$$2\mu + \nu - 1 = 0 \tag{20}$$

$$\mu = -b/a^2 \alpha \tag{21}$$

and

$$\mu = 6b^2/a^4 \alpha^2 \tag{22}$$

In this case, the conditions for proper diffusivity and good convergence cannot be simultaneously met, because they require μ to be negative and positive, respectively. Thus, the condition for good convergence should be dropped. The condition for proper diffusivity is met by any negative μ and the condition for unbiasedness is met on straight line with slope of -2, passing through (0,1) in the μ - ν plane. Let A be a point (0,1). Then on the semi-infinite segment of the straight line through A with slope -2 in $\mu \leq 0$ half plane the conditions of stability, unbiasedness and proper diffusivity are all satisfied. This is also a known result regarding the unconditional stability of the implicit scheme. Thus we choose

$$\mu = -b/a^2 \alpha \quad \text{and} \quad \nu = 1 + 2b/a^2\alpha \qquad (23)$$

However, the cost of unconditional stability is the non-satisfaction of the condition for good convergence, which also was satisfiable for the explicit scheme.

CONCLUSION

The method suggested above for the pedagogically preferable derivation of the finite difference schemes has other advantages. Conditions of stability i.e. (7) and (17) imply that no sinusoidal error pattern should be amplified when the solution advances from one time instance to the next. But they allow some error patterns to persist without attenuation. The conditions can be strengthened by replacing the limits ±1 by ±f, where f is a positive fraction. Then certain compromises would have to be tolerated in the conditions for unbiasedness, proper diffusivity and good convergence. The method explicitly brings out the nature of these compromises. Secondly, the method is very general. For example, to determine partial differential equations (9) and (19) simulated by the finite difference schemes of Eqs. (6) and (16) respectively, the classical Taylor's series expansion has been used. Superior alternatives to it exist and have been proposed (Moharir 1984, 1986, Moharir et al 1988). They could be used to derive the equations simulated and the nondefeatist procedure could still be followed. This procedure should be followed in view of its advantages brought out above.

REFERENCES

1. Hitch, C.J., 1953. Suboptimization in operations problems. J.Op. Res. Soc. Am. V. 1, pp. 87-89.

2. Hofstadter, D.R., 1986. Metamagical themas: questing for the essence of mind and pattern. Bantam Books. Toronto.

3. Machol, R.E., and Miles, R.F.Jr., 1973. The engineering of large scale systems, In Machol, R.E., and Miles R.F. Jr. (Eds), Systems concepts. John Wiley. New York. p.39.

4. Moharir, P.S., 1984. Finite difference ground water modelling. Internat. Workshop Rural Hydrogeol. Hydraul. Fissured Basement Zones. Roorkee. pp 201-207.

5. Moharir, P.S., 1986. Approximation - theoretic schemes for groundwater modelling. Reg. Workshop Groundwater Model. Roorkee. pp. 153 - 174.

6. Moharir, P.S., 1987. Relevance, success, limitations and dangers of mathematics. Math. Ed. V.3, pp. 25 - 32.

7. Moharir, P.S., Jain, M.K. and Khattri, K.N., 1988. Non-Taylor discretization. J.Inst. Electron. Telecomm. Engrs. V. 34, in press.

8. Noye, B.J., 1978, Numerical simulation of fluid motion. North Holland. Amsterdam. pp. 1 - 112.

HYDROGEOLOGY OF THE MIDDLE CHELIFF BASIN (ALGERIA) SIMULATION OF THE GROUNDWATER FLOWS

M. BOUZELBOUDJEN* et J. MANIA**

* Centre d'Hydrogeologie - Universite de Neuchatel (Suisse)
** Laboratoire de geologie structurale et appliquee - Universite de Franche-Comte, Besancon (France)

ABSTRACT

The hydrologic analysis of alluvial plain of the middle Chellif basin permits the successive definition of the boundaries of the alluvial reservoir and the associated sandstone formations, the aquifer geometry and the hydrologic budget. After a description of the aquifer stratas the hydrogeologic boundaries are discussed. It appears that the aquifer deposits are thick (100-180 m) on the right bank of the Cheliff river with the presence of sedimentary deposition gaps (Pliocene-Miocene) in the meridional part of basin. The utility of the mathematical model based on the finite element method, with steady state conditions, can justify the work hypothesis with an improvement of the knowledge of hydraulic caracteristics of the aquifer and the calculation of a complete hydrologic balance. It's shows that the effective precipitations and the surface waters of the Cheliff stream represent 40% and 22% of the inputs respectively. A second simulation based on the finite difference method, with unsteady conditions, leads to the evaluation of the values of specific yield and the determination of the impact of irrigation on the piezometric evolution of water table. The total water requirement (37 hm3) with potable water (16.6 hm3), with agricultural water (18.4 hm3) and with industrial water (2 hm3) in the year 2000 cannot be satisfied by pumping only from the aquifer.

GEOGRAPHICAL SITUATION

The Cheliff basin appears like an internal mountainous depression length of 500 km from East to West and width of 30 km to 80 km, in the north-west part of Algeria. The studied sector in the middle part of the plain (with an average elevation of 240 m) is about 50 km long between in the south the Doui (1039 m), the Ouarsenis (1985 m) and the Temoulga (530 m) mounts and in the North the Dahra and the Bou Maad mountains (Figure 1). Two narrowings of valleys limit the plain at West with the step of Ponteba and at East with the turning of Doui. Agriculture is the principal economic activity of the country which as 500,000 people. A statistical approch on the 1968-85 period shows total precipitations of 420 mm (maximum in February and minimum in July-August with 75 mm), an interannual average temperature of 18.5 C (minimum in January with 10.7 C and maximum in August with 28.5 C) and a potential evapotranspiration of 385 mm.

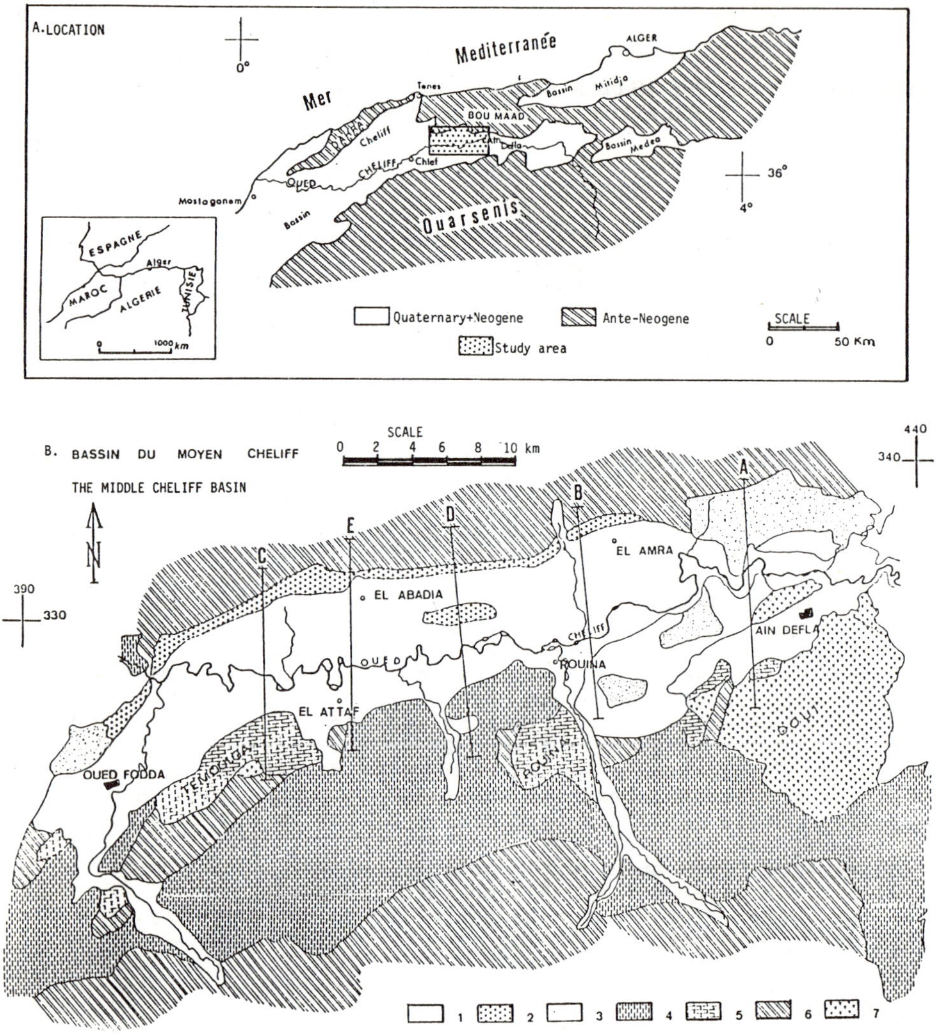

Figure 1: Location of the study area and regional geological map
Legend:-1:Quaternary(alluvial deposits),-2:Pliocene(sandstones),-3:Upper Miocene (sandstones with clays),4:Lower Miocene(clays),-5:Jurassic(limestones),-6:Cretaceous(marls),-7:Paleozoïc(schists)

GEOLOGICAL SITUATION

The Cheliff plain is located in a subsident unstable paleogeographical furrow (Glangeaud, 1955) bounded by two different folded and faulted domains where the tectonic is actually in activity like the example of the earthquake of El Asnam in 1980 (Meghraoui, 1982). The north boundary of the plain is affected by a tangential tectonic near the cretaceous of the Dahra mount with overturned folds towards the axial part of the plain (Lepvrier and al, 1970 - Lepvrier, 1971). The southern boundary (paleozoic of the Doui mount and jurassic of Rouina and Temoulga mounts) is disturbed by a polyphasic faulted tectonic (Mattauer, 1958 - Lepvrier, 197). The valley has been filled by about 3000 m of detritic sediments after the orogenic distention phases (Perrodon, 1957).

GEOMETRY OF THE AQUIFER RESERVOIR

The multi-stratas aquifer recognized by drilling is constituted by three aquifer layers:

- quaternary alluvial material (Figure 2a) composed by clays, pebbles and gravels with a big thickness down-stream (130 to 175 m) and little thickness up-stream (15 to 40 m). Lateral variations of facies has been found from up-stream to down-stream of the plain, this last sector being more detritic,

- pliocene sands and sandstones which present on the northern boundary of the plain (maximum thickness 80 m) disappear with chamfered edge fashion in the central part,

- miocene sandstones (20 to 80 m thick) with many clay intercalations transform in the down-stream of the plain to clays.

The substratum of the aquifer is made up of blue marls from the Helvetian about over 100 m depth with possible extension continuing on the northern central part of the basin. On the meridional boundary of the plain the reservoir formations lean against the primary and the secondary mountainous formations.

The right bank of the Cheliff wadi shows the maximum depth of the reservoir, the three effective aquier stratas (without impervious formations) being added up, between 150 to 180 m depth near El Abadia and El Amra (Figure 2b). Upward this last town the quaternary alluvial formation is lying on the Pliocene or on the Miocene. In the southern part on left bank, these formations can lie directly on the marly substratum of the Helvetian.

THE HYDROLOGIC OUTLINES OF THE CATCHMENT

The northern boundary of the plain (Figure 3) participates to the replenishment of the water table by the bias of little wadis draining the calcareous and stoneless outcrops and by the surface runoff on the reliefs of the Dahra and Bou Maad mounts.

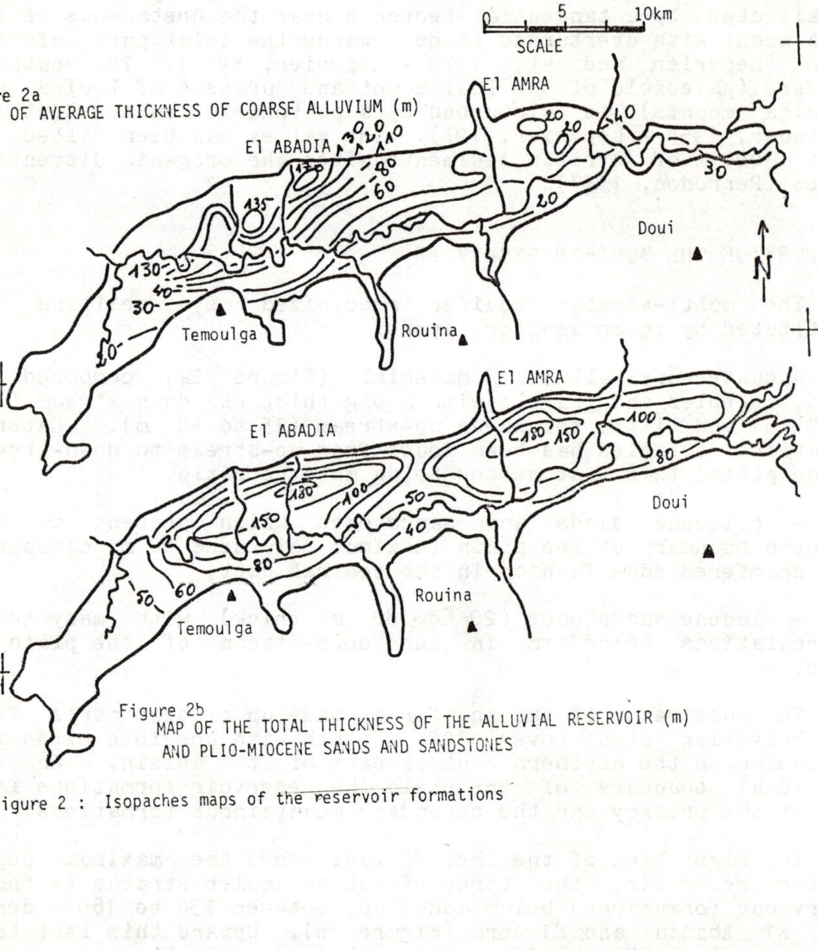

Figure 2a
MAP OF AVERAGE THICKNESS OF COARSE ALLUVIUM (m)

Figure 2b
MAP OF THE TOTAL THICKNESS OF THE ALLUVIAL RESERVOIR (m)
AND PLIO-MIOCENE SANDS AND SANDSTONES

Figure 2 : Isopaches maps of the reservoir formations

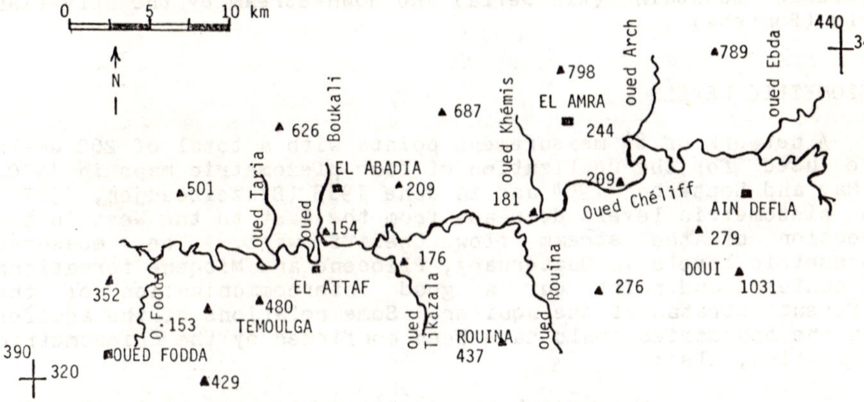

Figure 3:Hydrographic network and location of the topographical points (m) .

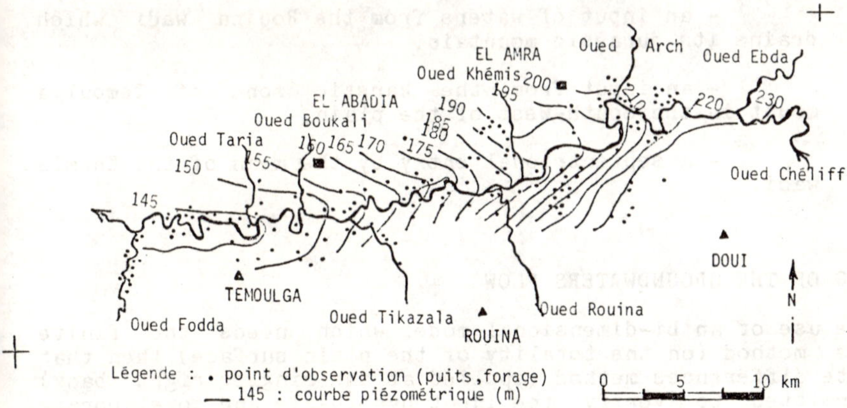

Figure 4:Piezometric map for the flood period of the year 1970 (m) .

The south boundary is impervious with runoff flow on marly formations outcrops in direction of the Cheliff wadi. Two steps bound the catchment up-stream at the crossing of the schistose paleozoic mountain (Ain Defla) and down-stream by the helvetian marls (Ponteba).

PIEZOMETRIC LEVELS

A network of 84 measurement points with a total of 200 wells were used for the realization of four piezometric maps in 1970, in May and September 1984 and in June 1985 (Bouzelboudjen, 1987). The piezometric levels decrease from the East to the West in the direction of the stream flow Cheliff wadi. The measured piezometric levels in Quaternary, Pliocene and Miocene formations are confused and point out a good intercommunication of the different stratas of the aquifer. Some relations of the aquifer with the boundaries could have been confirmed by the piezometric examination, also:

1. - the hydrologic participation of the paleozoic mountain with a draining forward in the North-West,
2. - an input of waters from the Rouina wadi which drains its jurassic mountain,
3. - an input from the karstic zone of Temoulga mount in the South-West of the plain,
4. - a septentrional entry by the mean of the Khemis wadi.

MODELING OF THE GROUNDWATERS FLOW

The use of an bi-dimensional model which needs the finite elements method (on the totality of the plain surface) then that of finite differences method (applied at the single right bank) has permitted to verify the input hypothesis and to elaborate some exploitation schemes with steady state conditions. The unsteady state has been chosen for the restitution of specific yield such as the simulation of the irrigation process according to mensual time step.

F.E.M. Modelisation

From a surface of 756 km2 (48x12) the catchment has been divided in 541 nodes and in 233 elements (Figure 5) of which 51 are mono-dimensional (with 3 nodes), 182 are bi-dimensional with 77 triangular elements (3 nodes) and with 105 quadrilaterial elements (with 8 nodes). The program has been optimized by Kiraly (1985). Several approaches have been tested:

1- the internal coherence of the model has been verified with the aid of various fields of transmissivity (sectors of E1

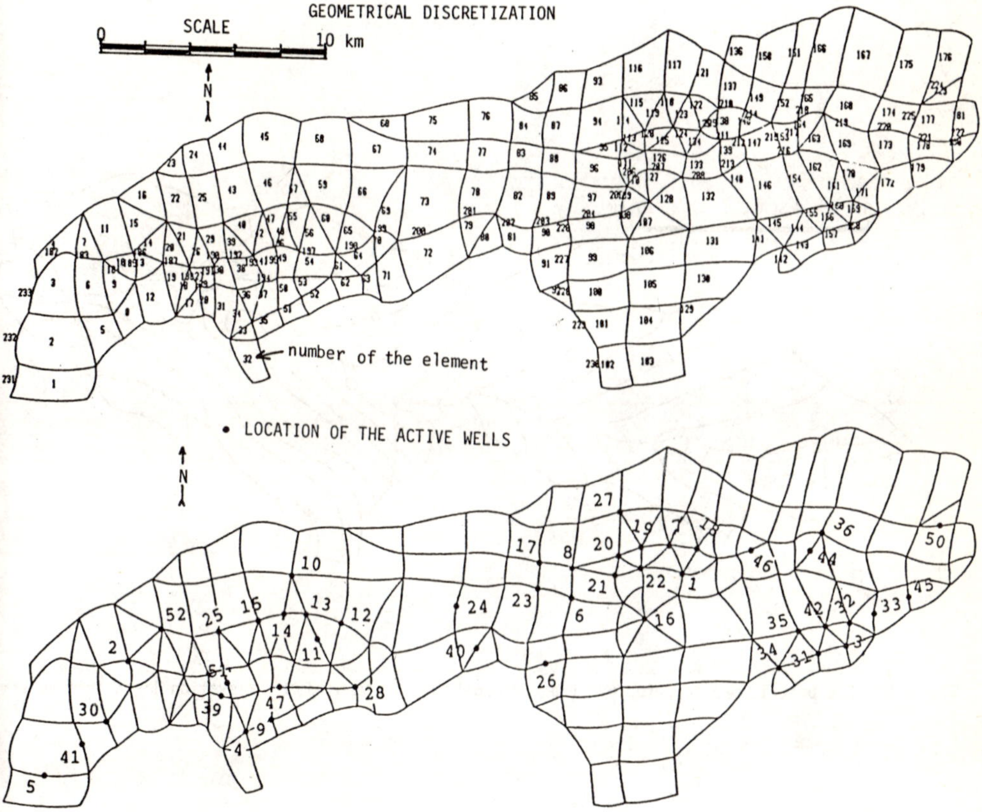

Figure 5 : Finite element grid of the Cheliff aquifer (upper case) and location of the active wells on some nodes

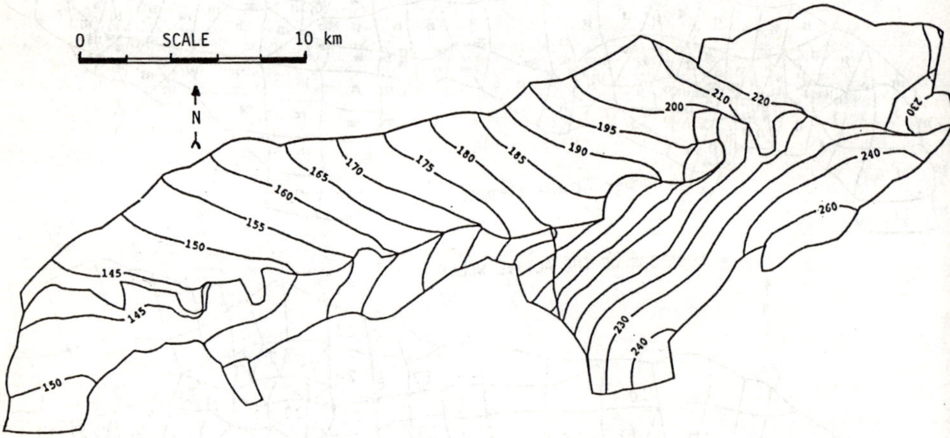

Figure 6 : Piezometric map computed by the FEM model for the year 1970 - Elevation lines in meters(m)

Abadia and El Amra and the left bank) which oscillate between 6.8 10 and 1.3 10 m2/s near El Abadia, between 9.4 10 and 1.5 10 m2/s near El Amra and between 2 10 and 3 10 m2/s on the catchment boundaries,

2- some hydraulic potentials have been specified on the north limit of the model (from 142 to 230.5 m),

3- a decrease of the draining effect of the Cheliff wadi is obtained by the drop of the values of the transmissivity,

4- an inflexion of the effective infiltration (from 9.5 10 to 4.17 10 m/s) and of constant potentials on the tributaries wadis of the Cheliff wadi.

The results obtained with steady state conditions have permitted a good reproduction of isopiezometric lines for the period of flood : of the year 1970 (Figure 6) and for the drought period of the year 1984 in which the hydrologic balance is given in Table 1. It's noted for inputs (0.7849 m3/s), the importance of effective infiltrations coming from the seepage loss of irrigation (39.4%) followed by the infiltrations from wadis (27.1%) in the recharging zones and by the meridional mountain groups (23.7%) and at last the inputs on the north and the south limits (9.8%). The outputs of the aquifer system are represented essentially by the draining zones of the wadis with 67.65% before the pumping rates flow (25.48%) and finally by the exchanges on the limits (3.55%) and on mounts (3.32%). The period in the year 1975 constitute an attractive point of the hydrologic budget of the aquifer before its over-exploitation (Table 2) where the wells abstractions reach only 4.6% of the total outputs, the wadi drains 83.1%. The inputs percentage of various units of the aquifer system are comparable with those of the year 1984.

F.D.M. Modelisation

This type of model with unsteady conditions and polygonal cells (Boonstra and al., 1981) has been chosen for its facility on personal computer. The right bank of the catchment has been discritized by 81 polygonal cells with variable size, the center of each cell is located on a well (Figure 7). The Cheliff wadi has been considered as a constant potential in the south part of the domain and in the North and Dahra mount constitute another boundary. The originality of this modelisation by comparison with the previous one is the introduction of hydraulic conductivities and specific yields. The time step of the simulation has been fixed for a period of one month in order to follow the irrigation effects from the waters dropped in the Cheliff wadi by the emptying of regional dams between February and June. The main results are the following:

1- the specific yield varies from 0.1% on the north limit, to 15% in the sector of El Amra and to 25% near El Abadia in central and downstream parts of the basin,

2- the higher hydraulic conductivities are observed

INPUT	Q m3/s	fract.%	total %	OUTPUT	Q m3/s	fract.%	total %
Infilt. by wadi				Exfil. by wadi			
*Chelif	.146	18.6		*Chelif	.420	53.51	
*Fodda	.004	0.51		*Fodda	.028	3.57	
*Rouina	.017	2.17		*Rouina	.034	4.33	
*Ebda	.046	5.86		*Ebda	.049	6.24	
			27.10				67.65
Limits exchan.				Limits exchan.			
*North	.0565	7.20		*North	.0209	2.66	
*South	.0204	2.60		*South	.007	.89	
			9.80				3.35
Effect. Infil.				Wells Supp.w +			
*intern	.309	39.37	39.40	Irrig.w	.200	25.48	25.48
Mounts Infil.				Mounts Exfil.			
*Doui	.096	12.33		*Doui	.005	.64	
*Temoul	.046	5.86		*Temoul	.004	.51	
*Rouina	.044	5.61		*Rouina	.017	2.17	
			23.70				3.32
TOTAL INPUTS	0.7849	100.00	100.00	TOTAL OUTPUTS	0.7849	100.00	100.00

Table 1: Hydrologic budget of the Cheliff basin after F.E.M modelling - Steady state conditions for the 1984 drought

INPUT	Q m3/s	fract.%	total %	OUTPUT	Q m3/s	fract.%	total %
Infilt. by wadi				Exfil. by wadi			
*Chelif	.283	24.40		*Chelif	.921	79.40	
*Fodda	.027	2.33		*Fodda	.0004	.034	
*Rouina	.004	0.35		*Rouina	.025	2.15	
*Ebda	.007	0.60	27.70	*Ebda	.018	1.55	83.10
Limits exchan.				Limits exchan.			
*North	.0560	4.85		*North	.0006	.052	
*South	.0180	1.55	6.40	*South	.00007	.006	.06
Effect. Infil. *intern	.506	43.62	43.60	Wells Supp.w + Irrig.w	.053	4.57	4.60
Mounts Infil.				Mounts Exfil.			
*Doui	.144	12.40		*Doui	.104	8.97	
*Temoul	.111	9.57		*Temoul	.012	1.03	
*Rouina	.004	0.35	22.30	*Rouina	.026	2.24	12.24
TOTAL INPUTS	1.160	100.00	100.00	TOTAL OUTPUTS	1.160	100.00	100.00

Table 2: Hydrologic budget of the Cheliff basin after F.E.M modelling - Steady state conditions for the 1971 flood

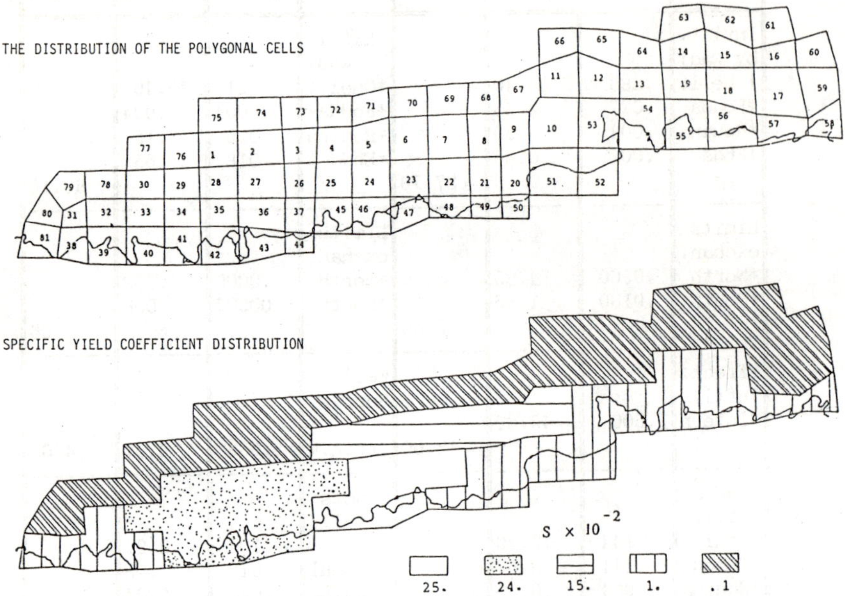

Figure 7 : Difference element grid of the right bank of the Cheliff wadi and geographical position of specific yield after the wedging of the model

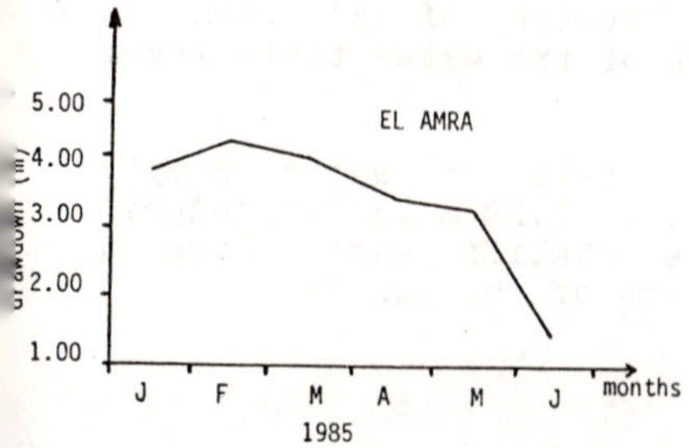

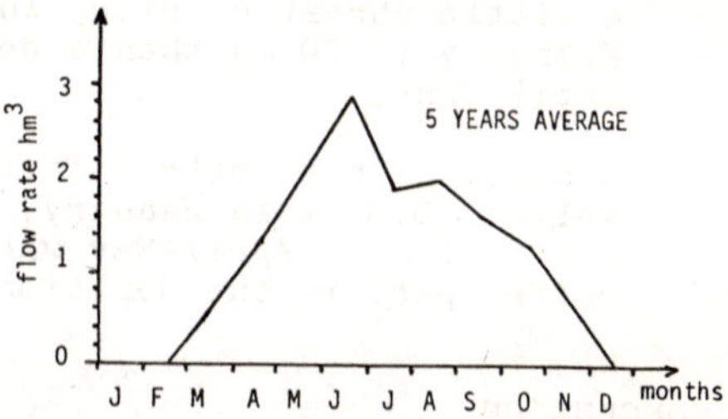

Evolution of the flow rates irrigation groundwaters

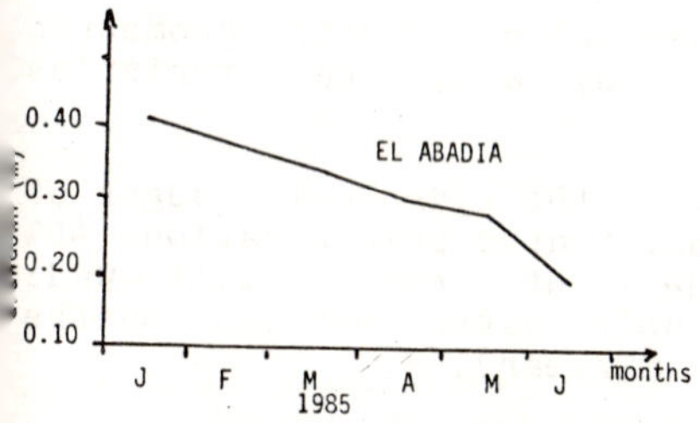

Figure 8 : The evolution of computed drawdown of the water-table during irrigation downstream (E1 Abadia zone) and are equivalent to the steady state simulation results,

3- the water table recharge by precipitations is null between May and October. Twelve wells are utilized for the irrigation (Flow rate of 2 10 m3/month; Figure 8) and the wells for the water supply distribution (0.24 10 m3/month) in the central part of the plain. A return flow from irrigation taken into consideration is about 40%.

4- the mensual step simultion on the period time of pumping January - June 1985 shows :

a little phreatic rise in the sector of E1 Amra in February (0.20 m) then a decrease of the water table level until June,

a continue piezometric fall of levels on water supply wells: 0.51 m in January, February, 0.24 m in March-April and 0.10 m in April-May where the Cheliff wadi plays a buffer part in the depltion process of the aquifer.

CONCLUSION

The hydrologic analysis and the support of mathematical simulation of groundwaters flows has shown :

- the great geographical variation of the geometric caracteristics (thickness) and the hydraulic caracteristics (conducitivity, specific yield)

- the important participation to the hydrologic budget of the natural replenishment by the infiltrated precipitation (40% or 9.9 10 m3/year, 1984) and from the inputs from Cheliff wadi (20% or 4.7 10 m3/year) and runoff water entries on the meridional mountains (24% or 5.96 10 m3/year),

- the flow rate of wells can reach 6.4 10 m3/year and the Cheliff wadi drains, in flood period, 13.4 10 m3/year,

- the mensual evolution of piezometric levels which is simulated with an mensual time step during irrigation shows a continual decreasing of levels with a buffer effect by Cheliff wadi which assumes a good regularisation of groundwaters.

REFERENCES

Boonstra, J., and De Ridder, N.A., 1981. Numerical modelling of groundwater basins, I.L.R.I. Publ., 29, Wageningen, p.226.

Bouzelboudjen, M., 1987. Hydrogeologie et bilan de la nappe d'E1 Amra - E1 Abadia par modeles mathematiques (bassin du moyen Cheliff, Algerie). These Doct. Ing., Univ. Franche-Comte, Besancon (France), p. 186.

Glangeaud, L., 1955. Les deformations plio-quaternaires de l'Afrique due Nord. Geol. Rundschau, band 1, p.181-186.

Kiraly, L., 1985. FEM 301 :a three-dimensional model of ground water flow simulation. Bull. Cent. Hydrol. Neuchatel (Suisse), p. 96.

Lepvrier, C., Magne, J., and Sigal, J., 1970. Donnees stratigraphiques et structurales sur les deformations telliennes d'un partie du Tell septentrional (secteur compris entre Cherchell, Miliana, El Asnam et Tenes-Algerie). Bull. Soc. Geol. France, t.22, ser.5, p. 794-804.

Lepvrier, C., 1971. Donnees relatives a la schistosite et au metamorphisme dans les massifs du Cheliff et de Bou Maad (autochtone Nord et mesotellien). C.R. Acad. Sc. France, t,273, p. 284-286.

Lepvrier, C., 1978. Les plis couches synschisteux des massifs du Cheliff (Tell algerois, Algerie). Rev. Geogr. Phys. Geol. Dynam., 2, v. 20, fasc.1, p. 123-136.

Obert, D., and Lepvrier, C., 1976. Paleotectonique en Afrique du Nord: l'exemple des Babors et de l'ensemble Bou Maad, massifs du Cheliff (Algerie). 4eme R.A.S.T. Paris, p. 308.

Mattauer, M., 1958. Etude geologique de l'Ouarsenis oriental, Algerie. Publ. Serv. Carte Algerie, 17, These, Univ. Franche Comte, Besancon (France).

Megghraoui, M., 1982. Etude neotectonique de la region Nord-East d'El Asnam:relation avec le seisme du 10 octobre 1980. These Univ. de Paris VII (France), p. 182.

Perrodon, A., 1957. Etude geologique des bassins neogenes sub-littoraux de l'Algerie Nord Occidentale, Publ. Serv. Carte Algerie, 12. These, Univ. de Paris VI (France).

INPUT	Q m3/s	fract.%	total%	OUTPUT	Q m3/s	fract.%	total%
Infilt.				Exfil.			
b wadi				by wadi			
*Chelif	.146	18.60		*Chelif	.420	53.51	
*Fodda	.004	0.51		*Fodda	.028	3.57	
*Rouina	.017	2.17		*Rouina	.034	4.33	
*Ebda	.046	5.86		*Ebda	.049	6.24	
			27.10				67.65
Limits				Limits			
exchan.				exchan.			
*North	.0565	7.20		*North	.0209	2.66	
*South	.0204	2.60		*South	.007	.89	
			9.80				3.35
Effect.				Wells			
Infil.				Supp.w			
*intern	.309	39.37		+			
			39.40	Irrig.w	.200	25.48	25.48
Mounts				Mounts			
Infil.				Exfil.			
*Doui	.096	12.33		*Doui	.005	.64	
*Temoul	.046	5.86		*Temoul	.004	.51	
*Rouina	.044	5.61		*Rouina	.017	2.17	
			23.70				3.32
TOTAL				TOTAL			
INPUTS	0.7849	100.00	100.00	OUTPUTS	0.7849	100.00	100.00

Table 1: Hydrologic budget of the Cheliff basin after F.E.M. modelling - Steady state conditions for the 1984 drought

INPUT	Q m3/s	fract.%	total%	OUTPUT	Q m3/s	fract.%	total%
Infilt. b wadi				Exfil. by wadi			
*Chelif	.283	24.40		*Chelif	.921	79.40	
*Fodda	.027	2.33		*Fodda	.0004	.034	
*Rouina	.004	0.35		*Rouina	.025	2.15	
*Ebda	.007	0.60		*Ebda	.018	1.55	
			27.70				83.10
Limits exchan.				Limits exchan.			
*North	.0560	4.85		*North	.0006	.052	
*South	.0180	1.55		*South	.00007	.006	
			6.40				.06
Effect. Infil.				Wells Supp.w +			
*intern	.506	43.62	43.60	Irrig.w	.053	4.57	4.60
Mounts Infil.				Mounts Exfil.			
*Doui	.144	12.40		*Doui	.104	8.97	
*Temoul	.111	9.57		*Temoul	.012	1.03	
*Rouina	.004	0.35		*Rouina	.026	2.24	
			22.30				12.24
TOTAL INPUTS	1.160	100.00	100.00	TOTAL OUTPUTS	1.160	100.00	100.00

Table 2: Hydrologic budget of the Cheliff basin after F.E.M. modelling - Steady state conditions for the 1971 flood

THE ROLE OF VERTICAL FLOW IN SEDIMENTARY AQUIFERS

K.R.Rushton, Department of Civil Engineering, University of Birmingham, UK

M. Thangarajan, NGRI, Hyderabad 500 007, India

Abstract

This paper reviews the conditions in three sedimentary aquifers from which substantial quantities of water have been abstracted. Evidence is considered which demonstrates the importance of vertical flows through the low permeability zones that overlie the main aquifers. Methods of estimating the magnitude of these vertical flows are discussed.

INTRODUCTION

Sedimentary aquifers are often important sources of water for irrigation, industrial and domestic purposes. Before the advent of powered pumps, water was usually drawn from large diameter dug wells by hand or using animaldriven bucket wheels. However, powered pumps allowed larger quantities of water to be drawn from the aquifer into the wells. This usually resulted in falling water tables and required the deepening of wells to maintain the yield. Falling pumping levels also occurred due to the increasing density of the wells. The next step in the aquifer development was usually the drilling of tube wells to greater depths within the aquifers. Aquifer performance tests frequently showed high yields. Therefore, instead of using dug wells, the deeper aquifers were exploited either by drilling tubewells or by constructing bores in the bottom of existing dug wells. Good yields were initially obtained but the experience in many aquifers is that pumping levels have fallen by tens of metres, the water table have continued to fall and it is often necessary to drill new wells to maintain the yield.

In attempting to understand the aquifer responses as described above it is essential to recognize that in these sedimentary aquifers both *horizontal* and *vertical* flows can occur. So much of groundwater analysis is dominated by the Theis equation and further more complex analytical solutions which assume that the flows in the aquifer are predominantly horizontal. However, in practice the groundwater will follow the easiest path.

This paper examines three case studies in India of deep aquifers from which substantial quantities of water are abstracted. The historical development of the aquifer is described in each case and from these three examples methods of identifying the presence of both vertical and horizontal flows are indicated. Then, methods of quantifying the vertical flows are introduced.

EXAMPLES OF VERTICAL FLOW

Mehsana Aquifer

The Mehsana alluvial aquifer in Gujarat covers an area of approximately 3000 km^2. In the south-west the thickness exceeds 300m but in the north-east, where the ground elevation is about 60m higher, the thickness is less than 40m. In the south-west there are roughly equal proportions of sand and clay but in the north-east there is little clay present.

Large diameter dug wells have been used for many centuries to supply a limited amount of water for irrigation purposes. However, the advent of powered pumps from the 1930's onwards resulted in a fall in the water table. This required the deepening of wells and the lowering of the surface centrifugal pumps; Fig. 1 illustrates the various stages in the development of the aquifer. Due to the declining water levels, exploratory boreholes were drilled deeper into the aquifer and at 100m or more below the ground surface a more permeable aquifer was tapped and the piezometric heads were found to be higher than the water table elevation. Pumping tests in the exploratory boreholes indicated a good yield for relatively small drawdowns.

From this evidence it was concluded that an alternative source of water had been found; it was assumed that the water in the confined zone originates from the higher ground in the north-east where there is little clay present. This area was called the common recharge zone (UNDP 1976). Rapid exploitation of these deeper aquifers has occurred using both tubewells with submersible pumps and dug-cum-bore wells with centrifugal pumps. This has led to a notable decline in both the groundwater heads in the deeper aquifers and in the water table elevation, Fig. 1. Severe falls in the water table elevations and the pumping levels continue to occur and new and deeper wells are required to maintain the output.

Important information about the Mehsana aquifer can be gained from an examination of groundwater hydrogrphs (Rushton 1986). Figure 2 shows the groundwater heads at three different depths within the aquifer. The groundwater heads in the shallower piezometer, P, reflect the changes in the water table elevation with rises during the monson season (for the timing of the monsoons see the lower part of Fig. 2) whereas the deeper piezometers, C_1 and C_2, begin to rise in February or March and the fall commences in October or November. Clearly the groundwater heads in the deeper aquifers do not respond to the onset of the monsoons; instead they are governed primarily by the abstraction for irrigation which builds up from September or October and falls to a much lower level in February or March following the harvest of the second crop.

Madras Aquifer

The Minjur-Panjetty aquifer to the north of the city of Madras is also an alluvial aquifer extending over an area of 500 km^2. The aquifer system consists of a coarse to medium grained sand formation resting on top of the impervious Gondwana clay; above the aquifer there is an aquitard consisting mainly of clay and silt with an unconfined formation of silt and sand with

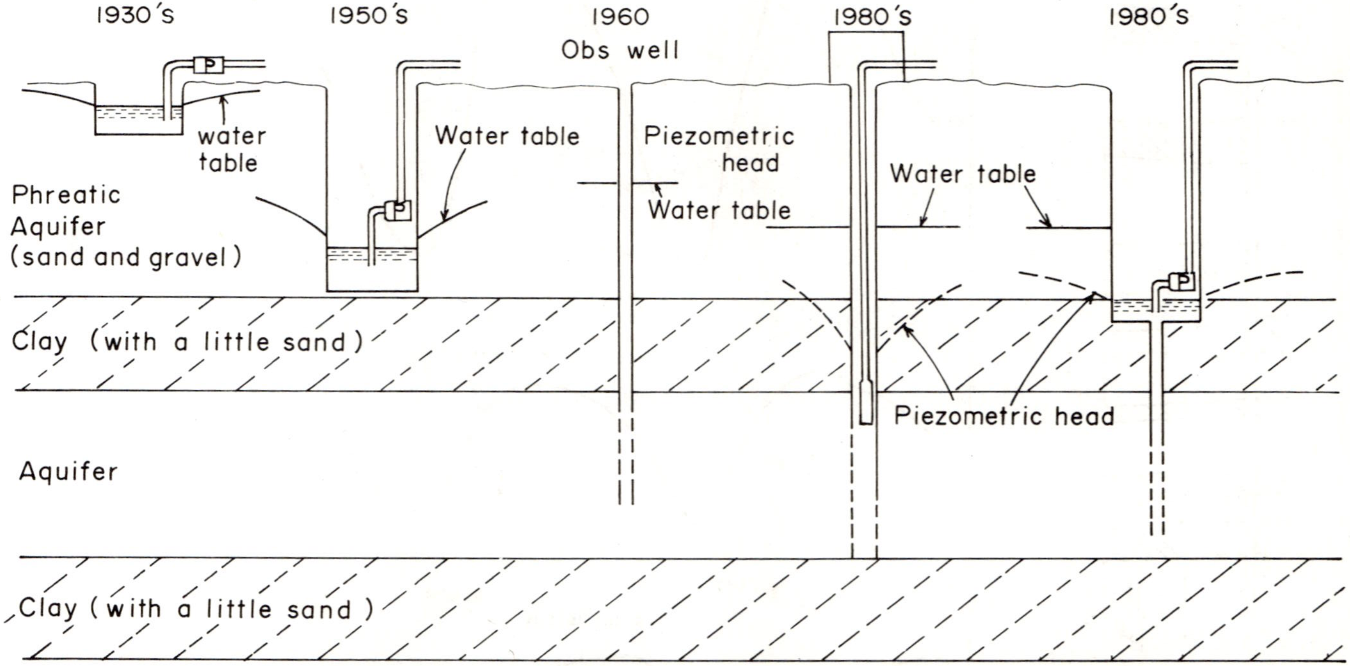

Fig. 1 Historical development of conditions in the Mehsana aquifer

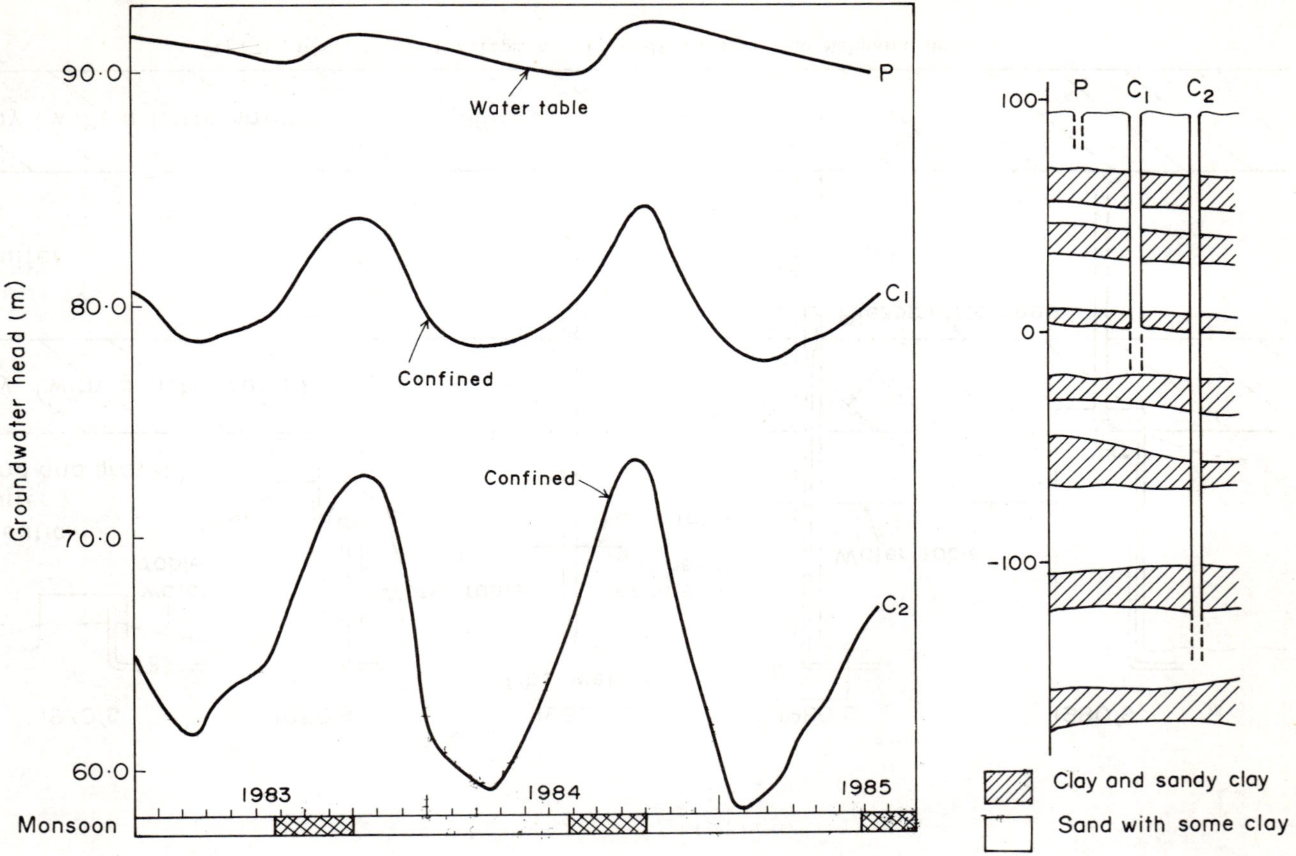

Fig. 2 Typical groundwater hydrographs for the Mehsana area.

clay lenses lying above the aquitard. The upper unconfined formation receives recharge from rainfall, return flow from irrigation and limited inflows from the major rivers. Initially the aquifer was exploited from dug wells and a number of dug wells are still in use. Subsequently the potentialities of deeper tubewells were identified and well fields were constructed for industries and augmentation of the Madras city water supply. This was followed by the exploitation of the deeper confined aquifer by farmers using tubewells or dug-cum-bore wells. Two crops are grown over most of the area with rice being the main crop.

As with the Mehsana aquifer, the development of this aquifer has led to severe declines in the pumped levels in the deeper tubewells. In one location it has been necessary to redrill deeper wells three times. However, the decline in the water table elevation has not been so severe. Figure 3 (which is redrawn from Krishnasamy and Sakthivadivel 1986) shows the response of a shallow and a deep well. The water level in the shallow well fluctuates due to the variability of the recharge. However, the deep observation well shows a serious decline. Of particular importance is the fact that in 1981 and 1982 the groundwater heads fell below the bottom of the confining layer. Another consequence of the falling heads in the confined region is that they are now below sea level causing saline water to move inland.

Neyveli Aquifer

The development of the Neyveli aquifer is somewhat different to the other two examples. Prior to 1961, there were domestic wells over much of the area, but in addition there were areas with numerous overflowing wells as indicated in Fig. 4. These wells typically penetrated 100-150 metres to tap a confined aquifer which occurs below overlying low permeability deposits of sandstones, lignite and clays.

Lignite deposits were identified above this confining aquifer and in 1961 pumping commenced to depressurise the confined aquifer so that open cast mining of the lignite could commence (GSI 1969, Gowrisankran et al 1980). The depth to the bottom of the lignite is between 70 to 120m below ground level. The lignite is used in thermal power stations and as a result of the increased availability of power, a rapid increase has also occurred in the number of wells used for irrigation.

Consequence of the increased pumping from the aquifer include:

- the cessation of flows from most of the overflowing boreholes,
- changes in conditions at the rivers Vellar and Gadilam from perennial to non-perennial conditions and
- falls in the groundwater heads in shallow and deep wells.

For example, Fig. 5(a) indicates that a fall in water table of about 4m has occurred at Karuveppilankurichi which is between the rivers Vellar and Manimukata. This fall in head is primarily due to the increased agricultural pumpage. A second hydrograph of the response of the confined aquifer, Fig. 5(b), shows that at Thambipettaipalayam a fall in excess of 12m has occurred; this is typical of the response of the confined aquifer.

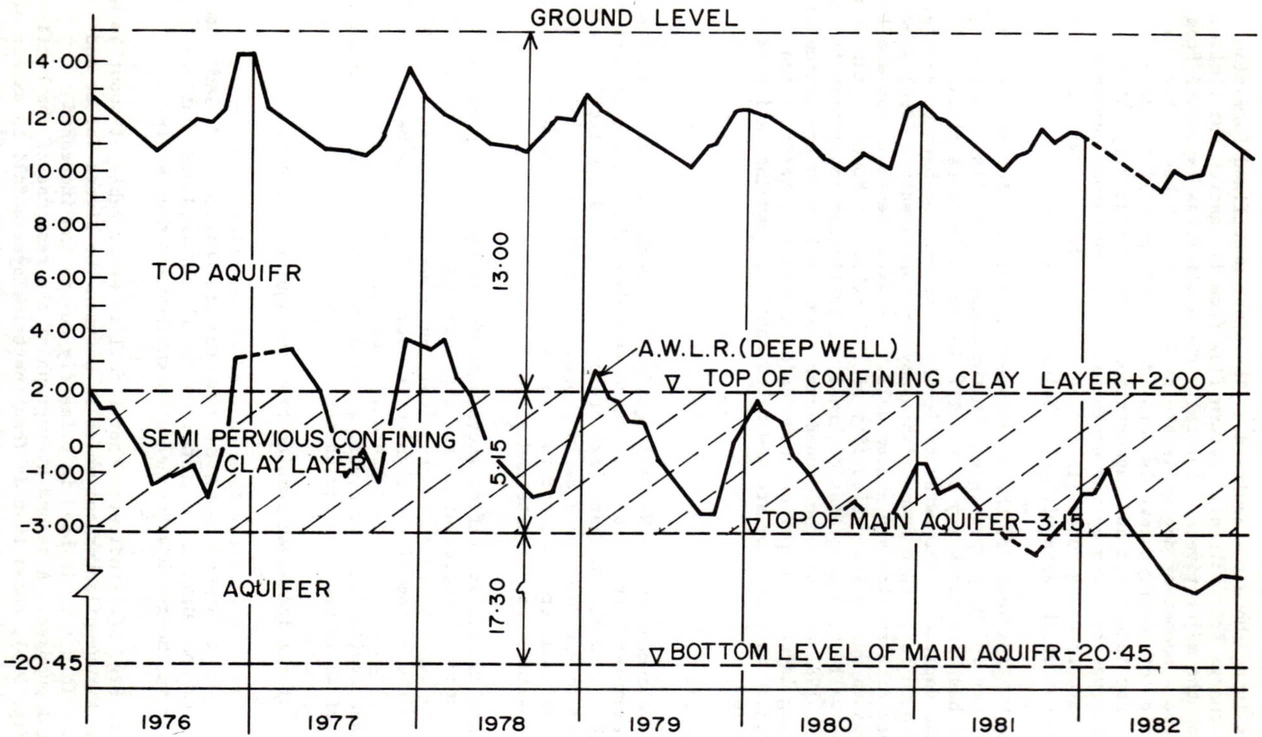

Fig. 3 Typical well hydrographs for the Madras aquifer.

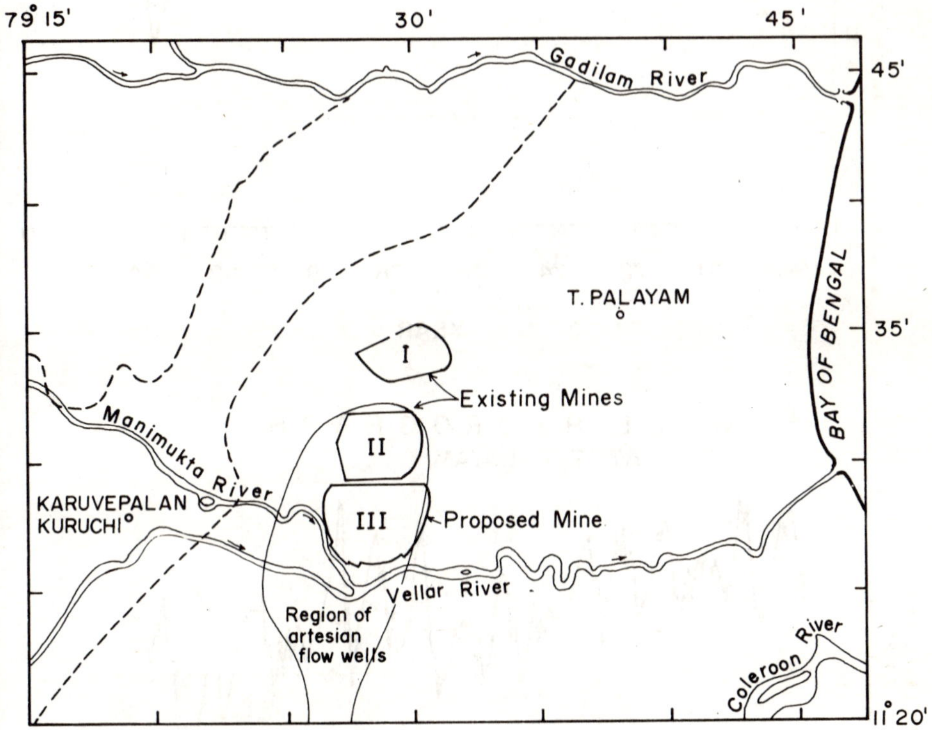

Fig. 4 Plan of the Neyveli aquifer.

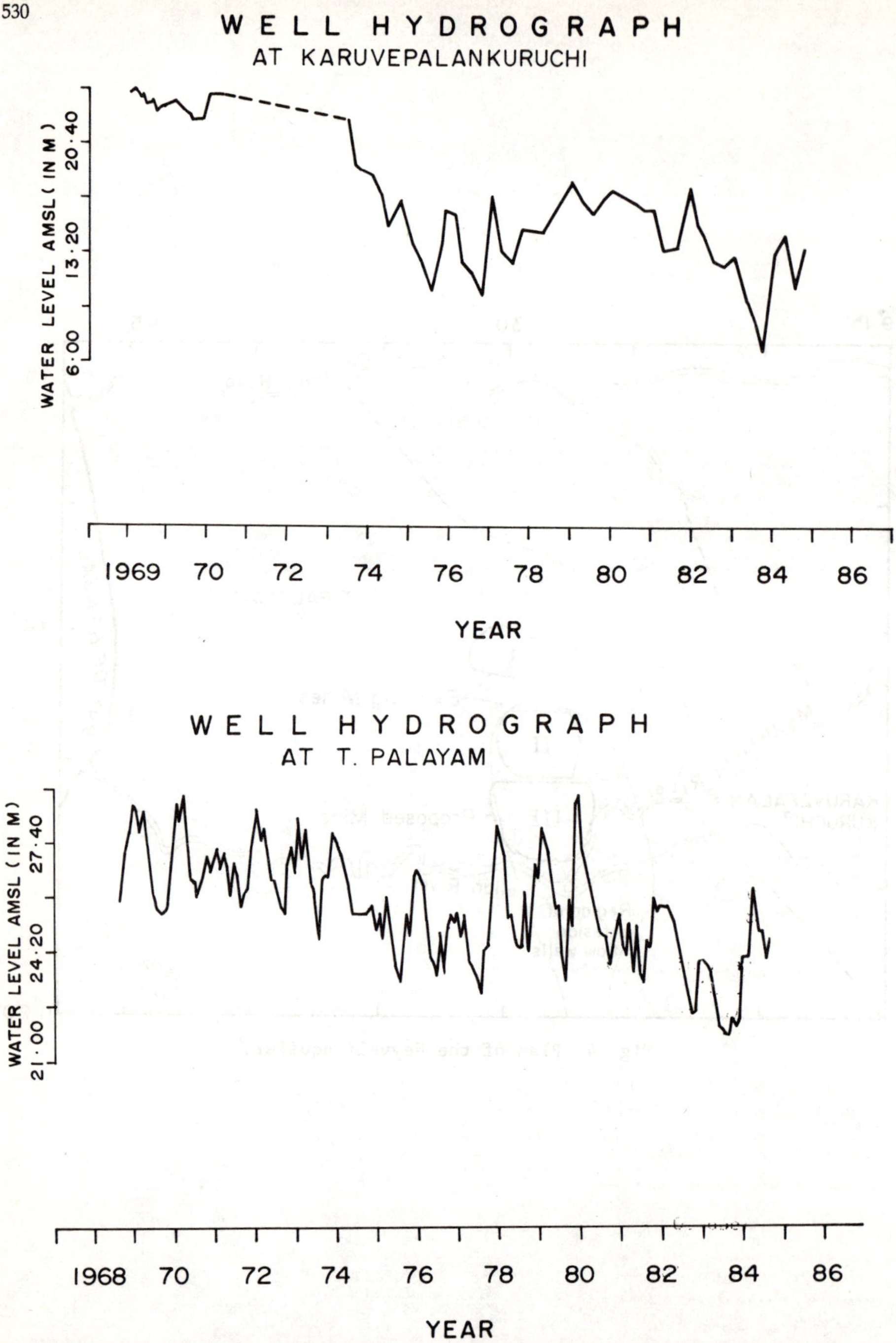

Fig. 5 Groundwater hydrographs for the Neyveli aquifer: (a) dug well response, (b) deep tubewell response. The location of the observation wells is indicated in Fig 4.

IDENTIFICATION OF VERTICAL FLOWS

The above examples can be used to illustrate the importance of vertical flows, yet in each case the existence of the vertical flows was not identified until a much later stage in the development of the aquifers. Therefore it is important to consider how it is possible to identify whether vertical flows are likely to occur. Five indicators of the occurrence of vertical flows are described below.

Overflowing wells and springs or high piezometric heads occurring prior to the exploitation of the aquifer. In the Neyveli aquifer there were extensive areas of springs and overflowing wells before the exploitation of the confined aquifer by deeper wells, Fig. 4. These overflowing wells occurred because the heads deeper within the aquifer were above ground level. When a well penetrates to the confined aquifer it by-passes the overlying lower permeability zones. There is also a substantial vertical gradient across the low permeability zones but, due to the low permeability, the actual quantity of water moving upwards will be of the order of 0.1 mm/d. Such a low flow cannot easily be identified; in fact it is because of the restriction in the upwards vertical flow that the groundwater heads build up in the lower aquifer. The higher piezometric heads in the confined regions of the Mehsana aquifer prior to exploitation (Fig. 1) occur for the same reason.

Falling piezometric heads and reduction in flow in perennial rivers. Abstraction from these deeper aquifers invariably leads to a decrease in the piezometric heads; this results in the reduction or cessation of flow from overflowing wells. In addition, a significant decrease in river flows usually occurs. Figure 6(a) indicates that when there are only a few wells tapping the confined aquifer, the yield can be obtained by intercepting lateral flows. Hence the pumped drawdown is small. However, as the number of wells increases, the demand on the confined aquifer increases such that it cannot be supplied by the lateral flows. Significant increases in the pumped drawdown do occur causing some increase in the lateral flows but the main result is that the vertical head gradients and therefore the vertical flows increase to meet the abstraction demand. This situation is illustrated in Fig. 6(b); note that the piezometric heads are now much lower than before abstraction commenced.

Increasing pumped drawdowns due to increasing abstraction. Figure 2 indicates that in the Mehsana aquifer the differences between the water table head, which are indicated by the upper line, and the piezometric heads in the lower aquifers, C1 and C2, double between October and March. The vertical flow components are proportional to this head difference. In October, the total pumpage is small and therefore a smaller vertical gradient is required, whereas in March when the crop demand and hence the total abstraction is far higher, larger vertical head gradients are required.

For the Neyveli aquifer, the annual cycle for irrigation wells is that significant abstraction occurs between January and July, this is reflected in a fall in the the piezometric heads, Fig. 5(b). In addition. during periods of low rainfall such as 1973-1976, abstraction increases to compensate for the low rainfall and this leads to a continuing decline in the piezometric heads. However, in the good rainfall year of 1979 and

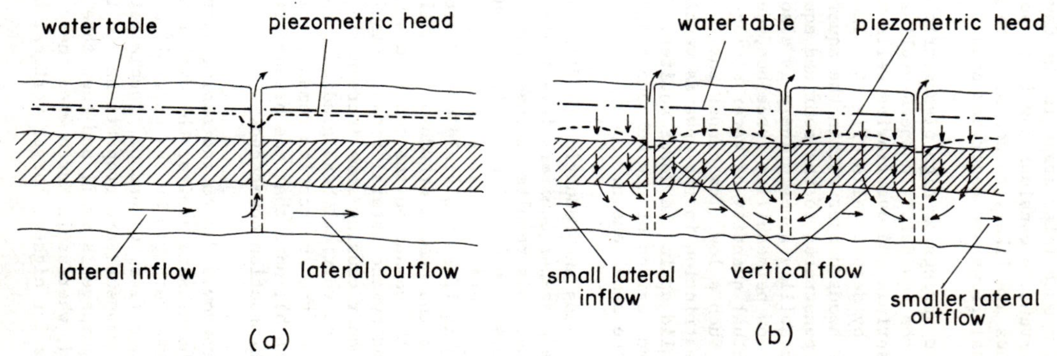

Fig. 6 Effect of different numbers of wells tapping the confined zone.

1983-84, the agricultural abstraction was less and rises occurred in the piezometric heads. Note that the piezometric head fluctuation does not vary due to the recharge but it does reflect the change in abstraction resulting from the rainfall distribution.

The Madras aquifer illustrates a further important feature. Since rainfall occurs due to the north-east monsoon, the abstraction during October to January is low and recoveries occur in piezometric heads, Fig. 3; during the remainder of the year the peizometric heads tend to fall. The average abstraction from the confined aquifer is increasing so that the piezometric heads exhibit a steady annual decline. In 1981 and 1982 the piezometric heads fell below the base of the semi-pervious layer. This means that there is no longer a fully saturated connection between the confined and unconfined aquifers and partial dewatering of the confined aquifer commences. Figure 7 illustrates that this leads to a severe reduction in the efficiency of the well and requires the drilling of a deeper well with solid casing extending to greater depths.

Continuing long term fall in the water table elevation. An examination of the water table hydrographs at Neyveli, Fig. 5(a) and Madras, Fig. 3 indicates that the water table does respond to the recharge resulting from monsoon rainfall. In the Neyveli aquifer the recharge was so high in the later months of 1979 that the water table recovered by almost 5.5m. However an examination of the long term trend indicates that there is a steady fall of about 0.3m per year which was reversed only during the period 1977-79. This steady decline is due to the vertical flow resulting from abstraction from the confined aquifers.

In the Madras aquifer the fall occurs at a slower rate. This is due to the lower hydraulic conductivities of the unconfined aquifer which limit the rate at which water can move downwards. This limited downwards movement of water from the unconfined aquifer through the semi-pervious zone is insufficient to meet the pumping requirement from the confined aquifers. Hence the development of dewatered zones in the confined aquifer.

Tracers can provide clear evidence of the occurrence of vertical flows. When tracers are injected in the unsaturated zone above the water table, the downwards movement depends on two factors:

- -the recharge from the soil zone
- -the vertical flows from the saturated water table through the less permeable zones to the confined aquifers.

In the unsaturated zone it is difficult to distinguish between the influences of these two factors but the continuing downwards movement of the tracer throughout most of the year supports the concept of vertical downwards flows from the water table.

QUANTIFICATION OF VERTICAL FLOWS

There are a number of methods of quantifying the vertical flows.

(i) Using Darcy's Law. Figure 2 indicated that the average vertical head difference betweeen the water table and the piezometer C_1 is 12m. Between

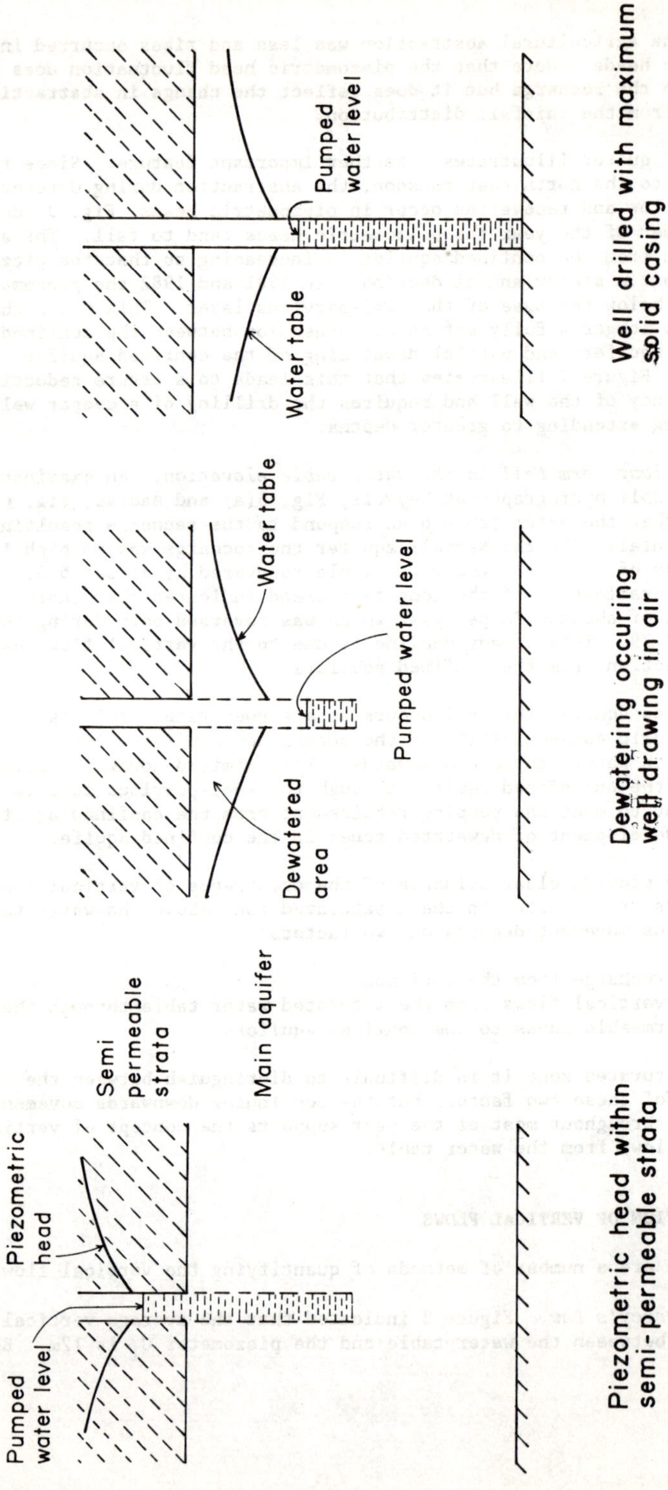

Fig. 7 Effect of the partial dewatering of the Madras aquifer.

the water table and the open portion of piezometer C_1 there are clay layers with a cumulative thickness of 32m. Typically the vertical hydraulic conductivity of these layers is 0.003 m/d and therefore from Darcy"s Law the downwards vertical velocity is

$$v_z = 0.003 \times (12/32) = 0.001125 \text{ m/d}$$

This is equivalent to 1125 m^3/d over an area of 1 km^2, which would provide sufficient water for three wells each pumping at 750 m^3/d for 12 hours each day for every day of the year.

(ii) Examination of typical cross section The simple calculations of the above section have indicated that the vertical flows can be significant. However, they do not provide much information about the relative magnitudes of the horizontal and vertical flow components. Indeed, when the typical cross section of the Mehsana aquifer is examined, Fig. 8, it appears that it is far easier for water to flow horizontally through the more permeable sand zones than to move vertically through the clay zones of low hydraulic conductivity (Rushton and Tiwari 1989).

The danger of the above assertion is that the distortion inherent in Fig. 8 is not recognized. If the cross-section is plotted to true scale, the horizontal dimensions would need to be multiplied by a factor of fifty. If this were done the horizontal sand layers would correspond to long narrow tubes.

When a well is pumped, the groundwater heads in the vicinity fall and water is drawn in both horizontally and vertically. Consider a well screen at the position X in Fig 8. Water can either be drawn in

-*horizontally* through a narrow channel of sand less than 20m in depth from the common recharge zone which is some 40 km distant with a hydraulic gradient of less than 1:1000. Furthermore, this well would be competing with many other wells tapping the same strata.

-*vertically* through clay layers having a cross-sectional area associated with the well of 0.33 km^2, a cumulative clay thickness of 32m with a vertical head gradient of 1:2.67

From the above description it is apparent that the resistance to flow in the vertical direction is often significantly less than that in the horizontal direction. Consequently, the greater part of the discharge is met by vertical flows. For the position X, vertical flows account for more than 90% of the discharge.

(iii) Use of Numerical Models Numerical models have been used to examine and understand the conditions in each of the aquifers. The model for the Mehsana aquifer is discussed by Rushton and Tiwari (1989) and a description of the Madras aquifer model is presented by Krishnasamy and Sakthivadivel (1987). For the Neyveli aquifer a series of models are being developed by the National Geophysical Research Institute.

The purpose of the first model was to confirm the importance of the vertical flow components; the model represents the main aquifer which lies beneath the low permeability zone and the lignite. Provision is made for

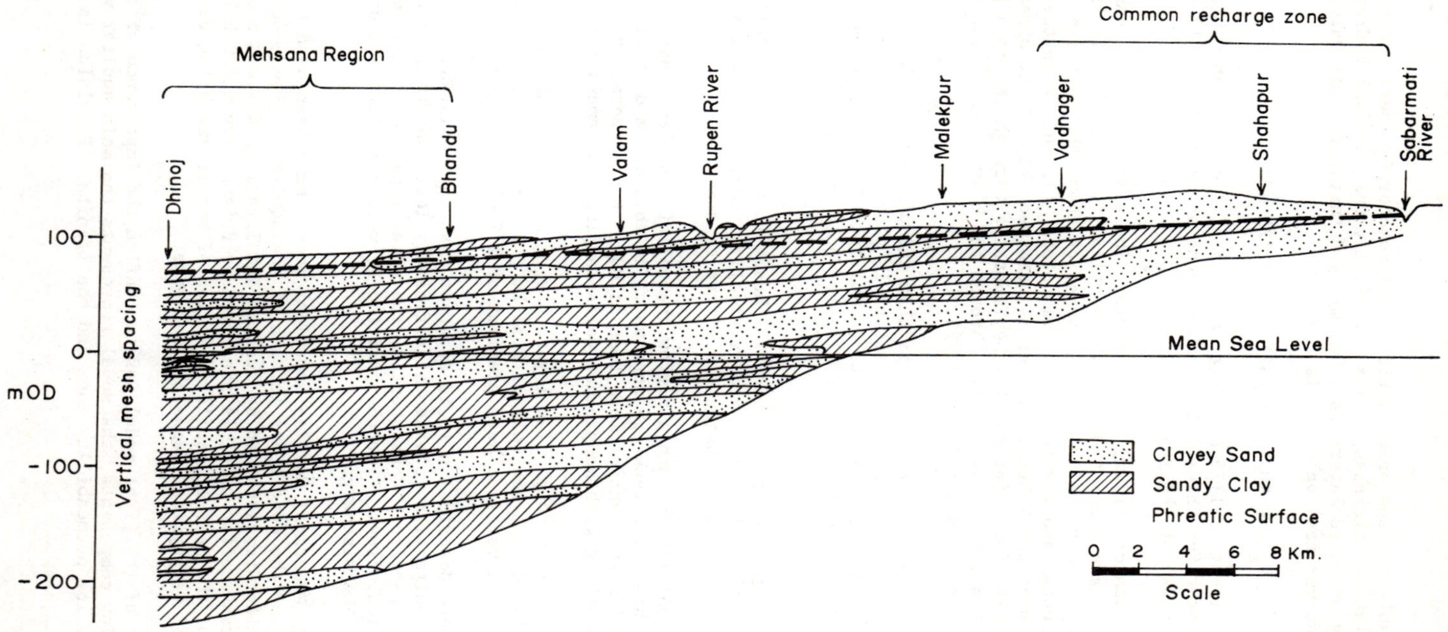

Fig. 8 Typical cross-section of the Mehsana aquifer.

the representation of vertical flows between the water table and the main aquifer over the whole of the study area; the magnitude and direction of the flow depends on the difference in groundwater heads between the water table and the main confined aquifer and on the effective vertical hydraulic conductivity. The study covers an area of 2200 km^2, the area is divided into a finite difference grid with mesh intervals varying from 4 km to 500m. The following features are incorporated in the model:

- head dependent vertical flows
- transmissivity varying with saturated thickness
- change between the confined and unconfined states
- river-aquifer interaction

One of the most important parameters in the model is the effective vertical permeability. As yet there is no means of directly determining the magnitude of this parameter. However, from the lithology, a value of 0.0022 m/d was selected; detailed pumping tests are currently being used to identify the magnitude of this parameter more accurately.

The model simulation commences in 1956 and continues to 1986. Satisfactory agreement has been obtained between field and modelled values of groundwater heads both in the recharge area in the west and elsewhere in the deeper aquifers. The cessation of flows from certain overflowing boreholes has also been simulated. The numerical model results indicate that during this period the vertical components of flow increased more than fourfold so that in 1986 40% of the water abstracted from the aquifer is provided by vertical flows. Therefore the recharge area to the west is not the only source of water. Currently a multi-layered model of the aquifer is under development.

CONCLUSIONS

This paper has examined the role of vertical flows in three aquifers in India. In each case the original concept was that groundwater originated from an up-dip recharge area. In the undisturbed state, this was a valid assumption about the conditions within the aquifer. However, heavy exploitation of the aquifers meant that the horizontal flows from the recharge zones were no longer sufficient to meet the demands and vertical flows became an important feature.

Therefore the main conclusion of this paper is that the possibility of vertical flows must be examined in all groundwater investigations.

References

Cookey, E.S., Rathod, K.S. and Rushton, K.R. 1987. Pumping from an unconfined aquifer containing layers of different hydraulic conductivity. Hydo. Sci. Jnl. v. 32, 43-57.

Gowrisankaran, S., Meyappan, M. and Radhakrishnan, M. 1980. Lignite deposits and groundwater development of the Neyveli Area. Bull. ONGC vol 17, no 1.

GSI 1969. Geology and groundwater aspects of the Neyveli Lignite Field, South Arcot District, Tamil Nadu. Memoirs of the Geological Survey of India, vol 94.

Krishnasamy, K.V. and Sakthivadivel, R., 1986. Regional modelling of non-linear flows in multiaquifer system. UNESCO Regional Workshop on Groundwater Modelling, Roorkee, 85-105.

Rushton, K.R. and Tiwari, S.C. Mathematical modelling of a multi-layered alluvial aquifer. Institution of Engineers, India.

Rushton, K.R., 1986. Vertical flow in heavily exploited hard rock and alluvial aquifers. Ground Water. v. 24, 601-608.

UNDP, 1976. Groundwater surveys in Rajasthan and Gujarat. Technical Report United Nations, New York.

ESTIMATING GEOHYDROLOGIC PARAMETERS FOR GROUNDWATER MANAGEMENT AND LAND SUBSIDENCE CONTROL IN KANTO PLAIN, JAPAN

N. SHIBASAKI
National Geophysical Research Institute
Hyderabad - 500 007

ABSTRACT

Detailed geohydrological investigations have been carried out in Kanto Plain, Central Japan for a basin-wide groundwater management and to evolve appropriate strategies for prevention of land subsidence. Geological and hydrogeological studies/analyses have been carried out to construct computer simulation models. Geohydrologic parameters have been estimated by statistical analyses using pumping tests/routine production tests of water wells and soil tests data. The calculated groundwater level and land subsidence by the vertical two-dimensional multi-aquifer model using finite element method have been satisfactory agreement with the observed values. The water balance and land subsidence studies which have based on accurate geological and hydrogeological studies/analyses have greatly helped evolve optimal exploitation schemes in this area.

INTRODUCTION

Kanto Plain is one of the largest plain in Japan and has been urbanized since 1950s. In 1950s decline of groundwater table and land subsidence occurred in eastern part of Tokyo where heavy pumping of groundwater was done. In 1960s the area of lowering of artesian heads and subsequent land subsidence were spread and minimum artesian head was recorded below -60 meters from the mean sea level. After enforcement of regulations for controlling groundwater the artesian heads in central to southern Kanto Plain have rised up, but the area of artesian heads lowering and land subsidence have been spread in nothern Kanto Plain.

Many hydrogeological investigations and computer simulations were carried out by several organizations in this area (Research Group for Water Balance, 1973, 1976). These studies, howerver, were carried out in small area and geological and geohydrological informations were not sufficient. In 1980s several observation wells have been constructed by core boring which has provided detail information of geology and geohydrology. Therefore, detailed geohydrological investigations have been carried out in nothern Kanto Plain for basin-wide groundwater management and prevention of land subsidence.

GEOLOGICAL STUDY

The geology of Kanto Plain consists of Tertiary and Quaternary sediments which are clay, silt, sand and gravel beds. These sediments are divided into the Kazusa Group, Shimofusa Group and Alluvial sediments in ascending order. The Kazusa Group is Plio-Pleistocene sediments and about 2,000 m in maximum thickness. The Shimofusa Group which thickness is about 300 m unconformably overlies the Kazusa Group. Alluvial sediments which mainly consist of soft clay, silt and peaty clay beds are distributed in central to northern Kanto Plain and bury so-called "fossil valleys".

This stratigraphy was established in the margin area of Kanto Plain so that subsurface geology in Central Part of Kanto Plain was not made clear. Then, the basic stratigraphy of subsurface geology has been established at the beginning of this study. The analyses of facies, diatom, foraminifera, pollen and volcanic ash etc., have been carried out by using core samples of five observation wells as the result of this new ten formations have been distinguished and named (1), (2), ..., (10) in descending order. After establisthment of basic stratigraphy, several geological sections have been made.

HYDROGEOLOGICAL STUDY

After making geological sections hydrogeological sections have been made to construct the vertical two-dimensional multi-aquifer models. In these sections four aquifer units have been classified and facies boundaries have been drawn. The correltion between geological formations and aquifer units is given below:

Geological Formation	Aquifer Unit	
(1)	H	
		Unconformity
(2), (3), (4), (5)	A	
		Unconformity
(6), (7)	B	
		Unconformity
(8), (9), (10)	C	

After dividing each colomunar section of water well into aquifer units, iso-depth contour maps and isopach maps of each aquifer unit have been made to construct the quasi-three-dimensional multi-aquifer model. Facies maps of each aquifer unit have also been made to calculate transmissibility and storage coefficient in this area. In these facies maps the facies have been expressed in content ratio of clay, sand and gravel beds which have computed as follows:

$$(\text{content ratio of each facies}) = \frac{(\text{total thickness of each facies bed})}{(\text{total thickness of each aquifer unit})} \times 100 \,(\%)$$

These isopach maps and facies maps have been divided into square shaped finite element mesh 2 km X 2 km in size and thickness and content ratio of each facies have been given in each mesh.

ESTIMATING GEOHYDROLOGICAL PARAMETERS

In simulation model of groundwater transmissibility and storage coefficient are very important. But, in fact, the number of accurate pumping test data are not so many that statistical analyses have been employed to use routine production tests of water wells.

The relationship between transmissibility and specific capacity was given by Logan (1964) as follows:

$$T = 1.22\ Sc$$

where T = transmissibility (m^2/day)
Sc = specific capacity (m^3/day/m)

Specific capacity and permeability are given as follows:

$$Sc = Q/s$$
$$K = T/b$$

where Q = discharge from the well (m^3/day)
s = drawdown of groundwater lebel (m)
K = permeability (m/day)
b = thickness of aquifer (m)

Using above formulas, permeability of aquifer where strainers are located can be calculated easily.

The storage coefficient which is obtained from pumping test is not so reliable (Research Group for Water Balance, 1976) that soil test data have been employed. From the compaction test. specific storage and storage coefficient are given as follows:

$$Ss = \frac{\gamma w . mv}{10}$$

$$S = Ss.b$$

where Ss = specific storage (1/m)
S = storage coefficient (dimensionless)
γw = density of water (g/cm^3)
mv = volume compressibility (cm^2/kg)
b = thickness of aquifer (m)

Besides it is impossible for gravel facies to carry out compaction test so that the presumed specific storage which was given by Domenico and Mifflin (1965) has been used. Permeability of clay and silt can be given from comaction test.

Then statistical analyses of permeability and specific storage have been done and average value and logarithmic standard deviation of these parameters in each facies have been computed. These results are shown in Fig.1 to Fig.3. It has been made clear that permeability is mainly controlled by facies but specific storage is controlled by facies and geological age of the sediments. Especially in clay and silt, their specific storage of each aquifer are different others.

In the vertical two-dimensional model, transmissibility (T) and storage coefficient (S) of each mesh have been calculated by

$$T = k.b$$
$$S = Ss.b$$

where k = permeability of facies (m/day)
Ss = specific storage of facies (1/m)
b = thickness of finite element mesh (m)

In the quasi-three-dimensional model, transmissibility (T) and storage coefficient (S) of each mesh in each aquifer unit have been calculated by

$$T = b.Rsg.Ksg/100$$
$$S = b.Rsg.Sssg./100$$

where b = thickness of aquifer unit (m)
Rsg = content ratio of sand and gravel (%)
Ksg = permeability of sand and gravel (m/day)
$Sssg$ = storage capacity of sand and gravel (1/m)

Apparent leakance (k'/b') of each mesh in each aquifer unit has been calculated by

$$k'/b' = k'c/(b.Rc/100)$$

where $k'c$ = permeability of clay (m/day)
b = thickness of aquifer (m)
Rc = content ratio of clay (%)

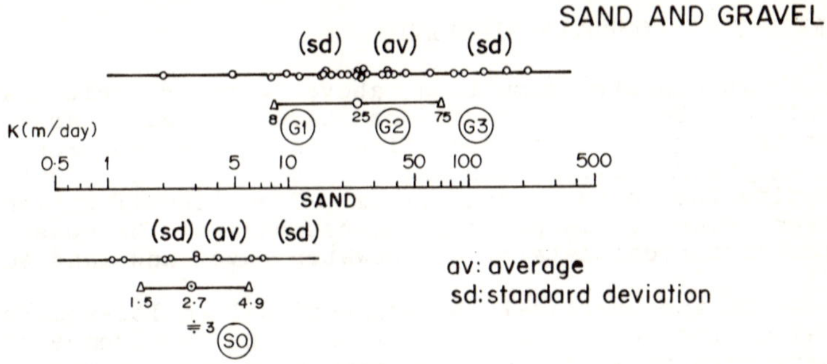

FIG. 1 ESTIMATED PERMEABILITY BY LOGAN'S METHOD

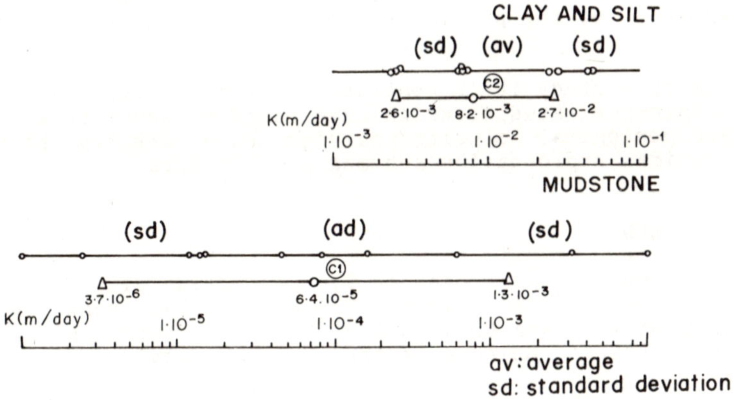

FIG. 2. ESTIMATED PERMEABILITY BY SOIL TEST

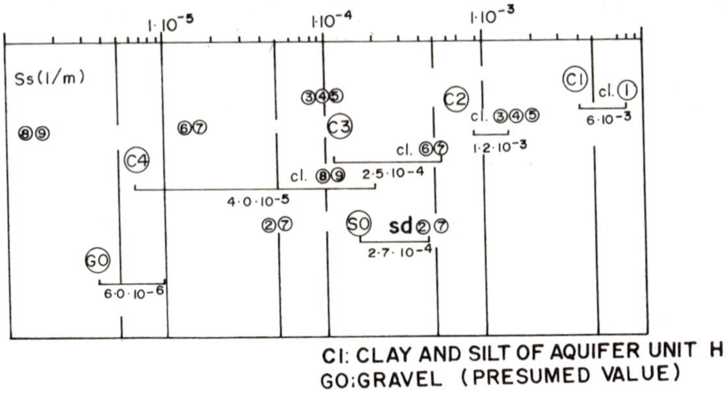

FIG. 3 ESTIMATED SPECIFIC STORAGE BY SOIL TEST

RESULT OF COMPUTEP SIMULATION

In computer simultion above geohydrologic parameters and initial boundary conditions have been given as a intial model. At the beginning of identification of geohydrologic parameters, long term calculation has been done without input of pumpage to modify the boundary conditions. Then identification of model has been done by inspecting the response of the model compared with the historical data of groundwater level and land subsidence.

In the vertical two-dimensional multi-aquifer model the calculated groundwater level and land subsidence of intial model have been satisfactory agreement with the observed values. In the quasi-three-dimensional multi-aquifer model the calculated values of intial model were different from observed values because apparent lakance was not proper. After modifying the lakance within the logarithmic standard deviations the calculated values have better than that of initial model. After fixing the groundwater level the parameters about land subsidence have been fixed.

From this study it is concluded that the water balance and land subsidence studies which have based on accurate geological and hydrogeological studies/analyses have greatly helped to evolve optimal exploitation schemes in this area.

ACKNOWLEDGEMENTS

The author thanks Professor T. Shibasaki (JICA Expert) for helpful advice and discussions. The author also thanks Professor M. Horiguchi (Saitama University), Mr. T. Kobayashi and Mr. K. Nakai (Earth Science Centre) who gave advice on geological studies.

REFERENCES

Domencio, P.A., and M.D. Milffin, 1965. Water from low permeability sediments and land subsidence, Water Resou. Res., v.4.

Logan, J., 1964. Estimating transmissibility from routine production tests of water wells. Ground water, v.2, no.2.

Research Group for Water Balance, 1973. Groundwater Resources Resh. (in Japanese). Kyoritsu Shuppan, Tokyo, pp. 397.

Research Group for Water Balance, 1976. Groundwater Basin Management (in Japanese). Tokai Univ. Press, Tokyo, pp. 242.

OPTIMAL PUMPING POLICY AND GROUNDWATER BALANCE FOR THE BLUE LAKE AQUIFER INVOLVING NON-LINEAR GROUNDWATER HYDRAULICS

A. K. Rastogi
Department of Civil Engineering
Indian Institute of Technology, Bombay
Bombay - 400 076.

ABSTRACT

Blue Lake Aquifer is a unconfined sloping aquifer in Northern California over run by Mad River which is regulated by Ruth Reservoir. High pumpage from the existing six wells of the order of few MGD operated by HBMWD causes appreciable drawdown compared to the saturated thickness of the aquifer. HBMWD also proposed another set of ten wells for meeting an anticipated daily demand of 23 MG. Therefore reduction of the governing nonlinear groundwater flow equation into a linear equation was not considered appropriate. An existing explicit scheme coded by previous workers on this project was found most suitable for the solution of inherent nonlinear equation. However many modifications to the existing algorithm were required before a groundwater balance was obtained for the Blue Lake Aquifer. HBMWD wanted to minimise the pumping cost for lifting the proposed 23 MGD from ten wells. This nonlinear optimisation problem was solved using the most recent Modified Embedded approach incorporating the appropriate constraints based upon the real system and on the optimisation requirements. It was estimated that HBMWD could have a annual saving of \$ 20,000 by implementing the annual optimal pumping policy (involving monthly planning periods as listed in Appendix I) compared to non-optimised policy (i.e. equal pumping of 2.3 MGD from ten wells).

INTRODUCTION

Blue Lake Aquifer (BLA) lies in Northern California in Humboldt Count y (Fig. 1-a). The aquifer has considerable slope (.21 %) towards the pacific ocean (west). It encompases an area of about 20 sq miles. It is a unconfined aquifer consisting of unconsolidated sand and gravel and over run by Mad River. The aquifer is underlain by Franciscan basement formation. Six ranney wells are operated by Humboldt Bay Municipal Water District which are a major source of supply to the Humboldt County. To meet the anticipated industrial, agricultural and municipal demand of water (23 MGD), ten additional wells were selected by HBMWD in the higher hydraulic conductivity area of the BLA Fig (1-b). This study was carried out first to examine a groundwater balance and the influence of additional pumping on head distribution in the BLA for one year planning period (June 1987-May 1988). Second to find an annual optimal pumping policy for the proposed wells by simulating a hydraulic management model of the BLA system.

Fig. 1-a : STUDY AREA LOCATION

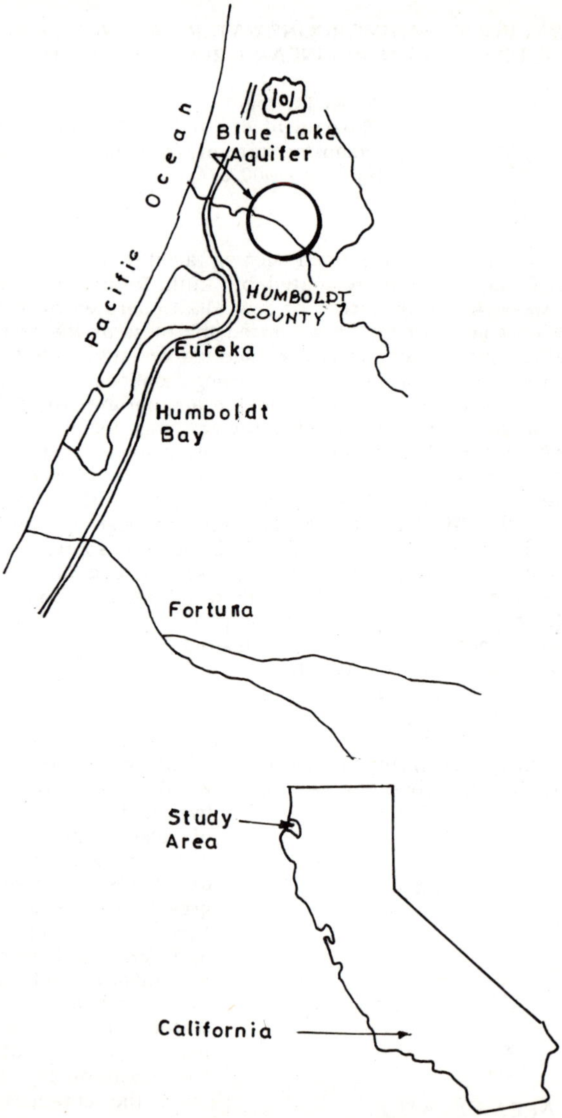

FIG. 1-a. STUDY AREA LOCATION

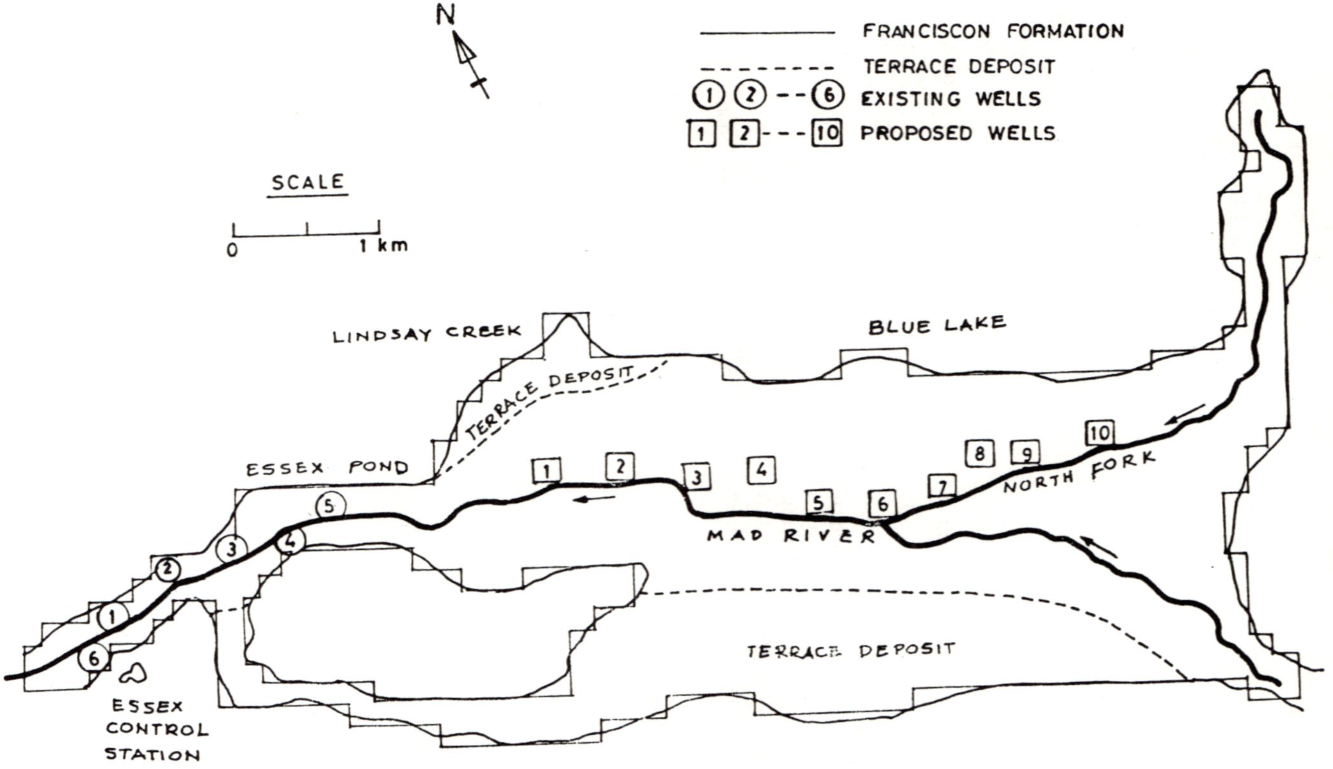

FIG. 1-b. BLUE LAKE AQUIFER

PREVIOUS WORK

Considerable amount of work on BLA was done by Willis and Chu (1981) and Chu and Willies (1984). This was a desirable feature to make maximal use of the existing work. Chu and Willis (1984) justified the use of a newly coded forward in time central in space, explicit finite difference scheme (FTCS) for the solution of the vertically averaged governing equation

Fig. 1-b : BLUE LAKE AQUIFER

$$\partial/\partial x \; K_x (h-\eta)\, \partial h/\partial x + \partial/\partial y \; K_y (h-\eta)\, \partial h/\partial y + \sum_{i \in \Omega} \delta (x - x_i)(Y - Y_i)\, N = S_y\, \partial h/\partial t \quad .....(1)$$

which described groundwater flow in unconfined, hetrogeneous, isotropic, sloping BLA system.

In eq. (1) $h(x,y,t)$ is the hydraulic head (ft), $Sy\,(x,y)$ is the specific yield, $K(x,y)$ is the hydraulic conductivity (ft/d) in the principal axes direction, $\eta\,(x,y)$ is the elevation of aquifer bottom (ft), and $N(x,y,t)$ is the recharge and/or pumping discharge (ft/d) in the domain Ω; δ is the Dirac delta function. Spatial distribution of the η values in the BLA is shown in Fig. (2).

The initial and boundary conditions for the problem were given as $h(x,y,0) = H(x,y)$ for all $x,y \in \Omega$*

$h(x,y,t) = HQ\,(x,y,t)$ for $x,y \in \partial\Omega$

$$|\partial h / \partial n|_{\partial\Omega_2} = 0$$

H is the initial groundwater head in the aquifer, HQ is the head along the Mad River, n is the normal direction to the impervious boundary. Appropriate free surface boundary condition is already incorporated in the non-linear hydraulic eq (1) (Bear 1978 ; pp 121). Fig. (3) shows the initial Mad River head in the BLA.

Fig. 2 : BLUE LAKE AQUIFER BOTTOM ELEVATION (Min = 3 ft. and Max = 66 ft.) (Impervious boundary is zero)

The finite difference formulation of eq (1) using the FTCS can be written as

$$S_{yij} \cdot (h^{t+1}_{ij} - h^{t}_{ij}) / \Delta t = 1/\Delta x \; [\overline{K^{x}_{ij}} \; \overline{(h^{t}_{ij} - \eta_{ij})^{x}} \cdot (h^{t}_{i+1\,j} - h^{t}_{ij}) / \Delta x - \overline{K^{x}_{ij}} \; \overline{(h^{t}_{ij} - \eta_{ij})^{x}} \cdot (h^{t}_{ij} - h^{t}_{i-1j}) / \Delta x] + 1/\Delta Y \; [\overline{K^{y}_{ij}} \; \overline{(h^{t}_{ij} - \eta_{ij})^{y}} \cdot (h^{t}_{ij+1} - h^{t}_{ij}) / \Delta Y - \overline{K^{y}_{ij}} \; \overline{(h^{t}_{ij} - \eta_{ij})^{y}} \cdot (h^{t}_{ij-1}) / \Delta Y] + N_{ij} \quad .....(2)$$

* where Ω is the aquifer region , $\partial\Omega_1$ is the impervious boundary region and $\partial\Omega_2$ represents the river region.

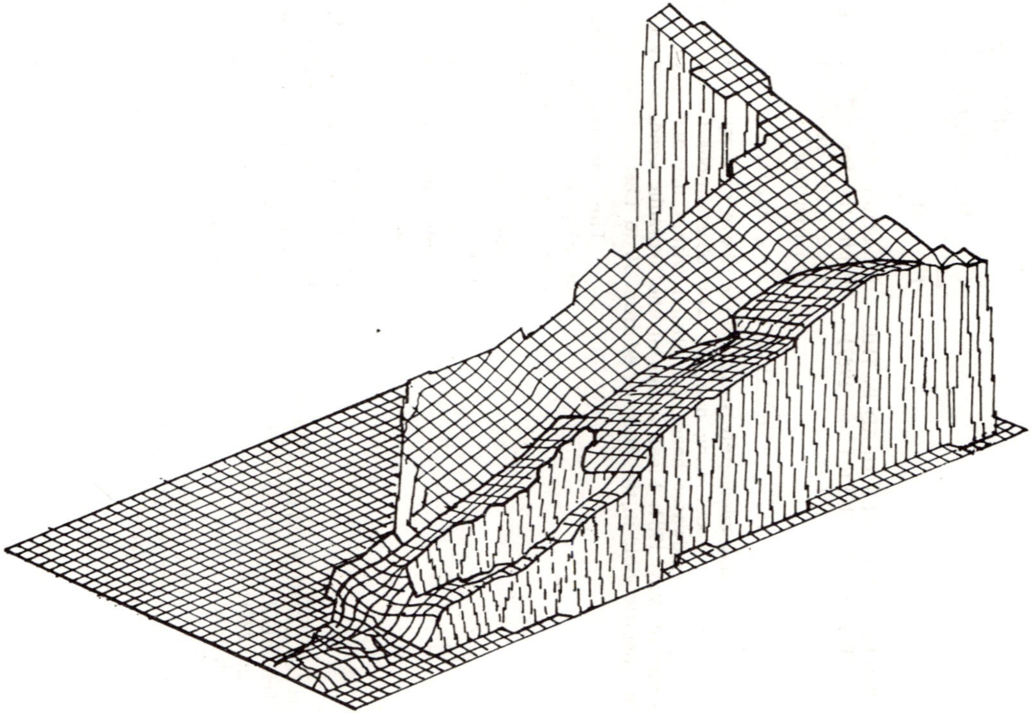

FIG. 2. BLUE LAKE AQUIFER BOTTOM ELEVATION (Min. = 3 ft. and Max. = 66 ft.)
(Impervious boundary is zero)

where the average operator for a given function (say f) is given as

$$\left.\begin{aligned} \overline{f^x} &= (f_{i+1j} + f_{ij})/2 \\ \overline{\overline{f^x}} &= (f_{ij} + f_{i-1j})/2 \\ \overline{f^y} &= (f_{ij+1} + f_{ij})/2 \\ \overline{\overline{f^y}} &= (f_{ij} + f_{ij-1})/2 \end{aligned}\right\} \quad .....(3)$$

Fig. 3 : INITIAL MAD RIVER HEAD IN THE BLUE LAKE AQUIFER (June 1987) Min = 30, Max = 131 ft.

Equation (2) can be solved explicitly for h_{ij}^{t+1}. The model stability constraint further required that

$$\Delta t \cdot \overline{k(h - \eta)}/\Delta x \cdot \Delta y \cdot Sy \leq 0.25 \quad .....(4)$$

A time step of 0.5 day adequately satisfied the above constraint. Chu and Willis (1984) calibrated the model using pumping well observations for one month and worked out the groundwater head distribution within the BLA system for one month period of May 1981. Final values of the hydraulic conductivity are shown in Fig. (4). Specific yield in the BLA varied from 10 to 15%.

Fig. 4 : HYD. CONDUCTIVITY VARIATION IN THE BLUE LAKE AQUIFER Min=2, Max.=270ft/d. (Impervious boundary Hyd. cond. is shown as zero.)

The model using FTCS developed by Chu and Willis (1984) (referred as earlier model) was adequate in the sense that it did not require linearisation of the Boussinesq eq (1) as discussed by Bear (1972). It was also less complicated and time consuming compared to numerical solutions reported for parellel problems by Hornberger et al (1970), Lin (1972), Yoon & Yeh (1975), Pinder & Gray (1977) and Ligget & Liu (1982). The above model was good only for small period of 50 days or less and could not be used for longer period of one year or more because river was assumed as a constant head boundary which is not adequate for simulating the annual flow dynamics of the BLA system involving significant variation in the Mad river stages. However the model was useful in giving the aquifer head values which were taken as initial values for the following modified model.

MODIFICATIONS INVOLVED

A significant modification involved the simulation of the Mad river as a dynamic flux boundary and not as a constant head boundary. The flux from or to the Mad river was worked out as

$$Q_{ijt} = K' / B \; (HRIVER_{ijt} - H_{ijt}) \quad .....(5)$$

where Q_{ijt} is the river flux (ft/d) per unit nodal area at any time t, K'_{ij} is

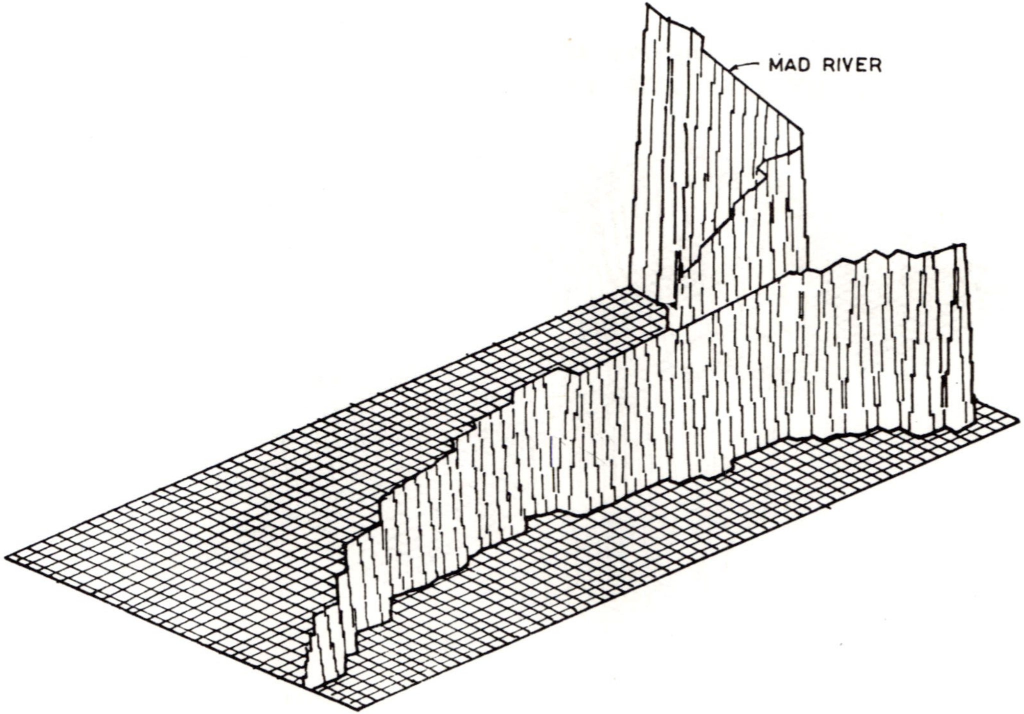

FIG. 3 – INITIAL MAD RIVER HEAD IN THE BLUE LAKE AQUIFER (June 1987)
Min = 30, Max = 131 ft.

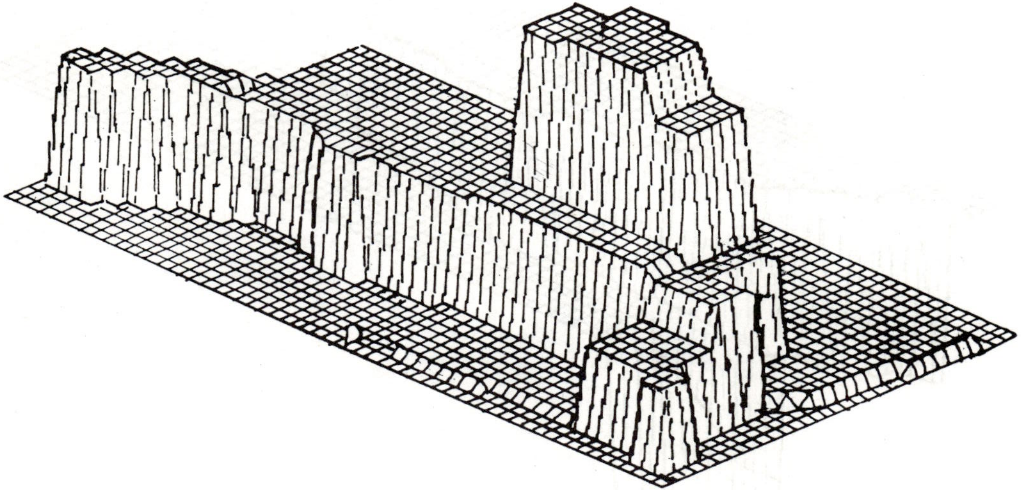

FIG.4- HYD. CONDUCTIVITY VARIATION IN THE BLUE LAKE AQUIFER
Min = 2 , Max. = 270 ft/d. (Impervious boundary hyd. cond. is shown as zero.)

the hydraulic conductivity (ft/d) of the Mad river bed, B_{ij} its thickness (ft), $HRIVER_{ijt}$ is the river head (ft) and H_{ijt} is the groundwater head (ft) vertically below the Mad River at node (i, j) at the same time (t). Equation (5) required a forecast of the river head at all the nodes simulating the Mad River for the planning period of one year. In the BLA 109 nodes simulate the Mad River length over the aquifer and the river head at all the nodes for one year period was obtained from the hydraulic model* of the Mad River. It was based on the solution of St. Venant's mass balance equation (Viesmann et el 1972, pp-190)

$$H.\partial v/\partial x + v\,\partial H/\partial x + \partial H/\partial t = Q(x,t) \qquad .....(6)$$

where H is the river head per unit width (ft), v is the down stream velocity (ft/s) in the (x) direction and Q denotes the lateral inflow (ft/s). Finite difference form of equation (6) was solved implicitly using the Newton - Raphson scheme of solution. The river velocity was known from the river discharge and the sectional area, the former was known from the Ruth reservoir releases which govern the flow to the Mad River. For the initial known values of the river head and the lateral inlows, river head at half day time step (required for the reservoir release consideration 50 miles further up stream) was predicted for the one year planning period by the hydraulic model. Fig. (5) shows the Mad river head forecast from the hydraulic model for one year period at a node (58, 32).

Thus for the first time step the initial groundwater head and river head values were used to work out the river flux term. Flux for the next time step was worked out using the river head predicted by the hydraulic model and groundwater head by the simulation model at the end of the first time step. This process continued for the total period of one year.

Fig. 5 : ANNUAL MAD RIVER HEAD VARIATION AT NODE (58, 32)

Surface recharge terms to the BLA were not considered relevant in the earlier model while the modified model included these values. Surface recharge data for the planing period was predicted from a rainfall-runoff model**. Fig. (6) shows the surface recharge to the BLA as worked out from the rainfall-runoff model. Monthly average values of the surface recharge were considered which was assumed uniformly distributed over the aquifer area. Earlier model used the average transmissivity $\{K.(h-\eta)\}$ values for simulating the flow at a node. This was appropriately changed to harmonic mean of transmissivity values which suitably made the transmissivity outside the impervious boundary zero and therefore the no flow condition was accuretely imposed without making the head across the boundary equal.

Fig. 6 : ANNUAL VARIATION OF SURFACE RECHARGE IN BLUE LAKE AQUIFER
1 cubic feet = 7.48 US GALLON

---

* developed by Prof. Mac Mckee, ** developed by Dr. Brad A Finney, both, Dept. of Env. Resour. Engg., Humboldt State Univ. California, USA.

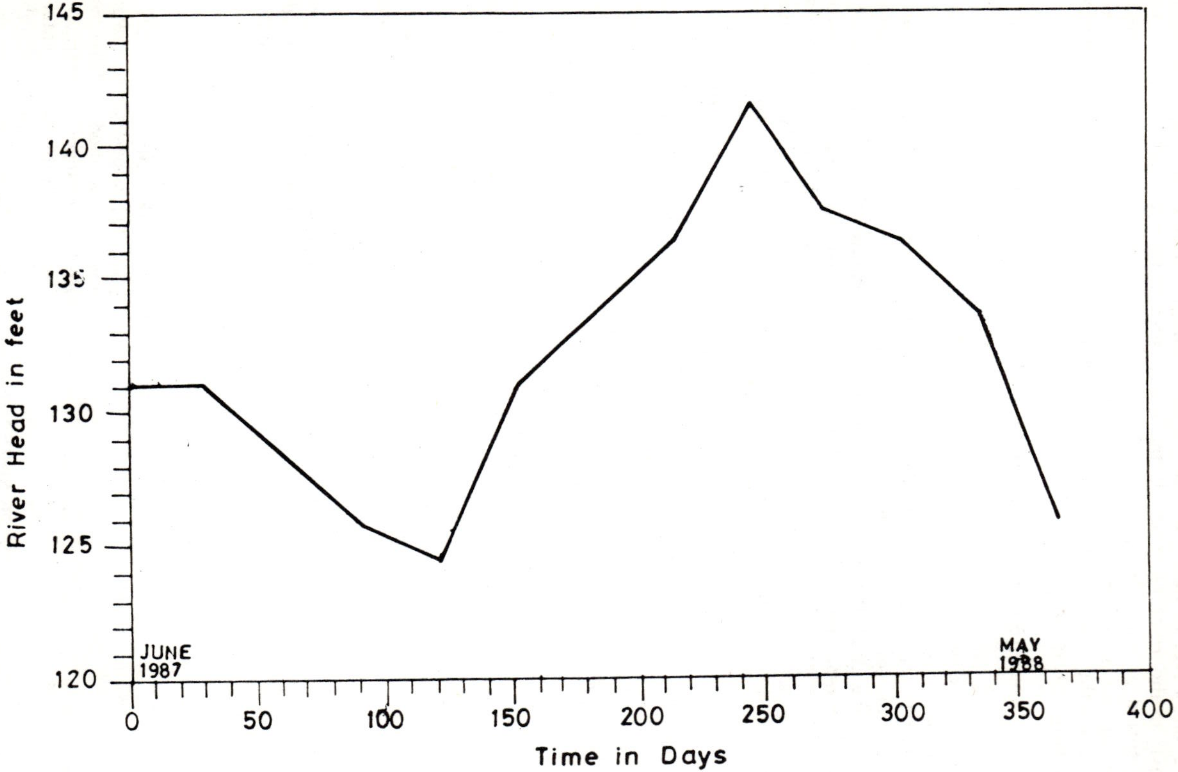

FIG. 5 - ANNUAL MAD RIVER HEAD VARIATION AT NODE (58, 32)

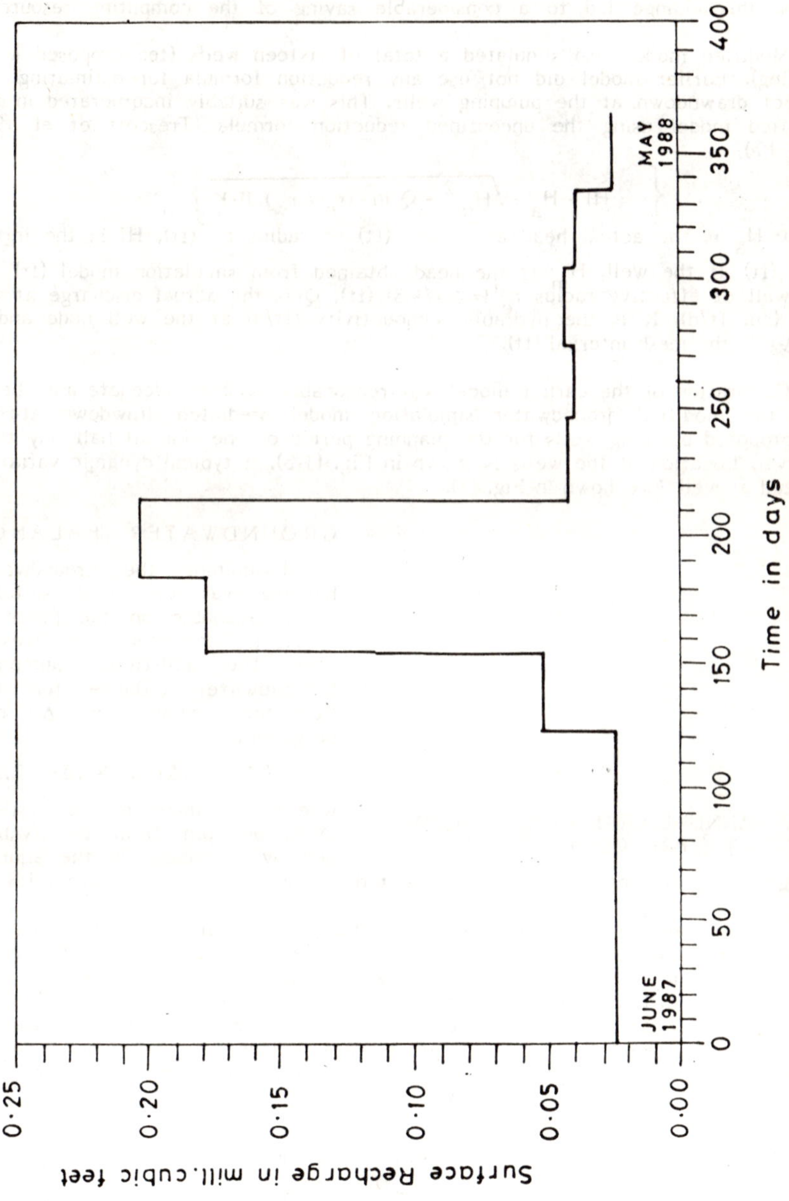

FIG. 6 - ANNUAL VARIATION OF SURFACE RECHARGE IN BLUE LAKE AQUIFER
1 cubic feet = 7·48 US GALLON

This was advantageous in simulating the long impervious boundary of the BLA as the algorithm was considerably shortened. More importantly, since the later optimisation model called the groundwater simulation model over two thousand times, this change led to a considerable saving of the computing resources.

Modified model also simulated a total of sixteen wells (ten proposed + six existing). Earlier model did not use any reduction formula for estimating the correct drawndown at the pumping wells. This was suitably incorporated in the modified model using the unconfined reduction formula (Trescott et el 1976 - pp. 10)

$$\left\{ s = HI - H_a = \sqrt{H_m^{\,2} - Q \ln \left(r_e / r_w\right) \pi \cdot K} \right\}_{ij}$$

where H_a is the actual head at a well (ft) of radius r_w (ft), HI is the initial head (ft) at the well, H_m is the head obtained from simulation model (ft) for the well of effective radius r_e ($= \Delta x/4.8$) (ft), Q is the actual discharge at the well (cu. ft/d), K is the hydraulic conductivity (ft/d) at the well node and $\Delta x = \Delta y$ is the mesh interval (ft).

Calibration of the earlier model was reasonably assumed adequate and therefore the modified qroundwater simulation model predicted drawdown at the ten proposed pumping wells for the planning period of one year at half day time interval. Location of the wells is shown in Fig. (1-b). A typical dynamic variation of head at well 7 is shown in Fig. (7).

GROUNDWATER BALANCE

Examining the groundwater balance was the most suitable check available on the proposed model to account for the stresses from the additional pumping. Groundwater balance for the BLA for a time step Δt can be given as

$$(I - 0) .\Delta t = Sy.\Delta t \qquad ..(7)$$

where I is input to the system, O is out put from the system and Sy is change in the aquifer storage during the interval Δt (as reflected by the changes in the water level).

Fig. 7 : ANNUAL VARIATION OF HEAD AT A PROPOSED WELL 7

The input to the system was surface recharge (evaporation/evapotranspiration losses were accounted during the surface recharge evaluation) and the river flux terms while the out-put was the total discharge from the sixteen pumping wells in the BLA. Change in storage was worked out by multiplying head difference at time interval Δt with aquifer area and the specific yield. After above modifications the BLA model satisfied the groundwater balance for the planning period of one year. Main components of the annual groundwater balance worked out from the model are presented in TABLE 1.

TABLE 1 (all units are in Million Gallon)

1. Net annual Mad River flux (MRF) = $\sum_{\Delta t=1}^{730} (MRF)_{i \in \partial\Omega_1}$ = 15400
2. Net annual surface recharge to aquifer (SR) = $\sum_{\Delta t=1}^{730} (SR)_{i \in \Omega}$ = 33.7

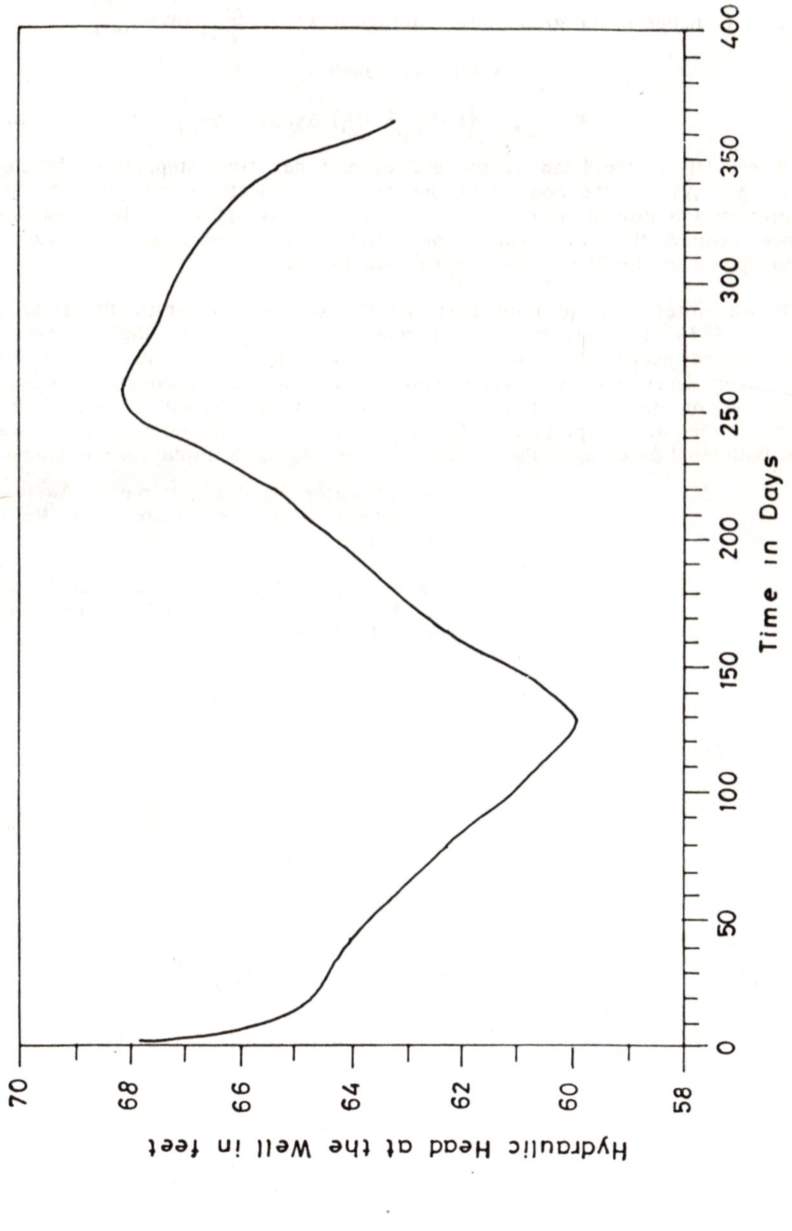

FIG.7 – ANNUAL VARIATION OF HEAD AT A PROPOSED WELL 7

cont. of table 1

3. Yearly pumping discharge from the 16 wells (PD) $= \sum_{\Delta t=1}^{730} \sum_{i=1}^{16} (PD)_i$ = 12987

4. Hence the annual change in the aquifer flux (CAF) $= \sum_{\Delta t=1}^{730} (CAF)_{i \in \Omega}$ = 2446.[illegible]

5. Yearly change in the groundwater storage (GWS) $= \sum_{\Delta t=1}^{730} (GWS)_{i \in \Omega}$ = 2447

where for a node i

$$(GWS)_{i,\Delta t} = \left\{ (HF_{i\Delta t} - HI_i) . \Delta x . \Delta y . \; Sy_i \right\} \qquad .....(8)$$

Where HF is the head at the end of half day time step, HI is the initial head, $\Delta x . \Delta y$ is the nodal area and Sy is the nodal specific yield. Since 4 ≈ 5 therefore the groundwater balance equation 7 was satisfied. The groundwater balance assured that a dynamic correlation betwen the aquifer stresses and their response on the BLA was correctly established.

It was interesting to note that for the six existing wells, the river flux was only 6836 MG/year. However increased withdrawl from the proposed ten wells induced additionally over 100% river flux into the Blue Lake aquifer system. This suggested a very good hydraulic continuity and adequate river-aquifer interaction for useful groundwater withdrawl as earlier implied in the preliminary reports of Bechtal Corp. (1960), Trotter & Assoc. (1960) and Cal. State Water Reso. Bull(1965) based upon the localised studies. Fig. 8 displays some components of groundwater balance of the BLA. It can be observed from the figure that the net aquifer storage of the BLA is a linear function of the Mad River flux. A typical three dimensional computer plot showing groundwater head distribution in the BLA at the end of month 5 (Oct. 1987) is shown in Fig. (9)

Fig. 8 : ANNUAL VARIATION OF TOTAL PUMPING(A) AQUIFER STORAGE(B) AND MAD RIVER FLUX (C) IN BLUE LAKE AQUIFER

Fig. 9 : HEAD DISTRIBUTION IN BLUE LAKE AQUIFER AT THE END OF OCT. 1987 (Value at impervious boundary is shown as zero) Min = 26 ft. and Max = 140 ft.

OPTIMISATION

HBMWD established an additional requirement of 23 MGD for the Humboldt County for which the authorities selected 10 tubewell sites. With this given condition it was required by the HBMWD to obtain an optimal pumping policy whereby the annual pumping cost to meet the above demand was minimised. The cost of pumping involved the product of pumping rate, the cost of energy and the lift. Cost of energy being constant, the minimum pumping cost was

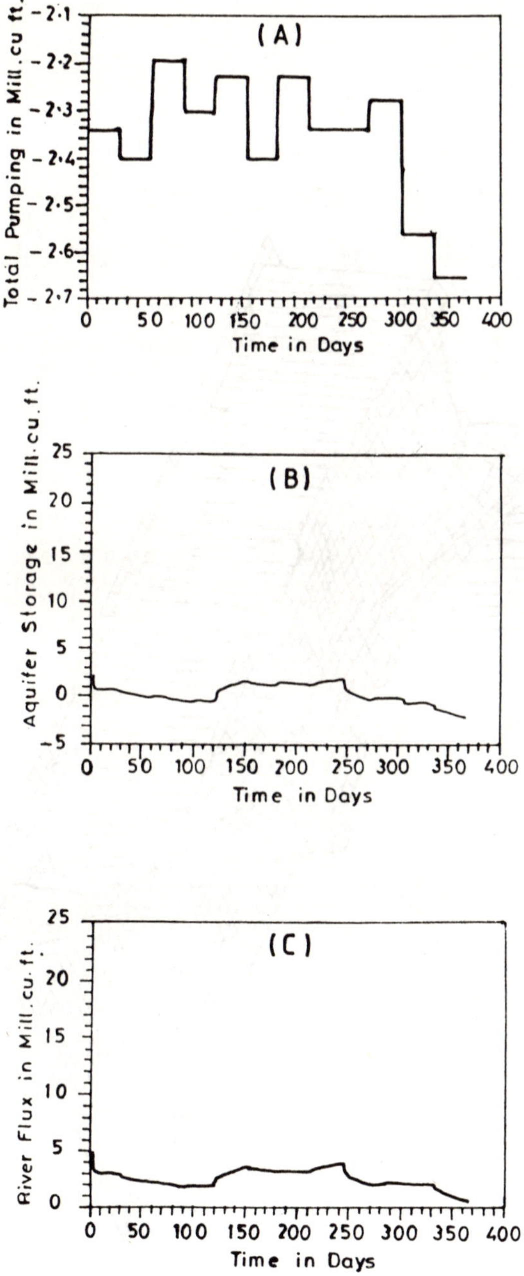

FIG.8 – ANNUAL VARIATION OF TOTAL PUMPING (A) AQUIFER STORAGE (B) AND MAD RIVER FLUX (C) IN BLUE LAKE AQUIFER

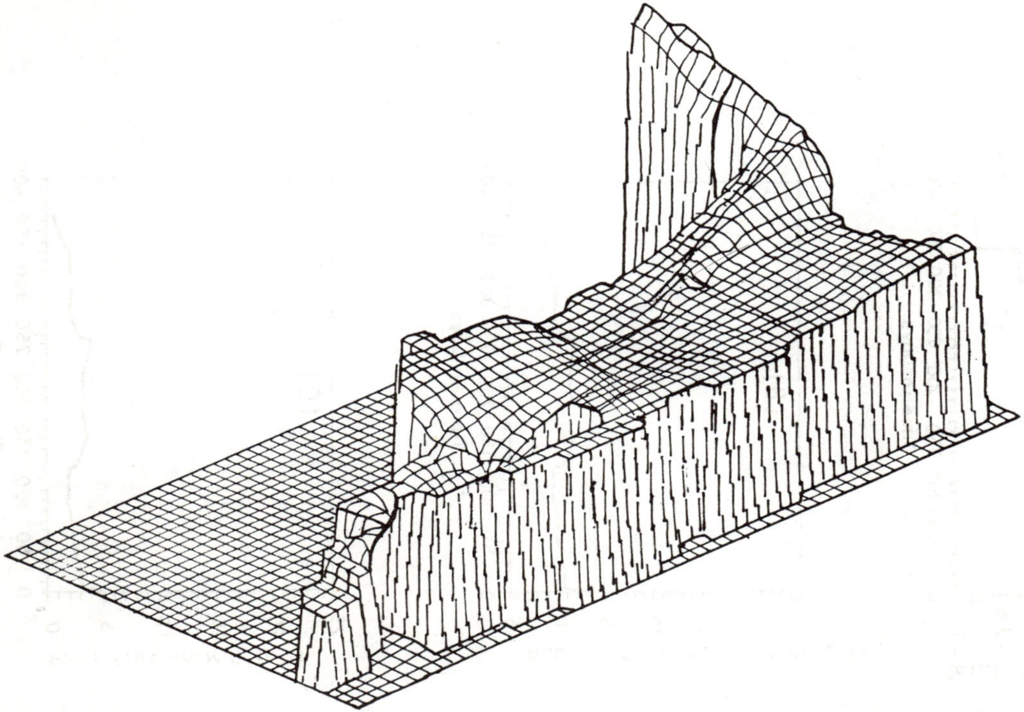

FIG. 9_ HEAD DISTRIBUTION IN BLUE LAKE AQUIFER AT THE END OF OCT. 1987
(Value at impervious boundary is shown as zero)
Min. = 26 ft. and Max. = 140 ft.

equivalent to finding the optimal (max) pumping rate and the (minimum) drawdown combination. The capital investment for pump installation was not considered here as their number was predetermined by HBMWD.

Principles of nonlinear optimisation theory were used to determine the annual optimal pumping policy. A very brief review of the recent hydraulic management model is given before use of the most appropriate technique is discussed.

Effect of aquifer stresses (pumping or recharge) can be adequately responded by the aquifer model and these simulation models have been used as groundwater management model, based on "what if" type approach (Wanakule et el 1986). It involves analysing the aquifer response to a large no. of possible variations in pumping and recharge policies and selecting the most appropriate technique. For the present problem involving pumping bounds from 1 to 3 MGD and 120 control variables might involve almost infinite number of aquifer simulations.

Response Matrix Approach (RMA) for linear and non-linear optimisation was used by Maddock (1974), Morel Seytoux et el (1980), Heidari (1982) and Willis (1984). This technique involved the generation of a unit response matrix by solving simulation modal several times to determine the response of unit pumping or recharge at selected points in the aquifer. Superposition was used to workout the cumulative influence of pumping or recharge. Wanakule et al (1986) reported that this method was exact only for confined aquifer or for small drawdown problems. The RMA was not considered appropriate for the Blue Lake Aquifer optimisation as large drawdowns were noted at the pumping zones.

Aquado (1974), Aquado and Remson (1980), Willis and Newman (1977), Remson and Gorelick (1980) and Willis and Liu (1984) used the Embedding Approach (EA) for solving the groundwater hydraulic management models. In EA the finite difference or finite element discretisations of the corresponding groundwater flow equatien were treated as part of the constraint set of the optimisation model. Therefore decision variables were the hydraulic heads at each node, pumping from the wells and recharge policy decisions (control variables). It therefore involved a very large number of dicision variables which was dependent on the product of the number of planning periods and the spatial discretization of the aquifer. As Gorelick et el (1984) observed "for computational economy and avoidance of numerical difficulties EA should currently be restricted to small steady state problem". For the BLA involving 61 x 31 nodes EA was also not considered appropriate.

A modification in the EA (Modified Embedding Approach MEA) was suggested by Gorelick et el (1984) and applied by Wanakule et el (1986). MEA expressed the state variable (head) as an implicit function of the control variable (pumping), thereby conceptually eleminating the model constraints and yielding a smaller reduced problem involving only the control variables. Overall problem could therefore be viewed as one of discrete time optimal control with reduced constraints. MEA was considered suitable for examining the optimal pumping policy of the Blue Lake Aquifer. For using MEA, the BLA optimisation problem was posed as :

$$\text{MIN. } F = \sum_{t=1}^{12} \sum_{i=1}^{10} C.\ Q(i,t).\left\{L(i) + \overbrace{HI(i) - HF(i,t)}^{\text{Draw down}}\right\} \qquad .....(9)$$

with the constraints

$$\Sigma\ Q(i, t) \geq D(t) \quad \text{where } D(t) = 23 \text{ MGD} \qquad .....(10)$$

$$u \geq Q(i, t) \geq l \quad \text{where } u = 3.5\text{MGD}, \; l = 1\text{MGD} \qquad .....(11)$$

$$HF(i, t) \geq 0.5\ HI(i) \qquad .....(12)$$

and t = 1, ... 12, i = 1, ...10

where obj function F represents the total cost of lifting 23 MGD from 10 wells for 12 operational period, t is the operational period (of one month each), i is number of the well, C is the cost (cents) of lifting unit MG of water by one foot, Q(i, t) is the average daily discharge (MGD) from well i during operational period t (month), HF(i, t) is the head (ft) at well i at the end of operational period t, HI(i) is the initial head at pumping well i, L(i) is the distance (ft) from the ground surface to initial water level HI(i) at each well location i ; u and l are the upper and lower bounds on the pumpage which are based upon real system (for upper bounds) and optimisation requirements (for lower bound). Thus upper bound on Q is the pumping discharge at which the head approaches critical (18 ft) similarly lower bound on pumping was selected 1 MGD to give a good start to the optimisation problem. It was chosen after the intial runs which showed that all optimal pumping values were above 1 MGD (refer Appendix-I). Equation (12) specify the lower bounds on the final head at well i at the end of each operational period (t). This bound on head is the critical head value for efficient pumping.

The objective function is non-linear having 120 nonlinear variables in Q. In equation (9) the term in bracket can be represented as an implicit function of Q. The groundwater simulation model was used to solve this implicit function value.

This nonlinear problem was optimised using the software package MINOS 5.1. MINOS (modular incore nonlinear optimisation system) is the most recent commercially available software package in USA to solve the nonlinear problems of optimisation. It was developed by Murtagh and Saunders (1987) utilizing the long experience of the faculty of the Dept. of Operations Research, Stanford University, California. MINOS uses the Reduced Gradient Method (Wolfe 1962) and a Projected Lagrangian Method (Robinson 1972 ; Rosen and Kreuser 1972). MINOS has the ability to estimate all gradient vectors numerically. An user written subroutine called FUNOBJ was developed for computing nonlinearities in the objective function. A line search tolerance of .01 was considered adequate for the present problem. Demand Constraints and all the upper and lower boounds on the head and well discharge values were suitably incorporated in the later section of a data input file (SPECF) which follows an MPS (mathematical progrmming systems) format for entering the data. Earlier section of the SPECF sets various run time parameters that describe the nature of the problem. MINOS assumed initial pumping values which satisfied the lower and upper bounds on pumping for each well and the total daily pumping demand constraints of 23 MGD. MINOS chooses this initial solution vector by arbitrarily keeping some pumping values equal to their upper (3.5 MGD) and the rest equal to their lower bounds (1 MGD). MINOS then called simulation model and passed on these initial pumping values to the simulation model. Simulation model returned the head values at these pumping wells for twelve operational periods to MINOS. MINOS worked out the objective function, checked for the optimality requirement and if the objective function was not optimal, passed these pumping values back to the simulation model. Again the simulation model used these pumping values and returned groundwater head values to MINOS. This process continued until the objective function was minimised to the specified tolerance limit. A flow diagram of the hydraulic management model is shown in Fig. (10).

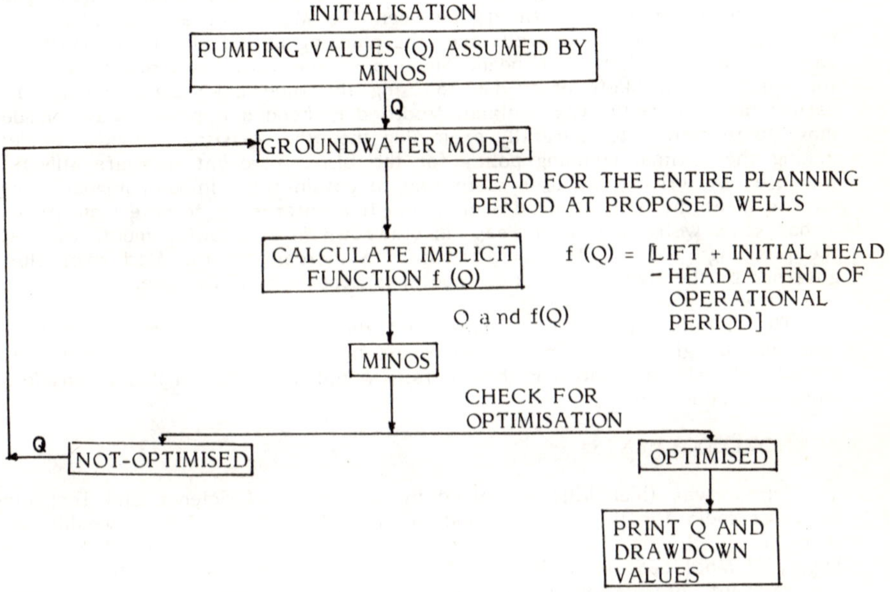

Fig. 10 - FLOW DIAGRAM FOR THE HYDRAULIC MANAGEMENT MODEL

MINOS called the BLA model 2500 times and took 30 hrs of computing time on a slow machine cyber-720 (more recent computers can find the solution in less then one hour). The minimum cost for pumping the additional 23 MGD for one year was given as $55, 300. However, if the wells were pumped at 2.3MGD each (not-optimised) for all the operational periods then the cost to deliver the same quantity of water would be $75,300. Therefore a net benefit of $20,000/ year was anticipated. The optimal yearly pumping policy and the corresponding drawdown are given in Appendix I for each operational period of one month.

CONCLUSIONS

Blue Lake Aquifer model involved the nonlinear governing flow equation. Requirement of half day time steps justified the use of explicit FTCS scheme, because explicit method is much faster (computationally) compared to implicit (iterative) methods and especially an implicit scheme is not superior to an explicit scheme for solution of inherent nonlinear equation (Richtmyer and Morton, 1967). As a matter of fact for the economy of computing resources it was extremely important presently to choose the least-time groundwater model since the hydraulic management model called the groundwater model several hundred times before the optimal pumping values were ascertained. Many modifications to the existing FTCS scheme were made. More importantly, the use of harmonic mean of transmissivity reduced the size of the algorithm and implemented the no flow boundary condition more accurately. Simulation of Mad River as a dynamic flux boundary replaced the existing constant head boundary concept of simulating a river.

This however demanded an advance knowledge of the river heads for the annual planning period which were provided from a hydraulic model of the Mad River. Simulation model responded well to the stresses from ten additional wells as was reflected by a correct groundwater balance for the Blue Lake Aquifer system. Due to good hydraulic continuity between the Mad river and aquifer, dynamic head variation at pumping wells over the planning period closely followed annual variation in the Mad river head. Since only the drawdown results were used for optimisation, likely error due to using the Blue Lake aquifer head of the earlier model initially was minimal. Modified Embedded Approach was considered most appropriate after carefully examining the other existing methods, for determining the optimal pumping policy for the BLA. A recent software MINOS 5.1 was used by writing the user subroutine to obtain the minimum annual pumping cost to meet the daily demand over a year. It is interesting to note from Appendix I that some wells show a decrease in drawdown in a following month even when the pumping goes up. This is attributed to increase in the Mad river flux as governed by the river head in that month. It's opposite is also true.

The details on hydraulic model and the rainfall-runoff model of the BL area will be given later in a paper/report by their developer. Presently author considered that incorporating their details would shift the emphasis considerably from the main issue.

ACKNOWLEDGEMENTS

Author was financially supported by the Dept. of Science and Technology, Govt. of India for the entire period of his stay in USA. Author would like to thank Profs Robert Willis, Mac Mckee and Brad Finney, Dept. of Env. Reso. Engg., Humboldt State Univ., California, for providing all the technical data and relevant information on the Blue Lake Aquifer. Prof. Willis also extended the use of MINOS to the author.

REFERENCES

Aguado, E., I. Remson, M. F. Pikul, and W. A. Thomas, 1974. Optimum puming for aquifer dewatering, J. Hydraul. Divn., Am. Soc. Civ. Eng., 100, 860-877.

Aguado, E., and I. Remson, 1980. Groundwater management with fixed charges, J. Water Resour. Plann. Manage. Divn., Am. Soc. Civ. Eng., 106, 375-382.

Bear, J., 1972. Dynamics of Fluid in Porous Media, Elsevier, New York, Ny, 361-438.

Bear, J., 1978. Hydraulics of Groundwater, Mc Graw Hill, New York.

Bechtel Corporation Consulting Engineers, 1960. Report on hydrogeological survey Humboldt Bay Municipal Water District, Eureka, California. (unpublished)

CaliforniaState - Dept. of Water Resour., Bull No. 142-7, 1965. Water resources and future water region, North Coastal Hydrographic Area, Vol. I.

Chu, W. S. and R. Willis, 1984. An explicit finite difference model for unconfined aquifers, Groundwater, 22(6), 728-734.

Gorelick, S. M., C. I. Voss, P.E. Gill, M. Murray, M.A. Saunders, and M.M. Wright, 1984. Aquifer reclamation design : The use of contaminant transport simulation combined with nonlinear programming, Water Resour. Res., 2o(4), 415-427.

Heidari, M., 1982. Application of linear systems theory and linear programming to groundwater management in Kansas, Water Resour. Bull., 18(6), 1003-1012.

Hornberger, G.M., J. Ebert, and I. Remson, 1970. Numerical solution of the Boussinesq equation for aquifer - stream interaction Water Resour. Res., 6(2), 601-608.

Knowles, T., 1981. GWSIM III - Groundwater simulation program, Program Document and User's Manual, Tex. Dep. of Water Resour., Austin.

Ligget, J. A., and P. L-F. Liu, 1982. The Boundary Integral Equation Method for Porous Media Flow, Allen Unwin, Winchester, MA, pp. 272.

Lin, C. L., 1972. Digital simulation of the Boussinesq equation for a water table aquifer, Water Resour., Res., 8(3), 691-698.

Maddock, T., III, 1974. Nonlinear technological functions for aquifers whose transmissivities vary with drawdown, Water Resour. Res., 10(4), 877-881.

Morel Seytoux, H.J., G. Peters, R. Young, and T. Illangasekare, 1980. Groundwater modelling for management, International Symposium on Water Resources Systems, Water Resour. Dev. Train. Centre, Univ. of Roorkee, India.

Murtagh, B.A., and M.A. Saunders, 1987. MINOS 5.1 user's guide, Systems Opti. Lab., Dept. of Opera. Res., Standord Univ., Standord, California.

Pinder, G. F., and W. G. Gray, 1977. Finite Element Simulation in Surface and Subsurface and Subsurface Hydrology, Academic Press, NewYork, 126-199.

Remson, I., and S.M. Gorelick, 1980. Management models incorporating groundwater variables, in Operation Research in Agriculture and Water Resources, edited by D. Yarn, and C.S. Tapiero, North Holland, Amsterdam.

Richtymer, R. D., and K. W. Morton, 1987. Difference Method for Initial Value Problem, 2nd Ed., Interscience.

Robinson, S. M., 1972. A quadratically convergent alogorithm for general nonlinear progrmming problems, Math. Prog. 24, 55-69.

Rosen, J.B., and J. Kresuer, 1972. A gradient projection algorithm for nonlinear constraints in Numerical Methods for Non-Linear Optimisation, edited by F.A. Lootsma, Academic Press, London & New York, 297-300.

Trescott, P.C., G.F. Pinder, and S.P. Larson, 1976. Finite difference model for aquifer simulation in two dimensions with results of numerical experiment, Tech. of Water Resour. Invest. of the U.S. Geolog. Surv., Chapter Cl.

Trotter, R. M., and Associates, 1960. Report on preliminary hydrogeological survey for city of Arcata, California. (unpublished)

Viesmann, JR, W., T.E. Harbaugh, and J.W. Knapp, 1972. Introduction to Hydrology Intext Educational, NewYork/London.

Wanakule, N.L., L.W. Mays, and L.S. Lasdon, 1986. Optimal management of of large scale aquifers : Methodology and application, Water Resour. Res. 22(4), 447-465.

Willis, R., 1984. A unified approach to regional groundwater management, in Groundwater Hydraulics, Water Resour, Monogr. 9, edited by J.S. Rosenshein and G.D. Benett, AGU, Washington, DC.

Willis, R., and W-S Chu, 1981. Water resources systems investigation of the Mad River Basin, Final report to Humbouldt Bay Municipal Water District, Eureka, California. (unpublished)

Willis R., and P. Liu, 1984. Optimisation model for groundwater planning, J. Water Resour. Plann. Mange. Div., Am. Soc. Civ. Eng., 110, 333-347.

Willis, R., and B.A. Newman, 1977. Management model for groundwater development, J. Water. Resour. Plann. Manage. Div., Am. Soc. Civ. Eng., 13, 159-171.

Wolfe, P., 1962. The reduced gradient method, unpublished manuscript, RAND Corporation, Santa Monica, California.

Yoon, Y. S., and W. W-G Yeh, 1975. The Galerkin method for nonlinear parabolic equation of unsteady groundwater flow, Water Resour. Res., 11(5), 751-754.

APENDIX I
(A)

WELL NO	1	2	3	4	5	6	7	8	9	10	
	2183742.	1287997.	2238806.	2078427.	2069962.	3100000.	3196239.	2259451.	2317950.	2215341.	1
	3370492.	2821725.	2523722.	1452630.	1452630.	1452630.	1452630.	3459899.	3500000.	1452630.	2
	2139291.	1774902.	1964886.	2620981. *	2597911.	3184454.	3252587.	1968981.	2043353.	1452630.	3
	2160475.	1590936.	1883656.	2799490.	2799482.	3188448.	3470727.	1842382.	1811749.	1452630.	4
	2057655.	2108993.	2712301.	3051484.	2866145.	3382878.	1452630.	1452630.	1952630.	1052630.	5
	3500000.	3500000.	3500000.	2663155.	1452630.	1452630.	1952630.	1452630.	1852630.	1452630.	6
	3500000.	3500000.	2663155.	1452630.	3500000.	1452630.	1652630.	1952630.	1452630.	1452630.	7
	3500000.	1452630.	1652630.	1452630.	1652630.	1452630.	3500000.	3373937.	3343883.	1452630.	8
	2235010.	1954183.	2309657.	2663015.	2544216.	3500000.	3266157.	1452630.	1452630.	1452630.	9
	2580308.	1452630.	1452630.	1452630.	3241156.	1452630.	3500000.	2970498.	3071183.	1452630.	10
	2146530.	1784859.	1972302.	2623521.	2600093.	3148474.	3251485.	1973379.	2046702.	1452630.	11
	3500000.	3500000.	1452630.	1152630.	1752630.	1452630.	1452630.	3377295.	3343271.	1452630.	12

Optimal pumping policy for the ten additional wells in Blue Lake Aquifer for 12 operational periods.

* average daily pumping in gallons during a month.

(B)

WELL NO.	1	2	3	4	5	6	7	8	9	10	MONTH
	10.89	18.88	17.92	11.62	10.89	18.23	17.75	16.05	15.33	13.15	1
	17.92	24.45	21.31	18.96	18.47	15.57	15.08	10.17	19.20	14.12	2
	15.25	21.79	20.82	16.71 **	16.46	12.83	12.83	10.65	19.20	17.75	3
	16.95	22.52	22.03	20.09	19.85	15.25	16.22	12.59	11.14	10.17	4
	14.04	23.73	22.27	17.43	16.46	19.93	16.54	14.84	14.36	13.63	5
	15.74	21.33	24.45	11.14	15.81	11.21	11.21	10.00	18.79	18.31	6
	12.83	20.94	16.46	13.63	10.41	17.82	17.09	16.13	14.43	14.19	7
	16.54	18.47	15.81	18.06	16.61	12.49	17.58	13.71	12.98	12.11	8
	13.39	10.41	19.68	13.39	11.94	18.79	16.85	12.74	10.56	10.17	9
	16.29	19.44	17.75	10.97	15.33	16.61	10.48	18.31	16.61	12.74	10
	17.26	13.07	11.86	17.26	16.29	12.42	11.94	19.52	17.58	15.88	11
	20.58	22.81	16.95	18.96	18.47	16.29	14.84	11.86	19.44	15.57	12

Corresponding drawdown for the ten additional wells in Blue Lake Aquifer for 12 operational periods.

** total drawdown at the end of the month in ft.

DETERMINATION OF ANISOTROPIC PERMEABILITY OF AN UNCONFINED AQUIFER IN EGYPT USING PUMPING TEST DATA OF A PARTIALLY PENETRATING WELL

Dr Engr Mamdouh Shahin
International Institute for Hydraulic
and Environmental Engineering
Oude Delft 95 - P.O. Box 3015, 2601 DA Delft
The Netherlands

ABSTRACT

The paper.discusses a number of methods used for determining the permeability of an aquifer. There is one group of methods that considers the horizontal permeability only. The first type of methods in this group is based on mechanical analysis of the aquifer material. Such a type requires pumping-test data only for calibrating the regression constant. The second type includes the analysis of test data from wells fully penetrating the aquifer. It's application to partially penetrating wells yields small permeability values. Estimation of the adequate correction factor is not always feasible and sometimes it does not produce a satisfactory improvement. The second group of methods takes into consideration the anisotropy of the aquifer permeability. Two methods based on the steady-state and one method on the unsteady state, both of groundwater flow towards the pumped well, are presented. All these methods have been applied to a pumped well situated east of the Nile Delta and the results obtained are reported and discussed.

INTRODUCTION

The increasing demand on water in Eqypt and the marked rise of the watertable in several parts of the country have urged the authorities to consider some of the technical and economic questions concerning the exploitation of groundwater aquifers. A proper exploitation of these resources can help satisfy some of the irrigation, drainage and water supply requirements. The potentials and limitations of the groundwater resources in Egypt have been presented by Shahin (1987a), and so is the feasibility of vertical drainage of the irrigated land (1987b).

Several references, amongst them El-Kashef (1983), have reported on the possible extent of saltwater intrusion in the aquifer underlying the Nile Delta as a consequence to pumping of groundwater.

STATEMENT OF THE PROBLEM AND OBJECTIVE OF THE PRESENT STUDY

The wells installed for pumping groundwater in Egypt, like many other countries, are provided with short screens which do not penetrate the full thickness of the aquifer from which water is pumped. The penetration-ratio barely exceeds 0,20. The results obtained from pumping test data are often regarded as uncertain because field conditions are mostly not in agreement with the assumptions underlying the traditional analysis. The vertical component of groundwater flow to a pumped well partially penetrating an aquifer should be considered in the analysis and not ignored.

The present study has thus been carried out with the following objectives:

i - Presenting some recent techniques for analyzing pumping test data, in which anisotropy of the aquifer material is taken into consideration. Those techniques are applied to the data obtained from a test well located east of the Nile Delta area.

ii - Comparing the results obtained in (i) to those obtained from the application of the traditional methods of analysis to the same pumping test data.

EXPERIMENTAL SET UP

The pumped well exists at a site called Abu-Shalaby. It is installed in the area east of the Nile Delta (see location map, Figure 1). The aquifer thickness is estimated at say 300 m. The total depth of the well is 64 m, of which the last 20 m are the screen length. This gives a penetration ratio of less than 0.07.

There are 5 observation wells A_1, A_2, A_3, A_4 and A_5 at distances of 0.75, 13.5, 27.5, 41.5 and 57.5 m respectively, from the centre line of the discharge well. These observation wells terminate at the same level as the bottom of the screen of the pumped well. The second set of observation wells form a battery of 5 wells; A_{28}, A_{52}, A_{75}, A_{100}, and A_{120}. The battery is located at 5 m distance from the centre line of the pumped well, and the code number refers to the depth in m below the static watertable. The discharge of the pumped well is led through a channel to the so-called spreading fan, therefrom to a depression which acts as a large recharge well 10 m in diameter.

GENERAL CONSIDERATIONS

Basic Assumptions Underlying All Methods

The basic assumptions in all methods used here are:
- the aquifer is homogeneous and of infinite areal extent,
- the water table before pumping is horizontal over the area influenced by pumping,
- the aquifer is pumped at a constant rate, and
- the storage in the pumped well can easily be neglected.

Correction of Drawdown Measurement

Since the discharge of the pumped well is recharged back to the aquifer not far from the well site (see Figure 1), the observed drawdowns are definitely affected by the recharge. The rise in the piezometric head has been calculated using the formula developed by Hantush (1964). This formula reads

$$s' = \frac{Q_i}{8\pi k_h . l} F\left(U, \frac{l}{r}, \frac{z}{r}\right) \quad \ldots (1)$$

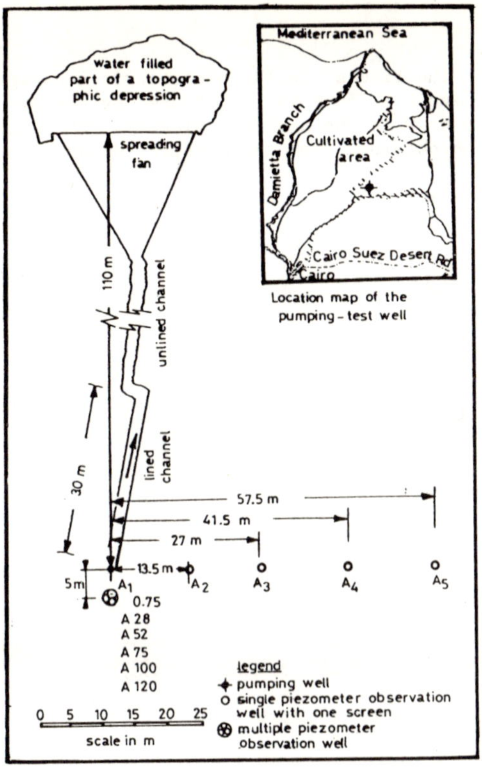

Figure 1 - Plan of the experimental site and location of the pumping well.

where,

s' = rise of piezometric level at a distance r from the test well and at depth z after a time t from the beginning of the test,

Q_i = rate of infiltration = Q_w = discharge of pumped well, equal to 4680 m^3/d,

k_h = horizontal permeability estimated at 60 m/d,

1 = depth of infiltration well below the groundwater table, taken as 5 and 25 m, and

$$F\left(U, \frac{1}{r}, \frac{z}{t}\right) = M\left(U, \frac{1-z}{ar}\right) + M\left(U, \frac{1+z}{ar}\right)$$

where,

$a = K_v/k_h$ and k_v = vertical permeability. Since k_v is taken as $k_h/4$, a becomes 0.5.

The function M {U, $(1 \pm z/ar)$} is an infinite integral, for which numerical values are given in the literature (e.g. ILRI, 1976).

As an example, the rise s' in the piezometric head against t for r = 100 m, z = 100 m and D = 150 m is illustrated by the two curves in Figure 2, for l = 5 and 25 m. D is the aquifer thickness. The drawdown s caused by pumping alone can be obtained from the measured drawdown s_m as

$$s = s_m - s' \qquad \ldots (2)$$

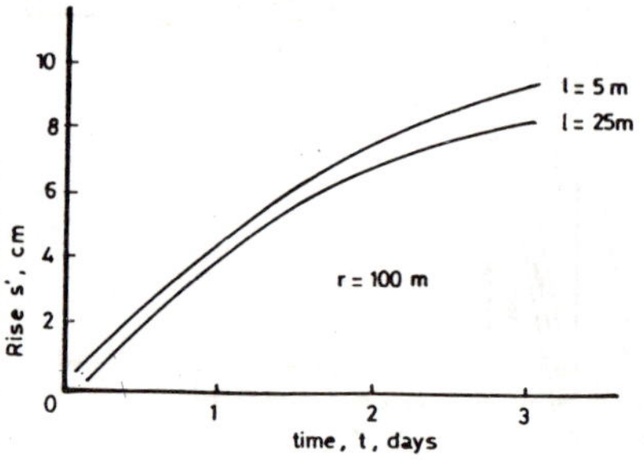

Figure 2 - Rise s' versus time t for r = 100 m, and l = 5 and 25 m.

METHODS OF ANALYSIS BASED ON AQUIFER ANISOTROPY

Steady-state of Groundwater Flow

a) A recent formula developed by RIGW/IWACO (1984), in which the aquifer is assumed to be infinitely deep can be written as

$$s\ (r,\ z) = \frac{Qw}{4\pi k_h\ (b-d)}\ F\ (\frac{ar}{b},\ \frac{z}{b},\ \frac{d}{b}) \qquad \ldots (3)$$

where d and b are the depths to the top and bottom faces, respectively, of the screen of the pumped well below the static water table level. Other notations have been already defined in the previous section.

The functional relationship $F\ (\frac{ar}{b},\ \frac{z}{b},\ \frac{d}{b})$ has been found by RIGW/IWACO (1984) as

$$F\ (\frac{ar}{b},\ \frac{z}{b},\ \frac{d}{b}) = \text{arcsinh}\ (\frac{z+b}{ar}) - \text{arcsinh}\ (\frac{z-b)}{ar}) +$$

$$+ \text{arcsinh}\ (\frac{z-b}{ar}) - \text{arcsinh}\ (\frac{z+b)}{ar}) \qquad \ldots (4)$$

The procedure of analysis is based on matching a plot of the field observed data and a set of type curves of (ar/b, z/b, d/b) versus z/b developed from eq. 4.
A curve representing s (r , z) = 0.28 versus z/b as obtained from the test data was found to be matched quite well by the type curve for x = 0.047 (see Figure 3). If F in eq. 3 is put equal to 1, k_h can be given by

$$k_h = \frac{Qw}{4\pi l s(r,\ z)} \qquad (3')$$

From eq.3' k_h can be found at 66 m/d. Moreover, as $k_v = k_h \cdot (xb/r)^2$, k_v can easily be found at 24 m/d.

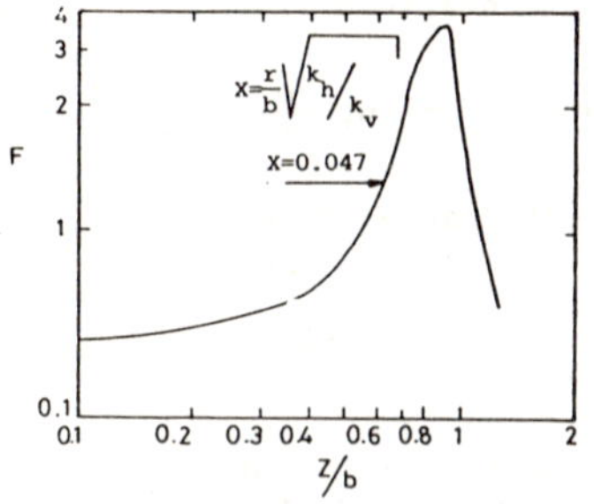

Figure 3 - Type curve used in the analysis of steady-state flow of groundwater towards a pumped well using the drawdowns in a battery of observation wells with constant r and variable z.

b) Another method based on the steady state of flow like the one in (a), except that the type curve is prepared for F (r, z) versus ar/b. The curve of s (r, z) versus z/b, already obtained from the field data is matched to the type curve. If the drawdowns are measured in observation wells where z is constant and r is variable there will be one type curve only.

From Figure 4 it can be seen that by using the above procedure one gets x = 1 and r/b = 1.9. Since x = ar/b the ratio a = kv/kh is 0.526 and kv = 18 m/d.

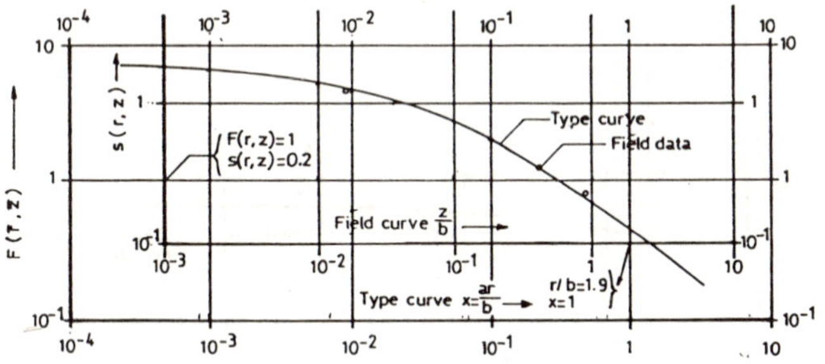

Figure 4 - Analysis of steady-steady flow of groundwater towards a pumped well using the drawdowns in observation wells with constant z and variable r.

Unsteady-state of Groundwater Flow

This method can be used when the state of the groundwater flow towards the pumped well is unsteady. This is the case when the time from the beginning of the pumping test is rather short that

$$t < sD \{1-(+z)/2D\}^2/5\ k_v \qquad \ldots (5)$$

Hantush (1964) showed that within the time range given be eq. 5 the time-dependent drawdown s (r, z, t) can be approximated by

$$s\ (r,\ z,\ t) = \frac{Qw}{8\pi k_h (b-d)}\ F\ (u,\ \frac{b}{r},\ \frac{d}{r},\ \frac{z}{r}) \quad \ldots (6)$$

where the functional relationship $F\ (u,\ \frac{b}{r},\ \frac{d}{r},\ \frac{z}{r})$ can be written as

$$F = M\ \{u,\ (b+z)/ar\} - M\ \{u,\ (d+z)/ar\} + M\ \{u,\ (b-z)/ar\} - M\ \{u,\ (d-z)/ar\} \quad \ldots (7)$$

M {u, (b or d ± z)/ar} is an infinite integral for which are available in the literature (e.g. ILRI, 1976).

Like in the previous methods, the procedure of analysis is based on the matching of a graphic plot of the field data and a set of type curves for the function F. In this method the drawdown in a single observation well is all what is needed.

Figure 5 shows the plot of the field data of s (r, z, t) versus the time t matched to the type curve of F versus 1/u, corresponding to x = 0.05. The drawdown curve for observation well A_{52}, where r = 5 m and b = 64 m, gave at the matching point 0.12 m for s and F = 1. With these figures substituted in eq. 3' the horizontal permeability k_h can be found at 78 m/d, which upon further substitution in x gives k_v at 32 m/d. The drawdown data of the observation well A_3 gave k_h at 85 m/d and k_v at 85 and 13 m/d, respectively. So, the average permeabilities obtained by this method of analysis are k_h = 83 m/d and k_v = 25 m/d.

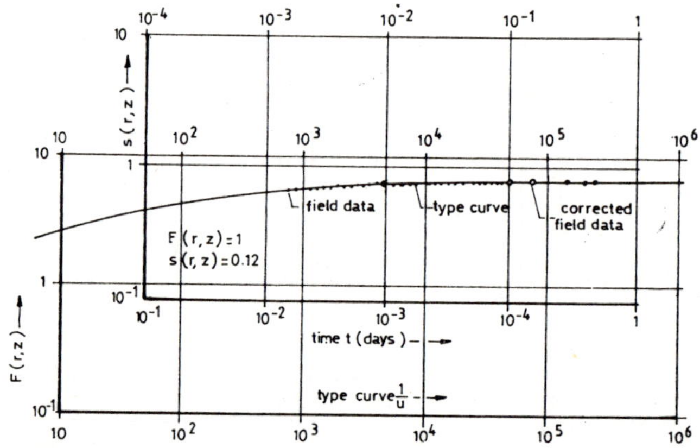

Figure 5 - Analysis of unsteady-state flow of groundwater towards a pumped well using the drawdown data of a single observation well.

ESTIMATION OF AQUIFER PERMEABILITY FROM THE GRAIN SIZE DISTRIBUTION

Disturbed samples of soil have been taken out while drilling the aquifer for installing the battery of observation wells. The upper 120 m of the aquifer consists of rather coarse sand with some gravel. Thin clayey lenses have been found at depths from the surface of 10-11 m, 22-24 m and 73-75 m, whereas fine sands predominate between 20 and 30 m, and between 70 and 85 m, below the surface. The grain size distribution of the sample taken at a depth of 57 m below the surface is shown in Figure 6.

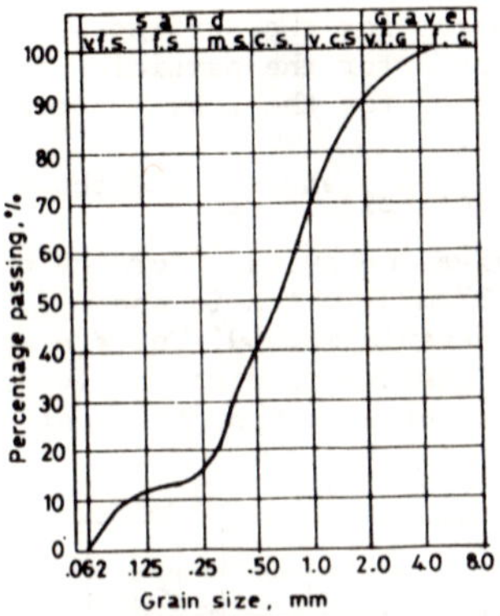

Figure 6 - Grainsize distribution curve of the aquifer material at a depth of 57 m below the surface.

a) Hazen's Method

The relationship between the permeability and the grain-size distribution can be written as

$$k = c\, d^2_{10}\ (0.7 + 0.03\ t) \qquad \ldots (8)$$

where,

k = permeability, m/d,

c = constant depending on the uniformity coefficient u, where u = d_{60}/d_{10}, i.e. the grain size, mm, corresponding to 60 and 10 % percentage passing (see Figure 6), and

t = temperature in degrees centrigrade

By carrying out the mechanical analysis at temperature t equal to 20° C, eq. 8 reduces to

$$k = 1.3\ c\ d^2_{10} \qquad \ldots (8')$$

This expression has been recommended by Farid (1980) whenever the uniformity coefficient is in the range between 1 and 5. For the sample shown in Figure 6, u = 5, the value of c is close to 800 and d_{10} is about 0.17 mm. Substituting these values in eq. 8' the value of k can be found at 30 m/d. The average k for the zone extending from the ground surface to 120 m deep is thus equal to 81 m/d (IWACO/RIGW, 1986).

b) ILRI's Method

ILRI (1976) developed an empirical relationship between the

permeability and the surface area of the sand fraction after having it corrected. The formula can be written as

$$k = 54000\ U^{-\frac{1}{2}}\ C_{so}\ C_{cl}\ C_{gr} \qquad \text{... (9)}$$

where, other than defined earlier,

U = specific surface of the main sand fraction,
C_{so} = correction factor for the sorting of the sand,
C_{cl} = correction factor for the particles finer than 16 μg
C_{gr} = correction factor for the gravel content in the sample.

The correction factors C_{so}, C_{cl} and C_{gr} are shown graphically in Figure 7.

The ILRI's method gives me average for the permeability of the top 120 m of the aquifer of 87 m/d, which is about 7.5 % higher than the average figure given by Hazen's method. The average figure of 83 m/d obtained for k using the grain-size distribution methods, is the same as the average figure for k_h obtained from the unsteady-state drawdown method.

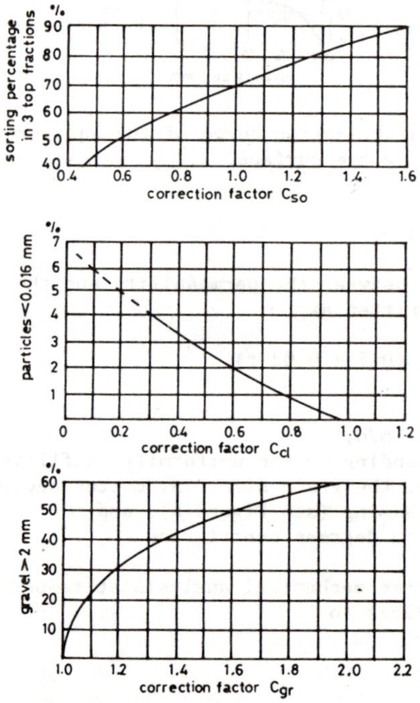

Figure 7 - Correction factors for estimating permeability of sands from their grain-size distribution.

ESTIMATING THE PERMEABILITY ASSUMING THE AQUIFER FULLY PENETRATED BY A PUMPING WELL (Thiem's Method).

The transmissivity T of an unconfined aquifer can be computed from

Thiem's formula

$$T = \frac{Q_w}{2\pi\ (h_2 - h_1)}\ \ln\ (r_2\ /\ r_1) \qquad \ldots (10)$$

where,

h_1 = water level in observation well, say no. 1,
h_2 = water level in observation well, say no. 2,
r_1 = distance between observation well no. 1 and pumped well, and
r_2 = distance between observation well no. 2 and pumped well.

Substituting in eq. 10; Q_w = 4680 m^3/d, $(h_2 - h_1)$ = 0.67 m and (r_2/r_1) = 36 one gets T at about 4000 m^2/d. If the whole depth of the aquifer, which is in the order of 200 m, is assumed to contribute to the flow towards the well, the permeability k will be in the order of 20 m/d. Since this assumption does not really hold, a correction has to be introduced. An estimate of the effective depth of the aquifer is the depth to the bottom of the screen of the pumped well added to it a distance equal to (1.0 - 1.5) times the length of the screen.

In the case of the aquifer in question the length of the solid part of the well is 44 m and the screen length is 20 m. This means that the effective depth of the aquifer is between 84 and 94 m, bringing the permeability of the aquifer material to 42 - 48 m/d. These figures are still small compared to those obtained from the methods based on the anisotropy of the aquifer material.

DISCUSSION AND CONCLUSIONS

The previous sections have shown clearly that the different methods for computing or estimating the aquifer permeability yield different results. The range of values obtained can be large indeed.

The assumption that the whole depth of the aquifer is contributing to the flow towards the pumped well, whereas the penetration ratio is too small is not correct and gives small values for the permeability coefficient. The introduction of some reduction factor to the thickness of the aquifer definitely improves the result though to a limited extent.

The methods based on the grain-size distribution of the aquifer material are too empirical. Unless proper constants are used, the results obtained could be misleading. This type of methods like the previous one does not take into consideration the fact that the aquifer permeability is anisotropic. This reduces the vertical permeability to zero. The computed permeability k is considered identical to the horizontal permeability, k_h.,

Finally the methods based on the anisotropy of the aquifer material, where both horizontal and vertical permeabilities are considered, seem to give more realistic and satisfactory results. The two steady-state methods, where the drawdown is measured in one of them in a vertical section and in the other in a horizontal section give alsmost the same values, especially for k_h. The unsteady state-method gives slightly high values for k_h and k_v.

To this end one can fairly recommend the use of the group of methods based on aquifer anisotropy for the analysis of the pumping test data. To overcome the difference in results between one method and another the average of all methods in the said group has to be used.

REFERENCES

Hantush, M.S.,1964. Hydraulics of wells. Advan. Hydrosc. Vol. 1, p 686-699.

ILRI, 1976. Drainage principles and applications. Publ. 16, Vol. 1, International Institute for Land Reclamation and Improvement, Wageningen, the Netherlands.

IWACO-RIGW, 1986. Development and management of groundwater resources in the Nile Valley and Delta, 1983-1985. 2nd Interim report of the Research Institute for Groundwater, Cairo, Egypt.

El-Kashef, A.I., 1983. Salt water intrusion in the Nile Delta. J. Ground water no. 2, p 160 - 167.

RIGW-IWACO, 1984. Hydrogeological inventory study of the Eastern Nile Delta Region. Progress report of the Research Institute for Groundwater (RIGW), Cairo, Egypt.

Shahin, M.M., 1987 a. Groundwater resources in Egypt: potentials and limitations. From "Water for the Future": Hydrology in Perspective (Pro ceedings of the Rome Symposium), IAHS Publ. no. 164, p 179-192.

Shahin, M.M., 1987 b. Technical feasibility of vertical drainage of irrigated lands in Egypt. Proc. International Conference on Groundwater and the Environment, p B.16-B.25, Kuala Lumpur, Malaysia.

LIST OF NOTATIONS

a = k_v/k_h
b = depth to the bottom face of the well screen
c = constant
d = depth to the top face of the well screen
d_{10} = diameter of particle corresponding to 10% passing on the size distribution curve
d_{60} = diameter of particle corresponding to 60% passing on the size distribution curve
h = elevation of water level in an observation well
k = coefficient of permeability
k_h = horizontal permeability
k_v = vertical permeability
l = length of well screen or recharge well
t = distance from centre of pumped well
s = drawdown caused by pumping alone
s' = rise of piezometric level due to aquifer recharge
s_m = measured drawdown
t = time
t^o = temperature
z = depth below the static groundwater table
C_{cl} = correction factor for the clay fraction in the sample
C_{gr} = correction factor for the gravel content in the sample
C_{so} = correction factor for the clay fraction in the sample
D = depth of aquifer

Q_i - recharge

Q_w - discharge of the pumped well

T - transmissivity of the aquifer

GROUNDWATER BASIN SIMULATION OF MAHI RIGHT BANK CANAL COMMAND AREA, GUJARAT, INDIA

S.K.Sondhi
Sr.Irrigation Research Engineer
Deptt. of Soil and Water Engg.
PAU, Ludhiana, India.

P.B.S.Sarma
Senior Hydrologist
Water Technology Centre
IARI, New Delhi, India

ABSTRACT

A groundwater basin simulation model was used to predict the dynamic behaviour of the water table in response to groundwater pumpage and net recharge in part of MRBC (Mahi Right Bank Canal) command area, Gujarat, India. The model is based on finite difference approach and uses iterative alternating direction implicit technique as the solution procedure. The model was calibrated and validated by using historical input data. Some statistical parameters such as the mean, standard deviation, the coefficient of correlation and a non-parametric statistical test were used to study the extent of agreement between the observed and the predicted water table depths. No significant difference was found between the distributions of the observed and the predicted water table depths at the 95 per cent confidence level. Net recharge to groundwater basin was calculated and used for determining additional groundwater potential. The study reveals that the additional groundwater potential available for development in the area is about 265 MCM. Sufficient scope exists for substantial increase in groundwater development.

INTRODUCTION

During the last three decades, there has been gradual change in the landuse pattern of the Mahi Right Bank Canal (MRBC) command area in the Gujarat State of India. This change was made possible with the introduction of canal irrigation in 1958. Prior to introduction of canal irrigation groundwater was the major source of irrigation. It continues to be so even after three decades of introduction of canal irrigation. Inspite of this, the water table in the command area has been rising continuously at an ever-increasing rate. This is due to the fact that there has been increase in groundwater recharge from conveyance and distribution systems of the canal irrigation project. Prior to introduction of canal irrigation, in about 88 per cent of the command area, the water table was below 9 metres from the ground surface. Areas with water table within 1.5 metres depth were non-existent, except some localised pockets. After about 22 years of irrigation practices (1980) the condition of high water table extended to almost the entire northern boundary of the command area. This situation calls for more attention to the need for better understanding of groundwater behaviour in the area. This in turns require an accurate groundwater simulation model. The model so developed can be used for predicting and evaluating the impact of any proposed changes in irrigation practices or management techniques in groundwater behaviour.

The specific aim of the present study was to first simulate the dynamic behaviour of water table in the MRBC area with the help of a digital model. Next, to use some appropriate parameters and a statistical test for evaluating performance of the model. Finally, to make an estimate of groundwater potential available for development.

DESCRIPTION OF THE MRBC COMMAND AREA

The command area of Mahi Right Bank Canal lies in the Kaira District of Gujarat State, India (Fig. 1). The geographical area covered by this project is 3,16,000 ha of which about 2,13,000 ha are cultivable. The area selected for this study is the region within the project area bounded by the rivers Mahi and Shedi and the Alang drain. It comprises approximately 2,95,200 hectares of land or about 93 per cent of the geographical area covered by the project.

The soils of the MRBC area are deep and are formed through alluvial deposits carried by Mahi and Sabarmati rivers. The soil texture varies from coarse to fine. Almost 60 per cent of the area has coarse textured soil. The soil are deep ($>$90 cm) with moisture holding capacity in the range of 38 to 50 per cent.

The topography of the area is generally plain and the area slopes from north-east towards south-west with an average gradient of about 1 to 1600. The land slope is steep in eastern parts and gentle in the central parts.

The climate of the area is semi-arid with an average annual rainfall of 823 mm. About 96 per cent of the rainfall is received during the south-west monsoon period (June-September) and there is substantial variation in monthly and annual rainfall.

Canal System

The Kadana Reservoir and Wanakbori Weir, both located on the Mahi river, constitute the headworks of the irrigation system (Fig. 1). The water distribution system comprises a main canal, six branch canals, a number of distributaries and outlets. The main canal and branch canals are lined. Below the outlet the water distribution is mainly field to field. The irrigation system has a conveyance efficiency of 63 per cent (Water Technology Centre, 1983).

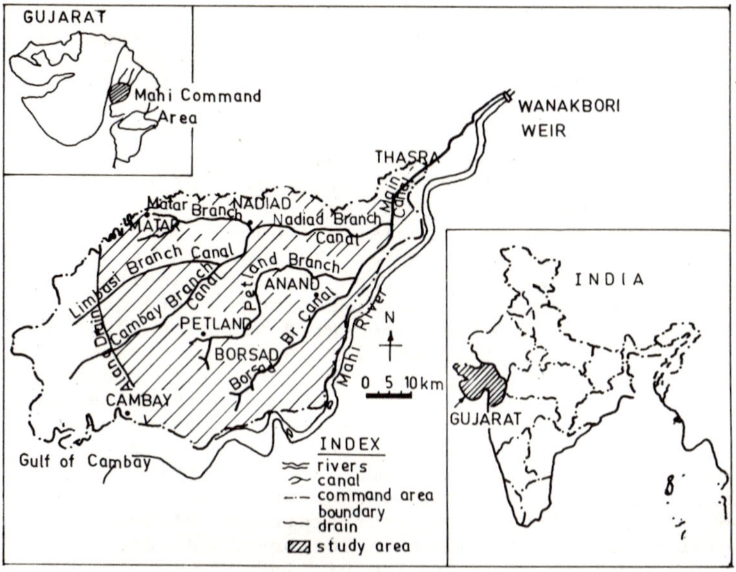

Fig. 1. Location map of the study area.

Geohydrological Conditions

The aquifers are mainly gravel and sand with an average thickness of 30 to 60 m. There is a distinct water table aquifer of 10 to 15 m thickness overlying a main confined aquifer. In the north eastern portion and in some areas to east of Cambay, the aquifer is unconfined. Over the remaining portion, leaky confined conditions are common. The groundwater quality is good in a major portion of the area.

Analysis of pumping tests conducted in several locations in the command area by Gujarat Water Resources Development Corporation indicate that the aquifer transmissivity ranges from 196 to 6830 m^2/day with the higher values occurring in the eastern portion. The storage coefficient varies between 0.00015 to 0.00417 and the leakage coefficient between 0.0016 to 0.006/day.

DIGITAL SIMULATION MODEL

The computer model used in this study solve the two dimensional flow equation given by:

$$\partial^2 h/\partial x^2 + \partial^2 h/\partial y^2 = (S/T)\,\partial h/\partial t + Q \qquad (1)$$

where, T = transmissivity of the aquifer, S = storage coefficient of the aquifer, h = groundwater potential, t = time, Q = net groundwater withdrawl rate per unit area, and x,y = rectangular coordinates.

The model was based on a digital computer model developed by Prickett and Lonnquist (1971). It allows for simulation of both leaky confined and unconfined zones of the aquifer. Essentially, the procedure consists of numerical solution of the flow equation by finite difference approach using iterative alternating direction implicit technique. The aquifer was discretized by superimposing a two dimensional rectangular grid over the region with 143 block centered nodes in 13 column (I) and 11 rows (I). In all 82 nodes lie in the study area (Fig.2). At each node, the initial groundwater potential, transmissivity, storage coefficients, average annual pumping periods at nodes with wells, leakage factor (in case of nodes representing leaky confined areas), elevation of water level in source bed, hydraulic conductivity, elevation of bottom of aquifer (in case of nodes representing unconfined areas) and type declaration, declaring a node as representative of leaky confined or unconfined area, were input to the model. A detailed description of the model is given by Sondhi (1984).

The computer output presents the groundwater potentials or drawdowns as a result of pumping of groundwater at different locations in the command area as indicated by the value of discharge assigned to the nodes. The pattern of distribution of final potentials is with reference to the initial groundwater potential at the nodes which were supplied to computer as data input. The output does not, however, reflect the influence of recharge to groundwater basin. Thus in order to use these results to compare with the historical data, the net recharge to be superimposed was estimated. The data for the period 1976-77 to 1979-80 were used for the study.

Assessment of Net Regional Groundwater Recharge

The net regional recharge (R_n) to groundwater was estimated as below :

$$R_n = R_g \pm Q_g - Q_p \qquad (2)$$

where, R_g = gross recharge to groundwater, Q_g = groundwater inflow/outflow to neighbouring areas, Q_p = groundwater extraction, through wells.

The gross recharge (R_g) to groundwater was estimated by using the following equation:

$$R_g = R_p + R_c + R_d + R_{ci} + R_{wi} \qquad (3)$$

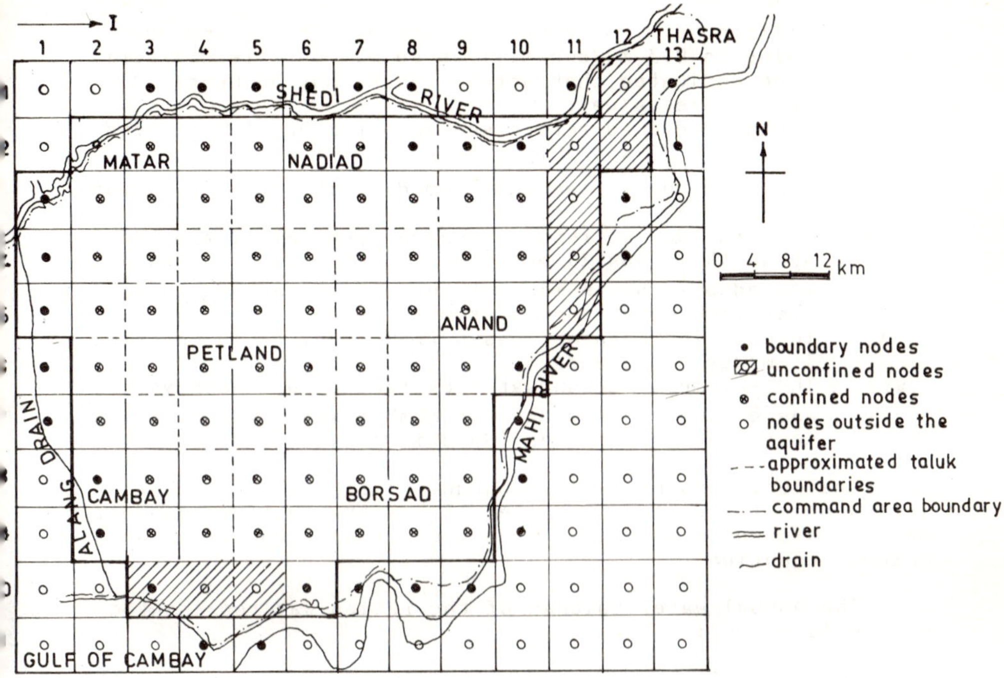

Fig. 2. Fixed, block-centered grid configuration for MRBC command area aquifer model.

where, R_p = recharge due to rainfall, R_c = recharge due to seepage from main canal and branches, R_d = recharge due to seepage from distributaries, R_{ci} = recharge due to percolation losses from areas irrigated by canal water, R_{wi} = recharge due to percolation losses from areas irrigated by wells.

The net groundwater recharge in the region was estimated using equations 2 and 3. The recommendations of the Groundwater Estimation Committee of the Ministry of Irrigation (1984) with some modifications based on field observations were used for estimation of recharge from rainfall, canal systems and groundwater irrigated areas. The following specific percentage allowance for various recharge components were used.

i) Recharge from rainfall (R_p) was taken as 18 per cent of the annual rainfall.
ii) Recharge due to seepage loss from the lined main and branch canals (R_c) were calculated as 0.6 m^3/sec/

million sq.m. of wetted perimeter (The recommended value is 0.3 m^3/sec/million sq.m., but field experience indicated that actual losses may be nearly twice this value).

iii) Recharge due to return seepage from canal irrigated fields (R_{ci}) was 30 per cent.

iv) Recharge due to seepage from distributaries (R_d) was 7 per cent.

v) Recharge due to return seepage from well irrigated areas (R_{wi}) was 15 per cent (This relatively low value was adopted as in many of these areas an underground pipe line conveyance system is used).

vi) In addition seepage losses from paddy field contribute 3 mm/day for 100 days.

vii) The volume of groundwater extracted (Q_p) each year was estimated from data of number and type of wells operating in the area (Sondhi, 1984).

The last component of recharge, namely, the sub surface inflow or outflow of groundwater across the aquifer boundaries was calculated as a residue from the annual (crop year) water balance equation.

The annual water balance of the region was represented by the equation :

$$(P+I) = (ET + E + Q_g + Q_p + Q_s \pm \Delta S_m \pm \Delta S_s \pm \Delta S_g) \qquad (4)$$

where, P = precipitation, I = irrigation water applied over the region by the canal system (Q_c) and groundwater (Q_p), ET = evapotranspiration, E = soil evaporation from uncultivated areas, Q_s = direct surface run off from the area, ΔS_m = change in soil moisture storage, ΔS_s = change in surface water storage, and ΔS_g = change in groundwater storage.

All the above components were expressed in volume units MCM (Million Cubic Metres) for the area of 2,95,200 ha. Since MRBC system remained practically dry at the beginning of June every year during 1976 to 1980, the change in surface water storage was assumed to be zero. Similarly, it was assumed that there is no significant change in soil moisture storage on annual basis. Since all the terms of the water balance equation except Q_g were known, this was determined as a residual of eqn. 4 (table-1). The details of how each one of these quantities were measured or estimated are given by Sondhi (1984) and Sondhi et al (1989). The net annual recharge and the corresponding water table rise was estiamted using eqns. 2 and 3 (table-2). Comparison with the observed water table rise, during the four years 1976-77 to 1979-80 showed (table-2) that the estimates of net regional groundwater recharge were realistic.

Table-1 Annual water balance of the MRBC command area

Sr. No.	Water balance component	Year 1976-77	77-78	78-79	79-80	Average annual
Input						
1.	Rainfall (P)	4440	3041	2453	2497	3108
2.	Irrigation (I) :					
	(a) Canal supplies (Q_c)	849	1244	1470	1448	1253
	(b) Wells (Q_p)	148	156	168	192	166
3.	Total input	5437	4441	4091	4137	4527
Output						
4.	Evapotranspiration (ET)	1180	1169	1282	1463	1274
5.	Evaporation (E)	615	553	487	466	530
6.	Pumpage (Q_p)	148	156	168	192	166
7.	Surface run off (Q_s)	2412	1461	1003	941	1454
8.	Change in ground water storage (ΔS_g)	806	593	156	425	495
9.	Total output	5161	3932	3096	3487	3919
10.	Sub-surface outflow from the area (Q_g) (3)-(9)	276	509	995	650	608

The net regional groundwater recharge obtained above was distributed amongst the different nodes using recharge distribution coefficients (Sondhi et al, 1989).

MODEL CALIBRATION

The model was calibrated by comparing the observed and predicted groundwater potentials after one year. The main variables to be calibrated were transmissivity, storage coefficient and leakage coefficient. Initially, the program was run by using initial values of the hydrologic variables and observed potential for June, 1976 as input. However, the initial results were not convergent with observed values of June, 1977. Therefore, a sensitivity analysis was carried out by varying the individual variable to study their effect on the output. Some of the values were revised after re-examining the field data. The final value of leakage coefficient was of higher magnitude than that which was initially inferred. Using adjusted values of the hydrologic variables, simulation runs were performed to predict the groundwater potentials. The predicted groundwater potentials thus obtained were used to plot the maps of depths to water table (Isobath)

Table-2 Net annual groundwater recharge in the project area (MCM)

Sr. No.	Source of Recharge	Recharge 1976-77	77-78	78-79	79-80	Average
1.	Rainfall (R_p)	799	547	442	450	560
2.	Seepage from main canal and branches (R_c)	33	33	31	39	34
3.	Seepage from distributaries (R_d)	57	85	101	99	85
4.	Returnflow from canal irrigation (R_{ci})	393	525	661	640	555
5.	Return flow from well irrigation (R_{wi})	22	24	25	29	25
6.	Groundwater extraction (Q_p)	148	156	168	192	166
7.	Groundwater outflow (Q_g)	276	509	995	650	608
8.	Net Recharge (R_n) = (1)+(2)+(3)+(4)+(5)-(6)-(7)	880	549	97	415	485
9.	Equivalent water table rise (m)	1.99	1.24	0.22	0.94	1.10
10.	Observed water table rise (m)	1.82	1.34	0.35	0.96	1.12

for the period 1976-77 to 1979-80. A typical observed and predicted isobath map for June 1979 is shown in Fig. 3. The percentage of area under different water table depth ranges predicted by using the model and the corresponding observed values are given in table-3. It may be seen from table-3 that there is close proximity between the corresponding values during June, 1979. The results obtained for the years 1976-77, 1977-78 and 1979-80 also were similar in nature.

A check was also made to see the distribution of simulation error. The simulation error equals the absolute difference between simulated water levels and the measured water levels. The average absolute deviation between the observed and the predicted water levels during the years 1976-77 to 1979-80 was 1.16,0.97, 0.91 and 0.91 m, respectively. The average deviations of this magnitude or more have been considered to give a satisfactory match (Kisiel et al, 1972; Karanjac et al, 1977; Klempt et al, 1979 and Boonstra and DeRidder, 1981).

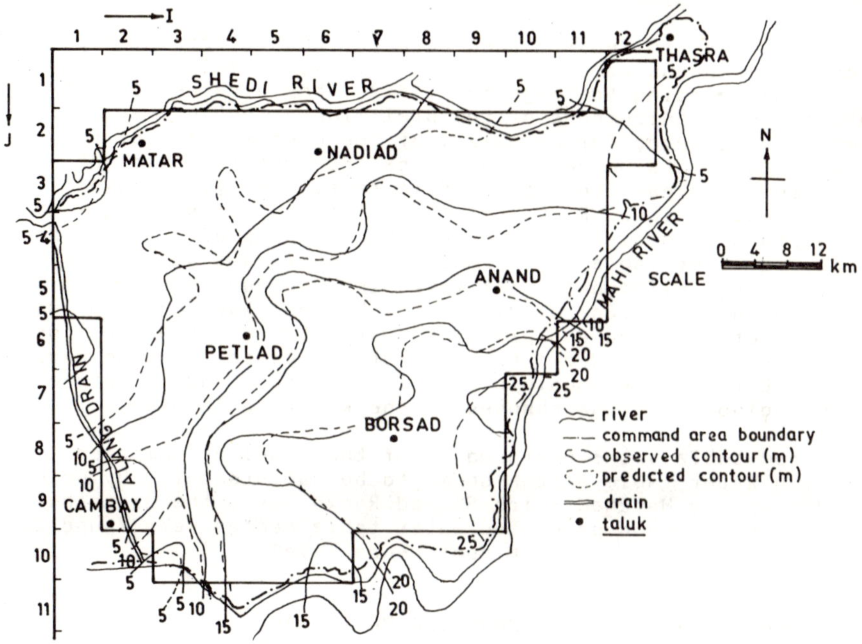

Fig. 3. **Map of observed and predicted Iso-baths (pre-monsoon, 1979) of MRBC command area.**

Table-3. Percentage of area under different depths to water table (range) in the month of June, 1979

Range of depth (m)	Percentage of area	
	Observed	Predicted
< 5	18.70(552)	21.21(626)
5-10	28.86(852)	26.12(771)
10-15	17.51(517)	16.36(483)
15-20	18.77(554)	19.95(589)
20-25	12.03(355)	12.16(359)
>25	4.13(122)	4.20(124)

Figures in parentheses refer to area in square kilometres.

Statistical Analysis Using Observed and Predicted Water Table Depths

To gain more insight into the performance of the groundwater basin simulation model, some statistical parameters of the distributions of the observed and predicted water table depths were evaluated. An appropriate statistical test was modeled to determine if there is any significant difference between the observed and predicted water table depths. The statistical parameters of the distribution and the results of the tests are given in table-4.

From the limited recorded data for these years, the coefficient of skewness were found to be 0.3315, 0.4141,0.3807 and 0.4716. According to Yevjevich (1972), these values do not satisfy the requirements of the preliminary test criterion for symmetry and hence for normality of the distribution of the data. Thus, tests on some statistical parameters could not be performed. Instead a non-parametric test, namely, Wilcoxon's Matched Pairs Signed Rank Test for related samples (Siegel,1956) was performed to discover whether the samples of observed and predicted water table depths had any significant difference in any respect. Wilcoxon's Test takes into account both the direction and the magnitude of difference between two series of data. The pair of observed and predicted water table depths for the same month was considered to be related because the same causative factors are responsible both in the model and in nature for bringing about the changes in the water table depth.

From table-4 the performance of the model as compared with observed water table depths seems to be satisfactory. By using the Wilcoxon's Matched Pairs Signed Ranks Test, the distribution of the observed and predicted water table depths were found to be identical at 95 per cent confidence level.

ASSESSMENT OF REGIONAL GROUNDWATER POTENTIAL

The average net annual groundwater recharge for the period 1976-77 to 1979-80 was 485 MCM (table-2). The average rainfall during this period was 1053 mm. However, the long term average rainfall for the region is only 823 mm. Thus the net recharge which when converted to long term mean rainfall was 380 MCM. About 70 per cent of this value was considered utilizable (Ministry of Irrigation, 1984). Hence, the additional groundwater potential of the region was estimated at 265 MCM.

The availability of additional groundwater resources at each node was estimated using recharge distribution coefficients. The additional potential available for development was not uniform at all nodes. It varied from 0.04 MCM in the north-east near the Shedi river, to 9.5 MCM in the central portion of the project area. In general, the potential for groundwater development was higher in the central and southern portion of the project area than in the western and northern portions.

SUMMARY AND CONCLUSIONS

The complex groundwater behaviour of MRBC command area was simulated with a digital simulation model based on finite

Table-4. Statistical properties of the distributions of observed and predicted water table depths in MRBC area

Criteria of comparison and/or Test and Hypothesis	1977		1978		1979		1980	
	Observed	Predicted	Observed	Predicted	Observed	Predicted	Observed	Predicted
Mean, m	13.10	13.18	11.89	11.70	11.52	11.45	10.58	10.35
Standard deviation, m	7.49	7.18	7.29	7.36	7.10	7.17	6.61	6.87
Coefficient of correlation	-	0.9634	-	0.9851	-	0.9848	-	0.9680
Coefficient of skewness	0.3315	-	0.4141	-	0.3807	-	0.4716	-
Wilcoxon's Matched Pairs Signed Ranks Test :								
H_0: No difference between observed and predicted water table	Do not reject H_0 at 0.05 α level		Do not reject H_o at 0.05 α level		Do not reject H_o at 0.05 α level		Do not reject H_o at 0.05 α level	
H_a: Observed and predicted water table are different								

Note : α = Size of type 1 error.

difference approach using iterative alternating direction implicit technique. Four years of data were utilized for the calibration of the model. Good agreement was obtained between observed and predicted groundwater potentials. To increase the confidence on the model performance, statistical parameters such as the mean, standard deviation, the coefficient of correlation were evaluated. Wilcoxon's Matched Pairs Signed Ranks Test was used and the distributions of the observed and predicted water table depths were found to be identical at the 95 per cent confidence level. It was, therefore, felt that the groundwater simulation model is capable of simulating the historic past reasonably well and can be used as a tool for predicting the behaviour of groundwater under different management plans.

The net recharge to groundwater from different sources was quantified to determine the available groundwater potential in the area. The available annual groundwater potential was estimated to be 265 MCM, which is about 1.6 times the existing level of groundwater development. Thus sufficient scope exists for further development of groundwater resources in the MRBC command area.

REFERENCES

Boonster,J. and De Rider,N.A., 1981. Numerical modelling of groundwater basin. ILRI Pub.No. 29, Wageningen, The Netherlands.

Karanjac,J.M., Altunkayank, and Ovul,G., 1977. Mathematical model of Ulcova Plain, Turkey a training and management tool. Ground Water, Vol. 15, No.5, p.348-357.

Kisiel,C.C., Supkow,D. and Tetley,W., 1972. Role of digital computer models of aquifers in water resources planning: case study in Tucson, Arizona. Int.Symp. on the Planning of Water Resources, Mexico City.

Klempt,W.B., Knowlees,T.R., Elder,G.R., and Sieh,T.W., 1979. Groundwater resources and models applications for the Edwards (Balcones Fault Zone) Aquifers in the San Antonio Region. Report 239, Texas Dept. of Water Res. Austin, Texas, USA.

Ministry of Irrigation, 1984. Groundwater estimation methodology. Rept. of the Groundwater Estimation Committee, Govt. of India, New Delhi, India.

Prickett,T.A. and Lonnquist,G.G., 1971. Selected digital computer techniques in groundwater resource evaluation. Bull.55, Illinois State Water Survey,Urbana, Ill.USA

Siegel,S.,1956. Non parametric statistics for behavioural sciences. McGraw Hill Book Co., New York, N.Y., USA

Sondhi,S.K., 1984. Groundwater basin simulation for conjunctive use in canal command areas. Ph.D. thesis, Division of Agricultural Engineering, IARI, New Delhi, India.

Sondhi,S.K., Rao,N.H. and Sarma,P.B.S., 1989. Assessment of groundwater potential for conjunctive water use in a large irrigation project in India. J.Hydro. (In press).

Water Technology Centre, 1983. Resource Analysis and Plan for efficient water management - A case study of Mahi Right Bank Canal Command Area, Gujarat. Res. Bull. 42, IARI, New Delhi, India.

Yevjevick,V., 1972. Probability and statistics in Hydrology. Water Resour. Publications, Fort Collins, Colorado, USA.

NUMERICAL SIMULATION OF GROUNDWATER FLOW IN FUHRBERG FIELD, HANOVER

S.N. Rai[1] and B. Hoffmann[2]

1. National Geophysical Research Institute, Uppal Road Hyderabad-500007, India.

2. Institut für Wasserwirtschaft, Hydrologie und Landwirtschaftlichen Wasserbau, Appelstr.9, Universität Hannover, 3000 Hannover 1, F.R. Germany

ABSTRACT

A mathematical model is presented to simulate dynamic behaviour of water table in response to recharge, pumpage and interaction with a river in Fuhrberg field near Hanover, F.R. Germany. Water balance analysis is presented for a period of 14 years. This model is used to predict spatio-temporal distribution of water table in response to proposed pumping and artificial recharge schemes.

INTRODUTION

Groundwater is the main source of water supply to Hanover City. A major part of it is obtained from Fuhrberg field. For the proper management of groundwater resources of this region, a detailed analysis of groundwater flow conditions and prediction of dynamic response of the aquifer system for proposed pumping and artificial recharge schemes are very essential. With this purpose a mathematical model for simulation of groundwater flow of this region is developed which would be helpful in decision making for the proper use of groundwater.

Fuhrberg field lies in the north of Hanover . The total area under investigation is 327 Km^2 with maximum extension of 26 km in east-west direction and 18 km in north-south direction. The northern boundary of the study area runs roughly along the Aller river from North of Hambühren to Schwarmsted. The western boundary runs along the Leine river from Schwarmsted to the north of Helstorf. The south-western boundary lies along the foothills of Brelinger Berge. The southern boundary extends from Bissendorf to the north of Kleinburgwedel. The eastern boundary runs parallel to the western boundary and passes through the east of Hambühren (Figure 1).

The Wietze river traverses the study area from south to north and joins the Aller river just out side of northern boundary. Six tributaries namely Wulbeck, Tiefenbruchgraben, Hengstbeeke, Bruchgraben, Muhlengraben and Rixfordergraben are

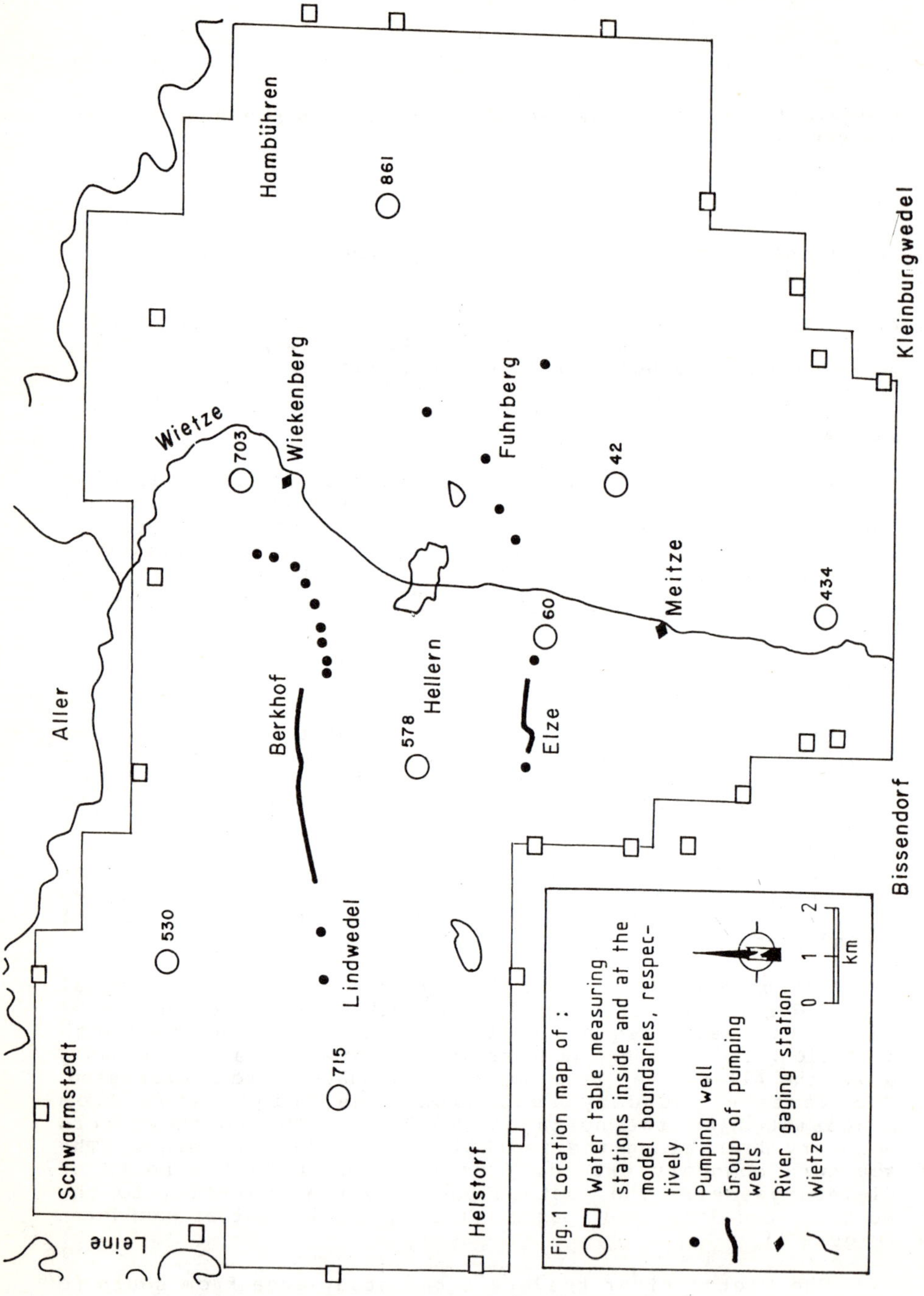

Fig. 1 Location map of :
○□ Water table measuring stations inside and at the model boundaries, respectively
● Pumping well
Group of pumping wells
River gaging station
Wietze

flowing into the Wietze. One tributary, Grosse Beeke, is flowing into the Leine river. However, historical data necessary for simulation of river flow is available for Wietze only. So the flow of Wietze is simulated in this study.

HYDROGEOLOGY

The aquifer is composed of quarternary sand and gravel. Layers of silt and sedimentary loam with varying thicknesses are also present locally within the aquifer. The aquifer is covered by relatively less permeable layers at some places and is underlain by tertiary bed rock. The bed rock is slightly dipping towards north-east direction and is very less permeable. Therefore it is treated as impermeable and the flow system is considered as a single layer unconfined aquifer. Its thickness varies from place to place. It varies between 25 to 35 meters in the basins of Aller and Wietze and reaches upto 70 meters at some places in the eastern and northern regions (Hoffmann et al., 1980). Water table in the southern region is comparatively at higher level and is continuously decreasing towards north direction. As a result the movement of groundwater is from South to North. Water table heights are measured at various stations located along the boundaries and inside of the study area. Locations of these stations are shown in Figure 1.

Precipitation is the main source of the groundwater recharge. In addition to it seepage from the Wietze, return flow from irrigation and inflow from adjacent areas also contribute to groundwater recharge. Precipitation occurs throughout a year and is more in winter season (from October to March) than in summer season (from April to September). The six months period of each season is taken as time increment in this study. Discharge of the groundwater is due to pumping, evaporation, evapotranspiration, seepage in to the Wietze and outflow across the boundaries. Pumping wells are located in Lindwedel, Berkhof, Elze and Fuhrberg regions. The groundwater of this region is also used by farmers for irrigation. However, historical data of pumped out groundwater for this purpose is not available. So the contribution of irrigated water is not considered. In order to simulate interaction between groundwater and the Wietze , water level of the river is measured at two stations located at Meitze and Wiekenberg.

MATHEMATICAL MODELLING

The study area is discretized into a system of square grid of 1 km length as shown in Figure 2. The simulated positions of water table height measuring stations, pumping wells, the Wietze and the river gazing stations are shown in this figure. A computer program " GW3MOD " developed at the Institut für Wasserwirtschaft, Hydrologie und landwirtschaftlichen Wasserbau (IWHLW), Univ. Hannover is used to compute varying water table

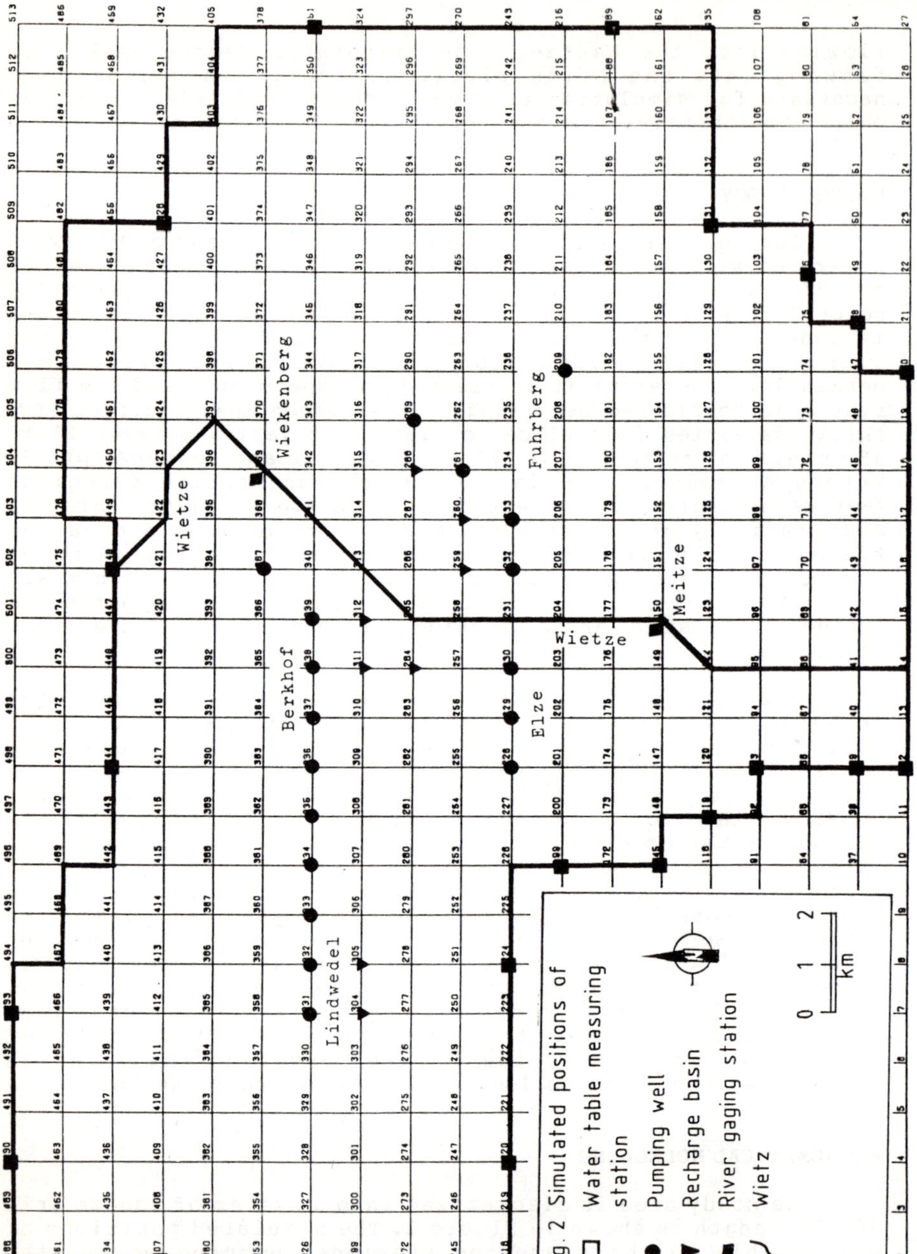

Fig. 2 Simulated positions of

heights at each node inside the model boundaries. The following water balance equation is used in this program.

$$\sum_{i=1}^{6} Y_{i,o} (h_i^{t+\Delta t} - h_o^{t+\Delta t}) + A_o q_o - Q_o$$
$$= A_o S_o (h_o^{t+\Delta t} - h_o^{t}) / \Delta t \qquad (1)$$

in which $Y_{i,o}$ = harmonic mean of transmissivity between nodes o and i; h_o = water table height at node o; h_i = water table height at node i; t = time; Δt = time increment, A_o = area of the element with node o; q_o = rate of recharge per unit time and per unit area; Q_o = pumping rate per unit time; and S_o = specific yield. This equation represents change in water table height at node o due to net flow from the six surrounding elements with nodes denoted by i = 1 to 6. In case of impermeable base there will be no inflow from the bottom. Seepage through a cross sectional area $B \Delta X$ of the river bed is computed using following expression.

$$Q = B \Delta X \, L \, (h_w - h) \qquad (2)$$

in which h is water table height, h_w is height of water level in the river, B is width of the river bed, ΔX is length of an element and L is the leakage factor defined as:

$$L = K_r / t_r \qquad (3)$$

K_r and t_r are hydraulic conductivity and thickness of the river bed, respectively.

Most of the input data used in this simulation study are taken from Hoffmann and Schulz(1984). These include water table heights on April 1968 to use as initial condition, elevations of the aquifer base, river bed and groundsurface, water table heights at the model boundaries for each time increment to use as boundary conditions, water level heights and width of the Wietze, average values of recharge and pumping rates for each time increment. Heights are measured from mean sea level. Water table heights at the boundaries are available for those nodes only where positions of the measuring stations are simulated. Water table heights for nodes lying between two successive measuring stations are determined by interpolation. Local values of hydraulic conductivity and specific yields are obtained from other previous works (Hames, 1972; Lillich et al, 1973; Klenke, 1975).

NUMERICAL RESULTS AND DISCUSSION

Calibration of the model is carried out in two stages: (i) Verification of the model and (ii) Prediction of spatio-temporal distribution of water table for the proposed rcharge and pumping schemes. Numerical results of both stages are

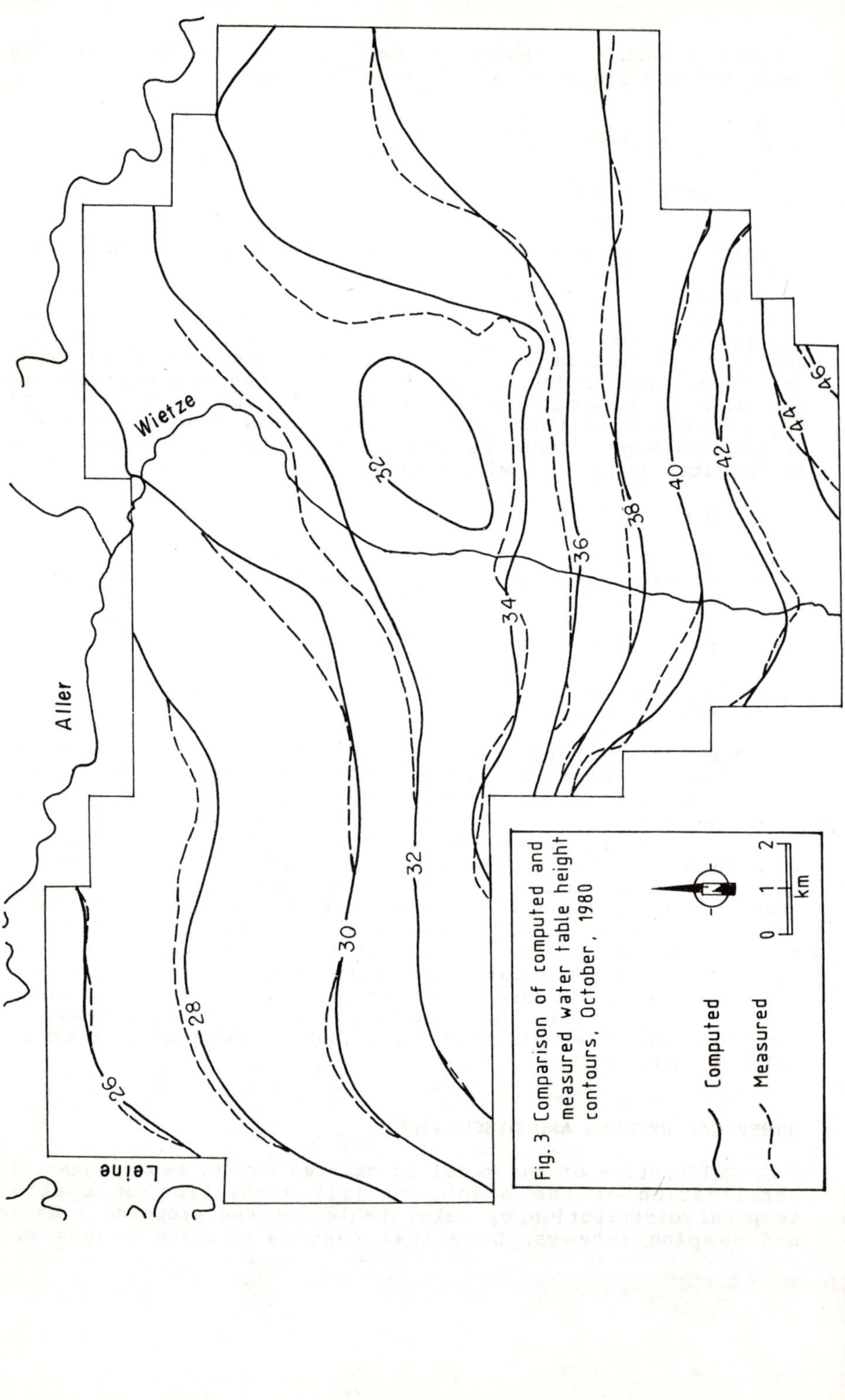

Fig. 3 Comparison of computed and measured water table height contours, October, 1980

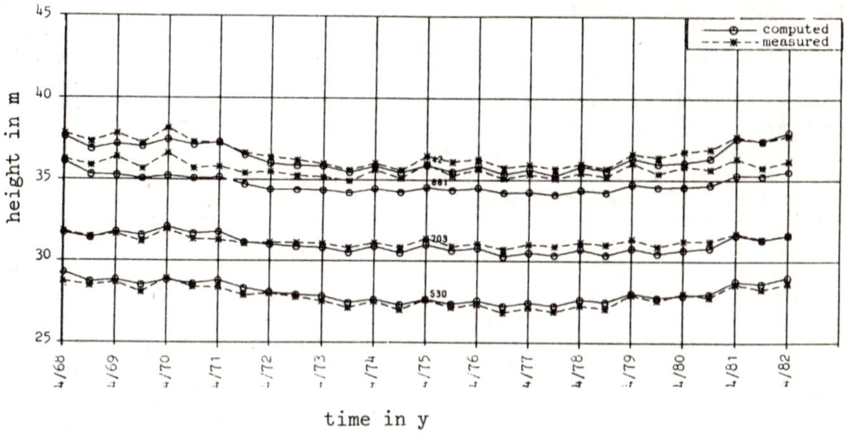

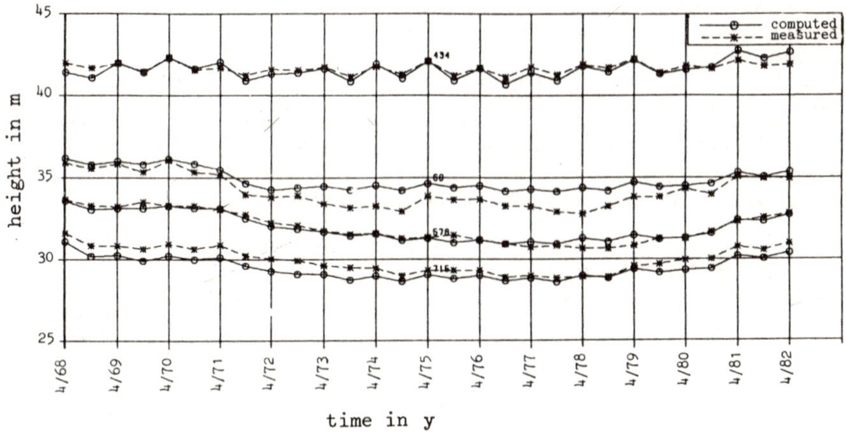

Fig. 4 . Comparison of computed water table heights with measured water table heights at various water table measuring stations.

discussed below.

(a) MODEL VERIFICATION

The verification of the model proceeded by trial and error adjustments of hydraulic conductivity, specific yield and leakage factor until a reasonable agreement between computed and measured water table heights was obtained. For these values of system parameters , water table height contours for each time increments of six months from April 1968 to April 1982 are computed and compared with the corresponding measured water table height contours (Rai and Hoffmann, 1984). An example of comparison for October 1980 is shown in Figure 3. The comparisons indicate good agreement between computed and measured water table height contours in most part of the study areas. In general discrepancies are found more near pumping wells. In further support of the verification of the model, computed and measured water table heights at eight stations are compared in Figure 4. Positions of these stations with numbers 42, 861, 703, 530,434, 60, 578 and 715 are shown in Figure 2. The figure shows satisfactory agreement between computed and measured water table heights. These results confirm validity of the model.

Computed rates of total groundwater recharge, pumping, seepage of the river, flow through boundaries and changes in storage are plotted in Figure 5. for each time increment. Positive values indicate inflow to the aquifer and negative values indicate outflow from the aquifer. It can be seen from the figure that recharge is more during winter season (October to March) and is lesss during summer season (April to November). An increase in the recharge value is associated with an increase in the storage and vice versa, as expected. The average change in the storage is only -0.58 Million m^3 /year. It shows that the system is nearly in a state of balance.

Table 1. Proposed rates of pumping and artificial recharge (in Mil m^3/Y).

	Lindwedel	Berkhof	Elze	Fuhrberg
Pumping rate	10	26	6.9	25
Artificial recharge rate	7.45	7.45		11.1

(b) PREDICTION

Purpose of the prediction is to investigate the combined effects of increase in pumping rates and locally surface applied artificial recharge on dynamic response of the system for average groundwater flow conditions. An analysis of water

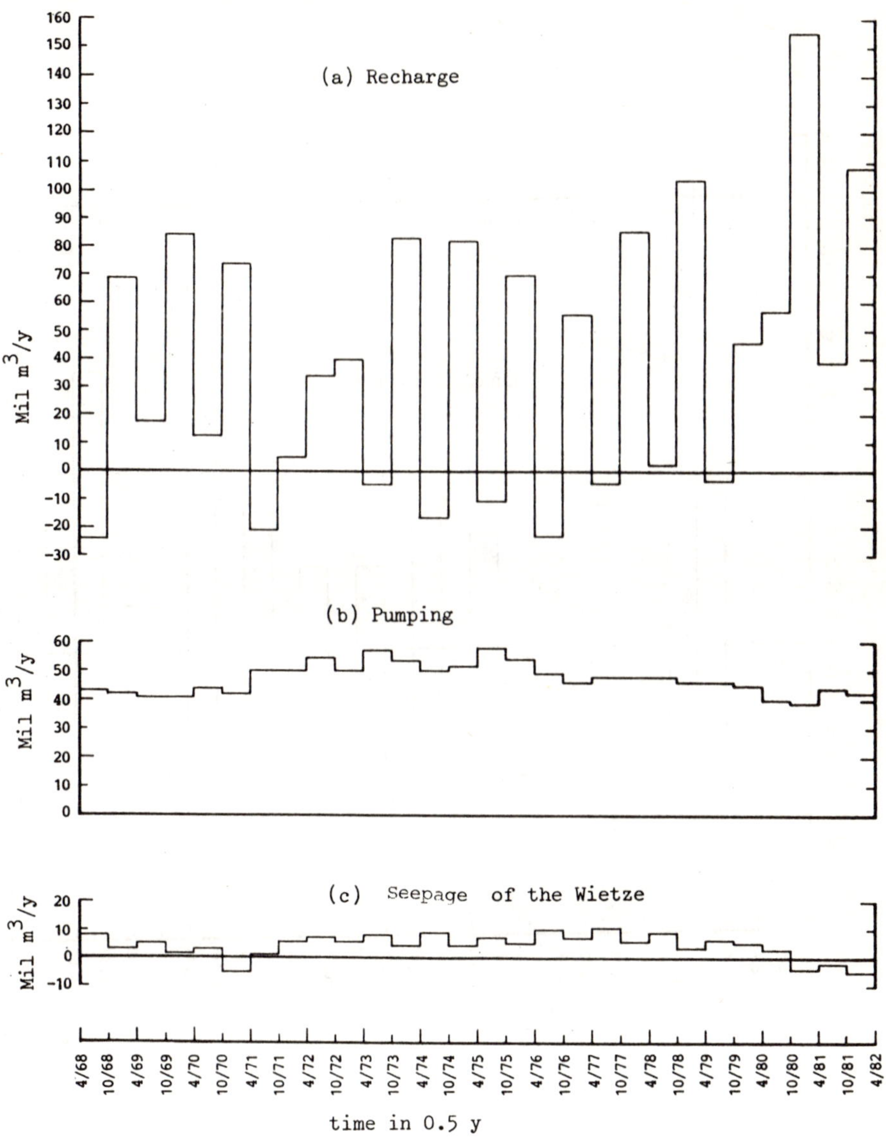

Fig. 5. Rates of total (a) groundwater recharge, (b) pumping and (c) Seepage of the Wietze.

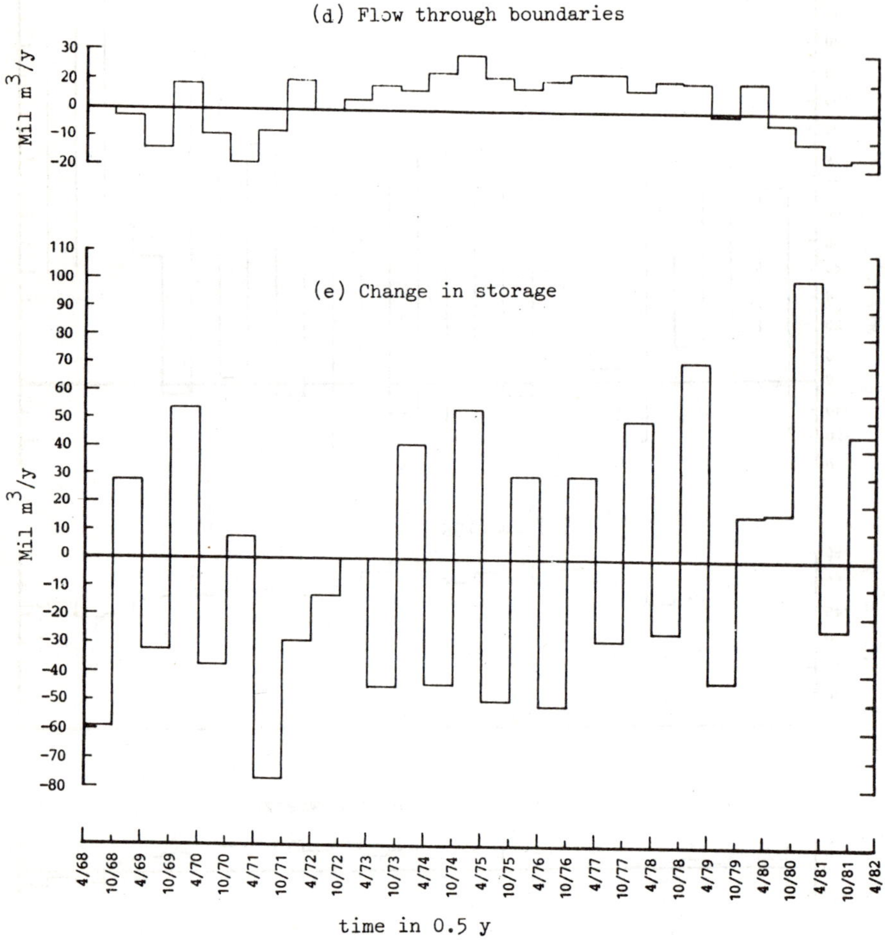

Fig. 5. Rates of total (d) flow through boundaries and (e) change in storage.

table fluctuations at the boundaries shows that the water table heights on October 1980 represent average water table heights for the entire period of investigation. So , the water table heights on October 1980 is taken as intial condition. The proposed rates of pumping and artificial recharge for different regions are given in Table 1. The location of artificial recharge basins are shown in Figure 2. The remaining input data are taken for the period from April 1978 to Oct. 1980 for six time increments each of six months duration. Water table height contours for each time increments are computed(Rai and Hoffmann,1984). Predicted water table heights at eight stations are plotted in Figure 6. The figure shows that a rise in the water table height is followed by a decline up to fourth time increment. For fifth and sixth time increments water table is continuosly rising. It is because an increase in the recharge value for the fifth time increament is followed by another increase in the sixth time increament.

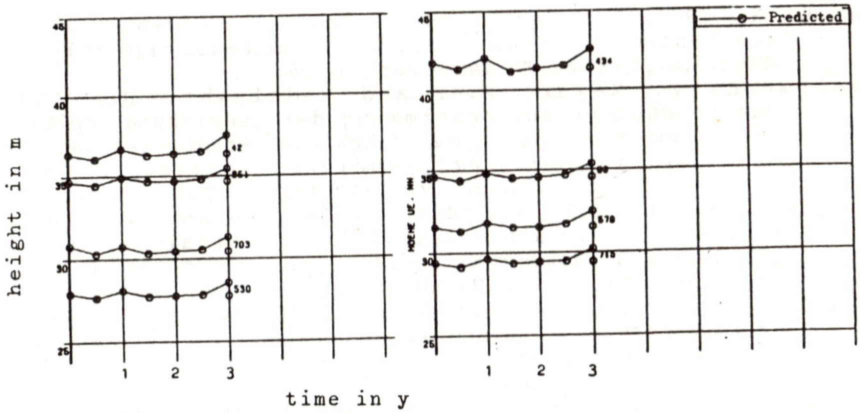

Fig. 6. Predicted water table heights at various water table measuring stations.

CONCLUSIONS

A mathematical model for simulation of the groundwater flow in Fuhrberg field is presented. The following conclusions are drawn from the present study.

(a) Good agreement between computed and measured water table heights confirms validity of the model for simulation of groundwater flow of the study area.

(b) Groundwater flow during period of investigation is found in a state of balance.
(c) Predicted results indicate an increase in the aquifer storage which leads to the rise of the water table .

ACKNOWLEDGEMENTS

Authors are thankful to Dr. H. Schulz, Messrs. R. Knoop, R. Foremny and Haring for valuable discussions. The first auther is grateful to Dr. R. N. Singh for his encouragement and to Prof V. K. Gaur, Director, NGRI, Hyderabad for his kind permission to publish this work. Thanks are due to Mr. G. Ramakrishna Rao for typing the manuscript. Support of the German Academic Exchange Service for this study is gratefully acknowledged.

REFERENCES

Hames, H., 1972. Beitrag zur Berechnung von dreidimensionalen, stationären und instationäre grundwasserproblemen. Mitt.a.d.,IWHLW, TU Hannover, H. 25.

Hoffmann, B. Meyer, H.H. and Radelfahr, H., 1980. Untersuchungen zur bestimmung der Auswirkung geplanter förder und anreicherungmassnahmen der Stadtwerke Hannover auf die grundwasserspiegelverhältnisse im raum Fuhrberger Feld., unpubl. report, IWHLW. Univ. Hannover.

Hoffmann, B and Schulz, H., 1984. Grundwasseranreicherung im Fuhrberger feld als modell für die optimale nutzungvon grundwasservorkommen im norddeutschen Flachland. IWHLW, Unpbl. report, Univ. Hannover.

Klenke, M., 1975. Einsatz eines hybriden Rechnersystems bei grundwasserhaushaltsuntersuchungen- Ein Beitrag zur lösung instationärer grundwasserströmungsproblem. Mitt. a.d. IWHLW, TU Hannover, H.32.

Lillich, W., Kuckelkorn, K.F. and Hofmann, W., 1973. Untersuchungen zum grundwasserhaushalt im repräsentativen lockergestinsgebiet Fuhrberger Feld bei Hannover. Bilanzjahre 1967 und 1968. Beiheft z. geol. jb.,107, Hannover.

Rai, S.N. and Hoffmann, B. 1984. Mathematical modelling of groundwater flow in Fuhrberg field. Unpublished report, IWHLW, Univ. Hannover.

SIMULATION OF REGIONAL HYDRODYNAMICS IN THE NEYVELI AQUIFER

C.P. GUPTA*, M. THANGARAJAN* and V.V.S. GURUNADHA RAO*
N.K. BALASUNDAR+ and S. MUTHURAM+

* National Geophysical Research Institute, Hyderabad-500 007
+ Neyveli Lignite Corporation, Neyveli-607 801

ABSTRACT

The groundwater flow regime in Neyveli basin has been perturbed due to groundwater withdrawals from the lignite mines and a large number of irrigation wells in the area. The available geohydrological data was thoroughly analysed for the conceptualization of the complex aquifer system and a mathematical model was constructed. The aquifer modeling studies helped evolve a first order quantitative understanding of the vertical flows as well as the river-aquifer interactions in the basin. It has been found that the leakage from the phreatic aquifer into the groundwater regime below the lignite seam significantly contributes towards controlling the hydrodynamics of the region. The model studies have also helped design field experiments for better quantification of spatio-temporal variations of geohydrological parameters.

INTRODUCTION

The Tertiary formations of Cuddalore sandstones comprise a complex artesian aquifer system in the Neyveli basin in South Arcot district of Tamilnadu, India. Since 1961, the piezometric surface of this aquifer has been lowered by about 60-70 m (in the lignite mines) due to depressurizing operations necessitated for safe mining of lignite occurring in the basin. Besides, the number of energized irrigation wells in the area has increased manifold due to easy availability of electrical power from Neyveli Thermal Power Station and increased agricultural activity in the basin. As a result, the total groundwater abstraction has considerably gone up causing an overall decline of regional piezometric heads in the artesian aquifer. A regional groundwater simulation study was taken up for a quantitative understanding of the regional hydrodynamics of the aquifer as well as to prognose some remedial measures.

HYDROGEOLOGY

Initially, an area covering about 2,300 sq km between Lat 11° 20' N to 11° 45' N and Long 79° 15'E to 79° 45' E was taken up for modeling studies (Fig. 1). This area was chosen on the basis of the observed piezometric declines during the past three decades.

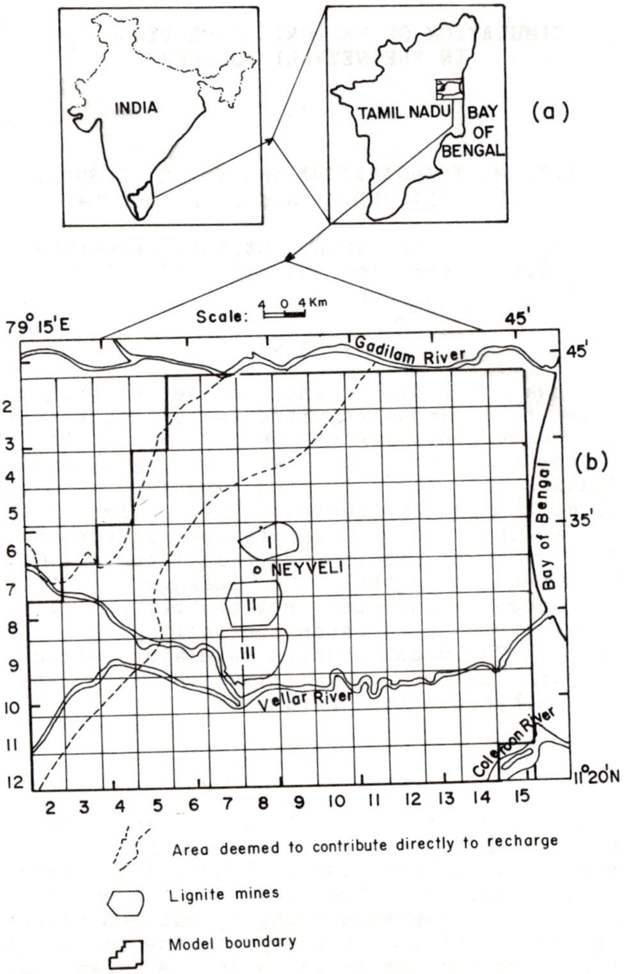

Fig. 1. (a) LOCATION MAP OF THE NEYVELI BASIN
(b) GRID PATTERN FOR TWO-DIMENSIONAL MODEL OF GROUNDWATER REGIME

The Cuddalore sandstone formation containing sands, gravel, pebbles, clays, and lignite is the main aquifer system in the basin. Two vertical sections along west-east and north-south are shown in Fig. 2.

The hydrogeologic set up in the basin is extremely complex. In the northwestern, western, and southwestern regions, water table conditions exist upto different depth ranges. In the remaining area, this aquifer is under semi-confined to confined conditions interspersed with low to very low permeability zones. The composition, texture, thickness, and areal extent of aquifer zones widely vary laterally as well as transversely.

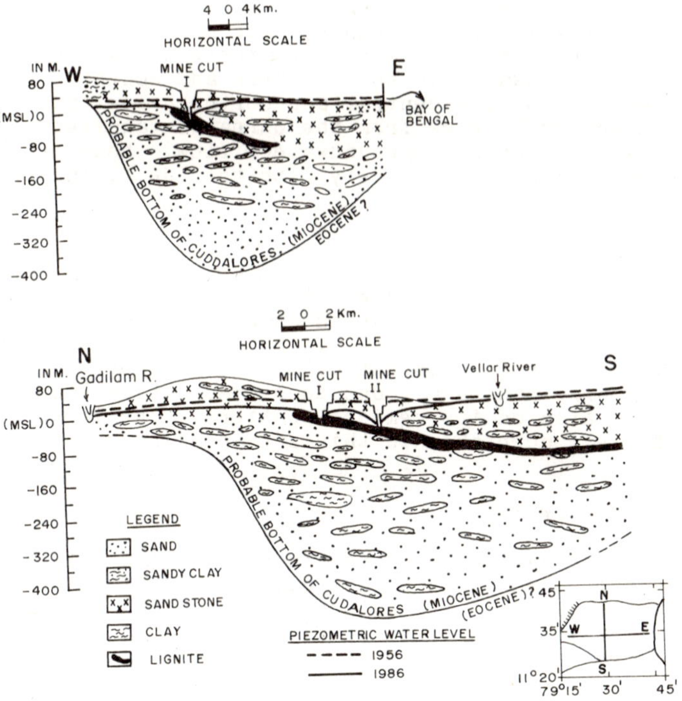

Fig. 2. INFERRED HYDROGEOLOGICAL SECTIONS

INPUT/OUTPUT STRESSES

Broad features of the groundwater flow regime and input/output stresses as inferred from the available information/data are as follows:

i. The regional gradient of piezometric water levels is east- and southeast-ward resulting into an outflow towards the Bay of Bengal.

ii. A subsurface inflow to the aquifer across the southwestern boundary of the area and an outflow across the northern boundary existed in 1956 i.e., before the onset of the mining activity. However, with the progressive lowering of piezometric heads, the flow direction has reversed across most parts of the northern boundary.

iii. Initially, the piezometric water levels in the southern and southeastern regions were higher than the water table while in the central and northern regions reverse was the case. As a result, there was an upward movement of groundwater in the southern and southeastern regions, while in the central and northern regions, there was a downward leakage from the phreatic aquifer. The area having upward groundwater flow, however, has been shrinking progressively with the declining piezometric heads.

Based on the lithological information and water level fluctuations, it was presumed that an area of about 420 sq km (Ref: Fig. 1) could be regarded as directly contributing to the recharge of the aquifer at a rate of 16.6% of the annual rainfall. Of course, this estimate required revision because the actual recharge has been found to depend not only upon the total rainfall but also on its temporal pattern. The injected tritium tracer studies undertaken in this area confirmed this view.

CONCEPTUAL FRAMEWORK

The conceptualization and quantification of the Neyveli aquifer have been beset with a number of difficulties. Neither the hydraulic conductivity values (vertical and horizontal) nor the transmissivity values in different areas and for different aquifer zones are reliably known. Notwithstanding, as a first order approximation, the vertical hydraulic conductivity was assumed to be 2.5×10^{-8} m/sec for the overburden (comprising the clayey layers and lignite). The transverse flows were calculated based on the differences between the interpolated/extrapolated phreatic and piezometric levels, the inferred values of vertical hydraulic conductivity, and thickness of the intervening layers.

An average horizontal hydraulic conductivity of 0.0005 m/sec was assumed on the basis of lithological data and some laboratory measurements, for the aquifer zones in the area deemed to be contributing directly to the recharge. A lower value of 0.0002 m/sec was taken for the aquifer zones in the remaining area. The transmissivity was estimated from the inferred spatial distribution of hydraulic conductivity values and the saturated thickness as derived from drilling logs. An abstracted map of the transmissivity values is given in Fig. 3.

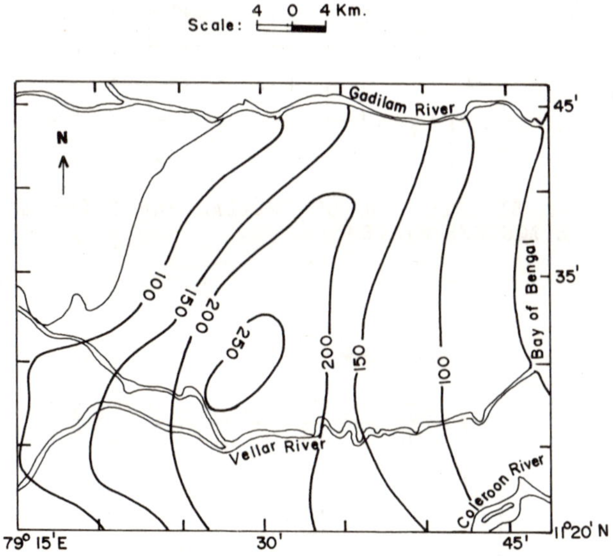

Fig. 3. TRANSMISSIVITY (in m^2/hr) CONTOURS

The mean annual input/output quantities for the steady state as obtained after a number of iterations during the calibration process, are depicted in Figs. 4 and 5.

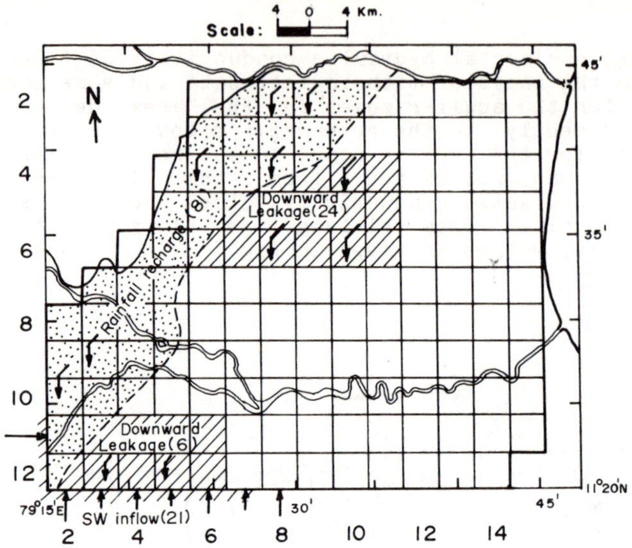

Fig. 4. SPATIAL DISTRIBUTION OF INITIAL INPUT (in mcm/year) TO THE AQUIFER SYSTEM

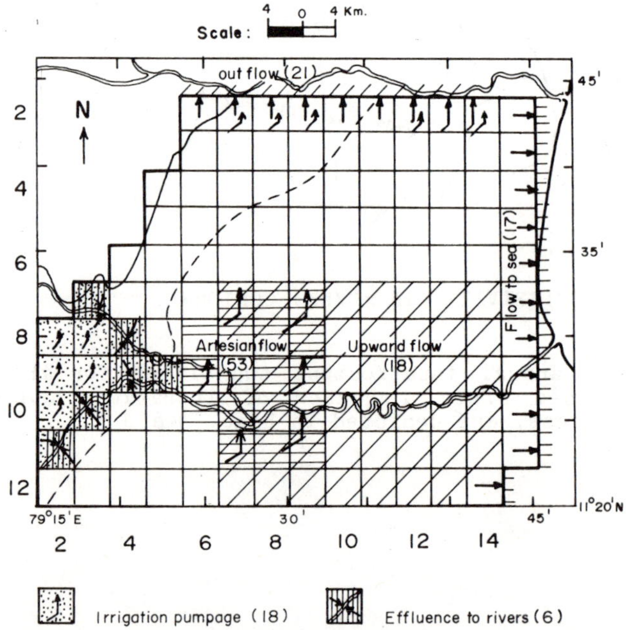

Fig. 5. SPATIAL DISTRIBUTION OF INITIAL OUTPUT (in mcm/year) FROM THE AQUIFER

AQUIFER MODELING

(i) Two-Dimensional Modeling:

Initially, a coarse grid two-dimensional model was constructed. The entire area was divided into 136 square grids of 16 sq km each (Ref: Fig. 1). A comparison of field and model values of water levels for June 1956 are presented in Fig. 6. Barring some localised mismatch, the general trend of piezometric contours may be seen to have been reasonably simulated.

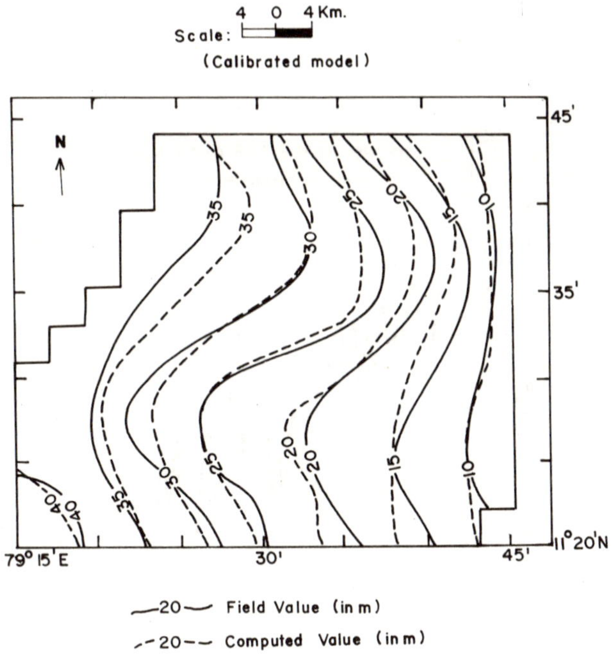

Fig. 6. COMPARISON OF COMPUTED AND FIELD VALUES OF PIEZOMETRIC WATER LEVELS (1956)

Dynamic calibration of the model was carried out for the period 1956-1986 through a large number of iterations and progressive adjustment of various parameters. A typical hydrograph (Fig. 7) shows the comparison of model results with field observations of piezometric heads.

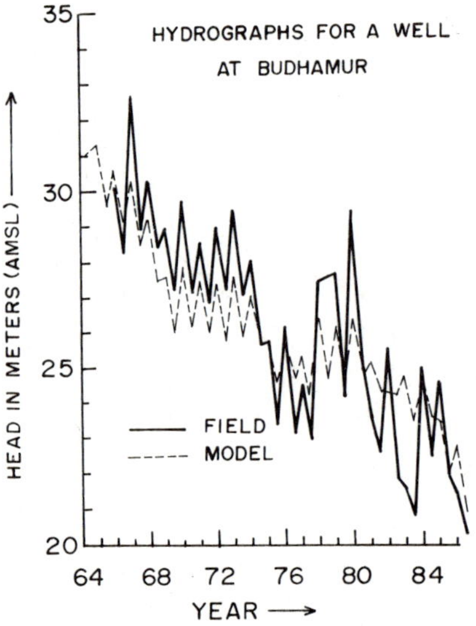

Fig. 7. COMPARISON OF COMPUTED AND FIELD HYDROGRAPH AT A TYPICAL OBSERVATION STATION

An important inference of this study is the realization of the significance of vertical flows in the aquifer system. Since the fluctuations in water table has been more or less only seasonal without a significant permanent declining trend and the piezometric heads have been progressively going down the difference between phreatic and piezometric water levels has been increasing (Fig. 8). As a result, the estimated downward leakage from the phreatic aquifer has increased from 30 mcm/yr in the pre-mining phase of 1956 to about 95 mcm/yr in 1985-86.

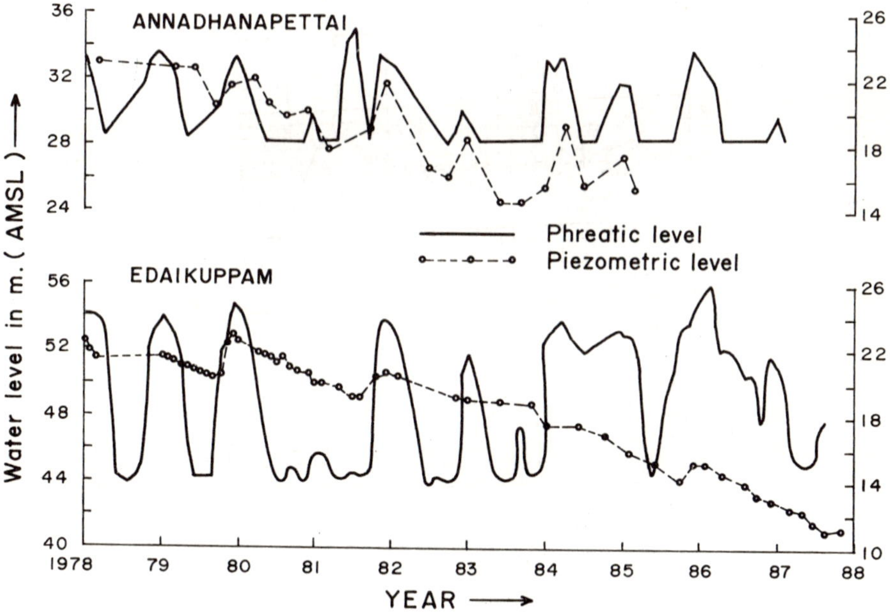

Fig. 8. HYDROGRAPHS SHOWING INCREASING GAP BETWEEN PHREATIC AND PIEZOMETRIC WATER LEVELS

(ii) Three-Dimensional Model:

As the vertical flows in the saturated zone have been found to be quite significant in the basin, a three-dimensional model of the aquifer was also constructed with a nodal spacing as shown in Fig. 9. There are 17, 18 and 11 nodes in X-, Y-, and Z-directions, respectively. The area under study has been extended in the east and south and covers about 3,000 sq km.

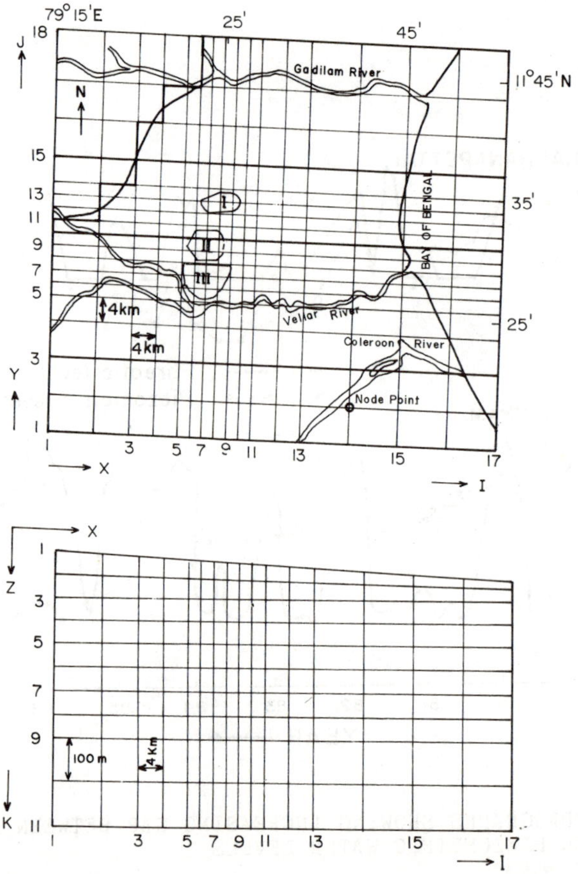

Fig. 9. SPATIAL DISCRETIZATION FOR THREE-DIMENSIONAL MODELING

The level of the top surface of the aquifer representing the water table has been assumed to be invariant during the steady state simulation. The hydraulic potential and flow lines along three vertical sections viz., J=3 (south), J=10 (through the mining area), and J=15 (in the north) as computed through the three-dimensional modeling are shown in Fig. 10. The vertical section of piezometric potentials and flow lines help visualize the vertical movement of groundwater more clearly.

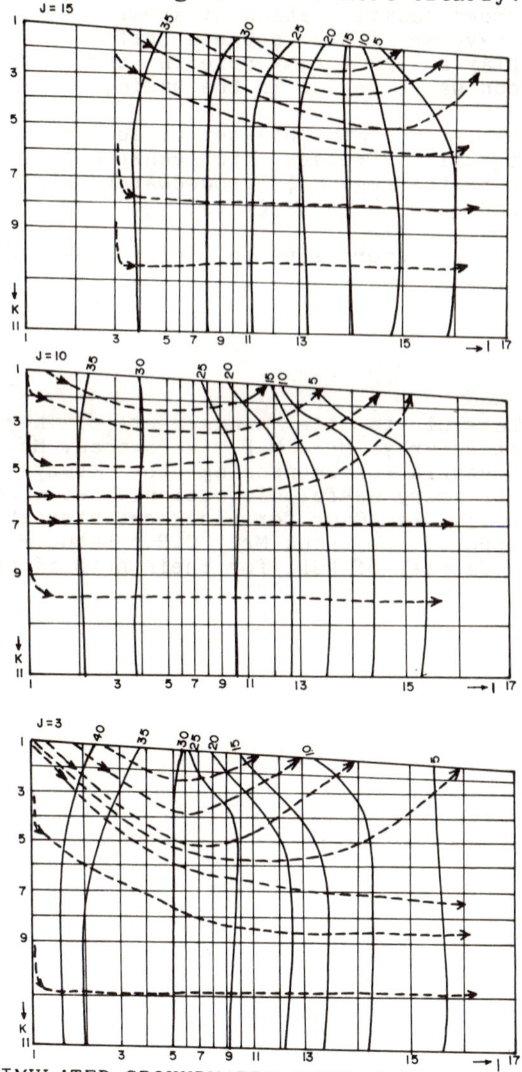

Fig. 10. SIMULATED GROUNDWATER POTENTIALS AND FLOW LINES (in 1956) ALONG THREE SECTIONS (Ref: Fig. 9)

—— 10m Hydraulic potential lines

– – – – Groundwater flow lines (inferred)

CONCLUSIONS

The modeling study has helped

i. gain a better insight of the hydrodynamics in the Neyveli basin than what was hitherto possible through field investigations alone,

ii. first order quantification of areal extent and magnitude of transverse flows between different aquifer zones which play a significant role in the hydrodynamics of the groundwater regime in the basin,

iii. characterize a compatible set of aquifer parameters and hydrological stresses to construct a model which represents the regional groundwater regime reasonably well, and

iv. design field experiments to improve the geohydrological data base.

ACKNOWLDGEMENTS

We are grateful to the management of Neyveli Lignite Corporation particularly Mr. M.P. Narayanan, Mr. M.S. Mani, and Mr. S.R. Sengupta for sponsoring the study and Mr. S. Gowrisankaran for very useful discussions. We also express our gratitude to Prof. V.K. Gaur for his keen interest in this work and continued support. We thank Mr. G.R. Babu, Mr. P. Sunder Rao and Mr. J. Gabriel of NGRI for their help in preparation of the manuscript.

COMPARATIVE MODEL STUDY FOR GROUNDWATER RESOURCES EVALUATION IN KELANTAN, MALAYSIA

Part 1: Model Evaluation

S. Awadalla, I. M. Noor and I.Ibrahim
Universiti Kebangsaan Malaysia,
43600 Bangi, Selangor, Malaysia.

Abstract

The increasing demand for water supply in Kelantan state, Malaysia has created a greater need for evaluation, development and management of groundwater resources. On the basis of recent hydrological investigation and available data on groundwater exploitation in Kelantan a two dimensions numerical groundwater model was developed to simulate the present situation and the possible future development of groundwater resources in the area. An optimum set of input data and parameters are used for a period of 11 years, to obtain the simulated head values. These values are then compared with the actual heads and a good agreement obtained.

Introduction

In Malaysia, Kelantan state is the only state that greatly relies on groundwater for its domestic water supply. It has the best potential for groundwater development in the country. Groundwater is mainly abstracted from the aquifer in Coastal Plain which lies on the northeastern part of the Kelantan State (Fig. 1). It covers an area of about 2000 km2. Flowing northward through the plain is the Kelantan River, the third biggest river in the Peninsular.

Water demand in the State is increasing every year, and by the year 1990 the population of the state is expected to reach 1,261,000 (3% annual growth rate). Along with the increase of domestic water demand, developments in agriculture and industry will inevitably put more emphasis on the groundwater resources. Thus, it is necessary to establish an evaluation and management scheme for groundwater optimum abstraction without any adverse effects such as groundwater undermining, subsidence, sea water intrusion, etc.

Digital groundwater model for the Kelantan Coastal Plain aquifer has been developed and it has been used for simulation and environmental impact study of the future water demand in the area (Noor, 1979). The model has also been used in the management of the Kelantan aquifer (Noor, 1985). In the present study a similar model was developed base on Prickett and Lonnquist (1971) model. Similar criteria and input parameter has been used for this present model. The purpose of this exercise is to compare the performance of the two models.

Geology and Hydrology

Most of the area is covered by Quaternary alluvium and topographically 90% of the coastal plain has elevation less than 75m above mean seea level. The plain is underlained in most part by the Mesozoic granites and they outcrop on both sides of the Kelantan River Fig. 2. The granites are also found outcropping as isolated hills in the coastal plain. Between the eastern and western granitic masses there are various types of sedimentary and metamorphic rocks of Paleozoic age. The sedimentary rocks include shales, sandstones, conglomerate, quartzite, limestone, siltstone and mudstone. Shale and quartzite are the predominant rock types.

The Quarternary alluvium has a thickness varying from a featheredge at the valley margin to more than 200m near the coast. The statigraphy of the alluvium is complex. This is because of the varying conditions under which the sediments were deposited. The alluvium consists of intermixing of gravel, sand, silt and clay. The percentge of mixing varies from place to place but generally the percentage of coarse materials increases with depth.

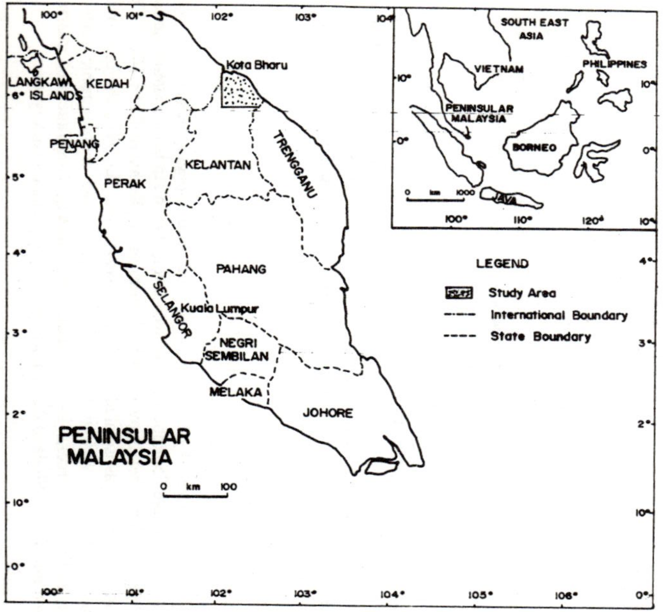

FIGURE 1. LOCATION OF STUDY AREA.

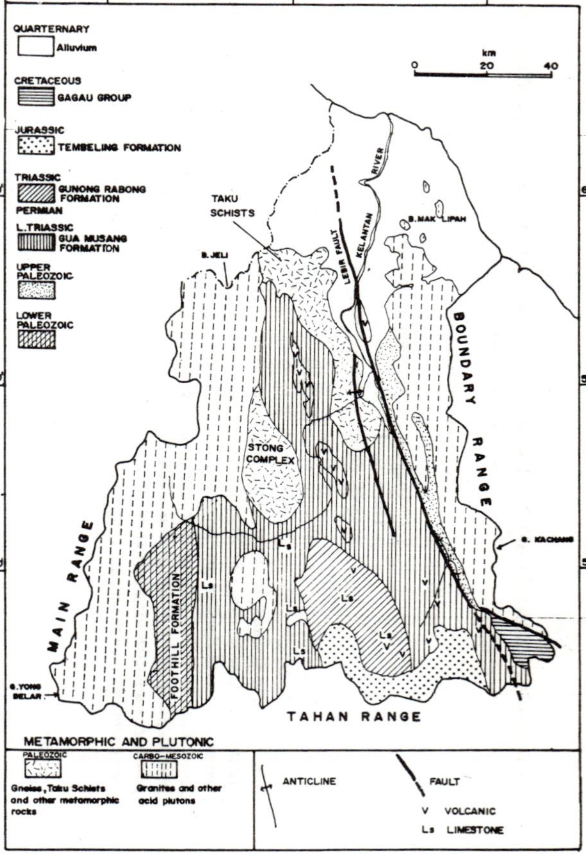

FIG. 2: Geology of Kelantan State.

The alluvium can generally be divided into two prominent aquifer zones, the upper and lower aquifer. The upper aquifer consists mainly of fine and medium sand and normally its thickness is less than 15m. The lower aquifer is much thicker and constitute a large groundwater reservoir. The top and lower aquifer are hydraulically connected. For the purpose of management the top and lower aquifer are considered to constitute as one huge water reservoir.

The average annual rainfall in the Kelantan coastal plain is about 2400 mm. About 12% of the total rainfall infiltrates into the aquifer as direct recharge. The total yearly groundwater outflow from the aquifer is about 178 x 10^6 m^3.

Most of the wells in the area are tapping the shallow top aquifer. These wells ranges in depth from 2m (village dugwell) to about 15m (drilled well in pumping station). There are only a few wells that exceed 15m in depth. Generally, well yield are high in the northern central part of the basin and low in the south eastern coastal area.

Hydraulic characteristics of the aquifer was determined through pumping tests of several wells and the pump test data were analysed by using Theis' (1935), Cooper and Jacob's (1946) and Boulton's (1963) methods. The average permeability and the specific yield of the aquifer are 197 m/day and 0.11 respectively.

Recharge to the aquifer is directly from the rain. The lower aquifer is being recharged through the percolation of water from the upper aquifer. However, boundary recharge also takes place through the surface runoff from the granite hills and the impermeable rocks located on both sides of the Kelantan River Valley.

The Kelantan River flow's across the coastal plain in a northwardly direction. The maximum daily flow of the river ever recorded is 17240 m/s and the average minimum flow is about 250 m/s. The river respond linearly to the rain falling in the catchment area and for this reason the area around the river is highly susceptible to flooding during the Northeast monsoon season.

The river also recharge the aquifer during normal high flow. This occurs normally when the flow exceeds the critical flow which is in the region of about 400 m/s.

Digital Model Simulation

The digital model is a computer program designed to simulate the response of an unconfined aquifer to pumpinng from one or more wells. This program can also be used to simulate the response of a confined aquifer in which the transmissivity does not vary significantly during the period of analysis. The groundwater reservoir may be irregular in shape and nonhomogeneous with infiltrations from lakes and streams.

Prickett and Lonnquist (1971) developed a ditigal model to simulate groundwater flow in an aquifer. The model solves the governing partial differential equations numerically by using iterative alternating direction implicit (IADI) method.

The partial differential equation governing the nonsteady, two-dimensional flows of groundwater in an artesian, nonhomogeneous and isotropic aquifer is

$$\delta\{Tx(\delta h/\delta x)\}/\delta x \quad + \quad \delta\{Ty(\delta h/dy)\} \; = \; S\delta h/\delta t \; + \; Q \qquad \text{Eq. 1}$$

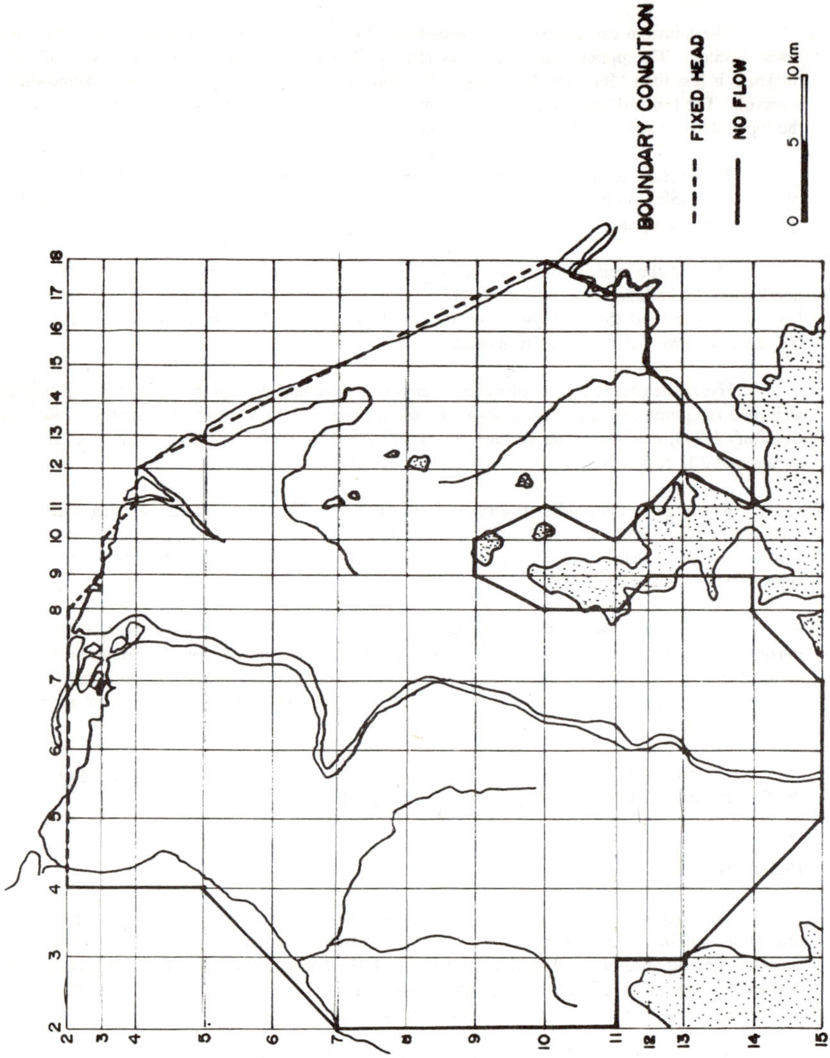

FIG 3. BOUNDARIES DELIMITING THE MODELLED AREA

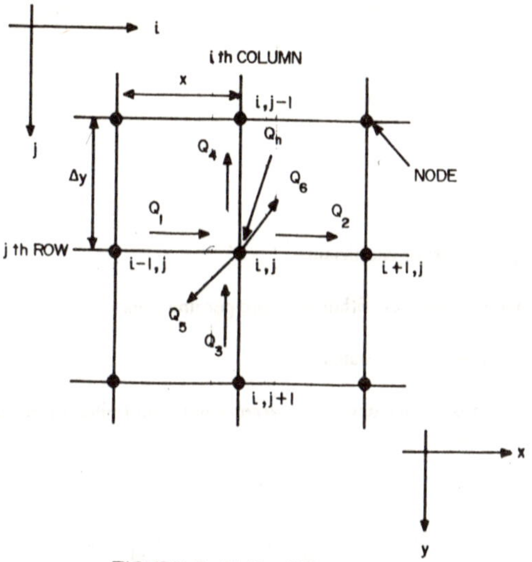

FIGURE 4. Finite difference grid.

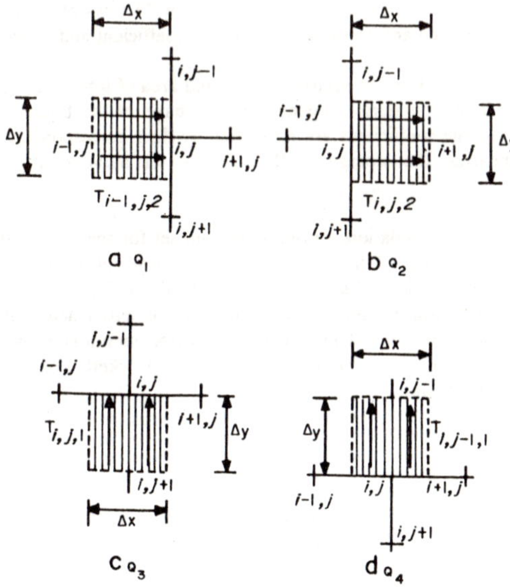

FIGURE 5. Vector volumes for node-to-node flow rate terms.

where

T = aquifer transmissivity

h = head

t = time

s = aquifer storage coefficient

Q = net groundwater withdrawal rate per unit area

x,y = rectangular coordinates.

This equation is further developed, according to the finite difference grid shown in Figure 4 to obtain the following.

$$T_{i-1,j,2}(h_{i-1,j} - h_{i,j})/\Delta x^2 + T_{i,j,2}(h_{1+1,j} - h_{i,j})/\Delta x^2 + T_{i,j,1}(h_{i,j+1} - h_{i,j})/\Delta y^2$$
$$+ T_{i,j-1,1}(h_{i,j-1} - h_{i,j})/\Delta y^2 = S(h_{i,j} - h\phi_{i,j})/\Delta t + Q_{i,j}/\Delta x\Delta y - Qn/\Delta x\Delta y \text{................Eq. 2}$$

As illustrated in Figures 5 and 6, Equation 2 is written in two forms, one to solve the node equations by columns and the other by rows. This includes the calculation of B and G arrays that are then solved for the model head values. The process is then repeated for several iterations until the values are within a certain predetermined tollerance; for one time step increment.

In applying this model to the problem at hand, the Kelantan aquifer is divided into a grid mesh as shown in Figure 3 where there are 17 columns by 14 rows of nodes. At each node the basic input parameters needed are transmissivity, storage coefficient and recharge.

At the 'no flow'boundries the vector volumes and area of influence are taken into account. The treatment of'no flow' boundaries depends on the shape of the boundries(Figure 7). The transmissivities and storage coefficients are multiplied by their corresponding shape factors. For constant head boundaries, the storage coefficients are set at very large values to simulate continuous recharge.

Before starting the simulation, an optimum data set for the aquifer that includes aquifer characteristics, abstraction rates of pumping stations, recharge rates, and river discharge versus time is obtained. This is a calibrated set developed by Noor (1979) for an 11 year period for the region based on actual measurement. An estimated set of initial head values is also used to facilitate the simulation of the digital model for the period mentioned. With this established, several modification and development are made on the Prickett Lonquists model so as to accomodate the problem at hand.

Analysis

With the calibrated optimum set of input values, a 11 year run is invoked on the digital model for the area as discretized by the grid mesh of Figure 3. The period covers the years 1967 through 1976.

Figurue 8.1 and 8.3 show the values of groundwater head for the two digital models and the actual case. There is a good agreement between the computed and actual heads where the difference are quite minor. However, at certain nodes or areas of the aquifer, the computed values have distinct differences compared with the actual measured values in the field. Since the grid size for a particular node is quite large (5km), the representation of the aquifer conditions at the

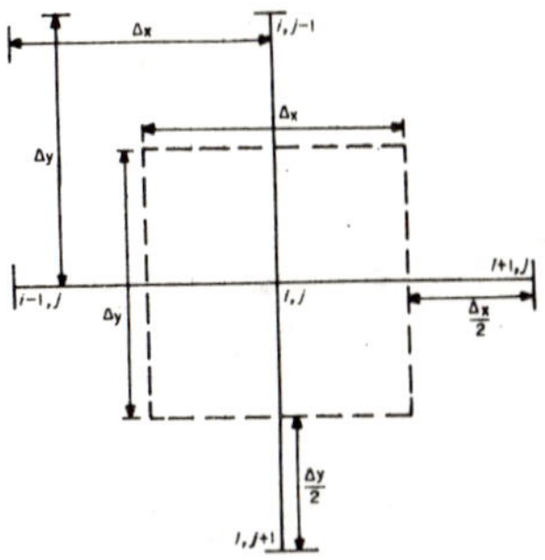

FIGURE 6. Vector volumes for flow rat terms $Q_{\cancel{x}}$, Q_{e} and $Q_{\cancel{y}}$.

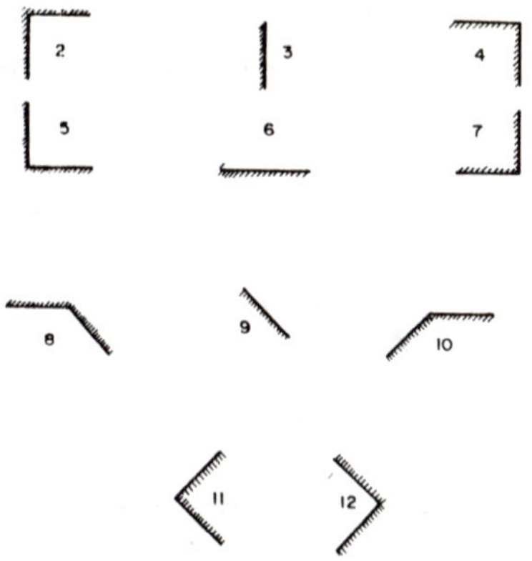

FIG 7: POSSIBLE SHAPES OF NON-FLOW BOUNDARY.

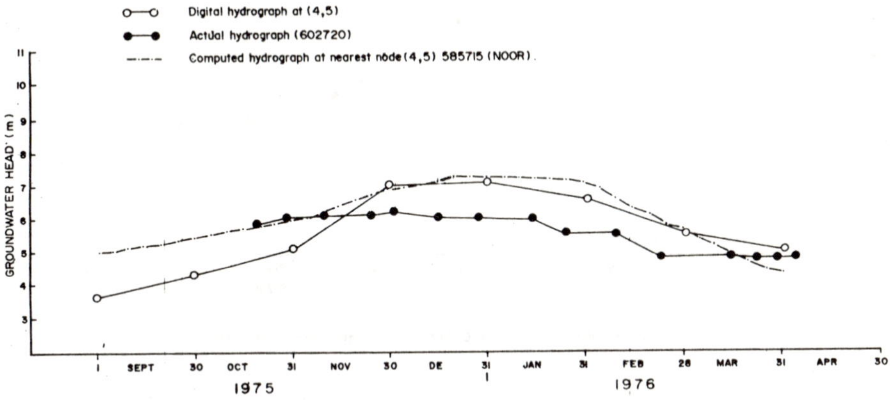

FIG: 8.1 COMPARISION OF OBSERVED AND COMPUTED HYDROGRAPHS AT BH 26.

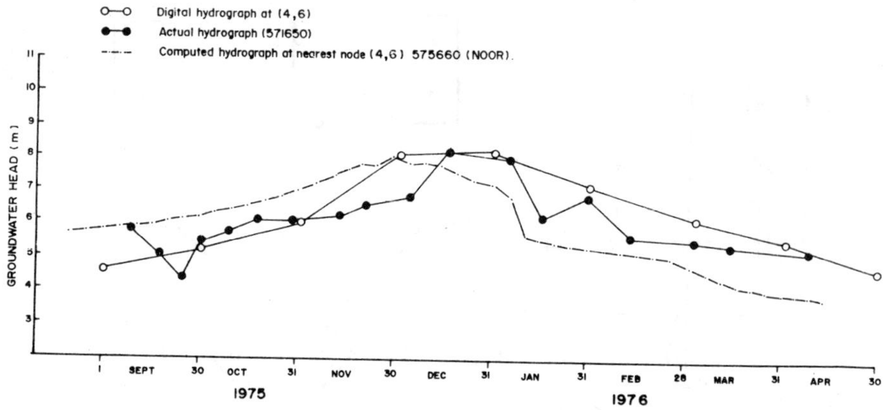

FIG 8.2: COMPARISION OF OBSERVED AND COMPUTED HYDROGRAPHS AT BH20

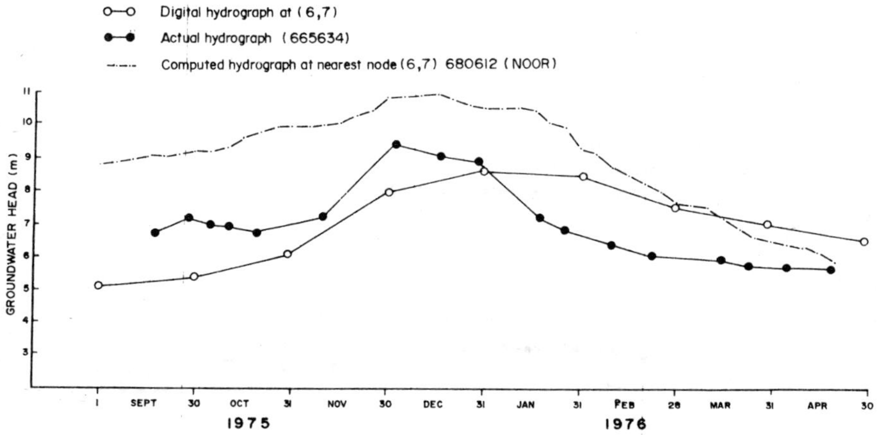

FIG 8:3: COMPARISION OF OBSERVED AND COMPUTED HYDROGRAPHS AT BH13

observation wells is restricted. The average parameters and data represented may not be accurate enough to predict the heads at the wells. Nevertheless, the computed values compare quite well with the observed heads ,and the trends are much alike.

Conclusion

This research was designed to provide a digital model capable of describing the responnse of an aquifer to a wide range of hydraulic stresses and to apply this digital technique to selected field problems.

Model results were compared with the field data and a good agreement between the computed and actual heads was achieved. This model can be a useful tool to handle future management study.with the field data

References:

Bolton,N,S.,1963. Analysis of data from non-equilibrium pupmping test allowed for delayed yield from storage, *Institute of Civil Engineer Proc.(British), v.28 p.603-610*

Cooper, H.H.,and Jacob, C,E., 1946. A generalised graphical method for evaluating formation constants and summerising well field history. *Transaction American Geophysical Union, v. 27(2) p.526-534.*

Noor , I,M.,1979. *Prefeasibility study of potential groundwater development in Kelantaan, Malaysia.* Unpublished Ph.D thesis, University of Birmingham.

___________, 1985, Groundwater management of Kelantan Aquifer. *Proceeding of Groundwater and Mineral Exploration Conference, June 1985,Bangkok.*

Prickett, T,A. and Lonnquist, C., 1971. Selected digital computer techniques for groundwater resources. *Illonois State Water Survey, Urbana, Bulletin 55.*

Theis, C,V.,1935. The relation between the lowering of piezometer surface and the rate and discharge of a well using groundweater storage. *Transactions American Geophysical Union 16thnn Annual Meeting, Part 2.*

MODELLING THE EFFECTS OF PROPOSED PUMPING FROM THE TAMBRAPARNI RIVER BASIN, TAMILNADU, INDIA.

A.BALASUBRAMANIAN[1] AND J.C.V.SASTRI[2]

[1]Department of Geology, V.O.Chidambaram College, Tuticorin, India.

[2]Department of Geology, University of Mysore, Mysore, India.

ABSTRACT

A two-dimensional finite-difference computer code has been developed to model the groundwater flow of Tambraparni river basin (Lat. 8°25' to 9°13'N &Long. 77°10' to 78°10'E), Tamilnadu and used as a predictive tool for forecasting the distribution of groundwater potentials under (i) normal, existing conditions and (ii) with a limited stress of 1% increase in the amount of extraction per year, upto 1991. The zones of depletion due to the proposed stress into this crystalline aquifer system, its ultimate effects on the groundwater exploitation schemes have been inferred. Unwarranted mining of groundwater changes the entire effluent river system into an influent one by 1991. It has been concluded from the results of this model that over exploitation of any groundwater resource should be stopped in order to "Save Ourselves" in future.

INTRODUCTION

A numerical groundwater flow model of Tambraparni river basin (Lat. 8°25' to 9°13'N & Long. 77°10' to 78°10'E; 5382 Sq.km. in area) has been developed in this study and utilized as a predictive tool for forecast. The study area is mainly composed of crystalline Archaean gneisses and charnockites except for a small linear strip of Tertiary sediments along the coastline. The basin lies in the arid to semiarid climatic zone receiving an average annual rainfall of 88 cms. Groundwater occurs mainly under unconfined conditions in the weathered and fractured zones of the crystalline formations. River Tambraparni is perennial in nature at present.

GOVERNING EQUATION

The digital simulation model developed in this study computes the head changes in the aquifer system at any specified time as a function of the various input aquifer characteristics. These changes are obtained by solving the equation for flow through porous media. The two-dimensional partial differential equation (PDE) governing the flow of groundwater in a non-homogenous, anisotropic aquifer, where the vertical variation in hydraulic conductivity is small to be neglected (Prickett and Lonnquist, 1971; Rushton and Redshaw, 1979) is

$$\frac{\partial}{\partial x}\left(T_x \frac{\partial h}{\partial x}\right) + \frac{\partial}{\partial y}\left(T_y \frac{\partial h}{\partial y}\right) = S \frac{\partial h}{\partial t} \pm W(x,y,t) \quad \ldots\ldots 1$$

Where h is the hydraulic head; T_x and T_y are the transmissivity tensors in x and y directions; S is the appropriate storage coefficient; W is the fraction of recharge or discharge with reference to space and time and t is the amount of time increment.

FINITE-DIFFERENCE APPROACH

The general solution of equation 1 does not exist. The approximate solution of the equation is obtained by numerical methods like finite-difference or finite-element approaches. Finite-difference method has been adopted to in this study.

In this method both time and space variables are treated as discrete parameters. For this, the region covered by the function has been subdivided into 21 columns and 16 rows amounting 336 nodes in which 51 are boundary nodes and 161 active nodes, lying within these boundary nodes (Map 1). As the basin is formed by an aquifer of extensive areal extent, a grid spacing of $\Delta x = \Delta y = 5000$ m has been used in this study.

SOLUTION STRATEGIES AND COMPUTER CODING

The four solution strategies available for solving equation 1 are (1) Forward Difference (Explicit), (2) Backward

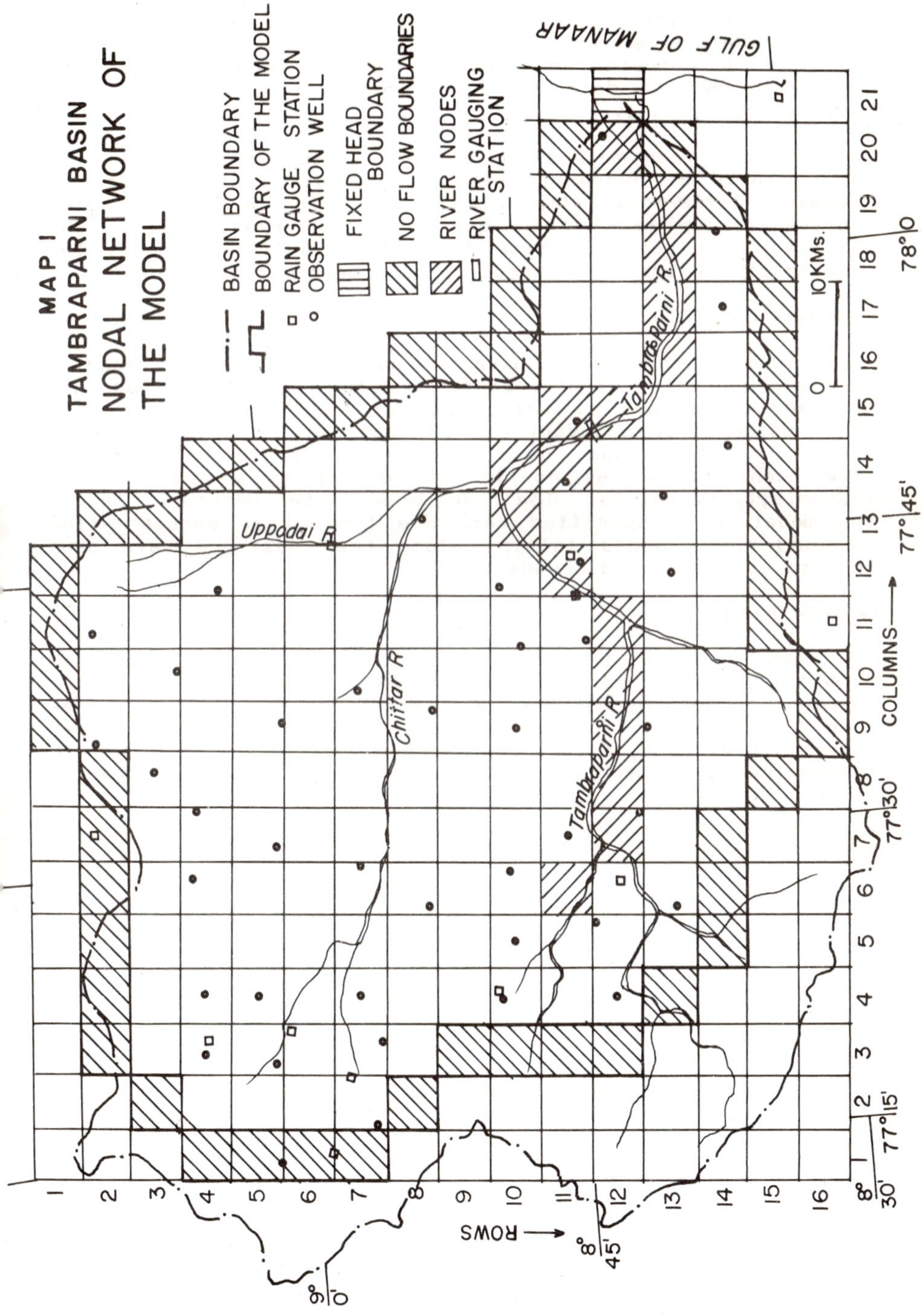
MAP I
TAMBRAPARNI BASIN
NODAL NETWORK OF
THE MODEL
BASIN BOUNDARY
BOUNDARY OF THE MODEL
RAIN GAUGE STATION
OBSERVATION WELL
FIXED HEAD BOUNDARY
NO FLOW BOUNDARIES
RIVER NODES
RIVER GAUGING STATION
GULF OF MANAAR
Uppodai R.
Chittar R.
Tambraparni R.
Tambraparni R.
10KMs.
0
ROWS
COLUMNS
77°15'
77°30'
77°45'
78°0
8°30'
8°45'
9°0'

Difference (Implicit), (3) Alternating - Direction Explicit (ADE) and (4) Iterative Alternating Direction Implicit (IADI) methods (Remson et. al. 1971; Rushton and Redshaw, 1979). A composite computer program in FORTRAN language has been developed and the data has been processed with the TDC 316 Computer of the University of Mysore.

All the four methods have been tested and they gave the same results of hydraulic heads. The study shows that the Backward Difference (Implicit) formulation with Successive Over-Relaxation (SOR) technique is better than Explicit and utilizes less computer time than ADE and IADI methods. The results of the Implicit procedure have been discussed here.

BOUNDARY CONDITIONS

Boundary conditions have a real and important meaning in the preparation of groundwater models. The three important boundary conditions of this model are (1) no-flow boundary, (2) constant-flow boundary and (3) constant head (fixed head) boundary. Impervious mountain fronts and the basin divide have been treated as no-flow boundaries by assigning zero transmissivity values. The zones where water enters or leaves the model at a specified rate are treated as constant-flow boundaries. Nodes falling along the coastline have been treated as fixed-head boundary.

HYDROGEOLOGIC INPUT

Nodal aquifer transmissivity values have been evaluated from the pumping test data of dug wells (Map 2). The aquifer basement topography has been evaluated from the study of bore-hole litholog data and geophysical soundings (Map 3). After several trial runs, a very low storage coefficient of 0.01 to the weathered formations gave the best results. Nodal recharges

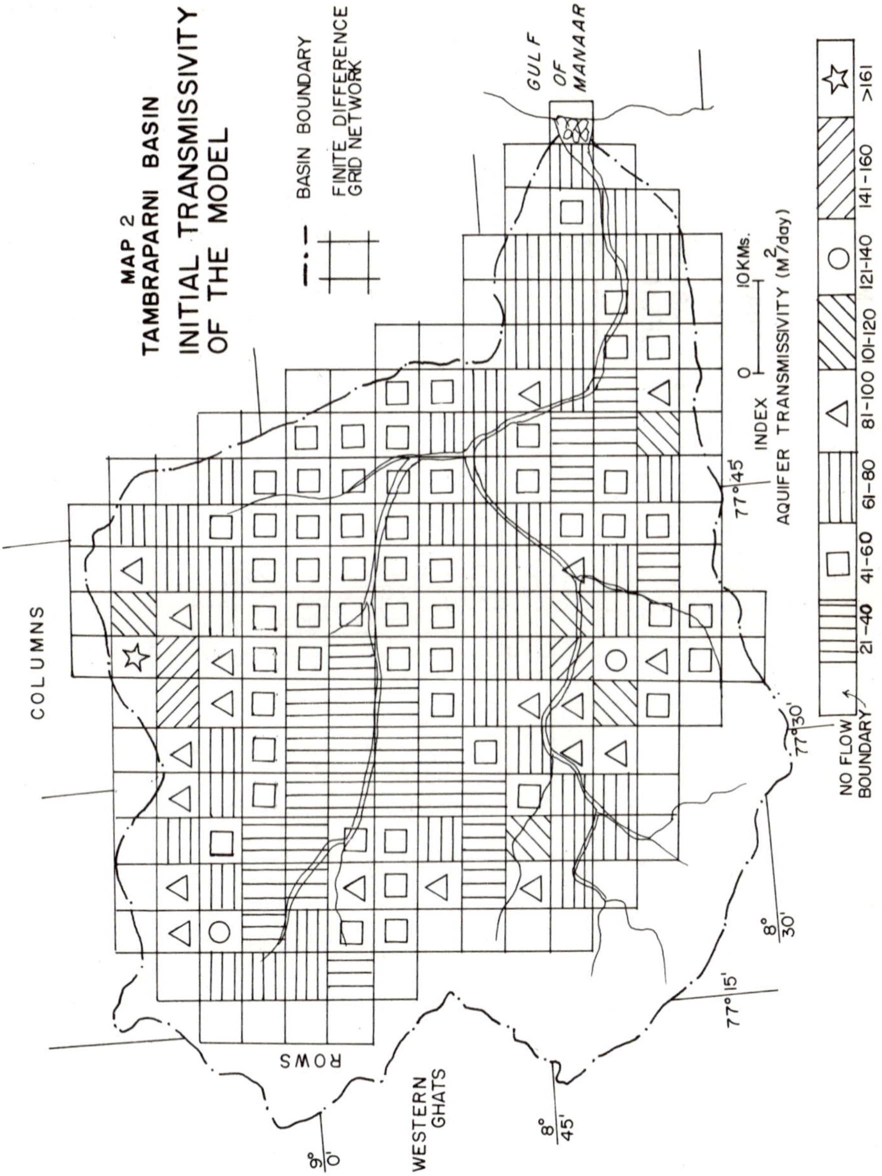
MAP 2
TAMBRAPARNI BASIN
INITIAL TRANSMISSIVITY OF THE MODEL
BASIN BOUNDARY
FINITE DIFFERENCE GRID NETWORK
GULF OF MANAAR
COLUMNS
ROWS
WESTERN GHATS
9° 0'
8° 45'
8° 30'
77° 15'
77° 30'
77° 45'
0
10 KMs.
INDEX
AQUIFER TRANSMISSIVITY (M²/day)
NO FLOW BOUNDARY
21-40
41-60
61-80
81-100
101-120
121-140
141-160
>161

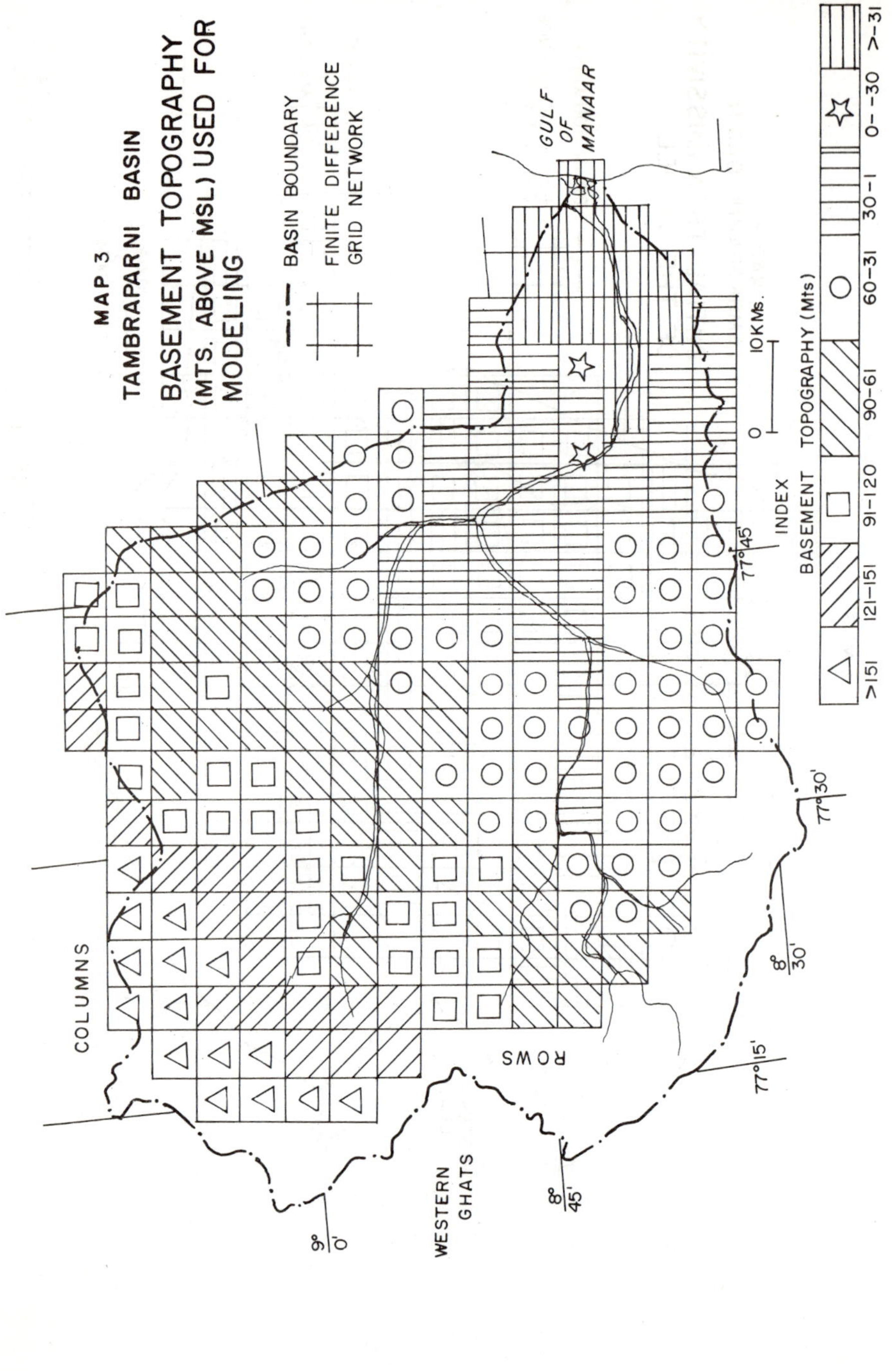
MAP 3
TAMBRAPARNI BASIN
BASEMENT TOPOGRAPHY
(MTS. ABOVE MSL) USED FOR
MODELING
BASIN BOUNDARY
FINITE DIFFERENCE
GRID NETWORK
GULF
OF
MANAAR
10 KMs.
0
INDEX
BASEMENT TOPOGRAPHY (Mts)
>151
121-151
91-120
90-61
60-31
30-1
0--30
>-31
77°45'
77°30'
77°15'
8°30'
8°45'
9°0'
COLUMNS
ROWS
WESTERN
GHATS

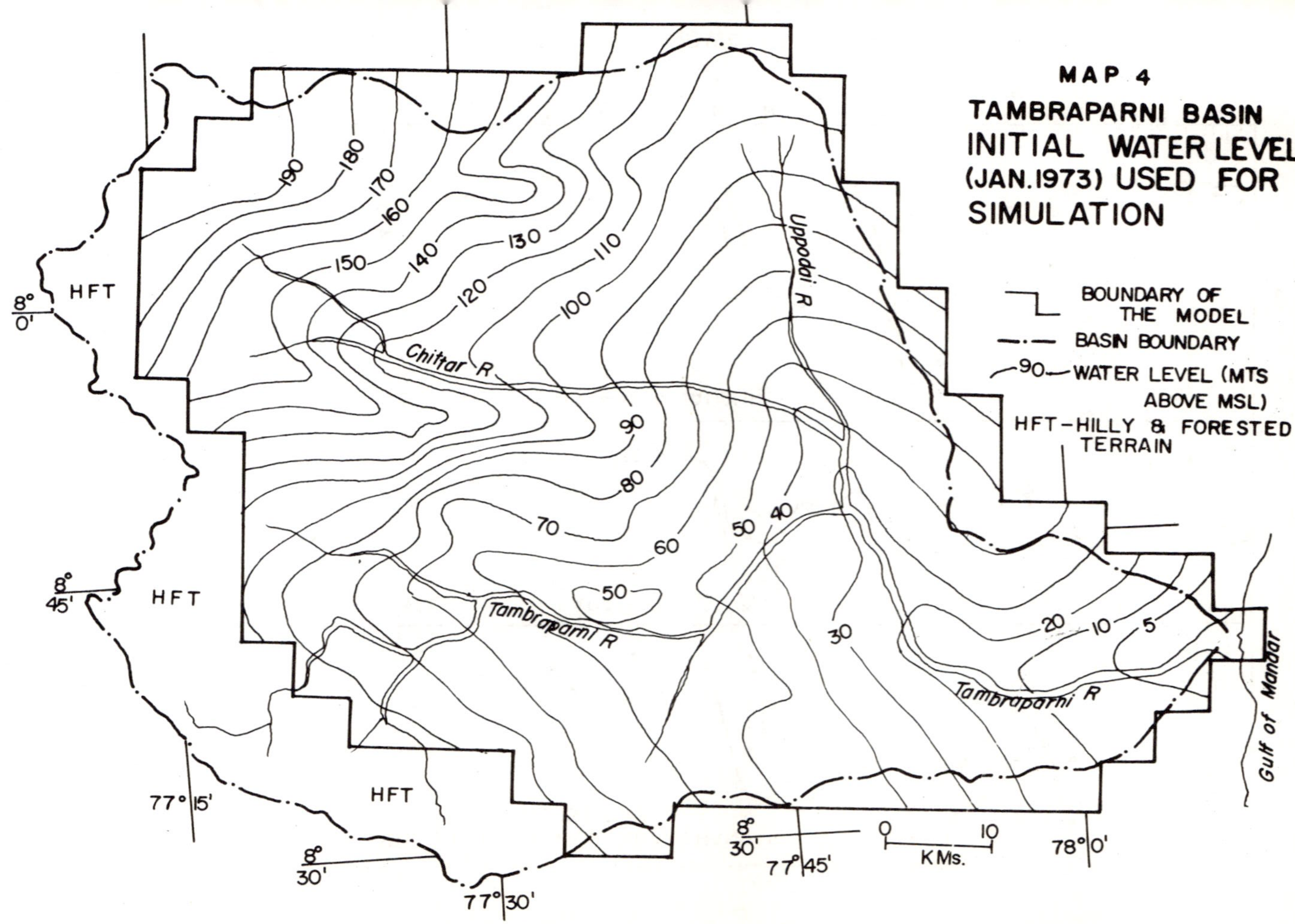

MAP 4
TAMBRAPARNI BASIN
INITIAL WATER LEVELS (JAN. 1973) USED FOR SIMULATION

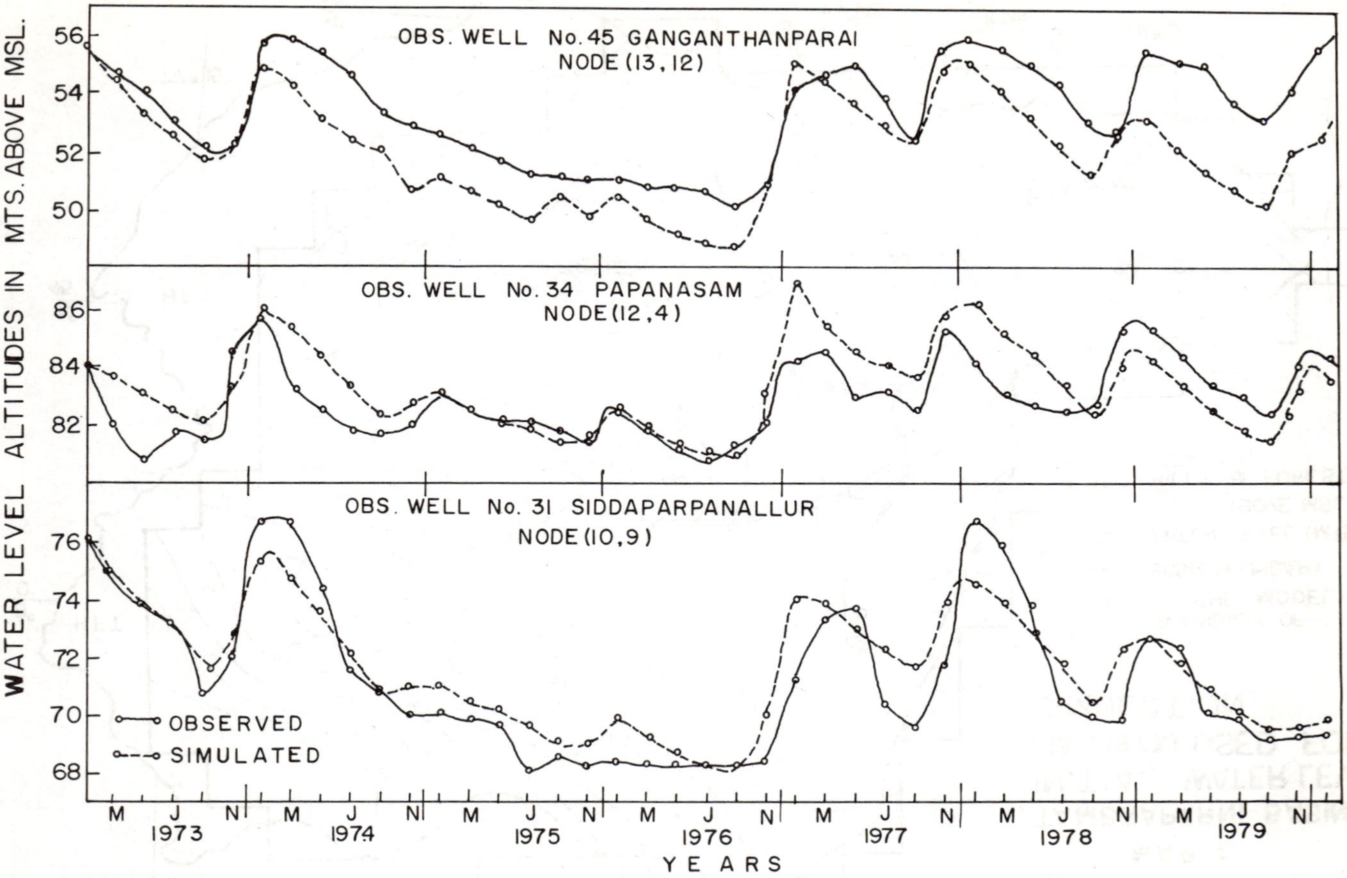

FIG. 1 SIMULATED AND OBSERVED GROUNDWATER LEVEL HYDROGRAPHS

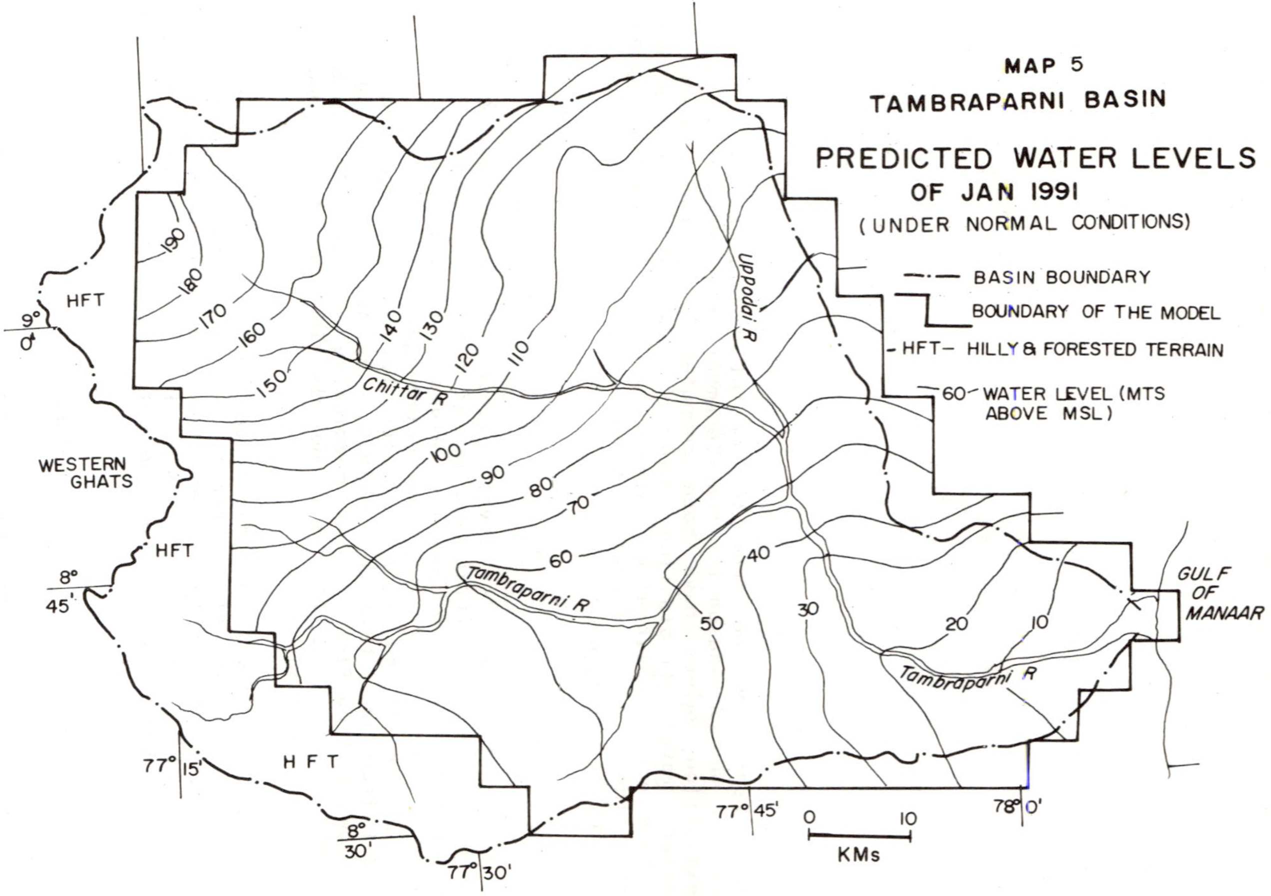

MAP 5
TAMBRAPARNI BASIN
PREDICTED WATER LEVELS
OF JAN 1991
(UNDER NORMAL CONDITIONS)

have been computed from the monthly precipitation volumes of the monsoon season (October-December) and the monthly nodal extraction values have been computed using the systematic water level fluctuation approach and also the method of subjective calibration (Biswas, 1976). It has been found that the average annual groundwater extraction amounts to 247 Mm^3 for this basin.

INITIAL CONDITIONS, CALIBRATION AND PREDICTION

January 1973 water level altitude above MSL (Map 4) has been used as the initial head map of the model. Minor modifications in the transmissivity values have been done for some of the nodes during the calibration phase, and Fig.1 shows the predicted and observed water level hydrographs of the model from Jan. 1973 to Jan.1979. By assigning the average inflows and outflows, the predictive simulation has been extended upto Jan. 1991 and the distribution of groundwater potential is shown in Map 5.

As the demand of groundwater increases year by year, a nominal increase of 1 percent per year to the present pumpages has been proposed and the same model has been used to predict the groundwater potential from the current year till Jan.1991. Map 6 shows the output of the model with the proposed stress into the system.

THE FUTURE

It is clearly seen in Map 6 that the effluent characteristics of the stream has disappeared. The river Tambraparni becomes an influent stream by 1991, if there is even 1 percent

MAP 6

TAMBRAPARNI BASIN

SIMULATED WATER LEVELS
OF JAN 1991
(UNDER PROPOSED STRESS)

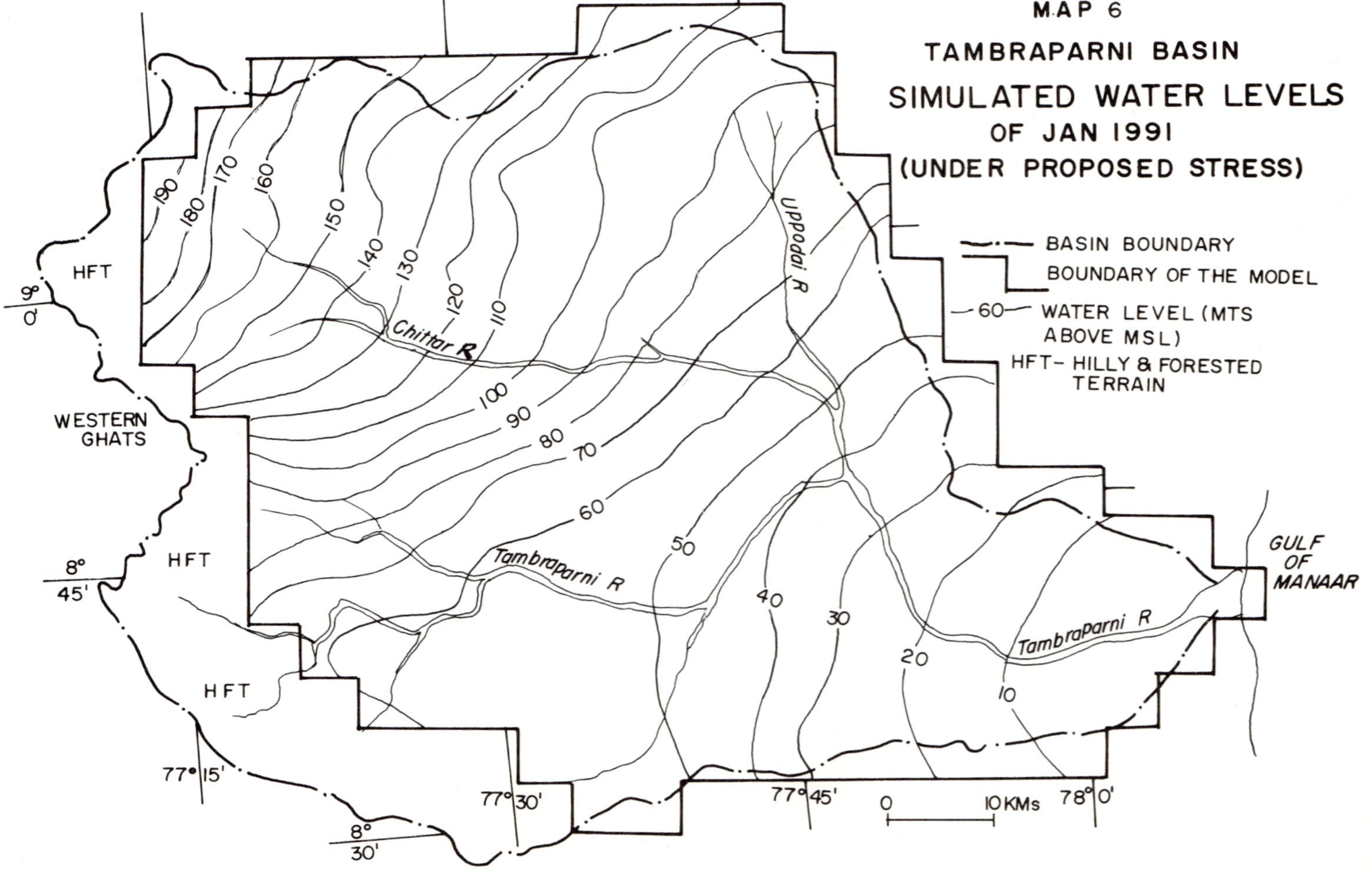

increase in the present day pumpages per year and the water levels are lowered to the extent of 5 mts by 1991.

As the population of this study area mainly depend on the surface and groundwater resources of this basin, care must be taken in the demand-supply schemes of the district in order to avoid over exploitation and to save ourselves in future.

ACKNOWLEDGEMENTS

The authors are thankful to thier parent organisations for providing the necessary facilities during the tenure of this work.

REFERENCES

Biswas, K. Asit., 1976 Systems Approach to Water Management, McGraw Hill Inc., 429 p.

Prickett, T.A., and Lonnquist, C.G., 1971 Selected Digital Computer Techniques for groundwater resources evaluation, Illinois State Water Survey, Bull. 55.

Remson, I., Hornberger, G.M., and Nolz, F.J., 1971 Numerical Methods in Sub-surface Hydrology, Wiley-Interscience, New York.

Rushton, K.R., and Redshaw, S.C., 1979 Seepage and Groundwater Flow, John Wiley and Sons Ltd., Chichester, 339 p.

*****

EXPLORATION AND DEVELOPMENT OF GROUNDWATER IN THE CRYSTALLINE ROCKS

G. Krishna Rao*, Central Department of Geology, Tribhuvan University, Kathmandu, Nepal

ABSTRACT:

Crystalline rocks which constitute a major part of the Peninsular India account for nearly 40 percent of the total aquifer extent in the country. Further irrigation potential of the country depends much on the understanding of these complex aquifers.

Field observations in different parts of Andhra Pradesh and Karnataka states, India, revealed that the behaviour of different rock types felsic to mafic, to tectonics and weathering is different. Fracture intensity is observed to be the highest in quartz veins followed by granites and granitic gneisses and related rocks. Mafic rocks are generally massive with sporadic fracturing. Similarly, the nature of weathered residue from the felsic rocks is relatively more permeable than from the mafic rocks. Petrological significance of the rocks with reference to fracture and granular porosity in these rocks is discussed.

The aquitard-aquifer setting of the weathered zone - fracture zones in the crystalline rocks is explained and mode of optimum groundwater development from them is suggested. In a given area, groundwater exploration and development is recommended in the quartz veins and felsic rocks in preference to mafelsic and mafic rocks.

INTRODUCTION:

Hydrogeologists include all the crystalline rocks in one group, viz., 'Hard Rocks' from groundwater point of view. Little attempt has been made to study in detail the rock types vis-a-vis their groundwater potential. The crystalline rocks which include gneisses and granulites of metamorphic and igneous origin range in their composition from felsic to mafic. These rocks which are generally 'aquifuges' become aquitards or aquifers due to secondary porosity developed by weathering and/or fracturing. It is therefore necessary to understand the behaviour and response of these rocks to tectonics and weathering.

JOINTS AND FRACTURES:

Joints and fractures are discontinuities in the rock masses that develop due to certain stresses or cooling. This study confines to those related to tectonic forces. Joints are in

* Permanent Address: Geology Department, Andhra University, Visakhapatnam, India-530003.

general systematically disposed about folds and faults and their pattern exhibit a rather uniform and consistent nature with respect to a particular stress system over large areas. Intensity of jointing is an important factor for groundwater potential rather than their orientation. Intensity depends to some extent on the vicinity of the formations to the disturbed zone. But this study aims to focuss the response of different formations in the same area subjected to the same tectonic forces.

It is observed during field studies in different parts of Andhra Pradesh and Karnataka states that quartz veins wherever present are in general intensely fractured either in the form of cleaving with their spacing as close as a few centimetres [plate. 1a & b] at some places or in criss-cross pattern depending on the nature of the stresses. Mafelsic gneisses situated close to such formations are observed to be massive with sporadic fractures. A fine grained mafic rock in the same area is found to be very massive with indistinct, short and discontinuous fractures [plate.2]. It is a general observation that felsic rocks like granites, granitic gneisses or even khondalitic type of rocks exhibit definite sets of joints which vary in their spacing from a few centimetres to a few metres and are continuous over large areas laterally and vertically. In contrast to them, mafelsic and mafic rocks are less fractured and remain mostly as massive bodies. It appears therefore that within a given area, and subjected to the same tectonic stresses, quartzites or quartz veins are more susceptible for fracturing followed by felsic and mafelsic rocks. Mafic rocks are the most resistant to fracturing. Ries and Watson [1958] describe granites as the one with the best development of joints and gabbros lacking any regular jointing and often showing great toughness.

Literature is scanty on the study of fracture intensity in relation to the rock type and the rock properties that control it. Under atmospheric pressure and temperatures, most rocks are brittle and fail by rupture at their elastic limit while they undergo plastic deformation above the elastic limit under high temperature and confining pressure. But this distinction cannot be applied to the rock types existing in the same area which have been subjected to the same PT conditions. All ruptures may be classified as tension fractures or shear fractures depending on their pattern and the forces acted upon. The compression, tensile and shear strengths of different rock types are given below:

ROCK TYPE	COMPRESSION [IN HOEK AND BROWN, 1980]	COMPRESSION [IN HABIB, 1982]	TENSILE	SHEAR [In Vutukuri et.al.,1966]
	MPa			
Dolerite	294-572			
Gabbro	205-351			
Gneiss	235-254			680-760
Granite	116-344	60-180	6-15	970-1020
Quartz diorite	243			
Quartzite	227-327	80-300	7-20	610

The above figures show that there is no significant variation in the strengths of different rock types and in fact quartzites are more or equal in resistance to compression and tensile stresses with granites and mafic rocks. The tensile strength of a rock which is responsible for the ruptures is considered to be 5 to 10% of its compression strength. The fact that there is a wide range of values for compression and tensile strengths even for the same rock type suggests that these parameters depend on the fabric of the rock related to mineralogy, grain size, porosity and inherent weak planes. It is significant that the shear strength of quartzites is relatively less than that of granites and gneisses. It infers that shear forces may be responsible for the differential fracture intensities in different rock types in a given area. Other things being equal, the crushing strength of igneous and metamorphic rocks is dependent on the interlocking of the minerals [Ries and Watson, op.cit.]. It suggests that the size and number of minerals in a rock influence its fracture intensity. It is a common observation that granulitic and fine grained rocks are tougher than banded or foliated and coarse grained rocks. Secondly, in quartzite which is a monomineralic rock, the fractures are more or less even and closely spaced since the rock is homogeneous. Pegmatites which are mainly a two mineral aggregate with much coarser in grain size are the second weakest for fracturing. Granites and other rock types are in general three or more mineral component rocks. Fractures in these rocks is in general wide. The fabric may play a dominent role on their fracture intensity. The physical properties of the individual minerals of the rock such as hardness and toughness may not be significant for the susceptibility of the rock to fracturing since most of the rock forming minerals are brittle and possess almost the same hardness at 5-6. In fact, quartz is harder than the other minerals with 7. However, presence of mica minerals in abundance makes the rock fissile. But micas usually occur as accessory minerals in the crystalline rocks.

Groundwater Conditions:

Groundwater potential of the crystalline rocks depends more on the joint and fracture systems and their intensity as they are known to be water bearing down to about 100 m. and more in exceptional cases. These deeper zones are tapped by bore wells mostly for domestic water supply. Their adoptation for irrigation purpose has not picked up in view of their low discharges and frequent failures. A study of bore well yields in the gneisses and khondalites of Visakhapatnam district [Subba Rao and Krishna Rao, 1984] has shown that they range from 1.4 to 14 m^3/hr. and the average is about 4 m^3/hr. In the gneisses and granites of Karnataka state, the yields are reported to vary from 2.2 to 4.5 m^3/hr. in many cases and exceptionally upto 23 m^3/hr. [Radhakrishna, 1971]. The mean yields in various regions of the hard rock aquifers in the world are most commonly between 2.8 and 7 m^3/hr. and 5 to 20 percent failures are common on an average [Davis and De wiest, 1966]. Though many rock types in different parts of the country have been explored and developed for ground water under various programmes, the performance of the wells are not correlated with rock type. Hence a detailed assessment of the potentiality of the individual rock types in the country is not available. However, such a study for some of the African countries [Anon, 1973] as given below throws some light on this aspect.

COUNTRY	LOCATION	GEOLOGY	YIELDS/WELL $m^3/hr.$
Zambia	Kalma, Choma	Quartz veins	8-12
Dahomy	Parakon	Fractured granites in tectonic depressions	7-8
Ivory Coast	Yamonssoukro	Fractured granites	6
Upper Volta	various parts	Granite gneisses	1-4
Mozambique	various parts	Granites, paragneisses, ortho-gneisses	4-8
	Pero Amela Vila Pery	- do -	12-20 [upto 25]
Togo	Dapango	Alkaline granite gneisses	0.04 - 0.2
	Palime	Granites and granodiorites with amphibolites and quartz	0.2 - 0.8
Uganda	Karamoja	Acid gneisses	5-50
Mauritania	south-east	Diorites	0.5
Namibia	Namaqualand	Gneisses and quartzites	1-20
Angola	south	Metamorphic rocks [fractured]	0.6
Mauritania	Fort Gourand	Mica schists and gneisses with pegmatite dikes and sills	0.8
Togo	Kanda	Chlorite schists with quartz veins	7-12
Madagascar	various	Weathered gneisses	0.4 - 1.2

It is evident from the above table that groundwater potential is high in quartz veins with the yields 8-12 m^3/hr. In the granites the yields range from less than 1 to 8 m^3/hr. and exceptionally upto 50 m^3/hr. in acid gneisses. The yields are very meagre in the granites associated with granodiorites and in diorites. It is significant to note that metamorphic rocks and schists record less than 1 m^3/hr. but the yields increase to 7-12 m^3/hr. when the schists are associated with quartz veins. In a sum up of the performance of bore wells in the African continent, it is stated that 5 m^3/hr. is a good yield for granites and granite gneisses and 1 m^3/hr. is appreciable for schists. Better yields are obtained in quartzose zones. It is a common observation during exploratory drilling in the granite and gneissic terrains that the yields of the bore holes suddenly increase whenever they pierce quartz or pegmatite veins. Detailed study of the bore well records in relation to the rock type as available in the records of different exploratory programmes of different organisations in the country would throw light on this aspect.

WEATHERED ZONE:

Weathered zone in the crystalline rocks assume maximum importance in groundwater development as they are exploited through shallow dug wells since ancient times. Weathering depends on the environmental factors like climate, topography and vegetation apart from rock type. But in a given area where the environmental factors are the same, the rock type decides the nature of the weathered residue. The stability of minerals during weathering is considered to be the opposite to the sequence of their crystallisation i.e., quartz is the most resistar while pyroxenes and calcic plagioclase are the least resistant. All the silicate minerals except quartz alter finally to clay minerals during weathering. Therefore quartz poor mafic rocks or basic rocks give rise to purely clayey substance under weathering while in the felsic or acidic rocks, quartz remains studded amidst such clay. In the latter case, the weathered zone resembles a sandstone with argillaceous matrix but with less induration while the former behaves as a clay. The weathered zones in the granites and granitic gneisses therefore possess certain degree of permeability depending on the quartz content in the rock and qualifies as a water table aqutard if not an aquifer. In the quartz poor rocks, the weathered zones may be similar to aquicludes.

Degree of weathering is another factor which influences the permeability of the disintegrated rock. In the completely weathered rock the primary minerals are usually replaced by clay minerals and hence the residue would be almost clayey. In the semi-weathered rocks, the primary minerals undergo collapse in their structures and partly replaced by clay minerals or some other intermediate products and the minerals and the rock retain their skeletal structures with voids due to leaching of certain chemical constituents. Therefore semi-weathered zones are relatively more permeable than the completely weathered zones.

Groundwater Conditions:

Weathered zones in the crystalline rocks form shallow aquifers [water table aqutards] with limited potentiality. While topography plays a role in the crystalline terrains in deciding the depth of water table and its fluctuation, the nature of weathered zone determines the inflow characteristics of the wells. Yields of dug wells in different parts of the states have a wide range from less than 30 m^3 to about 150 m^3/day obviously depending on the nature of the weathered zones. Bore wells in the weathered zones are in general considered to be failures with meagre yields. As shown in the above table, bore wells in the weathered gneisses of Africa show the yields in the range of 0.4 to 1.2 m^3/hr. This is due to very low permeability of the weathered zones. Therefore dug well is a popular structure in these zones which provides storage for the slowly percolating groundwaters. The pooled-up water in these wells is pumped once or twice in a day depending on the rate of recuperation.

MODE OF OPTIMUM GROUNDWATER DEVELOPMENT:

The above descriptions show that the crystalline rocks possess two distinct water bearing zones-weathered zone at the surface predominantly with granular porosity and fracture zones below. It has been well established now that these two zones form an aquitard-aquifer setting with the latter under semiconfined to confined conditions depending on the nature and thickness of the weathered zone and depth of the fracture zones. A study of bore wells in Visakhapatnam district [Subba Rao and Krishna Rao, op. cit.] revealed that the piezometric head in the fracture zones ranges from 1.5 to 25 m. with an average at 7.5 m. While the weathered zone is exploitable by dug wells, the deeper fracture zones can be approached only by bore wells. Restricting groundwater exploitation to any one particular horizon among these two, leaves the other untapped. Though it is considered that vertical leakage from weathered zone into the fracture zones under pumping conditions is quite significant [Sridharan et.al., 1987], the area of influence of such pumping would be minimum. Maximum groundwater flow in the weathered zones can be only 20 to 100 m. in an annual cycle i.e. post-monsoon to pre-monsoon period [assuming a hydraulic gradient of 1.0 and 'k' in the range of 0.1 m/day corresponding to silt, clay and their admixtures with sand, to 0.5 m/day for the laminated sandstone, Anon, 1981]. Therefore overdevelopment situation in the weathered zones of an area cannot be extended farther. Velocity of groundwater through fracture zones, on the other hand, depends on the fracture pattern and their intensity and would be much higher than in the weathered zones. Therefore for an optimum development of crystalline aquifers, it is suggested that dug wells and bore wells should be constructed in different parts of an area avoiding intereference. Since dug wells are generally clustered close to the villages, bore wells should be located far away from these areas to exploit the untapped deep groundwater resource. By this, bore wells can also draw deep horizon waters from the dug well areas without affecting the latter significantly since the influence of bore wells on vertical leakage diminishes with distance proportional to the cones of depression in the piezometric surface.

CONCLUSIONS:

Quartz content in the crystalline rocks plays vital role in making them potential aquifers. The weathered residue of quartz rich felsic rocks would be relatively more permeable than the quartz poor mafic or basic rocks. Similarly, quartz veins or quartzites and felsic rocks are more brittle than the mafic rocks and hence develop more fracture intensity which is ideal for high groundwater potential.

Accordingly, it is suggested that in a given area, groundwater exploration and development should be taken up in the quartz veins, quartzites and felsic rocks in preference to mafelsic and mafic rocks.

Optimum groundwater development in the crystalline rocks can be achieved by exploiting the weathered zones and fracture zones through dug wells and bore wells respectively but by constructing them in different parts of an area minimising their intereference.

REFERENCES:

Anon, 1973. Groundwater in Africa. United Nations Publication, p. 170.

Anon, 1981. Groundwater mannual. A water resource technical publication, U.S. Department of the Interior, p. 480.

Davis, S.N., and De Wiest, R.J.M., 1966. Hydrogeology. John Wiley and Sons Inc., New York, p. 463.

Habib, P., 1982. An Outline of Soil and rock Mechanics. Cambridge university Press, p. 149.

Hoek, E., and Brown, E.T., 1980. Underground Excavations in rock. Institute of Mining and Metallurgy, London, p.527.

Radhakrishna, B.P., 1971. Problems confronting the occurence of Groundwater in hard rocks. Proc. seminar on Groundwater Problems in Hard Rock Areas of India, Bangalore, p. 27-44.

Ries, H., and Watson, T.L., 1958. Engineering Geology. John Wiley & Sons, Inc., London, p. 750.

Sridharan, K., Mohan Kumar, M.S. and Lakshmana Rao, N.S., 1987. Aquifer-Water Table Aqutard model for hard rock areas. J.Geol. Soc. Ind., V.29, p.145-157.

Subba Rao, N., and Krishna Rao, G., 1984. Dug Wells and Bore Wells in hard rock aquifers of Visakhapatnam district, Andhra Pradesh - A Comparative Study.

J. Assoc. Expl. Geophy, V.4, No.4, p. 11-18.

Vutukuri, V.S., Lama, R.D., and Saluja, S.S., 1974. Handbook on Mechanical properties of Rocks, V.1, Trans. Tech. Publications, p. 280.

Plate 1a (top left) : Fracturing in quartz vein

Plate 1b (bottom left) : Same as 1a, close-up view shows brittleness of the rock

Plate 2(above) : Fracturing in a mafic rock

MANAGEMENT OF GROUND WATER RESOURCES IN COMPLEX AQUIFER SYSTEM OF RAJMAHAL TRAPS OF SAHEBGANJ DISTRICT, BIHAR.

BY

S.H.A. KHAN AND JAWED RAJA

SCIENTIST 'B' ASSTT. HYDROGEOLOGIST

CENTRAL GROUND WATER BOARD

ABSTRACT

Sahebganj district of Bihar being chronically drought prone area, the development of ground water resources for assured and continuous water supply has assumed a significant preposition.

Hydrogeological survey coupled with exploratory drilling in the Gumani river Sub-basin of Sahebganj district has revealed that major part of the area is underlain by Rajmahal trap of upper Gondwana, having a huge thickness of basaltic flow. The Rajmanal traps is comprised of seven basaltic flows with white clay and chert separating intertrapean beds.

In this paper an attempt has been made to examine the possibility of ground water in the typical trap country of Sahebganj district, Bihar.

It has been proved that the basalts too are productive when either whethered or fractured. Fractured Zone in their lateral as well as in depth has been delineated. The existance of potential fractured zone at 30-35 m and 70-100 m range has been established In general the Rajmahal traps shows vertical to horizontal joints trending E-W and N-S. At places columnar joints are also well developed. These rocks are Weathered with Characteristic Spheroidal Ex-foliation which give rise to large rounded boulders as the out crop.

Tensile fracture which caused the development of secondary porosity in these, otherwise impermeable rock, provides suitable reservoir for gound water storage.

The quantitative interpretation of aquifer parameters have given a positive sign of presence of productive fractured zone in space and time.

§§§§§§

INTRODUCTION

The Sahebganj district of Bihar State Spread over an are of 3405.6 Sq.Km. and lies between N latitude 24°15' to 25°22 and E longitude 87°30' to 87°55'. The district covering vas tract of land in basaltic terrain is traditionally inhabitate by tribals. From hydrogeological point of view Laterites an Basalts are by far the most important rock in the district. Th lawa flows are weathered to varying degree and has imparted th secondary porosity increasing the permeability of the basalts Apart from Cooling joints in individual flows, a system o fractures and shear cuts accross a number of flow layers. Th intensity of fractures varies in different region. These fracture have developed secondary porosity in these otherwise impermeabl formation and provides suitable aquifer condition for groundwate storage.

Decomposition and disinitegration of Rajmahal traps ha made this aquifer system hetrogeneous and complex. So far prope management of ground water resources in this district has no been done. This study reveals that a substantial quantum of groun water lies untapped which can be utilised to meet the domesti as well as to some extent irrigation and Industrial needs.

GEOMORPHIC REATURES

The area is characterised by hilly undulating topograph covered by lateritic soil and Rajmahal basaltic flows with fla valleys separated by topographic highs. Originally this elevate area existed as a more or less regular plateau formed o successive flows of basaltic traps, sub-aerial denudation ha however very much modified its primary forms and rivers actin through long period of time have excavated wide valleys. Th basalts are Weathered, Chracterised by Spheroidal exfoliatio giving rise to large rounded boulders as the out crops.

Forest abound in the rolling ranges with gradual slop exhibiting oval to domal crests and piercing Valleys.

GEOLOGY

Major part of the area in the district is covered by Rajmahal basaltic lava low of jurassic period which is under lained by sedimentry Gondwana rocks represented by Dubrajpur sand stone. The trap is having a huge thickness (450-600 m) of basaltic lava flows. It has been found that there are at least seven successive flows of basalt in the area, each flow ranging in thickness between 20 m and 75 m. However flows less than 20 m thickness do occur and there are also hills of traps having thickness of 150-200 m. These lava flows are interbedded by the sedimentary inter trappean beds. The traps are mainly composed of basalts and dolerites of tholeite type. The vesicular basalt is dark coloured, porphyritic and amygdaidal rock with the amygdules filled with beautiful Chalcedony and varities of Calcite, Zeolite, and other secondary minerals. The non-Vesicular basalts are hard and tough, Compact and medium to fine grained breaking with Conchoidal fractures. The vesicular type are Comparatively soft and break easily showing vertical and horizontal joints.

Like the Deccan traps, the Rajmahal traps appear to have been largely erupted through fissures though a Centralised focus of eruption is known near simra (Lat. 21°-1°-32", Long 87°-20'-45").

The type area is the Rajmahal hills at the head of the Ganges delta near the border of West Bangal and Bihar.

HYDROGEOLOGICAL CONDITIONS

Ground Water in this traps country occur under unconfined and semi-confined to confined state of disposition depending upon lithological, topographical and structural controls. Decomposition and disintegration of Rajmahal traps has made this aquifers system heterogeneous and complex.

From hydrogeological point of view laterites and basalts

are the most important rocks in the area. The traps are devoid of intergrannular porosity and their storativity and transmissivity depends mainly on secondary porosity imparted by weathering and fractures.

On the basis of detail hydrogeological Survey coupled with exploratory drilling four ground water worthy 30nes have been decifered and aquifers have been broadly divided as following:-

(a) Potential Ground Water exists in the pisolitic laterites and vesicular basalts.

(b) The joints forms potential aquifer at shallow to moderate depth.

(c) Aquifer of substantial values is found at Contact Zone of two basaltic flows.

(d) The fractures forms the conduit for the ground water movement and also acts as a reservoir forming potential aquifer.

a) At many places basalts are weathered forming pisolitic laterites (10-25 m) in thickness. These laterites are having very good effective porosity which provides a productive gound water shallow aquifers. The vesicular basalts when decomposed is yet another good zone of shallow gound water aquifer. Weathering imparts them secondary porosity.

The recouperation test conducted on dug wells (Pararia Longitude 87°-44'-45"; Lat 25°-1'-30"), at Adro (Lat.25°-13'; Long.87°-40'-50") tapping these aquifers shows that the discharges varies from 2-5 m^3/hr for a d.d. ranging from 0.5 to 1.0 m; the Specific Capacity from 20-40 m^3/day/m/d.d., (Papadopulous-Cooper method). Transmissivity from 90 to 125 m^2/day and the rate of inflow from 0.9 to 1.67 m^3/hr. The above test data shows that these aquifers are having good potential to Cater the domestic need of the area by means of dugwells & shallow bore wells if properly exploited.

b) The joints have also been developed at shallow depth due to the Cooling action of the lava-flow. General attitude of joints in basaltic flow is E-W & N-S. At places Columnar joints are well developed (on the Bansloi River bank near Amrapara). These joints provide the important function of Transmitting infiltrating rain fall from weathered zone in to the deeper fractured system.

The recouperation test conducted on dug wells tapping this zone has given an encourging result. The discharge ranged from 5-10 m^3/hr for a draw down of 0.75 to 1.25 m for pumping of 200 mts. The Sp. Capacity ranged from 25 to 50 m^3/day/m/d.d. Transmissivity ranged from 125 to 150 m^2/day and the rate of inflow varied from 1.75 to 2.00 m^3/hr. The data of recouperation test indicates that these zones are forming good repository of ground water in the area.

c) At the contact zone of two lava flow (near Brindavan) potential aquifer has been found at moderate depth. The exploratory drilling has confirmed the presence of potential contact zone.

The aquifer parameters are appended in table I.

d) Apart from the cooling joints in individual flows a system of fractures and shears cut across a member of flow layers. The sub-vertical deep seated fractures occurs as lineaments. The intensity of fracture varies in different region being higher in Barharwa, Borio and Berheit and Litipara regious than else where. However in the Barharwa region the density of fracture is much higher. In this part the trend of fracture varies from N 15° W-S 15° E, and N-S, where as in Pakur and Amrapara the trend is N 15° E-S 15° W,. The fractured zone with NW - SE trend is likely to be tensile fracture as it is more productive where as N-S and NE - SW trending fractures appears to be congugate shear fracture. These fracture systems has caused the development of secondary porosity in these other wise impermeable rocks which forms the conduit or passage for the movement of ground water and at the same time acts as a reservoir. This has been confirmed by exploratory drilling by CGWB.

The Quantitative assessment of the aquifers have given a positive sign of presence of these productive fractured zone (See Table I).

MANAGEMENT TECHNIQUES

In view of the complex nature of the aquifer and their intricate aquifer geometry the management of gound water resource is essential in the context of proper assessment and scientific planning.

Till recently the development activities in this district have proceeded in an unplanned manner resulting in the failure of ground water structures at places. The key of the problem lies in the efficient utilisation of ground water by suitable structures. So, the following methodology is suggested to harness and develop ground water resources which is so scare in this trap country.

In Sahebganj district mostly shallow ground water aquifer is harnessed by means of dug wells, which fails to solve even the domestic requirement particularly during summer as traditional wells are generally dug by means of simple hand excavation techniques which seldom penetrate more than a few meters below water table and seasonally often go dry with the decline of water table during the dry season. So to augment the yield of dug wells, these wells need extention of 5 to 10 m depth either through vertical extension bores or angular drilling methods so as to pierce the entire thickness of the weathered zone.

Shallow bore wells are now being widely constructed to ensure drinking water by the state Govt. Here also during the lean period due to the lowering of the water table these shallow bore wells become inoperative. These shallow bore wells too need extension of 5 to 10 m depth below the position of the water table during dry season (Pre-monsoon) which will become a reliable source of ground water for drinking as well as domestic purposes. For this regular monitoring of the ground water level should be carried out before the shallow bore wells are extensively constructed.

More over the joints and fractures developed at shallow depth should be delineated and site for dug wells (across the joints) and shallow bore well (across the fractures) should be selected which will provide copious supply of ground water to Cater the domestic need of the district.

Large scale development of shallow ground water resource should be confined to valleys and topographical depressions only. There are a few deep bore wells in the Sahabganj district, hence the deep seated fractured zones have not been extensinety tapped which if properly located will ensure the perinnial supply of ground water for domestic as well as agricultural and industrial uses.

The deep seated fractured zones should be delineated by study of aerial photographs on (1 160,000 scale) and satellite imageries on 1 : 2,50,000 scale). Ground Checking of these lineaments and elimenating the doubtful one from the local geological, structural and hydrogeological condition should be carried out. Then suitable surface geophysical methods should be employed to Pin Point the sites for drilling.

The deep seated fratured zones should be developed by means of drilled wells. It has been confirmed that drilled wells which draw water from the fracture system generally have much higher yield (Barharwa 51.6 m^3/hr), Borio (21.0 m^3/hr) than those which tap only from weathered layer Rajmahal (6.12 m^3/hr). Also the yield of such wells tapping fractured zones tend to be more stable and less subject to the Vagaries of drought for lack of reacharge.

Due to undulating nature of the terrain construction of sub-surfacr dams across small valleys and extracting water by Jack wells on the up stream side of this dam shall be a very efficient method of ground water development. Areas where the thickness of weathered mantle is of the order of few metre is ideal place for such membrance structure. Similar Structure across small streams would also be very effective to arrest- ground water run off (baseflow) which could be tapped by large diametre (10-12 m) dug wells.

Inter - montane valley fills along the vicinity of river forming river terraces should be developed for ground water extraction (as most of the rivers are effluent in nature) through sump well/radial wells.

Spring should also be harnessed by constructing check weirs for Copious supply of ground water.

The estimated ground water resources of the Sahebganj distt. is 241.95 MCM. Against this the net ground water draft is only of the order of 12.48 MCM. As such a balance of 228.47 MCM is available for ground water development through various ground water structures.

So, proper ground water management technique coupled with Scientific methodology judeously implemented, will ensure the regular supply of ground water for domestic. Agricultural and industrial uses of this Chronically drought prone tribal area.

ACKNOWLEDGEMENT

The authors are thankful to S/Sh. N.Kittu, Director, E.R. CGWB and Sh. S.K. Sinha, Scientist 'D' Officer in charge, B.S.U. Office, CGWB, Patna, who have been the driving forces in writing this paper. The authors are also thankful to their collegues particularly Mr. P.N. singh and Mr. K.K. Singh in critically discussing this paper.

REFERENCES

1. Krishnan M.S. — Geology of India & Burma (1971).

2. Saha, D.B. — Systematic Hydrogeological studies in Bhagirathi-Gumani - Torai Sub-basin, North eastern parts of Santhal Pargana's District, Bihar (1983).

3. Sinha Ray, S.P. & Dutta. S. (Mrs.) — Hydrogeology and Ground Water resources of Tribal district, Sahebganj, Bihar (CGWB, ER) (1987).

4. Trivedi M.K. — Geology and Mineral Resources of Santhal Parganas Distt. Bihar (Unpublished report) Ground Water Deptt., Govt. of Bihar, Dumka.

TABLE

AQUIFER PARAMETER OF THE EXPLORATORY BORE HOLE AT SAHEBGANJ DISTRICT, BIHAR.

Sl. No.	Location	Aquifer Material	Depth Drilled (m)	Zone encountoud (in m. b.g.l)	Static Water Level (m. b.g.l.)	Discharge (m^3/h)	Draw Down (m)	Transmi- (m^2/day)	Storage Ceffi-cient.
1.	Pakur	Weathered and fractured trap	100.60	ABANDONED AS DISCHARGES WAS NIL					
2.	Barharwa	Fractured 1 Trap	100.00	45-1000	9.30	51.60	8.92	333	$1.4x10^{-4}$
3.	Litipara	Vescicular Trap (Fractured)	98.35	30-50	11.36	9.24	13.19	86	-
4.	Amrapara	Trap (Weathered and fractured)	98.00	43-62	1.55	1.08	19.65	28	$1.3x10^{-2}$
5.	Brindavan	Trap (Weathered)	92.50	33-35 42-50	2.17	6.120	7.05	29	$6.1x10^{-5}$
6.	Berhait	Vescicular Trap (Fractured)	90.65	33-70	7.65	26.10	15.34	45	
7.	Gaurpara	Massive Trap	80.00	ABANDONED AS DISCHARGE WAS NIL					
8.	Borio	Highly fractured Trap	75.20	39-54	5.33	21.0	9.65	38	$1.7x10^{-5}$
9.	Rajmahal	Vescicular Trap (Weathered)	74.45	33-36	6.90	6.12	3.67	90	$7.3x10^{-3}$

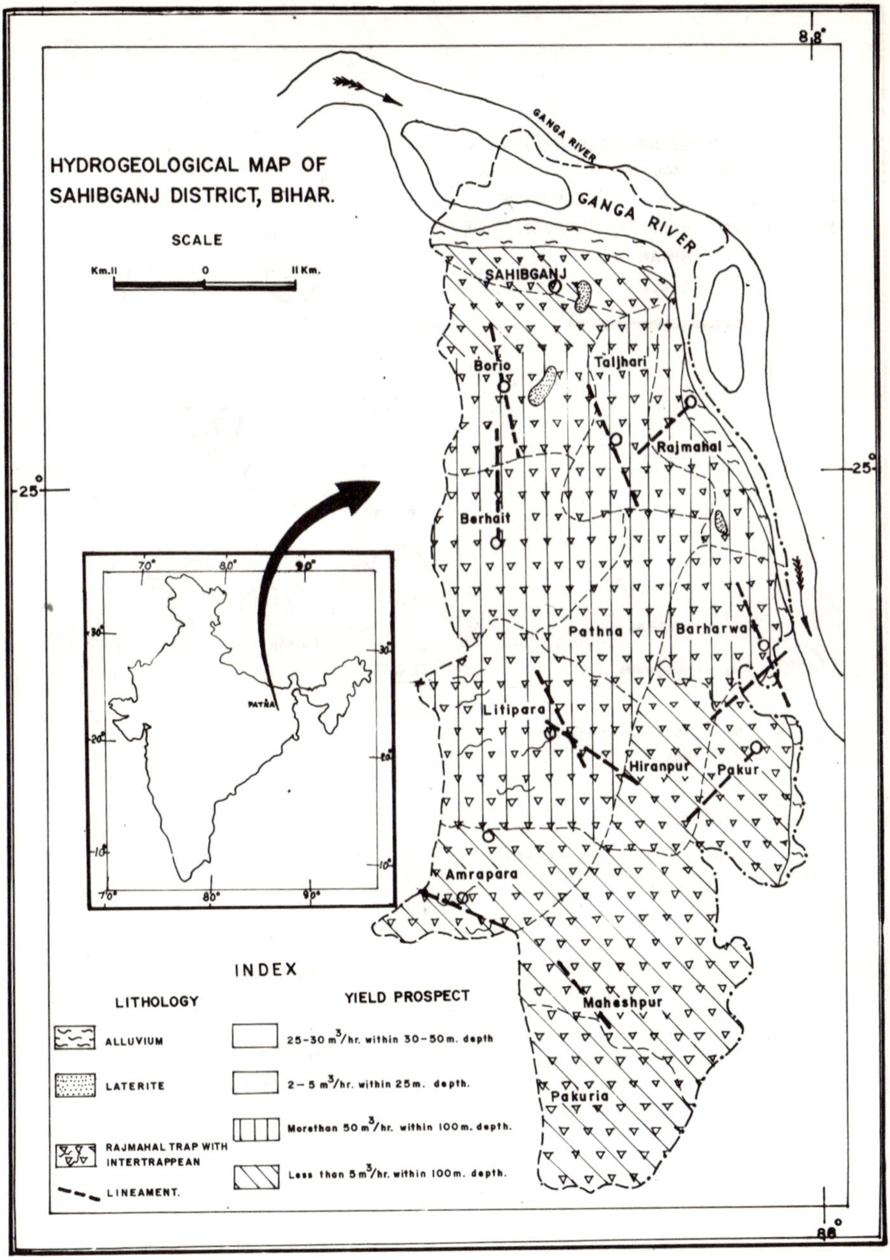
HYDROGEOLOGICAL MAP OF
SAHIBGANJ DISTRICT, BIHAR.
SCALE
Km.11
0
11 Km.
GANGA RIVER
GANGA RIVER
SAHIBGANJ
Taljhari
Borio
Rajmahal
Berhait
Pathna
Barharwa
Litipara
Hiranpur
Pakur
Amrapara
Maheshpur
Pakuria
88°
25°
25°
PATNA
INDEX
LITHOLOGY
ALLUVIUM
LATERITE
RAJMAHAL TRAP WITH INTERTRAPPEAN
LINEAMENT.
YIELD PROSPECT
25-30 m³/hr. within 30-50 m. depth
2-5 m³/hr. within 25 m. depth.
Morethan 50 m³/hr. within 100 m. depth.
Less than 5 m³/hr. within 100 m. depth.

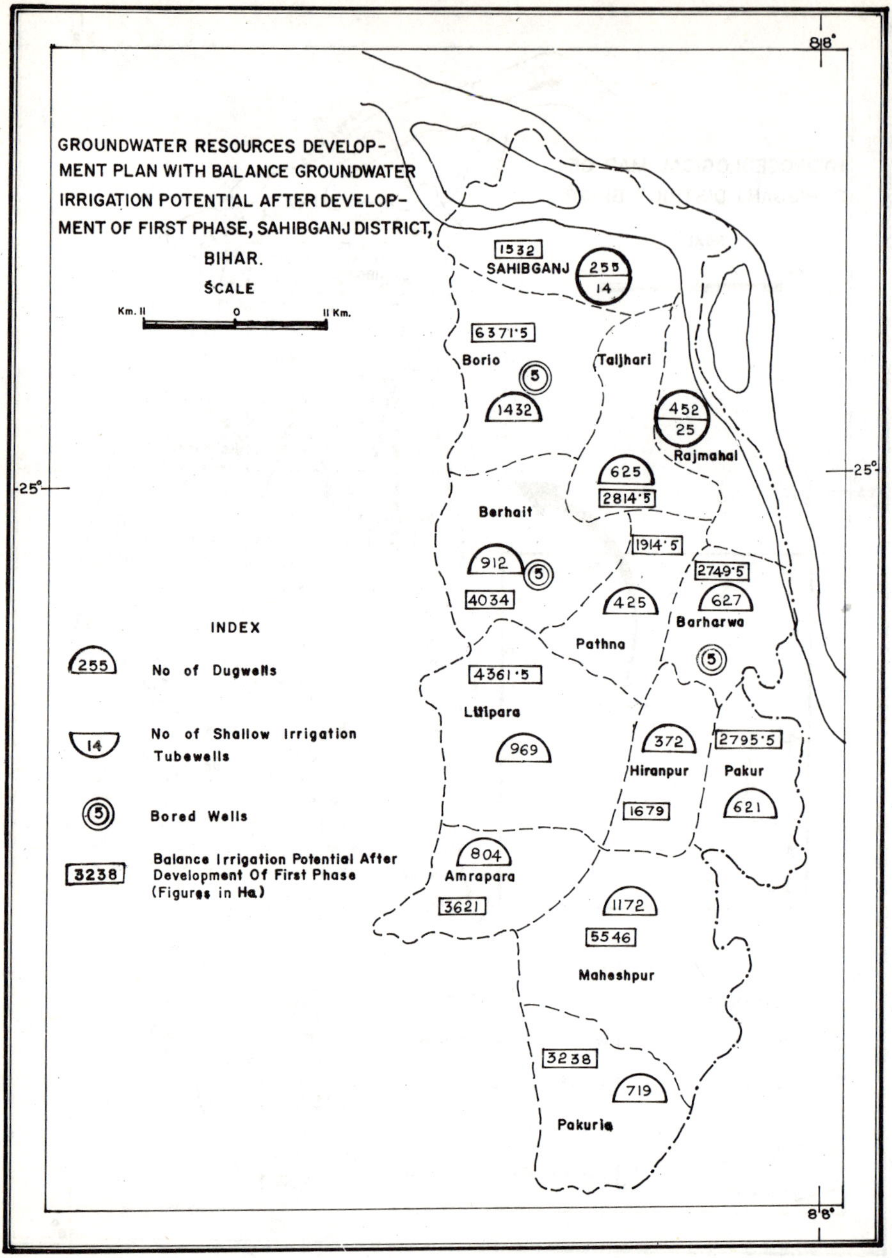
GROUNDWATER RESOURCES DEVELOPMENT PLAN WITH BALANCE GROUNDWATER IRRIGATION POTENTIAL AFTER DEVELOPMENT OF FIRST PHASE, SAHIBGANJ DISTRICT, BIHAR.
SCALE
Km. 11 0 11 Km.
88°
25°
1532
SAHIBGANJ
255
14
6371·5
Borio
5
1432
Taljhari
452
25
Rajmahal
625
2814·5
Berhait
1914·5
912
5
4034
425
2749·5
627
Barharwa
Pathna
5
4361·5
Litipara
969
372
Hiranpur
1679
2795·5
Pakur
621
804
Amrapara
3621
1172
5546
Maheshpur
3238
719
Pakuria
INDEX
255
No of Dugwells
14
No of Shallow Irrigation Tubewells
5
Bored Wells
3238
Balance Irrigation Potential After Development Of First Phase (Figures in Ha)

GROUNDWATER ASSESSMENT IN THE SEDIMENTARY BASINS OF EASTERN KENYA

George Okoye Krhoda
University of Nairobi
P.O. Box 30197, Nairobi
Kenya

ABSTRACT

This paper assesses the groundwater resources of the sedimentary basins of Eastern Kenya. The sediments were laid between Palaeozoic and Quaternary periods. The area is generally drought-striken and comprises nearly two-thirds of the total area of the country. The mean rainfall decreases while the rate of potential evaporation increases northwards and eastwards. Aridity of the area, therefore, results from high temperatures accompanied by lack of vegetation cover. The entire region is drained by sandrivers except for the Tana River that runs along the southern boundary.

Several geophysical methods have been used, with differing success in assessing the groundwater potential of the region. Because of the remoteness of the area, well logging has been identified as the most useful method in reconstructing the stratigraphy of the level. Preliminary investigation on the nature of the lithological logs shows that three types of geologic terranes have potable water within the sedimentary basins in the area. The first geologic terrane is the Pliocene and Quaternary deposits comprising of unconsolidated or poorly consolidated sand and gravel commonly interbedded with clay. The second terrane consists of semi-consolidated and consolidated anglomerate, sandstone or shale formations of which are often joined and possess intergranular and fracture porosity and permeability. The third geologic terrane comprises of limestone and gypsum beds which generally contain calcium sulfate. Further research is needed to quantify the water resources of the area.

Introduction

Finding and developing water resources is a vital prerequisite for the development of Kenya. Experience has proved that supplying clean water to an individual is the foundation on which rests the health and economic progress of a community. Higher incomes, improved health and increased leisure are spin-offs that accrue as a result of a reduction in the time and energy previously used in collecting water as well as an improved organization of the community. In turn, better community organization will increase labour quality, improve hygiene and reduce pressure that now exists on health and social services.

This paper assesses the groundwater resources of the sedimentary basins of eastern Kenya. Specific emphasis will be placed on techniques that have been used in data gathering and interpretation. The area has a semi-arid to arid climate and consists of a homogenous geological formation which makes it possible to generalise the present findings and also help other investigators working in the same environments to adapt some of the techniques used here. There is no single groundwater basin in the area.

The major water supply sources in the area includes the perennial Tana River that starts in the eastern Kenya highlands and flows into the Indian Ocean south of Lamu. The other important rivers in the region include Rivers Ewaso Ngiro and Daua. The Ewaso Ngiro waters becomes sub-surface from Habaswein eastwards while the flows of River Daua, an international river running along the border between Kenya and Ethiopia, has low flows during the dry seasons. The dry rivers are called laghas and these supply considerable water during the wet seasons. Most of the water supply is from boreholes, shallow wells and pans. The latter, probably, supply more than 95 percent of the total water supply in the project area. Although the supply is generally interrupted by breakages of water pumps, bursts of pipes and siltation of pans and dams, there appears to be a need to improve on the designs and plan conjunctive water use in a manner that allows ground and surface water interaction.

The Problem

The sedimentary rocks underlie nearly two-thirds of the country and is situated in the northeast and eastern regions which have low and variable rainfall. The area of the present study is approximately 180,000km^2 constituting of Northeast, Coast and Eastern Provinces (Table 1).

The general geology of the area comprises of sedimentary deposits ranging in age from Palaeozoic to Quaternary periods. These rocks include the Mansa Guda formation (Triassic), Daua limestone series and Mandera series (Jurassic) and Marehan series (Cretaceous) and Quaternary sediments.

Table 1: The Districts That Comprise the Study Area

Province	District	Area (Km^2)	Population (1979)
North Eastern	Mandera	26,470	105,610
	Wajir	56,501	139,319
	Garissa	43,931	128,857
Coast	Tana River	38,694	92,401
	Lamu	6,506	42,299
	Kilifi	12,523	430,986
Eastern	Kitui	31,099	464,283
	Isiolo	26,605	43,478
	Marsabit	78,078	96,216

The mean annual rainfall range between 700mm and 200mm. The rainfall is low, variable and usually occurs in the form of short duration storms. The high evaporation rate, about 2300mm, accompanied by unreliable rainfall makes the study area prone to droughts. The frequency of such droughts range between 4 and 7 year intervals.

The major short-coming in the study area is lack of adequate hydrometeorological data (Table 2). The well-watered areas have more stations than the areas most affected by drought. Most of the stations have short records and therefore are less useful for hydrogeological analysis.

There are two distinct rainy seasons in eastern Kenya. The long rains start in March and ends in May except in exceptional wet years when the rain ends in June. The short rains start from mid-October to mid-December.

An average rainfall for the basin is 535mm. Tana River basin, with an area of 132,000km^2 (about 22.7 percent of area of the country) contributes nearly 25 percent of the total runoff of the country. The other important surface water body of interest is River Ewaso Ngiro which has its headwater from Mt. Kenya and Aberdare Range and disappears at the Lorian Swamp. Only during exceptional floods does Ewaso Ngiro and Lagh Dera, as the river is known towards the east, reach the Somalia border. The Ewaso Ngiro river has an area of more than 15,000km^2 at Archer's post where the annual average discharge is about 1.5 million m^3.

A small area in the northwest lying between Finno and Derkali drains east and northward to River Daua. Although Daua River is large, it has a seasonal flow and forms an international boundary with Ethiopia. The water supply to Rhamu, Mandera town and Army barracks are from the river. When rainfall is adequate, the drainage ways of dry river beds become extensive,

Table 2: The Distribution of Rainfall Stations in the Study Area

District		Stations	Km^2/Station
North Eastern			
	Mandera	13	2036
	Garissa	7	6276
	Wajir	15	3766
Eastern			
	Isiolo	17	1565
	Kitui	107	291
Coast			
	Kilifi	69	181
	Tana River	27	1433
	Lamu	18	362
	Total	301	1061

Source: Various records at the Meteorological Department and the Ministry of Water Development, Nairobi.

broad, shallow and ill-defined water bodies. In such circumstances, groundwater resources remain the major reliable supply in the area.

GROUNDWATER INVESTIGATION

Geophysical techniques may be advantageously employed at many stages of groundwater resource development. Besides their speed and economy, many techniques fail to deliver useful results unless the right techniques are used and the conditions are favourable. Surface geophysical and well logging methods have been used in the present project.

Resistivity technique was used in the area to detect and map the extent, depth and thickness of aquaferous zones in the area. Although the technique could be used for determining the fresh-salt water interfaces, for different reasons, the emphasis was on measurement of depth and stratigraphic sequence of materials within the sedimentary basins. However, the use of resistivity analysis was abandoned because of both technical and interpretational problems in an area in which geology is little known.

Using a terrameter model SAS 300, a survey was carried along Dif-Wajir road. The readings of the terrameter were erratic when probing depths of 25m and 32m at several sites.

The cause of the erratic readings were due to the presence of dry sands in the area. The problem was erectified at some points by using current busters, or waiting to carry out the exploration after the rains or by carrying several thousands of gallons of water to wet the sands around the electrodes to improve the contact. In a desert environment, the cost of carrying out resistivity survey were greater than collecting information from the current boreholes that have been drilled in the study area.

The geohydrology of the area has been reconstructed using lithological logs which are records of stratigraphic units traversed during borehole drilling. The records are taken for every 2m depth traversed and sent to the Ministry of Water Development for filing. Out of 654 boreholes in the study area, 130 do not have logs at all while 110 have poor logs that are difficult to interprete. The problem becomes appreciable when it is realised that more than 40 percent of the lithological logs in Marsabit and Isiolo are poor. The table below gives a summary of the nature of lithological logs.

Table 3: Assessment of Groundwater Resources in Sedimentary Basins - Eastern Kenya - Summary of Borehole Physical Data

T.D. = Total Depth; W.S. = Water Struck; W.R.L. = Water Rest Level

District	Total No. B/H	Dry No. B/H	No. B/H	Successful					
				Max	Min	Mean	Mean	Mean	Mean
Marsabit	85	39	46	356	19	110	72	65	5.6
Isiolo	59	13	46	233	11	93	61	52	4
Mandera	52	13	39	300	19	142	95	49	4
Wajir	127	47	80	304	14	114	84	72	10
Garissa	88	32	56	231	8	118	103	98	4
Kitui	39	12	27	197	26	102	32	22	7

Source: Krhoda, G.O. Assessment of Groundwater Resources in Sedimentary Basins, Eastern Kenya: Progress Report, 1987

The Triassic sediments are represented by the Mansa Guda Formation which is constituted of conglomerates and sandstones deposited as a delta during transgression of the sea. It has good shallow acquifers, however at great thicknesses, the formation is overlain by siltstones and shales.

Table 4: Assessment of Groundwater Resources in Sedimentary Basis - Eastern Province

Summary of Lithological Logs

District	No. B/Hs.	B/Hs without Logs	B/Hs with Logs	
			Total	No. with Good Logs
Marsabit	85	23	62	27
Isiolo	59	13	46	24
Mandera	52	15	37	27
Wajir	127	15	112	85
Garissa	88	10	78	76
Kitui	39	4	35	23

Source: Krhoda, G.O. Assessment of Groundwater Resources in Sedimentary Basins, Eastern Kenya: Progress Report, 1987.

Recurrent downwarping in the Lower Jurassic period caused another sea transgression and Bur Mayo formation was deposited. The thickness of Bur Mayo Formation is 2400m (Ayer, 1952) and it is constituted of grey and brown limestone. The rocks dip gently 2° to 3° toward the east, southeast, south and southwest. Groundwater yield is variable, but range from 6 litres to 12 litres per minute in Wajir Bor, Talebba and Wal Giris to about 50 litres to 75 litres per minute north of Finno.

After regression, from middle Jurassic period the Cretaceous continental Maheran series were deposited. The series are constituted of Danissa Beds and Marehan Beds. The Danissa Beds, however, have variable lithology, constituting of lenses of interbedded limestone, shale, siltstone and sandstone. This type of formation indicates a change of the environment from marine to estuarine, to deltaic and to terrestrial conditions. The Danissa Beds have a thickness of 116m.

The Upper Maheran Beds weather to a reddish colour and are of continental origion. The beds have a gentle dip, 2° to 3° toward the east, southeast, south and southwest. They are constituted of sandstone, shale and siltstone to a thickness of 107m. The Upper Marehan Beds forms better aquifers especially in sandstone of white, pink or reddish colour, of fairly rounded quartz grains, and friable fine to medium-grained texture. Around Finno, the yields range from 40 litres to 90 litres per minute.

The Pliocene Merti Beds were laid during alternating pluvial and interpluvial periods. As a result the lithology of the beds are of varied character (Joubert, 1963). The beds are red, white, grey or tan in colour and include friable to consolidated gravel, grit, sand, silt and clay generally intercalated in lenticular beds. There are also scattered, concretionary zones commonly less than 1 metre thick of silt, sand and ground cemented by calcium carbonate.

The sequence of Quaternary deposits varies considerably over short distances which makes it difficult to recognize and correlate the different stratigraphic units (Swarzenski and Mundorff, 1977). These deposits are represented by Wajir Beds, characterized by limestone and gypsum beds. The Wajir Beds forms a shallow aquifer with yields ranging between 40 litres and 80 litres per minute.

Conclusion

A picture of the geohydrology of the sedimentary basins is beginning to emerge from the analysis of well logs of the area. Preliminary investigation on the characteristics of the lothological logs shows that there are three types of geologic terranes in which potable water is obtained in the sedimentary basin. First is the Quaternary and Pliocene deposits of unconsolidated or poorly consolidated sand and gravel commonly interbedded with clay. The others consists of semi-consolidated and consolidated anglomerate, sandstone or shale formations of which are often joined and possess intergranular and fracture porosity and permeability. The third geologic terranes is the limestone and gypsum beds which generally contain saline water (C3850) largely calcium sulfate. Further research is needed to demarcate these and other geologic terranes in the country.

REFERENCES

Ayers, F.M., 1976. Geology of the Wajir-Mandera District, Northeast Kenya. Government Printers.

Bestow, T.T., 1960. Northeastern Province Ground Survey. Esplanatory note on Sheet 13 - Lower Ewaso Ngiro.

Government of Kenya, 1962. An Investigation into the Water Resources of the Ewaso Ngiro Basin, Kenya. Ministry of Works, Hydraulic Branch Report 5. Government Printers.

Government of Kenya, 1977. The National Water Master Plan of Kenya. government Printers.

Government of Kenya, 1973. Sectional Study and National Programming for Community and Rural Water Supply, Sewerage and Water Pollution Control. Report 7. Groundwater Resources in Kenya WHO, Brazzaville.

Krhoda, G.O., 1986. The Groundwater Resources in semi-arid and arid parts of Eastern Kenya, in Geohydrology of Drought-Prone areas of Africa Common Science Council, Technical Publication Series No. 202, London.

Krhoda, G.O. 1987. Assessment of Groundwater Resources in Sedimentary Basins, Eastern Kenya: Progress Report.

Krhoda, G.O. Water Supply in Kenya Today and the year 2000 AD, in S.H. Ominde (ed.), Issues in Population Growth and Resources in Kenya by the year 2000 AD, Chapter 16, in press.

Krhoda, G.O., 1987. Water Resources Constraints and Prospects for Agricultural and Industrial development toward the year 2000 AD. Paper presented at the Workshop on Kenya's Industrial and Agricultural Straregies Toward 2000 AD, Kenya Economic Association, September 21-25, 1987.

Mason, P., 1955. Geology of the Meru-Isiolo Area. Government Printers.

Swarzenski, W.V. and M.J. Mundorff, 1977). Geohydrology of Northeastern Province, Kenya. USGS Water Supply Paper 1757 - N.

GONDWANA ARTESIAN BASIN IN WEST GODAVARI DISTRICT, ANDHRA PRADESH, INDIA

B. Jaya Kumar, K.V.S. Sastry and
K.V.J.R. Krupanidhi
Central Ground Water Board

INTRODUCTION

The Central Ground Water Board has carried out ground water exploration, during the period 1967 to 1972, in the Gondwana basin and identified potential aquifers down to a depth of 300 m in the West Godavari district, Andhra Pradesh, India. Based on the above results, this potential artesian basin is extensively developed by ground water abstraction structures to meet the irrigation requirements. With the increase in demand from irrigation sector for integrated water management, the CGWB extended its activity to the adjacent areas to explore the possibility of locating deeper Gondwana aquifer system underlying the younger formations. During the period 1977 to 1984, the exploration was carried out down to a depth of 600 m and proved the existance of deeper potential aquifers of Gondwana formations in the district.

DISPOSITION OF GONDWANA BASIN

The Gondwana rocks (Lower Triassic to Middle Jurassic age) occupying an exposed area of 1050 sq.km in the northwestern part of the district, (Fig.1) comprise thick sandstone beds, pebble beds, conglomerates, grits and shales dipping gently towards southeast. Locally, these rocks are classified into four divisions namely, Chintalapudi Sandstones, Gollapalli Sandstones, Raghavapuram Shales and Tirupati Sandstones. To the southeast of the exposed area, these rocks are overlain by younger formations, comprising basalts (Upper Cretaceous Eocene age), Rajahmundry Sandstones (Miocene-Pliocene age) and Alluvium (Recent age). In the northeastern part of the district, the khondalites and charnockites, gneisses and other associated rocks (Archaean age) are exposed and form the basement of the Gondwana sediments.

The Gondwana basin is characterised by graben - horst structures. There are a number of faults in the area, which are identified through the study of aerial photos and space images. The major lineaments trend in NE-SW direction and extend upto tens of kilometres. The major faults identified are traversing in a NE-SW direction. The occurrence of trap flows in a step like nature towards south from outcrop area, as evidenced by borehole data, confirms the presence of a series of parallel faults in this area. The other major set of faults occurs parallel to the trend of the main Gondwana basin along NW-SE direction. They are well marked by their influence in river courses.

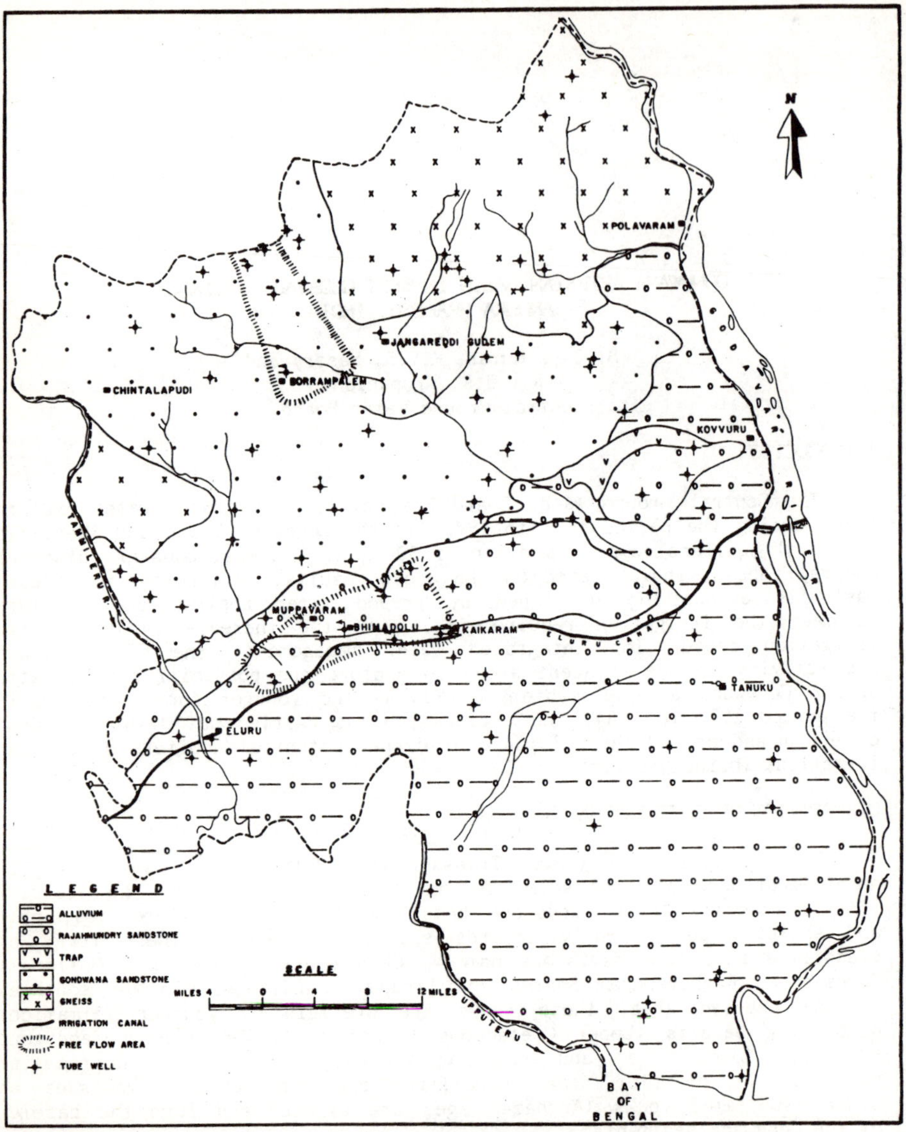

Fig.1 Geology and location of tubewells in West Godavari district, Andhra Pradesh, India

AQUIFER SYSTEMS

The ground water exploration carried out during the first phase has established a potential artesian multiaquifer system down to a depth of 300 m in the coastal Gondwana Group of formations in the district. The thickness of granular zones range between 32 and 234 m indicating wide lithological variation within the basin. The maximum thickness of the granular zones is observed along the axis of the basin. The yields of the wells vary from 700 to 2650 lpm for a drawdown of 12 m. The permeability of these aquifers varies from 1.2 to 27 m/day. Free flow zone is observed in the northern part of the district along the axis of the basin where thick clays act as confining beds.

Based on the above encouraging results of the Gondwana aquifer system and the demand to supplement the canal water irrigation in the southern part of the area, exploration was carried out down to a depth of 600 m in the central part of the district where the Gondwana aquifers are concealed by younger formations. During this deep exploration the entire thickness of younger formations had been penetrated and production wells were constructed tapping exclusively the Gondwana sandstones at certain places and at other places tapping the aquifers in both the younger Rajahmundry (Tertiary) sandstones, and the Gondwana sandstone. The Deccan traps form an ideal confining layer for the underlying Gondwana formations resulting in free flowing conditions in respect of the wells constructed tapping deeper Gondwana aquifers.

The deep exploration has brought to light the disposition of the Gondwana aquifer system (Fig.2) as well as its hydrological characteristics. The Gondwana sediments are encountered at different depths in the area mainly due to the two fault systems described earlier. The occurrence of the top of these sediments range from 46 to 383 metres. Below these depths, these sediments consist of granular zones varying in thickness from 100 to 260 m down to the explored depth of 600 m. It is expected that these granular zones extend beyond 600 m depth which is to be confirmed by further deep drilling. The yields from these granular zones range from 3700 to 6000 lpm for varying drawdowns of 19 to 25 m. The average permeability of these granular zones is 10 m/day. The elevation of potentiometric surface of these aquifers varies from 4 to 25.5 m above ground level, with a gradient of 1 in 200 towards southeast.

DEVELOPMENT PERSPECTIVE

The Gondwana artesian basin in West Godavari district is extensively developed by tubewells in outcrop portion during the past decade. However, the deeper Gondwana aquifers, below 300 m in the central part of the basin have not been exploited. These deeper highly potential aquifers extending from Eluru to Tadepalligudem can be tapped by a battery of heavy duty tubewells for integrated water management with the already existing canal water supply in the area to increase the irrigation potential particularly in the Rabi season.

The free flowing area lying between Muppavaram and Kaikaram, adjacent to the tail end portion of the Eluru irrigation canal, can be taken up for conjunctive use of surface and ground waters. Ground water potential of 83 MCM is estimated to occur in this area, which can be exploited by constructing 150 tubewells tapping the Gondwana aquifers below 300 m. These wells may be constructed along the canal as battery of clusters.

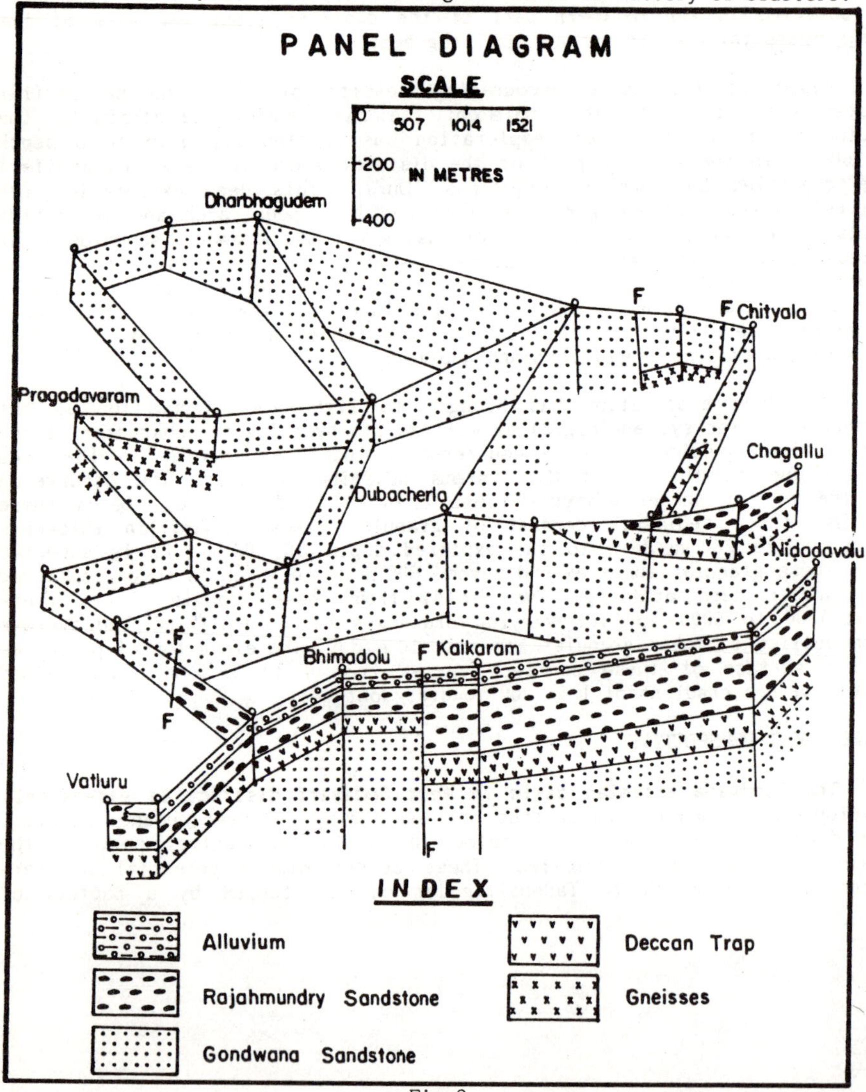

Fig.2

GROUNDWATER DEVELOPMENT FOR DROUGHT RELIEF IN CHANDRAPUR DISTRICT, MAHARASHTRA

M.SANKARAN, S. MARWAH, T.TALUKDAR,
M.R.K. REDDY AND A.K.MADHUKAR

CENTRAL GROUND WATER BOARD, NAGPUR

INTRODUCTION

During 1987-88 when Maharashtra was reeling under drought, the Central Government had advised the Regional Directorate to offer all possible help and co-operation in the field of groundwater development to combat the drought situation in the State.Two rotary rigs were diverted on priority for construction of tubewells for rural water supply. The services of hydrogeologists of Central Ground Water Board (CGWB) were utilised for selection of borewell and tubewell sites. The logger units of Central Ground Water Board were deployed for electrical logging of boreholes for the proper design of tubewells. The analysis of existing borewell data in relation to the structure and lineament in the hard rock areas were studied in detail. This was followed up by careful site selection and scientific design of tubewells, borewells and other structures in the district. Construction of tubewells in the sedimentary areas was taken up by the Central Ground Water Board and construction of borewells in the hard rock areas by the Groundwater Surveys and Development Agency,(GSDA), Government of Maharashtra.

Site Selection for Constructing Wells for Rural Water Supply in the Drought Affected Villages :

The CGWB deployed their hydrogeological personnel for selecting sites for bore/tubewells in the drought affected villages of the six taluks out of 10 taluks in the district. Based on the available data on lineaments, geology,hydrogeology of the district and on the actual study of hydrogeological conditions in each village, sites were selected for suitable groundwater structures depending upon the water requirements and water availability. The sites selected were shown to the local officials and detailed site selection reports incorporating all the details were sent to GSDA and Chief Executive Officer, Zilla Parishad, Chandrapur for implementation of the recommendations. This task of site selection for the district was completed by April,1988 before the peak summer so that remedial measures could be taken well ahead. The map of Chandrapur District is given in Figure 1.

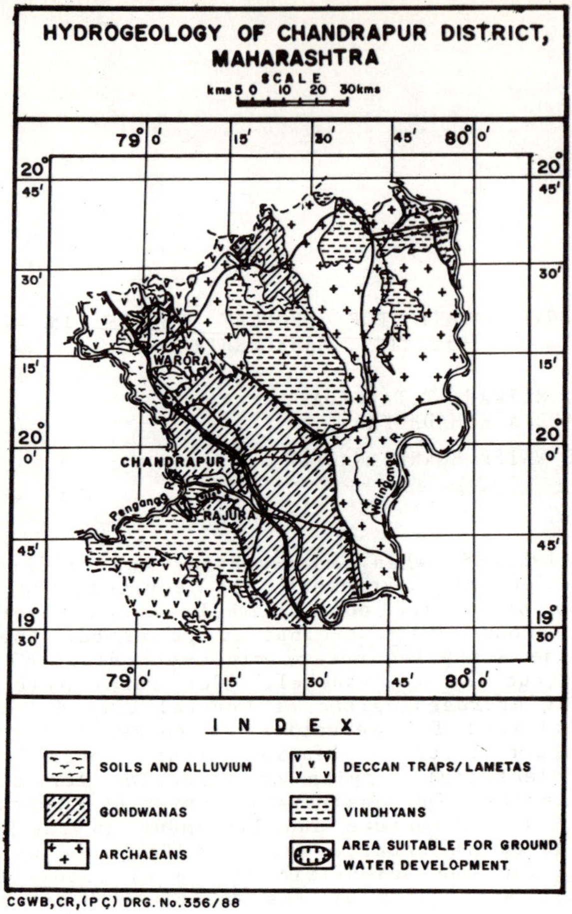

Figure 1 : Map of Chandrapur District

Out of 121 sites selected, 26 tubewells were constructed by CGWB and 24 borewells were drilled by GSDA. One borehole drilled by CGWB at Dongargaon was abandoned. Out of 24 borewells drilled by GSDA, one borewell at Devai was abandoned. 3 borewells were dry. The yield ranges from 0.14 to 5.2 l/s in the 21 borewells drilled so far.

Construction of Tubewells :

Two rotary rigs were deployed in Chandrapur district for construction of tubewells in the sedimentary areas for provision of drinking water supply to the drought affected villages. The priority villages were indicated by the State Government. After detailed surveys, the sites were selected and pinpointed for drilling. Drilling was normally done down to 70 metres. On completion of pilot borehole drilling, electrical logging was carried out and based on the study of electrical log, granular zones were demarcated.The construction of tubewells were carried out under the supervision of hydrogeologists. The tubewells were developed with air compressor. In view of quick need for provision of drinking water, the wells were fitted with hand pumps to be of immediate use. The quality of groundwater from the tubewells is uniformly good (except at Sidur). At Sidur, high chlorides and sulphates are reported.

The results of construction of tubewells under the drought relief work indicated good scope for development of groundwater in Kamthi, Barakar and even Talchir formations for drinking water supply to the rural population. Out of 27 boreholes drilled, 26 were converted into production tubewells. Only in the case of Dongargaon in Rajura taluk,the borehole was abandoned. The rocks of the Lower Gondwana group comprising of Talchir, Barakar and Kamthi formations occur over a wide area and are affected by a number of fault systems. The thickness of granular zones varies laterally and vertically. Kamthi formations which extend down to about 40 metres is the main aquifer, besides the Barakar and Talchir formations which are comparatively less yielding. The yield ranges from 0.04 to 9.24 l/s in the above three formations. The construction of tubewells in Talchir formations with gravel packing proved to be successful. Though the yield of these tubewells are rather low, they are sufficient to meet the domestic water supply needs of the villages.

CONCLUSIONS

In this district, the groundwater resource is the only major source meeting the needs of all water supply schemes in the rural areas and it has many advantages as far as potable water is concerned.

Based on the available data and by proper planning, the programme of construction of tubewells in the Gondwana formations and site selection in the 6 taluks of Chandrapur district was chalked out well ahead of summer months and the drought relief work was completed in time. Out of 27 sites selected in Gondwana formations, 26 tubewells were constructed successfully. The construction of tubewells in Talchir formations which had not been attempted earlier in this district proved successful in providing domestic water supply.

Out of 121 sites selected in Chandrapur district, 51 sites were selected for construction of borewells in the hard rock areas. Out of these, borewells were drilled at 24 sites by GSDA, Government of Maharashtra. At Devai, one borehole was abandoned and three borewells happened to be dry at the time of construction. A periodical checking and development of borewells is also necessary, as in certain cases, development and blasting, if necessary removes the choking and they are found to produce water. Similar planned groundwater studies in future will help in tackling the domestic water supply needs of the villages by providing permanent sources from groundwater which are free from pollution and contamination.

ACKNOWLEDGEMENTS

The authors are grateful to the Chairman, the Chief Hydrogeologist & Member, Central Ground Water Board and the Director, CGWB, Central Region for their kind permission and encouragement for publishing this paper.

ADVANCING FRONTIERS OF LATHI-BARMER FRESH WATER

MESOZOIC FORMATIONS OF BARMER DISTRICT

P.G. Adyalkar, National Environmental Engineering Research Institute, Nagpur - 440 020

While earliest geological work in the West Indian Thar desert was done by Blanford (1877), the first stratigraphical sequence was established by La Touche (1902) followed by extensive work by Sinha (1961). However, detailed account of the geology and hydrogeology of Barmer-Jaisalmer districts was presented by Adyalkar (1964 & 1971), and later summarised by Karanth et al (1981).

In this the Mesozoic Super Group is identified in Barmer district by two characteristic groups of fresh water formations in the form of older Lathi sandstone of Jurassic epoch and younger Barmer sandstone of Cretaceous epoch. Of these, the Lathi formation was hitherto identified in the extreme northern part by its alignment along the Bhiyar-Kanasar-Undu-Keshula line.

Then the scattered outcrops of the Lathi formation are seen near Harwecha, Jhanfli, Antra, Agoria, Kotara, Balasar and Siwala along the western margin, but the precise boundary of the Lathi formation in the western, southern and eastern sectors very much remained unidentified.

So also the Barmer formation was till very recently identified by its outcrops in the temple hill of Barmer town and the quarry of Lunu khurd with no precise boundary in any direction. This is due to the fact that the entire terrain including its geological setting is masked by sand dunes, as a typical effect of desertification of the terrain.

Recent study by satellite imagery has advanced the frontiers of these twin formations. Lathis of Barmer have now a well defined northern, western and southeastern border. South of the Pokaran-Sankra-Bhainsra water divide line the Barmer patch is separated from that of the Jaisalmer composite patch by a well defined NE-SW fault. With another NW-SE fault, it has a well defined western border along the Dharvi Khurd-Bhadka line, except for isolated occurrences of Antra-Siwala-Agoria. Then the southeastern boundary defined by the pediplain runs along the sutureline connecting Keshula with Bhadka. It has a southerly spur towards Bhimda from the mainland, and a widely separated north-south patch extending along the Gadra Road-Asadi line, together covering an area of over 3,000 km². Barmer formation also similarly extends along the NE-SW line from Barmer to Kawas along the pediments and buried pediments with an areal coverage of about 500 km². In fact, the shape of these twin formations roughly coincides with the twin landforms independently arrived at by the Central Arid Zone Research Institute together representing (i) flat and sandy undulating buried pediments, and (ii) flat aggraded older alluvial plains.

HYDROGEOLOGICAL MAP OF LATHI-BARMER FORMATIONS OF BARMER DISTRICT, RAJASTHAN

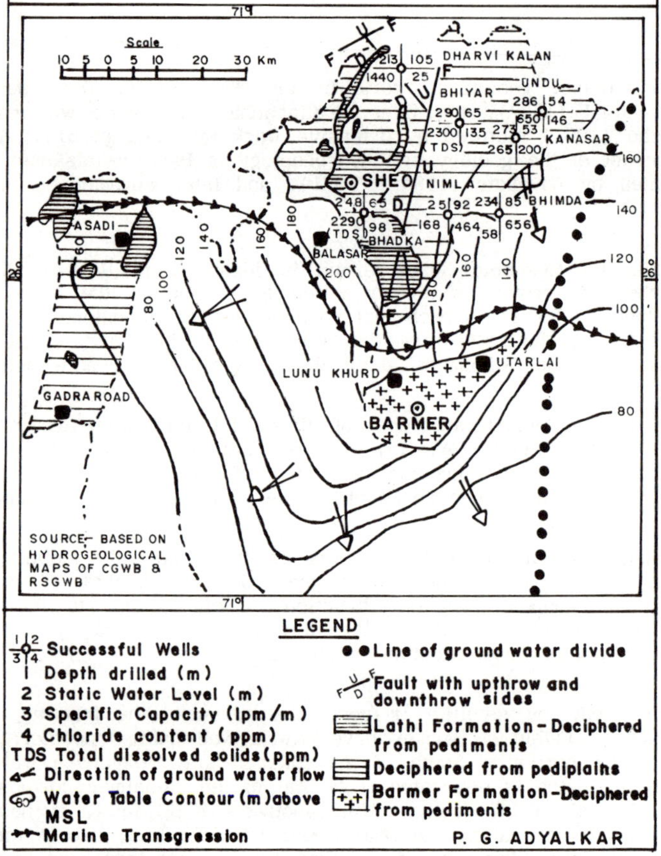

The Lathis comprise variegated sandstones with occasional shales, and represent near-shore deposits. These are medium-grained and moderately cemented, with the upper horizon in the form of silt and shale being relatively finegrained. On the contrary, the younger Barmers are in the form of white to grey silicified sandstone as exposed at Bhadres, Bisala, Botiya, Lunu Khurd, Nosar and Sonri. These are in the form of lacustrine and fluviatile fresh water deposits.

While it is difficult to locate fresh water lakes in the northern part of Western Rajasthan, it is equally true of southern, eastern and extreme western parts of Barmer district apart from there being famous salt water lakes of Pachpadra, Thob and Sanwarla. To explain this, Adyalkar (1971, p.9) had extended the southern arm of the Eocene sea with its crest along the Sundra-Jhinkali-Akli-Chidiya-Chibi line of Sheo-Barmer tahsils, extending through Pachpadra-Thob to Kaparda salt playa between Jodhpur and Ajmer. This marine transgression was more in the form of inundation of the low-lying areas, including those of Lathi and Barmer formations, to explain occasional salinity over the fringe areas, defying their precise identification in the ground water exploration programme.

As a result of the intensive work carried out from 1952-64 and beyond, Lathi-Barmer formations have been singularly recognised as the most dependable formations of the district. Ground water occurs in them under perched, phreatic and confined water conditions, the latter varying in depth, aggregate thickness and coarseness of the formation. While the phreatic surface ranges from 27 - 75 m b.g.l., the piezometric surface has a range of 32 - 92m The discharges obtained from the boreholes tapping 18 - 40 m thickness of granular zones range from 6 - 20 lps for a drawdown range of 2 - 15m. While the specific capacity ranges from 25 - 750 lpm/m of drawdown, the transmissivity ranges from 90 - 1170 m^2/d and hydraulic conductivity from 5 - 32 m/d.

Presently, their ground water resources have been conservatively estimated at 4500 MCM, but with advancing frontiers there lies tremendous scope for further augmenting the same.

Adyalkar, P.G., 1964. Geology and ground water conditions in parts of Barmer and Jaisalmer districts of Western Rajasthan, Expl. Tubewells Orgn., Unpublished Report.

Adyalkar, P.G., 1971. Palaeogeography, sedimentological framework and ground water potentiality of the arid zone of Western India, Proc. Symp. on Problems of Ind. Arid Zone, pp. 1-14.

Blanford, W.T., 1877. Geological notes on the Great Indian Desert between Sind and Rajputana, Rec. Geol. Surv. Ind., Vol. 10, pp. 10-21.

Karanth, K.R., Hussain, A. and Rao, M.K.M., 1981. Ground water resources and development potential of Barmer district, Rajasthan, CGWB Unpublished Report.

La Touche, T.H.D., 1902. Geology of Western Rajputana, Mem. Geol. Surv. Ind., Vol. 35, pt. I, pp. 1-116.

Sinha, B.P.C., 1961. Detailed geohydrological studies on the area covered by exploratory drilling in Barmer district, Rajasthan, Progress report for F.S. 1959-61, Geol. Surv. Ind., Unpublished Report.

INDRAVATI-KOLAB BASINS OF KORAPUT AS POTENTIAL GROUND WATER RESERVOIRS OF THE PURANAS IN ORISSA

P.G. Adyalkar,
National Environmental Engineering Research Institute,
Nagpur - 440 020.

The Indravati-Kolab (Sabari) basins of Koraput district in Orissa are in continuation with the identical basins of the same name downgradient in Bastar district of the Chhattisgarh Region of Madhya Pradesh. These are, therefore, simulating the largest basin (i.e. Mahanadi) of Chhattisgarh with generally similar lithology of the Purana formations of East-Central India, overlain with laterites of the Quaternary era and micro-alluvial valleys of the Recent age. They are thus the successive diminutives of the largest Purana basin with identical lithofacies.

Thus while the Mahanadi basin has an areal extent of 37,000 km², that of the Indravati is 4,800 km². The thickness of the Purana lithosediments of the Mahanadi basin is 1,200 m and that of the Indravati basin is 650m, with further deterioration in thickness of the Kolab basin. In Koraput district these twin basins occupy over 1,000 km² (Sharma et al, 1981).

The study of litho-sequence suggests that the Lower Tirathgarh Stage, 270 m in thickness, comprises shaly sandstones, sub-arkoses and orthoquartzites with a basal conglomerate. This is overlain by the Kanger limestones, shales and sandstones with again a basal conglomerate (140 m). This sequence is then topped by that of Jagdalpur dolomites, magnesian limestones and shales with stromatolites (150 m). They are studded with laterites of Quaternary era, occurring prominently both north and south of the Indravati river. The micro-alluvia are strewn in the twin valley belts in the low-lying areas. Here while the shales, limestones and micro-alluvial bodies occupy the pediplains and the valley fills, the sandstones, sub-arkoses, ortho-quartzites and conglomerates are represented by the pediments along with buried pediments and residual/structural hills (Adyalkar et al, 1972).

The sandstones, quartzites and conglomerates of the Purana sequence are poor and problematic aquifers with occasional network of 4 or more sets of joints, including a master-joint. Such network of joints extends in depth upto 30 - 60 m. They are often covered with weathered mantle, not exceeding 30 m in depth with thin or thick soil mantle. In this case it is advisable to follow the idea of hydraulic trough in morphological depressions. By following this principle one is likely to get ideal locations for constructing potential dug wells, often of large diameter to achieve best results. In all such cases, thicker the weathered mantle and/or fracture zone richer will be the potential of individual dug wells; and by following

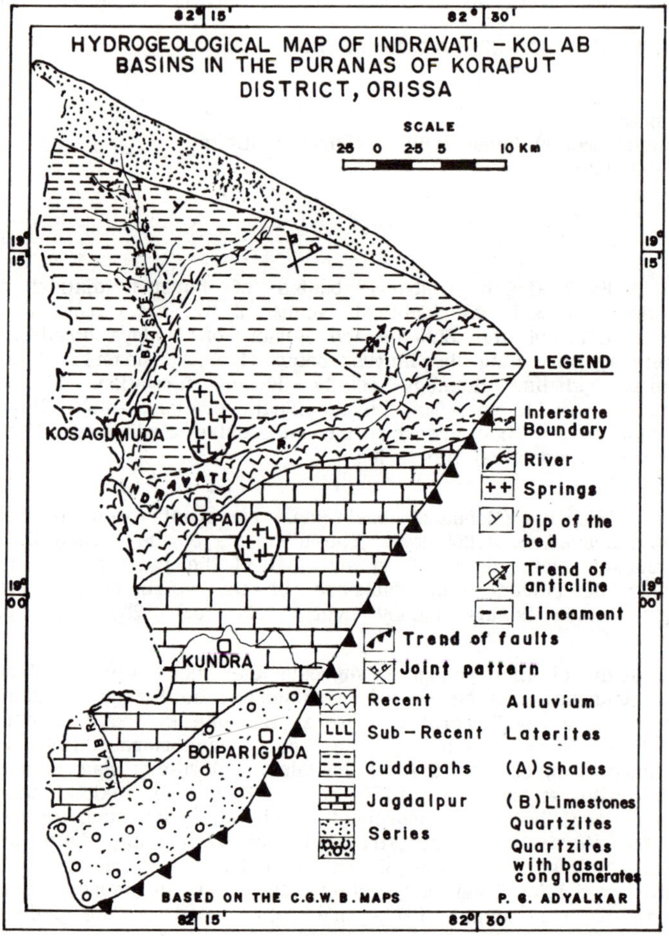
HYDROGEOLOGICAL MAP OF INDRAVATI - KOLAB BASINS IN THE PURANAS OF KORAPUT DISTRICT, ORISSA
SCALE
25 0 25 5 10 Km
82° 15'
82° 30'
19° 15'
19° 00'
KOSAGUMUDA
KOTPAD
KUNDRA
BOIPARIGUDA
INDRAVATI R.
BHASKEL R.
KOLAB R.
LEGEND
Interstate Boundary
River
Springs
Dip of the bed
Trend of anticline
Lineament
Trend of faults
Joint pattern
Recent
Alluvium
Sub-Recent
Laterites
Cuddapahs
(A) Shales
Jagdalpur
(B) Limestones
Series
Quartzites
Quartzites with basal conglomerates
BASED ON THE C.G.W.B. MAPS
P. G. ADYALKAR

the ideal conditions one may achieve subdued parameters of the Gondwana sandstones amongst them (Adyalkar, 1984).

The Kanger calcic limestone is flaggy but the Jagdalpur magnesian stromatolitic limestone is heterogeneous, with the result that there is classic development of karst in the Indravati-Kolab basins of Koraput district. In these gently dipping formations there are varieties of grikes and clints with multitude of sinkholes with benches of karstification extending to greater depths in all their grandeur as witnessed from the caves and caverns of Kotamsar and Gupteswar (Adyalkar, 1983) with development of giant sinkholes like those of Dhamtari (Adyalkar et al, 1973) with gaining and losing streams and occurrence of multitude of resurgences and pertes strewn all over. Indeed these twin valleys of the Purana sequence are the field museums of grand statuaries of karsts of their own type with the vertical walls furrowed with flutings and lapies. The whole sequence is again overlain with mutually coelescing micro-alluvial valleys, and in the pediment zone and residual hills of the Indravati basin developed in the form of lateritic flats of great significance. These triple formations in the form of karst, laterites and micro-alluvial valleys have rendered added significance to the valley fills, pediplains and structural hills from the point of view of their rich ground water potential as identified from the satellite imageries (Adyalkar, 1984).

With high average rainfall in the range of 170 - 190 cm, the wells in the micro-alluvial valleys and karsts are often limited to 12 m in depth. Individual tubewells, 90m deep, are capable of yielding 5 - 8 lps for a drawdown of 6 m, the piezometric level being 4.5 - 15 m b.g.l. The specific capacity ranges from 50 - 150 lpm/m of drawdown and yield potential 150 - 300 m^3/d for a drawdown of 5 - 10 m, when developed under strict hydrogeological control. The hydraulic conductivity ranges from 5 - 10m/d and the specific yield is 3 - 8% with a spacing pattern of 250 m. So also laterites though limited in extent are of great significance from the point of view of the occurrence of oozes and springs in the highland area and richness of their ground water potential (Adyalkar et al, 1978).

Adyalkar, P.G., 1983. Ground water survey and exploration techniques in karstic regions, Sem. Ass. Dev. Manag. Ground. Res., Vol. on Keynote Addresses and Spl. Pap., pp. 45 - 56.

Adyalkar, P.G., 1984. Hard rock hydrogeology in Indian scenario, Proc. 71st Ind. Sci. Cong., pt. II, Presidential Addresses, pp. 1-52.

Adyalkar, P.G., Phadtare, P.N. and Ramanna, K., 1973. Buried sinkhole of Dhamtari in the Charmuria limestone of Raipur district in Madhya Pradesh and its hydrogeological significance, Curr. Sci., Vol. 42, No. 4, pp. 131 - 133.

Adyalkar, P.G., and Radhakrishna, T.S., 1972. Karstic phenomenon in the carbonate rocks of the Indravati Series of Madhya Pradesh with their ground water possibilities, Proc. Ind. Nat. Sci. Acad., Vol. 38, pt. A, Nos. 5 & 6, pp. 133 - 141.

Adyalkar, P.G. and Sharma, S.K., 1978. Laterite as a potential source of ground water in parts of South Goa, Jour. Ind. Assn. Hydrol., Vol. II, Nos. 1 - 2, pp. 13 - 16.

Sharma, R.N. and Chougula, S.V., 1981. Hydrogeology and ground water potential of Koraput district, Orissa, Cent. Ground. Board Mimeo. Report.

GROUNDWATER MANAGEMENT PLANNING OF ALAPAKKAM AREA, TAMILNADU, INDIA

Dr. M.Y.A. Baig,
Reader in Civil Engg.,
S.V.U.College of Engg.,
Tirupati, A.P. (INDIA)

A.Chandrasekhar Rao,
Janathapet,
Kavali, A.P. (INDIA)

Introduction

The evaluation of groundwater potential is generally done with a view of possible exploitation to go either on its own or in conjunction with surface water resources. But on some specific occasions, particularly in coastal tracts, the evaluation is done to have management planned to avoid sea water intrusion.

A specific area located in South Arcot district, along east coast, India, has been chosen for systematic groundwater management studies. The investigations consisted of collecting extensive data on farm practices, groundwater consumption rates, groundwater levels, precipitation, soil particulars, pumping tests for evaluating aquifer characteristics, evapotranspiration etc. The study confirmed that the water from this coastal aquifer is not completely free from seawater contamination.

Physical and Geological Setting

The study area, shown in Fig.1 below is almost a plain country with a maximum elevation of about 38m and a minimum of about 15m in the tertiary formation area and less than 15m in the alluvium area. The tertiary formation includes lateritic sandstone, pebbles, sands, clayey sands, clays, etc. whereas the alluvium consists of materials which might have deposited due to weathering action of wind, river, etc. The general slope of the study area is towards south-east.

Climate

The climate of Alapakkam area is considered tropical and temperatures are highest (40°C) in the months of May and early June while December and January constitute the coolest period (17°C) of the year.

The region receives rainfall due to both south-west monsoon spreading from early June to about the end of September and north-east monsoon occuring from October to December, the latter monsoon being predominant. Systematic study of the meteorlogical data available for more than 50 years reveals that November experiences a maximum of 11 rainy days whereas February a minimum of 0.9 day. Average annual precipitation is 1322mm.

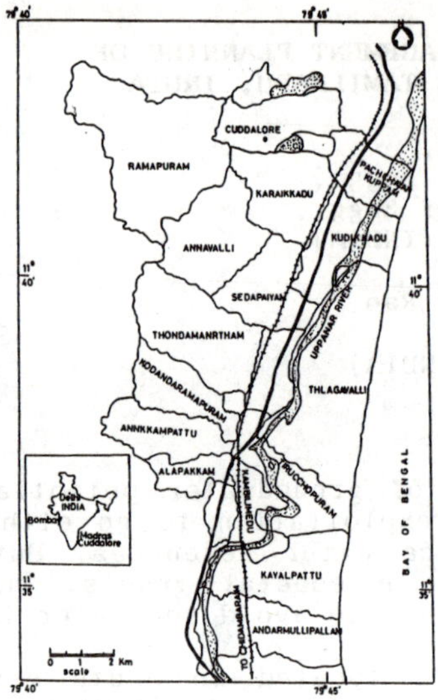

Fig.1. Physiographical Map of Alapakkam Area.

Hydrogeology

Groundwater occurs in Alapakkam area under water-table conditions. The average thickness of the phreatic aquifer is 34 metres with porosity 40-45% and specific yield 10-20%. Analysis of the pump test data from a few wells using Theis recovery method (2,4) yielded an average value of 1759 m^2/day for transmissivity and 51.53 m/day for hydraulic conductivity. The same data was also analysed by Jacob's method and the values obtained for transmissivity and permeability were found to be 1772 m^2/day and 51.9 m/day (4).

The specific capacity, obtained from the step draw down test data (1,7) from a typical well yielded a value of 2000 litres/minute at a drawdown of 5 metres. Yield factor is worked out to be 11.8 which indicates good yielding capacity of the area.

It is further computed that about 55,960 hectare metres of groundwater is reserved in the area. Average natural recharge of the area is worked out to be 1.14 m^3/s (natural recharge due to precipitation 0.938 m^3/s + natural recharge from surface water bodies 0.203 m^3/s). Average annual discharge of the groundwater due to evaporation alone using modified Thornathwaite method (8) is worked out to be 0.207 m^3/s.

Static and Dynamic Groundwater Reserves

The groundwater resources available for development can broadly be categorised as dynamic reserves and static reserves. The dynamic groundwater reserve represents the long term average annual recharge under conditions of maximum groundwater use. Static reserve is the groundwater contained within the permanently saturated zone of groundwater reservoir.

The average annual precipitation of the area is estimated to be 1322mm from which the natural recharge is worked out to be 269.5mm. This is about one fifth of the average annual rainfall.

Since the average thickness of the aquifer is 34 metres, assuming an average value of specific yield of 0.15, the aquifer has in static storage 5100mm as against the average annual recharge of 269.5mm. This static storage is about 4 years' total rainfall or 19 years' recharge (8).

Safe Yield

Although groundwater is a renewable resource, it is not inexhaustible. If the groundwater supplies are to be maintained perenially, the recharge must balance total discharge. The groundwater withdrawal rate is limited by safe yield which is defined as the amount of water that can be withdrawn annually from a groundwater basin without producing an undesirable result (7).

Of the several methods of determining safe yield, the one suggested by Kazmann, which is based on specific yield and average rise in water-table has been adopted in the present study. The safe yield is worked out to be 1.52 m^3/s whereas the total withdrawal from the area is of the order of 1.65 m^2/s (4). This has resulted in overdraft from the basin causing sea water intrusion in certain regions.

Groundwater Quality

Groundwater samples from a few control wells have been analysed once in every six months since 1980 and the variation of pH (6.4 to 8.6) of groundwater revealed that it is of tolerable degree to be used for both municipal and irrigation purposes in some regions.

The chloride-bicarbonate ratio is recommended by Revelle (5) to evaluate sea water intrusion. This ratio computed for one of the control wells about 8 Km away the coast line increased steadily from 0.5 to 2.9 during the period of observation 1980-86 (4). It is concluded that at this rate the entire belt is likely to contaminated by seawater intrusion in due course if the pumping pattern is not regulated in a programmed manner.

Recommended Method to Control Seawater Intrusion

In the present case, construction of a subsurface physical barrier is identified to be a potential method compared to other methods (7) on the basis of the experience in small reaches along the same coast. This physical barrier should be constructed near the coast to an optimum design depth using relatively impervious material. A trench backfilled with compacted silty clay mater may serve the purpose.

Acknowledgements

The authors wish to thank Prof. M.P.Chockalingam and Prof. Achutan of Annamalai University for their help and encouragem and to the personnel of the Groundwater Division, Cuddalore, Regio Meteorological Centre, IMD, Madras, for poroviding with the necess data required in connection with these investigations.

References

1. Abdel Aziz Ismail Kashef, 1986. Groundwater Engineering. McG Hill Book Co., U.S.A.

2. Baig, M.Y.A., 1986. Groundwater Potential Evaluation of a Mo Island in East Coast, India. Paper presented at the National Se nar on 'Grundwater Development in Coastal Tracts' conducted Trivandrum, Jan 29-30.

3. Brown, R.H., et al., 1972. Groundwater Studies - An Internatio Guide for Research and Practice, UNESCO.

4. Chandrasekhar Rao, A., 1988. Hydrogeological Investigations Groundwater Management Studies of Alapakkam Area, South Ar District, Tamilnadu - M.Tech. Thesis (unpublished) submitted Annamalai University, Annamalainagar, India.

5. Revelle, R., 1941. Criteria for Recognition of Seawater in Grou water. Trans. Amer. Geophysical Union, Vol.22, p.593-597.

6. Satish Chandra and Saxena, R.S., 1975. Water Balance Study Estimation of Groundwater Resources. J. CBIP, Special Issue. V 32. No.4. October

7. Todd, D.K., 1959. Groundwater Hydrology. John Wiley & Sons, In New York.

8. Wilson, E.M., 1969. Engineering Hydrology. The Macmillan Pr Ltd., London.

GROUND WATER FLOW IN SOUTHERN PART OF KOSI ALLUVIAL FAN AND ITS EFFECT ON WATERLOGGING, WITH POSSIBLE REMEDIAL MEASURES, DISTRICT SAHARSA, BIHAR, INDIA

P.N. SINGH
ASSISTANT HYDROGEOLOGIST,
CENTRAL GROUND WATER BOARD, PATNA

ABSTRACT

Remarkable ground water flow along the ground slope and waterlogging are common in the Kosi alluvial fan area. Its shallow depth to water level with phraetic condition within the sandy sub-surface horizon (aquifer) have surplus reserve of ground water.

Based on studies carried out in field, water table map and depth to water level map has been prepared which reveals the ground water flow and the causes of waterlogging in various parts of Saharsa district, Bihar (southern part of Kosi alluvial fan). The established facts about gentle hydraulic gradient and relatively steeper ground slope reveal the seepages condition which are ultimately giving birth to new river or accumulates water in low lying area in the form of waterlogging. The continous ground water flow is maintained by the influent condition of Kosi river towards its upper reaches. The depth to water level data for 7 years does not indicate any appreciable rise or difference in water level, even during year of scanty rainfall, no marked difference in water level has been noticed. Only rise of depth to water level has been recorded in the area alsdong unlined canal which is one of the major cause of waterlogging in the area, locally.

Besides the known remedial measures, some possible recommendation has been discussed on the basis of hydraulic gradient. Firstly, the construction of a number of fishing pond in waterlogged area creates concentric hydraulic gradient (locally) and will act as a decanting of ground water from the adjoining area and consequently lowering of depth to water level. Secondly, the same phenomena has also appealed here for Kosi river bed by desiltation, it will convert the position into efluent condition and consequently the lowering of water table as well as facilitates in the flood control.

1. INTRODUCTION

The study of water table behaviour and depth to water level in parts of Kosi alluvial fan has generated the idea to establish the facts about water logging problem in Saharsa district of Bihar. Regarding remedial measures of waterlogged area some possible suggestions have been discussed here through mechanism of hydraulic gradient. This methedology would also be suitable to adopt in the entire Kosi alluvial fan area, in which some three others district Purnea, Madhepura and Katihar also fall.

2. KOSI RIVER

River Kosi, known as 'sorrow' of Bihar, is notorious for its vagaries and has prefered to change its bed very often. The Kosi river was flowing through, Araria, Purnia about 125 years ago. The river Kosi has catchment area of 59000 Sq.Km which includes the highest peak of the world, Sagar Matha (Mt. Everest) and Mt. Kunchenjunga. Hence 5704 Sq. Km. of its Catchment basin are under glaciers ensuring perpetual flow and natural fall of water through out the year even during the dry summer season. During the rainy season the Kosi river and its tributaries always over flows and inundate large portion of low lying area, which later converts into waterlogged area/swamps/marshes.

3. KOSI ALLUVIAL FAN

Probably it is the world famous Alluvial fan deposited in front of Himalayas, its apical part starts from Chatra in Nepal and periphery extended along the western and Southern border of Saharsa district and Southern border of Madhepura and Katihar. It's eastern limit is bounded upto Forbisganj Araria and Purnia town. The total area of Alluvial fan is about 11000 Sq.Km. (including area in Nepal). The sediment laid down in this entire area is of recent age and of fine to coarse grained sands with silt. The present study area falls in middle and southern periphery of Kosi alluvial fan and along the river course. The slope of the alluvial fan varies from 305 m above MSL in north near Chatra (Nepal), to 38 m above MSL in the south near Sonharsa Raj over a distance of 170 Km.

Within the alluvial fan deposition is complicated due to frequent changes in the course of river. Some what sorted pervious material is deposited in the course while emerging from the hills. As the deposition proceeds the course of the principal channel is shifted many times and the abondaned channels get covered up with the heterogenous material. Towards the periphery of the alluvial fand the coarse clastic material gradually changes into fine grained aggregate and then into sand and loam. During the growth of alluvial fan the sediments are carried farther and farther out into the plain so that they overlap the sediments deposited earlier.

4. STUDY SETUP AND DISCUSSION

During the course of hydrogeological investigation of the area, the author has given main emphasis on water table behaviour, its effect on waterlogging and its remedial measures.

4.1 Depth to Water Level

4.1.1. The entire area was covered by establishing 114 dugwells as monitoring stations from which the data of depth to water level were recorded to draw the appropriate map (Fig.1). The depth to water level in the area is generally very shallow, ranging from 0.64 m below ground level to 3.72 m. b.g.1. in January 1988, with the exception at Pipra where it shows 4.30 m.b.g.1.

4.1.2. For the convenience of study, the area has been devided into three parts on the basis of latitudes, these are upper middle and lower part (See the table). The waterlogging area is calculated in percentage which shows increasing trend towards lower part. For example 58.3 % of the area of lower part or say periphery of alluvial fan is waterlogged (depth of water level is less than 1 m) where as it is 52.4% and 41.6% in middle and upper part respectively (see the table).

TABLE

Classified Area	Gradient		Percentaqge of Area with deff. Depth to water level				
	Ground slope m/Kms	Hydra-ulic m/Km	Lake	0 to 1 m	1 to 2 m	2 - 3 m	above 3 m
Upper part 26930' to 26°15'N	0.85	0.79	6.9	34.7	34.7	23.6	0.0
Middle part 26°15' to 26°00'N	0.30	0.32	7.4	45.0	30.4	14.8	2.12
Lower Part 26°00' to 25°40'N	0.17	0.10	10.5	47.8	20.9	15.0	5.5

4.2 Hydraulic Gradient and Surface Gradient

4.2.1. Water table contour was prepared (Fig. 2) on the basis of 50 reduced level data of water table from the established monitoring stations. The contour interval maintained in the map is 5 m. The contour values itself indicating that the ground water flow in this area is North to South.

4.2.2. The detailed survey was carried out along the Kosi embankment at Piprahi, Basawanpatti (see fig. 3 A) Satanpatti, Simri and Baptiyahi in which the reduced level of water table is more or less same as that of the level of water in Kosi river (see inset of Fig. 2)

4.2.3 From the behaviour of water table it is infered that Kosi river is influent in nature in the upper reaches.

4.2.4. The hydraulic gradient of upper, Middle and lower part has been calculated (see the table) that is 0.79 m/Km, 0.32 m/Km and 0.10 m/Km respectively. The gradient is decreasing towards south or at the perphery of Alluvial fan. Therefore the Velocity of ground water flow is more in the Upper part than in the lower part.

4.2.5. A sketchy coss-section (North to South) was prepared (Fig. 3 B) on the basis of reduced level data of water table as well as water level of water logged area. This sketch shows that the level of water in waterlogged area (Lake) is nothing but the water table.

4.2.6. The gradient of ground slope has been also calculated (see the table) that is 0.85 m/Km in upper part, 0.30 m/Km in middle part and 0.17 m/Km in lower part. The value shows that the gradient is steeper in the northern part where as gentle in the southern part.

4.2.7. Area of Seapages : The low lying area has developed because of the dried up old channel, ox-how lake and by means of man made excavation. These are the places where ground slope is steeper than hydraulic gradient and consequently the intersection of two is taking place in the form of Seapages. (Fig. 3 B). The water is actually oozing out of the surface to make the area waterlogged.

4.2.8. One of the aims of the construction of Embankment was to check the flow in the distributaries but virtually due to the heavy percolation of water through the sediments on the river bank, these distributaries are revived again. Besides that a number of Channels have developed with new courses. The intersection of the gentle hydraulic gradient and steeper ground slope in the old channels of Kosi "Dhara" give birth to these streams.

5. CAUSES OF WATERLOGGING

5.1 All the low lying area within the Kosi alluvial fan turned into waterlogged area because of its very low depth to water level condition as well as porous and permeable condition of the sub-surface matérial. Actually, all low lying area is intersected by the water table of Kosi alluvial fan.

5.2. The Kosi embankment has been constructed over many of the old channel for example, at north of Bhatabari (Nepal), Simari, Baptiyahi. Where it is observed that enormous underground flow is prevailing. And whereever such flow is facing the low lying area, it is oozing out to form the waterlogged area. This is seen all along the eastern side of the embankment.

5.3. The peripheral part of the alluvial fan is having very wide stretch of low lying area in comparision to apical part. This is the Characteristic of Alluvial fan deposits. Thus the large area (58.3%) has gone under waterlogging. (less than 1 m D.T.W.).

5.4. Probably the construction of unlined canal has been taken up by using earth material from just adjacent to the Canal, which has created low lying area all along the Canal, is now turn into the waterlogged area by receiving Groundwater as well as seepages from Canal. The same condition is also prevailing all along the road side as well as along the rail route side.

6. POSSIBLE REMEDIAL MEASURES

Surface and hydraulic gradients would play an important role in checking the waterlogging problem of Kosi alluvial fan. However it is very difficult to lower the water table as it remains recharged throughout the year due to the continuous supply of water from Kosi river in its upper reaches. Even then some of the possible remedial measures which would be economical too may be taken up to remove the waterlogging problem in this area.

6.1 Fishing Pond

All existing waterlogged area has to be considered for the construction of fishing ponds by excavating the earth material into each of 100x100x8 m size. According to the size, shape & extent of the waterlogged area the series of ponds may be constructed so as to accomodate the whole area.

The idea behind it is that the pond constructed upto 6 to 8 m depth, will create void in the sub-serface locally. In turn the hydraulic gradient of water table will become concentric towards pond and consequently the adjoinging aquifer will be de-watered (Fig. 3 E). Hence the water table will be lowered to some extent. Moreover this will make the adjoining area comparatively higher by spreading the excavated earth material from the pond. This land will have deeper depth to water level, hence free from waterlogging which will be additional cultivable land to the farmers.

This will be a job oriented planning and would also ease the waterlogging problem of the adjucent land. This work would have to be sponsered by Government by providing subsidy to the public.

6.2 Desiltation of Kosi river bed

The heavy load (about83,200 acre ft. of sediment load annually) brought by the Kosi river is deposited in its channel as soon as the gradient become gentle, causing siltation to a great degree. Because of continuous siltation the river bed is becoming shallower and shallower almost to the ground level, making the river influent.

If the desilting of the river bed is carried out in a planned way (Fig 3 F) the river bed will become deep, hence more volume of water will flow in the channel and probably it will convert the influent condition of the river into the effluent condition, following the same phenomena as shown in the fishing pond. It will certainly decline the water table along the adjoining areas of the river and will ease the waterlogging problem.

Like wise other tributaries of the Kosi river has to be desilted for greater base flow within the area.

6.3 Lining of Canel

The existed unlined Canal in the area has to be provided with the lining to check the seepages from canal.

CONCLUSION

Kosi alluvial fan deposits have high permeability and transmissivity. Most of the surface precipitation which is absorbed here percolates down to the water table and is swept down the gradient. The hydraulic gradient is from north to south. The influent condition of Kosi river towards upper reaches facilitating the tremendous ground water flow in the subsurface horizon, Causing waterlogging problem in low lying area of peripheral part of Kosi alluvial fan. The Remedies of water logging has been suggested by constructing fishing ponds in the waterlogged area and desiltation of Kosi river and its tributaries. The lining of Canal has also been suggested.

ACKNOWLEDGEMENT

The author is grateful to Mr. S.K. Sinha, Officer in Charge, Central Ground Water Board, Bihar State Unit, Patna and Mr. N. Kittu, Director, C.G.W.B., Calcutta for permitting to work on the present problem and for providing all possible help during field investigations.

REFERENCES

Uppal, H.L., 1972. Waterlogging in Punjab and the suspected burried ridge in the sub-alluvium. Proce. INSA, New Delhi. V. 40A, Nos. 5 & 6; p 366 - 382.

Williams, V.S., 1982. Tectonic tilting of Mountain-front alluvial fans near the Sapta Kosi gorge, eastern Nepal. Himalaya; land forms and processes p. 115 - 132.

FIG. 1.

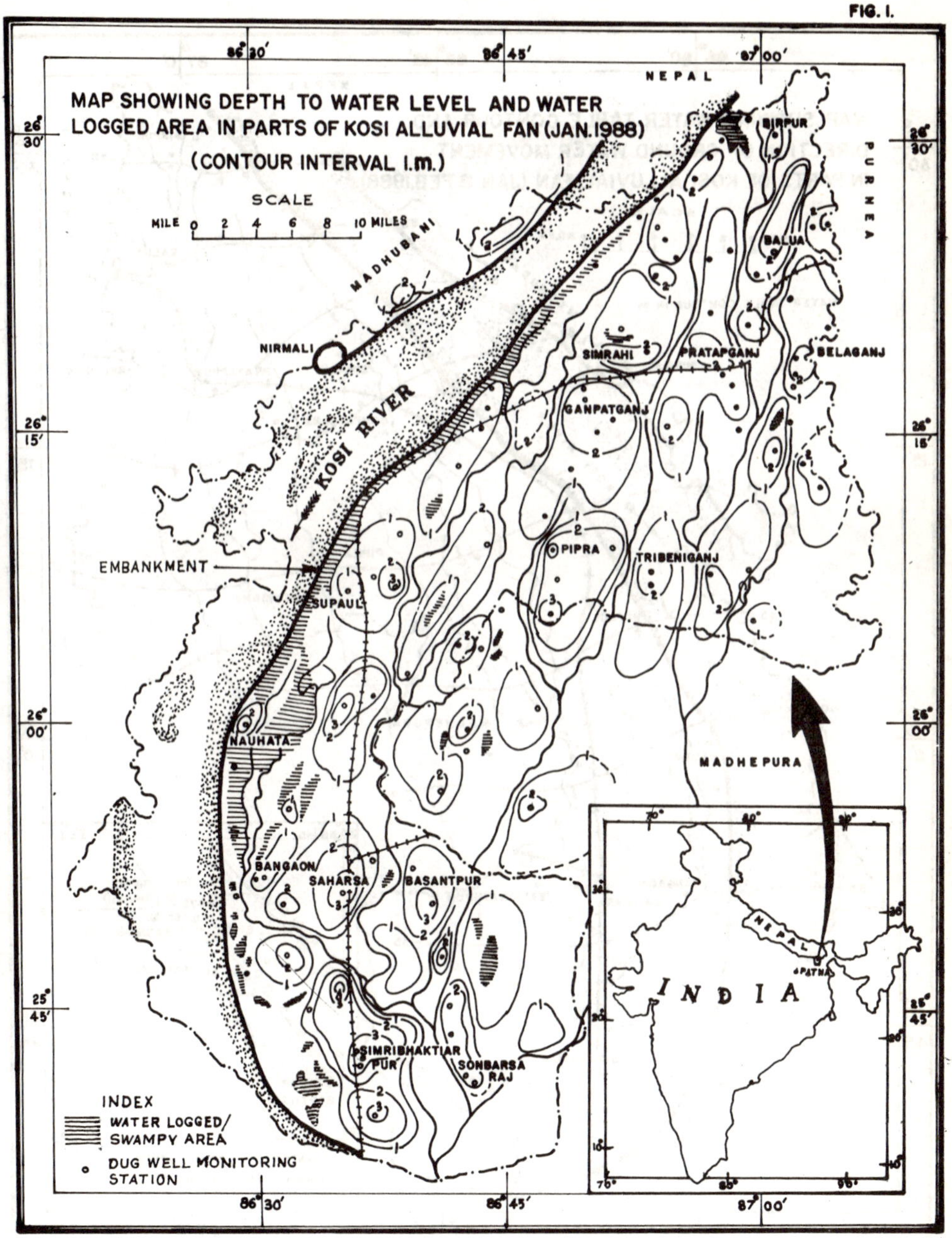
MAP SHOWING DEPTH TO WATER LEVEL AND WATER LOGGED AREA IN PARTS OF KOSI ALLUVIAL FAN (JAN.1988)
(CONTOUR INTERVAL 1.m.)
SCALE
MILE 0 2 4 6 8 10 MILES
86°30′
86°45′
87°00′
26°30′
26°15′
26°00′
25°45′
NEPAL
PURNEA
MADHUBANI
NIRMALI
KOSI RIVER
EMBANKMENT
BIRPUR
BALUA
SIMRAHI
PRATAPGANJ
BELAGANJ
GANPATGANJ
PIPRA
TRIBENIGANJ
SUPAUL
NAUHATA
MADHEPURA
BANGAON
SAHARSA
BASANTPUR
SIMRIBHAKTIAR PUR
SONBARSA RAJ
INDEX
WATER LOGGED/ SWAMPY AREA
DUG WELL MONITORING STATION
INDIA
NEPAL
PATNA

FIG. 2.

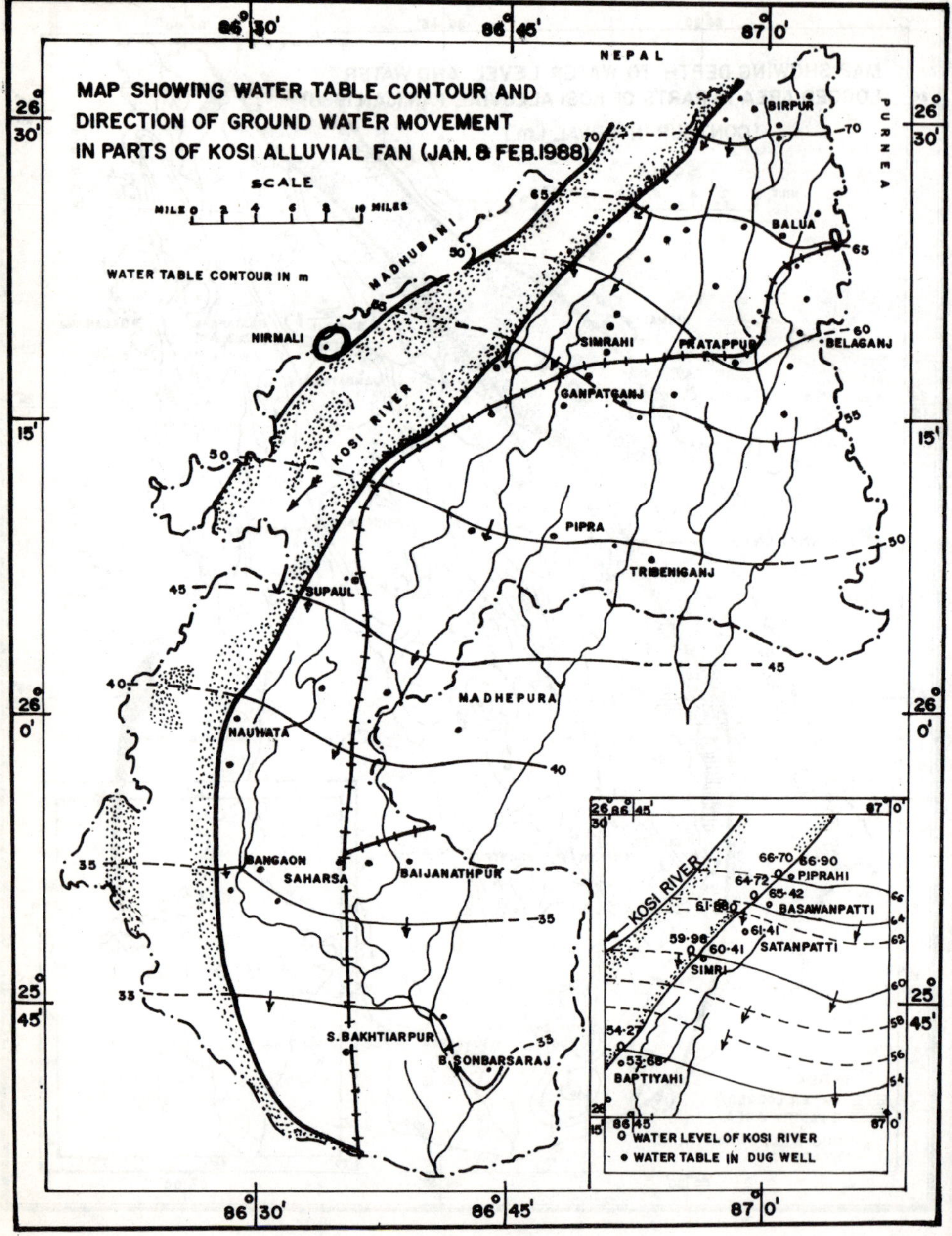
MAP SHOWING WATER TABLE CONTOUR AND
DIRECTION OF GROUND WATER MOVEMENT
IN PARTS OF KOSI ALLUVIAL FAN (JAN. & FEB.1988)
SCALE
MILES 0 2 4 6 8 10 MILES
WATER TABLE CONTOUR IN m
NEPAL
PURNEA
MADHUBANI
KOSI RIVER
BIRPUR
BALUA
NIRMALI
SIMRAHI
PRATAPPUR
BELAGANJ
GANPATGANJ
PIPRA
TRIBENIGANJ
SUPAUL
MADHEPURA
NAUHATA
BANGAON
SAHARSA
BAIJANATHPUR
S. BAKHTIARPUR
B. SONBARSARAJ
86° 30'
86° 45'
87° 0'
26° 30'
26° 15'
26° 0'
25° 45'
70
65
60
55
50
45
40
35
33
66·70
66·90
PIPRAHI
64·72
65·42
61·96
BASAWANPATTI
61·41
59·96
60·41
SATANPATTI
SIMRI
54·27
53·68
BAPTIYAHI
66
64
62
60
58
56
54
WATER LEVEL OF KOSI RIVER
WATER TABLE IN DUG WELL

FIG. 3

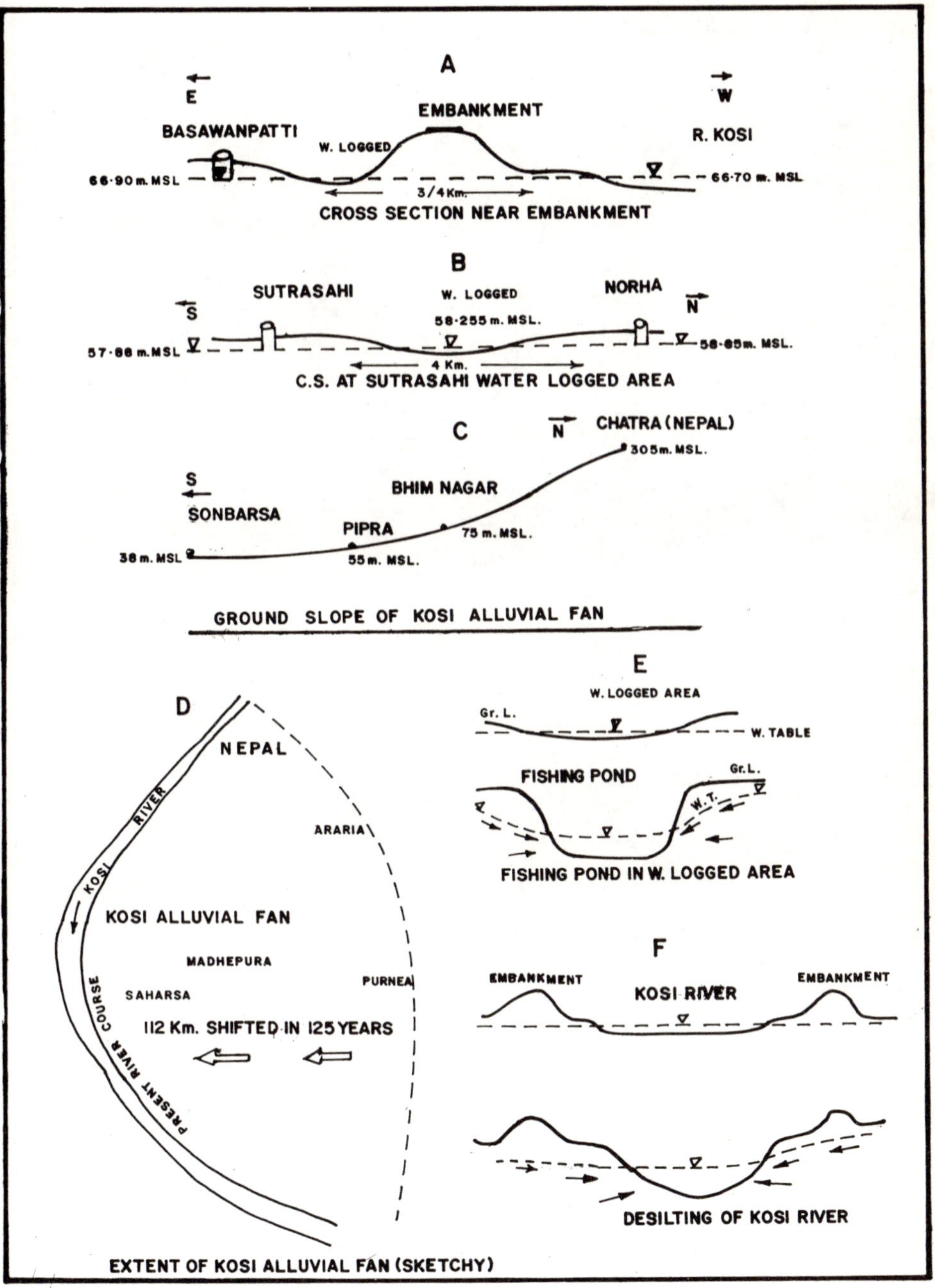
A
E
W
EMBANKMENT
BASAWANPATTI
W. LOGGED
R. KOSI
66·90 m. MSL
66·70 m. MSL
3/4 Km.
CROSS SECTION NEAR EMBANKMENT
B
SUTRASAHI
W. LOGGED
NORHA
S
N
58·255 m. MSL.
57·88 m. MSL
58·65 m. MSL.
4 Km.
C.S. AT SUTRASAHI WATER LOGGED AREA
C
N
CHATRA (NEPAL)
305 m. MSL.
S
BHIM NAGAR
SONBARSA
PIPRA
75 m. MSL.
38 m. MSL
55 m. MSL.
GROUND SLOPE OF KOSI ALLUVIAL FAN
D
NEPAL
KOSI RIVER
ARARIA
KOSI ALLUVIAL FAN
MADHEPURA
PURNEA
SAHARSA
112 Km. SHIFTED IN 125 YEARS
PRESENT RIVER COURSE
EXTENT OF KOSI ALLUVIAL FAN (SKETCHY)
E
W. LOGGED AREA
Gr. L.
W. TABLE
FISHING POND
Gr. L.
W. T.
FISHING POND IN W. LOGGED AREA
F
EMBANKMENT
KOSI RIVER
EMBANKMENT
DESILTING OF KOSI RIVER

SESSION S-5

POLLUTION OF AQUIFERS : GROUNDWATER QUALITY MONITORING ; MODELING OF MIGRATION OF NON-REACTIVE/REACTIVE CHEMICALS, RADIOACTIVE NUCLEAR WASTE ETC. ; REMEDIAL MEASURES

SELECTING EFFECTIVE METHODS FOR GROUNDWATER PROSPECTING AND EXPLORATION IN THE MEKONG DELTA

BUI XUAN PHAM
General Department of Mine and Geology, S.R. Vietnam

ABSTRACT

In Mekong delta, groundwater for domestic and industrial uses could be exploited from the young sedimentary formations of Keinozoic age. Hydrogeologically, these sedimentary formations are very much complexed with irregular hydrochemical zone distribution. The fresh water bearing layers inter-related with the saline ones, especially in the region lying between the Tien (Mekong) River and the Hau (Bassac) river.

Formerly, the main method for groundwater prospecting and exploitation in this region follows the "prospecting - preliminary exploration - detailed exploration" sequences. As regard the main technical measures for the research were drilling and pump-test projects : lines of boreholes to determine the boundary of fresh and saline waters, and with systems of boreholes (of various depths) at the same location to determine different formations. This method is very expensive and at times not fully satisfied the technical requirements.

To overcome certain weaknesses of the above said method as well as lower the expenses for this task, it is necessary to change the study method (i.e. omiting certain stages) and rearrange the projects' networks. As regard the technical measures, it is necessary to intensify the combination of various methods i.e. groundwater regime observation, hydrogeological chemistry, geophysics.

Among the above mentioned methods, the geophysic one will be the sole factor in studying the fluctuation law as well as drawing out the salt-fresh waters boundary according to the area and depth, aiming at minimizing the drilling and pump-test projects.

In Mekong delta, the groundwater prospecting and exploration aiming at serving the domestic and industrial objectives with the requirements of great reserves and centralized exploitation have been commenced only from the beginning of 1980. Up to now, dozens of regions with average area of 400-600 km2 were explored in which hundreds of boreholes with total length of nearly 100,000 m were drilled in certain regions. Through these tasks, a question is raised : in this plain, what research methods and technical measures are appropriate for the highest affect (satisfying the technical requirements with minimum cost) in groundwater exploration?

The above question was raised since the hydrogeological conditions in the plain are rather complicated in general, of which, worthy of note is the hydro-chemical zoning in many regions of the plain is not following a regular law, the salt-water layers intercalated with the fresh-water ones and rapidly changed in space. The water bearing formations with great reserves and good quality are lying rather deep underground. These characteristics of the region, the present research method as well as the present technical measures relying mainly on the drillings and pump tests to determine the hydrogeological parameters and to outline the water-bearing formations... cost this task immensely and sometimes not satisfy the technical requirements.

As an expert in hydrogeological field and rely on the groundwater exploration experiences in the plain, the author set forth some outlines of the hydrogeological conditions in the plain along with proposed research methods and technical measures to be applied in the groundwater exploration in this region.

I. GENERAL HYDROGEOLOGICAL CONDITIONS OF THE PLAIN

Geologically, the Mekong delta plain is typical of a complex artesian depression, formed by the loose sediments with great thickness that filled up the depression of Kainozoic age. The depression's basement structure is cut by a fault system, therefore, blocks were formed with uneven raising and sinking regimes. Hydrogeologically in general, the plain could be devided into the following water-bearing units.

(1) Holocene water-bearing formation - The holocene sediments are widespread over area of the delta. With a thickness of upto 50 meters. It consists of alluvial silt, clay, sand, peat with discontinuous water table aquifers. We distinguish two aquifer sequences, an alluvial aquifer and a littoral dune aquifer.

The alluvial aquifer has a static water level from 2 to 4, 5 m and yield from 0,01 to 0,09 1/s with total dissolved solids from 0,59 to 1,4 g/1. The main water type is C1-HCO -Na. Along the coast, the sandy dunes are crescent shaped and contain both fresh and brackish water with the water table varying from 1 to 2,5 meters. The yield of dug wells is from 0,1 to 0,5 1/s and the TDS value is from 0,35 to 3,9 g/l. Water type is Cl-Na.

These water are being used by the population (even the brackish water) because there is no other source of the fresh water. The main recharge is from rainfall so that the hydraulic and quality component is seasonally changing.

(2) The Pleistocene water-bearing formation - The Pleistocene formation consists of silt, silty clay, silty sand. Its hydrogeological characteristics are very complicated, depending on the hydrogeological regions.

(a) In the eastern region, the sedimentary structure is not uniform. In the plain of reeds (eastern part of the area), Pleistocene sediments cover the elevated areas, and the water table in these wells is 0,8 to 4 meters and the yield less than 0,1 l/s. The water type is g/l. In the southern part of the area, the Pleistocene formation dips to 40 m, with a thickness from 80 to 150 m. Water table changes from 0,9 to 1,7 m, the total dissolved solid value is increasing from northwest to southeast, from 0,7 to 6,7 g/l and the water type Cl-Na-Ca.

(b) In the region between Tien river and Hau river, the Pleistocene formation dips southeast-ward reaching a depth of 45 to 70 m with thickness of 130 to 180 m. In the upper part, the clay and silt beds are of 5 to 40 m thick. It is possible to divide this area into two zones based on their different hydrogeological natures as follows:

- The eastern area - The water is mostly saline to different degrees, in general, the total dissolved solids increase toward the sea coast, from 1,9 to 8 g/l.

- In the coastal area of Cuu Long province, the Pleistocene formation possesses a high artesian head with the static level of some drill holes rising up to 0,3 m M.S.L. Specific capacity of drill holes is 0,3 to 0,6 l/sm. Water type is Cl-Na but TDS value is low 0,7 to 0,9 g/l, so the water is suitable for domestic purpose and processing sea products.

(c) In the west region, Pleistocene sediments are slightly shallower, 40-50 m below the surface, 80-140 m thick and dip southeast-ward. The region is also divided into different hydrogeological areas.

- In the northern area, the water is artesian with a static level of 0,5 to 1,5 meters - TDS values are 0,5 to 1,1 g/l, water type is Cl-Na and Cl-ClCO_3-Na.

- In the Ca Mau peninsula area, the groundwater is also of high artesian head, with a static level of 0,5 to 1,2 meters, and a free flowing yield of 0,2 to 0,5 l/s - TDS value are less than 1 g/l. Thousands of well are using groundwater in the region.

(d) R ch Gia Gulf region. In the northern area, the groundwater is saline, water type is Cl-Na with TDS value 1,5 to 4,5 g/l. In the southern area, water is potable but when exploitation exceede the safe yield, the water could become saline in time.

(3) The Pliocene water-bearing formation - The Pliocene formation is wide spread across the Mekeong plain. In the eastern region, it appears on the surface and generally dips toward the central part of the plain to a depth of 200 meters. This formation can be devided into two aquifers ($N_2^a$and N_2^b).

(a) Eastern Region: In the northern area only the upper aquifer (N_2^b) exists. Its thickness varies from 60 to 120 meters. Water is under high pressure with static level 0,8 to 2,0 m. The aquifer is high yielding and good quality with TDS values less than 1 g/l.

- In the southern area, there are two aquifers. In a south-western direction approaching the Tien river, the upper aquifer (N_2^b) becomes saline in many places. The lower aquifer (N_2^a) a depth of 180 to 240 meters has thickness from 100 to 160 meters. The aquifer is saline with TDS values around 4 g/l.

(b) Central Region: The Pliocene water-bearing formation lies at the depth of 100 to 240 meters. These two aquifers N_2^a and N_2^b are totally saline with TDS values varying from 1,5 to 28 g/l.

(c) Western Region: The Pliocene formation is widespread over the western of the plain, with the depth from 80 to 200 meters. It is a high pressure water formation with static level from 0,05 to 10 meters. It is possible to divide this zone into two areas as follows:

- North area - the aquifer is totally saline
- South area - both aquifers N_2^a and N_2^b are potable

TDS value from 0,75 to 1 g/l. Water type is HCO - Na and HCO_3 -Cl-Na.

(d) Rach Gia Gulf Region: The Pliocene water-bearing formation is distributed over the depth range 60 to 100 meters. In the north area the lower aquifer (N_2^a) is absent, but in the southern area aquifers N_2^a and N_2^b exist: Thickness of the aquifer varies from 30 to 80 meters. Artesian water conditions exist in many holes. TDS are 1 to 2,5 g/l. The lower aquifer N_2^a is seem in the southern area at a depth from 160 m. It is nearly 100 meters thick. Permeability is low and its water quality is also not good.

(4) The Miocene water bearing formation - The Miocene water formation lies deep below the Mekong plain from 350 to 400 meters. This is an artesian aquifer of high pressure, with static level from 0,1 to 2 meters. The water is of good quality and high temperature of 39,5°C. Test pumping gave a yield 8,5 to 9,5 l/s. With this result, we can hope for a water supply of high volume, but continued detailed exploration is necessary.

Vertical hydrochemical characteristics in lower Mekong Plain: Hydrochemical characteristics of groundwater in the Mekong plain are complicated in chemical composition and total dissolved solids values. According to water sample analyses taken over the depth of drill-holes across the plain, four separate situations are seen, as follows:

- Total dissolved solids values increase harmoniously with well depth, but the chemical structure of the water does not change.

- Total dissolved solids values decrease harmoniously with well depth, chemical structure changes from Cl-Na or Cl-Na-Mg to Cl-Ca-Mg-Ca and finally to HCO_3 -Na or HCO_3 -Ca-Mg-Na and HCO_3 -Na-Ca.

- Total dissolved solids values increase gradually to

a certain depth, where water becomes abruptly fresh. Chemical structures change from Cl-Mg-Ca-Na; Cl-Na-Ca to Cl-Na, and other to HCO -Na, HCO_3 -Cl-Na-Ca and HCO_3 -Cl-Na.

- Total dissolved solids values increase to a definitive depth, the chemical structure does not change (Cl-Ca-Na-Mg), but when decreases but with a slight chemical structure change (but the same Cl-Na-Ca-Mg type).

II. SELECTING THE EFFECTIVE METHODS FOR GROUNDWATER Prospecting and Exploration in the Plain

From the above said characteristics of the plain and for the optimum results, we propose the following methods and technical measures to be applied in the tasks of groundwater exploration.

(1) Procedures and stages: Two research stages - prospecting and exploration - are more appropriate than the presently applied three stages (prospecting, preliminary exploration and detailed exploration). The prospecting stage comprises mainly of the outlining of the distributed fields of the water-bearing units and preliminary determination of the water quantity and quality. The exploration stage comprises mainly of the kprecise analysing for the field proposed to be installed with water exploitation project (water reserve determination along with the factors that ensure the reserve to be stable quantitatively and qualitatively throughout the exploitation process).

This procedure is not going against the assessment on the complex hydrogeological conditions of the plain, but very much appropriate on the basis of the completed hydrogeological map of the scale of 1:500,000 of the plain and the compiling one of the scale of 1:200,000. The region's general structure, the water-bearing units classified to the complex system, water-bearing formations and layers... were determined by the results of the map compiling tasks. Therefore, the prospecting stage is not confined to the task of pointing out the prospected water supply units, but clarified it at a higher level. As regard the preliminary exploration, if its task is to find out the prospected water-supply areas so as to be selected for detailed exploration only - it is not practically appropriate in the plain, since the selected area should be closely linked with a certain actual water supply project (not to mention many cases where water supply locations were decided prior to the prospecting and exploration deeds - therefore, the task of discovering the promising field in the preliminary exploration stage lost its significance). Therefore, after the completion of the prospecting stage, it is logic to shift to the detailed exploration stage without going through the preliminary exploration one. This proposal was proven by the experiences in establishing a number of groundwater exploration projects in the plain.

(2) Projects' Networks (mainly regarding boreholes and pump tests) - During the prospecting stage: In the plain, and within

this prospecting stage, the boreholes are distributed mainly for the sake of clarifying the water bearing units pointed out in the hydrogeological map (other than aiming at discovering the new water-bearing objects). Therefore, during this stage, it is better to confine the borehole arrangement according to the method of even distribution on the prospecting area, including the areas of complex hydrochemical conditions as the one lying between the Tien river and the Hau river. The best method is the combination of drilling, pump-test and geophysic measures to determine the boundaries of various water-bearing formations as well as boundaries of salt and fresh waters... For each prospecting area, only some lines of boreholes are required to have sufficient comparative and correlated data. The boreholes are arranged according to the single-borehole form at the places where geological and hydrogeological data were not sufficient and the water quality is still doubtful, and chiefly where groundwater exploration and exploitation is promising. At the same time, these boreholes should be located to serve both prospecting purpose and be used for small scale water supply needs. During this stage, groups of boreholes are only arranged in cases of hydraulic relations and salt-fresh waters boundaries determinations are required.

- During the detailed exploration stage: Arrangement methods and neworks of boreholes are determined depending on the reserve calculation method. As regard the water-bearing complex Q_{I-III} lying at the east and west of the plain, the reserves are mainly calculated by hydrodynamic method, therefore, groups of test boreholes are required. As regards the deep water-bearing formations, with complex hydrochemical conditions, the hydraulic method is the best one for reserve calculation. Locations of the group of boreholes drilled for the reserve calculation should be distributed on the fields intended to install the water exploitation projects. Besides, depending on the actual conditions of the explored area, the boreholes should be suitably arranged to make clear certain conditions of the water supply boundary, hydraulic relations... to guarantee the calculated water reserves.

(3) Increasing the application of the geophysic method: The geophysic method plays an important role in the Mekong delta's groundwater prospecting and exploration. It not only makes the data more precise, be the basis for the borehole networks and pump-tests... but also solves a number of tasks that other technical methods could not thoroughly fulfilled i.e. determining boundaries of various aquifers, salt-fresh waters... What's more important in the plain is the application of the geophysic method will minimize the number of boreholes, chiefly among the system of boreholes. Up to now, to study a vertical hydrogeological section, we have to make use of various systems of boreholes. At a certain location, sometimes 3 - 4 or even more boreholes are required since the number of boreholes in the system depend on the number of the available aquifers and the detail degree of study required (up to water-bearing formations or layers). It's clear that this method is very much expensive since in many cases, high precisions are not required : information on water abundance, water quality (fresh-salt)...

through preliminary research can be obtained by the geophysic method with only one borehole.

(4) Thoroughly exploit and utilize the observation data on surface water and groundwater movements for the purposes of determining the regime of various aquifers, the hydraulic relation between the aquifers as well as between the surface water and groundwater, calculating the water reserves, forecasting the appropriate groundwater exploitation, and other purposes that could not be fulfilled by other technical methods.

In the plain, the task of movement observation of groundwater should be carried out systematically and synchronously for the optimum results. Along with the regional observation networks, there should be the national one covering the whole delta. The dleta's sediments are of the same forming condition of loose rocks, therefore, that's the favourable condition for the arrangement and correlation of data among the observatories. The national observatories are distributed according to certain regions of different movement regimes. As for the regions where groundwater explorations are carrying out, these observatories be distributed according to the prospecting objectives - mainly according to various aquifers and tide-effected areas. The national observatories on the delta will be the standard stations and guide the regional ones in data comparison and correlation.

(5) The hydro-geochemical method be applied in studying the changing law of groundwater quality, determining the boundaries of various aquifers and salt-fresh groundwaters. As regards the plain with complex hydro-chemical conditions, this research method plays an important role : it not only makes the research objectives more precise, but also lowers the cost of the prospecting and exploration tasks. However, to obtain the above said objectives, the technical measures should be strictly observed in sampling for water chemical analysis : samples should be collected precisely in space, time and timely analysed.

Experiences from the groundwater prospecting and exploration during the recent time in the delta show that a great number of water samples were analysed, but due to insufficient implementation of the methods and technical measures, the effectiveness was low. Also from the last experiences in studying the delta's hydro-geochemical conditions, the number of surface samples for analysis could be decrease drastically (samples collected from the surface water, the outcrops, chiefly where almost no possibility of having any hydraulic relations between the surface water and groundwater).

(6) The aerial photography method be utilized in analysing the region's general structure as well as the smaller structures, aiming at forecasting the groundwater distribution law according to area and depth. As regards the Mekong delta, where canals, ditches, swamps are plenty, the application of aerial photography method is not only help us in forecasting precisely the groundwater distribution law, but also overcome the desperate costs of equipment transportation. The above said is a number of

methods and technical measures to be applied in the groundwater prospecting and exploration in the Mekong delta. Since the delta's hydrogeological conditions are complex and our experiences are thin, its sure that our proposals should be supplemented mostly with methods and technical measures of ever progressive the world over.

IMPACT OF INTENSIVE IRRIGATION ON GROUNDWATER REGIMEN IN PARTS OF OSMANSAGAR AREA, R.R.DISTRICT, A.P., INDIA

By
G.Shankar Narayana, Professor & Head, Geology Department, Osmania University Hyderabad; D. Vidyanath, Research Scholar, Geology Department, Osmania University, Hyderabad; and Dr. J.V.S. Murty, Director, Institute of Professional Groundwater Scientists, Hyderabad.

ABSTRACT

Groundwater regimen around Twin Cities has been polluted because of the impact of intensive irrigation and industrial effluents. The impact of intensive irrigation on groundwater regimen in parts of Osmansagar catchment area lying to the west of Twin Cities of A.P. in southern India has been investigated. The area spreading over 200 sq.km. forms part of the extensive peneplains of Peninsular Gneisses of Archaean age and occupies moderately elevated terrains of the Deccan Plateau, the altitude of which ranges between 544 and 615 m above MSL.

Groundwater occurs mainly under Phreatic conditions as well as confined conditions in wells tapping fracture zones. Vineyards, orchards and vegetable gardens are much in vogue near the Twin Cities. Normal Annual Rainfall being 798 mm, groundwater forms the only means for intensive irrigation. Further, the use of groundwater has exhorbitantly increased on account of the vagaries of rainfall and recurrence of droughts over the last two decades.

The quality of water, which was uniformly palatable for drinking in 1960s, has now deteriorated to a large extent, especially in the intensively irrigated tracts because of heavy dosages of fertilisers and pesticides; and recirculation of waters. The values of specific electrical conductance of groundwater in intensively irrigated area range from 2 850 to 5 000 m.mhos./cm. at 25°C. The concentration of sulphates is also very high.

The paper analyses the impact of intensive irrigation on the quality of groundwaters and suggests total abandoning of intensive irrigation in low rainfall regions of fissured rocks lest the soil regimen as well as geo-hydrochemical environment may be damaged beyond irreparable limits.

INTRODUCTION

Over the last two decades, the Twin Cities have touched the headlines in newspapers regarding the harmful pollution levels because of the industrial effluents and sewage. However, it is not common knowledge that agricultural pollution is high in large areas. In one of the field sojourns of the authors, it has been observed that the ground waters are highly affected because of the intensive irrigation in Osmansagar catchment of the district. A research programme on the pollution in intensively irrigated area is taken up in a small area of 200 sq.km. and the findings are presented herein.

Continued irrigation accumulates salts from vegetative matter, manures, fertilisers and pesticides in groundwater. Groundwater being the main source of irrigation in R.R.District, salts add up through the process of residual concentration consequent to the evaporation of waters. The accumulation is further accentuated through the recirculation of the groundwater with local depletion of the water table by heavy pumping.

Location

The study area lies to the west of the Twin Cities falling between the latitudes 17° 18' and 17 °20' N and longitudes 78° 08' and 78°20' E in the Survey of India topographical sheets No.56 K/3 and K/7. It forms a part of the catchment of Osmansagar dam (Fig.)

Physiography

The investigated area is a part of the Deccan Plateau, underlain by the Pre-Cambrian peninsular gneissic complex. The altitude of the area ranges between 544 and 645 m MSL with granitic and basaltic hillocks of 10 - 15 m height. The area is characterised by undulating topography and drained by fourth order stream courses enjoining the Osmansagar. The drainage is parallel and dendritic. The terrain is dotted by tanks, while seepages and springs have dwindled because of the intensification of the droughts over the last two decades.

The normal annual rainfall is 798 mm (IMD 1901 - 1950) as recorded at the Gandipet Station. But, it has decreased substantially over the last two decades with the vagaries of rainfall. The winter temperatures vary from 13°- 32° while the summers touch 44°C.

Quality of Groundwater

Nearly one hundred water samples have been collected keeping in view geomorphological and hydrogeological environments and intensive irrigation pockets. The results of analysis of the representative water samples are given in the table.

The ground waters are alkaline in nature, the pH values being 7.10 - 8.35. The waters are classified as Bicarbonate type. The general quality of groundwater excepting near irrigated pockets is very good, the specific electrical conductance ranging between 500 and 1 400 m.mhos/cm at 25°C.

Methodology

The quality of groundwater generally depends on residual concentrations which are influenced by lithological composition water, bearing layers, geomorphological conditions, groundwater movement, evaporation effects. For understanding

Fig.

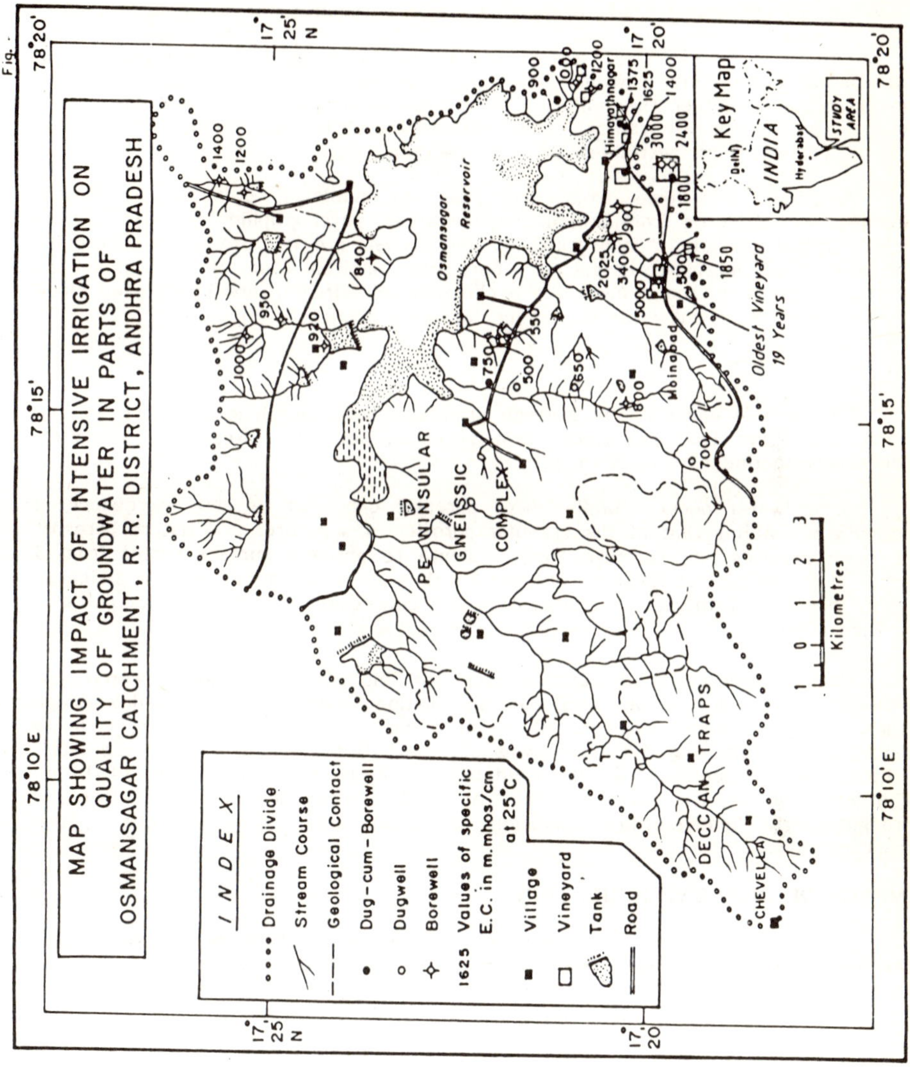
MAP SHOWING IMPACT OF INTENSIVE IRRIGATION ON QUALITY OF GROUNDWATER IN PARTS OF OSMANSAGAR CATCHMENT, R. R. DISTRICT, ANDHRA PRADESH
INDEX
Drainage Divide
Stream Course
Geological Contact
Dug-cum-Borewell
Dugwell
Borewell
1625 Values of specific E.C. in m.mhos/cm at 25°C
Village
Vineyard
Tank
Road
Osmansagar Reservoir
PENINSULAR GNEISSIC COMPLEX
DECCAN TRAPS
CHEVELLA
Himayathnagar
Moinabad
Oldest Vineyard 19 Years
Key Map
INDIA
Delhi
Hyderabad
STUDY AREA
Kilometres
78°10'E
78°15'
78°20'
17°25' N
17°20'

the levels of pollution and establishing a base for protracted studies in a region, one has to select an area where the involved parameters are minimal in number. Appreciation of the general groundwater conditions, heads and the geohydrochemical environment of hydrogeological regime requires the systematic collection of water samples and analyses, well inventory and pump test analysis. The chemical analysis results of waters are to be interpreted in the light of the influencing parameters. From the ionic concentration of the salts, it could be established whether the pollution is because of irrigation. Further protracted analysis of the specific radicals and salts shall establish the cause of the pollution as to fertilisers, pesticides, or other factors.

A small area free from industrial and municipal pollution is chosen. Samples are collected from drainage divide areas, physiographic lows and intensively irrigated pockets. The results of the analysis are interpreted to appreciate the effect of intensive irrigation. For the complrehension of the impact of intensive irrigation, the values of the chemical constituents are interpreted along the direction of groundwater movement following a micro-watershed.

HYDROGEOLOGY

The area is underlain by gneissic granites where saprolitic members, jointed fresh rocks and fracture zones form potential aquifers. Weathered saprolitic zones vary in thickness upto 18 m.

Groundwater occurs mainly under phreatic conditions in the interconnected porosity resembling that of the granular porosity of saprolites and fissure porosity of fresh granites and fracture zones. Depth to phreatic surface varies from 3 -to 5 m below ground level. It lowers to 15 - 20 m in the intensively irrigated tracts because of the heavy exploitation and consequent lowering of the phreatic surface locally.

Heavy withdrawals of groundwater for intensive irrigation lowered the phreatic surface locally as confirmed by shallow levels both up-and-down-slope sides. Large diameter dugwells of 10 m radius and 18 - 20 m depth, yield as much as 450 cu.m./day for drawdowns of 4 m. Borewells of 150 mm diameter, drilled down to depths of 60 m, yield upto 10 cu.m./h for drawdowns of 18 m. The levels of phreatic surface as well as yield of wells decrease in summer and drought years. Semi confined conditions are locally observed in borewells tapping fracture zones.

IMPACT OF INTENSIVE IRRIGATION

A critical analysis of the results of chemical analysis of ground waters revealed that the ground waters in vineyards record EC values exceeding 1 500 m.mhos/cm at 25 °C. Ground waters of wells in vineyards of more than 10 years of irrigation record EC values beyong 3 000 m.mhos. Older the vineyard, higher the EC value. The influence of groundwater movement within a vineyard is significant as observed from the ionic concentrations of the well waters in the background of the morphometric slope (Table).

TABLE: RESULTS OF CHEMICAL ANALYSIS OF REPRESENTATIVE WATER SAMPLES

Well	pH	Spe.Ele.Cond. m.mhos/ cm.at 25°C	CO_3^{-2}	HCO_3^-	Total Hardness	Ca^{+2}	Mg^{+2}	Na^+	K^+	F^-	Cl^-	SO_4^{-2}
			ppm									
1	7.45	900	Nil	60	420	120	73	62	83	1.1	265	349
2*	7.15	1000	40	350	260	160	24	95	80	2.5	350	265
3**	7.80	1200	15	100	540	140	97	65	81	1.4	240	518
4**	7.92	1375	30	140	320	60	63	71	81	3.7	300	315
5**	7.65	1625	20	310	240	140	24	91	81	1.7	300	261
6**	8.05	1400	40	270	220	40	44	68	81	1.5	225	249
7	7.40	900	20	200	220	140	19	65	83	1.4	220	230
8*	7.52	3000	10	70	700	260	107	95	82	1.3	362	649
9*	7.45	2400	Nil	70	740	420	78	72	81	1.2	340	637
10*	7.62	1800	Nil	60	1060	260	194	77	83	1.4	354	879
11*	7.70	1850	Nil	80	540	40	121	65	83	1.7	320	476
12	7.99	500	40	170	480	315	100	91	83	270	196	
13*	7.72	5000	10	90	2240	1020	297	76	85	1.6	300	1997
14*	7.10	3400	10	170	1660	560	267	75	82	1.5	290	1501
15*	7.40	2025	10	230	620	80	131	78	81	1.4	235	560
16	7.50	800	15	250	220	105	28	55	75	1.2	250	176
17	7.60	650	20	230	280	205	50	57	75	1.4	200	192
18	7.45	500	20	150	260	190	45	61	76	1.	250	180
19	7.6	750	25	200	260	210	63	58	74	1.2	255	241
20	7.3	550	30	170	240	40	49	46	72	1.4	270	10
21	7.9	1400	10	10	300	140	69	50	73	1.5	285	322
22	7.6	1200	15	140	255	180	18	45	70	1.3	260	82
23	7.6	1000	10	160	220	167	13	57	68	1.6	245	70
24	7.55	840	Nil	180	200	80	29	46	65	1.	280	60
25	7.6	1000	Nil	200	230	122	26	50	60	1.2	270	89
26	7.5	950	15	155	240	130	27	48	55	1.4	265	85
27	7.4	920	Nil	140	210	162	11	46	50	1.4	275	62

N.B.: 8* Samples from well located in vine-yards; 4** Water from wells in abandoned vineyards

The analysis following a micro-watershed concept clearly indicates a definite impact of the intensive irrigation. For an example let us take Moinabad micro-watershed. Commencing from the drainage divide the EC values recorded along the slope of the watershed are 500, 2 400, 3 000 and 2 025 (Fig.). The lowest value 500, corresponding to 300 ppm of TDS and indicating very good quality even for drinking purpose, is recorded along the drainage divide area. This can be taken as background value of the ground waters. Within a few hundred metres, the quality of groundwater in vineyard deteriorates to values of beyond 2 400. An important aspect is that the quality improves downslope side because of higher influx of groundwater through high pressure heads of phreatic surface at lower levels. This fact confirms also the local impact of intensive irrigation. The adjacent western vineyard records 5 000 value as it is older by 5 years and the phreatic surface is lower than that of the eastern vineyard due to the cumulative effect of the heavy pumping.

An increase in the sulphate and chloride contents in ground waters from irrigated tracts has been reported by Vogel way back in 1913. With discontinuation of the use of chloride fertilisers, the question of chloride concentration naturally does not arise. The sulphate concentration in the waters of vineyards ranges between 637(9) 1996(13) ppm, while the waters of other than vineyards r-record only values below 500. This effect could be ascribed to heavy dosage of fertilisers superphosphate and NPK, being the only fertilisers in vogue in the vineyards. The role of pesticides is yet under study.

CONCLUSIONS

Studies carried out in a type area of the Osmansagar catchment revealed that the quality of groundwater in vineyards deteriorated because of the intensive irrigation and large dosage of the fertilisers. Use of groundwater for intensive irrigation in low potential groundwater regimens of peninsular basement gneissic complex should be stopped. Otherwise the groundwater quality in the intensively irrigated tracts shall not only be polluted to irreparable levels but the soil regimen shall also be effected thereby rendering vast irrigated tracts into waste lands.

REFERENCES

Brown Robert: Groundwater. 2nd Elsevier Applied Sciences Publishers. pp D-203 - 206. 1986.

Elridge, E.F.: Irrigation as a source of water pollution. Jour. Water Pollution Control. Fed. 35,5. pp.614 - 625. Washington D.C. 1963.

Todd, D.K.: Groundwater Hydrology. 2nded. John Willey & Sons. pp. 332. 1984.

Vidyanath, D., Shankar Narayana, G., & Murty, J.V.S.: Environmental impact analysis of irrigation on groundwater regimen in Osmansagar Area, R.R.Dist., A.P., India. Int. Conf. EIA of Developing Countries. 1988.

Vogel, J.H.: Die Abwasser aus der Kallindustrie, Ihre Beseitigung Sowie Ihre Einwirkung in und an den wasser laufen. pp.591. Berlin. Bertraeger. 1913.

GROUNDWATER QUALITY OF AN ALLUVIAL AQUIFER

L Elango, S Manickam, M V Somasundaram, J H Tellam*
Centre for Water Resources, Anna University, Madras 600025, India.
*School of Earth Sciences, University of Birmingham, PO Box 363, Birmingham, UK.

ABSTRACT

This article describes the preliminary results of a study attempting to evaluate the long term water quality situation in a heavily exploited alluvial aquifer in Southern India. The aquifer consists of an upper low permeability unconfined part which acts as a semi confining layer to the the lower high permeability part. The system is heavily pumped for supply and local irrigation, and flow modelling suggests that much of the pumped water in deep wells is supplied by vertical leakage from the upper layer. Water quality is very variable spatially and appears to reflect the importance ot vertical flow; some correlation with land use is observed. Quality varies seasonally, some wells showing increase in salinity as water levels rise in response to the monsoon. This appears to be due to flushing of evaporation-concentrated salt in the unsaturated zone. Long term data suggest the system is generally chemically stable, possibly because of the volumetric importance of rainfall recharge. In some locations, however, quality is cnanging, and preliminary salt balance calculations suggests the system can under certain circumstances be sensitive to changes in chemical inputs (especially fertilizers).

INTRODUCTION

In India, as in many parts of the world, considerable effort has been expended with some success in attempting to understand the long term flow balances of heavily exploited aquifer systems. A great deal of effort has also gone into water quality surveys. However, fewer studies have been able to combine the two types of study in order to assess the long term water quality implications in areas of heavy irrigation. This article describes the results obtained so far for an investigation of an area of intensive agriculture in Tamil Nadu, South India; the aim of the work is firstly to understand the water quality data in the context of the hydrogeological processes, and secondly to use this understanding in evaluating the long term water quality implications for the aquifer.

The Arani-Korttalai Aquifer is an alluvial aquifer located just north of Madras. It is heavily exploited, the water being used for irrigation, public, and industrial supply. It has been the subject of several studies, including detailed mathematical modelling (Krishnasamy, l987). In addition, water quality data have been collected for some years by the Public Works Department (PWD) amongst others (Somasundaram, l988). The present study commenced 3 years ago (Elango, l985) with the initial aim of understanding the chemical variation in the aquifer, and subsequently of producing some simple salt balance capable of indicating the chemical stability of the system. Todate, results from detailed sampling and existing chemical data have been interpreted in the context of the modelling studies and preliminary salt balance calculations have been carried out.

HYDROGEOLOGY

The 500 km^2 aquifer lies 20 km north of Madras between the rivers Arani and Korttalai which flow for only a few days each year (Figure l). It consists of alluvium up to 50 m in thickness which is underlain by low permeability rocks of pre-Quaternary age.

The alluvium comprises sands and clays, with the finer grained beds becoming dominant upwards. The permeability of the lower part of the sequence is thought to be considerable (50-200 m/d), and that of the upper part significantly less. The aquifer has therefore been interpreted as a leaky system (Rushton, l986; Krishnasamy, l987) comprising a "lower (semi-confined) aquifer" and an "upper (unconfined) aquifer", the latter being roughly l0 m thick; hydrographs strongly support this interpretation (Figure 2), and regional flow modelling has been successfully undertaken using this conceptual model (Krishnasamy, l987). However, as expected in an alluvial sequence, the lithological variation is considerable (Figure 3), and locally the flow system may be considerably more complex.

Abrstraction originally occurred from shallow dug wells in the upper, lower permeability part of the succession. As water levels fell, boreholes were drilled to tap the lower aquifer. Water levels continue to fall, the system being still heavily exploited. From the flow modelling work, vertical leakage is thought to be a very considerable source of water in the vicinity of pumping boreholes as shown on Figure 2 (Rushton, l986), with each borehole drawing water from the upper aquifer within its catchment.

Rainfall is around l200 mm/a, 35% falling in the southwest monsoon (June - September) and 60% in the northeast monsoon (October - December). Potential evapotranspiration is estimated at l30-230 mm/month and recharge at 200 mm/a (about 70% from rainfall) (Krishnasamy, l987). 80% of the land area is cultivated.

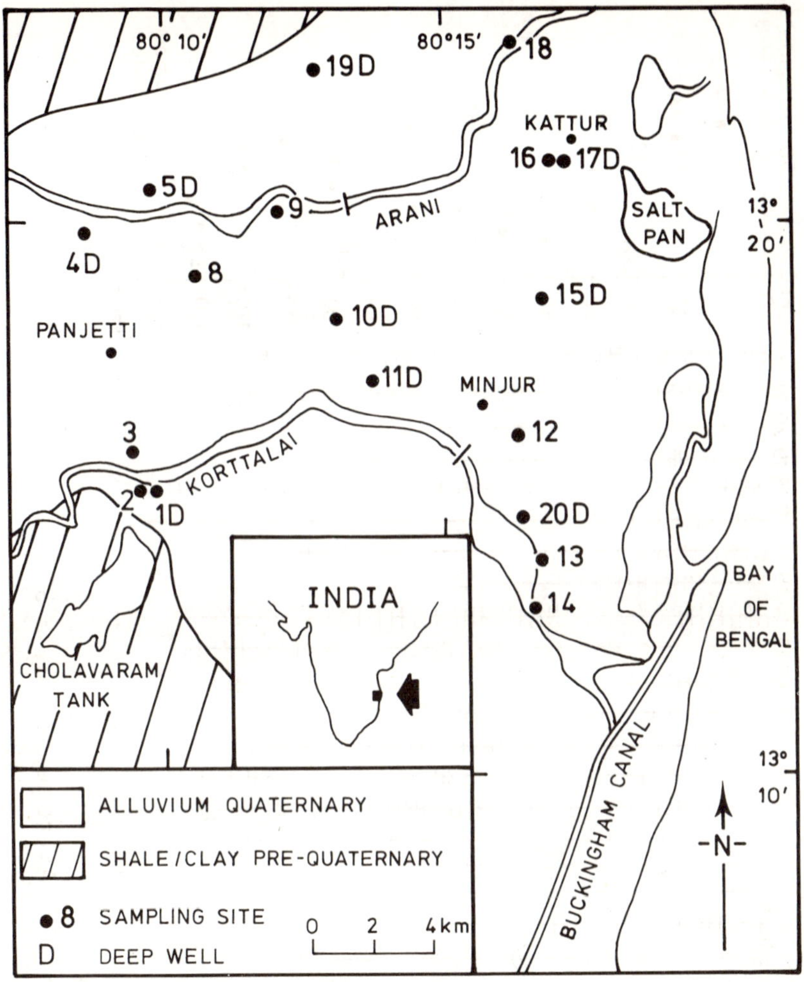

FIGURE 1. LOCATION MAP

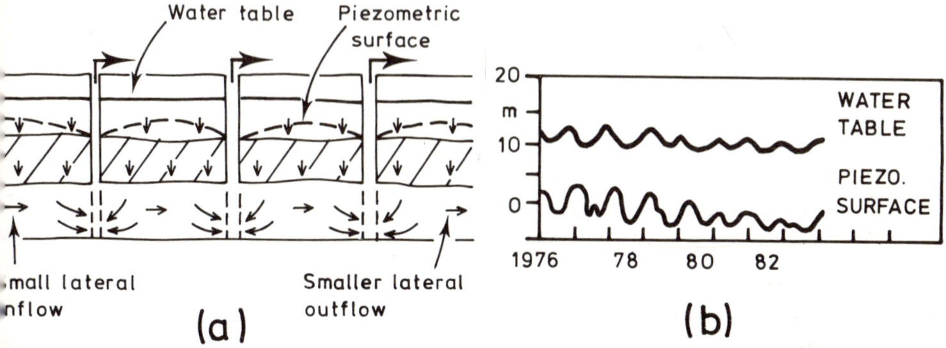

IGURE 2. (a) HYDROGEOLOGICAL MODEL (after Rushton, 1986) (b) HYDROGRAPHS

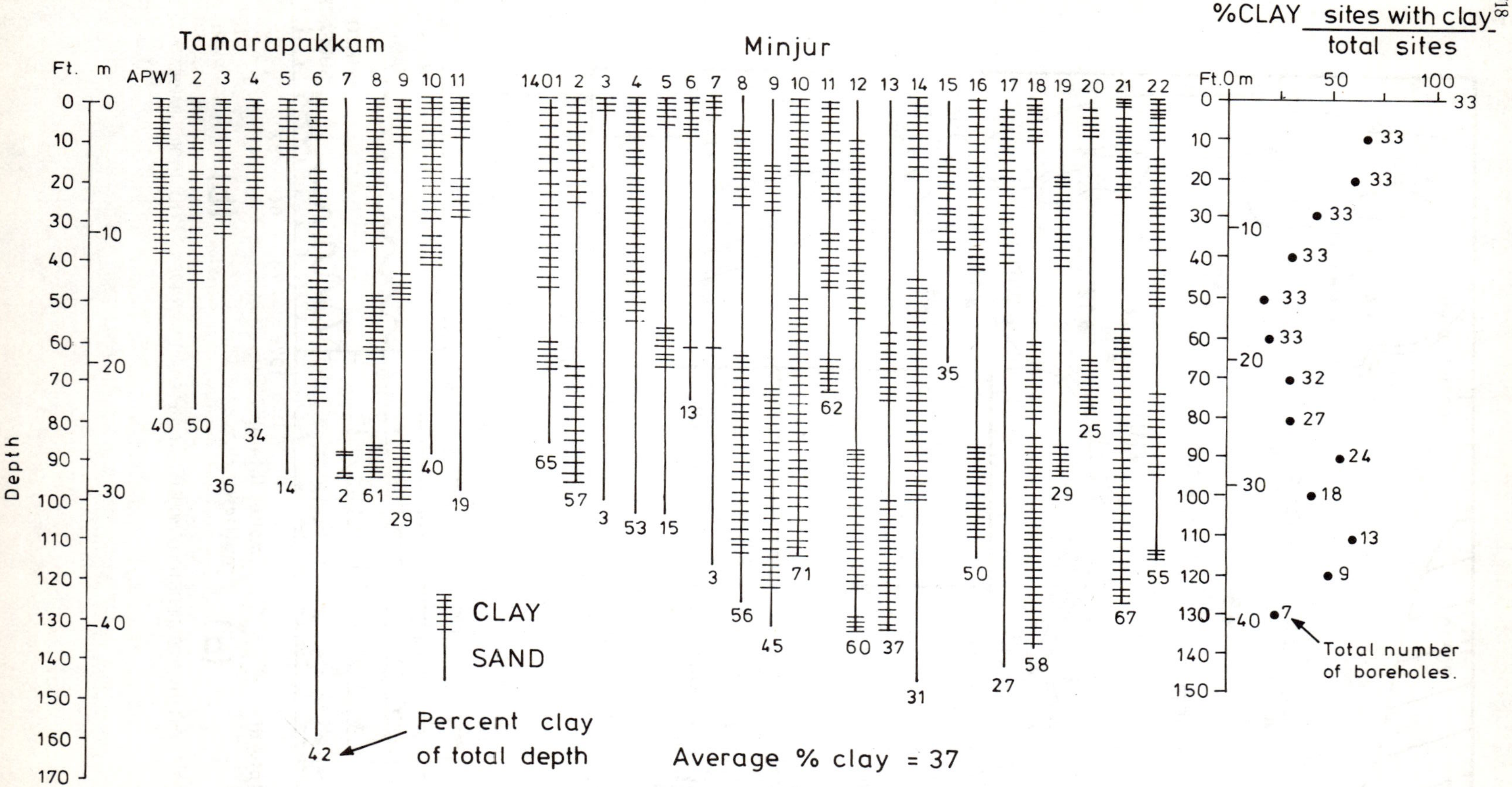

FIGURE 3. LITHOLOGICAL VARIATION

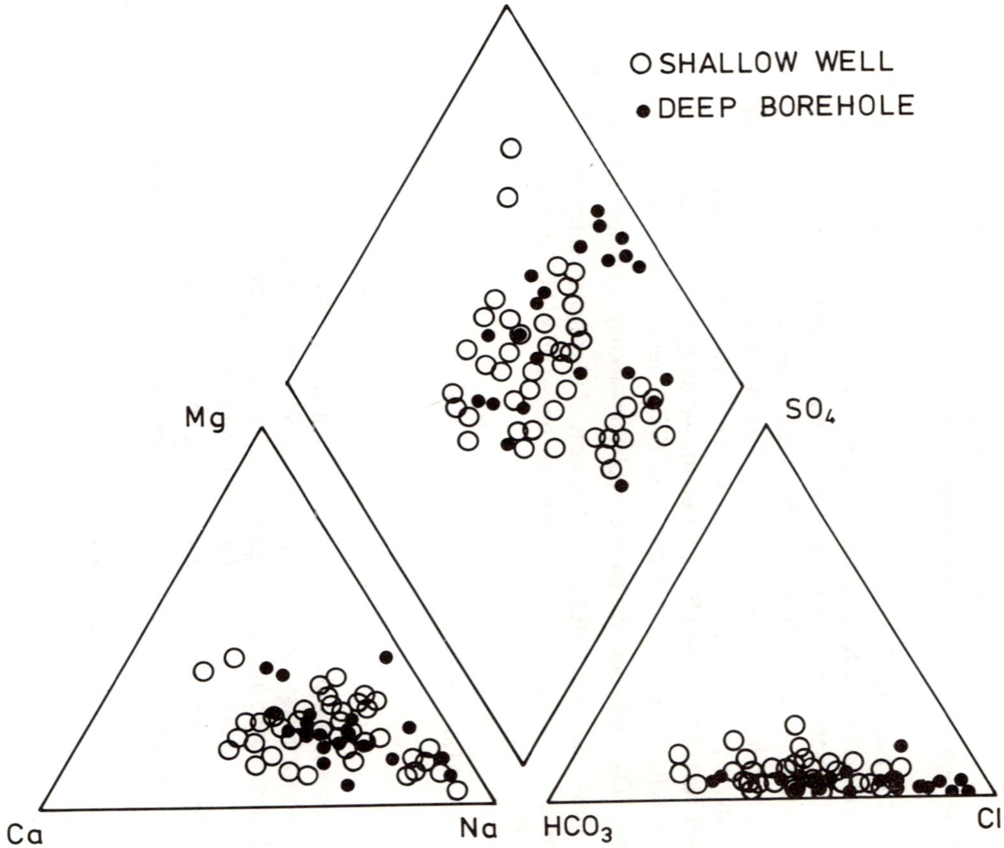

FIGURE 4. COMPARISON OF DEEP AND SHALLOW WELL CHEMISTRY

SAMPLING AND CHEMICAL ANALYSIS

Biannual sampling has been carried out by PWD since 1975. Each sample was analysed for major species (Na^+, K^+, Ca^{2+}, Mg^{2+}, Cl^-, SO_4^{2-}, HCO_3^-, and sometimes NO_3^-), and samples were taken pre and post main monsoon at a few sites. In addition UNDP data have been collected during 1967-1969, again at pre and post monsoon times.

The existing PWD and UNDP data suggested a marked seasonal variation in water quality, and thus for the present study it was decided that monthly samples would be taken from about 16 wells at various depths and associated with various) land uses. The sites are shown on Figure I, and details are given in Table I. Analysis was carried out at Anna University: temperature, pH, and electrical conductivity were measured in the field; Na^+, K^+, Cl^-, and F^- were determined using ion selective electrodes (Orion 901); Ca^{2+}, Mg^{2+}, and HCO_3^- were determined by titration; SO_4^{2-} was determined spectrophotometrically. Results have been processed using CHEM27, a University of Birmingham program incorporating an aqueous speciation model based on WATEQF (Plummer et al., 1976).

RESULTS AND INTERPRETATION

Introduction

In order eventually to investigate potential changes in chemistry due to exploitation, it is first necessary to understand the relationship between chemical variations in time and space, and the flow system. As a result, the data collected and collated during this study will be discussed in relation to the general hydrogeological model outlined above: the following sections therefore deal with a general description of the water quality, vertical chemical variations, lateral variations, seasonal variations and long term variations; a final section concludes and discusses possible modelling work.

General Chemistry

Table I gives typical analyses for each site sampled in the present study; the chemistry, as discussed below, varies considerably with season. The waters in general are oxidizing and calcite undersaturated; salinities increase towards the coast with maximum Cl^- concentrations of about 1200 mg/l, most waters having less than a few 100 mg/l Cl^-. $SO_4$2- concentrations are moderate to low (most < 40 mg/l), and HCO_3^- is usually a couple of hundred mg/l. The maximum

Site	Use	Date	EC	CONCENTRATIONS (mg/l)										
				Ca	Mg	Na	K	Cl	SO_4	HCO_3	pH	F	SH	NaH
1D	Irr	14/9/85	537	25	16	74	2.3	69	13	189	6.9	0.24	C2	S1
2	Irr	14/9/85	410	31	5	45	2.2	47	14	110	6.4	0.18	C2	S1
3	Irr	5/12/85	1039	54	24	133	1.1	185	37	298	7.0	0.29	C3	S1
4D	Irr	10/10/85	1190	73	29	153	1.8	325	26	338	7.1	0.21	C3	S1
5D	Irr	8/8/85	1042	42	23	141	2.3	196	19	195	6.9	0.36	C3	S1
8	Dom	3/1/86	1014	18	12	95	105	205	9	192	6.9	0.08	C3	S1
9	Dom	25/6/85	183	15	4	15	3.2	13	8	70	6.6	0.06	C1	S1
10D	Irr	8/8/85	1256	89	33	121	2.7	278	20	232	6.8	0.13	C3	S1
11D	Irr	5/12/85	959	51	37	71	2.0	167	12	320	7.0	0.19	C3	S1
12	Dom	14/9/85	1282	40	43	174	4.3	351	22	274	6.9	0.18	C3	S1
13	Irr	7/11/85	4770	221	105	935	25.1	684	84	326	7.4	0.22	C4	S4
15D	Irr	7/1/86	1790	98	62	261	4.2	715	20	238	7.2	0.12	C3	S2
16	Dom	7/1/86	1707	41	29	361	229	774	35	483	7.5	0.01	C4	S3
17D	Dom	10/10/85	2890	53	91	469	193	1010	43	597	7.0	0.03	C4	S3
19D	Dom	9/1/86	1577	91	48	223	2.5	476	33	364	7.6	0.10	C3	S1
20D	Irr	5/12/85	2720	164	67	514	7.1	1170	19	174	6.9	0.28	C4	S3

D = Deep (>10m) well/borehole
Dom = Domestic Supply
Irr = Irrigation Supply
EC = Electrical Conductivity (mg/l)

SH = Salinity Hazard
NaH = Sodium Hazard
(US Dept Agriculture classification)

Table 1 Example Analyses

recorded F^- concentration is 0.55 mg/l. NO_3^- concentrations range from < l mg/l to > l65 mg/l, but typical maximum values are about 40 mg/l. In irrigation terms the waters vary from good quality to poor quality (Table l): the presence of poor quality water makes it particularly important to understand the system; is the poor quality due to irrigation? Will the good quality water become poor quality and the poor quality water become unuseable?

A simple calculation suggests that rainfall evaporative concentration probably cannot explain the range of concentrations observed. Using the average recharge rate suggested by Krishnasamy (l987) of 0.23 m/a, and making several assumptions including that all recharge comes from rainfall, the rainfall Cl^- concentrations can be estimated by (recharge rate) x (groundwater concentration)/(rainfall rate). Calculations of this sort suggest rainfall concentrations of an average 73 mg/l (range 6.5-l33 mg/l) for the wells remote from the coast: rainfall data are few, but concentrations are in the range 4-l6 mg/l Cl^-, and therefore at many sites sources of solutes additional to rainfall are required.

Vertical Variation in Chemistry

Since hydraulic data suggest the system to consist of two aquifers - an upper unconfined aquifer and a lower (semi) confined aquifer - it is of interest to see if this is reflected in the chemistry.

Figure 4 shows typical comparisons of shallow (<l0m) and deep well data. It is clear that no systematic variation with depth is present; there is at least as much chemical variation laterally as there is vertically.

The similarity of water chemistry in shallow and deep parts of the aquifer system may reflect lack of chemical sources or lack of reactivity. But equally well it accords with the conceptual hydraulic model, which implies that a significant proportion of deep well water is derived from local vertical flow from the overlying shallow aquifer.

Lateral Variation in Chemistry

Figure 5 shows concentration distributions across the area. Near the coast, concentrations increase due to sea water intrusion, salt pan leaching, and possibly evaporation. Although important, the slightly saline water has been the subject of other publications (Ramalingam, l980; Rao et al., l987, Elango et al., l988), and will not be pursued here.

recorded F^- concentration is 0.55 mg/l. NO_3^- concentrations range from < l mg/l to > l65 mg/l, but typical maximum values are about 40 mg/l. In irrigation terms the waters vary from good quality to poor quality (Table l): the presence of poor quality water makes it particularly important to understand the system; is the poor quality due to irrigation? Will the good quality water become poor quality and the poor quality water become unuseable?

A simple calculation suggests that rainfall evaporative concentration probably cannot explain the range of concentrations observed. Using the average recharge rate suggested by Krishnasamy (l987) of 0.23 m/a, and making several assumptions including that all recharge comes from rainfall, the rainfall Cl^- concentrations can be estimated by (recharge rate) x (groundwater concentration)/(rainfall rate). Calculations of this sort suggest rainfall concentrations of an average 73 mg/l (range 6.5-l33 mg/l) for the wells remote from the coast: rainfall data are few, but concentrations are in the range 4-l6 mg/l Cl^-, and therefore at many sites sources of solutes additional to rainfall are required.

Vertical Variation in Chemistry

Since hydraulic data suggest the system to consist of two aquifers - an upper unconfined aquifer and a lower (semi) confined aquifer - it is of interest to see if this is reflected in the chemistry.

Figure 4 shows typical comparisons of shallow (<l0m) and deep well data. It is clear that no systematic variation with depth is present; there is at least as much chemical variation laterally as there is vertically.

The similarity of water chemistry in shallow and deep parts of the aquifer system may reflect lack of chemical sources or lack of reactivity. But equally well it accords with the conceptual hydraulic model, which implies that a significant proportion of deep well water is derived from local vertical flow from the overlying shallow aquifer.

Lateral Variation in Chemistry

Figure 5 shows concentration distributions across the area. Near the coast, concentrations increase due to sea water intrusion, salt pan leaching, and possibly evaporation. Although important, the slightly saline water has been the subject of other publications (Ramalingam, l980; Rao et al., l987, Elango et al., l988), and will not be pursued here.

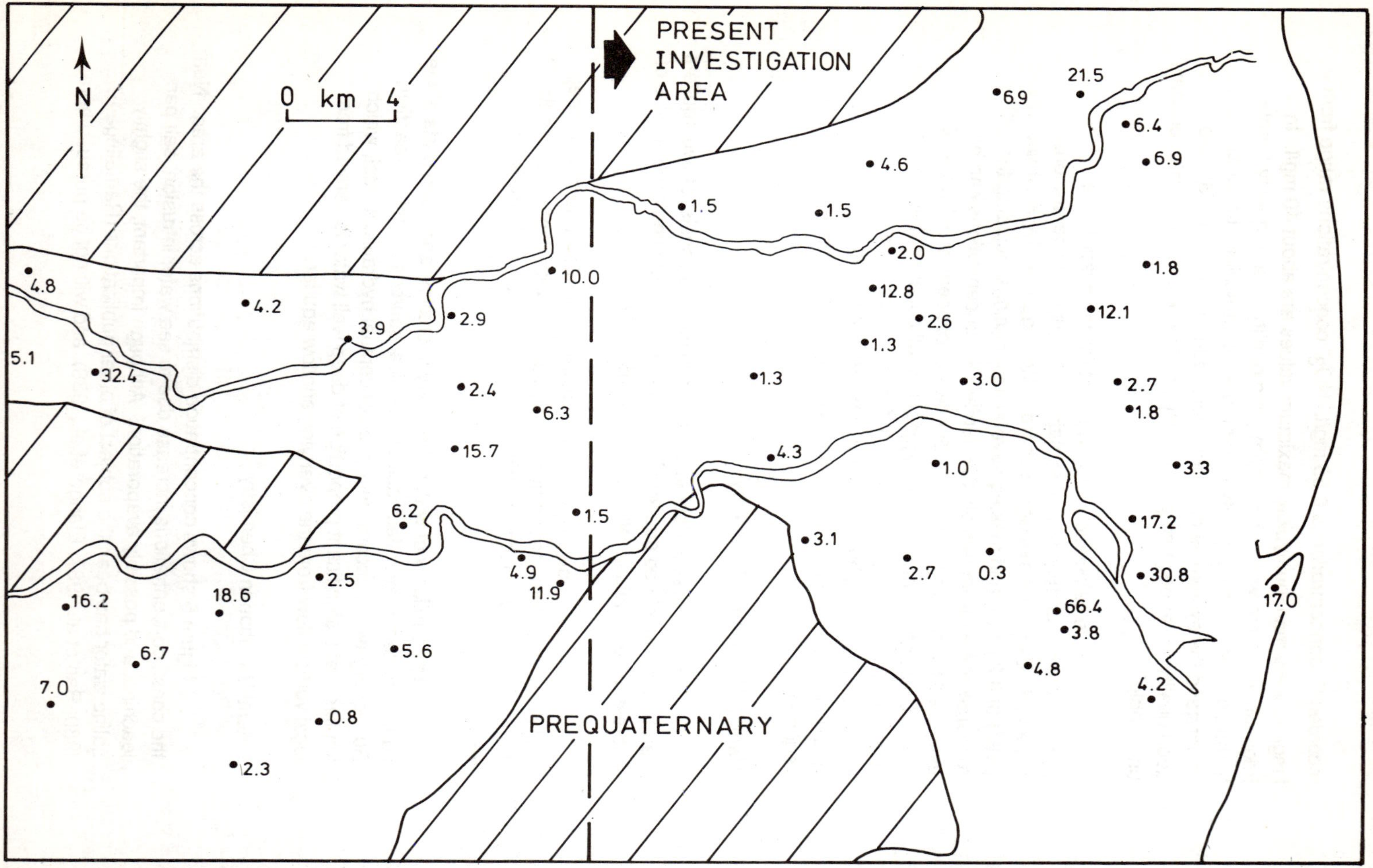

FIGURE 5 - REGIONAL Cl^- CONCENTRATIONS OCTOBER 1967 (meq/l) (UNDP Data)

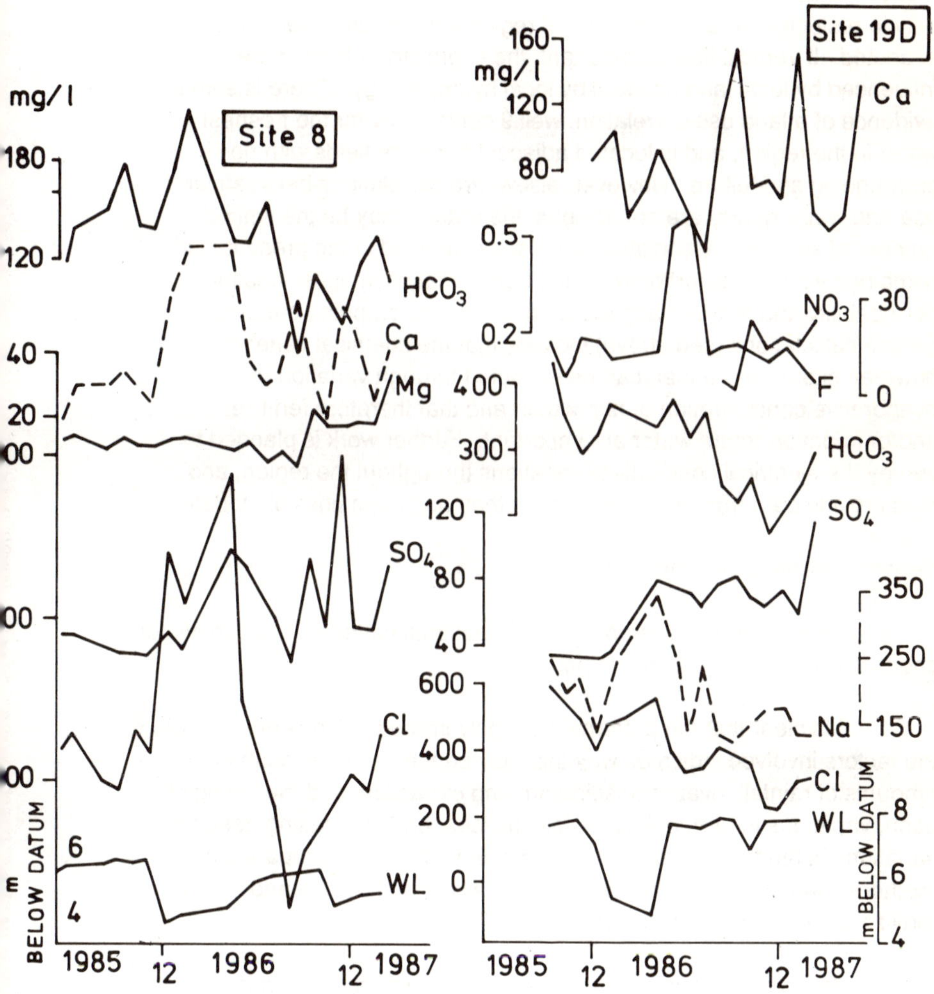

FIGURE 6. SEASONAL CHEMICAL VARIATION

Remote from the coastal areas the freshwater concentrations are highly variable, with no obvious patterns emerging in the poor and good quality water. This is again in accord with the general hydrogeological model in that if lateral regional flow were predominent with recharge only in restricted areas, some type of regional chemical pattern would be expected. If vertical flow is important, the chemistry is likely to be influenced by local land use and by local hydrogeology. There is some evidence of a land use correlation: well 9 contains by far the freshest water in the region, and is located adjacent to sports fields with no surrounding agriculture. However, elsewhere correlations between land use and water quality are not obvious: the reason may be the limited number of sites and associated land uses examined in the present study combined with the distortion resulting from variation in upper aquifer permeability and fertilizer application practice and cropping patterns. Simple calculations (see above) indicate that the chemical variation between sites is more than can be expected through variation in evaporative concentration of rain water, and that therefore fertilizer and/or irrigation return water are important. Further work is planned to survey the electrical conductivity variations throughout the region, and thus provide data from many more sites than has been possible todate.

Seasonal Variation in Chemistry

Figure 6 shows examples of the seasonal variation in groundwater chemistry over the investigation period.

It is clear that the patterns are highly irregular. Consideration of the factors involved indicates why this may be the case: variation in amounts of rainfall, evapotranspiration, and extraction, and the timing of each; variations in permeabilities, thicknesses, and well casing designs; variations in land use, irrigation amounts and timing, soil preparation methods, and crop types; variation in rainfall concentrations and fertilizer practice; and degree of mixing.

However, in several sites there is a general increase in total dissolved solids with water level and/or at monsoon onset (eg site 8, Figure 6). Similar correlations have been observed elsewhere (Sarma et al., l979). In the current area the increase in salinity appears to be mainly due to increase in Cl^- and associated cations, although occasionally alkalinity also increases. The simplest explanation is that recharging water flushes out evaporation - concentrated water from the unsaturated zone; in a chemically stable system the excess concentration in the early monsoon recharge would be balanced by a lower concentration in the late monsoon recharge. The fact that only some ions increase in concentration may be due to the kinetics of the dissolution reactions (Drever and Smith,

1978) and the fact that Cl^- will stay in solution to extremely low moisture contents because of the high solubility of its salts, whereas Ca^{2+}, HCO^-, Mg^{2+} and $SO_4{}^{2-}$ are all liked to be precipitated out far earlier. Some work has been carried out to model these effects using the computer model of Plummer et al. (1975).

The correlation of change in water level with change in concentration and the lack of lag time between monsoon and these events, might imply very rapid transport of recharge from groundlevel to the water table. However, this is not in accord with studies such as the tritium studies carried out by the National Geophysical Research Institute which imply slow movement through the unsaturated zone. At least two explanations are possible. Firstly, the dominent transport mechanism may be via routes which by-pass the slower intergranular flow; at the end of the dry season, many fields are heavily cracked and burrowed. Secondly, recorded water level rise at some sites is partly due to decrease in abstraction at the onset of the monsoon; as water levels rise, they will come into contact with the higher concentrations present in the porewaters of the unsaturated zone. This second mechanism implies that the unsaturated zone porewater concentrations are in excess of those of the saturated zone and could make unnecessary the flush-out idea.

Some evidence suggests that the first explanation (by-pass flow) is at least locally important. Well 3 data for 1986 was characterised by unusually high K^+ peaks in June and September. On inquiry it was found that the application of potassium fertilizer on the surrounding fields occurred in these two months. Further work on such local correlations is necessary, though not that easy.

Long Term Variation in Chemistry

Figure 7 shows typical long term chemical data collected by PWD. In general there is no indication that water quality is getting generally worse, despite continued irrigation. Some sites have stable chemical concentrations, others show increases, and other decreases with time. These observations support the contention that each site is dependant on its own local conditions, and that any general observations are likley to be inaccurate at the local scale. Where water quality appears stable in time, a certain degree of lateral inflow is implied, even if vertical supply is dominant.

The general apparent stability of the water quality in the area is very encouraging and in general supports the flow modelling conclusion that rainfall recharge comprises a substantial proportion of total recharge. However, water quality at some sites is deteriorating, and as

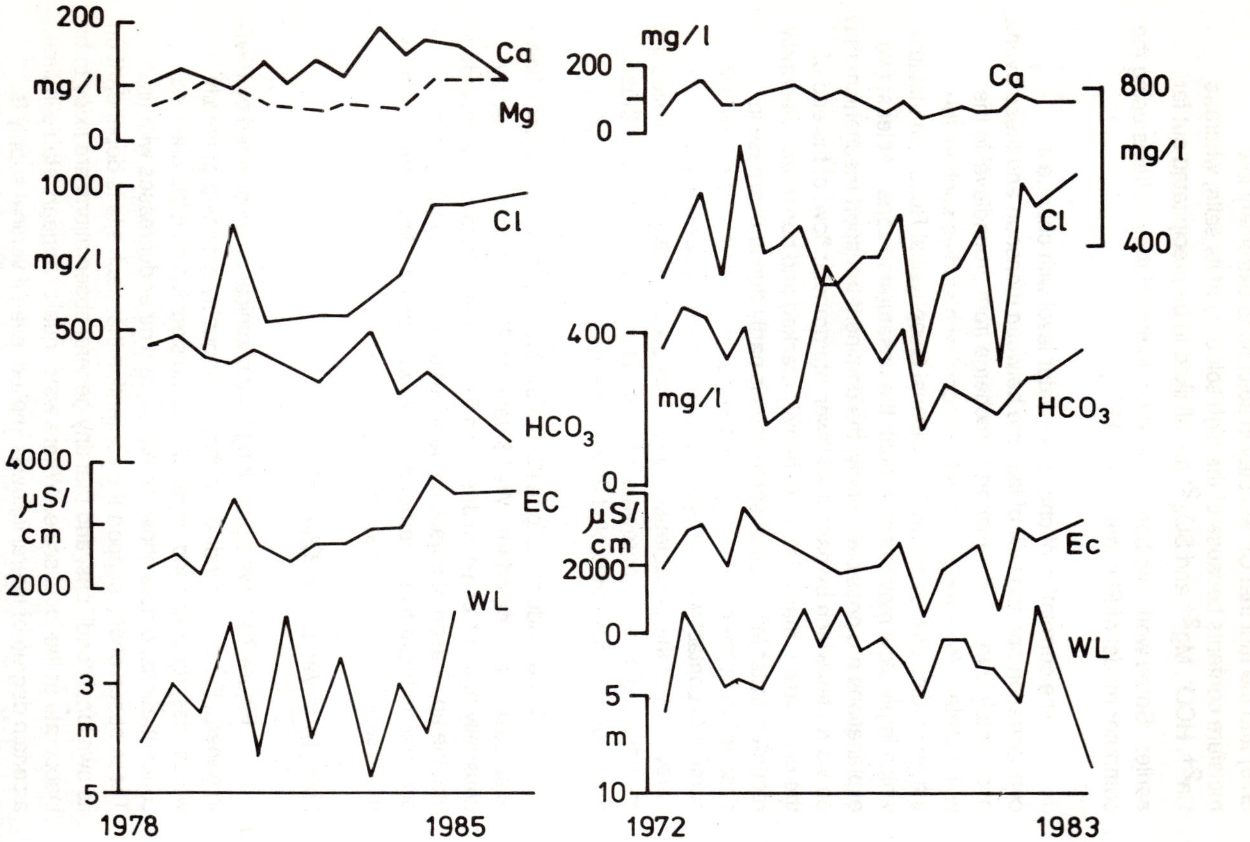

FIGURE 7. LONG TERM ½ YEARLY SAMPLES (PWD Data)

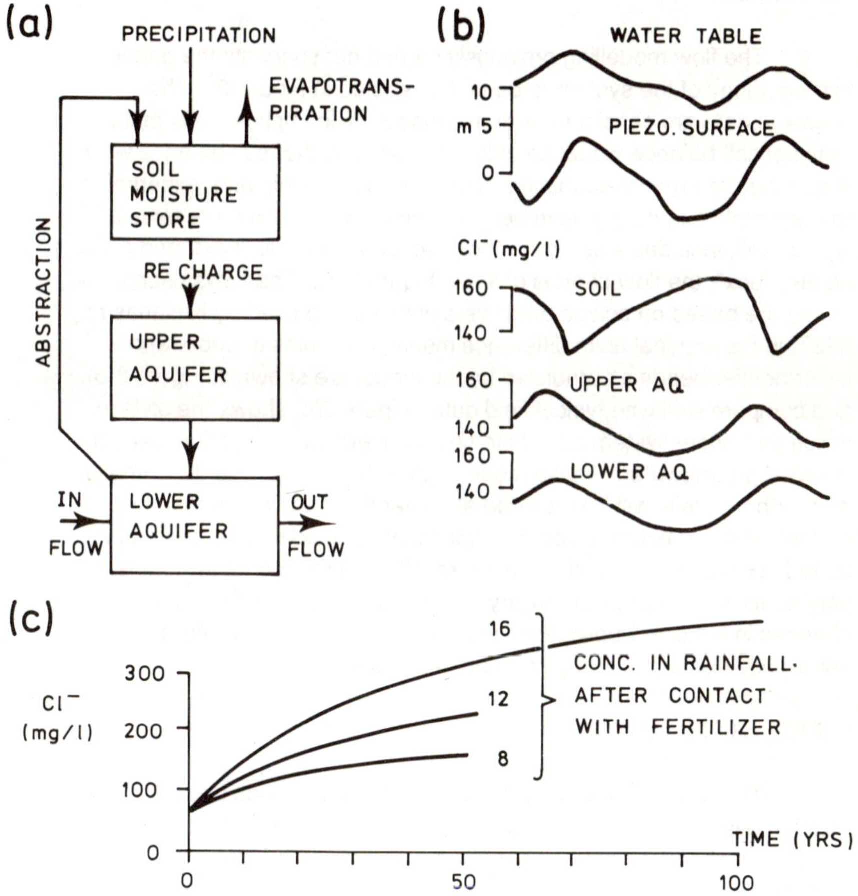

FIGURE 8. PRELIMINARY SALT BALANCE MODEL

yet the sensitivity of the system to changes in pumping rates, fertilizer applications, or cropping patterns is unknown: many wells yield only moderate quality water, and a small deterioration would be potentially serious. The investigation of the sensitivity of the system is therefore a major aim of the present study.

DISCUSSION

The flow modelling previously carried out suggests the general hydrogeology of the system to be that illustrated in Figure 2. The chemical data are consistent with this model. The implication is that a regional salt balance would be of limited use, and that salt balance work should centre around examining "typical" wells. To this end, experiments with a simple box model have begun. The model, which represents a typical well, includes a soil zone "box", an upper aquifer "box", and a lower aquifer "box"; the flow chart is given in Figure 8(a). The flows used in the model are based on scaled down versions of those used by Krishnasmay (I987) in his regional finite difference model; examples of upper and lower aquifer heads as predicted by the model are shown in Figure 8(b), and compare well with typical field data. Figure 8(c) shows the style in which the model system adjusts to commencement of irrigation using the pumped groundwater. These model studies have only recently commenced and further details will be provided at a later time; as yet, neither the mathematical or hydrogeological aspects of the model have been fully tested. However, it is clear from the results that the model system is very sensitive to rainfall chemistry and fertilizer input, and therefore changes in fertilizer applications may well disturb the equilibrium apparently existing in many parts of the aquifer.

CONCLUSIONS

The groundwater quality of an alluvial aquifer has been examined in order to evaluate future water quality variations. It has been possible to explain the chemical distributions in time and space by reference to the general hydrogeological mechanisms deduced from regional flow modelling. These (qualitative) explanations are specific to the aquifer, since the flow mechanisms are also specific. Although further testing of the qualitative model is necessary, it is very likely that despite the high permeability of the main (lower) aquifer, that future salt balance studies should consider the local scale rather than the regional scale, principally because of the importance of vertical flows from the upper to the lower aquifer system. Salt balance models, based on single well catchments, have been constructed; preliminary work suggests that water quality equilibrium can easily be disturbed by changes in chemical input such as brought about by changes in fertilizer application.

Future work will involve continuation of monitoring, electrical conductivity surveys to help in interpolation of results, development of the salt balance calculations, and examination of statistical approaches (Somasundaram, 1988; Elango and Manickam, 1987). Further data will come from a nitrate and pesticide study being undertaken jointly with the British Geological Survey; this will involve monitoring of specially drilled piezometer nests and should provide more information on the recharge mechanisms.

ACKNOWLEDGEMENTS

We would like to thank Professor K R Rushton for many very useful insights into the hydrogeology of this aquifer. We also thank Dr K V Krishnasamy and Dr R Sakthivadivel for sharing their knowledge of the system with us. Financial support is provided by the UGC of India, the British Council, and Anna University, and is gratefully acknowledged. Mr N J Shaheen carried out the salt balance modelling.

REFERENCES

Drever, J.I, and Smith, C.L., 1978. Cyclic wetting and drying of the soil zone as an influence on the chemistry of groundwater in arid terrains. Am. J. Sci., v.287, p.1448-1454.

Elango, L. 1985. Hydrochemical studies of groundwater. Unpubs. Post Graduate Diploma Project Rep, Centre for Water Resources, Anna University, Madras, India.

Elango, L. and Manickam, S. 1987. Weighted pair averaging cluster analysis in hydrogeochemistry. Proc. 6th Ind. Geol. Congr., Roorkee, 21-24 Feb. 1987.

Elango, L. Manickam, S. and Sakthivadivel, R. 1988. Saline groundwater in heavily exploited coastal aquifer a case study in Madras. Nat. Seminar on Ground Water Develop. in Coastal Tracts, Trivandrum, Jan. 1988.

Krishnasamy, K.V. 1987. Regional modelling of non-linear flows in multi aquifer systems. Unpubs. PhD thesis, Centre for Water Resources, Anna University, Madras, India.

Plummer, L.N. Jones, B.F. and Truesdell, A.H., 1976. WATEQF - a FORTRAN IV version of WATEQ, a computer program for calculating chemical equilibrium of natural waters. U.S. Geol. Surv. Water Resources Invest., 76-13, 63 pp.

Ramalingam, A.E., 1980. Studies on seawater intrusion in a coastal aquifer north of Madras in Tamil Nadu. Proc. 48th Research Sess. of CBIP, Hyderabad, Andra Pradesh, v.II, p.133-146.

Rao, S.H. Jain, S.K. Navada, S.V. Nair, A.R. Shivanna, K., 1987. Isotopic studies on seawater intrusion and inter-relationships between water bodies: some field examples. Isotope Tech. in Water Resources Develop., IAEA-SM-299/76.p. 403-426.

Rushton, K.R., 1986. Vertical flow in heavily exploited hard rock and alluvial aquifers. Ground Water, v.24, p.601-608.

Sarma, V.V.J. Prasad, N.V.B.S.S and Rajendra Prasad, P., 1979. The effect of hydrogeology on variations in the electrical conductivity of groundwater fluctuations. J Hydrol., v.44, p.81-87.

Somasundaram, M.V., 1988. Hydrochemical time series interpretation of overstressed Araniyar Kortalayar Aquifer. Visit Research Report, Centre for Water Resources, Anna University, Madras, India. 129 pp.

DROUGHTS AS MEANS OF DELINEATION OF AREAS PRONE TO SEA WATER INTRUSION

B.S. SUKHIJA, D.V. REDDY
NATIONAL GEOPHYSICAL RESEARCH INSTITUTE,
HYDERABAD - 500 007.
&
I. VASANTHAKUMAR REDDY
PONDICHERRY AGRO SERVICES & INDUSTRIAL CORPORATION, LIMITED, PONDICHERRY - 605 009.

...

ABSTRACT

Groundwater studies along the coast are most important to protect groundwater reserves from the sea water encroachment. Over-exploitation of groundwater along the coast results in lowering of water levels which may even cause reversal of hydraulic gradients. In order to prevent such a situation, long term groundwater input-output studies are required. However, prolonged drought periods in any area simulate conditions of over - exploitation of groundwaters, which can be effectively utilised for the study of such problems. In this paper, it has been demonstrated that by analysing water levels, the changes in flow directions and hydraulic gradients etc. of groundwater in the coastal region of Pondicherry, corresponding to normal rainfall, excessive rainfall and drought years, it is possible to delineate areas prone to sea water intrusion.

INTRODUCTION

Over-exploitation of groundwater along the coast may result in declining water levels accompanied by the reversal of hydraulic gradient. Over-exploitation in an area occurs because of ever increasing agricultural activity based on groundwater with limited recharge potential. Over-exploitation may also occur for a limited period due to the prevailing severe drought conditions extended over two or three years in an area. During drought period surface water resources get scarce, recharge is meagre and groundwater resource gets depleted heavily.

The purpose of this paper is to demonstrate that the comparative study of groundwater levels, groundwater flow and the gradients corresponding to normal rainfall, drought and excessive rainfall provide us a means to delineate areas prone to seawater encroachment. Present work has been undertaken at Pondicherry and its environs. The study period encompasses normal rainfall years (1977 to '79), years of low rainfall (1980 to '82) and excessive rainfall year of 1985. During the low rainfall and drought periods the agricultural activity was maintained (not allowed to be reduced) using large scale withdrawal of groundwater.

DESCRIPTION OF THE AREA

Investigated area comprised Pondicherry and its environs situated, between 11° 46' and 12° 03' north latitudes, and 79° 35' and 79° 53' east longitudes, along the coast of Bay of Bengal. Except for some hillocks in the eastern portion, the investigated area is mostly flat. General topographical gradient of the area is towards east and locally towards rivers. Physiographically the area can be divided into three parts drained by the rivers - the Gingee, the Malattar and the Ponnaiyar (Fig. 1). Apart from the rivers, there are many small and two big irrigation tanks in

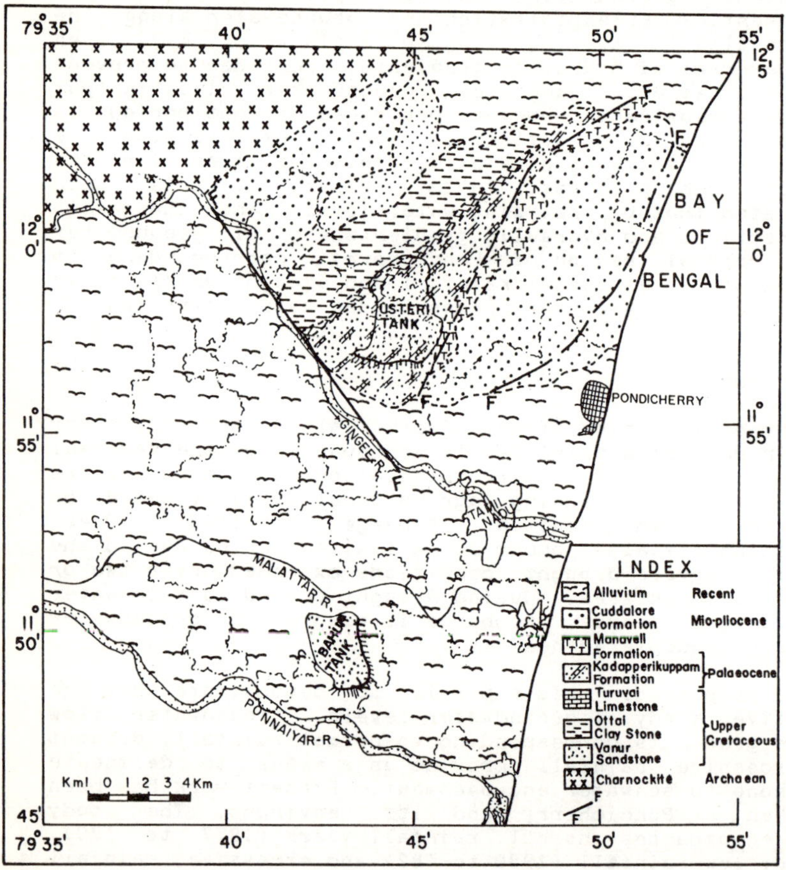

Fig. 1. Geological map of Pondicherry.

the region. The study area has coastal climate with temperature ranging between 15° C (minimum) and 43° C (maximum). The annual average rainfall from the two monsoons is 1150 mm, the south-west monsoon provides about 30% of total rainfall during June to Septemper, and north-east monsoon provides about 65-70% of the total during October to December. There is a considerable variation in the mean annual rainfall as can be seen in the Fig. 2. During last forty years, severe droughts have occurred in 1952,68,79, 80 & 82, when the rainfall was much less than the normal rainfall.

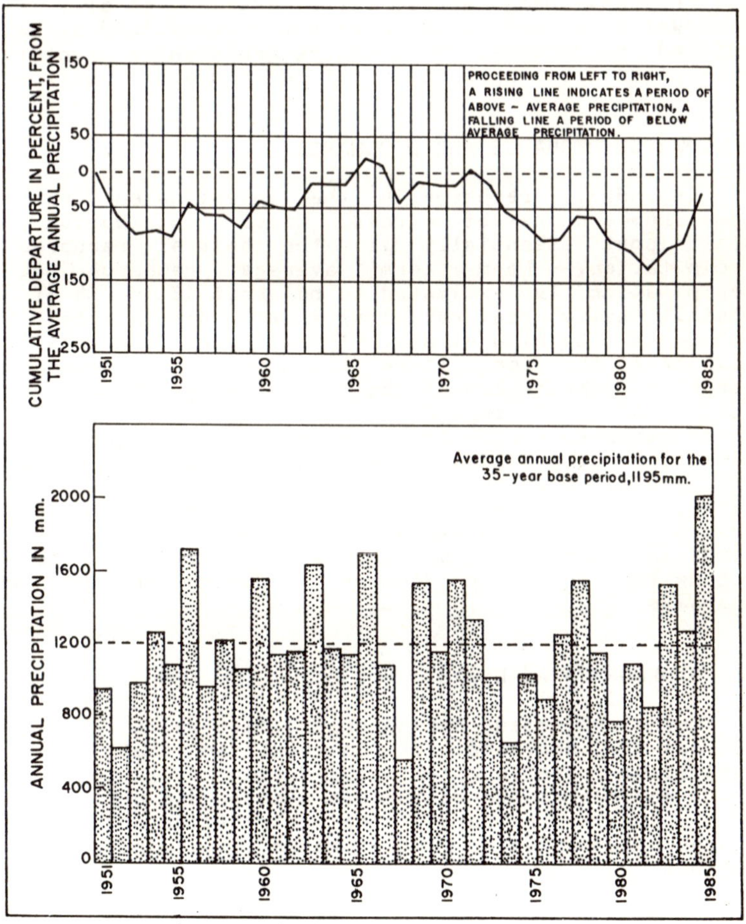

Fig. 2. Average annual precipitation and cumulative departure from average precipitation at Pondicherry (1951 to 1985).

Geologically Pondicherry and its environs is covered with cretaceous, tertiary and recent sediments on Archaean basement. Older formations are exposed only in the north of Gingee river and recent alluvium covers the southern part (Fig.1). The Central Groundwater Board of India has carried out extensive geohydrological studies in the area (Raghava Rao et al., 1976 & 1980). Groundwater occurs in eight hydrostratigraphic units. Alluvium deposits and Cuddalore sandstone are the most potential aquifers followed by the Vanur and the Ramnathopuram formations, and the Kadapperikuppam and Manaveli formations have marginal resources. Further groundwater studies related to individual aquifers have been carried out by Sukhija et al., (1987). Groundwater occurs in the area under unconfined as well as confined conditions. Most of the groundwater exploitation takes place from upper alluvial and cuddalore aquifers.

GROUNDWATER RECHARGE AND DISCHARGE

Groundwater recharge in the investigated area takes place from a number of sources. Major contribution to the total input essentially is the natural recharge from precipitation. Long term average recharge due to precipitation is estimated to be about 16 +/- 3% of local rainfall using environmental as well as injected tracer techniques (Sukhija et al., 1988). Apart from this, return flow from the agricultural lands forms the second major source as about 50% of the total geographical area is cultivated, and seepage from the existing tanks and canals is the other source. Important outflow of groundwater is extraction of groundwater from shallow and deep wells for agricultural purposes as there is very little flow towards rivers. Because of continuous drought during 1980 - 1982 consumption of groundwater increased tremendously with increasing number of new tube wells drilled by the State Groundwater Unit, Pondicherry, during drought relief programme.

WELL HYDROGRAPH ANALYSIS

Data collected for 35 observation wells from 1977 to 1982 has been analysed. It is observed that there is no significant rise in water table during July to September, which means that the south-west monsoon did not contribute to groundwater recharge. The contribution of north-east monsoon is observed as rise in water level during October to December.

A large variation in water level increase (1-8 m) is observed during normal rainfall years at different places. But during a low rainfall year rise in water table is very minimal.

DYNAMIC BEHAVIOUR OF GROUNDWATER

Dynamic behaviour of groundwater of Pondicherry was studied in space and time including the typical years of 1979 and 1983, corresponding to wet and drought conditions.

According to the rainfall data of Pondicherry; (Fig.2) 1974 was the year of severe drought followed by high rainfall (1550 mm) in 1978 below normal during 1980 and 1982, and was maximum (2030 mm) in 1985. In general, following the rainfall variation, groundwater level was observed to be high during 1979 and minimum during pre-monsoon of 1983. To study the dynamic behaviour of groundwater of Pondicherry, water level data of 35 dug wells as well as 35 deep wells were utilised to prepare potentiometric contour maps for the shallow and deep wells; which were studied as mentioned below to understand the effect of drought.

GROUNDWATER CONDITIONS DURING POST MONSOON OF 1979.

Fig.3. shows the potentiometric surface based on the data of shallow and deep observation wells during post-monsoon season of 1979. The study of water level maps indicates that there exist two easily distinguishable areas located in the north and south of Gingee river.

In the northern area, an elongated mound was observed parallel to the coast at a height of about 45 m (a msl) with a NE-SW groundwater divide. The flow of groundwater was observed towards all the directions from the mound with high hydraulic gradients. The gradients were about 13-20 m/km in the east, about 7.5 m/km towards west, and about 16 m/km towards Usteri tank. There was yet another groundwater divide in the upper reaches of Usteri tank with flow towards the Gingee river. In the north-western region hydraulic gradient was around 6 m/km.

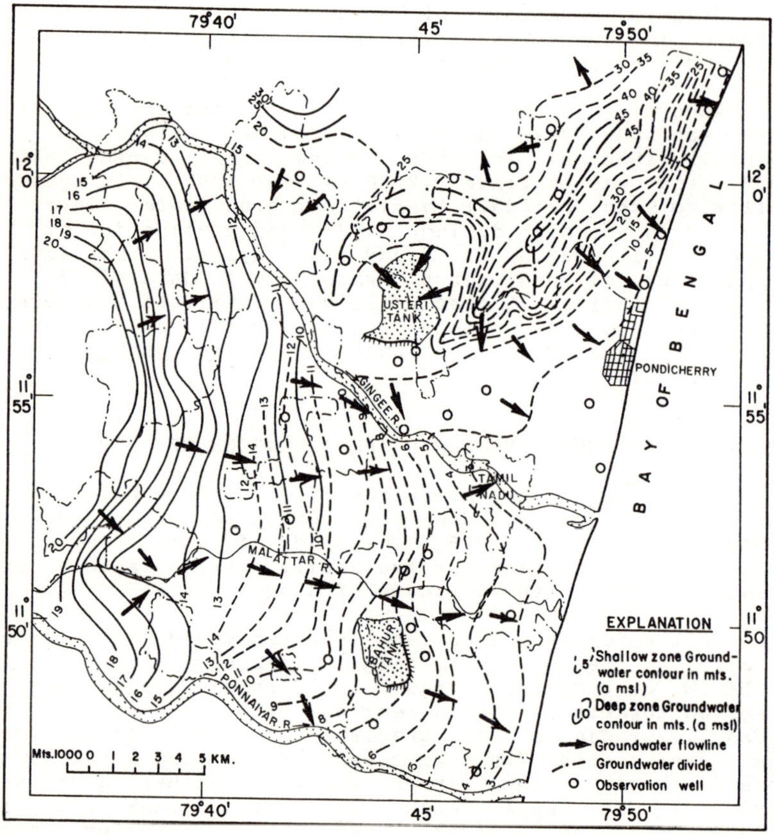

Fig. 3. POTENTIOMETRIC SURFACE OF GROUNDWATER DURING POST-MONSOON, AFTER HEAVY RAINS, OF 1979.

Water level contours in the south of Gingee river for the shallow as well as deep wells were approximately parallel to the coast line indicating east-ward flow. But however, it was observed that near the river course, the contours became concave indicating the effluent conditions for the rivers Gingee and Ponnaiyar. Except for upper reaches the parts of Malattar river also reflected effluent character.

GROUNDWATER CONDITIONS DURING PRE-MONSOON OF 1983.

Due to the drought during 1982- '83, there were significant changes in the groundwater regime (Fig. 4). As compared with 1979 levels, the height of the groundwater mound in the north-east was reduced to about 30m. The hydraulic gradients towards different directions were also reduced. The flow directions however remained unchanged. In the NW region the hydraulic gradient was reduced from 6 to less than 2 m/km. However, a large change in the groundwater flow direction was observed in the area south of the Gingee river. The study of phreatic groundwater contours indicated an oval shaped groundwater depression (3 to 6m below msl) around the Bahur tank. Because of the depression there was convergence of groundwater flow into the depression with hydraulic gradient of 3 to 4 m/km. In the west of the Bahur tank contours were parallel to the coast line.

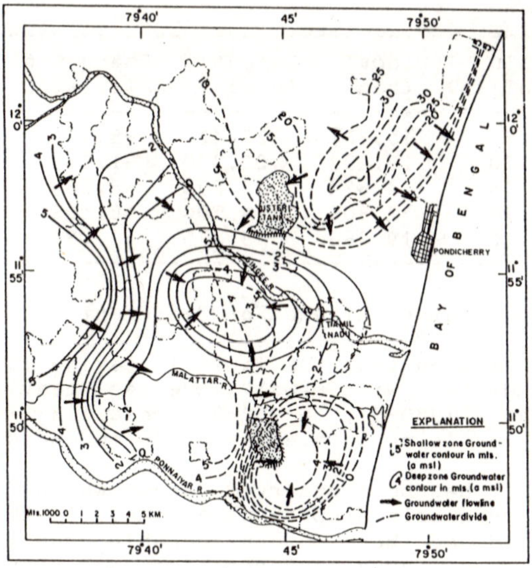

Fig. 4. POTENTIOMETRIC SURFACE OF GROUNDWATER DURING PRE-MONSOON(1983), AFTER DROUGHT (1980-1983).

A considerable variation in the flow direction of the piezometric surfaces was also observed during the drought period. The study of deep piezometric surface indicate a big groundwater depression (-5 m) in the central region also. However, this area does not warrant special attention as being away from the coast line.

DISCUSSIONS

Effect of drought on water levels in the area was two folds, firstly during drought recharge is reduced considerably, further as the agricultural activity was maintained (not allowed to decrease) and the requirement of irrigation water was met from groundwater as the surface water resources were meagre, this resulted in large exploitation of groundwater for a short duration of time. During this time appreciable drop in the mean groundwater level was observed.

Table. 1. shows the groundwater budget calculations for different years. Input, in the table, refers to annual recharge, return flow from agricultural fields, seepage from tanks and canals etc. Output is the total draft from different types of wells for different agricultural, domestic and industrial activities. The change in storage is calculated by the input-output difference. The calculated values of changes in the storage is cross-checked with observed yearly difference of groundwater level in the study area using a specific yield of 0.05 (Sukhija et al., 1987).

Period	Rain fall(m)	Input (ham)	Output (ham)	Change in Storage(ham) Calculated	Observed
1980-81	0.90	15,381	23,427	-8,046	- 10,978
1981-82	1.00	19,492	20,730	-1,238	+ 3,050
1982-83	0.84	14,624	22,893	-8,269	- 7,965
1983-84	1.90	22,906	14,500	+8,406	+ 7,313
1984-85	1.10	19,266	18,175	+1,091	+ 480

Table 1. Groundwater Balance For Different Years.

As can be seen from the Table 1; taking 1979 as a base year, groundwater withdrawal increased heavily (1.5 times input) during 1980-81, & 82-83 resulting in decline of groundwater levels on an average of 7m in the study area.

Thus, it has been observed, as shown in Fig. 4., that there was appreciable drop in water table in the area during drought period and there is considerable change in hydraulic gradient as well as directions of flow. During drought period the flow got completely modified. As a result, in the south of Gingee,flow broke down into smaller sub-systems with small mounds one each in the north and south, and a large depression. The maximum effect by way of lowering water level and reversing hydraulic gradient is observed in the south-east; thus, the area maximum prone to sea water intrusion is delineated (Fig. 4.).

CONCLUSION

It can be concluded that the large groundwater withdrawal in drought conditions simulate the over-exploitation of aquifers (which may occur in future), and the analysis of comparative dynamic behaviour of groundwater for normal rainfall and drought periods provides us a means to identify areas where hydraulic gradients reverse and there exist a possibility of seawater intrusion.

ACKNOWLEDGEMENTS:

We are grateful to Prof. V.K. GAUR, Director, NGRI, for his keen interest in the work and kind permission to publish the paper; to the Director of Agriculture, Pondicherry, for sponsoring the project to NGRI, Hyderabad. We are also thankful to our colleagues in the laboratory who helped in the preparation of this paper.

REFERENCES:

Raghava Rao, K.V., Krupanidhi, K.V.J.R., Balasubramaniyan, G., Subramaniyan, P.R., and Natarajan, P., 1976. Report on groundwater survey and exploration in the Union Territory of Pondicherry and its environs. CGWB Report, India.

Raghava Rao, K.V., Krupanidhi, K.V.J.R., Balasubramaniyan, G., Sankaran, M., Subramaniyan, P.R., and Natarajan, P., 1980. Groundwater survey and exploration in the Union Territory of Pondicherry and its environs. Bull. ONGC. v. 17, no. 1-16.

Sukhija, B.S., Reddy, D.V., and Vasanthakumar Reddy, I., 1987. Report on Study of groundwater resources of Pondicherry and its environs. T.R.NO: NGRI-87-SP-25. Category 'B'.

Sukhija, B.S., Reddy, D.V., Nagabhushanam, P., and Chand, R. 1988. Validity of the environmental chloride method for recharge evaluation of coastal aquifers, India. J. Hydro. v. 99, p. 349-366.

GEOCHEMISTRY OF SHALLOW GROUNDWATER AT KIGAMBONI PENINSULA ALONG DAR ES SALAAM COASTAL STRIP TANZANIA

H.H. NKOTAGU
DEPARTMENT OF GEOLOGY
UNIVERSITY OF DAR ES SALAAM,
TANZANIA

ABSTRACT

The paper presents types of groundwater, hydrochemical facies and chemical genesis of shallow groundwater. Sodium bicarbonate, sodium sulphate, sodium chloride, calcium bicarbonate and magnesium bicarbonate type of waters are identified.

Hydrochemical facies change from calcium-sodium facies near the shore to sodium-calcium facies landward for cations and from chloride-sulphate bicarbonate to bicarbonate-chloride sulphate facies, respectively for anions.

Water samples collected in plastic bottles from shallow wells were analysed using standard laboratory techniques. Modified stiff, piper trilinear diagrams and trend surface maps were constructed for ascertaining processes generating the chemical character of the shallow groundwater.

Dissolution of dolomitic limestone, cation exchange and mixing of fresh recharge water with interstitial marine water in addition to dissolution of airbone salts are the major processes responsible for the chemical character of the water.

However the shallow groundwater has been reviewed and found to be suitable for general domestic and irrigation purposes.

INTRODUCTION

The study area is located on the West coast of the Indian Ocean, between $6^{o}49'$ and $6^{o}50'$S latitudes and between $39^{o}18'$S and $39^{o}23'$E longitudes (Fig.1). The area is characterized by sea shore equatorial climate hot and humid with a mean daily temperature of 26°C and mean annual rainfall of 1000mm.

Geologically, the area is characterized by beach ridge, sands, sand dunes and beach deposits between Ras Rongoni and Ras Mji Mwema along the coast. Clays, silts and salt Marsh are sandwitched between the raised reef limestone of probable pleistocene to Recent. Superficial white buff sands with no index fossils dominate landward covering the rest of the area (Fig.2.0).

The chemistry of groundwater is normally determined by processes taking place in the unsaturated zone during recharge. Processes including cation exchange, dissolution of weathered material and mixing with marine interstitial water modify the natural recharging water. At Kigamboni, magnesium-bicarbonate ions dominate close to the shore while sodium sulphate ions dominate landward.

Samples were collected from productive shallow ring wells during the dry season. The ring wells whose depth range between 5.00 - 11.00mm and spaced at 2-3.0km apart are fitted with handpumps. Standard laboratory techniques as reported by Greenberg et al 1985 were used to analyse the samples immediately after field collection. Electrical conductivity values were measured in the laboratory and results are believed to represent field conditions with only minor changes if any. The water has been recommended for domestic and irrigation purposes.

METHODOLOGY AND MATERIALS USED

Sampling Techniques

Samples were collected and stored in one litre plastic bottles, during the dry season. Sample bottles were carefully cleaned with conc HCl then rinsed strongly with tape water finally with distilled water. Contamination was further avoided before sample collection at the sampling site by rinsing sample bottles, with the water to be sampled. The sample water temperature was measured directly during sampling, by use of a thermometer.

A total number of 27 samples were collected from 9 different shallow wells.

Laboratory Work

Both physical and chemical parameters were analysed using standard laboratory techniques (Greenberg et al 1985). The analytical results along with the methods used to determine them are given in Table 1.

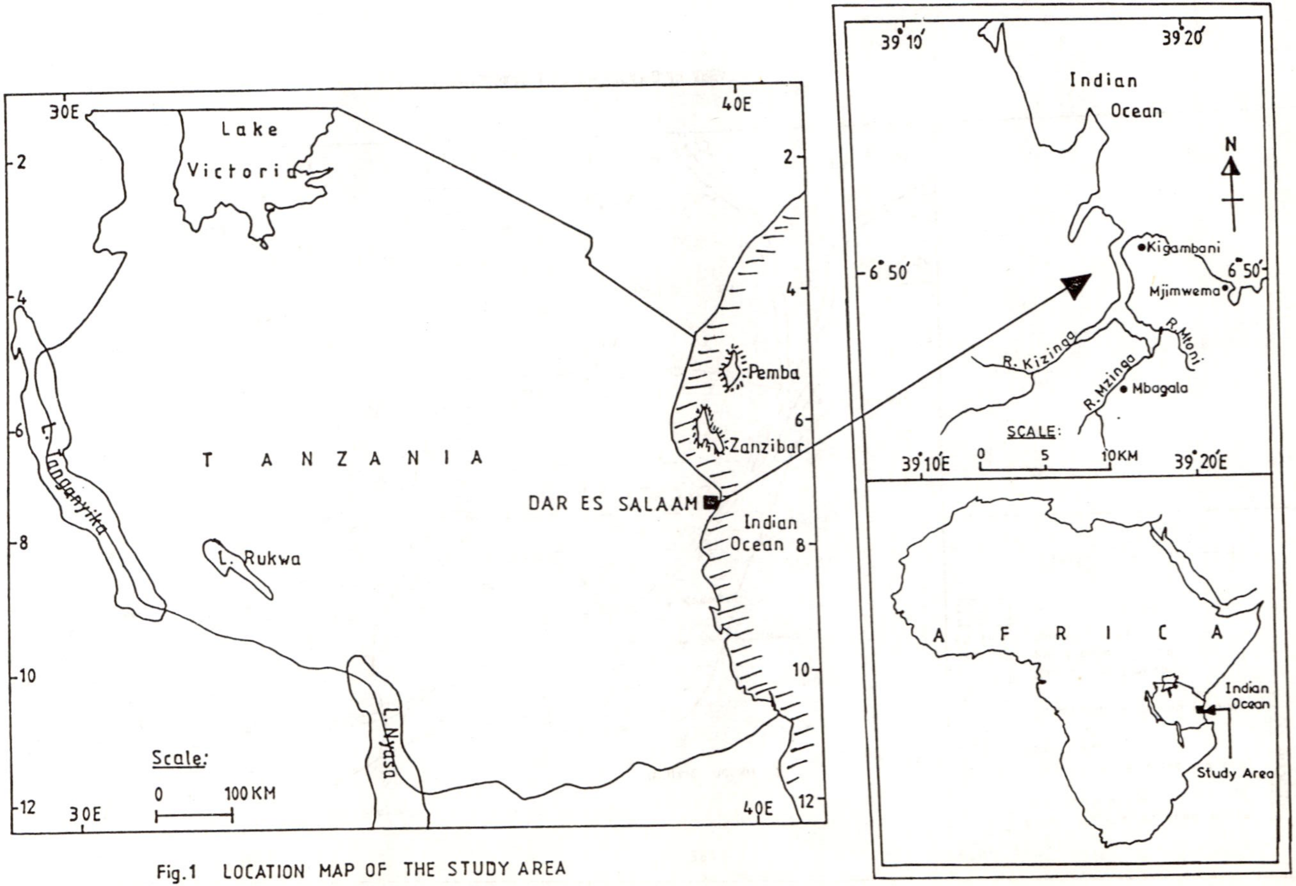

Fig.1 LOCATION MAP OF THE STUDY AREA

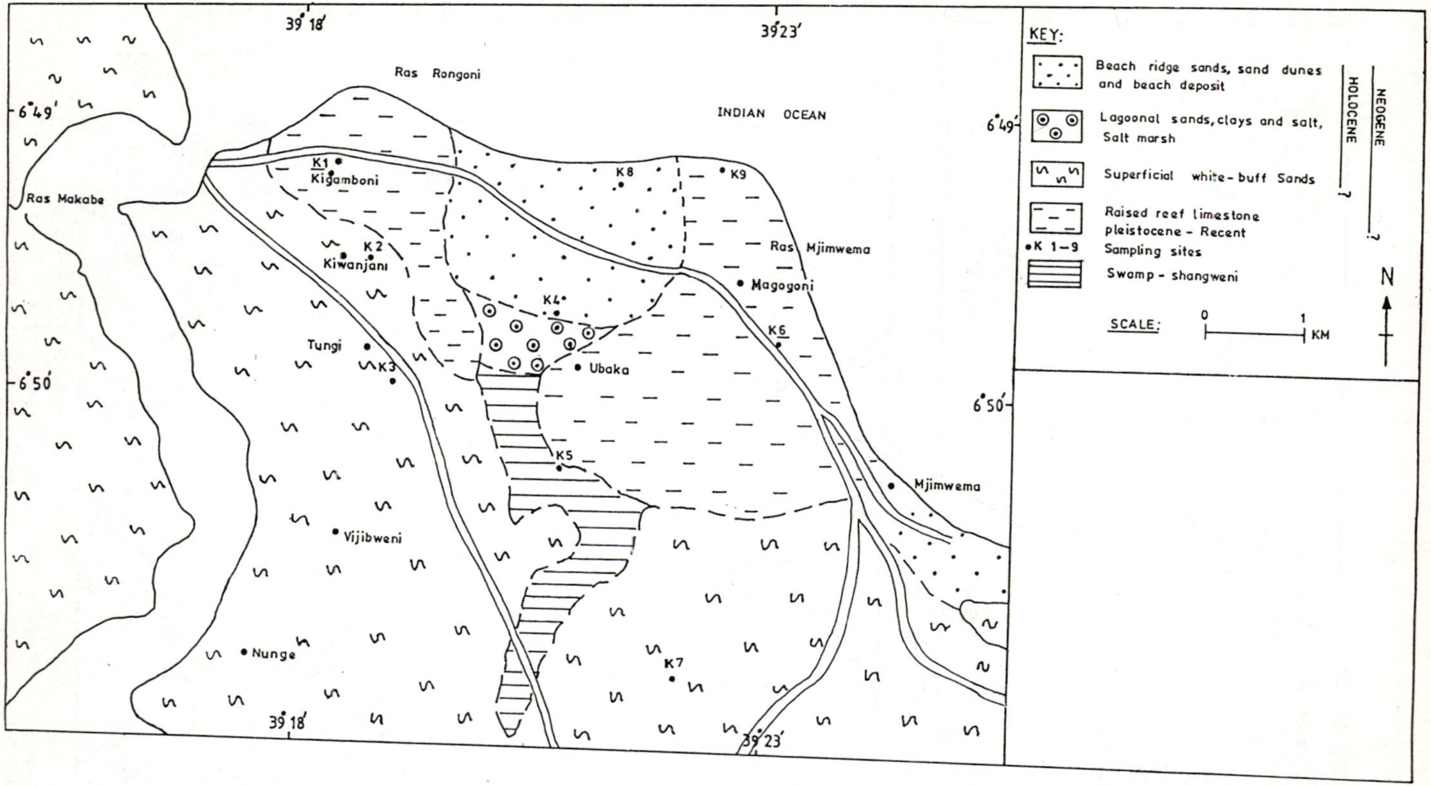

Fig. 2 GEOLOGICAL MAP OF KIGAMBONI AREA

TABLE I: CHEMICAL ANALYSES RESULTS :

SAMPLE NO	* PPM	I Na^+	1 K^+	1 Mg^{2+}	1 Ca^{2+}	1 Fe^{2+}	2 F^-	2 NO_3^-	3 SO_4^{2-}	4 CO_3^{2-}	4 HCO_3^-	4 Cl^-	5 TDS	6 CONDUCTIVITY	4 TOTAL ALKALINITY	4 TOTAL HARDNESS	7 TURBIDITY	8 COLOUR	9 PH	TOTAL CATIONS	TOTAL ANIONS	% ION BALANCE
K1	PPM	29.54	9.94	39.85	28.5	1.0016	0.0016	5.38	120	3.5	161.16	42.54	366	570	132.0	158.0	10	25	7.6	6.29	6.44	1.1
	EPM	1.29	0.3	3.28	1.42	0	0	0	2.6	0	2.64	1.2										
K2	PPM	163.10	10.37	22.64	23.60	1.0017	0.145	3.19	80	NIL	439.4	52.54	573	890	360	86.0	15	30	7.5	10.43	10.34	0.4
	EPM	7.12	0.28	1.86	1.17	0	0	0	1.66		7.2	1.48										
K3	PPM	193.70	6.23	16.03	12.8	NIL	0.064	0.34	194.8	NIL	269.6	54.2	861	1300	221.0	98.00	10	35	8.00	10.59	9.96	3.0
	EPM	8.46	0.17	1.32	0.64	0	0	0	4.36		4.4	1.5										
K4	PPM	471.3	15.90	60.62	41.56	NIL	0.11	4.54	675	NIL	322	219.79	1713.5	2750	510.0	68.00	10	25	7.90	27.97	30.45	4.2
	EPM	20.5	0.41	4.99	2.07	0	0.154	0	14.06		10.19	6.2										
K5	PPM	96.63	7.85	26.6	32.93	0.0159	0.15	2.79	95	NIL	287.91	58.14	461	710	236.0	140.0	9	20	7.50	8.32	8.30	0
	EPM	4.22	0.27	2.19	1.64	0	0	0	1.98		4.72	1.6										
K6	PPM	116.3	10.23	16.06	30.40	NIL	0.376	3.96	158	NIL	61.04	56.01	470.2	730	132.0	142.0	10	30	7.4	8.18	7.50	4.3
	EPM	5.08	0.26	1.32	1.52	0	0	0	3.29		2.64	1.57										
K7	PPM	151.59	8.60	13.48	45.78	0.036	0.078	1.25	155	NIL	158.6	156.62	690.9	1050	130.0	116.0	.5	10	7.00	10.23	10.14	0.4
	EPM	6.62	0.22	1.11	2.28	0	0	0	3.23		2.5	4.41										
K8	PPM	99.08	8.03	14.94	21.48	NIL	0.064	3.12	102.5	NIL	175.68	63.81	486	750	144.0	112.0	10	15	7.60	6.82	6.82	0
	EPM	4.31	0.21	1.23	1.07	0	0	0	2.14		2.88	1.8										
K9	PPM	122.1	12.95	44.58	166.00	NIL	0.058	1.02	144	NIL	314.76	166.62	716	1101	258.0	300.0	15	15	7.20	12.78	12.61	0.6
	EPM	4.25	0.44	1.62	6.147	0	0	0	2.99		5.16	4.7										

* Analytical methods used:

1. Atomic Absoption Spectrophotometer
2. Specific Ion electrode
3. Turbidimetric method
4. Titration
5. Volumetric method
6. Conductivity meter
7. Absorptometric method
8. Colourimetric method
9. PH meter

INTERPRETATION AND DISCUSSION OF RESULTS

Geochemical classification of waters

The geochemistry of shallow groundwater in the study area was achieved by use of trilinear diagrams. (Fig. 3.0).

The following water types are identified.

(i) sodium bicarbonate type at K_2 K_3 K_5 and K_8

(ii) sodium sulphate type at K_4 and K_6.

(iii) magnesium bicarbonate type at K_1

(iv) calcium bicarbonate type at K_9

(v) sodium chloride type at K_7

In addition the trilinear diagram indicates that, shallow groundwater at K_3, K_4, K_6, K_7 and K_8 is composed of primary salinity ie K^+Na^+ and Cl^- ions dominate while no cation-anion pair exceed 50% for samples K_1, K_5 and K_9. Sample K_2 is characterized by primary alkalinity ie Na^+, K^+, $HCO_3^=$ dominate.

Hydrochemical facies

Hydrochemical facies have been determined to denote the diagnostic chemical aspect of water solutions, occuring in the hydrologic system of the study area. The facies also reflect the responce of chemical processes in the lithologic framework and the pattern of water flowing in it. The facies have been designated according to Back W. (1966). Two anion hydrochemical facies have been identified as dominant chloride sulphate bicarbonate facies and bicarbonate-chloride sulphate facies. Similarly two cation hydrochemical facies have been analysed as dominant sodium-calcium facies and calcium sodium facies.

Complementary studies:

Additional studies including cation ratios, iso-chemical maps were conducted to complement the above results. It is noted that Ca/Mg ratios evolve from a high initial value to a much lower value than one as Ca^{2+} is lost through calcite precipitation. In the studied area all samples except K_6, K_7 and K_9 show dolomite dissolving at the expense of calcite precipitation and vice versa for sample K_9. While sample K_6 indicate calcite and dolomite dissolution to be at equilibrium. Sample K_8 reflect calcite super saturation with probable dolomite dissolution(table 2.0).

Trend studies indicate that chloride, total dissolved solids (TDS) and electrical conductivity increase towards the sands silts and lagoonal sand formation sandwitched between raised reef limestone. The observed increase of these parameters could be attributed to mixing of marine water with fresh shallow groundwater or dissolution of airbone salts by the recharging rain water. Location of samples K_3, K_4, K_6, K_7, and K_8 on the right vertex corner of diamond shaped field of the trilinear diagram supports this view(Fig. 3.0).

Table 2 Selected Major Ion ratios in epm at 25°C

Sample No.	Ca^{2+}/Mg^{2+}	$Na^{+}+K^{+}/Mg^{2+}$	$Na^{+}+K^{+}/Ca^{2+}$	$Cl^{-}/Na^{+}+K^{+}$	Cl^{-}/Mg^{2+}	Cl^{-}/Ca^{2+}	Interpretation
K_1	0.43	0.48	1.1	0.75	0.36	0.84	Dolomite dissolution with calcite precipitation.
K_2	0.63	3.98	6.32	0.2	0.79	1.26	"
K_3	0.48	6.54	13.48	0.17	1.11	2.34	"
K_4	0.41	4.19	10.10	0.29	1.2	2.99	"
K_5	0.75	2.05	2.74	0.35	0.73	0.97	"
K_6	1.15	4.04	3.51	0.29	1.18	1.03	Calcite and dolomite dissolution at equilibrium.
K_7	2.05	6.07	2.96	0.65	3.97	1.93	Calcite dissolution with probable dolomite under-saturation
K_8	0.87	3.67	4.22	0.39	1.46	1.68	dolomite dissolution with Calcite supersaturation.
K_9	3.99	2.89	0.72	1.0	2.90	0.72	Calcite dissolution with dolomite precipitation.

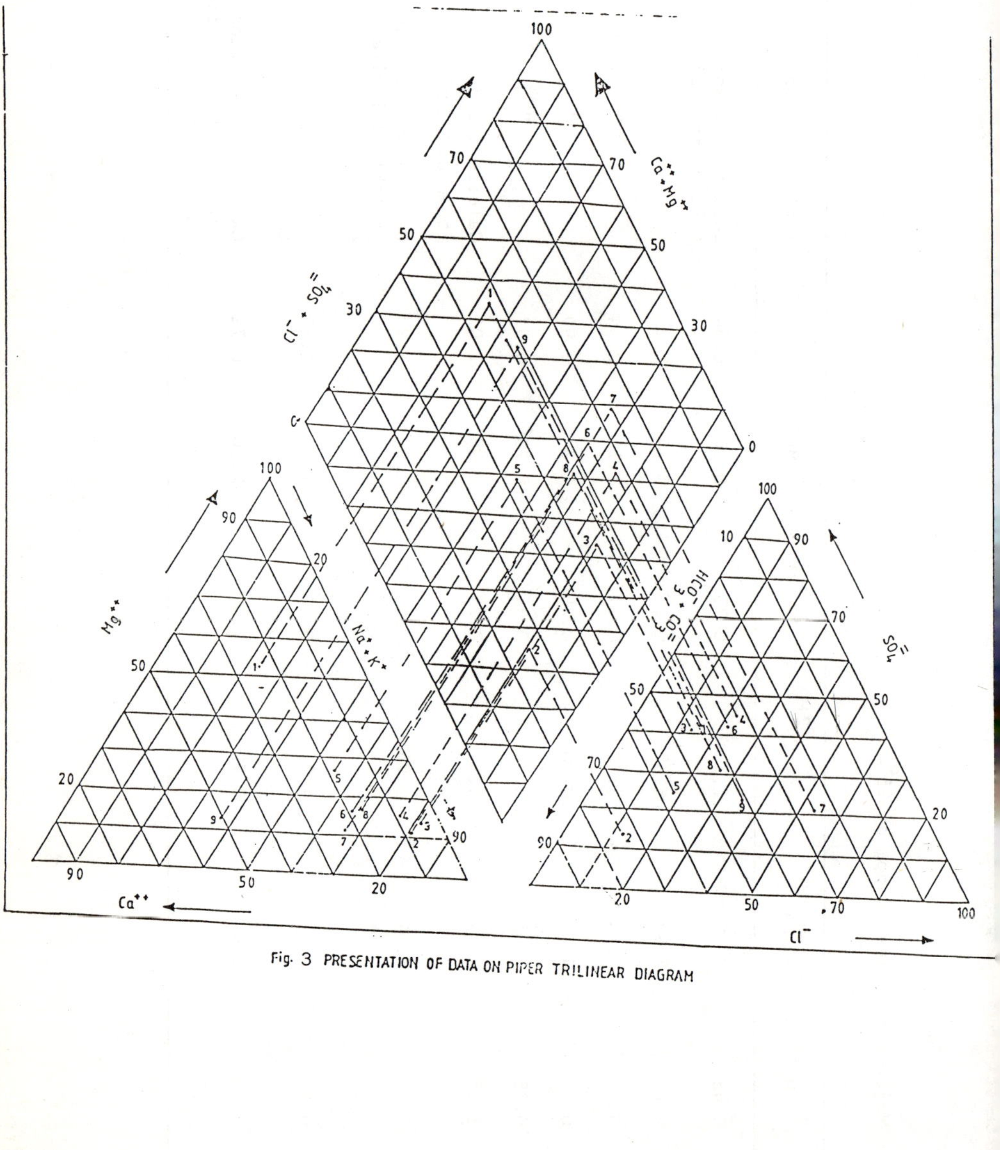

Fig. 3 PRESENTATION OF DATA ON PIPER TRILINEAR DIAGRAM

Water quality

The quality of shallow groundwater in the area has been reviewed for domestic and irrigation purposes. The guidelines as reported by WHO (1963) and US salinity diagram (HEM 1986) have been compared for recommending the suitability of the water for the purposes. It has been observed that the shallow groundwater is suitable for both purposes with the exception of K_4 where high sulphate concentration beyond allowable limits by WHO may be undesirable.

CONCLUSION

The area is dominated by soft and hard waters close to the beach sands and raised reef limestone respectively. The beach sands are sandwitched between the reef limestone. Dissolution of dolomitic limestone, cation exchange, and mixing of fresh recharge water with interstitial marine water may be the probable processes determining the chemical character of the shallow groundwater. Finally results show that the water is suitable for both domestic and agricultural purposes.

REFERENCES

Back,W., 1966 Hydrochemical facies and groundwater flow patterns in northen part of Atlantic coastal plain U.S. Geol. Surv. Prof. Paper 498-C, 16p.

Back,W., and Freeze, R.A., 1983 Chemical Hydrogeology. Benchmark papers in Geology 73 Hutchinson Ross Publ. Comp 416p.

Bargman, D.J., 1970. The climate of Dar es Salaam, Tanzania Notes and Rec. 71p. 56 - 64.

Bartholomew, H., 1963. Quarter degree sheet 186 Dar es Salaam Geological Survey of Tanganyika.

Canadian International Development Agency, 1979., Coast and Dar es Salaam Water Master Plan Vol. 1. IPI - 98.

Freeze, R.A., and Cherry, J.A., 1979. Groundwater Prentice Hall Inc. 603p.

Greenberg, A.E., Trussel, R.R., Clesceri, L.S., (Ed) 1985. Standard methods for the examination of water and wastewater, 16th edition 162p.

Hem, J.D., 1986. Study and interpretation of the chemical characteristics of natural water. 3rd edition. United states Geological survey Water supply paper 2254. United states Government Printing Office Washington 264p.

Koert W. and F. Tornau 1910. Zur geologie and hydrologies Von Dar es Salaam Und. Tanga (Deutch-Ostafrika) Abh. Preuss geol Lande Saust. 63.

Mnzava, L. Joel.,1986. Distribution; Chemistry and Origin of saline groundwater in Dar es Salaam, M.Sc. thesis (Unpub.) Department of Geology,University of Dar es Salaam.

Captions to Figures

Figures	Explanation
I	Location map of the study area.
2	Geological map of the study area.
3	Piper trilinear diagram.
K_1-K_9	Sampling sites
K_1	Kigamboni ferry
K_2	Kigamboni Msikitini
K_3	Lutheran Church
K_4	Kiwanjani
K_5	Amani Klab
K_6	Ubaka
K_7	Tungi
K_8	Madizini
K_9	Magogoni

GROUNDWATER QUALITY MONITORING-STATISTICAL MODELLING

Dr.K.S.V.Basivi Reddy,
Principal
Kakatiya Institute of Technology & Science, Warangal
India 506 015

Dr.M.Pandu Ranga Rao
Professor of Engineering Geology
Regional Engineering College, Warangal
India 506 004

SYNOPSIS

Water quality data obtained from Warangal Urban agglomeration which is a hard rock terrain was subjected to multiple regression analysis. 58 wells were monitored for 5 seasons and each sample was analysed for 13 parameters.

This analysis has been used to suggest models for predicting water quality. On the whole, it appears, that sodium either independently or in association with potassium, magnesium, calcium and bicarbonates are the causal variables for the determination of total dissolved solids. It is also seen that chlorides, either independently or in association with bicarbonates and magnesium appears to be important variables for the contribution of TDS. With the help of these models it is possible to predict the water quality in any water given one predictor value such as specific conductivity which in turn clearly indicates TDS.

A regression problem considers the frequency distribution of one variable when another is held fixed at each of several levels. A correlation problem considers the joint variation of two measurements, neither of which is restricted by the experiment. Correlation is a process by which the degree of association between samples of two variables X and Y is defined. The correlation coefficient is a mathematical definition of that association. The end product of the process of correlation is the correlation coefficient; it is not an equation.

1.MULTIPLE LINEAR REGRESSION

The 'regression' model assumes that some variable 'Y' responds to changes in other 'X' variables. The 'Y' variable is the quantity under study and is known as 'response' or 'dependent' variable. The X variables are those which exhibit a causal effect on the value of the Y and are known as the 'explanatory' or 'independent' variables.

The model is expressed by

Y = bo+b1x1+....bkyk

Where bo, b1 .... bk are the least squares estimators of the unknown parameters, bo is the intercept, while b1.... bk are regression coefficients. They are chosen in such a way as to minimise the squared sum of the residuals or deviations from the estimated line.

The major issues in the development of this model are;

a) The identification of those variables which have significant and separte effects on the dependent variables.
b) The model must not only provide good statistical fit to the present day data but must also be of a logical and meaningful form.
c) The variables must be meaningful in explaining the dependent variable behaviour.

With such an equation developed, it is possible to develop future levels of the dependent variable given future predictor indicators. The adequacy of the model can be tested by the Analysis of variance approach.

The total sum of squares (SST) is decomposed into regression sum of squares (SSR) and error sum of squares (SSE)

SST = SSR + SSE

The multiple correlation coefficient R.

R = SSR/SST

This indicates the degree of association between independent variable and the dependent variable. It varies between 0 and 1. Closer to 0 is worse but closer 1 is better. The significance of R is that its square R2 is approximately the decimal fraction of the variation of 'Y' accounted by independent variables xi, i.e. if R2 =0.941 then 94.1% of the total variance in the data is explained by the model.

2.'F'TEST

The regression sum of squares can be used to give some indication concerning whether or not the model is an adequate explanation of the true situation. One test is the F ratio.

$$F = \frac{SSR/k}{SSE(n-k-1)}$$ at k, n-k-1 degrees of freedom

From F tables at prescribed confidence level the value of F can be known. F calculated must be more than F tables in which case the variation in Y is explained and is not by chance.

3.'t'TEST

The t-statistic indicates the significance (or not) of the regression coefficient of each independent variable. Independent variables which have a 't' value of less than the table 't' value at the degrees of freedom do not have a significant relationship with the dependent variable and therefore, contribute nothing to the equation. If t3 calculated for a parameter3 is 2.543 and t at 90% level at (n-k-1) degrees of freedom is 3.36 from tables, i.e. t3 calculated is less than t tabulated, the coefficient does not significantly differ from zero. Hence, variable a3 can be dropped and other combinations tried.

PRECAUTIONS:

The following precautions are to be taken in the development of linear regression models.

a) Independent variables should not be intercorrelated
b) All variables should be capable of clear interpretation and measurable
c) The size of regression intercept in relation to the mean dependent variable (y) is to be small.
d) Signs must be logical

4. DEVELOPMENT OF LINEAR REGRESSION MODELS

In the present investigation the water samples were analysed to determine 13 different chemical properties. Sodium, Calcium, Potassium, Magnesium, Chlorides, Bicarbonates & Carbonates, Sulfates, Nitrates, Fluorides, hardness, total dissolved solids (TDS), sum of ions and pH.

Since TDS is the single parameter which could reflect the influence of all the dissolved constituents it is desirable to develop a model by means of which TDS could be predicted or computed given all or a few of the twelve independent chemical constituents. In other words it is hypothesised that

$$TDS = fé(Na, Ca, K, Mg, Cl, (HCO_3+CO_3), SO_4, NO_3, F)¿$$

There could be several types of functional forms but multiple linear regressions is the most powerful and easily explainable model available in the literature. For successful development of such a model the causal variable, must be truly independent of each other as explained earlier. For this purpose factor analysis was performed on the test battery and it was found that there are four basic dimensions which are truly uncorrelated.

It is now proposed to utilise the result in selecting such variables which are truly uncorrelated with a view to develop a number of regression models for possible selection. All the possible combinations were tried in this case. Regression models were developed with the help of a FORTRAN package and the models were examined for R2, F, T and intercept statistics. Those models which do not satisfy any of these statistics are considered as poor and hence rejected.

It was interesting to note that none of the models which contained Nitrates, Sulphates, pH were found to be satisfactory. As a result of this experience it was decided to repeat the programme with a new set of independent variables. The tables containing these programmes are lengthy and are not included in view of the space restrictions in the paper. The recommended models based on final Multiple Regression Analysis are as follows;

June 1981: (Premonsoon and Summer)

1.TDS = 4.54 Na + 406 say 4.5 Na + 400
2.TDS = 3.-3 Cl + 379 say 3 Cl + 380
3.TDS = 4.56 Na + 35.3 K + 203 say 4.5 Na + 35 K + 200
4.TDS = 4.52 Na + 21.6 Mg + 7.8 say 4.5 Na + 22 Mg + 8
5.TDS = 4.4 Na + 6.28 Ca + 105 say 4.4 Na + 6.25 Ca +100
6.TDS = 2.97 Cl + 1.5 HCO3-72.8 Say 3 Cl + 1.5 HCO3-75

October 1981: (Postmonsoon)

1.TDS = 1.86 Na + 17.86 Mg + 416 say 1.80 Na + 18 Mg + 400

December 1981: (Winter)

1.TDS = 3.28 Na + 414 say 3.3 Na + 400
2.TDS = 2.15 Cl + 457.7 say 2.15 Cl + 460
3.TDS = 3.04 Na + 4.31 Ca + 243 say 3 Na + 4.3 Ca + 250
4.TDS = 2.06 Cl + 1.2 HCO3 + 160 say 2 Cl + 1.2 HOC3 + 160
5.TDS = 3.22 Na + 14.22 Mg + 185 say 3.2 Cl + 14 Mg + 190

February 1982 : (Late Winter)

1.TDS = 3.9 Na + 280
2.TDS = 2.4 Cl + 380
3.TDS = 2.16 Hardness + 300
4.TDS = 3.75 Ca + 17.5 Mg + 40
5.TDS = 3.5 Na + 5.5 Ca

May 1982: (Summer)

1.TDS = 4.150 Na + 380
2.TDS = 2.5 Cl + 435
3.TDS = 3.75 Na + 5.5 Ca + 7
4.TDS = 3.75 Na + 5.5 Ca + 7
5.TDS = 2.5 Na + 1.6 HCO3-60

In order to predict water quality throughout year, attempts are made for conducting regression analysis on the data from 290 samples collected throughout the year over five seasons and the following models have been found satisfactory.

TOTAL ANNUAL DATA:

1.TDS = 3.25 Na + 480
2.TDS = 2.25 Cl + 480
3.TDS = 3.2 Na + 18.2K + 340
4.TDS = 3.00 Na + 5.75 Ca + 140
5.TDS = 6.40 Mg + 2.20 Cl + 380
6.TDS = 2.20 Cl + 0.35 HCO3 + 380
7.TDS = 3.15 Na + 18.5 Mg + 175
8.TDS = 6 Mg + 21.5 Cl + 0.334 HCO3 + 295

On the whole, it appears, that sodium either independently or in association with K, Mg, Ca and bicarbonates are the causal variables for the determination of TDS variable. Alternatively chlorides, either independently or in association with bicarbonates and Mg also appear to be the important variables for the contribution of TDS in the water with the help of these models. It is now possible to predict TDS in any water given one predictor value .

One important application of this analysis is that the entire water quality can be predicted through a single simple test like specific conductance which in turn clearly indicates the TDS.

DIGITAL MODELLING OF THE HYDROGEOCHEMICAL FACIES OF GROUNDWATERS IN TAMILNADU, INDIA.

[1]A.BALASUBRAMANIAN, [1]R.THIRUGNANA SAMBANDAM, [1]R.CHELLASAMY & [2]J.C.V.SASTRI

[1]Department of Geology, V.O.C.College, Tuticorin-8, India.
[2]Department of Geology, University of Mysore, Mysore-6, India.

ABSTRACT

Identification, classification and interpretation of hydrogeochemical facies are normally attempted through graphical representations. Any developmental planning of groundwater resources, needs a complete picture/model of the major ion characteristics for deriving all the quality parameters. Digital modelling of hydrogeochemical facies involves both numerical and statistical computations of hydrochemical data. Factor, Principal component and multiple regression analysis have been adopted to develop the deterministic and predictive hydrochemical models of groundwaters of Tamilnadu. The deterministic model could show the principal chemical components of aquifers, their pattern of distribution and the residence time of groundwaters in aquifers. Results of the predictive model could reduce the cost of chemical analysis of sequential water samples normally being carried out in a groundwater system.

INTRODUCTION

The identification, classification and comparison of hydrogeochemical facies of groundwater have so far been done by means of graphical representations (Piper, 1944; Stiff, 1951; Back, 1961; Zaporozec, 1972). Trilinear plots have been extensively used to classify the groundwaters (Back, 1960, 61) and show the regional trends in groundwater quality (Back and Hanshaw, 1965) of one or more groups of samples. Any developmental planning of groundwater resources needs a complete picture or model of the major ion characteristics of water. This could be obtained only through modelling the hydrogeochemical facies.

Recently, numerical analysis of hydrochemical data have been attempted to determine the geochemical patterns of groundwater by Lawrence and Upchurch (1976, 82) and Balasubramanian (1986). In this study, attempts have been made to develop the deterministic and predictive models for two river basins of Tamilnadu using digital data processing techniques.

FIG.1 FLOW CHART OF NUMERICAL DATA PROCESSING TECHNIQUES

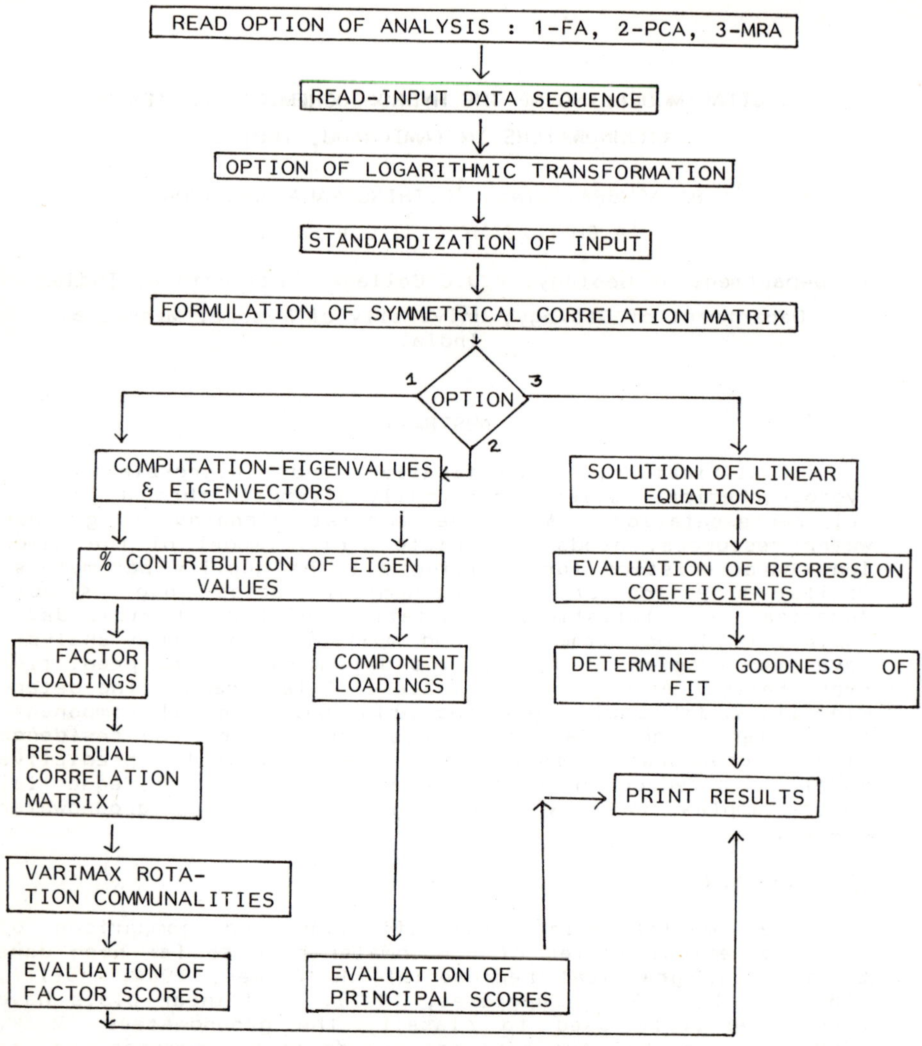

STUDY AREAS

The hydrogeochemistry of groundwaters collected from two river basins, viz., (1) Tambraparni and (2) Vaigai river basins of Tamilnadu have been utilized in this study. Tambraparni river basin is located between Long. 77°10' to 78°10'E and Lat. 8°25' to 9°13'N (5382 Sq.km. area), in the Tirunelveli and Chidambaranar districts. Major portion of the basin exposes charnockites and gneisses with a small patch of Tertiary sandstones and shell-limestones in the coastal belt. Groundwater occurs under watertable conditions in the crystalline terrain and under confined conditions in the coastal belt.

The Vaigai river basin, bounded between Long. 77°15' to 79°0'E and Lat. 9°15' to 10°25'N, lies in parts of Madurai, Ramnad and Anna districts of Tamilnadu. This basin also comprises of charnockites and Khondalites of Archaean age except for the coastal track where sedimentary rocks of Gondwana age overlies the Archaeans.

PROCEDURES

Factor Analysis (FA), Principal Component Analysis (PCA) and Multiple Regression Analysis (MRA) are the widely used statistical or numerical concepts in parametric classifications and modelling studies. Detailed mathematical descriptions of these methods could be seen in Davis (1974).

A general outline of the data processing operations performed during this study has been shown as a flowchart in Fig.1. The details of hydrogeochemical data taken for the present study are listed below.

Variables considered	Number of samples	Processing Analysis	River Basin
Ec, pH, Ca, Mg, Na+k, HCO_3, CO_3, Cl, NO_3, SO_4	53	Factor Analysis	Tambraparni
-do-	-do-	Multiple Reg. Analysis	-do-
Ec, Ca, Mg, Na+k, HCO_3+CO_3, Cl, SO_4	67	Multiple Reg. Analysis	Vaigai
Ca, Mg, Na+k, HCO_3+CO_3, Cl, SO_4	-do-	Principal Component Analysis	-do-

In FA and PCA, the dominant factors and principal components are extracted from the loadings matrix and the number of factors and components to be considered are determined based on an acceptable Eigen value, which normally is greater than 1.0. The scores (Factor scores in FA; Component Scores in PCA) scored by individual samples with reference to the extracted factors or components can give a clear picture about the dominant chemical ions of groundwaters over a basin. Models developed using the positive scores could be used for deterministic purposes.

In groundwater samples, Ec (in Mmhos) is mainly dependent on the concentrations of major ions like Ca, Mg, Na+k, HCO_3, CO_3, Cl, NO_3 and SO_4. The functional relationship could be taken as $Ec = f(Ca, Mg, Na+k, HCO_3, CO_3, Cl, NO_3, SO_4)$ and the resulting regression coefficients of a data matrix could be used for the development of a model. A simple conversion step has been adopted to in this study to compute the chemical parameters of a known value of the dependent variable, i.e., Ec.

RESULTS AND DISCUSSIONS

The dominant factors and principal components of groundwaters of the two river basins, extracted from the loadings matrix are

Basin	Extracted Parameters	Chemical Parameters	Cumulative % of variance
Tambraparni	FACTOR I	Cl, Ec, Ca and TDS	61.03
	FACTOR II	CO_3, pH, SO_4 and (Na+k)	75.47
	FACTOR III	HCO_3, (Na+k) and TDS	85.06
Vaigai	P.Component I	Cl, Mg, (Na+k)	52.50
	P.Component II	SO_4, HCO_3, Mg	72.75
	P.Component III	HCO_3, (Na+k)	87.69

Chebotarev (1955) has given the major ion evolution sequence of groundwater as:

Travel along the flow-path -->

$HCO_3 \rightarrow HCO_3+SO_4 \rightarrow SO_4+HCO_3 \rightarrow SO_4+Cl \rightarrow Cl+SO_4 \rightarrow Cl$

Increasing Age -->

These changes are expected when water moves from shallow zones of active flushing, through intermediate zones where the flow path is very sluggish and the water is old (Freeze and Cherry, 1979). It is very evident from the above table that the dominance of chloride ions in both the basins of crystalline terrain over the other variables of groundwater is due to the prolonged stay of water in the aquifer for a longer time.

The model developed for Tambraparni river basin has been shown as Map 1. The hydrogeochemical model of the Vaigai river basin could be seen in Chellasamy and Balasubramanian (1988).

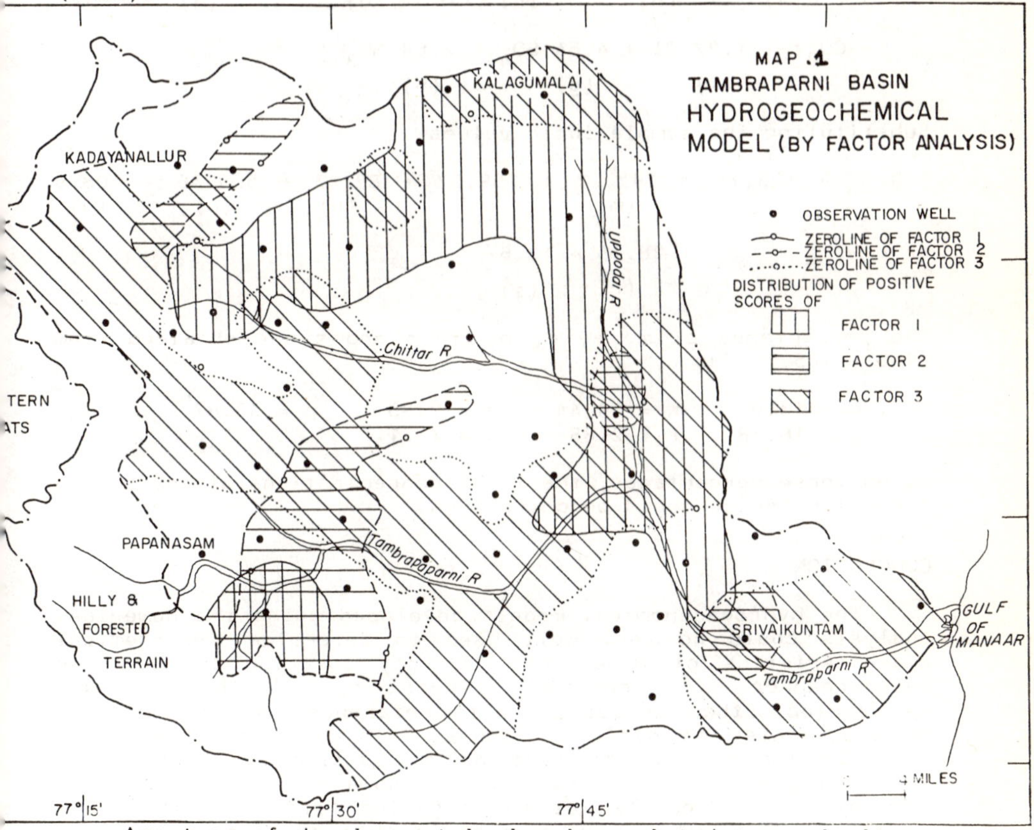

Any type of developmental planning related to agriculture or public water supply could be attempted using these maps.

The results of the multiple regression analysis are shown below.

Vaigai basin : (EC-in Micromhos)

$$EC = -105.75 + 2.22\,Ca + 1.84\,Mg + 1.12\,(Na+k) + 0.94\,(HCO_3 + CO_3) + 1.92\,Cl + 4.54\,SO_4 + 2.64\,NO_3.$$

Goodness of fit (r) = 0.99.

Considering a known value of Ec and its component concentrations, an inverse computation could be made to find out the percentage contribution of each ion as shown below:

$$Ec = -\underset{(a)}{105.75} + \underset{(b)}{2.22\,Ca} + \underset{(c)}{1.84\,Mg} + \underset{(d)}{1.12\,(Na+K)} + \underset{(e)}{0.94\,(HCO_3 + CO_3)} + \underset{(f)}{1.92\,Cl} + \underset{(g)}{4.54\,SO_4} + \underset{(h)}{2.64\,NO_3}$$

Substituting the sample ionic values

$$940 = -105.75 + \underset{(m)}{142.37} + \underset{(n)}{109.57} + \underset{(o)}{82.43} + \underset{(p)}{565.36} + \underset{(q)}{108.9} + \underset{(r)}{28.15} + \underset{(s)}{12.67}.$$

The percentages of m, n, o, p, q, r and s are obtained from (940 + 105.75) as,

m = 13.6 n = 10.44 o = 7.85 p = 53.87
q = 10.38 r = 2.68 s = 1.21.

Using these percentages, the ionic concentrations of any sample for which the Ec could be determined.

CONCLUSION

The hydrogeochemical models developed using the numerical analysis could be very well used to determine the chemical characteristics of groundwater at any location in which an enterpreauner is interested. These could further be used to develop the agriculture with reference to the chemistry of groundwater of different zones. Multiple regression analysis model could be used as a positive predictive tool in finding out the chemistry of groundwater if the dependent variable Ec is measured at any location (either in the field or lab). The methodology adopted in this study could be used for any groundwater system in order to derive better water management practices.

ACKNOWLEDGEMENT

The authors thank their organisations for providing the necessary facilities to carryout this work.

References:

Back, W., 1961. Calcium Carbonate saturation in groundwaters from routine analysis. USGS Wat. Sup. Pap. -1535-D, 14p.

Back, W., 1963. Preliminary results of a study of Calcium Carbonate saturation of groundwaters of Central Florida, Int. Assn. Sci. Hydrol., V.8(3).

Back, W., and Hanshaw, B.B., 1965. Chemical Geohydrology, Adv. Hydroscience, V.1., p.49-109.

Balasubramanian, A., 1986. Hydrogeological investigations in the Tambraparni river basin, Tamilnadu. Unpublished Ph.D. Thesis, University of Mysore, Mysore, 349 p.

Chebotarev, I.I., 1955. Metamorphism of natural waters in the crust of weathering. Geochim. et. Cosmochim. Acta. V.8., p. 22-48, 137-170, 198-212.

Chellasamy, R., and Balasubramanian, A., 1988. Hydrogeochemical studies of Vaigai River basin, Tamilnadu. Proc. Vol. Int. Sem. Hydrol., Andhra University, Waltair, (In Press).

Davis John, C., 1973. Statistics and data analysis in Geology, John Wiley & Sons, New York, 550p.

Freeze, R.A., and Cherry, J.A., 1979. Groundwater. Prentice Hall. Inc., New Jersey, 604p.

Lawrence, F.W., and Upchurch, S.B., 1976. Identification of geochemical patterns in groundwater by numerical analysis, Adv. Groundwater. Hydrol. AWWA Pub., p.199-214.

Lawrence, F.W., and Upchurch, S.B., 1982. Identification of recharge areas using geochemical factor analysis. Groundwater, V.20(6), p.680-687.

Piper, A.M., 1944. A graphic procedure in the geochemical interpretation of water analysis, Am.Geophys.Union. Trans, V.25, p.914-923.

Stiff, H.A.Jr., 1951. The interpretation of chemical water analysis by means of patterns. Jl.Petrol.Tech., 3(10), Techn. Note - 84, p.15-16.

Zaporozec, A., 1972. Graphical interpretation and water quality data. Groundwater, V.10(2), p.32-43.

ANALYSIS OF BREAKTHROUGH CURVES FOR ASSESSING MOVEMENT OF CONTAMINANTS

S.K.Gupta

Central Soil Salinity Research Institute,Karnal-132001 (India)

ABSTRACT

Transport of chemicals in porous media has been a subject of considerable importance. In the last few decades, it has received much attention, as well documented studies on pollution of ground water resources through non-point sources in agriculture have been released. Movement of chemicals, in ground water pollution problems is mostly studied utilizing breakthrough curves. A refinement to such presentation is made on the basis of analysis of breakthrough curves of linearly adsorbing ions. The theory is developed to present dimensionless leached fractions and parameter identification equations. Examples from field and laboratory studies have been used to assess the theory proposed in this study. It is opined that a switchover to leached fractions is possible and worthwhile. Mathematical models suggested in the paper could refine predictions needed to evaluate the ground water quality problems. These are better tools to analyze and predict the quantity of solutes entering the water resource system. They provide a disciplined format for assessing the relevant coefficients for use in future studies on modelling. Finally, the approach is consistant with existing equations and can be easily incorporated in the modelling of ground water quality monitoring and evaluation programmes.

INTRODUCTION

Use of chemicals in agricultural production programmes has come to stay. It is now understood fairly well that the use of such chemicals might not always be beneficial in the absence of proper management of the applied chemicals. Studies have established that chemicals used in agriculture are potential agents of ground water pollution and are considered non-point source. One of the major goals of any water quality evaluation and monitoring programme is to develop and/or verify models. This should provide a basis for technical approach for both the problem identification and solution development phase of the programme.

The modelling of chemicals has been subject of numerous investigations. Although these models elucidate processes and provide insight into the solute transport, yet major factors correlated with contaminant transport particularly which display dispersion and other mass transport mechanisms are not fairly well understood (Pandey and Gupta,1984; Seyfried and Rao, 1987). Sometimes breakthrough curves hide more than what these reveal. Therefore, systematic analysis of breakthrough curves to get maximum information on solute transport behaviour and processes is desirable.

Objective of the present paper is to provide a method to analyse breakthrough curves so as to predict total salt/contaminant moving to ground water and to estimate solute transport parameters. Paper also describes the applications and concludes that the modified equations could be adjusted in existing modelling framework for operational use.

THEORY

One dimensional dispersion convection equation for the movement of an adsorbing solute in a homogeneous porous media at constant moisture content is given by

$$D\frac{C}{x} - V\frac{C}{x} - Q = \frac{C}{t} \qquad (1)$$

here D is the hydrodynamic dispersion coefficient (L/T); C is the instantaneous concentration of the effluent; x is the space (L) and t is the time co-ordinate respectively; V is the average pore water velocity (L/T) and Q is the adsorption term. Equation(1) for an equilibrium controlled process and for a solute which gets adsorbed linearly on the solid matrix is written as:

$$D1\frac{C}{x} - V1\frac{C}{x} = \frac{C}{t} \qquad (2)$$

in which D1 =D/(1+k); V1= V/(1+k); k = BS/f; B is the bulk density and S is the adsorption coefficient which is the slope of the plot between C and C1; C1 is the concentration of the solute adsorbed on the solid matrix. The term (1+k) is often reffered to as the retardation factor and is denoted by R.

The solution for the boundary value problem for the step input case is given by Lapidus and Amundson (1952) as

$$C(x,t) = \frac{Ci}{2}\ \mathrm{erfc}\left(\frac{x-V1\ t}{2\ D1\ t}\right) + \exp\left(\frac{V1\ x}{D1}\right)\mathrm{erfc}\left(\frac{x-V1\ t}{2\ D1\ t}\right) \qquad (3)$$

while equation(4) gives the instantaneous concentration of the solute in the solution phase in space and time, a index of the total solute passing through a given point in space can be obtained by integrating equation(3) with time. Application of Laplace transform to equation(3) yields

$$Si = {}_{0}^{t}\ C(d,t)dt = \frac{Ci}{2}\ (t-d/V1)\mathrm{erfc}\left(\frac{d-V1\ t}{2\ D1\ t}\right) + \exp\left(\frac{V1\ d}{D1}\right)(t+d/V1)\ \mathrm{erfc}\left(\frac{d+V1\ t}{2\ D1\ t}\right) \qquad (4)$$

where Si is an index of total salts drained below depth d.

The fraction of the solutes leached is obtained by dividing equation (4) with the total solutes applied at the soil surface. Thus equation(4) after dividing by (Ci t) becomes

$$La = \frac{1}{2}(1-1/Pl)\,erfc\left(\frac{1-Pl}{2}\left(\frac{B}{Pl}\right)\right) + \frac{1}{2}\exp(B)(1+1/Pl)\,erfc\left(\frac{1+Pl}{2}\left(\frac{B}{Pl}\right)\right) \quad (5)$$

Here Pl is the number of pore volumes drained and are given by Vt/d(1+k) and B is the Peclet number and is given by Vl d/Dl.

To evaluate solute transport parameters dispersion and adsorption in this case, we select two points on the breakthrough curve. For large Pl, we have

$$k = (1-La)P - 1 \quad (6)$$

such that P=Vt/d. The parameter k can also be obtained by the relation

$$k = LA\ P \quad (7)$$

in which LA=La - La at P pore volumes.

Similarly for the second point which is taken at Pl=1, equation(5) is simplifide to

$$La = \exp(B)erfc(\ B) \quad (8)$$

For known La at Pl=1, the value of B can be obtained using trial and error. For large values of B, however, equation(7) can be written in a simple way by expanding the comlimentary error function term and retaining only the first term of the expansion. Thus

$$B = 1/\ \ La \quad (9)$$

MATERIALS AND METHODS

To verify equation(5) and the subsequent parameter identification equations, a two way approach was followed. The first approach consists of carrying out a laboratory experiment in plexiglass columns with chloride and sulphate as the test ions. It is expected that for the soil (Table 1) and chemical in question SO will behave as a weak adsorbent whereas chloride will act as a non-interacting ion.

Laboratory Experiment

An experiment was conducted in plexiglass columns of 45 cm length and 7.5 cm dia. The columns were uniformly packed to 40 cm length. Four columns were prepared to have two replications of each experiment on chloride and sulphate ions. The use of finite lengths of columns in these tests is justified because it is now well established that the diffrences in predicted concentrations with finite and semi-infinite boundary conditions are not significant (van Genuchten and Alves,1982; Gupta,1986). After packing, the columns were saturated with distilled water from the lower end to avoid any air entrapment. Distilled water was passed through the top of the column so as to drain nearly 4 pore volumes to ensure that the columns are solute free. The effluent water was also tested to eliminate any doubt about the presence of ions in the column. A 50 meq/l concentration solution of calcium chloride and calcium sulphate was passed through these columns separately and effluent collected in a beaker which was kept over a magnetic stirrer so as to shake the effluent continuously. Water samples from the stirred solution were taken at frequent intervals and the particular ions were measured. Chloride was estimated with titration against silver nitrate and sulphate through barium sulphate precipitate as described by Richards(1954).

Table 1. Physico-chemical characteristics of the soil used in the study

Mechanical analysis	Sand(%)	56.0
	Silt(%)	22.5
	Clay(%)	21.5
Texture	Sandy loam	
pH (1:2)	7.95	
EC(1:2)dS/m	0.29	
CEC(meq/100 gm of soil)	12	

Application Of The Theory To Data From Literature

Because the experiments were conducted only for a single weakly adsorbing ion. the data thus collected was not sufficient to firmly establish the validity of the theory. In order to test the theory more rigorously, data from several experiments involving different ions was collected and used for verification. The ions in question are sulphate (Wood,1978), chloride (Chan et al.1980), phosphate (Gupta and Greenkorn,1974) and Boric acid (Wild and Mazaheri,1979). The only care in selection was to ensure that the ion adsorption isotherm should be linear. The dimensionless concentration curves were used to obtain leached fractions (Fig.1).

RESULTS AND DISCUSSIONS

The rationale behind the use of the leached fraction equation for

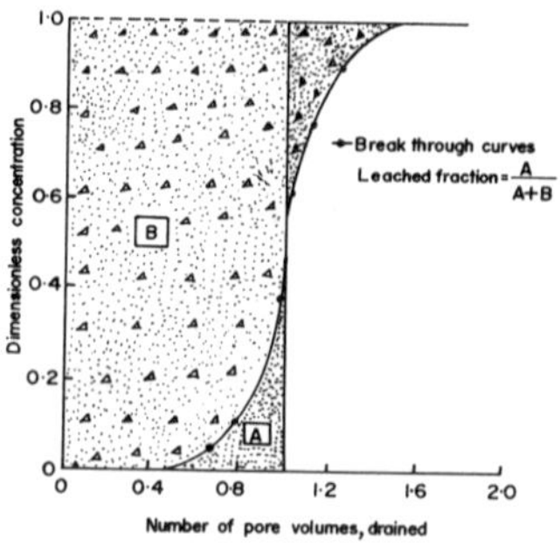

Fig 1. Definition sketch showing the evaluation of leached fractions using breakthrough curves

prediction purposes, apart from its ready applicability in solute balance and ground water pollution studies is highlighted in the definition sketch (Fig.2). It may be seen that whereas the solute concentration curves for adsorbing and non-adsorbing solutes merge at some pore volume, the leached fraction curves remain seperate at all pore volumes. It is because of the adsorption of the part of the solute on the solid matrix. At higher pore volumes the solutes in the solution phase for both the interacting and non-interacting ions are similar in concentration. It is also evident that at higher pore volumes the leached fraction becomes independent of the Peclet number and is a function of pore volume and the adsorption parameter k only. This fact forms the basis of deriving parameter identification equations.

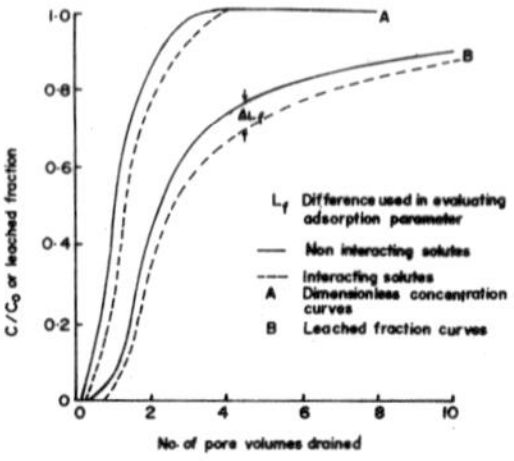

Fig 2. Schematics of dimensionless concentration and leached fraction curves for non-interacting and interacting ions.

Leached fraction curves for the chloride, sulphate and phosphate ions as predicted by equation(5) are reported in Fig.3

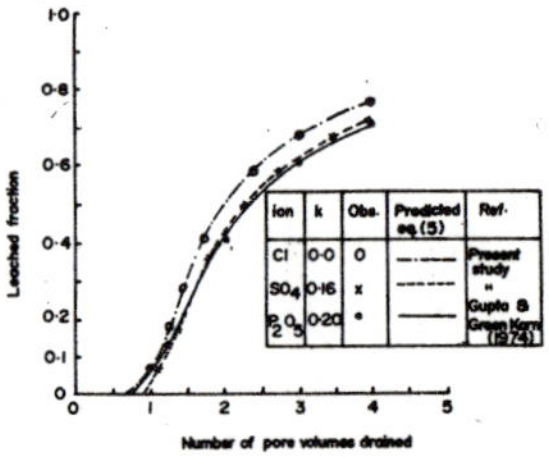

Fig 3. Observed and predicted leached fractions for non-interacting and interacting ions.

The values of the retardation factor and the hydrodynamic dispersion coefficient were obtained using equations (6) and (9) respectively for the chloride and sulphate ions. For phosphate ions the value reported by Gupta and Greenkorn (1974) were used. Leached fractions for the chloride and sulphate ions were obtained by direct measurement of the outflow while for the phosphate ion it was obtained utilizing the breakthrough curve. The close agreement between the predicted and observed values validates the applicability of equation (5) for use in solute transport studies.

Equations (6) and (7) are used to evaluate the adsorption parameter, k, for various ions. The data presented in Table 2 establishes the validity of the procedure. Out of seven values reported in the Table, five values closely agree with the reported values obtained through other procedures mainly through trial and error such that breakthrough curve is predicted well. To test the other two cases breakthrough curves were predicted utilizing the k values. A close agreement between the observed curve and the predicted curve is apparent (Fig.4). The value of hydrodynamic dispersion coefficient used was the one reported by Chan et al (1980)

Once the value of k is known, any of the procedures adopted in evaluating dispersion coefficient in the non-interacting solute transport can be modified to estimate the dispersion parameter. For example equation (3) has been used to suggest that D can be evaluated by estimating the slope of the breakthrough curve by drawing a tangent at the point given by (1+k) pore volumes(Gupta and Pandey,1978). However,if

Table 2 Observed and predicted values of the adsorption parameter

Test ion	La	P	Value of k		Reference
			Predicted	Reported	
Sulphate	0.713	4.0	0.148	0.150	This paper
Phosphate	0.700	4.0	0.200	0.196	Gupta & Greenkorn(1974)
Sulphate	0.060	2.676	0.161	0.160	Wood(1978)
Boric acid	(1)0.566	3.6	0.922	0.90 to 1.0	Wild and Mazaheri(1979)
	(2)0.322	3.6	1.439	1.45 to 1.60	
Chloride	0.420	2.0	0.016	*	Chan et al. (1980)
Chloride	0.460	2.0	0.080	*	Chan et al. (1980)

*Test in Fig.4

the leached fractions at (1+k) pore volumes are known use of equations (8) or (9) can directly be made to estimate this parameter. A comparative statement of the Peclet numbers obtained using differnt

Table 3. Comprison of predicted and reported values of the Peclet number for linearly adsorbing solutes

Test ion	Peclet number			Reference
	Trial and error	Slope method	Method of leached fraction	
Chloride (1)	106.1	-	106.4	Chan et al. (1980)
(2)	108.0	-	113.2	
Sulphate	73.0	71.1	74.7	This Paper
Phosphate	21.5	20.8	22.1	Gupta and Greenkorn(1974)

methods is given in Table 3. A close agreement again confirms that the theory proposed in this paper is promising.

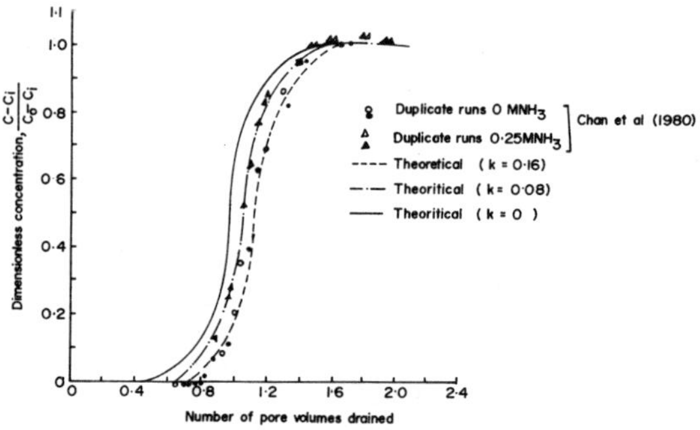

Fig 4. Observed and predicted breakthrough curves utilizing predicted adsorption parameter.

APPLICATIONS

Equation (5) is a variation of the solution of the dispersion convection model most often derived and used as the basis for ground water quality models. In the new form, an estimate of threats to ground water could be directly evaluated as it gives the fraction of the total solute that will contribute to ground water loading. Significantly, lack of comprehensive data base currently limits the application of models to field problems. It is in this direction that present improvements are significant as the true parameters can directly be evaluated from field data. The tests revealed that these equations estimated true pore volume which is more than the apparent pore volume (pore volume at which dimensionless concentration is 0.5) and is less than the absolute pore volume (estimated by independently measured porosity). Moreover, two parameters can be estimated simultaneously and no additional tests are needed to estimate adsorption coefficient. The method also dispenses with the need to assemble a breakthrough curve in field studies for parameter identification. The difficulties in assembling a breakthrough curve are too well known.

More important amongst many advatages which the leached fraction offer are that proposed changes in no way force any major change in the existing framework of field management models applied to ground water quality and monitoring programmes. It appears that it is a formidable tool and should find wide applications in future programmes.

REFERENCES

Chan, K.V., Goering, H.R., and Devey, G.B., 1980. Movement of chlorides in a soil with variable charge properties and chloride systems. J. Environ. Quality. Vol.9, p. 679-82.

Fried, J.J., and Combarnous, M.S., 1972. Dispersion in porous media. In Advances in Hydro science. Vol. 7, p. 169-82

Gupta, S.K., 1986. Salt movement and dispersion-convection equation: material balance concept. Bull. No. 9. Central Soil Salinity Research Institute Karnal. p. 48.

Gupta, S.K. and Pandey, R.N., 1978. Discussion and note on determining dispersion and adsorption coefficients. Ground Water. Vol. 16, 453-54.

Gupta, S.P. and Greenkorn, R.A., 1974. Detarmination of dispersion and non-linear adsorption parameters for flow in porous media. Water Resous. Res. Vol. 10, 839-46.

Lapidus, L. and Amundson, N. R., 1952. Mathematics of adsorption in beds. VI. The effect of longitudinal diffusion in ion exchange and chromatographic columns. J. Phy. Chem. Vol. 56, 984-88.

Pandey, R.S. and Gupta, S.K.,1984. Analysis of breakthrough curves: effect of mobile and immobile pore volumes. Aust. J. Soil Res. Vol. 22, P.23-30.

Richards, L.A.(Ed.), 1954. Diagnosis and improvement of saline alkali soils. U.S.D.A. Handbook 60, P.160.

Seyfried, M.S. and Rao, P.S.A., 1987. Solute and transport in undisturbed columns of an aggregated tropical soil: Preferential flow effets. Soil Sci. Soc. Am. J. Vol. 51, 1434-44.

van Genuchten, M.Th. and Alves, W.J., 1982. Analytical solutions of the one dimensional convective- dispersive solute transport equation. U.S.D.A. Tech. Bull. 1661. p.151

Wild, A. and Mazaheri, A., 1979. Pradiction of leaching rate of boric acid under field conditions. Geoderma. Vol. 22, 127-36.

Wood, W.W., 1978. Use of laboratory data to predict sulphate sorption during artificial ground water recharge. Ground Water, Vol. 16, 22-31.

DISPERSION OF POLLUTANTS IN POROUS MEDIA FLOW
ONE-DIMENSIONAL APPROACH

by
D. G. Paygude
College of Engineering, Pune

B. Vasudeva Rao
and
S. G. Joshi
Department of Civil Engineering
Indian Institute of Technology
Powai, Bombay - 400 076

SYNOPSIS :

The work presented herein deals with the numerical solution of the One-Dimensional form of the Convective-Dispersion Equation for the case of saturated homogeneous isotropic porous media flow using the Galerkin's Finite Element Formulation. The results obtained by using the numerical solution procedure are compared with those of the available analytical solution. Also the stability of these numerical results has been studied in terms of the dimensionless parameters Courant Number and Peclet Number.

INTRODUCTION :

With the rapid growth in urbanaization and industrialization, the possibilities of the contamination of both the surface water and ground water sources are rapidly increasing. The ground water sources which were being considered as relatively free from contamination, are gradually, becoming degraded as a result of transport of soluble chemicals via the rainwater percolating into the subsoil and ultimately meeting the ground water. Disposal of solid wastes and sewage on land, agricultural activities, disposal of liquid wastes originated from industries and disposal of radioactive waste are the major sources of ground water contamination. Remedial meassures are absolutely essential to prevent this contamination. If not detected in time, it may result in the damage of the aquifers beyond repair.

PROCESS OF CONTAMINANT TRANSPORT :

Various contaminants polluting the ground water sources are in the form of soluble chemicals or dissolved substances called solutes. There are other forms of solutes namely the natural constituents or minerals and artificial tracers. The transport of these solutes occur through an aquifer by convection and hydrodynamic dispersion. Convection or advection is the process by which the solutes are transported by the bulk motion of the flowing ground water. Hydrodynamic dispersion involves mechanical or hydraulic dispersion and molecular diffusion. As a result of convection and dispersion, the solute is transported by the ground water with varying degree of concentration, throughout the domain of flow. The differential equation describing the physical processess of convection and hydrodynamic dispersion known as Convective-Dispersion Equation is basically obtained from the statement of principle of conservation of mass as given under :

NET RATE OF CHANGE OF MASS OF SOLUTE WITHIN AN ELEMENTAL VOLUME = FLUX OF SOLUTE OUT OF ELEMENT - FLUX OF SOLUTE INTO THE ELEMENT +/- LOSS OR GAIN OF SOLUTE MASS DUE TO INTERNAL REACTIONS (1)

Convection and hydrodynamic dispersion are physical processes controlling the influx and efflux of the solute, whereas adsorption and radioactive decay are chemical or biochemical processess that cause the loss or gain of the solute mass.

OBJECTIVES OF THE PRESENT INVESTIGATION :

Looking into the need for the control of the ground water quality, it is necessary for a ground water hydrologist to study the engineering aspects of the problem of contaminant transport such as distribution of solute concentration in the aquifer, estimation of the level of contamination and planning of suitable remedial measures to prevent further contamination. From this point of view, the objective of the present study has been the solution of the differential equation describing the process of solute transport, that is, the solution of the dispersion equation, for the case of saturated, homogeneous isotropic porous media. Because the straight forward analytical solution for the most general form of this equation is not possible, a numerical solution using the Finite Element Method (FEM) has been attempted and the digital model of the equation is developed for the case of one-dimensional form. As the analytical solution is available for this case, the results from the digital model have been compared with the analytical solution. To simplify the problem, adsorption and radioactive decay were not considered in this study.

The problem of contaminant transport has been studied extensively during the last three decades. In the earlier part of this period, the investigations were mainly concerned with the theoretical development and the experimental work, whereas in the later part, attention seems to be focussed mainly on the numerical solution of the problem using numerical techniques. Bear [2] and Scheldegger [11] presented the general theory of dispersion in the porous media flow. Bachamat and Bear [1] described general equations governing the hydrodynamic dispersion of a homogeneous fluid in a homogeneous and isotropic medium. Dagan [4] solved the one-dimensional dispersion equation using perturbation technique. Guymon [7] gave an equivalent variational principle to the governing partial differential equation of one-dimentional diffusion-convection and developed a finite element solution requiring approximation in space domain only. The analytical solutions were developed by Marino [8] for two longitudinal dispersion problems in saturated porous media.

BASIC DIFFERENTIAL EQUATION GOVERNING CONTAMINANT TRANSPORT :

The basic differential equation describing the process of solute transport in saturated, homogeneous isotropic porous media and accounting for the effects of convection, hydrodynamic dispersion adsorption and radioactive decay can be derived from the principle of conservation of mass, Equation 1 and written as [1, 6] :

$$n\frac{\partial C}{\partial t} + (1-n)\frac{\partial S}{\partial t} = \frac{\partial}{\partial x_i}\left[nD_{ij}\frac{\partial C}{\partial x_j} - u_i C\right] - \lambda\left[nC + (1-n)S\right] + q C^* \quad \cdots (2)$$

The flow through the porous medium is assumed to be steady and obeys Darcy's law. Assuming linear equilibrium adsorption, the concentration in the solid phase can be expressed as

$$S = k \cdot C \qquad (3)$$

As a result of the above assumptions, Equation 2 takes the following form

$$R\frac{\partial C}{\partial t} - \frac{\partial}{\partial x_i}\left[D_{ij}\frac{\partial C}{\partial x_j} - \frac{u_i}{n}C\right] + \lambda R C - \frac{q}{n}C^* = 0 \qquad (4)$$

In which R, the retardation factor is given as

$$R = [\, 1 + k\,(1-n)/n \,] \qquad (5)$$

For the case of one-dimensional dispersion, the Equation 4 takes the following form

$$R\frac{\partial C}{\partial t} - D_{xx}\frac{\partial^2 C}{\partial x^2} + \frac{1}{n}\frac{\partial}{\partial x}(u_x \cdot C) + \lambda R C - \frac{q}{n}C^* = 0 \quad ..(6)$$

FINITE ELEMENT FORMULATION OF BASIC DIFFERENTIAL EQUATION GOVERNING CONTAMINANT TRNSPORT :

Let Equation 6 be written as

$$L\,(C) - P = O \qquad (7)$$

and the trial solution of Equation 7 be written as

$$C'(x,t) = [N]\{C\} = \sum_{i=1}^{n} N_i\,(x)\,C_i\,(t) \qquad (8)$$

where $N_i(x)$, $(1=1,2,3,........n)$ are the interpolation functions satisfying the boundary conditions and C_i (t) are the unknown coefficients which become the nodal values of cocentrations. The approximate functions C' (x,t) will be exact solutions of Equation 7 only if the error or the residual ϵ given as

$$\epsilon = L\,(C') - P \qquad (9)$$

Is equal to zero. This condition is equivalent to the requirement of orthogonality of the error ϵ to the interpolation functions N_i (x), that is,

$$\int N_i\,(x)\,[L(C') - p\,]\,dx = 0 \qquad (10)$$

Substituting the operator L and simplifying, the finite element formulation of Equation 6 takes the form :

$$\int_D \Big[R N_i \frac{\partial}{\partial t}\sum_{j=1}^{n} N_j C_j + D_{xx}\frac{\partial N_i}{\partial x}\frac{\partial}{\partial x}\sum_{j=1}^{n} N_j C_j - \frac{u_x}{n}\frac{\partial N_i}{\partial x}\sum_{j=1}^{n} N_j C_j$$

$$+ \lambda R N_i \sum_{j=1}^{n} N_j C_j\Big]dx = D_{xx}\frac{\partial C}{\partial x}N_i + \int \frac{q}{n}C^* N_i\,dx \;;(i=1,n) \quad ..(11)$$

Equation 11 can be written in matrix form as

$$[P]\dot{C} + [K]C = \{F\} \quad (12)$$

in which [P] and [K] are nxn square matrices and {F} is nx1 column vector such that

$$p_{ij} = \int_D R\, N_i\, N_j\, dx \quad (12a)$$

$$k_{ij} = \int \left[D_{xx} \frac{dN_i}{dx}\frac{dN_j}{dx} - \frac{u_x}{n}\frac{dN_i}{dx} N_j + \lambda R N_i N_j\right] dx \quad (12b)$$

$$f_i = \left[D_{xx}\frac{\partial C}{\partial x} - \frac{u_x}{n}\right] N_i + \int_D \frac{q}{n} C^* N_i\, dx \quad (12c)$$

The time derivative $\partial C/\partial t$ denoted as $\{\dot{C}\}$ in the above equation and can be approximated by using an Implicit Finite Difference Scheme as

$$\{\dot{C}\} = [\{C\}^{t+\Delta t} - \{C\}^t]/\Delta t \quad (13)$$

Also for better accuracy of results, the column vector {C} in the term [K] {C} can be expressed as an average of its values at time steps t and t +Δt as

$$\{C\} = [\{C\}^{t+\Delta t} + \{C\}^t]/2 \quad (14)$$

Substituting for {C} from equations 13 and 14 respectively, the Equation 12 takes the form

$$\left(\frac{1}{\Delta t}[P] + \frac{1}{2}[K]\right)\{C\}^{t+\Delta t} = \left(\frac{1}{\Delta t}[P] - \frac{1}{2}[K]\right)\{C\}^t + \{F\}^t \quad (15)$$

Equation 15 thus represents the general assembly of the set of Finite Element Equations. Solution of Equation 15 after incorporating the initial and boundary conditions, yields the unknown values of concentration C at the nodal points at any time instance when the values at the earlier time instance are known. In order to obtain the global assembly of the set of FEM equations, for the entire domain as given by Equation 15, the element matrices are first assembled for individual elements and then properly superposed. In the present study, the element matrices are obtained by using the methods of numerical and direct integration schemes and the global assembly was done in a compact form by using Sparse Matrix Techniques.

INITIAL CONDITIONS :

For the disperstion problem, the initial concentration distribution over the entire domain under consideration is assumed to be zero. Mathematically, the condition can be written as

$$C(x,o) = 0, \text{for } x >= 0 \quad (16)$$

BOUNDARY CONDITIONS :

Following forms of upstream boundary conditions given by Marino [8] are assumed in the present study.

Case I	:	$C(o,t) = C_o\,;\ t>0$	(17)
Case II	:	$C(0,t) = C.\ \mathrm{Exp}\,(\gamma t);\ t>0$	(18)
Case III	:	$C(o,t) = C\,[1 - \exp(-\gamma t)]; t>0$	(19)

SOLUTION OF SYSTEM OF EQUATIONS :

The set of simultaneous equations in compact matrix form was solved by using Gauss-Seldel iterative scheme. This method is particularly suitable to the unsymmetric matrices when stored in compact form using Row-Oriented method.

ANALYTICAL SOLUTIONS :

Marino [8] has given the following form of analytical solution to the one-dimensional dispersion for the above cases of upstream boundary conditions.

Case I:

$$C(x,t) = \frac{C_0}{2}\left[erfc\left(\frac{x-ut}{2\sqrt{Dt}}\right) + exp\left(\frac{ux}{D}\right) erfc\left(\frac{x+ut}{2\sqrt{Dt}}\right)\right] \quad \ldots(20)$$

Case II:

$$C(x,t) = \frac{C_0}{2} \cdot exp(\gamma t)\left\{exp\left[\frac{x(u-\psi)}{2D}\right] erfc\left(\frac{x-\psi t}{2\sqrt{Dt}}\right) + exp\left[\frac{x(u+\psi)}{2D}\right] erfc\left(\frac{x+\psi t}{2\sqrt{Dt}}\right)\right\} \quad \ldots(21)$$

Case III:

$$C(x,t) = \frac{C_0}{2}\left\{erfc\left(\frac{x-ut}{2\sqrt{Dt}}\right) + exp\left(\frac{ux}{D}\right) erfc\left(\frac{x+ut}{2\sqrt{Dt}}\right) - exp(-\gamma t)\left[exp\left(\frac{x(u-\phi)}{2D}\right) erf\left(\frac{x-\phi t}{2\sqrt{Dt}}\right) + exp\left(\frac{x(u+\phi)}{2D}\right) \times erfc\left(\frac{x+\phi t}{2\sqrt{Dt}}\right)\right]\right\}$$

in which $\psi = (u^2 + 4D\gamma)^{1/2}$ and $\phi = (u^2 - 4D\gamma)^{1/2}$ (22)

In the above solutions, the adsorption and radioactive decay are not taken into consideration.

DISCUSSION OF RESULTS :

The domain under study is 600 meters long has been descretised as shown in Fig. 1 by using a simple line element as shown in Fig 1a with linear interpolation function. The results have been obtained for all the possible combinations of the element size and time step in order to study the variation of FEM results in terms of the dimensionless parameters, Courant Number and Peclet Number.

Figs. 2 to 4 show the graphs of relative concentration against distance for a typical element length of 6.25 m and for various time steps and Figs.5 to 7 shown the similar graphs for a typical time steps of 1 day and for various element sizes in respect of the three boundary conditions stated above in Equations 17 to 19. The results of the analytical solution have superimposed on these graphs.From the graphs, it is observed that the FEM results compare well with those of the analytical solution. For oscillation free numerical solution with regard to temporal and spatial distribution respectively, the criteria for Courant Number and Peclet Number in respect of one-dimensional dispersion are specified by Daus and Frind [5] as given below :

$$C\,N_x \leqslant 1.0 \quad \text{and} \quad P\,N_x \leqslant 2.0$$

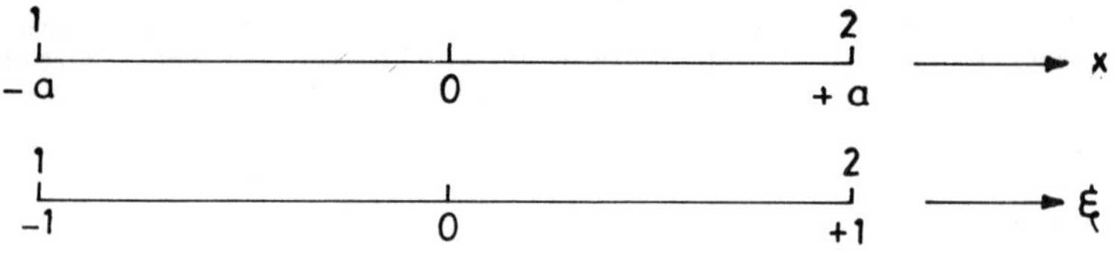

FIG 1.a – LINE ELEMENT

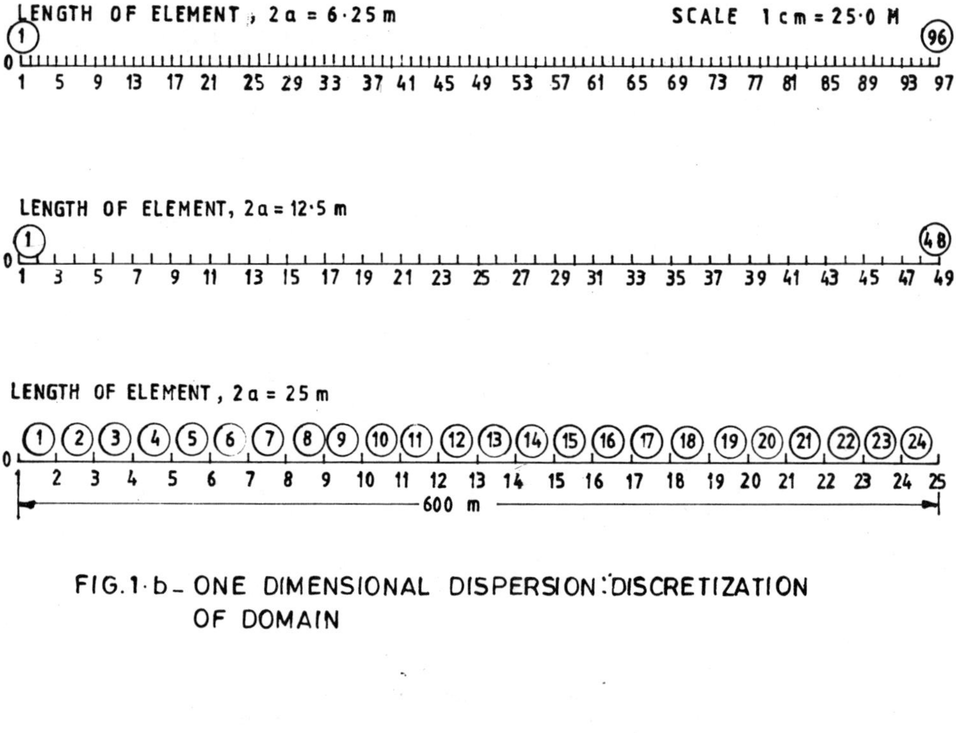

FIG.1·b – ONE DIMENSIONAL DISPERSION: DISCRETIZATION OF DOMAIN

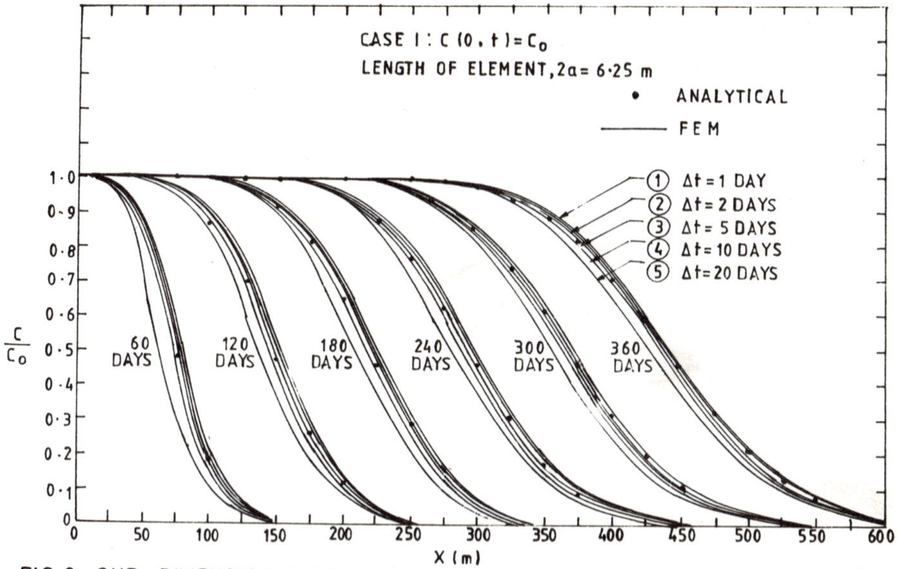

FIG. 2 _ ONE-DIMENSIONAL DISPERSION : COMPARISON OF FEM AND ANALYTICAL RESULTS (TIME STEP VARIATION)

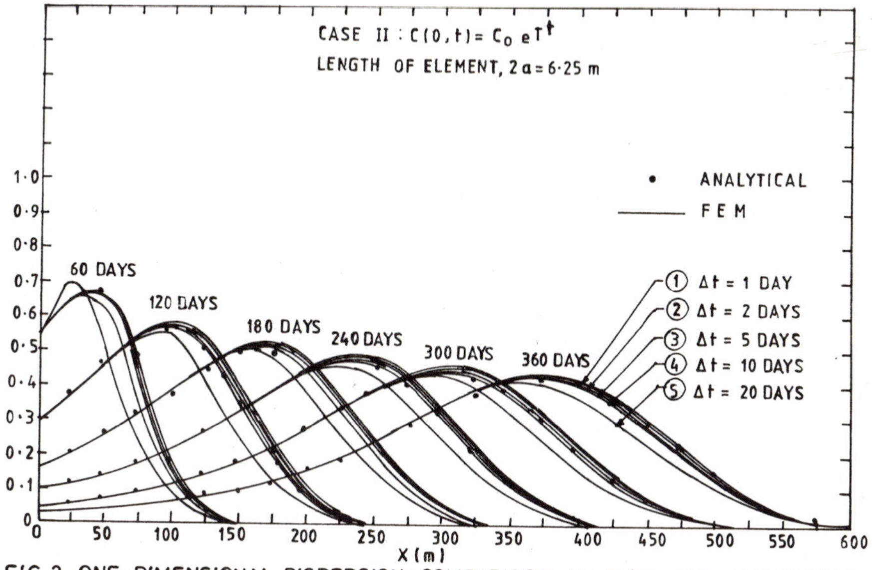

FIG. 3 _ ONE-DIMENSIONAL DISPERSION : COMPARISON OF FEM AND ANALYTICAL RESULTS (TIME STEP VARIATION)

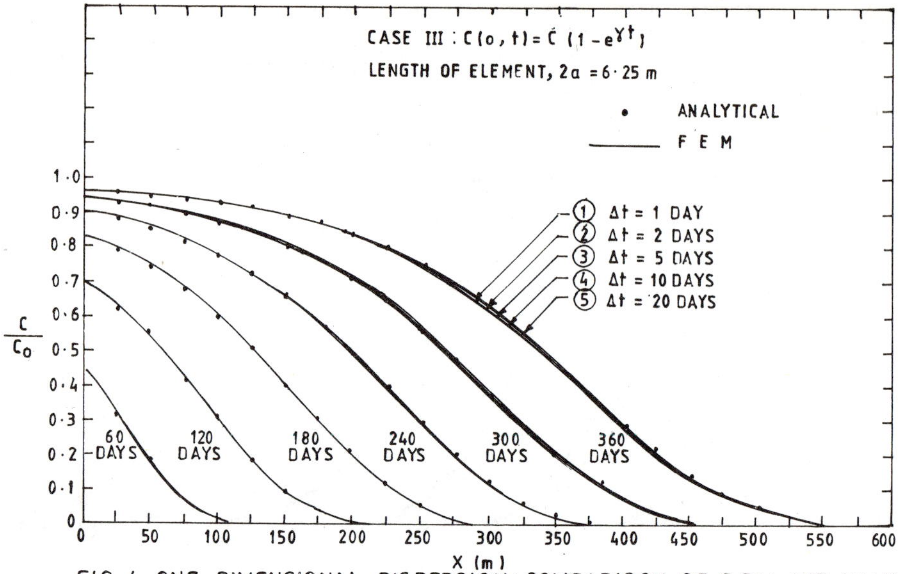

FIG. 4 _ONE-DIMENSIONAL DISPERSION: COMPARISON OF FEM AND ANALYTICAL RESULTS (TIME STEP VARIATION)

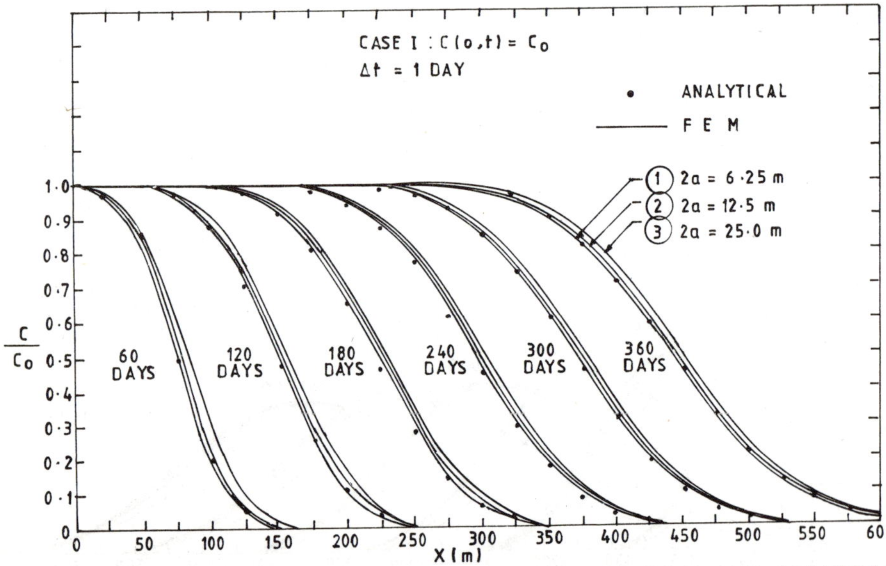

FIG. 5 _ONE DIMENSIONAL DISPERSION: COMPARISON OF FEM AND ANALYTICAL RESULTS (ELEMENT SIZE VARIATION)

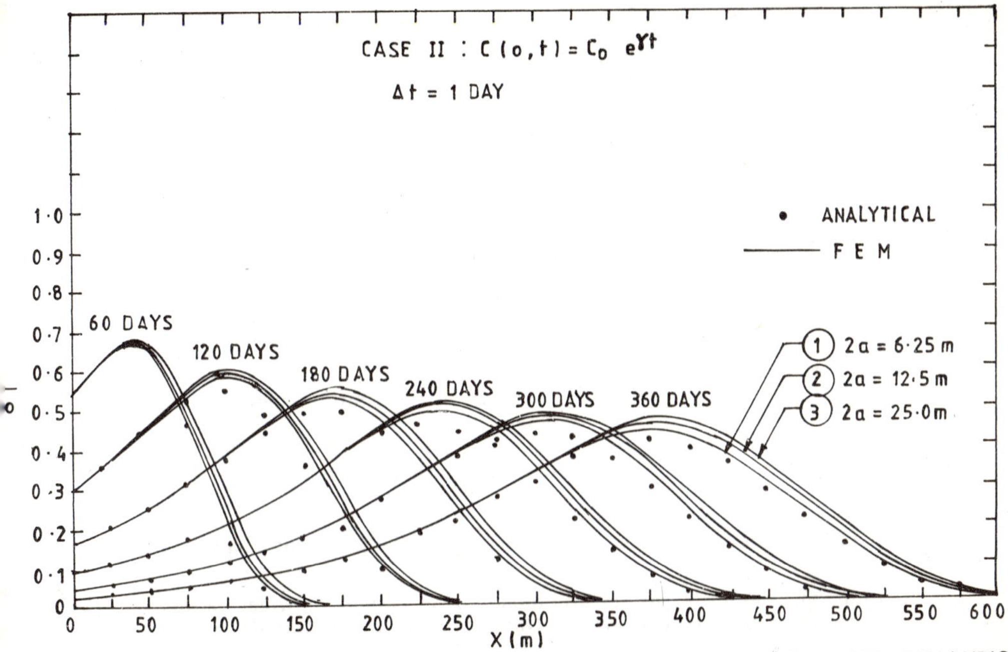

FIG. 6_ ONE-DIMENSIONAL DISPERSION: COMPARISON OF FEM AND ANALYTICAL RESULTS (ELEMENT SIZE VARIATION)

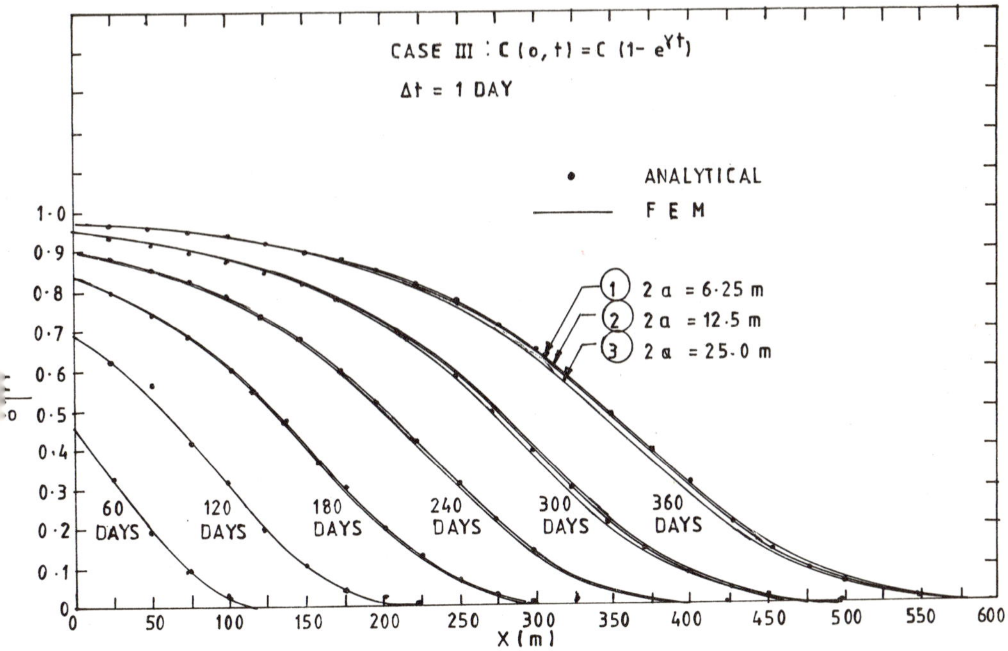

FIG. 7_ ONE-DIMENSIONAL DISPERSION: COMPARISON OF FEM AND ANALYTICAL RESULTS (ELEMENT SIZE VARIATION)

where $C\ N_x$ and $P\ N_x$ are defined by the following equations :

$$\text{Courant Number, } C\ N_x = u_x\, \Delta t / \Delta x \qquad (23)$$

$$\text{Peclet Number, } P\ N_x = u_x\, \Delta x / \ D_{xx} \qquad (24)$$

In the present results, the Courant Number ranges from 0.05 to 3.9 and Peclet Number ranges from 1 to 4. Even for this wider ranges of numbers, the FEM results are found to be stable and oscillation free. The FEM results closely follow the analytical solution results when the Courant Number is of the order 0.5.

CONCLUSIONS :

The major aim of the study was to develop the finite element digital model for the solution of the problem of one-dimensional dispersion in saturated porous media flow and compare the FEM results with those of the analytical solution. As a result of the investigation to this effect, following conclusions can be drawn based on the numerical results.

1. The FEM results are in close agreement with those of the analytical solution.
2. One-dimensional FEM results are found to be stable for wider ranges of Courant Number and Peclet Number than those stated earlier.

In addition to the above conclusions, following observations have been made :

1. There was about 70 to 80 % saving in the memory core of the computer when a compact storage scheme to store the Sparse matrix was used.
2. There was a considerable saving in the CPU time when an element matrix was formulated byusing direct integration instead of numerical integration.

ACKNOWLEDGEMENT

The authors gratefully acknowledge the financial support form the Department of Atomic Energy, Government of India, to carryout this work as part of the Research Project entitled "Some Studies on the Dispersion of Chemical Nuclides in the Saturated Porous Media Flow Using Finite Element Method".

NOMENCLATURE :

[A]	-	Global assembly matrix
[A1]	-	1 th element of element matrix
a	-	Half element length
C_*	-	Concentration of dispersing mass in liquid phase
C^*	-	Concentration of source fluid
C_o	-	Input concentration
C/Co	-	Relative Concentration
{C}	-	Column vector of nodal values of C
$\{\dot{C}\}$	-	Column vector of nodal values of $\partial C/\partial t$

CN_x	-	Courant Number
D,DL,Dxx	-	Longitudinal dispersion coefficient
Dij	-	Hydrodynamic dispersion tensor
$\{F\}$	-	Right Hand Side column vector or force vector
f_l	-	l th element of the force vector
$[J]$	-	Jacobian matrix
$[K]$	-	Stiffness matrix
K_{ij}	-	Element of stiffness matrix in the ith row and jth column
k	-	Constant of proportionality between concentrations in solid and liquid phases
Ni	-	Interpolation functions or basis functions
n	-	Porosity of the medium, also the number of nodal points
$[P]$	-	Capacitance matrix
P_{ij}	-	Element of the capacitance matrix in ith row and jth column.
PNx	-	Peclet Number
q	-	Discharge rate of the source fluid
R	-	Retardation factor
S	-	Concentration of dispersing mass in solid phase
t	-	Time elapsed since start of computations
u, ux	-	Darcy velocity
u_i	-	Darcy velocity component in the ith direction
x_i x_j	-	Spatial coordinates
x,y,z	-	cartesian coordinates
α_L	-	Longitudinal dispersitivity
Δt	-	Time step
ϵ	-	Tolerance criterion in the computations
γ	-	Coefficient in connection with boundary conditions
λ	-	Radioactive decay coefficient
ϕ	-	Field variable
ξ	-	Local coordinate
erfc(x)	-	1 - erf(x)
erf(x)	=	$\frac{2}{\sqrt{\pi}}\int_0^x \exp(-y^2)\,dy$

REFERENCES

1. Bachamat, Y. and Bear, J. 1964. The General Equation of Hydrodynamic Dispersion in Homogeneous, Isotorpic Porous Mediums, journal of Geophysical Research, Vol. 69, No. 12, June 15, pp. 2561-2567.

2. Bear, J. 1961. "On the Tensor Form of Dispersion in Porous Media", journal Geophysical Research, Vol. 66, No. 4, April, pp. 1185-1197.

3. Bear, J. 1972. "Dynamics of Fluids in Porous Media", American Elsevler Environmental Science Series, New York.

4. Dagan, G. 1967. "Hydrodynamic Dispersion in a Non-homogeneous Porous Column, Journal of Geophysical Research, Vol. 72, No. 16, August, pp. 4075-4080.

5. Daus, A.D. and Find, O.E. 1985. "An Alternating Direction Galerking Technique for Simulation of contaminant Transport in Complex Ground Water Systems", Water Resources Research, Vol. 21, No. 5, May, pp. 653-664.

6. Freeze, R.A. and Cherry, J.A. 1979. "Groundwater", Prentice Hall Inc. Englewood Cliffs.

7. Guymon, G.L. 1970. "A Finite Element Solution of the One-Dimensional Diffusion-Convection Equations", Water Resources Research, Vol. 6, No. 1, Feb. pp. 204-210.

8. Marino, M.A. 1974. "Longitudinal Dispersion in Saturated Porous Media", journal of the Hydraulics Division, ASCE, Vol. 100, No. HY1, January, pp. 151-157.

9. Paygude, D.G. 1986. "Dispersion of Pollutants in Saturated Porous Media Flow", Ph.D. Thesis, Department of Civil Engineering, Indian Institute of Technology, Bombay.

10. Rao, B. Vasudeva; Joshi, S.G. and Paygude, D.G. 1986. "Some Studies on the Dispersion of Chemical Nuclides in the Saturated Porous Media Flow by Using Finite Element Method", Report Submitted to the Board of Research in Nuclear Studies, Department of Atomic Energy, Government of India, Bombay from the Department of Civil Engineering, Indian Institute of Technology, Bombay.

11. Scheidegger, A.E. 1961. "General Theory of Dispersion in Porous Media", Journal of Geophysical Research, Vol. 66, No. 10, October, pp. 2373-3278.

12. Zienkiewicz, O.C. 1977. "The Finite Element Method", McGraw-Hill Book Co. New York.

NUMERICAL MODELLING OF MASS TRANSPORT EQUATION

M. Banerjee,

Department of Geophysics,
Banaras Hindu University,
Varanasi 221 005 (U.P.)

ABSTRACT

A numerical finite difference model has been presented for solving the convective-diffusive-dispersion equation for mass transport through a rigid isotropic and homogeneous porous medium. The set of finite difference equations for the transport equation and boundary conditions have been presented together with the stability criteria. Computer experiments have been carried out for salt disposition within a profile of a porous material.The results for mass concentration within a profile during infiltration and redistribution processes have been presented, which exhibit a lag in salinity front and salt peak. The dispersion effect is found to be more pronounced during redistribution event.

INTRODUCTION

The general vertical mass transport equation for a porous medium involves convection, molecular diffusion and hydrodynamic dispersion; this is a cumbersome hydrodynamic problem which needs to be solved only by a numerical technique (Bear, 1972; Fried and Combarnous, 1971).

In the past, many experiments on salt movement through porous media have shown assymetrical effluent distributions (Biggar and Nielsen, 1967) which can be predicted with the convective dispersive equation. The non-linear adsorption studies also showed such assymetrical behaviour (Green et al., 1972 and Davidson et al., 1972). A complex modelling system for unsteady state movement of non-reactive salt described by Bresler and Hanks (1969) and Warrick et al. (1971) which considers varying velocities, and a mathematical system described by Dagan (1967) which takes a varying density in one direction into consideration. Generally, in most unsteady state studies the redistribution phase has been neglected. Swanson and Dutt (1973) pointed out that there is a need for hysteresis numerical models.

In the present paper the detail of the finite difference numerical model for mass transport through a rigid (incompressible), isotropic and homogeneous porous medium has been presented, alongwith computer experiments for mass disposition within a soil column of 3.5 m length.

MASS TRANSPORT EQUATION

The flow equation for mass transport under unsteady state conditions may be given:

$$\frac{\partial C}{\partial t} = \frac{\partial}{\partial Z}\left[D_z . \frac{\partial C}{\partial Z} \right] - \frac{\partial (V_z . C)}{\partial Z} \tag{1}$$

where: t = the time, C = the mass concentration, D_z = the combined dispersion coefficient, V_z = the average pore velocity, and Z = the depth (positive upwards).

The expression for D_z, V_z and D_L may be written as follows:

$$D_z = D_m + D_L \tag{2}$$

$$D_L = a . V_z \tag{3}$$

and

$$V_z = \frac{q}{\theta} \tag{4}$$

where: D_m = the molecular diffusion coefficient, D_L = the longitudinal dispersion in the direction of flow, q = the volumetric flux of water, θ = the volumetric fluid content, and a = the dispersivity.

Combining the equations (2) and (3), we obtain

$$D_z = D_m + a . V_z \tag{5}$$

The surface flux (F_s) of the mass can be written as

$$F_s = - D_z \left(\frac{\partial C}{\partial Z}\right) + V_z . C \tag{6}$$

NUMERICAL METHOD

Finite Difference Approximations :

The eqn (1) together with the boundary and initial conditions in eqn (6), and eqns (23) through (26) can be solved numerically using a finite difference model of Crank-Nicholson's type (Smith, 1965).

The Taylor series expansion in forward difference form can be written as:

$$C_{j,k+1} = C_{j,k} + \Delta t.\frac{\partial C_{j,k}}{\partial t} + \frac{(\Delta t)^2}{2!}.\frac{\partial^2 C_{j,k}}{\partial t^2} + O(\Delta t)^2 \qquad (7)$$

where: j's and k's indicate the space and time steps, respectively.

The discretization of t and Z has been achieved in a manner such that

$$Z = j.\Delta Z; \qquad j = 0, 1, 2, \ldots, N; \qquad (8)$$

and

$$t_k = \sum_{s=1}^{k} (\Delta t)_s \qquad k \geqslant 1; \qquad (9)$$

For flow to be convective type, the eqn (1) is expressed as

$$\frac{\partial C}{\partial t} = -\frac{\partial(V_z . C)}{\partial Z} \qquad (10)$$

Differentiating eqn (10) w.r.t. space and time and we get

$$\frac{\partial^2 C}{\partial t^2} = V_z^2 . \frac{\partial^2 C}{\partial Z^2} + (V_z . \frac{\partial V_z}{\partial Z} - \frac{\partial V_z}{\partial t}) . \frac{\partial C}{\partial Z} \qquad (11)$$

Combining eqns (1), (7) and (11), we obtain the linearized form of the finite difference expression as

$$S_j . C_{j-1,k} + T_j . C_{j,k} + U_j . C_{j+1,k} = W_j \qquad (12)$$

where:

$$S_j = X_{j,k} + Y_{j,k} \cdot \frac{\Delta Z}{2} \qquad (13)$$

$$T_j = \frac{1}{r} - 2.X_{j,k} \qquad (14)$$

$$U_j = X_{j,k} - Y_{j,k} \cdot \frac{\Delta Z}{2} \qquad (15)$$

$$W_j = \frac{1}{r} \cdot C_{j,k+1} \qquad (16)$$

where:

$$X_{j,k} = (D_z)_k + \frac{\Delta t}{2} \cdot (V_z)^2_{j,k} \qquad (17)$$

$$Y_{j,k} = (V_z)_{j,k} - \frac{\Delta t}{2} \cdot (V_z)_{j,k} \cdot \frac{(V_z)_{j+1,k} - (V_z)_{j-1,k}}{(2.\Delta Z)} - \frac{\Delta t}{2} \cdot \frac{(V_z)_{j,k+1} - (V_z)_{j,k-1}}{(2.\Delta t)} \qquad (18)$$

and

$$r = \frac{\Delta t}{(\Delta Z)^2} \qquad (19)$$

In eqn (1) other derivatives $(\frac{\partial C}{\partial Z})$ and $(\frac{\partial^2 C}{\partial Z^2})$ are obtained using central difference approximations. The coefficient of the set of simultaneous equations for a given time increment for a tridiagonal matrix. For this type of matrix, the technique of matrix inversion known as Gauss elimination can be set up as simple, rapidly accomplished recursive computation (Ricktmyer and Morton, 1967; Peaceman and Rachford, 1955; and others).

Therefore, in general the solution of eqn (12) is

$$h_i = P_i - Q_i \cdot H_{i+1} \qquad (20)$$

where:

$$P_i = \frac{W_i - S_i \cdot P_{i-1}}{T_i - S_i \cdot Q_{i-1}} \tag{21}$$

and

$$Q_i = \frac{U_i}{T_i - S_i \cdot Q_{i-1}} \tag{22}$$

where: $i = 1, 2, \ldots, N-1, N$.

Initial Condition:

The initial condition for mass distribution in the profile may be expressed as:

$$C_{(Z,0)} = C_j(Z) \tag{23}$$

where: $C_j(Z)$ is the initial mass concentration at various depths.

Boundary Conditions:

The flux of mass during infiltration event is written as:

$$F_{(0,t)} = q_{(0,t)} \cdot C_0\ (t) . \frac{1}{\theta}; \quad \text{for} \quad 0 \leqslant t < T \tag{24}$$

where, C_0 is the mass concentration at $Z = 0$, and T is the time period during which flux was applied. Now, during the absence of flux of mass (during redistribution event),

$$F_{(0,t)} = 0; \qquad \text{for} \quad T \leqslant t < 2T \tag{25}$$

Generally, the mass concentration does not change with time after some depth, say $Z = L$. Therefore, the mass concentration at various times becomes

$$C_{k+1} = C_{k-1} \qquad \text{for} \quad t > 0 \tag{26}$$

Now, for the derivation of boundary condition preparation equation, we can write using eqns (1), (6), (7) and (11) as:

$$\frac{C_{2,k+1}}{\Delta t} = \frac{C_{2,k}}{\Delta t} + \frac{1}{\Delta Z}.\ [(D_z)_{2,k}.\frac{C_{3,k} - C_{1,k}}{2.\Delta Z} - V_z.C_{2,k} - F_s]$$

$$+ \frac{\Delta t}{2}.\ [V_z^2.\ \frac{C_{3,k} - 2C_{2,k} + C_{1,k}}{\Delta Z^2} + (V_z.\ \frac{(V_z)_{3,k} - (V_z)_{1,k}}{2.\Delta Z}$$

$$- \frac{(V_z)_{2,k+1} - (V_z)_{2,k-1}}{2.\Delta t}).(\frac{C_{3,k} - C_{1,k}}{2.\Delta Z})] \qquad (27)$$

Arranging eqn (27) in the form,

$$C_{1,k} = A.\ C_{2,k} + B.\ C_{3,k} + D \qquad (28)$$

where:

$$A = \frac{1}{2M.\Delta Z}.\ [\ \frac{2}{r} - 2.\Delta Z.\ V_z - 2.\Delta t.\ V_z^2\ ] \qquad (29)$$

$$B = \frac{1}{2.M}.\ [\ \frac{2}{\Delta Z}.\ E_{2,k} + \Delta t.\ G_{2,k} - ((V_z)_{2,k+1} - (V_z)_{2,k-1})] \qquad (30)$$

$$D = -\frac{1}{M.\Delta Z}.\ [\ \frac{1}{r}.\ C_{2,k+1} + \Delta Z.\ F_s\ ] \qquad (31)$$

and

$$E_{2,k} = (D_z)_{2,k} + \frac{\Delta t}{4}.\ V_z.\ [(V_z)_{3,k} - (V_z)_{1,k}] \qquad (32)$$

$$G_{2,k} = \frac{V_z^2}{\Delta Z} + \frac{(V_z)_{2,k+1} - (V_z)_{2,k-1}}{2.\Delta t} \qquad (33)$$

$$M = \frac{1}{2.\Delta Z}.\ E_{2,k} - \frac{\Delta t}{2}.\ G_{2,k} \qquad (34)$$

The first tridiagonal equation is

$$S_2.\ C_{1,k} + T_2\ .\ C_{2,k} + U_2.\ C_{3,k} = W_2 \qquad (35)$$

Substituting eqn (28) we have

$$S_2.[A.C_{2,k} + B.C_{3,k} + D] + T_2.C_{2,k} + U_2.C_{3,k} = W_2 \tag{36}$$

When rearranged eqn (36) becomes

$$(T_2 + S_2.A).C_{2,k} + (U_2 + S_2.B).C_{3,k} = W_2 - S_2.D \tag{37}$$

From eqns (35) and (37), the boundary condition preparation equations are

$$T_2 = T_2 + S_2.A \tag{38}$$

$$U_2 = U_2 + S_2.B \tag{39}$$

$$W_2 = W_2 - S_2.D \tag{40}$$

and the solution equation is given by eqn (28).

Stability :

To ensure a stable solution of mass transport equation, two necessary conditions have been obtained following a Neumann stability analysis (Neumann and Richtmyer, 1950). The conditions are:

$$\Delta t \leqslant \frac{\Delta z^2}{2.X} \tag{41}$$

and

$$\Delta t \leqslant \frac{2.X}{Y^2} \tag{42}$$

PHYSICAL STATE OF THE SYSTEM

System Description:

The porous material used in this study is sand, the characteristics of which have been given by Banerjee and Watson (1984). The saturated water content value of sand is

2.92×10^{-1} m^3/m^3 and the corresponding value of the hydraulic conductivity is 4.8×10^{-1} m/hr. A semi-infinite profile of sand of 3.5 m length is used in the analysis with an initial water content of 5.5×10^{-2} m^3/m^3. It has been assumed that initially the salt concentration is absent in the profile. A surface flux condition has been assumed with the infiltration water being applied at 3×10^{-1} m/hr. The salt concentration is of 10 mg/litre. The value of hydrodynamic diffusion coefficient (D_m) for salt movement in sand has been assumed as 3×10^{-8} m^2/hr.

The various scanning hysteresis soil-water characteristic curves have been defined by using the -model of capillary hysteresis (Banerjee and Watson, 1984).

Flux Pattern :

Three cycles of intermittency have been initiated at the surface with the time periods of each infiltration and redistribution events as 25 and 75 min, respectively, thus giving a total simulation period of 300 min (5 hours) in all.

SOLUTION PROCEDURE

The numerical solution for concentration versus depth has been obtained by computing the equations (12) through (22) for appropriate boundary conditions. The boundary condition preparation and solution equations are described by the equations (28) to (34) and equations (38) to (40). The value of D_z has been calculated by eqn (5). The stability criteria given in eqns (41) and (42) have been found to satisfy for smaller time steps and nodespacing of 0.03 m. The programming part is still in a developmental stage and needs more refinement.

RESULTS AND DISCUSSION

The mass concentration-depth profiles at cummulative times of 10 and 25 min (of first cycle) and 225 min (of third cycle) during infiltration event, and the profiles of 50 and 100 min (of first cycle) and 250 and 300 min (of third cycle) during re-distribution event, have been shown in Fig. 1. The figure shows that thesalt, in the semi-infinite sand column, is slowly displaced downward during each event. During infiltration event the salinity front shows a slight steepening of the concentration profiles (the profiles of 10, 25 and 225 min).However,the

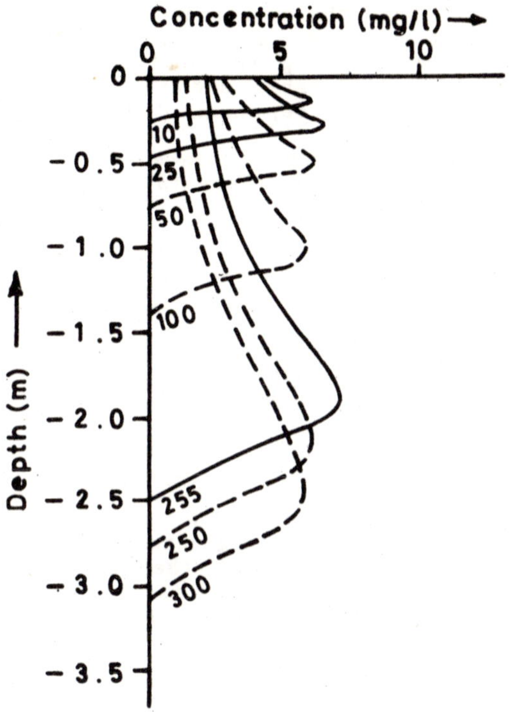

Fig. 1 - Salt concentration-depth profiles during three cycles of intermittency.

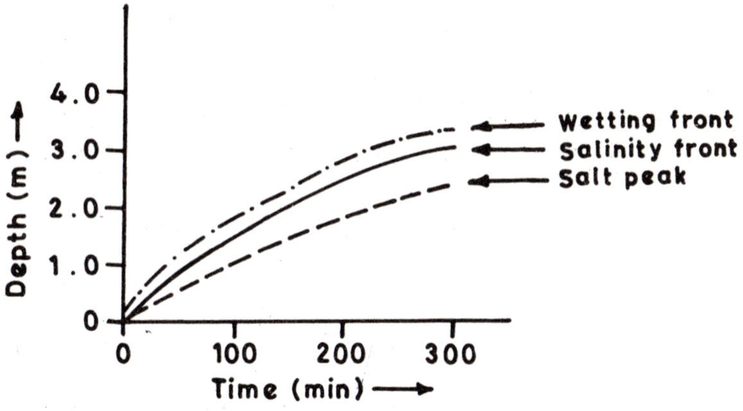

Fig. 2 - Relative disposition of wetting front, salt peak concentration and salinity front with time.

225 min profile shows the magnitude of salt peak increased which may signify that the salt accumulation is taking place during infiltration event. During redistribution event the magnitude of the salt peak diminishes and the front appears at deeper horizons with further steepening of the concentration profiles (see the profiles of 50, 100, 250 and 300 min). The steepening of profiles and occurrence of peaks of decreasing magnitude at greater depths signify that there is a dispersion of salt.

Further, elucidation of the dispersion of salt as it passes through the sand column is facilitated in Fig. 2. For smoothing of the curves the bicubic spline has been employed. The figure clearly indicates that with the increase of time the lag of salt disposition increases. Further it is to be noted that at a particular instant the depth of salinity front is observed relatively at a shallower depth. Similarly the salt peak concentration occurs still at a shalllower depth.

Eventual inference may be that the intermittent mass transport study of this type may give a detail and clear picture of mass distribution alongwith water within the vertical section of a porous medium which may be controlled from the surface by changing the flux pattern. This also enables to have a control over the rate of wetting front and salinity front movements which is of great significance in relation to the recharge of soil water storages and aquifers. Ultimately all these controls would have direct bearings in the modification of soil water and groundwater qualities.

CONCLUSIONS

The studies of the salt movement under intermittent condition indicate the most of the salt is being transported by convection, but dispersion is found to be more significant during each cycle of intermittancy. Also it is found that the salt peak concentration and salinity front have always been lagging behind the wetting front which again increased with time. The shape of the concentration profiles gets altered due to increasing dispersion effect. The dispersion is more pronounced during redistribution events, thereby signifying the importance of numerical model (with the accounting of hysteresis effect) in the studies of leaching of salt. Finally, for the environmental consideration, we can predict that such models may provide guidelines for setting the control between the surface supply of salt with water and the changes in the quality of soil water and ground water during sprinkler irrigation, etc.

ACKNOWLEDGEMENTS

The financial assistance received from U.G.C. under D.R.S. Project no. P-o1/200B is gratefully acknowledged.

REFERENCES

1. Banerjee, M., and Watson, K.K., 1984. Numerical analysis of soil water movement under conditions of rapid intermittency of water application, Water Resources., v. 20, no. 1, p. 119-125.

2. Bear, J. 1972. Dynamics of fluids in porous media, American Elsevier Publication Co., New York

3. Biggar, J.W., and Nielsen, D.R., 1967. Miscible displacement and leaching phenomena, Irrigation of Agricultural lands, Agron. 11 ed. R.M. Hagan et al., Amer. Soc. Agron., Madison Wis., p. 254-274.

4. Bresler, E., and Hanks, R.J., 1969. Numerical method of estimating simultaneous flow of water and salt in unsaturated soil, Soil Sci. Soc. Amer. Proc., v. 33. p. 827-832.

5. Dagan, G., 1967. Hydrodynamic dispersion in a non-homogeneous porous column, Journal of Geophysical Research, v. 72, p. 4075-4080.

6. Davidson, J.M., Manasell, R.S., and Baker, D.R., 1972. Herbicide distribution within a soil profile and their dependence upon adsorption-desorption, Soil and Crop Sci. Soc. Florida Proc., v. 32, p. 36-41.

7. Fried, J.J., and Combarnous, M.A. 1971. Dispersion in porous media, Advn/ Hydrosci. v. 3, p.170-280.

8. Green, R.E., Rao, P.S.C., and Corey, J.C., 1972. Solute transport in aggregated soil : tracer zone shape in relationto pore velocity distribution and adsorption, Proc. 2nd Sym. Fundamentals of Transport Phenomena in Porous Media, IAHR-ISSS, Guelph, August 7-11, 1972, v. 2, p. 732-752.

9. Peaceman, D.W. and Rachford, H.H., Jr. 1955. The numerical solution of parabolic and elliptic differential equations, J. Soc. Ind. Appl. Math. v. 3, p. 28-41.

10. Richtmyer, R.D., and Morton, K.W. 1967. Difference methods for initial value problems, 2nd ed. Wiley (interscience), New York.

11. Smith, G.D. 1965. Numerical solution of partial differential equations. Oxford Univ. Press, London and New York.

12. Swanson, R.A., and Dutt, G.R., 1973. Chemical and Physical processes that effect atrazine movement and distribution in soil systems, Soil Sci. Soc. Amer. Proc. v. 37, p. 872-876.

13. Von Neumann, J., and Richtmyer, R.D., 1950. A method for the numerical calculations of hydrodynamic shocks, J. of Applied Physics, v. 21, p. 232-257.

14. Warrick, A.W., Biggar, J.W., and Nielsen, D.R., 1971. Simultaneous solute and water transfer for an unsaturated soil, Water Resour. Res., v. 7, p. 1216-1225.

A SINGLE-PHASE/TWO-PHASES NUMERICAL MODEL USING NESTED SQUARE MESHES

E. LEDOUX[x], S. SAUVAGNAC[xx] and A. RIVERA[x]

[x] Centre d'Informatique Géologique, Ecole des Mines de Paris, Fontainebleau, France
[xx] National Geophysical Research Institute, Hyderabad, India

A numerical model (NEWVAR) to simulate both the regional groundwater flow and the transient movement of a discrete interface between freshwater and saline-water or any other fluid has been developed. NEWVAR is designed to allow the analysis of a regional two-dimensional groundwater flow in coastal aquifers and permits the prediction of regional freshwater level (mono-phase) and two-dimensional freshwater/saline-water interface location (two phases) using nested square meshes.

The numerical solution is based on the finite difference method. After an optimal renumbering search, the Gauss-Jordan direct method or the Gauss-Siedel iterative method, through a predictor-corrector scheme, are used for solving steady-state non-linear equations in the two-phase mode, successively for the freshwater and saline-water terms. In transient state the system of equations is simultaneously solved through semi-implicit method. Different original procedures are used to avoid difficulties, generaly found in such two-phase models in tracking the transient position of the interface toe. The use of nested square meshes allows both a fine resolution in certain areas alongwith the possibility to reach hydrogeological boundaries which are far away using a minimum of meshes.

The numerical solution was tested against analytical ones for the cases of an advancing-retreating interface and of a floating freshwater lens over saline-water. These tests showed good agreement, thus verifying the FD approximation and the original procedures used to overlay the toe tracking dilemma.

INTRODUCTION

The numerical modeling of groundwater flow and pollution has become a very widely used tool for simulating the hydrodynamic behaviour of aquifer systems. In general, a numerical model permits the prediction of the response of an aquifer to stress, and presents alternative suggestions for use of this aquifer.

Given the increasingly rapid development of coastal areas, costal aquifers constitute a very important, sometimes the only, source of freshwater for domestic use, industry or agriculture. However, the quantitative evaluation of coastal aquifer is very complex because of the proximity of the sea and the contact between the freshwater and the saline-water which requires special attention and special modeling techniques.

Many researchers have made quantitative analyses of saline-water/freshwater systems attempting to describe mathematically the physical system and its important mechanisms (Cooper et al. 1964, Bear and Dagan, 1964, Pinder and Page, 1977, etc.). Freshwater and saline-water are miscible fluids with a density difference and therefore there is always a contact or a transition zone between them caused by hydrodynamic dispersion. Under certain conditions, the width of this zone is small, compared to the thickness of the aquifer, so that the zone of gradual transition from freshwater to saline-water may be approximated as a sharp interface (Sauvagnac, 1987).

In this paper we insist on the use of a model of practical interest and we make use of the discrete interface approach to propose a finite-difference regional two-dimensional model (NEWVAR) to simulate the feshwater/saline-water interface motion. An issue of this work is the compatibility of the single-phase multi-layered model NEWSAM (Marsily et al., 1978 and Girard et al. 1981), with an extension to two-phase model presently described, both using nested square meshes. This model has been applied to a real coastal aquifer in Mexico (Rivera, 1987).

1) CONCEPTUAL MODEL

Figure lb shows a plane representation of our conceptual model for a regional analysis and figure la shows a schematic representation in a vertical cross-section with the nomenclature to be used in our model.

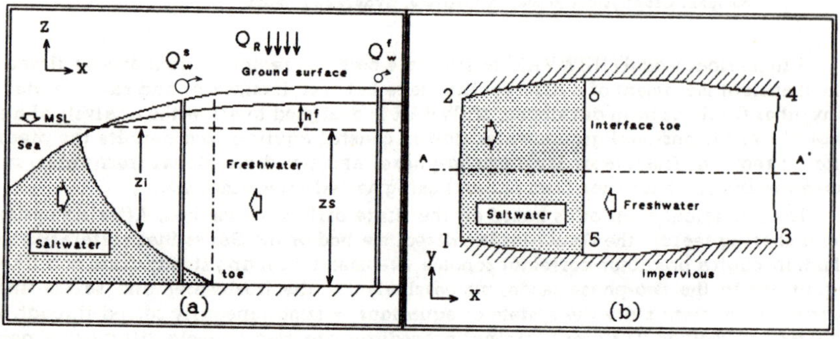

Figure 1 : Conceptual model

The x axis is taken positive in the horizontal plane inland. The boundary 1-2 represents the coast line and it is the source of the saline-water intrusion into the aquifer. The boundary 3-4 is an arbitrary flow limit where the aquifer receives a certain amount of freshwater inflow (Qinj). The boundary of the line 5-6 in figure lb is the downward limit of the saline-water intrusion in the aquifer and it will represent the interface toe in the x-y plane in our model. Line 1-5-3 and 2-6-4 are given impervious boundaries of the whole aquifer system in x-y plane.

The datum of our model is the sea level, so that the z-axis will be taken vertically upward as positive. QR is the recharge of the aquifer, say natural replenishment from recharge and QWf and QWs are the pumping from the freshwater zone and from the saline-water zone respectively.

2) BASIC EQUATIONS

In order to construct the conceptual mathematical model for a two-phase flow we have to start from the most general basic equations (terms are defined in the nomenclature section).

2.1 General equations for two fluids 1 and 2

The basic equations for flow through porous media are:

- **Darcy's equation,**

$$U_1 = -\frac{kk_1}{\mu_1}(\text{grad } p_1 + \rho_1 g z)$$
$$U_2 = -\frac{kk_2}{\mu_2}(\text{grad } p_2 + \rho_2 g z) \qquad (1)$$

Where U_1 and U_2 are Darcy's velocities for fluids 1 and 2.

- **The state equations of fluids considering isothermal conditions,**

$$\rho_1 = \rho_1(p_1) \qquad \mu_1 = \mu_1(p_1)$$
$$\rho_2 = \rho_2(p_2) \qquad \mu_2 = \mu_2(p_2) \qquad (2)$$

- The fundamental **mass balance equations** for an incompressible rock matrix,

$$\begin{aligned} \mathrm{div}(\rho_1 U_1) + \varepsilon \frac{\partial(\rho_1 s_1)}{\partial t} = 0 \\ \mathrm{div}(\rho_2 U_2) + \varepsilon \frac{\partial(\rho_2 s_2)}{\partial t} = 0 \end{aligned} \qquad (3)$$

- The relationship between **capillary pressure** pc and the pressures pl and P2 which in general is a non unique relationship,

$$p_1 - p_2 = pc\,(s_1) \qquad s_1 + s_2 = 1 \qquad (4)$$

Particular case: **Incompressible fluids** with the use of the hydraulic potential or hydraulic head as,

$$h_1 = \frac{p_1}{\rho_1 g} + z \qquad h_2 = \frac{p_2}{\rho_2 g} + z \qquad (5)$$

$$\begin{cases} U_1 = -\dfrac{kk_1\rho_1 g}{\mu_1}\,\mathrm{grad}\,h_1 = -K_1 k_1 .\,\mathrm{grad}\,h_1 \\ U_2 = -\dfrac{kk_2\rho_2 g}{\mu_2}\,\mathrm{grad}\,h_2 = -K_2 k_2 .\,\mathrm{grad}\,h_2 \end{cases} \qquad (6)$$

2.2 The freshwater/ saline-water interface

It can be shown that there is an identity between a miscible fluid formulation and a two-phase formulation if :

1) The relative permeabilities are linear in s

2) Dispersion is independant of velocity (diffusion)

Neglecting dispersion we can simplify the problem as the capillary pressure becomes zero. We can then write h_2 as a simple function of h_1 using f and s indexes for fresh and saline-water, respectively as:

$$\begin{aligned} pc = 0 \rightarrow p^f = p^s \rightarrow \rho^f g(h^f - z) = \rho^s g(h^s - z) \\ \rightarrow h^s = \frac{\rho^f}{\rho^s} h^f + \frac{\rho^s - \rho^f}{\rho^s} z \end{aligned} \qquad (7)$$

General equations

The general equations, neglecting the variation of ρ^f and ρ^s in grad h^f and grad h^s are as follows:

$$\begin{cases} \mathrm{div}\,(\rho^f K k^f .\,\mathrm{grad}\,h^f) = \varepsilon \dfrac{\partial(\rho^f s^f)}{\partial t} \\ \mathrm{div}\,(\rho^s K k^s .\,\mathrm{grad}\,h^s) = \varepsilon \dfrac{\partial(\rho^s s^s)}{\partial t} \end{cases} \qquad (8)$$

$$h^s = \frac{\rho^f}{\rho^s} h^f + \frac{\rho^s - \rho^f}{\rho^s} z \qquad s^f + s^s = 1 \qquad (9)$$

Two-dimensional case

For the two-dimensional case we assume that: (1) a sharp interface of elevation Zi separates freshwater and saline-water in the aquifer (Fig. la), (2) Dupuit's assumption of no vertical flow is valid and (3) permeability is constant along the vertical.

Integrating along the vertical (z) we obtain equation (10) :

$$\rho^f\left[\frac{\partial}{\partial x}\left(\int_{zs}^{h^f}(K^f k^f)_x\,dz\,.\,\frac{\partial h^f}{\partial x}\right)+\frac{\partial}{\partial y}\left(\int_{zs}^{h^f}(K^f k^f)_y\,dz\,.\,\frac{\partial h^f}{\partial y}\right)\right]=\int_{zs}^{h^f}\rho^f q^f dz+\frac{\partial}{\partial t}\int_{zs}^{h^f}\varepsilon\,\rho^f S^f dz$$

$$\rho^s\left[\frac{\partial}{\partial x}\left(\int_{zs}^{h^f}(K^s k^s)_x\,dz\,.\,\frac{\partial h^s}{\partial x}\right)+\frac{\partial}{\partial y}\left(\int_{zs}^{h^f}(K^s k^s)_y\,dz\,.\,\frac{\partial h^s}{\partial y}\right)\right]=\frac{\partial}{\partial t}\int_{zs}^{h^s}\varepsilon\,\rho^s S^s dz$$

Where $\partial/\partial t$ is the total derivative in the deformable control volume from ZS to h^f.

$Q^f = \int_{zs}^{h^f} q^f dz$ Making $\int_{zs}^{h^f}(K^f k^f)_x\,dz = K_x^f(h^f - Zi)$ (11) $Q^f = \int_{zs}^{h^f} q^f dz$

the transmissivity of the freshwater with $Q^f = \int_{zs}^{h^f} q^f dz$ which is the freshwater flow per unit area and assuming $s = 0$ for $ZS < z < Zi$, $s = 1 - s$ we obtain our final system (12):

$$\frac{\partial}{\partial x}\left[K_x^f(h^f - Zi)\,\frac{\partial h^f}{\partial x}\right]+\frac{\partial}{\partial y}\left[K_y^f(h^f - Zi)\,\frac{\partial h^f}{\partial y}\right] = Q^f + S\,\frac{\partial h^f}{\partial t} - \varepsilon\,\frac{\partial Zi}{\partial t}$$

$$\frac{\partial}{\partial x}\left[K_x^s(Zi - ZS)\,\frac{\partial}{\partial x}\left(\frac{\rho^f}{\rho^s}h^f + \frac{\rho^s - \rho^f}{\rho^f}Zi\right)\right]$$
$$+\frac{\partial}{\partial y}\left[K_y^s(Zi - ZS)\,\frac{\partial}{\partial y}\left(\frac{\rho^f}{\rho^s}h^f + \frac{\rho^s - \rho^f}{\rho^f}Zi\right)\right] = Q^s + \varepsilon\,\frac{\partial Zi}{\partial t}$$

Boundary conditions

The boundary conditions consist of the usual Dirichlet condition where h^f and Zi are prescribed, and the Neumann condition where the normal derivative of h^f, $\partial h^f/\partial n$ is used to prescribe the freshwater or saline-water inflow across a boundary.

3) NUMERICAL SOLUTION

The NEWVAR model solves the set of differential equations by integration in a finite mesh using the Finite-Difference Method. Space and time are discretized for solving the functions $h^f(x,y,t)$ and $Zi(x,y,t)$. Searching a solution for the previous functions we have used the functions $H(x,y)$ and $ZI(x,y)$ for each considered time-step on each element of the domain.

One of the principal characteristics of NEWVAR is that it allows the use of nested square meshes. 10 different size are accepted, providing that two adjacent meshes have at most a factor of 2 of difference in size. This technique allows more precision in some areas of the domain (e.g. pumping wells or places near the shore in our case) with a minimum of computer memory without using unnecessary meshes, for further details see Marsily et al. 1978.

3.1 Integration of the diffusivity equations of both fluids over a mesh

On each mesh i we define:

- a constant function Hi at the mesh center
- a constant function ZIi at the mesh center
- The same horizontal permeability K for each fluid.

Saltwater flux Q^s and freshwater flux Q^f between two adjacent meshes 1 and 2 through element's face ABCD (Fig. 2) are approximated by:

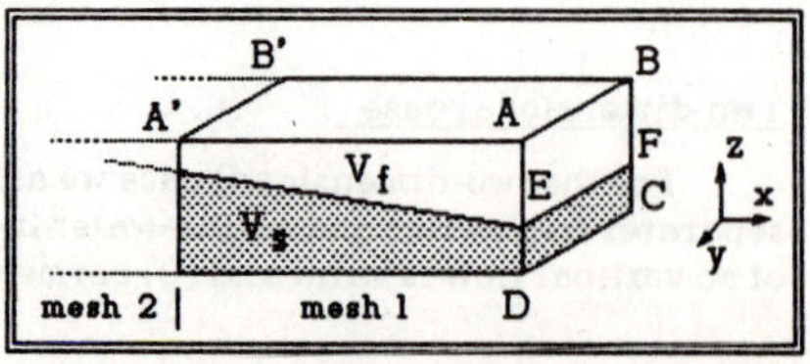

Figure 2 : Numerical approximation

$$Q^f = \int_{ABEF} K_x \left(\frac{H_2 - H_1}{a}\right) dydz = \left(\frac{H_2 - H_1}{a}\right) . \int_{ABEF} K_x \, dydz = \left(\frac{H_2 - H_1}{a}\right) . \int_{AB} T_x^f \, dy = T_{1,2}^f \, (H_2 - H_1)$$

$$Q^s = \int_{EFCD} K_x \left(\frac{H_2^s - H_1^s}{a}\right) dydz \qquad (13)$$

$$= \frac{\rho^f}{\rho^s} \left(\frac{H_2 - H_1}{a}\right) . \int_{CD} T_x^s \, dy + \frac{\rho^s - \rho^f}{\rho^f} \left(\frac{ZI_2 - ZI_1}{a}\right) . \int_{CD} T_x^s d$$

$$= T_{1,2}^s (H_2 - H_1) + C_{1,2}^s (ZI_2 - ZI_1)$$

The average transmissivities $T_{1,2}^f$, $T_{1,2}^s$ and $C_{1,2}^s$ between two meshes are formulated as the harmonic mean of the two horizontal transmissivities of elements 1 and 2 as it follows for the case of two meshes of the same size.

$$T_{1,2}^f = 2 \, \frac{K_2(H_2 - ZI_2) \, . \, K_1(H_1 - ZI_1)}{K_2(H_2 - ZI_2) + K_1(H_1 - ZI_1)}$$

$$T_{1,2}^s = 2 \, \frac{\rho^f}{\rho^s} . \frac{K_2(ZI_2 - ZS_2) . K_1(ZI_1 - ZS_1)}{K_2(ZI_2 - ZS_2) . K_1(ZI_1 - ZS_1)} \qquad (14)$$

$$C_{1,2}^s = \frac{\rho^s - \rho^f}{\rho^f} . \frac{\rho^s}{\rho^f} . T_{1,2}^s$$

3.2 Solution of the set equations

3.2.1 Steady state

The steady-state finite-difference form of the system is expressed in matrix form as follows:

$$\begin{cases} \bar{\bar{T}}^f . (\bar{H}) = (\bar{Q}_w^f) \\ \bar{\bar{T}}^s . (\overline{ZI}) = (\bar{Q}_w^s) \end{cases} \qquad (15)$$

where :

$\bar{\bar{T}}^f$: Matrix composed with the coefficients $T_{1,2}^f$

$\bar{\bar{T}}^s$: Matrix composed with the coefficients $T_{1,2}^s$, $C_{1,2}^s$ and H^f

$\bar{H}$: Vector containing the unknown freshwater head Hi for all meshes not containing a constant head condition

$\overline{ZI}$: Vector containing the unknown interface level for all meshes not containing a constant interface level condition

$\bar{Q}_w^f, \bar{Q}_w^s$: Vectors containing the known freshwater and saline-water inflow and outflow in all the meshes

In this system T^f and T^s depend of H and ZI. The system is solved using Gauss-Jordan direct method or Gauss-Siedel iterative method successively in the freshwater and the saline-water with iterations over the values of the intermesh average transmissivities until H and ZI become stable (e.g. predictor-corrector method).

3.2.1 Transient state

The transient state FDM of the system is solved simultaneously using a semi-implicite method. The coefficients of the matrix which depend on H and ZI are defined at the beginning of the time step t and the unknown values of ZI and H are defined at the end of the time step $t+\Delta t$. This semi-implicite method is stable for any value of the time step. Generally we let the program start with a critical time step which represents the duration of the fastest transfer on the smallest mesh of the model. It seems to be a correct value to discretize the time after a perturbation occurs in the aquifer. This time step will follow a geometrical progression until the entire period of the simulation is achieved.

3.2.3 The movement of the interface

The above equations represent an appropriate description of the system behaviour, however their numerical solution entails some difficulties if one wishes to simulate the interface movement in transient condition. Physically, the seawater term of equation (12) becomes null for meshes inland of the toe, preventing the resolution of the equation. Some authors, to avoid the difficulty of the moving boundary at the interface discontinuity, ignore the existence of an impervious bedrock, or at least consider it very deep to avoid such discontinuity. This procedure assumes a constant transmissivity for the saline-water and then considers the freshwater part as a lens.

Others (Shamir and Dagan, 1971) have circumvented this problem in one dimension by tracking the toe regenerating a finite difference grid each time step such that the toe coincide with the edge of a grid block. However, for a 2-D areal flow this technique involves prohibitive computational time.

Some other authors (e.g. Mercer et al., 1980) have evaluated the use of explicit corrections to prevent mass balance errors. These techniques however were not successful in a problem with an advancing saline-water wedge. Sa Da Costa and Wilson (1979) proposed a technique to track the movement of the toe using Finite Element method with the Gauss Quadrature points.

In order to overcome the toe dilemma, which presents difficulties in the numerical solution of a two-dimensional areal flow, the NEWVAR code adopts a stratagem consisting of the following procedures.

The meshes located along and inland the toe have, at the beginning, a zero-flow (transmissivity) term for saline-water T^s. Normally the code uses the harmonic mean to calculate the average transmissivities $T^s_{1,2}$ and $C^s_{1,2}$ between two adjacent nodes 1 and 2 (Fig. 3). This gives a zero transmissivity term to seawater in equation (12) upstream of the wedge, preventing the movement of the interface (Fig. 3a). Therefore the mesh 1 situated inland of the toe is assigned by the code a minimum seawater transmissivity T^s_{min} which does not allow non-zero flow terms (Fig. 3b and 4).

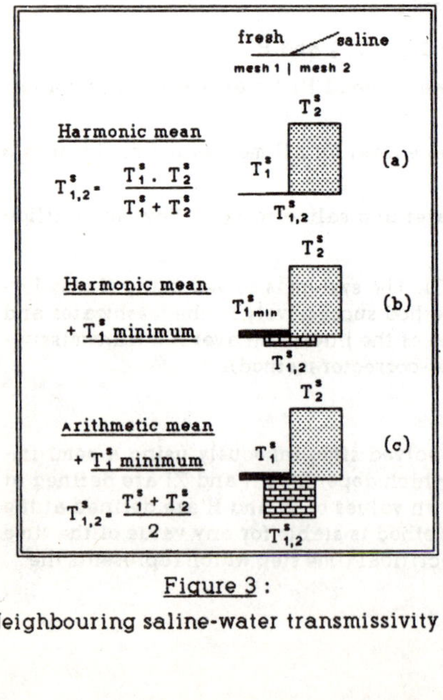

Figure 3 :

Neighbouring saline-water transmissivity $T^s_{1,2}$

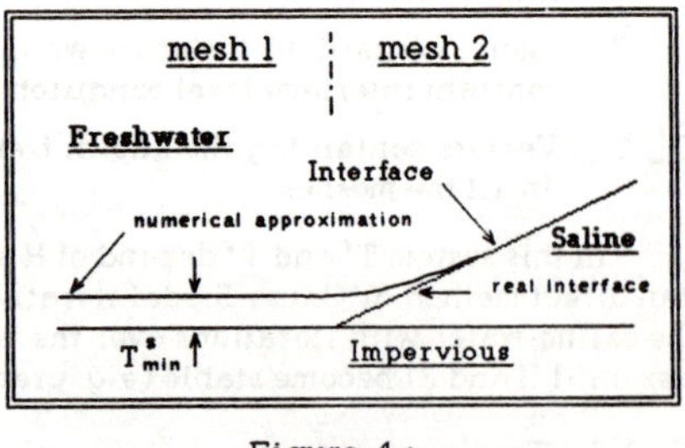

Figure 4 :

Numerical approximation for saline-water transmissivity inland of the toe

To calculate the relatively fast motion of the interface toe not necessitating prohibitive number of time-step, another procedure is added: For the meshes 1 and 2 situated along and inland of the toe, NEWVAR uses nó longer the harmonic mean but the arithmetic mean to determine the average transmissivity $T^*_{1,2}$ between the two neighbouring meshes (Fig. 3c).

Finally, the code uses a displacement porosity ε 100 times smaller inland of the toe of the wedge (mesh 1 in our example) to minimize the saline-water term in the equation (12) and to obtain a negligible storage term for seawater in balance with the above corrections in the saline-water inter-meshes transmissivies $T^*_{1,2}$ and $C^*_{1,2}$.

4) MODEL VERIFICATION

In order to verify that the finite-difference approach is adequately approximating the partial differential equation (12), we carried out several tests of comparison with existing analytical solutions. First we tested the location and movement of the interface for both steady and transient conditions using the analytical solution by Isaacs and Hunt (1986). Another set of tests were also made by comparing the development of a freshwater lens over saline-water with Hantush's (1968) analytical solution.

One must note that actually no analytical solution can give exact representation of the transient movement of the interface, then we must keep in mind the different assumptions involved to obtain the final expressions.

4.1 Advancing interface

Isaacs and Hunt (1986) proposed an approximate analytical solution to calculate the transient movement of the saline-water/freshwater interface in a coastal aquifer.

Figure 5 shows Isaacs and Hunt's conditions used in their solution. L(t) is the length of the transient position of the toe of an advancing interface due to the change in the freshwater inflow from Q_0 to Q_1. Q_0 is the freshwater inflow used before starting the transient simulation, Q_1 is the final freshwater inflow starting at time t=0. Their solution involves physical and mathematical simplifications as follows :

Physical assumptions : (1) an isotropic and homogeneous aquifer, (2) incompressible medium, (3) abrupt interface, (4) Dupuit principle which are essentially the same as the ones used in our model. These leads to a non-linear differential equation same as the one described in Bear and Dagan (1964).

Mathematical assumptions : The terms of the previous equation are scaled and the one with the smaller order of magnitude is discarded to obtain a non-linear first order equation which will be then linearized and integrated to give the final expression.

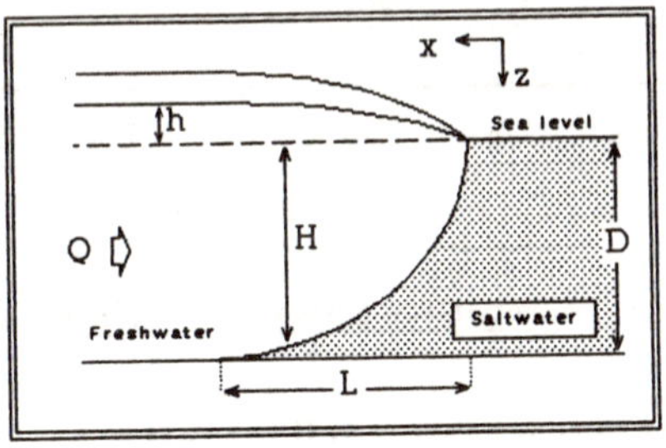

Figure 5 : Isaacs and Hunt's model

Their general solution is :

$$L(t) = \frac{E.K.D^2}{2.Q_0} + \frac{Q_0 - Q_1}{n.D} \qquad \text{for } 0 < t < \frac{n.E.K.D^3}{2.Q_0^2}$$

$$L(t) = \sqrt{\frac{2.E.K.D.t}{n}} - \frac{Q_1.t}{n.D} \qquad \text{for } \frac{n.E.K.D^3}{2.Q_0^2} < t < \frac{n.E.K.D^3}{2.Q_1^2}$$

$$L(t) = \frac{E.K.D^2}{2.Q_1} \qquad \text{for } t < \infty$$

Where n is the porosity and E the relative densities difference (≈ 0.025)

This solution was compared with the well known Shamir and Dagan's numerical solution (1971) for finding a maximum difference in the toe position of 10 %.

NEWVAR used for this comparison a grid of 92 nested square meshes of sizes are 12.5, 25. and 50 meters.

Figure 6 shows the comparison of Isaacs and Hunt's solution with the NEWVAR model for the interface toe motion. We have chosen the previously defined critical time-step which is estimated by the program. Keeping in mind the uncertainities of the analytical solution, we can see from the figure that the numerical solution is in good agreement with it. Both solutions give the same positions for the initial and final steady-state. Isaacs and Hunt's solution tends to overestimate the interface's movement as estimated by the authors. The maximum deviation was estimated as 10% from Isaacs and Hunt's analytical model as compared with Shamir and Dagan's numerical solution. In our comparison, the maximum difference found is 7% at the toe position.

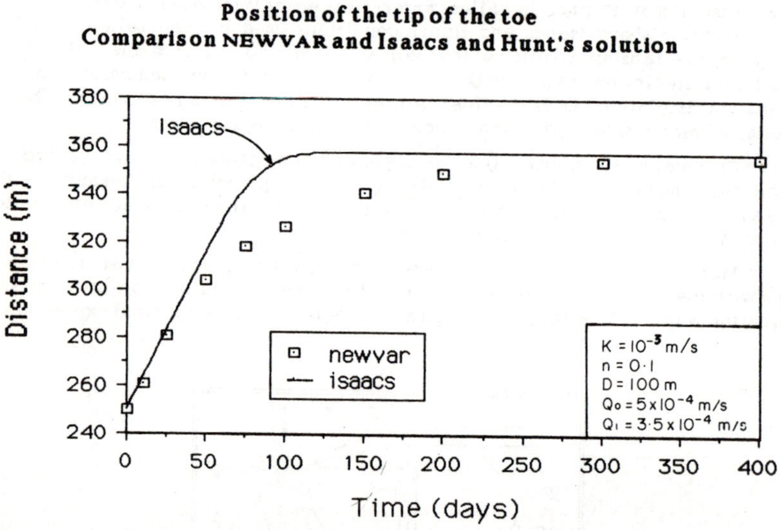

Figure 6 : Comparison NEWVAR with Isaacs's solution

4.2 A floating freshwater lens over saline-water

NEWVAR can also be used to simulate the transient development of a freshwater lens over saline-water. Such situation can be found in case of artificial recharge in a saline aquifer to reject forward the saline-water.

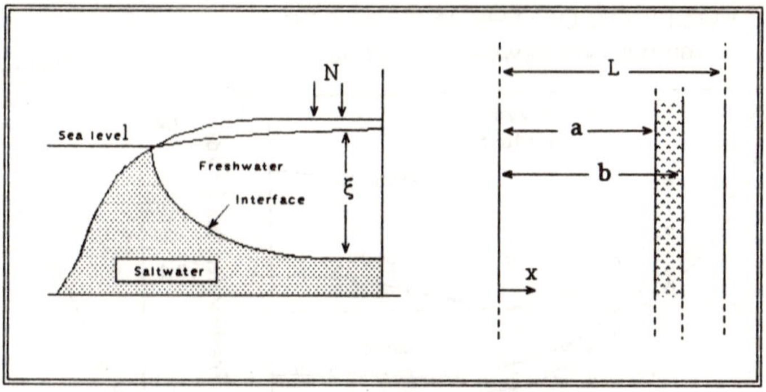

Figure 7 : Hantush's analytical model

An approximate analytical solution was proposed by Hantush (1968) for thick aquifers near a sea-coast. Figure 7 shows the conditions used by Hantush in a coastal aquifer of finite lenght L, receiving uniform recharge N along an infinite strip parallel to the coast. Both analytical and numerical solutions assume the same approximations, except that the analytical one considers an infinite saltwater permeability (e. g. an infinitly deep aquifer).

Hantush final equation is :

$$\xi^2 = 2.C.\left[2.(b-a).x - f(x,t)\right] \qquad \text{for } 0 \leq x \leq a$$

$$\xi^2 = 2.C.\left[2.(b-a).x - (x-a)^2 - f(x,t)\right] \qquad \text{for } a < x < b$$

$$\xi^2 = 2.C.\left[b^2 - a^2 - f(x,t)\right] \qquad \text{for } b \leq x \leq L$$

$$f(x,t) = \left| \frac{32.L^2}{\pi^3} . \sum_{n=0}^{\infty} Rn.\mathrm{Sin}\, \frac{(2.n+1).\pi.x}{2.L} \right|$$

$$Rn = \frac{1}{(2.n+1)^3} \left[\mathrm{Cos}\, \frac{(2.n+1).\pi.b}{2.L} - \mathrm{Cos}\, \frac{(n+1).\pi.a}{2.L} \right] . \exp\left(- \frac{(2.n+1)^2.\pi^2}{4} . \tau \right)$$

$$C = \frac{N}{2.\delta.(1+\delta).K} \qquad \tau = \frac{K.\delta.\bar{\xi}.t}{ne.L^2} \qquad \delta = \frac{\rho^s - \rho^f}{\rho^f}$$

where ne : effective porosity

ξ : depth of the interface

$\bar{\xi}$: average depth of the interface

ρ : density for each fluids

The numerical model uses a grid of 61 nested square meshes ranging from 2.5 to 5 meters in size. Figure 8 shows the comparison between the two solutions. Sa Da Costa and Wilson (1979) have shown that Hantush's solution does not conserve mass at the beginning of the simulation. Our comparison shows good agreement specially after time = 50 days. The differences found at the beginning are related with the above deviation of the analytical solution.

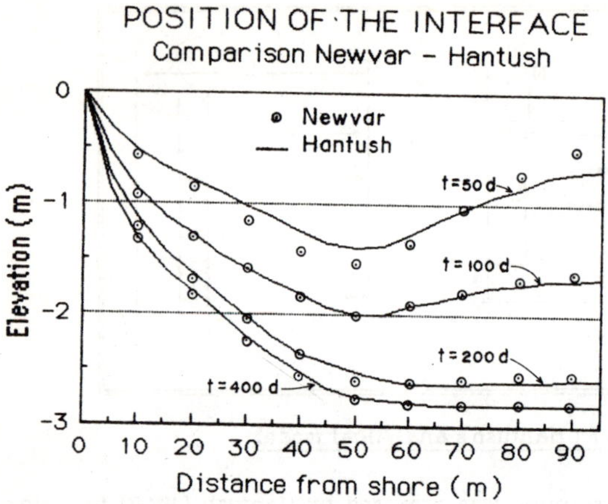

Figure 8 : Comparison NEWVAR - Hantush

Position of the interface

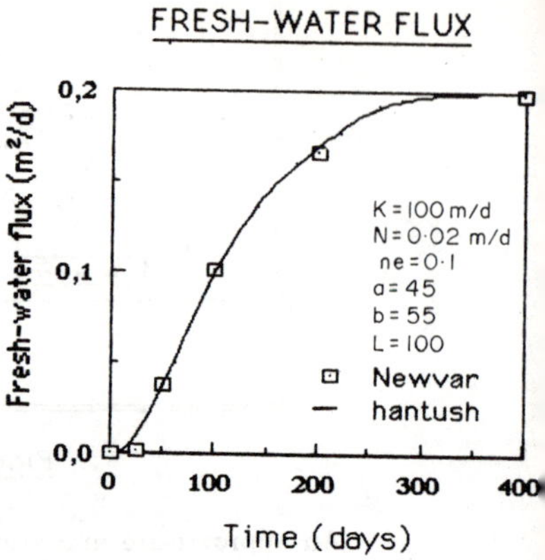

Figure 9 : Freshwater outflow

Another test performed using Hantush's solution was the calculation of freshwater outflow to the sea. Figure 9 shows the results of both solutions with an almost perfect match. The values of the parameters used for these simulations are presented in the figure 9.

CONCLUSIONS

NEWVAR has demonstrated that it is an accurate and efficient model to represent seawater intrusion in coastal aquifers.

The stratagems adapted to overcome the toe movement dilemma give correct results in transient state.

The toe tracking stratagems show a dependance on the time step. The choice of the critical time step generally give best results.

An interesting issue is the compatibility of NEWVAR with the single-phase multi-layered model NEWSAM which uses also nested square meshes of 10 different sizes.

Our model can also be used for any two-phase fluids system, The program considers different values of density and dynamic viscosity in the mathematical expressions.

In case of a confined aquifer, NEWVAR changes, in the above defined inter-mesh transmissivities, the parameter h' by the top of aquifer level Z_T. We include also in the saline-water part of equation 12 the variation of storage using S^* (storage coefficient for saline-water).

In further developments, NEWVAR will be able to treat Multi-layered aquifers to be fully compatible with NEWSAM. The physical phenomenon is not well known and then more physical studies have to be done before attempting to model it.

NOMENCLATURE

g	gravitational acceleration
k	intrinsic permeability
k_1, k_2	relative permeability for fluids 1 and 2
K	hydraulic conductivity (Darcy permeability)
μ	dynamic viscosity
U	Darcy's velocity
p	pressure
ρ	density
ε	porosity
Z	elevation
s	saturation
pc	capillary pressure
h^f, h^s	freshwater and saline-water hydraulic head
Q_{inj}	freshwater injected into aquifer
Q_R^f	freshwater recharge
Q_w^f	freshwater withdrawal
Q_w^s	freswater withdrawal
Zi	interface elevation
ZS	elevation of the bottom of the aquifer
S	storage coefficient of the aquifer
ε	displacement porosity between freshwater and saline water

REFERENCES

Bear J. and Dagan G. , 1964
Moving interface in coastal aquifers, Hydraul. DIV., Proc. Am. Soc. Civil Eng., 90(HY4): 193-213.

Cooper H.H., Kohout H.R. and Glover R.E., 1964
Seawater in coastal aquifer., U.S. Geological Survey, Water Supply pap. 1613-c, 84 pp.

Girard G., Ledoux E. and Villeneuve J.P., 1981
The modèle Couplé - A jointly simulation of surface runoffs and watergroud flows on a hydrological watershed. Cahier ORSTOM, série Hydrol., XVIII, N° 4, pp 195-279

Hantush M.S., 1968
Unsteady movement of fresh water in thick unconfined saline aquifers., Bull. Int. Assoc. Sci. Hydrol., Vol. XIII (2), pp 40-60

Issacs L.T. and Hunt L., 1986
A simple approximation for a moving interface in a coastal aquifer., J. Hydrol., 83 : pp 29-43

Marsily G. de, Ledoux E., Levassor A., Poitrinal D. and Salem A., 1978
Modelling of large multilayered aquifer systems: theory and application., J. Hydrol., 36 : pp 1-34

Mercer J.W., Larson S.P. and Faust C.R., 1980
Simulation of salt-water interface motion., Ground Water Vol. 18, n° 4, pp.374-385

Pinder G.F. and Page R.H., 1977
Finite element simulation of saltwater intrusion on the South Fork of Long Island., In: Finite Elements in Water Resources Proc. of the first Int. Conf. on Finite Elements in Water Ressources, London, pp 2.51-2.69

Rivera, A., 1987
Etude par modélisation numérique d'un aquifère côtier au Mexique., Mém. de D.E.A., Paris School of Mines, 115 pp.

Really, T.E. and Goodman A.S., 1985
Quantitative analysis of saltwater freshwater relationships in groundwater systems - A historical perspective., J. Hydrol., 80 : pp. 125-160

Sa Da Costa A. and Wilson J.L., 1979
A numerical model of seawater intrusion in aquifers., Cambridge R.M. Parsons Lab. for Water Resources and Hydrodynamics, MIT report n° 247

Sauvagnac S., 1987
Les intrusions salines, approche analytique, comparaison avec un modèle numérique., Mém. de D.E.A., Ecol. Nat. Sup. des Mines de Paris, 123 p.

Shamir V. and Dagan G., 1971
Motion of seawater interface in coastal aquifer : a numerical solution., Water Resour. Res. 7 (3), pp. 644-657

WATER TREATMENT TECHNOLOGY FOR REMOVAL OF EXCESS FLUORIDE

W.G Nawlakhe and K.R. Bulusu

National Environmental Engineering Research Institute, Nagpur - 440 020.

ABSTRACT

Removal of excess fluoride from drinking water is necessary to prevent the sufferings of village population from fluorosis. NEERI has developed a cheap and simple technique for defluoridation which is called as Nalgonda Technique. The technique is easily adapted by villagers. It is adaptable at domestic as well as community levels. The defluoridation operations are performed on fill-and-draw basis and also on continuous operation basis. Fill-and-draw type operations are discussed here for domestic, small community and rural water supply systems.

Domestic defluoridation can be performed in a bucket. A small hand-operated fill-and-draw unit of 200 litres capacity for domestic use is also designed which can be easily operated and maintained by the villagers. For population of 200 villagers, similar unit is designed. The details of both these units are given. A defluoridation scheme for population upto 5000 is discussed. Its cost estimates for a population of 2000 are also included. Six fill-and-draw defluoridation plants have been commissioned in various parts of the country based on Nalgonda Technique.

INTRODUCTION

The village population mainly depends upon ground water for drinking and other domestic uses. Ground water sources contain metallic and non-metallic substances and they cause impurities to the water when present in excess of certain limits. Drinking water standards have been framed by World Health Organisation. In India, Indian Standard Institution and Indian Council of Medical Research have also prescribed standards for various parameters present in drinking water. The water is safe for drinking when the substances present are below the prmissible limits. It needs treatment when any one or more of the parameters are in excess of these limits.

Government of India has set up a Technology Mission on Drinking Water to provide safe water to rural population. The Mission has identified brackishness, fluoride, iron and other chemical and bacteriological contamination as problem parameters. The needs for development of effective technologies for treatment of water to drinking water standards has been stressed. NEERI has been entrusted with the responsiblity to develop and provide technologies for defluoridation, iron removal, disinfection and slow-sand filtration. Technology on defluoridation is discussed in this research paper.

OCCURRENCE OF FLUORIDE

Fluoride is so highly reactive that it is never encountered in its elemental gaseous state except in some industrial processes. It occurs in the earth´s crust notably as fluorspar, cryolite and fluorapatite. These fluoride minerals are nearly insoluble in water. Fluoride is encountered in water when conditions favour their dissolution from the fluoride bearing minerals.

Occurrence of excess fluoride bearing waters was reported by many research workers in Andhra Pradesh, Bihar, Gujarat, Haryana, Karnataka, Madhya Pradesh, Maharashtra, Orissa, Punjab, Rajasthan, Tamilnadu and Uttar Pradesh. There are an estimated 8700 villages in India which have problem of excessive fluoride in water affecting 2.5 million people.

HEALTH EFFECTS OF FLUORIDE IN WATER

Fluoride is beneficial to certain extent when present in concentration of 0.8 - 1.0 mg F/L for calcification of dental enamel especially for the children below 8 years of age. But, it causes dental fluorosis if present in excess of 1.5 mg F/L and skeletal fluorosis beyond 3.0 mg F/L if such water is consumed for abot 8 to 10 years.

As per WHO Guidelines for Drinking Water Quality (1982) and Water Technology Mission of the Government of India, the permissible limit for fluoride in drinking water is 1.0 mg F/L. It can be extended to 1.5 mg F/L if there is no alternative source in the village. The water needs treatment above this limit.

REMOVAL OF EXCESS FLUORIDE FROM WATER

Since it became known that excess fluoride in drinking water caused mottling of teeth, many methods have been suggested from time to time for removal of excess fluoride.

Materials like clays, minerals, ion-exchange resins, activated carbon, activated alumina, sulphonated coals and serpentine were tried by NEERI for removal of excess fluoride from water. In-situ, chemical treatment with lime, magnesium salts, were also studied. Defluoron-2, activated alumina and magnesia processes were successful in removing fluorides.

All these processes suffer from one or more drawbacks, such as high initial cost, lack of selectivity for fluorides, poor fluoride removal capacity and complicated or expensive regeneration of medium. In order to overcome this lacuna, alum coagulation method has been revived and a technology evolved in NEERI which is popularly referred as Nalgonda Technique[1]

NALGONDA TECHNIQUE

Nalgonda Technique comprises addition in sequence of (i) bleaching powder, (ii) lime and (iii) aluminium sulphate or aluminium chloride or a combination of these two. Addition of these chemicals is followed by flocculation, sedimentation and filtration. The process is so simple and adaptable that even a layman can make use of it at village level.

Aluminum salt is responsible for removal of fluoride from water. Lime helps in increasing the floc density and rapid settling. Lime addition has also the advantage of maintaining the alkalinity of the raw water if it is less for formation of flocs. Bleaching powder is used for disinfection. The doses of aluminium salt depend upon the fluoride and alkalinity in the raw water. Table 1 indicates the doses of alum required to obtain permissive limit of fluoride in water at various fluoride and alkalinity levels. Lime dose is 1/20th of the alum dose. The dose of bleaching powder is so adjusted that residual chlorine in the defluoridated water will be 0.2 mg/L.

Table 1 : Approximate Alum Dose (mg/L) Required to obtain Permissive Limit (1 mg F/L) of Fluoride in Water at Various Alkalinity and Fluoride Levels

Test water Fluorides, mg F/L	Test Water Alkalinity, mg $CaCO_3$/L							
	120	200	300	400	500	600	800	1000
2	143	221	273	312	351	403	468	520
3	221	299	351	403	507	520	585	767
4	*	403	416	468	559	598	689	936
5	*	*	507	598	689	715	884	1010
6	*	*	611	715	780	936	1066	1209
8	*	*	*	*	988	1118	1300	1430
10	*	*	*	*	*	*	1508	1690

Note : * To be treated after increasing the alkalinity with lime or sodium carbonate.

Nalgonda Technique can be applied at domestic level as well as community level. The defluoridation operations are performed on fill-and-draw basis and also on continuous operation basis. Fill-and-draw type operations are discussed in this paper.

DOMESTIC DEFLUORIDATION USING BUCKET

Domestic defluoridation can be performed using a bucket of 60 L capacity with a tap 3 - 5 cm above the bottom of the bucket. The villager has to know the doses of chemicals for the raw water to be defluoridated. The doses are fixed by testing the water for fluoride and alkalinity and by referring the Table 1. After fixing the doses, the villager has to be informed how much volume of 10% alum solution be used for defluoridation operation. Approximate volumes of alum solution required to be added in 40 litres of raw water to obtain permissive limit of fluoride at various fluoride and alkalinity levels are given in Table 2.

Table 2 : Approximate volume of alum solution (millilitre) required to be added in 40 litres test water to obtain permissive limit (1 mg F/L) of fluoride in water at various alkalinity and fluoride levels.

Test water fluoride mg F/L	Test water alkaliniy, mg $CaCO_3$/L							
	125	200	300	400	500	600	800	1000
2	60	90	110	125	140	160	190	210
3	90	120	140	160	205	210	235	310
4		160	165	190	225	240	275	375
5			205	240	275	290	355	405
6			245	285	315	375	425	485
8					395	450	520	570
10							605	675

Bleaching powder and lime solutions are added first and mixed well with water. Alum solution is then added and the water stirred slowly for 10 minutes and allowed to settle for one hour. The supernatant which contains permissible amount of fluoride is withdrawn through the tap or decanted, if there is no tap, into another container for consumption. The settled sludge is discarded and the same bucket is used for another operation. Domestic operation is shown in Fig.1.

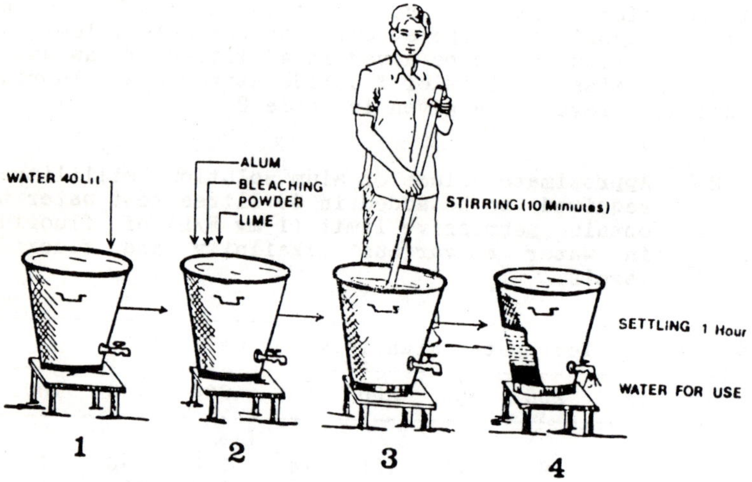

Fig. 1 : DEFLUORIDATION AT DOMESTIC LEVEL

DOMESTIC DEFLUORIDATION UNIT

This is also a fill-and-draw unit having 200 litres capcity. It comprises a cylindrical vessel with 1 m depth equipped with a hand-operated stirring mechanism as shown in Fig. 2. The unit is filled with raw water and the same defluoridation operation is performed as in bucket. The settled sludge is withdrawn through the valve at the bottom of the unit.

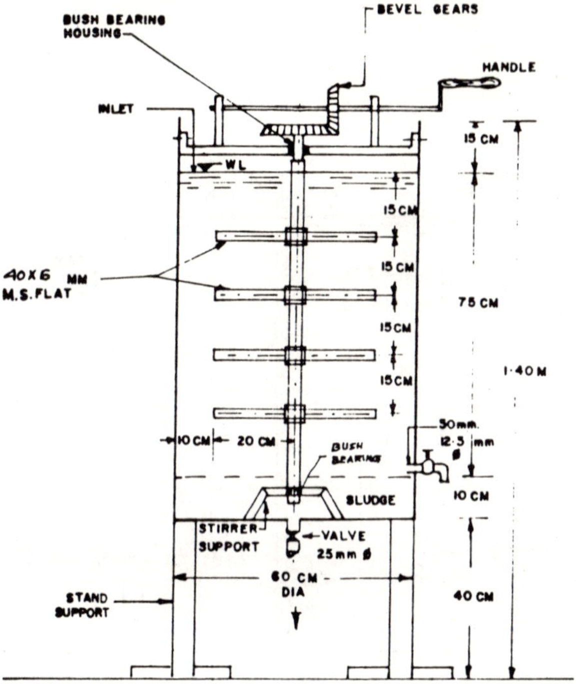

Fig. 2: FILL AND DRAW TYPE DEFLUORIDATION UNIT
CAPACITY - 200 LITERS

FILL-AND-DRAW DEFLUORIDATION PLANT FOR SMALL COMMUNITY

This plant can be used by communities upto 200 population. The plant comprises a hopper-bottom cylindrical tank with a depth of 2 m equipped with a hand-operated or power driven stirring mechanism as shown in Fig. 3. The raw

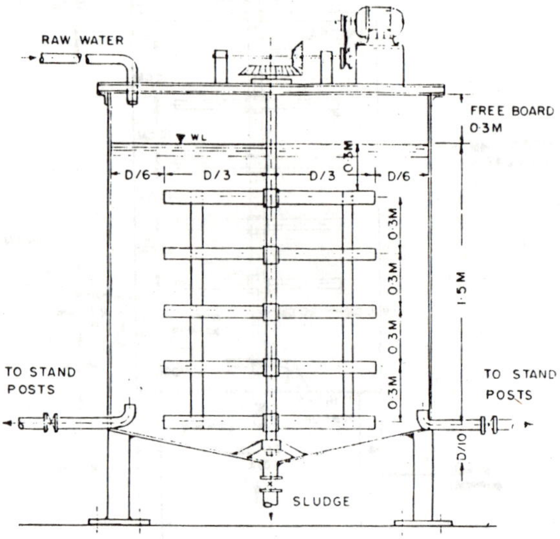

Fig. 3 : FILL-AND-DRAW DEFLUORIDATION PLANT

BASIS : 40 lpcd DOMESTIC WATER
(FOR POPULATION UPTO 200)

D = DIAMETER OF PLANT

Water Depth	=	1·5 M
Depth of sludge cone	=	D/10
Shaft Diameter	=	50 mm
Free Board	=	0·3 M

water is pumped or poured into the tank and defluoridation operation is performed. The flocs are allowed to settle for 2 hours. The defluoridated supernatant is withrawn through the tapp and supplied through the standposts. Plant dimensions are given in Table 3.

The notable features of the plant are :

(i) With a pump of adequate capacity the entire operation is completed in 2-3 hours and a number of batches of defluoridated water can be obtained in a day.

(ii) The accessories needed are a few and these are easily available. These include 16 L buckets for dissolving alum, preparation of lime slurry, bleaching powder and a weighing balance.

(iii) The plant can be located in the open with precautions to cover the motor.

(iv) Semi-skilled labour can perform the operation independently.

Table 3 : Plant diameter (meters) for population upto 200 on the basis of 40 lpcd defluoridated water through standposts.

Population	Water volume (M^3)	Plant diameter (M)	Suggested H.P. for motor
50	2	1.30	1.0
100	4	1.85	2.0
200	8	2.60	2.0

Alum required to be added per <u>each</u> batch of treatment (Grams, alumina ferric, IS : 299 - 1962)

= (Water volume, M^3) x (Alum dose for that particular water mg/L)

Fresh Bleaching Powder (gram per <u>batch</u>)

= 3 x (Water Volume, M^3)

FILL-AND-DRAW DEFLUORIDATION PLANT FOR RURAL WATER SUPPLY

The fill-and-draw defluoridation plant for rural water supply comprises cylindrical tank of 10 m^3 capacity with a dished bottom, inlet, outlet and and sludge drain. The cylindrical tank has the sturdy railings. Each tank is fitted with an agitator assembly consisting of (i) 5 HP drip proof electric motor : 3 phase, 50 Hz, 1440 RPM with 415 V ± 6% voltage fluctuation and (ii) gear box for 1440 RPM input speed with reduction ratio 60 : 1 to attain an output speed of 24 RPM, complete with downward shaft to hold the agitator paddles. The agitator is fixed to the bottom of the vessel by sturdy, suitable stain-less steel supporting bushings.

The defluoridation scheme comprises tanks of 10 m^3 capacity each, a sump well and an overhead reservoir as shown in Fig.4. Raw water is pumped into the units and treated by Nalgonda Technique. The detention period for

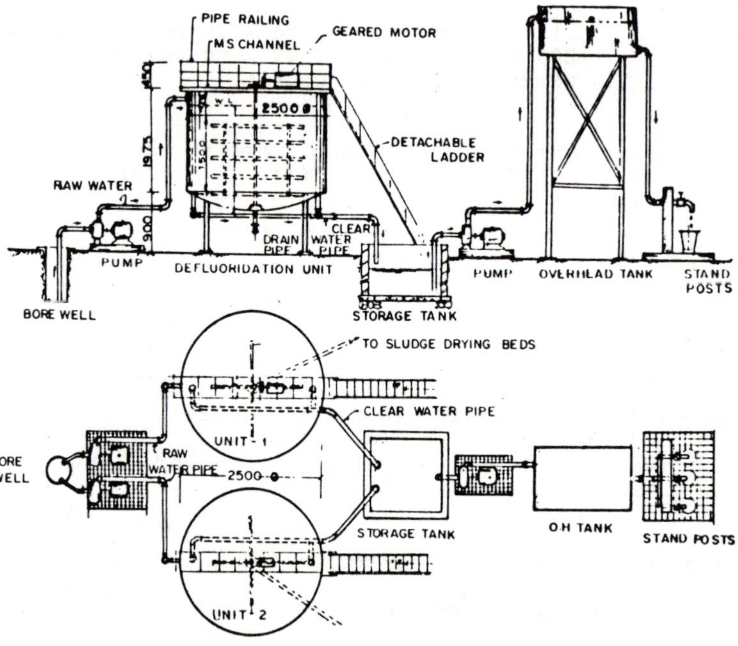

Fig.4 FILL-AND-DRAW DEFLUORIDATION SYSTEM FOR RURAL WATER SUPPLY

flocculation is 30 min and that for sedimentation is 2 hours. The treated water is collected into the sump and is pumped into the overhead reservoir from where it is supplied through stand posts. The entire operation is completed in 3-4 hours and the same tank can be reused for subsequent operations for a number of times. The number of tanks are decided depending upon the number of times each tank is reused and the water supply need of the village population. The scheme is recommended upto 5000 population for rural water supply. Based on Nalgonda Technique, six fill-and-draw defluoridation plants have been installed in the country. Out of these, three plants are installed in Andhra Pradesh, two in Gujarat and one in Haryana under Water Technology Mission Programme. The plants are functioning satisfactorily and the villagers are supplied with defluoridated water for drinking and other domestic purposes.

TECHNOLOGY AWARENESS CAMPS

Technology Awareness camps were held for demonstration of Nalgonda Technique to field engineers, social scientists and villagers. A National Defluoridation Camp was held at Amereli (Gujarat) and two Regional Camps at Palkudoddi (Andhra Pradesh) and Mehsana (Gujarat) under Water Technical Mission Programme. The Camps were aimed at training the representatives of the participating agencies of various states in the country afflicated with the problem of excess fluoride in drinking water. The activities of the camp programmes included (i) lecture and discussion sessions, (ii) demonstration of defluoridation technology at the camp site, (iii) visit to fluorosis affected villages, (iv) demonstration of Nalgonda Technique under field conditons, (v) evaluation of water quality and fixing the dose of chemicals for the identified sources and (vi) public meeting and demonstrations at important places.

COST ESTIMATES FOR FILL-AND-DRAW DEFLUORIDATION PLANT FOR RURAL WATER SUPPLY

The running cost of the defluoridation varies between Re 1.00 and Rs 5.00 per m^3 depending upon fluoride and alkalinity of the raw water. A typical example of cost estimates for 2000 population is given below :

Population	:	2000 @ 40 lpcd
Water Supply Need	:	80 M^3 per day
No. of operations of each tank per day	:	Two
No. of tanks required	:	Four
Capital cost of the the plant	:	Rs. 5,00,000

Water Quality :

Fluoride	-	5.0 mg F/L
Alkalinity	-	400 mg $CaCo_3$/L

Chemical doses :

Alum	-	600 mg/L
Lime	-	30 mg/L
Bleaching powder	-	5 mg/L

Total Annual Cost :

-	Depreciation at 5% p.a.	Rs. 25000
-	Interest at 12% p.a.	Rs. 60000
-	Personnel	Rs. 61200

Chemist 1 at Rs 1500 pm		Rs 1500
Helpers 4 at Rs 900 pm		Rs 3600
	Total	Rs 5100
	p.a.	Rs 61200

-	Maintainance at 5% p.a.	Rs	25000
	Total cost p.a.	Rs	171200
	per day	Rs	469

Chemical Cost :

-	Alum consumption, 48 kg/d at Rs 2/kg	Rs	96
-	Lime consumption, 2.4 kg/d at Re 0.70/kg	Rs	1.68
-	Bleaching powder, 0.4 kg/d at Rs 1.50/kg	Rs	0.60
	Total chemical cost per day	Rs	98.28

Power Cost :

Power consumption, 35 units per day at Re o.50/unit	Rs	17.50

Operational and Maintenance Cost :

Total Operational Cost/d	=	Rs 469 + 98.28 + 17.50
	=	Rs 584.78
Operational Cost/M^3	=	Rs 7.31

Running Cost :

Total Running Cost	=	Rs 98.28 + 17.50
	=	Rs 115.78
Running Cost/m^3	=	1.45

Annual running cost of defluoridation per capita is estimated as Rs 27.17.

REFERENCES

1. Nawlakhe, W.G., Kulkarni, D.N., Pathak, B.N. and Bulusu, K.R., 1974. Defluoridation of Water with Alum. Indian J. Env. Hlth. V. 16, p. A-F.

2. Nawlakhe, W.G., Kulkarni, D.N., Pathak, B.N. and Bulusu,K.R., Defluoridation of Water by Nalgonda Technique. Indian J. Env. Hlth. V. 17, p. 26-65.

3. Bulusu, K.R., Sundaresan,B.B., Pathak, B.N., Nawlakhe, W.G., Kulkarni, D.N. and Thergaonkar, V.P., 1979. Fluoride in Water, Defluoridation Methods and their Limitations. J. Inst. Engrs. (India), Env. Engg. Div., V. 60, Pt EN1, p 1-26.

MEASURES TO OVERCOME THE SALINITY PROBLEMS OF THE COASTAL AREAS IN GUJARAT

J.F.Mistry
Secretary, Water Resources Department
Government of Gujarat
Gandhinagar - 382 010

ABSTRACT

Misuse of land and groundwater in the vicinity of coastal area of Saurashtra region in Gujarat State has created manmade ecological imbalance resulting into a classical problems of persistent ingress of saline sea water in the coastal belt. This has affected the quality of groundwater in about 120 villages where 2,80,160 people reside. The paper highlights the causes of the problem, remedial measures planned for implementation and monitoring of the salinity ingress affected area. The achievements in reduction in total dissolved salts, improvement in land and crop etc. are also discussed.

INTRODUCTION

The State of Gujarat has the longest sea coast line i.e. about 1/3rd (1600 km) of the total coastal line of India (4800 km). Out of 1600 km long sea coast of Gujarat, Saurashtra and Kachchh cover a coastal length of about 1125 km. This coastal area which is a very fertile and productive tract is affected by salinity ingress due to manmade ecological imbalance created by continuous heavy withdrawals of groundwater for the purpose of irrigation and water supply combined with decreased recharges and poor rainfall conditions.

MOST CHRONIC AREA

The most heavily salinity affected region lies within a coastal strip of 160 km length between Una and Madhavpur (Fig.1). In this region, the fertile land got converted into saline land in about 35,000 ha in 1971 and process continued till about 1,00,000 ha of land became saline in 1977. The presence of salt content in groundwater beyond the acceptable limits in 120 villages posed problems for drinking water as well as irrigation. About 12,562 wells have gone out of commission. The population affected in the villages is about 2,80,160 souls. The other reaches affected by salinity ingress are Una-Bhavnagar, Madhavpur-Maliya and Maliya-Lakhpat in Saurashtra and Kachchh regions.

FIG. 1 SALINITY INGRESS IN THE COASTAL AREA OF SAURASHTRA & KACHCHH, GUJARAT STATE (INDIA)

PROBLEMS

The problem causing concern in this area is of highly cavernous and permeable nature of limestone in the coastal region where the saline water intrudes inland. The groundwater being the main source of irrigation besides the rainfall in the area, the effect of saline water intrusion is ruinous. Except a few check dams and minor tanks, no medium irrigation project has been undertaken so far to provide a permanent solution or as an alternative source for meeting the agriculturists' demand of irrigation water in view of the extensive well irrigation already existing.

With the development in agricultural techniques, the agriculturists switched over from conventional techniques of drawing water from wells to modern devices such as electricall operated pump sets of high efficiency. This concentrated local pumping resulted in lowering the water table, and it in turn paved the way for easy intrusion of saline water inland on account of reversal of hydraulic gradient. The result is the deterioration in the quality and reduction in the quantity of groundwater. The increase in the total dissolved salts further affected severely the agricultural output as also the income of the farmers.

In the last few years, inadequate rains have resulted in very low recharge to the groundwater body. The situation further aggravated when the farmers started utilising water from the wells which were adversely affected due to higher withdrawals and poor recharge and which were the only source of irrigation available to them. The use of the waters from these affected wells with the high percentage of total dissolved salts caused certain deterioration in the quality of agricultural soils. The persistent withdrawals from these wells caused persistent ingress of saline water inside the coastal belt and has now created the problem of inadequate drinking water supply in this area.

In the coastal regions of Saurashtra and Kachchh, salinity is found to have been caused due to different reasons. There are vast areas along the coast between Bhavnagar and Una which have turned saline due to inherent salinity in that saline area. These areas seem to have remained under the sea for quite a long time and the saline water held in the clay and silty strata in these regions with higher water table has caused salinity in the agricultural lands. In the reach between Madhavpur and Dwarka, it is observed that the southwest winds from the sea blow landwards almost at right angles. The winds bring saline water vapour and affect the coastal agricultural lands upto about 15 to 20 km inside. Similarly, the winds also carry saline silt and sand particles on to the agricultural lands. In the reach between Dwarka and Malia, several open salt pans have been responsible for spreading of salts in areas around these pans and so increasing in salinity of agricultural lands.

STEPS TO SOLVE THE SALINITY PROBLEM

The State Government was well aware of this grave situation and had formed a recharge committee in 1972. To find ways and means of combating the salinity ingress problem in the coastal area of Saurashtra and Kachchh, the Government of Gujarat appointed a High Level Committee-I for the reach from Madhavpur to Una in 1976. The Second High Level Committee was appointed in 1978 to suggest measures for the coastal reaches of Bhavnagar to Una, Madhavpur to Dwarka, Dwarka to Malia and Malia to Lakhpat. Both the Committees were headed by the Chief Secretary of the State.

These two High Level Committees studied the problem of salinity ingress in depth and recommended various remedial measures to arrest the ingress of salinity, such as

A Management Techniques

- i Change in the cropping pattern
- ii Regulation of groundwater extraction by groundwater legislation

B Recharge Techniques
- i Check Dams
- ii Recharge Tanks
- iii Recharge Wells
- iv Spreading Channels
- v Afforestation

C Salinity Control Techniques
- i Tidal Regulators and Bandharas
- ii Fresh water barrier
- iii Extraction barrier
- iv Static barrier

The High Level Committee-I suggested construction of more than 75 check dams, 7 recharge tanks, 200 recharge wells, 7 tidal regulators, 10 bandharas, spreading channels of 60 km length, static barrier of 67 km length, afforestation of 10,000 ha and also change in the cropping pattern. These recommendations are outlined in Figure 2.

The two Committees submitted their reports for the four reaches as under:

	Reaches of Saurashtra and Kachchh	Estimated cost (Rs. in crores)
I	Una-Madhavpur	64
II	Una-Bhavnagar	168
III	Madhavpur-Malia	370
IV	Malia-Lakhpat	186
	Total	788

A comprehensive statement showing the details of various measures, recommended by the two committees in each of the four reaches is given in Annexure I.

PROGRESS FOR THE HIGH LEVEL COMMITTEES-I AND II AREAS

In the first phase, the works costing Rs. 35.87 crores including 5 tidal regulators, 9 bandharas, 150-200 check dams, 1 recharge reservoir and 1000 nala plugs and 10,000 ha afforestation were planned to be taken up in 1980 with the aid of the World Bank. The World Bank while appreciating the technical recommendations of the High Level Committees, approved works of Rs. 29.7 crores.

Uptill now, 2 tidal regulators (Vadodara Zala and Vasoj), 7 bandharas (Sil, Sheriaj, Sangawada, Zankharwada, Lati, Badalpur, Barda), 20 recharge wells, 5 recharge tanks and 138 check dams are completed and 22 check dams are in progress. Moreover, the works of Meghal, Medha Creek, Muldwarka tidal regulators, Sodam bandhara and Vrajmi recharge reservoir are in progress. The work of 1516 nala plugs and afforestation in 2732 ha is completed.

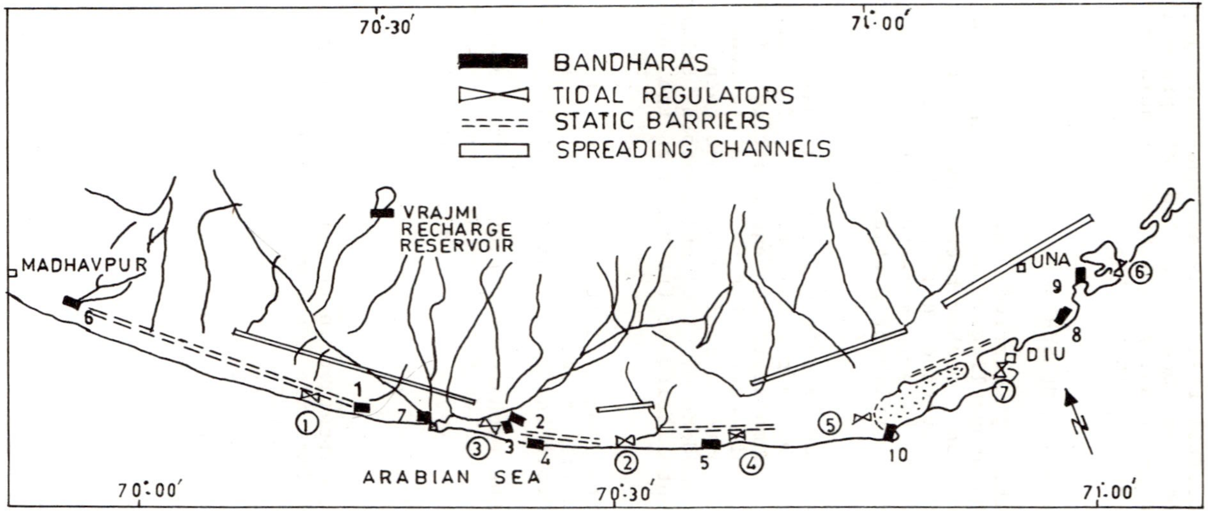

LIST OF BANDHARAS

1. ADRI 2. BADALPUR 3. LATI
4. HIRKOT FORT 5. BARADA 6. SIL
7. VERAVAL 8. JAKHARWADA 9. KHADA
10. SODAM

LIST OF TIDAL REGULATORS

1. MEGHAL 2. VADODARA ZALA
3. SOMNATH 4. MUL DWARKA
5. PIPALVA 6. SAYAD RAJPARA
7. VASOJ

FIG. 2

SALINITY CONTROL SCHEMES RECOMMENDED IN UNA MADHAVPUR REACH

Over and above the work taken up as stated above, 82 check dams in other coastal reach between Una-Bhavnagar and Madhavpur-Malia-Lakhpat have been completed at present and 100 check dams are in progress. Also 1 tidal regulator viz. Rukmavati in Kachchh area is in progress.

RESULTS OF SALINITY CONTROL MEASURES

As per the monitoring carried out in 1984 in Una-Madhavpur reach on the above lines, it is found that the salinity has been remarkably pushed back towards the coast between Chorwad and Una covering an improved area of about 18,000 ha out of 1,00,000 ha area affected. The quality of groundwater has improved (fully or partially) in 35 out of 120 villages.

With the construction of nearly 220 check dams in the affected area along the whole coastal strip of Saurashtra and Kachchh regions, a cumulative storage capacity of about 145 million cubic feet of sweet water has been created uptill now and the irrigation potential of 1740 ha has been created through lift irrigation done by the cultivators themselves during the intermittent filling of the basin on the upstream of the check dams.

In addition to the above lift irrigation facilities, there is a direct benefit of recharge of groundwater aquifers which ultimately controls the extent of salinity.

The details of the recharge techniques and estimation of recharge in each technique worked out in case of Malia-Lakhpat reach of Kachchh region of the Gujarat State are described below.

Check Dams

Check dams are constructed in the bed of rivers with regulation device to store the monsoon flow, which otherwise passes to the sea. The structures are provided with Godbole gates, having automatic operation facilities to store water during the recession of each flood. During prolonged stream flow periods, the gates remain open so that the flood flows down the stream channel unrestricted. Through the repeated filling of the pool behind the gates, the water will percolate through the wetted lake areas into the ground. Some water will also be available for supplementary irrigation to Kharif crops by lift irrigation, which in turn will add to the recharge of soil in the command. It is proposed to construct 740 check dams in the entire reach of Malia-Lakhpat.

The reservoir spread of the check dams on an average is taken as 10 ha. Recharge rate is taken as 25% of the seepage loss of 2.4 cumecs/Mm^2 (8 cusecs/Million sq. ft.) as in the case of unlined channels as per Indian Standards. This assumption is based on the fact that most of the check dams will be constructed on rocky formation. The recharge rate would work out to be 0.6 cum/Million sq. metre of the submerged area.

Considering the effective period for the recharge through check dam beds as 50 days in a year, the total groundwater recharge from 740 check dams works out to be

$$\frac{740 \times 10 \times 0.6 \times 3600 \times 24 \times 50}{10^6}$$

$$=193 \ Mm^3/Year$$

Recharge Tanks

Some check dams are proposed to be used for diverting surplus water to recharge tanks in nearby low depression areas. The water spread is assumed to be 10 ha for each tank. For two months as effective recharge period, the recharge from 25 recharge tanks in the area, at a rate of 0.6 cumecs/Million sq.m of spread on tank bed works out to be

$$\frac{25 \times 10 \times 0.6 \times 3600 \times 24 \times 60}{10^6}$$

$$= 8 \ Mm^3/Year$$

Recharge Wells

Recharge wells are dug like open well with 3 m to 4 m diameter. They will then be filled with rubble and sand filter at top. These wells will be useful to recharge pervious aquifers located at some depth below ground level. The recharge rate considered for recharge wells is of the rate of 4500 litres/ha. The period considered for recharge is only 20 days, assuming a frequency of 4 floods with 5 days duration on an average during every monsoon.
Considering 150 recharge wells in the region, the recharge capability of recharge wells works out to be

$$\frac{150 \times 4500 \times 24 \times 20}{100}$$

$$= 0.30 \ Mm^3/Year$$

Gully Plugs

Total 20,000 gully plugs are proposed to be constructed in the region. Impounding of water behind these plugs will be 3675 cu m. The recharge rate is worked out to be 20% of the total water impounded which is worked out as under

$$\frac{20000 \times 3675 \times 0.2}{10^6}$$

$$= 1.5 \text{ Mm}^3/\text{Year}$$

Annexure-II shows the total groundwater recharge by various schemes and Benefit/Cost in each reach of Saurashtra and Kachchh areas of the Gujarat State.

ESTIMATION OF EXTRACTION RATE

The analysis of the data of quantum of withdrawal and the quantum of natural recharge made for Una-Madhavpur reach of Saurashtra region for the year 1977, which was a good year, shows that the drawal was 102 Mm^3 of water which was more than the natural recharge. Against this, 83 Mm^3 of water can be added as a result of implementation of various methods of recharge. Thus, inspite of these schemes there will remain over withdrawal to the extent of 102-83=19 Mm^3. The twofold measures against the excess withdrawal are

i to enforce groundwater legislation for future expansion of construction of wells, putting engines, electrification etc. and

ii to take steps to direct the farmers to reduce their present withdrawals to achieve groundwater balance.

BENEFIT COST RATIO

The normal method of working out benefit cost analysis applicable to irrigation project will not be applicable for recharge of salinity control schemes, as such schemes have in addition to direct benefits many more indirect benefits. The cost benefit ratio has therefore been worked out on direct and indirect benefits from flood protection works supplemented by direct benefits of lift irrigation.

i	Total cost of the project (Madhavpur-Una reach)	Rs.	19.54 Crores
ii	Total benefit (direct & indirect)	Rs.	32.60 Crores
	Therefore the overall benefit/cost		1.67

MONITORING OF SALINITY INGRESS

In order to study the extent of salinity ingress areawise and qualitywise, special monitoring cell has been created. It carried out the macro-level geophysical and geohydrological investigation along the coastal tract through the following reachwise numbers of observation wells

	Reach	No. of observation wells
1	Bhavnagar to Mahuva	169
2	Mahuva to Una	153
3	Madhavpur to Okha	215
4	Okha to Malia	156
5	Coastal tract of Kachchh	245
6	Una-Madhavpur	257

The scope of monitoring covers the following

- i Observation of water level fluctuation through a network of observation wells during premonsoon and postmonsoon every year to determine the seasonal variation.
- ii Observation of chemical quality of groundwater to monitor seasonal and long term variations.
- iii Conducting repeat geoelectrical soundings on oriented profiles every year to determine movement of saline/ fresh water interface, if any.

Evaluation and corelation of geoelectrical and geohydrological data gives a comprehensive picture of variation in the areas and subsurface horizon.

Results of Study

The large amount of data so obtained from macro-level studies was processed and evaluated with a view to projecting any movement in saline/fresh water interface and variations in the quality of groundwater. The results donot indicate any drastic change in five reaches except Una-Madhavpur, because the reversal of the effects of the salinity is very slow process. Moreover, the effectiveness was arrested due to failure of monsoon for consecutively three years in the state.

However, due to the partial implementation of the scheme (138 check dams, 2 tidal regulators, 7 bandharas, 5 recharge tanks, 20 recharge wells and 1516 nala plugs) in the Una-Madhavpur reach as well as the good rains in 1988 monsoon have reactivated the recharge process in the area. The fact is established from the postmonsoon monitoring of the water table and water quality (October, 1988) which indicates the rise in the water table and improvement in water quality. This interim set of observations reveal that improvement in the water quality (TDS $\angle$ 2000 PPM) is observed in an area of approximately 25000 ha after obliterating the reverse situation existed in May 1988.

CONCLUSION

We will need all the scientific expertise to evaluate critically what we have done so far. We must also bring to bear the results of research in science and technology towards evolving more effective and economical ways of containing salinity in the coastal area in Gujarat and elsewhere in the country. Sundervan in West Bengal and some parts of Kerala State in the country are also suffering from salinity ingress problems. The Water Resources Department of Gujarat State has tackled the problem effectively in Saurashtra and Kachchh regions and the Government of Gujarat will be happy to help other States to solve the problems of the salinity ingress in their coastal areas.

REFERENCES

Mistry J.F., July, 1983. Salinity Ingress in the coastal area of Saurashtra - Gujarat State (India), ICID Bulletin, Vol. 32, No.2, p 52 to 60.

Report of High Level Committee-I,1977 and Report of High Level Committee-II, 1983 & 1984. To examine the problems of salinity ingress along the coastal area of Saurashtra and Kachchh, Government of Gujarat.

ANNEXURE-I COMPREHENSIVE STATEMENT SHOWING THE DETAILS OF VARIOUS WORKS/MEASURES RECOMMENDED BY THE HLC-I AND HLC-II IN THE SAURASTRA AND KACHCHH COASTAL AREA

Sr.No.	Description	Unit	Una-Madhavpur HLC-I	Una-Bhavnagar HLC-II	Madhavpur-Malia HLC-II	Malia-Lakhpat HLC-II
1	Length	Km	160	180	425	360
2	Affected area	Ha	1,00,000	1,40,000	4,60,120	3,65,200
3	Affected population	Souls	2,80,160	2,20,200	5,79,373	2,50,000
4	Affected Wells	Nos.	12,562	8,800	11,388	10,478
5.	Affected villages	Nos.	120	166	248	245
6	Estimated cost	Rs.in crores.	64	168	370	186
7.	Land Reclamation	Ha.	-	10,000	15,500	10,000
8.	Irrigation potential	Ha.	19,850	33,113	65,640	37,545
9.	Check dams	Nos.	75	300	460	740
10.	Tidal Regulators	Nos.	7	8	23	15
11.	Bandharas	Nos.	10	29	48	40
12	Afforestation	Ha	50,000	15,750	25,000	14,000
13	Nala plugs	Nos.	10,000	20,000	45,400	20,000
14	Recharge wells	Nos.	200	450	680	150

ANNEXURE = II

TOTAL GROUNDWATER RECHARGE

AND BENEFIT/COST IN VARIOUS REACHES BY VARIOUS SCHEMES

Sr. No.	Schemes	Recharge in MCM/Years			Benefit/Cost		
		Bhavnagar-Una	Madhavpur-Malia	Malia-Lakhpat	Bhavnagar-Una	Madhavpur Malia	Malia-Lakhpat
1	Check dam	155.50	238.46	191.80			
2	Recharge Tanks	9.33	18.66	7.77		Overall benefit / Cost	
3	Spreading Channel	6.98	30.79	5.87	1.27	1.23	1.55
4	Recharge well	2.16	1.46	0.32		Benefit	
5	Gully Plugs	-	33.37	1.47		Cost	
6	Infiltration recharge from applied irrigation through lift.	14.00	97.70	67.07	1.61	1.64	1.45
	Total	187.97	420.44	247.30			

Safeguards for Groundwater Pollution in India

Rameshwar D.Verma
Department of Civil Engineering
Malaviya Regional Engineering College,Jaipur 302017(India).

ABSTRACT

Groundwater usage has increased at an accelerating rate in recent years and indications are that the trend will continue. Groundwater pollution has now become a major issue in the development of water resources, management of natural resources and the protection of the environment. Since groundwater is an important reliable source of drinking water in India, protection of its quality should become a high priority management goal.

Many cases of groundwater pollution in India have been reported in the literature. These have been discussed and reviewed. Important causes of groundwater pollution as identified are: haphazard urban development without adequate attention to sewage and waste disposal, rapid industrialization without proper treatment and disposal of waste products, insanitary dumping of refuse and other solid wastes near aquifers, excessive use of fertilizers and pesticides for agricultural development and poor drainage in agricultural soils. Future trends of groundwater pollution in India are projected. New uses of groundwater reservoirs for waste storage and artificial recharge, if not properly planned and designed, would seriously impair groundwater quality.

Experiences of developed countries for safeguarding their groundwater are presented and discussed.Presence of organic chemicals, especially solvents in groundwater supplies are reported in the United States and other industrialized countries. The hazardous wastes generated every year in the developed countries, when not properly disposed, caused severe groundwater pollution. These experiences are analyzed in the context of industrial development and water usage in India. If adequate longterm safeguards are not imposed to prevent groundwater pollution in the country, the quality of groundwater may become an alarming limiting factor for the use of this precious resource.

****

SAFEGUARDS FOR GROUNDWATER POLLUTION IN INDIA

RAMESHWAR D.VERMA
DEPARTMENT OF CIVIL ENGINEERING
MALAVIYA REGIONAL ENGINEERING COLLEGE
JAIPUR 302017 (INDIA)

INTRODUCTION

During the last decade groundwater pollution has become a major issue in the management of natural resources and the protection of the environment. As nearly one fifth of all water used in the world is obtained from groundwater resources and in many areas groundwater is the only fresh source, protection of its quality has become high priority management goal.

In the past, water resources planning was devoted largely to development of supplies to meet quantitative demands. Increasing attention is now being devoted to quality of water and its management. It is generally recognized that both aspects of water resources - quantity and quality - are important and inter-dependent. New uses of groundwater reservoirs for waste storage, artificial recharge, gas storage, and other purposes, raise difficult questions about groundwater quality and its protection. A contaminated surface water reservoir may be restored to health in a few years or less; parts of a groundwater reservoir may be rendered useless for centuries by indiscriminate introduction of foreign, poor quality substances.

This paper reviews some case studies of groundwater pollution in India. Causes of groundwater pollution have been described. Future trends of groundwater pollution in India are projected. Safeguards to prevent pollutiom are identified and discussed.

CASE STUDIES OF GROUNDWATER POLLUTION

Many cases of groundwater pollution in India are reported in literature. Some of these are discussed below(Anand Lal - 1984):

Groundwater quality of twenty-seven wells in Jaipur was investigated during 1969-1971 by Olania and Saxena (1977). Located in 20 m to 6 km proximity from selected six(6) open refuse dumps, these wells exhibited the pollution of groundwater by the presence of high concentrations of dissolved solids, chlorides, iron and COD as a results of leachates from the refuse dumps.

In another investigation Olania et al (1978) reported high MPN count of total coliforms, faecal coliforms and faecal streptococcus, and high concentrations of chlorides, nitrites, nitrates and sulphates and COD in water of wells located at 30 to 60 metre (average) distance from sewage farms at Jaipur.

Studies conducted by Subha Rao and Venkata Rao (1973) revealed that water supplied from an infiltration gallery constructed in the bed of non-perennial river SARDA (Ankapalli Town - Andhra Pradesh) exhibited severe taste, odour, colour and filamentous growth which was a result of indiscriminate discharge of untreated sugarcane distillery wastewater into the river bed, as reported by them.

Sridhar and Pillai (1973) reported the pollution of groundwater at Bangalore. They conducted field and laboratory investigations of water samples collected from wells located in the neighbourhood of (1) the sewage channels on the outskirts of Bangalore City, (ii) Septic Tanks on the campus of the Indian Institute of Science and (iii) the septic tanks of four industrial establishments. Results of investigations revealed that all the wells exhibited sewage pollution.

In a geohydrological investigation in northern and south-western Haryana localised rise in nitrate concentration from 43 mg/l at Singhani to 1920 mg/l in Narnaul groundwater was observed by Kakkar (1981). In the absence of any geological source of nitrate in the area and non-uniform distribution of nitrate along the direction of groundwater movement, nitrate pollution is likely to be caused by sewage and agricultural waste.

From the field studies in Ludhiana in Punjab, Kakkar and Bhatnagar (1981) concluded that groundwater at shallow depths of aquifer near few bicycle factories had been polluted by haxavalent chromium and cyanides in concentrations ranging from 3.5 to 12.9 mg/l and 0.05 mg/l to 0.98 mg/l respectively. Other trace elements such as copper, Zinc, Cadmium, Cobalt, Molybdenum, Strontium, Lithium and Silver were also detected in groundwater in different concentrations, although below the permissible limits for drinking water.

In another investigation analysis of water samples drawn from the hand pumps installed in Industrial Area of Ludhiana, Punjab revealed that groundwater had been contaminated by haxavalent Chromium, Copper, Zinc and Iron in concentrations ranging from 0.05 to 1.75 mg/l, 0.04 to 0.16 mg/l, 0.05 to 0.17 mg/l and 0.71 to 4.0 mg/l respectively. (Goyal et al 1981)

Well water quality monitoring was carried out by Tanwar (1981) for four years. He observed continued increase in concentration of salts (such as chloride and sulphates of calcium and magnesium) as a result of increased use of fertilizers (4 to 57 kg per hectare within 12 years) in Haryana. Author suggested that in case no preventive step is taken the groundwater pollution could become a serious problem.

Quality of groundwater in Warangal Andhra Pradesh has deteriorated because of land disposal of untreated waste water from tannery and textile mills which found access to groundwater. Fortunately the pollution had travelled only 5 metres from the disposal site and wells located in sandy loam and rocky formations at 150 metres from disposal site were found unaffected (Naram 1981)

In Tamilnadu also, the tannery waste is posing problem to ground water quality. Krishnaswamy and Haridass (1981) reported the results of ground water quality investigations along a seasonal river Palnar that receives organic/inorganic wastewaters from tanneries located at Vaniyambadi and Ambur (close to the origin of river). A significant rise in total dissolved solids, salinity and hardness in wells located near factories led the authors to conclude that preventive measures must be instituted immediately to control groundwater pollution.

Handa et al. have reported cases of groundwater pollution in south-western. Uttar Pradesh. The investigators noted Nitrate concentrations between 112 mg/l to 694 mg/l at fourteen different places (1982). In other investigation (1983a) authors noted high concentrations of Chromium (21,000 microgram Cr/l from handpump near National Chemicals Factory Kalpi Road, Kanpur, 2000-2270 microgram Cr/l from another hand-pump installed nearby, and 31,000 microgram Cr/l in water of a tubewell in the compound of Orient Packing Limited Faridabad). Authors could not notice any other chromium source near water sampling point in these cases. In yet another investigation the same authors noticed (1983b) Potassium concentration over 200 mg/l in eight dug wells and 1050 mg/l in a well at Agra. In the absence of any geological source for potassium - deposites in Uttar Pradesh, authors suggested that such high values owe their origin to the excessive and unscientific use of potassium fertilizers coupled with poor water resources management. The Commission of European Community has limited potassium concentration to 10 mg/l in potable water. Detailed dscription and review of ground water pollution cases throughout the world are given by Todd and Mc Nulty (1976).

CAUSES OF GROUNDWATER POLLUTION

Pollution may be caused due to spreading of the pollutant on the ground surface or by the direct injection of the pollutants into the aquifer system. Degree and extent of the pollution larely depends on the quality and concentration of the pollutants and the porosity, permeability, hydraulic gradient and storage capacity of the aquifer, besides the time factor. Specific causes of pollution are:

(i) One of the most important cause of groundwater pollution is haphazard urban development without adequate attention to sewage and waste disposal. Sewage disposal without adequate treatment is a widespread phenomena in ill-planned or even well planned development of urban conglomerations, which may result in entry of bacteria, nitrogen,phosphorus and other pollutants such as ABS.

(ii) Rapid industrialization without provision of proper treatment and disposal of waste and effluent is another source of pollution. Industries such as sugar factories, textile mills, distilleries, tanneries, mineral and metal processing industries and chemical industries may generally cause pollution. The waste products of industries may contain hazardous substances such as arsenic, haxavalent chromium, cadmium and cyanide in toxic concentrations.

(iii) Excessive use of fertilizers and pesticides for agricultural development may pollute groundwater resources. Many of the pesticides and herbicides are cumulative poisons.

(iv) Another factor of groundwater pollution is due to some geological formations, mineral deposits and mineral benefaction plants releasing excessive dissolved substances.

(v) Waste products of industries may be a potential source of ground water pollution. Oceanic and atmospheric nuclear explosions and waste products of nuclear power plants may cause the contamination of ground water due to seepage of polluted surface water into the ground.

(vi) Pollution may also occur due to subsurface disposal of wastes through pits, mines, wells and septic tanks. Artificial recharge if not properly planned and designed may pollute groundwater. Saltwater intusions due to excessive pumping of fresh water in coastal aquifers may be another cause of ground water pollution.

EXPERIENCES OF DEVELOPED COUNTRIES AND FUTURE PROJECTIONS FOR INDIA

(i) Hazardous and Toxic Wastes:

In the past, the main thrust of the United States water pollution control programme was to abate traditional pollution - i.e. Biochemical oxygen demand (BOD), total suspended solids (TSS) and heavy metals. Now the U.S. Environmental Protection Agency (EPA) is striving to control hazardous and toxic pollutants in industrial discharges and groundwaters. After the passage of the Resource Conservation and Recovery Act in 1976, the industries in the United States are pre-occupied with how to manage their hazardous waste (a list of these wastes have been promulgated by EPA in 1980). A typical item on this list would be : wastewater treatment sludges from electroplating operations or air-pollution control scrubber sludges from coke-ovens and blast furnaces. There are about 100 different waste streams on this list including wastes from such industries as wood preserving, organic chemicals, pesticides, explosives, petroleum refining, leather tanning and finishing, iron and steel, primary copper, primary lead and primary zinc. Beyond this, EPA has another list of toxic chemicals or components that are often found in these industrial wastes. Thus, any firm discharging a waste stream containing one or more of these components is considered to have a hazards waste. A waste can be considered hazardous if it is : ignitable , corrosive, reactive, or toxic as determined by the Extractive-procedure Toxicity Test (Dallaire 1981). Morrison (1981) reports that organic chemicals, expecially solvents are showing up in groundwater all over the United States and in many other industrial countries as well. As local drinking water problems multiply , and ways of treating them continue to develop, more communities may choose - or be forced - to treat their well water. Krill and Sonzogni (1986) reported presence of volatile organic chemicals in wisconsins groundwater in the United States. Screening of 1174 community wells and 617 private wells showed that 65 community wells had detectable levels of volatile organic chemicals (VOC) and that 5 community wells had concentrations above the recommended health advisory level. Also 82 private wells had detectable levels of VOC, 14 of which had concentrations above the health advisory.

With the increased industrial and agricultural development in India in the future, the waste products would find access into groundwater of the country. Unless strong preventive measures are taken, our ground water resources would be polluted of hazardous and toxic wastes.

(ii) Agricultural Impact:

Contamination of groundwater in suffolk country Newyork (USA) by agricultural chemicals (fertilizers, pesticides, herbicides, nematocides and fungicides) has been reported by Baier et al (1987). During this testing, 101 agricultural or organic compounds were evaluated, 41 were found in the ground water (four of which were in concentrations greater than the permissible limits). Nitrates from fertilizer applications were also present in quantities exceeding the primary drinking water standard. In another study Lau and Mink (1987) reported that Ethylene dibromide (EDB) and 1,2-dibromi-3 chloropropane (DBCP), which are organic chemicals used to kill nematodes in pineapple fields, were found in the groundwaters of Hawaii. The presence of EDB and DBCP means an agricultural impact on ground water. Both EDB and DBCP are known to cause cancer and various other toxic effects in test animals.

India has already seen a green revolution. With the increased demand of food and fibre in the country we would have to further modernize our agricultural system. This would require increased use of fertilizers, pesticides and herbicides which may find access into our groundwaters. The likelihood of pesticides contamination depends in part on an areas hydrologic characteristics.

(iii) Radio-active Wastes:

The injection of low and intermediate level radioactive wastes into deep geologic formations is practiced at many places. These wastes may pollute the groundwater aquifers.

Todd (1976) described and reviewed the cases of groundwater pollution due to radioactive wastes. Distribution of radionuclides in groundwater has been reported from underground nuclear explosions. Migration characteristics of various radionuclides through the sediments of Hanford plant (USA) gathered through an extensive network of monitoring wells and sophisticated monitoring equipment have been reported by Brown (Todd-1976).

The use of atomic energy for peaceful purposes is going to increase in India. Many nuclear power plants are likely to be constructed to meet the growing shortage of energy in the country. Unless the wastes generated from the power plants are disposed properly, there is a risk that the wastes may find access into our groundwaters.

(iv) On-site sanitation and Waste Disposal System:

There is risk of groundwater pollution due to septic tank effluents, landfill leachates and on-site sanitation such as pit or bore-hole latrines and other unsewered waste disposal systems. The problem due to septic tank effluent has been described in detail by Kreiss(1978). Lewis et al (1982) have discussed and reviewed the risk of groundwater pollution by on-site sanitation in developing countries. They have concluded great need for more research into the problem and have found the unsewered waste disposal systems to cause nitrate contamination of groundwater in may hydrogeological environments. Ostendorf(1985) has described a case of unconfined aquifer contamination by waste lagoons.

There is already ground water pollution in India due to the above effects. Many of these have been reported in this paper under 'Case studies of groundwater pollution'. The situation may further worsen due to unplanned urban development and unsatisfactory waste (solid as well as liquid) disposal systems which might occur due to population pressures.

SAFEGUARDS TO PREVENT GROUNDWATER POLLUTION

(i) Wastewater standards:

Treated or untreated wastewater should not be allowed to be discharged on land, particularly if there is an aquifer below it, unless it is ascertained that it will not contaminate the groundwater resources. Effluents from septic tanks, leachates from landfills and on-site sanitation systems should be carefully investigated so as to meet minimum water quality standards for the prevention of groundwater pollution. Such standards should be formulated for different wastewaters and for different soil, geological and hydrogeological parameters. Verma and Mathur (1985) have described a methodology to develop such standards. No new sanitary landfill or septic tanks sludge disposal site should be allowed within the aquifer semsitive area.

(ii) Hazardous and Toxic Wastes:

Ground water protection standards must be established for each substance that has already been detected in groundwater or that has a reasonable probability of entering the groundwater. For each substance an 'enforcement standard' and a 'preventive action limit' should be set. According to the U.S. Environmental Protection Agency, Enforcement standards define when a violation has occured and subject the activity, practice or facility that is the source of the substance to immediate enforcement action. The preventive action limit represents a lesser concentration of the substance than the enforcement standard. The preventive action limit serves two purposes. First, it is used as a criterion in the design of facilities or management practices. Second, it serves as the trigger for instituting remedial actions to correct contamination problems before they progress to the stage that enforcement standards are attained or exceeded.

The disposal of solid or liquid wastes, especially hazardous and toxic industrial wastes, should not be allowed underground unless their long term effects on groundwater have been investigated to prevent pollution.

(iii) Agricultural practices:

Excessive use of fertilizers, pesticides and herbicides should be avoided particularly from agricultural areas adjacent to sensitive aquifers. Irrigation return flows should preferably be allowed to join a surface water source through proper drainage system or be suitably treated before being allowed to join the groundwater table.

(iv) Legislation:

There is need of a groundwater protection act in the country. The existing Environment (Protection) Act 1986 doe not fully consider the interests of groundwater resources. The main provisions of the groundwater protection act should include establishment of groundwater standards, compensation for owners of contaminated private wells, expanded groundwater monitoring and sampling programs. The act should provide stringent actions against industries which discharge hazardous and toxic wastes without adequate treatment into groundwater. There should be provisions of land-use management and pesticides control as longterm means to protect groundwater sources.

(v) Water Quality Monitoring:

Regional or long-term changes in water quality should be monitored by establishing a few observation points throughout the area covered by the plan. Where the objective is simply to monitor trends in water quality change, annual sampling may be adequate. More closely spaced data points and more frequent sampling would be needed for special water quality problems, such as sea or other saline water intrusion, upwelling of connate saline water, or artificial recharge of an aquifer with water quality inferior to that which is naturally recharged or in areas where there is potential pollution as a result of liquid or solid waste disposal.

The objective of the monitoring may be to show the extent of contamination, not necessarily the sources of contamination. Groundwater movement is exceedingly slow, and pollutants disperse considerably. This means traces of contaminants migrate far in advance of the bulk. Thus, warning of groundwater pollution is possible if monitoring capability at very low concentrations (ng/l) and standards are available.

(vi) Public Awareness and institutional Measures:

In many cases of groundwater pollution or potential pollution, remedies or prevention of contamination are institutional instead of technological. If people can be convinced that an aquifer is a precious resource in need of safequarding, local political processes can be mustered to prevent degradation. Protection of an aquifer is environmentally and economically more desirable than curing an already contaminated body of groundwater. However, such preventive actions do not come without a significant investment of local funds, effort and coordination. Communication with citizens and the news media on technically uncertain topic that involves human health may prove to be difficult in a time of crisis. Pre-educate the public. Assist the news media with interpretations of technical materials and develop a protocol to verify reports. correct errors and stop rumours and false or misleading stories. Confirm laboratory reports before acknowledging contamination. Educate the public about the assessment and management of risk to human health posed by polluted groundwater.

CONCLUSIONS

Many cases of groundwater pollution in India and elsewhere have been reported. Causes of such pollution are discussed. In the light of the experiences of the developed countries future trends of groundwater pollution in India are projected. For safeguarding the quality of our groundwater resources we need to develop wastewater standards for municipal and industrial wastes alongwith hazardous and toxic wastes, should avoid excessive use of fertilizers, pesticides and herbicides in our agricultural operations through legislation amd should develop water quality monitoring system and create public awareness to the problem.

REFERENCES

1. Anand Lal "Waste water standards for preventing groundwater pollution in Semi-arid-land". August 1984. M.E. Dissertation presented to the University of Rajasthan under the guidance of Dr. R.D. Verma.

2. Baier, J.H; B.W. Lykins Jr; C.A. Fronk and S.J. Kramer, 'Using Reverse osmosis to Remove Agricultural Chemicals From Groundwater'. Journal, American Water Works Association. August 1987. Vol. 79 , No.8, p. 55-60.

3. Dallaire, G, "Industry Challenges EPA on whether certain wastes are hazardous or not." Civil/Engineering, ASCE. September 1981, p. 79-81.

4. Goyal ,M.R ; O.P.Abroi and A.K.Vohra, 1981. "Pollution of upper aquifer in punjab ". Proceedings of an international symposium held in the Netherlands (March 23-27, 1981), p. 105-110.

5. Handa, B.K ; D.K.Goel ; A.Kumar and T.N. Sondhi, 1982 pollution of Groundwaters by Nitrates in Uttar Pradesh. Technical Annual IX journal Indian Association for water pollution control, p.95-103.

6. Handa, B.K. ; D.K.Goel ; A.Kumar and T.N.Sondhi 1983 a. Pollution of Groundwater by chromium in Uttar Pradesh. Technical Annual X, journal Indian Association for water pollution control, p.167-176.

7. Handa, B.K. ; D.K.Goel ; A.Kumar and T.N.Sondhi 1983 b. Pollution of Groundwater by potassium in Uttar Pradesh. Technical Annual X, journal Indian Association for water pollution control, p.177-186.

8. Kakkar, Y.P. 1981. 'Nitrate pollution of Groundwater in southern and South Eastern Haryana, India'. Proceedings of an international symposium held in the Netherlands, March 23-27, 1981. p.125-129.

9. Kakkar, Y.P. and N.C.Bhatnagar, 1981 Groundwater pollution due to industrial effluents in Ludhiana, India. Proceedings of an international symposium held in the Netherlands (March 23-27, 1981), p. 265-272.

10. Kreiss. J.F. 1978. Management of small waste Flow . Small Scale Waste Management Project. University of Wisconsin, U.S.A. U.S. Environmental Protection Agency Report No. EPA - 600/2 - 78-173.

11. Krill, R.M and W.C. Sonzogni, " Chemical Monitoring of Wisconsins' Groundwater". Journal, American Water Works Association, September 1986, vol 78, No. 9 p.70-75.

12. Krishnaswamy, R and G. Haridass 1981. " Groundwater pollution by tanneries in Tamil Nadu (India). Proceedings of an Inernational symposium held in the Netherlands (March 23-27, 1981), p.287-290.

13. Lau, L.S. and J.F. Mink, " Organic contamination of Groundwater : A Learning Experience." Journal, American Water Works Association, August 1987. Vol. 79, No.8, p. 37-42.

14. Lewis, W.J., S.S.D Foster, B.S Drasar, 1982. 'The Risk of Groundwater Pollution by on-site Sanitation in Developing Countries.' Report No.1. International Reference Centre for Waste Disposal, Switzerland.

15. Morrison, A. ' If your citys' well water has chemical pollutants, then what?' Civil Engineering, ASCE. September 1981. p. 65-67.

16. Naram, K.R. 1981. "Groundwater pollution in Warangal Town, Andhra Pradesh, India". Proceeding of an international symposium held in the Netherlands (March 23-27, 1981) p.163-168.

17. Olania, M.S and K.L.Saxena, 1977. Groundwater Pollution by open Refuse Dumps at Jaipur. Journal Environmental Health (India) vol 19 (3) p. 176-188.

18. Olania, M.S; W.G.Nawlakhe, H.C.Sharma and K.L.Saxena, 1978 pollution studies of well waters in sewage farms at jaipur, journal Environmental Health (India), vol 20 (4), p 398-412.

19. Ostendorf, D.W. 1985, Unconfined Aquifer Contamination by Waste Lagoons,' Proceedings of the National Water Well Association Conference on Practical Applications of Groundwater Models, Ohio State University, August 19-20, 1985 P.278-289.

20. Sridhar, M.K.C and S.C. Pillai, January 1973. pollution of Groundwater at Bangalore, Proceedings of the symposium on Environmental pollution published by NEERI Nagpur, India (1973) p. 59-68.

21. Subba Rao, P and Venkatarao, 1973. Groundwater pollution by industrial wastes : A case study of iron infilteration gallery waters, journal Environmental Health (India) vol. 15 (1), p. 14-20.

22. Tanwar,B.S. 1981 " Groundwater pollution and its protection in Haryana, India ". Proceedings of an international symposium held in the Netherlands (March 23-27, 1981), p.1097-1100.

23. Todd, D.K. and D. McNulty " Polluted Groundwater - A review of the significant literature " 1976. Water information center, Inc Huntington, Newyork.

24. Verma, R.D. and A.L. Mathur 1985 " Wastewater standards for Preventing Groundwater Pollution in semi-arid lands." Proceedings of the National Water Well Association Conference on Practical Applications of Groundwater Models. Ohio State University, August 19-20, 1985. p. 311-325.

****

Hydrochemical Classification of the Groundwater of the Wadi al Bih Wellfield in the Northern Part of the United Arab Emirates

Faig ElSchami and Abdul Rahim Hamdan

Department of Geology, U.A.E. University (P.O.Box 15551 Al Ain)

ABSTRACT

The wadi al Bih wellfield is considered one of the most important fresh water wellfields in the northern Emirates. Hydrochemistry of the groundwater is discussed in detail to show the variability areal distribution of the major groundwater chemical types and the change of the groundwater quality with time. Nine major groundwater types are identified and delineated. The overall chemistry of wadi al Bih groundwater indicates that alkalis exceed alkaline earths and strong acids exceed weak acids. The salinity of the production water wells ranged from 360 to 3670 mg/l in 1987, and from 435 to 4140 mg/l in 1988.

INTRODUCTION

The United Arab Emirates is predominantly an aried country. Fresh groundwater is considered a matter of prime importance for domestic use and for irrigation. The wadi al Bih wellfield is one of the most important fresh water wellfields in the Emirate of Ras Al Khaimah. It is located in the Ruus al Jibal mountains in the northern part of the United Arab Emirates (Fig.1). Since 1980 the Ministry of Electricity and Water has drilled 23 production water wells for the purpose of drinking water utilization. The wellfield is presently supplying

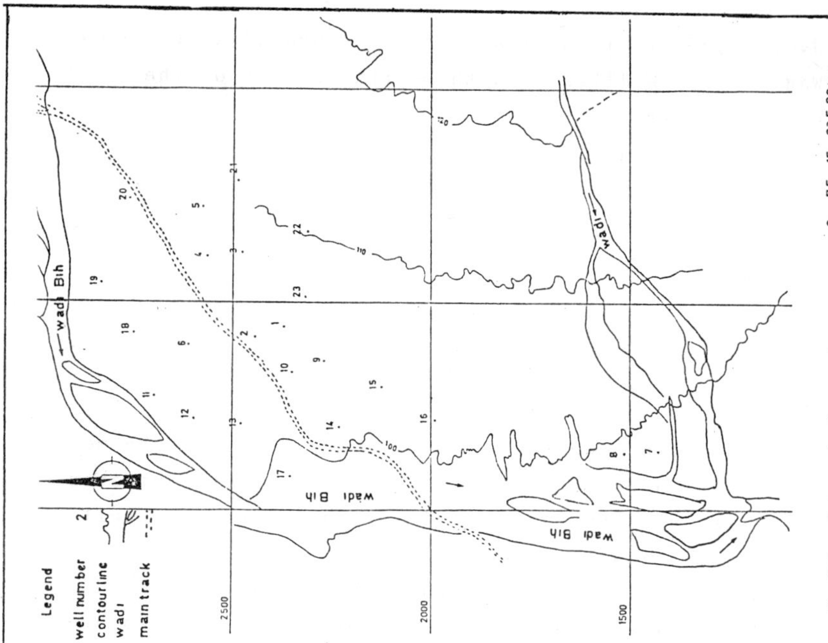

Fig. 2 Distribution of watersamples

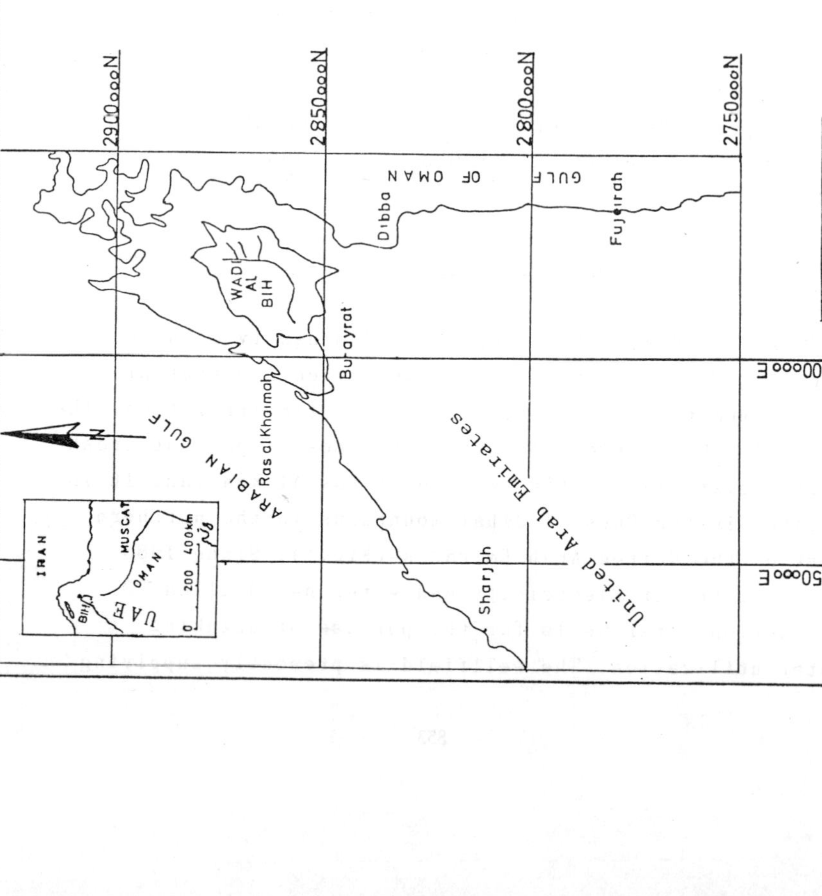

Fig. 1 Location map of wadi al Bih

over 4 million gallons per day to the city of Ras Al Khaimah The groundwater is contained in the valley fill which consists of carbonate blocks, boulders, cobbles and pebbles derived from the Ruus al Jibal Group, Elphinston Group. and Musandam Group. This stratigraphic succession ranges in age from the Permian to the Lower Cretaceous.

Data on chemical analyses during 1987 were obtained from the Ministry of Electricity and Water, while chemical analyses of 1988 were carried out by the author. This study intends to classify the groundwater of wadi al Bih, to delineate the areal distribution of the groundwater groups and the horizontal change of water quality with time.

AREA and TOPOGRAPHY

Wadi al Bih is a deeply incised valley located in the Ruus al Jibal mountains. The total catchment area of this valley is about 475 sq.km. The valley has an average length of over 41 km and a series of tributaries. It has an average width of more than 20 km over much of its length.

CLIMATE

The climatical condition of the wadi is characterized by an arid climate by high mean annual temperature (26°C) and high potential evaporation (2000 mm). The mean annual precipitation in wadi al Bih is of the order of 135 mm.

GENERAL GEOLOGY, LITHOLOGY and STRATIGRAPHY

Ruus al Jibal and the catchment area of wadi al Bih are almost exclusively composed of carbonate rock.

Limestone predominates but dolomite constitutes a significant proportion of the sequence. The limestone is slightly argillaceous, stratified, fossiliferous often fractured and porous. The carbonate succession in Jabal Hagab which overtook wadi al Bih attains a thickness of about 3500 m.

The sequence belongs to the Hajar Super Group which comprises the Ruus al Jibal Group (Permian to early Triassic), the Elphinston Group (late Triassic) and the Musandam Group (Jurassic to Lower Cretaceous).

Wadi al Bih has a box-shaped valley charaberized with deep alluvial fill composed of poor cemented limestone blocks, boulders, cobbles, pebbles and coarse sands. The thickness of the unconsolidated deposits ranges from 100 m to 129 m. The variability of thickness is in part dependant on the buried topography of the underlying bedrock.

AQUIFER OCCURRENCE

The groundwater occurs in the Quaternary valley fill and the fractured and Karstic underlying limestones. The groundwater level at wadi al Bih wellfield is about 80 m below the surface. Thus the aquifer system can be considered basically to be made of two parts - an upper part of nearly 10 m saturated valley fill and the lower part of at least 100 m of Karstic limestone. According to Khan (1986) groundwater occurs in the water table as well as the semi-confined conditions as revealed by the geological setting and the analysis of pumping test data. The confined layer is formed by locally cemented gravels or by a mixed zone of gravel and clay at the bottom most portion of the alluvial deposits. The transmissivity (T) of aquifer ranges between 588 and 3045 $m^3/d/m$ with a mean value of 1500 $m^3/d/m$. The storage coefficient (S) values range between 0.001 and 0.05. This confirms the unconfined to semi-confined nature of the aquifer.

The yields of the wells in the area range from 27.3 to 90.9 m^3/h and the specific capacity ranges between 1.28 and 61.76 $m^3/h/m$.

HYDROCHEMISTRY

The understanding of the hydrochemistry of groundwater in wadi al Bih is based mainly on the interpretation of groundwater analyses of 23 wells (Fig.2) sampled early in 1988, as indicated in table 1. The results of the analysis of groundwater in the wellfield in wadi al Bih show that the groundwater is rich in chloride and sodium. The interpretation of groundwater salinity has shown that in 1987 it ranged from 360 to 3670 mg/l, whereas in 1988 it increased from 435 to 4140 mg/l (Fig. 3 and 4). The figures show that the salinity increases from the SW to the N direction then it decreases to the NE direction. This pattern may be attributed to the extensive pumping in the drought season and the high production rate. This conclusion is substantiated by the increase in salinity at the center of the wellfield.

CLASSIFICATION and CHARACTERISTICS of the GROUNDWATER

Several methods are available for graphical representation of water analyses. These methods allow the direct correlation of samples, and the identification of similar and different groundwater, mainly based on the ionic ratios and concentrations. Some of these methods use a pattern representing the analysis in a way producing shapes or diagram, reflecting the different ionic concentrations of the major ions in distinguishing diagrams. Methods of this type may use the bar-graph (Reistle, 1927) or the patterns of Stiff (Stiff, 1951) or the nomograph (Schoeller, 1935). The correlation of many samples may be achieved by using the trillinear plotting system suggested by Piper in 1944.

Table 1. Chemical Analysis of Wadi Al Bih Groundwater, 1988. (Units of concentration in epm)

Well No.	T.D.S.	Ca	Mg	(Na + K)	Cl	SO_4	HCO_3
1	2410	4.19	6.71	29.73	35.31	2.81	2.59
2	3200	5.3	7.9	40.3	49.4	2.6	2.5
3	1800	4.3	5.1	20.7	26.1	1.9	2.1
4	1315	4.1	4.7	13	18.3	1.6	2
5	760	2.1	1.1	7.4	7	1.9	2.5
6	4140	6.66	8	55.6	64.9	3.1	2.5
7	1435	3.5	2.5	16.5	16.6	3.1	3
8	1170	2.7	2.9	13	14	2	2.5
9	1310	2.9	2.9	16.6	14.3	3.8	2.5
10	2520	5.9	5.9	30	37	2.2	2.5
11	2135	5.9	4.9	24.8	31	2.5	2.3
12	2740	5.7	6.9	33.8	44.8	2.5	2.3
13	2320	6.1	4.1	28.3	30.8	2.2	2.5
14	1455	3.7	3.5	16.5	26.6	2	2.5
15	1360	3.5	2.9	15.4	26.2	2	2
16	1285	3.3	3.1	14.2	15.9	2	2.7
17	1480	3.9	3.3	16.7	19	2	3
18	1030	4.1	3.7	8.9	13.1	1.7	2.2
19	435	2.1	1.7	2.7	4	0.7	2
20	545	1.8	1.4	4.8	4.7	1	2.4
21	800	2.5	1.7	7.9	7.6	1.8	2.7
23	2035	4.9	4.6	24.2	28.9	2.2	2.7

Table 2. Summary of the Groundwater Chemical Groups for the Period 1987 - 1988.

Year	Group Symbol	No. of samples in each Group	T.D.S. (ppm)	Water type	Remarks
1987	3331111(1)	5	3168	NaCl	Cl > (Na +K) $Cl > SO_4 > HCO_3$
	3131113(2)	4	1077	NaCl	Cl > (Na +K) $Cl > SO_4 > HCO_3$
					$Cl > HCO_3 > SO_4$
	3133112(3)	1	835	NaCl	Cl > (Na +K) $Cl > HCO_3 > SO_4$
	3331113(4)	4	2130	NaCl	Cl > (Na +K) $Cl > HCO_3 > SO_4$
	3333113(5)	1	1360	NaCl	Cl > (Na +K) $Cl > HCO_3 > SO_4$
	3132111(6)	1	1240	NaCl	Cl > (Na +K) $Cl > SO_4 > HCO_3$
	3131111(7)	3	1093	NaCl	(Na +K) > Cl $Cl > SO_4 > HCO_3$
	3231111(8)	1	1495	NaCl	Cl > (Na +K) $Cl > SO_4 > HCO_3$
	3231112(9)	1	1350	NaCl	Cl > (Na +K) $Cl > SO_4 > HCO_3$
	3133113(10)	1	645	NaCl	Cl > (Na +K) $Cl > HCO_3 > SO_4$
	3233222(11)	1	805	NaCl	Cl > (Na +K) $Cl > HCO_3 > SO_4$
1988	3331111(1)	3	3096	NaCl	Cl > (Na +K) $Cl > SO_4 > HCO_3$
	3231111(2)	1	3200	NaCl	Cl > (Na +K) $Cl > SO_4 > HCO_3$
	3131111(3)	5	1216	NaCl	Cl > (Na +K) $Cl > SO_4 > HCO_3$
	3133113(4)	3	928	NaCl	Na > Cl $Cl > HCO_3 > SO_4$
					Cl > (Na +K)
	3331113(5)	1	1435	NaCl	Cl > (Na +K) $Cl > SO_4 > HCO_3$
	3331112(6)	2	2725	NaCl	Cl > (Na +K) $Cl > SO_4 > HCO_3$
	3131112(7)	1	2520	NaCl	Cl > (Na +K) $Cl > SO_4 > HCO_3$
	3131113(8)	5	1625	NaCl	Cl > (Na +K) $Cl > HCO_3 > SO_4$
	3133111(9)	1	2035	NaCl	Na > Cl $Cl > HCO_3 > SO_4$

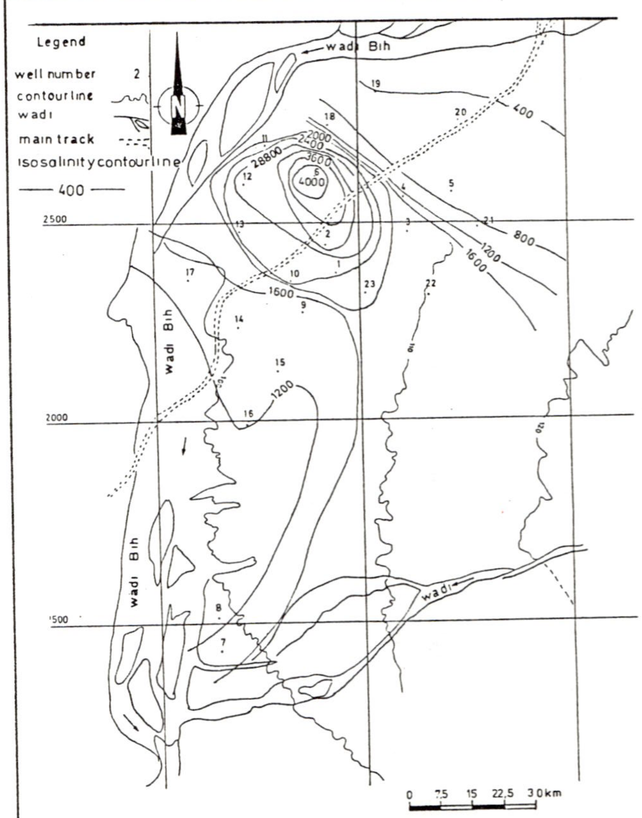

Fig. 3 Areal distribution of groundwater salinity 1987

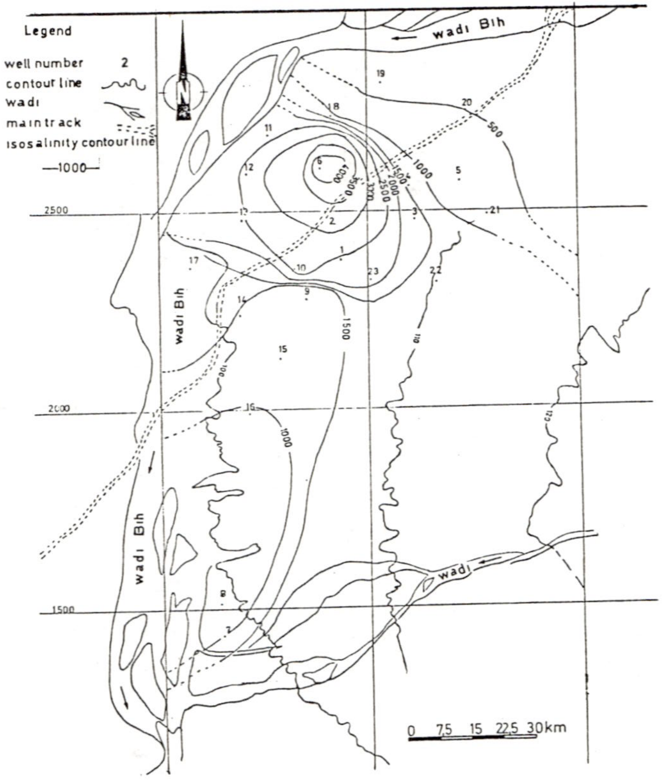

Fig. 4 Areal distribution of groundwater salinity 1988

In this study two methods have been used for the study and representation of the groundwater analyses of the well-field in wadi al Bih. The first is the Piper diagram (1944) (Fig. 5). The projected points in the Piper diamond show that alkalis exceed alkaline earths and strong acids exceed weak acids.

In the second method the interpretation and presentation of the groundwater analyses is achieved by the nomograph of Schoeller (1935). Five major groups exist in the wadi al Bih well-field. The five groups are characterized by descending values of both anions and cations arranged as follows:

First group: $Cl > SO_4 > HCO_3$; $Na > Mg$ Ca; $Cl > (Na + K)$
Second group: $Cl > SO_4 > HCO_3$; $Na > Ca$ Mg; $Cl > (Na + K)$
Third group: $Cl > HCO_3 > SO_4$; $Na > Ca$ Mg; $Cl > (Na + K)$
Fourth group: $Cl > HCO_3 > SO_4$; $Na > Ca$ Mg; $Na > Cl$
Fifth group: $Cl > HCO_3 > SO_4$; $Na > Mg$ Ca; $Cl > (Na + K)$

The third and fourth groups are rather similar so far as with respect to the relationship between cations and anions. However, the separate cation and anion relationships are considered but they differ with respect to Cl/Na ratios where the fourth group shows Na to be greater than Cl. Such difference may be due to exchange or base exchange phenomena affecting waters of the two groups. The average composition of groundwater chemical groups is shown in Schoeller nomograph which displays the relative concentration of cations and anions in each group (Fig.6).

NUMERICAL APPROACH to the CLASSIFICATION of GROUNDWATER

This method is derived by Haddad (1981). All methods of classification depend on the variation of ionic ratio and ionic percentage. Generally the classification takes into consideration the relationship between major cations and anions. The ratios are arranged and given symboles as follows: Cl/HCO_3; SO_4/HCO_3; Cl/SO_4; $(Na + K)/Cl$; $Ca/(Na + K)$; $Mg/(Na = K)$ and Ca/Mg.

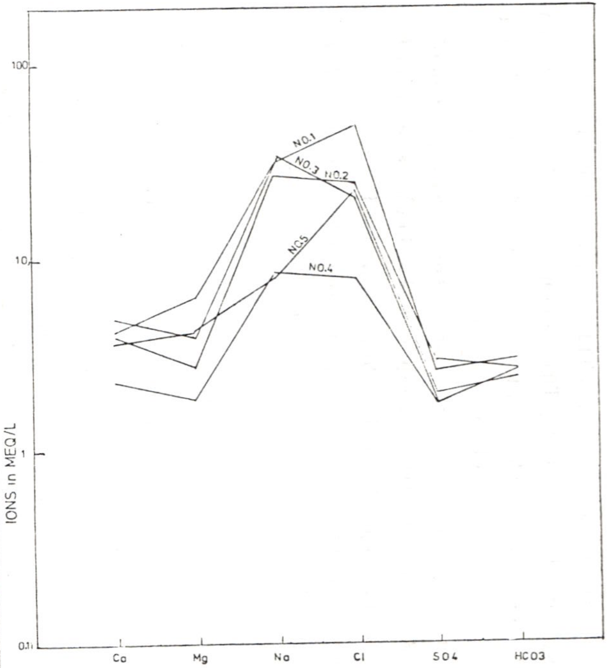

Fig. 5 Piper's diagram for groundwater analysis of Wadi al Bih

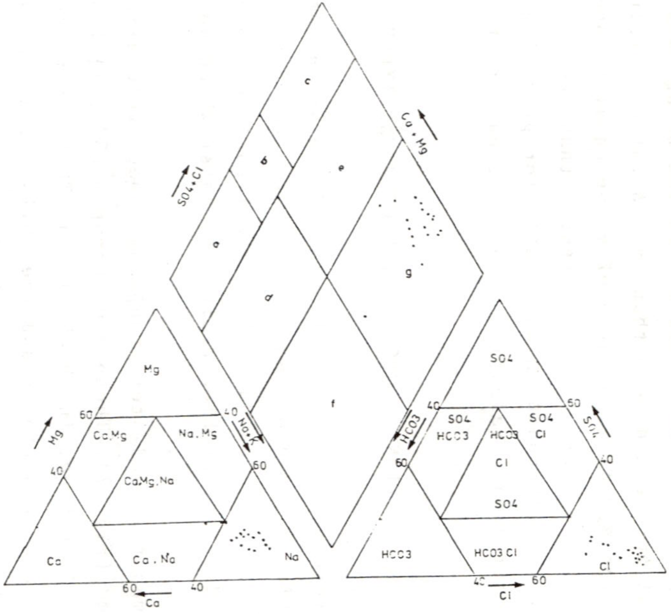

Fig. 6 Average chemical characters of groundwater groups for the year 1988

Three possibilities exist: the ratios being less than one, equal to one and more than one. A ratio of value less than one is given symbol 1, a ratio of value equal to one is given the symbol 2, a ratio of value greater than one is given the symbol 3. This approach is applicable for providing rapid classification and evaluation of a large number of symbols.

AREAL DISTRIBUTION of GROUNDWATER CHEMICAL TYPES

The hydrochemical classification of the groundwater in wadi al Bih is based on samples collected and analysed in June 1988, as well as data of chemical analyses obtained from the Ministry of Electricity and Water which were carried out in 1987.

Nine major groundwater groups have been recognized and delineated (Fig. 7). These are: group 1 (3331111), sodium chloride water type, in which Cl Na. This water type occurs in the northwest strip around the wells W1, W6 and W12. The salinity increases in these wells due to the concentration of chloride ions. Group 2 (3231111) is a sodium chloride water type in which Cl (Na +K). This group is only concentrated around the well 2 on the center of the wellfield. Group 3 (3131111) is a sodium chloride water type in which Cl (Na + K) and is located in a north-south strip around W3, W4, W8, W15 and W19. Group 4 (3133111) is a sodium chloride water type in which Na Cl and Cl (Na + K) and it centers in north-east strip around W5, W20 and W17. Group 5 (3331113) is a sodium chloride in which Cl (Na +K) and is restricted only around W7. Group 6 (3331112) is a sodium chloride water type in which Cl (Na +K). This group is only recognized in W8 and W11. Group 7 (3131112) is a sodium chloride water type in which Cl (Na +K) and is situated around W8 and W11. Group 8 (3131113) is a sodium chloride water type in which Cl (Na +K). This group covers the major part of the wellfield and it includes W13, W14, W16, W18 and W23.

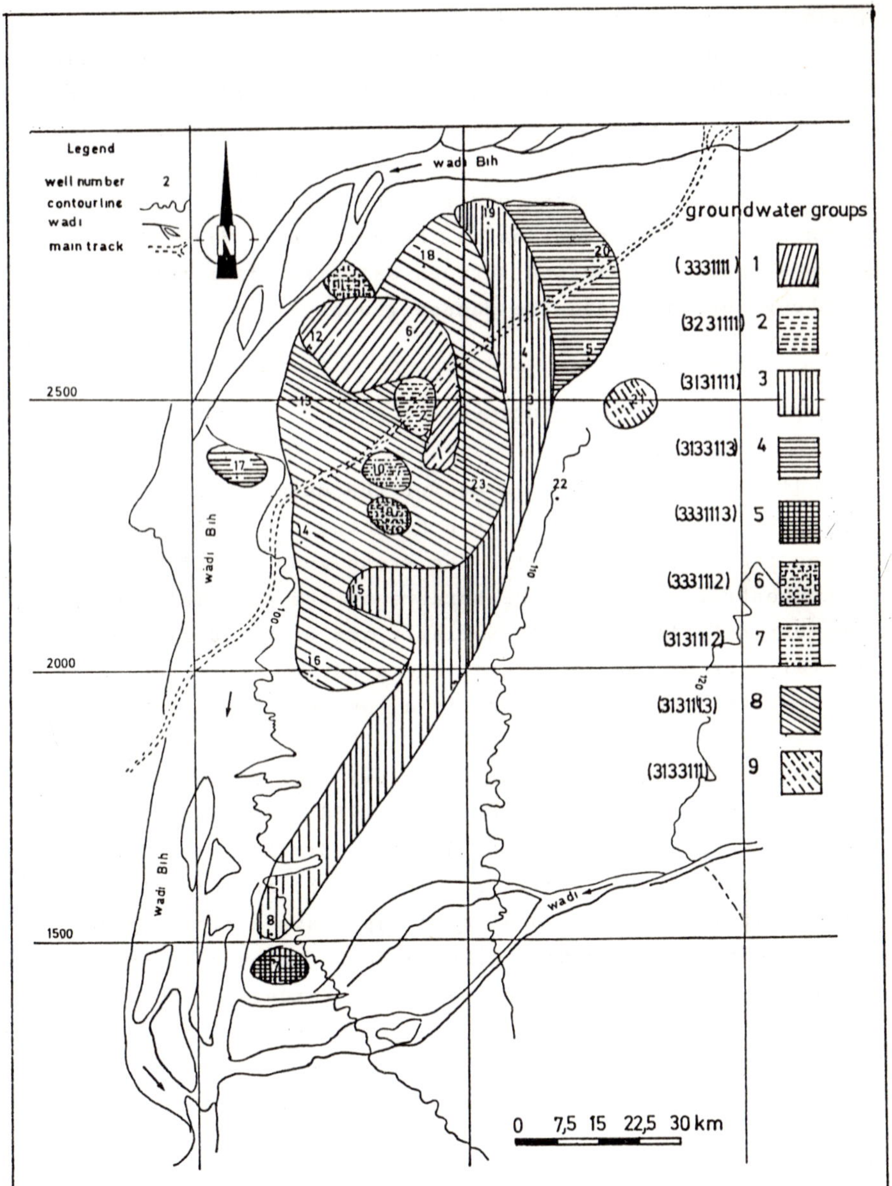

Fig.7 Distribution of groundwater groups for the year 1988

Group 9 is a sodium chloride water type in which Na Cl and it centers around W21 in the east. A summary of the groundwater chemical groups is shown in table 2. Table 1 represents the chemical analysis of wadi al Bih water in 1988. The close correspondence of the groundwater groups with the total dissolved solids mapped for the year 1988 (Fig.4), which has limited variation in the horizontal direction, shows a clear correlation between salinity distribution of the groundwater and the mapped groups.

CONCLUSIONS

Wadi al Bih wellfield is one of the main fresh groundwater resources in the northern Emirates. The salinity ranges from 500 - 4000 mg/1. The fresh groundwater body is concentrated towards the borders of the wellfield, while the most notable change in salinity has occurred along the central part of the wellfield. The study of the groundwater qualitytrend shows that there is a considerable change in salinity with time.

Groundwater chemical analysis shows that there has been nine groundwater chemical groups in 1988. These are sodium chloride (3331111), (3231111), (3131111), (3133113), (3331113), (3331112), (3131112), (3131113) and (3133111). Areal distribution of isosalinity map shows a clear agreement with the distribution of groundwater groups where the fresh water is limited towards the borders of the wellfield. Water wells show that alkalies exceed alkaline earths and strong acids exceed weak acids.

REFERENCES

Al-Ruwaih, F.M. 1985. Hydrochemical classification of the groundwater of Umm Al-Aish. Kuwait, J. Univ. Kuwait (Sci) 12.

Haddad, R.H. 1981. A proposed numerical approach to the classification of groundwater. First Symposium on the Future Development of Water Resources in the Gulf and Arabian Peninsula. Kuwait 3 - 5 March 1981. J. Gulf Arab. Penin. Studies (Kuwait University), Suppl. pp.47 - 61.

Hem, J. D. 1970. Study and interpretation of the chemical characteristics of natural water. U.S. Geologic Survey, Water Supply Paper, no. 1473, 2nd. ed.

Hudson, R.G.S., McGugar, A. and Morton, D.M. 1954. The structure of the Jabal Hagab Area, Trucial Oman, Quart. J. Geol. Soc. London, 110, pp. 121 - 152.

Khan, J.S. 1986. Groundwater conditions in the wadi al Bih in the northern part of the United Arab Emirates. Report for the Ministry of Electricity and Water. Dubai, U.A.E. (unpubl.).

Piper, A.M. 1944. A graphic procedure in the geochemical interpretation of water analyses. Geophys. Union Trans. Vol. 25, pp. 914 - 923.

Reistle, C.E., Jr. 1927. Identification of oil-field waters by chemical analysis. U.S. Bur. Mines Techn. Paper 404.

Schoeller, H. 1935. Utilite de la notion des exchanges des bases pour la comparison des eaux souterraines. Soc. Geol. Comptes Rendus, Sum. et Bull. 5, vol. 5, pp. 651 - 657.

Stiff, H.A., Jr. 1951. The interpretation of chemical water analysis by means of pattern. J. Petr. Techn., vol.3, no. 10, pp. 15 - 17.

SEMI-ANALYTIC SOLUTION OF ONE-DIMENSIONAL DISPERSION CONVECTION EQUATION

Sita Ram Singh and Ashok Kumar Saini
Department of Soil and Water Engng.
Punjab Agricultural University Ludhiana

S.K.Kamra
Div.of Agril.Engng.
CSSRI Karnal

ABSTRACT

Dispersion convection equation adequately describes the transport of solute in porous media. The space discretisation of the one-dimensional equation by Galerkin's finite element method yields a system of time dependent inhomogeneous ordinary differential equations. The solution of such an inhomogeneous matrix-vector differential equation is obtained with the help of the matrix exponential in exactly the same way as if one was dealing with a scalar equation. The eigen value - eigen vector method of Euler is employed for computing the matrix exponential. The resulting solution is continuous in time. Results from the aforesaid model are compared with an analytic solution of the dispersion convection equation which ascertains that the model is accurate for low Peclet numbers.

INTRODUCTION

The dispersion convection equation adequately describes solute transport in unsaturated and saturated subsurface flow fields as well as rivers, lakes and oceans.It is also widely used to model transport of pollutants in the atmosphere. In one-dimensional subsurface flow domain the equation is as below.

$$\theta\ R_f\ \delta C/\delta t = \delta/\delta y\ (\theta D\ \delta C/\delta y - qC) - S(y,t) \qquad (1)$$

in which C is solute concentration in liquid (ML^{-3}); θ is volumetric water content($L^3\ L^{-3}$); q is specific discharge (LT^{-1});D is longitudinal dispersion coefficient (L^2T^{-1});$R_f$is retardation factor;S is rate of withdrawal (or production) of solute per unit volume of the porous medium ($ML^{-3}T^{-1}$); y is space coordinate positive downward (L) and t is time (T). When the porous medium is fully saturated like domain beneath the water table, θ is replaced by the porosity of the medium, n. In case of solute production S becomes (-S).

When D is large in relation to q, dispersion dominates and the partial differential equation (1) is parabolic. However, when q is large, convection dominates and the equation becomes hyperbolic. Thus in a non uniform unsteady flow domain the differential equation may either be parabolic or hyperbolic in space and time.

Numerous analytic solutions of (1) or its variants for constant values of the parameters θ, D, q and S, and several boundary and initial conditions have been developed. Van Genuchten and Alves (1982) have reported these solutions along with many computations and computer programmes. Amongst the several numerical methods, the Galerkin finite element-finite difference technique has been widely employed in developing general solutions to one and two-dimensional dispersion convection equations. However, the application of straight forward finite element approximations immediately encountered difficulties in situations where convection dominated the transport. In order to overcome numerical dispersion in solution, several investigators either developed finite element solutions to modified dispersion convection equation or they employed special techniques.Van Genuchten(1987) added and subtracted $[qt/(6\theta^2 R_f)]$ to D in (1) at times t and (tt), respectively. Heinrich and Envia (1982) theoretically derived the addition of $\underline{D} = 1/2\ |V|h\alpha$ to D in the steady state form of (1), where h is local grid size and α a parameter given below.

$$\alpha = \coth |V|h/2D - 2D/|V|$$

Adey and Brebbia (1974) proposed a mixed Euler-Lagrange algorithm for two-dimensional transport equation assuming a fixed fluid mass and a known velocity field in each element in which the dispersion was allowed to take place at each time step but convection was calculated afterwards.

Neuman and Sorek (1982) decoupled advection from dispersion in one dimensional dispersion convection equation The resulting advection problem was solved by the methods of (i) reverse streak lines, and (ii) continuous particle tracking. Dispersion was handled by implicit finite element on a fixed grid. Botha et al. (1982) proposed a moving mesh finite element method for one dimensional dispersion convection equation.

PURPOSE OF STUDY

In all the preceding algorithms time was discretised using conventional finite difference technique which demands marching through all the intermediate time steps to develop solution at a desired time.The purpose of this study was to develop finite element solution continuous in time so that computations at any time could be made by simply substituting the value of time in the solution. It was achieved by discretising the space coordinate using Galerkin's finite element method and solving the resulting system of time dependent ordinary differential equations by Euler's eigen value technique.

A continuous in time eigen value solution to the spatially discretised two-dimensional solute transport equation in steady ground water flow using complex matrix analysis has been developed by Hwang et al. (1984). A similar analysis but employing real arithmetic has been developed by Umari and Gorelick (1986). Even though the latter authors used real arithmetic, their treatment is difficult to understand and programme on computer.The proposed algorithm also uses real arithmetic after the evaluation of eigen values and eigen vectors, but it is much simpler to comprehend and programme on digital computers.

AUXILIARY CONDITIONS

In order to solve equation (1) one needs to specify initial and boundary conditions. The flow domain beneath the land surface upto the basis layer is shown in Fig.1.

Boundary Condition at the Land Surface

Since the flow of water in soil profile is assumed steady,therefore, specific discharge q at every point of the profile is independent of time.In absence of sources and sinks the magnitude of q equals the rate of infiltration or evaporation.However, when sources or sinks are present,q vary in space.

The principle of mass conservation requires that normal components of the solute flux entering and leaving every point of the land surface are equal. Let R be the rate of infiltration (-R for rate of exfiltration) of water having concentration C_e and **J** be a solute flux vector leaving the land surface. The outward normal to the land surface is **1n**. When the land surface is horizontal,**1n** is in the direction of (-**1y**). The boundary condition at the land surface becomes:

$$-\mathbf{RC_e} \cdot \mathbf{1n} = -\mathbf{J} \cdot \mathbf{1n} \qquad (2)$$

where $\mathbf{RC_e} = RC_e\mathbf{1y}$

The minus sign in (2) is to reverse the direction of **1n**. Substituting equations for **J**, **1n** and $\mathbf{RC_e}$ in (2), one gets:

$$-RC_e\,\mathbf{1y} \cdot (-\mathbf{1y}) = -\,(-\theta D\,\delta C/\delta y + qC)\,\mathbf{1y} \cdot (-\mathbf{1y})$$

or,

$$RC_e = -\,\theta D\,\delta C/\delta y + qC \qquad (3)$$

An equation like(3) could be derived at any other boundary of the flow domain as well.Sometimes instead of assigning general flux type boundary condition given by (3), one assigns Dirichlet type boundary condition

$$C(0,t) = C_e(t) \qquad (4)$$

at the land surface. Here $C_e(t)$ is the concentration of solutes in the inflowing water.

Solute concentration, C_e, in water vapour is zero. Thus, when evaporation takes place from the land surface, boundary condition (3) becomes:

$$- \theta D \, \delta c/\delta y + qC = 0 \qquad (5)$$

Evaporation causes increase in solute concentration at the evaporating site. However, as the soil solution gets saturated, the salts precipitate out of the solution phase.

Boundary Condition at the Water Table and Impervious Basis Layer

The downward flowing water through the soil profile picks up salts in the profile and carries to the water table. One of the aims of solving the one-dimensional dispersion convection equation is to determine the concentration C_{in}, of the ground water recharge at the phreatic surface. We substitute C and q in place of C_e and R respectively, in (3) to get the following boundary condition beneath the soil profile at the water table :

$$\delta c/\delta y = 0 \qquad (6)$$

When there is evaporation from the water table we substitute the concentration of the ground water, C_g, in place of C_e in(3) to get the following boundary condition at the bottom of the soil profile:

$$- \theta D \, \delta c/\delta y + qC = RC_g \qquad (7)$$

Some authors (Van Genutchen, 1987) use the concentration of ground water, say C(L,t), as the boundary condition at the soil aquifer interface. Thus one gets:

$$C(y,t)|_{y=L} = C_g(t) \qquad (8)$$

where L is the depth of ground water below the land surface.

When the basis layer is impervious to flow, q = 0. Its substitution in (3) yields equation (6) as the boundary condition.

Initial Condition

The concentration of solutes in the soil profile at t =0, as obtained from sample measurements, is specified as the initial condition for the differential equation (1); i.e.

$$C(y,0) = C_o(y) \qquad (9)$$

Thus equations (1), (3), (4), (5), (6), (7) and (9) complete the formulation of the boundary value problem of solute transport through soil profile. Both the downward and upward fluxes with infiltrating and evaporating water has been considered. It may be noted that during upward flow of water, R should be replaced by (-R) in all equations.

FINITE ELEMENT DISCRETIZATION

Derivation of Ordinary Differential Equations

Consider the following linear operator for the one-dimensional solute transport in partially saturated porous media:

$$L(C) = \theta R_f\ \delta C/\delta t - \delta/\delta y[\theta D\ \delta C/\delta y - qC] \qquad (10)$$

Thus (1) becomes

$$L(C) + S(y,t) = 0 \qquad (11)$$

We approximate the solution C(y,t) by $C^*(y,t)$ given by

$$C^*(y,t) = \sum_{j=1}^{NN} C_j(t)\ N_j(y) \qquad (12)$$

where $C_j(t)$ are unknown coefficients to be determined, and $N_j(y)$ are one dimensional basis functions whose total number is NN. As C^* is an approximator to the solution, its substitution in (11) leaves a residual:

$$L(C^*) + S(y,t) \qquad (13)$$

If the approximate solution were the exact solution, the residual would vanish. In Galerkin's finite element method, we make the residual orthogonal to each basis function. It forces the residual zero in an average sense. Thus

$$\langle\ [\ L(C^*) + S(y,t)\ ]\ ,\ N_i(y)\ \rangle = 0 \qquad (14)$$
$$\text{for } i = 1,2,3,\ldots,\ NN.$$

Substituting (10) and (12) in (14) and integrating term by term and simplifying one gets:

$$[AM]\ \{dC/dt\} = [\ [DM] + [EM]\ ]\ \{C\} + \{F\} \qquad (15)$$

where [AM],[DM] and [EM] are square matrices of size (NNxNN) and {dC/dt},{C} and {F} are column vectors of size (NNx1). Typical elements of the matrices are given below.

$$AM_{ij} = \sum_{e=1}^{NE} \int_e \theta R_f\ N_i\ N_j\ dy \qquad (16)$$

$$DM_{ij} = - \sum_{e=1}^{NE} \int_e \theta D \, (dN_i/dy)(dN_j/dy) \, dy \qquad (17)$$

$$EM_{ij} = \sum_{e=1}^{NE} \int_e q(dN_i/dy) \, N_j \, dy \qquad (18)$$

$$F_i = F_i - F_{wi} \qquad (19)$$

$$F_{si} = -[(-\theta D \, \delta c/\delta y + qc) \, N_i(y)]\Big|_{y=0}^{y=L} \qquad (20)$$

$$F_{wi} = \sum_{e=1}^{NE} \int_e S(y,t) \, N_i(y) \, dy \qquad (21)$$

Equation (20) gives the solute flux at the boundaries of the flow domain; Eq.(21) is due to salt removal by any means from the flow domain.

Application Of Boundary Conditions

Effect of all the boundary conditions is incorporated in Eq.(20). When the flux of water is downward, the boundary condition (3) is of flow-in type(concentration known) and boundary condition (6) is of flow-out type(concentration unknown). Substituting (3) and (6) in (20) one gets:

$$\begin{aligned} F_{si} &= - [(-\theta D \, \delta C/\delta y + qC) \, N_i(y)]\Big|_{y=L} \\ &\quad + [(-\theta D \, \delta C/\delta y + qC) \, N_i(y)]\Big|_{y=0} \\ &= RC_e \, N_i(0) - q(L)C(L,t) \, N_i(L) \end{aligned}$$

$$\text{for } i = 1,2,3,\ldots, NN \qquad (22)$$

Since the basis function is nonzero only at the node with which it is attached and is zero at all other nodes; therefore, F_{si} becomes:

$$F_{s1} = RC_e \qquad (23)$$

$$F_{s2} = F_{s3} = \ldots = F_{sNN-1} = 0$$

$$F_{sNN} = - q(L)C_{NN}(t) = [BM]\{C\} \qquad (24)$$

where [BM] is of (NNxNN) size with all entries zero except the entry in the NNth row and NNth column, which is (-q(L)). Thus after incorporating the boundary conditions, one gets:

$$[AM] \, \{dc/dt\} = [\, [DM]+[EM]+[BM] \,] \, \{C\} + \{F\} \qquad (25)$$

in which AM_{ij}, DM_{ij}, and EM_{ij} are given by (16), (17) and (18) respectively; and $F_1 = F_{s1} = RC_e$ as given by (23), and $F_2=F_3 = \ldots = F_{NN} = 0$.

When the flow of water is upward, the boundary condition at the water table is of flow-in type and that of the land surface is of flow-out type. With the help of (5) and (7) equation (20) becomes:

$$F_{si} = -[RC_g\ N_i(L)] + [0\ N_i(0)] \quad \text{for } i = 1,2,\ldots, NN$$

or, $F_{s1} = F_{s2} = \ldots = F_{sNN-1} = 0$

and $F_{sNN} = -RC_g$ (26)

In this case the system of ordinary differential equations is described by (15) where F_{si} is defined by (26).

SELECTION OF BASIS FUNCTIONS AND COMPUTER PROGRAMME

Here, out of several choices of basis functions, linear Lagrange polynomials are associated with each node. The basis function has the value unity at the node with which it is associated, and is zero at all other nodes, as shown in Fig.1.

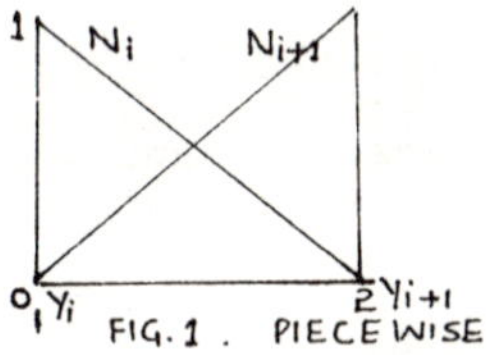

FIG.1. PIECEWISE LINEAR BASIS-FUNCTIONS

Fortran 77 computer programme of the one dimensional solute transport model is in two parts. The first part is the finite element programme to generate coefficient matrices. When ground water flow is steady, q and θ are independent of time. Consequently for each discretization the coefficient matrices are generated only once. Thus matrix [H] becomes time invariant. The second part of the model computes all the eigen values and associated eigen vectors of the matrix [H] using driver subroutine RG of Smith et al. (1976, pp 7.1-9 to 7.1-234). The computations of eigen values τ_i for i = 1,2,..., NN, eigen vectors stored in matrix [Z], and $[Z]^{-1}$ are performed only once. A subroutine FUNDA written by the authors computes the fundamental matrix solution:

$$[C(t)] = [Z]\ [QD] \tag{27}$$

at any time t. Since [QD] is a function of time, therefore the fundamental matrix [C(t)] is computed at each time, t. Substitution of [C(t)] yields the matrix exponential at any time t. Once the matrix exponential is obtained, the computation of solute concentration at each node and time, t, is straight forward.

COMPARISON OF THE MODEL RESULTS WITH ANALYTIC SOLUTIONS

The accuracy of the model was studied by comparing its results with an analytic solution which describes distribution of inflowing solutes into a soil column of finite length. The analytic solution is for the convection dispersion equation (1) with $S(y,t) = 0$, constant q and θ, and constant auxiliary conditions:

$$C(0,t) = C_e \tag{28a}$$

$$\delta C(L,t)/\delta y = 0 \tag{28b}$$

$$C(y,0) = C_0 \tag{28c}$$

The solution derived by Cleary and Adrian (1973) is as below (Van Genuchten and Alves, 1982; p.11).

$$C_a(y,t) = C_0 + (C_e - C_0)\,A(y,t) \tag{29}$$

where

$$A(y,t) = 1 - \sum_{m=1}^{\infty} [2\,\beta_m\,\mathrm{Sin}(\beta_m y/L)\,\exp\{(vy/2D) - (v^2t/(4DR_f)) - (\beta_m{}^2Dt/(L^2R_f))\}]/[\beta_m{}^2 + (vL/2D)^2 + (vL/2D)] \tag{30}$$

with $v=q/\theta$, the intertitial velocity; and where the eigen values β_m are the positive roots of the equation:

$$\beta_m\,\mathrm{Cot}(\beta_m) + vL/2D = 0 \tag{31}$$

The deviation of finite element cum eigen value solution from the analytic one is due to the space descretization of the differential equation (1). L_2 and Tchebycheff norms (Prenter, P.M.; 1975, p 6-11) were employed to measure the goodness of the numeric solution. The two norms are as below.

L_2 norm:

$$1/\sqrt{L}\,||C_a - C||_2 = [\int [C_a(y,t) - C(y,t)]^2 dy]^{0.5}(1/\sqrt{L}) \tag{32}$$

It gives average distance of the numeric solution $C(y,t)$ from the analytic solution $C_a(y,t)$ at time t.

Tchebycheff norm:

$$||C_a(y,t) - C(y,t)|| = \max_{0\le y\le L} |C_a(y,t) - C(y,t)| \tag{33}$$

It describes the maximum difference between analytic and numeric solutions in the domain $0\le y\le L$ at time t.

The integration in L_2 norm (32) was performed using trapezoidal rule. (Burden, Faires, and Reynolds; 1981, p 149).

The accuracy of the model results were studied for different values of dispersion coefficient, D, and specific discharge, q combined through a dimensionless Peclet number:

$$P_e = |v| L_c/D \quad (34)$$

Here, in addition to terms defined earlier, v is pore water velocity, q/θ, and L_c is a characteristic length, which has been taken as unity.

In all the computations C_e, C_o, L and θ were assigned the values of 1.0,0.0, 10m, and 0.4, respectively. Only the values of specific discharge q and dispersion coefficient D were varied to achieve different Peclet numbers. The flow domain was divided into 10, 20 and 40 linear elements and computations were performed for Peclet numbers of 0.25 to 5.0. Fig.2a,b shows the plot of L_2 and Tchebycheff norms against pore volume (qt/θL) for 20 elements and Peclet numbers 0.25, 1.0, 1.5, 2.0, 2.5, 3.0, 3.5, 4.0, 4.5 and 5.0. It is evident from the figure that at the low Peclet numbers the deviation in semi-analytic solution from the analytic one is negligibly small. It is clear from the figure that the error reduces with increase in time at all Peclet numbers. At low Peclet numbers the trend is uniformly decreasing but at high Peclet numbers it is decreasing with oscillation.

REFERENCES

Adey, R. , and Brebbia, C., 1974. Finite element solution for effluent dispersion. Numerical Methods in Fluid Dynamics,(Brebbia and Connor,editors).Pentch Press, London,pp.325-354.

Bellman,R.E., 1970 . Introduction to Matrix Analysis, Second Edition. McGraw-Hill , New York.

Botha,J.F., Herbst,B.M., and Schoombie,S.W., 1982. On the numerical solution of the diffusion-convection equation. Finite Element in Water Resources,(Holz et.al, editors),Springer-Verlag, Berlin,pp 14-3 to14-8.

Braun, M.,1976. Differential Equations and Their Applicatioins, Second Edition. Springer-Verlag,New York.

Cleary, R.W., and Adrian, D.D., 1974. Analytical solution of longitudinal convective-dispersive equation for cation adsorption in soils. Soil Science Society of America Proceedings,Vol.39: pp. 197-199.

Heinrich, J.C., and Envia, E., 1982. Finite element techniques in transport phenomena. Finite Element in Water Resources,(Holz et. al editors),Springer-Verlag, Berlin, pp.14-27 to 14-40.

Hwang, J.C.,Cho, W.C., and Yeh, G.T., 1984. An eigenvalue solution continuous in time to the spatially discretized solute transport equation in steady groundwater flow. Water Resources Research, V.20, no. 11.pp.1725-1732.

Neuman, S.P., and Sorek, S., 1982. Eulerian-Lagrangian methods for advection-dispersion. Finite Element in Water Resources,(Holz et. al editors),Springer-Verlag,Berlin, pp.14-41 to 14-68.

Prenter, P.M., 1975. Splines and Variational Methods. John Willey and Sons, New York.

Smith, B.T., Boyle, J.M., Garbow, B.S., Ikebe, Y., Klema, V.C., and Moler, C.B., 1976. Matrix Eigensystem Routines- EISPACK guide: Springer-Verlag, New York.

Umari, A.M.J., and Gorelick, S.M., 1986. The problem of complex eigensystems in the semi-analytical solution for advancement of time in solute transport simulations: A new method using real arithmetic. Water Resources Research, V.22, no.7. pp.1149-1154.

Van Genuchten, M.Th., and Alves, W.J., 1982. Analytical Solution of the One-Dimensional Convection-Dispersive Solute Transport Equations. U.S. Department of Agriculture, Agricultural Research Service, Technical Bulletin no. 1661.

Van Genuchten, M.Th, 1987. A numerical model for Water and solute movement in and below the root zone: Research report no.121, United States Department of Agriculture, Agricultural Research Service, U.S. Salinity Laboratory River side, California.

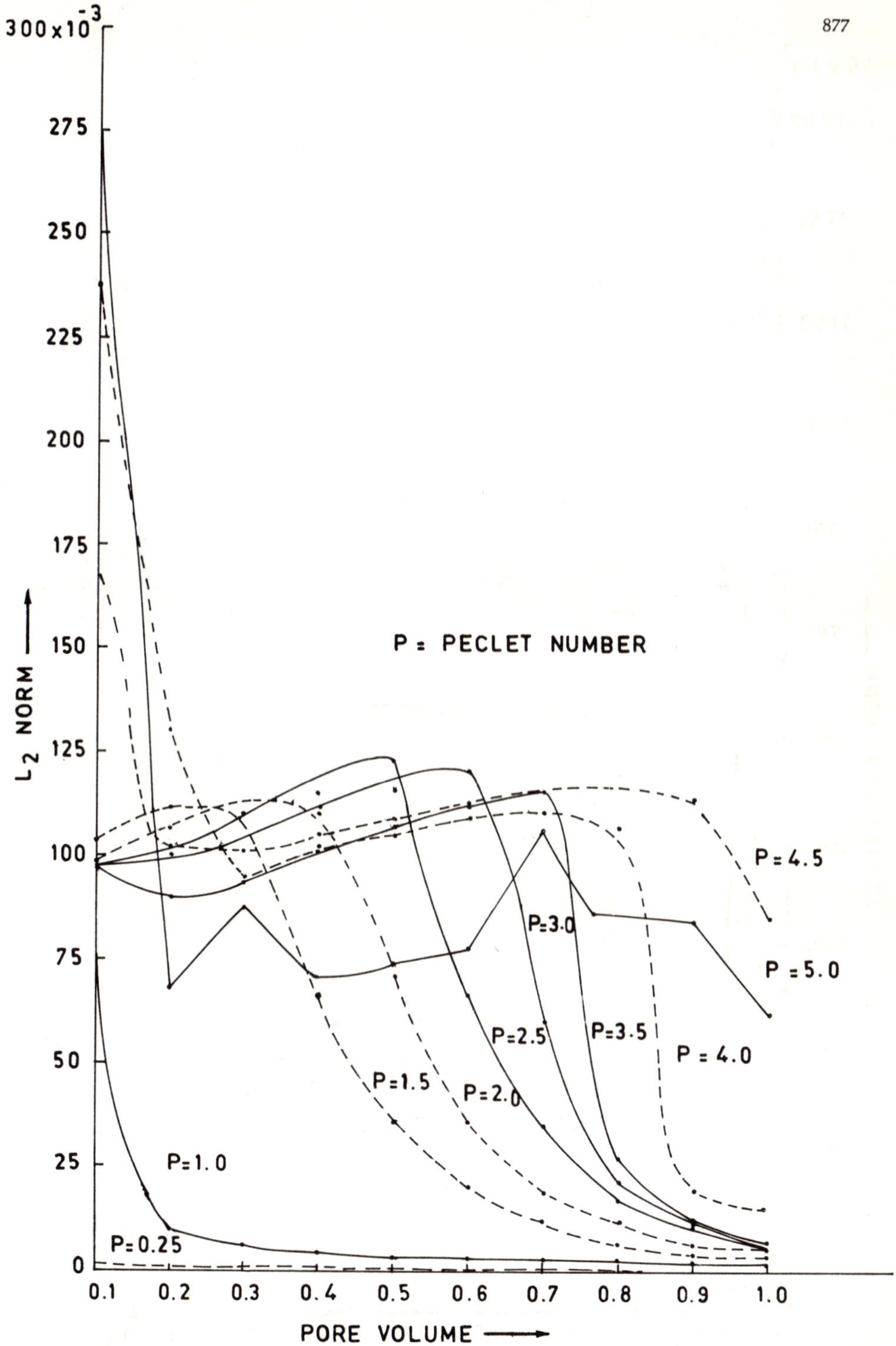

FIG.2a. AVERAGE ERROR IN SEMI-ANALYTIC SOLUTION AT DIFFERENT PECLET NUMBERS

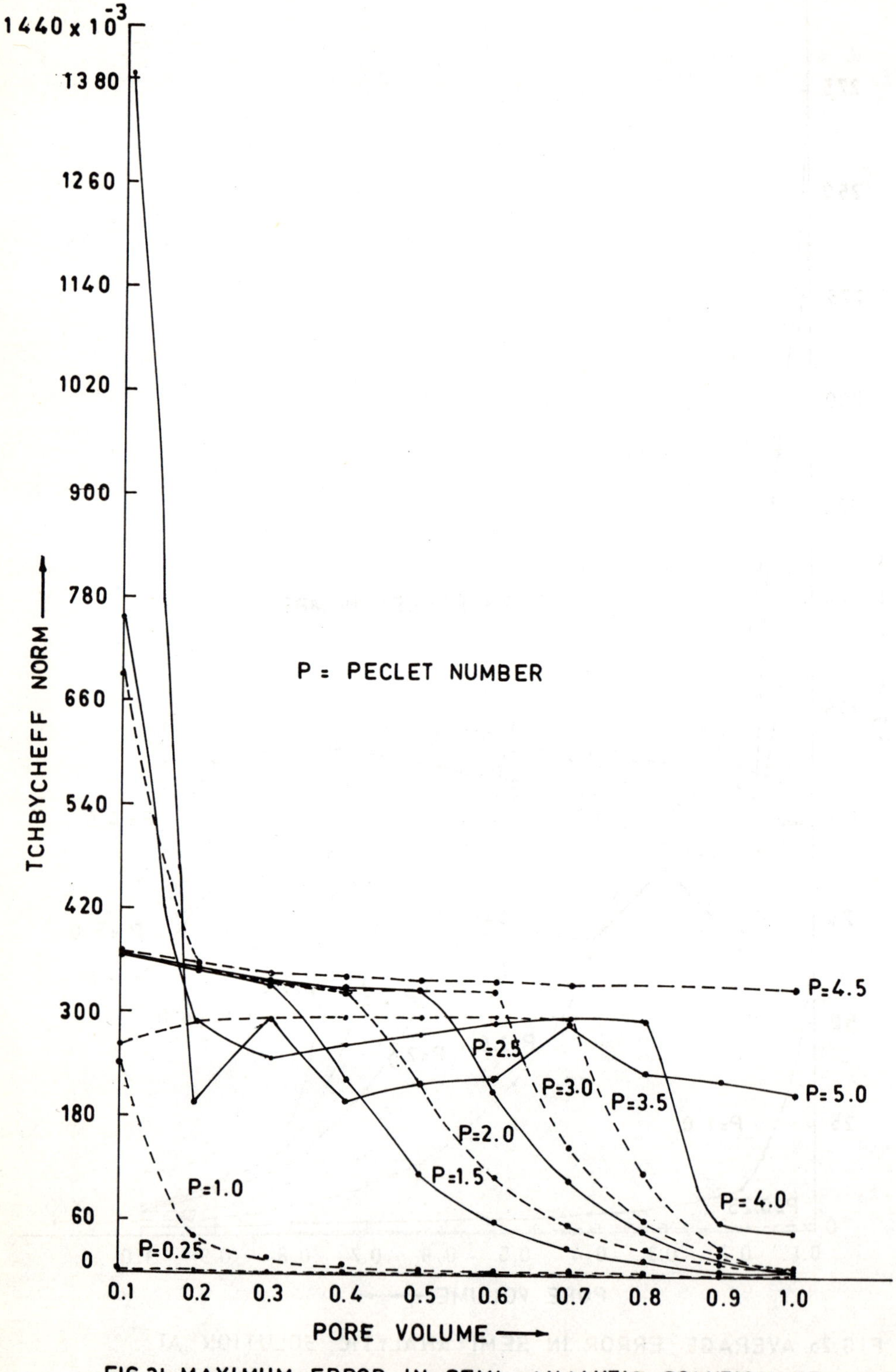

FIG.2b. MAXIMUM ERROR IN SEMI-ANALYTIC SOLUTION AT DIFFERENT PECLET NUMBER

NITRATE POLLUTION OF GROUNDWATER IN LUCKNOW AREA, U.P

by
V.K. Sahgal, R.K. Sahgal, Y.P. Kakar
Central Ground Water Board, Lucknow

Abstract

With the establishment of direct relationship between nitrate concentrations in drinking water and incidence of mathemoglobinemia which can prove fatal to infants, the study of nitrate in water supplies has assumed great importance. Consequent to observance of high concentrations of nitrate in ground water of Lucknow area, detailed study was undertaken which has revealed nitrate levels several times higher than the prescribed limit of 45 mg/l.

The study area which lies in the sub-basin of Central Ganga alluvial/ plain is underlain by thick alluvial deposits of Quaternary age comprising clay, kankar, and sand. Two groups of aquifers exist upto a depth of 183.18 meters, the top aquifer is in the range upto about 107 metres b.g.l. interspersed with thin clay beds and lenses and second aquifer is in the range of 107 to 145 meters b.g.l.

The present study indicates that nitrate concentrations of ground water are over 100 mg/l at several places and concentration as high as 650 mg/l has been observed in shallow groundwater. Monitoring of nitrate concentrations at deeper levels has revealed that pollutants have adversely affected groundwater quality in the depth range of 100-135 meters with nitrate concentrations to the extent of 145, 168, 590 mg/l in some of tubewells. Preliminary investigations about the sources of nitrate pollution indicate that leachates from soak pits, septic tanks, unlined sewage drains and abondoned wells used for dumping wastes are responsible for degradation of groundwater quality.

Apart from monitoring nitrate levels in shallow and deep aquifer groups, interrelated topics such as mobility of nitrogen in the unsaturated one, temporal variations and statistical relationships have been discussed. In view of deleterious affects of nitrate, the study highlights necessity of regular surveillance of these ions along with methodology to check pollutants in order to ameliorate groundwater pollution to safe-guard the health of infants and growing children in the area under study.

1. INTRODUCTION

Nitrate is the end product of oxidation process of nitrogen and its high concentration in water is often an indicator of pollution from sewage or fertilisers. Nitrate in surface and groundwater may be contributed by bacteria and algae, decomposition of organic matter in the soil, leaching of fertilisers, human and animal excreta. Industrial effluents of nitrogenous industries may also contribute in increasing the nitrate of natural waters.

As a result of ingestion of water with high concentration of nitrate, infants may suffer from cyanosis (mathemoglobinemia) or blue baby disease. Nitrate is converted to nitrate and then to ammonia by bacterial reduction in the human system. Excess of nitrite is absorbed by hemoglobin and converted to mathemoglobin as a result of which oxygen transporting capacity of hemoglobin molecule is impaired. The reason that only infants and children are effected by this disease and not adults has been attributed to under-developed metabolic enzyme, greater reactivity of fetal hemoglobin and relatively small blood volume in comparison to fluid intake.

Consequent to observance of nitrate in groundwater of Lucknow area in concentrations higher than the maximum permissible limit of 45 mg/l, systematic study was undertaken during 1987-88. In this study, objectives were to find overall distribution of nitrate in groundwater, temporal variations in comparison to investigation conducted during 1981, changes in nitrate concentration at deeper levels, sources of nitrate concentration and to suggest methods to control nitrate pollution of groundwater.

II. MATERIALS AND METHODS

For the study conducted during 1987-88, 46 groundwater samples were collected from shallow levels upto the depth of about 30 meters and 15 samples were collected from deeper levels in the range of 155 meters. In the earlier study conducted during 1981, 52 and 13 samples were collected from shallow and deeper levels, respectively. Collection of water samples and chemical analysis was carried out by standard methods given in literature (Brown et al, 1972; A.P.H.A., 1976; Handa, 1976). For determination of nitrate, mostly chromotropic acid method was used and high values were cross-checked with distillation method.

III. THE AREA, GEOLOGY AND HYDROGEOLOGICAL CONDITIONS

Being the capital of Uttar Pradesh and center of commercial activities, Lucknow is thickly populated with a population of 10.07 lakh according to 1981 census which at present is likely to be much higher. It has an area of 79 sq. Km. with tropical climate and annual normal rainfall of approx. 1015 mm.

The area lies in the Gomti sub-basin of central Ganga alluvial deposits of Quarternary age comprising clay, kankar and very fine to medium grained sand.

Studies carried out by Singh (1988) in the area around N.E. Railway City Station, Lucknow have revealed that there are two groups of aquifers down to a depth of 183 meters. Depth to water level varies from 2.40 to 11.53 b.g.l. and the depth of dug well structures ranges from 10.72 to 29.30 meters b.g.l.

The first or top aquifer occurs in the depth range of 30 and 107 meters b.g.l. and is interspersed with thin clay beds with range in thickness from 2 to 40 meters. The second aquifer which underlies the clay bed, varies in thickness from 6 to 37 meters and it occurs in the depth range of 107 to 145 meters b.g.l. Ground water occurs under water deeper aquifer appears to be extensive and in a confined state of disposition due to presence of extensive clay beds seperating the first and second aquifers.

IV. MOBILITY OF NITROGEN TO THE SATURATED ZONE

The organic matter component in the soil present initially or added subsequently is decomposed by bacteria and fungi to aminoacids which in turn are degraded to carbon dioxide, ammonia, nitrate etc. Urea is hydrolysed to ammonia by various species. Clay minerals such as illite, montmorillonite etc. which have expandable clay lattices, can fix ammonium ions in non-exchangeable forms. However, under tropical moist and aerobic conditions, oxidation of ammonia to nitrate occurs particularly in neutral to alkaline soils. In acidic soils, ammonium ions are directly absorbed by the plant roots. Soil micro-organisms also tend to use nitrogen present in the soil for growth. In general, the nitrogen present or added to the soil is removed by various processes such as (i) uptake by micro organisms, (ii) uptake by plants, (iii) volatalisation of ammonia, (iv) decomposition of nitrites in acid soils and (v) denitrification.

Nitrates which are not removed by these processes may be leached to the saturated zone resulting in increase of nitrate concentration of ground water. The behaviour of nitrate ions has similarity with those of chloride ions. Nitrate salts usually have high solubility and are not absorbed by the clays present in the soils and strata. Walter et al. (1975) have observed that irrespective of the form in which nitrogeneous fertilisers are applied, these are changed to mobile nitrate by natural processes. Nitrate which is not taken up by aforementioned processes may eventually find its way to the groundwater system depending upon the recharge water moving through the profile.

V. DISCUSSION OF RESULTS

As pointed out earlier this study is based on chemical analysis of 52 groundwater samples from shallow zone (mostly less than 30 meters), 13 samples from deeper zones (aprox. 100-150 meters) collected during 1981 and 46, 15 samples collected from shallow and deeper zones, respectively during 1988. The sampling stations along with nitrate ranges with the help of symbols are depicted in Fig. 1.

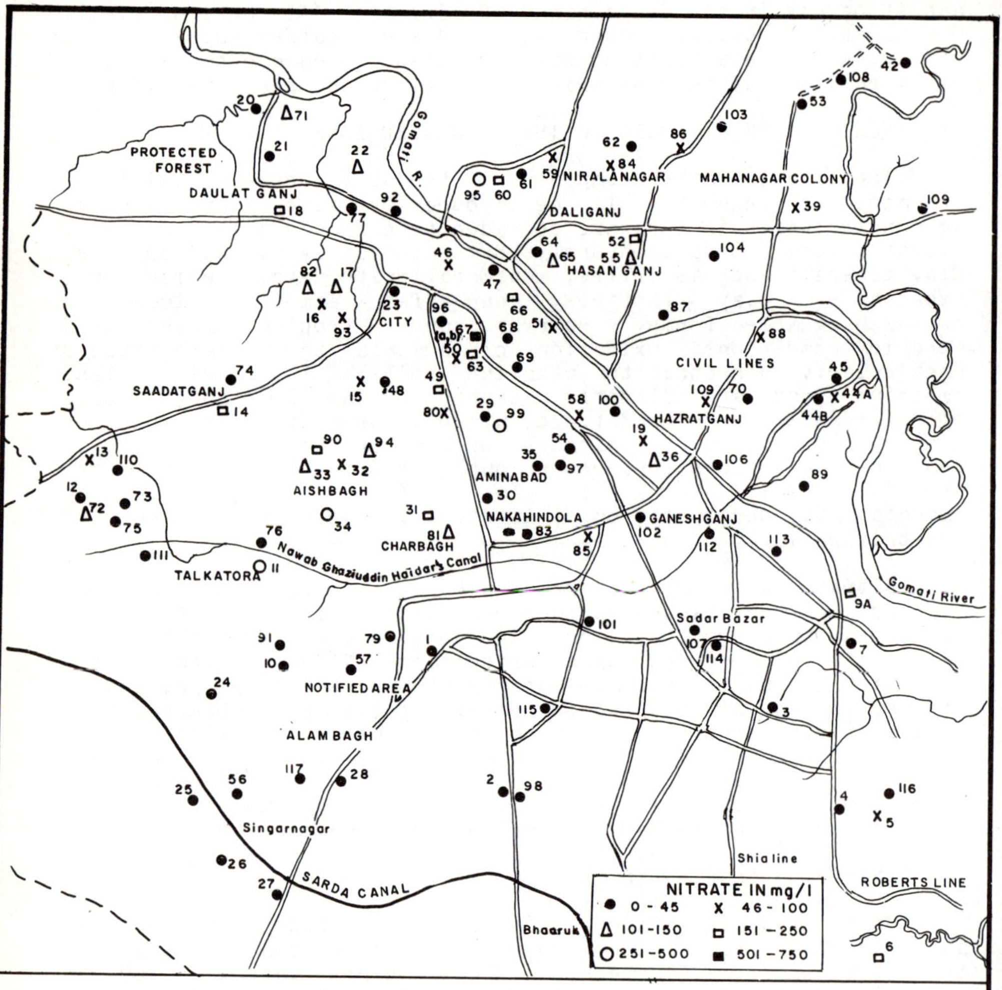

Fig. 1 LOCATION OF SAMPLING STATIONS IN LUCKNOW AREA ALONG WITH NITRATE RANGES OF GROUNDWATER.

SCALE

Mts. 500 0 500 1000 1500 Mts.

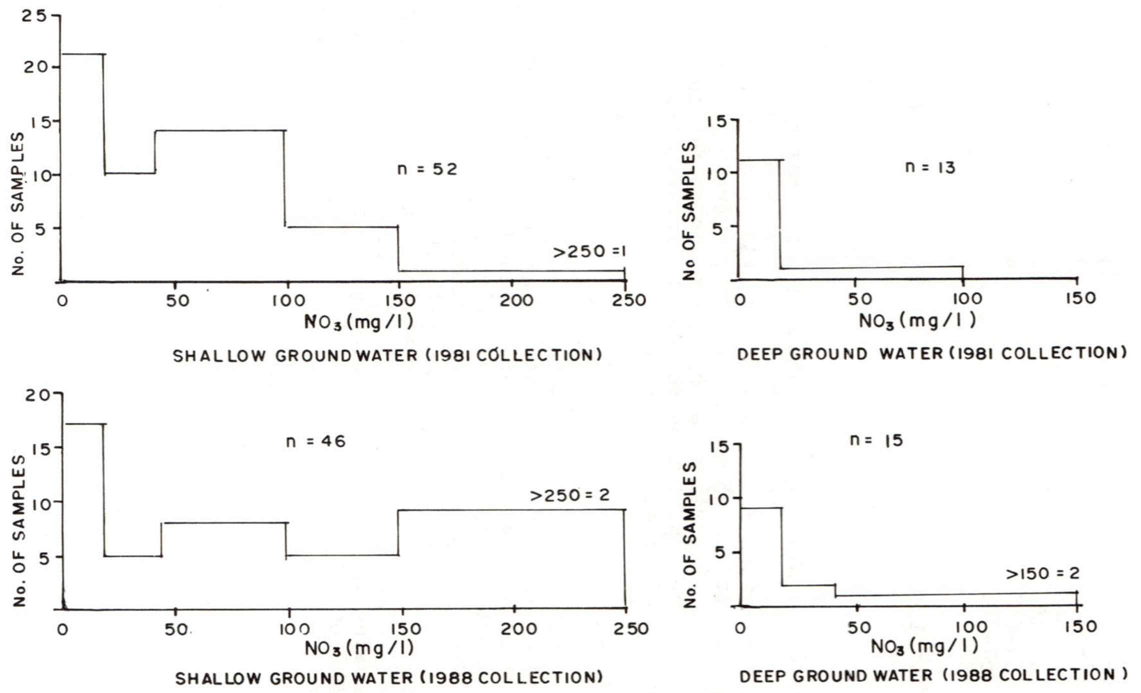

FIG.-2. NUMBER OF SAMPLES IN DIFFERENT NITRATE RANGES IN SHALLOW AND DEEP GROUD WATER OF LUCKNOW DURING 1981 & 1988,

Table 1. Distribution of number of samples from shallow and deep groundwater in different nitrate ranges.

Nitrate range (mg/l)	year-1981		Year-1988	
	Shallow	Deep	Shallow	Deep
0-20	21	11	17	9
21-45	10	1	5	2
46-100	14	1	8	1
101-150	5	-	5	1
151-250	1	-	9	1
251-500	1	-	2	-
501-750	-	-	-	1
Total	52	13	46	15

The data of nitrate distribution in shallow groundwater (Table 1) indicates that out of 52 wells monitored in 1981, nitrate concentration in 21 wells (40.4%) was more than the permissible limit of 45 mg/l. In 1988, there were 24 out of 46 sampling stations (52.2%) more than this concentration indicating increase in nitrate pollution during the intervening period. As the samples are distributed throughout the area and are not limited to a particular locality, this increasing trend is likely to have a fair amount of reliability.

The distribution of nitrate in groundwater of the study area (Fig.1) indicates that concentrations are high in several parts. Apart from city area which is characterised by dense population, some of the outskirts such as village Neelmatha in south-western part and Pump Station in north-western part have high concentrations of nitrate. In general, the central and northern parts of the study area have comparatively higher concentration of nitrate in groundwater. Though effect of fertilisers may be there in causing increase of nitrate levels in the outskirt rural areas, high concentrations in the thickly populated areas may be due to sources other than those related to agricultural activities.

There are clusters of sampling stations near City Station (Fig.1) with high nitrate concentrations to the extent of 590, 330, 233, 192 mg/l. On northern side also, there are several places with nitrate concentrations in groundwater above 100 mg/l. It is evident that patches of groundwater with high nitrate concentration are present throughout the sutdy area. As it appears to be a case of point pollution, contours have not been drawn as these would give erroneous interpretations. There are several places with high concentration of nitrate surrounded by groundwater with low concentration indicating localised affect of pollution. For this reason, nitrate ranges in groundwater are depicted with the help of symbols in Fig. 1.

Table 2. Chemical analysis results of selected constituents present in ground water with nitrate concentrations more than 45 mg/l (1988 collection)

S.No.	Location	Map Ref No.	E.C. (micro-siemens)	pH	concentration in mg/l. HCO_3	Cl	NO_3	Ca	Mg
	Shallow (upto approx 30 mts)								
1.	Neelmatha	6	1374	7.50	445	103	184	56	63
2.	Pipraghat Village	9(a)	1539	7.75	378	163	185	32	80
3.	Alamnagar	13	1543	7.30	268	261	46	37	73
4.	Matadin Road	14	1939	8.00	433	255	192	22	117
5.	Muszid Ibrahim	15	565	7.50	201	44	52	24	35
6.	Chaupatiya	17	828	7.50	195	99	124	24	49
7.	Aishbagh Road	31	1361	7.50	256	149	240	44	70
8.	Motijheel	32	2086	7.50	610	284	76	84	39
9.	Lakshaman Ganj	33	798	7.50	207	71	102	20	46
10.	Madayganj	60	1361	8.00	464	71	175	98	62
11.	Talkatora Road	11	1366	7.30	165	181	270	24	66
12.	H.A.A. Road	16	926	7.30	366	57	84	52	68
13.	Thakurganj	18	1558	7.00	488	121	232	14	86
14.	Raniganj	22	1593	7.50	409	145	125	32	74
15.	Pande Talab	34	1161	7.50	220	103	262	36	53
16.	J.P.Krishna Lane	36	880	7.50	244	74	140	16	51
17.	Mahanagar	39	771	8.00	183	89	93	28	44
18.	Tilak Marg	44(a)	1243	7.40	464	74	60	78	52
19.	Teri Bazar	49	1326	7.60	390	103	192	118	58
20.	City Station (area)	50	814	7.70	293	46	50	62	53
21.	I.T. Crossing	52	1074	8.00	250	113	160	74	67
22.	University Road	55	1977	7.50	671	181	136	184	63
23.	Sitapur Road	59	1570	8.00	457	131	100	136	90
24.	City Rly. Station I	63	1197	8.15	162	135	233	69	63
25.	City Rly. Station II	67a	3230	7.15	651	387	650	148	159
	Deep (appx. range 100-140 mts)								
1.	NE Rly. Station	67b	1914	7.90	287	237	590	164	141
2.	Lumbeshwar Temple	65	1387	7.90	256	140	145	75	71
3.	River Bank Colony	66	931	8.00	329	74	168	62	61
4.	Medical College	46	943	7.50	354	60	100	60	55

Such high levels of nitrate in groundwater in several parts of the study area are a health hazard particularly for the reason that at most of these places, water is being used for drinking purposes.

VI. SOURCES OF NITRATE IN GROUNDWATER

The main sources of nitrate to groundwater include:

i. precipitation,

ii. geological deposits,

iii. industrial effluents,

iv. fertilizers,

v. sewage, septic tanks and animal wastes.

Precipitation: It is a minor source of nitrate to groundwater. Studies carried out by Yalon (1964) have revealed ammonium concentration in the range of 0.5 to 28 mg/l in rainfall over Israel with nitrate concentration nearly one-fourth of ammonium ions. Studies carried by Handa et. al (1986) have shown that nitrate concentration in rainfall over Lucknow is mostly less than 5 mg/l indicating that contribution from precipitation is quite small in the study area.

Geological Deposits: In absence of any major reported geological deposits of nitrate in the study area, the contribution from this source is not likely to be significant.

Industrial Effluents: The effluents of industries manufacturing nitrogenous fertilisers, distillaries, breweries, butcheries, meat packing etc. contain high concentrations of nitrate. Though a brewary is present in the study area, the effluents are discharged into the river and its effect on groundwater is likely to be minimal. Butcheries and slaughter houses operate on a small scale and the effect of leachate from these sources is also not likely to be conspicuous.

Fertilisers: Use of fertilisers and other agricultural activities such as barn yard, silo wastes, disposal of crop residues can contribute towards enrichment of nitrate levels of groundwater. Studies carried out by Handa (1983), Kakar (1985) have revealed that fertilisers may be one of major sources of high concentrations of nitrate in several parts of the country.

In the study area, though fertilisers may be contributing nitrate to groundwater in some of rural areas in the outer skirts of the city or as a result of horticultural activities in the city area, the overall contribution is not likely to be very significant. This is coroborrated by the fact that high nitrate levels in groundwater have been observed in thickly populated areas where use of fertilisers is not significant.

Sewage, Septic Tanks, Animal Wastes: Domestic sewage when discharged on land without proper treatment is one of the main sources of nitrate contamination of groundwater particularly in areas with high permeability of soils and shallow water table. In the study area, most of sewage is being discharged through drains and sewer pipes into the river Gomti. Seepage from unlined drains is likely to contribute nitrate to the ground water system. In addition to this, septic tanks are extensively used as mode of sewage disposal in several areas of Lucknow. Leakage from septic tanks can be one of the major sources of increase in nitrate levels of groundwater in these areas.

Apart from direct contact of waste water from septic tanks with shallow groundwater, these can also cause pollution of deep groundwater. The leachate of septic tank may find its way into the tubewell through the angular space between the borehole wall casing and gravel being permeable medium may permit the effluent to move vertically down into the aquifer system. Singh (1985) has attributed high concentration of nitrate in deep groundwater in parts of Lucknow area to this reason.

Animals such as cows, buffaloes etc. produce large amount of nitrogeneous wastes. As animals in dairies are confined to comparatively small places, the excreta gets accumulated which on leaching increases the nitrate content of groundwater. Morris (1966) has reported that nitrogen content of animal manure slurries expressed as N varies from 0.13 to 0.42% for cattle, 0.3 to 0.9% for hogs and 0.41 to 1.7% for poultry. The nitrogen concentration of waste water from cow-sheds and milking paralors has been found to be 4200 mg/l (Loehr, 1970).

Schmidt (1972) has reported that there is comparatively high concentration of chloride in groundwater as a result of sewage pollution. For the chemical analysis data of study area, an attempt was made to find statistical correlations between chloride and nitrate concentrations of shallow ground water collected during 1988. The regression equation between these two variables chloride (Y) and nitrate (X) based on 46 samples has been found to be $Y = 0.77 X + 1.43$.

The value of correlation coefficient between chloride and nitrate ions has been observed to be + 0.77. This high positive correlation coefficient coroborrates the finding observed during field work that percolation of sewage into the groundwater system is likely to be one of the major sources of nitrate contamination of ground water in Lucknow area.

Statistical relationships for groundwater at deep levels could not be determined due to insufficient data.

VII. TEMPORAL VARIATIONS

The changes in chemical characteristics of groundwater at a place with passage of time depend on several variable factors. These include extent of recharge, changes in vegetation cover on the land surface, reactions in soil and vadose zone and contribution from activities of man. The net change in chemical characteristics of groundwater may be due to cumulative effect of these factors.

In the study area, nitrate concentrations in groundwater from the same 11 sampling stations were monitored during the years 1981 and 1988, the results of which are given in Table 3.

Table 3. Temporal variations in nitrate concentration of groundwater.

No.	Map Ref. No.	Source of ground water	Location	NO (mg/l) 1981	1988
1.	1	Dug Well	C.P.S. Colony	not detected	7.0
2.	3	"	Nehru Road	not detected	2.5
3.	5	"	Race Course	57	42
4.	15	"	Masjid Ibraham	40	52
5.	33	"	Lakshman Ganj	64	102
6.	35	"	Aminabad Thana	10	20
7.	58	"	Aminudaula Library	52	38
8.	51	Hand Pump	Gandhi Bhawan	66	23
9.	55	"	University Road	15	136
10.	66	Tubewell	River Bank Colony	72	168
11.	7	"	Dilkhusha Garden	2.5	11

From the data given in Table 3, there is increase in nitrate concentration of groundwater at 8 out of 11 sampling stations and decrease at the remaining places. Out of the samples in which nitrate concentrations have increased, at three places, the increase is less than 10 mg/l. At the remaining places, the increase is of higher order and is quite significant at S.No. 5, 9 and 10. As already pointed out, the temporal changes in chemical characteristics are as a result of cummulative effects of several factors. The most important out of these in the study area is likely to be pollution effect of sewage discharge on groundwater which goes on varying. The decrease in nitrate concentration in groundwater at 3 places may be attributed to lesser sewage discharge or greater dilution effect. In general, nitrate pollution of groundwater in the study area indicates increasing trend with passage of time.

VIII. CHANGES IN NITRATE CONCENTRATION WITH DEPTH

In order to study the changes in nitrate concentrations of groundwater with depth, samples were collected from both shallow and deeper levels. In cases, where deep well was not available at same point, the monitoring was carried out from deep well available in the nearest vicinity. The results along with location, depth details are given in Table 4.

Table 4. Changes in nitrate concentrations of groundwater with increase in depth.

S.No.	Map Ref. No.	Location	Ground water source	Depth (meters)	NO (mg/l)
1(a)	63	City Station I.	H.P.	15.0	233
1(b)	67	NE Rly. Station	T.W.	102.04	590
2(a)	51	Gandhi Bhawan	H.P.	12.19	66
2(b)	69	Dufferin Hospital	T.W.	130.00	25
3(a)	52	I.T. Crossing	H.P.	30.48	160
3(b)	65	Lumbeshwar Temple	T.W.	131.11	145
4(a)	47	Jawahar Nagar	H.P.	Less than 30.48	9
4(b)	64	Daliganj	T.W.	128.34	40
5(a)	47	Jawahar Nagar	H.P.	12.80	9
5(b)	66	River Bank Colony	T.W.	132.63	168
6(a)	25	Kanausi	H.P.	30.48	1.8
6(b)	28	Chander Nagar	T.W.	200.00	18
7(a)	84	Nirala Nagar	D.W.	9.48	56
7(b)	62	C.S.I.R. Colony	T.W.	150.17	6.8

As a result of activities of man, the concentration of nitrate is usually higher in groundwater at shallow level which decreases with increase of depth. This trend is evident in samples at S.No. 2, 3 and 7. However, the opposite trend of higher concentration of nitrate in groundwater at deeper levels is evident in samples at S.No. 1, 4, 5 whereas the change is not significant in case of groundwater at S.No. 6.

It is likely that at these places leachate from septic tanks may enter through the annular space between the borehole well and casing.

IX. MEASURES TO CHECK GROUNDWATER POLLUTION

In view of high concentration of nitrate in groundwater of the study area at several places, there is immediate necessity to take steps to control and prevent pollution. Some of the measures which can help in this direction are:-

i) Unlined drains through which sewage flows from various parts to the river should be lined to prevent seepage and consequently degradation of ground water quality.

ii) The septic tanks through which leakage is expected should be examined and cement grouting may be carried out on the sides to make these insulated to prevent seepage of sewage to groundwater system.

iii) Raw sewage should be treated before it is discharged on land.

iv) Abandoned wells should be closed from top to prevent being used for dumping wastes.

v) Fertilisers should be used according to optimal requirements of soils and crops to prevent leaching of fertilisers to groundwater.

vi) To prevent pollution from wastes of dairy animals, excreta of animals should be accumulated in closed places to prevent leaching to ground water.

X ACKNOWLEDGEMENTS

The authors are grateful to Shri B.P.C. Sinha, Chief Hydrogeologist and Member, Central Ground Water Board, New Delhi for his help, encouragement and permission to publish this paper.

REFERENCES

A.P.H.A. 1976. Standard methods of analysis of water and waste water. American Public Health Assoc., USA, 14th Edition, pp 33-362.

Brown E, Skougstad M.W., Fishman M.J. 1970. Methods for collection and analysis of water samples for dissolved minerals and gases. U.S.G.S., Book V, Chapter AI, pp 1-394.

Handa B.K., 1975. Technical manual on modern methods of analysis. Technical Manual No. 3, Ministry of Irrig. and Agri., pp 1-160.

Handa B.K., 1983. Effect of fertiliser use on groundwater quality in India. National Seminar on groundwater development. A perspective for the year 2000 AD.

Handa B.K., Adarsh Kumar, Goel D.K. 1986. Rainfall composition in Lucknow in 1982 and its significance to man, Mausam, 37, 2, pp 245-254.

Kakar Y.P. 1985. Nitrate pollution of groundwater in Haryana. Seminar on Water Quality and its management. Central Board of Irrig. and Power, New Delhi. pp. 59-60.

Loehr R.C. 1970. Control of nitrogen from animal waste waters. Proc. 12th Sanitary Engg. Conf. Urban, Illinois, pp 153-169.

Morris W.H.M. 1966. Economics of liquid manure disposal from confined livestock. Proc. on Natural Symp. on animal waste management. A.S.A.E. Publ. SP-0366, pp 126-131.

Schmidt K.D. 1972. Aquifer protection in rehabilitation, Bull Session No. 4, GroundWater, Jan-Feb. 1972, pp 73-88.

Singh B.K. 1988. Report on groundwater pollution in an area around north-eastern Railway Station, Lucknow, CGWB, NWR, Lucknow pp 1-22.

Walter M.F., Bubenzer G.D., Converse J.C. 1975. Predicting vertical movement of manurial nitrogen in soil Trans. ASAE 18, pp 100-105.

Yalon D.N. 1964. The concentration of ammonia in rain water over Israel in relation to environmental factors. Tallus 16(2), pp 200-204.

STUDY OF GROUNDWATER POLLUTION AT TIRUMALA, TIRUPATI, INDIA.

B.S. Sukhija, D.V. Reddy and V.K. Saxena

National Geophysical Research Institute,
Hyderabad - 500 007.

...

ABSTRACT

This paper relates to the investigation carried out with the objective of determining the degree, extent, nature and source of groundwater pollution in and around TIRUMALA TEMPLE AREA, TIRUPATI. Towards this end, about 60 wells including the four important wells in the temple campus were selected to monitor water levels and water chemistry. Water level data collected so far has indicated the existence of groundwater mounds around the temple area and the movement of the Pushkarni water towards the well in the temple area. Groundwater chemistry has indicated excess of nitrates and suspended solids in most of the wells. The nitrate contents of the Pushkarni are close to those of two near by wells supporting influent condition of the Pushkarni. The high content of suspended solids, fat and food material were detected in one of wells. A simple tracer experiment designed and carried out in the temple area has established the large kitchen to be one of the important sources of pollution besides other sources mentioned above.

INTRODUCTION

The present study relates to an investigation of pollution at the Tirumala temple at Tirupati in India (Sukhija et al. 1988). The Tirumala temple at Tirupati is one of the biggest pilgrim centres of India, the temple being richest in India and perhaps second in the world. The study was undertaken to determine the nature, extent and delineate possible sources of pollution of groundwater in the temple with special reference to four wells known as the Bangaru Bavi, the Viraja Nadi, the Pottu Bavi and Poola Bavi; each one being used for specific purposes. The study is mainly based on the regional groundwater flow dynamics, hydrochemistry along with simple tracer experiment conducted in the temple premises.

DESCRIPTION OF THE STUDY AREA

The study area encompasses the Tirumala town situated at an elevation of about 850 mtrs in the Nagari hills, Chittoor District of Andhra Pradesh (Fig. 1). The average annual rainfall is about 950 mm. Physiographically the temple is situated in the lowest portion of a basin like

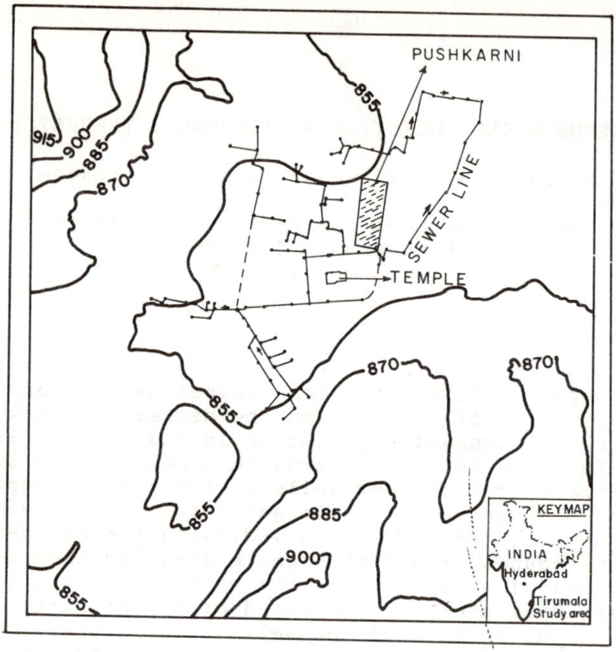

Fig.1. Location of Tirumala Temple, Topography and Sewer Lines Around It.

structure which is surrounded by steep hillocks rising to 950 mtrs. In the north of the temple, there is a big artificial water reservoir called Pushkarni in which thousands of devotees take bath before entering the temple. There exist a number of dense residential complexes surrounding the temple, where sanitary conditions are deplorable. The sewage is lined as well as unlined and it runs all along the four streets of the temple.

Geologically the study area is situated in the extreme southern end of the Cuddapah Basin and the area is occupied by quartzites which rest directly on the Archaean gneiss. The great Epharchean Unconformity is well known geological feature to be seen around. Groundwater occurs in the area under unconfined conditions and recharge is expected to take place through lateral flow from the surrounding hillocks as well as through direct infiltration.

RESULTS AND DISCUSSIONS

In order to gain the information about regional movement of groundwater and pollutants, water levels and hydrochemical data were generated from the sixty selected observation wells. The analysis showed there was a high degree of pollution in the temple wells. The wells showed high concentration of suspended solids (90 ppm), nitrates (100-350 ppm) and high colliform bacteria, high fat contents etc. Highest nitrate concentration (108 ppm) is observed in the Bangaru Bavi water. The nitrate concentration in the wells located in the thick residential complex just out-side the temple range from 40 to 70 ppm. In and around Pushkarni this concentration is 45 ppm. However, it was noticed that nitrate concentration of water samples in deep bore well outside the temple premises was negligible. The Viraja Nadi well, inside the temple premises is the most polluted giving off an obnoxious odour. Figs.2&3. show the iso-concentration plots for nitrates and suspended solids.

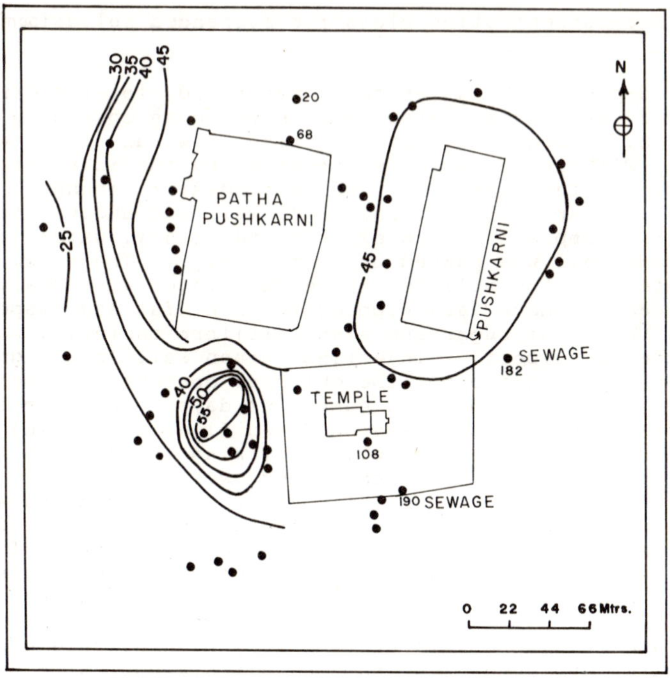

Fig. 2. Iso-concentration plots for nitrates(ppm).

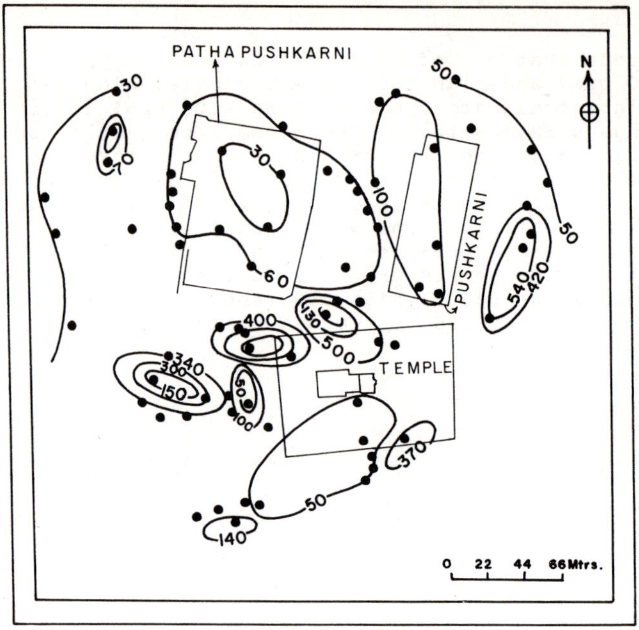

Fig. 3. Iso-concentration plots for suspended solids(ppm).

The depth to the water table was found to vary from 2 to 9 m and groundwater flow was inferred as coverging from the north, south and west directions towards the temple (Fig. 4). With-in the area, encompassed by the four streets surrounding the temple, two groundwater mounds were detected; one around the Viraja Nadi well and another to the west of the temple. In an experimental study, water in the Viraja Nadi well was pumped out completely. The water that accumulated in the well two-three days later was analysed. It was found to have very high contents of fat and suspended solids, traces of sugar and also colliform bacteria, making the kitchen a suspect. Injecting common salt in the ground below the kitchen floor, the chloride content of water in the well was monitored for a number of days. An increasing trend after the injection was observed finally reaching five times the normal concentration, confirming the kitchen to be one of the sources. For the mound west of temple the suspected source is residential complex in which groundwater is at higher elevation. At the time of investigation water in Pushkarni was at a higher elevation than that of the general water table around creating influent conditions for the wells close by. Almost similar nitrate concentration in the Pushkarni, Poola Bavi and Pottu Bavi further corroborate this observation. The very high nitrate content of Bangaru Bavi suggested that the nitrate content might be high due to leakage from sewage.

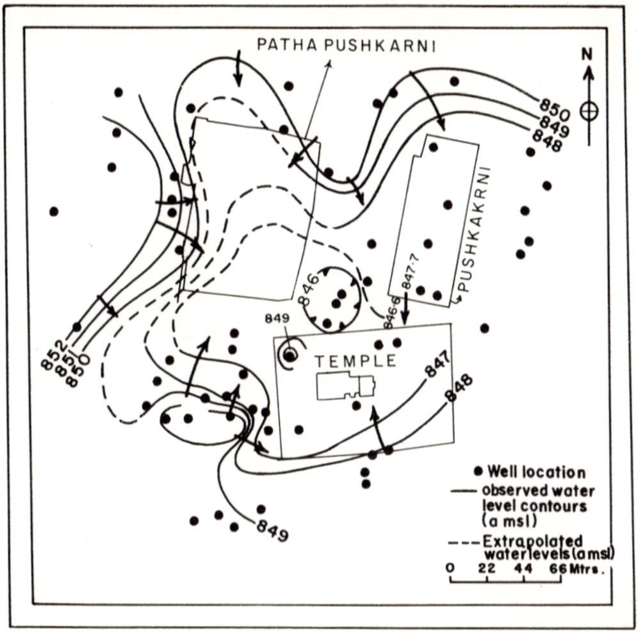

Fig. 4. Groundwater movement surrounding Tirumala Temple.

CONCLUSIONS

From the detailed regional geohydrological, hydrochemical and tracer study in and around the Tirumala temple, Tirupati it has been possible to elucidate different pollution sources such as private dense residential complexes, large kitchen of the temple, the sewer line net work and the Puskarni water reservoir.

ACKNOWLEDGEMENTS

Authors are grateful to Prof. V.K. Gaur, Director, NGRI for his keen interest in the work. This work was sponsored by the TTD authorities. They have very kindly provided all the facilities for carrying out studies. Water level data and some of the chemical data obtained after painstaking efforts were generously provided by them and made use of in this paper is gratefully acknowledged.

REFERENCE:

Sukhija, B.S., Reddy, D.V., and Saxena, V.K., 1988. Pollution of groundwater at the Tirumala temple, Tirupati - A preliminary study.

A CASE STUDY OF INTEGRATED EXPLORATION PRACTICE FOR SHALLOW FRESH WATER AQUIFERS IN SOUTHERN PARTS OF RAMNAD DISTRICT (TECHNOLOGY MISSION PROJECT)

J. SRINATH, A. RAJA MOHOMMED AND K.M. NAMBOODIRI
AFPRO GIT UNIT III, COIMBATORE

The average annual rainfall in the Mudukulathur block of Ramanathapuram district (Tamil Nadu) vary from 400 to 700 mm which is mainly received through the north-east monsoon. The sub-surface water is mostly saline and at few places it is saltier than sea water. The drinking water ponds, irrigation tanks and occassional shallow sweet water pits are the main drinking water structures. The pits and shallow wells dug out in the tank beds support the drinking water need in summer and the quality of water very often changes even within a day. AFPRO GIT Unit III selected about 100 square kilometres in parts of Mudukulathur and Kadalady blocks which are the most problematic area of Ramnad district (the shaded part in Fig.1) under the CAPART Technology Mission Project, for detailed surface water and ground water studies.

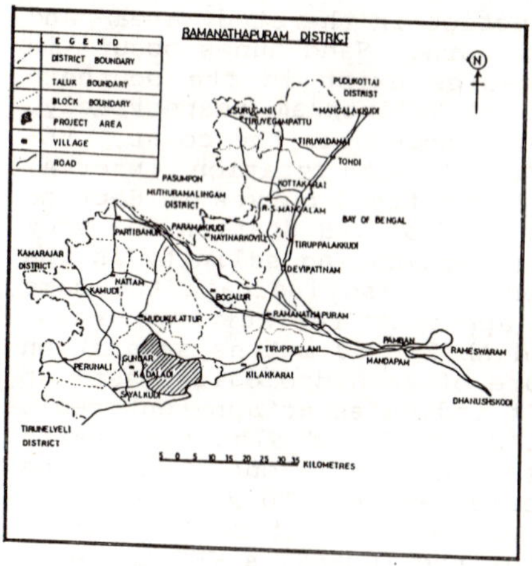

FIG. 1: LOCATION OF THE PROJECT AREA.

The area is underlain by fluvio-marine deposits (sand and clay), sand dunes, terrisand and alluvium deposits (Fig.2). Numerous irrigation tanks recharge the shallow groundwater body. (Fig.3). CGWB NGRI, State Government departments and other agencies conducted several scientific studies to unearth the hidden sweet water in this area. However, their study mainly corresponds to deeper horizons. Few blast holes drilled by ONGC and the information given by locals indicate availability of sweet water in pockets/channels at shallow depth. Hence an attempt to investigate the shallow sweet water bearing pockets were taken up in the project so as to provide shallow handpumps/infiltration wells.

The remote sensing data indicated fringes of burried channel near Kumarakkurichi Village. Vertical Electrical Soundings and water sample analysis (EC and pH) were carried out at known shallow sweet water bearing sites, near saline water wells and near the wells where the quality changes with time. Based on the informations thus collected, the area was classified into four major categories for planning the detailed field work. (Field layout is shown in Fig.3).

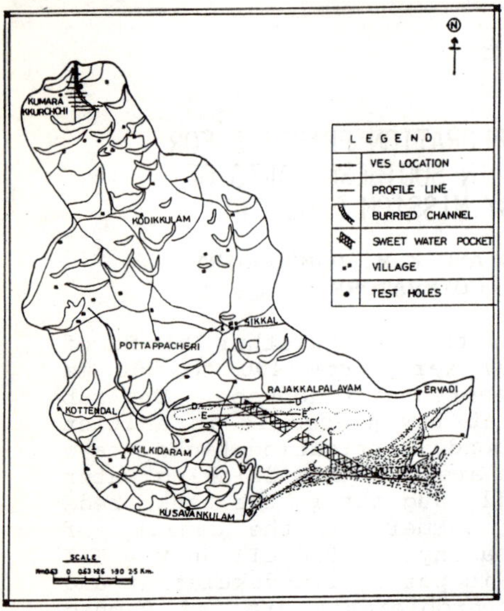

FIG. 2: MAP SHOWING BURRIED CHANNELS, GEOLOGICAL AND STRUCTURAL FEATURES

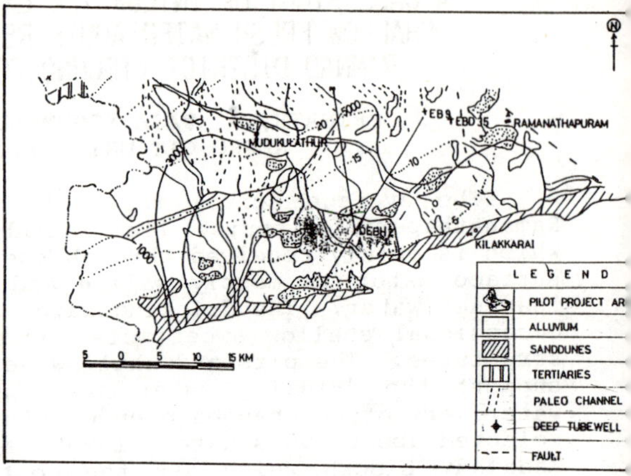

FIG. 3: PHYSIOGRAPHY AND FIELD LAYOUT MAP

The coast runs in east-west direction in the study area and near Senankuruchi it takes a southern turn. Sand dunes occur in the eastern part and sandstone exposures exist in the southern and western part near Senankuruchi and Valinokkam respectively. To study the possibility of sweet water body near the coast, VES traverse (AA', BB', and CC' in Fig.3) with station interval of 300 metres were conducted. Most of the field curves in Section AA' (Fig.4) and the locations beyond east of VES station 20 show very low order of apparent resistivity indicating salinity in the area. The VES curves in the section BB' also indicate the same feature. Fig.5(a) shows the VES apparent resistivity cross-section for AB/2 = 10, AB/2 = 25 and AB/2 = 50 metres separation which probably may indicate the nature of geohydrological set up in the traverse of shallow depth. It indicates an apparent resistivity high for the VES station No.12 to 15 and also the nearby surface indications revealed occurrence of sand stone, which has more or less north west-south east orientation. To study the continuity of these sandstone patches and also to study the variation and nature of the aquifer towards the in land, a VES traverse, perpendicular to the strike of the sand stone/coast (CC' in Fig.2) was conducted with a station interval of 300 metres. The qualitative nature of the curve shows more or less a same trend except increase in the order of apparent resistivity at few locations. Fig.5(b) shows the vertical section of this traverse for AB/2 = 10, 25 and 50 metres. It indicates an apparent resistivity high near VES station No.5 to 7 and this high coincides with the extention of the sandstone strike from the coast.

Series of VES profiles from east to west direction along the stretch of the belt (DD', EE', FF', in Fig. 2) were conducted to confirm or otherwise the continuity of the fresh water channel into the inland (terisand belt). Mostly KQ and KH type curves are obtained. The maximum apparent resistivity varied from 50 ohm-m to 100 ohm-m and the least was around 5 to 10 ohm-m. The Fig. 5(c) to 5(e) show the vertical section of the apparent resistivity along the traverses (DD', EE', FF' in Fig.2). It can be seen from the 3 profile sections that there was a remarkable apparent resistivity high along NW-SE direction. The comparison with the field observation indicated occurrence of fresh water bearing wells along this zone. When the profile sections along the coast, perpendicular to the coast and 3 terisand profile sections were collaterally analysed with hydrogeological data they indicated a continuous patch of sweet water bearing aquifer body along NW - SE direction from the coast. This was later confirmed by test drilling and production filter points. The quality of water in few wells analysed in this area and the surrounding area corraborates these findings.

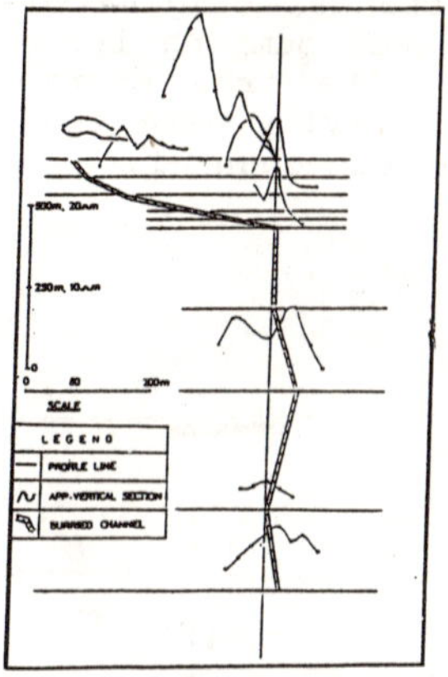

FIG. 4: FIELD VES CURVES-
COASTAL AREA
(SECTION AA')

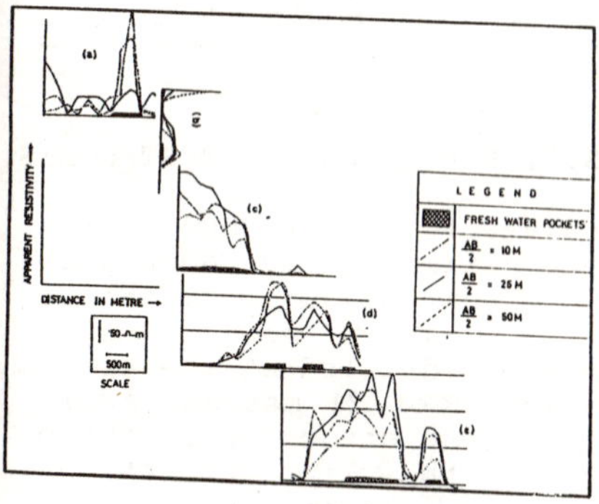

FIG. 5: APPARENT RESISTIVITY VERTICAL SECTION - COASTAL BELT AND TERRISAND AREA

The remote sensing data indicate an abrupt end of a burried channel near Kumarakkurichi village. VES traverse were conducted perpencular to the suspected channel (Fig.3). Fig.6 shows the detailed plan of the VES traverse. The lateral traverse distance was 20 metres and the traverse were conducted with 250 metres distance apart first and when the field analysis had indicated a change in trend of the channel, it was restrictedto 25 metres to 50 metres. The same figure contains the amplitude variations of each profile. The peak to peak correlation yield the course of the channel. The collatteral analysis indicated that this particular burried channel originated from Gundar river and was getting recharged. Few pumping irrigation filter points and shallow open wells in the area corraborate these findings. The quality and the resistivity data indicated deteriorated nature of the water away from the channel.

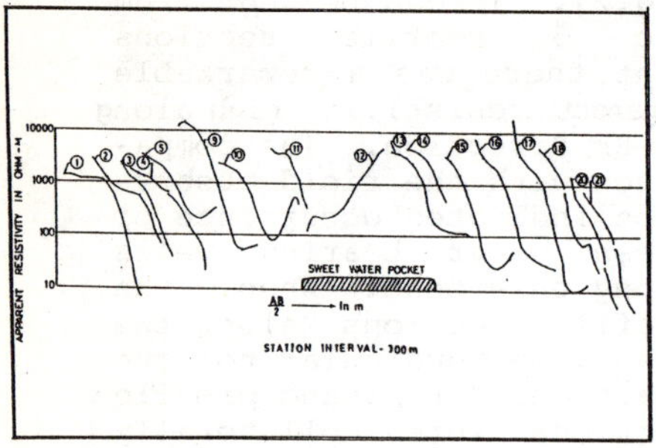

FIG. 6: KUMARAKKURICHI BURRIED CHANNEL

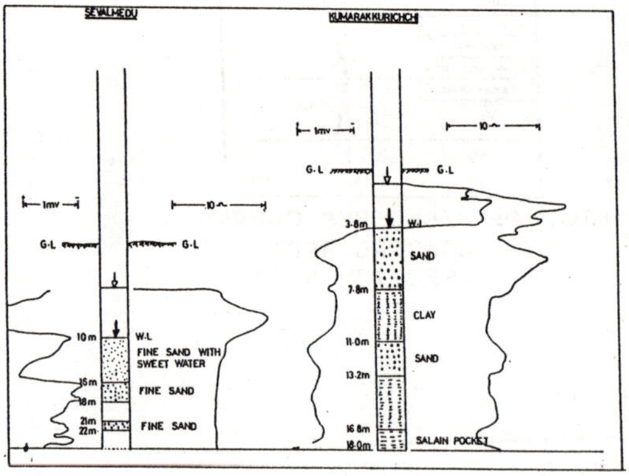

FIG. 7: S.P. & P.R. LOG OF TEST HOLES

In order to further understand and verify the hydrogeological and geophysical studies, test holes were drilled using the Hydra-Drill equipment, a portable direct rotary drilling machine. First test hole was drilled at Sevalmedu terri sand area to a depth of 24 metres and the second hole was drilled at the central part of the burried paleo stream to a depth of 21 metres. Self potential and single point resistance log were conducted in these test holes. The Fig.7 shows these logs and their interpretations. The analysis indicated the existence of fresh water upto a depth of 18 metres and 13 metres respectively.

Based on the analysis, few production wells were constructed by the Hydra-Drill equipment at (i) Kumarakkurichi, (ii) Sivalinga puram, (iii) Rajakkalpalayam and (iv) Thothamangalamwadi to a depth of 10.6, 10.8, 13.8 and 4.9 metres respectively. All the wells are fitted with India Mark II hand pumps and are yielding potable water.

This study showed that combined with remote sensing and quality data measurements, VES traverses can be used to locate the fresh water pockets in this type of geohydrological set up. It is also useful in tracing the paleo channels. The fresh water pocket is limited even in terri sand area though it is elevated to 30 metres above the sea level. In such environment, pumping scheme may have to be abandoned.

HYDROGEOCHEMICAL MODELING

V. NARAYANAN NAIR
Department of Geology, University of Kerala
Kariavattom P.O., Trivandrum-695 581

Groundwater, which is always in motion, except under certain conditions while travelling through aquifers interacts with the sub surface environment. During this movement, it may dissolve, transport and deposit mineral matter. Therefore the major ion composition of groundwater reflects the overall effects of hydrology, mineralogy, and weathering. In addition to this anthropogenic sources of metals and other trace elements entering natural waters represent a social risk. It is all the more important in the Indian context because of the indiscriminatory disposal of industrial wastes into surface and subsurface sites. Moreover, the nation's nuclear power programme is going to take a big leap, which aims at establishing a total generating capacity of 10,000 MW by the turn of the century. Naturally scientists are at present engaged in a search for suitable, deep geological formations which can be used as repositories for the increasing volumes of radioactive wastes. It is in this context, that we understand the importance of hydrogeochemical modeling studies in achieving a sufficiently thorough and reliable aqueous speciation of waters for the evaluation of social risks.

A hydrogeochemical model is the integration of mathematical, thermodynamic and kinetic relationships on which the geochemical calculations are based. These calculations are facilitated by the use of modern electronic computers. The basic data required is the chemical analysis of water samples, which should include all of the major elements (cationic and anionic) and important trace elements. The reliable calculation of the distribution of a dissolved metal among its numerous aqueous species requires careful and systematic sampling and the analysis of the water. In addition to this, data on mineralogy, hydrology, isotopic data and other relevant informations are also necessary.

Hydrogeochemical modeling is largely a mental exercise, but is facilitated by three types of calculations: (1) Speciation calculations, (2) Mass balance calculations, and (3) calculations of mass transfer and reaction path (Plummer, 1984).

The equilibrium speciation calculations determine the State of saturation of the water with respect of various minerals, based on a thermodynamic model and observed water quality data. These calculations can tell us whether a particular mineral would tend to dissolve on precipitate in a given aqueous environment.

In mass balance calculations, the amounts of mineral reactants and products which must dissolve and/or precipitate between the recharge and discharge points of a flow path are determined in order to account for the observed water quality. Mass balance calculations can also be used to solve problems of mixing of waters.

The mass transfer calculates the amounts of phases entering or leaving the aqueous phase required to account for the observed changes in composition between two water samples along a flow path. The mass transfer calculations are based on a thermodynamic model of the mineral-water system and depends on the solubility of minerals, precipitation, starting water composition, assumed set of irreversible reactions, absorption and mixing of water. In general, mass balance relations are used to define the mass transfer during the chemical evolution of one water to another.

More than 50 computer codes for these models have been prepared by various workers in the field (Nordstrom and Ball 1983). These computerized models are developed for different purposes and there is no single code, which can be used for all of the applications. In aqueous geochemistry, calculations to solve an equilibrium problem was first carried out by Garrels and Thompson (1962). This led to the development of various other computerized chemical models. The first generation program of water-mineral equilibria model is that of Truesdell and Jones (1973, 1974). This program, WATEQ, is written in PL/1. Later on it has been modified and translated into FORTRAN by Plummer et al. (1976), known as WATEQ. The WATEQ program contains selected thermodynamic data base for use with low-temperature natural waters. In WATEQ manganese speciation is also included WATEQ 2 (Ball et al., 1979, 1980) was written with the expansion of the data base to include complexes of Br and I, several metastable solids and several ion pairs of major constituents. The computer code also has been reorganized. WATEQ 3 (Ball et al. 1981) is a related program, with uranium species added. These programs are suited for treating both major and minor elements in the groundwater environment. For calculations at geothermal temperatures, the programs SOLMENEQ (Kharaka and Barnes, 1973) and EQ 3 NR (Wolery, 1983) are highly useful (Plummer, 1984).

Mass balance calculations can be carried out by the program BALANCE (Parkhurst et al., 1982), a FORTRAN computer program designed to help define and quantity chemical reactions between groundwater and minerals. This program also solves problems involving mixing of end-member waters, oxidation-reduction reactions and simple isotope balance.

In mass transfer calculations, the first reaction path code, PATHI, was developed by Helgeson et al. (1970). Later on relatively faster codes based on the reaction path concept have been developed of which EQ 3/6 (Wolery, 1984) and PHREEQE (Parkhurst, et al. 1980) seems to be the most important ones. A companion program to PHREEQE viz. PHRQINPT have been written for interactive construction of PHREEQE input data sets (Fleming and Plummer, 1983).

Recently Ball et al. (1987) developed the code WATEQ4F, adapted to execute on an IBM PC or compatible microcomputer. This is a FORTRAN 77 version of WATER 2 and the calculation procedure is the same. The programs PHREEQE and BALANCE are also now available in the modified from to run on an IBM PC or compatible (Ball et al., 1987).

Although more than 50 computer codes are now available, only few have been discussed here, There are various limitations, even in the use of these, especially in the thermodynamic data base used by the various models. The results obtained can be used to establish base line geochemical conditions for potential repository sites and to simulate the geochemical behaviour of transported radio nuclides.

REFERENCES

Ball, J.W, Jenne, E.A, Nordstrom, D.K., 1979. WATEQ2-A computerized chemical model for trace and major element speciation and mineral equilibria of natural waters in Jenne, E.A.Ed. Chemical Modeling in Aqueous Systems. American Chemical Society Symp. Ser. No.93, p.815-835.

Ball, J.W., Nordstrom, D.K. and Janne, E.A., 1980. Additional and revised thermochemical data and computer code for WATEQ2-a computerized chemical mode for trace and major element speciation and mineral equilibria of natural waters. U.S.G.S. Water-Resources Investigations 78-116,109 p.

Ball, J.W., Jenne, E.A. and Cantrell, M.W, 1981. WATEQ3-A geochemical model with uranium added. U.S.G.S. Open file report 81-1183, 81 p.

Ball, J.W., Nordstrom, D.K., and Zachmann, D.W., 1987. WATEQ4F-A personal computer FORTRAN translation of the geochemical model WATEQ2 with revised database. U.S.G.S. Open file report 87-50., 108 p.

Fleming, G.W, and Plummer, N.L., 1983. PHRQINPT-An interactive comppputer program for constructing input data sets to the geochemical simulation program PHREEQE. U.S.G.S. Water-Resources Investigations Report 83-4236., 108 p.

Garrels, R.M. and Thompson, M.E., 1962. A chemical model for sea-water at 25 c and one atmosphere total pressure. Am. J.Sci. U. 260,p.57-66

Helgeson, H.C, Brown, T.H, Nigrini, A, and Jones, T.A., 1970 calculation of mass transfer in geochemical processes involving aqueous solutions. Geochim. et cosmo. Acta, V. 34, p. 569-592.

Kharaka, Y.K, and Barnes, I, 1973. SOLMNEQ-Solution-mineral equilibrium computations. NTIS Tech. Report PB 214-899, Springfield, Virginia, 82p.

Nordstrom, D.K, and Ball, J.W., 1983. Chemical models, computer programs and metal complexation in natural water. Int. symp. on trace metal complexation in Natural Water.

Parkhurst, D.L., Thorstenson, D.C., and plummer, N.L., 1980. PHREEQE- A compputer prgram for geochemical calculations. U.S.G.S. Water Resources Investigations 8096. 210 p.

Parkhurst, D.L., Plummer, N.L., and Thorstenson, D.C., 1982. BALANCE-A computer program for calculating mass transfer geochemical reactions in groundwater U.S.G.S. Water Resources Investigations 82-14. 29 p.

Plummer, N.L, 1984. Geochemical modeling: A comparison of forward and inverse methods. Proc. First candadian/American conference on Hydrogeology- practical Applications of Groundwater Geochmistry, Banff, Alberta, Canada, Plummer, N.L, Jones, B.F, Truesdell, A.H., 1976. WATEQF-A FORTRAN IV version of WATEQ, a computer program for calculating chemical equilibria of natural waters. U..S.G.S. Water Resources Investigations 76-13. 61p.

Truesdell, A.H., and Jones, B.F., 1973. WATEQ-A Computer program for calculating chemical equilibria of natural waters. NTIS - PB2 - 20464, 77p.

______, 1974. WATEQ-A computer program for calculating chemical equilibria of natural waters. U.S.G.S. Jour. Res. Vol.2, p. 233-248.

Wolery, 1983. EQ3NR-A computer program for geochemical aqueous speciation solubility calculations: User's guide and documentations UCRL-53414. Lawerence Livermore National Laboratory.

Wolery, T.J., 1984. EQ6-A computer program for reaction-path modeling of aqueous geochemical systems : User's guide and documentation. Lawerence Livermore National Laboratory.

SIWANA-BARMER FRESH WATER BOWLS AS A PANORAMA OF MICRO-ALLUVIAL VALLEYS IN RAJASTHAN DESERT

P.G.Adyalkar and V.K. Bhagdikar,
National Environmental Engineering Research Institute,
Nagpur - 440 020.

Micro-alluvial valleys are essentially tiny and individually less significant alluvial bodies strewn in the midst of the Malani group of rocks with very limited extent. While they rarely exceed individually 500 km² in areal extent, in vertical extent they have a further serious limitation. In arid region like Barmer the thickness naturally exceeds as the water table recedes in depth to a serious limitation in rainfall, frequent incidents of drought with overexploitation and general desiccation of the terrain.

In this report, the studies carried out by Adyalkar (1962), Dutt et al (1962), Mani (1972 & 1973), Phadtare (1969) and Adyalkar (1973) in Raipur district of Madhya Pradesh, Bhir, Wardha and Yavatmal districts of Maharashtra and Jhunjhunu-Sikar districts of Rajasthan are of particular significance. By virtue of their occurrence all these valleys are classed as micro-alluvial valleys (Adyalkar, 1984).

Hence the productivity of a shallow and/or semi-confined heterogeneous aquifer in micro-alluvial valley is never in doubt, though the particular sections remaining granular may remain indeterminate. In a simple valley-fill, deposited from the adjacent hills the material will be althrough granular and continuous, with the bottom section often with boulders, cobbles, pebbles and gravels. In case of relict valleys of the Deccan Peninsula the material will be more evenly granular with a relatively better aquifer althrough. In the faulted lacustrine basin the occurrence of granular zone, though undoubted, may have serious limitation in extent.

As evident in respect of all alluvial valleys, coarser the formation and more equigranular it is higher will be its porosity, permeability and yield potential, should it be saturated with ground water. So also larger the admixture of clayey and silty material lesser will be porosity, permeability and yield potential. Interconnection of voids with facility to transmit ground water is more important in permeability determinations, with the result that soft clay, silt and tuff with very high order of porosity have relatively very low permeability.

In this light, in a recent study of Barmer district two potential fresh water bowls have been identified, one in the southeastern sector covering the major part of Siwana tahsil, and the other west of Barmer in an amphitheatrical region of Barmer tahsil.

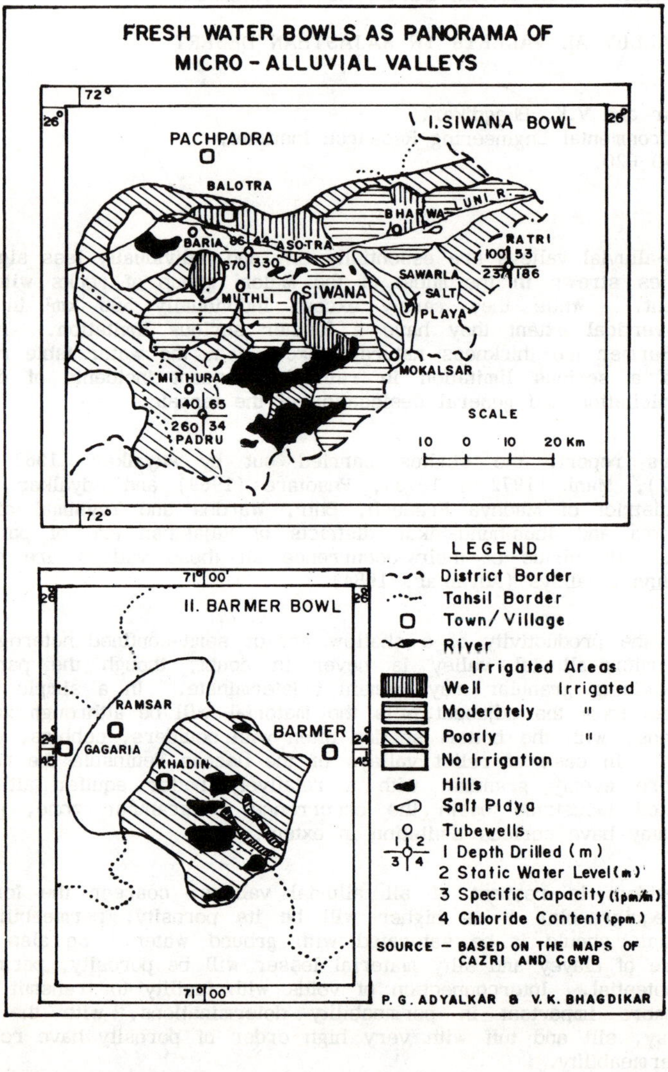
FRESH WATER BOWLS AS PANORAMA OF
MICRO - ALLUVIAL VALLEYS
72°
26°
I. SIWANA BOWL
PACHPADRA
BALOTRA
BHARWA
UNI R.
RATRI
100 53
237 186
BARIA
86 44
570 330
ASOTRA
MUTHLI
SIWANA
SAWARLA
SALT
PLAYA
MOKALSAR
MITHURA
140 65
260 34
PADRU
SCALE
10 0 10 20 Km
LEGEND
District Border
Tahsil Border
Town/ Village
River
Irrigated Areas
Well Irrigated
Moderately "
Poorly "
No Irrigation
Hills
Salt Playa
Tubewells
1 Depth Drilled (m)
2 Static Water Level (m)
3 Specific Capacity (lpm/m)
4 Chloride Content (ppm.)
SOURCE — BASED ON THE MAPS OF CAZRI AND CGWB
71° 00'
II. BARMER BOWL
RAMSAR
GAGARIA
KHADIN
BARMER
P. G. ADYALKAR & V. K. BHAGDIKAR

The Siwana bowl has an areal extent of about 1,500 km^2, with an average annual rainfall of 364 mm. It has two principal relict channels in the form of Ver and Luniwala rivers with several ephemeral streams, all of which disappear within short distances, sinking the entire drainage in the sandy expanse to enrich the shallow aquifer body limited to a maximum of about 150 m b.g.l., overlying the rhyolite-granite basement with very high drainage density. Indeed the drainage density in this part is 0.1 to 0.5 km/km^2, which is the highest in the district. The Barmer bowl has an areal extent of about 600 km^2 with rhyolite or Barmer sandstone as the basal rock limited to a shallower depth of 60 m b.g.l. in an area with an average annual rainfall of 242 mm.

In the Siwana bowl, ground water occurs under phreatic and shallow confined condition in sand and gravel beds, which are mutually interconnected. While there is a centripetal drainage in the southern part, that in the northern part is northerly towards the Luni river; in the western part it disappears in the sand dunes. In Barmer, the formations comprise silt, sand, kankar and gravel. They typically represent the panorama of mutually coelescing micro-alluvial valleys, masked with sand dunes.

The discharge of the tubewells in the Siwana bowl varies from 10 - 20 lps for a drawdown of 6 - 20 m tapping 40 - 65 m of granular zones in a depth range of about 100 m. The specific capacity varies from 35 - 200 lpm/m of drawdown with matching high transmissivity and hydraulic conductivity. On the contrary, dug wells in both the bowls are capable of yielding upto 100 m^3/d with variable phreatic surface. With an areal extent of about 2,000 km^2 the static ground water resources have been estimated at 2,000 MCM, the major part of which remains to be tapped.

The quality of water even in the arid tract of Siwana in Barmer is very good, with (i) total dissolved solids ranging from 310 - 1,080 mg/1, (ii) total hardness from 70 - 260 mg/1 as $CaCO_3$, (iii) chlorides in the range of 50 - 220 mg/1, and (iv) sulphates from 40 - 120 mg/1. However, the quality of water in the phreatic zone is relatively superior. Then there are over a dozen springs located in the southern part of the area, where the quality of water is still superior, being of "Excellent" quality conforming to the International Standard. In case of Barmer area the quality of water is relatively inferior. Along the fringe of the Siwana area as also in the area close to Lake Sawarla the fresh water floats over saline water.

According to Nefedov and Popova (1972, pp. 24-26) it is possible to precisely identify intermontane micro-alluvial valleys with shallow phreatic surface, studded with bogs, oozes, marshes, springs and mochazhinas in non-arid regions, or when brooklet, stream or river disappears over sandy desertic area in the form of blind delta from small, medium or large scale (1 : 12,500) aerial photographs. In the latter case it is possible to identify them by locating suffosion funnels and chocolaks covered with phreatophytes (Prosopis cineraria, Tamarix articulata, Saccharum bengalense, etc.) . Indeed in Barmer area they have been identified from the satellite imageries.

Adyalkar, P.G., 1962. Report on groundwater investigation for water supply to the Khetri Copper district of Rajasthan, Expl. Tube. Orgn. Mimeo. Report.

Adyalkar, P.G., 1984. Hard rock hydrogeology in Indian Scenario, Proc. 71st Ind. Sci. Cong., pt. II; Presidential Address, pp. 1 - 52.

Adyalkar, P.G., Phadtare, P.N. and Ramanna, K., 1973. Buried sinkhole of Dhamtari in the Charmuria limestone of Raipur district in Madhya Pradesh and its hydrogeological significance, Curr. Sci., Vol. 42, No. 4, pp. 131 - 133.

Dutt, D.K., Karanth, K.R. and Srivastava, J.P., 1962. Geology and groundwater conditions in the Singhana river basin, Jhunjhunu district, Rajasthan, Geol. Surv. Ind. Mimeo. Report.

Mani, V.V.S., 1972. Report on systematic groundwater studies in parts of Wardha sub-basin, Maharashtra, Cent. Ground. Board. Mimeo. Report.

Mani, V.V.S., 1973. Report on systematic hydrogeological studies in parts of Wardha sub-basin, Yeotmal district, Maharashtra, Cent. Ground. Board, Mimeo. Report.

Nefedov, K.E. and Popova, T.A., 1972. Deciphering of groundwater from aerial photographs, Translated from Russian by Dr. V.S. Kothekar, Amerind Publ. Co. Pvt. Ltd., New Delhi.

Phadtare, P.N., 1969. Report on groundwater investigation for rural water supply in Gevrai tahsil of Bhir district, Maharashtra, Cent. Ground. Board, Mimeo. Report.

DISTRIBUTION OF FLUORIDE IN SHALLOW AQUIFERS IN AND AROUND DISTRICT ROHTAK, HARYANA STATE

Sunil Kumar and Azhar Syed
Department of Geology, Panjab University
Chandigarh 160 014, India

ABSTRACT

Fluoride concentration from 0.6 to 1.2 mg/l is essential in potable water for substantial protection against dental caries, while a higher concentration can cause mottling of teeth and a still higher concentration may lead to fluorosis.

The present paper depicts the geochemical factors affecting the occurrence and distribution of fluoride in shallow aquifers in and around district Rohtak, Haryana, India. The analytical results reveal that high fluoride concentration is a major problem in the area studied. Fluoride concentration exceeding the desirable limit (0.6 to 1.2 mg/l) is a major health hazard as shown by 36 samples. 14 samples showed it to range between the desirable limits; however, 5 samples had a value below 0.6 mg/l. High values can be attributed to the use of phosphatic fertilizers and the presence of fluor-apatite, which are being leached down to the saturated zone by return irrigation flows.

INTRODUCTION

Fluoride is an element which is essential in trace amount for all human beings and is one of the normal constituents of all diets. It is a minor constituent of the natural water, but plays an important role in assessing the quality of water for domestic as well as for agricultural uses. Deleterious effects of fluoride on human systems are well known. Evidences have shown that its concentration in potable water affects the condition of human teeth and bones. High fluoride values have been observed in groundwater from large part of the area but the problem is more severe in Jhajjar, Madina, Chuckawas, Machhrauli, Chirya, Kasni, Silana, Bahadurgarh, Bawan and Kalanaur.

SIGNIFICANCE OF FLUORIDE

Fluoride acts acts as a two-edged sword: high doses (i.e., above 1.2 mg/l) of it lead to tooth and bone fluorosis and low doses (i.e., below 0.6 mg/l) can cause dental caries (tooth decay). Researches carried out on the effect of fluoride concentration around 1 mg/l in potable water reduces dental caries by replacing hydroxy group in hydroxy apatite from dental enamel isomorphically. This reduces the solubility of the enamel in acid and acts as an inhibitor of the bacterial enzymes responsible for the production of acid which attacks the enamel in caries. However, a higher concentration, viz. 1.5 mg/l or above, results in staining of the tooth enamel varying from yellow to black brown, while at still higher levels, viz. 5 to 10 mg/l of fluoride, pathological changes like stiffness of the back and difficulty in performing natural movement and results in chronic fluorosis take place on the continuous use for long periods.

The desirable limit for fluoride in drinking water is 0.6 to 1.2 mg/l (ISI, 1983). If the limit is below 0.6 mg/l, water source should not be rejected but suitable public health measures should be taken. The maximum limit may be extended to 1.5 mg/l if no alternative source is available. The highest desirable and maximum permissible limits for fluoride as prescribed by Indian Council of Medical Research are 1.0 and 1.5 mg/l respectively.

The optimum fluoride level (Table 1) for a given community depends on climatic conditions because the amount of water (and consequently the amount of fluoride) ingested by children is primarily influenced by the air temperature. This relationship was first studied and reported by Galagan (1953), Galagan and Lamson (1953) and Galagan and Vermillion (1957), but has been further investigated and reported by Richards et al. (1967).

Table 1 (Modified after EPA,1976)

Annual average of maximum daily air temperature °C	Recommended Control Limits fluoride concentration in mg/l			Approval limit mg/l
	Lower	Optimum	Upper	
9.9 - 12.0	1.1	1.2	1.3	2.4
12.1 - 14.6	1.0	1.1	1.2	2.2
14.7 - 17.6	0.9	1.0	1.1	2.0
17.7 - 21.4	0.8	0.9	1.0	1.8
21.5 - 26.2	0.7	0.8	0.9	1.6
26.3 - 33.4	0.6	0.7	0.8	1.4

METHODOLOGY

Water samples were collected in well-cleaned and pre-treated double-stoppered polythene bottles of two-litre capacity from the existing dugwells and tubewells. Suspended matter in the samples was removed by filtering through Whatman No.542 filter paper prior to analysis. EC and pH were determined at the time of collection of water samples. During the course of detailed and partial chemical analysis of water samples, fluoride was determined specifically by Spectrophotometeric Eriochrome Cyanine R method (Rainwater and

Thatcher, 1960). Optical density was measured at 540 nm.

RESULTS AND DISCUSSIONS

The analytical results reveal that the fluoride concentration in the shallow aquifers in some of the areas is well within the permissible limits (0.6 to 1.2 mg/l, ISI, 1983). At Samargopalpur, Chamaria, Bedwa, Seman, Meham and Sisarkhas, marginal low values (i.e., below 0.6 mg/l) have been obtained; while at Jhajjar, Chuckawas, Kasni, Silana, Bahadurgarh, Bawan, Khera, Bahu, Rohad, Machhrauli, Palri, Sankol, Chirya, Madina and Kalanaur, isolated patches of moderately high to very high fluoride concentration (i.e. above 1.2 mg/l and up to 4.10 mg/l) have been obtained. The distribution of fluoride in shallow aquifers has been shown in Fig.1.

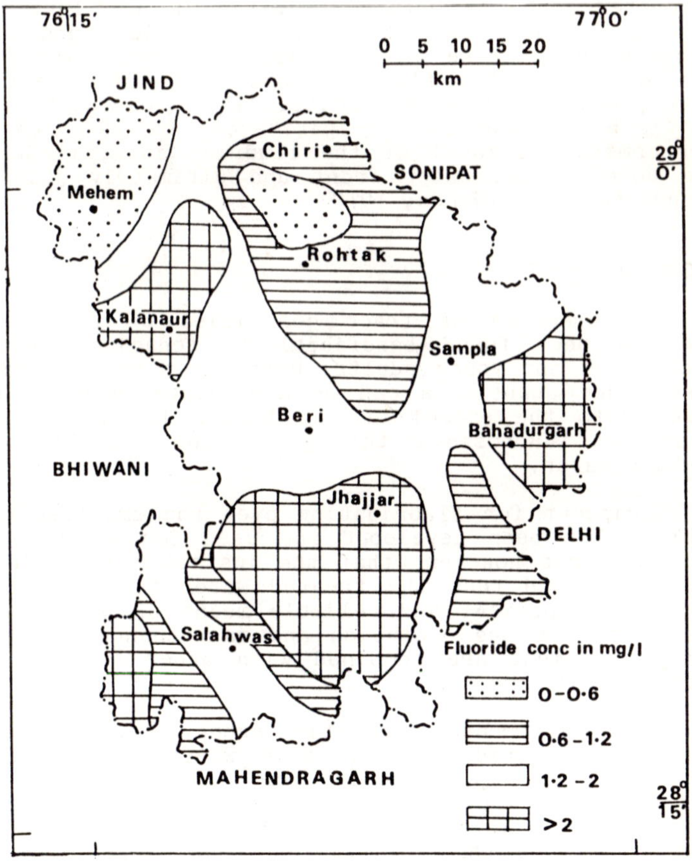

Fig.1 : DISTRIBUTION OF FLUORIDE IN SHALLOW AQUIFERS IN AND AROUND DISTRICT ROHTAK, HARYANA

On computation of analytical results, it was found that there exists a negative correlation of fluoride ion with the total hardness (i.e., an increase in calcium ion concentration results in the decrease of fluoride ion and vice-versa).

A comparative study of the available chemical analysis of the water samples collected during the pre-monsoon and post-monsoon periods shows that there is a remarkable change in the concentration of various dissolved constituents while no change was observed in the fluoride concentration. This indicates that the origin of fluoride in shallow aquifers may be due to the weathering of rocks or the continuous use of phosphatic fertilizers.

The fluoride concentration above 1.2 mg/l may be attributed to the use of phosphatic fertilizers or fluor-apatite, which are being leached down to the main ground water body (saturated zone) by return irrigation flows. Irrigation with water containing small amounts of fluoride ions would tend to concentrate these ions in soil. If the calcium ions are removed as calcite, fluoride ions will either get absorbed or co-precipitated with calcite. If the carbonate concentration is high enough to precipitate calcium as calcite without a concomitant removal of the fluoride ions, these ions may move down to the water front and ultimately join the main groundwater body (Handa, 1979).

RECOMMENDATIONS

The presence of trace quantities of fluoride in the diet has been found beneficial in reducing dental caries in children and young adults. Water is a media through which these trace quantities can be assimilated through body processes into the enamel of the teeth. It is very important to prevent fluorosis b the use of low fluoride content of drinking water.

Equipment for fluoridating even the smallest home water supplies has been developed and used for several years. It is recommended however, that the installer maintain the home fluoridator and test the treated water for fluoride levels. It is an economical and reliable means of providing fluoridated water if the operation and maintenance of the fluoridating equipment are combined with other home water supply sources.

The Environmental Protection Agency, USA, has recommended 1 mg/l of fluoride in irrigated water for continuous use but upto 15 mg/l of fluoride for short term use (20 years) on fine soils. It is therefore advisable to take adequate care in the use of phosphatic fertilizers while using groundwater with more than 1 mg/l, so that there is no unwanted excessive accumulation of fluoride ions in soil from these combined sources. In areas of low fluoride levels, fluoride-rich foods like sea-fish, cheese and tea should be used.

E.P.A., 1976. National interim primary drinking water regulation. EPA-570/9-76-003.

Galagan, D.J., 1953. Climate and controlled fluoridation. J.Am.Dent.Assn. 47:159.

Galagan, D.J., and Lamson, G.G., 1953. Climate and endemic dental fluorosis. Pub.Health Rep. 68: 497.

Galagan, D.J., and Vermillion, J.R., 1957. Determining optimum fluoride concentrations. Pub. Health Rep. 72: 491.

Handa, B.K., 1979. Effect of return irrigation flows from irrigated land on the chemical composition of groundwater from shallow unconfined aquifers. Prog.Water Tech., 11: 337-349.

I.C.M.R., 1975. Manual of standards of quality for drinking water supplies, 2nd edition, Special Report Series No.44. Indian Council of Medical Research, New Delhi.

I.S.I., 1983. Indian Standard specification for drinking water. IS: 10500- Indian Standards Institution.

Rainwater, F.H., and Thatcher, L.L. 1960. Methods for collection and analysis of water samples. U.S. Geol. Survey. Water Supply Paper No.1454.

Richards, L.F., et al., 1967. Determining optimum fluoride level for community water supplies in relation to temperature. J.Am.Dental Asn. 75.

NITRATE POLLUTION IN THE GROUNDWATER OF SANGAREDDY AREA, ANDHRA PRADESH, INDIA.

G.SHANKAR NARAYANA, V.SUDARSHAN, C.NARSIMULU
Department of Geology, Osmania University, Hyderabad-7.

ABSTRACT

The hydrogeochemical investigations of the groundwater from parts of the Manjira river basin revealed two main zones of nitrate facies. The nitrate concentrations in the northern region of the river are below 45 ppm the safe limit prescribed by WHO for potability of water and interestingly on the southern region of the river near Sangareddy township the concentrations are varying from 45 to 165 ppm. The nitrate concentrations are measured by Orion EATM 940 Expandable Ionanalyser which is a highly sophisticated micro-processor controlled PH/ISE instrument.

INTRODUCTION

The presence of high nitrate concentrations would normally indicate pollution of groundwater at some stage of its history. Since presence of excess nitrate ions is deleterious to health, their occurrence in high concentrations in natural water is a matter of great concern to environmental scientists. The leaching of nitrates from agricultural land has been a major research focus in recent years. Although commercial fertiliser is a suspected major source of nitrate in groundwater, researchers have also identified natural organic nitrogen, livestock, septic tanks and atmospheric inputs as contributing to contamination of groundwater (Hill, 1982). The study area is composed of peninsular granites and gneisses with a few patches of alluvium. Most of the area is occupied by red-sandy soil and low-lying areas (along the river belt) are covered by black-cotton soil. Groundwater in hard rocks occurs under water-table conditions in weathered and fractured zones. Depth to water table is varying from 4.5 to 17.4 meters. Groundwater in alluvium occurs under water-table conditions in saturated sand. The depth of the wells ranges between 7.0 and 8.5 meters.

METHODOLOGY

A total of 42 representative samples were collected from dug wells, dug-cum-borewells, borewells and filter points in May 1988 in one litre polyethylene bottles for chemical analysis. The model 93-07 Orion Nitrate electrode and 90-02 double junction reference electrode are used with Orion EATM 940 Expandable Ionanalyser to measure nitrate concentrations. Inner chamber of the reference electrode is filled with Orion cat. No. 900002 solution saturated with Ag Cl. Outer chamber filling solution is prepared by diluting 2 ml Ionic Strength Adjuster ($2M(NH_4)So_4$) to 100 ml with distilled water.

Electrode is soaked in distilled water for 15 minutes and in a standard nitrate solution (e.g.,100 ppm) for one hour before the analysis. For 100 ml of all standards and samples 1 ml of preservative solution (6.2 g. reagent grade boric acid in 100 ml boiling water) is added to prevent biological degradation of the solutions. The Nitrate Interference Suppressor solution is mixed in an equal volume with samples and also with standards for the removal of a variety of interfering anions present in samples. 2 ml Nitrate Ionic Strength Adjuster is added to 100 ml of standards and samples to adjust ionic strength.

Direct measurement technique is employed for measurement of concentrations which involves calibrating the Ionanalyser with one to five standards of known concentrations in ppm. Unknown sample concentrations are then read directly from the display in ppm.

RESULTS AND DISCUSSION

Studies carried out on the nitrate content have shown that in all the cases, except in two samples, the concentrations are above 10 ppm and they are varying from 9.0-165 ppm (Fig.1). The concentrations in about 14 samples are found to be more than 45 ppm. The PH values of the samples are of alkaline range. The E.C. Values are varying from 300 to 2400m.mhos/cm. The F concentrations are ranging from 0.4 to 1.5 ppm except in few samples where it is exceeding 2.0 ppm. The K concentrations are within the range of 10 ppm (0.9 to 5.9 ppm) except in two samples where the values have gone upto 190 ppm. The C1 values range from 9.9 to 430 ppm and Ca concentrations are varying from 1.1 to 130 ppm. It is important to note that the hydrochemistry of the study area indicates two main zones of nitrate facies. The nitrate concentrations in the northern region of the Manjira river is less than 45 ppm and interestingly on the southern region of the river i.e. in and around Sangareddy township the concentrations are more than 45 ppm.

It is reported that the water containing more than 45 ppm of nitrates causes methemoglobinemia in infants and various toxic effects in ruminants (Todd, 1980).

There are numerous sources of nitrogen compounds present in natural waters (Handa et al. 1982), the most important being i. Atmosphere ii. Geologic sources iii. Soils iv. Atmospheric nitrogen fixation and v. Anthropogenic sources which include a.Industrial wastes containing N-compounds b. Human and animal wastes and c.Agricultural activities viz., green manuring, addition of manure or compost, use of synthetic N-fertilisers. The anthropogenic activities have modified the nitrogen cycle both on the micro and macro scale. The human excreta in big towns finds its way into the septic tanks and some times into open fields. Similarly animal excreta in the villages generally soaks into the soil.

The use of synthetic N-fertilisers like Urea, Calcium ammonium nitrate, Ammonium phosphate, Ammonium phosphate, Ammonium sulphate, Ammonium nitrate, Ammonium chloride and Urea ammonium phosphate is on the increase which is likely to be of importance.

The leaching of nitrates to groundwater may be explained in two broad groups, the decomposition of organic matter containing nitrogenous matter into amino acids, which are further degraded into ammonium sulphate etc., (aerobic proteollysis) or into ammonia amines etc., (anaerobic proteolysis).The other group is the hydrolysis of urea to ammonium and its subsequent oxidation to nitrate. These nitrates in so for as they are not taken up by the plants tend to leach down to the saturated zone and degrade the quality of groundwaters.

According to Walter et al (1975) all nitrogenous fertilisers regardless of the form in which applied are ultimately changed to mobile nitrate by natural processes. Amount of water mixing through the profile determine rate and extent of nitrate movement.

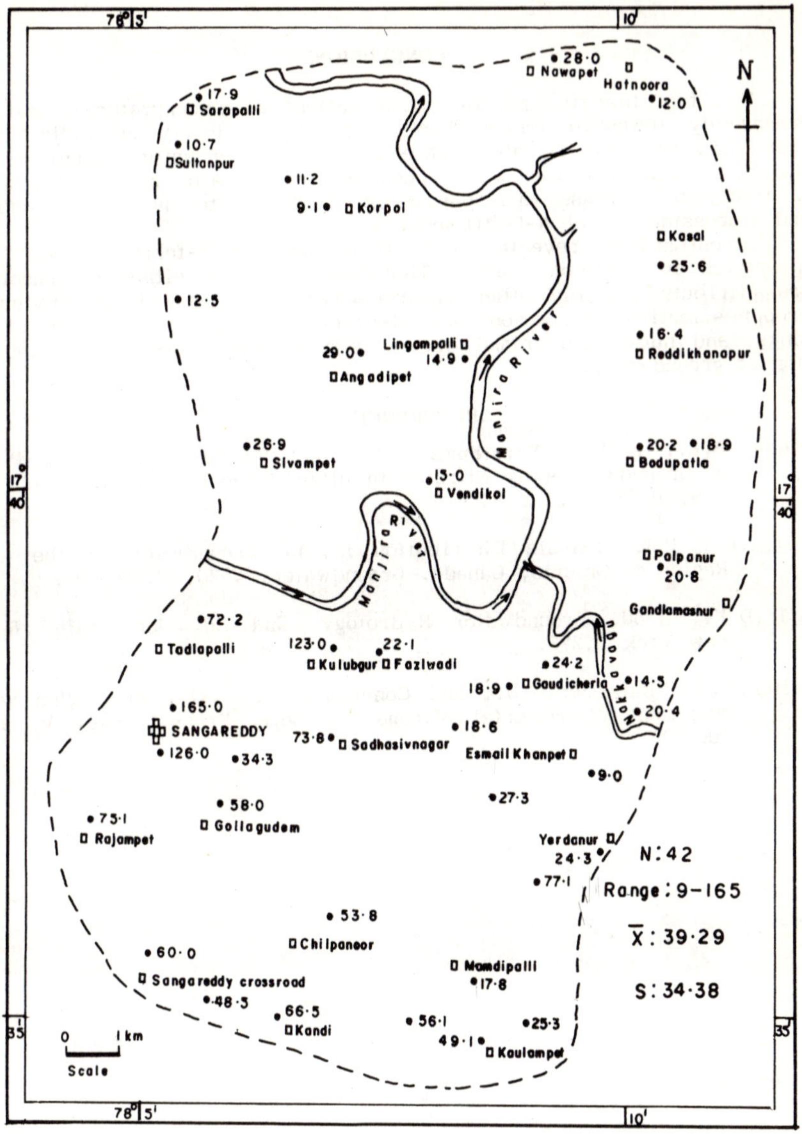

Fig. 1 Spatial distribution of Nitrate concentrations (in ppm) in Groundwater samples.

Fig.1 Spatial distribution of Nitrate concentrations (in ppm) in Groundwater samples.

CONCLUSIONS

The fact that the nitrate concentrations of groundwater in and around Sangareddy township (more than 45 ppm) is clearly an indication that there has been human interference in the form of anthropogenic processes which is polluting the water occuring in phreatic zone. This tendency is likely to increase with passage of time due to increase in population and increasing use of N-fertilisers.

Investigations revealed that there are no N-fertiliser and related industries in the study area. Therefore the high nitrogen concentrations are attributed to the other anthropogenic activities. It is obvious that proper scientific management of soil-water, N-fertilisers, location of septic tanks, and human and animal wastes is essential to prevent the contamination of groundwater.

REFERENCES

Handa, B.K., goel, D.K., Kumar, A., and Sondhi,T.N., 1982. Pollution of Groundwaters by Nitrates in Uttar Pradesh. IAWPC. Tech. Annual. V.9, P.95-103.

Hill,A.R., 1982. Nitrate Distribution in the Groundwater of the Alliston Region of Ontario, Canada. Groundwater, V.20, P.696-702.

Todd,D.K., 1980. Groundwater Hydrology. 2nd Ed. Joh. Wiley & Sons. New York P.275.

Walter,M.F., Bubenzer,G.D., and Converse,J.C., 1975. Predicting vertical movement of manurial nitrogen in Soil. Trans. ASAE, V. 18. P. 100-105.

****

SESSION S-6

AUGMENTATION OF WATER RESOURCES : ARTIFICIAL RECHARGE OF AQUIFERS ; RAINWATER HARVESTING ; CONSTRAINTS AND LIMITATIONS

Underground Storage of Surplus Flood Waters in Gangetic Plain

RAMA
Tata Institute of Fundamental Research
Homi Bhabha Road, Bombay - 400005

Abstract:-

Generous rainfall during the months of July-August-September causes floods in Gangetic basin, followed by shortage of supply during dry months. The physiography and topography of the region provides a very limited scope for surface storage.

There exists the possibility of storing some of the surplus run-off of the rainy season in underground reservoirs, for use during the dry period. The scheme envisages emptying the existing natural reservoirs which are normally always full and which are likely to fill by themselves after they are emptied, without any extra engineering. The artificial part of the scheme consists in generating productive consumption, by intensive irrigation, for the large quantity of water withdrawn. The emptied reservoir is expected to absorb a part of the flood pulse also. The scheme is likely to have maximum potential in the Gangetic plain. The most promising reservoirs, meeting the above criteria, appear to be the sandy aquifers connected with the beds of seasonal streams.

Outline of Underground Storage Scheme

Major part of the annual precipitation in the alluvial plains of India comes during three or four monsoon months, i.e. June, July, August and September. There are occasions when rainfall is intense and widespread, with the result that streams (nadis) and rivers overflow their banks and cause vast devastation. During the eight dry months, there is immense demand for irrigation which is partly met by groundwater withdrawal. The exploitation of groundwater has now reached a level close to the recharge capabilities in several regions. The increasing irrigation demands now require conservation of flood waters. But the scope for construction of dammed reservoirs in the Himalayas is very limited. Construction of surface reservoirs in the alluvial plains is also impractical for reasons of cost, land submergence and desiltation effort. Conventional underground reservoirs would be even more impractical.

Here, a proposal is made to create sufficient pore space right under the beds of non-perennial monsoonal streams (nadis) by deliberate excessive pumping of the underground aquifer. This pore space is expected to soak a sizable fraction of the run-off during the monsoon season. The basic premise is that the sandy beds of the streams and the adjoining aquifers are highly permeable; a conservative estimate of infiltration rate of a few cms of water per hour should suffice.

It may be remarked here that the proposal envisages pumping around seasonal streams only i.e. the streams which become active during monsoon but run dry during the rest of the year. It is a common observation that the aquifers under their beds stay usually full all through the year, with the consequence that there is no storage available under the bed. Sufficient storage can however be created by large scale intensive pumping. The economics of the proposal becomes favourable when the large amount of water envisaged to be pumped during the dry season is usefully consumed locally. There is sufficient demand for this water since the farmers are beginning to practise intensive cultivation in this region.

The proposal for replenishable underground reservoirs is illustrated here with the help of a number of diagrams. They however do not indicate any particular operational policy which will be decided only after investigating several factors like stream discharge, permeability of the bed and adjoining aquifer, topography and land use practices.

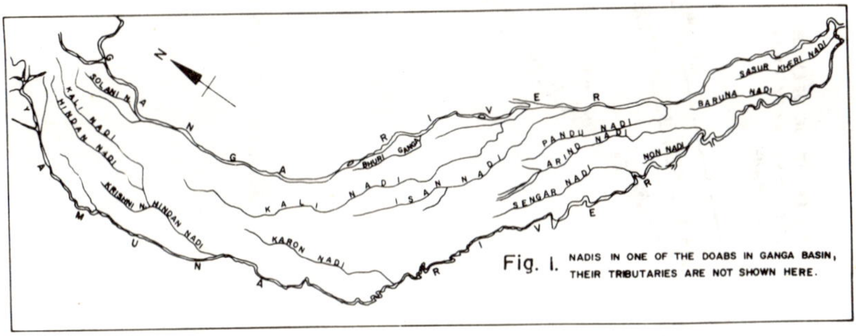

Fig. 1. NADIS IN ONE OF THE DOABS IN GANGA BASIN, THEIR TRIBUTARIES ARE NOT SHOWN HERE.

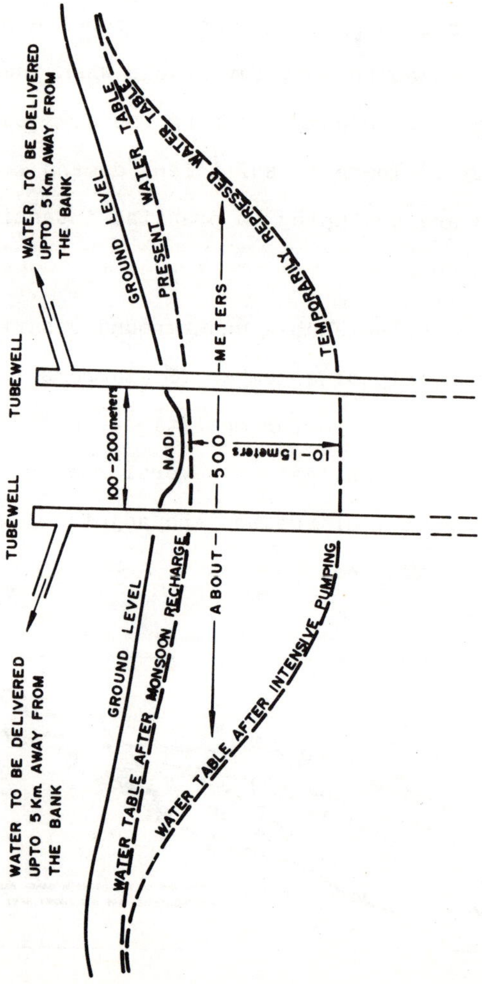

A SCHEMATIC DRAWING OF PROPOSED UNDERGROUND RESERVOIR

FIG. 2

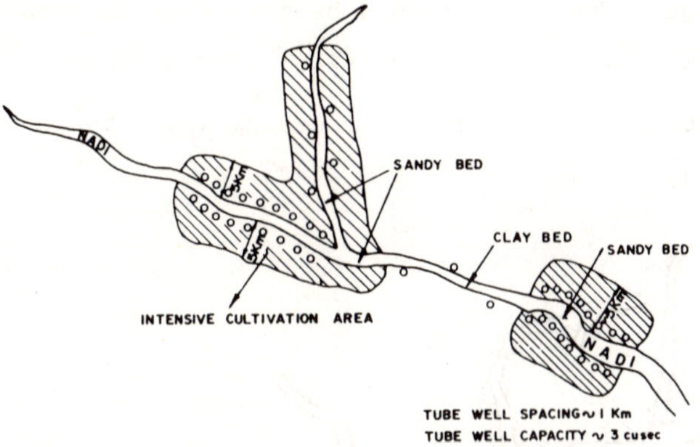

FIG. 3 A SCHEMATIC DIAGRAM FOR INTENSIVE CULTIVATION BY TUBEWELL IRRIGATION

An attractive aspect of the above scheme is that installation of even a single pump begins to yield returns immediately. Significant results on flood mitigation will however become perceptible only when the scheme is employed extensively.

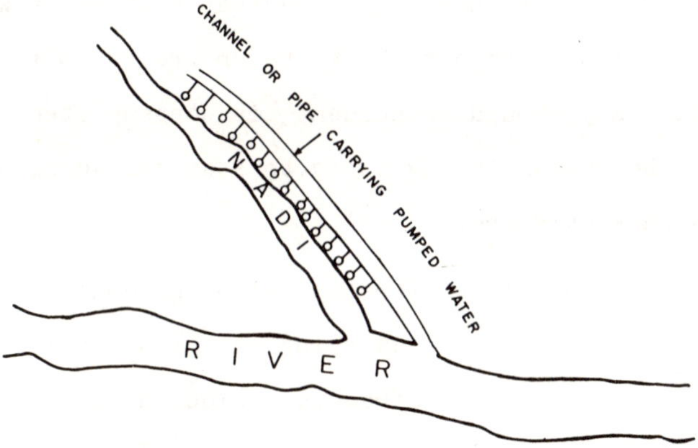

AUGMENTING RIVER DISCHARGE (CANAL SUPPLY) WITH WATER PUMPED FROM NADI BED DURING DRY SEASON

FIG. 4

The proposal can be (in fact must be) carried out gradually, in steps. The draft should be gradually increased to match the actual limit of bed infiltration of monsoonal run-off. The problem of continuously receding water table must never be permitted.

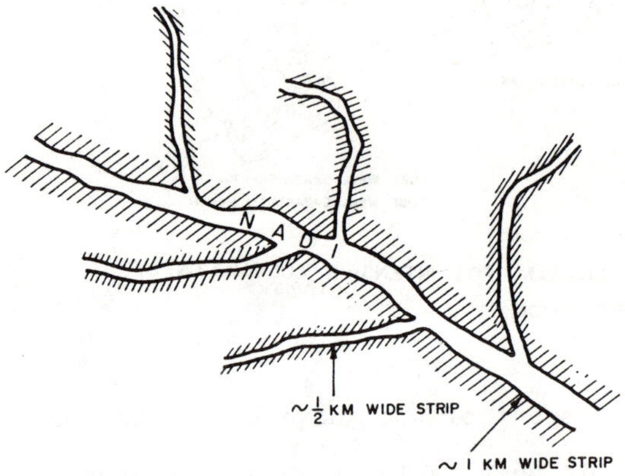

AREAS SUITABLE FOR INTENSIVE CULTIVATION BASED ON LOCAL PUMPING BY TUBEWELLS (NO PIPED TRANSPORT INVOLVED)

FIG. 5

It may further be pointed out here that neither it will be possible nor it is desirable to soak the entire run-off, only a sizable fraction, say around 20 percent, is contemplated (and that is a lot). The system is envisaged to stay self-desilting as these streams currently are.

Three sites, about 10 Km stretch, along Hindon (a tributary of Yamuna) have been selected for carrying out pilot projects. Surface and subsurface survey required to decide on installation of appropriate pumpage are underway.

Only one site (near Saharanpur) has been investigated as yet. About sixty boreholes made to a depth of 50 meters in the project area encountered some clay lenses which were not extensive. About thirty boreholes were converted into observation wells for measuring the water table depth. While the design of the pumping complex was being discussed, an unprecedented drought in 1987 led to lowering of water table by 5-6 meters in the project area. During the 1988 monsoon, a few flood episodes led to almost full recovery of water table in the project area. The preliminary analysis of the data appears very promising, and plans are now afoot to exploit this resource. The work on the pilot project is undertaken by the Government of Uttar Pradesh.

AUGMENTATION OF WATER RESOURCES IN CANAL COMMANDED AREA

Dr. A.S.Chawla
Director

Sri A.S.Ansari
Senior Research Fellow

Water Resources Development Training Centre
University of Roorkee, Roorkee

ABSTRACT

Augmentation wells are constructed along the canal to augment its supplies during the period of shortage of the surface water. When these augmentation wells operate,the induced seepage increases with decrease in the distance of the wells from the canal, decrease in coefficient of storage and increase in duration of pumping.

In case, the augmentation wells are located too close to the unlined canal, the recirculation of canal water will increase during pumping, due to increased induced seepage resulting in increased cost of the unit volume of net water pumped from ground water reservoir. On the other hand if the wells are located far away from the canal, the cost of conveyance system and the energy cost for pumping water will be more. Moreover, closer the wells, the more will be drawdown due to interference and hence energy cost of the pumping will be more. On the other hand if the spacing is more, the cost of power transmission line will be more. In the present study optimum distance of the wells from the canal and spacing among the wells have been worked out for the minimum unit cost of net pumped water. The effect of various aquifer and operational parameters have been studied. The optimum distance of wells from the canal increases with decrease in storage coefficient, increase in the value of the transmissibility and increase in duration of pumping.

INTRODUCTION

Conjunctive use of surface and ground water is necessary to meet the increasing demand of irrigation water for intensive irrigation. Advantages of the conjunctive use of surface and ground water, in addition to availability of increased total water supply, include smaller surface distribution system, smaller evapotranspiration losses and improvement of power load and pumping plant use factors. The conjunctive use requires planned use of ground water reservoir and success of a conjunctive use project in a basin depends on the availability of space in ground water basin to store the recharged water to be used when needed. One of the principal advantages of the use of ground water reservoir is the economic value of the aquifers as a transmission system in partial replacement for surface distribution system.

Unlined or lined canals are constructed to supply water for irrigation. From unlined canals, water percolates into the adjacent aquifer which can be pumped for irrigation. The wells can be located in the fields and operated directly for irrigation. An alternative is to have wells located near the canals, which pump water from the aquifer and deliver into the canal. Such wells are called 'Augmentation Wells'. Commands of state-owned direct irrigation wells are normally outside the canal command. State owned direct irrigation wells normally have a capacity of 0.04 cumecs or more. Private wells owned by farmers, normally of smaller capacity averaging about 0.01 cumecs may be inside the canal command for supplemental irrigation or outside the canal command for meeting the full irrigation demand.

There are some areas where either ground water is not fit for irrigation or the water table is too deep to pump water out economically. To irrigate these areas, large capacity augmentation wells can be constructed in the areas having plenty of ground water fit for irrigation, along the existing distribution network of canals to augment their supplies during the period when surface supplies are short. In the system of conjunctive use of surface and ground water, the augmentation wells are reported to be more efficient than direct irrigation wells as augmentation wells have several advantages such as separate and exclusive electric feeders can supply uninterrupted power supply and they use the existing distribution network of canals. Cost analysis has shown that the large capacity wells are more economical as compared to small capacity wells. Large capacity direct irrigation wells present the problems of management whereas large capacity augmentation wells have no such problems.

Augmentation wells may be installed in a row parallel to the canal at some distance from it. When these wells are operated, a part of pumped water will be resident water but a part will also be drawn from canal due to induced seepage. In case, the augmentation wells are located very close to the unlined canal, the recirculation of canal water will increase during pumping, resulting in the wastage of energy used for pumping the seepage water. On the other hand, when the wells are moved away from the canal, the induced seepage is reduced but the cost of the conveyance structure from the well to the canal increases. Moreover, closer the wells, the more will be the drawdown due to interference and hence energy cost of pumping will be more. On the other hand, increase in spacing among wells will increase the cost of power transmission line. So, the optimum distance of the wells from the canal and their spacing should be such that the cost of the unit volume of net pumped water (volume of pumped water minus induced seepage) from the ground water reservoir is minimum including the cost of energy for pumping water and transmitting it to the canal and the capital cost of pipeline from well to the canal. These wells are operated by the central authority responsible for canal supply rather than by the local farmers. Thus there is an opportunity to optimize the number, location and operation of these wells.

Jindal and Khepar (1980) have studied the problem of optimum spacing of augmentation tubewells. They assumed that the induced seepage from the canal to the well is uniform during the pumping period and the drawdown in the well is independent of the well location. They minimize the total annual cost rather than the cost of unit volume of water drawn from the ground water reservoir. The cost of the pipeline as conveyance system from well to the canal was calculated for the induced recharge, whereas it should have been calculated for the total water pumped from the well. They also considered the induced seepage equivalent to that at the end of the pumping period and not the average during the period.

Chawla and Hazarika (1982) studied the problem for a single well, in which the drawdown in the well was related to the well location and the duration of continuous pumping. The average induced seepage was obtained by integrating the induced seepage over the pumping period instead of considering it uniform. Moreover, instead of minimizing the total annual cost (as done by Jindal and Khepar), the cost of pumping the unit volume of net water drawn from the ground water reservoir was minimized.

Normally for the augmentation of canal water supplies, a finite number of wells are provided. Kalita (1982) optimized the distance of the wells from the canal and the spacing among the wells.But it is valid for the infinite number of wells.

Present study gives the method to find out the optimum spacing among the finite number of wells installed in a row parallel to the canal and the distance of the well from the canal such that the unit cost of net water pumped from the ground water reservoir is minimised. The optimum values of these two parameters have been obtained for different values of other parameters which affect them. P.V.C. pipes have been considered for conveying water from wells to canal.

INDUCED SEEPAGE

The canal is assumed to be hydraulically connected to the aquifer and is, therefore, considered to be a recharge boundary. The resistance to flow from canal caused by silt deposited on its bed is also neglected. Initially the entire pumped water is drawn from the ground water reservoir and as the cone of depression reaches the canal boundary some part of pumped water is drawn from the canal. The fraction of the pumped water drawn from the canal due to induced seepage then increases with time. Eventually an equilibrium may be obtained when the entire discharge is drawn from the canal.

For a single well located at a distance 'a' from the canal, the ratio of induced seepage from the canal and the pumped water from the well (Theis, 1941) is given by

$$P_r = \frac{2}{\pi} \int_0^{\pi/2} e^{-F_f . \sec^2\theta} d\theta \qquad \text{..........} \qquad (1)$$

in which,

$$F_f = \frac{a^2 S}{4Tt} \qquad \text{...........} \qquad (2)$$

S = storage coefficient

T = coefficient of transmissibility

t = time from the beginning

$\theta = \tan^{-1}\left(\frac{x}{a}\right)$

x = distance along recharge boundary measured from perpendicular joining the real and image wells.

Now consider a battery of augmentation wells (2n + 1) located along the canal at a distance 'a' from the canal with a spacing 'd' among the wells (Fig. 1). All the wells are assumed to be operating simultaneously at a uniform pumping rate. In this case induced seepage from the canal will depend upon the hydraulic properties of the aquifer, distance of the wells from the canal, spacing among the wells, duration of continuous pumping and the number of the wells.

The ratio of induced seepage Q_r from the canal to the total pumped water due to pumping from (2n + 1) wells at the rate of Q is given by

$$P_r = \frac{Q_r}{(2n+1)\,Q}$$

$$P_r = \frac{2}{(2n+1)\pi} \sum_{j=-n}^{n} \int_{\tan^{-1}(jp)}^{\pi/2} e^{-Ff.\sec^2\alpha}\, d\alpha \quad \ldots\ldots \quad (3)$$

in which

$$\alpha = \tan^{-1}(x^1+jp) \quad \ldots\ldots\ldots \quad (4)$$

$$x^1 = \frac{x}{a}$$

$$p = \frac{d}{a}$$

x = distance along the line of recharge canal measured from the perpendicular joining central real and image wells.

If t_o is the continuous duration of pumping, the ratio of the volume of induced seepage from the canal and the pumped water is given by:

$$P_{r(av)} = \frac{2}{(2n+1)\ t_o} \int_{o}^{t_o} \sum_{j=-n}^{n} \int_{\tan^{-1}(jp)}^{\pi/2} e^{-Ff.\sec^2\alpha}\, d\alpha\, dt \quad ..(5)$$

The net water pumped by the battery of the wells from the ground water reservoir in a year is given by:

$$Q\ (net) = 3600\ (2n + 1)\ Q.ta\ [1-P_{r(av)}] \quad \ldots\ldots\ldots\ldots \quad (6)$$

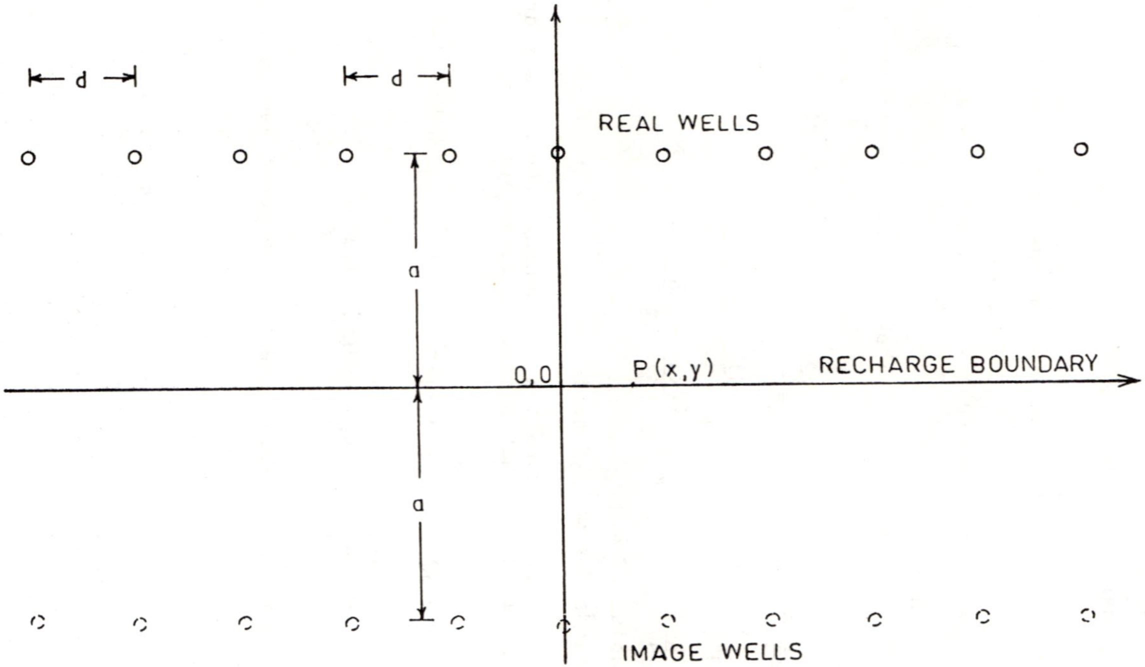

FIG.1 - BATTERY OF WELLS NEAR A RECHARGE BOUNDARY

in which,

ta = annual pumping hours

Q = discharge of a well in cumecs

Drawdown after time t from the beginning of the pumping in jth well of a battery of (2n+1) wells from the central well is given by:

$$s_j = \frac{Q}{4\pi T}\left[W(U_1) - W(U_2) + \sum_{i=1}^{n+j} \{ W(U_{3i}) - W(U_{4i})\} + \sum_{k=1}^{n-j} \{W(U_{5k}) - W(U_{6k})\} \right] \quad \text{.........} \quad (7)$$

in which

$$U_1 = \frac{r_w^2 . S}{4Tt} \qquad U_2 = \frac{(2a)^2 . S}{4Tt}$$

$$U_{3i} = \frac{(id)^2 . S}{4Tt} \qquad U_{4i} = \frac{\{(id)^2 + (2a)^2\} . S}{4Tt}$$

$$U_{5k} = \frac{(kd)^2 . S}{4Tt} \qquad U_{6k} = \frac{\{(kd)^2 + (2a)^2\} . S}{4Tt} \quad \text{........} \quad (8)$$

r_w = radius of well

W stands for well function

If t_0 be the continuous duration of pumping, the average drawdown at jth well during that period is given by:

$$(s_{av})_j = \frac{Q}{4\pi T t_0} \int_0^{t_0} \left[W(U_1) - W(U_2) + \sum_{i=1}^{n+j} \{W(U_{3i}) - W(U_{4i})\} + \sum_{k=1}^{n-j} \{W(U_{5k}) - W(U_{6k})\} \right] dt \quad \text{...........} \quad (9)$$

If j = 0, the expression will give the average drawdown at the central well.

COST FUNCTION

Determination of optimum location of augmentation wells requires minimization of the unit cost of net water drawn from the ground water reservoir. It requires determination of cost of well construction, cost of pipeline for transporting the pumped water from well to the canal, energy cost of pumping the water from the well and transmitting it to the canal and the cost of transmission lines for feeding power to the wells.

The cost of unit volume of net pumped water from the ground water reservoir is given by:

$$C_u = \frac{(2n+1)\ C_W + C_2\ (2n+1)a + C_p + T_n.2n.d}{3600\ [1-P_{r(av)}]\ (2n+1)Q.t_a} \quad \ldots\ldots. \quad (10)$$

in which

C_u = cost of unit volume of net pumped water

C_W = annual cost of construction of a well

$(2n+1)$ = number of wells

C_2 = annual cost of delivery pipe per metre length

C_p = annual energy cost for pumping water and for transmitting the same through the pipeline between the pumping well and canal.

$$= C_1 . \frac{9.8\ t_a\ Q}{\eta}\ [(2n+1)\ h + (s_{av})_{\text{central well}} + 2 \sum_{j=1}^{n} (s_{av})_j + (2n+1)a\ F_L] \quad \ldots\ldots.(11)$$

η = efficiency of the pumping units

C_1 = cost of energy per KWH

Q = discharge of a well in cumecs.

h = depth of the water table or piezometric surface below ground surface (m)

F_L = friction loss per unit length of the pipeline.

For PVC pipe, the friction loss was computed by the expression (9).

$$F_L = 0.0009\ \frac{Q^{1.773}}{D^{4.773}} \quad \ldots\ldots\ldots\ldots. \quad (12)$$

T_n = Annual cost of power transmission line per metre length.

D = diameter of the delivery pipes

OPTIMAL SPACING

The unit cost of net water pumped from the ground water reservoir (Eq. 10) is a function of aquifer parameters, cost of well, conveyance system and energy, number of wells, operational parameters such as continuous running hours and total annual operating hours and position of wells in relation to canal and spacing among the wells. For a given set of aquifer conditions there will exist a unique combination of all other variables for which C_u will be minimum. The difficulty in solving the problem as such is that the operational parameters and cost parameters will be subjected to constraints specific to each site condition. For example, the operation parameters will depend on the type of rotational system of irrigation operating in the scheme, schedule of power availability, total quantity of water to be pumped etc. The cost parameters will depend on the type of conveyance system. The diameter of the pipe for the conveyance system is a function of discharge and has been determined by the computer program developed by Chawla and Hazarika (1982). For the study to be useful, the problem has been formulated as an unconstrained problem and then solved for various operational conditions to find out the optimal combination of spacing among the wells, d^* and distance of the wells from the canal, a^* for a finite number of wells installed in a row parallel to canal. Powell's method (7) has been adopted to minimise the unit cost of net pumped water. It is one of the direct search methods. For this method a computer program developed by Powell (6) has been used. The results have been plotted in the form of graphs to see the effect of various parameters on the optimum location of wells i.e., on the values of a^* and d^*.

RESULTS AND DISCUSSIONS

Equations 5 and 9 have been integrated numerically using Gauss legendre quadrature method. Induced seepage from the canal due to pumping from a battery of wells located along the canal depends on the aquifer parameters, spacing, location and number of wells and duration of pumping. The ratio of average induced seepage to the total pumped water, $P_{r(av)}$ has been plotted against various parameters to bring out their effect,(Fig. 2 to 4). A perusal of Figs. 2 and 3 indicates that with increase in the values of a and S, the value of $P_{r(av)}$ decreases. It is seen that the value of d and number of wells have no significant effect on the value of $P_{r(av)}$. A perusal of Fig. 4 indicates that the value of $P_{r(av)}$ increases with the increase in the value of piping duration, t_o.

Optimal values of distance of wells from the canal, a^* and spacing among wells, d^* are functions of aquifer parameters, pumping duration, well discharge and cost parameters. The values of a^* and d^* have been plotted against various parameters to study their effect. A perusal of Fig. 5 indicates that the value of a^* and d^* decrease with increase in the value of coefficient of storage. However, the decrease in the values of a^* and d^* is very sharp for low values of S. The values of a^* and d^* increase with increase in the continuous duration of pumping, t_o (Fig.6). It is seen from Figs. 7 and 8 that the value of a^* increases with increase in the value of coefficient of transmissibility, T and the value of d^* increases with the increase in the value of annual pumping hours. The number of wells do not affect the optimal spacing and distance. It is also seen from Figs.5 and 6, that the pumping capacity of the well affects the optimal spacing, d^* significantly. The sensitivity of the values of a^* and d^* to the capital cost of wells, electric transmission system and the conveyance system has also been studied. Spacing among the wells is seen to be very sensitive to the cost of the transmission system (Fig.9). The capital cost of the wells and conveyance system affect slightly the optimum distance of wells,a^*.

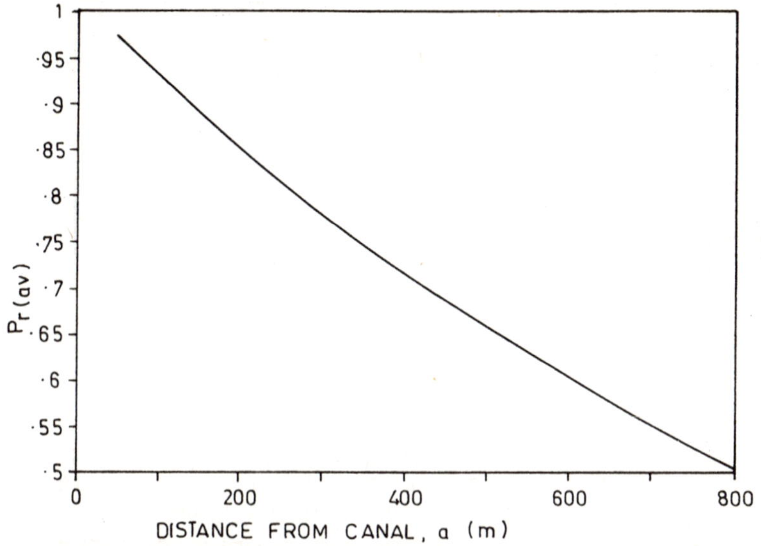

FIG.2-EFFECT OF DISTANCE OF WELLS FROM CANAL

(T= ·033 m^2/sec., S= ·01 , t_o = 168 hrs.)

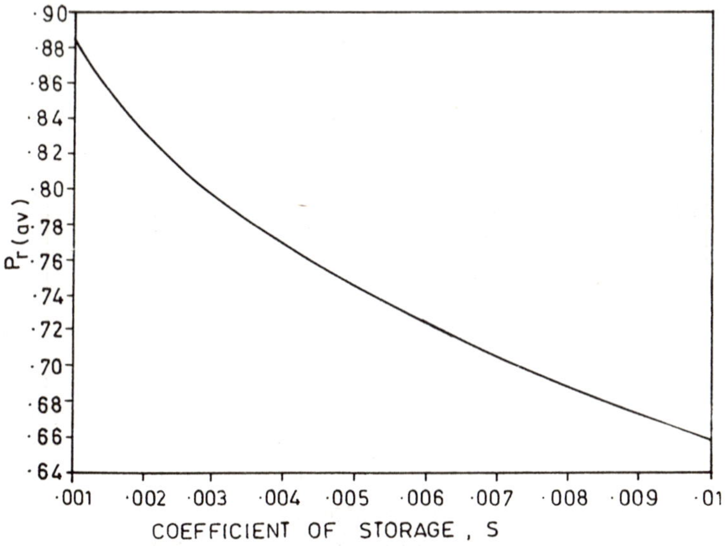

FIG.3-EFFECT OF COEFFICIENT OF STORAGE

(T= ·033 m^2/sec , t_o = 168 hrs , a = 500 m)

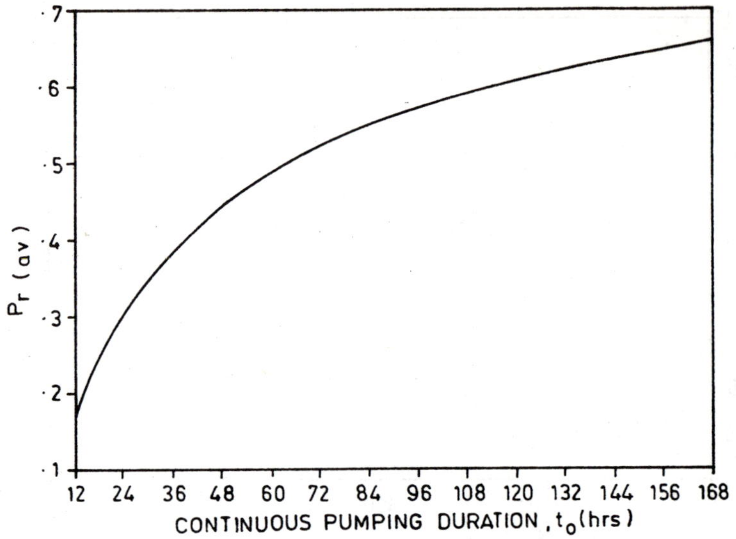

FIG. 4-EFFECT OF CONTINUOUS PUMPING DURATION
($T = \cdot 033\ m^2/sec$, $S = \cdot 01$, $a = 500 m$)

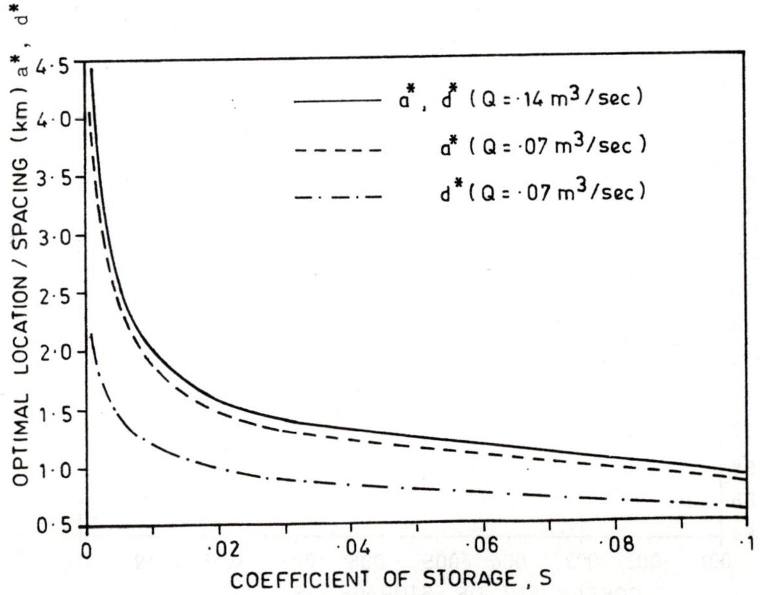

FIG. 5-EFFECT OF COEFFICIENT OF STORAGE
($T = \cdot 033\ m^2/sec$, $t_o = 168$ hrs, $t_a = 5000$ hrs, T_n = Rs. 2·51)

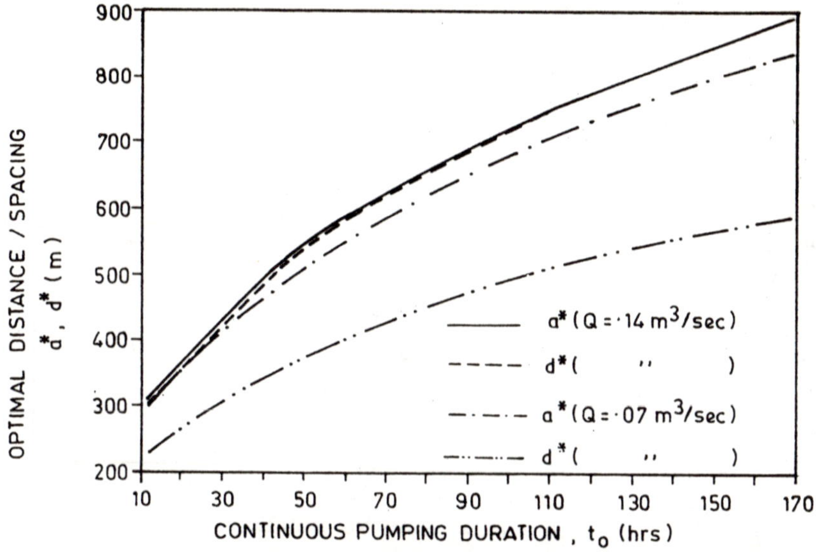

FIG.6 - EFFECT OF CONTINUOUS PUMPING DURATION

($S = 0{\cdot}1, T = {\cdot}033\ m^2/s, t_a = 5000$ hrs, T_n = Rs. 2·51)

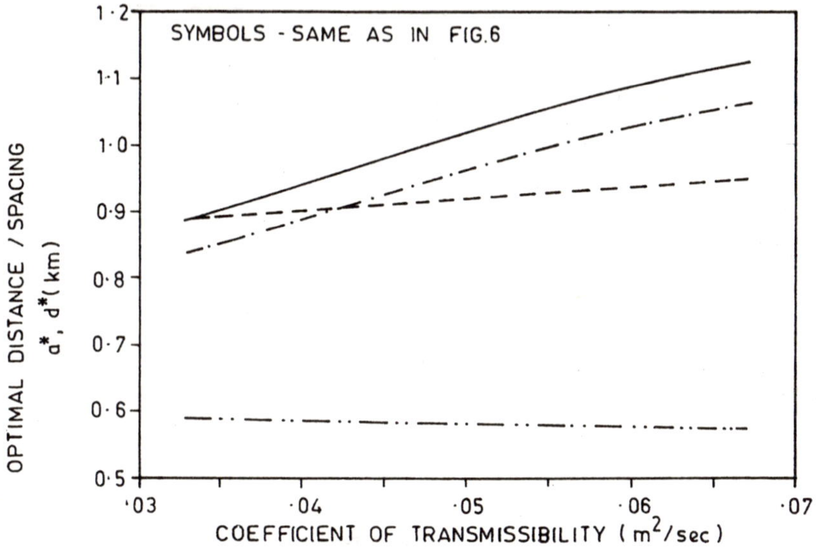

FIG.7 - EFFECT OF TRANSMISSIBILITY

($S = 0{\cdot}1$, $t_o = 168$ hrs, $t_a = 5000$ hrs, T_n = Rs. 2·51)

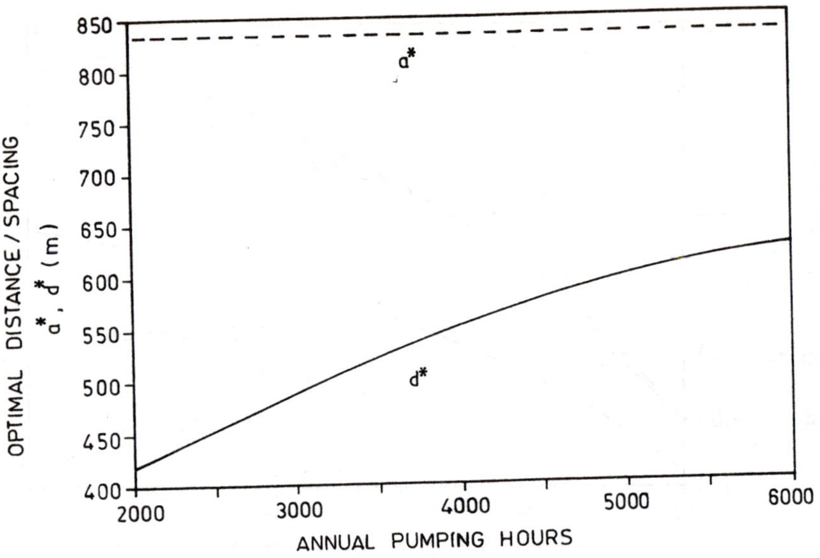

FIG. 8 - EFFECT OF ANNUAL PUMPING HOURS
($T = \cdot 033 m^2/sec$, $S = 0 \cdot 1$, $t_o = 168$ hrs, $Q = 0 \cdot 07 m^3/sec$, $T_n =$ Rs. 2·51)

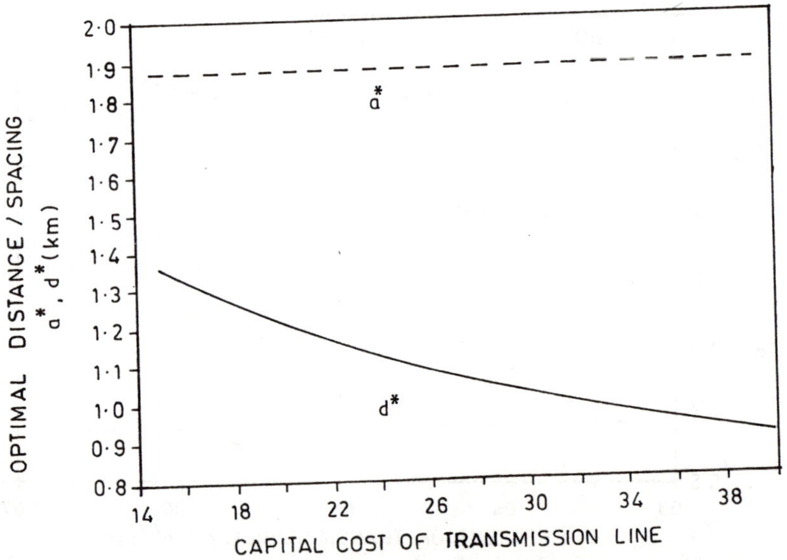

FIG.9 - EFFECT OF COST OF ELECTRIC LINE
($T = \cdot 033 m^2/sec$, $S = \cdot 01$, $t_o = 168$ hrs, $t_a = 5000$ hrs, $Q = \cdot 07$ cumec)

CONCLUSIONS

The average induced seepage from the canal due to pumping from a system of wells placed along the canal increases with decrease in the value of a and S and with increase in the continuous duration of pumping. The number and spacing among the wells do not affect the induced seepage.

The optimal distance of wells from the canal and spacing among the wells increase with decrease in the values of S and increase in continuous pumping duration. The value of a^* increases with the increase in the value of T and the value of d^* increases with the increase in the value of t_a. But the value of a^* is not affected by the change in the value of t_a. The number of wells do not affect the optimal distance and spacing. Optimal spacing among the wells is affected by the discharge of the wells.

REFERENCES

1. Chawla, A.S. and Hazarika, S (1982), "Optimal Location of Augmentation Wells". Proceedirgs of All India Seminar on Water Resources, Its Development and Management held at Chandigarh.

2. Chawla, A.S., S.Hazarika and Kalita, K.C.(1983), "Economics of Canal Augmentation Wells", Proceedings of Seminar on Assessment, Development and Management of Groundwater Resources", Central Ground Water Board, New Delhi.

3. Hantush, M.S.(1964), "Hydraulics of Wells", in Chow,V.T. "Advances in Hydrosciences", Academic Press, New York.

4. Jindal, P.K., and Khepar S.D.(1980), "Optimum Spacing of Augmentation Tubewells from Canal", Journal of the Institution of Engineers (India), Civil Engg.Div. Vol. 61, Pt. C1.

5. Kalita, K.C.(1982), "Optimum Location of Augmentation Tubewells", Dissertation submitted to the University of Roorkee in Partial Fulfilment of the requirement for the award of M.E. Degree.

6. Kuester, J.L. and Mize, J (1973), "Optimization Techniques with Fortran". Mc Graw Hill Book Company.

7. Rao,S.S.,(1984), "Optimization Theory and Application", Willey Eastern Limited Press, IInd edition.

8. Theis, C.V. (1941), "The Effect of Well on the Flow of a Nearby Stream", Trans. of Amer. Geoph. Union, Vol.22, Pt. 3.

9. Wavin PVC Pressure Pipes. A World Guide-Wavin Overseas, Zwolle, Netherlands.

EFFECT OF DRAINAGE CHANNEL ON EVOLUTION OF WATER TABLE DUE TO RECHARGE FROM TWO PARALLEL CANALS

D.N. Bhargava, Chief Engineer, Irrigation Design Organisation, Roorkee
G.C. Mishra, Scientist, National Institute of Hydrology, Roorkee
Satish Chandra, Director, National Institute of Hydrology, Roorkee

ABSTRACT

A canal-aquifer-drainage channel interaction problem has been solved using discrete kernel approach. The time of activation of a drainage channel which runs parallel to two continuously running ridge canals has been predicted. The return flow entering to the drainage channel at different times after its activation has been quantified. The evolution of water table prior to and after the activation of the drainage channel has been predicted. It is found that the time of activation depends on the aquifer parameters, seepage rate from the canals, distance of the drainage channel from the canals, and the elevation of the drainage channel. If a drainage channel is located nearer to the canals, it receives return flow at a higher rate. The water table between two continuously running canals takes the shape of a plateau and it is marginally affected by the presence of the drainage channel.

Introduction

The irrigation canals are constructed in a basin where a natural drainage system generally exists. A part of the recharge due to seepage from the canals will be taken away from the basin through the natural drainage system. If a drainage channel exists in the vicinity of the canals, the water table may come up near the bed level of the drainage channel in course of time due to seepage from the canals. The drainage channel activates only when the water table reaches its bed. The return flow to the drainage channel will be governed by the difference in potentials at the periphery of the drainage channel and in the aquifer below its bed. The time of activation of the drainage channel and return flow to it are governed by:

i) the seepage losses from the canals,
ii) the spacing between the canals,
iii) distance of the drainage channel from the canals,
iv) hydraulic conductivity, storage coefficient, and
v) initial saturated thickness of the aquifer.

In this paper, the evolution of water table in the presence of a drainage channel which runs parallel to a canal system, and the performance of the drainage channel have been studied. The study has been made for parallel canals which are hydraulically unconnected with the aquifer. Such a situation may arise in case of ridge canals. The rate at which the water will be drained by the drainage channel is controlled by width of drain, depth of water in it, and its bed slope. In the present study, the width of the drain has been taken into consideration. The depth of water in the drain, after it receives water, has been assumed to be nominal and its variation has not been considered in the present analysis. The boundary condition that has been imposed at the drainage channel is that the drainage channel after receiving water acts as a constant head boundary.

Statement of the Problem

Two parallel canals have been constructed on high ridges in a homogeneous and isotropic porous medium of infinite areal extent. There exists a drainage channel in the vicinity of the canals. The dimensions and locations of the canals and the drainage channel are as shown in Fig.(1). On account of large difference in the elevation of bed level of the ridge canals and the water table underneath, the ridge canals are not hydraulically connected with the aquifer. Therefore, the seepage occurs at constant rate from the ridge canals. The drainage channel situated at a lower level gets activated when the piezometric surface rises above its bed level. Thereafter, it gets hydraulically connected to the aquifer. The rate of return flow to the drainage channel is controlled by the potential difference between the channel and the aquifer below the bed of the channel. The permeability of the aquifer is K. The initial saturated thickness of the aquifer is $\bar{H}$ and the storage coefficient is ϕ . It is required to determine the time at which seepage from the ridge canals enters the drainage channel, the temporal and spatial variation of water table rise, and the rate of return flow.

Analysis

The assumption made to carry out the analysis are as follows:

i) The hydraulic properties of the aquifer remain constant with respect to time and space.
ii) The flow due to seepage from ridge canals is vertically downward until it reaches the water table.
iii) Dupuit's assumptions are valid.
iv) The time span is discretised by uniform time-step. Within each time-step the return flow to the drainage channel is constant, but it varies from step to step.
v) The return flow to the drainage channel is linearly proportional to the difference in the potentials at the periphery of the drainage channel and in the aquifer below its bed.

Let the return flow entering to the drainage channel from unit length during time-step 'n' be $Q_r(n)$. According to the linear relationship the return flow is given by:

$$Q_r(n) = \Gamma_r[\sigma_r(n) - S_r(o,n)] \qquad ..(1)$$

where $\sigma_r(n)$ is the drawdown to the water level in the drainage channel

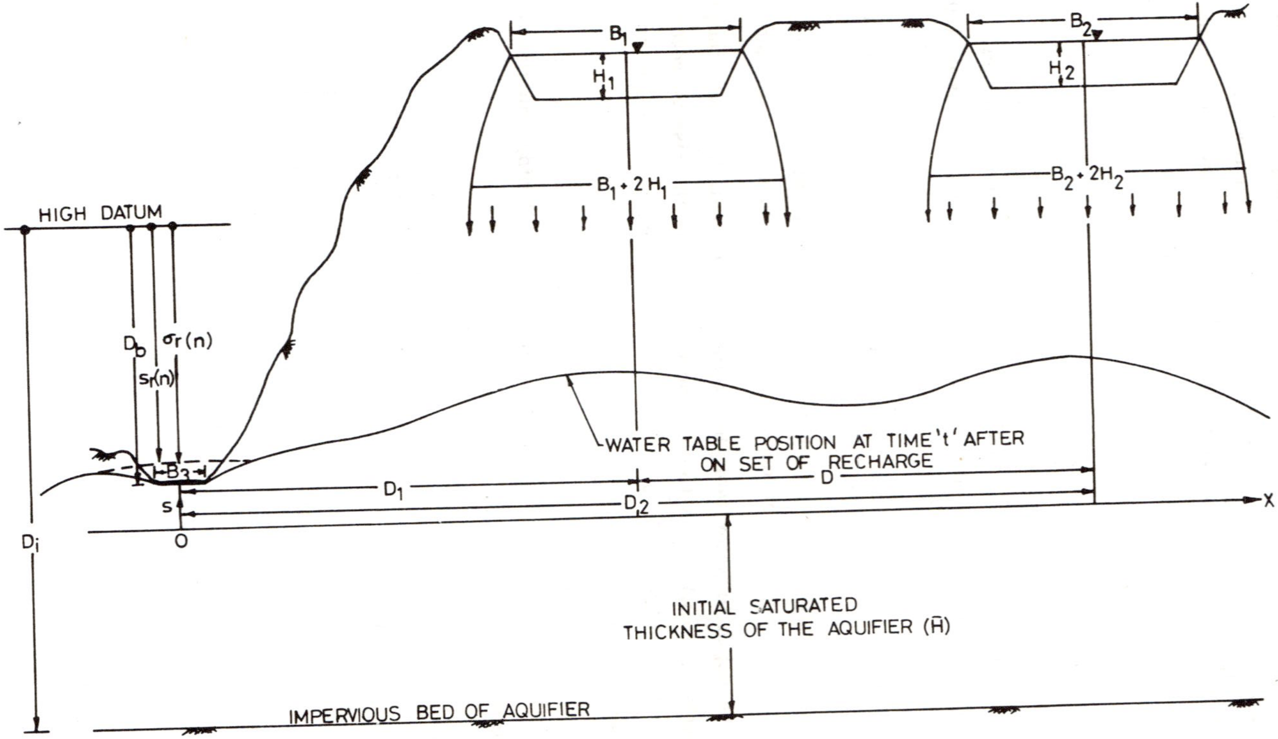

Fig.1 Schematic section of two parallel ridge canals and a drainage channel

measured from a high datum during n^{th} time-step and $S_r(o,n)$ is the drawdown in the aquifer under the channel measured from the same datum during n^{th} time-step. According to Herbert (1970) the reach transmissivity coefficient, Γ_r, for the drainage channel is given by the following formula:

$$\Gamma_r = \pi K/\log_e(0.5\pi e/W_p) \qquad ..(2)$$

in which e is the saturated thickness below the drainage channel and W_p is the wetted perimeter of the channel. Γ_r is the reach transmissivity for unit length of the drainage channel.

Let the longitudinal bed slope of the drainage channel be so steep that water in the drainage channel is disposed of quickly and a small depth of water is maintained in the channel. Under such situation, $\sigma_r(n)$ can be assumed to be equal to the depth to channel bed from the high datum (D_b).

The depth to water table below the channel bed, $S_r(o,n)$, comprises two parts, $S_1(o,n)$ and $S_2(o,n)$, where $S_1(o,n)$ is the rise on account of seepage from ridge canals and $S_2(o,n)$ is the decrease in height due to irrigation return flow.

The component of the water table rise at any time due to seepage from the ridge canals can be obtained using the following equation given by Poluborinova-Kochina (1962):

$$S_1(x,t) = \frac{K\alpha t}{2T}\,\mathrm{Erf}\left(\frac{x+B/2+H}{\sqrt{4\alpha t}}\right) - \frac{K\alpha t}{2T}\,\mathrm{Erf}\left(\frac{x-B/2-H}{\sqrt{4\alpha t}}\right)$$

$$+ \frac{K}{4T}(x+B/2+H)^2\,\mathrm{Erf}\left(\frac{x+B/2+H}{\sqrt{4\alpha t}}\right) + \frac{K\sqrt{\alpha t}}{2T\sqrt{\pi}}(x+B/2+H)\,e^{-\frac{(x+B/2+H)^2}{4\alpha t}}$$

$$- \frac{K}{2T}\frac{\sqrt{\alpha t}}{\sqrt{\pi}}(x-B/2-H)\,e^{-\frac{(x-B/2-H)^2}{4\alpha t}} - \frac{K(x-B/2-H)^2}{4T}\,\mathrm{Erf}\left(\frac{x-B/2-H}{\sqrt{4\alpha t}}\right)$$

$$- \frac{K\sqrt{(x^2)}(B+2H)}{2T} \qquad ..(3)$$

$$= F(x,B,H,t) - \frac{K\surd(x^2)(B+2H)}{2T}$$

in which B is the width of canal at water surface, H is the maximum depth of water in the canal and x is the abscissa measured from the canal centre. The above expression for water table rise is based on the assumptions that seepage from a ridge canal is K(B+2H) and width of the recharging strip at the time it joins the water table is (B+2H). The component of water table rise under the drainage channel due to seepage from the ridge canals at the end of time-step 'n' is given by:

$$S_1(o,n) = F(-D_1,B_1,H_1,n) - K \surd(-D_1)^2(B_1+2H_1)/2T$$

$$+ F(-D_2,B_2,H_2,n) - K \surd(-D_2)^2(B_2+2H_2)/2T \qquad ..(4)$$

where D_1 and D_2 are the distances from the centre of the ridge canals to the centre of the drainage channel, D_2 being equal to (D_1+D). The first two terms on right side of Equation (4) represent water table rise due to the first ridge canal and the last two terms represent the rise due to the second ridge canal.

Let water enter the drainage channel for the first time at the end of time-step, m.

Therefore,

$$D_i-D_b = \bar{H} + F(-D_1,B_1,H_1,m) - K\surd(-D_1)^2(B_1+2H_1)/2T$$

$$+ F(-D_2,B_2,H_2,m) - K\surd(-D_2)^2(B_2+2H_2)/2T \qquad ..(5)$$

The time-step, m, during which water enters the drainage channel, can be known by an iteration from this equation. The lowering in water table at the centre of the drainage channel, at the end of nth unit time-step, $S_2(r,n)$, due to return flow which has entered the channel upto n^{th} time-step, is given by

$$S_2(o,n) = \sum_{\gamma=m+1}^{n} Q_r(\gamma)\ \partial(o,B_3,n-\gamma+1) \qquad ..(6)$$

in which,

B_3 = width of the drainage channel.

The discrete kernel coefficients $\delta(o,B_3,n)$ are defined by the following equations:

$$\delta(x,B_3,n) = F(x,B_3,n) - F(x,B_3,n-1) \qquad \text{for } n \geq 2 \qquad ..(7)$$

$$\delta(x,B_3,1) = F(x,B_3,1) - \frac{1}{2T}\sqrt{(x)^2} \qquad \text{for } x \leq -\frac{B_3}{2} \text{ and } x \geq \frac{B_3}{2} \qquad ..(8)$$

$$= F(x,B_3,1) - \frac{1}{2B_3T}\left(x^2 + \frac{B_3^2}{4}\right) \qquad \text{for } -\frac{B_3}{2} \leq x \leq \frac{B_3}{2}$$

where,

$$F(x,B,t) = \frac{\alpha t}{2BT} \text{ Erf}\left[\frac{x+B/2}{\sqrt{(4\alpha t)}}\right] - \frac{\alpha t}{2BT} \text{ Erf}\left[\frac{x-B/2}{\sqrt{(4\alpha t)}}\right]$$

$$+ \frac{1}{4BT}(x+B/2)^2 \text{ Erf}\left[\frac{x+B/2}{\sqrt{(4\alpha t)}}\right] - \frac{1}{4BT}(x-B/2)^2 \text{ Erf}\left[\frac{x-B/2}{\sqrt{(4\alpha t)}}\right]$$

$$+ \frac{\sqrt{(\alpha t)}}{2BT\sqrt{\pi}}(x+B/2) \exp\left[-(x+B/2)^2/(4\alpha t)\right]$$

$$- \frac{\sqrt{(\alpha t)}}{2BT\sqrt{\pi}}(x-B/2) \exp\left[-(x-B/2)^2/(4\alpha t)\right] \qquad ..(9)$$

Splitting the temporal summation appearing in Equation (6) into two parts,

$$S_2(o,n) = \sum_{\gamma=m+1}^{n-1} Q_r(\gamma)\partial(o,B_3,n-\gamma+1) + Q_r(n)\partial(o,B_3,1) \qquad ..(10)$$

Referring to Fig.(1), the expression for depth to water table from the high datum at the centre of the drainage channel can be written as

$$S_r(o,n) = D_i - \bar{H} - S_1(o,n) + S_2(o,n) \qquad ..(11)$$

Substituting for $S_r(o,n)$ in Equation (1),

$$Q_r(n) = \Gamma_r[\sigma_r(n) - D_i + \bar{H} + S_1(o,n) - S_2(o,n)] \qquad ..(12)$$

Substituting $S_1(o,n)$ and $S_2(o,n)$ from Equations(4) and (10) respectively in Equation (12) and rearranging,

$$Q_r(n) = [\frac{1}{\Gamma_r} + \delta(o,B_3,1)]^{-1} \cdot [\sigma_r(n) - D_i + \bar{H} + F(-D_1,B_1,H_1,n)$$

$$- K\surd(-D_1)^2(B_1+2H_1)/2T + F(-D_2,B_2,H_2,n) - K\surd(-D_2)^2(B_2+2H_2)/2T$$

$$- \sum_{\gamma=m+1}^{n-1} Q_r(\gamma)\delta(o,B_3,n-\gamma+1)] \quad \text{for } [n \geq (m+1)] \qquad ..(13)$$

In particular for the $(m+1)^{th}$ time-step the return flow rate is given by:

$$Q_r(m+1) = [\sigma_r(m+1) - D_i + \bar{H} + F(-D_1,B_1,H_1,m+1)$$

$$- 0.5K\surd(-D_1)^2(B_1+2H_1)/T + F(-D_2,B_2,H_2,m+1)$$

$$- 0.5K\surd(-D_2)^2(B_2+2H_2)/T][\frac{1}{\Gamma_r} + \delta(o,B_3,1)]^{-1} \qquad ..(14)$$

Results and Discussion

Results have been presented for a drainage channel which is located to the left of the first canal. The time at which the drain gets activated, the rate at which the return flow enters the drainage channel, and loci of the water table at different time have been presented. The non-dimensional times, $Kt/(2\phi\bar{H})$, at which water first enters into the drainage channel, have been predicted for $B_1/\bar{H} = B_2/\bar{H} = 0.03$ and 0.06, $H_1/\bar{H} = H_2/\bar{H} = 0.003$, $B_3/\bar{H} = 0.02$, for various values of D_1/H and $D/\bar{H}$. It is seen that with larger spacings between the parallel canals, the time at which the drainage channel activates, is delayed. For example, with $B_1/\bar{H} = B_2/\bar{H} = 0.03$, and $D_1/\bar{H} = 0.3$, the non-dimensional time for activation of drain is 0.028 if spacing between the ridge canals, $D/\bar{H}$, is 0.08. For $D/\bar{H} = 2.0$, the corresponding time is 0.042. If the distance of the drain from the first canal, i.e., $D_1/\bar{H}$, is increased, the drain activates at a later time, other dimensions remaining same. For example, if $D/\bar{H} = 0.5$ and $D_1/\bar{H} = 0.3$, the drain activates at non-dimentional time = 0.0395, whereas it activates at non-dimensional time = 0.0875, if $D_1/\bar{H} = 0.6$. It is also seen that for larger width of parallel canals, a drain would activate at earlier time, other dimensions remaining same. For example, for $B_1/\bar{H} = B_2/\bar{H} = 0.03$, $D_1/\bar{H} = 0.3$ and $D/\bar{H} = 0.5$, the drain activates at non-dimensional time=0.0395, whereas if $B_1/\bar{H} = B_2/\bar{H} = 0.06$, the corresponding time is 0.025. The time of activations for different sets of canals are presented in Table 1.

Table 1 - Non-dimensional time at which the water enters the drainage channel evaluated for $D_i/\bar{H}$ = 1.1, $\sigma(n)/\bar{H}$ = 0.098 and $B_3/\bar{H}$ = 0.02; the drain lies to the left of the first canal

Width of the canals $B_1/\bar{H}, B_2/\bar{H}$	Depth of water in the canals $H_1/\bar{H}, H_2/\bar{H}$	Spacing between the canals $D/\bar{H}$	Distance of the drain from the first canal $D_1/\bar{H}$	Non-dimensional time at which the drainage channel activates $Kt/(2\phi \bar{H})$
0.030	0.003	0.08	0.30	0.0280
		0.18		0.0320
		0.50		0.0395
		1.00		0.0415
		2.00		0.0420
		0.08	0.60	0.0675
		0.18		0.0735
		0.50		0.0875
		1.00		0.0955
		2.00		0.0965
0.06	0.003	0.08	0.30	0.0190
		0.18		0.0215
		0.50		0.0250
		1.00		0.0255
		2.00		0.0256
		0.08	0.60	0.0490
		0.18		0.0535
		0.50		0.0620
		1.00		0.0650
		2.00		0.0651

The variation of return flow to the drainage channel with time after the drain gets activated, is presented in Fig.2 for $B_1/\bar{H} = B_2/\bar{H} = 0.03$, $H_1/\bar{H} = H_2/\bar{H} = 0.003$, $B_3/\bar{H} = 0.02$, and $D_1/\bar{H}$ = 0.3 and 0.6, for various values of $D/\bar{H}$. Similar results for canal of large width with $B_1/\bar{H} = B_2/\bar{H} =$ 0.06 are presented in Fig.3. It is seen from these figures that water enters the drain at an increasing rate. For example, for $B_1/\bar{H} = B_2/\bar{H} =$ 0.03, $D_1/\bar{H}$ = 0.3 and $D/\bar{H}$ = 0.5, the non-dimensional flow $[Q_r(n)/(KH_1)]$ entering to the drainage channel per unit length is 1.2553 at non-dimensional time = 0.1395. The flow rate to the drainage channel at non-dimensional time = 0.3995 is 3.5337. The predicted return flow to the drainage channel presented in the figures are valid only if the depth of water in the drainage

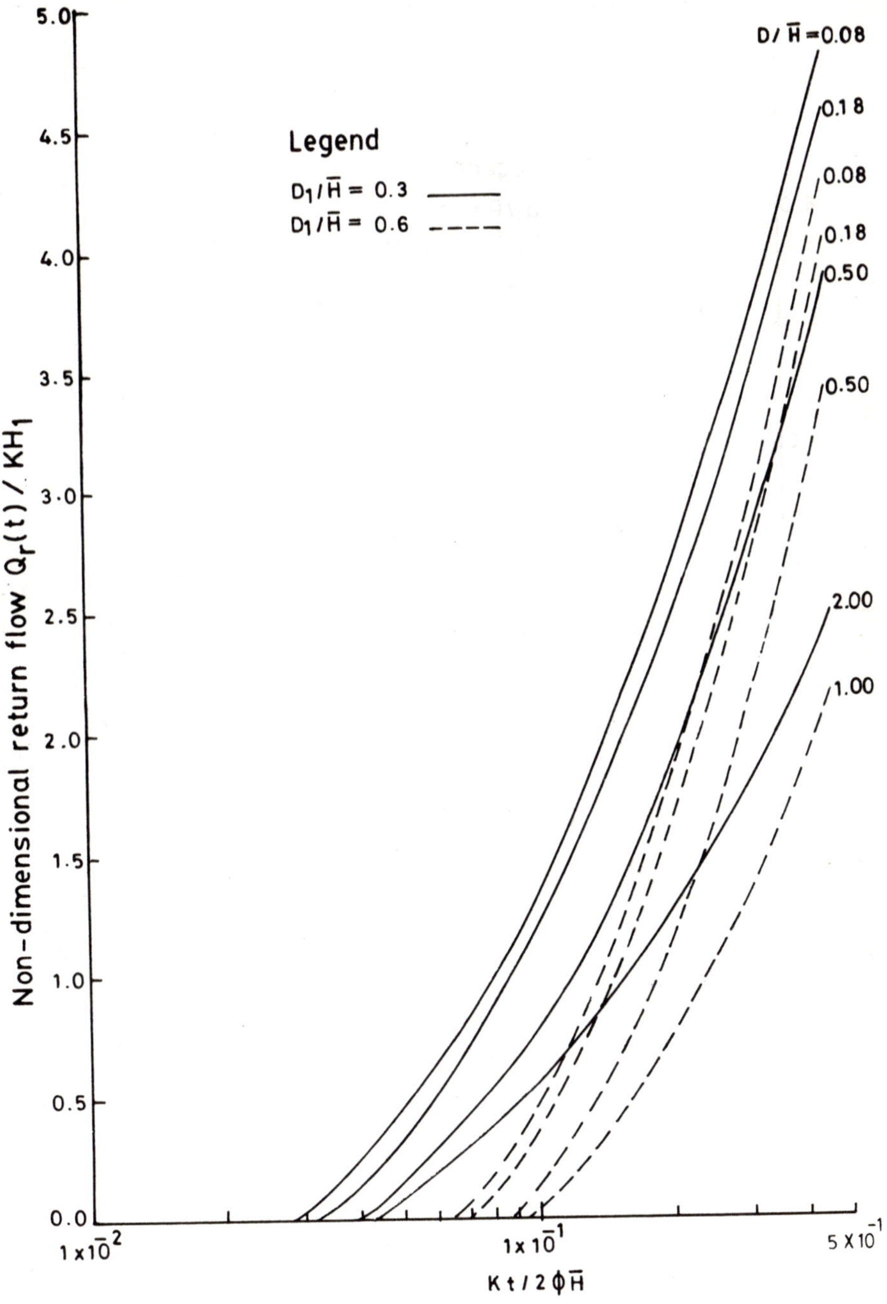

Fig. 2 Variation of the return flow to the drainage channel with time, evaluated for $B_1/\bar{H} = B_2/\bar{H} = 0.03$, $H_1/\bar{H} = H_2/\bar{H} = 0.003$, $B_3/\bar{H} = 0.02$, $\sigma^1(n)/\bar{H} = 0.098$, $D_i/\bar{H} = 1.1$, & and $D_1/\bar{H}$=0.3 & 0.6

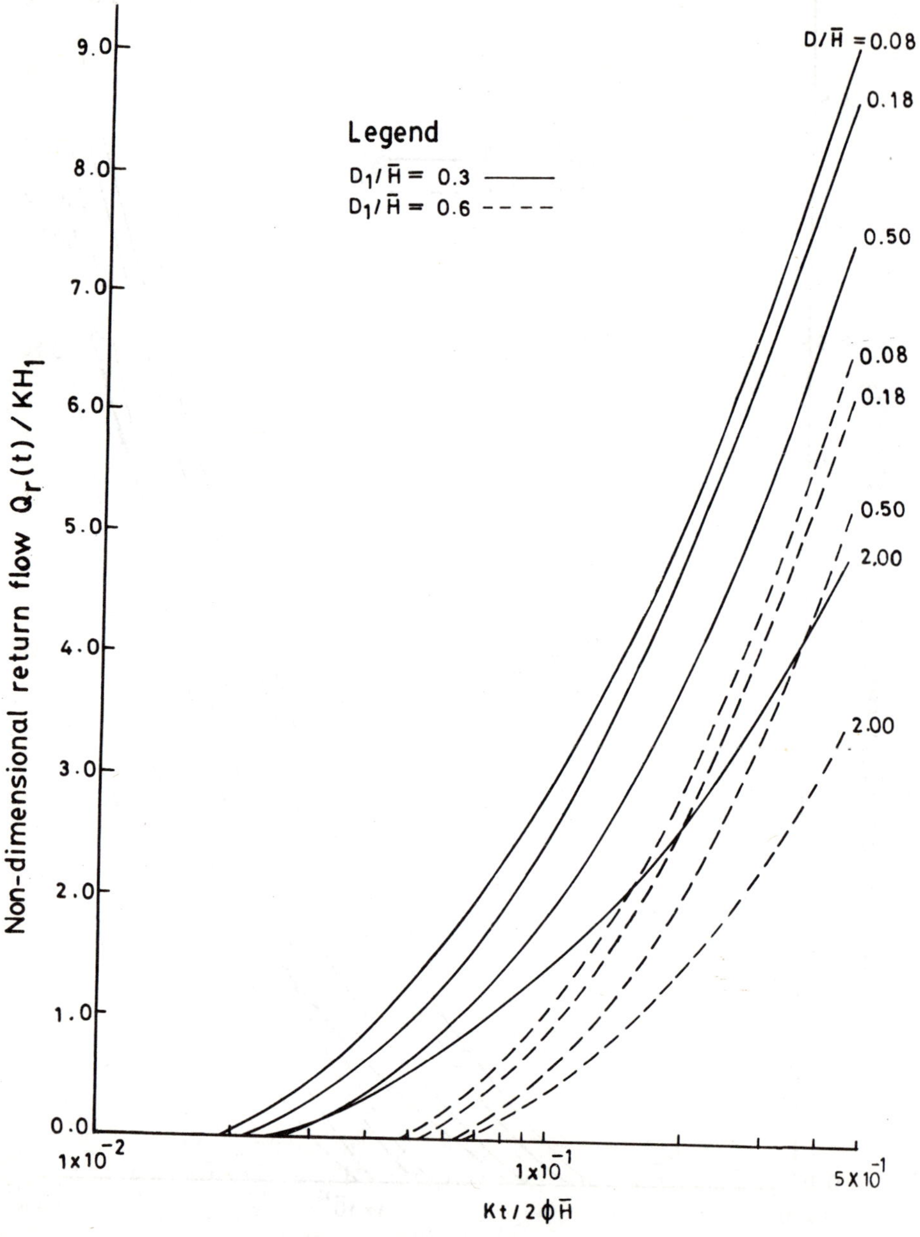

Fig. 3 Variation of the return flow to the drainage channel with time, evaluated for $B_1/\bar{H} = B_2/\bar{H} = 0.06$, $H_1/\bar{H} = H_2/\bar{H} = 0.003$, $B_3/\bar{H} = 0.02$, $\sigma(n)/\bar{H} = 0.098$, $D_i/\bar{H} = 1.1$, & $D_1/\bar{H} = 0.3$ & 0.6

channel is very nominal as has been assumed. It is inferred from the slope of the graphs presented in Figs.2 & 3 that with decrease in $D/\bar{H}$, the return flow rate is increased.

Typical results of evolution of water table with time, due to seepage from parallel ridge canals, in the presence of drainage channel, are depicted in Fig.4 for $B_1/\bar{H} = B_2/\bar{H} = 0.06$, $H_1/\bar{H} = H_2/\bar{H} = 0.003$, $D/\bar{H} = 0.18$, & $D_1/\bar{H} = 0.3$.

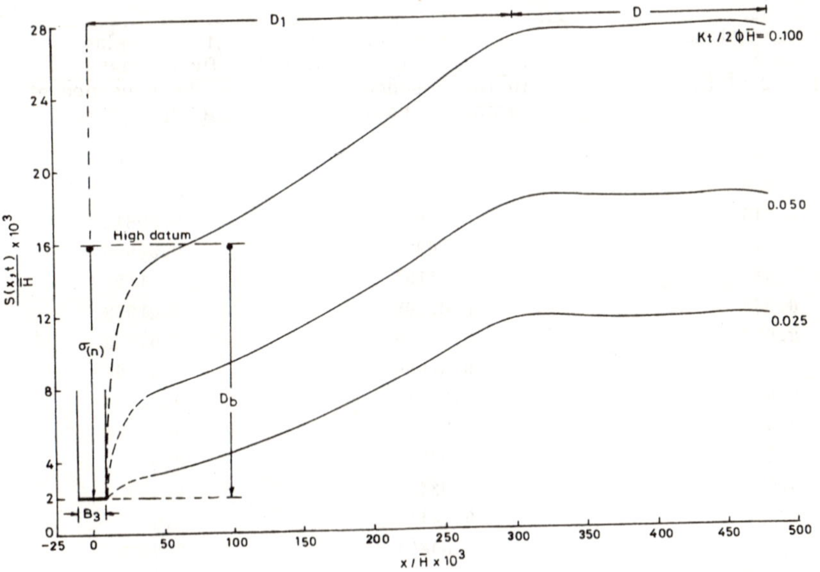

Fig.4 Typical water table positions of different times after the drainage channel gets activated, evaluated for $B_1/\bar{H} = B_2/\bar{H} = 0.06$, $H_1/\bar{H} = H_2/\bar{H} = 0.003$, $D/\bar{H} = 0.18$, $D_1/\bar{H} = 0.3$, $\sigma(n)/\bar{H} = 0.098$, and $D_i/\bar{H} = 1.1$

The water enters the drainage channel at the end of non-dimensional time = 0.0215. The water table evolutions shown in the figure are after the drainage channel has been activated. It could be seen that at non-dimensional time = 0.050, the water table rise, $S(x,t)/\bar{H}$, under the first canal, which is nearer to the drain, is 0.01798. Had there been no drainage channel nearby, the non-dimensional water table rise under this canal would have been 0.01803. [Table 2]. Thus, due to the presence of the drainage channel, the water table evolution under the canal has not been influenced appreciably upto non-dimensional time = 0.050. The results of water table rise at various times below the centre of the first canal, with and without the presence of drainage channel, are showin in Table 2.

Table 2 Rise of water table below the centre of first canal with and without drainage channel nearby, evaluated for $B_1/\bar{H} = B_2/\bar{H} = 0.06$, $B_3/\bar{H} = 0.02$, $H_1/\bar{H} = H_2/\bar{H} = 0.003$, $D/\bar{H} = 0.18$, $D_1/\bar{H} = 0.30$, $\sigma(n)/\bar{H} = 0.098$, and $D_i/\bar{H} = 1.1$

Non-dimensional time $Kt/(2\phi\bar{H})$	Water table rise $[S(x,t)/\bar{H}]$ below the centre of the first canal	
	In the absence of drainage channel	In the presence of drainage channel
0.005	0.00368	0.00368
0.010	0.00609	0.00609
0.015	0.00812	0.00812
0.020	0.00990	0.00990
0.025	0.01151	0.01151
0.030	0.01299	0.01299
0.035	0.01436	0.01436
0.040	0.01565	0.01564
0.045	0.01687	0.01684
0.050	0.01803	0.01798
0.100	0.02751	0.02707
0.150	0.03481	0.03379
0.200	0.04110	0.03924
0.250	0.04661	0.04387
0.300	0.05160	0.04791
0.350	0.05619	0.05151
0.400	0.06047	0.05476
0.450	0.06449	0.05772

It could be seen that at non-dimensional time = 0.2, the water table rise below the centre of the first canal is 0.03924 when the drainage channel exists nearby. The water table rise without the drain would have been 0.04110.

Conclusions

The following conclusions have been drawn from the study presented above:

i) A drain activates at earlier time, for canals, having larger width, other parameters remaining the same.

ii) If the spacing between the parallel ridge canals is increased, the time for activation of drain is delayed.

iii) A drain activates at earlier time if its distance from the parallel canals is decreased.

iv) With the presence of the drainage channel, the maximum water table height under the left canal is marginally affected upto certain initial period.

References

Herbert, R., 1970. Modelling partially penetrating rivers on aquifers model. Ground Water. v. 8, p. 29-36.

Polubarinova - Kochina, P.Ya, 1962. Theory of ground water movement. Princeton University Press, U.S.A.

WATER CONSERVATION AND ARTIFICIAL RECHARGE TECHNIQUES IN HARD ROCK AREAS

Saleem Romani & P.N. Sharma
Scientist "C" Scientist "D"

Central Ground Water Board
North Central Region
BHOPAL (M.P.)

ABSTRACT

Water conservation and artificial recharge of groundwa- is gaining importance as one of the strategies to fight droughts. The need is felt all the more in hardrock areas where the water supplies are essentially dependent upon groundwater but which are frequently visited by droughts or where, because of the terrain conditions much of the rainwater is lost as flash or instantaneous floods and the local streams are for much of the year dry or atleast intermittent in nature.

Jhabua district in Madhya Pradesh presents a situation where barren hard rocks with steep gradients provide conducive factors for heavy runoff of surface water within short period after rainfall. The drought prone tribal district was taken up for studies under the Technology Mission on "Drinking Water in villages and Related Water Management". The paper describes the techniques adopted for water conservation and artificial recharge in the hard rock terrain of Jhabua district. Case histories have been given for two sites in the Deccan trap and two sites situated in the Bagh beds. The techniques of lateral recharge, vertical recharge and recharge by submergence have been described for areas selected on the basis of hydrogeological surveys.

INTRODUCTION

Droughts are a recurring feature in India. With the severity of droughts the debate on the ways to combat the situation comes to acquire dimensions. Water conservation and artificial recharge of groundwater are gaining importance as few of the strategies to fight droughts. There is presently a wide gap between theoritical planning of the strategies and their field implementation. The State of the art techniques and methodology in the field of water conservation and artificial recharge techniques, specially in hardrock areas, are in their infancy in India. It is so because the required importance to the area-specific-hydrogeological-situation is often not given and projects are designed without adequate measures to meet the actual field conditions. The feedback from such projects even field conditions. The feedback from such projects even if available does not add much to the knowledge resulting in slow pace of development in techniques and methodologies.

The drought prone tribal district of Jhabua in Madhya Pradesh was taken up for studies under the Technology Mission on "Drinking Water in Villages and Related Water Management". The district is representative of typical hard rock situation where barren hills with steep gradients provide conducive factors for heavy runoff of surface water within a short period after the rainfall. The situation offered scope to design low cost structures in hardrock areas which would help in harnessing the surface and ground water in a planned and coordinated way. An attempt has been made in the paper to introduce some techniques adopted for water conservation and artificial recharge of grou -ndwater at sites under different hydrogeological set-ups in the Jhabua district. The methodology can be easily replicated in other parts of the country with similiar hydrogeological situations.

HYDROGEOLOGICAL SETUP OF JHABUA DISTRICT

Jhabua district is located in the western part of the Madhya Pradesh State, covering an area of about 6792 sq.km. between the north latitudes 21° 55'and 23° 15' and and the east longitudes 74°01' and 75° 01'. Narmada river forms the southern boundary of the district whereas the Mahi river forms the northern and north eastern boundary (Fig-1). The district lies within the basins of these major rivers. There is a chain of hills known as "The Vindhyachal" running across and extending northwards towards Udaipur in Rajasthan. The main physiographic units are (i) the Hilly terrain (ii) the Plateau area and (iii) the Undulating plains. The hilly terrain and plateau area of northern, north western and southern parts of the district are occupied by the Deccan trap lave flows of the Cretaceous to the Eocene age. The flows are mainly of the Aa type with their characteristic feature of usually an exceedingly rough fragmentery top surface. Vesicular horizons are generally absent or limited in thickness thereby reducing the chances of the presence of good aquifers for groundwater storage. The weathered zones,joints, fractures and vesicular zones, of whatever thickness though, form the main water bearing horizons.

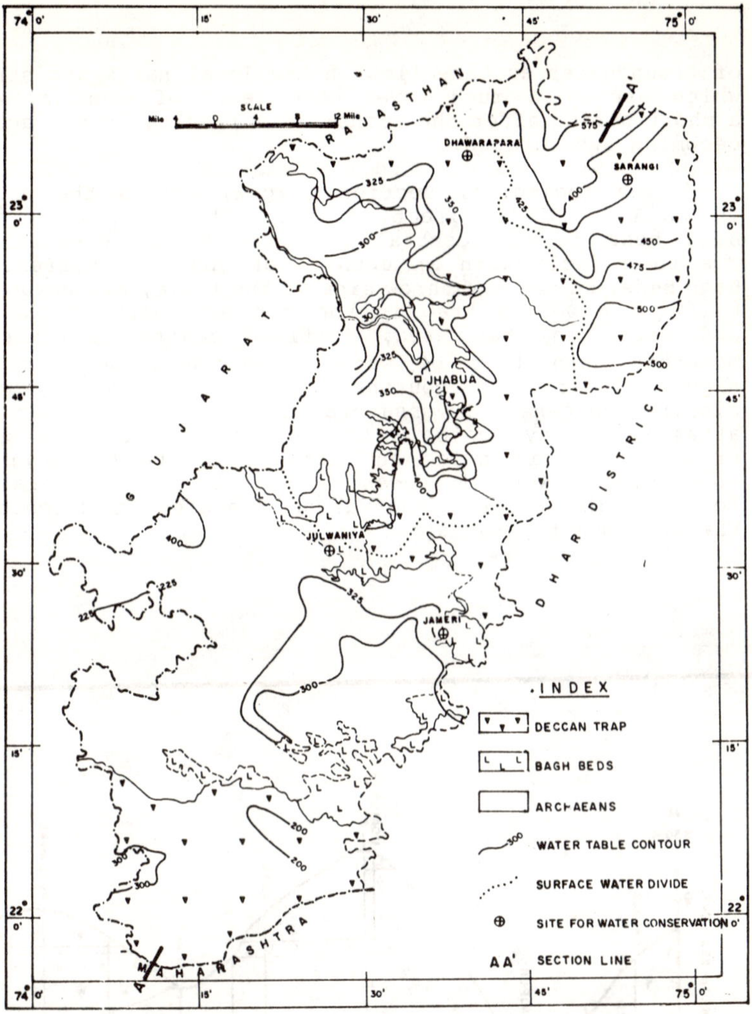

Fig.1 Hydrogeological Map of Jhabua District

The underlying Bagh beds of the Upper Cretaceous age are exposed at the fringe of the Deccan Trap area in the southeastern and central parts of the district. The sandstones and limestones belonging to the Bagh bed series are generally about 12 to 20 metre thick in the central parts but exceed even about 90 metres in the southeastern parts. In the central parts these rocks generally lie along the surface water divide of the Narmada and the Mahi basins occupying hilly terrain. The geographical location of these rocks at high elevation causes the draining out of groundwater as base flow in the local nalas and streams, which are at a much lower base level of erosion, within a short period after the rains: this results in the aquifers becoming unsaturated.

The geological section (Fig-2) across the district in S30°W - N30°E direction show that the presence of the Bagh beds in the central parts at higher elevations is due to a fault with an upthrow of about 20 metres. The Bagh beds, on the downthrow side of the fault, are encountered below the Deccan trap in the northern part of the fault and form, in that area, confined aquifer/(s) yielding moderate quantities of water. The Bagh beds occupying higher elevation in sourthern part of the fault, (upthrow side)remain generally unsaturated as the infiltrated water, after the rainy season, flows out in the form of springs or base flow in nalas and streams in the Archaean rocks occupying lower altitudes. The situation offers scope for adopting water conservation measures as discussed in the subsequent paragraphs.

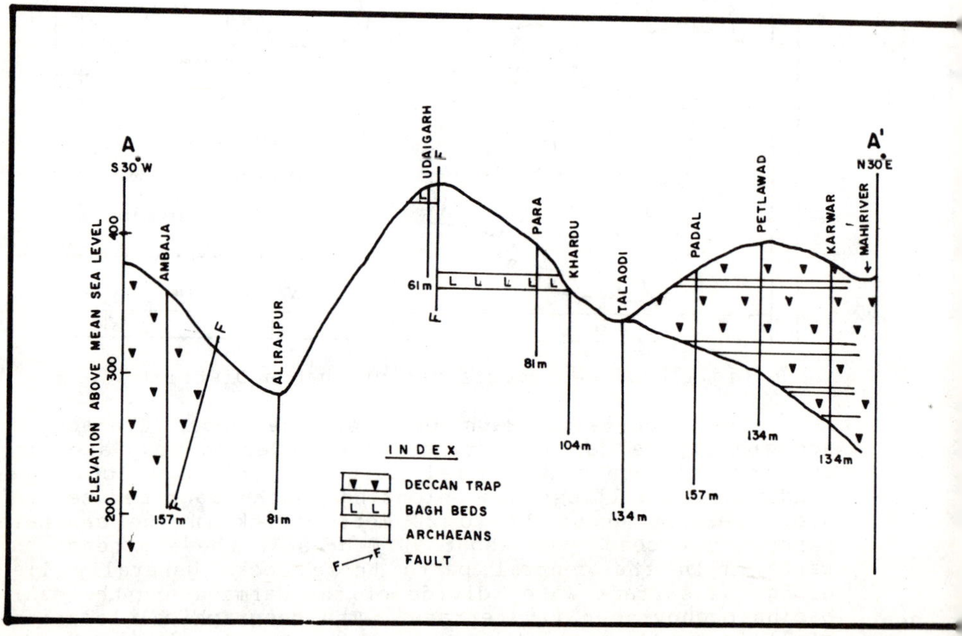

Fig.2. Geological Section Along AA', Jhabua District(MP)

The central and northwestern part of the district is covered with the Archaean group of rocks (the Aravallies and the Banded Gneissic complex). The rock types show wide variations comprising granites, gneisses, basic rocks, schists, phyllites and quartzites. These rocks are generally impervious and poor in groundwater storage. Groundwater, in them, occurs under phreatic conditions in the weathered, jointed and fractured horizons of the different rock units.

WATER CONSERVATION AND MANAGEMENT TECHNIQUES IN JHABUA DISTRICT, MADHYA PRADESH.

Background :- Jhabua district is a drought prone tribal district of M.P. The average annual rainfall of the district is about 829 mm. During 1983, the district recorded the lowest rainfall since 1947 with the district average of about 420.2mm. At the peak of drought the Prime Minister visited 3 villages of Jhabua and received petitions related to the provision of drinking water. The district was included under the Technology Mission on "Drinking water in villages and Related Water Management" and Central Ground Water Board was asked to provide the necessary scientific input for all future works. The typical scenerio of the district is that of the barren hard rock hills with steep gradients, a factor conductive to the making of heavy runoff of surface water within a short period after rainfall. This necessitated investigation of scope for water conservation and the measure to be adopted for the same. Studies were undertaking at a few sites on experimental basis to develop a model for water conservation and artificial recharge in a hard rock terrain, such as this, so that it may be replicated in other parts of the country with similar hydrogeological situations. The techniques adopted for water conservation and artificial recharge for a proper management of surface and groundwater have been dealt with for two sites in the Deccan trap and two sites situated in the Bagh beds.

The case history of the sites in the Deccan Trap demonstrate how hydrogeological mapping of lava flows has helped in the proper location of low cost structures like gully plugs and check dams for water conservation and artificial recharge. In view of the fact that minor irrigation tanks in other parts of the district with reservoir wholly in Bagh beds, showed heavy vertical seepages and the tanks could not hold surface water beyond 3 to 4 months after every rainy season, the sites selected for artificial recharge into the Bagh beds were located at the contact of the Bagh beds with the underlying Archaeans to ensure no subsurface leakage. The details of the methodology adopted is discussed below -

Lateral Recharge :

Site : Sarangi, Block Petlawad (23°04';74°54'; 46 I/16):

The area is covered with massive basalts of the Deccan Trap formation. The village with a population of 2128 is a problem village with an inadequate availability of drinking water. Out of the ten boreholes drilled by the P.H.E.D. to meet the local water supply needs, only four have yielded limited quantity of water which can sustain only handpump supplies. A nala passing through the village flows westwards and carries the runoff during rainy season to the Larki Nadi. The maximum relief of the catchment area and the nala bed exceeds about 30 m within a distance of about 1 km. Thus,due to the massive nature of the basalt most of the surface water flows out to the Larki River without adding much to the local groundwater storage.

Hydrogeological surveys of the area brought to light the contact of the massive basalt with the underlying vesicular basalt rock in the nala section at a distance of about 1 km.west of the village. This vesicular basalt horizon,about 2 m in thickness, was found to be mostly unsaturated with only the bottom few centimeters forming the phreatic aquifer. Small pools of water in the nala represented the surface manifestations of this saturated or phreatic level (Fig-3).

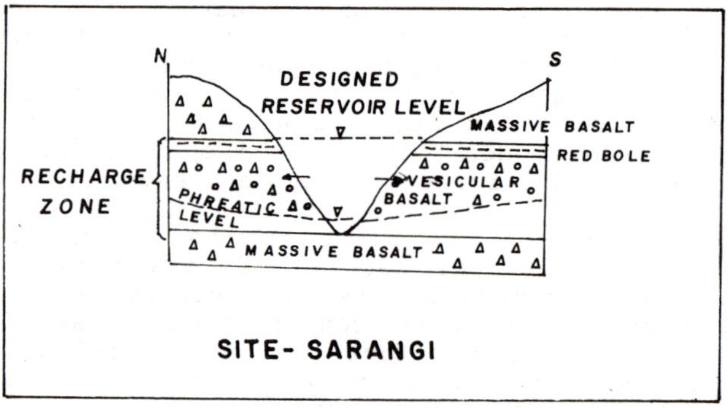

Fig.3- Lateral Recharge

The site offers scope for lateral recharge of groundwater in the unsaturated part of vesicular basalt horizon. On the recommendations of C.G.W.B. the Irrigation Department of M.P.Govt. has initiated the construction of a stop-dam across the nala. The masonary dam is proposed to be 3.1 m high so that the full thickness of vesicular basalt is submerged under the reservoir water. This will, when the reservoir is full, provide the necessary head for the lateral flow of water to artificially recharge the vesicular basalt aquifer.

The main dam would have a masonary length of 23.75 m length which will check any leakage of the reservoir water downstream. The main dam would also be provided with proper puddle so that the stored water of reservoir does not leak across the dam and the flow of water underground takes places only or mostly laterally.

Nine piezometers on both sides of the nala at variable distance have been installed to observe the extent of lateral recharge and the temporal variations in the groundwater levels vis-a-vis the level of water in the reservoir.

Vertical Recharge :

Site : Dhawarapara, Block Thandla ($23^{o}05'$; $74^{o}30'$;46 I/12);

The site is located in northern part of the Jhabua district, covered with the pahoehoe type of lava flows of the Deccan Trap. The area exhibits typical rolling topography with smooth hammocky barren hillocks covered with highly jointed massive basalt. A number of small gullies and nalas bedeck the area and meet northwards with the Undwa nala. The main nala runs in the SW direction and drains into the river Pat. The maximum altitude of the hillocks is 360 m a.m.s.l., and they overview the narrow stretch of the Undwa nala at an average altitude of about 340 m.a.m.s.l. Thus due to steep gradient, the water in the catchment area flows rapidly to the nala and within a short period all the surface water is lost to the Pat river.

In the Undwa nala section a vesicular basalt horizon of about 2 m thickness is seen exposed under a cover of about ½ to 1 m. thick massive but jointed basalt. The underlying vesicular basalt is unsaturated partly for the reasons that it is covered by massive impermeable basalt rock and partly becuase the slopes are steep and the surface runoff cannot take much advantage of the joints for infiltration or percolation its residence time during the surface-flow over the massive rock being very short.

The site offers scope for vertical recharge of the vesicular basalt horizon by construction of percolation tanks either by taking advantage of the existing joints and fractures or by excavating fully the massive rock bed and thus exposing the vesicular basalt layer at the bottom.

Forest Department of the M.P. Govt. has undertaken work of water conservation and artificial recharge at this site (Fig.-4). A small catchment area of 417 ha. has been selected for the work. To conserve the surface water flowing out to the main drainage about 125 gully plugs have been constructed at various elevations in the catchment. Four small check-dams have also been constructed at higher elevations with a small percolation tank so that the surface water in catchment can be recharged downwards through the jointed basalt. A part of the surface runoff during the rainy season can be thus recharged to groundwater. A part of the overlying massive basalt has also been removed in the reservoir area of these tanks for vertical seepage. The main earthen dam across the nala has been constructed at the lowest elevation of the nala with proper foundation treatment. The full thickness of vesicular basalt under the earthen dam foundation has been sealed by clay puddling to check sub-surface leakage of groundwater across the dam. The height of the dam is 5 m with a length of 26m.

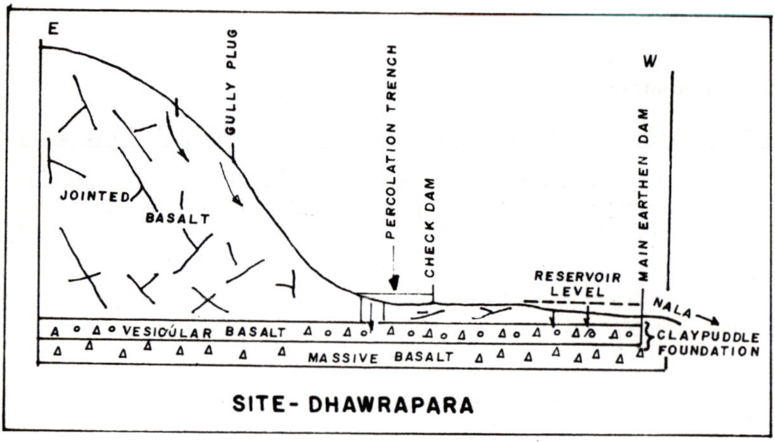

Fig-4. Vertical Recharge - Site Dhawarapara.

The reservoir of the dam would not only store the surface water coming from the spillways of small check dams upstream but also would recharge the underlying vesicular basalt horizon.

Recharge by Submergence :

Site : Jameri, Block Jobat (22°26';74°39'; 46 J/11);

The hillocks in the Jameri Reserved Forest area are occupied by limestone and sandstone of the Bagh Formation at an altitude of about 380 m.m.s.l. A small nala cutting across these rocks runs down to an altitude of 340 m.m.s.l. and joins another westerly flowing higher order nala which finally meets the Dohi river. Due to steep gradient the surface runoff flows rapidly through the Bagh beds to the granitic terrain at lower elevations. A sizeable thickness of the Bagh beds remains unsaturated and much of the runoff is lost to the impervious granite area. A micro catchment area of 0.05 sq.km. has been selected near Jameri for artificial recharge of the Bagh beds (Fig-5).

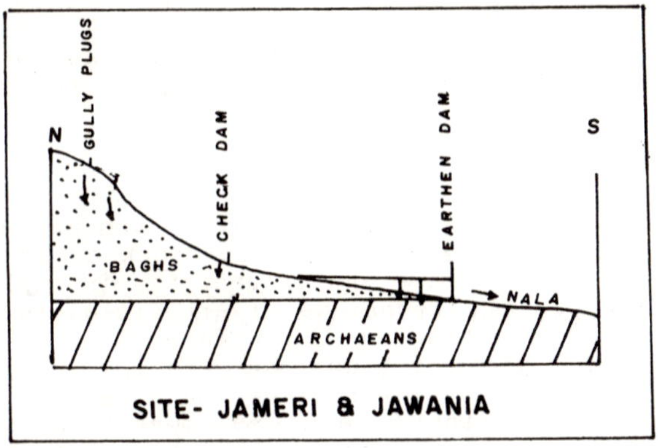

Fig.5-Recharge by Submergence

About 30 gully plugs have been made in the catchment area at various elevations to check the surface water run-off. At lower elevation (345 m.m.s.l.) a small earthen dam has been constructed to recharge the Bagh beds and also check the flow of water to the main earthen dam down-stream. A main earthen dam of 6.5 m height with a reservoir area of about 0.723 ha. has been constructed at the contact of the Bagh beds and the granite exposed in the nala bed. As the foundation of dam is on a fresh massive granite rock there is no possibility of sub-surface leakage. It is expected that an area of 0.25 ha. would be submerged in the reservoir which is covered by the permeable Bagh beds which would thus get artificially recharged. The number of small check dams at various points at higher elevations upstream of the nala would provide the necessary time for water to percolate underground.

Site : Jawania, Block Bhabra (22°32';74°27'; 46 J/6):

The area presents a situation similar to that at the Jameri site. The Bagh beds occupy hilly region with a maximum elevation of 420 m.a.m.s.l. A nala flowing south-wards through the Bagh beds comes down to an elevation of about 380 m.m.s.l. where phyllites of the Archaean complex are exposed. The Bagh beds are mostly unsaturated due to steep topographic slopes which cause heavy runoff of surface water finally to the Archaean terrain. A micro-catchment area of 0.31 sq.km. has been selected for water conservation. Rubble check-dams of 1 m. height have been constructed at various altitudes to plug the gullies and thus recharge the Bagh beds. At lower elevation (385 m.m.s.l.) a small earthen dam of 1.5 m height has been constructed in the Bagh beds to check the flow of surface water and recharge the same. The main earthen dam of 3.6 m height has been constructed at the contact of the Baghs and the Archaeans.

The creation of reservoir along with concomitant recharge of the Bagh beds by submergence in the reservoir area would help in conjunctive use of surface and groundwater for afforestation programme in the Reserve Forest.

CONCLUSIONS

The studies in Jhabua district have given an insight for designing low cost structures in hardrock areas which help in harnessing the surface and groundwater in a planned and coordinated way. The technique of artificial recharge of groundwater by creating small surface reservoirs in favourable hydrogeological set-ups would help in optimal utilization of total water resources of the area. However, the wide gap between planning and implementation can be bridged with profit only if the techniques devised are such as to take into consideration the specific hydrogeological situations. The role of hydrogeology in identifying the areas for water conservation and management are often neglected resulting in poor results. The case hist-ories discussed in the paper demonstrate how the selection and adoption of the right methods can bring about the best interaction between the surface and groundwater in planning water conservation programmes. The models if replicated in other parts of the country with similiar hydrogeological situations would prove an effective tool to fight droughts in hardrock areas of our country on a long term basis.

ACKNOWLEDGEMENTS

The authors wish to place on record their sincere thanks to Shri B.P.C.Sinha, Chief Hydrogeologist & Member, Central Ground Water Board for his permission to present the paper.

REFERENCES

Romani, S (1988.) On the techniques of water conservation and artificial recharge adopted in the drought affected Jhabua district of M.P. BHUJAL NEWS V. 3 No.1 pp. 19-21.

:::

ROLE OF DRIP IRRIGATION IN GROUND-WATER MANAGEMENT IN BASALTIC TERRAIN : A CASE STUDY OF SOLAPUR DIST.

M.K.ZAMBRE
D.B.F.Dayanand College of
Arts & Science,
SOLAPUR-413 002

***

INTRODUCTION:

The strategy of efficient management of limited groundwater resources aquires vital importance in our country. Maximum development of groundwater resources for beneficial use involves scientific planning in terms of entire groundwater basin. The management objectives must be selected in order to develope and operate these basins. These involves not only geologic and hydrologic conditions but also economic, legal, political and financial aspects. Optimum economic development of water resources in an area requires an integrated approach that coordinates the use of both surface and groundwater resources.

The assessment of groundwater in Deccan volcanies is indeed a difficult task. This is because of groundwater in these rocks is not uniformly distributed in the subsurface, as the weathering pattern and lateral extention of hard and jointed structures of basalt rocks are unpredictable. The water potential in a chronically drought prone area like Solapur district, depends upon the annual precipitation, geomorphological setup and hydrogeological characters of the basin. The food production is always affected because of ground-water potential, as the average water table goes 70 to 80 metres below ground level.

Now days it is very expensive to explored such deeper aquifers. Under this circumstances drip irrigation system proved to be most successful technology in groundwater management.

REVIEW OF LITERATURE:

The idea of drip irrigation came into existence in 1860 in Germany. Few irrigation scientists have developed their fields through the pipe lines which were made up of earthen ware. In the true sence, Prof. O.E.Roub of Mishigaon State University has demonstrated the irrigation through the screened pipes made up of baked earthen wares. Soon after the second world war, there was a large demand for plastic pipes. In 1952, S.Bloss, a engineer, has taken a patent of cocks for drip irrigation and showed howmuch growth of a plant is possible if the water is supplied to its root zone area. In 1962, Dr.K.Dartor and his associates, established the procedure of plastic pipes having drip facilities. The first international conference, on drip irrigation system was held in Israel in 1971, where about 20 research papers were presented (Magar,1988). In 1985, third international congress on drip irrigation was held where more than 250 research papers were presented. In India Drip irrigation has very recent origin. Experiments are going on in various research institutions, e.g. water technology centre or IARI and other agricultural institutes. In Maharashtra, Forest Research stations have taken leading step by installing a number of drip systems in Teak seed orchard and the results obtained are quite satisfactory. In Maharashtra, the farmers are using this system since 1985 for irrigation their lands.

HYDROGEOLOGICAL SETUP OF SOLAPUR DISTRICT:

The Solapur district of Maharashtra State (Fig.1) covering an area of about 15,000 Sq.Kms. is situated in Bhima basin and geographically divided into eleven talukas, out of which Madha, Mohol, Karmala and Sangola are known to be drought prone. The district receives an average annual precipitation 640 mm. Over greater part of the district the topography is undulatory with altitude ranging from 410 to 470 mtrs. above M.S.L. In the western part of the district hill range trending in north-West-south-east direction are present and higher elevation is marked by planner surface at 600 mtrs. above mean sea level. The plains in the district are drained by the rivers Bhima, Sina, Man, and Bori and their tributaries.

Geologically, the district is constituted of Deccan Volcanics of Cretaceous - Eocene age. The volcanics are represented by basaltic flows having an almost horizontal disposition, though easterly dipping flows are met with in Akkalkot taluka (Adyalkar and Mani 1972) The flows are totally devoid of hypabyssal intrusions and intertrappean bed. However, through out the district there are layers of red bed formations On the basis of field observations, the basaltic flows are grouped into following types, namely -

1) Hard and massive, fine grained of dark grey color.

2) Vesicular and amygdaloidal flows of pink to grey color in which the amygdules are filled with green earth or zeolitic material. This type is more susceptible for weathering.

3) Red bed formation which is composed of iron rich clayey material, generally brick red color and which vary in thickness from few centimetres to six meters. At places first two types are found to be traversed by closely spaced and fractures.(Zambre and Thigale,1980)

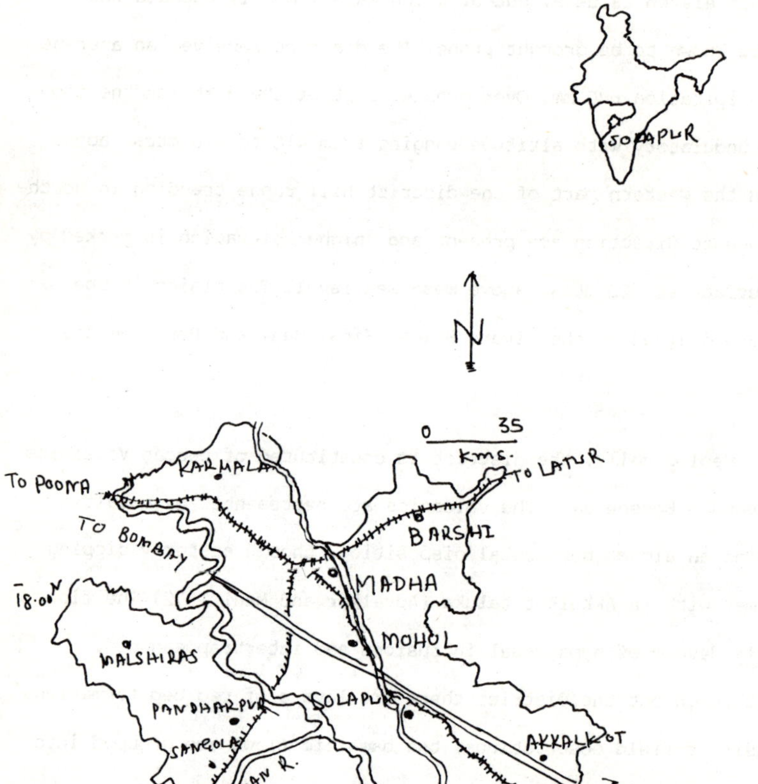

FIG. 1. LOCATION MAP OF SOLAPUR DISTRICT.

The weathered and jointed or fractured and vesicular horizons of the flows are usually considered to be ideal site for ground-water development. (Deshpande, 1975, Radhakrishnan Nair et .al.1979). The groundwater in such horizons occur under both water table and confined condiation. The porosity in such horizons was found to be very high. However, lateral and vertical distribution of these water potential formation is erratic and their recognition and delineation are the main problems in prospecting groundwater.

The use of large diameter wells (dug wells) to withdraw ground-water for irrigation and domestic use is a common pratice in india. In drought prone area, because of low recharge, these dug wells either yield very poorly or get dry in summer season. As a result crop failure is frequent. It is very important to manage the optimum utilistion of the available ground water potential for irrigation purpose when there is acute shortage of water. Under this circumstances there is no alternative to go for drip irrigation system for increase the average food production.

DRIP IRRIGATION SYSTEM:

Drip irrigation is an efficient method of providing water directly into the soil at the root zone of the plant. And it permits the irrigation to limit watering closely to the consumptive use of the plant. Thus drip irrigation minimises the conventional losses such as deep purcolation, runoff and soil evaporation. (I.Pai Wu and Harris M. Gitlin, 1974).

This method of irrigation (Fig.2) is accomplished by using small diameter plastic lateral lines with devices called emitters or drippers, at selected spacing so as to deliver water to the soil surface near the base of the plants. The lateral can be of small plastic tube

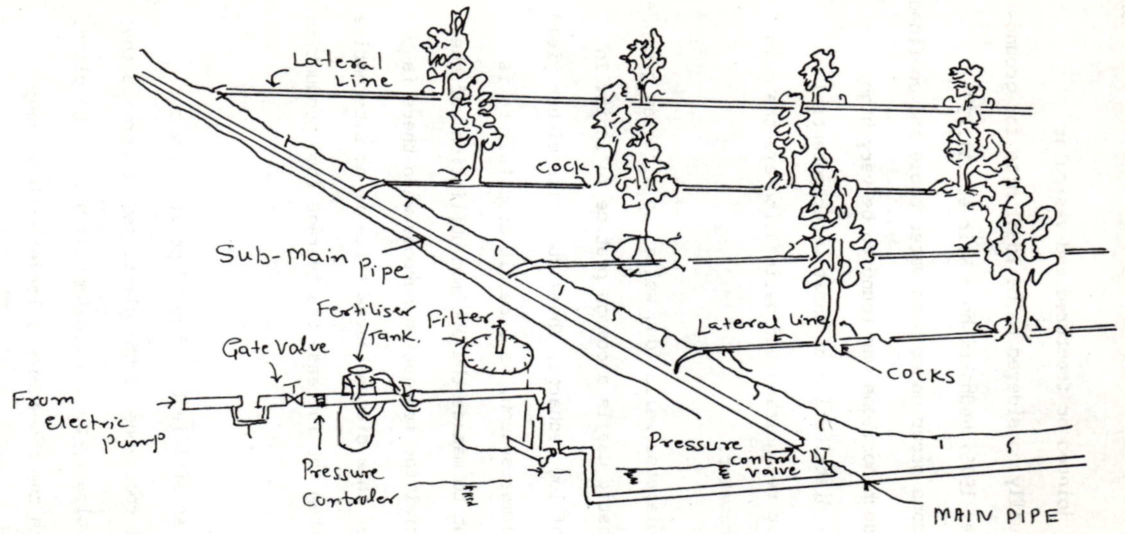

FIG. 2. LAYOUT OF DRIP IRRIGATION SYSTEM

combined with emitters or simply a small thin walled plastic tube with orificies. The laterals are designed for distributing water into the field with exceptional degree of uniformity. The sub main acts as a central system which can be adjust water pressure in order to deliver the required amount of flow into each lateral. It is also used to control the irrigation time to individual field.

The main line serves as conveyence system for delivering the total amount of water for the Drip Irrigation System. The emittters, laterals, sub-main lines are considered as principle parts of drip irrigation system. There are supporting parts such as filters, flushing units, pressure regulators, pressure gague, fittings, valves,fertilizer injectors etc. which are used to serve different purposes in a drip irrigation system.

PROCEDURE OF DRIP IRRIGATION:

From the main, feeder lines are run across the field and laterals are inserted on the same. Low density polythelene pipes are laid along the plants or tree rows with outlets or drippers inserted at appropriate intervals. These are designed to supply water at low rates (1 to 10 litres per hour) directly to the soil. Low pressure ranging from 1/3 to 3 Kg/Cm^2 are sufficient to work. Pressure control valves are also fitted. Usually 3.75 to 7.00 Cm dia are used for the main and 2.5 to 3.75 cm are used commonly for sub-mains in the drip system. For laterals 6 mm, 9 mm, 15 mm and 18 mm diameter pipes are used to suit various situations. In hilly areas where pressure variations become excessive, sub-mains with control valves are used to feed laterals. Centrifugal pumps are the most ideal one to supply water at constant pressure and adjustable discharge. Once an even flow from each outlet is achieved the system should not be altered and water

TABLE-1

Area Irrigated by Drip System in Solapur Dist,1987

Sr.No.	Talukas	No. of Farmers	Area under Drip in Hectares	Crop Cultivate
1.	Akkalkot	29	87	Grapes
2.	Barshi	21	63	Grapes
3.	Karmala	01	03	Grapes
4.	Madha	14	42	Grapes
5.	Malshiras	04	12	Grapes
6.	Mangalwedha	05	15	Fruits
7.	Mohol	21	63	Fruits
8.	Pandharpur	20	60	Fruits
9.	Sangola	09	27	Fruits
10.	Solapur N.	78	234	Fruits
11.	Solapur S.	26	78	Sugarcane
	TOTAL:	228	684	

Source: District Agriculture Officer,Solapur.

The drip irrigation system has been adopted by the farmers in Solapur District early in 1986. By the end of 1987, about 684 hectares land come under drip irrigation system for cultivation of grapes, and sugar canes. No scientific data however is available regarding water potential requirement. But it has been observed that there is saving of water a of about 55 percent, when the drip system is adopted.

In South Solapur taluka drip system has been used for the cultivation of sugar cane. And it is observed that there is saving of groundwater of about 60 percent and increase of production of about six percent.

CONCLUSION

Erratic and irregular rainfall and frequently occuring drought situation is a common phenomenon for a farmer in the state. Out of total cultivable land of 20.8 million hectares, at the most 26 percent land comes under the surface irrigation system. By the application of groundwater resources the figure goes to 40 percent. But if the drip irrigation system is used, then this percentage would goes to 60 to 70. This drip irrigation system would proved to be the back bone of state economy. Optimum utilisation of groundwater will definately bring the additional output in food producttion.

ACKNOWLEDGEMENT

The author would like to thank the Principal, D.B.F.Dayanand College of Arts & Science, Solapur for his continued support and encouragement in completing this paper. He also thanks to the district Agriculture Officer, Solapur for supplying the necessary data.

REFERENCES

Deshpande,B.G.,(1975), Some Aspects of Hydrogeology of the Deccan Traps, Proc.Indo-German Workshop, N.G.R.I., Hyderabad, PP 105-114

I.Pai Wu and Harris M.Gitlin,(1974),Drip Irrigation System Design in Metric Units, Cooperative Extention Service, University of Hawaii, Miscellaneous Publication.144.

Magar,S.S.,(1988), Drip Irrigation System (in Marathi) Continental Publication,P.128

Radhakrishna Nair,P.V., Pendharkar,A.S., Navada,S.V., and Rao,S.,(1979) Groundwater Recharge Studies in Maharashtra, Isotope Hydrology, I.A.E.A. Vienna,V.2 PP 803-826

Sivanappan,R.K., and O.Padmakumari,(1980),Drip Irrigation, College of Agriculture Engineering, Tamil Nadu Agriculture University, Goimbatore

Zambre,M.K., and Thigale,S.S.,(1980),Geophysical Investigation for Groundwater in Solapur District, Maharashtra, Proc.3rd.Ind.Geol. Cong.Poona. PP 435-443.

****

POSSIBILITIES OF INDUCED RECHARGE AND DEVELOPMENT OF DEEPER AQUIFERS FOR AUGMENTING WATER SUPPLY TO AMBALA URBAN AREA, HARYANA STATE

T.G. Firozuddin, G. Suryanarayana
D.S.S. Murthy & K.V.J.R. Krupanidhi
Central Ground Water Board

ABSTRACT

There is mounting concern regarding the problem of developing adequate water supplies to Ambala urban area as the locally available ground water supplies through tubewells will not be able to meet the growing demands. In the present paper the results of the studies carried out to delineate the water table aquifer along the riverine tract of Dangri river, the geometry and the productivity of the aquifer, its capacity to hold the required amount of water and possibilities of induced recharge for the river have been examined. It is estimated that a quantity of 2.66 MCM can be pumped out from the aquifer during the dry season. The estimated induced recharge from the river to the aquifer is about 0.833 MCM during the monsoon season. A possibility of development to the extent of 7.30 MCM of the deeper aquifers occurring below 150 m depth has also been discussed as an additional source to augment the water supply.

INTRODUCTION

In recent years ground water supplies using induced recharge by surface water into the water table aquifer have been extensively developed. In a broad perspective the induced recharge means altering hydrological conditions in an aquifer such that more than normal seepage takes place from a perennial water body. This can effectively be done by increasing the hydraulic gradient between gaining and losing water bodies provided the capacity of the gaining body is sufficient to accommodate the excess or induced recharge. In case of ephemeral water body such as Dangri river this can effectively be achieved by replenishing water table aquifer by flood waters. For this condition, over-development of the aquifer to cause substantial depletion in the ground water storage has to be done during summer (dry spell) preceding the floods usually in monsoon. In the present paper, the studies conducted on exploring the possibilities of creating storage space for induced recharge into the water table aquifer in the riverine tract of Dangri river to augment water supply is discussed.

Presently Ambala urban water supply requirement is about 29.36 MCM per day which is met through tubewells constructed down to a depth of 150 metres bgl and is likely to become doubled in near future. Most of these tubewells have started showing decline in discharge and in some of these tubewells the discharge rate is reduced to such a low level that pumping from these wells is proved to be uneconomical. The present paper also deals with the study of ground water developmental potentialities of deeper aquifers below 150 m depth as an additional source to augment water supply to Ambala urban area.

HYDROLOGICAL SETTING

The mean annual rainfall is 800 mm and about 90% of the annual precipitation is received during the South-West monsoon period between mid-June and September. A study of the actual rainfall for the period 1953-78, reveals that more than half the number of years recieved rainfall between 810 mm and 1270 mm. This suggests that the normal years are more common than the abnormal years, the ratio being 11:3.

The main drainage of the area is Dangri river which is ephemeral in nature and flows from NNE to SSW direction. A major component of flow in it is provided by the rainfall during monsoon season with base flow for a short period immediately afterwards. The Dangri river in the upstream area records average discharge of 78.81 cumec during monsoon (June-October), out of which 75.01 cumec discharge is recorded between July and September months.

HYDROGEOLOGY

The area forms a part of Indo-Gangetic alluvial plain and consists of interbedded and lenticular deposits of clay, sands of various grades, silt and gravel and also kankar intermixed with clay in various proportions. The most permeable water bearing zones in the alluvial deposits are gravel and sand beds. These beds have been tapped by open wells and tubewells. The depth of the dug wells generally ranges from 7.0 to 13.0 m bgl and cavity wells with an average depth of 22.0 m bgl. The urban water supply demands of Ambala town are mainly met with by tubewells tapping aquifers down to a depth of 150 m bgl. The movement of ground water in the phreatic aquifer is towards SW and also WSW directions with a hydraulic gradient varying from 0.6 m to 2.0 m per kilometre (Fig.1). More or less similar pattern of ground water movement is noticed in deeper aquifers also.

GROUND WATER EXPLORATION OF SHALLOW AQUIFERS

Detailed hydrogeological studies followed by 32 electrical soundings using Schlumberger method were carried out along the riverine tract of Dangri river in an area of 67 sq.km. Based on the results of the soundings, nine slim holes were drilled down to a maximum depth of 75 m bgl to confirm the results of the vertical electrical soundings. Subsequently three sites were selected for construction of well fields at Kojkipur, Babial and Khelan located near slim hole nos. 6, 3, & 5 respectively (Fig.2) where the sand predominant zones are encountered which are considered suitable for induced recharge.

Aquifer Geometry

Three distinct granular zones were observed down to a depth of 75 m bgl in the study area. The top layer which is of 7 m thickness is silty followed by medium to very coarse sand. The sand beds are generally thicker towards southern parts around south of Boh and Ramgarh and gradually thin down towards east and north. The thickness of the

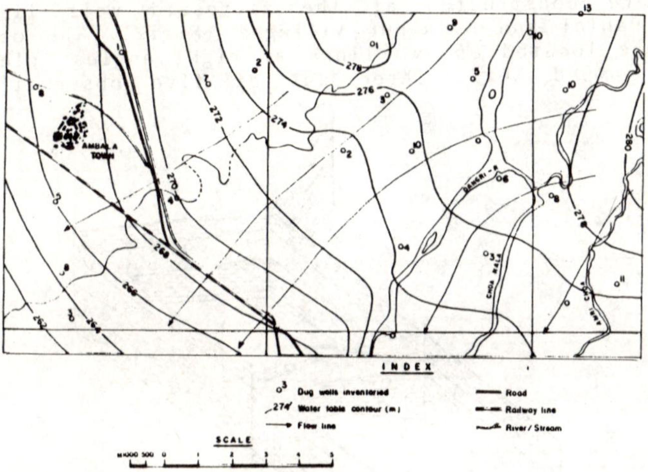

Fig.1. Water-table contour map (August 1978)

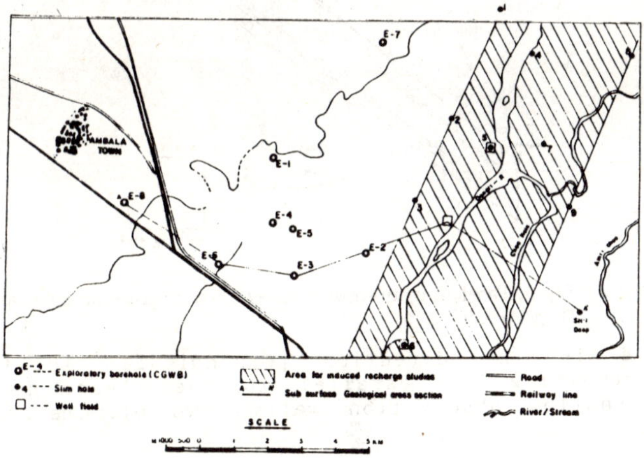

Fig.2. Map showing area of induced recharge studies and deep ground water exploration in Ambala area, Haryana State.

granular zone varies from 6 to 14 m and occurs down to a maximum depth of 29.5 m bgl(Fig.3). This zone is hydraulically connected with the surface water body i.e., the Dangri river and hence of much importance for the purpose of induced recharge. Second granular zone occurs at a depth between 24.66 and 47.50 m with a maximum thickness of 20 m and the third granular zone occurs between 56 and 60 m bgl.

Well Field Design

Five observation wells each, tapping the top aquifer zone, were constructed at the aforesaid well fields near Khelan, Babial and Kojkipur villages (Fig.2). The observation wells were located on two lines at right angles interesecting at the pumped well. Three out of five observation wells

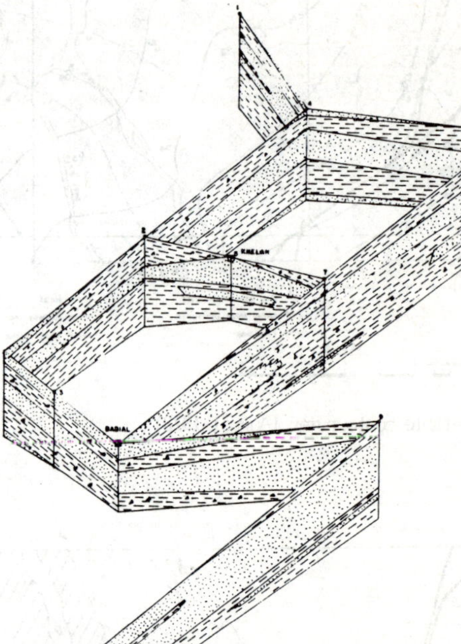

Fig.3. Panel diagram showing disposition of granular zones along Dangri river

constructed in the direction perpendicular to the river and the remaining two parallel to the river. In addition to these shallow observation wells, two piezometers tapping second and third aquifer zones were also constructed close to the pumped well to know the possible hydraulic interconnection between the different zones.

Hydraulic Properties of Shallow Aquifers

The quantity of surface water that can be induced to recharge the top (water table) aquifer varies with the amount and proximity of surface water, the hydaulic conductivity, transmissivity of the aquifer, permeability of stream bed and hydraulic gradient created by pumping. Three long duration (5 to 9 days) aquifer performance tests were carried out on the wells to evaluate the aquifer parameters. In table-1 the hydraulic properties of the aquifer determined by various methods at three sites are given. The highest hydraulic conductivity of the shallow alluvial aquifer is 25.4 m/day at Kojkipur where well sorted medium to coarse sand makes up the water table aquifer. In the other two sites i.e. at Babial and Khelan, the hydraulic conductivity is 20.44 m/day and 17.8 m/day respectively where the aquifer material is medium grained sand. No drawdown is observed in piezometers which tapped relatively deeper aquifer zones indicating different aquifer systems not connected with the top water table aquifer.

Table-1. Hydraulic properties of the shallow aquifers

Test site	Pumping rate (m^3/day)	Boulton			Jacob			Minkin	
		T (m^2/day)	K (m/day)	SY %	T (m^2/day)	K (m/day)	SY %	T (m^2/day)	K (m/day)
Khelan	916	143	17.8	10	108	13.5	10.7	186	23.25
Babial	916	184	20.44	3.4	159	17.6	3.18	274	30.52
Kojkipur	346	178	25.42	10	188	26.8	2	-	-

Dangri river, being ephemeral in nature, water table aquifer may be replenished only by flood waters during rainy season. Under these conditions it is economically advantageous to create an overdraft during the dry season when demands are also at the maximum and permits use of the water table zone to the maximum of its capacity. In such cases the most favourable lay out of the well field will be along a river bank at a distance of about 100 m from the river bed, keeping a distance of 400 m between two wells. Hence, it is safe for the purpose of calculations that well field will operate without any recharge during summer period, where the water resources will improve during rainy season on account of monsoon rains and then the same aquifer will have one positive boundary, the river, to replenish the ground water draft partly by induced recharge.

The aquifer parameters determined by pumping tests have been utilised for designing of the well field which could be taken into account for the calculation of ground water resources from the shallow aquifer. In such cases the lowering of water levels has to be worked out in the well field. The actual drawdown in the centre of the well field at an average discharge of the individual well is expressed by Musket formula without any hydraulic boundaries. The formula reads:

$$S = H - \sqrt{H^2 - \frac{Q}{\pi K}\left(\ln - \frac{\lambda}{2\pi r_o} + \frac{3.55\sqrt{at}}{\lambda}\right)}$$

Where H = Thickness of aquifer in metres
Q = Average discharge in each well m^3/day
K = Average hydraulic conductivity in m/day
λ = Distance between each well in metres
a = Water level conductivity in m^2/day
t = The time in days during which the well field is being pumped continuously without recharge during monsoon period
r_o = Radius of well intake section

Based on the above formula, the drawdown estimated in the centre of the well field during the dry season is 7.77 m with an average discharge of 726 m^3/day at each well. This drawdown is within the permissible limits (i.e. 9.8 m for the central well). The well field is based on 15 wells, each at a distance of 400 m to cover the length of 6 km on the western bank of Dangri river. With the aforesaid rate of pumping during 245 days of the year the well field would work without any recharge and the ground water draft during the dry season will be 10890 m^3/day. These resources can be utilised for augmenting the water supply to Ambala urban area in dry season before implementation of induced recharge under natural conditions.

The quantity of induced infiltration into the water table aquifer is estimated from a graph of Glover and Balmer (1954) as quoted by Karanth (1987) which works out to 0.833 MCM during the monsoon period of the year under natural conditions (Fig. 4).

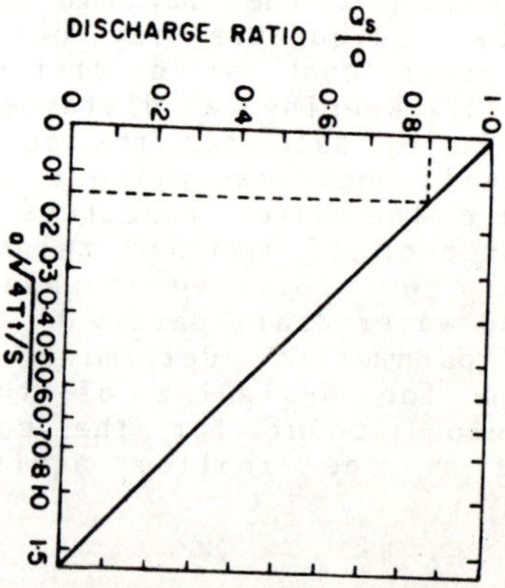

Fig. 4 Portion of well discharge contributed by a recharge source (Glover and Balmer, 1954)

GROUND WATER DEVELOPMENTAL POTENTIALITIES OF DEEPER AQUIFERS

Seven deep tubewells tapping the aquifer zones between the depth range 150 and 450 m bgl at locations marked E1 to E5 and E7 and E8 in Fig.2 were constructed with observation wells in the urban area of Ambala town and the results of the pumping tests are given in table-2.

Deep ground water exploration in the area has revealed that the aquifer system between 150 m and 450 m bgl (Fig.5) is hydraulically connected throughout most of the region either by interbedding meander belt deposits or stream scour and subsequent back fill. This may be considered as multilayered unit with possible interconnections and

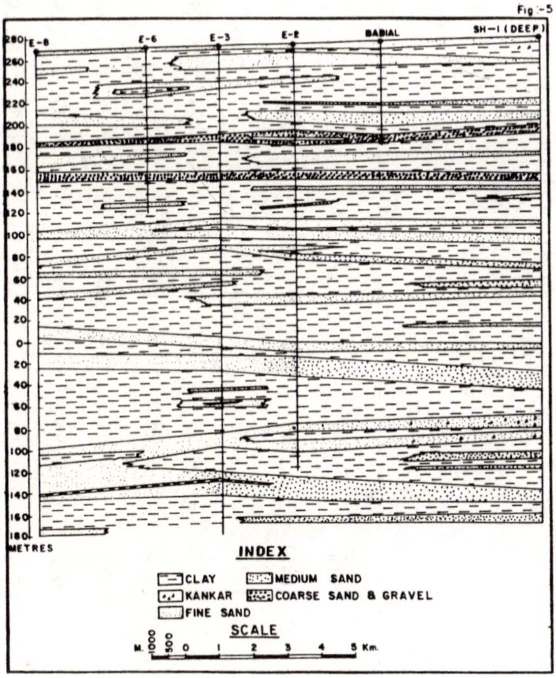

Fig.5. Sub-surface lithological cross-section along A-A' showing disposition of granular zones down to 450 m.

Table-2

Hydraulic properties of deeper aquifers in Ambala area

Bore-hole No.	Depth of tubewell (m.bgl)	SWL (m.bgl)	Discharge (m^3/day)	Drawdown (m)	Transmissivity (m^2/day) kb	Hydraulic conductivity (m/day) k	Storativity S	Hydraulic conductivity of semi-pervious confining layer (m/day) k'	Leakage Coefficient k'/m'
E - 1	426.0	17.9	914	6.71	1436	12.7	1.4×10^{-3}	-	-
E - 2	445.0	16.11	648	11.47	227	2.34	-	-	-
E - 3	420.0	16.14	1924	4.82	2514	24.89	8.7×10^{-3}	-	-
E - 4	447.0	18.15	2507	9.09	587	16.77	1.24×10^{-4}	2.3×10^{-4}	1.17×10^{-5}
E - 5	345.0	24.72	1978	4.47	1020	15.0	-	-	-
E - 6	150.0	24.89	1570	7.89	833	29.22	2.8×10^{-3}	3.2×10^{-2}	2.077×10^{-3}
E - 7	441.5	13.71	916	10.93	304	4.37	5.2×10^{-4}	-	-
E - 8	425.0	14.88	2191	4.54	1246	21.3	3.14×10^{-3}	-	-

leakage. The computed transmissivity of the deeper aquifer system varies from 227 m^2/day to 2188 m^2/day. High values in the western and southern parts of the area indicate high permeable aquifers which can sustain high capacity (heavy duty) tubewells. Hence the most favourable lay out of the well field will be in these zones from the court area (E8) in the west upto Dangri river passing through the areas where E3 & E5 are located covering a distance of 10 km (Fig.2). It is estimated that if a well field is developed along this zone by deep wells keeping 1000 metres distance between two wells with a pumping rate of about 2000 m^3/day, about 7.30 MCM water per annum can be withdrawn from this zone without any artificial recharge measures. The drawdowns at the centre of the well field of 10 deep wells in 10 km length is estimated to be 8.9 m based on F.M. Bochever formula (1969) for confined aquifer without hydraulic boundaries. This is far less than the maximum permissible drawdown i.e. 70% thickness of the aquifer tapped plus the pressure head above the top of the aquifer.

The formula reads

$$s = \frac{Q\Sigma}{4\pi Km} \ln \frac{2.25\,at}{R^2_{bw}}$$

Where $Q\Sigma$ = Total discharge of the well field in m^3/day
Km = Transmissivity of the aquifer in m^2/dayr
a = Water level conductivity in m^2/day
t = The time in days during which the well field is being pumped continuously without recharge i.e. period of operation of well field without recharge in days t = 10,000 days.
R_{bw} = Radius of one big well which will represent the entire well field and which is given by R_{bw} = 0.1 m where length of the well field 10,000 m in this case. Therefore R_{bw} = 0.1 x 10,000 = 1000 m.

HYDROCHEMISTRY

The concentration of chemical constituents in the ground water from shallow and deep aquifers are remarkably low and are well within the standards recommended by World Health Organisation (1984) for drinking water. The specific conductance varies between 316 and 523 and 566-944 microsiemens/Cm at 25°C in shallow and deep ground waters respectively. The fluoride content varies between 0.3 and 0.79 ppm and is within the permissible limit for drinking purposes.

The ground water from shallow aquifer is classified mostly as $Ca-HCO_3$ facies, whereas ground water from deeper aquifers is classified as $Na-HCO_3$ and $Na-Cl-HCO_3-SO_4$ facies.

CONCLUSIONS AND SUGGESTIONS TO AUGMENT WATER SUPPLY

The results of the investigation showed that the highly permeable granular zones, which can act as the most favourable zone for induced recharge, occur in the area around Khelan, Babial and Kojkipur along the riverine tract of Dangri river with average saturated thickness of about 14 metres. It is suggested to construct a well field in this area, with an average discharge of 726 m^3/day from each well. The well field should be laid parallel to the river on its western bank at a distance of 100 m from the river keeping spacing of 400 m between two shallow tubewells of 30 m depth. The corresponding drawdown in the centre of the well field is calculated to be 7.77 m. About 10890 m^3/day of ground water can be withdrawn during summer months which can be utilised for augmenting the water supply to Ambala urban area. This amount of water can be withdrawn without desaturing the zone tapped in shallow water table aquifer. It is expected that with intensive development, water levels will go down in these areas as the system tends to be in dynamic equilibrium thereby creating additonal storage space for induced infiltration. It has been estimated that a quantity of 0.833 MCM can be recharged by induced infiltration during monsoon season from the Dangri river. Further augmentation of water supply to Ambala town to the tune of 7.30 MCM of water per annum can be attempted by tapping the deeper aquifers in the depth range of 150 to 450 m bgl through a well field of 10 tubewells with a pumping rate of about 2000 m^3/day at each well.

REFERENCES

Firozuddin, T,G., 1980; Deep ground water exploration and possibilities of induced recharge from Dangri river, Ambala area (CGWB unpublished report).

Karanth, K.R., 1987; Ground water assessment, development and management, tata mac graw Hill

Singh, R.N. and Tandon, A.K., 1978; Geophysical investigation for induced recharge studies in Ambala Cantonment area - Ambala district, Haryana State and in parts of Patiala district, Punjab State (CGWB Unpublised report).

UNDP., 1983; Ground water studies in the Ghaggar river basin in Punjab, Haryana and Rajasthan - Artificial Recharge Studies, Vol.III.

INDUCED RECHARGE BY PERCOLATION TANKS IN PARTS OF MAHARASHTRA

Y.J. PARDHASARADHI
CENTRAL GROUND WATER BOARD

ABSTRACT

Many percolation tanks were constructed in Maharashtra state by impounding minor water courses. The average catchment area is 3.5 sq.km.with average gross storage of 200 TCM-both with a high coefficient of variation. There are about 475 such tanks in the Sina and the Man basins where a representative study is made by selective monitoring of tank levels, contents, ground water levels in wells, and evaporation rate. Evaporation on an average accounts for 10.7% of total depletion of tank contents. An average percolation rate of 45 mm/day was estimated for post monsoon period which may include the visible seepage appearing in water courses downstream. Those wells which are influenced by percolation show increased yield of 10 to 30%, by taking into consideration water levels and time factor. Some of the aspects of percolation tanks are outlined in the light of the findings.

INTRODUCTION

Large tracts of land in the Maharashtra state receive insufficient rainfall. Construction of percolation tanks had been taken up by the state agencies in such areas with a view to increase the groundwater potential and better performane of groundwater structures in respective command areas. Runoff from small individual catchments of about 3.5 sq.km is impounded by constructing suitable weirs and embankments.These reservoirs store on an average 200 TCM each of water. Such stored waters induce additional recharge for a prolonged period even after the monsoon activity ceases till the water is exhausted due to combined influence of percolation and evaporation. The quantitative aspects of the induced recharge needs continuous appraisal as considerable money has been invested in this endeavour.

An opportunity to study the percolation tanks in certain areas of the Maharashtra state arose as part of the Sina Man Project studies undertaken by the Central Ground Water Board between 1976 and 1980 (Pardhasaradhi 1980).

SINA AND MAN BASINS

Sina basin (11970 sq.km) and Man basin (4710 sq.km) form part of the upper catchment of the river Bhima in the Maharashtra state. These basins are included between N. latitudes 17°00' and 19°52' and E. longitudes 74°22' and 76°06'. Geologically the two basins are comprised of Deccan trap basalts. Successive subhorizontal flows of basalts are encountered which vary in thickness from 9 to 88 meters The sequence of each flow generally starts with a massive base and ends with a vesicular top. Weathered top of preceeding flow is presumed to be baked by the immediate successive flow resulting in the clayish red bole beds.

The basins experience semi arid climate receiving rainfall mainly between June and October every year from the southwest monsoon. The average annual rainfall was seen tobe 616 mm over the Sina basin and 520 mm over the Man basin (Rao etal. 1980).

Fig-1 Location of study percolation tanks in Sina and Man basins

332 percolation tanks in the Sina basin and 145 in the Man basin were in existence spread within the districts of Solapur, Sangli, Satara, Ahmednagar, Beed and Osmanabad. Runoff from small catchments of an average 3.5 sq.km. are impounded to fill reservoirs of average storage of 200 TCM. There is however a great variability in the catchment area as well as storage as reflected by the coefficients of variation which are 0.99 and 0.73 respectively. Recharge to groundwater storage is accentuated by, among other processes, percolation of stored water in tanks. The effect of induced recharge needs quantitative assessment both for the volume of additional water that percolates and the time span over which the percolation is effective. An attempt has been made to study some of these aspects of 7 percolation tanks in these basins (Fig.1). These are at villages Hotagi, Kasegaon, Shelgaon, Agalgaon, Babardi, Shivane and Mhaswad.

Method of study

Storage changes in tanks are effected by many factors - from surface runoff of catchment areas, by direct addition from rainfall,

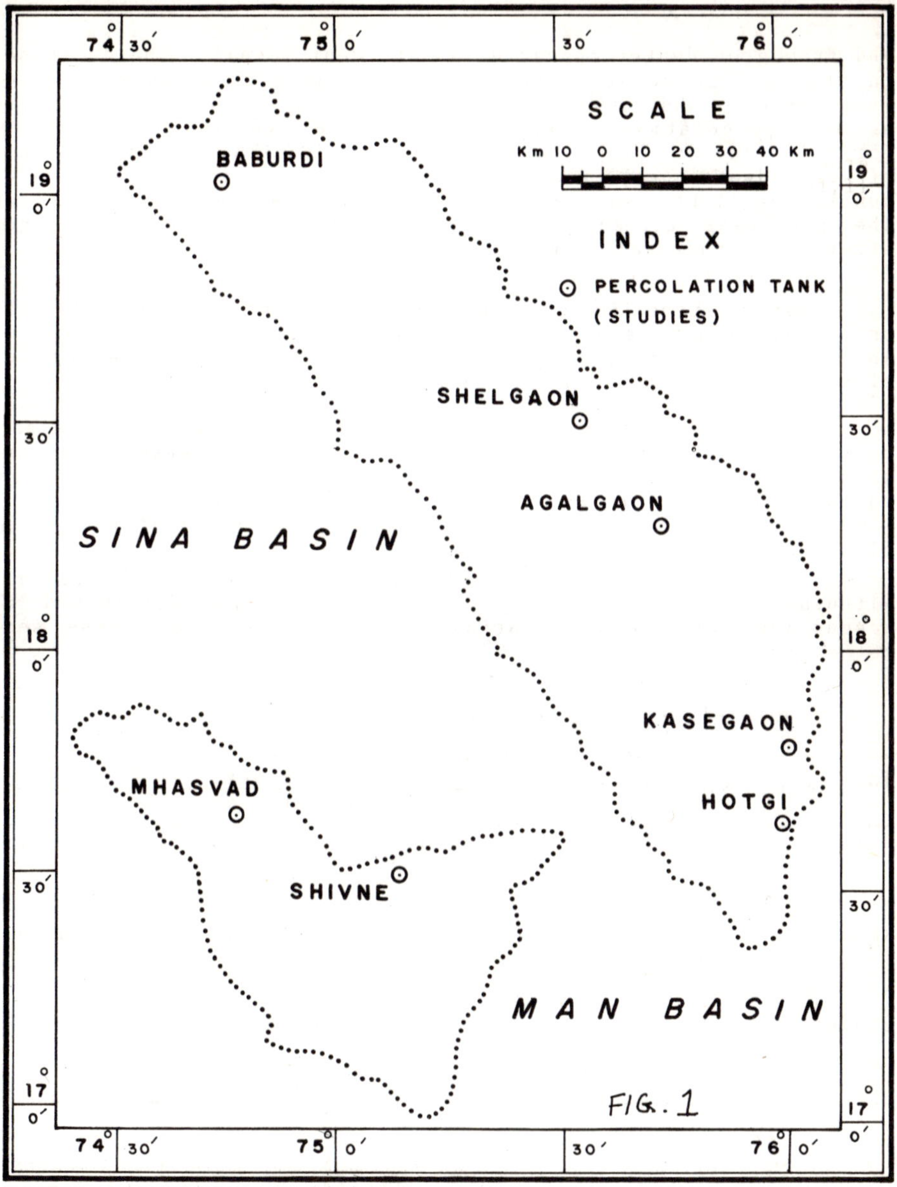

C. G. W. B., C. R. (Y. K. C.) Drg. No. 1 (a) / 89

and from groundwater appearing as effluent seepage. The decrease in storage can be accounted for as recharge to groundwater bodies, evaporation, and other usage through distribution works.Insofar as the percolation tanks are concerned the loss in storage is entirely due to percolation and evaporation. A certain amount of groundwater may reappear as visible seepage or groundwater runoff,for different lengths of time, downstream of the tanks. The processes of evaporation and percolation are more vigorous after the withdrawal of monsoon for obvious reasons.

Water stage in the tanks is recorded periodically either on the embankment slope or on the staff gauges specifically erected in the tank. Storage volumes are calculated from capacity curves of the tanks' survey.

Evaporation from reservoir surface is the other factor to be determined. Pan evaporation data from the nearest meteorological station were taken as representative. Tank evaporation is taken as 0.7 times of Pan evaporation rate observed.

FINDINGS

Tank stages are observed for different periods after the withdrawal of the monsoon. Seepage rates are determined after accounting for the evaporation. Ranges of some of these seepage rates are represented under table 1.

Table - 1
Percolation rates in selected tanks

Tank	Period of observation	Range in mm/day		
		Total loss	Evaporation	Seepage
Hotagi	14.8.78 to 31.3.79	44.3 to 65.7	3.8 to 7.6	36.8 to 61.7
Kasegaon	6.12.78 to 25.5.79	23.1 to 81.1	4.0 to 8.1	15.0 to 76.0
Shelgaon	21.12.78 to 19.5.79	47.1 to 62.1	4.0 to 7.8	42.5 to 54.3

Percolation rates are seldom uniform even within the same tank. Depiction of a set of percolation rates observed from different tanks in the Sina basin shows that the percolation rates are dominant particularly between 40 and 60 mm/day (Fig.2). The median percolation rate can be taken as 45 mm/day.

Fig-2 Percolation rates in selected tanks and
Fig-3 Percolation pattern in two tanks at Hotagi and Kasegaon

Figure-3 depicts the pattern of percolation volumewise for the tanks at Hotagi and Kasegaon. This includes that part of the percolation also which may reemerge downstream as effluent seepage.

Percolation influence

Increase in groundwater storage is reflected in the fluctuations in water table in respective command areas.The areal extent and magnitude of such fluctuations to be observed are dependent upon, among others, the volume of water seepage, hydraulic conductivity, extent, inter-relation of various aquifers concerned, irrigation practices and topographic factors like slope and occurrence of water courses.

A periodic study of fluctuations in water levels in selected observation wells vis-a-vis storage factor in tanks bring out the influence of the storage water over groundwater reservoir in space and time. The complex physiographic setting and hydraulic nature of aquifers involved, however, make no two areas similar in response to such a regime obtaining. Nevertheless a random approach of water level fluctuations in selected wells to establish, comparing these fluctuations with that of an independent well nearby, whether or not the response in the wells is appreciably greater than that in wells located outside the influence of the tanks.

FREQUENCY OF PERCOLATION RATES

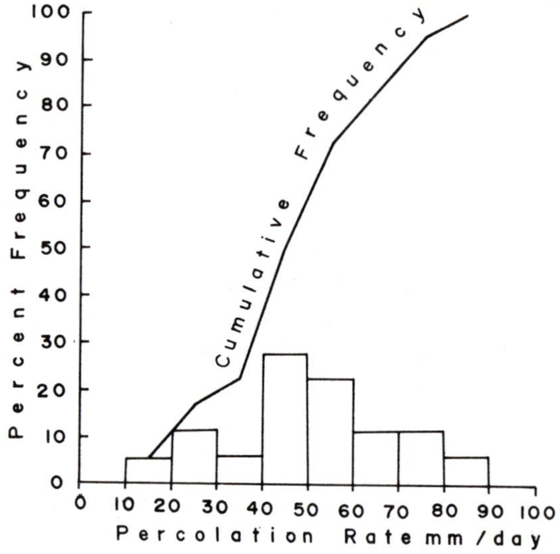

INFILTRATION PATTERN IN PERCOLATION TANKS

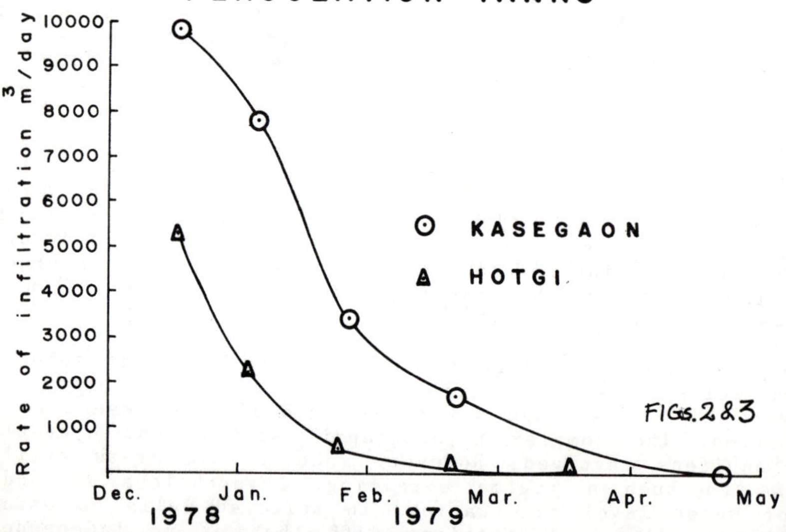

FIGS. 2&3

C.G.W.B. C.R. (Y.K.C.) Drg. No. 1(b)/89

Assuming that the specific yield is homogeneous within the vicinity of a tank, the vertical fluctuation in water levels and time factor involved directly suggest the difference in response in those wells. This aspect can be assessed by planimetering and comparing the areas contained within the hydrographs of individual wells. Representative values for some wells under Hotagi and Kasegaon percolation tanks are given under Table -2.

Table -2

Percolation tank influence on wells

	Planimetered area	Percentage increase over observation wells	Water level range BMP in meters.
Hotagi P.T.			
Hotagi O.W.	347	-	12.10 to 4.85
P.T.well-1	350	+0.9	7.55 to 1.35
P.T.well-2	457	+31.7	9.1 to 0.5
P.T.well-3	237	-31.7	5.5 to 0.7
Kasegaon P.T.			
Ule O.W.	542	-	11.75 to 3.40
P.T.Well 1	849	+56.6	9.35 to 0.20
P.T.well-2	600	+10.7	6.5 to 3.7

It may be seen that there is vast difference in the amount of additional groundwater avaiability even in wells within the command of a single tank.In some wells even there is absolutely no influence of the expected additional percolation.It however remains that hydrographs offer a valuable tool for such comparative studies. Feasibility and selection of suitable sites shall be attempted in the light of the above observations.

SUMMARY

Direct measurements of water recharged from percolation tanks is possible by accounting for evaporation from the total depletion. Study of seepage volumewise may not be representative in view of the high variability in storage capacities.An equally valuable alternative is to find out the column of water recharging for limited periods of time. In this method only the vertical fluctuations of the tank level are accounted for either as recharging or as evaporating. Within such a period it can be assumed that the surface spread will not vary to a significant level.

The percolation is observed to be generally less during the monsoon period and immediately thereafter. Reason for this could be the saturated condition of the aquifers and the invisible sub-surface inflow to the tank from the catchment.

Clogging of the pores in the tank bed by silt and clay and increased evaporation during the dry summer periods retard the recharge from the percolation tanks from March onwards.

Selection of percolation tank sites above viable aquifers and regular removal of silt from the tank bed before onset of each monsoon shall improve the performance of a percolation tank. Constructing a series of check dams across the stream course shall be an alternative to percolation tanks. This way submergence of useful land under reservoir is precluded while the effect of additional recharge is achieved.

ACKNOWLEDGEMENTS

The author is grateful to the Chairman, the Chief Hydrogeologist, and the Director (Central Region) of the Central Ground Water Board for their kind permission.

REFERENCES

Pardhasaradhi, Y.J.,1980. Hydrological studies in Sina and Man basins, Maharashtra. CGWB Unpub.Report.

Rao, C.V.V.S., Malhotra, V.J. and Gopal Prasad,1980. Hydrometeorological studies of Sina and Man basins, CGWB Unpub. Report.

ARTIFICIAL GROUNDWATER RECHARGE IN DROUGHT PRONE AREAS

SUBRAHMANYAM, K.M.[1], PRAKASAM, P[2]

Artificial recharge of groundwater has assumed significance in the various drought prone areas of Andhra Pradesh, India during the last decade. Most part of these drought prone areas have limited surface water potential compelling heavy dependence on the groundwater resources to meet various requirements. These areas are mostly underlain by hard rocks which form shallow aquifers of limited storage capacity and this coupled with erratic rainfall, characteristic of these areas, makes artificial groundwater recharge imperative.

Construction of percolation tanks has been carried out extensively in these areas at suitable locations selected based on scientific hydrological and geohydrological investigations. These tanks have been found to be economically viable and are of immense utility in augmenting groundwater resources and have gained such popular appreciation that the public in these drought prone areas demand for conversion of many of the existing minor irrigation tanks into percolation tanks. A series of check dams constructed can substantially augment groundwater recharge besides conserving soil and runoff in the watersheds.

I. INTRODUCTION

The Economy of Andhra Pradesh State is essentially dependent on agriculture. Since a major part of the area of this State is subject to erratic rainfall distribution both in time and space, harnessing of water resources for irrigation has been intensified since independence. Though the State is endowed with some major river systems, there are vast zones which suffer from paucity of surface water resources and as such groundwater is the only source from which the various requirements are to be met. Thanks to the increased availability of institutional finance, introduction of deepwell pumpsets and rapid progress in rural electrification, groundwater development has been intensified during the last two decades.

The ultimate irrigation potential of the State using surface water resources has been estimated as 8.5 million hectares. At the beginning of the First Five Year Plan, the total irrigated area in the State was 2.7 million hectares of which groundwater irrigation was in an extent of 0.3 million hectares. The total irrigated area now has an extent of

---

1. Director.
2. Senior Hydrologist

Ground Water Department, Hyderabad, India.

4.34 million hectares, while groundwater accounts for 1.1 million hectares.

Nearly 85 per cent of the State is underlain by hard rocks like granites, gneisses and other consolidated rocks. These are incidentally the areas which have limited surface water potential and where demand for groundwater development is maximum. The weathered, fractured and jointed zones in these rocks form aquifers but they are shallow and of limited storage capacities. Owing to the above, the recent intensfication of groundwater development, especially in drought prone areas of the State, has resulted in continuous decline of groundwater levels rendering a huge number of open wells in these areas dry. The farmers are thus compelled to deepen the wells and drill deep borewells and in-well-bores, but this can only offer a temporary relief, the recharge to groundwater basins remaining constant. The Government of Andhra Pradesh recognised the urgent need to reverse this unhappy trend and took up on a massive scale, schemes to accomplish artificial groundwater recharge through construction of percolation tanks and checkdams during the last decade. Further, the National Water Policy, which was adopted by India during 1987 lays emphasis on this important programme by stating that "Groundwater recharge projects should be developed and implemented for augmenting the available supplies".

II. ARTIFICIAL RECHARGE TECHNIQUES

Artificial groundwater recharge projects aim at augmenting the natural infiltration of precipitation or surface water into underground formations by some method of construction, spreading of water or by artificially changing natural conditions. Schemes contemplating artificial recharge as the sole objective can be planned or in the alternative artificial recharge can be realised as an objective incidental to certain other developmental programmes. For instance, soil conservation practices, which aim at conservation of fertile top soil and improved agricultural practices intended to maintain proper soil structure incidentally aid in reduction of surface runoff and maintain high infiltration rates and thus augment groundwater recharge.

Artificial recharge can be accomplished through various spreading methods, through injection wells or through pits and shafts. Spreading methods include flooding method, natural channel method etc. The local field conditions including the physiography and hydrogeology of the area have to be carefully considered before selecting a method for artificial recharge. The observations made and experiences gained from the artificial recharge schemes implemented in Andhra Pradesh are discussed briefly in this paper.

III. RECHARGE THROUGH PERCOLATION TANKS

A percolation tank is a small tank with storage capacity

in the range of 0.14 to 0.57 million cu.m. and formed by the construction of an earthen bund across a stream at a suitable location upstream of the zone which needs to be recharged. Selection of the site for formation of a percolation tank is preceded by extensive hydrological, geohydrological and hydrogeological investigations. Some of the criteria for preliminary selection of sites are as follows.

1. There should be adequate scope to accommodate, in the aquifer, the augmented recharge from the proposed percolation tank. This implies that the groundwater levels in the influence zone are deep enough and should not be less than 3 metres below ground level during post-monsoon season and weathering extends to a fairly good depth and is extensive.

2. There should be substantial number of working wells in the influence zone and the well density should atleast be 2 wells per sq.km.

3. There should be no springs and seepage zones in the influence zone of the tank.

4. Soils in the water spread area should have good structure to permit reasonable percolation rates.

5. There should not be extensive outcrops of hard massive rock, intrusive bodies like dykes across the general direction of seepage immediately down stream of the bund.

6. The influence zone of a percolation tank may be approximately taken as 6.4 km. along the stream draining the area and 0.8 km. across for preliminary study. However, the effective influence zone shall be delineated during detailed hydrogeological study of the area. The influence zone may be limited to the area with topographic gradient not exceeding 3 per cent, apart from the hydrogeological considerations such as existence of adequate thickness of weathered zone and fractured zones. The presence of massive outcrops, change of formation, intrusive bodies and streams of higher order across the general direction of seepage, abrupt changes in ground slope, springs and seepage zones should be taken into account to limit the influence zone.

As a part of the geohydrological and hydrogeological study for clearance of the sites to form percolation tanks, complete inventory of the existing wells is made and the geological and geomorphological mapping of the area is carriedout. Infiltration tests at representative locations in the water spread area of proposed tank are carriedout to assess the infiltration capacity of the soils. Soil samples at these test sites should be collected for laboratory analysis and pumping tests on some representative wells in the influence zone should be conducted to estimate the aquifer characteristics, specific capacity and recouperation rates of the wells. A minimum recouperation rate of 0.0014 cum. per second is considered desirable for hard rock aquifers.

It is necessary that a preliminary estimate of the possible augmentation of recharge that can be effected from the proposed percolation tank is made so that the most ideal sites are takenup for construction in needy areas. Indiscriminate construction of percolation tanks will not only result in poor returns on the investments made but also cause certain adverse effects such as water logging, deterioration of groundwater quality, environmental degradation etc.

The Ground Water Department investigated so far 830 sites for formation of percolation tanks and cleared 638 sites. The Irrigation Department constructed, during the last 6 years alone, 316 percolation tanks and 691 checkdams in the Statefor accomplishing artificial recharge. More than 80 per cent of these structures were constructed in the drought prone areas of Rayalaseema Region.

IV.RECHARGE FROM MINOR IRRIGATION TANKS CONVERTED INTO PERCOLATION TANKS

People of these drought prone areas have recognised the utility of these structures so much that there is a popular demand to form more such structures in these areas and to convert the existing minor irrigation tanks into percolation tanks. However, conversion of minor irrigation tanks into percolation tanks has to be takenup with due caution owing to the following reasons.

1. The bunds of minor irrigation tanks are generally founded on hard massive rock outcrops or on shallow basement or along massive dykes to prevent seepage losses and to ensure better foundation stability. Further, the cutoff provided below the bund inhibits seepage. This problem can be overcome by construction of one or two small percolation ponds or a series of checkdams in the influence zone at favourable locations and periodically release water into them from the main percolation tank during non-monsoon periods.

2. If the silt deposit in the water spread area is thick and comprises of fine grained material, seepage rates into the influence zone will be greatly reduced. To ensure reasonable percolation rates, it is recommended to make trenches in the water spread close to the bund to penetrate into the weathered aquifer zone and then fill them with gravel. It is necessary that the top layer of the gravel in these trenches is replaced periodically. Periodical desilting of the percolation tanks is essential to maintain the infiltration rates. As desilting is an expensive affair, the farmers should be encouraged to use the silt in the water spread area for manufacture of bricks, modification and enrichment of the soils in the agricultural fields etc., and some incentives in the form of concessional transport or mechanised excavation may be provided.

Theoretically speaking, potential recharge rate at any given moment in the ayacut area of a minor irrigation tank should be more compared to that in the influence zone of a percolation tank constructed at the same location. This is for the simple reason that recharge will be effected not only from tank storage as underseepage from the bund but also from applied irrigation over the entire command.

However, in practice, in shallow hard rock areas, groundwater build-up in the ayacut will take place within a short time after the tank gets filled in and this potential recharge cannot be mobilised in the absence of a scientific conjunctive use programme implemented in the command. It is hard to expect, in the prevailing system, the farmer to voluntarily resort to conjunctive use when adequate canal water is available. What is needed is a self-imposed discipline by the local community, by reducing the use of tank water and make it up by pumping groundwater. This demands active involvement of the farmers concerned who need to be educated properly of the utility of these artificial recharge programmes and the need to conserve the precious water resources through judicious water management practices, adopting modern irrigation systems, switching over to less water intensive crops etc.

The ultimate objective should be to treat and operate most of the existing minor irrigation tanks ideally as 'minor irrigation-cum-percolation tanks'.

V. RECHARGE THROUGH CHECKDAMS

A series of checkdams constructed in cavernous limestone areas at appropriate locations in the recharge zones based on detailed hydrological and hydrogeological investigations aid greatly in groundwater recharge. Checkdams are also effective in other hardrock areas with steep slopes and they serve the twin purpose of groundwater recharge and soil conservation. However, in zones with gentler slopes, percolation tanks generally perform better than checkdams and are also cost effective.

VI. PERCOLATION CANALS

A bold and successful artificial recharge experiment conducted during 1979 and repeated periodically since then, in Mutchukota valley in eastern parts of Anantapur district in the State, involved release of water through an irrigation canal and a net work of natural stream courses.

The Mutchukota valley is situated between two hill ranges and bounded by River Pennar in the north and river Chitravati in the south. The valley comprises of two water sheds, which drain into Pennar and Chitravati Rivers. The area is underlain by Tadipatri shale of Cuddapah Group. The groundwater levels in the area during pre-project period were very deep and open wells, which are as deep as 30 metres are not uncommon. The area is characterised by a semi-arid climate and rainfall is highly erratic. The valley is ideally suited for growing citrus, which is a highly remunerative perennial crop but which needs a reliable irrigation supply during certain critical periods. Water was let out for the first time during 1979 into Distributary No.29 of the Tungabhadra Project High Level Canal Stage-I and into the network of natural stream courses in the valley from the storage in the Subbarayasagar balancing reservoir, which is a part of the above project. It was observed that the groundwater levels rose substantially and to the extent of about 12 metres and the citrus gardens in an extent of about 400 hectares in the valley which were about to be lost due to water shortage were saved. Encouraged by this recharge experiment, numerous additional wells have came up in the area and artificial recharge is being accomplished during summer seasons since then. the hydrographs of two observation wells in this area shown below demonstrate the extent to which groundwater build-up can be effected through artificial recharge.

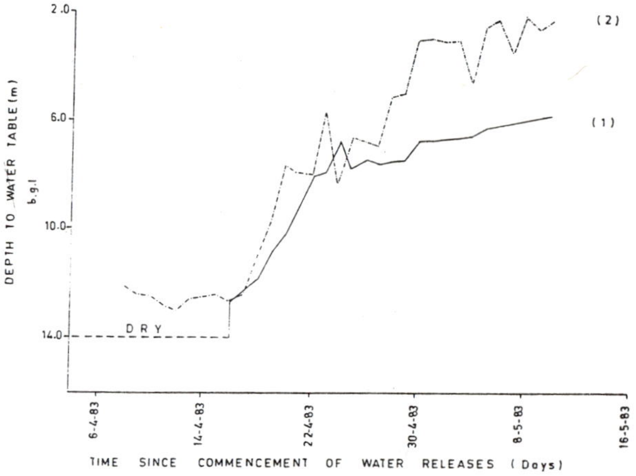

Fig.1 Groundwater Build-up due to recharge from Percolation Canal, Anantapur district.

VII.ARTIFICIAL RECHARGE IN THE TAILEND AREAS OF IRRIGATION PROJECT COMMANDS.

There are extensive zones in the tailend reaches of the Tungabhadra Project Complex Command area in Cuddapah and Anantapur districts in Andhra Pradesh, where chronic deficiency of water supply or even non-receipt of canal supplies is experienced. The farmers in these zones are thus compelled to go in for extensive groundwater use in these areas. As a result of the above and owing to the fact that these areas are located in chronically drought prone areas, there has been a constant decline in groundwater levels. The hydrographs presented below pertain to the observation wells in a shallow alluvial aquifer along river Pennar in Cuddapah district.

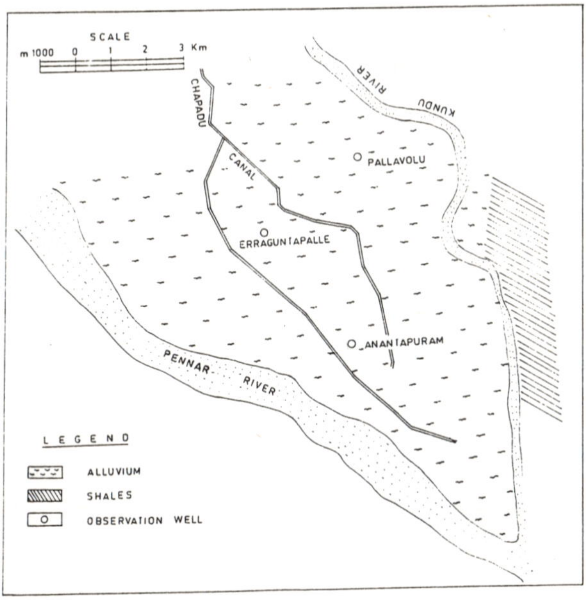

Fig.2 Location map of tail end reaches in command area, Cuddapah district.

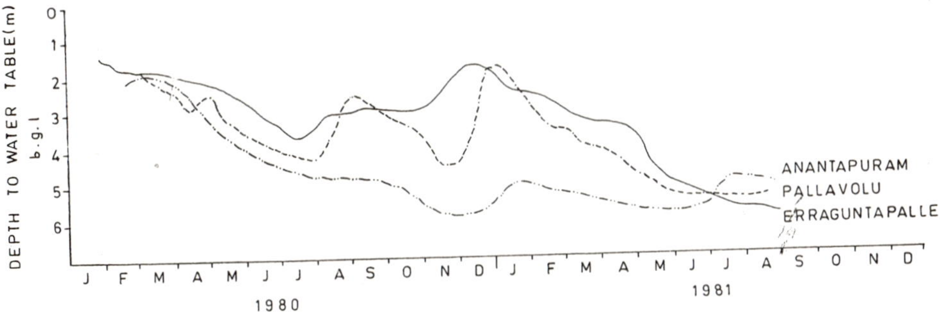

Fig.3 Groundwater decline in Tail end reaches of command, Cuddapah district.

After a detailed geohydrological and hydrogeological study of the area it was recommended to utilise the existing canal system in an extent of 1600 hectares as recharge canals and feed them once in a fortnight solely for groundwater recharge. This recommendation is under consideration.

VIII. CONCLUSIONS

Artificial recharge of groundwater has been accomplished successfuly at numerous sites in the drought prone areas of the State of Andhra Pradesh during the last decade by means of percolation tanks and checkdams. These areas are underlain by hardrock aquifers with limited storage capacities and are subject to erratic rainfall and experience high potential evaporation rates. Recharge through percolation tanks and checkdams has proved to be appropriate under such conditions. Percolation canals have also served well under certain conditions, to recharge the aquifers in the State.

REFERENCES

Subrahmanyam, K.M. et al, 1982. Percolation tank as an aid to groundwater recharge: A review of geohydrological studies. Proceedings of Seminar on Hydrological investigations during the last 25 years in India, Association of Hydrologists of India, May, 1982, P.205-209.

Todd, D.K., 1959. Groundwater Hydrology, John Wiley and Sons Inc.

1979, Manual on formation of Percolation Tanks in Andhra Pradesh, Government of Andhra Pradesh.

HYDRAULIC FRACTURING TEST FOR GROUNDWATER SIMULATION IN A KARSTIC LIMESTONE AQUIFER - HATTINGEN/BADEN-WEURTTENBURG, WEST GERMANY

R. JUNG* and S. SANKARAN+

* Bundesanstalt fur Geowissenschaften und Rohstoffe
Stilleweg 2, Postfach 510153, 3000 Hannover 51, West Germany

+ National Geophysical Research Institute, Hyderabad-500 007

ABSTRACT

In areas of low permeable rocks, groundwater is mainly transported through bedding planes, natural joints and fracture zones. Because of the uncertainity to locate such potential zones boreholes drilled in such areas are often poor yielding. Application of hydraulic fracturing in low yielding boreholes is a potential tool in creating new fractures and connect it to the water zones around the borehole, if any.

The principle of hydraulic fracturing is essentially in applying high hydraulic pressure to the walls of selected zones in a borehole until failure of the surrounding rock occurs. The experiment is continued in order to create new, highly permeable flow channels which are intended to lead to increase the yield of the well.

The hydrofracturing tests was carried out at a depth of about 100 m in a borehole near Hattingen located in karstic limestone. The injection unit has a pump with a maximum pressure of 300 bars and a maximum flow rate of 400 l/min. The fractures were initiated by injecting water at low flow rates followed by higher for fracture propagation. The hydraulic parameters viz., transmissivity, storativity, fracture width and fracture area of the induced fractures were studied after the end of each injection phase by observing the pressure decline during the respective shut-in periods.

From the water level measurements before and after the test it could be concluded that it is possible to connect a non productive borehole in karstic limestone to water carrying joints by inducing hydraulic fractures. The analysis of the data show that the size of the fracture and the length of the intersection between the induced fracture and the natural joints could be increased after establishing the connection by injecting water with higher flow rates. The maximum pressure required to initiate fractures was 38 bars. A maximum width of 0.55 mm at fracture propagation was achieved with a flow rate of 6.64 l/min. The maximum transmissivity achieved was 14.2 DM (Darcy Meter).

The exponential decay of the fluid pressure after shut-in was analysed to estimate the hydraulic parameters. The estimated residual transmissivity and fracture width were too low. This limited the specific yield to a value too low for large scale groundwater supply. To achieve higher transmissivity and fracture width, acidizing the fracture or use of gels with or without propping materials are possible solutions.

INTRODUCTION

In areas with low permeable rock groundwater is mainly transported in natural joints and fracture zones. Because of the uncertainity to locate high permeable joints boreholes drilled in such areas are often poor yielding. Application of hydraulic fracturing in dry or low yielding boreholes is a potential tool in creating new fractures and connect it to the water bearing joint system (WILLIAMSON, 1982).

The present hudroofrac-test was an attempt to stimulate the borehole near Hattingen and connect it to a water bearing fault suspected in the neighbourhood of the borehole. The objective of the stimulation was to investigate the extent by which the specific yield could be increased using water as stimulant, and to possibly use this borehole for long term monitoring of water level.

LOCATION, GEOLOGY AND EXPERIMENTAL SETUP

The borehole near Hattingen is located 1 km SE of a series of sinkholes along the river Donau (Fig. 1). River water flows through these sinkholes (698 m NN) and discharges to the Aach well (475 m NN) about 5 km SE of the borehole at a rate of 10 m /s. In contrast to the observed water level of about 90 m (bgl) in other boreholes, the water level in Hattingen was only 23 m (bgl) which is mainly due to the poor connection of the borehole to the karstic aquifer system.

The major geological formation is jurrasic limestone and the investigated section of the borehole belongs to the oxford limestone series of karstic nature. The lithology of the borehole is shown in Fig. 2. The total depth of the borehole is 150 m. Fig. 3 is a schematic representation of the experimental setup. The test intervals were sealed off by a straddle packer inflated with a fluid pressure of about 100 bars. To allow packer inflation and to change from inflation to injection a pressure actuated valve installed below the top packer was used. The packers were hydraulically connected to the surface by a steel tubing with a maximum pressure of about 500 bars and an inner diameter of 50 mm. A flexible tank of 50 m served as water reservoir. The pump used for fracture propagation and hydraulic testing has a maximum pressure of 300 bars, its maximum flowrate being 400 l/min. The flowrate was controlled by a flow controlling valve between 0-400 l/min, and measured by magneto-inductive flow meters with a resolution of 1. Fluid pressure was measured at the wellhead and in the test interval. The pressure sonde placed in the test interval had a piezo resistive pressure transducer with an accuracy of 0.5%. The wellhead pressure was measured by straingauge pressure transducers with an accuracy of 0.25%. All data were continuously recorded by 6-channel tape recorder and by a computer controlled data acquisition system, consisting of a data logger, a computer and a plotter. More details are given by REICH et al., (1986).

HYDRAULIC TESTS AND RESULTS

Two series of hydrofracturing tests were performed at intervals between 87.1 m - 88.7 m and 101.8 m - 103.3 m. In both intervals fractures were initiated by injecting small volumes of water at low flow rates and propagated further during subsequent injection periods with higher flow rates and volumes. The hydraulic parameters of the induced fractures were studied after the end of each injection phase by observing the pressure decline during the respective shut in periods. A low rate injection test was conducted between these two fracture tests to conform the results of the first test series.

Figs. 4 and 5 show the pressure and flow rate recordings of the two frac test series. Several parameters were derived from the shut-in periods of these records as illustrated in Fig. 6. In this gigure fluid pressure is plotted in logarithmic scale as function of time. The exponential decline of the fluid pressure observed during shut-in and the low water level at the end of the shut in period indicates that a hydraulic connection between borehole and karstic aquifer was achieved. The exponential pressure decline during shut in were analysed by using the model described in the Appendix 1. The transmissivity of the fracture, Tc at frac-extension and TH at hydrostatic pressure, their respective fracture widths (wc and wH) and the fracture area (A) were calculated by using equations given in Appendix 2. The results of the fracturing test between 87.1 m - 88.5 m are listed in Table 1. The shut in pressure Psi of about 26 bars measured after each injetion phase is equal to the normal stress acting on the fracture plane. It can be observed that the storage capacity (dV/dp)1 of the fracture and the fracture area increased during each injection phase. This indicates that due to the increased flow rate in the later injection periods the fracture size increased even after achieving a hydraulic connection to the karstic aquifer system. In contrast to the transmissivity Tc at frac-extension pressure, which increased with increasing flow rate, the residual transmissivity TH at hydrostatic pressure remained constant. A similar phenomena is observed for the fracture widths.

In Fig. 7 the variation of the residual specific yield with flow rate is depicted. Similar to the transmissivity of the fracture the specific yield C2 of the intersection between fracture and borehole remains constant. The specific yield C1 of the intersection between fracture and karstic aquifer increases with increasing flow rate indicating an increase in the length of the intersection.

A low rate injection test with two different flow rates was carried out to evaluate the specific yield of the well after the first series of frac experiments. The pressure record of this test is shown in Fig. 8. The fluid injection was continued until steady state conditions were reached. The specific yield obtained from this test as given in Table 1 is about two times higher than the specific yield values from the shut in periods. Though the reason for this discrepancy is not clear, nevertheless it is attributed to the error in the absolute values derived from the shut in curves. This does not affect the conclusions arrived at.

th ath anh and and the dip of the induced fractures were evaluated from the fracture imprints (Figs. 9 and 10) obtained by an impression packer after the frac experiments. Inspite of several natural joints vertical fractures were induced in both cases striking almost N-S and NNW-SSE respectively. This coincides with the direction of the maximum horizontal stress in that area (RUMMEL and JUNG, 1975).

SUMMARY AND CONCLUSIONS

It could be proved that it is possible to connect non productive borehole in karstic limestone to water carrying joints by inducing hydraulic fractures. The analysis of the data show that the size of the fracture and the length of the intersection between the induced fracture and the natural joints could be increased after establishing the connection by injecting water with higher flow rates. Accordingly the interconnection between induced fracture and natural joints could be improved. On the other hand the residual width and transmissivity of the fracture were not increased. This limited the specific yield to a value which is by at least one order of magnitude too low for water supply purposes. Further investigations should concentrate on methods to achieve higher fracture widths and fracture transmissivities. Possible solutions are acidizing or to use gels with or without propping materials.

ACKNOWLEDGEMENTS

The experiments were carried out in collaboration with the Geologisches Landesamt Baden-Wurttemberg. Our sincere thanks are due to Dr. Villinger and Dr. I. Stober of BGR. Also we thank the Wasserwirtschaftsamt rottweil (Mr. Haupt) for their help and for long term water level recording. We are grateful to Prof. Dr.H.-J. Durbaum for his encouragement and keen interest. Financial support was given from BGR funds, Energie and Rohstoff-versourgung No.53 208. The evaluation and analysis of the above experiment was a part of work carried out by Mr. S. Sankaran during his study period as DAAD stipendiat in Federal Republic of Germany.

REFERENCES

JUNG, R., 1981: Propagation and Hydraulic Testing of a large unpropped Hydraulic Fracture in Granite. - Geol. Jb., E 39: 37-65, Hannover.

REICH, W., HENNING, H., PIECHOTTA, G., and JUNG, R., 1987: Technical facilities at the Falkenberg Test Site. - Geol. Jb. E 39: 29-36 Hannover.

RUMMEL, F., and JUNG, R., 1975: Hydraulic Fracturing Stress Measurements near the Hohenzollern Graben-structure SW Germany. - Pageoph. v. 113.

WILLIAMSON, W.H., 1982: The use of hydraulic techniques to improve the yield of bores in fractured rocks. - Australian water resources council, conference ser. no. 5.

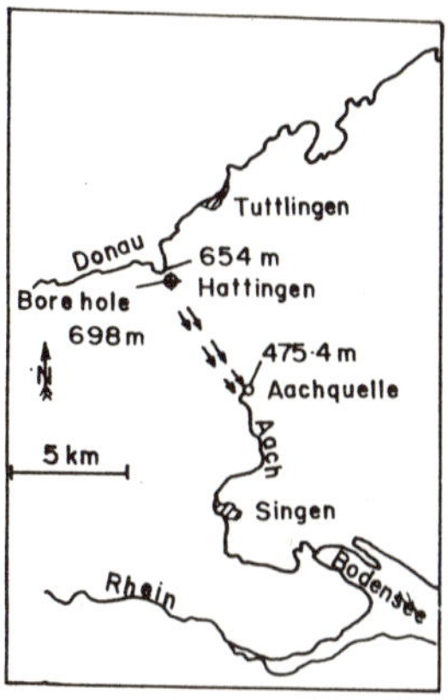

Fig.1: Location and hydrogeological condition

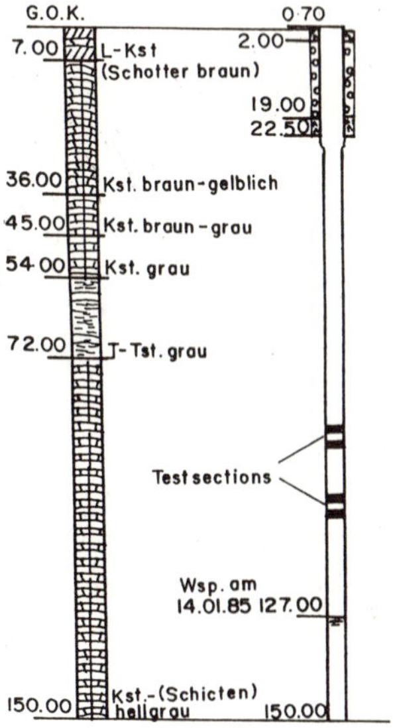

Fig. 2. Lithological section and test intervalls
Borehole diameter 150mm (open clay)
Kst. Limestone, T: Clay, Tst: consolidated hard clay

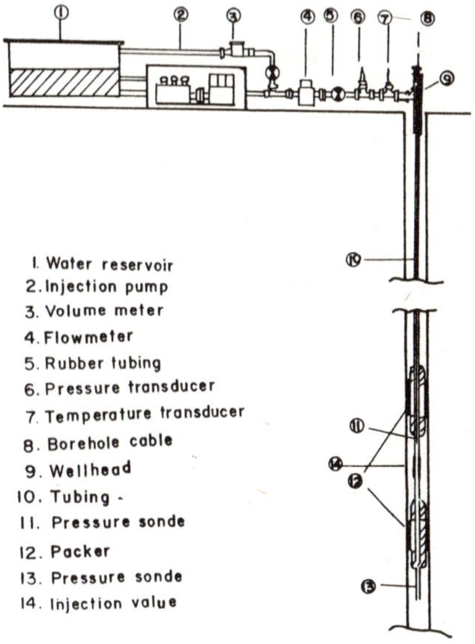

Fig. 3. Experiment set-up

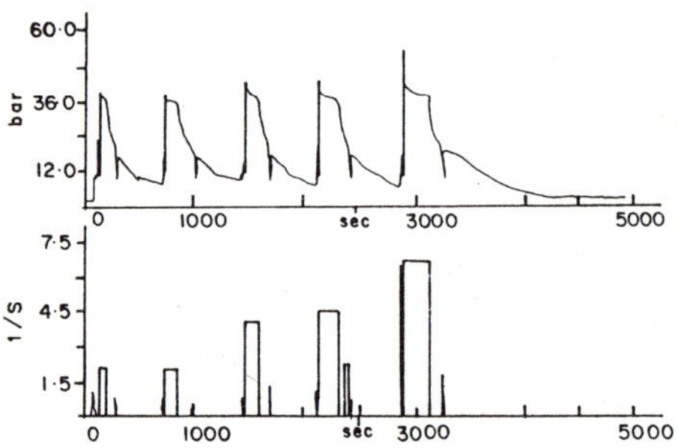

Fig.4 Records of fracture propagation test 86 OCT 16 at interval 87·1 m — 88·5 m

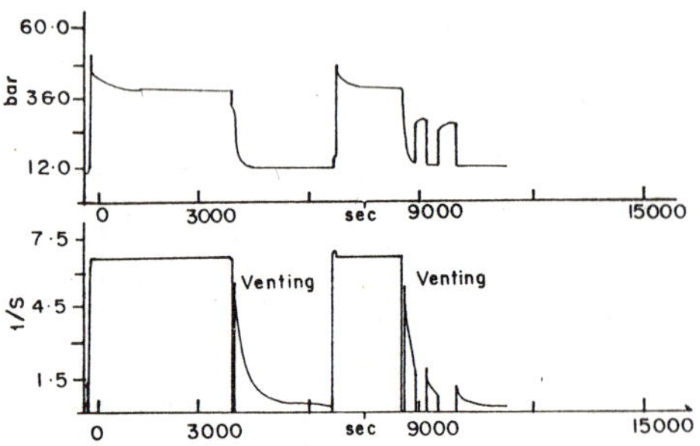

Fig. 5 Records of fracture propagation test 86 OCT 20 at interval 101·8 m — 103·3 m

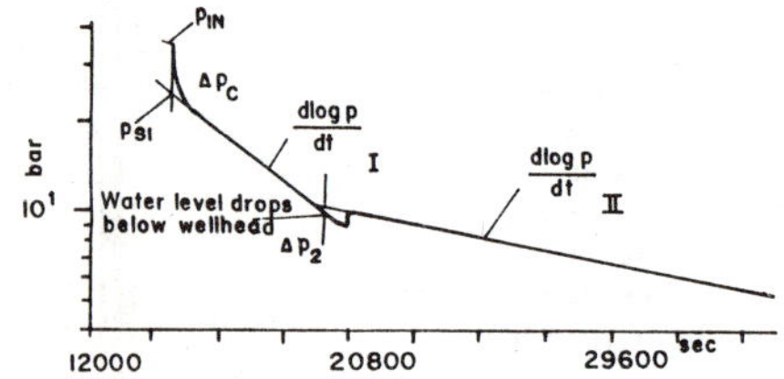

Fig. 6 Half-logarithmic plot of the down-hole pressure during shut-in.

P_{IN}: Iinjection pressure, p_{si} : shut-in pressure

Δp_2: pressure difference between borehole and fracture (S.App.1)

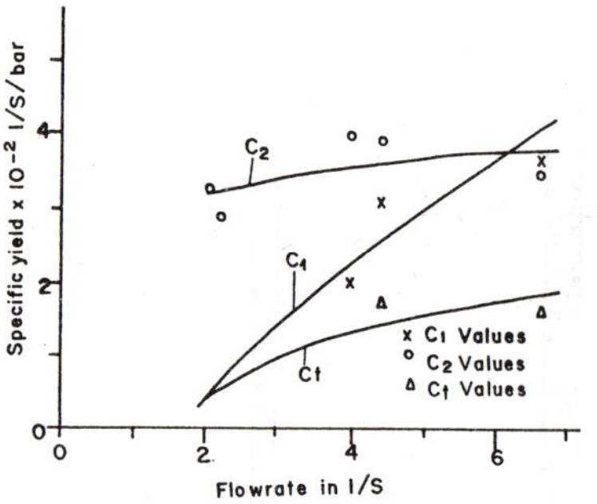

Fig.7: Variation of specific yield with flowrate

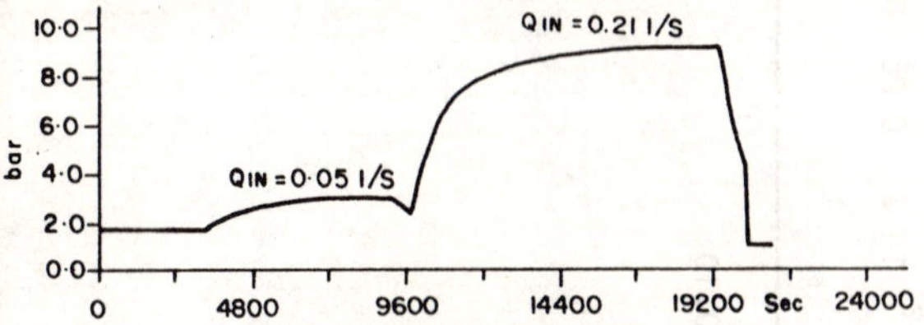

Fig. 8 Record of dowhole pressure for low-rate injection experiment.

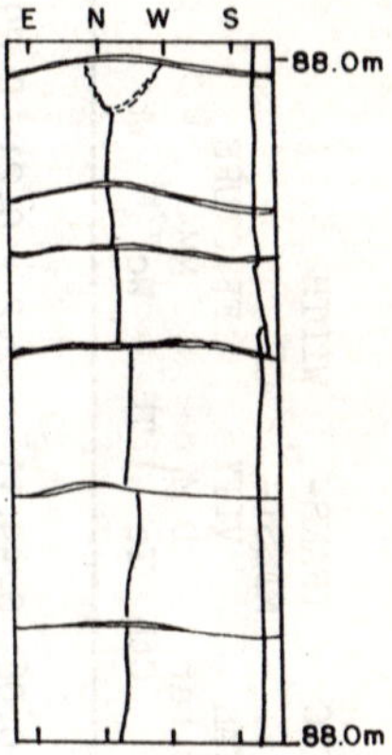

Fig. 9 Fracture imprint (first test series) Horizontal lines are natural joints (filled with clay)

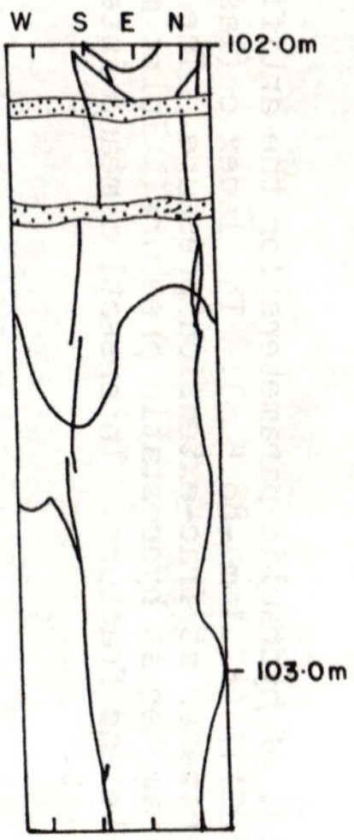

Fig. 10 Fracture imprint (second test series). Horizontal layers are two major joints filled with clay.

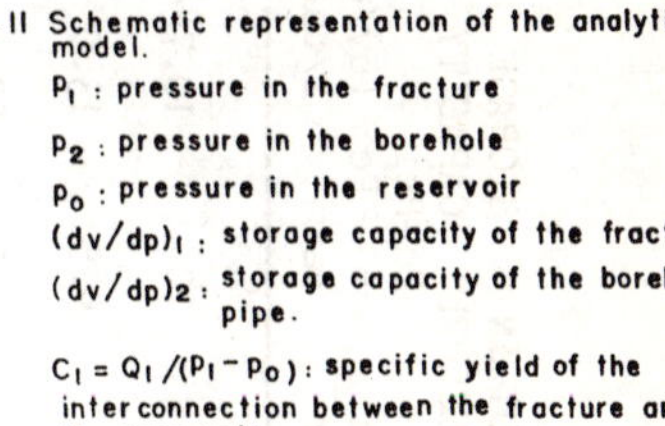

Fig. 11 Schematic representation of the analytical model.

p_1 : pressure in the fracture

p_2 : pressure in the borehole

p_o : pressure in the reservoir

$(dv/dp)_1$: storage capacity of the fracture

$(dv/dp)_2$: storage capacity of the borehole pipe.

$C_1 = Q_1/(p_1 - p_o)$: specific yield of the interconnection between the fracture and the reservoir

$C_2 = Q_2/(p_2 - p_1)$: specific yield of the interconnection between the borehole and the fracture

PHASE	FLOW RATE Qin l/s	TOTAL VOLUME Vl l	INJECT. PRESSURE Pc bar	SHUT IN PRESSURE Psi bar	Pc-Psi = Pc bar	STORAGE CAPACITY (dv/dp)1 l/bar	FRACTURE AREA F M 10	SPECIFIC YIELD INTERSECTION l/s.bar Clc	C1h	SPECIFIC YIELD INLET l/s.bar C2c	C2h	SPECIFIC YIELD TOTAL l/s.bar Ctc	Cth	TRANS-MISSI-VITY D.M Tc	Th	WIDTH OF FRACTURE MM Wc	Wh
1	2.14	1370	36	26	10	10.7	2.2	0.09	$0.51.10^{-2}$	0.21	$2.9.10^{-2}$	0.06	$0.44.11^{-2}$	2.2	0.21	0.29	0.14
2	2.0	2240	36	26.5	9.5	13.0	2.6	0.09	$0.45.10^{-2}$	0.21	$3.3.10^{-2}$	0.06	$0.40.10^{-2}$	2.1	0.24	0.29	0.14
3	4.0	5120	36	26.5	9.5	50	10.0	0.18	$2.0.10^{-2}$	0.42	$4.0.10^{-2}$	0.12	$1.30.10^{-2}$	6.4	0.29	0.42	0.15
4	4.41	7760	35.8	26	9.75	73	14.5	0.19	$3.1.10^{-2}$	0.45	$3.9.10^{-2}$	0.13	$1.70.10^{-2}$	7.4	0.29	0.45	0.15
5	6.64	15580	37.4	26.8	10.6	73 (assumed)	14.5	0.27	$3.7.10^{-2}$	0.63	$3.5.10^{-2}$	0.19	$1.60.10^{-2}$	14.2	0.26	0.55	0.15
6	0.05/0.21													2.90.10			

Table 1: List of hydraulic parameters for the artificial fracture from experiment 860CT17 (87.1 m -88.5 m). The index c (last columns) means value of the parameter at frac-extension pressure. The index H means value of the parameter at hydrostatic pressure. Inlet means penetration between borehole and fracture. Intersection means intersection between fracture and joint.

APPENDIX 1

Analytical solution for parameter estimation from the pressure decline during shut-in.

The analytical solution for the system response after hydrofrac test depicted as pressure time-plot in Fig. 6 can be derived by representing the hydraulic condition of the system as a model shown in Fig. 11. The exponential decline of the fluid pressure (Fig. 6) during shut-in can be explained by an analogy of an electrical model consisting of two capacitors each discharging through a constant resistor. The first capacitor representing the fracture has the capacitence analogous to the storage capacity as $(dV/dp)_1$. The second capacitor represents the storage capacity of the hydraulic system $(dV/dp)_2$ (pipe, pump and test interval). Due to the fact that the water level drops below the well head during shut-in the storage capacity of the hydraulic system has two different values.

For case 1 where the water level is at well head the storage capacity $(dV/dp)_2$ is given by the compressibility of the hydraulic system. For case 2 where the water level is below the well head the storage capacity $(dV/dp)_1$ is equal to the change of the fluid volume in the tubing for unit change in pressure. The two resistors represent the flow resistance between the fracture and aquifer (c_1) and between borehole and fracture (c_2).

Case 1: **Water level at well head.** $Q_2 \approx 0$

$$Q_1 = dV/dp \;.\; dp/dt$$

$$c_1\,(p_1 - p_o) = (dV/dp)_1 \;.\; dp/dt$$

$$\text{for } p_o = 0,\; p_1 = 1/c_1 \,.\, (dV/dp)_1 \,.\, dp_1/dt \qquad (1)$$

The solution for the above differential equation can be written as :

$$p_1 = p_{10} \cdot \exp\left(\frac{-c_1}{(dV/dp)_1}\, t\right)$$

$$\text{i.e.} \quad \frac{d\,(\log P_1)}{dt}\,\mathrm{I} = -\frac{c_1}{(dV/dp)_1}\log e \qquad (2)$$

Case 2 : Water level below well head. $Q_2 > 0$

$$Q_2 = dV_2/dt = (dV/dp)_2 \,.\, dp_2/dt \qquad (3)$$

$$Q_1 = dV_1/dt + dV_2/dt$$

$$= (dV/dp)_1 \,.\, dp_1/dt + (dV/dp)_2 \,.\, dp_2/dt \qquad (4)$$

also $\quad c_1 = Q_1/(p_1 - p_o) \quad$ and $\quad c_2 = Q_2/(p_2 - p_1)$

Hence equation (3) can be written as :

$$c_2(p_2 - p_1) = (dV/dp)_2 \,.\, dp_2/dt \quad \text{and}$$

equation (4) as :

$$c_1(p_1 - p_o) = (dV/dp)_1 . dp_1/dt + (dV/dp)_2 . dp_2/dt$$

At $p_o = 0$, $p_1 = 1/c_1 . (dV/dp)_1 . dp_1/dt + (dV/dp)_2 . dp_2/dt$

It is assumed thathe pressure difference between the bore-hole and fracture changes much slower than that between fracture and reservoir. Then $dp_2/dt = dp_1/dt$.

Hence $\quad p_1 = 1/c_1 . (dV/dp)_1 + (dV/dp)_2 . dp_1/dt \qquad (5)$

The solution of equations (5) can be written as

$$p_1 = p_{10} \cdot \exp\left(\frac{-c_1}{(dV/dp)_1 + (dV/dp)_2} \log e\right) \qquad (6)$$

By combining equation (2) and equation (6) we obtain

$$\frac{(d(\log p_1)/dt)_{I}}{(d(\log p_1)/dt)_{II}} = \frac{(dV/dp)_1 + (dV/dp)_2}{(dV/dp)_1}$$

i.e.

$$(dV/dp)_1 = (dV/dp)_2 \Big/ \left(\frac{(d(\log p_1)/dt)_{I}}{(d(\log p_1)/dt)_{II}} - 1\right) \qquad (8)$$

The specific yield at the hydrostatic pressure

$$c_1 = (d(\log p_1)/dt)_{I} . (dV/dp)_1 . 1/\log e \qquad (9)$$

The specific yield at the hydrostatic pressure

$$c_2 = Q_2/ (p_2 - p_1) = Q_2/\Delta p_2$$

At the end of the injection $Q_2 = Q_{IN}$ for steady state condition. Δp_2 is the pressure difference between the borehole and the fracture.

from case II

$$Q_2 = (dV/dp)_2 . dp_2/dt \quad \text{and for } dp_2/dt = dp_1/dt$$

and $\quad p_1 = p_{10} \cdot \exp^{-\lambda t} \quad$ wth $d(\log p_1)/dt = (d(\log p_1)/dt)_{II}$

$$c_2 = (dV/dp)_2 . (d(\log p_1)/dt)_{II} . p_1/\Delta p_2 . 2.3 \qquad (10)$$

Using equations 8, 9 and 10 the parameters $(dV/dp)_2$, c_1 and c_2 can be obtained from the slopes of the two linear parts of the shut-in curves and their offset.

APPENDIX 2

Calculation of transmissivity, fracture width and area.

Using the difference between injection pressure p_c and the instantaneous shut-in pressure the transmissivity T_c at the fracture initiation pressure can be obtained for turbulent flow condition (JUNG 1986).

$$T_c = \mu Q_{IN} / \left[2\pi \Delta p_1 \left(\frac{c_t}{c_1 (Q/Q_c - 1)} \right) + \ln \frac{r_e}{r_o} - \ln \frac{Q}{Q_c} \right] \qquad (1)$$

Where μ is viscosity of the fluid

Q_{IN} is the injection flowrate

Q_c is the critical flowrate for the onset of turbulent flow wich was assumed as 0.5 l/s.

The transmissivity values were calculated for

$$\frac{r_e}{r_o} = 100 \quad \text{and} \quad \frac{c_t}{c_1} = 1$$

For laminar flow condition the transmissivity at hydrostatic pressure T_H can be obtained from

$$T_H = \mu Q_{IN} / (2\pi \Delta p_2) \, . \, \ln \frac{r_e}{r_o} \qquad \text{(after JUNG 1986)} \qquad (2)$$

Here $C_2 = Q_2 / \Delta p_2$ hence we have

$$T_H = \mu / (2\pi c_2) \, . \, \ln \frac{r_e}{r_o} \qquad (3)$$

Where c_2 can be obtained using equ. (10) in Appendix 1.

The values of the storage capacity of the fracture (dV/dp), were used for estimates of fracture area in each phase using the relation

$$dV/dp = A \, . \, \frac{1}{K_n}$$

where A is the area of the fracture

K_n is the normal stiffness parameter

The width of the fracture (w_c) at the fracture initiation pressure and W_H at the hydrostatic pressure were calculated with the known values of T_c and T_H as

$$Tc = w_c^3/12 \qquad \text{and} \qquad T_H = w_H^3/12$$

A STIMULATION STUDY CONCERNING STIMULATION OF GROUNDWATER BORE WELLS BY HYDRAULIC FRACTURING

T.N. GOWD and M. GOVARDHAN
National Geophysical Research Institute, Hyderabad-500 007

ABSTRACT

Vertical hydrofracture growth in low permeable aquifers (permeability of 1 or 25 or 50 or 75 or 100 md) has been simulated on the basis of the Perkins-Kern-Carter two dimensional hydrofrac model. The results indicate that the frac penetration of 60% to 100% of the drainage radius (about 235 ft) of groundwater borewells and a frac width of about 10 mm can be achieved by injecting a viscous frac fluid (viscosity=50 CP) at a rate of 500 LPM for 40 minutes. On the consideration that 5 mm thick proppant pack of 20-40 mesh size sand can be created in the 10 mm wide hydrofractures, we have estimated the likely increase in well yields due to such a hydrofrac stimulation operation and found that the well yields increase 10-16 folds after the stimulation. These results have thus provided theoretical basis for designing hydrofrac stimulation equipment and stimulaion procedures.

INTRODUCTION

Hydrofrac stimulation technology is being used all over the world to increase productivity and life of oil and gas wells (Howard and Fast, 1970). The stimulation procedure involves creation of a large fracture by hydraulically fracturing the reservoir rock and propping the fracture with proppants (sand etc.) to keep them open to conduct fluids from deep within the reservoir into the borewells. The authors have carried out a computer simulation study (Gowd and Govardhan, 1987) to enquire into the possibility of application of the hydrofrac stimulation technology for enhancing yield of groundwater borewells in hard rock terrains and attempted to estimate optimal stimulation parameters (pumping rate, pumping time, frac fluid viscosity, sand pack thickness) in relation to aquifer rock properties (permeability, porosity, minimum horizontal principal stress), and likely increase in water well yields after hydrofrac stimulation. This simulation study has thus provided a theoretical basis for designing hydraulic fracturing stimulation equipment as well as stimulation procedures.

Some of the important results of this simulation study are presented in this paper.

HYDROFRAC AND AQUIFER MODELS

On the basis of the Perkins-Kern-Carter two dimensional hydrofrac model we have simulated the growth of vertical hydraulic fractures of constant height of 30 ft or 60 ft in low permeable aquifers due to stimulation operations at different rates of fracturing fluid injection. In this study, we have considered aquifers with permeability of 1, 25, 50, 75 or 100 md giving rise to water-well yields in the range of 10-300 gph (Table 1), and then studied the fracture growth as a function of different pumping rates (5-30 CFM) and pumping time (20 or 40 minutes). A fracturing fluid with a viscosity of 50 CP, and a differential pressure (injection pressure minus pore fluid pressure) of 1000 PSI were considered in this study. The differential pressure was estimated on the basis of in-situ stress magnitudes in the Hyderabad region (Gowd et al, 1986).

Yield (gph)	S (Sq.m/Day)	K (md)	Porosity
10-20	0.00285	1	0.05
60-70	0.0712	25	0.05
150	0.1420	50	0.05
250	0.2130	75	0.05

Table 1. Typical yield of low producing borewells, and transmissivity (S), permeability (K) of the corresponding aquifers.

RESULTS AND DISCUSSION

The results indicate that depending on aquifer permeability (K=1, 25, 50, 75, 100 md) hydrofractures of 470-260 ft length (Fig.1), about 10 mm width (Fig.2), and 30 ft height (assumed) can be created by injecting a viscous fluid (M=50 CP) at a rate of either 20 CFM (about 500 LPM) for a period of 40 minutes or 30 CFM (about 800 LPM) for a period of less than 40 minutes (Table 2). This amounts to achieving fracture penetration (half frac-length) of the order of 60% to 100% of the drainage radius which we have estimated to be about 235 ft on the consideration that drainage area of a groundwater borewells is about 5 acres only (Thangarajan, 1986). On the consideration that 10 mm wide hydrofracture can be propped with 5 mm thick sand pack of 20-40 mesh size sand, and that one foot thick 20-40 mesh size sand pack subjected to an overburden pressure of 1000 PSI will give rise to a fracture conductivity of about 10 darcy- foot (Dowell Publication , 1973), we have estimated the relative conductivities (Ratio of fracture conductivity to formation permeability) of the hydrofractures in the five different aquifers (K=1, 25, 50, 75, 100 md) to be in the range of about 460-4.5. From these values of frac penetration and relative conductivities, productivity increase index PI (ratio of productivity after stimulation to productivity before stimulation) has been estimated for the five different aquifers by using the standard curves published by McGuire and Sikora (1963). From this we have found that the water-well yields increase 10-16 folds after stimulation (Fig.3). This will, thus, result in increasing waterwell yields from 10-20 gph to 400 gph,

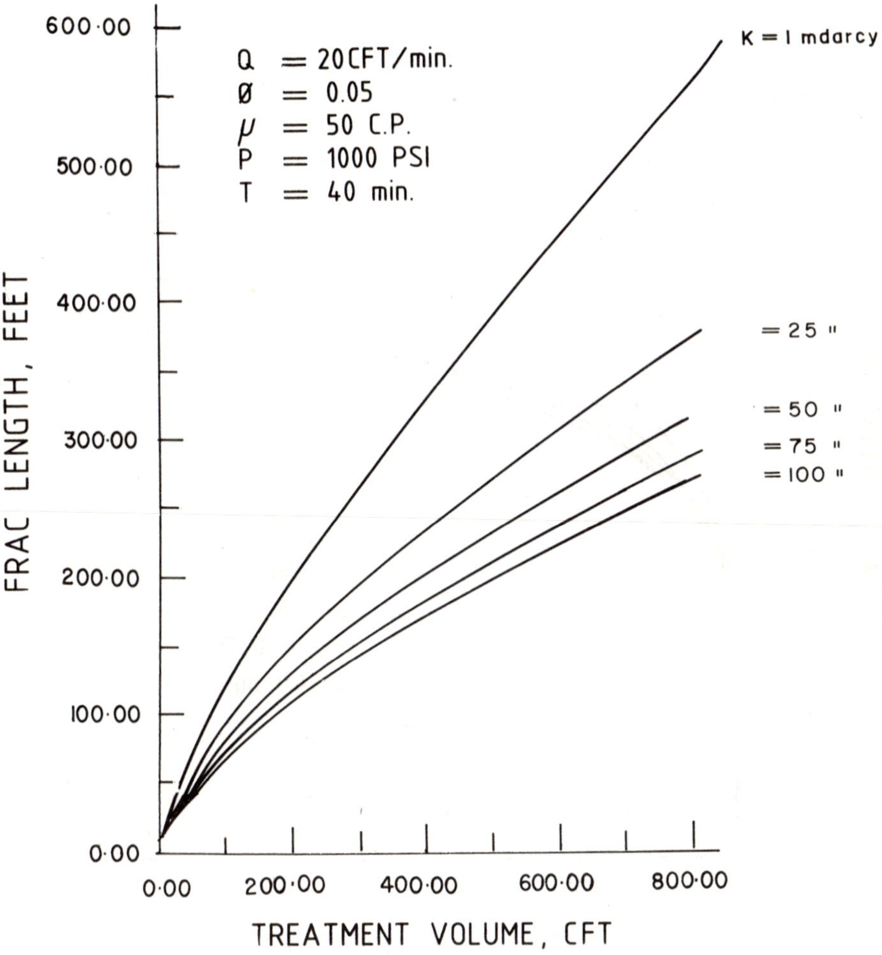

Fig. 1: Vertical Hydrofracture length as a function of treatment volume

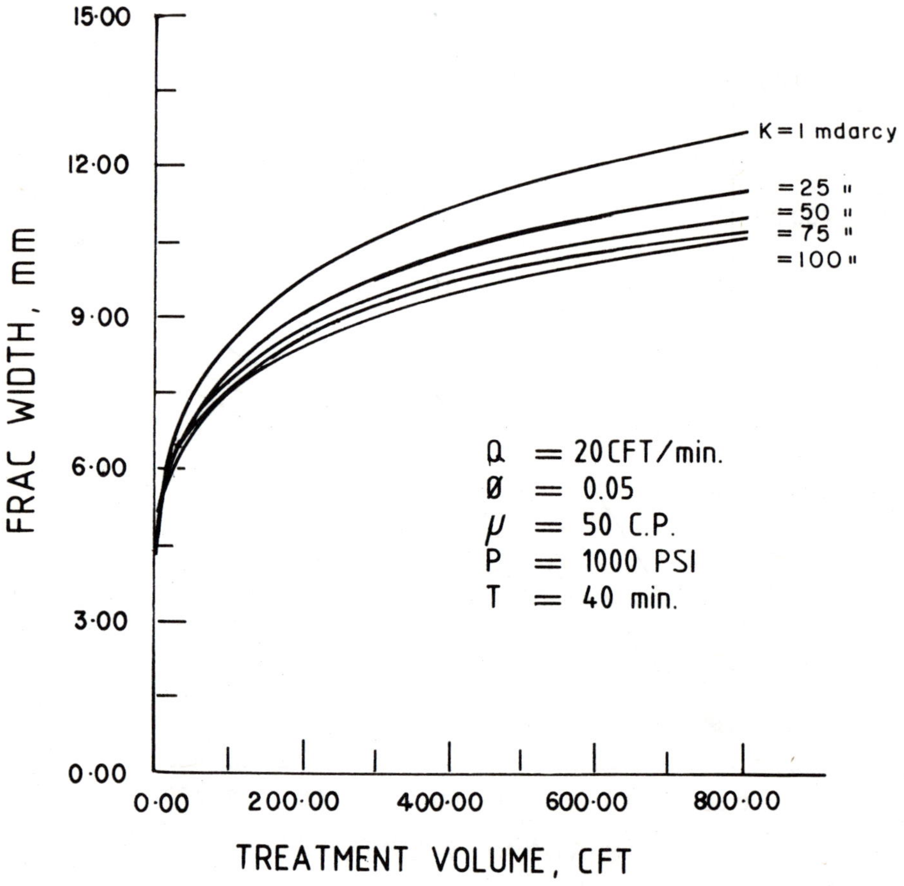

Fig. 2: Vertical Hydrofracture width as a function of treatment volume

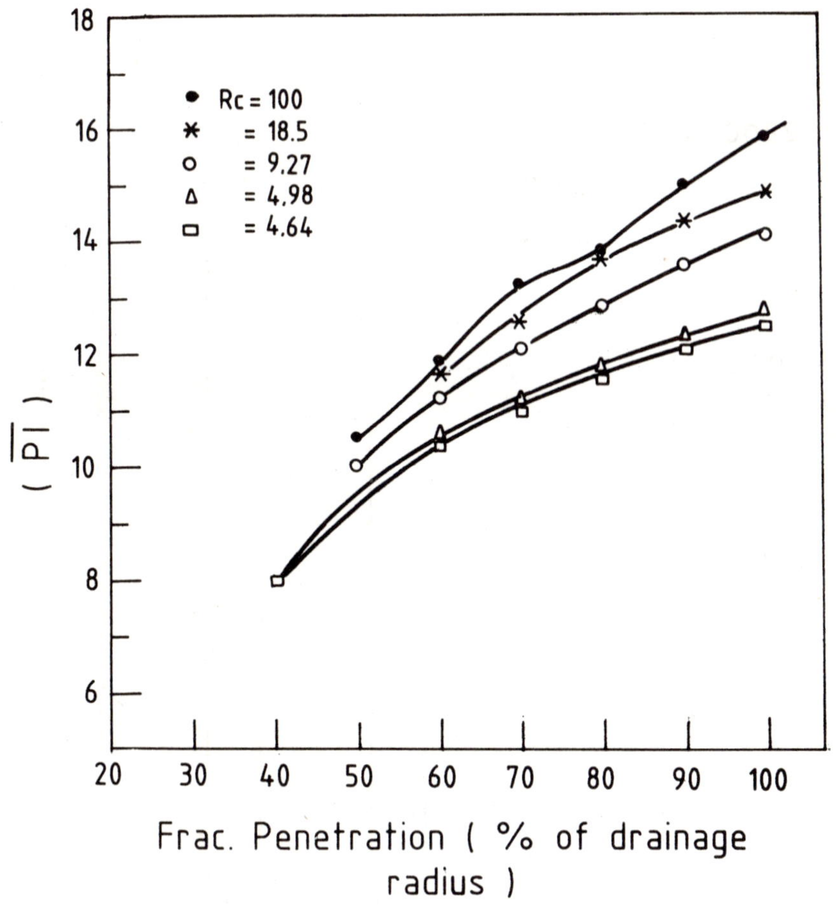

Fig. 3: Productivity increase index as a function of relative conductivity and penetration of vertical hydrofracture, estimated for groundwater borewells of 6" diameter and 5 acre drainage area (drainage radius of 235 ft) in five different low permeable aquifers having permeability of 1, 25, 50, 75 and 100 md.

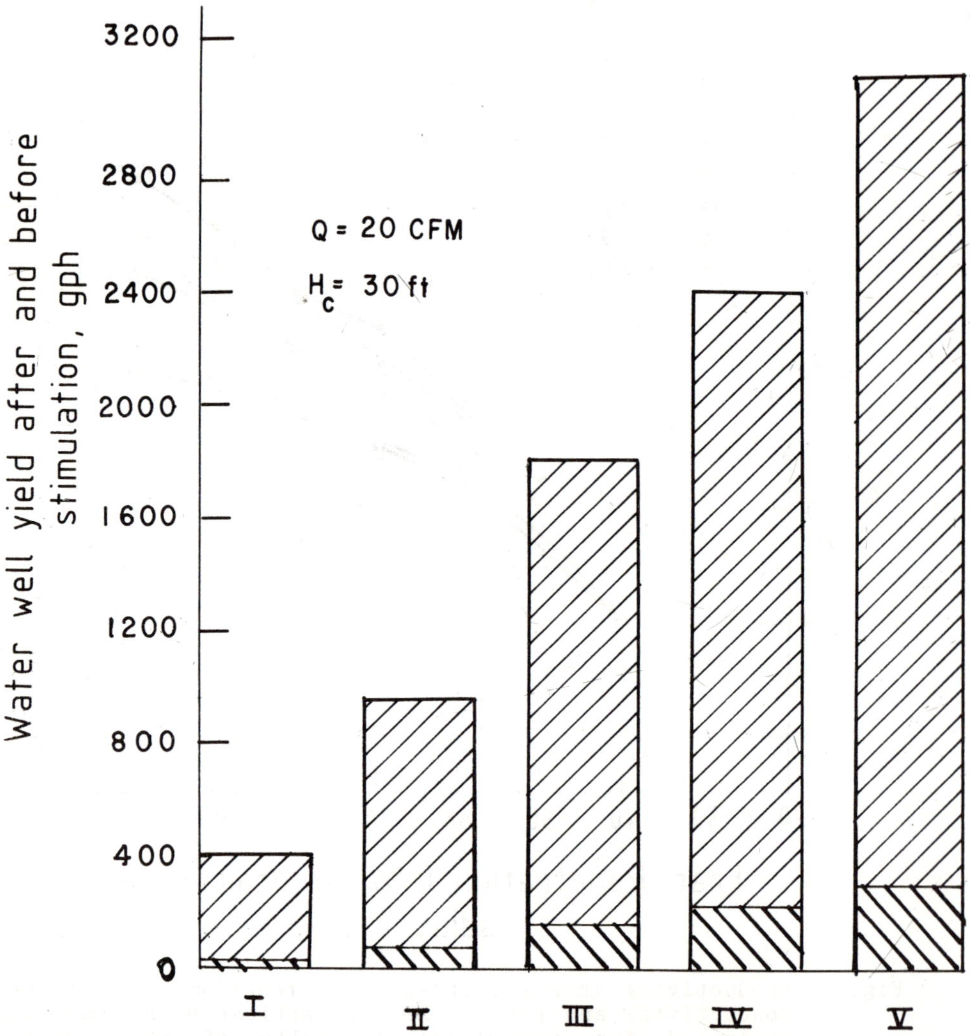

Fig. 4: Estimate of the yield of groundwater borewells after stimulation, in respect of different low permeable aquifers; borewell diameter=6", drainage radius=235 ft, aquifer permeability (K)=1(I), 25(II), 50(III), 75(IV), 100(V) md, Q=20 CFM, T=40 minutes, H=30 ft.

from 60-70 gph to about 1000 gph after stimulating extremely low permeable aquifers (K=1,25 md) and from 100-300 gph to 2400-3500 gph after stimulating low permeable aquifers (K=50, 75, 100 md), as shown in Fig.4 and Table 3.

Ex. No.	Q CFM	K md	length ft	width mm	PI
GBS 21	15	1	457.90	11.26	--
GBS 22	15	25	295.50	10.09	--
GBS 23	15	50	249.00	9.67	--
GBS 24	15	75	228.00	9.46	--
GBS 25	15	100	221.00	9.31	--
GBS 26	20	1	552.90	12.68	16.00
GBS 27	20	25	370.60	11.48	13.50
GBS 28	20	50	313.20	11.00	12.00
GBS 29	20	75	284.20	10.74	10.50
GBS 30	20	100	268.00	10.59	10.50
GBS 31	30	1	720.60	15.00	16.00
GBS 32	30	25	506.80	13.73	15.00
GBS 33	30	50	433.00	12.86	12.00
GBS 34	30	75	391.50		
GBS 35	30	100	365.40	12.66	11.50

Table.2 Values of the model parameters used to stimulate the growth of vertical hydrofractures caused due to water well stimulation operations at pumping rates of 15, 20 and 30 CFM for 40 minutes. GBS 21 through 35 are the simulation exercises. Length and width of the hydrofractures acheived in each of these exercises and the corresponding productivity increase indices (PI) are also given; Q is the pumping rate; K is the aquifer permeability;

Either by increasing sand pack thickness or by using coarse sand (10-20 mesh size) as propping material higher relative conductivities can be achieved and thus still higher yields can be obtained after stimulation.

K md	Yield before stimu-lation	Yield after stimulation		
		Q=30 CFM Hc=30 ft	Q=20 CFM Hc=30 ft	Q=30 CFM Hc=60 ft
1	10-20	200-400	200-400	135-270
25	60-70	900-1050	810-945	630-735
50	150	2125	1800	1500
75	250	3000	2625	2000
100	300	340	3150	2400

Table 3. Yield of groundwater borewells after stimulation.

CONCLUSIONS

1. A hydrofrac pump with a discharge rate of 500 LPM at a pressure of 1500 PSI should be employed for stimulating waterwells by hydraulic fracturing.

2. Gelled water with a viscosity of 50 CP should be used as frac fluid to achieve desired frac length and frac width and to transport sand proppant into the fracture.

3. 20-40 mesh size sand should be mixed with frac fluid in such a proportion that a 5 mm thick sand pack could be created in the fracture to achieve the desired relative conductivities and thus to enhance well yields 10-16 folds after hydrofrac stimulation job is completed.

ACKNOWLEDGEMENTS

The authors are grateful to Prof. V.K.Gaur, Director, NGRI, Hyderabad for his encouragement and for according permission to publish the paper.

REFERENCES

Carter, R.D. (attributed to) in Howard, G.C. and Fast, C.R., 1957. Optimum fluid characteristics for fracture extension. Dril. and Prod.Prac., API 216.

Gowd, T.N., Srirama Rao, S.V., Chary, K.B. and Rummel, F., 1986. Insitu stress measurements using hydraulic fracturing method. Earth and Plan. Sci. (Proc. Ind. Acad. Sci.) 95 (3), pp 311-319.

Gowd, T.N. and Govardhan, M., 1987. Stimulation of groundwater borewells by hydraulic fracturing - A simulation study. NGRI Technical Report T.R.No.NGRI-87-Environ-32, 112p.

Howard, G.C. and Fast, C.R., 1970. Hydraulic fracturing. Publ. Society of Petroleum Engineers of AIME (1970).

Perkins, T.K. and Kern, L.R., 1961. Width of hydraulic fracturing. J. Pet.Tech. pp 931-949.

Thangarajan, M. Personal Communication.

AUTHOR INDEX

International Workshop on

APPROPRIATE METHODOLOGIES FOR DEVELOPMENT AND MANAGEMENT OF GROUNDWATER RESOURCES IN DEVELOPING COUNTRIES

(February 28 - March 4, 1989)

VOLUME I

Chief Editor
C.P. GUPTA

Editors:
Shakeel Ahmed
V.V.S.G. Rao
M.T. Rajan

NATIONAL GEOPHYSICAL RESEARCH INSTITUTE
(C.S.I.R.)
HYDERABAD - 500 007, INDIA

OXFORD & IBH PUBLISHING CO. PVT. LTD.
New Delhi Bombay Calcutta

ISBN 81-204-0406-8 (For Set)
ISBN 81-204-0407-6 (Vol. I)
ISBN 81-204-0408-4 (Vol. II)

Published by Mohan Primlani for Oxford & IBH Publishing Co. Pvt. Ltd., 66, Janpath, New Delhi 110001 and printed at Rekha Printers, Okhla Industrial Area, Phase II, New Delhi 110020

1-F9-8.5

FOREWORD

Groundwater is the great hidden component of the hydrological cycle comprising the most important natural resource of our planet. The history of groundwater utilization goes as far back as the history of mankind itself. But the stupendous growth of population, agriculture, and industry has compelled an indiscriminate exploitation of this resource. As a result, a variety of environmental hazards such as declining water levels in the aquifers, reduction of groundwater effluence to rivers and streams, deterioration of water quality, land subsidence etc., have set in despite vast strides that the multidisciplinary science of geohydrology has taken over the last few decades.

Utmost care has, therefore, to be exercised in the exploitation and management of this precious resource. In order to evolve a pragmatic and scientific plan for the management of groundwater resources, one needs to quantify the characteristic hydrogeological, hydrometeorological, hydrological, hydrogeochemical and other relevant parameters; establish an optimal network for data acquisition, storage and retrieval; and adopt appropriate methodology for data interpretation and analysis. Already, a lot of work has been done to evolve and apply appropriate techniques related to these aspects, particularly in the developed countries.

It is heartening that an International Workshop is being organized to share the experiences and knowledge of experts from developed as well as developing countries. I hope that it would culminate into making concrete suggestions on such techniques and strategies as could be easily replicated in the developing countries with their limited resources of man, material, money, for optimal harnessing and proper management of their groundwater resources.

Hyderabad
January 6, 1989

HARI NARAIN

PREFACE

Water is essential for existence. An assured supply of wholesome water has, therefore, been the prime concern of all societies. In our own time the increasing stress on the supportive and assimilative capacities of water systems has led to a growing realization that it is only through a scientific approach to their management, that we can hope to sustain the quality and regular availability of this basic human requirement. The need for appropriate integrated studies in developing countries is all the more important in view of a burgeoning population, widespread aridity, scarcity of fiscal resources and vagaries of hydrometeorological conditions.

Since groundwater, surface water and land comprise a complex interlocked system, they must be considered and managed in their totality, towards making the most optimal use of this precious resource. With the advent of remote sensing and sophisticated data processing techniques, the speed and scale of investigations of visible resources such as land and surface water have greatly improved. An understanding of the invisible groundwater regimes and quantification of their characteristic parameters, however, continue to remain mostly inferential and indirect. Groundwater resources are now being increasingly tapped to meet societal needs on account of their possible availability in the vicinity of the consumer. However, it is important to recognize that although groundwater has ubiquitous occurrence and vast magnitude, its potential as a perennial source is subject to wide spatio-temporal variations. It is renewable but not inexhaustible. It is underground but not entirely protected from pollution.

Some of the related issues that the ensuing Workshop IGW-89 would address, include identification of the state-of-the-art techniques for geohydrological parameter estimation, acquisition of data, analysis of flow in the unsaturated zone, studies of contaminant migration and augmentation of groundwater. All these are essential pre-requisites for reliable estimation and rational management of this vital resource. Cost-effectiveness of these techniques is especially important for developing countries.

I sincerely hope that deliberations of this Workshop will go a long way in enhancing our collective technical competence, facilities and infrastructure on the one hand and socio-political awareness of these problems, on the other. For thus alone, can we successfully face the formidable challenge of meeting the ever-increasing demand for water.

Hyderabad
January 19, 1989

Vinod K. Gaur

SPONSORS FOR IGW-89

EMBASSY OF FRANCE
2, AURANGZEB ROAD
NEW DELHI-110 001, INDIA

INTERNATIONAL CENTRE FOR TRAINING AND EXCHANGES IN GEOSCINCES
103 RUE DE LILLE
75007 PARIS, FRANCE

AUSTRALIAN INTERNATIONAL DEVELOPMENT ASSISTANCE BUREAU
C/O THE AUSTRALIAN HIGH COMMISSION
1/50 G, SHANTI PATH
CHANAKYAPURI
NEW DELHI - 110 021, INDIA

COSTED
24, GANDHI MANDAPAM ROAD
MADRAS - 600 025, INDIA

ASSOCIATION OF GEOSCIENTISTS FOR INTERNATIONAL DEVELOPMENT
NATURAL RESOURCES DIV.
U.N. BUILDING, RAJADAMNOEN AVE.
BANGKOK 10200, THAILAND

DEPARTMENT OF RURAL DEVELOPMENT
MINISTRY OF AGRICULTURE
KRISHI BHAVAN
NEW DELHI - 110 001, INDIA

NEYVELI LIGNITE CORPORATION
NEYVELI - 607 801, INDIA

RAIL INDIA TECHNICAL AND ECONOMICS SERVICES
646, ANNA SALAI
MADRAS - 600 006, INDIA

CMC LIMITED
POSNETT BHAVAN, TILAK ROAD
HYDERABAD-500 001, INDIA

UNIVERSITY GRANTS COMMISSION
BAHADUR SHAH ZAFAR MARG
NEW DELHI - 110 002, INDIA

NATIONAL REMOTE SENSING AGENCY
BALANAGAR
HYDERABAD - 500 037, INDIA

TAMILNADU WATER AND DRAINAGE BOARD
CHEPAUK
MADRAS - 600 005, INDIA

UPTRON INDIA LTD
10, ASHOK MARG
LUCKNOW (U.P.), INDIA

PANCHAYAT RAJ ENGINEERING DEPARTMENT
SADHANA BUILDING, HYDERGUDA
HYDERABAD - 500 029, INDIA

NATIONAL BANK FOR AGRICULTURE AND RURAL DEVELOPMENT
P.B.NO.63520, WORLI
BOMBAY - 400 018, INDIA

AGRICULTURAL ENGINEERING DEPARTMENT
GOVT. OF TAMILNADU
NANDANAM
MADRAS-60035, INDIA

INTEGRATED GROUNDWATER SERVICES PVT. LTD.
71, HUDA COMPLEX, TARNAKA,
HYDERABAD - 500 007, INDIA

TERRA SCIENCE,
16-47, PRASANTHI NAGAR,
HYDERABAD -500 039, INDIA

STEERING COMMITTEE

PROF. V.K. GAUR DIRECTOR, N.G.R.I. HYDERABAD-500 007, INDIA	CHAIRMAN
DR. S. PLATTARD SCIENTIFIC ATTACHE EMBASSY OF FRANCE NEW DELHI-110 001, INDIA	MEMBER
DR. M. LELEU DIRECTOR CIFEG 103, RUE DE LILLE 75007 PARIS, FRANCE	MEMBER
MR. S.R. SENGUPTA GENERAL MANAGER (MINES) NEYVELI LIGNITE CORPORATION NEYVELI-607 803, INDIA	MEMBER
MR. G. GHOSH DIRECTOR, TECHNOLOGY MISSION DEPARTMENT OF RURAL DEVELOPMENT KRISHI BHAVAN NEW DELHI-110 001, INDIA	MEMBER
MR. K.N. JOHRY ADVISOR AND HEAD, I.S.C., CSIR NEW DELHI-110 001, INDIA	MEMBER
PROF. S. RADHAKRISHNA DEPARTMENT OF PHYSICS INDIAN INSTITUTE OF TECHNOLOGY MADRAS-600 036, INDIA	MEMBER
DR. SHAKEEL AHMED SCIENTIST, N.G.R.I. HYDERABAD-500 007, INDIA	JOINT SECRETARY
DR. C.P. GUPTA SCIENTIST, N.G.R.I. HYDERABAD-500 007, INDIA	SECRETARY

ORGANIZING COMMITTEE

PROF. V.K. GAUR DIRECTOR, N.G.R.I. HYDERABAD-500 007, INDIA	CHAIRMAN
MR. MOHD. INAMUL HAQ ADVISOR TECH. MISSION ON DRINKING WATER D.R.D., KRISHI BHAVAN NEW DELHI-110 001, INDIA	MEMBER
MR. K.M. SUBRAMANYAM DIRECTOR, A.P.STATE GROUNDWATER DEPT. SOMAJIGUDA HYDERABAD-500 463, INDIA	MEMBER
MR. T. HANUMANTHA RAO DIRECTOR GENERAL WALMATARI DIRECTORATE OFFICE SECRETARIAT BUILDINGS HYDERABAD-500 022, INDIA	MEMBER
MR. N.K. BALASUNDER DY. GENERAL MANAGER (HYDROLOGY) NEYVELI LIGNITE CORPORATION NEYVELI-607 803, INDIA	MEMBER
MR. K.V.J.R. KRUPANIDHI DIRECTOR, C.G.W.B. 3-6-291, HYDERGUDA HYDERABAD-500 029, INDIA	MEMBER
PROF. B.H. BRIZ KISHORE DIRECTOR (PLANNING AND CO-ORDINATION) J.N.T.U. HYDERABAD-500 028, INDIA	MEMBER
DR. P.V.L.N. RAO MANAGING DIRECTOR,NRDCS LIMITED P.B. NO. 1067, HIMAYATNAGAR HYDERABAD-500 029, INDIA	MEMBER
DR. A. APPA RAO SCIENTIST, N.G.R.I. HYDERABAD-500 007, INDIA	MEMBER
DR. M. THANGARAJAN SCIENTIST, N.G.R.I. HYDERABAD-500 007, INDIA	JOINT SECRETARY
DR. C.P. GUPTA SCIENTIST, N.G.R.I. HYDERABAD-500 007, INDIA	SECRETARY

SCIENTIFIC ADVISORY COMMITTEE

PROF. G. DE MARSILY
UNIVERSITY PARIS VI AND
PARIS SCHOOL OF MINES
PARIS, FRANCE
CHAIRMAN

PROF. K.R. RUSHTON
DEPARTMENT OF CIVIL ENGINEERING
UNIVERSITY OF BIRMINGHAM
BIRMINGHAM B15 2TT, U.K.
MEMBER

DR. T. POINTET
BRGM, B.P. 6009
45060, ORLEANS, FRANCE
MEMBER

DR. K. TRIPPLER
BGR, STILLEWEG 2
3000 HANNOVER 51
FEDERAL REPUBLIC OF GERMANY
MEMBER

PROF. B.B.S. SINGHAL
DEPARTMENT OF EARTH SCIENCES
UNIVERSITY OF ROORKEE
ROORKEE-247 667, INDIA
MEMBER

PROF. S. RAMASESHAN
DEPARTMENT OF CIVIL ENGINEERING
INDIAN INSTITUTE OF TECHNOLOGY
KANPUR-208 016, INDIA
MEMBER

DR. A.S. CHAWLA
DIRECTOR, W.R.D.T.C.
UNIVERSITY OF ROORKEE
ROORKEE-247 667, INDIA
MEMBER

PROF. P.G. SASTRY
DEPARTMENT OF CIVIL ENGINEERING
REGIONAL ENGINEERING COLLEGE
WARANGAL, INDIA
MEMBER

PROF. R. SAKTHIVADIVEL
DIRECTOR, CENTRE FOR WATER RESOURCES
GUINDY ENGINEERING COLLEGE
ANNA UNIVERSITY
MADRAS-600 025, INDIA
MEMBER

PROF. V. LAXMINARAYANA
DEPARTMENT OF CIVIL ENGINEERING
INDIAN INSTITUTE OF TECHNOLOGY
KANPUR- 208 016, INDIA
MEMBER

MR. S. GOWRISANKARAN DEPUTY GENERAL MANAGER NEYVELI LIGNITE CORPORATION NEYVELI-607 803, INDIA	MEMBER
DR. DEEPAK KASHYAP READER, INSTITUTE OF HYDROLOGY UNIVERSITY OF ROORKEE ROORKEE-247 667, INDIA	MEMBER
DR. SHAKEEL AHMED SCIENTIST, N.G.R.I. HYDERABAD-500 007, INDIA	MEMBER
MR. V.V.S. GURUNADHARAO SCIENTIST, N.G.R.I. HYDERABAD-500007	JOINT SECRETARY
DR. C.P. GUPTA SCIENTIST, N.G.R.I. HYDERABAD-500007, INDIA	SECRETARY

CONTENTS

VOLUME II

S-4: MANAGEMENT OF GROUNDWATER RESOURCES PARTICULARLY IN HARD ROCKS: PARAMETERIZATION OF GROUNDWATER FLOW REGIME; MODELING OF COMPLEX AQUIFERS; SIMULATION OF FLOW IN FRACTURED ROCKS; ASSESSMENT OF DYNAMIC POTENTIAL OF AQUIFERS

SESSION S-1A

GEOHYDROLOGIC PARAMETER ESTIMATION : RECENT ADVANCES IN CONVENTIONAL TECHNIQUES ; USE OF TRACER TECHNIQUES

PUMPING TEST IN A LOW PERMEABILITY CLAYSTONE AQUIFER EXECUTED WITH HIGH-RESOLUTION ELECTRONIC WATER LEVEL RECORDERS AND EVALUATED WITH THE VERSATILE COMPUTER PROGRAM HYDROPAR

K. Trippler, Federal Institute for Geosciences and Natural Resources, Hannover, Fed. Rep. of Germany.

Abstract

A long duration pumping test was carried out in October 1983 west of Hannover in a low permeability, highly fractured claystone aquifer system. Groundwater was pumped continuously for 120 hours (0.96 m^3/hour) from one well. The drawdown was automatically measured in 13 observation wells with high-resolution ultrasonic transducers, which function as both transmitter and receiver. The data are stored directly on magnetic tape. The data can be processed directly by computer using the versatile program HYDROPAR. The results of the pumping test showed that the aquifer is semiconfined (S=0.001) and has a low transmissivity (T=0.21 m^2/hour)

1. The measurement of the groundwater level in observation wells

An electronic transducer is installed below the groundwater level in an observation well. Initially the groundwater level in the well is measured manually. Subsequently, the relative movement of the groundwater level can be registered by measuring the time required for an ultrasonic pulse to travel from the transducer to the water surface and back. This is done with a device for measuring very short time intervals, having the resolution of 0.1 μs. This measurement of the relative position of the groundwater level can be achieved with a high resolution of 0.14 mm.

Groundwater levels in 30 observation wells can be measured simultaneously within 1.5 s. The digitized data are stored on magnetic tape for further processing.

Other parameters (pumping rate, precipitation, atmospheric pressure, temperature and conductivity of the groundwater) are also measured and stored together with the pumping test data.

A block diagram of the instrumental set up is presented in Figure 1 (Kohlmeier et al. 1985).

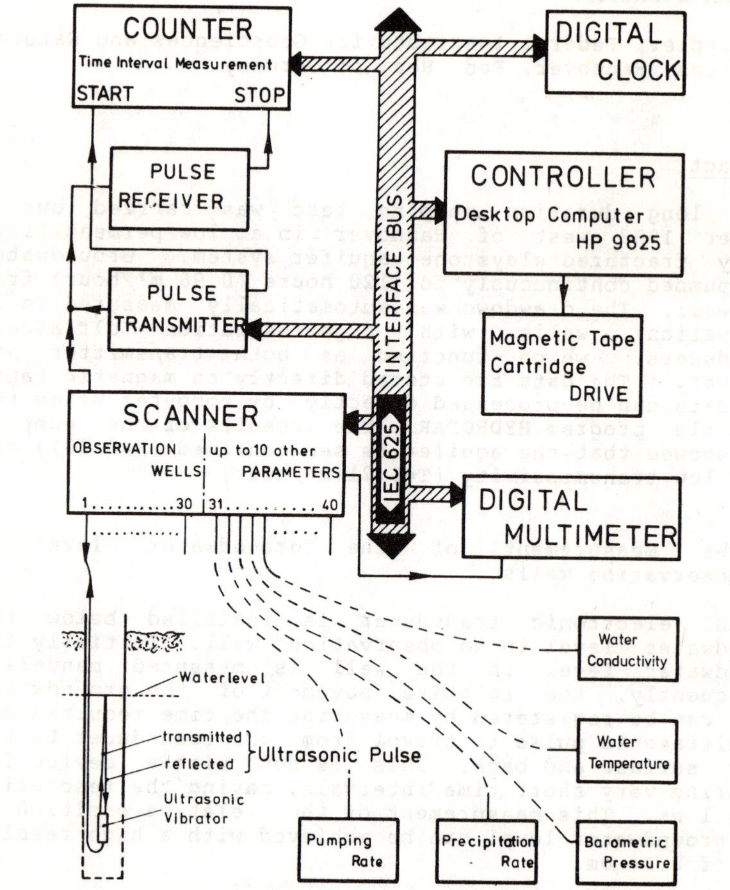

Fig. 1: Equipment for data acquisition during pumping tests; determination of water level by measurment of the transit time of ultrasonic pulses.

(after KOHLMEIER et al., 1985)

2. The computer program HYDROPAR

The program HYDROPAR (NIWS, initial release) for the evaluation of pumping test data is based on the formulas of Theis (1935), Hantush (1964), and others. A model of the aquifer is necessary for the evaluation of pumping test data. Four models are offered by the program HYDROPAR:

- confined aquifer ("Theis conditions")
- semiconfined aquifer ("Hantush conditions")
- semiconfined aquifer with partially penetrating wells
- multi semiconfined aquifer system.

Other options offered by the program are shown in Table I, e.g.

- multi- or single-well configuration,
- variable or constant pumping rate, and
- steady-state or time dependent conditions.

Apart from the raw data, the selected aquifer version, and the estimated parameters, together with their estimated upper and lower limits (Table I), are the input for the program. The aquifer and statistical parameters for the best-fit model curve are the output of the program. The confidence limits for pairs of aquifer parameters can be calculated from the statistical parameters.

The HYDROPAR program tries to find a solution within twenty iterative steps. If the program is not successful an output is given to that effect.

A major advantage of HYDROPAR with respect to the common manual evaluation of pumping test data is that the differences between the observations at two neighboring wells can be used for the evaluation. This procedure avoids the influence of local precipitation and evapotranspiration.

3. The pumping test

3.1 The hydrogeological environment

The wells used for the pumping test are located in a fractured claystone aquifer of low hydraulic conductivity in the plains of Lower Saxony, Federal Republic of Germany.

Table 1: Aquifer types, program options, necessary aquifer parameters, and arrangement of the input data

Aquifer type	Options: Well field	Options: Q	Options: Time	Necessary parameters	Arrangement of input data
1 confined aquifer	no	var	var	T S	h t r
	yes	var	var	T S	h t x y
2 semiconfined aquifer	no	var	con	T C	h r
	yes	var	con	T C	h x y
	no	var	var	T C S	h t r
	yes	var	var	T C S	h t x y
3 semiconfined aquifer & partially penetrating wells	no	con	con	k_h k_z C D	h r z
	yes	con	con	k_h k_z C D	h x y z
	no	con	var	k_h k_z C D S	h t r z
	yes	con	var	k_h k_z C D S	h t x y z
4 two semiconfined aquifers	no	con	con	T_1 T_2 C_1 C_2 C_3	h r 1 (or 2)
	yes	con	con	T_1 T_2 C_1 C_2 C_3	h x y 1 (or 2)
	no	con	var	T_1 T_2 C_1 C_2 C_3 S_1 S_2	h t r 1 (or 2)
	yes	con	var	T_1 T_2 C_1 C_2 C_3 S_1 S_2	h t x y 1 (or 2)

Remarks: a. Pumping tests with single well, cartesian coordinates may be used.
b. Input data: h – groundwater level in observation well
t – time of measurement
r,x,y,z – coordinates in space
1 (or 2) – aquifer number: 1 – upper, 2 – lower aquifer
c. Parameter: T – transmissivity
S – storage coefficient
C – hydraulic resistance (C=D'/k'); D' thickness & k' permeability of the semiconfined layer

The regional groundwater system is recharged in hills about 5 km east of the wells, which are about 50 m below the recharge area.

Detailed hydrogeological investigations are necessary to understand the movement of groundwater in the area as a whole and specifically through a waste disposal site.

3.2 The pumping test

The pumping test was carried out in October 1983. The test lasted 230 hours and consisted of three phases:

- measurement of the static groundwater level (38 hours);
- drawdown of the groundwater level by pumping at a rate of 0.96 m^3/h (120 hours);
- recovery of the groundwater level (72 hours).

The amount rainfall during the pumping test totaled 22 mm.

The locations of the pumping well and the observation wells are shown in Figure 2.

3.3 Evaluation of the pumping test data with HYDROPAR

It is assumed that the aquifer is semiconfined in the area of a waste disposal site. Version 2 (Table I) of HYDROPAR was selected for the evaluation of the pumping test. Some graphs produced by HYDROPAR are shown in Figures 3a – d.

The empirical values and the best-fit drawdown curve for observation well S5, 4.19 m from the production well, are shown in Figure 3a. The drawdown and the recovery curves for observation well S6 are shown in Figure 3b, together with the best-fit model curve. The traditional graphical evaluation of the data after Hantush (1964) is shown in Figure 3c. The confidence limits for the parameters S and T are shown in Figure 3d.

The results are given in Table II, together with technical data for the wells. Please note that the data for the observation wells more than 100 m from the production well could not be evaluated by HYDROPAR.

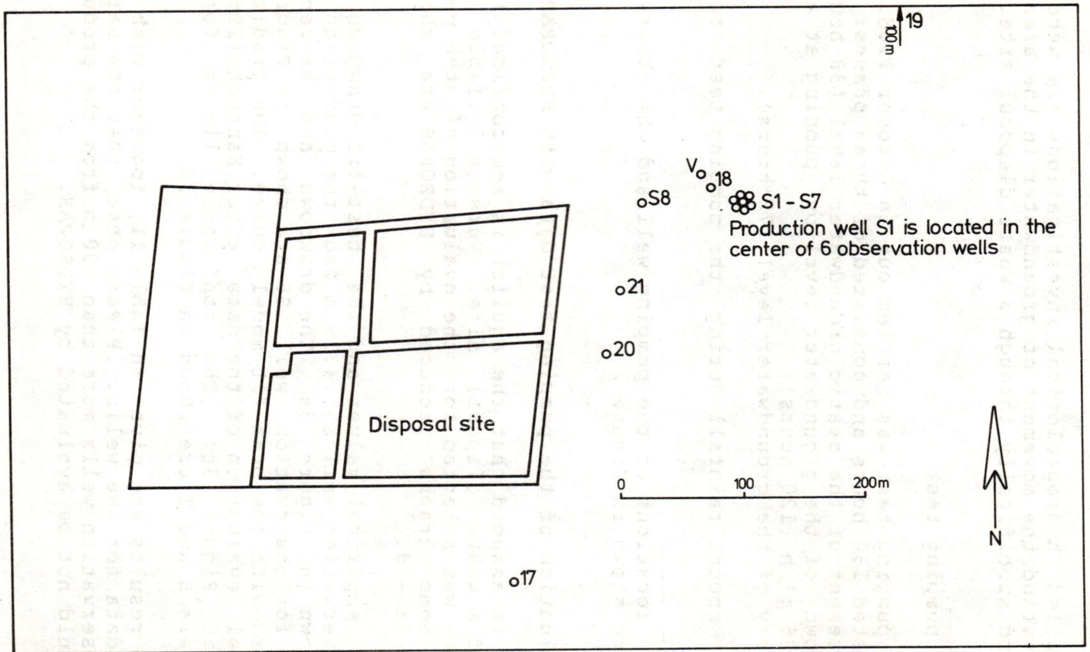

(after KILGER et al., 1985)

Fig. 2: Location of the wells involved in the test

Aquifer is semiconfined

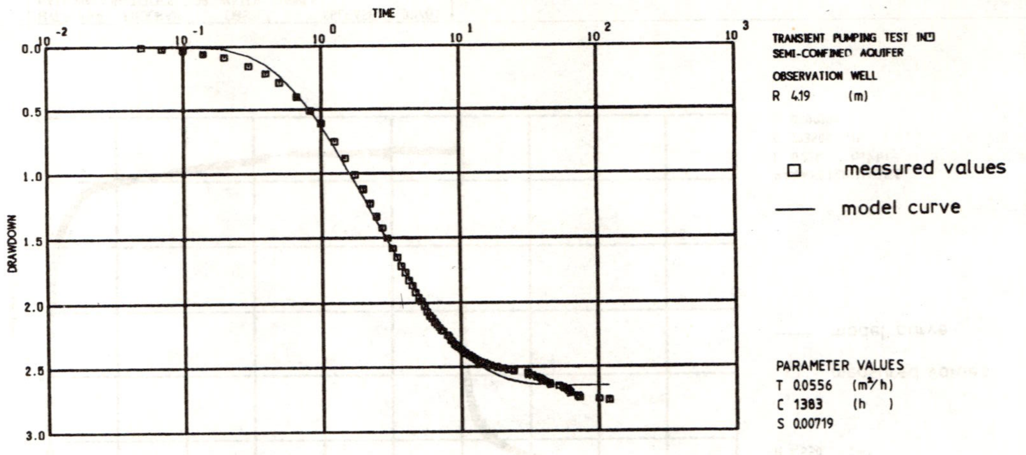

HYDROPAR PROGRAM OBSPLT RELEASE 82/01
NATIONAL INSTITUTE FOR WATER SUPPLY
P.O. BOX 150
2260 AD LEIDSCHENDAM
THE NETHERLANDS

Fig. 3a: Reaction of the water level in observation well S 5

(after KILGER et al. 1985)

Aquifer is semiconfined; drawdown & recovery in observation well S6

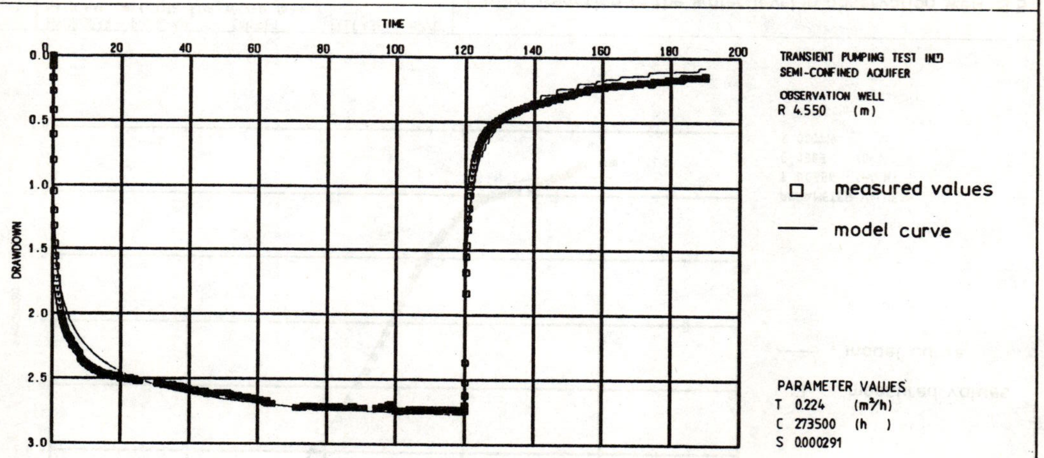

HYDROPAR PROGRAM OBSPLT RELEASE 82/01
NATIONAL INSTITUTE FOR WATER SUPPLY
P.O. BOX 150
2260 AD LEIDSCHENDAM
THE NETHERLANDS

Fig. 3b: Observation well S6; drawdown and recovery

(after KILGER et al. 1985)

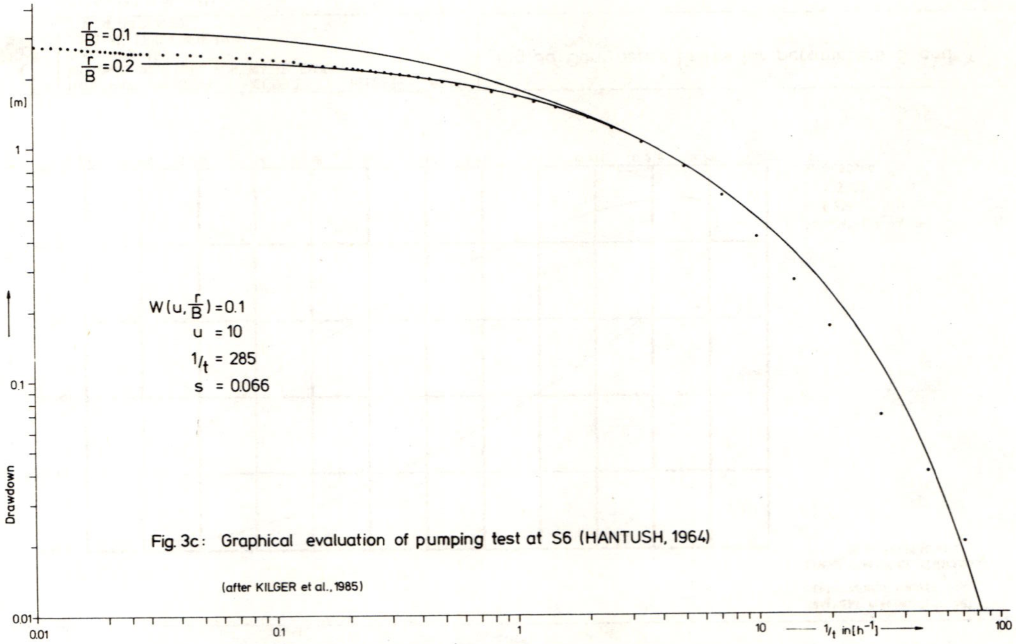

Fig. 3c: Graphical evaluation of pumping test at S6 (HANTUSH, 1964)

(after KILGER et al., 1985)

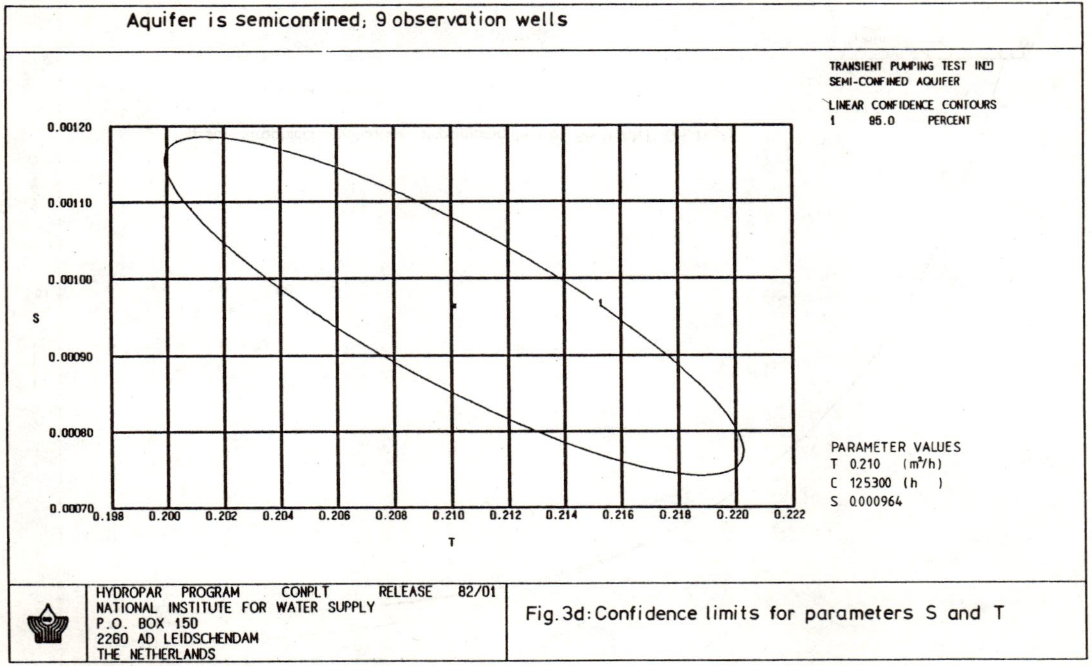

Fig. 3d: Confidence limits for parameters S and T

(after KILGER et al. 1985)

able 2: Pumping test: technical data and results

Well No.	Results			Technical data			Draw-down	Remarks
	T	S	C	Ø	filter	r		
S1				5	5.5 – 25.0	0.00	5.37	prod. well
S2	0.159	0.0028	6,580	5	20.5 – 25.0	3.95	2.24	obs. wells
S3	0.044	0.027	513	5	17.5 – 22.0	2.84	2.78	"
S4	0.080	0.0049	2,020	5	14.5 – 19.0	3.86	2.73	"
S5	0.056	0.0072	1,380	5	11.5 – 16.0	4.19	2.76	"
S6	0.155	0.0007	21,800	5	8.5 – 13.0	4.55	2.75	"
S7	0.139	0.0018	6,560	5	5.5 – 10.0	3.87	2.53	"
S8	0.2	0.0002	172,000	5	no filters	80.	0.80	"
17	–	–	–	–	5.0 – 25.0	310.	0.02	no result
18	0.207	0.0003	85,500	5	5.0 – 25.0	15.9	1.73	
19	–	–	–	5	5.0 – 25.0	273.	-0.03	no result
20	–	–	–	5	5.0 – 25.0	132.	-0.01	no result
21	–	–	–	–	5.0 – 25.0	173.	0.01	no result
V	0.11	0.002	9,500	4	1.0 – 5.0	23.2	0.95	
9 wells	0.21	0.001	125,000	–	–	–	–	S2 – S8, 18,V
S6	0.224	0.0003	273,500	5	8.5 – 13.0	4.55	2.75	drawd.& recov.
S6	0.115	0.0008	8,000	5	8.5 – 13.0	4.55	2.75	graph. eval.

Remarks: T = transmissivity (m^2/h) C = hydraulic resistivity (h)
S = storage coefficient Ø = diameter of the filters (inches)
r = distance from the pumped well (m)
Filter = depth range of the filters: from ... to ... (m)
Drawdown = maximum drawdown (m)

3.4 Results

The mean transmissivity T is about 0.2 m^2/hour and the storage coefficient S is about 0.001. The thickness of the aquifer is estimated to be m = 40 m. The hydraulic conductivity k (T=m*k) is 0.0000014 m/s. The mean hydraulic resistance C is 125 000 hour (C=D'/k'). The thickness D' of the top layer is about 1 m. The hydraulic conductivity k' of the top layer is 2.2×10^{-9} m/s.

The results confirm that the test was carried out in a semiconfined aquifer system of low to medium hydraulic conductivity. In arid and semi-arid areas such an aquifer can be used for domestic water supply.

References

HANTUSH, M., S.: Hydraulics of Wells. In: Advances in Hydroscience. Academic Press, Inc., New York 1964.

KILGER, B., MEINEN, M. & TRIPPLER, K.: Auswertung eines Pumpversuchs in einen Brunnen im Tonstein der niedersächsischen Unterkreide bei Münchehagen. Report of the NLfB, Hannover 1985.

KOHLMEIER, R., GIESEL, W. & STRAYLE, G.: Determination of water levels in observation wells by measurement of transit time of ultrasonic pulses and calculation of hydraulic parameters. Natural Resources and Development, vol. 21 1985.

NIWS: HYDROPAR — A program package for the statistical analysis of pumping test data. National Institute for Water Supply, The Netherlands, Initial release.

ANALYSIS OF PUMPING TEST DATA FROM BOREWELLS IN FRACTURED ROCKS OF KARNATAKA

S.K. MISHRA[1], B.B.S. SINGHAL[1] and D.C. SINGHAL[2]

[1] Department of Earth Sciences, University of Roorkee, Roorkee

[2] Department of Hydrology, University of Roorkee, Roorkee

ABSTRACT

Several methods exist for the computation of aquifer characteristics from pumping test data under different hydrogeological conditions. The conventional pumping test data analysis procedures developed for homogeneous media have many assumptions that are not applicable to field conditions in the heterogeneous media such as fractured rocks. In the present study, the aquifer characteristics and well characteristics are computed from data of pumping tests conducted in the Northern parts of Karnataka. An attempt is made to compute aquifer characteristics using double porosity model. The time-drawdown curves match to a greater extent with the type curves for double porosity model and results computed are consistent with the field observations.

The use of Rorabaugh's method for well characteristics in fractured formations discriminates between laminar, turbulent and transitional nature of flows. The specific drawdown versus discharge plots help in a better understanding of flow patterns in fractured formations.

INTRODUCTION

The exploration and development of groundwater resources is closely associated with advances in well hydraulics. In order to secure the supply of groundwater by efficient means, serious considerations have been given to the nature of fluid motion and therefore, much of the recent research has been concentrated on improving the understanding of the phenomenon of fluid flow towards pumped well.

In particular, recently, fractured formations have received much attention, mainly in engineering geological problems, in groundwater hydrology and in petroleum engineering, because of the development of hydraulic fracturing and the recognition of naturally fractured reservoirs as important producers of oil and water.

The choice of a particular method is greatly dependent on the appreciation of local geohydrological conditions as there are many similarities in type curves given for various aquifer types and well characteristics. Therefore, it is necessary to assess the applicability of a particular theoretical model to the field conditions. It is generally agreed that the conventional methods primarily developed for homogeneous reservoirs are inadequate (Streltsova-Adams, 1978; Mishra, 1986) and that new approaches are required. Detailed theoretical studies about the fluid flow behaviour in fractured formation have been made by Boulton and Streltsova (1975, 1977a, b, 1978), Streltsova-Adams (1978),

Bourdet and Gringarten (1980), Gringarten (1982) and others.

The step drawdown tests are of help in the proper selection of suitable discharge and also to know the degree of turbulence as well as status of development of wells (Jacob, 1947; Rorabaugh, 1953; Lennox, 1966 and Machie, 1982).

The present paper gives support for the applicability and validity of double porosity model for the computation of aquifer characteristics. Further, the Rorabaugh's method for computation of well characteristics, although developed for homogeneous media, shows applicability in fractured formations on account of the similar response patterns.

THEORETICAL CONSIDERATIONS

Double Porosity Model for Aquifer Characteristics

The concept of a double porosity model based on continuum mechanics as representative of the behaviour of a fractured reservoir was first introduced by Barenblatt et al. (1960). The model considers the rock mass broken up into blocks of irregular size and shape by fractures forming continuum media. To apply the method of analysis of infinitesimals to the fractured rock medium, Barenblatt et al. (1960) assumed that any infinitely small volume of rock consists of a large number of porous blocks, as well as a large number of randomly distributed, sized and oriented fractures. With this approach, the drawdown in hydraulic head for any distance from the pumped well and time t, in the fracture, denoted by s_1 and the corresponding drawdown, averaged over the thickness of the porous block, denoted by s_2, were regarded as the two values of drawdown at the point r of the double porosity medium.

The hydraulic parameters, using this model, for the fractured formations can be computed from following relations as given by Streltsova-Adams (1978).

$$B_1^2 = \frac{T_1}{\alpha_1 S_2} = \frac{K_1}{K_2} hH = \frac{K_1 h}{K_2/H} = \frac{T_1}{T_2'} \quad \text{....1}$$

$$T_1 = K_1 h, \; T_2' = \frac{K_2}{H} \quad \text{....2}$$

$$q = [\eta^2 (1+x^2)^2 - 4\eta x^2]^{\frac{1}{2}} \quad \text{....3}$$

$$v^2 = (\eta - 1)/\eta \quad \text{....4}$$

$$\eta = 1 + S_2/S_1 \quad \text{....5}$$

$$\theta = 4T_1 t/r^2 S_1 = 4c_1 t/r^2 \quad \text{....6}$$

The model with the delayed yield from block to fractures, is valid for finite H and h and only during the later time periods of pumping (Streltsova-Adams, 1978).

The variation in this model have been investigated by a number of workers. There is no particular agreement, perhaps because different results are possible for fractured formations which have varied heterogeneous aquifer characteristics. However, the basic concept and ultimate solutions are essentially identical, the apparent differences come from various parameters used in the derivation.

Rorabaugh's Method for Well Characteristics

Jacob (1947) formulated the drawdown in a well as consisting of two components :

$$s = BQ + CQ^2 \qquad ....7$$

where one component with first power of the discharge is the formation loss and the other component with second power of the discharge is the well loss. The equation (7) represents a linear pattern for specific drawdown (s/Q) versus discharge plot which is used to determine the well characteristics.

Rorabaugh (1953) modified the Jacob's equation for well characteristics after noting that specific drawdown/discharge plot showed concave upwards trends in a number of cases. He included the term for linear well losses and exponent for turbulent flow.

$$s = (B + C')Q + CQ^n \qquad ....8$$

The equation (8) representing a more general drawdown expression is useful to make a distinction between laminar flow, transitional flow between laminar and turbulent, and turbulent flow conditions. Although he suggested an upper limit to the exponent n around 2.5, the later workers (Lennox, 1966) show an upper limit as high as 3.5.

RESULTS AND DISCUSSION

The present study is based on pumping tests carried out at twelve well fields; the details of well fields are given in Table 1. The time-drawdown data curves from aquifer performance tests are characterised by two or three segments as shown in Fig. 1 and Fig. 2.

These segments correspond to the intrinsic parameters of the aquifer, which can be related to the response of the fractured medium, with storage coefficient S_1 and to the response of the porous medium with storage coefficient S_2. Therefore, for complete determination of the storage function of the fractured reservoir, it necessitates to obtain data for a long duration pumping test, which will show all segments of drawdown curve to calculate storage coefficient, S_1 and S_2.

Since the fractured aquifers represent heterogeneous media the concept of double porosity model for heterogeneous conditions was adopted

TABLE 1 : DETAILS OF WELL FIELDS

Well No.	Site	Diameter (mm)	S.W.L. (m)	Discharge (m^3/day)	Duration (min)	Formation
1	2	3	4	5	6	7
PW1	Anchatgiri, Dharwar	152	25.61	290.4	650	Gray-wacke
OW1	"	114	24.30	-	-	"
PW2	Yamakarni, Belgaum	152	33.63	154	1440	Basalt
OW2	"	114	32.825	-	-	"
PW3	Ambadgatti, Belgaum	152	36.50	393.6	1400	Gray-wacke
OW3	"	114	31.06	-	-	"
OW41	Sirsi, Karwar	114	4.34	227.1	1300	Laterite Gray-wacke
OW42	"	114	2.84	-	-	"
PW5	Kallapur, Dharwar	152	27.64	164	1300	Gray-wacke
OW5	"	152	30.29	-	-	"
OW61	Tondikatti, Belgaum	152	18.0	590.4	1410	Lime-stone
OW62	"	152	25.0	-	-	"
PW7	Srinivas, Saradgi, Gulbarga	152	1.915	820	3000	Basalt
OW71	"	152	1.695	-	-	"
OW72	"	152	2.06	-	-	"
PW8	Farhatabad, Gulbarga	152	8.45	475.2	1000	Lime-stone
PW9	Devargonal, Gulbarga	152	3.35	432	1000	Pink-Granite
PW10	Bamanbudni, Bijapur	152	2.82	151.2	1500	Dolomite
PW11	Khajjidoni, Bijapur	152	11.69	259.2	1400	Lime-stone
PW12	Lokapur, Bijapur	152	6.416	665.28	1500	Lime-stone

TABLE 2 : HYDRAULIC PARAMETERS COMPUTED FROM AQUIFER PERFORMANCE TESTS BY DOUBLE POROSITY MODEL

Well No.	Transmissivity of fractures $T_1 (m^2/day)$	Storage coefficient of fractures S_1	Storage coefficient of Blocks S_2	Drainage factor $B_1 (m)$	Characteristic of fracture block flow $\alpha_1 (day^{-1})$	Transmissivity of blocks in vertical direction $T'_2 (m^2/day)$	Hydraulic diffusivity $c_1 (m^2/day)$
PW1	2.63	-		-	-	-	-
OW1	16.50	6.2×10^{-3}	5.6×10^{-2}	303	3.2×10^{-3}	1.8×10^{-4}	2661
PW2	0.64	-	-	-	-	-	-
OW2	26.07	2.04×10^{-4}	1.84×10^{-3}	42.5	7.84	1.4×10^{-2}	127794
PW3	11.75	-	-	-	-	-	-
OW3	44.75	5.81×10^{-5}	5.76×10^{-3}	18.87	21.82	1.25×10^{-1}	768854
OW41	40.16	5.5×10^{-2}	4.9×10^{-1}	113	6.3×10^{-3}	3.1×10^{-3}	730
OW42	23.78	9.0×10^{-3}	8.1×10^{-2}	102	2.8×10^{-2}	2.2×10^{-3}	2642
PW5	3.11	-	-	-	-	-	-
OW61	117.45	8.2×10^{-2}	7.4×10^{-1}	49.75	6.4×10^{-2}	4.7×10^{-2}	1432
OW62	111.86	5.1×10^{-2}	4.6×10^{-1}	1119	1.9×10^{-4}	8.9×10^{-5}	2193
PW7	233.0	-	-	-	-	-	-
OW71	758.76	4.3×10^{-4}	3.8×10^{-3}	96	21.6	8.2×10^{-2}	1764558
OW72	741.52	7.11×10^{-4}	6.4×10^{-3}	82	17.2	1.1×10^{-1}	1042925
PW8	29.09	-	-	-	-	-	-
PW9	18.09	-	-	-	-	-	-
PW10	6.68	-	-	-	-	-	-
PW11	12.89	-	-	-	-	-	-
PW12	278.64	-	-	-	-	-	-

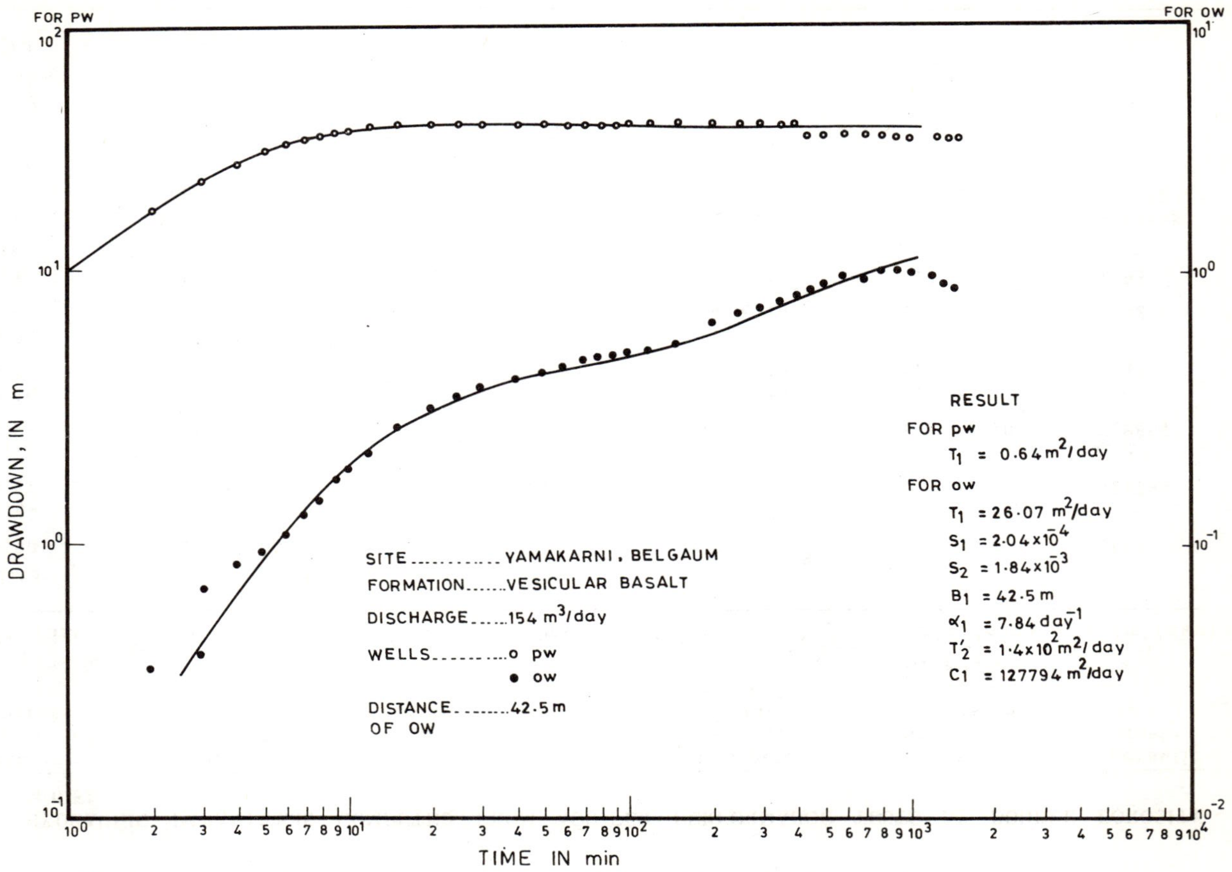

FIG. 1 : Time-Drawdown Curves for Well Field at Yamakarni, Belgaum.

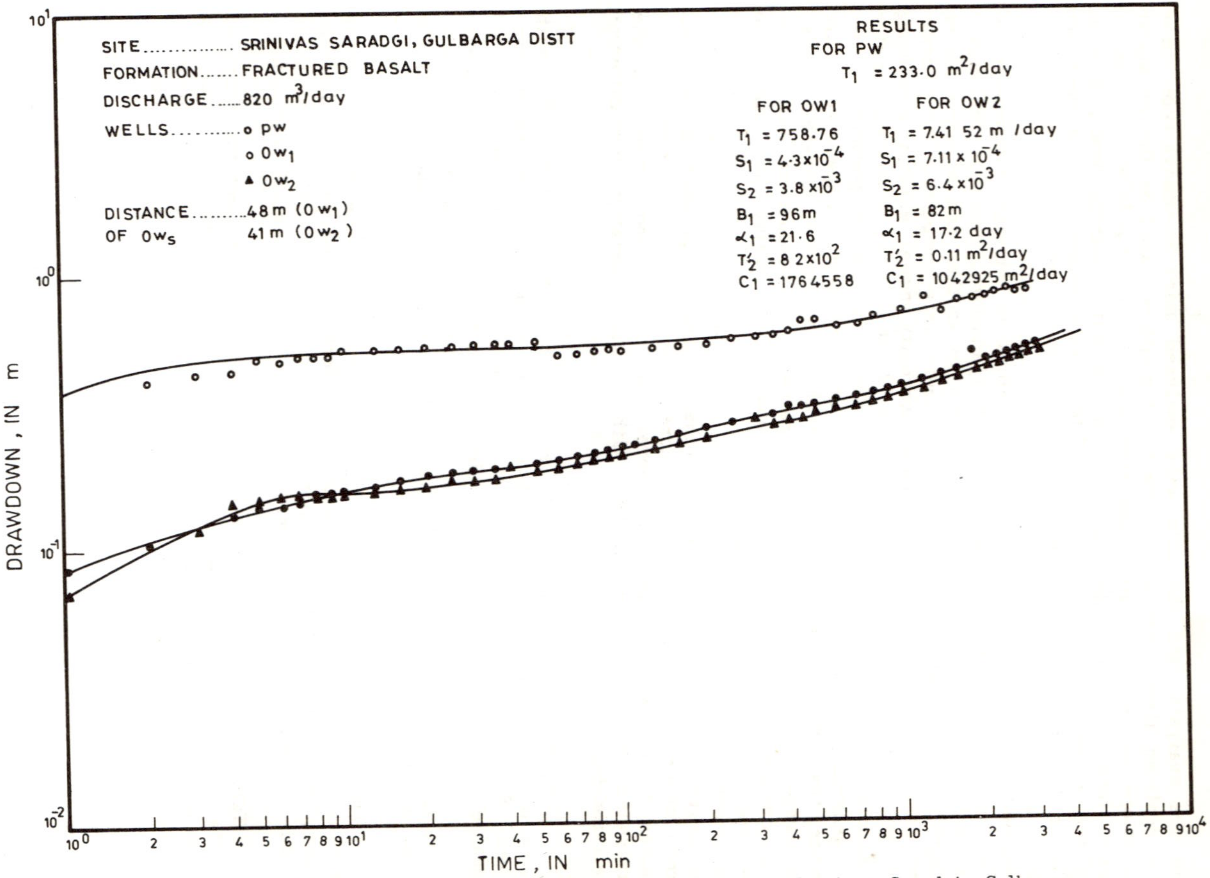

FIG. 2 : Time-Drawdown Curves for Well Field at Srinivas Saradgi, Gulbarga.

for the analysis, and the data were analysed using Streltsova-type curve for confined flow conditions in fractured aquifers based on the drawdown equations solved by Boulton and Streltsova (1977a, b, 1978). The static water levels in the wells and depth of the water bearing zones, in most of the cases indicated that the aquifers were under confined condition (Table 1). It was, also, evident from the characteristic shapes of drawdown curves which match to a greater extent with the type curve for confined condition and the computed values of storage coefficients of fractures fall within the range of confined aquifers. However, the degree of confinement varies from one well field to another and at some places slightly higher values of storage coefficients were computed. The hydraulic parameters computed using double porosity model exhibit convincing results for heterogeneous media. Table 2 shows the computed aquifer characteristic values for different well fields by double porosity model. In general, the transmissivity values obtained from pumped wells are lower than the transmissivity values obtained from observation wells. In pumped well, due to excessive drawdown on account of well losses and dewatering of fractures, the transmissivity values are likely to be lower. Therefore, the drawdown data from pumped well may not give representative values of transmissivity. Accordingly, the values of transmissivity, storage coefficient and other hydraulic parameters computed from observation well data are more dependable. The transmissivity values in graywackes are low whereas in basalts and limestones, the transmissivity values are higher with a wide variation for different well fields.

Comparison of storage coefficient values of the fractures, S_1, with those of blocks, S_2, indicate that the storage coefficient values of fractures, S_1, are lower than that of the blocks, S_2. This is expected because the coefficient of storage depends on the bulk-modulii of elasticity of the medium and since the compressibility of the fractures will be negligible as compared with that of blocks, the S_1 values are likely to be smaller.

The transmissivity values in the vertical direction in the blocks (T_2') are very low and are different in different formations. The drainage factor, B_1, and α_1 which is characteristics of fracture-block flow show a wide variation in their values for different hydrogeological conditions.

Table 3 shows the details of step drawdown tests and the computed results by Rorabaugh's method. The analysis provides the values of formation loss, well loss and the turbulency exponent, n. It is apparent that the formation loss and well loss values do not indicate any definite pattern. The well losses could be on account of resistance to flow offered at the well boundary by the irregular pattern of fractures. The factor, C, may be considered as the variable turbulence coefficient due to variable nature of fractures, the frictional factors and other geometrical factors (Machie, 1982). The turbulency factor, n, values are well within the limits suggested by Rorabaugh (1953). However, in one case the value is as high as 3.6 which indicate excessive losses due to a much greater turbulency.

The specific drawdown versus discharge plot provide a understanding of nature of flow in fractured formations (Fig. 3). The curves for well no. PW7 and PW12 show a sloping straight line which suggests the occurence of turbulent flow for the entire test period. The curves for well no. PW10 and PW11 show a concave downward trend which indicate that the well field was not properly developed prior to test and as test progresses, the development of the formation occurs. In such case, after development, retesting should be done. The curve for well no. PW8 shows a concave upward trend which may be due to occurrence of higher turbulence because of rough fracture geometry. It may also be possible that the pumping is being done of a limited fracture zone.

TABLE 3 : WELL CHARACTERISTIC VALUES OF RORABAUGH'S METHOD

Serial No.	Well No.	Rock type	Formation loss, BQ (m)	Well loss, CQ^n (m)	Turbulency exponent n
1.	PW7	Basalt	0.46	0.73	1.93
2.	PW8	Limestone	2.09	1.16	2.36
3.	PW10	Dolomite	7.97	0.28	1.94
4.	PW11	Limestone	3.15	4.65	2.18
5.	PW12	Limestone	0.15	0.66	3.6

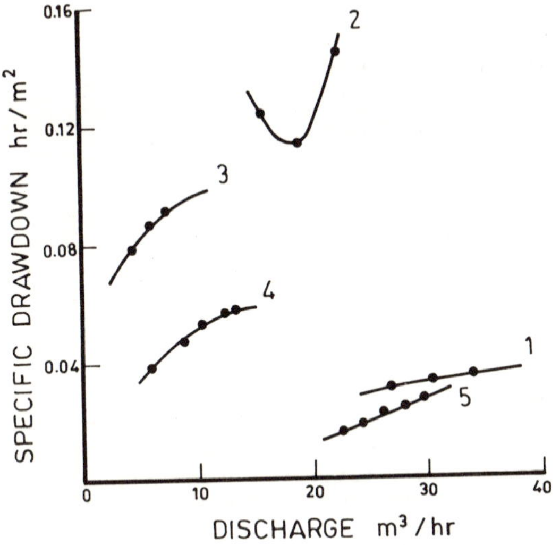

FIG. 3 : Well Characteristic Plot

CONCLUSIONS

In view of the pumping test data and field observations from the present work, it has been concluded that flow in fractured formations is complex. The double porosity model devised for heterogeneous media is the suitable method for the analysis of data from fractured formations. However, the use of this model is within certain limitations on account of departure from assumed conditions such as arbitrary fracture pattern, anisotropy etc.

The pumping test data analysed, make verification and application of the double porosity model. Moreover, the hydraulic parameters computed using this model provide transmissivity and storage coefficient values for fractures and blocks separately, the fracture-block flow relation and hence provide a better insight of the fractured aquifer.

The Rorabaugh's method for computation of well characteristics can be used for fractured formations since the aquifer response shows similar pattern as in case of homogeneous aquifers. It provides information about the flow nature and an exact turbulency exponent. The specific drawdown versus discharge plots help in understanding the flow mechanism in the aquifer. Moreover, it suggests about suitable discharge and proper development of well.

ACKNOWLEDGEMENTS

The authors are thankful to the National Bank for Agriculture and Rural Development (NABARD) for financial support through a consultancy project. The help and facilities received from various organisations viz. Action for Food Production, Central Ground Water Board, and State Groundwater Cell, Bangalore are gratefully acknowledged.

SYMBOLS

B	Linear formation loss coefficient
B_1	Drainage factor
C	Non linear well loss coefficient
C_1	Hydraulic diffusivity
C'	Linear well loss coefficient
h	Thickness of fractures
H	Thickness of block
K_1	Coefficient of permeability of fracture
K_2	Coefficient of permeability of block
n	Turbulency exponent
Q	Constant volume of water per unit time discharged from pumped well
r	Horizontal distance from pumped well axis to any point
s_1	Drawdown of hydraulic head at any point of fracture
s_2	Drawdown of hydraulic head at any point of block
S_1	Coefficient of storage for compressible fracture
S_2	Coefficient of storage for compressible block
t	Time since pumping started
T_1	hK_1; Coefficient of transmissibility of fracture
T_2	K_2/H; Coefficient of transmissibility of block in vertical direction
α_1	$\pi^2 c_2/4H^2$; Characteristics of fracture block flow
c_1	T_1/S_1

REFERENCES

Barenblatt, G.E. et al., 1960. Basic concepts in the theory of seepage of homogeneous liquids in fissured rocks. J. of Appl. Maths. and Machne. (USSR), V. 24(5), p. 1286-1303.

Boulton, N.S., 1963. Analysis of data from non equilibrium pumping tests allowing for delayed yield from storage. Proc. Inst. Civ. Engg., V. 26, p. 469-482.

Boulton, N.S., and Streltsova, T.D., 1977a. Unsteady flow to a pumped well in a fissured water bearing formation. J. Hydrol., V. 35, p. 257-269.

Boulton, N.S., and Streltsova, T.D., 1977b. Unsteady flow to a pumped well in a two layered water bearing formation. J. Hydrol., V. 35, p. 245-256.

Boulton, N.S. , and Streltsova, T.D., 1978. Unsteady flow to a pumped well in an unconfined fissured aquifer. J. Hydrol., V. 37, p. 349-363.

Bourdet, D., and Gringarten, A.C., 1980. Determination of fissure volume and block size in fractured reservoirs by type curve analysis. Presented at the SPE-AIHE, 55th Annual Technical Conference and Exhibition, Dallas, Texas, Sept. 21-24, Dallas. Soc. of Petrol. Engs. SPE-9293.

Gringarten, A.C, 1982. Flow test evaluation of fractured reservoirs. In: T.N. Narsimhan (Editor), Recent trends in Hydrogeology. Geol. Soc. Am. Special paper 189, p. 237-263.

Jacob, C.E., 1947. Drawdown test to determine effective radius of artesian wells. Trans. Am. Soc. Civ. Eng., V. 112, p. 1047-1070.

Jones, F.O. Jr., 1975. A laboratory study of the effects of confining pressure on fracture flow and storage capacity in carbonate rocks. J. Pet. Technol., V. 1, p. 21-27.

Lennox, D.M., 1966. Analysis and application of step drawdown test. J. Hydraul. Div., Proc. Am. Soc. Civ. Eng., V. 92, p. 25-48.

Machie. C.D., 1982. Multi rate testing in fractured formations. In papers of the groundwater in fractures rock conference 1982, Australian Water Resources Council Conference, Canberra.

Mishra, S.K., 1986. Analysis of pumping test data from bore wells in fractured formations of Karnataka State, India, M. Tech. Dissertation (unpublished), Department of Earth Sciences, University of Roorkee, Roorkee, p. 191.

Rorabaugh, W.I., 1953. Graphical and theoretical analysis of step drawdown test of artesian well. ASCE Proc. Separate No. 362, V. 79.

Streltsova, T.D., 1976. Hydrodynamics of groundwater flow in a fractured formation. Water Resour. Res., V. 12(3), p. 405-415.

Streltsova-Adams, T.D., 1978. Well hydraulics in heterogeneous aquifer formations, In : V.T. Chow (Editor), Advances in Hydrosciences, V. 11, p. 357-423.

ANALYSIS OF TEST PUMPING DATA FROM UNCONFINED AQUIFERS BY SIMULATION

Deepak kashyap*, P.K.Sarkar** and A.M. Khaled**

ABSTRACT

Application of Neuman's theory coupled with the principle of superposition is suggested for analyzing test pumping data from unconfined aquifers. The procedure, essentially oriented towards requirements of practising professionals, involves repeated intelligent guesses of the parameters till a close enough match between the observed and the computed time-drawdown curves is obtained. Such an analysis can account for partial penetration of wells, extra intermittent pumpings from the aquifer during the test and not high enough S_y/S ratio. The estimated parameters include storage coefficient, specific yield, transmissibility and vertical anisotropy (K_z/K_r). A case study is included.

INTRODUCTION

The type curve procedure based upon Boulton's theory (Boulton, 1954, 1963) is widely accepted for analysis of test pumping data of unconfined aquifers. However, the procedure incorporates a number of assumptions, which may be severely or mildly violated in most of the pumping tests on unconfined aquifers.

Boulton's theory attributes the delayed yield phenomenon to sluggish vertical flow above the falling watertable and quantifies it empirically by introducing a constant α, the reciprocal of the delay index. The role of unsaturated zone in the mechanism of delayed yield is a subject of controversy Neuman(1972).

The assumption of fully penetrating pumped and observation wells does not generally hold good in practice. The partial penetration may have a pronounced effect on

---

* Reader
** Ex-graduate students

Department of Hydrology
University of Roorkee
Roorkee, U.P. 247 667, India.

drawdowns monitored in the observation wells. Theoretically, it is possible to eliminate such an effect by placing observation wells sufficiently away from the pumped well. However, the drawdowns at such a distance may be too small to be measured accurately.

It is presumed in Boulton´s type curve analysis that the observed drawdowns are caused exclusively by the controlled abstraction of water from the pumping well. For this assumption to hold good, there must not be any other pumping from the aquifer in the hydraulic vicinity of the observation wells during the pumping test. However, in practice it may not be always possible to persuade the well owners not to operate their wells during a long duration pumping test. Such uncontrolled pumpings cause additional drawdowns in the observation wells and worsen the match between the type curve and the observed time-drawdown curve. It is possible to account for the additonal drawdowns by reconstructing the type curves provided all such pumpings commence at the same time as the test pumping and continue to occur at uniform rates throughout the duration of pumping test. However, in practice the uncontrolled pumpings are intermittent in nature. If left unaccounted, the additional drawdown lead to an underestimation of aquifer parameters.

Boulton´s theory is applicable for all values of (S_y/S); S_y and S being respectively the specific yield and storage coefficient of the aquifer. However, in order to avoid multiple type curves, Boultons´ type curves are constructed for S_y/S ratio tending to infinity. Such an assumption may hold provided S_y/S is atleast more than 30. For most of the good unconfined aquifers the ratio may be well over 30. But these days, with depleting water resources, it is becoming necessary to tap even not so good aquifers for which the ratio may be less than 30. The second segment of the time-drawdown curve of such aquifers is not horizontal. This may cause additional subjectivity in the analysis.

The paper presents a computer assisted numerical simulation procedure for analyzing pumping test data, from unconfined aquifer, by Neuman´s theory and the principle of superposition. The procedure accounts for partial penetration of wells, uncontrolled pumpings and low S_y/S ratio. The procedure is illustrated by analyzing data from a pumping test, conducted on partially penetrating wells. During the test, intermittent uncontrolled pumping from three private tubewells in the near vicinity are known to have occurred and their recorded relevant data were available.

THE PROPOSED PROCEDURE

The various features of the proposed procedure are described in the following paragraphs.

Neuman´s Theory

Neuman(1972,74) presented an analytical solution of two dimensional axi-symmetrical saturated flow equation which is capable of reproducing all the three segments of the time-drawdown curve observed in practice.Thus, he could rationally explain the mechanism of delayed yield (which he renamed as \`delayed watertable response´) by sluggish vertical transfer of water within the saturated zone i.e., without taking any recourse to unsaturated zone. He described the dimensionless drawdown (4π Ts/Q) in terms of seven dimensionless variables i.e., $\sigma = S/S_y$; $\beta = K_z/K_r (r/b)^2$; $t_s = Tt/r^2S$; $Z_{1D} = Z_1/b$; $Z_{2D} = Z_{2D} = Z_2/b$; $l_D = l_1 d/b$ and $d_D = d/b$, where s is the drawdown at a distance r from the pumped well at a time t since the commencement of pumpage at a uniform rate Q; K_r and T are the storage coefficient, lateral hydraulic conductivity and transmissibility of aquifer; Z_1, Z_2, d,b and l are the vertical dimensions of aquifer and the wells (Fig.1).

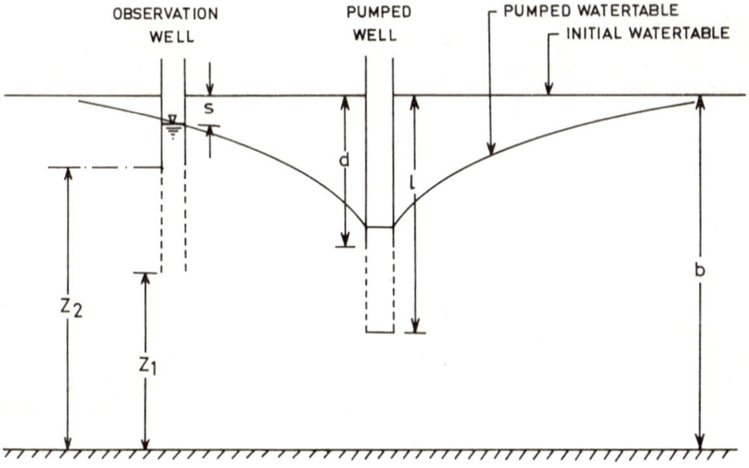

FIG. 1. - SECHEMATIC DIAGRAM OF AN UNCONFINED AQUIFER WITH PARTIALLY PENETRATING WELLS

Khaled(1984) developed a computer program in FORTRAN for calculation of average drawdown by Neuman´s theory.

Principle of Superposition

The contribution of the uncontrolled pumpings towards the measured drawdowns can be estimated by invoking the principle of superposition with respect to the discharge and drawdown.

The necessary linearity may nearly hold provided the drawdowns are small as compared to the average saturated thickness of the aquifer.

The intermittent uncontrolled pumpings or even time-variant test pumping can be mathematically replaced by a number of series of uniform pumpings (-ve or +ve) each starting at a different time but all continuing till the end of the test pumping. Thus, the resulting drawdown at any time can be estimated by summing up algebraically the individual drawdowns or build-ups, each computed by Neuman´s theory.

Let the uncontrolled pumpings be described by n series of time invariant pumpings. So, total drawdown in an observation well at a radial distance r from the test pumping well at time t since the beginning of test pumping is given by

$$s^T(r,t) = s(r,t) + \sum_{i=1}^{n} s(r_i, t-t_i^*) \qquad \ldots (1)$$

where, s(r,t) is the drawdown for the main pumped well; t_i^* is the time lag between the beginnings of the test pumping and the ith conceptual pumping (t_i^* is negative when the uncontrolled pumping starts earlier and the vice versa) r_i is the radial distance from the observation well to the ith conceptual well.

Simulation Procedure for Parameter Estimation

The time-drawdown curves can be computed for various values of four parameters namely, S, S_y, T and K_z/K_r. The parameter estimation essentially involves arriving at such values of the parameters, which yield the best reproduction of the observed time-drawdown curves. The classical visual approach (generally termed as the type curve procedure), though simplest, is not feasible if all the features discussed earlier are intended to be accounted for. The numerical approach involves minimization of a functional (generally sum of the squares) of the residues i.e., the difference between the observed and the computed drawdowns. The functional can be minimized by standard optimization techniques (Kashyap et al, 1988). However, the necessary expertise and mainframe computer facilities may not be generally available in practising organisations.

Simulation procedure is essentially a mix of the visual and the optimization techniques. The residue functional (i.e., sum of the squares of the residues) is computed for a large enough sets of parameter values. The choice of a subsequent set of trial parameter values can be based upon superposing the computed time-drawdown curve, in accordance with the preceding set of values, over the observed time-drawdown curve. The nature and extent of the deviations can enable the analyzer to 'improve' upon the parameter values; recapitulating that

i) Storage coefficient (S) affects only the first segment of the time-drawdown curve. Any increase in S decreases the computed drawdowns and vice versa.

ii) Transmissibility (T) influences all the three segments. Any increase in T decreases the computed drawdowns and vice versa in all segments. Further, the influence of uncontrolled pumping gets more pronounced and is reflected in the time-drawdown curve with a smaller time lag, as the transmissibility increases and vice versa.

iii) Specific yield (S_y) affects only the third segment. Any increase in S_y decreases the computed drawdowns and vice versa.

iv) Vertical anisotropy (K_z/K_r) influences the second segment. An increase in (K_z/K_r) decreases the computed drawdowns, causes an earlier onset of the second segment and shortens the span of the second segment and vice versa.

The associated improvement in the match can be quantified by computing the residue functional and comparing it with the preceding value.

This way the iterations can be continued till no further improvement in match is possible i.e., the visual match does not improve and the residue functional does not decrease, any further.

CASE STUDY

A pumping test of 5760 minutes duration was conducted by the Groundwater Investigation Organisation, U.P. on unconfined aquifer at Khajuri-Akbarpur,Distt. Saharanpur (Personal Communication, Er. Raj Pal Singh, Executive Engineer, GWIO, Roorkee,U.P. 247667, India). The discharge from the pumping well was 4620 m^3/day and drawdowns were monitored in two observation wells. The assembly diagram of pumped and observation wells were available. The pumping well was fully penetrating while the observation wells were partially penetrating.

The time-drawdown data reveal that the drawdowns displayed declining trends between 500 minutes to 900 minutes and at 2400 and 4000 minutes. The phenomenon occured in both the observation wells causing sudden kinks in the time-drawdown curves.

The appearance of such kinks on time-drawdown curves can be due to one or more of the following causes-time variation of test pumping discharge; sudden closure of pump due to mechanical failure; uncontrolled pumping from private tube wells, and their closures during the test in near vicinity of test pumping.

An enquiry revealed that in the present case the kinks are on account of intermittent uncontrolled pumping from three private tubewells in the near vicinity. Fortunately, the relevant data (Tables 1 and 2) and assembly diagrams of the wells were mostly available or could be subsequently collected.

Sl. No.	Description	Diameter mm.	Discharge m^3/day	l (refer Fig 1) m.	d m.	Distance from Obs. Well 1 m.	Distance from Obs. Well 2 m.
1.	Private Tube-Well 1	100	571.720	46.555	13.015	210.0	230.0
2.	Private Tube-Well 2	100	857.570	28.265	16.021	280.0	255.0
3.	Private Tube-Well 3	100	1000.500	46.560	13.020	260.0	210.0

Table 1 - Characteristics of Private Tube Wells

Sl.	Description	Time Since Beginning of Test Pumping					
		Period 1		Period 2		Period 3	
		From (min.)	To (min.)	From (min.)	To (min.)	From (min.)	To (min.)
1.	Private Tube Well 1	-425	+355	+2455	+3235	+4675	+5335
2.	Private Tube Well 2	-425	+355	+2455	+3235	+4675	+5335
3.	Private Tube Well 3	-425	+335	+2455	+3235	+4675	+5335

Table 2 - Schedule of Running Hours of Private Tube Wells

The intermittent uncontrolled pumpings were replaced mathematically by 18 pumping series (Sarkar,1988).

The data were analyzed by the procedure outlined in the preceding paragraphs. The results are given in Table 3. It can be seen that Boulton's type curve analyses have yielded lower estimates of transmissibility in case of well 1 and higher

Description	Method of analysis	Parameters Estimated				
		Transmi-ssibility (m^2/day)	Storage Coeffi-cient	Specific Yield	Vertical Aniso-tropy	day
Observation Well 1	Proposed Proceedure	1330.0	0.0008	0.101	0.013	Na
	Boulton's type curves	1050	0.0005	0.07	Na	0.23
Observation Well 2	Proposed Procedure	1364	0.001	0.054	0.025	Na
	Boulton's type curves	1671	0.0006	0.005	Na	1.32

Table 3 - Estimated Aquifer Parameters

estimates in case of well 2. The underestimation in case of well 1 could be due to lack of accounting of the uncontrolled pumpings,since the extra drawdowns are explained by lowering the parameter value.The overestimation in case of well 2 does not deserve much attention since the match of the time-drawdown curve of well 2, unlike of well 1, was very unsatisfactory (Sarkar, 1988). This is possibly because well 2 generally lies closer to the locations of the uncontrolled pumpings.

The reproduction of the observed drawdowns by the estimated parameters is shown in Figure 2. The improvement in the reproduction as a consequence of incorporating the uncontrolled pumpings can be seen by comparing the firm and the dotted lines.

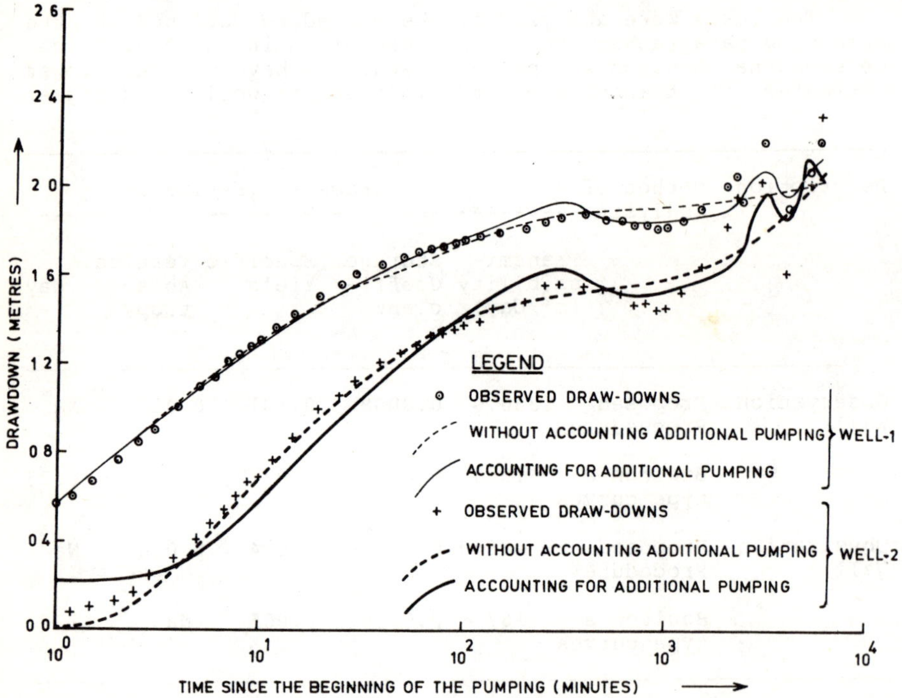

Fig. 2 - REPRODUCTION OF THE OBSERVED TIME DRAWDOWN DATA

CONCLUSIONS

The suggested procedure of test pumping analysis can account for, among others, time variant test pumping discharge, uncontrolled pumpings and partial penetration of wells provided adequate data are recorded during the test. Thus, the following suggestions are made:

(i) Periodic measurement and documentation of the test pumping discharge.

(ii) If there are some pumpings from wells, tapping the same aquifer in near vicinity of the test, just prior to or during the test pumping then discharges and running hours of such uncontrolled pumpings and distances of the wells from the observation wells should be measured and recorded.

(iii) The position of screen(with reference to unpumped water table level) and its length should be properly recorded for test pumping well, observation wells and the wells which may have been operated just prior to or during the test pumping.

Such an approach leads to more rational analysis of test pumping data, as reflected by a better reproduction of the observed time-drawdown curves.

ACKNOWLEDGEMENT

The authors are grateful to Er. Raj Pal Singh, Executive Engineer, Groundwater Investigation Organisation, U.P. for his help in carrying out the study.

REFERENCES

Boulton,N.S., 1954. Unsteady radial flow to a pumped well allowing for delayed yield from storage. Ass. Int. Hydrol.Sci.Rome Publ.37(2), p.472-477.

Boulton, N.S., 1963. Analysis of data from non-equilibrium pumping tests allowing for delayed yield from storage. Proc. Inst. Civil Eng.(London), 26, p.469-482.

Kashyap, D., Dachadesh, P., and Sinha,L.S.J.,1988. An optimization model for analysis of test pumping data. Groundwater, v.26(3).

Khaled, A.M., 1984. Analysis of test pumping data for unconfined aquifers. M.Eng. Thesis, Dept. of Hydrology, University of Roorkee.

Neuman, S.P., 1974. Effect of partial penetration on flow in unconfined aquifers considering delayed gravity response. Water Res. Res., 10(2), p.303-312.

Neuman, S.P., 1975. Analysis of pumping test data from anisotropic unconfined aquifers considering delayed gravity response. Water Res.Res., 11(2), p.329-342.

Sarkar, P.K., 1988. Analysis of test pumping data of unconfined aquifers by simulation. M.Eng. thesis, Dept. of Hydrology, University of Roorkee.

A CRITICAL EVALUATION OF PUMPING TEST DATA FROM FRACTURED CRYSTALLINE FORMATION

P.N. Ballukraya*, R. Sakthivadivel**, B. Rukmangada Reddi*.

* Dept. of Applied Geology, University of Madras, Madras 600 025, India.

** Centre for Water Resources, Anna University, Madras 600 025.

ABSTRACT

Conventional methods of estimation of aquifer parameters pose difficulties when applied to pumping test data of borewells in crystalline rock aquifers. The time drawdown curves obtained from pumping and observation wells reflect the variation in hydraulic conductivity both in space and direction. Formation of segments in the time-drawdown curves need not necessarily be due to barrier boundaries as .has been observed in alluvial aquifers, but may be the effect of varying hydraulic conditions in the aquifer itself with respect to pumping time. A multiple pumping test carried out in a crystalline formation was analysed and it is concluded that the data from the pumping well itself is a better indicator of Transmissivity characteristics than observation wells and that they can be estimated more accurately using a numerical method than by conventional methods.

INTRODUCTION

In a borewell drilled in crystalline rocks, water is transmitted due to the secondary porosity developed as a result of weathering, fracturing or jointing of the parent rock. The thickness of weathered and/or fractured zone varies within short distance resulting in an uneven thickness of both the aquifers. The pattern of fractures and joints systems introduce an element of directional variation in Hydraulic conductivity. Determination of aquifer parameters under such circumstances becomes highly ambiguous due to the nature of the joint pattern.

Further, majority of the borewells derive a large portion of water from the weathered rock zone which acts as a storage reservoir. This stored water is transmitted to the well through the fractures which receive their supply through vertical leakage from the weathered zone. Many a time, the weathered zone acts as unconfined aquifer and the fractured zone is mostly under confined condition. this interaction of confined and unconfined aquifers makes the mathematical treatment of hard rock aquifers very complex.

The common practice is to treat the entire depth, below the water level, upto the deepest water bearing horizon as a single homogeneous aquifer. The assumption is valid to the extent that this zone is made up of an interconnected system of fractures/joints, surrounding blocks of impervious rock. In deriving flow equations for such an aquifer, two basic approaches are made: 1. the discrete approach where equations of flow are separately written for each fracture (1,4) and 2. the continuum approach wherein, the entire system is treated on par with a porous medium.

HARD ROCK AQUIFERS

A study of closely spaced borewell logs indicates that the water bearing fractures in the relatively unweathered rock, lying below the weathered zone are more or less horizontal, or they are a subdued replica of surface topography, since most of them are tension release joints (1,2). It is also observed that these fractures contribute the largest share of water to a borewell yield, the rest coming from the weathered zone (1,8).

Through model studies, Snow (2) showed that the hydraulic conductivity of fractured rocks can be represented by anisotropic permeability tensors, determined by the number of joint sets present. The hydraulic conductivity is maximum along the near-horizontal sheet fractures than that in the inter-connecting sets of inclined/vertical joints present in the weathered rock. Since the borewell gets water from both weathered rock and the fractures, any pumping test data analysis should take into account the factor of simultaneous existence of unconfined and confined (or leaky) conditions within the aquifer. Further, casing pipe is generally lowered upto relatively harder bed rock, which is much below the static water level. The depth of blank casing, alters the time-drawdown characteristics of the pumping well. The observation wells, when present, may have different length of lining than the pumping well even if they are closeby, and this again affects the time-drawdown characteristics of the observation wells.

PUMPING TEST DATA ANALYSIS

Several attempts have been made in the past to find out an appropriate mathematical model for analysing the data from hard rock aquifers. However, it is still common to use the conventional methods such as Theis, Jacob's, Boulton's leaky aquifer and Neuman's unconfined aquifer methods. Rao, B.R. (2) gives several test cases using Jacob's straight line method to solve for Transmissivity in hard rock areas taking only the initial segment of drawdown-log.time curves, using the pumping well itself as observation well. Where unconfined conditions are indicated he employed Walton's method. Radhakrishnan and Venkateshwarulu (5) have used all the methods while evaluating aquifer response in an observation well constructed in ultrabasic crystalline rocks. Uhl Jr. and others (8) mainly employed the cooper-Jacob method (Non leaky confined aquifer) while detemining the aquifer parameters in several crystalline areas. Rushton and others have discussed the hard rock aquifer characteristics in a number of papers (6,7).

PUMPING TESTS AT ALATHUR (80°10'30"N. -12°41'30"E.)

In the present study four borewells were used, all of which are close to each other (fig.1). Two pumping tests were carried out, one in borewell no.1 and the other in no.4. Observations were made in all the four borewells during the tests, which were conducted over a time period of 1020 and 720 minutes respectively. Recovery was noted for about 150 minutes after the cessation of pumping. The hydrogeological details of the borewells are given in fig.1. The general lithology is about 6 meters of overburden (clay,soil), about 8 to 12 meters of weathered charnockite and fresh rock below with a water bearing fracture around 23 meters depth in all the four borewells. The static water level in the area is about 6 meters below ground level. The length of blank casing inserted in the borewells is 13.5m, 10.5m, 12.0m, and 8.0 meters respectively in wells 1 to 4.

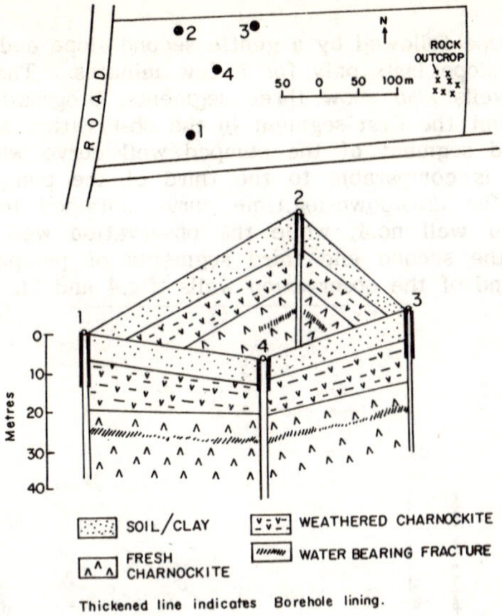

Fig.1 - Location map (top) and fence diagram showing subsurface geology.

Analysis of Data

The pumping test data obtained is quite interesting and attempts to analyse the data using conventional techniques proved difficult. For example, the drawdown-log.time curve for the pumped well no.4 (fig.2) shows three clear segments:

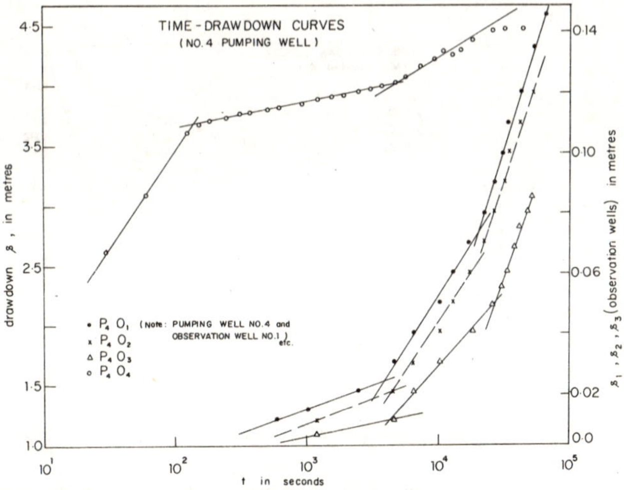

Fig.2. - Log.t. - drawdown curves for pumping well no.4 and observation wells 1,2, and 3.

an initial steep slope followed by a gentle second slope and a third steeper slope. The initial steep slope lasts only for a few minutes. The drawdown curves for the observation wells also show three segments, progressively steeper from the preceeding ones, but the first segment in the observation well curves is comparable to the second segment of the pumped well curve while the second in the observation wells is comparable to the third of the pumping well. When well no.1 is pumped, the drawdown-log.time curve obtained for the pumped well is similar to that of well no.4, while the observation well data shows only two segments, again the second and third segments of pumped well comparable to the first and second of the observation wells (fig.4 and 5).

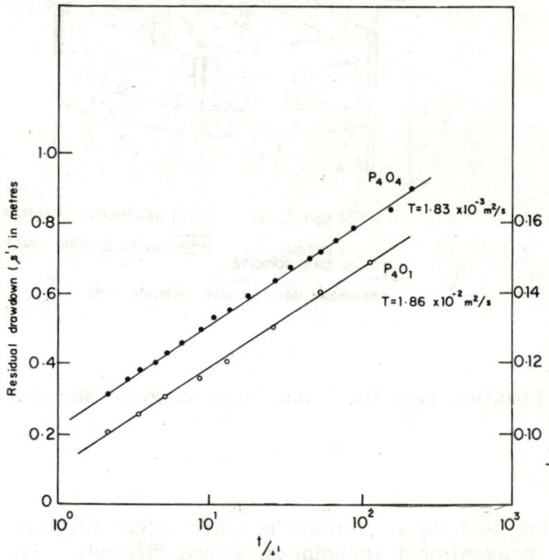

Fig.3. - Residual drawdown curves for pumping well no.4.

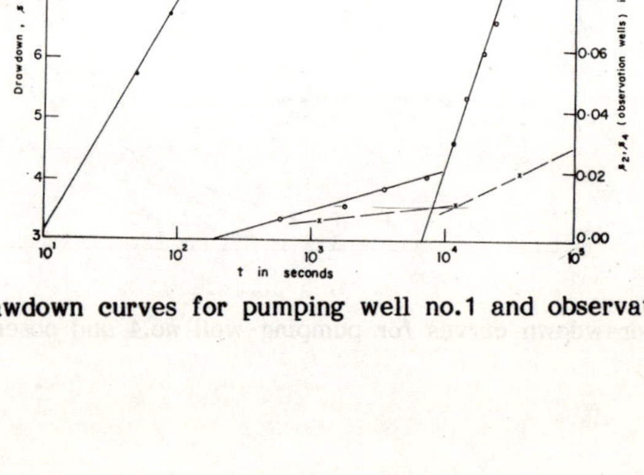

Fig.4 - Log.t. - drawdown curves for pumping well no.1 and observation wells 2 and 4.

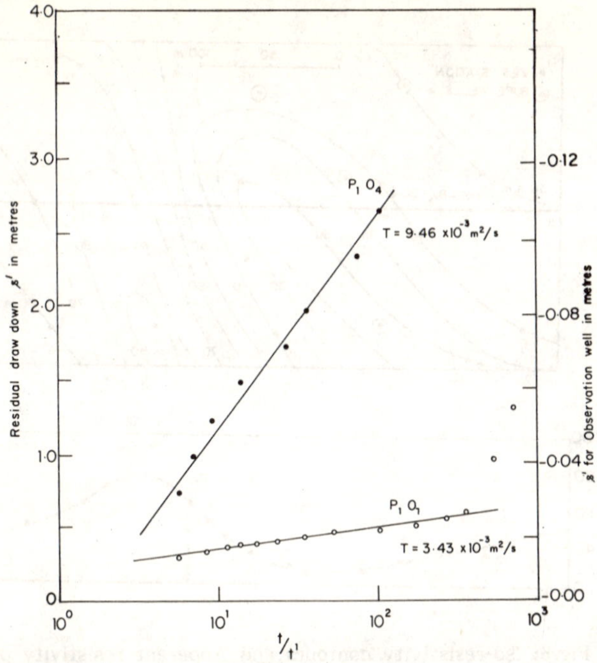

Fig.5 - Residual drawdown curves for pumping well no.1.

It is interesting to note that when well no.1 is pumped, the drawdown in wells 1,2, and 3, which are 98m.,76m. and 74 meters away from it are 0.135m.,0.12m. and 0.075 meters respectively. That is, the nearest well shows a smaller drawdown while the farthest well shows a much higher drawdown! However, when well no.1 is pumped, the observations wells 2 and 4 register 0.02m. and 0.13 meters drawdown while no.3 well remains unaffected (zero drawdown) even after 1020 minutes of pumping. This indicates the existence of abrupt hydraulic discontinuity in the area, particularly between wells 3 and 4 as compared to other zones.

For a homogeneous confined aquifer, the drawdown-log.time curve is a straight line. The formation of segments generally indicates boundary conditions, - in this case, impermeable boundaries if observation well data are considered and a recharge boundary followed by an impermeable boundary in case of pumped well data (2). In fact, the electrical resistivity surveys have indicated that the thickness of weathered rock zone decrease rather sharply to the east, at a distance of about 150 meters from well no.1 (fig.6). In the well field itself, there are minor variations in the weathered rock zone thnickness - from 14 to 18 meters as shown from drilling logs. This also suggests the possibility of an undulating bed rock topography. However, this may not affect the drawdown trend to any appreciable levels, since at no time during the test the water levels went below 7 meters in any of the observation wells. The possible barrier boundary, formed by the thinning of weathered rock layer, is also quite far away to cause large drawdowns, considering the fact that the maximum drawdown recorded is only 0.135 meters.

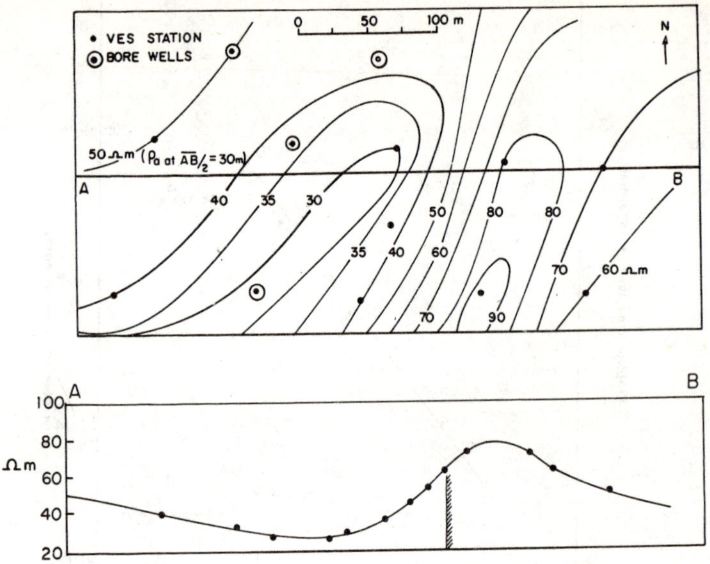

Fig.6: Iso-resistivity contours and Apparent resistivity profile

Various conventional techniques such as Cooper-Jacob's straight line method, Theis recovery method, Walton's leaky aquifer type curves and Neuman's type curves for unconfined aquifer were used to analyse the pumping test data. With Jacob's method, different T and S values are obtained for the various segments in the drawdown curves. Walton's and Neuman's type curves were used by considering the aquifer as leaky confined or unconfined. However, the observation wells data are not amenable for analyses with these methods as the log-log plots have a progressively increasing slope. The T and S values obtained by the various methods are given in Table-1. However, it is observed that the S values obtained for the pumping well data are incorrect because of the initial steep slope of the drawdown curves.

DISCUSSION

The initial steep rate of drawdown in the pumped well is due to the fact that blank casing has been provided upto 13.5m and 8.0 meters respectively in wells 1 and 4, which are quite below the static water table, thus shutting off a large portion of the saturated zone. Thus, in the first few minutes of pumping, because of the low transmissivity of the aquifer, the borewell gets de-watered and hence the steep drawdown. Once the pumping water level goes below the borewell lining, and a substantial portion of the aquifer is exposed to the well, the rate of drawdown appreciably decreases. This is clearly reflected from the fact that in the first 10 minutes of pumping, the drawdowns in wells 1 and 4 are 8.8 and 4.1 meters respectively, which compares with the length of blank casing in these wells - 13.5 and 8 meters. Therefore, under such conditions, the initial segment in the time-drawdown curve has to be neglected for the purpose of analysis. Unfortunately, because of this initial segment, the Storativity values cannot be determined accurately by conventional techniques, using pumping well data. The value obtained for the second segment will be absurdly low due to the steep slope of the first segment as well as the very small r^2 value in the equation $S = 4Tt_0/r^2$.

A. Jacob's straight line method :

	Segment 1		Segment 2		Segment 3	
	T	S	T	S	T	S
P_4O_4	-	-	0.00244	3.0×10^{-11}	0.000784	0.00163
P_4O_1	0.0323	0.00013	0.00832	0.60355	0.00387	0.00568
P_4O_2	0.0305	0.00412	0.00819	0.00667	0.00416	0.0105
P_4O_3	0.0457	0.00714	0.011	0.0119	0.0044	0.0185
P_1O_1	-	-	0.000915	1.1×10^{-7}	0.000211	0.0844
P_1O_2	0.11	0.000158	0.0261	0.0165		
P_1O_4	0.0422	0.0021	0.00416	0.00738		

B. Theis Recovery Method :

P_4O_4 : T = 0.00183 ; P_4O_1 : T = 0.0186 ; P_1O_1 : T = 0.00343 ; P_1O_4 : T = 0.00946

C. Walton's method for Leaky Aquifer Condition:

P_4O_4 : T = 0.0000145 ; S = 0.0147 ; P_1O_1 : T = 0.0000211 ; S = 0.0221

D. Neuman's method for Unconfined Condition:

P_1O_1 : $T_{(A)}$ = 0.0000636 ; $T_{(B)}$ = 0.000061 ; P_4O_4 ; $T_{(A)}$ = 0.000196 ; $T_{(B)}$ = 0.000179

Note: P_4O_4 means Pumping well is no.4, Observation well is also no.4.
P_1O_4 means Pumping well is no.1, Observation well is no.4, etc.,

Table 1: Values of Transmissivity, T (in m^2/sec) and Storativity, S.

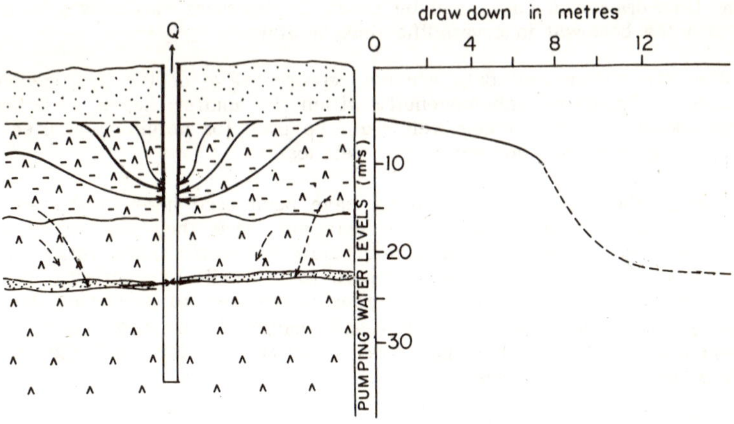

Fig.7: Groundwater flow into the borewell from the aquifers.

Vertical Flow Component

The presence of blank casing below the water table induces a very large vertical flow component in the initial stages of pumping (fig.7). Once the pumping water level goes below the lining, the rate of increase of head loss is comparatively small because of the availability of a large area for flow and hence, the log.t.-drawdown curve has a gentle slope. Gradually, the vertical flow component decreases with time and lowering of water surface. However, with the expansion of the cone of depression and further lowering of pumping water level, three factors come into play: (1) there is a vertical leakage to the fracture aquifer through the aquitard (unweathered rock), (2) the area of flow at the well face decrease and (3) contribution to the well discharge from the fracture aquifer increases. All these will result in greater loss of head as the area of flow is small, equal to the fracture thickness (0.1 to 0.2 m.) and the saturated weathered zone at the well face at any given time. The change in slope occurs (to a steeper gradient) when the pumping water level is at 15.9 m. in well no.1, when only 2.1 m. of saturated weathered rock zone is open to the well, out of the initial 12 meters. In well no. 4, the change in trend takes place once the pumping water level reaches 10.7m., leaving only 2.8 m. of saturated weathered rock out of the total 7.5 meters. In well no.1, initially, 4.5 m. of weathered rock (38%) is open to the well compared to 5.5 m. (73%) in well no.4. This is possibly the reason for a higher T value (for the second segments) for well no.4, and larger time for the increase in the slope of the curve to take place (4800 seconds as against 2300 seconds in well no.1). The T values obtained for the third segment is lower than that obtained for the previous segment, though the hydraulic conductivity (K_x) may well be higher than that of the weathered rock zone.

CONCLUSION

From the foregoing analysis it is concluded that the first segment in the drawdown curve (for the pumping well) is due to dewatering of storage water in the borehole because of blank casing below water table; the second segment represents flow predominantly from the weathered rock zone, while the third segment is the result of decrease in flow area and larger contribution from the underlying fracture aquifer which is under semiconfined condition. Therefore, the last two segments of the time-drawdown curve can be taken to represent two phases of ground water flow into the borewell in crystalline rock aquifers.

The observation well data are not representative of the hydrogeological conditions due to directional inhomogeneity within the aquifer; hence, it is suggested that the drawdown in the pumping well itself be used for obtaining aquifer parameters, after making necessary correction for well losses.

Though the conventional methods have been used here to get an approximate idea of the aquifer parameters, it is apparent that they are not adequate. The various factors such as borewell lining below the water table, large initial vertical flow component, anisotropism with regard to hydraulic conductivity, varying contribution from weathered and fractured zone in response to increasing drawdown, etc., all make the conventional techniques and analytical methods inadequate. The problem may be dealt with a numerical model such as the one suggested by Rushton which is being tried at present.

REFERENCES

1. Ballukraya,P.N., and R.Sakthivadivel., 1984, Analysis and Interpretation of Electrical Resistivity Data from Hard rock areas for Ground water Exploration, Report No.33, Central Board of Irrigation and Power, New Delhi.

2. Evaluation of Aquifer Parameters, 1982, Central Ground Water Board Manual, New Delhi.

3. Freeze, A.R., and J.A. Cherry, 1979, Groundwater, Prentice-Hall.

4. Kumaraswamy,P., and R. Sakthivadivel, Analysis of Pumping test data from shallow wells in hard rock aquifers.

5. Radhakrishnan,I., and T. Venkateshwarulu, 1980, Aquifer response in the ultramafic field area of Sukhinda valley - Orissa, (India) , Ground water Vol.18, pp.479-484.

6. Rushton,K.R., 1986, Vertical flow in heavily exploited hard rock and alluvial aquifers, Groundwater Vol.24, pp.601-608.

7. Rushton,K.R., and R. Sakthivadivel, 1989, Well Hydraulics with special reference to hard rock terrains, J1. of App. Hydrology, Vol.1, pp. 1-16.

8. Uhl Jr.V.W., K.Nagabhushanam and J.O. Johanson, 1979, Hydrogeology of Crystalline Rocks: Case studies of Two areas in India, Nordic Hydrology, Vol.10, pp.287-308.

COMPUTATION OF TRANSMISSIVITY FROM STEP-DRAWDOWN TESTS

M.SRINIVAS
Action for Food Production (AFPRO)
Geohydrological Investigation Team
Hyderabad - 500 007.

ABSTRACT

The total drawdown, s_w, in a pumped borewell is given by the equation $s_w = BQ+CQ^2$, where BQ and CQ^2 are drawdowns caused due to aquifer loss and well loss components. The main drawback in Transmissivity computations from Step-drawdown test data is the significant well-loss component.

In this paper, a method is suggested for estimating the Transmissivity, T, from pumped data of step-drawdown tests. Employing Jacob's method of analysis, well losses are computed and corrected for in the draw-down values for each step. The corrected drawdowns are used in the Logan's formula modified appropriately. Logan;'s formula, as such, uses maximum drawdown --s_{mw}-- values and the resulting T values were erroneous upto 50% or more, and hence were at best first-time-round-approximations.The very well with each other as also with T values computed from long-duration test data. A computer program in FORTRAN is also given for complete analysis and results by inputting just the field data.

The method suggested here enables cost-effectiveness and time-saving vis-a-vis the tedious, conventional long-duration tests in the determination of Transmissivity.

INTRODUCTION

As is well known, groundwater forms one component of the hydrologic cycle in nature.Though a replenishable resource, the accumulation of groundwater in the lithological formations, inter alia, is a function of time and hence the need for proper management of the precious resource. The fact that groundwater resources are not uniformly distributed in natures underlines the need for taking effective measures for groundwater exploration and management.

The coefficient of Transmissibility and Storage are two essential hydraulic parameters that define the aquifer characteristics. Several methods have been developed by various workers for computation of these two aquifer characteristics which have far-reaching significance.

The basic data collection for the purpose includes measurement of water- levels in a borewell corresponding to discharge rates for varying intervals of time. There are two types of tests usually conducted; i) constant discharge test-where the discharge rate is kept constant; and ii) variable discharge test - where the discharge rate is varied, usually deliberately. While the constant discharge test data is generally relied upon for calculating the coefficient of Transmissibility, or Transmissivity, T, and coefficient of storage, S, with the observation wells available, there are not many reliable methods for calculating these aquifer characteristics from step-drawdown test data.

In this paper, a simple method of calculating the Transmissivity values from step-draw down test data is discussed, which has been successfully tested on several data sets.

The essential methodology consists of analysing the step-drawdown test data by Jacob's method, correcting the drawdown of each step for the well-loss so as to obtain aquifer loss, and then substituting the discharge rate and aquifer loss in Logan's formula to obtain closely agreeing values of T for each step. The method is applicable to both steady-state and near-steady state conditions.

STEP-DRAWDOWN TEST

The step-drawdown test consists of pumping a borewell with atleast three discharge rates, preferably in increasing order, and measuring the corresponding pumping water levels with time. Initially, the water levels are recorded at closer intervals of time when the depletion is faster, and subsequently with increasing intervals of time.

RECOVERY:

Long duration test data could be utilised to determine the optimum pumping period and also calculate T values from both the pumping and recovery data as per standard methods. Recovery measurements are made for at least 100 minutes, i.e., 2 log cycles immediately after pumping.

The graph log (time) 't' versus recovery plotted and the time required for complete recovery is computed by extrapolation of this plot to zero drawdown level.

The optimum pumping period is determfined from the recovery data by substracting the recovery time period from 24 hours. For practical purposes, 95% of recovery, if not more, can be considered as complete recovery. It may be noted that complete recovery means recouperation to static water level with respect to the top of the aquifer as determ ined from driller's data/well log.

The optimum pumping period may also be determined by computerisation. However, it is preferable to do it manually so as to avoid errors. (Srinivas et al).

The basic formula for determing drawdown s in a borewell is

$$s = BQ + CQ^2 \qquad \text{-------(1)}$$

where B - aquifer loss coefficient

C = well loss coefficient

and D = the discharge rate of the borewell.

The term CQ^2, representing the well-loss component in equation (1), is normally difficult to evaluate as its behaviour is a function of both the discharge rate Q, and time t. In addition, the physical properties namely, the design, construction and development of the well, or deterioration of the screen, affect the well-loss.

JACOB'S METHOD OF ANALYSIS:

Jacob (Cooper and Jacob, 1946) have suggested a graphica method for determining the coefficients B and C so that the aquifer loss and well loss components are determined.

The field data obtained from step-drawdown test is plotted on a semi-log graph as drawdown s versus (log) time t.

The plots of each step are extrapolated to the optimum pumping period determined from the recovery graph and the extrapolated drawdowns noted. The ratios of extrapolated drawdowns (s_i) and the corresponding discharge rates (Q_i) are plotted on a linear graph paper with Q_1, as s_i/Q_i versus Q_i. i represents the step number.

A least square line is fitted through the points. The intercept of this line with time axis where Q is zero gives the value of B and the gradient gives the value of C.

From the C value obtained, the well-loss, CQ_i^2, can be calculated for each step and substracted from corresponding extrapolated drawdowns. This we term as Aquifer loss s_{AL} .

LOGAN'S FORMULA

Logan (1964) reasoned that if a well is pumped for such a long period that the flow is in steady-state, then an approximate estimation of the order of magnitude of the Transmissivity can be made using the Thiem formula for a confined aquifer which can be written as

$$T = 2.30\ Q \log (r_{max}/r_w)/\ 2\pi\, s_{mw} \quad ----(2)$$

Where r_w = radius of the pumped well

r_{max} = radius of influence (radius of cone of depression)

and s_{mw} = maximum drawdown in the pumped well.

Logan further stated that the accuracy of the calculation depends only on the accuracy of the measurement of s_{mw} (on which well losses may have a substantial influence) and on the accuracy of the ratio r_{max}/r_w. As r_{max}/r_w cannot be accurately determined generally, Logan opined that although the variations in r_{max} and r_w may be substantial, the variation in the logarithm of their ratio is much smaller. Hence, assuming average conditions of radii, he suggested a value of 3.33 for the log ratio as a rough approximation. Substituting this value in equation (2) above, we get Logan's formula:

$$T = 1.22\ Q/s_{mw} \quad -----(3)$$

where s_{mw} is the maximum drawdown in the pumped well.

Use of Logan's formula in the above form results in highly erroneous results of the order of 50% or more according to Krusemann and de Ridder, and can at best be only first-time-round approximations.

The author found that the error was as high as 100% in some cases.

MODIFICATION OF LOGAN'S FORMULA:

It may be appreciated that the high variation rate in the T values obtained by Logan's formula could be attributed to the influence of well losses on the drawdown values. Thus if well-losses could be corrected for from the maximum water level, s_{mw}, then we may be able to nullify the error. Hence Logan's formula was modified as:

$$T = 1.22\ Q_i/\ (s-CQ_i^2) \quad -----(4)$$

$$= 1.22\ Q_i/BQ_i \quad -----(5)$$

where i represents the steps.

Computing the well losses component as C x Qi_i, C having been obtained from Jacob's method described earlier, we can computer T_i, for each step of the step-drawdown test employing modified Logan's formula viz. equation (4). The T_i obtained for each step should match closely with any other step for a confined aquifer.

This is true theoretically also, since the drawdown now is composed of entirely the drawdown due to aquifer loss. The T_i therefore should be same for any discharge rate within the optimum time period and for steady state conditions.

COMPUTER PROGRAM:

A simple computer program in FORTRAN is given in Appendix which caclulates T for each step. The input values comprise essentially the field data of pumping water levels and time. The optimum time period is obtained from recovery.

EXAMPLE 1

Step-drawdown test was carried out on a borewell in the premises of National Institute of Rural Development, Rajendranagar. The pumping was conducted in 5 steps of 120 minutes duration each and with increasing rates of discharge. At the end of last step, recovery measurements were made for 100 minutes.

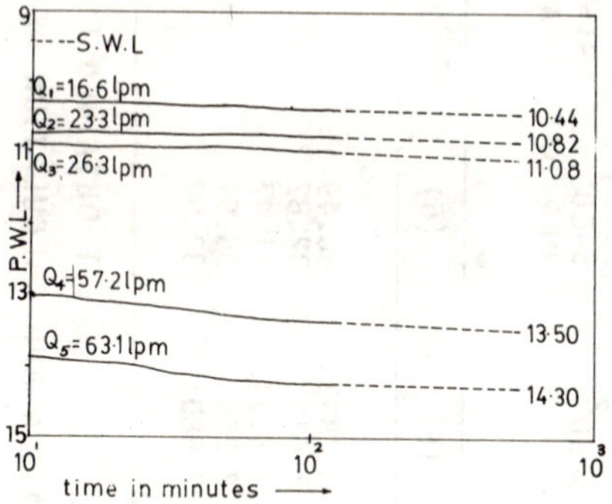

Fig.1 shows the pumping water level versus log time t plots for the 5 steps and the extrapolation to the optimum pumping period of 545 minutes determined from recovery data.

FIG.1: PLOT OF PWL Vs. DRAWDOWN AND EXTRAPOLATION TO OPTIMUM TIME PERIOD.

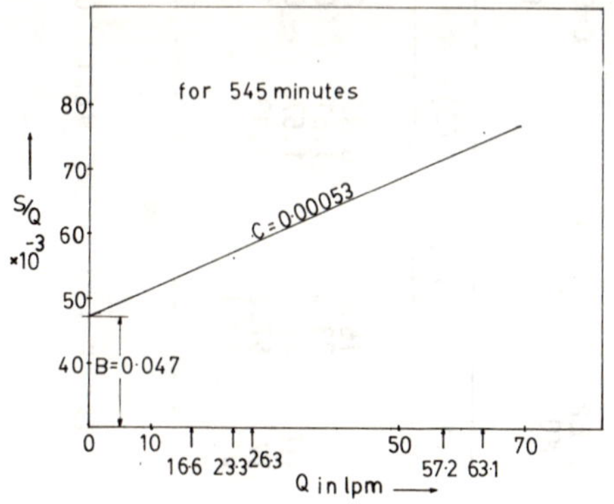

Fig.2 shows the s_i/Q_i versus Q_i and the intercept and slope values as shown in Table 1.

FIG.2: PLOT OF SPECIFIC DISCHARGE Vs. DISCHARGE RATES AND ESTIMATION B & C.

STEP NO.	Q lpm	EXTRAPOLATED P.W.L.IN Mtrs.	EXTRAPOLATED DRAWDOWN'S' IN Mtrs (3)-S.W.L.	S/Q
(1)	(2)	(3)	(4)	(5)
I	16.6	10.44	0.93	0.0560
II	23.3	10.82	1.31	0.0562
III	26.3	11.08	1.57	0.0597
IV	57.2	13.50	3.99	0.0698
V	63.1	14.30	4.79	0.0759

TABLE-1 SHOWS CALCULATION OF SPECIFIC DISCHARGE

FOR EXAMPLE-1

STEP NO.	Q in lpm	S	C=0.00055 CQ^2	$S-CQ^2$	$\frac{Q}{S-CQ^2}$ PER DAY	TML=1.22x1.44x $\frac{Q}{S-CQ^2}$ GRAPHICAL	TML FROM TRANSTEP
(1)	(2)	(3)	(4)	(5)	(6)	(7) (	(8)
I	16.6	0.93	0.14605	0.78395	30.49	37.20	40.62196
II	23.3	1.31	0.28773	1.02223	32.82	41.85	40.33938
III	26.3	1.57	0.36659	1.20343	31.47	38.39	39.96942
IV	57.2	3.99	1.7341	2.256	36.51	44.54	43.14519
V	63.1	4.79	2.110	2.680	33.90	41.36	38.47245

T OBTAINED FROM LONG DURATION TEST=43.9m^3/d/m

TABLE - 2 SHOWS WELL LOSS ESTIMATION AND CALCULATION OF T VALUE

FOR EXAMPLE 1

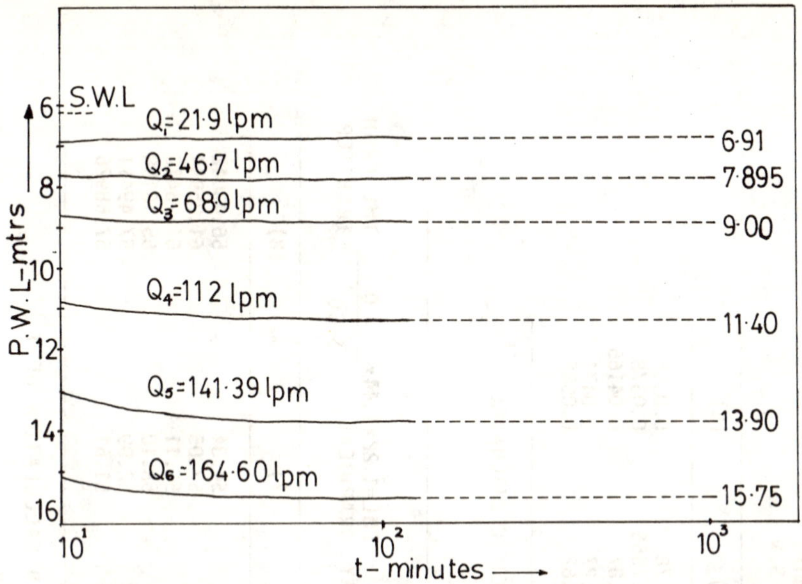

Fig 3 Plot of PWL vs Time and extrapolation to optimum time period

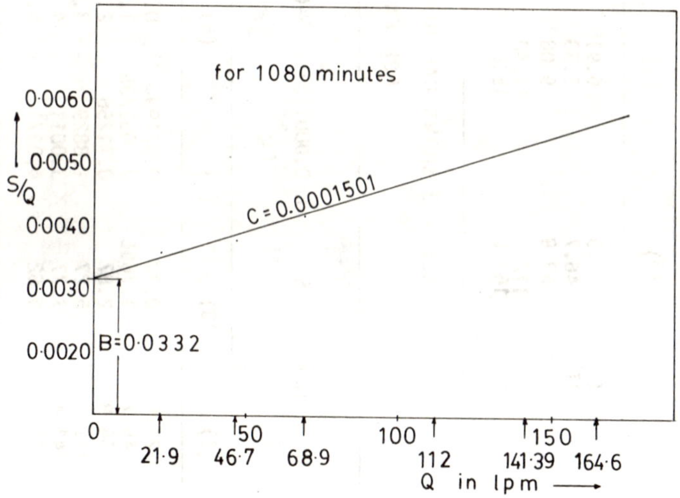

Fig 4 Plot of specific discharge vs discharge rates and estimation of B & C

STEP NO.	Q lpm	EXTRAPOLATED P.W.L. in mtrs	EXTRAPOLATED DRAWDOWN 'S' in mtrs (3) - S.W.L.	S/Q
(1)	(2)	(3)	(4)	(5)
I	21.9	6.910	0.78	0.0356
II	46.7	7.90	1.7655	0.0378
III	68.9	9.00	2.87	0.04165
IV	112.0	11.40	5.27	0.0471
V	164.0	15.8	9.62	0.0585

TABLE - 3 SHOWS CALCULATION OF SPECIFIC DISCHARGE FOR EXAMPLE - 2

STEP NO.	Q in lpm	S	C=0.0001501 CQ^2	$S-CQ^2$	$\frac{Q}{S-CQ^2}$ PER DAY	TML=1.22x1.44x $\frac{Q}{S-CQ^2}$ GRAPHICAL	TML FROM TRANSTEP
(1)	(2)	(3)	(4)	(5)	(6)	(7)	(8)
I	21.9	0.78	0.07199	0.7080	44.54	54.34	56.43042
II	46.7	1.7655	0. 032735	1.4382	46.76	57.05	61.25944
III	68.9	2.87	0.71256	2.1574	45.99	56.11	61.26472
IV	112.0	5.27	1.8829	3.387	47.62	58.10	65.16198
V	141.4	7.77	3.0011	4.7689	42.70	52.09	57.49411
VI	164.0	9.62	4.0371	5.5829	42.30	51.61	57.68926

TABLE-4 SHOWS WELL-LOSS ESTIMATION AND CALCULATION OF T VALUES FOR EXAMPLE-2

Column (7) in tables 2 & 4 give the T values obtained by graphical/analytical method for the two examples, while column (8) gives the T values obtained from Program TRANSTEP.

RESULTS

The Transmissivity values calculated for each step tallied very closely with each other and in turn with the T values computed from long duration test conducted on the same borewell.

Optimum pumping period can be defined as the period beyond which the aquifer properties may not hold good; implying that the drawdown pattern may fall in the overpumping range.As described elsewhere, it is preferable to arrive at the optimum pumping period from the recovery data.

It may be noted that one of the conditions for Jacob's method, that u should be small (< 0.01) will be satisfied in confined aquifers for moderate distances from the pumped well within an hour or less if radius r is relatively small (Krusemann & de Ridder, 1970). Hence, the method suggested in this paper for computation of T from step-drawdown test data is justified. However, it would be reasonable to obtain data for atleast 2 log cycles (100 minutes) of time or 120 minutes. It may be mentioned that the initial part of the time-drawdown graphs where there is faster drawdown is ignored and values after a lapse of time say 30 minutes may be considered for computational purposes.

The method has been tested for hard rock area of granite falling in and around Hyderabad district.The general depth of hard rock occurrence is 40 metres and borewells usually are drilled beyond this depth.Therefore, the aquifer or aquifer system is generally fully penetrated.

Another significant conclusion that can be drawn is the instantaneous calculation of T from the modified Logan's formula, which is written as

$T_i = 1.22\ Q_i/s\text{-}CQ_i^2$ in this paper.

In place of $(s\text{-}CQ_i^2)$ one can use the aquifer loss itself as calculated from aquifer loss coefficient B. Then the above formula can be written as:

$$T = 1.22 \times Q_i/BQ_i$$

$$= 1/B$$

Hence, in effect, the inverse of aquifer loss coefficient B multiplied by 1.22 gives the Transmissivity in M^2/day. The variation in T values calculated employing coefficient 'C' and 'B' could be explained due to the well loss component comprising of several losses.The well losses comprise of head loss in the well due to pumping, well loss due to flow through the gravel envelope or developed zone outside the well casing and through the screen or other openings in the casing, as well as flow inside the well, which is itself turbulent.Thus the well loss is expected to vary with some power of Q i.e., C $\propto Q^n$. Jacob (1947) considered n=2 used in this paper, while Rorabaugh (1953) concluded a value of 2.5 or more.

Further it should be mentioned that this method has been applied to borewells whose radius r is 6" or 6½". Hence 'well' in this paper refers to borewell only and not an open well.

ACKNOWLEDGMENTS:

The author is thankful to the Executive Director, Mr. R.M. Myles, and Dr.V.K.Dixit, Head WRD Department, AFPRO, New Delhi, for their permission and constant encouragement. Grateful thanks are also due to Mr.K.N. Dewangan, Sr. Hydrogeologist, for useful discussions.

REFERENCES

1) Kruseman G.P., and de Ridder, N.A.,1970. Analysis and Evaluation of Pumping Test Data.

2. Logan, 1964. Estimating Transmissivity from routine production tests of water wells. Groundwater, Vol.2, nr.1:35-37.

3. Bouwer Herman, 1978. Groundwater Hydrology, Mc Graw-Hill Kogakusha Ltd.

4. Srinivas, M., Arun Kumar, I. and Patangay, N.S. - A Computer Program for determining the optimum discharge rate from step-drawdown tests. Submitted to Indin JOurnal of Earth Sciences, Calcutta.

APPENDIX

```
C       PROGRAM TRANSTEP
C       COMPUTATION OF TRANSMISSIVITY FROM STEP DRAWDOWN
        TEST
        DIMENSION T(50), P(50,10),D(50,10),DSX(10)
        DIMENSION DDEX(10),B(10),C(10), Q(10)
        DIMENSION AL (10), AQL (10)
99      READ (4,10) AN, AM, SWL, TE
        N=AN
        M=AM
        WRITE (6,110)N,M,SWL,SDD,TE,TIT
110     FORMAT (2I3,3F10.5,A,/)
10      FORMAT (5F10.5)
        READ (4,120) (Q(J),J=1,M)
        WRITE(6,1120)(Q(J),J=1M)
        READ (4,120)(T(I),I=1,N)
120     FORMAT (10F10.5)
1120    FORMAT (5X,10F10.5,//)
        DO 222 J=1,M
222     READ (4,30)(P(I,J),I=1N)
30      FORMAT (10F8.3)
        DO 15 J-1,M
        DO 15 I-1,N
15      D(I,J)=P(I,J)-SWL
44      FORMAT (9F10.5)
        DO 25 I=1,N
25      WRITE (6,44)(T(I),(P(I,J),J=1,M))
        DO 35 J=1M
        ANN=AN-16
        STM=0.0
        SPL=0.0
        STM2=0.0
        SPL2=0.0
        STPL=0.0
        DO 45 I1=17,N
        STM=STM+T(I1)
        SPL=SPL+P(I1,J)
        STM2=STM2+(T(I1)*T(I1))
        SPL2=SPL2+(P(I1,J)*P(I1,J))
45      STPL=STPL+(T(I1)*P(I1,J))
        B1=STPL-((ANN*STM/ANN)*(SPL/ANN)
        B2=STM2-((ANN*STM/ANN)*STM/ANN)
        B(J)=B1/B2
        C(J)=(SPL/ANN)-(B(J)*STM/ANN
        DDEX(J)=((TE*B(J)+C(J))-SWL)/G(J)
35      CONTINUE
        SQ12=0.0
        SQ1=0.0
        SBQ1=0.0
        SBQ2-0.0
        SQBQ=0.0
        DO 70 K1=1,M
        SQ1=G (K1)+SGl
        SQ12=(Q(K1)*Q(K1)+SQ12
        SBQ1=SBQ1+DDEX(K1)
        SBQ2=SBQ2+(DDEX (K1)*DDEX (K1)
```

```
70      SQBQ=SQBQ+(Q(K1)*DDEX(Kl)
        AK1=M
        BQ1=SQBQ-(SQ1*SBQ1/AK1)
        BQ2=SQ12-(SQ*SQ1/AK1)
        BQ=BQ1/BQ2
        CQ=SBQ1/AKl-(BG*SQl/AKl)
        DO 333 J=1,M
        AQL(J)+CQ*Q (J)
        AL(J)=DSX(J)-(BQ*Q(J)*Q(J))
        T(J)=1 22*1.44*Q(J)/AL(J)
        TAQL=1 22*1.44*Q(J)/AGL(J)
        WRITE (6,159), T (J),
159     FORMAT (5X T(J)=',F10.5,/)
333     CONTINUE
        TYPE *,'GIVE (1) TO CONTINUE, (0) TO STOP'
        ACCEPT *,LKL
        IF (LKL, EQ.1) TO 99
        STOP
        END.
C       N=NO.OF SETS
C       M=NO.OF STEPS
C       TE=TIME PERIOD OF EXTRAPOLATION
```

APPROPRIATE WELL DESIGN FOR DEVELOPING FRESH WATER AQUIFER IN A MIXED QUALITY AQUIFER SYSTEM

K.V.J.R. Krupanidhi, V.V.S. Mani and
N. Varadaraj
Central Ground Water Board

ABSTRACT

The success of a ground water abstraction structure depends on its proper type, design and selection of optimum pumping system suitable for requirements. Incidents of failure of structures due to formation water quality deterioration, absence of sustained yield and faulty construction are common, particularly in the sedimentary tracts. The problems are manifold in the coastal areas.

The exploratory drilling operations carried out by the Central Ground Water Board since 1958 and the latest studies under Technology Mission in the sedimentary tract of Ramanathapuram district, Tamil Nadu have given very promising results in the form of a number of high yielding fresh water tubewell sources. The study of lithology, drilling time log and electrical log gives the nature of granular zones present in the borehole location and a first-hand information on the quality of water in various granular zones. The understanding of the regional hydrological conditions, performance of existing ground water abstraction structures in monsoon and summer periods, the behaviour of fresh water-saline water interface at select places gives an important guideline for selection of type of structures, design of well like diameter, depth, slotted length, screen diameter, slot size, pumping method and the likely yield and quality of water.

Select case histories of well design, their difference with the conventional method in the same environment and the success rate in quantity and quality aspects of ground water are highlighted in the paper.

INTRODUCTION

Ground water abstraction structures of various types, designs and depths are commonly used for drinking, irrigation and industrial purposes. Proper design of the abstraction structure is necessary in accordance with the purpose, prevailing hydrogeological setting and required yield. There are incidents of failure of structures due to inadequacy of yield, absence of sustained yield in summer months, quality deterioration due to over pumping, poor construction, corrosion of pipes and clogging of screens etc. The problems are manifold in the sedimentary rocks, particularly in coastal areas where fluvio-marine and marine sediments are tapped. There may be more than one group of granular sediments with varied chemical quality of formation water. Also wide variation in aquifer characteristics due to facies changes and structural disturbances are common. The occurrence of fresh water floating over saline water, fresh water aquifer underlying saline aquifers and sandwitched between saline aquifers are established in many parts of the country. The proper development of these aquifers without adversely disturbing the hydrostatic equilibrium requires suitable well design.

BACKGROUND INFORMATION FOR WELL DESIGN

The hydrogeological and hydrochemical details of the terrain are necessary for selecting suitable type of abstraction structures. In case of sedimentary tract with mixed quality formation water, the study of litholog, drilling time log and electrical log gives the nature of granular zones and to some extent the quality of water in various zones. The understanding of the regional hydrogeological setting, performance of existing ground water abstraction structures in monsoon and summer periods, the position of salt water-fresh water interface under static conditions, the behaviour of fresh water - saline water interface due to pumping and other hydrogeological data help in selection of the type of structures, design and also an idea of the likely quality and yield. The experience in the sedimentary tract of Ramanathapuram district has shown that the yield and quality of water of any tubewell is unpredictable. However, the integrated approach under Water Technology Mission, Ramanathapuram district in selection of sites and proper care in construction of tubewells with the background knowledge of exploration results has given very good results in the form of a number of high yielding fresh water tubewells.

SELECTION OF TYPE OF STRUCTURES

Based on the hydrogeological survey results, the ground water potential map of part of Pasumpon Muthuramalingam and Ramanathapuram districts has been prepared. The hydrogeological units identified (Fig.1) have been useful for selection of type of well viz. open well, infiltration well, Feeder well, Collector well, Shallow Tubewell, deep tubewell etc. The salient features of the various hydrogeological units and sectors identified in parts of Pasumpon Muthuramalingam and Ramanathapuram districts of Tamil Nadu are presented in Table-1.

TABLE-1
SALIENT FEATURES OF VARIOUS HYDROGEOLOGICAL UNITS AND SECTORS IDENTIFIED IN PARTS OF PASUMPON MUTHURAMALINGAM AND RAMANATHAPURAM DISTRICTS

Zone (With depth range)	Sector (with depth range)	Salient hydrogeological features and suitable well type
1. Shallow fresh water(6-60 m)	1.A Coastal Dunes (6-10 m)	Fresh water lense over saline aquifers. Moderate potential. Total dissolved solids (TDS) 100-500 ppm. Suitable for large diameter open wells/feeder well-Collector well system. Safe yield of open wells 20-30 m^3/day.
	1.B River flood plain(4-8 m)	Fresh to brackish water with fresh/ saline water aquifers at depth. Moderate potential. TDS 500-1500 ppm. Suitable for large diameter infiltration wells and galleries. Yield 50-100 m^3/day.
	1.C Crystalline rocks(10-60 m)	Fresh water to brackish water under water table and/or confined conditions. TDS 500-1500 ppm. Suitable for large diameter open wells and 60-100 m deep borewells. Yield 25-100 m^3/day for open wells, 20-100 lpm for borewells.
	1.D Palaeo-channels(10-25 m)	Fresh to brackish water under water table and semi-confined conditions. TDS 500-3000 ppm. Suitable for shallow tube-wells. Yield 100-300 lpm.

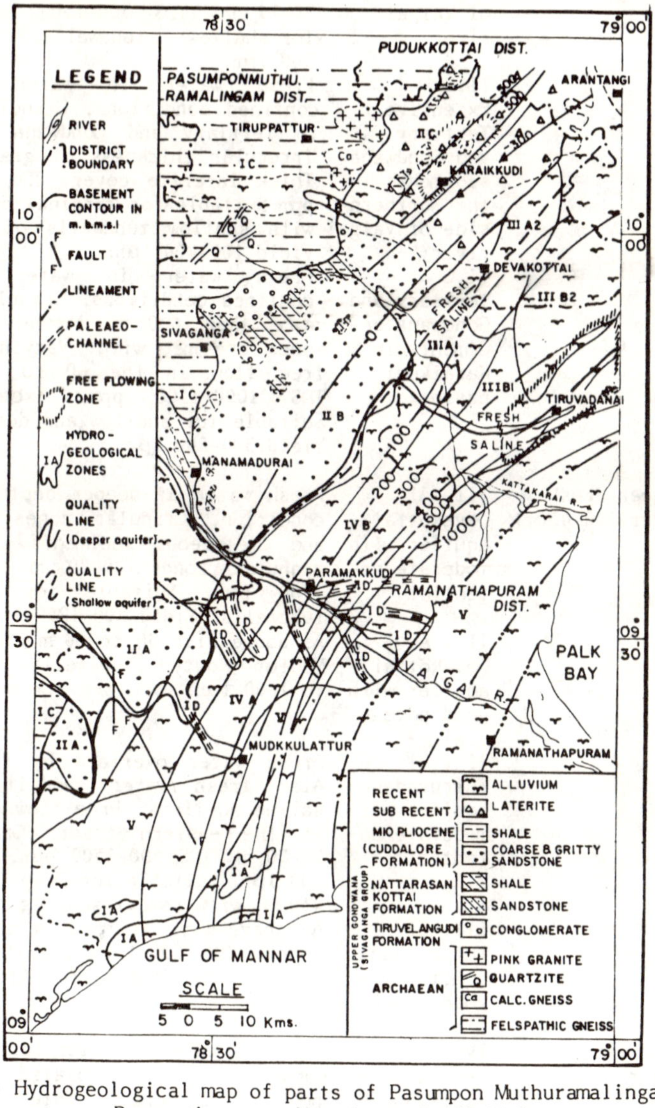

Fig.1: Hydrogeological map of parts of Pasumpon Muthuramalingam and Ramanathapuram districts, Tamil Nadu

II. Moderate depth fresh water zone (40-100 m)	II.A Isolated pockets in southern side of Vaigai river	Fresh water in Tertiary sandstones overlying the crystallines. Ground water potential in laterite is limited. Thickness of sediments around 30-75 m. TDS 500-1500 ppm. Suitable for shallow tubewells, yield 20-300 lpm.
	II.B Extensive Tertiary and Gondwana sediments in the Northern side of Vaigai river	Fresh water in water table and confined conditions. Granular sediments in Tertiary and Gondwana rocks. Very irregular thickness of granular zones. Thick laterite cover. TDS 500 to 1500 ppm. Suitable for marginal development with shallow tubewells of 40 to 100m. yield 100-400 lpm.
	II.C Tertiary and Gondwana sediments in Karaikudi basin .	Fresh water in water table and confined conditions. Granular sediments more than 600 m. Number of granular zones are met with. Varying thickness, free flow in the 60 to 100 m zone. TDS 1000-1500 ppm. Excessive Iron. Suitable for shallow and deep tubewells. Yield 300-800 lpm.
III. Deeper Fresh water(100-450m)	III. A1 Devakottai aquifer with moderate potential.	Fresh water at deeper depth in confined condition. Granular zones in Tertiary and Cretaceous sediments. Thickness of granular zones 20-30 m. TDS 700 to 1500 ppm. Suitable for Medium duty tubewells. Yield 300-1250 lpm.
	III. A2 Devakottai aquifer with good potential.	As above with 30 to 60 m granular zones. Medium duty tubewells. Yield 800 to 1600 lpm.
	III. B1 Thiruvadanai aquifer with moderate potential.	Fresh water overlain by saline water. Also fresh water sandwitched between saline aquifers. Free flowing conditions in the eastern tract. Granular zones 20-30 m. TDS 500-1500 ppm. Yield 400 to 800 lpm. Suitable for Deep Tubewells (300-450 m) with cement sealing technique.
	III. B2 Thiruvadanai aquifer with good potential.	As above with 30 to 60 m thick granular zones. TDS 500-1500 ppm. Deep wells of 300-450 m. Yield 800 to 1600 lpm.
IV. Brackish Water	IV.A Tertiary sediments South of Vaigai River.	Brackish water in water table and confined conditions with isolated fresh water pockets in perched water table condition. Thickness of sediments vary from 20 to 100 m. Few Paleo-channels and granular sediments form potential pockets. TDS 1500-3000 ppm. Suitable for shallow to deep tubewells. Yield 150 to 1600 lpm.

	IV.B Tertiary sediments North of Vaigai River.	Brackish water under water table and confined conditions. 3 to 4 granular zones of varied thickness, free flowing conditions at places. Fresh water pockets are limited and are not dependable sources. Suitable for shallow to deep tubewells of 60 to 400 m. Yield 400-800 lpm).
V. Saline Water	V. Coastal aquifers	Water table and confined conditions. Granular zones in Alluvium, Tertiary and older rocks. 7 to 10 granular zones identified. TDS 3000-30,000 ppm. Yield 400-1600 ppm. Not suitable for drinking and irrigation purposes.

The presence of a number of surface water bodies, (tank, Oorani) have resulted in isolated ground water pockets of fresh water which are not considered for the demarcation of zones. Such pockets are not found to be dependable sources due to limited ground water potential and the quality deteriorates in summer. These sources developed by ring wells and open wells dry up in summer and drought periods. The domination of marine environment during the deposition of the sediments and at present day has resulted in the adverse hydrochemical condition. The fluvial and marine environments, in time and space have resulted in complex hydrogeological set up and irregular distribution of granular zones. However, if a first hand information is available for any study area a suitable type of abstraction structure can be planned.

WELL DESIGN IN SAND DUNES AND FLOOD PLAINS

During the study of the existing wells in sand dunes and flood plain areas of Ramanathapuram district, the following factors are observed adversely effecting the yield and deterioration of the quality of formation waters:

i) Poor yield due to impervious lining of well, ineffective porous concrete in infiltration well, and provision of inadequate weepholes.

ii) Absence of sustained yield due to limited thickness of granular zones.

iii) Absence of sustained yield due to limited thickness of granular zones.

iv) Quality deterioration in summer due to upconing of underlying saline water during pumping as seen in infiltration wells at Karan and Mandapam areas.

v) Complete well failure due to saline intrusion.

The working of feeder well-collector well systems in some well fields at Mandapam Camp and Rameswaram island have indicated the choking of connecting lines between feeder wells and collector wells.

The understanding of fresh water-saline water interface due to pumping has to be clearly visualised before selecting the well site, its dimensions and pumping well pattern. For working out the design of an open well, the parameters like Transmissivity (T), Radius of influence (R), designed specific capacity (Q/s) and dimension (r) have to be considered. In Ramanathapuram area, the Transmissivity of coastal aquifers varies from 100 to 300 m^2/day and the radius of influence has been estimated to be of the order of 160 m. The specific capacity of wells varies from 200 to 500 lpm/m drawdown. The dimensions of proposed dug wells can be worked out from the following formula:

$$T = \frac{Q}{s} \times 529.7 \times \log \frac{R}{r} \times 10^{-3}$$

For an aquifer with transmissivity value of 300 m^2/day, the optimum diameter to get the specific capacity of 300 lpm/m drawdown should be 4 m. The dimension should be more for aquifers with less transmissivity and 4.5 m (diameter) is the general dimension. The drawdown created in a well tapping a fresh water lense underlain by saline water will result in upconing of the interface as per Ghyben-Herzberg relationship; the maximum rise being registered under the well where drawdown is maximum. Accordingly if the pumping rates are kept low, a new dynamic equilibrium may be established with the upconed interface occupying an intermediate position between the level at static conditions and the bottom of the well without entry of saline water into the well. The schematic section of undisturbed condition in a dune complex in Mandapam area with different abstraction structures and the likely effect of pumping on interface is shown in Fig.2. It can be seen that the large diameter open wells, with moderate extraction, are safe.

Large diameter infiltration well with radial arms can give maximum yield and pumping of well will result in less drawdown and less disturbance of the interface. In case of higher requirement to the order of 50 to 100 m^3/day, a well field with 10 to 15 feeder wells of 1.5 m dia with 50-60 m spacing and a collector well of 4.5 m dimension is suitable. The pumping of collector well having sealed bottom is safe. The feeder wells can be located in the slope of dunes and collector well at the topolow can help in ensuring natural gradient for collection of water. The well fields with feeder well-collector well system at old head works of Rameswaram Township and at Mandapam camp located in sand complexes are functioning satisfactorily for the past 40 years. The pumping level in an infiltration well should be fixed above mean sea level as far as possible. However, the water table fluctuation will determine the pumping level. When the pumping level is below mean sea level, the pumping rate should be as low as possible so that the drawdown will be gradual and small, which will avoid sudden upconing of interface. Periodical pumping of infiltration wells with safe yield determined from pumping tests is necessary.

Karanjac (1975) method can be used for determining the optimum yield of large diameter dug wells and infiltration wells in the coastal aquifers.

$$Q_R = Q_P \frac{t_P}{t_P + t_R}$$

Where Q_R is the optimum yield of the well, Q_P is the constant discharge for a known time t_p, Q_R is the equivalent average inflow rate from the aquifer during the time of well recovery t_R. If the well is pumped at the discharge of Q_R, the well will not go dry and on the otherhand, the chemical equilibrium of underlying interface will not be disturbed adversely.

The pumping test conducted at Sathakonevalasai infiltration well in Mandapam area for 300 minutes with a discharge of 258 lpm has resulted in 0.84 m drawdown. The near complete recovery was obtained within 100 minutes, and using Karanjac formula, the safe yield is estimated as

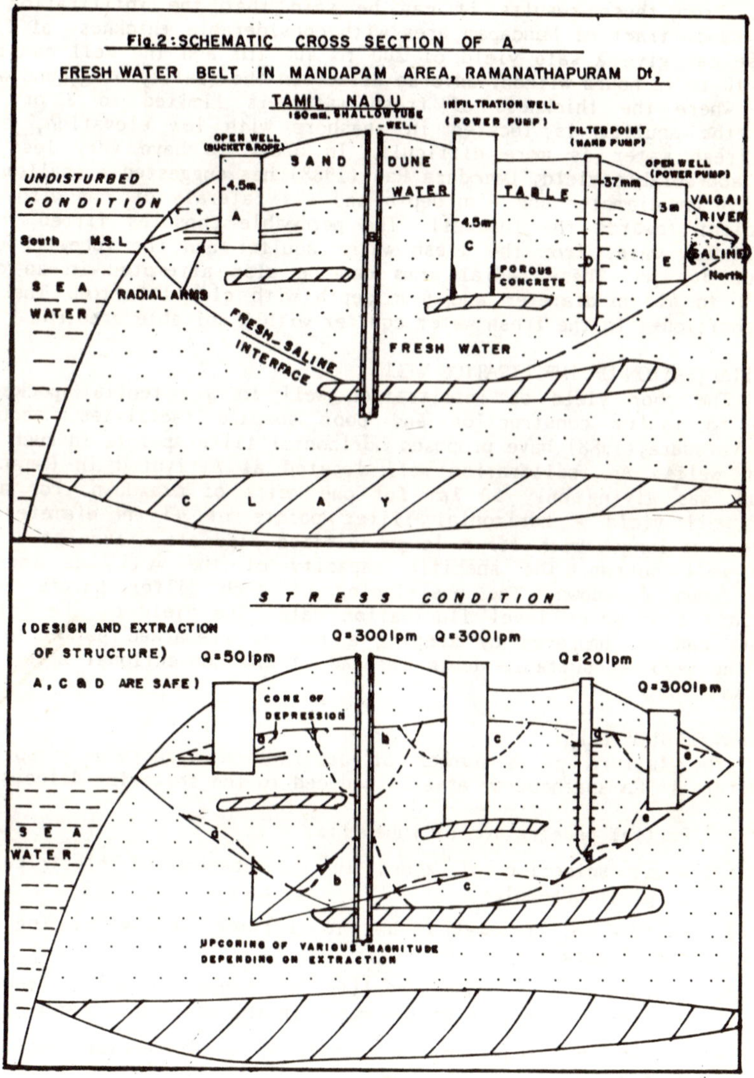

Fig.2: SCHEMATIC CROSS SECTION OF A FRESH WATER BELT IN MANDAPAM AREA, RAMANATHAPURAM Dt, TAMIL NADU

195 lpm. Another infiltration well in coastal dunes at Tuttivalasai tested for 120 minutes and the safe yield is estimated as 220.5 lpm. The safe yield of Pillaimadam infiltration well is estimated as 430 lpm. From these results it can be said that the infiltration well in the sandy tract of Mandapam area with considerable thickness of freshwater lense can give a safe yield of 200 to 400 lpm and the well can be pumped for 10 to 12 hours without adverse effect on the quality of ground water.

Where the thickness of fresh water is limited to 2 or 3 metres and the aquifer is located in seashore with low elevation, extraction of fresh water is more difficult. In order to have very less drawdown and appreciable yield, Damodara Rao (1988) has suggested a shallow tubewell of 150 mm diameter and 6 m depth and six laterals of 9 m with necessary gradient towards the tubewell. The permeable capsules fitted in radials will draw water from the fresh water aquifer and the tubewell can yield about 36 lpm. These radial arms can be also attempted in an open well of 3 to 4.5 m diameter and 6 m depth with closed bottom. The arms can be positioned in the fresh water aquifer with a suitable gradient.

REVITALISATION OF INFILTRATION WELLS

The poor yield of infiltration well in a potential pocket may be due to faulty construction and poor seepage facilities. Abdul Majeed and Varadaraj(1988) have proposed horizontal filter points in such infiltration wells. An infiltration well located at Attiyuttu in Ramanathapuram taluk was giving only 50 lpm for one metre of drawdown. For augmenting the well yield 4 horizontal filter points of 43 mm diameter and 4.5 m to 5 m length with 1.5 m length slotted pipe are driven at 1.5 m above the well bottom. The specific capacity of the well has increased to 240 lpm/m drawdown. The positioning of such filter points has to be decided from water level fluctuation data. The yield of the infiltration wells can be improved by keeping a row of prepacked permeable capsules in the well at suitable depth instead of the conventional porous concrete and weep holes.

DESIGN OF TUBEWELLS

The failure of a number of abstraction structures tapping deeper aquifers in Ramanathapuram area is deduced to the following defects:

i) Partial penetration of tubewells.

ii) Wrong selection of zones in the absence of electrical logging of pilot boreholes.

iii) Failure of tubewells due to silting during pumping (improper gravel packing).

iv) Failure due to poor quality of formation water (contamination of fresh water with saline water in the absence of cement grouting).

v) Failure due to poor yield inspite of thick granular zones (formation characteristics).

The results of a number of exploratory boreholes drilled by Central Ground Water Board in coastal areas of Tamil Nadu and identification of problems encountered in construction of production wells in the saline tracts have helped as guidelines for the development of proved area by construction of proper production tubewells.

ELECTRICAL LOGGING

The electrical logging of pilot borehole in a coastal environment with quality problems is found to be essential for every production well. The study of SP, Resistivity logs and Natural Gamma log helps in demarcating the fresh water bearing granular zones. The electrical logging of boreholes in Ramanathapuram district under Technology Mission has enabled to improve the success rate of shallow tubewells in respect of yield and quality of water of tubewells to 78.7 per cent in comparison to the 40 per cent success rate under regular production well programme. The combination of various geophysical log characteristics for a particular aquifer of interest is established by test wells. Based on the local key developed, the yield and quality of proposed tubewell can be inferred fairly accurately.

CEMENT SEALING (GROUTING)

The contamination of fresh water with poor quality formation water from different granular zones along the annular space of the tubewell is possible. Central Ground Water Board has perfected a cement sealing technique which arrests the flow of formation water from unwanted zones provided the different quality zones are separated by clay lenses of adequate thickness and extent. A slurry of cement is injected in the annular space so as to settle in the desired depth range and clay layer with adequate thickness. The fresh water zones to be tapped are confirmed by individual zone/quality tests. The typical well designs with single and double cement sealing are given at Mangalkudi and C.K.Mangalam tubewells constructed by CGWB under Technology Mission source creation work and the assembly details are given in Fig.3. The safe yield is estimated to be 1220 lpm and 400 lpm respectively from the estimation of aquifer and well characteristics. The quality of water is sweet and potable with total dissolved solids of 554 and 896 ppm, whereas the tubewell without cement grouting in the same area recorded a TDS of 750 to 2000 ppm. However in the absence of thick clay in between the fresh water and saline water aquifer system as seen in top aquifers of Paramakudi area, the cement sealing is not found to be effective. The vertical leakage of formation water from aquifer to aquifer has resulted in quality deterioration during pumping.

GRAVEL PACKING

Suitable size of gravel (4-6 mm) for Ramanathapuram area is shrouded in the annular space of borehole around the screen. In areas where selective tapping of fresh water aquifers is required, the gravel packing is given only for the zones tapped and the remaining length is packed with clay. In order to confirm the position of the gravel packing at desired depth, sounding of the annular space of the bore is done with a sounding rope. Also, a margin of gravel packing is given to accommodate the sinking of bridged gravels during development. This will ensure proper gravel packing against the aquifers tapped.

CONCLUSIONS

The tapping of fresh water in the mixed quality regime is practised commonly with non-scientific methods resulting in poor yield or failure due to quality deterioration of formation waters. The following guidelines in design of wells will be helpful in increasing the success rate of abstraction structures.

i) The aquifer wise ground water potential, elevation of water table, its fluctuation and behaviour of fresh water-saline water interface has to be visualised. The well design and pumping should be such that the drawdown is minimum. Large diameter infiltration well with radial arms is suitable for moderate extraction of 20 to 30 m^3/day. For higher yield requirement in the order of the 50 to 100 m^3/day, a well field

Fig.3: TYPICAL WELL DESIGNS WITH CEMENT SEALINGS RAMANATHAPURAM DISTRICT, TAMIL NADU

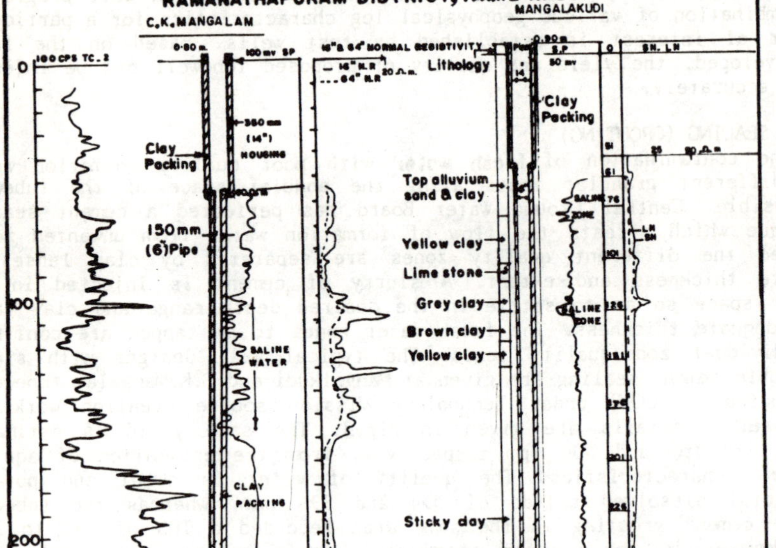

with 10 to 15 feeder wells and a large diameter collector well is suitable. Filter point is suitable for limited extraction in the order of 30 lpm.

ii) Revitalisation of poorly constructed infiltration wells can be made by horizontal filter points.

iii) Electrical logging of every borehole is necessary for proper construction of tubewells in quality problem areas. Developing of the regional key for various aquifer systems can be made from geophysical log characteristics, drilling time log, litholog and remote sensing techniques. The progressive changes in quality of formation water and thickness of granular zones have to be visualised while siting production wells.

iv) Tubewell construction should be made with proper slot size, screen length, gravel size etc. Tubewells should be developed till sand free water is obtained after adequate compressor development.

v) The selection of cement sealing depth, depth range of gravel packing, margin for gravel sinking during development, sounding at different stages of well construction to check the desired depth of gravel and clay packing as well as cement grouting are important factors for a good production tubewell. A sonic log enables us to check the cement grouting.

vi) The fresh water aquifers without sufficient thickness of confining layers are likely to receive significant vertical leakage from other saline layers during pumping. In a group of mixed quality aquifers, only fresh water zones with considerable thickness of confining layers have to be tapped.

REFERENCES

Abdul Majeed, N and Varadaraj, N.,	1988	Source finding committee report for Ramanathapuram district. Unpublished report of Technology Mission.
Damodara Rao, T.,	1987	Exploration of an unnoticed water source. The sandy coast. Proceedings of First Regional Seminar on Technology Mission, Gandhigram p 15-17.
Karanjac, J.,	1975	Brief note on testing the possibility of defining optimum yield of dugwells of large diameter. Ground Water Vol.13, No.4, p. 53-65.
Karanth, K.R.,	1987	Ground Water Assessment, Development and Management, Tata McGraw-Hill Publishing Co. Ltd., New Delhi.

RURAL WATER SUPPLY INVESTIGATIONS IN SCARCITY VILLAGES IN BHANDARA DISTRICT, MAHARASHTRA

P.R. SUBRAMANIAN
CENTRAL GROUND WATER BOARD, NAGPUR

ABSTRACT

Recurrent failure of monsoon from 1982 culminated in an alarming drought condition in many parts of Maharashtra, especially in Vidarbha region. Bhandara district was one of the severely affected districts. Most part of the district is underlain by rocks of Pre-Cambrian age belonging to Sausar and Sakoli series, comprising of mica-schists, quartz-mica-granulites and gneisses. The Sakoli group of rocks occur in a triangular fashion and comprise essentially of argillaceous sediments with arenaceous intercalations, metamorphosed to varying degrees. Granites, granitic gneisses and basic intrusives and extrusives occur associated with above rocks. The Central Ground Water Board carried out hydrogeological surveys in 315 scarcity villages and selected sites for borewells, tubewells etc. Drilling carried out so far details of which are available have indicated good yields ranging from 500 to 28,000 lph.

INTRODUCTION

During 1987 severe drought conditions existed all over the country due to failure of monsoon. The cumulative effects of the deficit rainfall in the previous years added to the crisis and an alarming situation had arisen. Similar drought conditions prevailed in many parts of Maharashtra State also, especially all over the Vidarbha region.

The Government of Maharashtra approached the Central Ground Water Board (CGWB) for assistance in the planning and execution of drought relief work. The CGWB extended immediate help by redeploying hydrogeologists from regular field investigations to the drought relief work and by giving all possible technical and equipment assistance. An action plan of hydrogeological surveys and site selections for water supply in drought affected villages was chalked out. The investigations were taken up in November 1987 and selection of sites for water supply in all drought affected villages were completed by April 1988, well before the peak of summer season.

Bhandara district was one of the severely drought affected districts in the Vidarbha region. Six field hydrogeologists and one geophysicist were deployed for the work in this district. This district is divided into three sub-divisions,viz; Bhandara, Gondia and Sakoli and comprises of 13 talukas. The normal monsoonal rainfall received in this district (June to October) is 1339.9 mm. The rainfall received in this district from June 1987 to middle of September 1987 is only 709.9 mm., almost 50% less than normal. An examination of the rainfall of previous years in this district indicates that deficit rainfall had occured during 1982, 1984 and 1985. Thus a cumulative deficit resulted which coupled with the abnormally low precipitation of 1987 created severe drought conditions.

Methodology adopted

On receipt of the list of scarcity villages from the State Government Agency, the villages were marked on toposheets. The area of work was demarcated and aerial photos of the same were studied.This study indicated the presence of lineaments, burried channels, drainage pattern and potential areas which were all marked on toposheets.This was followed by field work when every affected village was visited and all existing data was collected on an exhaustive format which was standardised earlier. Hydrogeology of the area was studied by inventory of the existing dugwells and borewells and by ascertaining their present yields. Combining these data with the lineaments, depth of weathered zone, topographic lows and occurrence of joints and fractures, suitable sites for dugwells or borewells for water supply were selected.In addition, resistivity surveys were conducted in 19 villages for confirmation of the sites selected. The sites were pin-pointed in the field and a detailed sketch map was made in the report for reference during actual construction. The finalised site selection reports were then duly scrutinised and sent to the Senior Geologist,Groundwater Surveys and Development Agency (GSDA),Bhandara and the Chief Executive Officer, Zilla Parishad,Bhandara for follow-up action and implementation of the recommendations.

Scarcity conditions

In Bhandara district altogether 315 scarcity affected villages were referred to the Central Ground Water Board for combating the existing drought. The total population in these 315 villages is about 3,22,000. The shortage of drinking water in these villages was upto an extent of 194.89 lpd.at the time of the surveys. The villages are more or less evenly distributed all over the district as shown in Fig.I.

Geology and Hydrogeology

The entire Bhandara district is underlain by Sausar and Sakoli group of rocks belonging to Pre-Cambrian age.

The Sausar Group of metasedimentary rocks forms an arcuate belt about 200 kms. in length and about 30 kms. in width. It extends from Chindwara district of Madhya Pradesh in the west where it has a NW-SE trend and passes through Nagpur and Bhandara districts with an approximately E-W trend to Balaghat district in Madhya Pradesh in the east where it assumes a NE-SW trend. The part of Sausar belt lying within Bhandara district is composed predominently of Gneisses, Quartz-Mica schists and phyllites.

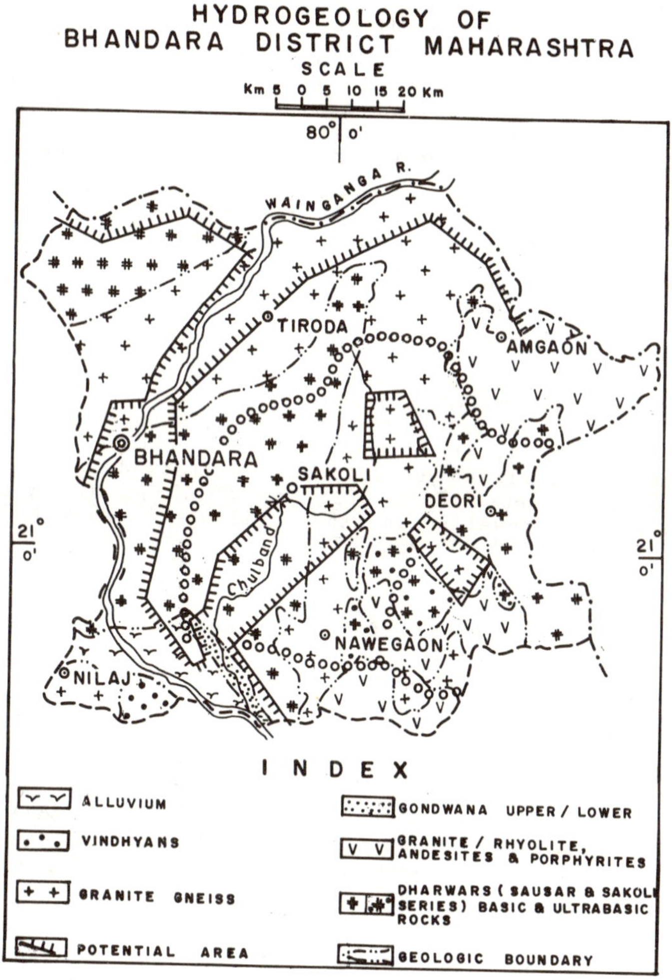

Fig. 1 Hydrogeology of Bhandara district, Maharashtra

The Sakoli group of rocks occupies a triangular shaped area within the district with its apex near Gondia and the base stretching towards SW and SE direction. It is known as the Bhandara Triangle. The rocks of this group comprise essentially of argillaceous sediments with minor arenaceous intercalations and are metamorphosed to varying degrees.

Associated with these rocks granites, granitic gneisses and basic extrusives and intrusives are found to occur. These are restricted to narrow patches in the eastern and south-eastern parts of the district.

A small occurrence of Vindhyan formations consisting of pink sandstones is seen in the southwest corner of the district. Rocks of Gondwana age consisting of sandstones, shales, conglomerates and clays occupy a narrow strip of area in the SW part of the district with a NW-SE trend.

Recent alluvium is restricted mostly to the banks of major river, Wainganga and its tributaries. It has a variable thickness, ranging from 6 to 20 metres.

Groundwater occurs in weathered portions and in the jointed and fractured zones in the massive rocks beneath the weathered zone. The phreatic aquifers are tapped by dugwells ranging in depth from 1.6 m. to 23.05 m.below ground level. The depth to water level in these wells ranged from 0.25 m. to 20.42 m.below ground level. However, in a major number of wells the depth to water level ranged from 6 to 8 metres. The deeper aquifers occuring under confined to semi-confined conditions are tapped by borewells which range in depth from 27.50 to 80.45 m. The yield of the existing borewells in the villages surveyed ranges from nil to about 30,000 lph.

Sites selected

Altogether a total of 315 villages had been surveyed and 276 sites were selected. In some villages sites were not selected for additional groundwater structures as it was found on investigation that the available water was sufficient to cope with the present needs and there was no scarcity. For computing the actual daily requirement of water for domestic needs of the village,the yardstick adopted was i) 40 litres per capita for human population and ii) 30 litres per capita for livestock. A total of 271 sites for borewells, one for tubewell and two each for dugwells and dug-cum-borewells. Of the above, the GSDA has conveyed the results of drilling in 146 sites. The results show that 138 sites are successful, 7 have been found dry and one has been abandoned by GSDA.Out of these 63 borewells have yields in the range of 500 to 2,000 lph; 48 have yields in the range of 2,000 to 5,000 lph; 27 borewells have turned out to be high yielding ones, with the discharge in 16 of them being from 5,000 to 10,000 lph; in 9 borewells the discharge ranges from 10,000 to 20,000 lph.and in 2 borewells it was of the order of 27,950 lph.The suitability of the methodology adopted during the field surveys especially for crystalline terrain, is indicated by the high percentage of success which is over 95%.

Acknowledgements

The author is grateful to the Chairman, the Chief Hydrogeologist and the Director (Central Region) of Central Ground Water Board, for their kind permission.

APPLICATION OF RADIOACTIVE TRACER TECHNIQUES FOR DETERMINING GROUNDWATER VELOCITY AND DIRECTION

K. SHIVANNA, U.P. KULKARNI, S.V. NAVADA AND A.C. EAPEN
ISOTOPE DIVISION
BHABHA ATOMIC RESEARCH CENTRE
TROMBAY
BOMBAY 400 085

ABSTRACT :

Single well tracer techniques for groundwater velocity measurement and multiple well technique for determining the direction of groundwater flow were employed in the following groundwater studies. (a) Leakage of reservoir water in a lateritic area in Sawantvadi in Maharashtra: Radiotracer studies were carried out at the poip dam site in Sawantwadi in Maharashtra to investigate the seepage of the reservoir water. It was observed most of the seepage occured from under the dam through the laterite soil. The direction of seepage water was towards the natural stream in the location. (b) Seawater intrusion in the Minjur aquifer near Madras in Tamil Nadu: A radiotracer study was carried out in the coastal Minjur aquifer to determine the intrusion rate of the salt water towards the well field. The intrusion rate of the salt water inland was found to be low of the order of few centimeter per day near the coast. (c) Impact of paper mill effluent on the groundwater body at Erode in Tamil Nadu: The radiotracer study carried out near M/s. Seshasayee Paper Mill in Erode helped in determining the rate at which effluents move towards the groundwater body.

INTRODUCTION :

Artificial radioactive tracers can be used in groundwater investigations via the single well and multiple well techniques. The single well dilution technique can be used to obtain an independent direct measurement of the filtration velocity in a water bearing formation under natural or induced hydraulic gradient. The filtration velocity interpreted in conjunction with other parameters can, under favourable conditions provide valuable information on the permeability, transmissivity and amount of water moving through an aquifer.

In multiple well technique the direction of groundwater flow could be determined. If the time required by the radiotracer to cover a certain distance is observed the transit velocity could also be determined. The paper describes the principles of the techniques used and a few case studies.

PRINCIPLES OF THE TECHNIQUES USED :

(a) Groundwater velocity measurement by point dilution method :

The point dilution method aims to relate the observed dilution of tracer introduced into a well-screen to the rate of the undisturbed groundwater flow in the aquifer. If there is a steady flow and the tracer is homogeneously distributed throughout the dilution volume at all times, the groundwater velocity could be computed using equation

$$V_f = \frac{(-V\ )}{(\alpha F t)} \cdot \ln \frac{C}{C_0} \text{ ----------- } (1)$$

where V is dilution volume.
F is Cross section of the dilution volume perpendicular to the direction of undisturbed groundwater flow.
Co is initial tracer concentration of the tracer.
C is tracer concentration at time t.
α is constant.

The value of α can be evaluated theoratically or can be determined experimentally at the laboratory. The value of α depends on bore hole geometry and on the permeabilities in area of bore hole (aquifer, filter gravel and filter tubes). Normally α values range between 1 and 3.

(b) Measurement of the direction of groundwater flow:

The direction of groundwater flow could be measured provided the stream flow is horizontal and vertical currents are negligible. A tracer is introduced into a segment of a bore hole and is carried away in the direction of flow due to horizontal flow. If the activity of the tracer cloud is measured later by a suitable, directionally oriented collimated detector, (Drost W. et.al,1968) the direction of maximum activity corresponds to the direction of flow.

In the present work multiple well technique was used to find out the direction of flow. In the multiple well method a tracer is introduced into the groundwater in a injection point and monitored in the downstream bore holes. The direction between the injection and the monitored bore hole gives the direction of flow. If the time taken by the tracer to flow from injected bore hole to the downstream bore hole is noted, then the distance velocity could be computed.

CASE STUDIES :

1. Seepage studies in lateritic area at poip minor irragation tank in Malvan valley of Konkan Region, Maharashtra:

Poip minor irragation tank was completed in the year 1974. The maximum height of the dam is 12.36 meter and length is 510 meters with waste wier of length about 200 meter on its right flank. The capacity of the reservoir at full reservoir level of 107.0 meter is 1.16 M. Cub.meter. A canal of length 1.2 km is provided there to irrigate 175 hectares of land. Every year the reservoir level goes down below sill level with in 3-4 months after the monsoon, though as per design, it is expected to hold water till completion of the rabbi season. As such there is practically no irrigation through the canal, but it is found that some land 2 km., down stream of the dam is benefited as wells in the area are getting abundant water even in summer season which was not the case before the construction of this tank. Hence the radiotracer work has been carried out to study the hydrogeological behaviour of the laterite and to delineate the groundwater flow direction and to determine the groundwater velocity in the vicinity of poip dam.

The geology of the area comprises of a thick laterite bed overlain by very thin loose lateritic soil cover. The upper hard portion of the laterite profile is concretionary which has hardened on exposure following erosion. This upper crust is rich in iron and kaolinite nodules and appears to have a cellular structure. The above cap is underlain by a zone comprising lithomarge and highly decomposed bed rock. The lithomargic clay here is yellow in colour, which means, it is ferugenous underlying this zone, and granulite is the parent rock.

EXPERIMENTAL DETAILS :

The radiotracer experiments attempted at the poip dam site were two fold;

(a) Determination of groundwater velocity using single well dilution techniques.

(b) Determination of groundwater flow direction using multiple-well technique.

In the single-well dilution technique about 200 uCi Bromine-82 as ammonium bromide was injected in the bore hole Jo and Mo (Location of the bore holes have been indicated in fig.2) on September 4, 1986. The water in the bore holes was continuously mixed and the concentration of the gamma tracer was monitor using portable scintillation detector. Average filtration velocity (V_f) for Jo and Mo computed using the formula (1) were 0.1 m/d and 1.0 m/d.

A few undisturbed gamma tracer scanning of bore holes were also carried out; from this it was observed that the water flow rates through the upper 5-6 meters of the soil layer at Jo and Mo was more compared to the flow rates below 6 meters. This could be either due to high permeability of the top layer or due to the high flow rate of seeping water from reservoir.

In multiple-well experiments tritium radioisotope as tritiated water (HTO) was used to study the direction of

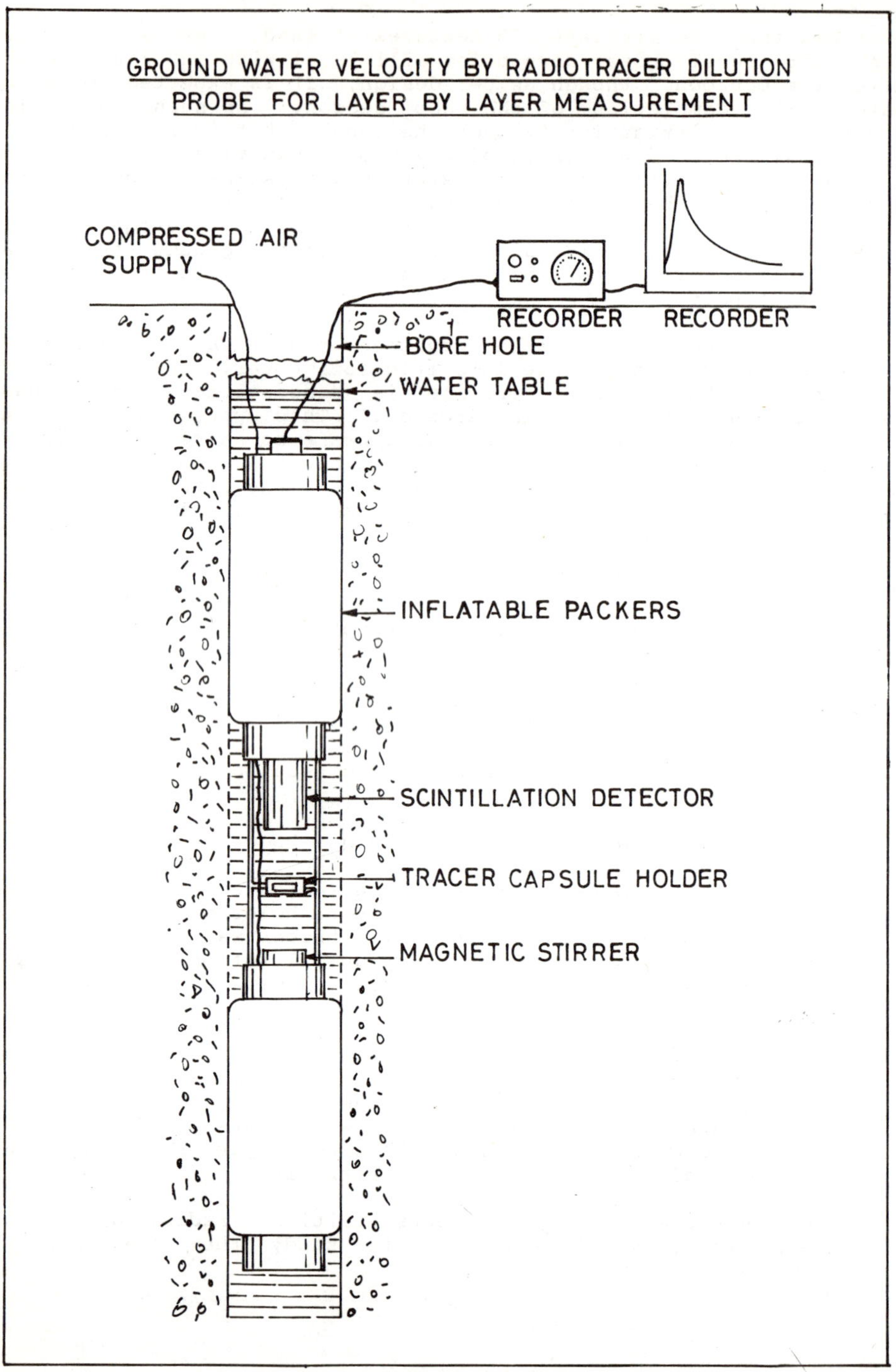
GROUND WATER VELOCITY BY RADIOTRACER DILUTION
PROBE FOR LAYER BY LAYER MEASUREMENT
COMPRESSED AIR SUPPLY
RECORDER
RECORDER
BORE HOLE
WATER TABLE
INFLATABLE PACKERS
SCINTILLATION DETECTOR
TRACER CAPSULE HOLDER
MAGNETIC STIRRER

Fig. 1

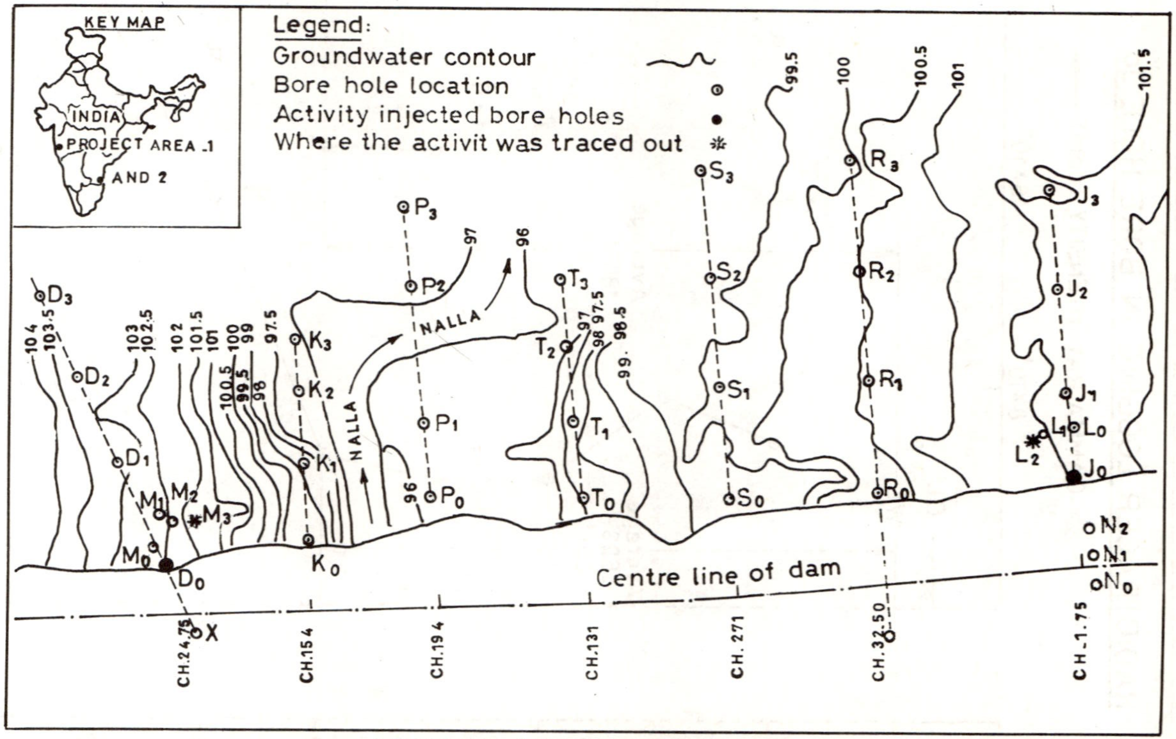

LOCATION OF THE POIP DAM SITE

Fig. 2

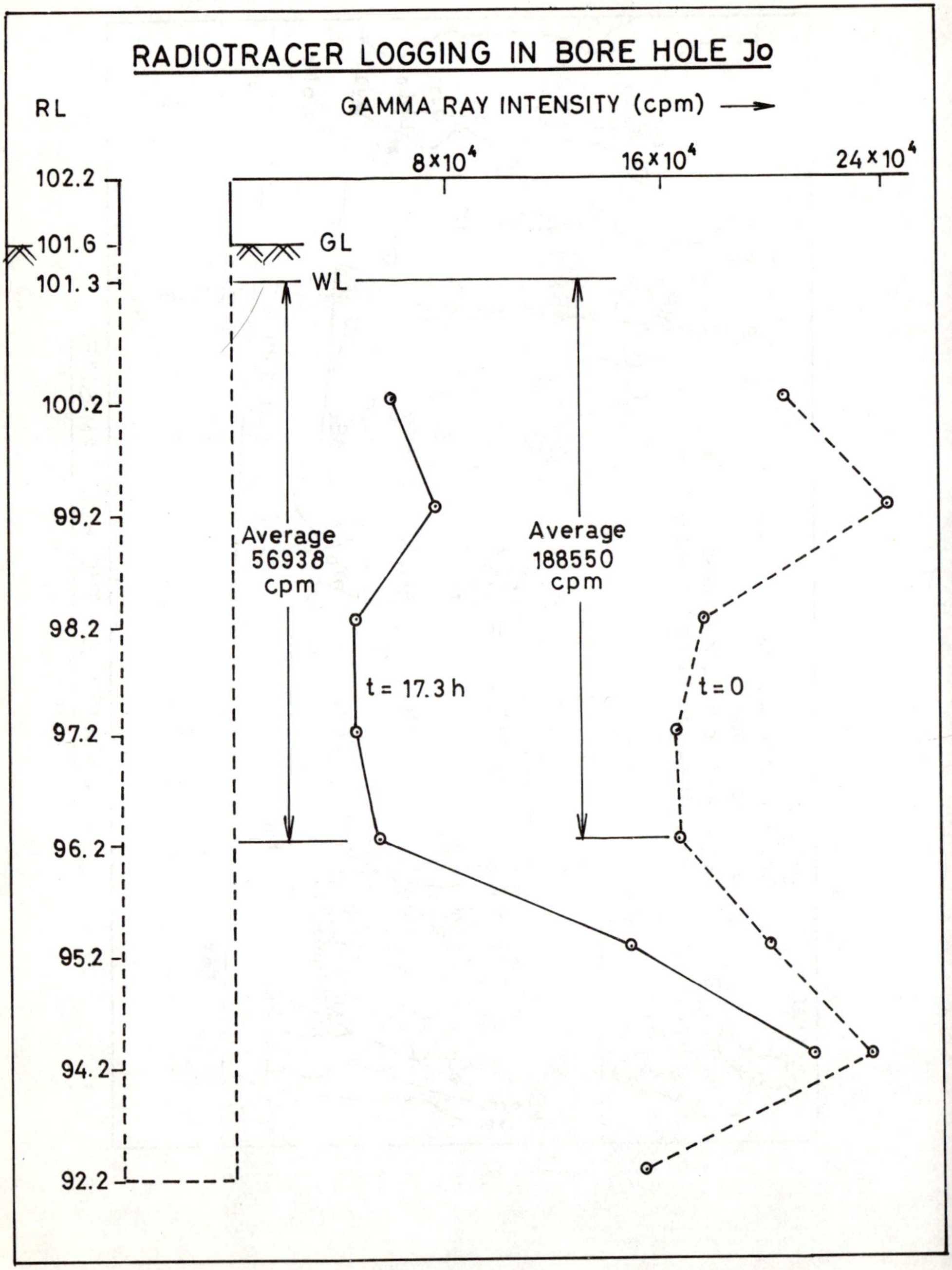
RADIOTRACER LOGGING IN BORE HOLE Jo
RL
GAMMA RAY INTENSITY (cpm)
8×10⁴
16×10⁴
24×10⁴
102.2
101.6
101.3
100.2
99.2
98.2
97.2
96.2
95.2
94.2
92.2
GL
WL
Average 56938 cpm
Average 188550 cpm
t = 17.3 h
t = 0

Fig.3

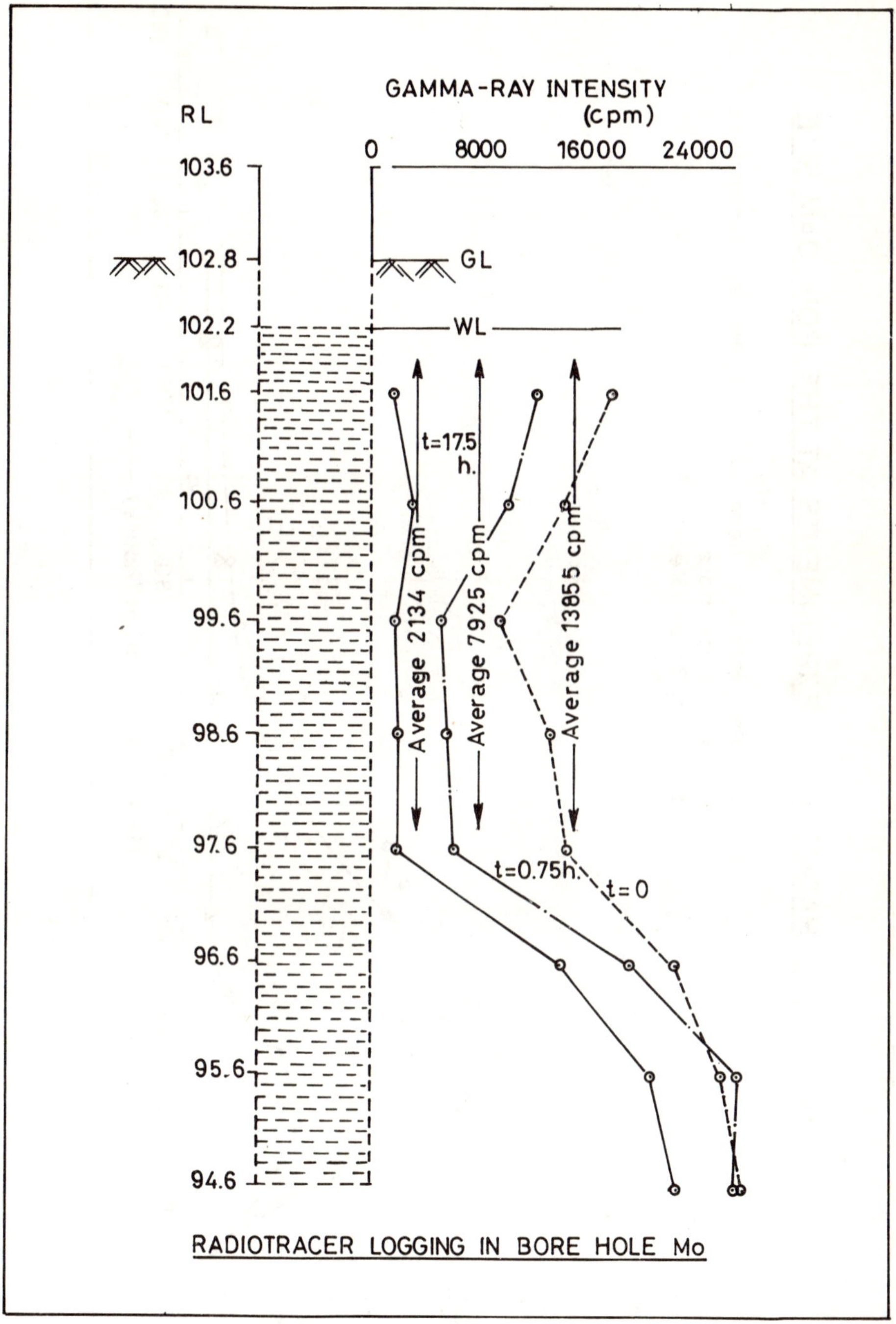
GAMMA-RAY INTENSITY
(cpm)
RL
0
8000
16000
24000
103.6
102.8
GL
102.2
WL
101.6
t=17.5 h.
100.6
99.6
98.6
97.6
Average 2134 cpm
Average 7925 cpm
Average 13855 cpm
t=0.75h.
t=0
96.6
95.6
94.6
RADIOTRACER LOGGING IN BORE HOLE Mo

Fig. 4

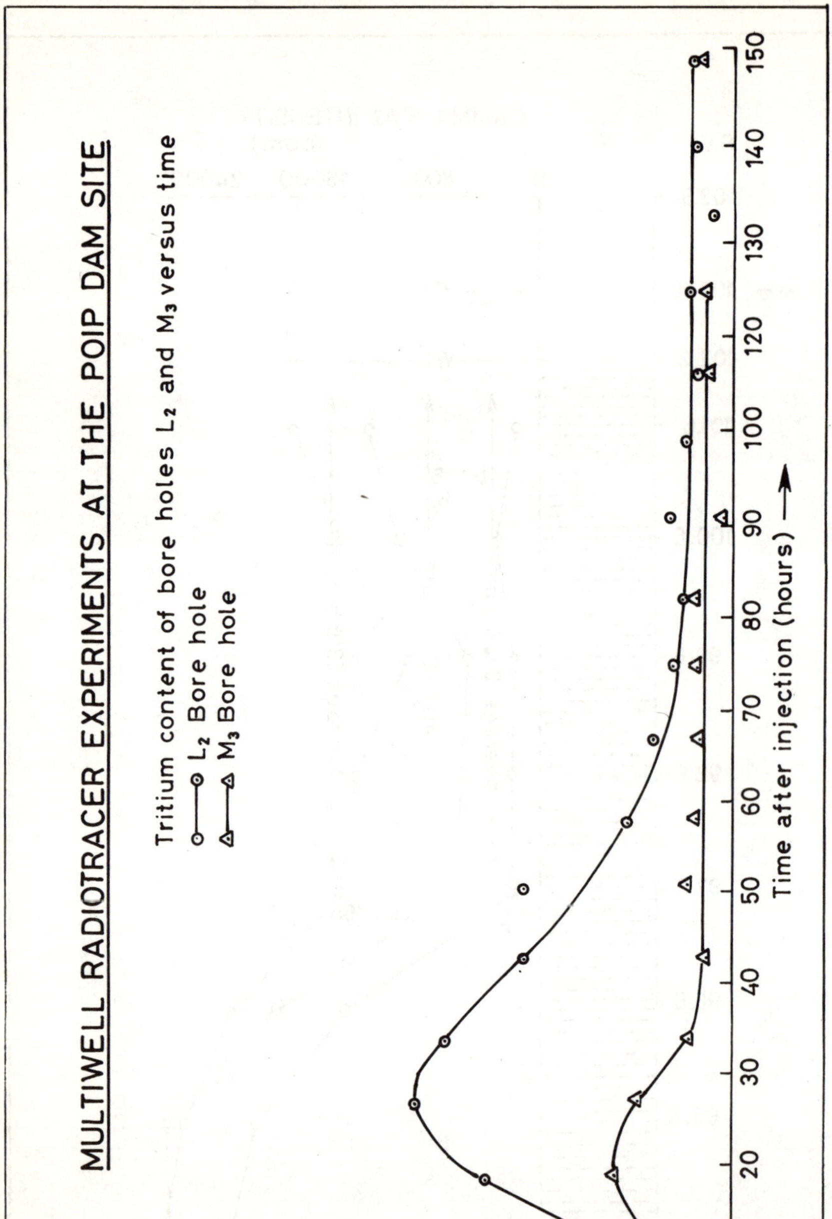
MULTIWELL RADIOTRACER EXPERIMENTS AT THE POIP DAM SITE
Tritium content of bore holes L2 and M3 versus time
L2 Bore hole
M3 Bore hole
Tritium content of samples (dpm/ml)
Time after injection (hours)
0
10
20
30
40
50
60
70
80
90
100
110
120
130
140
150

Fig. 5

groundwater flow. About 10 uCi of tritiated water was injected in bore holes Mo and Jo on September 5th, 1986. Water samples were collected from various bore holes on the downstream side and were analysed for tritium with a liquid scintilation counter. Fig. 5 shows the plot of tritium contents of two bore holes L2 and M3 with respect to time. It was observed that the water flow direction on the left and right flank of the poip dam site were from Mo to M3 and Jo to L2. Transit velocity (Vt) of groundwater near both the experimental bore holes have been computed using the formula

$$Vt = \frac{X}{\bar{T}} \qquad .....2.$$

Where X is the distance between the injected and monitoring bore hole and T is the average transit time of water flow (of tracer in this case). The transit velocity near Jo and Mo were 8.5 meter/day and 14.4 meter/day respectively. Fig. 3 & 4 show the mixed profile of Br-82 tracer, in the bore holes Jo and Mo. In this profile one can see that the groundwater velocity is more in the top 5-6 meters of water column in both the bore holes.

Vertical velocity measurements :

Vertical flow rates were measured using Bromine-82 tracer in a number of bore holes on the dam which tap various depths. It was observed that most of the seepage occured from under the dam through the laterite soil.

2. Seawater intrusion into the Minjur aquifer near Madras in Tamil Nadu.

Environmental isotope geochemical studies confirmed that the progressive increase in the salinity of the Minjur aquifer was due to sea water intrusion (S.M. Rao et.al 1987). To determine the intrusion rate of the salt water towards the aquifer well · field groundwater velocity measurements were carried out in a few piezometers in that area. Location of the piezometers have been shown in fig. 6.

In this experiment 6.5 cm. diameter point dilution probe was used (Fig.1). Bromine-82 as ammonium bromide kept in gelatine capsule was used as the radiotracer. The rubber inflators defining the measuring volume were inflated using a nitrogen cylinder kept on the ground surface. The built in battery operated magnetic stirrer mixed the tracer slowly and throughly in the measuring volume. Table-1 shows the results of measurements.

LOCATION OF PIEZOMETERS AND OBSERVATION WELLS IN MINJUR AREA

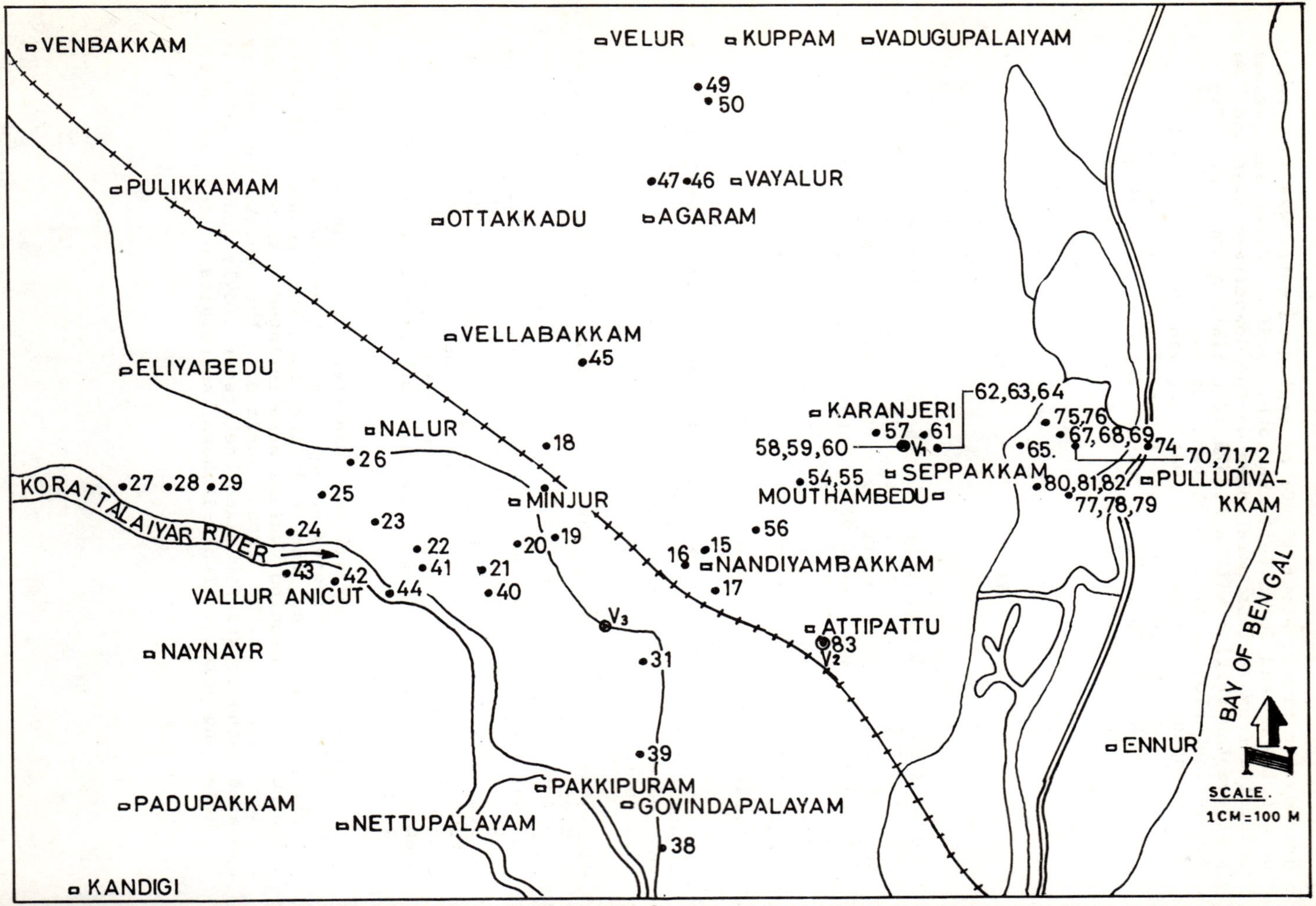

TABLE - 1

Groundwater velocity measurements:

Sl.No.	Piezometer	Reference (Fig.6)	Piezometer diameter (Cm)	Transit Velocity (Cm/d)
1.	11521C	V1	10	2
2.	11556C	V2	10	3
3.	11535C	V3	15	600

From the table it can be seen that piezometer near the salt pan area have a transit velocity of about 2 cm/d, which means that the flow velocity of saline water near the salt pan area is very small, indicating long residence time.

A piezometer in the western side of the aquifer near the production wells showed a higher groundwater velocity of a few meters per day. The higher transit velocity is obviously due to the proximity of the production well field. This latter information is of particular interest in controlling the aquifer exploitation.

3. Impact of paper mill effluent on groundwater body at Erode, Tamil Nadu.

Radiotracer experiments were carried out during March 2-3, 1988 to study the groundwater velocity near M/s. Seshasayee paper board limited. The experiment formed part of an investigation to study the impact of paper mill effluent on groundwater body, conducted by the Tamil Nadu Pollution Control Board.

Radiotracer experiments were conducted in eight bore holes (bore holes locations have been shown in Fig. 7) to determine the groundwater velocity using single well dilution technique with Bromine-82 as the radiotracer.

About 100 uCi of Bromine-82 as ammonium bromide was injected in each bore hole on March 2, 1988. The bore hole waters were initially mixed thoroughly and the concentration of the radioactive tracer in terms of count rate was measured with water proof scintillation probe connected to a count rate meter. Average filtration velocity has been calculated using the formula 1 and values given in the table 2.

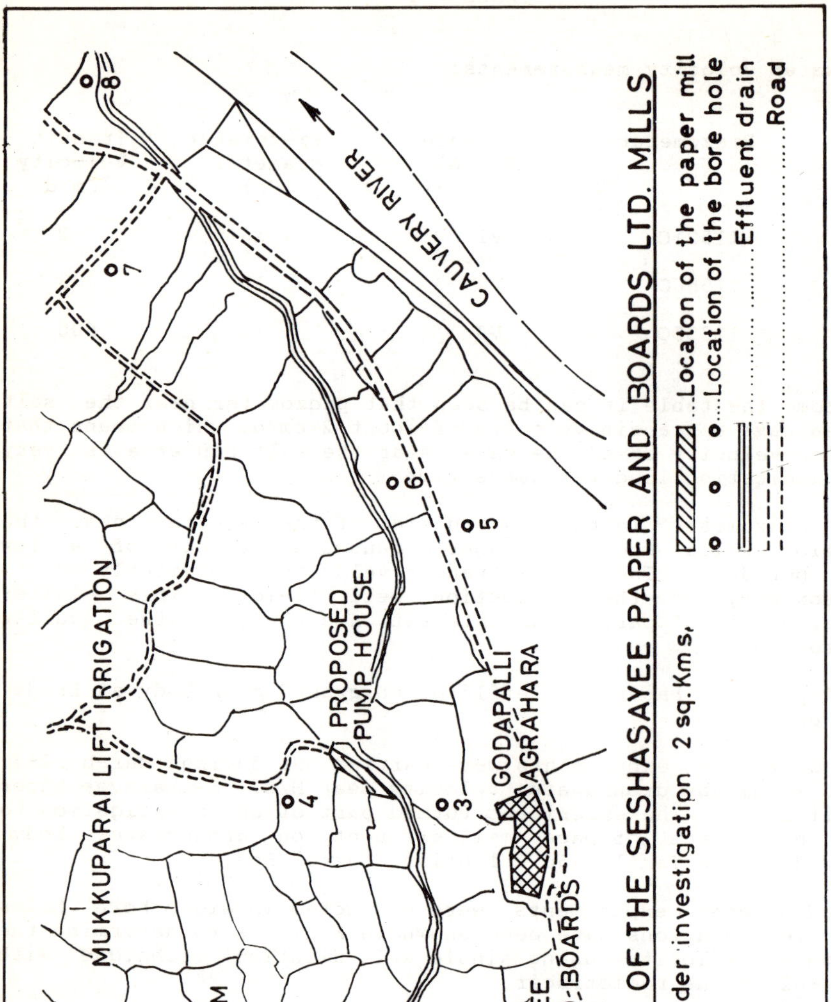
LOCATION MAP OF THE SESHASAYEE PAPER AND BOARDS LTD. MILLS
Area under investigation 2 sq.Kms,
Locaton of the paper mill
Location of the bore hole
Effluent drain
Road
N
PALLIPALAYAM VILLAGE
MUKKUPARAI LIFT IRRIGATION
PROPOSED PUMP HOUSE
GODAPALLI AGRAHARA
SESHASAYEE PAPER AND BOARDS MILLS LTD.
CAUVERY RIVER
1
2
3
4
5
6
7
8

Fig.7

TABLE - 2

Filtration velocity measurements:

Sl.No.	Bore hole No.	Filtration Velocity (m/d)
1	1	0.114
2	2	0.05
3	3	0.2
4	4	0.1
5	5	0.045
6	6	0.04
7	7	0.25
8	8	0.1

In eight bore holes average filtration velocity was in the range from 0.04 to 0.25 m/d. The data was further processed to determine the transit time of the discharge of pollutants in this location.

Conclusion:

The usefulness of groundwater velocity and direction measurements using radiotracer techniques in seepage studies, seawater intrusion and pollution studies has thus been illustrated.

Acknowledgements

Acknowledgements are due to Shri V.N.Yelgaonkar for help during poip dam tracer studies.

References:

1. Drost W., et.al., 1968. Point dilution methods for investigating groundwater flow by means of radioisotopes. Water resources research, 4,(1) P.125-145.

2. Guidebook on nuclear techniques in hydrology, 1983. Technical reports series no. 1, I.A.E.A., Vienna, P.223-253.

3. Moser H., et.al., 1983. New experiences with the use of radioactive isotopes in hydrology, Radioisotopes in Hydrology, I.A.E.A., Vienna, P.282-295.

4. Rao.S.M., et.al., 1987. Isotope studies on seawater intrusion and interrelations between water bodies, some field examples. Procedings of an international symposium on the use of isotope techniques in water resources development, IAEA-SM-299/76 P.403-423.

ENVIRONMENTAL ISOTOPE-AIDED STUDIES ON WATER RESOURCES IN THE REGION OF CHEJU ISLAND

J.S.AHN, J.W.Kim, and J.H.Kim

Korea Advanced Energy Research Institute, Daejeon, Korea

Z.K.U, C.C.Choung, and S.J.Song

Cheju National University, Cheju, Korea

Abstract

The type of groundwater yielded from existing deep wells situated within 2km from coastal line in U-do and Songsan area is classified as category 7, that is primarily salinity dominated by alkalies and strong acid such as saline water. The chlorine contents of the groundwater ranges from about 200ppm to 600ppm. Tritium, deuterium, and oxygen-18 contents of seawater are 5.74±0.49 TU, -4.6± 1.2 o/oo, -0.8±0.2 o/oo in maximum, and 8.67±0.45 TU, -24±0.18 o/oo, -3.6±0.12 o/oo in case of rainwater, and 15.11±0.61 TU, -43.4±1.12 o/oo, -7.2±0.5 o/oo in case of pure and non-affected groundwater by salt water encroachment. In general the groundwater bodies being highly affected by salt water intrusion are characterized by that content of tritium is similar or quite same with one of the pure-nonaffected groundwater and contents of stable isotope are higher than those.

1. Introduction

Quaternary basaltic rock in Cheju Island ranks first among the important water bearing rock in Southern Korea. Spread out in successive sheets over extensive areas, it forms the only source of water supplies throught large sections of Cheju Island. The water occurs in large joint openings and other cavities. It also occurs in the zones of vesicular and fragemental materials such as scorias and Pyroclastic deposits between successive lava flows. This rock is no generally traversed by large opening that it takes in the surface water very rapidly. Basalt gives rise to numerous large springs.[1] Many examples could be given of large springs discharges from basaltic lava on Cheju Island. Due to rapid growing economy of Cheju Island in the field of tourism, industries, agriculture etc, the demand of water continues to increase year by year. Therefore heavy pumpage of groundwater creates salt water encroachment problem into fresh groundwater bodies in some part of the island.

In order to provide effective counter measures to protect the groundwater body from sea water intrusion, isotope-aided technique and chemical analysis were applied in the areas of Songsan and Kujwa to ascertain whether the high salt content of existing well were caused by sea water intrusion.
-For the study,

1) Water samples were, collected seasonally from 15 different sites; 11 of groundwater bodies, 2 of spring waters and 2 of sea waters. (see table 1)
2) Temperature, PH, main anion and cation contents and electrical conductivity of the samples were measured[2], and the type of water was determined in accordance with Walton's classification.[3]
3) Tritium content analysis was counted at KAERI, and deuterium and oxygen-18 analysis was made by Australian Atomic Energy Commission.[4.5.6]

Table.1. Description of water samples, Cheju Island.

Sample No.	Location	Source of water	Distance from sea (km)	Well Depth (m)
G1	Kujwa up, Handong ri	Ground water	2.25	70
G2	Kujwa up, Pyongdal ri	Ground water	2.00	80
G3	Kujwa-up, Sehwa-ri	Ground water	1.50	45.5
G4	Kujwa-up, Sangdo-ri	Ground water	1.75	50
G5	Kujwa-up, Chongdal-ri	Ground water	1.85	35
G6	Kujwa-up, Songdang-ri	Ground water	9.50	200
G7	Kujwa-up, Hado-ri	Spring water	-	-
G8	Kujwa-up, Hado-ri	Sea water	-	-
G9	Songsan-up, Sihung-ri	Ground water	1.75	82
G10	Songsan-up, Susan-ri	Ground water	5.50	51
G11	Songsan-up, Susan-ri	Ground water	2.25	51
G12	Songsan-up, Sinyang-ri	Ground water	1	54.5
G13	Songsan-up, Ojo-ri	Spring water	-	-
G14	Songsan-up, Sinyang-ri	Sea water	-	-
G15	Songsan-up, Sinsan-ri	Ground water	4.6	
S1	Soguipo-shi, Songsan-dong	Ground water	0.23	60
S2	Soguipo-shi, Songsan-dong	Ground water	0.25	34
S3	Soguipo-shi, Sogui-dong	Ground water	0.15	62
S4	Cheju shi, Samdo 2-dong	Ground water	0.05	
U1	Udo, Choil ri	Ground water	1.0	38.5
U2	Udo, Choil ri	Ground water	0.75	38.0
U3	Udo, Choil ri	Pond water	0.40	
U4	Udo, Sogwang-ri	Sea water	-	-

4) 4 sampling sites were added at Soguipo-shi are (3 ground waters) and Cheju-shi (1 ground water) to confirm the interaction between fresh waters and sea water.

2. Geography and climate

Cheju Island being an oceanic volcanic island, is located 180km south of the mainland occupying 1,825 km^2 with a population of 474,000 in 1985. (see Figure-1) In the center of the island, Mt. Halla, the volcanic pile of Cheju, rises 1,950m above sea level. The rocks of Cheju are almost entirely of volcanic origin with a predominance of basalt, but including trachyte, trachyandesite, aphanitic andesite, and some volcanic ash and cinder. Most of the volcanic rocks are vesicular and are cut by numerous joints and fissures. Lava tunnels and scoriaceous contact zones of individual lava flows are common and provide conduits for groundwater movement.

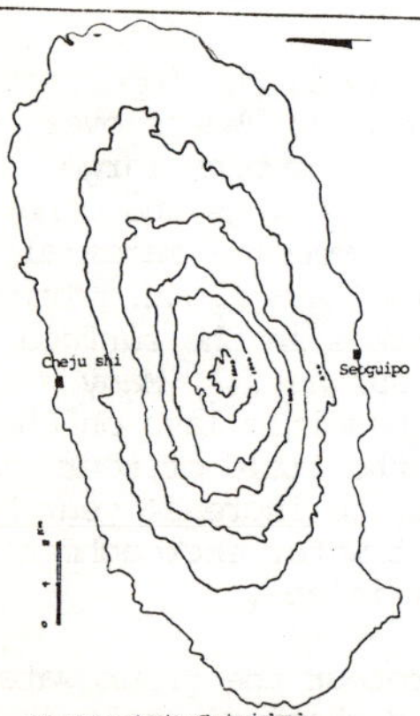

Fig.-1 Map showing Cheju island.

Annual rainfall averages about 1,440m/m. This total is undoubtedly exceeded at higher altitudes in the interior of the island. The heaviest precipitation occurs from June through September during the south-west monsoon season, when sixty percent of the annual total falls. But, despite the heavy rainfall, there is little sustained streamflow on the Island. Torrential runoff takes place during and immedately after a rain, but the high permeability of the volcanic rocks encourages very rapid infiltration. Streams are usually short and small, lack substantial base flow, and are very steep. The major water resource is therefore groundwater stored in the volcanic rocks.1)

3. Hydrogeology and Groundwater Occurrence

3-1. Hydrogeology

The basaltic rocks typical for an oceanic island contain extraordinarily permeable aquifers. The original lava congeals on the flanks of shield volcanos as thin layers of solid rock separated by equally thin layers of pyroclastic deposit. The solid rock layers are frequently broken by vertical fractures and horizontal openings such as lava tubes.

The clinker or scoria consists of an assortment of heterogeneous rock pieces mixed with finers. The combination of fractures and other opening in the connected large void spaces in the clinker and scoria give the rock mass outstanding aquifer properties. The combination of high infiltrability and permeability makes basalt aquifers the preferred groundwater resources on the island.

On Cheju Island, as along any coastal area, groundwater is classified into the two general categories. Aquifers having no contact with sea water carry "High Level Groundwater" because they normally occur at considerable elevation above sea level. The groundwater in aquifers in which fresh water is hydraulically connected to sea water is called "Basal Groundwater" when the entire aquifer below the fresh water portion is saturated with sea water. High level groundwater is normally not as voluminous as basal groundwater because ordinarily it is restricted to complicated geological areas in the island interior whereas basal aquifers are the subsurface reservoirs where practically all groundwater accumulates before being discharged into the sea. High level groundwater occurs appreciably above the basal groundwater table. Characteristically, all high level groundwater ultimately issues as surface springs or passes as subsurface seepage into basal aquifers. The basal groundwater as distinguished from the high level groundwater is a voluminous body of fresh water that occurs in a deep position beneath the land surface known as the "Ghyben-Herzberg Lens".

The rainfall filters into the ground through secondary porosities such as joints, lava tubes, scoria and the contact zones of layers. The water seeping into the ground reaches the contact planes, and will continue downward filtering of slope migration depending on the permeability of the rocks. When permeable geologic structures are connected with tunnels or have large cavities in the underground, the groundwater will form high level groundwater. But some of it continues seeping into deeper portions and will form the basal groundwater.[7)]

Along the southern coastal preimeter several springs form streams that flow for a few hundreds meters and then cascade over cliffs into the sea below. As mentioned above, most of the rainfall disappears rapidly into the permeable volcanic rocks. From higher parts of the island it generally moves downward thorugh joints and some pyroclastic layers and then laterally through sloping zones of scoria or similar permeable material, emerging at the surface locally as high or medium altitude springs that give rise to small streams. These small streams disappear again, sometimes within a few hundred meters, where rocks are highly permeable. Basalt layers dip toward the sea, whereas ash beds dip in various directions in conformance with local parasitic volcanic cones.

Substantial parts of the water supply filter into a basal fresh water lense that discharges at springs around the perimeter of the island and offshore. The discharge of fresh groundwater through springs located offshore and along the perimeter of the island is estimated about 525,000m^3/day through 185 springs.[8)] In addition, 206 water wells were drilled in order to supply agricultural purposes between 1972 and 1985. Total amounts of groundwater able to abstract from those wells are reported at about 470,000m^3/day.[8)]

3-2. Yield of Groundwater

Records were obtained of 171 water wells drilled in Cheju Island. Amongst the wells drilled on Cheju Island, 88 wells were located at an elevation between 10 and 50 m above the mean sea level, 42 wells between 50 and 100m, 29 wells between 100 and 150m, and the rest of 12 at higher elevations. Table 2 and 3 are summations of the average and the range of well depths, and the specific capacity of the wells drilled on Cheju Island.

Table 2. Average depth, yield, specific capacity and elevation of wells drilled on Cheju Island. (after Hahn.J.S. 1985)

Elevation (m)	Average elevation (m)	Average yield (m^3/day)	Average depth (m)	Average Sp.C (m^2/day)	No.of well
0 - 50	29.6	1,927.3	73.0	1,768.8	88
50 - 100	73.2	1,398.3	91.3	1,113.3	42
100 - 150	143.6	1,275.0	143.6	706.3	29
Above 150	204.2	991.6	192.3	427.0	12
Average		1,621.0	97.8	1,333.4	

Table 3. The range of depth, pumping rate, and specific capacity of the wells drilled on Cheju Island. (after Hahn.J.S. 1985)

Elevation (m)	Depth (m)	Yield (m^3/day)	Sp.C (m^2/day)
0 - 50	26 - 152	812 - 4,121	34.9 - 8,900
50 - 100	35.5- 126	855 - 2,445	37.3 - 3,800
100 - 150	104.5- 222	800 - 2,030	9.7 - 728.5
Above 150	92 - 300	300 - 1,500	12.3 - 3.609

As shown in the Table 3, wells drilled at the lowest possible place in the topography have more chances to encounter voluminous water-bearing zones: that is basal groundwater reservoir and yield more water than those drilled at higher places.

Of the 88 wells drilled at elevations below 50m, 36 wells(41% of the total) had pumping water levels lower than the mean sea level. For example, at a well drilled at Sinjo, Daejeong in South Cheju, the static water level was the same as sea level when the well was drilled 122m deep. When it was pumped at the rate of 2,000 m /day, the pumping water level dropped to 90m below the land surface which was 55m lower than sea level. If the extraction of groundwater is continued in this area, the hydraulic balance of the fresh water lense shall be broken and then salt water will intrude into the fresh water zone. Such phenomena of salt water encroachment is observed in this study area.

3-3. Temperature of the Groundwater

The temperatures of water form springs, water wells, as well as sea water were measured three times in 1983. The temperatures of groundwater from wells ranging in depth from 13 to 240m were recorded at 13°C to 16°C in April, 13.5°C to 18°C in May, and 14°C to 17°C in June. But groundwater temperatures from deep wells whose depths ranged from 90 to 240m were constant throughout the year, they measured from 13°C to 14.5°C. The temperatures of springs fluctuated with greater ranges than the water of deep wells and was quite similar to the temperatures of limestone springs.

4. Water Budget

The annual total precipitation of Cheju Island is estimated at about 3.38 billion metric tons per year. Amongst the total, 1,846 billion cubic meters of water are transevaporated and discharged into sea through stream channel, 1.54 billions being equavalent 45% of total precipitation infiltrated into the bedrock aquifer and stored as fresh groundwater bodies, 0.2 billion among the total infiltrated water into aquifer is discharged into the sea through 185 springs.

Therefore at least 0.9 billion cubic meters of fresh water is annually retained in basaltic aquifer.[7] The demand for water on Cheju continues to increase due to an expanding population, development of tourism and agriculture and fisheries. However there are constraints in the use of surface water and water well drilling has so far been haphazard. Water well are sometimes responsible for the intrusion of sea water into well-fields(see Figure 2). Therefore systematic development of underground water resources and their protection is of prim importance to fulfillment of the Cheju master plan. To protect wells from sea water intrusion and to ensure a stable water budget it is desirable to construct ditch tunnel-type water intake facilities.

5. Isotope Aided Studies on Salt Water Encroachment.

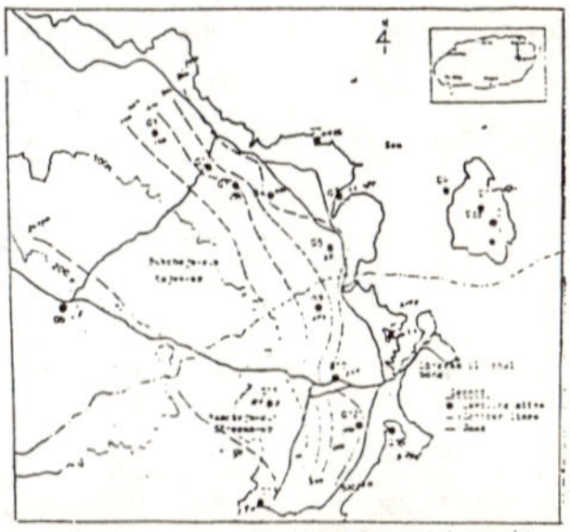

Fig.2 The contour map of [illegible] (Songsan, [illegible] Area) [illegible]

15 water samples from Songsan area, 4 samples from Soguiposhi and 4 other samples from U-do were taken seasonally or once in order to study the interaction between salt water and fresh groundwater bodies. Amonst 3 were sampled from sea water of coastal line and two from spring water located also at coastal line and I from pond and rest 17 from water wells drilled in basalt (see Figure 2). In Songsan area where the highest chlorine contents were observed, samples were collected twice in dry season of March and May, once on wet season of Sugust, and once in end of wet season of October and were analysed the main cations and anions, pH, and conductivity of each water to correlate with seasonal variation of isotope contents and to determine the interaction between salt and fresh water.[3,9] Besides of the study of seasonal variation of chemical components, of each water samples, water samples were collected 3 times with interval of I week after heavy rain of July, 16, 1986 in order to see any special fluctuation of chemical components due to recession of precipitation (see Table 4 and 5).[10]

Table 4. Seasonal variation of chemical component of water samples from Kujwa and Songsan areas.

Sample No.	Temp. (°C)	pH	Na (ppm)	K (ppm)	Mg (ppm)	Ca (ppm)	Cl (ppm)	NO_3 (ppm)	SO_4 (ppm)	CO_3+HCO_3 (ppm)	Conductivity (µmho/cm^{-1})
G1	14	7.6	57.4	9.4	13.0	9.4	115	1.18	15.6	38.0	568
	15	7.7	57.6	11.5	11.3	8.3	120	1.06	15.9	38.0	590
	15	7.5	42.1	5.8	10.5	7.7	98	1.54	12.2	29.5	445
		7.0	53.6	6.3	9.8	7.3	103	0.77	16.4	35.6	474
G2		7.6	137	17.0	27.8	22.5	303	1.60	44.1	37.4	1,173
	16	7.6	142	18.6	27.4	22.5	336	1.49	39.4	34.3	1,266
	16	7.6	81.2	9.5	46.1	35.9	206	1.74	25.0	31.5	854
	-	7.2	105	11.3	17.5	16.0	210	1.14	32.7	35.2	921
G3	15	7.6	92.5	10.0	14.5	14.5	190	1.45	26.1	40.8	798
	16	7.5	84.2	14.1	17.4	14.0	192	1.26	23.0	41.9	782
	15	7.5	92.8	10.9	23.0	14.3	348	1.80	37.2	35.2	1,135
		7.1	20.4	3.0	7.9	10.0	42.5	1.33	11.2	36.6	267
G4	15	7.4	229	33.4	46.4	27.8	589	0.77	98.0	48.4	2,224
	15	7.6	290	35.8	44.6	26.6	616	0.57	67.6	48.4	2,245
	15	7.5	192	15.1	34.3	19.4	459	1.42	57.0	46.7	1,780
		7.4	260	16.9	31.5	20.2	491	0.53	73.5	42.2	1,920
G5	15	7.4	145	19.2	31.9	19.6	328	0.70	18.0	40.5	1,280
	15	7.7	150	21.3	25.8	18.2	331	0.53	42.0	41.4	1,232
	15	7.5	92.0	9.3	18.1	11.8	245	0.90	32.6	26.7	1,010
		7.5	131	10.0	19.7	12.3	268	0.42	37.2	36.6	1,115
G6	14	7.5	8.2	2.0	3.5	4.4	12.4	1.80	3.3	31.8	113
	16	7.7	8.3	3.7	2.7	3.8	11.5	2.06	4.5	28.7	113
	15	7.6	8.6	2.9	3.5	4.0	14.5	1.74	6.1	21.9	112
		7.6	9.1	4.2	3.4	4.1	10.1	1.94	9.2	26.7	112
G7	14	7.1	445	49.4	69.2	31.3	936	1.02	123	36.6	3,343
	15	7.6	1,150	140	205	70.3	6,467	0.76	313	48.7	8,632
	14	7.4	663	49.4	110	32.2	1,427	0.78	140	38.5	5,015
		7.0	548	32	101.8	30.7	975	0.59	153	39.4	3,480
G8	12	8.1	8,970	220	1,120	470	19,321	0.47	3,102	133	56,515
	15	8.2	8,630	928	995	369	22,087	0.49	2,135	130	56,404
	25	8.2	7,450	545	1,135	608	17,017	0.39	1,860	126.6	47,100
	-	8.1	9,370	527	1,105	405	17,159	0.53	2,653	120	45,446
G9	15	7.8	179	16	25.7	4.2	271	0.90	45.7	38.0	1,016
	16	7.7	108	18.5	22.8	17.6	272	0.72	296.6	39.4	1,062
	17	7.7	67.0	6.5	12.9	10.2	136	1.00	18.6	29.5	602
	-	7.8	108	9.3	20.7	13.8	241	0.25	39.4	36.6	990
G10	15	7.5	251	22.8	38.5	20.3	511	2.99	65.3	38.0	1,088
	16	7.6	238	31.5	37.4	21.1	529	2.85	579	36.6	1,933
G11	16	7.6	37.1	6.3	11.8	9.5	81.5	1.19	12.2	35.2	394
	17	7.8	32.7	7.5	7.7	9.0	77.1	1.12	9.9	42.2	391
	15	7.4	28.0	3.7	7.9	7.0	51.8	2.99	9...	30.9	288
		7.6	32	4.2	8.7	7.5	58.5	1.21	26.5	36.6	316
G12	16	7.4	205	23.3	43.8	23.2	488	0.70	65.3	49.2	1,795
	16	7.7	201	26.4	36.8	21.9	490	0.56	62.3	43.0	1,787
G13	15	7.4	1,192	46	109	66	2,154	2.85	343	46.4	7,276
	15	7.5	1,050	98.3	134	54.1	2,195	2.79	257	44.2	7,398
	16	7.5	847	51.6	96.4	35.5	1,545	3.07	221	14.1	5,595
		7.5	2,248	129.2	322	86.5	3,970	2.35	510	53.5	12,874
G14	14	7.9	8,250	570	998	400	17,691	0.50	2,857	130	52,546
	16	8.2	8,910	933	1,040	380	18,746	0.40	2,570	127	59,118
	25	8.1	7,890	408	813	965	13,153	0.63	880	121	38,550
	..	7.9	9,740	552	1,718	393	17,159	0.39	2,551	101	47,489
G15	-	7.0	5.7	2.4	6.4	8.4	9.0	0.31	5.5	38.0	105
		7.1	8.7	2.7	4.5	4.5	8.2	0.27	8.3	39.4	116

First line : Data of March, Second line : Data of May,
Third line : Data of August, Forth line : Data of October.

5-1. Chemical Characteristics and Seasonal Variations of Ground Water

The pH of groundwater having average of 7.4±0.4 ranges from 7 to 7.8 while the one of

Table 5. Changes of chemical component after heavy rain in Kujwa and Songsan area.

Sample No.	pH	Na (ppm)	K (ppm)	Mg (ppm)	Ca (ppm)	Cl (ppm)	NO_3 (ppm)	SO_4 (ppm)	CO_3+HCO_3 (ppm)	Conductivity (μmho/cm^{-1})
G1	7.6	41.2	6.9	8.0	5.8	73.4	2.29	8.7	34.5	358
	7.6	44.6	7.9	8.6	6.4	96.6	1.87	8.6	33.2	376
	7.5	49.9	5.0	9.5	6.9	93.8	1.78	8.7	33.8	392
G2	7.6	78.8	6.1	13.0	10.2	147	2.48	19.7	34.6	623
	7.6	105	6.7	17.3	13.0	200	1.81	25.4	33.8	796
	7.6	110	7.7	18.6	13.8	214	1.76	29.7	35.2	846
G3	7.5	127	8.3	22.0	13.8	250	2.62	34.0	36.9	994
	7.5	128	8.7	19.4	11.8	244	2.34	34.0	38.0	937
	7.5	152	10.6	21.9	12.9	262	2.08	20.4	40.8	1,220
G4	7.5	201	10.4	28.3	17.3	371	2.27	26.9	38.3	1,373
	7.5	259	11.2	33.1	18.5	283	2.01	32.6	38.0	1,641
	7.5	317	13.7	41.5	21.0	547	2.27	64.0	39.4	2,034
G5	7.6	175	8.6	24.2	14.2	316	1.34	44.0	34.9	1,232
	7.6	179	9.8	22.9	14	326	1.14	45.4	39.4	1,225
	7.6	176	8.8	21.2	12.9	303	1.04	45.4	39.4	1,178
G6	7.5	10.1	3.7	3.6	14.7	6.2	2.37	1.9	34.9	117
	7.5	8.8	2.6	3.5	4.2	18.0	1.67	1.9	23.9	103
	7.6	8.6	2.8	3.6	4.2	10.6	1.75	1.9	25.3	104
	7.7	174	9.2	12.6	9.2	146	1.13	24.0	37.1	592
	7.6	174	8.1	11.6	8.9	137	1.02	29	36.6	559
	7.5	169	7.1	11.8	8.9	131	1.00	22.6	36.6	558
G11	7.6	23.6	2.7	6.1	4.9	34.2	4.68	8.9	28.7	219
	7.7	27.7	3	7.4	6.0	47.5	4.13	8.9	30.9	249
	7.7	295	3.9	8.1	6.4	47.2	3.22	8.9	32.4	262

First line ; One week after heavy rain.
Second line; Two week after heavy rain.
Third line ; Three week after heavy rain.

Table 6. Chemical data of water samples from Cheju-shi and Soguipo-shi area.

Area	Sample No.	Temp (°C)	pH	Na (ppm)	K (ppm)	Mg (ppm)	Ca (ppm)	Cl (ppm)	NO_3 (ppm)	SO_4 (ppm)	CO_3+HCO_3 (ppm)	Conductivity (μmho/cm^{-1})	Remark
Chejushi and Soguipo shi	S1	18	7.1	50.6	4.9	3.2	7.7	28.7	4.69	6.1	26.7	200	First line: Sampled at May
		-	6.8	88.5	10.5	4.4	4.7	26.1	1.77	7.9	24.4	201	
		-	6.9	41.4	3.6	5.0	9.7	26.9	7.39	11.2	36.6	220	
	S2	16	5.9	21.0	7.1	3.0	12.3	53.2	7.07	10.5	31.5	223	Second line : Sampled at August
		15.5	6.9	17.9	3.3	4.7	10.2	36.7	1.52	9.3	19.1	201	
		-	6.7	70.2	5.2	8.7	12.4	68.6	7.09	16.3	53.5	441	
	S3	18	7.9	18.5	5.3	3.9	17.3	23.9	5.34	5.0	69.5	247	
		17	7.6	21.7	3.8	4.7	8.2	22.2	1.26	7.1	64.7	224	
		-	7.7	15.2	4.8	5.8	15.7	14.5	4.81	9.4	17.5	240	Third line : Sampled at October
	S4	-	7.4	23.6	10.7	3.7	12.5	39.9	4.11	10.7	63.9	301	
		-	-	-	-	-	-	-	-	-	-	-	
		-	7.2	18.7	9.4	7.7	13.8	29.6	4.98	17.9	61.9	230	
U-do (August)	U1	7.4	7.3	132	10.7	16.5	8.2	221.0	1.23	34.9	8.4	952	- : not measured
	U2	7.7	7.7	226	13.3	31.2	19.5	417.5	0.64	64.0	70.3	1,550	
	U3	6.8	6.8	8.6	2.2	1.4	0.8	11.3	0.26	7.9	8.4	80	
	U4	8.1	8.1	9,500	535	1,135	398	18,125	0.36	1,980	122.4	48,150	

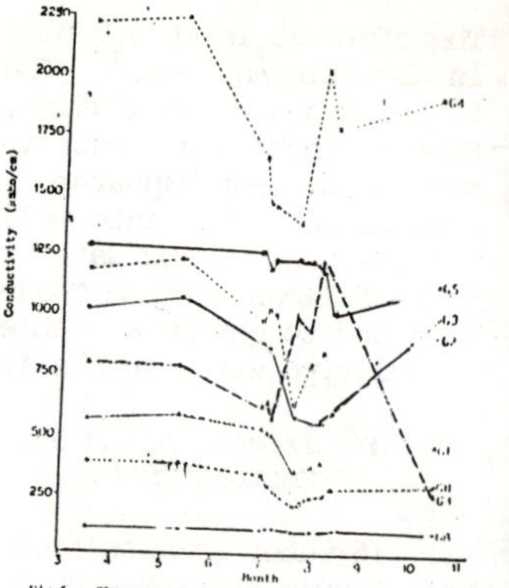

Fig.6. The seasonal changes of conductivity of water samples from Kujwa and Songsan areas.

seawaters are about 8.05+0.15 in average (see Table 4).

The concentration of potassium and chlorine, and conductivity of groundwater being highest values in March and May which belongs to dry season tends to decrease (dilute) in wet season (August) and again to increase slowly from October (end of wet season) upto the contents of early dry season, in Songsan, U-do, Cheju shi, and Soguiposhi area as shown in table 5,6. But the groundwater of G-6 well located about 9.5km far from coast line at the altitude higher than 200 meter above sea level shows quite constants values of potassium, chlorine and conductivity even in seasonal changes.[11] It is considered that the dilution effects of those contents of groundwater in wet season is due to the infiltration effect of rainfall intq aquifer body or the effect of ground water recharge near the well sites. In contrast, the increasing phenomena of those components in dry season is considered to be a fact that interaction between salt water and groundwater reserved specially in coastal area are taking place. (see Figure 6) Figure 2 shows on iso-chlorine map of Songsan area. As shown on the figure all groundwater except from G-6, G-11 and G-15 well which are located quite distance from coast line contains more than 150ppm of chlorine which is objectionable to drink. In other words, the groundwater sampled from G-4, G-10 and G-12 well located near the sea-coast contains more than 400 ppm of chlorine and very higher conductivities in comparision of other ground water. It is deemed that most of groundwater bodies occurred nearer sea coast are interacted or intruded by sea water owing to artificial abstraction of groundwater through wells.[12]

Table. 7. Environmental tritium, deuterium, oxgen-18 levels of water samples, Cheju Island.

Sample No.	March			May		
	Tritium (TU)	Deuterium (per mill)	Oxygen-18 (per mill)	Tritium (TU)	Deuterium (per mill)	Oxygen-18 (per mill)
G1	14.08±0.49	-38.6±1.70	6.9±0.20	16.06±0.45	-38.40±1.98	-6.59±0.02
G2	11.16±0.56	38.5±0.40	6.9±0.04	14.74±0.51	-38.60±0.69	-6.52±0.03
G3	11.68±0.53	40.8±0.02	6.8±0.01	16.97±0.63	-42.40±0.57	-6.48±0.01
G4	10.39±0.61	38.5±1.70	-6.9±0.01	13.05±0.58	-38.85±0.64	-6.51±0.01
G5	14.26±0.57	-40.5±0.03	-6.8±0.02	17.18±0.54	-39.85±0.63	-6.46±0.03
G6	11.69±0.48	43.4±0.20	-7.2±0.05	15.11±0.61	-39.60±2.26	-6.72±0.04
G7	10.74±0.59	40.7±1.70	6.9±0.04	14.44±0.63	-39.65±0.53	-5.86±0.02
G8	5.74±0.49	-2.40±0.02	-0.2±0.04	3.76±0.49	-3.20±0.99	-0.06±0.03
G9	13.22±0.53	-40.8±0.02	6.9±0.01	15.49±0.52	-41.20±2.65	-6.40±0.01
G10	13.24±0.53	44.3±0.80	7.0±0.03	12.99±0.56	-39.65±1.63	-6.51±0.02
G11	15.11±0.63	42.4±1.70	6.9±0.04	15.59±0.53	-39.40±1.56	-6.41±0.01
G12	14.87±0.58	-42.4±0.02	6.8±0.01	16.41±0.59	-39.65±0.64	-6.46±0.09
G13	11.23±0.53	38.8±0.02	6.8±0.02	10.72±0.47	-41.70±0.71	-5.98±0.01
G14	3.73±0.48	4.60±1.20	0.8±0.02	3.67±0.45	-3.00±1.79	-0.02±0.01
S1				14.00±0.57	-43.00±1.79	-7.05±0.08
S2				12.50±0.49	-43.90±1.75	-7.18±0.05
S3				5.51±0.43	-41.70±1.41	-6.92±0.09
S4				14.46±0.51	-49.97±1.17	-7.62±0.04
rain				8.67±0.45	-24.10±0.18	-3.61±0.12

Table. 8. Tritium levels of water samples collected in Kujwa and Songsan areas after heavy rain.

Sample No.	After one week	After two weeks	After three weeks
	Tritium (TU)	Tritium (TU)	Tritium (TU)
G1	16.71 ± 0.51	14.24 ± 0.47	14.29 ± 0.47
G2	10.05 ± 0.46	12.55 ± 0.53	12.89 ± 0.53
G3	13.15 ± 0.49	15.35 ± 0.48	13.93 ± 0.45
G4	11.53 ± 0.57	11.42 ± 0.46	11.07 ± 0.59
G5	16.43 ± 0.45	16.16 ± 0.53	16.72 ± 0.47
G6	16.66 ± 0.59	13.37 ± 0.47	12.93 ± 0.57
G9	16.48 ± 0.51	14.86 ± 0.55	16.66 ± 0.54
G11	11.92 ± 0.63	14.39 ± 0.56	13.02 ± 0.48

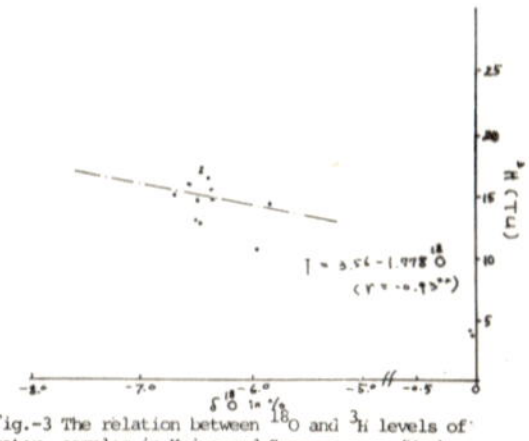

Fig.-3 The relation between ^{18}O and ^{3}H levels of water samples in Kujwa and Songsan area. (May)

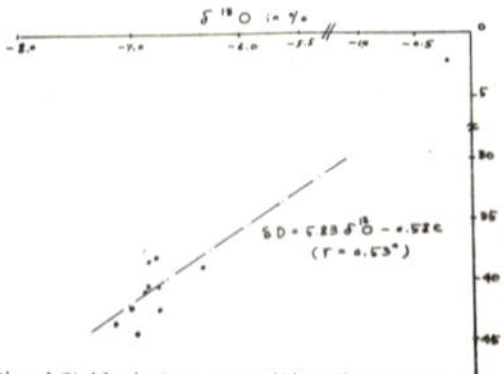

Fig.-4 Stable isotope composition of water samples in Kujwa and Songsan area (March).

5-2. Isotope Hydrology

a) Tritium contents and stable isotopes.[13]

Tritium contents of sea water during March and May was 3.67+0.45 TU and 5.74+0.49 TU, respectively[14] and the one of groundwater ranges from 10.37+0.62 TU to 17.18+0.54 TU which is almost 3 times of sea water and the one of rainfall(May) is recorded about 8.67+0.45 TU, and springs ranges from 10.74+0.59 to 14.44+0.3 TU as shown on table 7. According to the contents of tritium of each source of water, spring water are abundantly affected by groundwater and infiltrated rain water. The chemical and isotope contents of G-6 ground water can be used as standard value of ground water in this area because of its location and others. Tritium, Deuterium and Oxygen-18 contents of G-6 ground water in dry season(March) were 11.69+0.8 TU, -43.4+0.2, and -7.2+0.05 per mill respectively(see Figure 3,4). But the contents of G-4 ground water showing 589 ppm of chlorine were 10.39+0.6 TU, -38.5+1.7 per mill, and -6.9+0.01 in same period. The fact that the isotope contents of G-4 sample are considerably lesser than G-6 sample in same period is deemed to be a criteria of salt water encroachment into G-4 well field.[15] In general the groundwater bodies highly affected by salt water intrusion contains similar or same contents of tritium and higher contents of stable isotopes in Songsan area.[16] But the groundwater bodies not affected by salt water in Soguipo, U-do and Cheju area also shows very slight seasonal variation of chlorine contents whose maximum is less than 70 ppm(see Table 5). The isotope contents of each well (see Table 7) is quite constant in comparision with the contents of Songsan area,

After heavy rain of July, 16, the groundwater in Songsan area, were sampled 3 times with one week intervals each and measured the chemical components and isotope contents (see Table 8). Amongst water samples of Songsan area, the tritium contents of G-6 groundwater is decreasing time to time, that is, 16.66 TU after 1 week to 12.9 TU after 3 week due to recharge effect of rainfall or fresh groundwater. But the tritium of contents of G-4 groundwater which is deemed to be seriously affected by sea water intrusion show constant value during 3 weeks.[17)]

5-3. Classification of Waters.

In other to determine the type of waters occurred in the study area, Walton's classification is applied. All groundwater sampled and analysed in dry season except G-6, S-3, classified in category 7, in other words, non carbonate alkalie (primary salinity) exceeds 50%-that is, chemical properties are dominated by alkalies and strong acids such as saline water (see Figure 5)

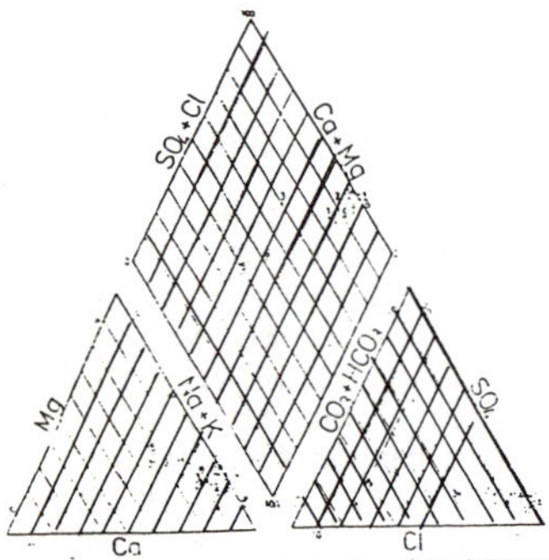

Trilinear diagram of water samples from Kujwa and Songsan area (October)

It is meant that all groundwater reserved and occurred near coastal line (say about within 2km from sea coast) are already seriously interacted or intruded by sea water. Some of the groundwater sampled at G-3, S-2 well changes their water type of NaCl(Category 7) in dry season to $CaCl_2$ and $CaHCO_3$ type(Category 9 & 5) at wet season by possible direct recharge of rainfall or indirect recharge of pure fresh groundwater in the well site.

6. Results and Recommendations

The salt content and electrical conductivity of groundwaters as well as spring waters in areas of Songsan and Kujwa were much higher than those of Cheju-shi area. In other words, 5-10 times in salt content and 30-60 times in electric conductivity were observed. Since the chlorine contents of groundwaters in these area were higher than 150ppm, the groundwaters are not to be suitable for drinking purpose but still usable for agricultural irrigation.

The fact that there were significant correlation among the salt (Na, Cl) contents, electrical conductivities and stable isotope contents, indicates the groundwaters close to the coastline were affected by sea water. The high ratio of $Cl/HCO_3 + CO_3$ in those groundwaters also provides the evidence of sea water intrusion.

Considering that the salt contents of groundwater body were much diluted by heavy rain-fall, the catchment basin seemed to be small and the mixing process might occur rapidly. Additional survey on the existing boreholes in U-do Island area showed that the groundwaters were not suitable for drinking but could used for the irrigation purpose. So recommendation on future groundwater development in Cheju Island is as followings;

1) Water wells showing sea water intrusing were located in general, below 50m altitude water level of 50 to 80m below ground surface. Accordingly, it is recommended to develope new water wells, at the higher altitude in order to secure of better quality groundwaters.

2) To meet the need of large water quantities in future, the ditch tunnel type-water intake system should be considered in aspect of water economy and the protect of fresh groundwater bodies from sea water intrusion.
3) A tentative hydro-geological survey, using environmental isotope aided technique, is necessary at the stage of feasibility study of siting and designing of ditch tunnel on Cheju. Consequently further study should be continued to investigate the western part of Cheju Island.

Acknowledgement

This work was initiated and supported by KAERI since 1984. First of all I deeply thank to Dr.Pil Soon Han, President, KAERI, whose heartfelt advices and continuous supports make this work successful. We also thank to Dr. G.E. Calf, Australian Atomic Energy Commission, who guided me to this work and provided technical and scientific advices. Sincere thanks are also extended to Dr. Wandowo, Center for the application of isotopes and radiation, Indonesia for his constant advices and stable isotope analysis.

Reference

1. Han.J.S. 1981, Ground Water of Bed Rocks in South Korea, Peninsula.J.Korean Asso. Hydrological, Scin. 14(4):73-81
2. Greenberg.A.E. Connors.J.J. and Jenkins.D. 1981, Standard Methods for the Examination of Water and Wastewater. Amm. Public Health Ass., Washington. pp.152-155, 270-271, 439-440
3. Piper.A.M. and Graphic.A. 1944, Procedure in the Geochemical Interpretation of Water Analyses. Am. Geophy. Union Trans. 25:914-923
4. Calf.G.E. and Seatonberry.B.W. and Smith.L.W. 1980, The measurement of Natural Levels of Tritium in Water. AAEC/E373
5. Alvarea.L.W. and Coming.R., Radioactive Hydrogen. A Correction Phys. Rev. 58(2):197
6. Hartley.P.E., Rapid.A. 1984, Mass Spectrometers Method for the Determination in Water at Natural Levels. Anal. chem. 52:2232
7. Han.J.S. and Ahn.J.S. 1984, Studies on Isotope Hydrology in South Korea. J. Geolo. Scl. 20(1):72-84
8. Ministry of Agri. 1984, Annual Report of Groundwater Development in Cheju pp11-33
9. Ahn.J.S., Kim.J.S. and Kim.J.S. 1980, Studies on Tritium Analysis and its Monitoring with Special Reference to Power Reactor Operation, KAERI/RR-204/80
10. Ahn.J.S., Kim.J.S., Han.J.S., Airey.P. and Payne.B. 1983, Environmental Isotope-Aided Studies on River Water and Ground Water Interaction in the Han River Basin. J.Korean Asso. Hydro. Scin. 16(4):245-252
11. Love.S.K. 1944, Cation-Exchange in Ground Water Contaminated with Sea Water Near Miami, Florida, Trans. Amer. Geophysical Union, Vol.25, pp.951-955
12. Todd.D.K. 1959, Ground Water Hydrology. John Wiley & Sons. Inc. pp293-294
13. Ahn.J.S., Kum.J.S, Airey.P. and Payne.B. 1981, Environmental Isotope-Aided Studies on River Water and Ground Water Interaction in the Region of Seoul. J.Korean Nucl. Soc.13(2):85-96
14. Ahn.J.S. et al, 1984, Environmental Isotope-Aided Studies on Water Resources in the Region of Cheju(II). J.Korean Asso. Hydro. Scien. 17(2):72-79
15. Revelle.R. 1941, Criteria for Recognition of Sea Water in Ground Waters, Trans. Amer. Geophysical Union, Vol.22, pp.593-597
16. IAEA, 1983, Guidebook on Nuclear Techniques in Hydrology. Tech. Report No.91 IAEA. Vienna, pp.276-278, 319-325, 351-356
17. Davis.G.H., Lee.C.K, El Bradley and Payne.B. 1970, Geohydrologic Interpretation of a Volcanic Island from Environmental Isotopes. Water Resources Research, 6(I):99-109
18. Payne.B.R., Qujano.L. and Latarre.D.C. 1979, Environmental Isotopes in a Study of the Origin of Salinity of Ground Water in the Mexicali Valley, J. Hydrol. 41:201-215

GROUNDWATER RECHARGE ESTIMATION BY TRITIUM TRACER TECHNIQUE IN SATHUPALLY AREA, ANDHRA PRADESH, INDIA.

*Ch.SUDARSANA RAJU, **C.S.MURTI and ***M.A.HALEEM

* Geology Department, Osmania University, Hyderabad - 500 007

** N.G.R.I., Hyderabad – 500 007. *** Central Groundwater Board, Nagpur.

ABSTRACT

Hydrogeological studies have been carried out in Sathupally area of Godavari Valley, A.P. The geological formations present in the area include Archaean gneisses and kamti sandstones. In the recent past there has been a spurt in tapping the under grounwater resources in the region. With a view to estimate the rainfall recharge to groundwater in Tammileru river basin of 517 sq.km. in Sathupally area during the hydrological cycle of 1985-86, 200 µCi tritiated water of 4 ml. quantity was injected at 25 selected sites at a depth of 80 cm. b.g.l. in the 1st week of June'85 before the on set of southwest monsoon. Assuming the piston flow model, it has been observed that the depth of recharge has varied from -9 cm. to +24 cm. at different sites in the area of investigation. The values are found to be positive in the recharge area and negative at the discharge portion of the basin. Average recharge depth has been found to be 6.4 cm and the total recharge in to the groundwater reservoir due to rainfall infiltration during monsoon of 1985-86 has been estimated to be 33 MM^3.

The tritium break through curves have been theoritically interpreted using diffusion model and a comparison of the data with that of the piston flow model is presented in the paper. The recharge estimation by piston flow model and the diffusion model are found to agree fairly.

* * * *

INTRODUCTION

Sathupally area of Khammam district in Andhra Pradesh forms a part of Godavari valley and comprises of Archaean gneisses and Kamti sand stones. It is a backward region and for the upliftment of the tribals in the area the Andhra Pradesh State Irrigation Development Corporation has taken up a program in 1975 to drill large diameter deep tube wells on the recommendation of Central Ground Water Board. As a part of detailed hydrogeological investigations in Sathupally area groundwater recharge estimation has been made in Tammileru river basin by applying tritium tracer technique. The tritium break through curves have also been theoritically interpreted using diffusion model. Tammileru is a euphemeral stream which starts in the north east corner of the area, flows southwest upto the middle, then to the south and at the end takes a southeastern course. The area under investigation lies in between lat.17°02'12" and 17°24'08"N and long.80°47'23" and 81°00'17"E and falls in toposheet No.65 C/15, C/16 and 65 G/3 and is 517 Km^2 in extent (Fig.1).

The Tammileru basin enjoys semiarid tropical climate with an average annual precipitation of 737 mm at Sathupally. About 75% of this is contributed by southwest monsoon (June-September), 20% by the northeast monsoon (October-December) and the rest by occasional showers during Jan.-May. Maximum temperature touches 47°C during summer, April-May, and minimum 16°C during winter, December-January, (Konala and Haleem, 1982).

Most of the precipitation in the area is drained out as run off due to elevation difference as it forms upper part of the basin and only a minor portion would go to recharge the unconfined aquifers through infiltration. To estimate this part of recharge contributed by precipitation, tritium tracer experiments were conducted at 25 selected sites in the basin.

In this paper a brief description of the technique is presented and the results are discussed.

ESTIMATION OF GROUNDWATER RECHARGE BY TRITIUM TAGGING METHOD

Tritium (^3H)[R] is a radioactive isotope of hydrogen which under goes decay emitting beta radiation with a half life of 12.26 ± 0.1 years and releases energy of 18 kev. (E_{max} = 18 kev, $T_{½}$=12.26 years).

Tritium is produced in the earth's atmosphere due to interaction of cosmic ray produced secondary neutrons with nitrogen nucleii (Libby, 1946). Tritrium is also produced artificially in nuclear reactors.

The environmental and injected tritium methods for estimation of groundwater recharge were developed based on the pistonflow model originally suggested by Zimmermann et al. (1967 a, b).

The model is based on the assumption that the soil moisture moves downwards in descrete layers, i.e. the rain water added at the ground surface forms a fresh layer of water in the top soil and percolates downwards by pushing down an equal amount of water from the succeeding layer of the unsaturated zone and this process would continue up to the water table, till an amount equivalent to the input is added to groundwater.

Accordingly, if the moisture of any horizontal plane is tagged with any suitable tracer like tritium, then with the tritiated water, (HTO)[R], the tracer would be carried along the moving water. At any given time subsequently the position of the tracer will give the transit length of the intially tagged moisture layer. To locate the position of the tracer depth-wise soil cores are collected from the tritium injected site and the tritium activities of the moisture distillate from each core are determined with the help of a liquid scintillation spectrometer.

The position of the tracer will be indicated by a maxima or peak

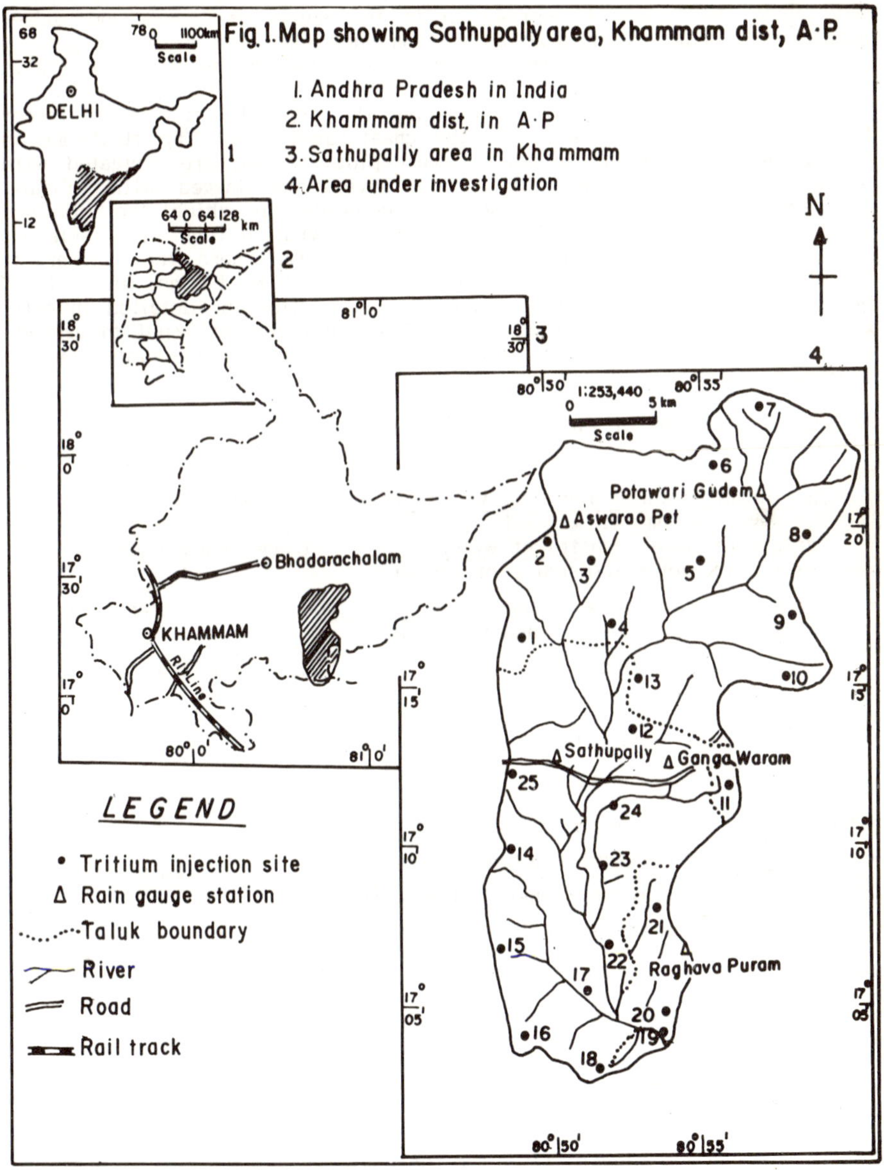
Fig. 1. Map showing Sathupally area, Khammam dist, A·P.
1. Andhra Pradesh in India
2. Khammam dist. in A·P
3. Sathupally area in Khammam
4. Area under investigation
N
DELHI
Scale
1100 km
64 0 64 128 km
Bhadarachalam
KHAMMAM
Rl. Line
1:253,440
0 5 km
Potawari Gudem
Aswarao Pet
Sathupally
Ganga Waram
Raghava Puram
LEGEND
• Tritium injection site
Δ Rain gauge station
Taluk boundary
River
Road
Rail track

in the tritium activity Vs depth profile. Molecular diffusion, stream line despersion, asymmetrical flow and other heterogeneties of the aquifer may cuase the tracer to spread in to the subjacent layers and consequently the peak may appear as a broadened one. The centre of gravity of the profile then corresponds to the tracer maxima as shown in Fig.2.

There are two methods of applying the artificial tritium as tracer for the study of infiltration in the unsaturated zone. The first one is the irrigation technique in which small plots of soil are irrigated with tritiated water such that the surface layers are tagged with tritium. The second method is the injection technique in which the moisture below the root zone soil layers is tagged with tritium at a selected site. The injection method is usually preferred where the number of investigation sites would be large, such as river basins, while the irrigation method is selected at one or a few sites for detailed investigations. The recharge or the amount of water added as input to groundwater per square centimeter area of soil at a site is calculated in the following way. :

Recharge = Fraction of water per unit volume x
(L^3) (tracer transit length x Unit area)

Recharge/unit area=

$$.k\ \frac{\text{Volume of water per unit volume of soil}}{\text{Volume of wet soil}} \times \text{tracer transit length}$$

$$= \frac{\text{weight of water/density of water}}{\text{weight of wet soil/density of wet soil}} \times \text{tracer transit length}$$

$$= \frac{\text{weight of water}}{\text{weight of wet soil}} \times \frac{\text{density of wet soil}}{\text{density of water}} \times \text{tracer transit length}$$

$$= \text{Wet weight fraction of soil moisture} \times \text{Density of wet soil} \times \text{Tracer transit length}$$

because the density of water is $\simeq$ 1

$$R_{PF} = M_w \cdot \rho_w \cdot h$$

where R_{PF} = recharge by pistonflow model

M_w = wet weight fraction of soil moisture

ρ_w = wet weight density of soil core

h = transit length of the tracer

FIELD WORK

200 µCi tritiated water of 4 ml quantity was injected at 25 selected sites, at a depth of 80 cm b.g.l. during the first week of June 85, before the onset of southwest monsoon. Out of the 25 sites 16 are nearer the boundary of the basin and the rest 9 are located in the interior. Holes of 14 mm diameter were made in the ground up to a depth of 80 cm by driving a tapered mildsteel rod. Tritiated water was then injected at 80 cm depth with the help of a hypodermic syringe and a copper pipe of 6 mm outer dia and 60 cm. length.

After allowing for a soaking time of about 5 minutes, the copper pipe was pulled out and the injection hole was packed up with the excavated soil. The site was left undisturbed through out the monsoon season. For exact relocation of the injected site, bearing distances with permanent markers were measured and recorded.

Depthwise soil core samples from 19 sites were collected during the first week of March, 86. C-Class steel pipes of 40 mm diameter of different lengths having tapered edges were hammered to obtain 30 cm. long soil cores upto a considerable depth below the injection level. These depths varied from 1.2 m to 2.1 m at different sites depending on the nature of soil, manual labour etc. Each soil core was double sealed in a polythene cover and all the packets of one site were again sealed in a polythene roll. The packets were brought to the laboratory for analysis.

LABORATORY STUDIES:

Each 30 cm long soil core is weighed in the laboratory and the bulk densities are estimated for each core from the known volume of coring pipe. From each core 100g of soil sample is used to estimate the moisture content by the oven drying gravimetric method. Moisture of the remaining soil fraction is distilled out in a vacuum distillation set up at a partial pressure of 4 to 6 cm.Hg. (Athavale et al. 1980).

4 cc of water sample distillate of every soil core is mixed with 6 cc of liquid scintillator (Instagel)(R) in a 20 cc low potash glass vial. Tritium concentrations of all the samples are counted in a L.K.B. Rack β automatic liquid scintillation spectrometer. The counting equipment has a routine background of 40 cpm. (counts per minute) in tritium channel and the counting efficiency is estimated at 17 percent, using standards.

TRITIUM BREAK THROUGH PROFILES:

The output data of all the parameters of a site are recorded in a tabular form showing the depth, wet bulk density, moisture content, tritium concentrations and the tritium activities. Plotting of a few tritium profiles are shown in Fig.2 for comprehension.

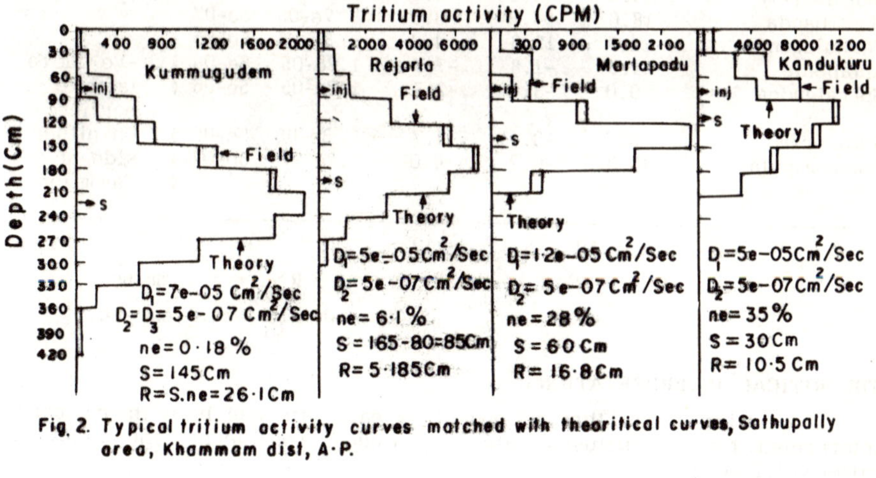

Fig. 2. Typical tritium activity curves matched with theoritical curves, Sathupally area, Khammam dist, A·P.

Recharge value for each site has been evaluated and the results are shown in Table No.1 along with a comparative statement of theoritically computed data.

TABLE – 1

TABLE SHOWING THE TRITIUM RECHARGE VALUES IN SATHUPALLY AREA KHAMMAM DISTRICT, ANDHRA PRADESH

Name of the Site	Porosity %	Recharge in Cm: Piston Flow Model	Recharge in Cm: Advection diffusion Model	Dispersion Co-efficient Cm^2/S: Longitudinal	Dispersion Co-efficient Cm^2/S: Transversal	Remarks
	n_e	R_f	R_{Th}	D_1	D_2	
Regulapadu	7.0	0.0	0.0	1.2-05	5e-07	
Chandraipalam	17.4	17.6	14.8	1.2e-05	5e-05	Recently irrigated
Kummugudem	18.0	23.5	26.1	7e-05	5e-07	
Rejarla	6.1	14.3	5.2	5e-05	5e-07	In contact with archaeans
Marlapadu	28.0	11.1	16.8	1.2e-05	5e-07	
Venkatapuram	24.0	6.3	6.0	1.2e-05	5e-07	
Kandukuru	35.0	9.3	10.5	5e-05	5e-07	
Parvatapuram	19.0	5.5	8.6	1.2e-05	5e-07	
Duddipadu	31.0	5.0	6.2	1.2e-05	5e-07	
Raghavapuram	14.0	9.0	12.6	1.2e-05	5e-07	
Tumburu	2.1	10.5	2.1	5e-05	5e-06	Low effective porosity
Chintampalle	12.7	3.0	3.8	5e-05	5e-07	
Gangavaram	19.0	4.9	3.8	1.2e-05	5e-07	
Mustibanda	18.0	4.9	3.6	7e-05	5e-07	
Parkalagundi	4.7	11.3	4.2	5e-05	5e-06	
Arlapenta	24.0	-1.8	-7.2	1.2e-05	5e-06	-Ve due to laterite barrier
Gurraigudem	9.0	-3.4	-2.7	1.2e-05	5e-06	
Sivapuram	6.7	-9.1	-4.7	5e-05	5e-06	On either side of Tammileru
Chennampeta	37.0	-0.7	0.0	1.2e-07	5e-07	

$n = 19$	19		
$\Sigma x = 121.2$	109.7		$R_{pF} = 6.4$ cm/sec.
$\bar{x} = 6.38$		5.77 and	$R_{Th} = 5.8$ cm/sec.

THEORITICAL INTERPRETATIONS:

The tritium break through profile of each site has been theoritically interpreted by the diffusion advection model, considering the following differential equations:

$$\frac{dc}{dt} = D\frac{d^2c}{dx^2} - V\frac{dc}{dx} \qquad \text{............} \qquad (1)$$

where c is the concentration of tracer.
D is the dispersion coefficient.
V is the mean velocity.
x,t are the space and the time variables.

There are a number of solutions available of the above equation for different initial and boundary conditions. For a point injection in an infinite medium, the commonly applied solution is given by the following equation:

$$C(x,t) = Co\,\frac{e^{-\frac{(x-vt)^2}{4\,Dt}}}{\sqrt{4\pi\,Dt}} \qquad \text{..............} \qquad (2)$$

with the initial conditions

$C\ (x,t) = Co\ .\ \delta r$

Where δr is the Dirac delta function, and the boundary condition:

$$\frac{dc}{dt}(x,t)\Big|_{r=0} = 0$$

$C\ (x.\ \infty\ ) = 0$

The complete solution for an orthogonal system with the flow along the x-axis, equation (2) will be written as (Lenda and Zuber, 1970):

$$C(X,Y,Z,t) = Co\ .\ \frac{e^{-\frac{(x-vt)^2}{4D_1t}}}{\sqrt{4\pi D_1t}}\ .\ \frac{e^{-\frac{Y^2}{4D_2t}}}{\sqrt{4\pi D_2t}}\ .\ \frac{e^{-\frac{Z^2}{4D_3t}}}{\sqrt{4\pi D_3t}} \qquad (3)$$

Where D_1 is the longitudinal dispersion coefficient and D_2, D_3 are the transverse dispersion coefficients.

Putting $D_2 = D_3$ and s=vt=transit length, equation (3) will be read as:-

$$C(X,Y,Z,t) = Co\ .\ \frac{e^{-\frac{(x-s)^2}{4D_1t}}}{\sqrt{4\pi\,D_1t}}\ .\ \frac{e^{-\frac{-Y^2}{2D_2t}}}{4\pi\,D_2t} \qquad \text{...............} \qquad (4)$$

Corrections for the following factors are to be applied in the above equation.

Radioactive decay	(λ)
Instrument efficiency	(η)
Core losses	(ϵ)
Effective porosity	(n_e)
Flow cross section	(F)

Equation (3) will now be written as:

$$C(X,Y,Z,t) = Co \frac{\lambda . \eta . \epsilon}{Fn_e} . \frac{e^{-\frac{(x-vt)^2}{4D_1 t}}}{\sqrt{4 \pi D_1 t}} . \frac{e^{-\frac{Y^2}{2D_2 t}}}{(4 \pi D_2 t)} \quad .....(5)$$

All the tritium profiles have been interpreted by equation (5)

The maximum activity at any time t can be obtained by putting X=0, Y=0, Z=0.

$\left(\frac{\lambda . \eta . t}{Fn_e} = 1 \right)$ which is given as

$$C(t) = Co \frac{1}{\sqrt{4\pi D_1 t}} . \frac{1}{4 \pi D_2 t} \quad .....(6)$$

An approximate value for D is estimated, under radial symmetry, by comparing the net activity of the profile with the equation No.6. Proper value for D_1 and D_2 are then chosen after 2 or 3 runs and their computations are performed such that $S_2 < S < S_1$. A few profiles are shown in Fig.2 and the theoritically estimated D,ne and estimated recharge values are given in Table 1.

Recharge for theoritical curves is calculated using the formula:-

$$R_{Th} = s.n_e \quad ..... (7)$$

Where s is water table fluctuation.
n_e is specific yield.

DISCUSSION:

The recharge values computed by the piston flow model are in the range of -9 cm to +24 cm with an average value of 6.4 cm; positive values representing recharging sites and negative values representing discharging sites.

The theoritically estimated average recharge has the range from -8.2 cm to +26.1 cm with an average of 5.8 cm. The average recharge of recharging sites has been found to be 6.4 cm by the piston flow model; and that calculated by theory is 5.8 cm. Similarly the average . loss at the discharging sites is -3.75 cm by piston flow model and that by the theoritical estimation is -4.87 cm.

It can be noted thus that the average of positive recharge in general agrees with the theoritically estimated values. The total quantum of input to groundwater through vertical infiltration of rainfall is calculated to be 33 MM^3 for the area of 517 km^2 with an average recharge value of 6.4 cm and rainfall of 737.2 mm during the study period of June 85-March, 86.

CONCLUSIONS AND RECOMMENDATIONS:

The average annual recharge to the phreatic aquifers of the Tammileru basin during the study period June 85 to March 86, has been estimated as 6.4 cm by piston flow method. The average rainfall of 5 rainguage stations during the corresponding period is 737 mm and thus the fractional amount of recharge is 8.6% of the rainfall.

The quantam of groundwater recharge to the unconfined aquifers through vertical infiltration has been estimated to be 33 MM^3 over the study area admeasuring 517 Km^2.

Theoritical interpretation of tritium break through curves was estimated by advection diffussion model and the following information was deciphered.

a) The longitudinal dispersion coefficient in the direction of flow is higher than the transverse dispersion coefficient. Further the longitudinal dispersion coefficient is higher in irrigated lands.

b) There is a good agreement of estimated average recharge by the piston flow model, and the advection diffussion model. However, the agreement between the models for individual sites is unsatisfactory. Thus studies are limited to the observations made by the authors in the field and the laboratory analysis.

For obtaining good results, seasonal, regular changes are to be considered and long term studies are recommended.

ACKNOWLEDGEMENT :

The authors thank Prof.V.K.Gaur, Director, N.G.R.I. for providing facilities for the work. Vacuum distillation of soil samples and liquid-scintillation counting of tritium activities were performed by Shri. G.R. Anjaneyulu and Shri.M.Dhanaraju, which is gratefully acknowledged. Mr.Haleem is thankful to the Central Ground Water Board and Shri.P. Subramanian, Director, C.G.W.B. Central Region, for permitting to carry out research work.

REFERENCES

Athavale, R.N.; Murti,C.S.; and Chand, R.; 1980. Estimation of recharge to the phreatic aquifers of the Lower Maner Basin, India, by using the tritium injection method. J.Hydro.V.45, p.185-202.

Konala,R.K.R.; and Haleem, M.A.; 1982. Hydrogeological conditions in Khammam District, A.P., District report, C.G.W.B., Southern Region, Hyderabad.

Lenda,A.; and Zuber,A.; 1970. Tracer dispersion in Groundwater Experiments, Isotopes in Hydrology, IAEA, Vienna, p.619-641.

Libby,W.F.; 1946. Atmospheric He^3 and radiocarbon from cosmic radiation, Phys.Rev. p.671-672.

Zimmermann,U.; Enhalt,D.; and Munnich,K.O.; 1967a. Soilwater movement and evapotranspiration: changes in the isotope composition of the water. Isotopes in Hydrology. IAEA, Vienna. p.567-586.

Zimmermann,U.; Munnich,K.O.; and Roether,W.; 1967b. Downward movement of soil moisture traced by means of hydrogen isotopes. Geophy. Monograph No.11, AGU, Washington, p.28-36.

ENVIRONMENTAL CHLORIDE METHOD FOR GROUNDWATER RECHARGE STUDIES: POTENTIALITIES AND LIMITATIONS

B.S. Sukhija, P. Nagabhushanam, and D.V. Reddy.

National Geophysical Research Institute,
Hyderabad - 500 007, India.

...

ABSTRACT

Study of groundwater recharge is a basic requirement in any water balance study. The application of environmental and other isotope tracer techniques for such purpose has been demonstrated by many workers in different hydrologic regimes. However, the potentialities of environmental chloride - a simple geochemical tracer, are being studied in depth only recently. The present paper discusses the potentialities and limitations of environmental chloride for estimating recharge in natural environmental as well as in changed land use conditions and for artificial recharge studies. For evaluation of recharge, currently two methodologies are in vogue: one using the chloride concentration of groundwater and soil water, and comparing with input chloride content; another using soil profiles wherein total deposition of chloride is studied. Average chloride concentration can be related to long term recharge and chloride concentration in rainfall based on chloride balance. Recent chloride studies have also elucidated the mechanism of recharge and possible environmental hazards as a result of deforestation. It is concluded that environmental chloride within its limitations is a simple geochemical tracer with large potential.

INTRODUCTION

Quantum of groundwater resources is inevitably dependent on the rate of recharge to the aquifer system. Groundwater recharge rate is a vital hydrologic parameter to be carefully evaluated and utilised for optimum groundwater exploitation. Study of recharge characteristics are important not only under natural conditions but also changed environmental conditions. Conventional methods of recharge estimation such as water balance studies, lysimeter utilisation etc. have intrinsic lacunae by way of a number of assumptions, insufficient data using empirical approaches etc.. Isotope techniques, for reliable recharge estimation, developed by Munnich and his co-workers (1967) were based on piston flow model of groundwater movement. Environmental radio isotopes of H-3, C-14, Si-32, Kr-85, and stable isotopes of H-2, O-18 and C-13 have found extensive application in groundwater recharge studies. Tritium tagging method(Zimmermann et al. 1967) and the environmental tritium method (Munnich et al. 1967) have been quite extensively used all over the globe for groundwater recharge estimates. Quite a sizeable number of researchers (Smith et al. 1970; Sukhija and Rama, 1973; Allison and Hughes, 1974; Dincer et al. 1974; Sukhija and Shah, 1976 etc.) have demonstrated the usefulness of the environmental tritium, present in the unsaturated zone and groundwater for different hydrologic regimes. Because of ever decreasing levels of environmental tritium over the years, artificial tritium has gained popularity (Datta et al. 1973; Athavale et al. 1980; Gupta and Sharma, 1984) for quick (but valid only for the experimental period) estimates of groundwater recharge.

Environmental chloride is yet another geochemical tracer having potential utility in hydrology. However, the potentialities of environmental chloride are being studied in depth only recently. The present paper discusses not only the potential utilities of environmental chloride for estimating recharge in natural environmental as well as in changed land use conditions and for artificial recharge studies but also the associated limitations of the methodologies.

METHODOLOGIES:

Environmental chloride is deposited on land by atmospheric deposition processes (rainfall plus dry fallout) and chloride fallout is concentrated near coast lines. If the chloride present in the unsaturated zone and groundwaters has surface environment (i.e. sea) as its source and there exists no other source or sink underground, it is possible to obtain chloride balance. The ratio of chloride concentration in rain water to that in percolated water is utilised to estimate recharge.

Currently two methodologies are in vogue: (1) Concentration method and (2) Flux method. Though basically both are based on chloride balance, the former one involves the comparison of chloride concentration of groundwater and soil water with input chloride content of precipitation, and total chloride content of soil profiles is balanced against total chloride concentration of precipitation. In the flux method recharge rate is obtained from turnover time, which in turn is obtained from the ratio of input flux and its total chloride abundance.

Many researchers have effectively utilised the environmental chloride present in the unsaturated zone and groundwater regimes for natural and artificial recharge estimates, salinity problems, soil water movement studies etc. As discussed in detail further in the paper, the comparison of chloride in the soil profile and that in groundwater has been utilised to elucidate soil water movement (Sharma & Hughes, 1988). Moreover, continuous monitoring of the chloride concentation and hydraulic head in space and time can reveal important information about change in groundwater recharge characteristic as a result of deforestation (Williamson et al., 1987). Similarly, spatial and temporal changes of chloride in groundwater can be linked to that of percolation tank, and can effectively be utilised for artificial recharge studies (Sukhija & Reddy, 1987).

POTENTIALITIES OF ENVIRONMENTAL CHLORIDE:

1. RECHARGE STUDIES:

Eriksson and Khunakasem (1969) have carried out a study making use of chloride concentrations in groundwater for estimating recharge rates in coastal plain aquifer in Israel. They have assumed that the matrix of the aquifer does not contain chloride, and in more permeable aquifers relicts of marine waters were washed out long ago, and the varying chloride concentrations in the profile is attributed to the changes in the regolith. For the two computational methods based on chloride and water balance equations for simple and complex groundwater flow patterns, the recharge values obtained were almost equal (92 and 95 mm /year respectively).

In correlative studies using different environmental isotopes, many investigators have opined that the environmental chloride could very well be utilised in hydrological studies (mainly for recharge evaluation, salinity problems). Sukhija and Rama (1973), in alluvial tracts of Gujarat, using limited chloride data available for groundwater and precipitation found recharge rates in fair agreement with those obtained by environmental tritium studies. In a comparative study involving chloride, deuterium and oxygen-18 (Sharma and Hughes, 1985) for recharge estimation in the deep coastal sands of Western Australia, the chloride balance study yielded an areal average recharge of approximately 15% of the average precipitation. Because of the small components of stable isotope enrichment due to evaporation of rainfall and the direct evaporation from the soil, the recharge values obtained by deuterium and oxygen-18 analyses are subjected to doubt. Variability, if any, in the chloride concentrations below 10 meters is ascribed to vertical non-uniformity of the soil and seasonal variability of chloride input, rainfall and evapotranspiration. Recharge values obtained by environmental chloride and tritium analyses of an unconfined aquifer in Gambier Plain of Southern Australia showed fairly good agreement between the two techniques though the mean annual recharge values varied between 50 and 270 mm for various hydrologic units (Allison and Hughes, 1978; Table 1).

Sl. No.	Hydrologic unit	Mean Annual Recharge(mm) estimated using Chloride	Tritium
1.	Sand over heavy clay	70	50
2.	Volcanic soils	100	100
3.	Sand over sandy clay	140	100
4.	Sand over thin sandy clay over lime stone	105	120
5.	Terra rossa over limestone	150	130
6.	Thin sandy loam over limestone	140	155
7.	Aeolianite	200	195
8.	Skeletal soils	250	270

Table 1. Estimates of mean annual recharge for the various hydrologic units (Allison & Hughes, 1978).

In semi-arid regions, combined geochemical and isotopic studies of both unsaturated zone and of shallow groundwaters can be effectively undertaken to determine not only the present day direct recharge component, but also a recharge chronology of immediate historic times, which may be important in the estimation of long term water resources (Edmunds and Walton, 1980). Artificial tritium and environmental chloride studies carried out recently by Sukhija et al. (1988) in Pondicherry area, which is a semi-arid coastal region, have established the usefulness of chloride ion in determining the long term recharge rates for various hydrological units. Results obtained by chloride analyses, for repeated two samplings at the sites, and artificial tritium study were fairly in good agreement (Table 2, and fig.1). The differences were attributed to the different time periods for which recharge rates were obtained.

		RECHARGE RATES OBTAINED BY		
Sl.No.	LOCATION	CHLORIDE	INJECTED TRITIUM	REMARKS
1.	L.R.PALAYAM	26.3	31.6	16% less by Cl^-
2.	MURATTANDI	16.1	19.8	18% less by Cl^-
3.	IDAIYANCHAVADI	14.6	33.0	55% less by Cl^-
4.	ARIYUR	16.1	12.5	28% high by Cl^-
5.	TIRUBHUVANAI	9.0	8.3	≈ equal

Table 2. Comparison of recharge rates (% of local rainfall) obtained by environmental chloride and injected tritium methods (Sukhija et al., 1988).

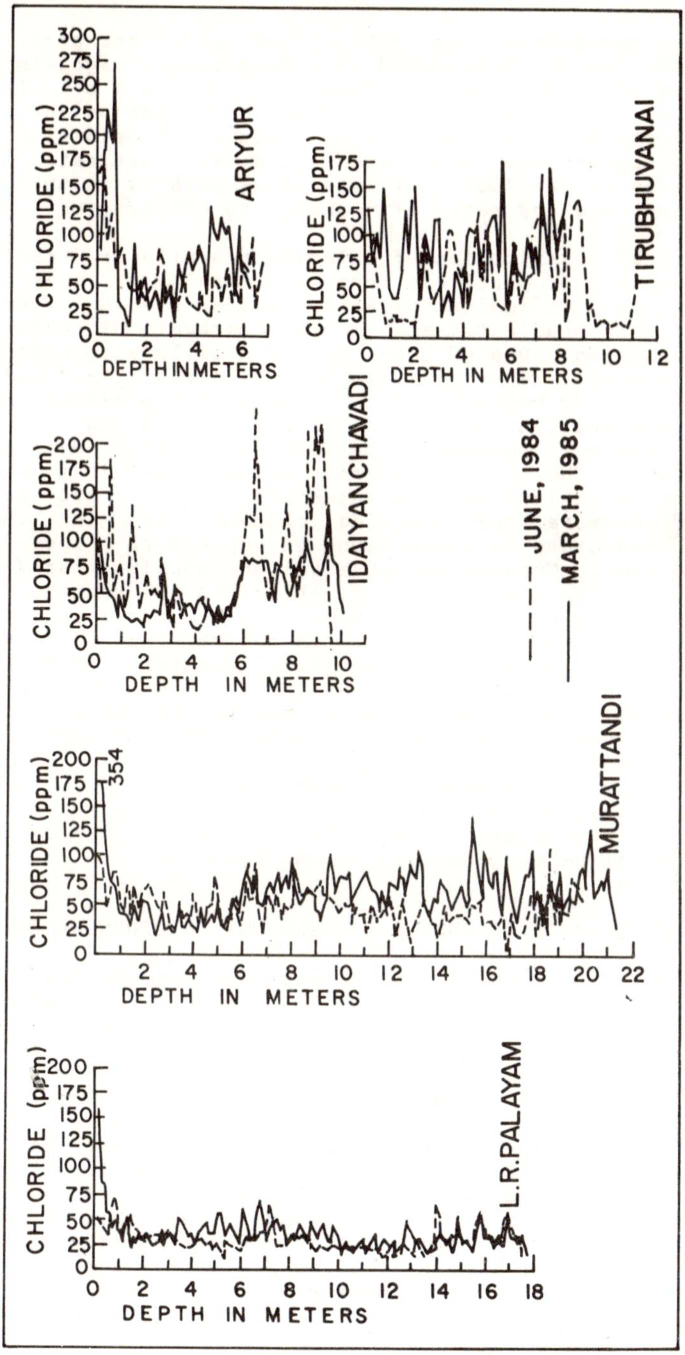

Fig.1. Depth variation of environmental chloride in soil moisture (Sampled in 1984 & 1985 ; Sukhija et al., 1988).

2. SOIL WATER DYNAMICS:

The mechanisms of water flow and recharge to groundwater were investigated in a deep clayey regolith in south-west Western Australia (Johnston, 1987 a & b). A 700 sq.m. area was intensively studied for a period of two years. Vertical distributions of natural chloride in 13 profiles upto 31m deep were used to estimate the distribution of vertical soil water flux density in the 16m unsaturated zone and rates of recharge to groundwater.

Vertical distribution of chloride essentially resulted in two types of profiles: one with a bulge or peak and others without it indicating marked difference in chloride concentration gradients and the total amount of chloride stored in the unsaturated zone. In the profile with chloride bulges, the soil water flux (infiltration rate 2 - 7 mm/yr) was reduced by 5m below ground. Some of the profiles show steep concentration gradients below this depth, which probably reflect boundaries between strata with different flow patterns. The profiles without bulges occur in a small area where large infiltration rates(50-100 mm/yr) are noted and further-more the chloride concentration in these profiles are close to those found in upper 4 to 5 m of the saturated zone. Groundwater observations confirm that areas of large infiltration rates contributed to the recharge of groundwater. The rates of recharge to groundwater estimated from chloride concentration in the saturated zone were as much as two orders of magnitude greater than the infiltration rate above the water table. Also Sharma and Hughes (1985) have observed that 50% of the total annual recharge occurs via movement of water through preferred pathways bypassing the soil matrix and propunding a bimodel system for water movement through the profile . Thus, chloride profiles revealed that the preferred pathways are likely to be associated with certain structural, textural and geological heterogeneities in the regolith.

3. STUDY OF EFFECTS OF CHANGE OF LAND USE ON GROUNDWATER:

Although clearing of the forest has been acknowledged as a precursor for salinity problems, there is a clear need for quantifying the mechanism or develop suitable model to predict the consequences of land use change . Environmental chloride in particular, and salt balance in general, can provide the vital information about the changing land use. Williamson et al., (1987) have studied 5 small forest catchments in the Collie River Basin in south-west Western Australia where the land use changed from forest to agriculture. Each catchment was instrumented to measure parameters of water and salt balance. The salt content determined as chloride ion of catchment waters was obtained by daily sampling of stream flow and periodical sampling of storage rainguage and groundwater. It was found that the stream flow and salt increased immediately after clearing. In the high rainfall catchment stream flow increased about 4 times to 48% rainfall, the salt load increased to 15 times the mean salt fall (measured as chloride ion) of 7 gm-sq.m./year. Direct runoff increased 4 fold to 16% of yearly stream flow. Absolute increase in stream water and salt output from partially cleared catchment in the lower rainfall location were about an order of magnitude less than for the higher rainfall catchment though the relative increase was similar.

An example of the large scale changes of chloride concentration and associated hydraulic head changes observed provided by a piezometer is shown in figure (2). The figure shows pre-and-post-clearing conditions. As a result of forest clearing, piezometer hydrograph has risen indicating increased recharge. The potentiometric surface has risen to 2.6 m/yr for three years indicating increased recharge rate.

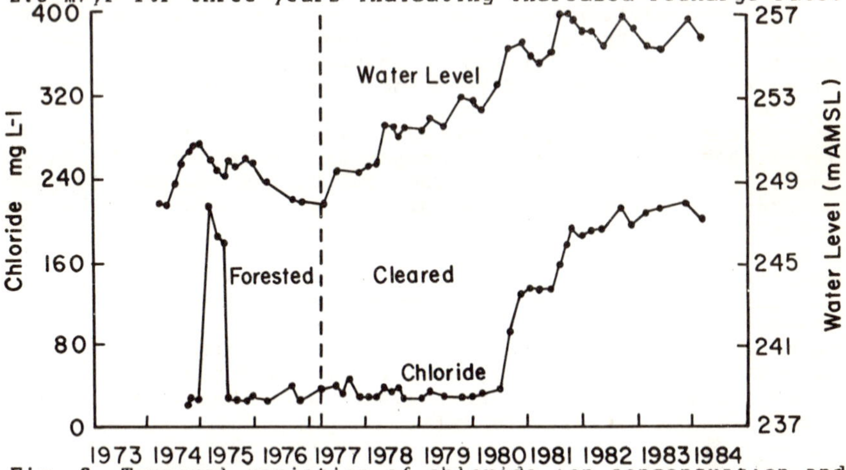

Fig. 2. Temporal variation of chloride ion concentration and water level in the pallid zone aquifer from piezometer (0863) in Wights catchment (Williamson et al., 1987).

The chloride concentrations vary significantly in time; in the high rainfall decreasing chloride concentration occur in the upper slopes where recharge appears to be high. In the pre-clearing conditions only once chloride concentration rose to 40 mg/l levels but, however, four years after the clearing the concentration was mostly 200 mg/l. According to Peck and Williamson (1987), the pre-clearing peak after the abnormally wet year of 1974 and post clearing increase in 1980 are related to water levels rising into more saline often found as a bulge in the salt profiles above the water table at forest side in this catchment. However, it is clear that there was substantial change in the chloride ion concentration as a result of forest clearing. Thus, it is demonstrated that the chloride ion is a simple tool to be used for the study of land use problems.

3.4 ARTIFICIAL RECHARGE STUDIES:

In studies pertaining to artificial groundwater recharge through percolation tanks, chloride analyses of tank water, groundwater and well waters have yielded encouraging results. An increasing trend of chloride concentration from the tank down the wells in command area is observed; which,obviously, established the movement of tank water towards the wells in the command area (Sukhija & Reddy, 1987). Thus, environmental chloride can prove useful in artificial recharge studies as well.

4.LIMITATIONS:

In groundwater recharge studies utilising environmental chloride as a tracer, the following information should be taken into account. Important assumptions about no source or sink of chloride should hold and chloride profiles should reflect steady state; and other important sources of chloride in addition to precipitation could be from applied fertilisers (particularly potassium chloride).and human waste etc.; average chloride fallout is assumed to be uniform in time, the assumption that chloride deposition (mass chloride per unit time) of present day rainfall may not represent the past period, as the rainfall depends on additional variables such as changing climatic conditions etc.,this aspect is of importance for arid zones; land use changes at the sampling sites; deep undisturbed soil samples to be obtained; because of chloride being electro -negative ion it may indicate early breakthrough and higher velocity (short travel time) by about 20% or more than water itself especially in clayey materials.

CONCLUSIONS:

It is concluded that environmental chloride, a simple geochemical tracer, within its limitations can not only provide realistic recharge rates in natural environmental conditions but also it can be effectively utilised in the associated environmental studies like soil water dynamics, recharge estimates in changed land use, artificial recharge schemes etc..

ACKNOWLEDGEMENTS:

The authors are grateful to Prof. V.K. GAUR, Director, N.G.R.I. for his keen interest in the work and his permission to publish this paper.

REFERENCES:

Allison, G.B., and Hughes, M.W., 1974. Environmental tritium in the unsaturated zone; estimation of recharge to an unconfined aquifer. Proc. Symp. Tech. Isot. Hydro. v. 1, p. 52-72, IAEA, Vienna.

Allison, G.B., and Hughes, M.W., 1978. The use of environmetal chloride and tritium to estimate total recharge to an unconfined aquifer. Aust. J. Soil Res. v. 16, p. 181-195.

Athavale, R.N., Murti, C.S. and Chand, R., 1980. Estimation of recharge to the phreatic aquifers of the Lower Maner Basin, India, by using the tritium injection method. J. Hydro. v. 45, p. 185-202.

Datta, P.S., Goel, P.S., Rama, and Sanghal, S.P., 1973. Groundwater recharge in Western Uttar Pradesh. Procd. Ind. Acad. Sci. v. LXXVII, Sect. A.I.

Dincer, T., Al-Mugrin, A. and Zimmermann, U., 1974. Study of infiltration and recharge through the sand dunes in arid zones with special reference to the stable isotopes and thermonuclear tritium. J. Hydro. v. 23, p. 79-109.

Edmunds, W.M., and Walton, N.R.G., 1980. A geochemical and isotopic approach to recharge evaluation in semi-arid zone; past and present. Proc. Symp. Arid Zone Hydrol., Invest. Isot. Tech., IAEA, Vienna.

Eriksson, E., and Khunakasem, V., 1969. Chloride concentration in groundwater, recharge rate and rate of deposition of chloride in the Israel coastal plain. J. Hydro. v. 7, p. 178-197.

Gupta, S.K., and Sharma, P., 1984. Soil moisture transport through the unsaturated zone: tritium tagging studies, in Sabarmati Basin, Western India. Hydrol. Sci. J. v. 29(2), p. 177-189.

Johnston, C.D., 1987a. Distribution of environmental chloride in relation to subsurface hydrology. J. Hydro. v. 94, p. 67-88.

Johnston, C.D., 1987b. Preferred water low and localised recharge in a variable regolith. J. Hydro. v.94, p. 129-142.

Munnich, K.O., Roether, W., and Thilo, L., 1967. Dating of groundwater with tritium and C-14. Isot. Hydrol. IAEA, Vienna. p.305-320.

Peck, A.J., and Williamson, D.R., 1987. Effects of forest clearing on groundwater. J. Hydro. v. 94, p. 47-65.

Sharma, M.L., and Hughes, M.W., 1985. Groundwater recharge estimation using chloride,deuterium and oxygen-18 profiles in the deep coastal sands of Western Australia. J. Hydro. v. 81, p.93-109.

Smith, D.B., Wearn, P.L., Richards, H.J., and Rowe, P.C., 1970. Water movement in the unsaturated zone of high and low permeability strata by measuring natural tritium. Isot. Hydrol. IAEA, Vienna. p. 73-87.

Sukhija, B.S., and Rama., 1973. Evaluation of groundwater recharge in the semi-arid region of India using the environmental tritium . Proc. Ind. Acad. Sci. v. 77(6), p. 279-292.

Sukhija, B.S., and Reddy, D.V., 1987. Study of artificial recharge through Potharam percolation tank by environmental chloride. Geophy. Res. Bul. v. 25, no. 1, p. 27-31.

Sukhija, B.S., Reddy, D.V., Nagabhushanam, P., and Chand, R., 1988. Validity of the environmental chloride method for recharge evaluation of coastal aquifers, India. J. Hydro. v. 99, p. 349-366.

Williamson, D.R., Stokes, R.A., and Ruprecht, J.K., 1987. Response of input and output of water and chloride to clearing for agriculture. J. Hydro. v. 94, p. 1-28.

Zimmermann, U., Munnich, K.O., and Roether, W., 1967. Downward movement of soil moisture traced by-means of hydrogen isotopes. Geophys. Monogr. Am. Geophys. Union. v.11.

SPECIFIC YIELD ESTIMATIONS BY RADIO-ACTIVE TRACER IN INDO-GANGETIC PLAIN

R.K.Raja
Research Officer

S.N.Rawat
Research Officer

S.S.Chhabra
Research Supervisor

U.P.IRRIGATION RESEARCH INSTITUTE, ROORKEE

ABSTRACT

Specific yield is an important hydrogeological parameter of an aquifer to determine change in ground water storage. Due to the variability of alluvial material, the evaluation of specific yield has long been recognised as being subject to large inaccuracies. It is usually determined either by laboratory or field measurements. Both type of measurements have distinct limitations and various practical difficulties. Over the past few decades tracer techniques have been developed which provides a comparatively better alternative to get somewhat reliable results of specific yield and describes the basic characteristics of an aquifer. A field study has, therefore, been carried out to estimate specific yield at Devla (Distt. Saharanpur), Rampur (Distt. Bijnor), Biralsi (Distt. Muzaffarnagar), Badaut (Distt. Meerut) and Chharra (Distt. Aligarh) lying in Indo-Gangetic plain of Western Uttar Pradesh using tritium as radio-active tracer. Pump test & grain size analysis methods have also been carried out to compare the results so obtained by tracer technique at the same test sites. The study indicates that the specific yield values obtained by tracer method vary from 8.18 to 23.48% and it varies from 7.40 to 23.23% by pump test for shallow unconfined aquifer in five districts of Indo-Gangetic Plain.

The details of the study and the results achieved therefrom are given in this paper. The tracer technique gives fairly acceptable values of specific yield and are in close agreement with the results so obtained by pump test method and differs to some extent with the values obtained by grain size analysis method.

0.0 INTRODUCTION

0.1 Rapid agriculture growth in Uttar Pradesh requires systematic groundwater planning to increase efficiency, save energy and to manage groundwater pumpage. Specific yield is an important hydrogeological parameter of an aquifer to determine change in groundwater storage. This can support more economic development of groundwater potential of a region Therefore, a better understanding of the specific yield is very important.

0.2 Meinzer defined the specific yield of rock or soil, as the ratio of the volume of water that will drain by gravity from a saturated rock to the total volume of the rock:thus ratio actually is expressed as percentage.Specific yield depends principally upon the properties of the media viz. specific retention, porosity, grain size distribution, consolidation of the media, and the shape and arrangement of the various particles.

0.3 Number of methods are in vogue to determine the specific yield. These are classified (i) laboratory methods and (ii) field methods. Both types of measurements have distinct limitations. The laboratory methods rather determine the specific retention and porosity.Specific yield is found out by the difference of porosity, η and the specific retention, Sr.

$$Sy = \eta - Sr \qquad ...(1)$$

Commonly employed laboratory methods are (i) sample saturation and drainage (ii) determination of specific yield from the particle size (iii) centrifuge moisture equivalent and (iv) moisture tension techniques. The field methods are (i) pumping test (ii) field saturation and drainage (iii) sampling after lowering the water table, (iv) recharge method and (v) tracer technique.

0.4 However all the methods have limitations but tracer technique is direct and independent for the determination of specific yield. The method is based on the assumption that the aquifer is homogeneous and isotropic, flow is horizontal and concentration of tracer is kept homogeneous during measurements. Moreover, artificial radio-active tracer can be measured in extremely low concentrations and often insitu, making possible the design of convenient and efficient field experiments.

1.0 OBJECTIVE AND SCOPE

1.1 The study has been carried out to determine specific yield of shallow aquifer at five test sites viz. Devla (Saharanpur), Rampur (Bijnor), Biralsi (Muzaffarnagar), Badaut (Meerut), and Chharra (Aligarh), all lying in Indo-Gangetic plain using radio-active tracer technique. The pump test and grain size analysis methods have also been carried out to compare the results of both techniques at the same test sites. The scope of the study include:

(i) to carry out field experiments using radio-active tracer technique to estimate specific yield of shallow aquifers.
(ii) to compare results and examine the suitability of the radio-active tracer technique.

2.0 SELECTION OF TRACER

2.1 The choice of the artificial radioactive tracer to be used depends on the features of the problem. In radio-active tracer technique where the water movement is traced directly, tritium in the form of tritiated water (HTO) is commonly used as an artificial radio-active tracer. Bromine- 82, Iodine-131 and Cromium-51 are the other radio-active tracers which can be used insitu to trace groundwater movement.

Tritium in the form of tritiated water has been used for the present study because of the following reasons:

(i) The tritiated water behaves chemically in the same way as ordinary water and therefore the tracer is "conservative".
(ii) It has a convenient half life of 12.50 years, long enough for the experimental work.
(iii) The detection sensitivity of the tracer is very high (about 3.0 x 10^{-5} μci/ml) so that the activity level can safely be kept below the health hazard levels.

3.0 METHODOLOGY

3.1 The principle of the method of specific yield determination in a saturated zone is based on the approximate equality of porosity and partial volume of water. The following two methods are generally used for estimating specific yield by radioactive tracer technique.

(i) Single Well Pulse Method.
(ii) Double Borehole Method.

1. Single Well Pulse Method

A tracer pulse is introduced into the aquifer through a well, while it is being recharged with sufficient pressure to avoid transit effects. The tracer pulse is then pumped out from the same well after a certain delay time (T) during which the tracer would have drifted away with the natural velocity in the direction of local gradient of flow. The activity of the pumped water is monitored and recorded. Thus method represents the flow in the aquifer over a larger distance and volume and is independent of vertical currents.

Consider a case of an aquifer of thickness b, and let its specific yield be Sy, under ideal conditions, the geometry of pumping is cylindrical. Suppose the tracer pulse (activity of tritiated water) has moved a certain distance 'r' in time T, when it is pumped back and monitored. If V represents the volume of water to be pumped back in order to obtain the peak activity.

$$V = Qt = \pi . r^2 . b . Sy \qquad ...(2)$$

Where Q is the pumping rate and t is the duration of pumping to obtain peak. The velocity of flow v is thus equal to r/T.

$$\text{or} \quad r = vT \qquad ...(3)$$

Substituting this value of r in Eq.2 we get

$$v = \frac{1}{T}\sqrt{\frac{Qt}{\pi . b . Sy}} \qquad ...(4)$$

This relationship can be made use of to determine the value of 'v' first and then the value of specific yield, Sy. However, in this method one is required to carry out a number of experiments with different delay times preferably at different points in the same area. Each time the activity of pumped water is monitored to obtain the peak activity.

The values of delay time T when plotted against corresponding values of $\sqrt{Qt/\pi b Sy}$ will give a straight line. The slope of this straight line will give the value of the distance velocity of flow 'v'. Thereafter using the so obtained value of v, the value of specific yield, Sy can be obtained from eq. (4).

2. Double borehole method

In this method a tracer is introduced into the groundwater at injection point (Doner Well). Detection is carried out in the direction of flow in another well (Acceptor Well).

The tracer concentration in the pumping well, resulting from an instantaneous injection in another well, is given by the formula,

$$C(r,t) = \frac{m}{V} \frac{1}{\sqrt{4\pi \frac{D}{vr}\left(\frac{t}{t_o}\right)^3}} \exp\left[\frac{1-\left(\frac{t}{t_o}\right)^2}{4\frac{D}{vr}\cdot\frac{t}{t_o}}\right] \qquad \ldots(5)$$

where m is the mass (activity) of the injected tracer, t_o is the mean transit time of water, $t_o = r/v$, r being the distance between the two wells and v being the mean interstitial velocity of the flow, and D is the dispersion coefficient. In the radial flow v is not constant, but the ratio of D/v remains constant, i.e., eq. (5) applies for the case of a pure hydrodynamic dispersion, volume V appearing in eq. (5) is the total volume of water in saturated pores in which the water movement takes place. For the case of a cylinderical aquifer of the thickness b and the mean specific yield, Sy, this volume is:

$$V = \pi . r^2 . b . Sy \qquad \ldots(6)$$

The above formula applies for the case when the involved volume of the depression cone is negligible or it does not reach a confined aquifer. If this is not so a more general formula has to be used:

$$V = Sy\pi \int_0^r 2\, r'.b(r')dr' = \pi . r^2 . b . Sy - Vd \qquad \ldots(7)$$

where Vd represents the volume of the involved part of the depression cone. It may also happen that the shape of the water bearing layer cannot be approximated by a simple cylinder minus the depression cone. In such cases the volume V has to be calculated as the volume of a corresponding geometrical body.

The volume V is, on the other hand, equal to the volume of water pumped out during the time t_o i.e.,

$$V = Q\, t_o \qquad \ldots(8)$$

where Q is the pumping discharge. Thus combining eq.(8) with eq.(6) or (7), one can calculate the specific yield.

The specific yield, Sy can thus be obtained from the above relationship. However, the essential conditions for the use of this method are:

(i) The radial velocities induced by pumping should be larger than the distance flow velocity.

(ii) The distance between two wells 'r' should be greater than the thickness of the aquifers.

(iii) The volume of drawdown cone at the pumped well should be smaller as compared to the volume of water extracted.

4.0 TEST SITES AND FIELD WORK

4.1 A field study has been carried out to determine specific yield of shallow aquifer using radio-active tracer technique at the following five test sites:

1. Devla (Distt.Saharanpur)
2. Rampur(Distt. Bijnor)
3. Biralsi(Distt.Muzaffarnagar)
4. Badaut (Distt. Meerut)
5. Chharra (Distt.Aligarh)

Their locations on the map of U.P. is shown in Fig.1. The test sites lies within the alluvial plains of river Ganga. The plains are made up of materials like clay,silt, sand and gravel and at places modules of kankar also appear within the layers of sand and clay.

4.2 To determine specific yield by double well method, the observation well were drilled upto 50 to 55 meters (shallow aquifers) and the distance between pumped to observation wells varies from 14.50 to 22 meters. The soil samples were also collected to know the grain size distribution at the change of strata during drilling. The details of tracer technique data is given in Table I.

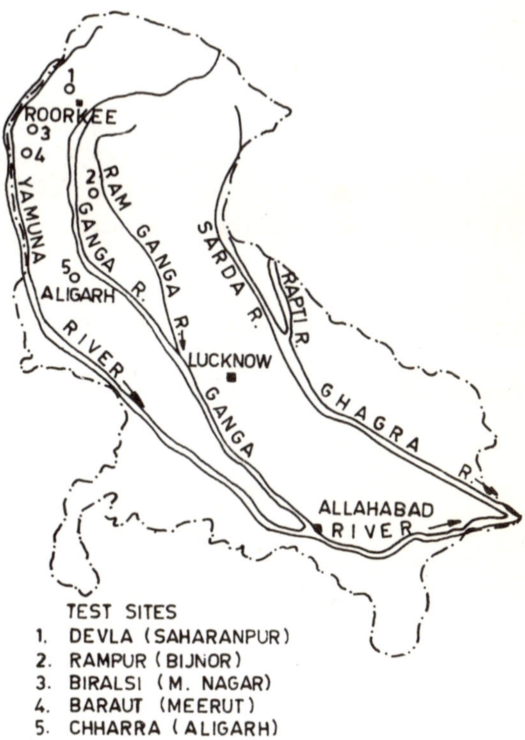

FIG.1 - LOCATION OF TEST SITES

Table - I

DETAILS OF TRACER TECHNIQUE DATA

Sl. No.	Name of test site	Name & Activity of tracer used (mci)	Date and time of injection	Duration of pumping (hours)
1	2	3	4	5
1.	Devla	220 (HTO)	5.3.86 at 20.00 hrs	77
2.	Rampur	220 (HTO)	20.8.86 at 14.00 hrs	100
3.	Biralsi	250 (HTO)	13.9.86 at 14.00 hrs	97
4.	Badaut	250 (HTO)	13.10.86 at 10.00 hrs	96
5.	Chharra	220 (HTO)	12.6.86 at 9.00 hrs	78

4.3 The discharge of pumped well was measured by area velocity method by constructing a drain and as well as by volumetric method at test sites.

4.4 Radio-active tracer studies

4.4.1 Tritium in the form of tritiated water (HTO) as radio-active tracer was injected into observation wells at all the sites. The water inside the wells were properly stirred to ensure proper mixing in the entire water column of the observation wells. Pumping at a constant discharge varying from 1040 m^3/day to 1440 m^3 /day was started in advance of the injection of the tracer so that flow conditions may be achieved properly. In all of the cases the duration of pumping was done upto 77 to 100 hours.

4.4.2 The radio-active water samples were collected in 20 cc polythene double lid samples bottles from pumped well at an interval of 15 minutes during pumping period to know the arrival of activity from observation to pumped well. These samples were brought to nuclear laboratory of the Institute for determining the tritium concentration. In laboratory, one ml. of the each collected water sample was mixed with 15 ml. of Dixone-Nephthalene cocktail liquid scientillator and then counted on Automatic Liquid Scientillation System (ECIL model LSS-34). The tritium activity of each sample at different timings enabled the plotting of C-t curve to determine the peak of arrival of tracer. Knowing the time of peak (t) and distance of both the wells (r), the specific yield was determined by the following relationship:

$$Sy = \frac{Q \cdot t}{\pi \cdot r^2 \cdot b} \qquad \ldots(8)$$

4.5 PUMP TEST

4.5.1 The pump tests were carried out side by side at the same test sites in the same borehole alongwith the tracer studies so as to have comparison of the test results by both methods. Various techniques are being practiced in analysing the pumping test data viz. Thies, Jacob, Chow, Hantush, Walton & boulton etc. for determining the aquifer parameters. In five districts of Indo-Gangetic plain, Boulton's delayed yield method has been used to estimate specific yield because of the fact that the shallow aquifers in Indo-Gangetic plain are generally phreatic aquifers (with delayed yield) or semi unconfined aquifers. The coefficient of transmissibility and specific yield has been estimated by using the following relationships:

$$T = \frac{Q}{4.\pi.s}.W(U_y, r/b) \qquad ...(9)$$

$$S_y = 4\ T\ t\ U_y\ /r \qquad ...(10)$$

4.5.2 For water level measurements, a simple devide, as suggested by Johnson Well Driller's was fabricated for recording the drawdowns in the observation well within an accuracy of 1 mm.

5.0 DISCUSSION

5.1 The lithological logs in respect of Devla experimental site for the pumped well and observation well drilled are shown in Fig.2 along with details like depth of boring, the position of strainer and the depth of water table. In the area round about experimental sites, there is no source of surface irrigation and as such there is likelihood of no influence on the observations. Fig. 3 gives the C-t curve for the Devla experimental site to know the transit time of arrival of radio-active tracer upto pumping well. The details of pump test carried out and Boulton's curve along with observation points are shown in Fig.4.

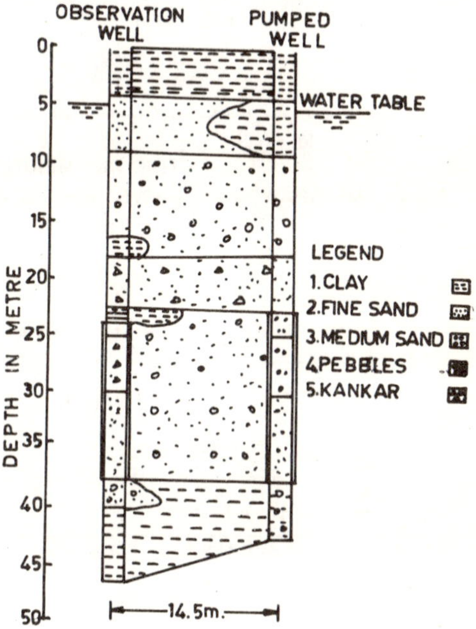

FIG.2 -STRATA CHART OF OBSERVATION AND PUMPED WELLS AT DEVLA EXPT. SITE

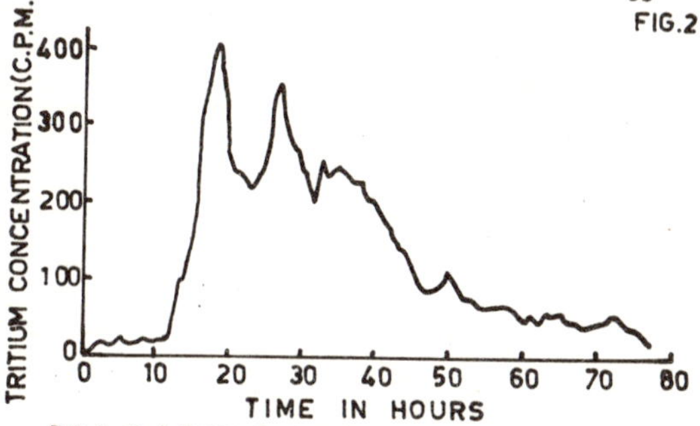

FIG.3-C-t CURVE FOR DEVLA EXPT. SITE

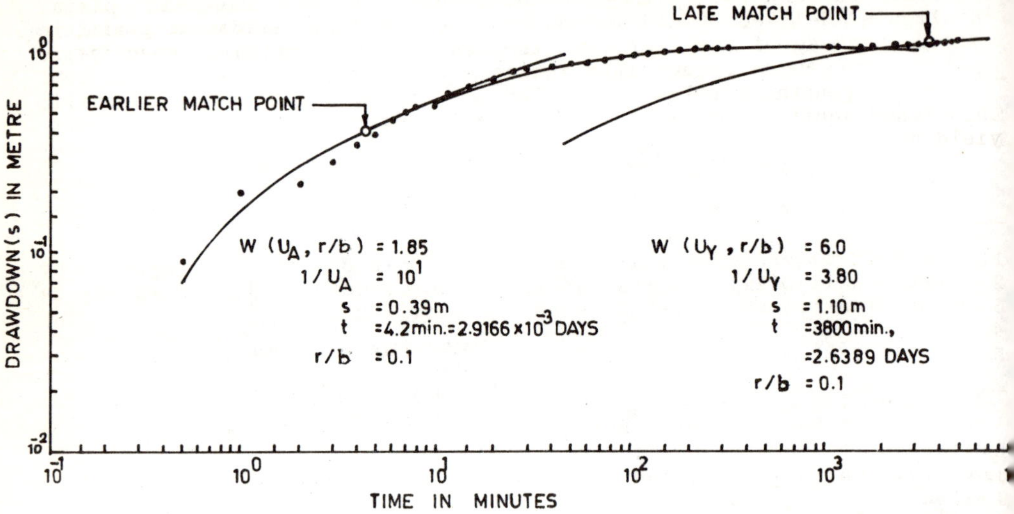

FIG.4 - TIME-DRAWDOWN CURVE FOR DEVLA EXPERIMENTAL SITE DISTT. SAHARANPUR

Table - II gives the results of specific yield by radio-active tracer technique.

TABLE - II

RESULTS OF SPECIFIC YIELD BY RADIO-ACTIVE TRACER TECHNIQUE IN INDO-GANGETIC PLAIN

Sl. No.	Name of test site	Discharge m^3/day	Time of arrival of peak days	Aquifer thickness metre	Distance between pumped and observation well (r)	Specific yield %
1.	Devla	1440	0.7500	15.00	14.50	10.91
2.	Rampur	1372	1.9375	16.25	20.00	13.01
3.	Biralsi	1310	2.9096	19.00	22.00	13.20
4.	Badaut	1040	3.6750	16.00	18.00	23.48
5.	Chharra	1130	1.5000	16.50	20.00	8.18

The results of pump test method are summarised in table III.

TABLE - III

RESULTS OF SPECIFIC YIELD BY PUMP TEST METHOD (BOULTON'S APPROACH) IN INDO-GANGETIC PLAIN

Sl. No.	Name of test site	Boulton's Constant W(Uy,r/b)	Boulton's Constant $\frac{1}{U_y}$	Boulton's Constant r/b	Drawdown (s) metre	Time (t) day	$Sy=\frac{4TtU_y}{r^2}$ %	$T=\frac{Q}{4\pi s}W(U_y, r/b)$ m^2/day
1.	Devla	6.00	380	0.1	1.10	2.6389	8.26	625.00
2.	Rampur	4.00	40	0.2	0.76	1.0416	14.97	575.00
3.	Biralsi	5.20	100	0.1	0.60	1.7361	12.97	903.93
4.	Badaut	5.40	100	0.1	0.33	2.0833	23.23	1354.95
5.	Chharra	5.60	100	0.1	0.52	0.7639	7.40	968.89

5.2 A perusal of Fig. 2 indicates the existence of clay layers some of which are lenticular and discontinuous within the matrix of sand and a few other extent laterly to fairly long distances. The soil samples collected during drilling were analysed by sieve and hydrometer. The percentage of different grain-size particles such as gravel coarse, sand, medium sand, fine sand, silt, and clay were found out. The specific yield values thus determined from a trilinear graph of textural classification (Fig.5) developed by Prill and Johnson for soil samples collected from Jan Joaquin and Antelope Valleys, California. The data are shown by lines of equal specific yield values on a trilinear graph of textural classification. The value of specific yield obtained by grain-size analysis collected from different test locations are given in Table IV.

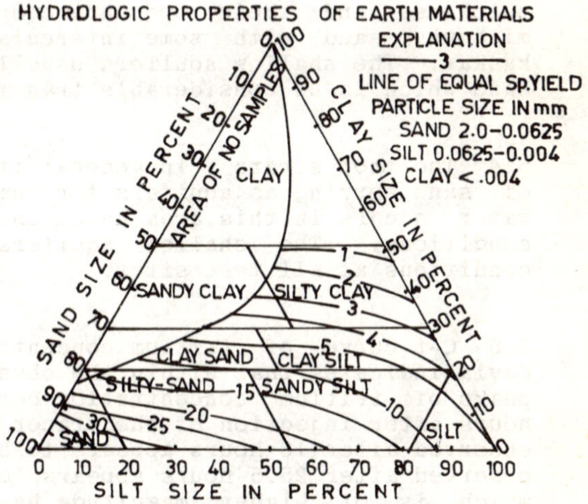

FIG.5-SOIL CLASSIFICATION TRIANGLE SHOWING RELATION BETWEEN PARTICLE SIZE AND Sp.YIELD

TABLE - IV

RESULTS OF SPECIFIC YIELD BY GRAIN SIZE ANALYSIS

Sl. No.	Name of test site	% of Grain Size						Specific yield %
		Gravel	Course	Medium	Fine	Silt	Clay	
1.	Devla	2.0	13.0	45.0	7.0	13.0	20.0	11.90
2.	Rampur		8.0	57.0	7.0	15.0	13.0	16.00
3.	Biralsi	-	18.0	58.0	3.0	4.0	17.0	16.00
4.	Badaut	-	11.0	59.0	8.0	14.0	8.0	24.50
5.	Chharra	2.0	8.0	48.0	5.0	15.0	22.0	10.00

5.3 It may be inferred from the borehole cuttings that the sand be forming the potential aquifer are fairly wide-spread with variabl thickness and their beds are composed of fine to medium graine micaous sand with some interculation of hard clay and clay wit kankar. The shallow aquifers usually contain fine and medium graine sand which is of considerable transmissibility and high permeability

5.4 The top strata in general is clay and below this exist laye of sand serving as aquifers for shallow and deep tubewells. Groun water occurs in this area under unconfined as well as semi-confine conditions. The shallow aquifers are generally under unconfine conditions at all test sites.

5.5 C-t curve of tritium concentration of pumped water in case c Devla indicates that within the observation period of 77 hours, tw peaks of tritium concentration corresponding to 18 hours and 25. hours after injection of the tracer were observed (Fig.3). The pea observed after 18 hours appears to be the actual peak while the pea observed after 25.5 hours appears to be secondary. The primary pea which is of higher magnitude has been considered for computir transit time for the peak activity and the specific yield has bee calculated using eq. 8. As a result of this experiment the specifi yield works out to 10.91% for Devla test site and the result of other sites are given in table II.

5.6 The time- drawdown data observed as a result of pump test wer analysed by Boulton's Delayed Yield Method (Fig.4) and the value c specific yield comes to be 8.26% by this method for Devla test site.

The details & results of other test sites are also given in thi paper.

6.0 RESULTS

6.1 The results of specific yield obtained by Radio-active tracer technique, pumping test & grain-size analysis methods are as below:

Sl. No.	Name of site	Specific Yield %		
		Radioactive tracer technique	Pump test method	Grain-size analysis method
1	2	3	4	5
1.	Devla	10.91	8.26	11.90
2.	Rampur	13.02	14.97	16.00
3.	Biralsi	13.20	12.97	16.00
4.	Badaut	23.48	23.23	24.50
5.	Chharra	8.18	7.40	10.00

6.2 Radio-active tracer technique provides a fairly acceptable values of specific yield and are in close agreement with each other but the values differs to some extent by the grain-size analysis method.

REFERENCES:

1. Halevy, E., Nir.A., 1962.The determination of aquifer parameters with the aid of radio-active tracers. J.Geophys.Res.67, p.2403.

2. Borewozk, M.Mairhofer,J.A.,1967. Single well pulse technique. Isotope in Hydrology (Proc. Sym. Vienna, 1966), IAEA, Vienna, p.507.

3. Meroado, A., Halevy, E. 1966. Determining the average porosity and permeability of a stratified aquifer with the aid of radio-active tracer. Wat. Resour, Res. 2.p.525.

4. Hazzara, I.B., Girgis, R.K., Sadd, K.F.,1965., A method for the determination of aquifer parameters using radioactive isotope. International journal of applied radiation and isotope, p.487-493.

5. International Atomic Energy Agency, 1968. Guide book on nuclear technique in hydrology. IAEA, Vienna.

LIST OF CAPTIONS

η = Porosity of soil medium
Sy = Specific yield
Sr = Specific retention
V = Volume of pumped water
v = Velocity of flow
Q = Discharge of pumped well
T = Delay time
t = Duration of pumping
b = Thickness of aquifer
r = Distance between pumped and observation wells
m = Activity of the injected tracer
t = Mean transit time of water
D = Dispersion coefficient
C = Tracer concentration
W(Uy,r/b),
1/Uy = Boulton's constants

JOINT SYSTEM ANALYSIS FOR THE EVALUATION OF FLOW IN FRACTURED AQUIFER

By

P. K. Das* and S. Basu Mallik**

* Central Ground Water Board, Unit-III Janpath, 24 Kharvelnagar, Bhubaneswar - 751 001, Orissa, India.

** Professor, Geological Sciences, Jadavpur University, Calcutta - 700 032, India.

ABSTRACT

An intensive study has been made of Joint System of Precambrian Crystalline rocks occurring at the township of Khatra in Bankura District, West Bengal. The area is mantled by saprolite derived from weathering of Phyllites, Micaschist, Granite, Amphibolites and Quartzites. The saprolite aquifer is considered to be isotropic in nature which may be evidenced from the water table contour maps prepared from close observation of wells. The saturated fractured rocks occurring just below the saprolite aquifer can be determined from electrical resistivity studies and borehole data. Analysis of the joint system of this aquifer can be made from the exposed rocks. The predominant joint sets have been determined by graphical techniques from the central tendencies of all the joint planes giving rise to a characteristic joint geometry of different sectors in the area. Study of different sets of dispersion shows that the medium is largely anisotropic in character and that the principal permeability direction of joint sets is contained along approximately SE directions. It can be shown that the courses of streams and creeks appear to bear some relation with the major permeability direction and also that the base flow of the river Kasai is maintained by the fractured aquifer.

Introduction :

Khatra in Bankura district is situated at the South Western part of West Bengal and is known to be drought prone area (Fig. - 1). The area is underlain by hard crystalline metamorphic rocks of precambrain age with undulating topography. The area is chiefly drained by Kasai river which originates in the neighbouring district of Purulia and flows down in the southern part of the area. The main tributaries of Kasai river in the study area are Kanria Jhor and Mura Jhor which flow towards the southern directions before their confluences. The tributaries begin to dry up from their upstream side in the summer months when Kasai river maintain its base flow only.

The objective of the present work is intended to study joint system for understanding and evaluation of flow in fractured aquifer.

The movement pattern of groundwater is controlled by various structural features in the bed rocks. Although groundwater circulates mainly in the weathered mantle, some of it moves into and through bed rock and drain into stream within and adjacent to the study area.

The analysis of flow in fractured media is very complex Two extreme cases may be realised in nature (i) When individual planner conductors such as faults in rock are independent and infrequent that each may be treated as a separate channel (ii) When aggregates of fracture as in fault breccia show resemble sedimentary pores that the medium may be assumed to be continuous (Muskat 1949). In the Individual conductors having low fracture density it is required to analyse the flow with non continuum approach based on Navier stokes equations (Wittke, 1973). In real situations there may however be wide departures from the two extreme cases depending on the number of sets and apertures of fractures.

Geological Setup of the Area :

The area forms a part of the extensive precambrain tract in eastern India (Dunn, 1929) and is capped by a mantle of varying thickness. The mantle

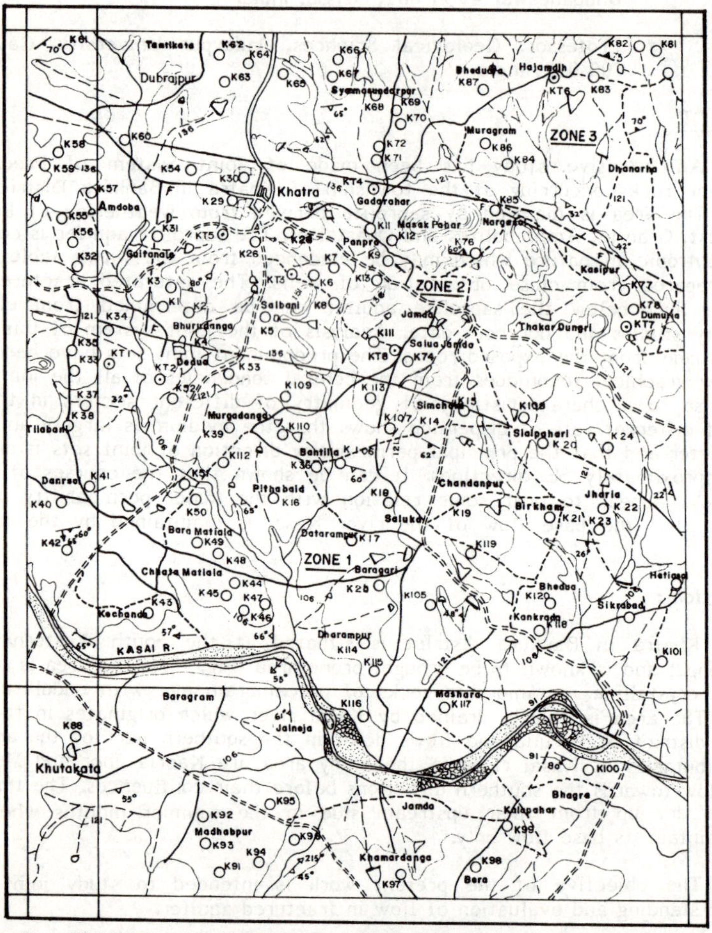

Fig 1. GEOLOGICAL MAP OF THE AREA WITH LOCATION OF OBSERVATION WELLS

M1000 0 1 2 3 Km.

N

SCHISTOSITY FOLIATION	SHEAR PLANE
FOLD-AXIS	STRIKE AND DIP FOLIATION RELATED TO SHEARING
FAULT PLANE	OBSERVATION WELL

consists of a soil zone of 1-2 m thickness and a zone of highly weathered rocks which can be referred to as the saprolite. Lithologically the area has been subdivided into three zones (Fig. - 1).

Zone I consists of phyllite micaschist, granite, garnet biotite-schist and hornblende-schist. The rocks of zone 2 are mostly amphilbolite occurring in the northern part of the area. The zone 3 consisting chiefly of granitic rocks is confined to the central part of the area. These rocks form a band-like appearance which tend to be parallel with the general schistosity plane of the rocks of the other zones. The structural features (Fig. - 1) in the bed rocks area are important in that some of them affect the direction and rate of groundwater movement. Although groundwater circulate mainly through bed rock and drains into the stream channels. Planner features in the bed rocks include bedding, foliation, cleavage, joints and faults along which the groundwater may move. Linear features and folds may also influence the movement of ground water generally to a lesser degree than the planner surface.

Ground Water Flow System Analysis :

The post-monsoon and pre-monsoon water levels have been measured in 120 open wells. The location of the wells as shown in Fig. -1 are well distributed that the contour lines could be drawn at 6 m interval through the point on the water table having same altitude to give a more or less congruous picture. The water level contour map of the post-monsoon period (Fig. - 2) may be used for the possible flow pattern in order to predict expectable phenomena or to interpret actual features that are associated with the natural distribution of groundwater in drainage basins. The analysis of groundwater flow in hard crystalline rocks is extremely difficult. The use of flow nets may however be one of the most powerful tools in understanding the possible flow patterns. It may be assumed that the unconfined saprolite aquifer is nearly a homogeneous and istropic system.

In ansotropic medium the construction of flow net is highly complicated because the relation of orthogonality between the equipotential lines and the flow lines is no longer valid. In this case the flow lines are determined from the inverse hydraulic conductivity ellipse about a point.

Subsurface Geology :

The electrical resistivity provides a general information on the nature of subsurface rock materials. The resistivity method can be successfully employed for groundwater investigations where a good electrical resistivity contrasts exist between the water-bearing formations and the underlying rocks (Zohdy etal, 1969).

In hard rock areas the weathered and fractured rocks is capable of yielding sufficient quantities of water. These methods are therefore suitable for groundwater locations in crystalline rocks.

.For this several electrical soundings have been made in order to make an estimate of the thickness of the weathered profile and the fractured zones. Three representative vertical electrical sounding curves are presented in fig. 3.

Interpretation of these curves indicate that fractured layer occurs at a depth range 5-11 mbgl. The fractured layer shows a range of resistivity 50-200 ohm m for a thickness of 15 m to 30 m (Basu Mallik, 1985). This data is also supported by drill core analysis.

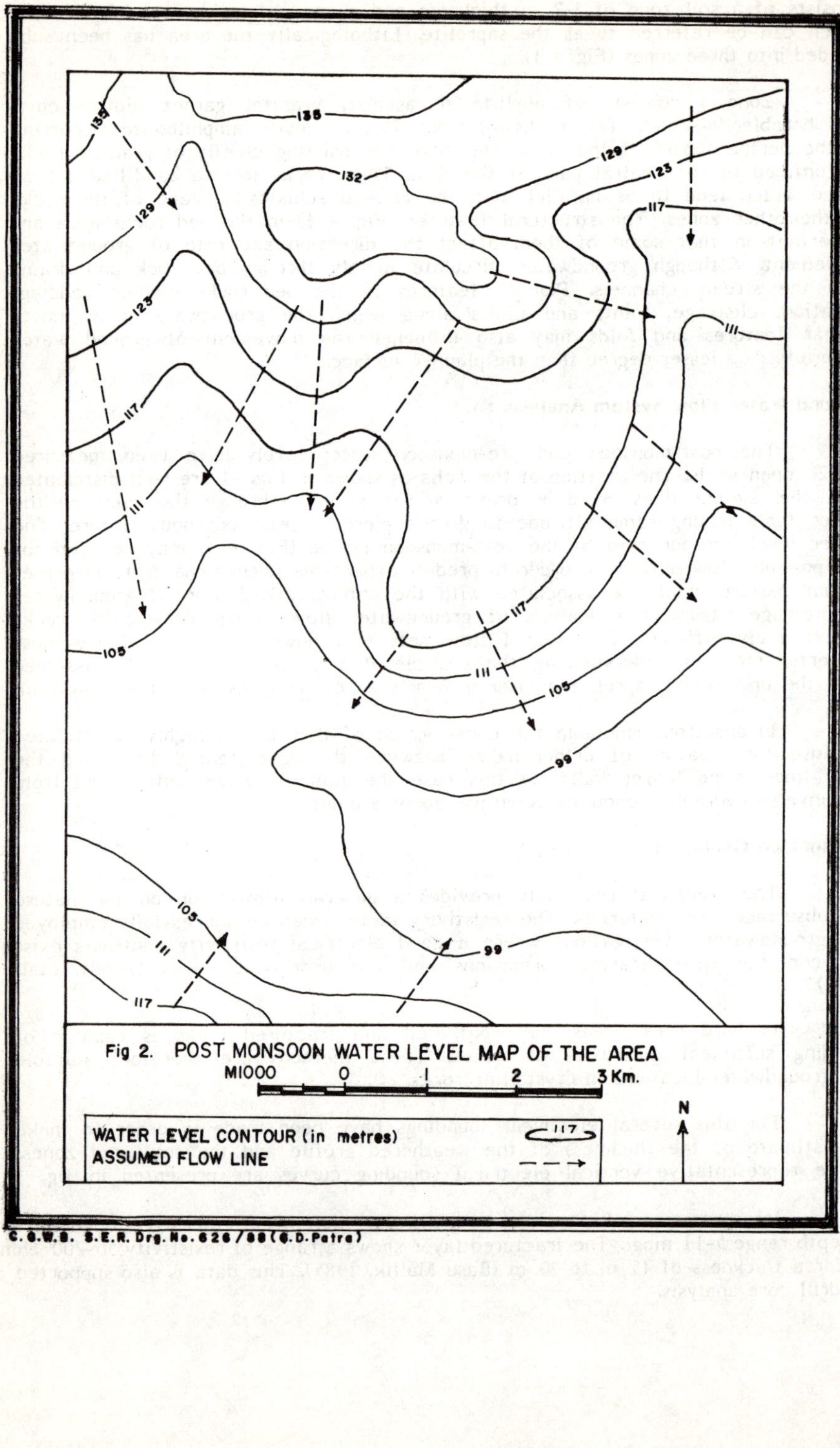

Fig 2. POST MONSOON WATER LEVEL MAP OF THE AREA

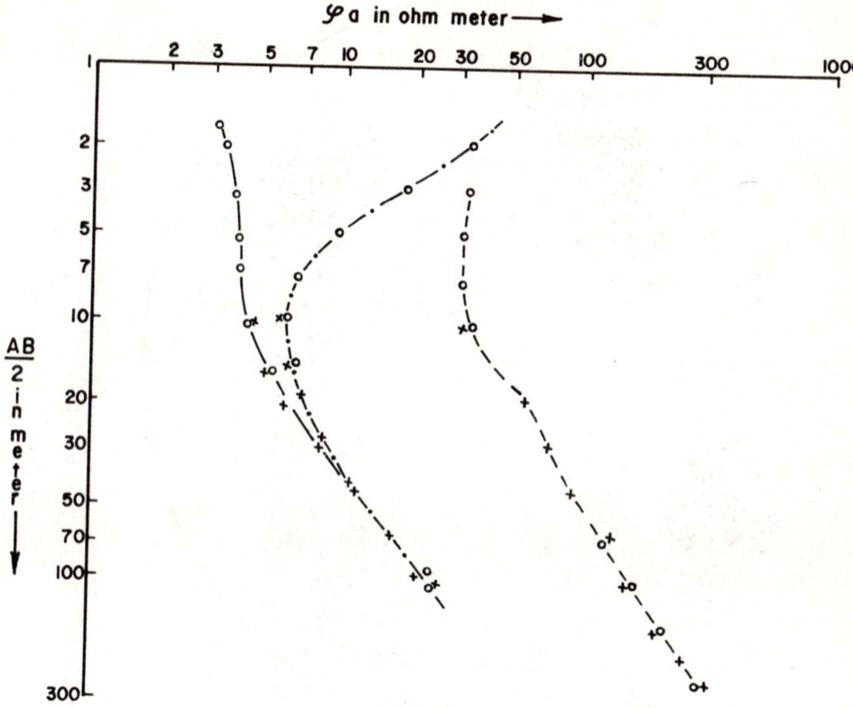

Fig-3. TYPICAL ELECTRICAL RESISTIVITY CURVES IN THE AREA

Joint System Analysis and Principal Permeability Directions of the Rocks :

Among various structural features the joints and fractures are considered to have profound influence on the occurrence and movement of groundwater in crystalline rocks. The rocks present in the area are everywhere jointed and otherwise fractured into fairly well defined sets. There is as well a multitude of minor joints and other fractures that display no recognizable pattern of orientation or attitude. Measurement of a large number of joints/fractures has been made in the rocks of the area, taking atleast three to four sets of dominant joint planes from each outcrop. The joints occurring in different sets can be measured with their frequencies and orientations. (Das, 1986). In weathered rock outcrop both the gentle and steeply dipping joint planes are developed but as seen in the drill cone specimens low dipping joint planes decrease significantly with depth. Hence the horizontal and subhorizontal joints are assumed to be due to release of pressure associated with weathering and erosion of rocks. Therefore on elimination of the inconsistent and horizontal fractures, the poles of the joint planes are plotted in a stereographic projection (Fig. - 4). For convenience of detailed analysis the joints have been considered separately from the three zones as described earlier. The poles of the joint sets are shown in figure 5 and from their central tendencies the predominant joint planes sets can be constructed (Fig. 6). The orientation and frequency of joint planes of the three zones are given in table 1 below.

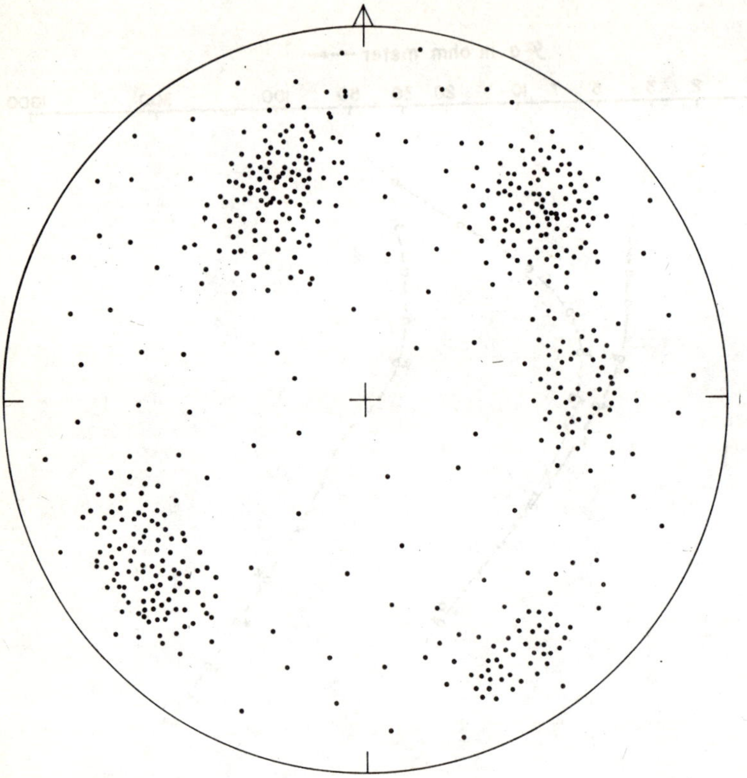

Fig. 4 Distribution of the poles of all the joint planes as measured in the rocks plotted on lower hemisphere.

Table - I

Zone - I :

Joint orientation	Frequency no. joints per meter
(i) 53° — 223°	15
(ii) 72° — 328°	8
(iii) 63° — 56°	12

Zone - II :

Joint orientation	Frequency no. joints per meter
(i) 63° — 56°	11
(ii) 52° — 156°	10
(iii) 50° — 263°	6

Zone - 3 :

(i)	70° — 52°	13
(ii)	66° — 222°	10
(iii)	57° — 160°	7

The fractured media such as these are very common in hard rock areas and their geometry lies beteween an individual planar conductor such as faults in the rocks and aggregate of fractures as in fault breccia. The method of non-continuum mechanics based on the Navier-Stokes equations is generally applied for flow analysis in individual fractures. It is however possible to carry out the analysis of flow in fractured rocks with continuum approach if the fracture frequency is sufficiently dense so that the fractured media behaves identical hydraulically with granular porous media. The important parameters governing the permeabilities of the fractured media are the orientation, frequency and openness of joint sets in rocks. Three principal sets of joints may be orthogonal or not. The orthogonal sets may be different in dispersions and in spacings. The principal permeability axes are parallel to the central tendency of the sets, the major axis being parallel to the normal of the weakest set of joints and the minor axis, parallel to the normal of the strongest. The medium may be cosidered inherently anisotropic if the frequency of fracture is different in one direction than it is in the other direction. Snow (1968, 1969) has demonstrated that fracture flow problems can be solved using standard porous media techniques utilising Darcy's law and anisotropic conductivity tensor. In the model studies of anisotropic permeability of fractured medium Snow (1969) determined the principal axis of permeability from different sets of fractures with assumed frequency.

In real situations there may be several departures from the model analysis of anisotropic permeability. However the results of model analysis may be useful guide for determination of possible directions of principal permeability. The three predominant joint sets of zone-1 have been plotted in fig.-6a. The three joint sets are different in disperion and in spacing (table - 1, Fig. 3a). The principal axes of permeability are parallel to the central tendencies of the sets and are denoted by P_1, P_2, and P_3. The figure 6a shows that the joint sets are nearly orthogonal and the major axis P_1 is parallel to the normal of the weakest set of joints and minor axis P_3 is parallel to the normal of the strongest joint set. P_2 is then the intermediate axis. This suggests that the joint sets striking NW-SE and dipping NE largely controls the groundwater movement in the rocks of zone-1.

In figure 6b the dominant sets of the rocks of the zone-2 are shown. Here also joint sets are nearly orthogonal. Therefore in determining the principal axes of permeability the frequency is more important than the orientation and disperions of the joint planes. On the basis of these reasonings the normal of the joint set striking N-S is parallel to the major axis, P_1. Thus P_2 and P_3 are respectively the intermediate and minor axis of permeability (Fig. 6b). The flow of groundwater in the rocks of the zone-2 is chiefly controlled by the joint set striking NE-SW.

The orientations of the joint sets developed in the zone-3 are represented in figure 6c. It is clearly evident that joint sets are nonorthogonal and the major axis of permeability lies along the direction P_1 given by the greatest number of joint intersections. It is difficult to determine the positions of the other two axes in cases similar to these. As the joint sets striking E-W is very close spaced and is dveloped rather strongly the minimum axis may be assumed to lie at P_3 normal to the P_1 and to roughly towards the normal direction of the joint set.

The intermediate axis can therefore lie along the NE-SW direction. Therefore the groundwater flow in the zone-3 is chiefly controlled by the joint intersections along the south-east direction.

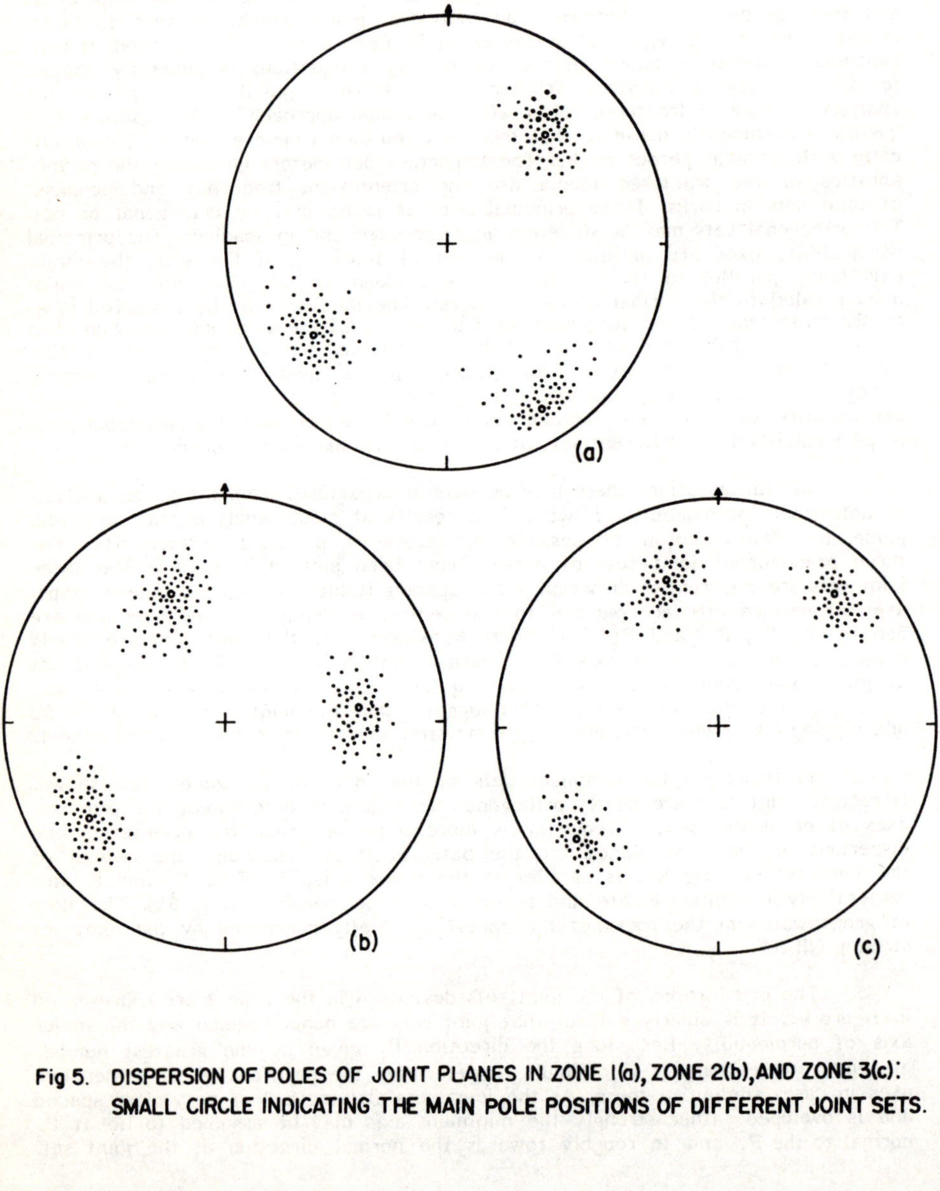

Fig 5. DISPERSION OF POLES OF JOINT PLANES IN ZONE 1(a), ZONE 2(b), AND ZONE 3(c): SMALL CIRCLE INDICATING THE MAIN POLE POSITIONS OF DIFFERENT JOINT SETS.

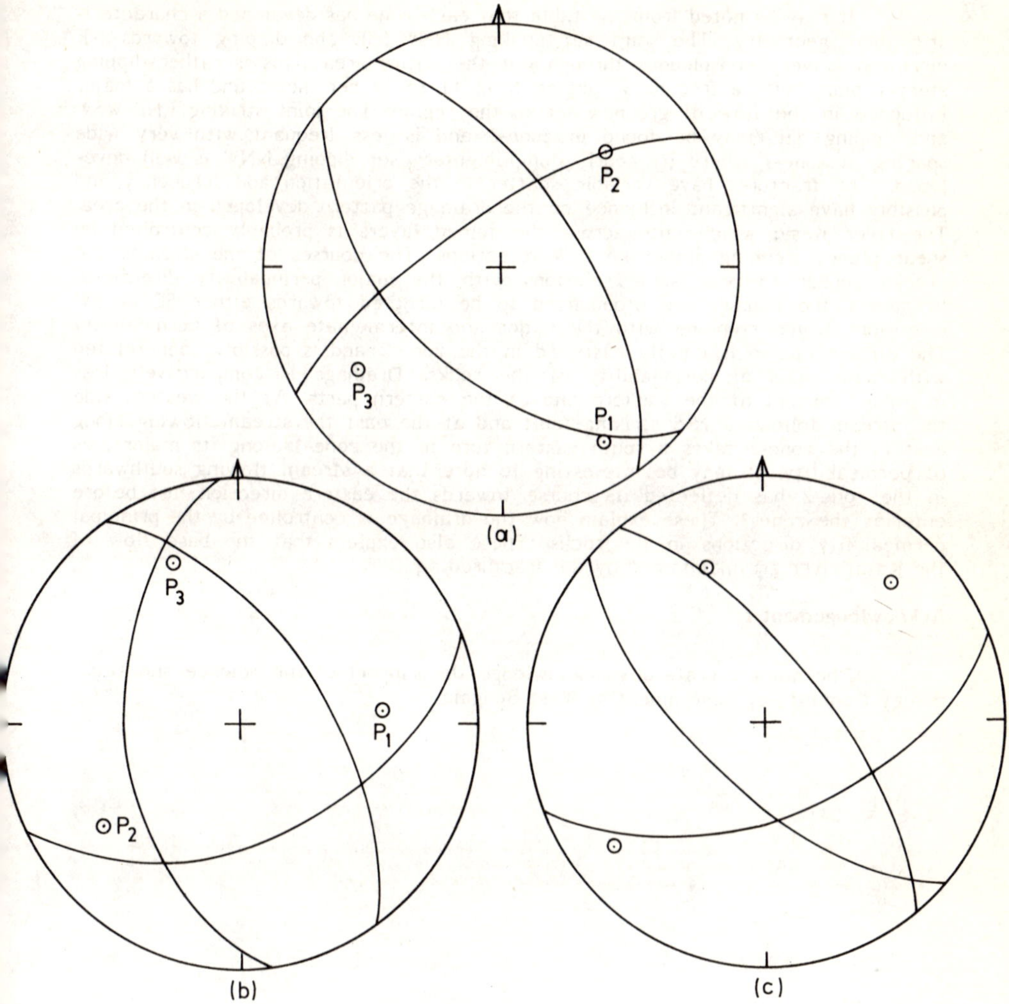

Fig. 6 Showing major joint systems and the principal permeability directions (P_1,P_2,P_3): a) Zone-1 b) Zone-2 c) Zone-3.

Results and Discussions :

It may be noted from the table that each zone has developed a characteristic joint geometry. The joint set striking NNW-SSE and dipping towards NE direction is very conspicuous through out the entire area. This is rather dipping steeply plane with a frequency ranging from 11 to 13 per meter and has a major influence in the flow of groundwater in the region. The joint striking ENE-WSW and dipping SSE may be found in zone-2 and is less frequent with very wide spacing in zone-1 where its nearly complimentary set dipping NNW is well developed. The fractures have variable scatter in the orientation and frequency and possibly have significant influence on the drainage pattern developed in the area. The river Kasai which cuts across the folded layers is probably controlled by shear planes trending along the E-W directions. The courses of the streams and creeks appear to bear some relations with the major permeability directions. In zone-1 the courses are pronounced to be directed towards either SE or SW directions which coincide with the major and intermediate axes of permeability The stream course is chiefly eastward in the zone-2 and is possibly been related with major axis of permeability of the rocks. Drainage is comparitively less in zone-3 except at the western and at the eastern parts. At the western side the stream follows a N-S striking fault and at the east the stream flowing along east in the zone-2 takes a south eastern turn in the zone-3 along its major axis of permeability. It may be interesting to note that a stream flowing southwards in the zone-2 has deflected its course towards the eastern direction just before entering the zone-3. These explain how the drainage is controlled by the principal permeability directions in the rocks. These also explain that the base flow of the Kasai river are maintained by the fractured aquifer.

Acknowledgement :

The authors gratefully acknowledge the support of the Science and Technology Committee, Government of West Bengal.

REFERENCES

Basu Mallik, S.,	1983	Behaviour of fractures in hard rocks-A study by surface geology and radial Ves method. Geoexploration, V 21, p 181.?
____________,	1985	Geoelectric investigations for groundwater resources at the trijunction of Purulia, Bankura and Midnapur district, West Bengal - report submitted to the Science & Technology Committee, Government of West Bengal.
Das, P. K.,	1986	Hydrogeologic Behaviour of some Crystalline Rocks of Bankura district, West Bengal. Unpublished Ph.D Thesis.
Dunn, J. A.	1929	The geology of north Singhbhum including parts of Ranchi and manbhum districts. Mem. Geol. Surv. Ind. 54, pt. 2, p64.
Muskat, M.	1949	The theory of potentiometric models, Trans. Amer. Inst. Min and Metal. Engrs., 179, p216.
Snow, D. T.	1968	Rock fracture spacings, openings, and porosities. J. Soil Mech. Found. Div., Proc. Amer. Soc. Civil Engrs., 94, p73.
____________,	1969	Anisotropic permeability of fractured media. Water Resources Res., V. 5, p1273.
Zohdy, A. A. R.	1969	The use of Schlumberger and equatorial soundings in groundwater investigations near EL Paso, Texas, Geophysics, V. 34, p 713.

MANAGEMENT AND CONSERVATION OF GROUND WATER RESOURCE THROUGH POINT DILUTION TECHNIQUE USING RADIOTRACER - SOME CASE HISTORIES

Dr. K.S. Pandey
Scientist'B'

S.K.Sinha
Scientist'D'

CENTRAL GROUND WATER BOARD

ABSTRACT

Management and conservation of ground water resources using cost effective techniques such as point dilution technique, has of late assumed greater importance in the wake of increasing cost of irrigational inputs which has a direct bearing on the day to day life of every individual. This point dilution technique involves the use of radiotracers for estimating seepage loss and filtration velocity from unlined canals directly and independently from the field areas. The water columns present in boreholes along the canal is utilised as media for injection of I^{131} as radiotracer and its detection of subsequent dilution, with time after injection, in next boreholes helps in finding out quantum of seepage loss and filtration velocity. Thus remedial and preventive measures (an outcome of this methodology) if adopted to check water losses from conveyance system can be of immense help in conservation of invaluable water resources and as well as preventionof water logging of fertile lands.

Experiments, adopting Point Dilution Technique using radiotracer I^{131} were conducted by UNDP/CGWB in the year 1977-78 along the banks of Ghaggar Branch of Sirhind Canal (Sangrur district-Punjab), Amloh minor at Haripur (Patiala district-Punjab) and water logged area near Baropal (Suratgarh-Rajasthan). The quantum of seepage loss of water has been evaluated by computing values of filtration velocity, wetted perimeter and hydraulic gradient.

This inexpensive method of determination of seepage loss using radiotracer is most suited for homogeneous and isotropic formation and its application should be encouraged in similar areas in the country.

INTRODUCTION

The single borehole dilution technique has been used by many workers for determining the localised filtration velocity of ground water flow. Filtration velocity called Darcian velocity is usually evaluated from Darcy's formulla through the measurements of hydraulic gradient and requires a knowledge of permeability of the aquifer. The filtration velocity interpreted with their parameters can provide valuable information on permeability, transmissivity and the quantum of water moving through the aquifers. The general principle of single borehole technique is that when a tracer is injected into the entire volume of water in a well uniformly, its concentration decreases with time due to the horizontal flow through the screened well. However, the volume of water in the well at a constant rate. The decrease of radiotracer concentration follows a simple exponential law.

It is well known that considerable water losses occur from the irrigation canals which are termed losses through conveyance system. The seepage losses in water courses (unlined) are larger since the ratio of the wetted area to the volume of water carried is higher. Seepage losses create water logging and soil salinity problems decrease of water flow in canals. On the other side deep seepage from unlined canal contribute to recharge of aquifers and improves ground water quality in saline tracts along the canals.

To measure seepage losses from canals, a number of experiments has been conducted by various workers (Vaidhianathan et.al,1937; Robinson and Rhower, 1957; Krishnamurthy and Rao, 1968; Garg and Chawla, 1970 and Sehadri, 1978) by means of single well dilution technique. As it is well understood that seepage from canals take place in the form of sub-surface streams flowing at different hydraulic gradients and different velocities at different depths, the filtration velocity tnus measured by taking any depth of water column in borehole will be due to the average velocity of the streams. Keeping this in mind a number of experiments were conducted using radiotracer I^{131} for injection into boreholes along the bank of Ghaggar Branch of Sirhind Canal at Akbarpur of Sangrur district, Amloh minor at Haripur of Patiala·district of Punjab and in water logged area at Baropal closed to the Ghaggar depression, Suratgarh (Rajasthan) by taking different depths of water level in boreholes and at different distances from the bank of canal.

The radiotracer technique is more economical from the point of view of time, cost and repeatability of the studies than the other methods. This method is best suited for homogeneous and isotropic formation. The tracer technique is relatively simple to use in deep canals with turbide and fast moving water, whereas canals with heavy weed growth create adverse bottom conditions.

The techniques employed earlier in determination of ground water flow using salts and dyes are being replaced by the use of radiotracers owing to their characters of higher sensivity in detection. The low concentration of tracer was used in experiments so that the effects of the tracer on the Chemical and Physical properties of the ground water would be minimised. A few advantages of using radiotracer for seepage losses measurements are as follows:-

i) Radiotracers are detected in the low concentrations which do not disturb the ground water flow pattern.

ii) These are readily detected by Scientillation gamma probe, and

iii)The radioisotopes used as tracers do not get precipitated or absorbed in soil.

The success and accuracy in determination of ground water flow depends to a large extent on the proper selection of the radioactive tracer.

Factors involved in selection of Radiotracer:

The following factors usually governed the choice of a radiotracer for seepage flow measurements;

a) The radioisotopes should be gamma emitting radiation to enable its efficient detection in water,

b) It should be available in the form of highly soluble water and should be stable in water medium i.e. it should not precipitated nor absorbed in the soil,

c) It should be easily detectable at the experimental sites,

d) It should have the maximum permissible concentration

e) Its half life should reasonably be high go that observation may be continued for reasonable period.

A greater number of radioisotopes had been tested in the laboratory as well as in the field, but only a few radioisotopes were formed to be useful in hydrogeological investigations, as which are tabulated below:-

Element	Mass number of radioisotopes	Half life	Radiation	MPS above natural background μci/ml in solution
Bromine	Br^{82}	36 hours	β^-, γ	$3x10^{-4}$
Chromium	Cr^{51}	27.8 days	β^-, γ	--
Iodine	I^{131}	8.04 days	β^-, γ	$2x10^{-6}$

EXPERIMENTAL DETAILS AND PROCEDURES

The experiment was conducted along Sirhind canal of Ghaggar Branch at RD 102,000 at Akbarpur in Sangrur district while the second experiment was carried out along Amloh minor canal at Haripur (RD 10,500) of Punjab state. Case histories of the experiments falling under Punjab state are dealt with in this paper. At each experimental sites four piezometers of 127 mm in diameter, drilled down to depth of 9.0 m were constructed through Handboring method along the left bank of Sirhind canal and Amloh minor canal respectively. The piezometers were thus designed by wrapping brass strainer over 1/16 inch slotted pipe for tapping saturated granular zones. The piezometers were constructed perpendicular to the canal alignment in a row at distances of 0.47 m, 1.0 m and 9.0 m respectively.

The length of screen portion was 2.0 metres and the area of openings was more than 10%. Suitable gravel pack according to aquifer material was used in the construction of piezometers, Iodine (I^{131}) as Sodium Thiosulphate solution of specific activity 3.7 μci/ml was injected into the piezometer located on the bank of the canal in different amounts according to the volume of water in borehole. After injecting Iodine-131 in borehole, the entire water column was thoroughly stirred with the help of bailer to make the concentration of radiotracer (I-131) homogeneous. Scientillation gamma probe was lowered into the borehole to measure the decrease of radioactivity in time in the form of CPS (Counts/100 seconds or Electric Pulses/100 seconds) for each borehole at different depths for calculation of filtration velocity (Table - 1). In the remaining two piezometers, periodical radioactive logging was done at the experimental sites to monitor the time of arrival of radiotracer. On the basis of these data, transit time and velocity of ground water flow were calculated. The measurements of canal's shape and size along with discharge were taken at the sites. In order to know the water losses by seepage, types of soils at the site were collected for detailed study in the laboratory. The lay-out of the experimental sites along with phreatic lines are shown in figures (Fig. 1,2).

RESULTS AND DISCUSSIONS

The filtration velocity (Vf) of the horizontal water flow in the absence of all other flows through the borehole was calculated with the help of the following formula.

$$Vf = -\frac{V}{\alpha Ft} \ln C/Co \text{ ------------- (i)}$$

where V = volume of water in which dilution of tracer in borehole takes place.

Table - 1 A. Details of Piezometers, depth to water level and observed radiotracer concentration with time at site Akbarpur of Punjab.

Sirhind Canal at Akbarpur Site.												
Piezometer No.	Distance of Piezometers from bank of Canal (mts)	Depth to water level (m bmp)	Radiotracer concentration at different depths									
			1.60 2.25 m bmp		2.25 m bmp		3.20 m bmp		4.90 m bmp		5.80 m bmp	
			t (Sec.)	C/Co	t (sec.)	C/Co	t (Sec.)	C/Co	t (Sec.)	C/Co	t (Sec.)	C/co
P-Inje	0.47	0.20	Injection Point									
P-I	1.0	0.27	18300	0.065	20700	0.275	25200	0.117	25200	0.198	-	
P-II	2.0	0.45	15590	0.273	15540	0.10	17400	0.20	17640	0.55	-	
P-III	9.0	4.95	-	-	-	-	-	-	19620	0.24	19740	0.56

Table - 1 B. Details of Piezometers, depth to water level and observed radiotracer Concentration with time at site Haripur of Punjab.

Piezometer No	Distance of Piezometers from level Canal (m)	Depth to Water level (m bgl)	Radiotracer Concentration at different depths (bmp)			
			6.58 m		7.50 m.	
			t (Sec.)	C/Co	t (Sec,)	C/Co
P - I	0.40	5.90	Injection Point			
P - 2	1.00	6.03	78720	0.128	124680	0.26
P - 3	2.00	6.21	9600	0.092	8650	0.124

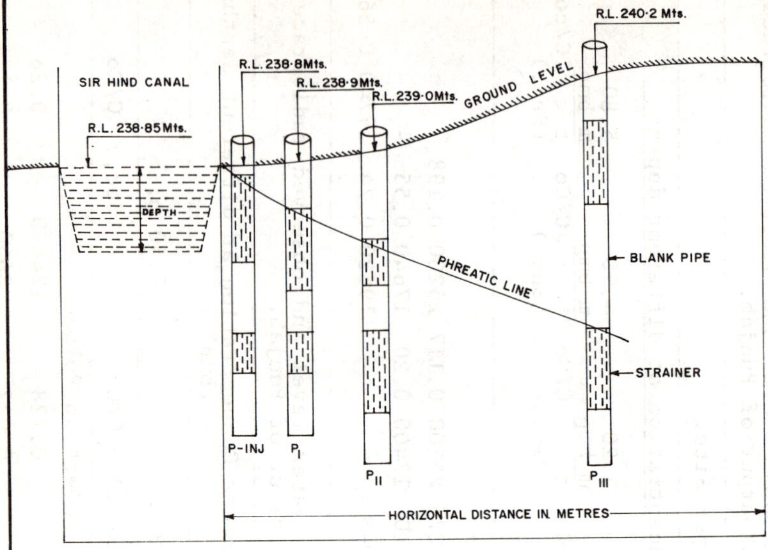

LAY OUT OF PIEZOMETERS AT EXPERIMENTAL SITE AKBARPUR (RD, 102·000) ALONG SIRHIND CANAL IN SANGRUR DISTRICT OF PUNJAB

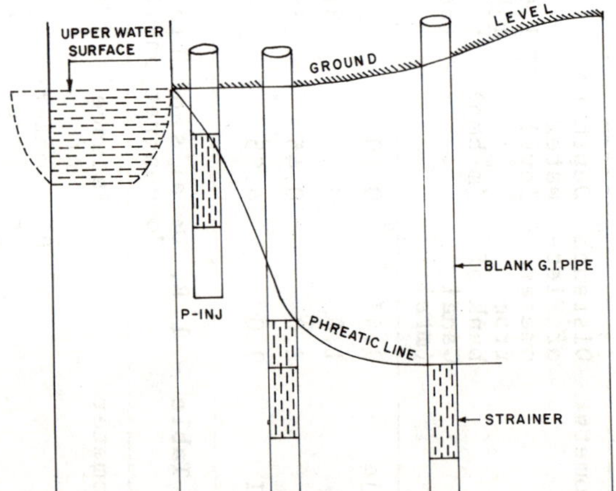

LAY OUT OF PIEZOMETERS AT HARIPUR (RD, 10·500) ALONG AMLAH MINOR CANAL IN PATIALA DISTRICT OF PUNJAB

F = Cross-section of the measuring volume perpendicular to the direction of the undisturbed ground water flows,

t = time, interval in which tracer concentration changes from Co to C,

Co = Tracer concentration at the beginning of measurement,

C = Tracer concentration at the end of measurement

α = Correctin factor which accounts for distoration of the flow lines due to the presence of borehole.

The value of α - factor can be estimated from the ratio

= Qb/Qf

Where Qb is horizontal flow rate in the borehole
Qf is the flow rate at the same cross-section in the aquifer.

When the borehole is constructed with the strainer, gravel pack with percentage of openings in the filter more than 10%, the value of α becomes 2.0. If the probe (detector) has a radius (r0) and response of detector is linear, then according to Borowczyk, Mairhofer, Zuber, Kots and Krolikowski (1964), the equation (1) becomes

$$Vf = - \frac{II\ (rl^2 - ro^2)}{\alpha\, dl \quad t} \ In \quad C/Co \qquad \text{-------(2)}$$

where dl - is inner diameter of a filter tube.

The decrease of radiotracer activity was determined by plotting time vs.cps on semilog paper (Fig.3) and then the filtration velocity was computed. It was observed that the concentration of radiotracer decreases with time due to the horizontal flow through the well. The highest filtration velocity was observed in medium sand whereas in clayey sand much lower value was recorded. Coefficient of permeability 'K' was calculated using Darcy's formula.

$$K = \frac{Vf}{I} \qquad \text{\_\_\_\_\_\_\_\_(3)}$$

Coefficient of permeability was estimated for each depth of filtration velocity measurements in all piezometer. An average coefficient of permeability was computed taking into account of the average of filtration velocity from

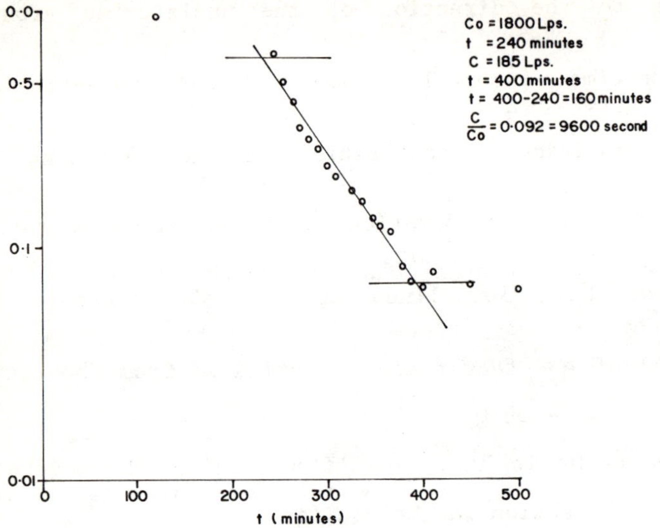

RELATIONSHIP OF C/Co AND TIME OF RADIOISOTOPE AFTER INJECTION IN THE PIEZOMETER AT AMLAH MINOR CANAL IN PATIALA DISTRICT OF PUNJAB

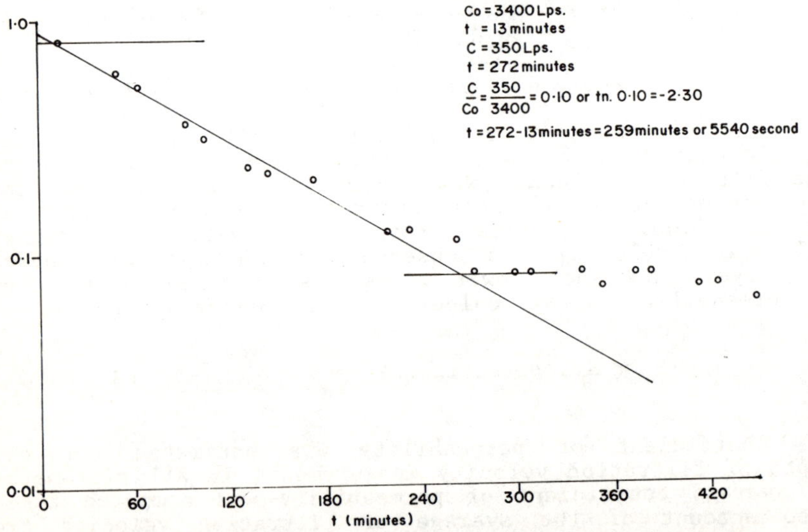

RELATIONSHIP OF C/Co AND TIME OF RADIOISOTOPE AFTER INJECTION IN THE PIEZOMETER AT SIRHIND CANAL AT AKBARPUR IN PATIALA DISTRICT OF PUNJAB

the test piezometer as well as hydraulic gradient (I) from the water level contour map. Seepage loss of water from canals was calculated from the following formula (Uppal & Singh):

$$S = K.\ P\ (Wa)\ .\ I \text{ ------------(4)}$$

where S is Seepage loss
P(Wa) is wetted area
K is coefficient of permeability
I is hydraulic gradient

The data computed from the equation (4) depict value of seepage loss of water from Sirhind canal i.e. S = 2.37 Cumecs/M Squ. metre of wetted area indicating thereby the characteristic of sandy loam, whereas the value of S = 0.0425 Cumecs/M Sq. metres of wetted area of Amloh minor canal at Haripur of Punjab denotes the characteristic of clayey sand. It is clear that the seepage from canal takes place in the form of sub-surface streams flowing at different gradients, therefore, the measurement of radiotracer dilution velocity may be affected by the length of the water column in the piezometer.

From the above findings, it has been observed that major percentage of losses occurs in water courses and field channels through seepage. The seepage losses in water courses are much greater since the ratio of the wetted area to the volume of water carried is higher. They are mostly contructed in filled soils and poorly maintained. In unlined earth channels, the water losses may vary depending upon the type of the soil through which the channel traverses.

As a result of seepage from unlined canals of the irrigation system many areas in central and western part of the Ghaggar basin have become water logged either permanently or seasonly. The lining of canals is a must, if extensive water logging is to be prevented in future. The seepage losses from the irrigation water conveyance and distribution system being the biggest contributor to ground water recharge, reduction of these losses by means of lining of different components of the system is expected to provide great relief. This will not only assist in control of water table, but also provide additional fresh water supplies for undertaking reclamation programme in saline areas. Realising the importance of this measure, the Govt. have already undertaken the lining work in a big way.

It is clear that the irrigation water is a costly input to a farmer, whether it is received from a public canal outlet or a tubewell or a dugwell. Water lost in channels is money lost. Water losses by seepage can be reduced by lining the channels.

Table - 2. Details of Canal data and estimated value of filtration velocity and seepage loss of water from canals.

		Sirhind Canal	Amlah minor canal
1.	Depth of Canal	= 6.45'	= 2.63'
2.	Width of Canal	= 60'	= 15'
3.	Slope of Canal	= 2.236'	= 1.45'
4.	Wetted perimeter at RD 102	= 74.42' or 22.68m^2	= 5.73 m^2
5.	Wetted perimeter between RD 67-103,000	= 2.6x10^5 m^2	= 213.35 m^3
6.	Length of wetted canal between RD 86-103,000	= 5181.6 m	= 1222.5 m^3
7.	Wetted area	= 117518.7 m^3	= 6 to 7 causec
8.	Discharge of Canal during the experimental period.	= 1090 cusec	= Clayey sand
9.	Type of Soil	= Sandy loam	= 0.802 m/day
10.	Filtration velocity	= 0.571 m/day	= 0.425 cumecs/ M Sq.mt.of wetted area.
11.	Seepage loss	= 2.37 Cumecs/M Sq. mt. of wetted area.	

Lining irrigation channels is expensive. It may cost to the farmer 2 to 10 times as much as constructing the unlined channels. In view of the high cost of lining channels, decision to use lining needs full justification. Which may be based on tangible benefits alone or both tangible as well as intangible. To show the economic feasibility, the present worth of the annual tangible and intangible benefits resulting from the installation of lining must be equal to or greater than the cost of the lining.

The radiotracer technique is best suited to homogeneous and isotropic formation and is also more economical because of limited time and cost of the studies than any other methods. This approach is direct and independent for the determination of aquifer parameters i.e. coefficient of permeability with the help of filtration velocity of ground water. Seepage loss of water from canals was determined with the help of filtration velocity, hydraulic gradient and wetted area which was calculated in the field. The radiotracer technique is relatively simple to use and can be applied for measurement of seepage losses of water from canal as well as to locate leakages in oil pipe line etc.

ACKNOWLEDGEMENT

The authors place their sincere gratitude to the Chief Hydrogeologist and member and Director, Central Ground Water Board, Eastern Region, Calcutta for their kind permission to utilise the departmental data and provide facility in preparation of this paper.

REFERENCES

Borowizyk, M., Mairhofer, J. , Zuber, A., Etal., 1964 Laboratory investigations on the determination of filtration Velocity by means of radioisotopes; Rep. No. 541/VI, Warsaw (NOV, 1968)>

Krishnamurthy, K., & Rao, S.M. 1969. Theory and experiments in canal seepage estimation using radioisotopes; Jour. of Hydrology. North Hall and Publishing Co.

Robinson, A.R. & Rahwar, C. 1957. Measurement of canal seepage. Trans. Am. Geo.Sec. of Civil Engineers, Vol. 122, Paper No. 2865.

Vaidhianathan, et.al.,1937. A new method of determining seepage from canals in areas of high water table : Research Pub. No. 7. Vol. V, Punjab Research Institute Lahore.

USE OF SINGLE WELL DILUTION TECHNIQUE FOR SEEPAGE MEASUREMENTS FROM CANALS USING RADIO-ACTIVE TRACER

A.N.BHARGAVA
RESEARCH OFFICER

P.R.SINGH
ASSTT.RES.OFFICER

S.S.CHHABRA
RESEARCH SUPERVISOR

U.P.IRRIGATION RESEARCH INSTITUTE, ROORKEE

ABSTRACT

Various parameters related to canal and borehole mainly the height of water column in borehole at different distances from canal bank play an important role for determining the sub-surface flow of water from canals. In the past, the earlier investigators have not considered this important factor while evaluating the seepage velocity by single well dilution technique. Keeping this aspect in mind, a field study has been carried out by Isotope Group, U.P.I.R.I., Roorkee from various earthen canals of Uttar Pradesh using radio-active tracer. The experiments have been carried out at various sections of canals and the variation of the tracer dilution velocity with various parameters related to canal and borehole has been studied. Knowing the seepage velocity, the seepage loss from canals by single well dilution technique has been estimated by the mathematical expression, $q = 2\ V_s \cdot d.\ \Theta \cdot \text{Cosec}\ \Theta$. The seepage losses at test sites have been found of the order of 1.83 and 2.16 $m^3/sec/10^6\ m^2$ of wetted surface for Upper Ganga Canal and Deoband Canal respectively.

0.0 INTRODUCTION

0.1 Canals are constructed and introduced in the areas to provide irrigation to crops for meeting the need of food for growing population[1] No doubt, these canals help in increasing the agricultural production but on the other hand excessive seepage from canals results not only in heavy loss of irrigation water which can otherwise be used profitably for irrigation purposes, but it creats water logging in the vicinity of canals.

0.2 Of the various factors which influence seepage loss from a canal, the most important are the boundary conditions of the flow domain and permeability of the material comprising the flow domain[1] Only after a correct assessment of the coefficient of permeability and boundary conditions, the seepage losses can be estimated either by numerical method or analytical technique. Needless to say that it is very difficult to determining the insitu coefficient of permeability of the porous medium. In order to avoid the difficult task of estimating the coefficient of permeability and the prevailing sub-surface boundary condition, the experimental techniques like ponding method and inflow-outflow method have been used for a correct estimation of seepage losses. Nuclear technique using suitable tracer has been used in recent years to estimate seepage loss from canals because of its easy operation in comparison to other experimental methods.

0.3 In nuclear technique, a single well dilution method using radio-active tracer is generally used to determine the seepage velocity from running canals. Various parameters related to canal and borehole mainly the height of water column in borehole at different distances from canal bank play an important role for determining the sub-surface flow of water from canals.

0.4 Keeping this aspect in mind, a field study has been carried out by Isotope Group of U.P.I.R.I., Roorkee at several sections of various canals of Uttar Pradesh using tritium as radio-active tracer to study the variation of the tracer dilution velocity with various parameters related to canal and borehole. A relation among the height of the water column in borehole, tritium dilution velocity, distance of borehole from the centre line of canal and the length of the curved surface of canal is established. Thereafter, a seepage formula viz. $q = 2.V_s. d. \Theta \cdot \text{Cosec}\ \Theta$ has been developed for correct assessment of seepage loss from unlined and lined canals. This formula is based on seepage velocity, gradient of phreatic line and distance of borehole from central line of canal. The losses estimated by the formula developed at U.P.I.R.I., Roorkee is found to agree with seepage losses measured by other experimental methods.

The details of derivation of the formula, studies carried out and the results are presented in the present paper.

1.0 SINGLE WELL DILUTION TECHNIQUE

1.1 Single well dilution technique is generally used for determining the seepage velocity from canals. In this technique, the boreholes are first made on the bank of the canal so as to extent to a depth well below the groundwater table and thus cover most of the probable flow zone. Tritiated water (HTO) is then injected into the groundwater contained in the borehole and the mixture is stirred thoroughly to make it uniform. To know the tracer dilution with time radio-active water samples are collected subsequently from the injection hole at different time intervals. The analysis of these water samples for their radio-active concentration enables the determination of seepage velocity and there-by the seepage discharge from the canal.

1.2 The tracer dilution is based on the principle that as with time fresh water due to seepage from the canal enters the well, and equal volume of concentrated water leaves the well uniformly in the direction of subsoil flow. Thus the water in the well gets diluted and the tracer concentration in the well goes on decreasing with time.

1.3 DERIVATION OF THE FORMULA:

The following assumptions are made while deriving the formula for estimating the seepage velocity from the canals:

(i) The canal is treated as line source.

(ii) The flow lines are radial and equipotential lines are circular.
(iii) The porous medium at any reach has constant permeability which may vary from reach to reach.
(iv) The plane of flow is normal to the canal axis.

Let the variation of concentration of the tracer with time is ıeasured in a vertical borehole which is located at a distance 'd' 'rom the centre of the canal and the diameter of the borehole is D .nd the borehole is taken to a depth h below the water table. The 'olume of water, V in the bore hole is equal to $\pi . D^2 . h/4$. Let : is the concentration of the tracer at any time t.

Assuming that the water entering to the borehole is free from he tracer used, the mass balance for a period from t to $t+\Delta t$ can be 'ritten as:

Total inflow of tracer to the borehole in time Δt = total utflow of tracer from the borehole in time Δt + change in oncentration of the solution in borehole x volume of solution in orehole.

If q is the total inflow of water into the borehole in unit ime, the mass balance equation can be written as:

$$0 = q C \Delta t + V . \Delta C$$

r

$$\frac{V}{q} \frac{dc}{C} = -dt \qquad \ldots(1)$$

Let the time is measured from the instant the tracer is added to he water in the borehole. Integrating t and applying the initial ondition that at t = 0, C = Co , the following expression can be erived:

$$t = \frac{V}{q} . \ln \frac{C_o}{C} \qquad \ldots(2)$$

Assuming that the distortions in the flow pattern due to resence of the borehole are occuring in plane normal to the borehole xis and that there is no distortion below the bottom of the orehole, the total flow rate q entering to the borehole can be pproximated to (Mc-Whorter etal, 1975)

$$q = \phi . D . V_s . h \qquad \ldots(3)$$

n which, Vs = Seepage velocity in a direction normal to the canal axis at large distance from the bore hole.
ϕ = a factor for hydrodynamic field deformation.

The above relation has been derived on the assumptions that ıter flows with a uniform gradient at large distance from the orehole and the aquifer has infinite areal extent. Replacing the value of q and V in eq. (2).

$$\ln \frac{C_o}{C} = \frac{4.\phi.V_s.t}{\pi D} \qquad ...(4)$$

$$V_s = \frac{\pi.D}{4.\phi.t} \ln \frac{C_o}{C} \qquad ...(5)$$

Thus when logarithm of concentration is plotted at various times, the resulting graph is a straight line.

Knowing ln C_o/C with time from C-t curve, the seepage velocity, Vs can be determined by eq.(5).

1.4 Once the seepage velocity is determined the seepage loss from the canal can be derived as follows:

An idealised configuration of stream lines is shown in Fig.1. Let θ is the angle made by the tangent to the phreatic line with the vertical at the borehole. θ can be approximately known by knowing positions of water table at two near - by points. In the absence of the borehole the water will move in the radial direction. Since Vs is the component of velocity along horizontal direction, the superficial velocity in radial direction is given by Vs . Cosec θ. The total segment circular area = 2 d . θ. Hence the total seepage loss from the canal per unit length is given by:

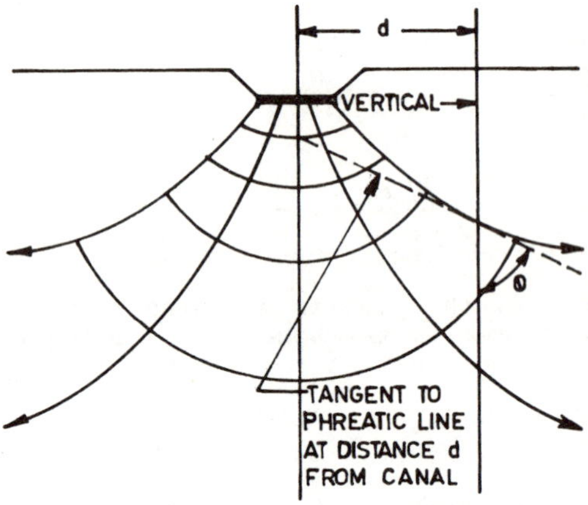

FIG.1 -IDEALISED FLOW CONFIGURATION

$$\begin{aligned} q &= \text{Velocity x Area} \\ &= V_s . \text{Cosec } \theta \times 2.d. \theta. \\ &= 2.V_s .d. \theta. \text{ Cosec } \theta. \qquad ...(6) \end{aligned}$$

2.0 EXPERIMENTAL DETAILS AND PROCEDURE

2.1 Several experiments have been carried out to estimate seepage loss from various canals of U.P. viz. Upper Ganga Canal, Lower Ganga Canal, Deoband Branch, Jhansi Canals, Banda Canals, Lalitpur and Jamni Canals using single well dilution technique.

Only two representative cases are reported in the present paper. The first experiment was carried out along Upper Ganga Canal at Km. 42.40 while second experiment was carried out along Deoband canal at Km. 6.4 in Distt. Saharanpur. The location of test sites is shown in Fig. 2.

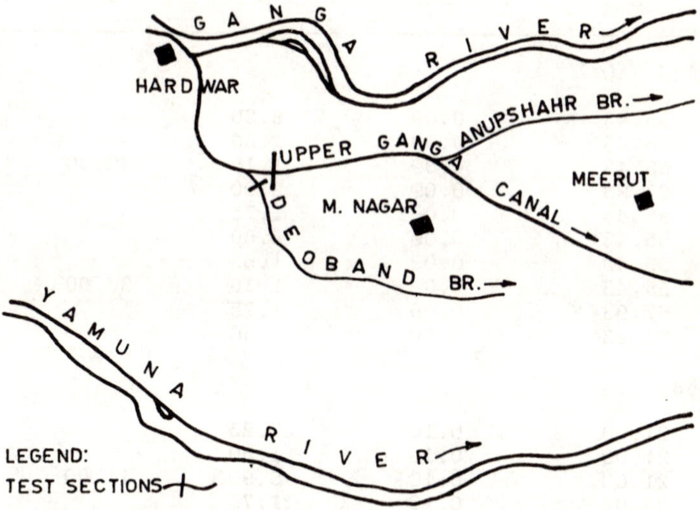

FIG. 2 - LOCATION OF TEST SECTIONS

Five to six boreholes each of 9 to 10 cm dia were drilled by hand auger on each bank of both the canals. The details of depth, distance of borehole and height of the water column are given in Table I. The boreholes were cased by the copper strainer of appropriate diameter in the lower part upto upper surface of water while blind pipe were used in the upper part upto the ground surface. The bore holes were dug upto different depths to know the effect of the height of water column in boreholes. The lay-out of the experimental sites along with phreatic lines are shown in Fig. 3 and 4.

TABLE - I

Details of Borehole Data

Borehole No.	Distance of borehole from centre line of canal. (d) m	Diameter of borehole (D) m	Height of water column in borehole (h) m	Length of one side wetted curved surface of canal (L) m	Angle (θ) radian
UPPER GANGA CANAL					
LA	55.43	0.09	2.20		
LA_1	55.43	0.09	1.50		
LA_2	55.43	0.09	1.15	36.00	1.5533
LB	67.93	0.09	2.00		
LC	85.43	0.09	1.75		
RA	55.43	0.09	2.00		
RA_1	55.43	0.09	1.65		
RA_2	55.43	0.09	1.10	36.00	1.5533
RB	67.93	0.09	1.75		
RC	85.43	0.09	1.50		
DEOBAND CANAL					
LN	21.00	0.10	2.25		
LN_1	21.00	0.10	1.50		
LN_2	21.00	0.10	0.90	11.00	1.5309
LO	28.00	0.10	1.75		
RN	20.00	0.10	2.50		
RN_1	20.00	0.10	2.00		
RN_2	20.00	0.10	1.25	11.00	1.5533
RO	29.00	0.10	2.00		

The boreholes LA,RA,LN and RN were used for single well dilution technique to estimate seepage loss while other boreholes were made for plotting phreatic line.

2.2 Tritiated water (HTO) of specific activity 2 μci/cc was injected in the boreholes in different amounts according to the volume of water in boreholes. After injecting tritium in boreholes the entire water column was thoroughly stirred to make the concentration homogeneous. Water samples were collected in the leak proof screw capped plastic containers from each borehole at different times and brought to Isotope Laboratory of U.P.I.R.I., Roorkee for tritium activity measurements. In order to know the type of soil at the test sites, the soil samples were also collected from the borehole.The analysis indicates that the type of soil is silty clay at both the test sites.

The tritium activity in each tritiated water sample was measured in the form of counts/minute for each bore hole with the help of Automatic Liquid Scintillation Counting System (LSS-34, ECIL Model).

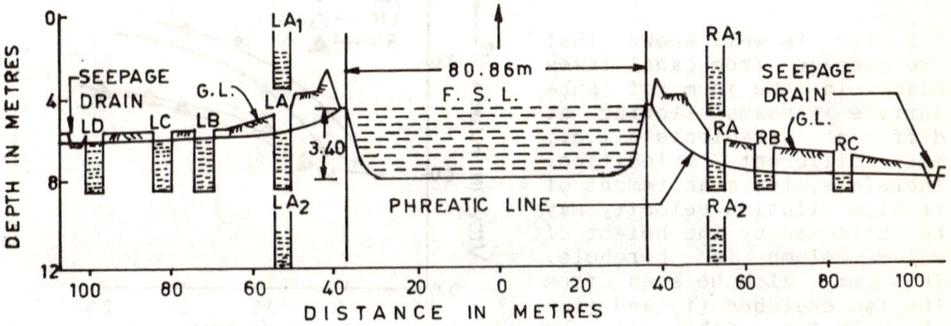

FIG. 3 - CROSS SECTION OF GANGA CANAL AT Km. 42.4

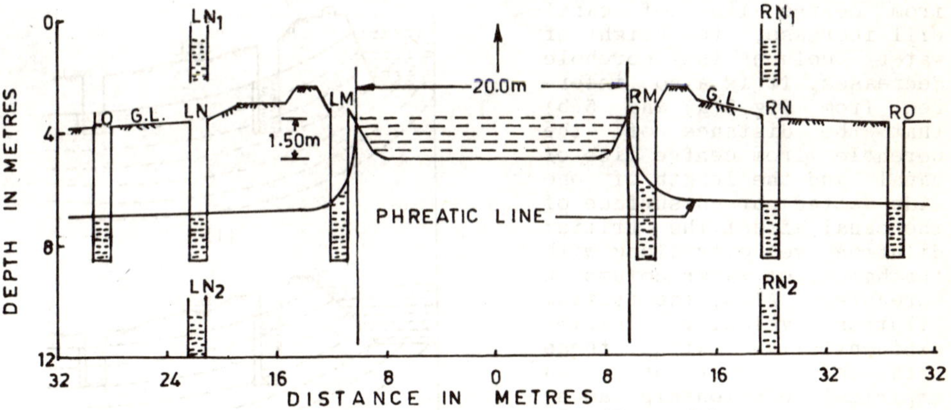

FIG.4-CROSS SECTION OF DEOBAND BRANCH AT Km. 6.4

3.0 DISCUSSION

3.1 The tritium dilution velocity and seepage velocity for each borehole of both the test sites evaluated by eq. (5) and (7) are summarised in table II. The graph between tritium dilution velocity and height of water column in boreholes is shown in Fig. 5(a). The perusal of Fig.5 (a) indicates that tritium dilution velocity increases with the increase of height of water column in boreholes. The shape of the graphs are similar to the curve of the charging of a condenser.

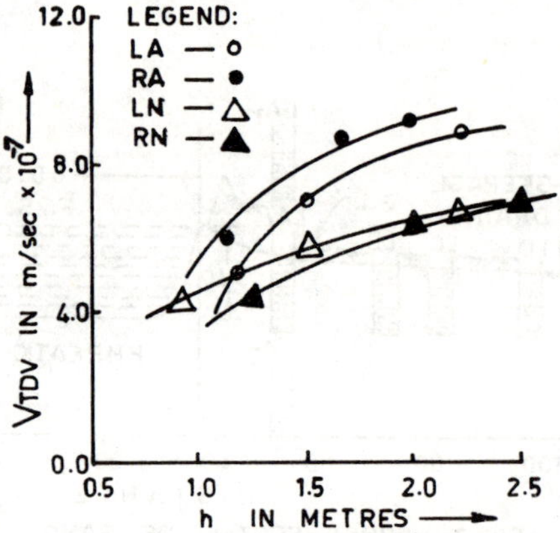

FIG.5 (a)-VARIATION OF V_{TDV} WITH h

3.2 It is well known that the seepage from canal takes place in the form of sub-surface streams flowing at different gradients i.e., with different velocities. Therefore, the measurement of tritium dilution velocity may be affected by the height of water column in borehole. This may also be seen from the two sketches (i) and (ii) shown in Fig. 5(b). If the length of the wetted curved surface of the canal will increase, the height of water column increases, and if the distance of the bore-hole from centre line of canal will increase, the height of water column in borehole decreases. It is also inferred from Fig.5(a) and 5(b) that the distance of the borehole from centre line of canal and the length of one side wetted curved surface of the canal affect the tritium dilution velocity along with the height of water column in boreholes. Also, the tritium dilution velocity varies exponentially with these parameters. Therefore, an empirical relationship among the height of water column in borehole and various other parameters of canal along with tritium dilution velocity is established and is given by:

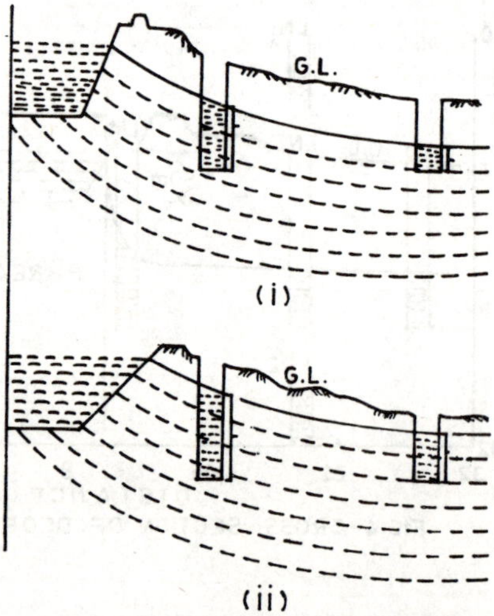

FIG. 5(b)- VARIATION OF h, d AND L.

$$V_{TDV} = V_{OTDV}\left[1-e^{-h.d/L}\right] \quad ...(7)$$

Where V_{TDV} , is the observed tritium dilution velocity.
V_{OTDV} , is the general tritium dilution velocity averaged over the different velocities of the existing water streams in borehole and is equal to Darcy's velocity.
h , is the height of the water column in borehole.
d , is the distance of borehole from the centre line of canal.
L ,is the length of one side wetted curved surface of the canal.

Table - II

Tritium Dilution Velocity , Seepage Velocity and Seepage Loss for Test Sites

Borehole No.	Tritium dilution velocity (V_{TDV}) $\times 10^{-7}$ m/sec	Seepage velocity (Vs) $\times 10^{-7}$	Average seepage velocity (Vs) m/sec	Seepage loss (q) m^3/sec/10^6 m^2	Average Loss (q)
UPPER GANGA CANAL					
LA	8.75	9.05			
LA_1	7.00	7.77			
LA_2	5.00	6.03	7.62	1.69	
LB	6.50				
LC	5.30				1.83
RA	9.20	9.64			
RA_1	8.80	9.42			
RA_2	6.20	7.60	8.89	1.97	
RB	7.00				
DEOBAND CANAL					
LN	6.80	7.60			
LN_1	5.80	7.83			
LN_2	4.40	7.41	7.61	2.28	
LO	6.00				
RN	7.00	7.37			2.16
RN_1	6.40	7.23			
RN_2	4.50	6.31	6.97	2.04	
RO	6.10				

3.3 Besides the above factors, the accuracy of the determination of seepage velocity by single well dilution technique depends entirely on the value of factor ϕ. The disturbances around the bore-hole during drilling will modify the permeability of the surrounding soil and consequently the factor ϕ. The maximum value of ϕ may be equal to 2 as mentioned by Mewhorter, D.B. Moreover, Gaspar, (1969) has reported that with high permeabilities of experimental soil and apparent velocity of the order of about 10^{-3} cm/sec, the use of a filter bed around the borehole should reduce the factor ϕ in the range from 0.5 to 0.75. The factor ϕ has also been determined in the field at both the test sites and ranges from 0.8 to 1.2 for silty clay soil. Thus, the value of factor ϕ may be taken as 1 while calculating seepage velocity for estimating seepage loss from canals having silty clay soil.

4.0 PRECAUTIONS AGAINST HEALTH HAZARDS

4.1 During the course of the study the following precautions were taken against health hazards:

(i) The radio-tracer was transported to the injection site in glass bottles packed in appropriate shielding complying with relevant Radiological Safety Regulations.

(ii) The people who collected and analysed the radio-active water samples were cautioned and trained properly. They were provided aprons and hand gloves to protect their clothes and hands.

(iii) The area surrounding the injection sites was declared restricted for all other than the radiation workers and was supervised properly. The warning signs were also displayed at proper places.

(iv) The quantity of radio-active tracer used was well within safe limits from radiological safety considerations and was of the order of 2.0μci/cc.

(v) The contaminated articles after experiments were brought back after being packed properly in polythene sheets and bags.

5.0 CONCLUSIONS.

5.1 The seepage losses from Upper Ganga Canal and Deoband Canal as observed by eq.(5), (6) and (7) are summarised in table III. These losses when worked out by Numerov's theoretical approach and other methods compare fairly well with the results obtained by using the developed relationship.

Table-III

Results of Seepage Losses

Sl.No.	Name of canal	Chainage	Seepage loss, $m^3/sec/10^6$ m^2	
			Single well Dilution Technique	Numerov's Theoretical Approach
1.	Upper Ganga Canal	42.40	1.83	1.55
2.	Deoband Canal	6.40	2.16	1.96

5.2 The height of water column in borehole, the distance of the borehole from central line of canal and length of one side wetted curved surface of the canal affect the seepage velocity. Thus, the emp rical relationship as developed can be used to get comparatively reliable results of seepage velocity from canals.

5.3 Single well dilution technique based on average seepage velocity and gradient of phreatic line appears to be reasonable for computing seepage losses from irrigation channels.

5.4 The experiments carried out on different canal systems in Uttar Pradesh indicates that the seepage loss varies from 0.50 to 7.00 $m^3/sec/10^6$ m^2 of wetted surface.

REFERENCES

1. Bhargava, D.N., Raja, R.K., Kumar, A. and Chhabra, S.S., 1983. Estimation of seepage losses from an unlined channel, a field study by nuclear technique. Proceeding Vol.II - Hydraulics, Fiftieth Annual Research and Development Session, CBIP, p. 45-60.

2. Garg, J.M., Kumar,D., Kumar, A. and Chhabra, S.S. 1981. Measurement of seepage losses from ganga canal by radio -tracer technique.Proceedings Vol. II-Hydraulics, Fortyninth Annual Research and Development Session, CBIP, p.59-81.

3. Gaspar, E., and Oncescu, M., 1969. Validity conditions of point dilution method, water in the unsaturated zone, Proceeding of the Wageingen Symposium IASH/AISH-Unesco, p.256-263.

4. McWhorter, D.B., R.K. Skogerboe, and G.V. Skogerboe, 1975. Water quality control in mine spoils. Upeer Colorado River Basin, Environmental Protection Technology Series, Environmental Protection Agency, Pub. 670.

5. Bhargava, A.N., Singh, P.R., and Chhabra, S.S., 1988. Development of tracer technique for seepage loss measurement from lined and unlined canals. Proceeding of Seminar of Canal Lining, I.R.D.O., Roorkee.

LIST OF CAPTIONS

D	=	Diameter of borehole
h	=	Height of water column in borehole
V	=	Volume of water in borehole
Co	=	Concentration at time, t = 0
C	=	Concentration at any time t
t	=	Time of dilution
q	=	Total inflow of water
Ø	=	Distortion factor due to presence of borehole
Vs	=	Seepage velocity
d	=	Distance of borehole from central line of canal
Θ	=	Angle which the borehole makes with the phreatic line
L	=	Length of one side wetted curved surface of canal
V_{TDV}	=	Tritium dilution velocity
V_{OTDV}	=	Observed seepage velocity by the developed formula
μCi	=	Micro Curie

AN APPROACH IN DETERMINATION OF AQUIFER GEOMETRY AND ESTIMATION OF AQUIFER PARAMETERS OF AN AREALLY EXTENSIVE UPPER KAMTHI SANDSTONE AQUIFER IN PARTS OF GODAVARI VALLEY, ANDHRA PRADESH

M. Veeranna and K.V.J.R. Krupanidhi
Central Ground Water Board

INTRODUCTION

Northern parts of Warangal and Karimnagar districts of Andhra Pradesh form a part of Pranhita-Godavari Valley. The area is underlain by the Gondwana Supergroup of sedimentary rocks. Gondwanas are divisible into Lower and Upper Groups. The Lower Group comprises Talchir, Barakar, Barren Measures and Kamthi Series. The Upper Group consists of Maleri, Kota and Chikiala Series. They are mainly sandstones with subordinate clays, siltstones and shales. The general trend of Gondwana formations is in NW-SE with dip ranging from 8 to 12° in NE direction. The dip of beds as determined from the bed elevation from borehole data is about 9°. The variations in strike and dip directions and changes in magnitude are commonly associated with numerous faults traversing the area. The major fault directions are NW-SE and NE-SW. The syn-depositional and post-depositional faulting activity in Gondwana grabben in NW-SE direction may be responsible for the greater width and consequent vast areal extent of Kamthi Series and repetitive occurrence of older formations towards centre of the valley.

METHODOLOGY

For determination of aquifer geometry and estimation of aquifer parameters of Gondwana aquifers of Godavari Valley, Central Ground Water Board has carried out drilling under, ground water exploration and first approximation resource evaluation studies in the lower reaches of Godavari Valley. The sites were selected after systematic hydrogeological surveys and deep resistivity spot sounding studies. Drilling, down to a depth of 300 m bgl was undertaken at 18 sites of which 5 sites were located in Upper Kamthi formation. After pilot hole drilling and electrical logging, aquifer zones were deciphered and zone tests were conducted. Exploratory wells were constructed tapping the aquifer horizons occuring beyond the depth of 60 m bgl. The wells were tested and aquifer parameters were determined.

KOYYURU WELL FIELD TEST

Coarse to gritty sandstones of nearly 170 m thickness have been encountered beyond a depth of 67 m bgl near Koyyuru village village (18°36'15":79°50'22"; 56 N/14) in Karimnagar district. A tubewell constructed tapping this aquifer has become an artesian flowing well with free-flow of 33 m^3/hr at a discharging head of 2.10 m agl (139.4 m AMSL) while the piezometric head was 7.54 m agl (144.848 m AMSL). Two observation wells OW-I and OW-II were constructed at distances of 60 and 120 m from this tubewell tapping the same aquifer horizons. Manometers of transparent rubber tubing were erected on observation wells and the fall and rise in piezometric surfaces were directly recorded during tests. A long duration pumping test of 4320 minutes duration was carried out starting from 31-12-76 at a pumping discharge of 2915 lpm. The head losses at test well and near and farther observation wells were 22.71, 8.99 and 6.99 m respectively. The observation wells data plots on double logarithmic paper showed two negative boundaries. The initial drawdown data was fitted with Theis type curve and the estimated transmissivity values of aquifer are found to be 668 and 776 m^2/d for OW-I and OW-II data respectively. The estimated storage coefficient values for the aquifer are 5.0×10^{-4} and 3.3×10^{-4}. Distance-drawdown graphs of Koyyuru well field area for different periods of pumping were constructed. The radius of influence after 1000 minutes of pumping at a discharge of 2915 lpm was determined as 10 km. Aquifer parameters were also determined at Koyyuru through the free-flow test method of Jacob and Lohman. The variable discharge data at constant drawdown after opening sluice valve and recovery of water level after closure of sluice valve were recorded. The transmissivity of the aquifer using recovery data is estimated to be 672 m^2/d.

YEDLAPALLI WELL FIELD TESTS

An exploratory borehole was drilled at Yedlapalli village (18°34'56":79°51'55"; 56 N/14) at a distance of 3.7 km from Koyyuru towards south-east. The same Upper Kamthi aquifer horizons were encountered below a depth of 133 m bgl. An observation well was constructed at a distance of 120 m tapping the same granular horizons. A long duration pumping test was conducted at Yedlapalli starting from 6-6-78 for a period of 5220 minutes at a discharge of 1627 lpm. Though the Yedlapali well did not turn out to be a flowing well its potentiometric surface was found to be 2.905 m bgl (144.645 m AMSL) which was above the piezometric head of Koyyuru well (142.00 m AMSL). The ground elevations at Yedlapaili and Koyyuru are 147.55 and 138.025 m AMSL respectively and the general land slope is from Yedlapalli to Koyyuru. The fall in potentiometric level from Yedlapalli to Koyyuru also indicates the direction of ground water flow. The Yedlapalli observation well data was analysed with Jacob's method and the transmissivity and storage coefficient values were estimated to be 291 m^2/d and 2.3×10^{-4} respectively. The test data did not show any boundary effects.

The piezometric surface at Koyyuru OW-I was observed during tests at Yedlapalli. The fall in water level was observed after 165 minutes of pumping at Yedlapalli. The total fall in the potentiometric surface at Koyyuru was 0.41 m at the end of pumping. There were no wells tapping this aquifer and the drawdown effect was entirely due to pumpage at Yedlapalli. The drawdown data showed two trends indicating two negative boundaries as obtained during Koyyuru well tests. The initial drawdown data was analysed with Jacob's method. The transmissivity and storage coefficient values of aquifer at Koyyuru were estimated to be 718 m^2/d and 1.0×10^{-4} respectively which were analogous with the values obtained from the tests at Koyyuru well field.

CONCLUSIONS

The Gondwana basin of Godavari Valley comprises sandstones, clays and siltstones striking NW-SE direction with gentle dip towards NE. The formations are traversed by numerous faults. It has become an intricate problem in deciphering the depth, thickness and extent of aquifers in view of the sloping and disturbed strata. The lithological and electrical logs of Yedlapalli and Koyyuru boreholes have indicated occurrence and extension of the potential confined Upper Kamthi aquifer. The observations at Koyyuru while pumping at Yedlapalli have enabled to know the presence of hydraulic continuity of this artesian aquifer though it is associated with structural disturbances. It has also helped in deciphering the location of negative boundaries beyond the Koyyuru well field and not in between Koyyuru and Yedlapalli well fields. In such situations a planned construction of small diameter observation wells at distance of 4 to 5 km would be of immense help in understanding the aquifer geometry and estimation of aquifer parameters in the region and thus save huge construction and testing costs of conventional well fields.

SUBMERSIBLE PUMP TESTS CONDUCTED ON BORES DRILLED BY D.T.H. COMPRESSOR RIGS AND THE COMPARISON OF TWO DISCHARGES

A.J. MERH *K.H. CHATROLA*

Gujarat Water Resources Development Corporation Ltd.
Ground Water Division, Opp. Jam Tower
Rajkot 360 001 - Gujarat - India

INTRODUCTION :

The shortage of drinking water in 1986-87 and 1987-88 in Gujarat State posed a challenge to hydrogeologists in locating suitable site of a bore hole in basaltic terrain. The Gujarat Water Resources Development Corporation Ltd. pressed all their field parties to do the ground water hunting in hard rock. Remote sensing, geohydrological survey, geophysical surveys interpretation, discussion and finally the pointing of site were done in great haste.

PROCEDURE :

Drilling was done by D.T.H. compressor rigs. Drilling sites of 150mm dia bores were attended by geologist and lithologs were prepared. Discharges during drilling were measured by 90° V notch method. A target of minimum 1 lps discharge suitable for submersible pump was kept before the geologist.

For the bores which had clear discharges of more than 1 lps, pumps were recommended. But for the bores which yielded less discharges during drilling, the lithologs were checked. Lithologs did show weathered zones and thus test pumping by submersible pump was initiated to compare the two discharges.

With great difficulty and persuation the Mechanical Engineer agreed to arrange for a submersible pump and generator unit. Firstly one 9 HP submersible pump was used for test pumping. The generator unit was assembled on a small truck and taken to the site. Five pump tests were conducted with 9 HP pump, and then 16 test pumpings were done with 5 HP submersible pump. One sluice valve was used on delivery pipe to throttle discharge rate. In most of the tests the throttle valve was first set on 1 lps discharge and the pump was run. The time gap between drilling and test pumping was about 2 months and therefore decrease in discharges were anticipated. Please see statement No. 1 for details.

OBSERVATIONS :

Comparison between two discharges show that out of 21 cases, 13 cases show increase in discharges, when the bore holes were pumped. In four cases there was no change and in 4 cases there was decrease in discharge. The decrease may be due to the time gap between drilling and test pumping. On the basis of these tests 6 cases prompted the use of submersible pumps, 5 deep well pumps were recommended and 6 hand pumps were suggested. In four cases the discharge obtained was very less for any use.

The above results indicate following points for consideration.

REMARKS :

1. Whenever discharge in compressor is less than 1 lps check litholog

and then do test pumping.

2. Be vigilant and accurate in identifying rock cutting while drilling.

3. Blasting by gelatine sticks can be tried to develop bore hole.

4. Fractured rock, calcite veins, highly weathered zone may cause air loss during drilling. Do test pumping.

5. Majority of lithologs show weathered zones. Correct locating the drilling point is important than a sorry figure. Prick on potential point only.

6. Repeated washing of bore hole while drilling is essential for developing.

ACKNOWLEDGEMENT :

The authors thank Shri G.D. Patel, Managing Director of Gujarat Water Resources Development Corporation Ltd., Gujarat. The authors also thank Shri S.C. Sharma, Chief Geohydrologist, Gujarat Water Supply and Sewarage Board for his guidance and encouragement.

STATEMENT No. 1 COMPARISON OF COMPRESSOR DISCHARGE TO SUBMERSIBLE PUMP IN BASALTIC COUNTRY

Sr. No.	Village	Taluka	Depth m	Comp. Dis. lps	Pump HP	Head m	Time min	Pump Dis. lps	Change	Install.
1	Khankotada	Kalavad	120.00	0.78	9	77.00	300	4.83	Increase	Sub. Pump
2	Jam-Vanthali	Jamnagar	170.00	0.78	9	97.00	150	4.33	Increase	Sub. Pump
3	N.Bhagedi	Kalavad	120.00	0.027	9	110.00	13	2.50	Increase	Dip W. Pump
4	Khandhera	Kalavad	132.62	0.38	9	97.00	30	1.66	Increase	Hand Pump
5	Theba	Jamnagar	152.40	0.38	9	70.10	150	0.58	Increase	Dip W. Pump
6	Dared	Jamnagar	60.96	0.027	5	39.62	285	2.75	Increase	Sub. Pump
7	Kanakpur	Kalyanpur	121.92	Dry	5	100.58	330	2.25	Increase	Sub. Pump
8	Bada	Jamnagar	100.58	0.83	5	94.50	400	1.20	Increase	Sub. Pump
9	Vinzalpur	Jamnagar	90.00	0.33	5	82.00	300	1.00	Increase	Sub. Pump
10	Kalyanpur	Kalyanpur	146.00	0.06	5	104.00	75	1.13	Increase	Dip W. Pump
11	Ishwaria	J.Jodhpur	100.58	Dry	5	97.54	200	0.50	Increase	Dip W. Pump
12	Masitiya	Jamnagar	120.00	0.38	5	103.63	210	0.33	No chang	Dip W. Pump
13	Chela	Jamnagar	100.58	Dry	5	97.54	60	0.33	Increase	Hand Pump
14	Shekhpat	Jamnagar	64.01	0.14	5	57.91	300	0.19	No change	Hand Pump
15	Vaniyagam	Jamnagar	119.78	0.38	5	91.44	300	0.30	Decrease	Hand Pump
16	Sapar	Jamnagar	60.96	0.38	5	43.00	300	0.28	Decrease	Hand Pump
17	Matva	Jamnagar	121.92	0.075	5	39.62	300	0.20	Increase	Hand Pump
18	Dhaturia	Kalyanpur	105.16	0.07	5	100.00	240	0.07	No change	No Recomm.
19.	Moti Khavadi	Jamnagar	60.96	0.075	5	34.00	150	0.055	Decrease	No Recomm.
20	Limbuda	Jodia	86.87	Dry	5	82.30	75	0.008	No change	No Recomm.
21	Lalparda	Jamnagar	90.00	0.16	5	70.00	7	Dry	Decrease	No Recomm.

AQUIFER PARAMETER ESTIMATION AND GROUNDWATER POTENTIALITY IN GRANITE-BASALT TERRAIN NEAR HYDERABAD, INDIA.

Ch.SUDARSANA RAJU AND M.MURALIDHAR
Geology Dept., Osmania University, Hyderabad-500 007, India.

The area under investigation is 200 sq.km to the southwest of Sadasivpet at the contact between Peninsular granite and Deccan basalts. It is about 70 kms to the northwest of Hyderabad on the Hyderabad-Bombay National Highway No.9. It falls in the Survey of India toposheet No.G/56 and lies between Long.77°50' - 77°57'E and Lat.17°37' - 17°38'N (Fig.1). The area is well-drained and most of the agricultural and domestic water demands are met by groundwater and there is still scope for further groundwater development.

Large diameter dug wells are the main sources for irrigation in the area. Well inventory data is collected from 80 wells. Pumping tests are conducted at 10 wells, 5 from each of the formations, at different locations (Fig.1). The pumping test data is analysed by using Papadopulos-Cooper (1967) (Fig.2) and Theis et al. (1963) recovery (Fig.3) methods to determine the aquifer characteristics. Slitcher's (Slitcher, 1906) method (Fig.4) is applied to estimate well characteristics (Table 1). Several others (Krupanidhi et al., 1973; Viswanathaiah et al., 1978) also used these methods.

Transmissivity (**T**) values determined by Papadopulos-Cooper method range from 12 to 35 m^2/day except in one case. At Momanpet the value is very high, 190 m^2/day. However, there is no marked difference between the T values of granites and baslats. Two of the tests conducted in close proximity in basaltic terrain have T values which are fairly close, 14 and 19 m^2/day. The T values decrease as the depth of the wells increased.

Storage coefficient (**S**) values in the area range from 0.002 to 0.15. As in the case of **T**, there is no marked difference or relationship in the values of **S** of the two formations.

Recovery data is collected and the T values are determined by Theis method. These values range from 266 to 992 m^2/day. They decrease as the depth of the wells increased. The values show that there is no significant difference in the **T** of the granitic area from that of the deccan trap area.

The values of **T** obtained by Theis recovery method are much higher than those obtained by Papadopulos-Cooper method using drawdown data. In granitic terrain, the minimum value of **T** obtained by Papadopulos-Cooper method is 12 (Well No.4) and the maximum is 35 m^2/day (Well No.1), whereas, in basalts the minimum and maximum values are 14 m^2/day (Well No.7) and

Fig. 1. Map showing the location and sites of pumping tests of the area.

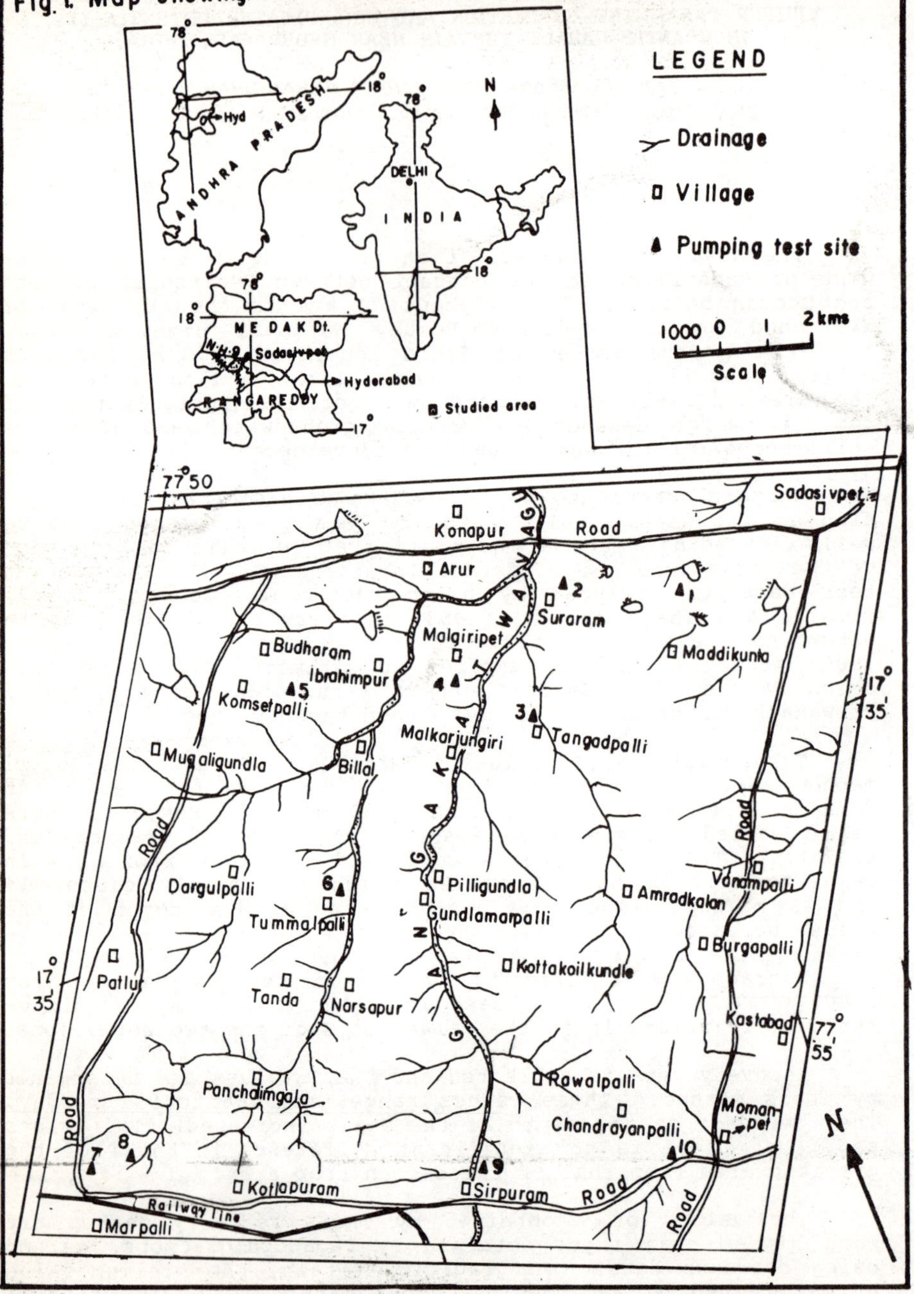

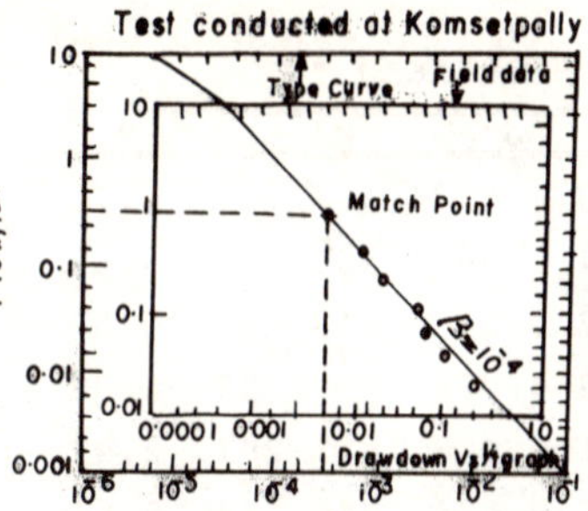

Fig. 2. Pumping test analysis of large diameter well no.5 by Papadopulos–Cooper method.

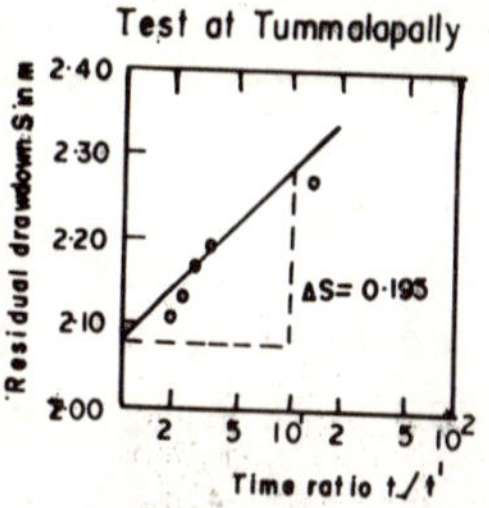

Fig.3. Analysis by Theis recovery method well no.6

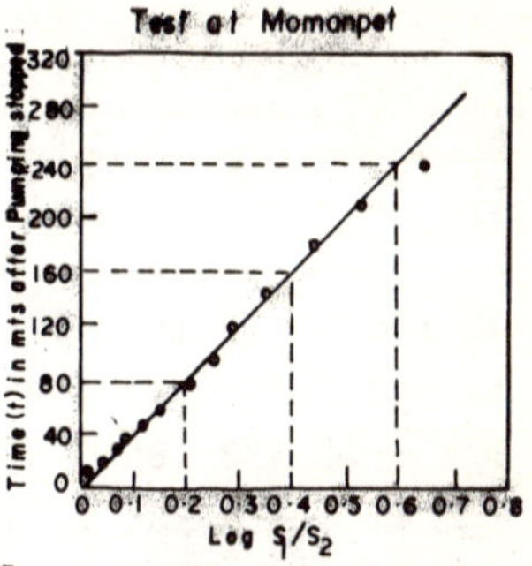

Fig. 4. Slitcher solution for determining specific capacity (c) at well no. 10.

Fig.1: Discharge Q=550 lpm: $r=5.25m$: $r^2=27.565$ From Type curve $F(U_w\beta)=0.31$: $U_w=3.1\times10^{-4}$: From field data $S_w=0.95m$: $1/t=0.0057$ = 8.208days: $T=\frac{QF(U_w\beta)}{4\pi S_w}=\frac{550\times1440\times0.31}{4\times3.142\times0.95\times1000}=20.566m^2/days=21m^2/$ days: $S=\frac{U_w\times4\times T\times t}{r_w^2}=0.0076$.

Fig.2: Q = 434 lpm: S=0.195m: $T=\frac{2.3Q}{2\pi\Delta S}=\frac{2.30\times432\times1440}{4\times3.142\times0.195\times1000}=586m^2/$ day.

Fig.3: A=113.143: $C=\frac{2.303\times A\times\log(S_1/S_2)}{t}=\frac{0.639}{1000}=639$ lpm/m/m.d.d.

Unit area Specific Capacity$=\frac{C}{A}=5.65$ lpm/m/m^2.

190 m^2/day (Well No.10). In the case of Theis recovery method the minimum value of T is 439 and maximum is 992 m^2/day, in granitic terrain and 270 and 822 m^2day in the basaltic region. However, similar trend is maintained in the results obtained by the two methods in both the formations. The decrease in the values of T as the depth of wells increased is perhaps attributable to the fall in degree of weathering.

Specific Capacity (C) of the wells is determined by Slitcher's formula. The C values range from 19 to 102 lpm/m and in one case it is 207 and in another 639. Formation-wise there is no distinct difference in the C values but to have uniformity, the unit area specific capacities of the wells are calculated. They vary from 0.5 lpm/m^2 to 2.39 lpm/m^2.

From the field studies and aquifer studies it is found that the area has vast untapped groundwater potential. Considering the local hydrogeological factors one should go for suitable bore or dug wells for judicious exploitation of the available groundwater resources to the maximum possible extent. There is also a need to go for more dependable methods for

TABLE 1: PUMPING TEST ANALYSIS DATA OF LARGE DIAMETER WELLS IN GRANITE-BASALT TERRAIN NEAR HYDERABAD.*

# A	B	C	D	E	F	G	H	I	J	K	L	M
1.	48.0	602	90	1.50	300	0.25	0.16	35	0.0174	992	35	0.73
2.	44.8	486	180	4.10	300	0.47	0.24	24	0.003	533	19	0.42
3.	145.7	272	180	1.50	300	0.37	0.16	17	0.01	448	83	0.57
4.	44.2	250	75	1.00	300	0.18	0.15	12	0.029	439	39	0.88
5.	86.6	550	180	0.90	180	0.22	0.22	21	0.008	659	207	2.39
6.	44.8	434	180	2.31	180	0.20	0.195	22	0.024	586	28	0.63
7.	38.5	240	100	1.45	240	0.52	0.254	14	0.010	662	71	1.85
8.	44.2	260	180	2.87	240	0.36	0.30	19	0.002	270	22	0.50
9.	56.8	243	50	0.59	300	0.19	0.24	18	0.033	266	102	1.80
10.	113.2	624	120	0.42	240	0.31	0.20	190	0.15	822	639	5156

#A=Well No., B=Area of Cross Section, C=Discharge(lpm), D=Pumping Time, E=Drawdown in mtrs. at the end of pumping, F=Recovery Time, G=Recovery(m), H=Recovery difference per log cycle of time(S), I,J=T.S. Papadopulos-Cooper method, K=T.Theis Recovery method, L=Specific Capacity(C)lpm/m, M=Unit area Specific Capacity.lpm/m/m.

*Well No.1 - 5 from Granitic Terrain and 5 - 10 from Basaaltic region.

analysing the pumping test data obtained from large diameter dug wells in hard rock terrains.

REFERENCES

Krupanidhi,K.V.J.R., Venkataraman,S., Matturia,D.P. and Murthy, K.N., 1973. Studies on specific capacities of wells in hard rock and alluvial formations of Mysore state. Proc.Int.Symp. Development of Groundwater Resources, Madras. V.1, P.II, pp.61-74.

Papadopulos,I.S. and Cooper,H.H., 1967. Drawdown in a well of large diameter. Water Resources, V.3, pp.241-244.

Slitcher,C.S., 1906. The Underflow in Arkanses Valley in Western Kansas, U.S.G.S. Water Supply, Paper No.153,p.61.

Theis,C.V., Brown,R.H. and Meyer,R.R., 1963. Estimating the transmissivity of aquifer from the specific capacity of the wells. In Bentall, Ray, Compiler. Methods of determining permeability, Eransmissivity and drawdown. U.S.G.S. Water Supply Paper, 1536-1, pp.331-340.

Viswanathiah,M.N. and Sastri,J.C.V., 1978. Specific capacity of wells in some hard rocks of Karnataka, J.Geol.Soc.Ind. V.19, No.9, pp.426-430.

ROLE OF GROUND WATER IN CONTROLLING ECOLOGY OF AREA A CASE STUDY OF GHANA BIRD SANCTUARY - DISTRICT BHARATPUR, RAJASTHAN

M. MEHTA

CENTRAL GROUND WATER BOARD

Introduction

The Ghana Bird Sanctuary, ditrict Bharatpur, Rajasthan is one of the famous bird sanctuary of the world and number of migratory birds migrate to the sanctuary in the winters. The sanctuary suffered acute water shortage in the year 1979 and thereby affecting the ecology and migration of birds. The area enjoys near sub-humid climate. The average annual rainfall is 670 mm. The rainfall in this area is very erratic and the rainfall variability is around 40 percent.Due to high rainfall variability, quite frequently the bird sanctuary suffer acute water shortage in lake thereby effecting the ecology of the area and it was feared that migration of birds will also be affected.

Relation between lake level and ground water level

In order to study the lake level and ground water level, a gauge in the lake and an observation well in the sanctuary was established and monitored periodically (Fig.1). It is observed that there is interconnection between lake water and ground water. The lake level is marginally higher than the ground water level. It is also observed that the average loss of water in the lake between October 1977 and May 1978 is 120 mm. per month of which seepage losses is around 40 mm. Any supply scheme to maintain the lake water level at any given stage over 1 sq.km. will require about 0.12 million m3 (4,000 m3/day); in addition to the quantity required for filling up the lake. It was estimated that around 0.3 million m3/day shall be required to maintain the lake level in limited areas of sanctuary.

Hydrogeology and aquifer characteristics

The main water bearing formation in and around the sanctuary is unconsolidated Quaternary sediments,the thickness of which is around 150 m. Based on exploration it was established that first aquifer with depth to water level between 3 to 5 m.bgl.is met between 0 to 40 m.The second and third aquifers are confined.The third aquifer from 90 to 140 is under hyperpiestic conditions with free flow in tubewells tapping it. From the detailed contour maps of piezometric surface and topographic configuration, it is deciphered that the entire santuary is under flowing well conditions. The exploratory well near the sanctuary at Mallah (77°30' : 27°11')gave a free flow discharge of about 30m3/hr. Pumping test for a period of 10485 metres was also conducted on Mallah tubewell with an observation well at a distance of 36 m. The tubewell was pumped at the rate of 151.74 m3/h. which created 14.03 and 3.395 m. of drawdown in pumping and observation wells respectively. The transmissivity of third aquifer (90.140 m) is 707.2 m2/d, the storage coefficient is 2.04 x 10 -4 and the permeability is 25.25 m/d. The electrical conductivity of water from deeper aquifer was 3312 m.mhos/cm.

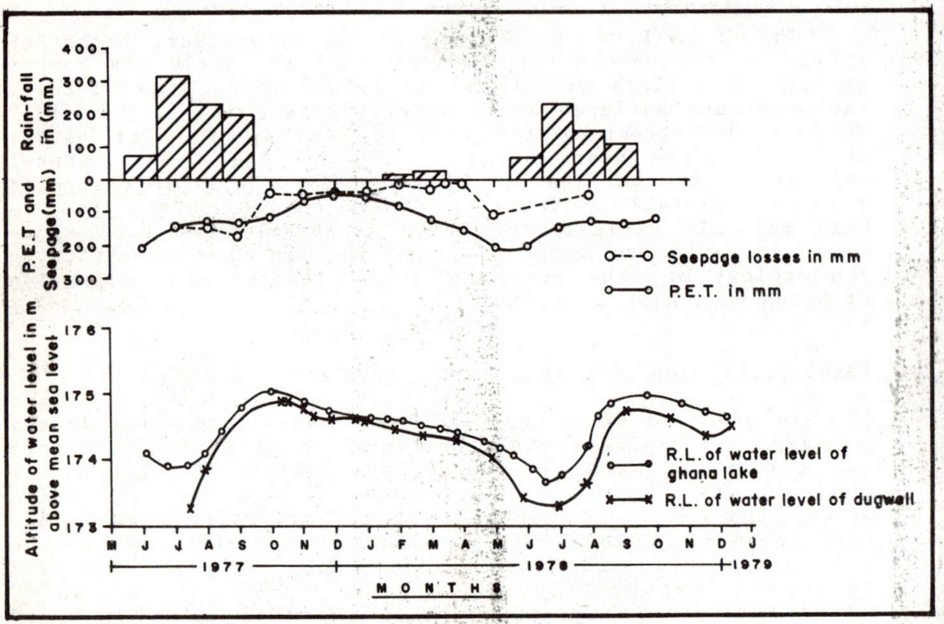

Fig.1 : Relation between Ghana lake water level,groundwater level, rainfall,seepage and evapotranspirational losses,Ghana Bird Sanctuary,District Bharatpur, Rajasthan

An attempt was also made to estimate the theoritical drawdown and interference effects both for pumping and free flowing conditions (Table 1).

Distance from the tubewell (m)	Theoritical drawdown in (m)	
	Due to 16 hrs.of pumping at the rate of 150^3/hr.	Due to free flow after 30 days with discharge of 720 m^3/d.
10	4.44	1.85
100	2.56	0.81
1000	0.74	0.44
2000	0.25	0.34
4000	0.02	0.22

Table 1 : Theoritical drawdown due to pumping of well and free flowing wells

Findings and results

The Ghana Bird Sanctuary is located in the saline tract of Banganga river basin and there is interconnection between the lake water and groundwater. Thus within the sanctuary area, it is not advisable to tap the shallow aquifer which although fresh, may result in tapping the lake water and merely recycling it instead of augmenting. The only alternative left to augment the water in lake is to develop the deeper aquifers (90-140 m) having moderate quality of water (E.C. 2500 to 3300 micromhos/cm.). The deeper aquifer is under artesian pressure and the piezometric head is about 2 to 3 meters above ground level within the entire sanctuary area. The expected free flow discharge from tubewells tapping this aquifer is about 30 m3/hr. If the wells are pumped, the discharge will be around 150 m3/hr. If the required quality of water in summer for aquatic life is to be met from free flowing wells, about 15 such wells will be required. The distance between such tubewells should be around 5 km. to prevent any significant interference.The required quality of water can also be obtained through 5 tubewells, if these are pumped at the rate of 150 m3/hr. The distance of pumped well should be around 600 m.to prevent any significant interference. The free flow from the five wells can be used round the year to maintain the water in the lake.

These findings were put to immediate use by the State Government. Three high capacity wells with free flowing conditions were constructed in the lake bed. This has contributed in no small measure to the continuance of ecological environment in the sanctuary due to filling up of water in lake by ground water, the birds migrated and settled in the lake as usual.

INSTRUMENTATION AND TECHNIQUE OF GROUNDWATER EXPLORATION (WITH NATURE OF AQUIFER) IN UNCONSOLIDATED SEDIMENTS (GANGETIC PLAINS) OF UTTAR PRADESH, NORTHERN INDIA

AVINASH Ch. NAUTIYAL

Department of Geology, Lucknow University
Lucknow-226 007, India

ABSTRACT

The instrumentation and technique of the hydraulic rotary drilling with reverse circulation (reverse rotery) and associated electric-submersible pump - as commonly and successfully employed in parts of Uttar Pradesh (e.g. districts of Lucknow, Barabanki, Bahraich, Gonda, Basti, Jaunpur, etc.) for the location of aquifers and recovery of sub-surface waters from the soft, unconsolidated sediments (Quaternary age) of the central Gangetic alluvial plains (or Ganga basin) - has been illustratively, comprehensively described and discussed. The rotary rig (Rig R.C. 600/27 type, Ludhiana Punjab made) - capable of drilling shallow holes upto 150 m depth and mounted on lorry trailers, to the giant machines with masts about 60 m high - is used for drilling underground water wells. Initially, a large diameter well (of 68.30 cm diamt hole) is drilled in dry unconsolidated sediments (using auger bits) upto a depth of 4.55 m With the rotary drilling in progress and slightly muddy fluids (or water) in circulation, the R.R. fish tail bits (of 60.72 cm, 68.30 cm diamt) are employed up to 31.91 m depth; and further drilling is continued up to 130 m, while using fish tail R.R. bits (60.72 cm diamt). Subsequently, the drilled hole is lined inside through a steel housing pipe (35.42 cm diamt) to a depth of 45.60 m. From this depth, the well is further narrowly lined inside (especially on aquifer horizons) through a special lining stell pipe (20.24 cm diamt) called screen (having unperforated and coarser slotted linings) to a depth of 130 m. Moreover, since the aquifer having fine sand particles, an envelope of coarse sand (1.50 mm size) and gravel (2-8 mm size), called a filter pack, is poured into the annular space between the aquifer and the screen. The filter retains the aquifer particles outside its outer boundary and the screen slot size is chosen to prevent the filter pack itself from passing through the slots. During the reverse circulation of the rotary method, the drilling fluid is pumped down the annulus and returns up the drill pipe, bringing sands, clay, clay with kankar, and seldom gravels. An electric-submersible pump (with five impellers) is placed inside a steel pipe (15.18 cm diamt), surrounded by a housing pipe, at a depth of about 33.44 m, being submerged within the shallowest first aquifer zone (between 24.32 m and 33.44 m depth). Some of the finest sand particles enter the well when it is first pumped (with 1,200 liters per minute maximum capacity) during the development process, but thereafter the well produces sand free waters fit for human consumption. A good number of shallow wells drilled in the Gangetic alluvial plains commonly revealed sandy aquifers of high potentiality of yielding groundwater. Two of such wells drilled (up to 130 m depth) in southeastern Lucknow's alluvial plains revealed distinct inter-layering of soft clay (grayish yellow), sand (light gray to light olive gray), clay and kankar (yellowish gray), clay with kankar and sandy fragments (very light gray with yellowish tinge), and four sandy unconfined aquifers (each 9 m thick) horizons (sands being yellowish gray to very light gray, fine grained, winnowed) were located, with these sediments collectively called Bhangar in the Ganges valley of northern India.

INTRODUCTION

The Indo-Gangetic basin (forming the alluvial plains) in Uttar Pradesh is underlain by a thick pile of alluvium (of fluvial sedimentation), spreading over 209,388 sq km in the state. It was developed infront of an orogenic mountain chain, with a stable peninusula on the southern side; and its configuration is controlled by the tectonic strike of the Himalayas. The structure and origin of the basin has been thoughtfully discussed by Wadia (1959), Krishnan (1961), Poddar (1962), Mathur and kohli (1963), Sengupta (1964), Mool Chand et al. (1964), Fuloria (1969), Singh and Rastogi (1973), and Singh (1977, 1988). Various interesting views have been expressed to its origin: the supposition of "fore-deep" of Suess, "Rift-valley" hypothesis of Burrard, and generally accepted concept of "sag" formed between the northward drifting Indian continent and the Tethyan basin.

The great alluvial tract of Ganga, Brahmaputra and Inus, spreading in 8,50,000 sq km area, is the largest alluvial tract of the world, and forms one of the three main physiographic divisions of the Indian subcontinent. The Ganga basin's tract in Uttar Pradesh is drained by the Ganga River system. Among the major tributaries to the Ganga River is the Yamuna, lying to the south and west of the former. Whereas, Ramganga, Gomti, Sharda, Saryu and Rapti Rivers occupy north of the Ganga River. The Ganga River is the most important river of the Gangetic alluvium. Originating from a glacial source in the Himalayas, it flows across the Himalayas (before descending into the alluvial plains), where it transforms its course parallel to the Himalayan axis. Its E-W flow in the alluvial plain is controlled by some weak zones in the Gangetic alluvium (Singh and Rastogi, 1973; Singh, 1977). Several Himalyan and Peninsular rivers join its eastward flowing course. About 60 per cent of the Indo-Gangetic rivers get waters from the Himalyan glaciers, which account for 17 per cent of the total Himalayan region. The Himalayan influential streams feed the under underground waters (as aquifers) in the unconsolidated sediments of the Central Ganga Plain.

The Ganga basin's alluvial tract in Uttar Pradesh is sub-divided into four sub-zones, viz., Bhaber, Tarai, Central Ganga Plain (Fig. 1) and Marginal Alluvial Plain. The vast alluvial tract - located south of the Tarai with southern limits close to Yamuna till its confluence with the Ganga and subsequently being the Ganga - forms the hydrogeological unit of the Central Ganga Plain in the state. The tract is characterised by a plain of low relief, having numerous fluvial depositional and erosional features. These features are suggestive of the various streams (of the Ganga River system) had occupied and abandoned several courses while the plain was under development. In the Central Ganga Plain, the record of various physical and environmental changes in the Ganga River system and the depositional history is preserved in the form of sub-surface lithological sedimentary units and configuration. This configuration is disclosed mainly by the lithological log of the boreholes drilled by the Central Ground Water Board, Uttar Pradesh Jal Nigam (Lucknow), and State Tubewell Construction Department (Lucknow) in Uttar Pradesh. In the central and southern parts of the plain, various types of bed rock were encountered at many a place within the depth range of 296 to 591 m (Khanna, 1988).

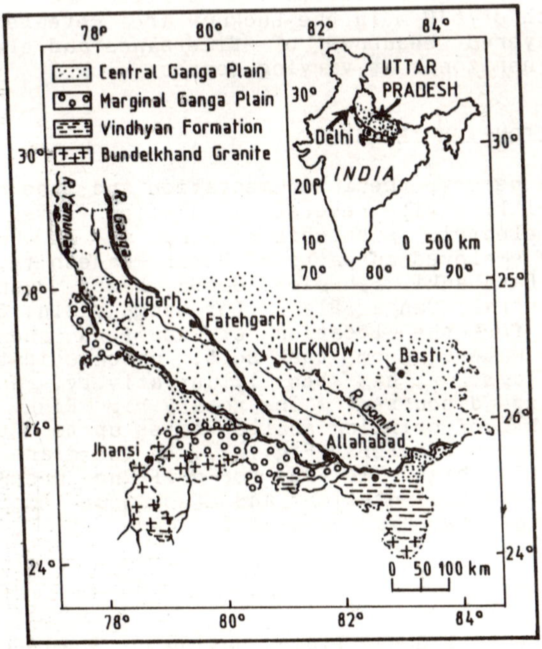

Fig. 1. Index map showing the location and geological setting of the Central Ganga Plain and the Marginal Gangetic Alluvial Plain and adjoining areas in Uttar Pradesh (modified after Banjpai et al., 1986).

However, the thickness of the overlying Pleistocene to Recent formations - as revealed through a few boreholes sunk and seismic surveys conducted in connection with exploration for oil - varies between 500-1,000 m near the Himalayan foot hill zone decreasing southwards.

The landscape of Lucknow district (centrally located in Central Ganga Plain) has generally resulted from the fluvial processes of Gomti River during the late Quaternary times. Rising from the Tarai region of Pilibhit district (Uttar Pradesh) and draining the higher alluvial surface of Ganga and Ghaggra Rivers, Gomti flows in a general south-easterly direction to confluence into Ganga near Saidpur in Ghazipur district. Geomorphologically, the major landforms of the Lucknow area comprise two distinct levels of paired terraces and the active flood-plain of Gomti, embody with an array of morphologic features prominent among which are point-bars, meander-scrolls, meander-cutoffs, ox-bow lakes, abandoned channels, natural levees, back-swamps, etc. Geologically, the sediments composing the terraces, comprise predominantly silt and clay with minor proportions of fine to very fine sand. Sub-surface data from boreholes drilled for tubewells to a maximum depth of 610 m in the Lucknow area revealed the presence of inter-layered sequence of fine sands and clays associated with kankar horizones at varying depths.

INSTRUMENTATION AND TECHNIQUE

In this report, the instrumentation and technique of the hydraulic drilling with reverse circulation (reverse rotary) and associated electric submersible pump - as commonly and successfully employed in parts of Uttar Pradesh (e.g. districts of Lucknow, Barabanki, Bahraich, Gonda, Basti, Jaunpur, etc.) or alluvial Central Ganga Plain (or Ganga basin, Subramanian et al., 1987) for the location of aquifers and recovery of underground waters from the soft, unconsolidated sediments (Quaternary age) - has ben illustratively, comprehensively described and discussed. The rotary rig (Ring. R.C. 600/27 type) - capable of drilling shallow holes up to 150 m depth and mounted on lorry trailers (Fig. 2), to the giant machines with masts about 60 m high - is used for drilling underground water wells. The instrumentation and technique employed may be sub-divided into two headings, as follows.

WATER WELL DESIGN

In the Central Ganga Plain, having unconsolidated sediments (sands, clays, clay and kankar formations) in shallower depths, the rotary rig (Rig R.C. 600/27 type, Ludhiana Punjab made) - capable of drilling shallow holes (of 68.30 cm diamt) up to 150 m depth and mounted on lorry trailers, to the giant machines with masts about 60 m high - is used for drilling underground water wells. Tall masts permit the use of long individual lengths of drill pipe, thus reducing the time spent in adding or removing lengths. They similarly speed up the process of installing casing. In unconsolidated sediments (sand, clay, clay and kankar, sand and sandstone fragments), the sand, particularly where it is saturated with water, seldom keeps collapsing into the hole (causing minor caving problems).

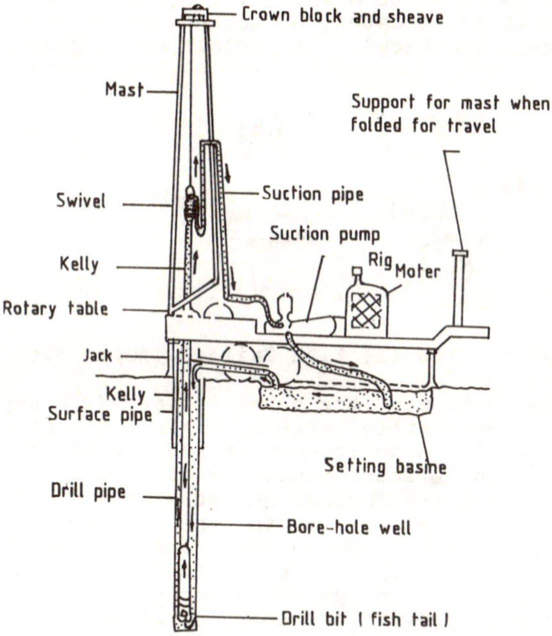

Fig. 2. Hydraulic rotary drilling with reverse circulation (reverse rotary). The main features of a trailer-mounted rotary drilling rig.

This problem, above the water table in a well, can be overcome through linging the shaft with concrete poured in place. Boreholes are usually supported by inserting lenghts of pipe, called casing or lining tubes, of diameter slightly smaller than the drilled diameter of the borehole. The space between the outside of the casing and the borehole wall is usually filled with a thin concreate called Cement grout. In addition to providing support, this prevents dirty or polluted surface water or soil water from running into the well outside the casing after heavy rain.

The basic parts of a conventional rotary drill (Fig. 2) are - a mast and hoist, a power operated revolving table that rotates the drill stem and bit, a rig pump and a power unit or engine. Drag bits of either the fish tail or auger bit (Fig. 3 A, B) are best suited with the drill pipes in an unconsolidated sand and clay formations. The slotted well screens (20 cm diamt, 3.04 m length; slots 5 cm long and 1.50 mm wide) are positioned at the four aquifer zones (between 24.32 m and 33.44 m, 52.91 m and 59.29 m, 82.68 m and 95.15 m, 102.44 m and 108.52 m depth) in a near zoo located, SE Lucknow water well of Uttar Pradesh Jal Nigam (Lucknow), to obtain a higher specific capacity.

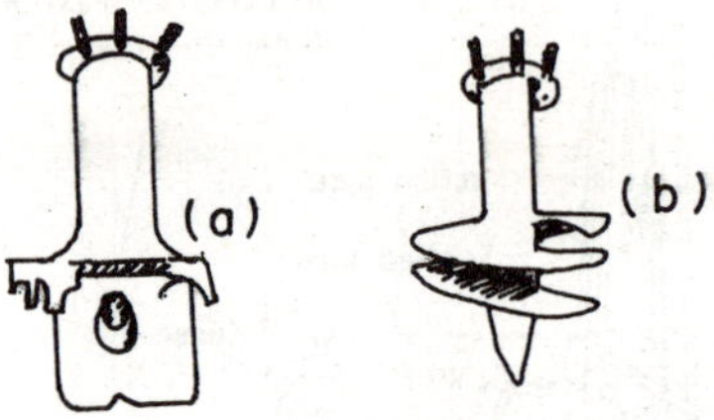

Fig. 3. Rotary drill bits (A, Fish Tail; B, Auger bit).

Moreover, from bottom of the borehole (at 130 m depth) and in continuation of the slotted screens, plain drill pipes (20 cm diamt, 3.04 m length) are positioned upwards up to 84 m. From this height, through a short connector pipe neck, the slotted screens and plain pipes are further joined with a housing pipe (35 cm diamt) for 45.56 m length, the free open end of the latter terminating slightly above the concrete plinth lying outside the borehole. Thus, below the water table, the insertion of unperforated linings would support the surrounding aquifer but would also prevent the entry of water. Furthermore, an electric-submersible pump (Figs. 4, 5) is placed inside a steel pipe (15.18 cm diamt.), surrounded by the abovementioned housing pipe, at a depth of about 33.44 m being submerged within the shallowest frist aquifer horizon (between 24.32 m and 33.44 m depth).

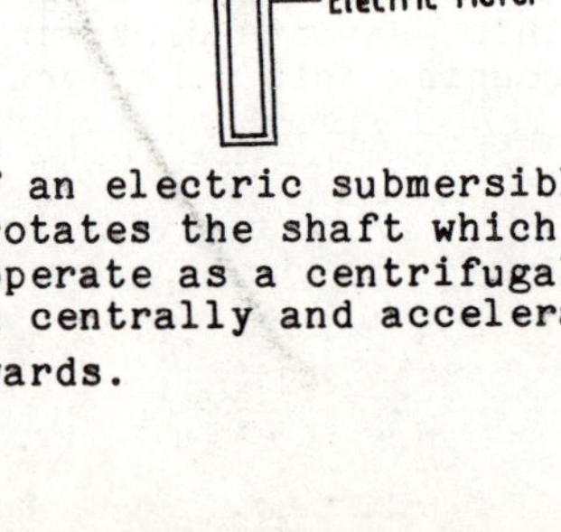

Fig.4. The principle of an electric submersible pump. The electric motor rotates the shaft which carries the impellers that operate as a centrifugal pump, drawing water in centrally and accelerating it outwards and upwards.

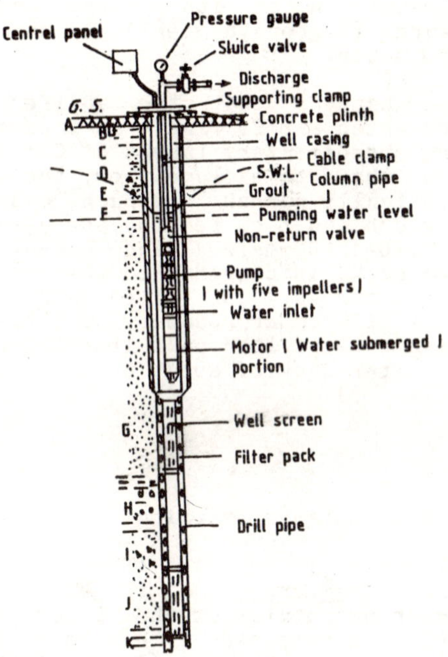

Fig.5. Well design, not to scale(with electric-submersible pump).

A well completion with fine sand, requiring a screen and a filter pack. litholog (of unconsolidated sediments) with two aquifer horizones: G, J.A, Soil (grayish yellow clay, humus); B, clay, yellowish gray; C, sand, very light to olive gray, fine grained; D, clay and Kankar, yellowish gray; E, clay with sandstone fragments, yellowish gray; F, clay, yellowish gray; G, sand, yellowish gray, fine grained, winnowed, representing First Aquifer; H, clay and kankar, yellowish gray, with sandstone fragments, very light gray; sand and sandstone fragments, very light gray with yellowish tinge; J, sand, yellowish gray, fine grained, winnowed, representing Second Aquifer; K, clay, yellowish gray, soft.

In the electric-submersible pump, a water proof electric motor is installed beneath the pump (where it is cooled by the water flowing past it) and drives it direct. Since it has to be raised more than 50 m from pumping water level (of first aquifer) to the surface, five impellers (or five stages) are used, one above the other (Fig. 4), each imparting energy to the water. The principle of an electric-submersible pump is such that the electric motor rotates the shaft which carries the impellers. Each impeller in effect operates as a centrifugal pump, drawing water in at its centre and accelerating it outsards and upwards (Raghunath, 1987). A seal prevents water from entering the motor.

It sometimes happens that the aquifer is composed of clastic particles which are so fine - as in central, eastern and northeastern part's sedimentary layers of Central Ganga Plain of Uttar Pradesh - that to exclude them, the screen slots would have to be impractically narrow. In such cases coarser slots (with 1.50 mm width and 5 cm length) are used, and an envelope of coarse sand (1.50-1/68 mmsize) and gravel (2-8 mm size), called a filter pack, is poured into the annular space between the aquifer and the screen. The filter pack retains the aquifer particles outside its outer boundary, and the screen slot size is chosen to prevent the filter pack itself from passing through the slots. A filter pack of about 60 to 90 mm width is placed in an annulus (Fig. 5).

TECHNIQUE

Reverse Circulation (Rotary) Drilling

In the rotary drilling rig (Rig. R.C. 600/27 type, mounted on lorry trailers, 2), the bit (fish tail and auger types) on this machine rotates at the lower end of a hollow steel tube, which is composed of lengths of drill pipe screwed together. Initially, an auger bit (55 cm width, 90 cm length) is attached (through bolts) to a drill pipe (15 cm diamt, 3.04 m length), which remains connected to the kelly and rotary table placed on the lorry trailer. And a large diameter well (of 68.30 cm diamt hole) is drilled in dry unconsolidated (overburden) soil (mixture of soft, grayish yellow clay and humus) up to a depth of 4.55 m. Subsequently, the bit is changed (through bolts) to fish tail type with the drilling pipe for further rotary drilling. In the reverse circulation (rotary) drilling, the direction of drilling fluid (or water) flow is reversed, i.e. from the annular space between the drill pipe and the wall of the hole through the bit into the hollow drill pipe upwards and discharged by the rig pump into the settling basin (along with sand, clay, clay and kankar samples). However, with the drilling in progress and slightly muddy fluids (or water) in circulation, the R.R. fish tail bits (of 68.30 cm and 60.72 cm diamt) are employed up to 31.91 m depth; and further drilling is continued up to 130 m depth, while using the R.R. fish tail bits (60.72 cm diamt). During the drilling in progress and rig pump in operation, the clear water from the settling basin returns to the borehole by gravity flow. Relatively high velocity of water in the drill pipe enables the unconsolidated sediments (sand, clay, clay and kankar, sand with sandstone fragments, seldom minor gravels) to be carried to the surface of the settling basin, where they are collected as fresh rock samples (and the remaining suspended sedientary particles are settled at the bottom of basin). Thereafter, they are dried in the sun and sampled either in cloth or plastic bags (marked with borehole depth numbers).

Interestingly, the boring in sedimentary layers is done without a casing and hydrostatic pressure is employed to support the walls of the borehole during construction. Moreover, water level in the borehole is kept about 2 m above natural level or at ground level. The settling basin constructed is about 6.08 X 12.16 m (4.55 m depth) size. Importantly, circulation of large amount of make-up water (2,500-5,000 lpm) by the use of a large capacity suction pump is needed and maximum development is wanted to remove the light filter cake formed. Furthermore, the diameter of the hole is kept large (68.30 cm) in relation to the drill pipe (15 or 20 cm diamt), so that the velocity of the descending water in annular space remains low (30 cm/sec or less). The bit along with drill pipe is rotated at speeds varying from 10 to 40 rpm (depending upon the degree of compaction or strength of the unconsolidated sedimentary layers). The large diameter favours completion of the well by artificial gravel (2-8 mm size) packing. Importantly, the suction end of the rig pump is connected through the swivel to the kelly and drill pipe. And the suction head of the pump limits the drill pipe lengths to 3.04 m; 20 cm drill pipes are commonly used so that kankar (along with clay or sand) of size up to 5 cm can be brought up through the bit (fish tail) hole to the drill pipe. Interestingly, reverse rotary is best suited to drill large diameter wells (of 68.30 cm diamt hole) in soft, unconsolidated sediments of the Central Ganga Plain of Uttaar Pradesh.

Ultimately, the borehole is lined inside through a steel housing pipe (35.42 cm diamt) from the ground surface to a depth of 45.60 m. An electric-submersible pump is placed inside a steel pipe (15.18 cm diamt) surrounded by the housing pipe, at a depth of about 33.44 m, being submerged within the shallowest frist aquifer horizon (between 24.32 m and 33.44 m depth).

Well Development:

The well development causes reversals of water flow through the screen openings, so that fine sediments are washed out. And the unconsolidated sedimentary particles are rearranged in the naturally developed well, forming a graded filter (reversed filter) with rings of increasing porosity and permeability towards the well in an artificially gravel packed (or filter packed) well. Ultimately, the well yields clear sand free water. This entire process of clearing borehole requires about two weeks' (with 18 hours each day) continuous operation. And the tendency of several small particles to bridge between large particles is overcome during the reversal of fluid flow around the screen. Indeed, some of the finest sand particles enter the well when it is first pumped through the electric-submersible pump (five stage), yielding water with maximum capacity of 1,200 liters per minute during the development processs, but thereafter the well produces sand free water fit for human consumption. It has been observed by the Uttar Pradesh hydrogeologiests/engineers that such type of underground water wells last for 12 to 15 years, while operating continiously in the Central Ganga Plain.

NATURE OF AQUIFER

In the Central Ganga Plain (Uttar Pradesh) four to eight aquifers (of unconfined, semi-confined to confined conditions) have beeen recorded in the unconsolidated sediments (called Bhangar) up to 440 m depth. Two of such wells drilled (up to 130 m depth) in southeastern Lucknow's (near zoo) alluvial plains revealed distinct inter-layering of soft clay (grayish yellow), sand (light gray to light olivegray), clay and kankar (yellowish gray), clay with kankar and sandstone fragments (very light gray with yellowish tinge); and four sandy, unconfined aquifer (each 9 m thick) horizons (occurring between 24.32 m and 33.44 m, 52.91 m and 59.29 m, 82.68 m and 95.15 m, 102.44 m and 108.52 m depths) were located. Whereas, the shallow wells drilled (up to 122 m depth) in northeastern Uttar Pradesh (e.g. Bahraich, Gonda and Basti districts) revealed up to eight sandy, unconfined aquifers (3 to 9 m thickness) horizons. On the contrary, the underground water wells drilled (up to 440 m depth) in western Central Ganga Plain (e.g. Aligarh distt.) disclosed four major sady aquifer horizons (e.g. shallowest first and third being unconfined type, second being semi-confined type, and fourth of confined condition) occurring in the sandy clay, silty clay and kankar-clay layers (Khanna, 1988). It appears that the western Central Ganga Plain (e.g. Aligarh-Etah unconsolidated sedimentary belt), being closed to the peninsular Sheled (with Archaean-proterozoic basement), yield varying underground water run off conditions. Whereas, the central and northeastern Central Ganga Plain (e.g. Lucknow-Basti sedimentary belt) - located closer to southern Himalayas (being the catchment areas for alluvial plains) - reveal high yielding potential of underground waters with greater run off conditions

ACKNOWLEDGEMENTS

The author is grateful to Mr. A.N. Verma (Engineer) and Mr. V.K. Gupta (Chief Driller) of Uttar Pradesh Jal Nigam (Lucknow), Mr. R.K. Misra (Engineer) of State Tubewell Construction Department (Lucknow), and to hydrogeologists of Central Groundwater Board in Uttar Pradesh for help, discussions (at well site areas) and encouragement during this study. Thanks are extended to Prof. S.K. Singh of Lucknow University (Head, Geology Department) for facilities to complete this work.

REFERENCES

Bajpai, V.N., and Gokhale, K.V.G.K., 1986. Hydrogeomorphic classification of the marginal Gangetic Alluvial Plain in Uttar Pradesh, India, using satellite imageries. J.Geol. Soc. Ind., V.28, p. 9-20.

Fuloria, R.C., 1969. "Selected lectures on Petroleum ExpLoration". Oil and Nat. Gas comm., p. 171.

Khanna, S.P., 1988. Hydrology of Central Ganga Plain, Uttar Pradesh. "Workshop on Gangetic Plain-Terra Incongnita", Luck. Univ. (Geol. Dept.), p. 1-4.

Krishnan, M.S., 1961. Mem. Geol. Surv. Ind., p. 1.

Mathur, L.P., and Kohli, G., 1963. 6th World Petroleum Congress, p. 633.

Mool Chand, Datta, A.N., Chellam, R.S., Ghosh, A.P., Awasthi, A.M., Awasthi, D.N., and Garg, V.C., 1964. Proceedings XXII Int. Geol. Cong., sec.2, pt. II, p. 260.

Poddar, M.C. (Editor), 1962. Oil and Nat. Gas Comm., Technical Publication, v. 1, p. 68.

Raghunath, H.M., 1987. Ground Water. Wiley Eastern Ltd., New Delhi, p. 563.

Sengupta, S., 1964. Report of XXII Int. Geol. Congress, New Delhi, v. 11, p. 334.

Singh, I.B., and Rastogi, S.P., 1973. Tectonic framework of Gangetic alluvium, with special reference to Ganga River in Uttar Pradesh. Curr. Sci., v. 42, no. 9, p. 305-307.

Singh, I.B., 1977. Bedding structures in a channel sand bar of the Ganges River near Allahabad, Uttar Pradesh, India. J.Sed. Petrol., v. 47, no. 2, p. 747-752.

Singh, I.B., 1988. Geological evolution of Gangetic Plain-What we know, and what we do not know ?. "Workshop on Gangetic Plain-Terra Incognita", Luck. Univ. (Geol. Dept.). p. 1-5.

Subramanian, V., Sitaswad, R., Abbas, N., and Jha, P.K., 1987. Environmental geology of the Ganga River Basin. J. Geol. Soc. Ind., v. 30, p. 335-355.

Wadia, D.N., 1959. Geology of India. Macmillan and Co., London, p.536.

SESSION S-1B

GEOHYDROLOGIC PARAMETER ESTIMATION : GEOPHYSICAL METHODS ; STOCHASTIC METHODS ; INVERSE TECHNIQUES

PARAMETER ESTIMATION FOR A QUASI-THREE-DIMENSIONAL FLOW MODEL

A. Das Gupta
Asian Institute of Technology
P.O. Box 2754, Bangkok, Thailand

A. Srikanthan Herath
Nippon Koei Co., Ltd.
Tokyo, Japan

Abstract: The parameter estimation procedure for a quasi-three-dimensional flow model is based on the sensitivity analysis approach. The flow model consists of a two-dimensional partial differential equation representing flow in individual layer and vertical leakage terms coupling the layers with transmissivities, storativities and leakage factors as unknown parameters of the model. An alternate direction implicit method is used to solve the equation numerically with specified initial and boundary conditions. The parameters are identified by improvement on a set of initially given parameters by the square of the difference between the observed and computed piezometric head values. Initially the model responses were evaluated by applying the algorithm for a hypothetical two-layer aquifer system. The possibility of application for a multi-layer aquifer system underlying the city of Bangkok was then investigated. The sensitivity analysis approach could be used to identify a set of parameters in a hydrologic model. However the main drawback in the method is the demand for excessive computation time.

INTRODUCTION

The groundwater is being increasingly used to meet the domestic, industrial and agricultural demand. Dependence on groundwater as a vital resource necessitates understanding of the dynamic behaviour of the system leading to a proper management of the resource. The use of simulative flow models of groundwater basins has become an established management tool. The model must be calibrated with historical records before applying it for prediction of aquifer response according to future operational policies. The calibration process usually requires adjustments in parameters till the model response duplicates field observation. This exercise is referred to as the inverse problem when the piezometric heads have been actually observed and it is required to solve the hydrodynamic equation of flow subject to specific initial and boundary conditions for parameters like transmissivity, storativity, leakage factor etc. There has been a substantial number of research papers dealing with the inverse problem in the last two decades. Yeh (1986) summarized many of the contributions reported in the literature. The main goal of the various approaches to the inverse problem is to obtain unbiased and consistent estimates of groundwater parameters based on measured values of the pertinent field variables.

This paper deals with the development and validation of an inverse technique for the identification of a set of parameters in a multi-layer aquifer system based on the sensitivity analysis approach. The quasi-three-dimensional flow model consists of a two-dimensional partial differential equation representing flow in individual layer and vertical leakage terms coupling the layers with transmissivities, storativities and aquitard leakage factors as the model's unknown parameters. Due to the difficulties associated with solving a system of non-linear partial differential equations and the unknown nature of intial and boundary conditions and sensitivity coefficients, the proposed technique is first validated using a hypothetical two-layer aquifer system with an unconfined aquifer overlying a semi-confined aquifer. Finally, an attempt has been made to apply the technique for possible improvements of the model parameter values used in the management study of the multi-layer groundwater basin underlying the city of Bangkok.

FLOW MODEL

The flow phenomenon in a multi-layer aquifer system is simulated using a quasi-three-dimensional numerical model. Though a three-dimensional model facilitates a realistic means of simulating a multi-layer system, its high core storage requirement and its high computation cost makes it uneconomical. Moreover, variations of aquifer parameters, as needed by the model, are not readily obtainable from field investigations. The physical configuration of the multi-layer aquifer system is shown in Figure 1(a). An aquitard and the aquifer immediately below it are considered as one stack and stacks are numbered from top to bottom. This system can be simplified, as shown in Figure 1(b) for modelling purpose. The flow of groundwater is considered to be two-dimensional in the aquifer plane with hydraulic leakage between the aquifers being provided through leakage flux incorporated at each model point. Following the derivation of Pinder and Bredehoeft (1968), the governing equation for the unsteady groundwater flow in a confined nonhomogeneous anisotropic aquifer of a multi-layer aquifer system can be written as

$$\frac{\partial}{\partial x}\left(T_x \frac{\partial h}{\partial y}\right)_\ell + \frac{\partial}{\partial y}\left(T_y \frac{\partial h}{\partial y}\right)_\ell - W_\ell = \left(S \frac{\partial h}{\partial t}\right)_\ell \qquad (1)$$

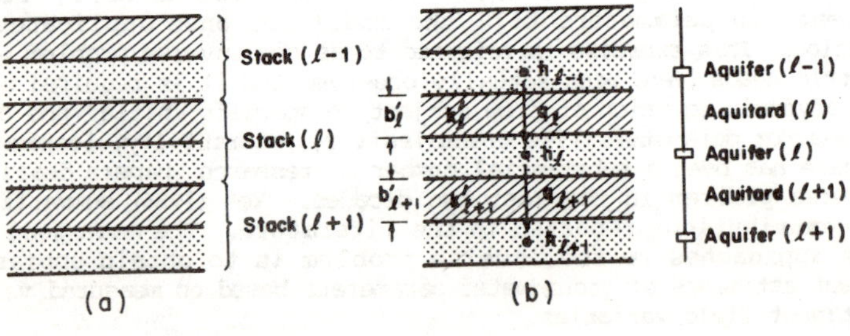

Figure 1 Multi-layer system and its model representation with specification of the leakage flux

where h is the hydraulic head; T_x and T_y are transmissivity values of the aquifer in respective coordinate directions x and y; S is the storativity of the aquifer; W stands for the volume flux per unit area of aquifer; and subscript ℓ indicates that all quantities are referring to the particular aquifer under consideration. The term W includes discharge (Q), recharge (R) and the leakage flux through confining layers above and below the aquifer under consideration and is expressed as (Figure 1b)

$$W = Q - R - q_\ell + q_{\ell+1} \tag{2}$$

The two leakage flux terms, q_ℓ and $q_{\ell+1}$, are proportional to the hydraulic gradient across the respective confining bed and take the form

$$q_\ell = \frac{k'_\ell}{b_\ell} (h_{\ell-1} - h_\ell) \tag{3}$$

$$q_{\ell+1} = \frac{k'_{\ell+1}}{b_{\ell+1}} (h_\ell - h_{\ell+1}) \tag{4}$$

where k'_ℓ and $k'_{\ell+1}$ are the vertical hydraulic conductivities of aquitard ℓ and $\ell+1$, respectively; b'_ℓ and $b'_{\ell+1}$ are thickness of the respective confining layer; and $h_{\ell-1}$, h_ℓ, $h_{\ell+1}$ are respectively the piezometric head in aquifer $\ell-1$, ℓ and $\ell+1$.

In the above partial differential equation, Equation 1, if the initial conditions, i.e.,

$$h(x,y,o) = h_o(x,y) \quad , \quad x,y \in \Omega \; ; \tag{5}$$

the boundary conditions, i.e.,

$$h(x,y,t) = h_1(x,y,t) \quad , \quad x,y \in \Gamma \text{ and } t > 0 \tag{6}$$

and the hydraulic parameters, i.e., T, S and R (= k/b) are known, then solution of Equation 1 subject to conditions (Equations 5 and 6) uniquely determines the hydraulic head distribution for $t > 0$, for any excitation of the aquifer. This is the direct problem in groundwater. To achieve a meaningful solution of the direct problem, it is necessary to have precise information on the initial and boundary conditions and the hydraulic parameters of the aquifer and the aquitard.

For numerical approximation of the governing equation, the aquifer area is discretized into several cells by two sets of line parallel to the x and y axis. The aquifer properties in each cell are assumed constant and are represented at the central point of the cell known as the node. Using difference approximation, the governing equation is transformed into a set of linear algebraic equation. A modified form of the alternating direction implicit method (ADI) of Peaceman and Rachford (1955) has been used to solve the resulting set of equations.

INVERSE PROBLEM

In inverse problem, the governing equation subject to initial and boundary conditions, defined by Equations 1 through 6, are solved for hydraulic parameters, T, S and R (= k/b) by comparing the computed hydraulic head distribution, h (x,y,t), with the observed one, $h^*(x,y,t)$. Hydraulic heads are invariably measured at discrete points in an aquifer which implies that the inverse problem is always discrete. The methods of solution to the inverse problem are either direct or indirect. The direct approach treats parameters as dependent variables in a formal inverse boundary value problem. However, the problem can be solved under certain restrictions. The indirect approach is essentially a trial-and-error procedure. The values of parameters are guessed and the groundwater model is solved knowing the aquifer geometry and the historical excitation. The output of the model is compared with the actually observed piezometric heads using a suitable error criterion. If the criterion is satisfied within certain limits, the procedure is terminated; otherwise, a new set of parameters is determined and the procedure is repeated until convergence is attained. Most often, the output error criterion used is the least square function in which the objective is to find T, S and R so that Z is minimized where

$$Z = [h(\bar{T}, \bar{S}, \bar{R}) - h^* (T, S, R)]^2 \tag{7}$$

For an aquifer area Ω and time interval t, Equation 7 may be written

$$Z = \int_t \int_\Omega [h(\bar{T}, \bar{S}, \bar{R}) - h^*(T, S, R)]^2 \, d\Omega \, dt \tag{8}$$

Since the observations are available at discrete points, Equation (8), for M number of observation over a time period T^* may be written as

$$Z = \sum_{t=1}^{T^*} \sum_{m=1}^{M} [h_{t,m}(\bar{T}, \bar{S}, \bar{R}) - h^*_{t,m}(T, S, R)]^2 \tag{9}$$

In addition to this least square criterion, a flatness criterion is used to keep the parameter values within certain bounds to insure the plausibility of results.

There are two types of errors associated with the inverse problem: (1) the system modelling error, as represented by the performance criterion, and (2) the error associated with parameter uncertainty. An increase in parameter dimension will generally improve the system modelling error, but will increase the parameter uncertainty and vice versa. The optimal level of parameterization depends on the quantity and quality of data.

SENSITIVITY ANALYSIS

The method of sensitivity analysis in parameter identification described by Vemuri et al. (1969) is employed here to develop an algorithm for identification of hydraulic parameters in a multi-layer aquifer system. Consider a physical system with piezometric head denoted by h^* is represented by the differential equation

$$\dot{h}^* = F(h^*, t, \pi) \quad ; \quad h^*(o) = h^*_o \tag{10}$$

where π represents parameter vector in the physical system and t is the time. The goal is to construct a model, characterized by

$$h = f(h, t, \mu) \quad ; \quad h(o) = h_o \tag{11}$$

such that the model responses h and the dynamic responses h^* of the actual system are close to each other in some acceptable sense. In Equation 11, μ represents the parameter vector in the model as against π representing the system parameters. The closeness of $h(t, \mu)$ to $h^*(t)$ is conveniently expressed by $E(t, \mu)$

$$E(t, \mu) = h^*(t) - h(t, \mu) \tag{12}$$

It is possible to minimize $E(t, \mu)$ be adjusting μ vector by $\Delta\mu$ such that

$$h^*(t) \approx h\ (t, \mu + \Delta\mu) \tag{13}$$

Since h is a smooth function of μ, it can be expanded by Taylor series about μ and by neglecting the higher order terms, Equation 13 can be expressed as

$$h^*(t) - h(t, \mu) = \frac{\partial h}{\partial \mu} \Delta\mu \tag{14}$$

where $\partial h/\partial\mu$ stands for sensitivity coefficient vectors and $\Delta\mu$ is the vector indicating the change in parameter values.

The sensitivity coefficients indicate how the hydraulic head changes when the aquifer parameters are adjusted; therefore, it is logical to use the sensitivity coefficients in developing an inverse process for groundwater modelling. If the region is homogeneous in μ, Equation 14 can be written for different time levels over a period of time and $\Delta\mu$ evaluated provided piezometric heads $h^*(t)$ are available at those levels. The general problem however is that the region is non-uniform in μ and measurements are available for any number of previous time periods. Equation 14 is then applied spatially for a particular time level. By considering the flow along the streamtubes it can be shown that there is a complete mathematical independence between parameters of two adjacent streamtubes; and if parameter values at every point in one streamtube is multiplied by the same number, then the piezometric head distribution along that tube will not change (Emsellem and De Marsily, 1971). Therefore the change in the parameter value for every reference nodal point is indicated as

$$\Delta\mu_{ij\ell} = \mu_{ij\ell}\ \delta_\mu \tag{15}$$

where δ_μ is a common multiplication factor for the considered region. The whole region under consideration can be subdivided into several subregions with constant parameters having enough data points.

The sensitivity coefficients (in this case $\partial h/\partial T$, $\partial h/\partial S$ and $\partial h/\partial R$) are obtained by differentiating Equation 1 with respect to T, S and R respectively for uniform parameter regions. The internal boundary conditions applying at discontinuities in transmissivity are that the specific discharge normal to the boundary and the hydraulic head along the boundary remain unchanged as the boundary is crossed. These conditions are satisfied by the discretized flow equation which satisfies mass balance at each grid and the selection of ΔT as $\delta_T\ T_{ij\ell}$ to ensure a constant flow ratio. The resulting set of equation obtained numerically for the sensitivity coefficients are solved simultaneously with the numerical solution of the flow equation. At the end of simulation period, Equation 14 is the following form

$$Y = X A \tag{16}$$

where

$$Y = [(h_1^* - h_1) \; (h_2^* - h_2) \; \ldots\ldots\ldots\ldots (h_N^* - h_N)]'$$

$$X = \begin{bmatrix} (\frac{\partial h}{\partial T} T)_1 & (\frac{\partial h}{\partial S} S)_1 & (\frac{\partial h}{\partial R} R)_1 \\ \vdots & \vdots & \vdots \\ (\frac{\partial h}{\partial T} T)_N & (\frac{\partial h}{\partial S} S)_N & (\frac{\partial h}{\partial R} R)_N \end{bmatrix} \tag{17}$$

$$A = [\, \delta_T \;\; \delta_S \;\; \delta_R \,]'$$

In Equation 17, N is the number of grid point; and the prime denotes a transpose.

If data are available over a number of time periods more than the number of parameters, it is clear that the system is overdetermined and that parameter values calculated directly over various time intervals may not agree due to error in the head values or changing sensitivities. The inverse procedures attempt to calculate the best values of parameters over the historical period of records by minimizing some error functional.

MODEL VALIDATION

The algorithm developed for the parameter identification has two basic components. One that predicts the flow field in a groundwater basin with known characteristics can be directly validated by applying the model to a system for which analytical solution exists. While there is no direct way to check on the accuracy of algorithm for the parameter identification. Any validation attempt has to presuppose that the proposed scheme is feasible.

Piezometric Head Computation

A two aquifer system with identical properties confined by aquicludes on either side and separated by an intermediate aquitard was selected for model validation. The transmissivity and storativity coefficient of aquifers were 0.72 m^2/hr and 1×10^{-3} respectively. The lower aquifer was centrally pumped at a rate of 1000 m^3/day. The leakage factor of the aquitard was 36×10^{-7}/hr and the aquitard storage was neglected in computation. A non-uniform grid network was used to cover a radial distance of 13,200 m from the well. Grid spacing varied geometrically with a factor of 1.5 from a value of 500 m at the center. The time steps used were also varied geometrically with a factor of 1.1 from an initial time step of 1 hr up to a maximum of 500 hr. The analytical solution for drawdown by Hantush (1967) was used to validate numerical results. The drawdown of piezometric surface along a radial line after 3000 hr. of pumping predicted by the model was in good agreement with the analytical results, as shown in Figure 2(a). Figure 2(b) shows the variation of drawdown with time at 500 m distance from the well. The intial deviation is partly due to truncation errors in small drawdown.

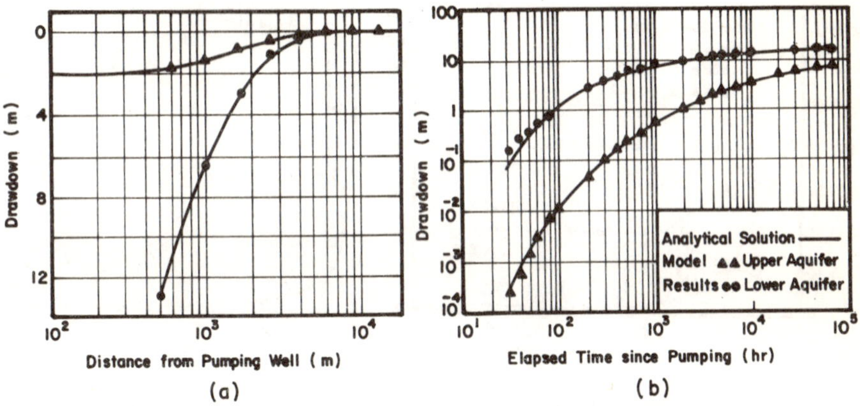

Figure 2 Comparison of model results with analytical solution : piezometric head distribution

Parameter Identification

Due to excessive computing time requirement even for a small system it was not possible to have several runs to determine whether the parameter values always converge to a unique solution for a given system. The criterion for terminating the computation process is either the least square error less than a specified value or the difference between computed and observed piezometric values less than a specified limit or simply by the number of iterations to be performed. The existance of the multiple solutions was tested by performing a foreward computation for a set of parameters. In the inverse computation the value of parameter R was restricted so that it could not reach the true value. The solution of this problem is summarized in Figure 3. The least square error was reduced to a nominal value beyond which there was no further improvement. If least square criterion is adopted there can be a number of combinational values for the parameters. However if a tolerance level is taken as a criterion there is a possibility of arriving at a unique solution. The selection of $\Delta\mu_i = \delta\ \mu_i$ imposes a restriction on the selection of the variation of μ^k.

If the intial estimate μ^k is greater than the true parameter value μ^{k+1} then δ_μ falls to a narrow band between zero and -1 as shown in

Figure 4. Even a slight variation of δ_μ has a marked influence on μ^{k+1} in this range. The oscillations of S value was due to such an overprediction at a stage in iterations. It may take several runs before the value is restored. Storativity S was the most vulnerable parameter to such oscillations probably due to greater difficulty associated with the solution of corresponding sensitivity equation. In such a case the minimum value of least square error corresponds to the best approximation of parameters. This also suggest that the initial estimate provided for the first run should be always a lower value to provide good results.

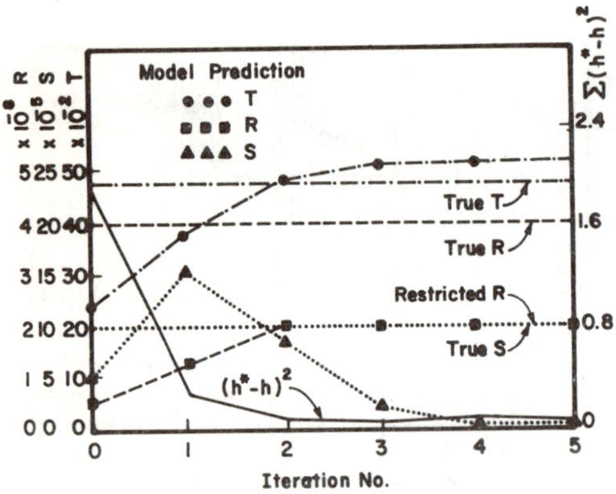

Figure 3 Parameter identification of centrally pumped single aquifer

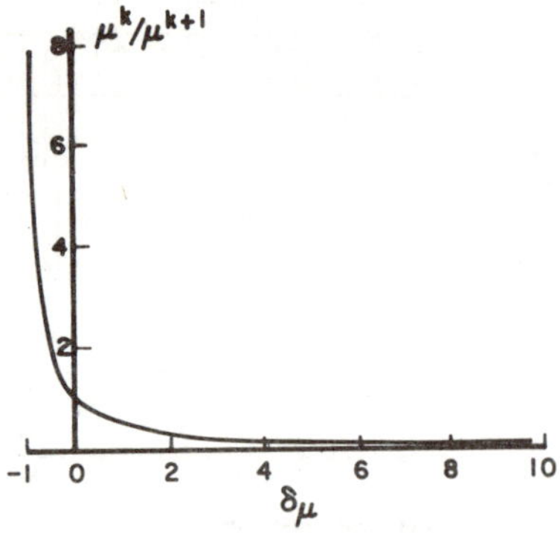

Figure 4 Variation of parameter with multiplication factor

The algorithm for parameter identification was tested by applying to the two aquifer problem considered for the validation of piezometric head computation. The piezometric head distribution was first obtained for leakage factor, storativity and transmissivity value of 36×10^{-7}/hr, 1×10^{-3} and 0.72 m^2/hr respectively for a period of 1000 hr. With these values taken as the observations the model was simulated for the inverse problem. Initial estimates of 36×10^{-8}/hr, 5×10^{-4} and 0.36 m^2/hr for R, S and T were given at the beginning of the simulation. Figure 5 shows the variation of parameter values and reduction of least square error for 5 runs. The leakage factor was recovered in the first run and the least square error was reduced to a value less than 1/5 of the initial figure. In this particular problem leakage factor was the most important parameter as the entire drawdown of the top aquifer was caused by leakage to the lower aquifer. After the fourth run the parameters estimated by the model were in good agreement with the initial given values. Table 1 gives the values of multiplication factor δ for parameter adjustment and $\Sigma(h^*-h)^2$ values for each run. While advancement of model predicted piezometric surface to the observed one with iteration is indicated in Figure 6. This application shows that with a large number of observation points the model behaviour is stable and reliable. The piezometric head computation for 1000 hours for the initial values to be taken as observations took approximately a CPU time of 75 seconds while for the inverse solution the CPU time requirement for 5 iterations was 23

CPU minutes.

MODEL APPLICATION

The model has been applied for updating the parameter values of aquifers for which a management study has been undertaken. Withdrawal of groundwater from multi-layer aquifers underlying the city of Bangkok, mainly for municipal supply and industrial use, is about 1.3 million m^3/day. Such quantity undoubtedly exceeded the natural recharge to the aquifers. The depletion of groundwater due to long-term heavy pumping has caused rapid decline in piezometric head from few meter below ground surface in 1955 to more than 50 m in 1982. The decline in artesian pressure has resulted in land subsidence as well as contamination of fresh water supply by salt water encroachment. A comprehensive investigation over a four-year period was undertaken with the objective of formulating certain management procedure whereby the high rate of decline of piezometric surface could be checked which in turn should restrict the rate of subsidence in Bangkok. Answer to this requirement was obtained through simulation study of the aquifer system to different pumping scenarios in future. Among the different simulated sequences, the temporal and spatial distribution of pumpage that satisfies the specified criterion on expected future response of the system is taken as the best operating policy of the system. The modelled area considered is same as that used for rehabilitation study by Piyasena (1982). The values of parameters used in this study were considered as initial estimates. The model included only three aquifers from the ground surface namely Bangkok, Phra Pradaeng and Nakhon Luang aquifer. The system was considered to be confined by an aquiclude below the third aquifer and a constant water table was maintained above the top aquifer. The model boundaries were selected sufficiently away from the main producing areas so that constant piezometric head could be assumed along the boundaries.

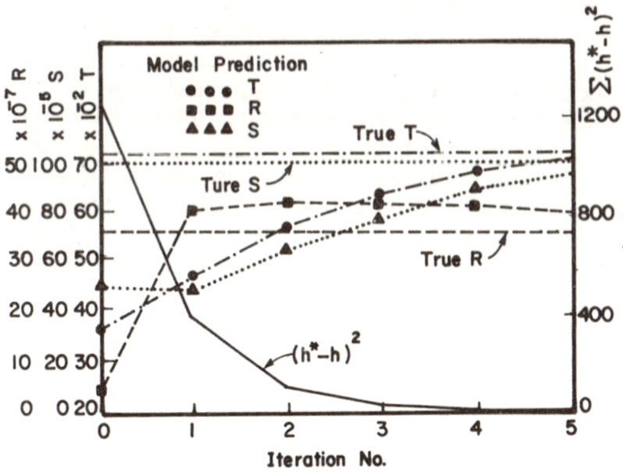

Figure 5 Parameter identification model validation for a two aquifer system

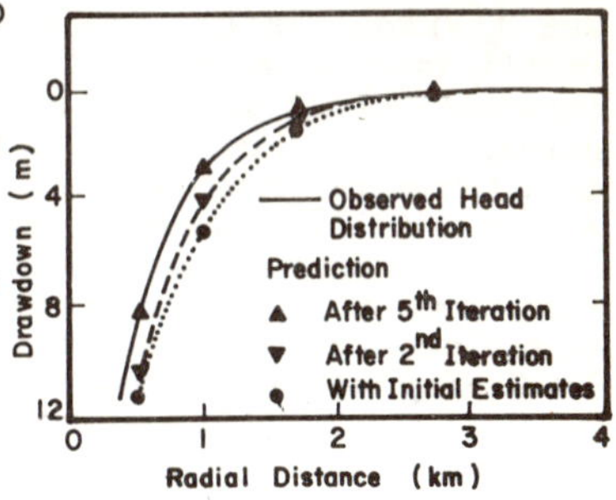

Figure 6 Improvement on prediction of piezometric surface with iteration

Table 1 Variation of Least Square Error and Parameter Improvement Factors with Each Iteration - Two Aquifer problem for Model Validation.

Iteration	$\Sigma(h^* - h)^2$	δ_S	δ_T	δ_R
1	1276.50	-.0464	.3831	13.22
2	373.03	.3980	.3447	.1190
3	99.36	.3212	.2160	-.6958
4	27.72	.2378	.1350	-.2132
5	4.39	.1428	.0696	-.3005
6	0.42	.0725	.0336	-.2931

Even though the parameter estimation model was validated for the constant parameters it was applied for a region of distributed parameters. The initial estimates for the model were taken as 30 times the storativity values and 10 times the leakage factor values corresponding to the early calibrated values. A period of two years was allowed during simulation without any discharge from the system for it to come to equilibrium state. The time step for the simulation was taken initially as 60 day geometrically increasing with a factor of 1.1 to a maximum of 120 day. Whenever there was a change of pumping pattern the time step was reduced to 60 day. Due to limitation on the available computer resources the iteration in each of the difference equations was restricted to four. The results of the first run showed surprising accuracy of the initial estimates. The changes required in the storativity and the leakage factor were insignificant with multiplication factor for the Transmissivity as .17 was estimated to cover a least square error of 20135. The time with this restricted solution scheme for the single simulation was 35 CPU minutes which made it virtually impossible to have more trials with improved accuracies. A second run was performed with these factors multiplied by 0.5, 0.8, 0.8 respectively for storativity, transmissivity and leakage factor. To reduce the CPU requirement sensitivity coefficients were computed only for the last three years. The figures obtained did not provide a significant improvement on the least square error. The CPU time required for this simulation was 10 CPU minutes, a significant reduction on the initial computing time. Variation of least square error and parameter improvement factors are tabulated in Table 2. The results obtained are promising though not as satisfactory as one would wish, however it should be remembered that during the computation of the sensitivity coefficients and the piezometric surface the iterations were terminated before the desired accuracy was attained.

Table 2 Variation of Least Square Error and Parameter Improvement Factors with Each Iteration - Bangkok Aquifer System

Iteration	$\Sigma(h^* - h)^2$	δ_S	δ_T	δ_R
1	23629.47	-.0269	.17	-.0056
2	20135.39	-.0311	.1317	-.0365
3	19773.85	-.0296	.0936	-.0288

CONCLUSION

The algorithm developed is capable of identifying a set of reliable parameters for a multidimensional aquifer system. Due to the particular nature of the partial differential equations governing the flow in a groundwater basin the parameters are extremely sensitive to small fluctuations in the input data, boundary conditions and the solution procedure adopted for the sensitivity equations, thus making it necessary to use the model with caution. The main drawback in the method is the demand for excessive computation time. While the piezometric head computation in a multi-layer aquifer system is tedious and time consuming the inverse problem requires solution of three such additional equations. This implies that in place of single simulation of algorithm for the inverse problem three test runs could be performed with trial parameter distributions. Depending on the number of iterations needed to arrive at satisfactory results in inverse process, model calibration through manual adjustment of parameters could be more practical and economical.

However for small aquifer system or for a system where piezometric measurements are available for more than at one time level sufficiently close so that the required simulation period is not so long this algorithm provides an easy way of estimating parameter values. It was noted that the influence of storativity in the final predicted piezometric surface is very small. While a change of storativity value affects the shape of time drawdown curve but this parameter does not show significant influence on the long term computation of piezometric surface. Large amount of computation involved in the solution for a quasi-three-dimensional system makes its practically impossible to have more than three or four iterations in the inverse computation. Therefore to increase the usefulness of the algorithm it is necessary either to find an appropriate convergence accelerator or to treat the system as an amalgamation of two dimensional systems.

REFERENCES

Emsellem, Y., and De Marsily, G., 1971. An automatic solution for the inverse problem. Water Resour. Res. v. 7, no. 5, p. 1264-1283.

Hantush, M.S., 1967. Flow to wells in aquifers separated by a semipervious layer. J. Geophys. Res. v. 72, no. 6, p. 1709-1720.

Peaceman, D.W., and Rachford, J.R., 1955. The numerical solution of parabolic and elliptic difference equations. J. Soc. Ind. Appl. Math. v. 3, p. 29-41.

Pinder, G.F., and Bredehoeft, J.D., 1968. Application of digital computer for aquifer evaluation, Water Resour. Res. v. 4, no. 5, p. 1069-1093.

Piyasena, W., 1982. Rehabilitation of a depleted aquifer system through artificial recharge - an application to Bangkok. Doctoral Dissertation, Asian Institute of Technology, Bangkok.

Vemuri, V., Dracup, J.A., and Erdmann, R.C., 1969. Sensitivity analysis method of system identification and its potential in hydrologic research, Water Resour. Res. v. 5, no. 2, p. 341-349.

Yeh, W. W-G., 1986. Review of parameter identification procedures in groundwater hydrology: The inverse problem. Water Resour. Res. v. 22, no. 2, p. 95-108.

INVESTIGATIONS OF GROUNDWATER RESOURCES AT SELECTED LEVELS IN DROUGHTPRONE AREA OF PAHARI BLOCK, DISTRICT MIRZAPUR,U.P.,INDIA

G.S. YADAV AND T. LAL
Department of Geophysics,
Banaras Hindu University,
Varanasi - 221 005, India.

ABSTRACT

Vertical Electrical Sounding was carried out at 28 locations using Schlumberger configuration for the investigation of groundwater resources in the droughtprone area of Pahari Block, District Mirzapur, U.P. The data were interpreted by partial curve matching and Automatic Iterative Methods of Resistivity Sounding Interpretation (AIMRESI). Based on the results of investigation, three vertical geoelectrical cross-sections representing the base of the alluvial cover, i.e. Vindhyan sandstone exposed in the central parts of the area, and the distribution of groundwater resources along the respective profiles have been presented. Its depthwise areal distribution throughout the region under study have been presented in the four horizontal geoelectrical cross-sections at selected levels. Apart from these sections, a depth contour map is also presented showing the variation of thickness of alluvial cover and fractured sandstones on the top of resistant substratum. Hydrogeological conditions present a complex picture especially due to lithological variations of the alluvial cover derived mainly from Vindhyan sediments. The study indicates that the exploitable groundwater is mainly confined to the sandy layers and sandy-clayey layers in the northern parts of the area under investigation whereas in the southern central parts the groundwater is confined only to the fractured rocks of Vindhyan sandstone.

INTRODUCTION

With the ever increasingdemand for water supply and the inadequate surface water especially in the droughtprone areas, attention is turned to groundwater resources mainly because it is replenishable by natural hydrological processes and could be even perenial provided caution is exercised to maintain a proper balance between the need and the supply. There have been a number of studies to explore the groundwater resources (Keller and Frischknecht, 1966; Zohdy and Jackson, 1969; Koefoed, 1979, etc.). But these were confined to study the vertical variation of the lithological changes along a particular profile. However, in the present paper, an attempt is made, for the first time, to study the horizontal geoelectrical cross-section at any given depth which would enable us to get a lucid picture of the areal distribution of different zones that in turn lead to infer the actual locations of exploitable groundwater. Pahari Block situated in east-south-east of Mirzapur city, is chosen for the present study. The actual area under study lies between 25° 02' N, 82° 38'E to 25° 10'N, 82° 45'E (Fig. 1). The annual rainfall of the region is around 100 cm of which 87% falls in the monsoon season only.

HYDROGEOLOGY OF THE AREA

The area under investigation falls on the northern slope of Vindhyan plateau which has maximum height of about 500 m. It gradually slopes down in this region towards north and is covered with a varying thickness of Pleistocene and Recent alluvial deposits of river Ganga and its tributaries in the northern parts of the area. The outcrops of country rocks in and around the area are the Kaimur sandstones. The lithological information of the area is available through shallow wells. The covered hard rocks are highly jointed and have developed fractures and cracks which permit percolation of copious amount of groundwater to greater depths.

In this area groundwater is mostly tapped from the alluvium from sandy layers and pockets in clay beds around Chatar Nadi and towards river Ganga. At some places, lenticular sand masses, entrapped in the alluvium represent good aquifers. The depth of water table varies from 6 to 8 m and differs from place to place during March-April. The area is recharged mainly through rain water during monsoon period. A considerable amount of moisture is lost because of increased evaporation and evapotranspiration in the summer season. The wells piercing through the alluvium down to the weathered or fractured rocks never get dried. However, wherever the hard rock is encountered at the bottom, the yield is poor and wells dryup.

MATERIAL AND METHODS

Twentyeight vertical electrical soundings were carried out using Schlumberger configurations. A d.c. resistivitymeter was employed. The maximum lengths of current electrode spread (AB) were kept between 300 to 600 m from place to place depending upon the accessibility of the area. Sounding curves suggest a few three-layer geoelectric sections of H- and A- type, a number of four-layer sections of HA- and KH-type and the rest five-layer sections of AKH- and HKH-type. All the field curves were interpreted with the help of auxiliary point charts and three-layer master curves of Rijkswaterstaat (1969). Accuracy of interpretation was checked by estimating the total longitudinal conductance for curves showing infinite resistivity for the last layer.

Layer parameters obtained by above partial curve matching technique were utilized to compute the multilayered sounding curves using digital linear filters of Ghosh (1971) and Yadav

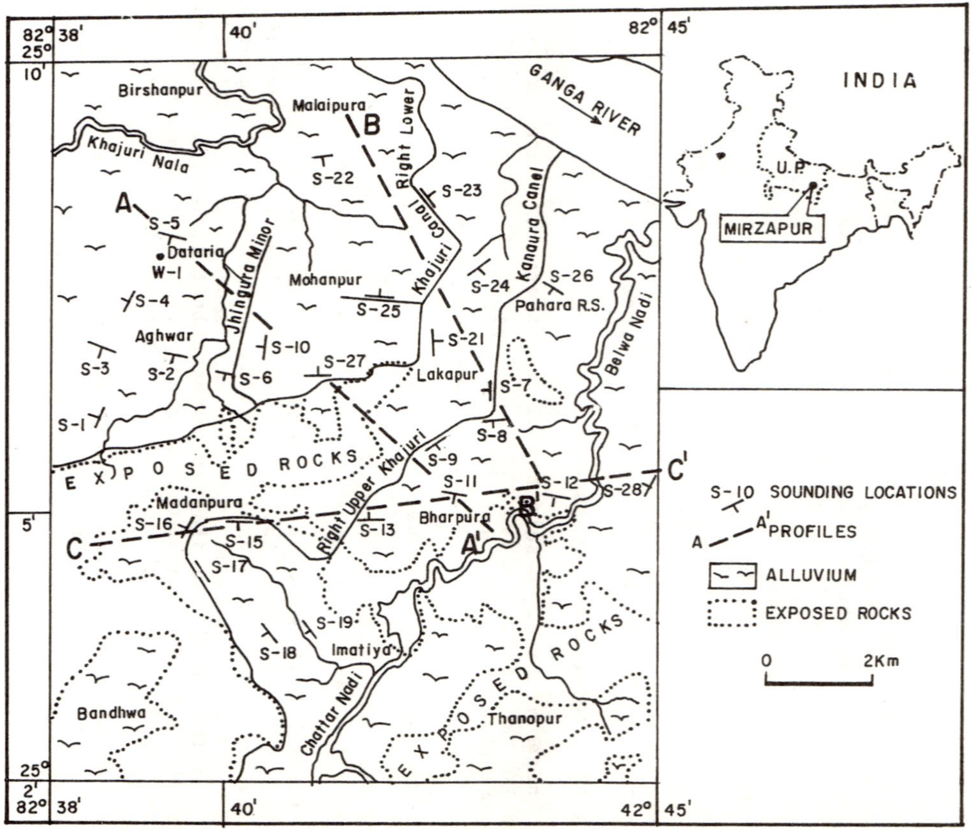

Fig. 1 - Location map of the area.

(1985) on ICL 1904S computer. For better accuracy, the interpretation obtained by curve matching has been modified as a result of theoretical computation of apparent resistivity curve using a computer program i.e. Automatic Iterative Methods of Resistivity Sounding Interpretation (AIMRESI) of Yadav(1982).

RESULT AND DISCUSSIONS

Detailed study of evaluated layer parameters has revealed fluctuations in resistivity values of the top layer due to the variation of moisture content and also due to change in surface conditions from place to place. Its thickness also varies from 1 to 9 m. The resistivity of the second layer shows lower range, between 3 to 17 ohm-m, indicating predominance of clay at many places. Its thickness varies from 4 to 90 m. The resistivity of the aquifers present in the northern parts of the area varies between 20 to 35 ohm-m whereas from the central southern parts the resistivity of water bearing zones varies from 60 to 195 ohm-m. The high resistivity values are indicative of bed rock, i.e. Vindhyan sandstone.

Vertical Geoelectrical Cross-sections:

Three vertical geoelectrical cross-sections are prepared based on the results of vertical electrical sounding data along the profiles AA', BB' and CC' as shown in the location map (Fig. 1). The shallow borehole information available in the area of investigation is also incorporated in the sections. The lithological information available from the existing borehole (W - 1) is indicated at its respective position on the geoelectrical cross-section AA'. For the sake of clarity, resistivity values are written in their respective positions.

Figure 2(a) shows a cross-section AA' (NW - SE) between Dataria and Bharpura based on the results of six vertical electrical soundings and one existing borehole data. The depth of the interface between the clay and sand layer as determined from geoelectrical sounding is in excellent agreement with that obtained from the borehole. A good thickness of sandy layer is present in the north-west side of the section. The thickness of sandy horizon decreases sharply towards exposure of Vindhyan rocks. This sandy horizon is underlain by a thick horizon of clayey bed between sounding locations S-2 and S-10 only. Vindhyan sandstones are exposed between sounding locations S-27 and S-9. Due to this reason, the soil cover near the exposed rocks is very thin which gradually increases in thickness away from the exposed rocks. On the south-east side of the Vindhyan exposure, thickness of the alluvial cover has shown gradual increase reaching upto a maximum of 65 m at S-11. Such good thickness of sandy horizon in the north-west side of the exposure and clay mixed with fine sands in the south-east side are expected to form prolific source of groundwater.

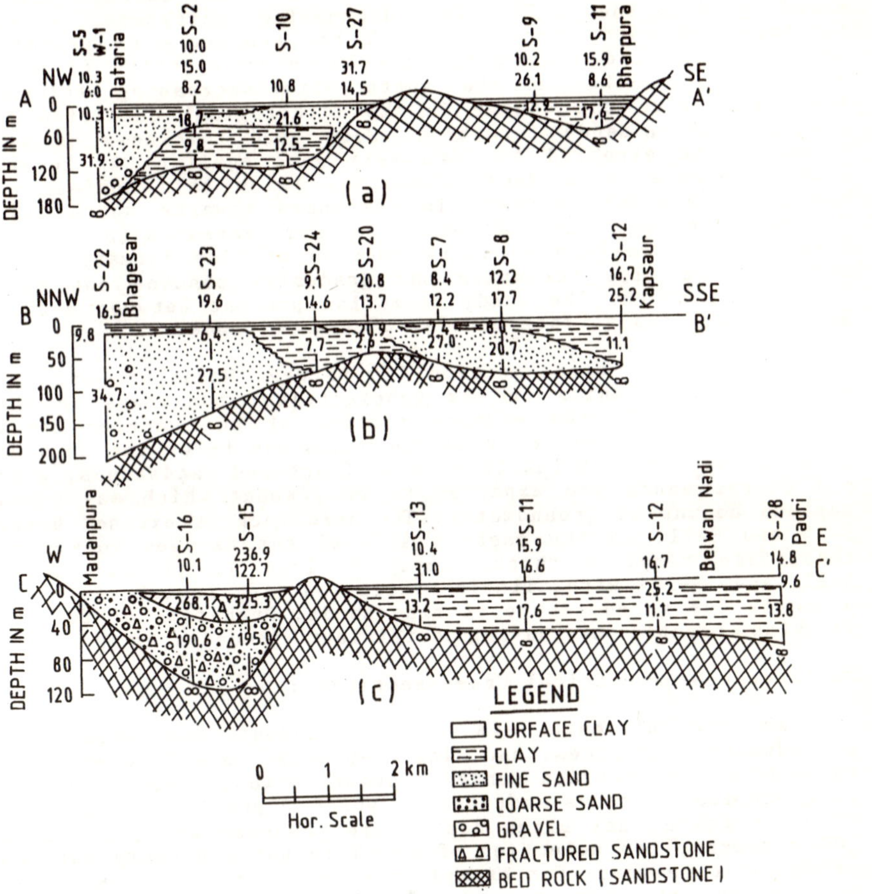

Fig. 2 - Vertical Geoelectrical Cross-sections along profiles : (a) AA', (b) BB', and (c) CC'.

A second cross-section BB' between Bhagesar to Kapsaur based on the results of evaluated layer parameters of seven geoelectrical soundings is shown in Fig. 2(b). This NNW - SSE profile lies in the similar geological setting not very far from the profile AA'. Although no borehole data is available along the profile, it is still possible to arrive at certain definite conclusions due to similarity in the geological settings of the area. Below the surface clay, a very thin clay layer is shown between geoelectrical soundings S-22 and S-23, the thickness of which has shown gradual increase further south, touching directly the bed rock before S-24 and extends towards sounding location S-20 with decreasing thickness. A good thickness of sandy formation with the admixture of gravel is present below the clay bed. This layer is expected to extend maximum upto S-24 in the south with decreasing thickness associated with the change in the lithological condition from gravel sand to sandy formation. Such good thickness of sandy formation is expected to form very good aquifer. At location S-20, presence of a sandy formation below thin surface clay has shown gradual increase in thickness towards south beyond S-8 extending upto bed rock which further decreases in thickness towards S-12 overlain by thick section of clay formation. This near surface clay formation has gradually pinched out beyond S-7 in the north. The sandy formation present between S-20 and S-12 is expected to form aquifer of significance.

A west-east cross-section CC' between Madanpura and Padri based on the results of six geoelectrical soundings is shown in Fig. 2(c). In the western portion of the section between S-16 and S-14, a lens of Vindhyan sandstone below the thin soil cover is detected. Below this lens, fractured sandstones, gravel and coarse sands are expected to be present which may contain copious amount of groundwater. The hard rock is exposed between S-14 and S-11. In the eastern side of the exposed rock, below thesurface clay, a thick clayey horizon is present between S-11 and S-28 with little variation in resistivity value probably due to admixture of fine sands. Such type of formation is also capable of having sufficient groundwater resources.

Horizontal Geoelectrical Cross-sections :

As the ultimate aim is to findout tne presence of groundwater resources and its depthwise areal distribution throughout the region under investigation, four horizontal geoelectrical cross-sections, which provide the areal distribution at any given depth, are prepared at 8, 25, 50 and 100 m depths, the selection of which is not arbitrary but based on specific reasons. For example the one at 8 m is chosen mainly to indicate the presence of groundwater table. However, mere presence of groundwater table will not suffice unless the areal distributions at the depths of exploitable groundwater are known. Accordingly the horizontal cross-sections are prepared at the above mentioned depths and the different zones are identified based on the resistivity values.

Figure 3(a) shows the horizontal cross-section at 8 m depth which reveals more or less similar characteristics as are observed at the ground level, with a few exceptions such as northward shift of the boundary between clay and hard rocks and the presence of fractured rocks in a small strip in the southern central parts. One can also notice the slight extension of small pocket of hard rock in eastern central parts. The

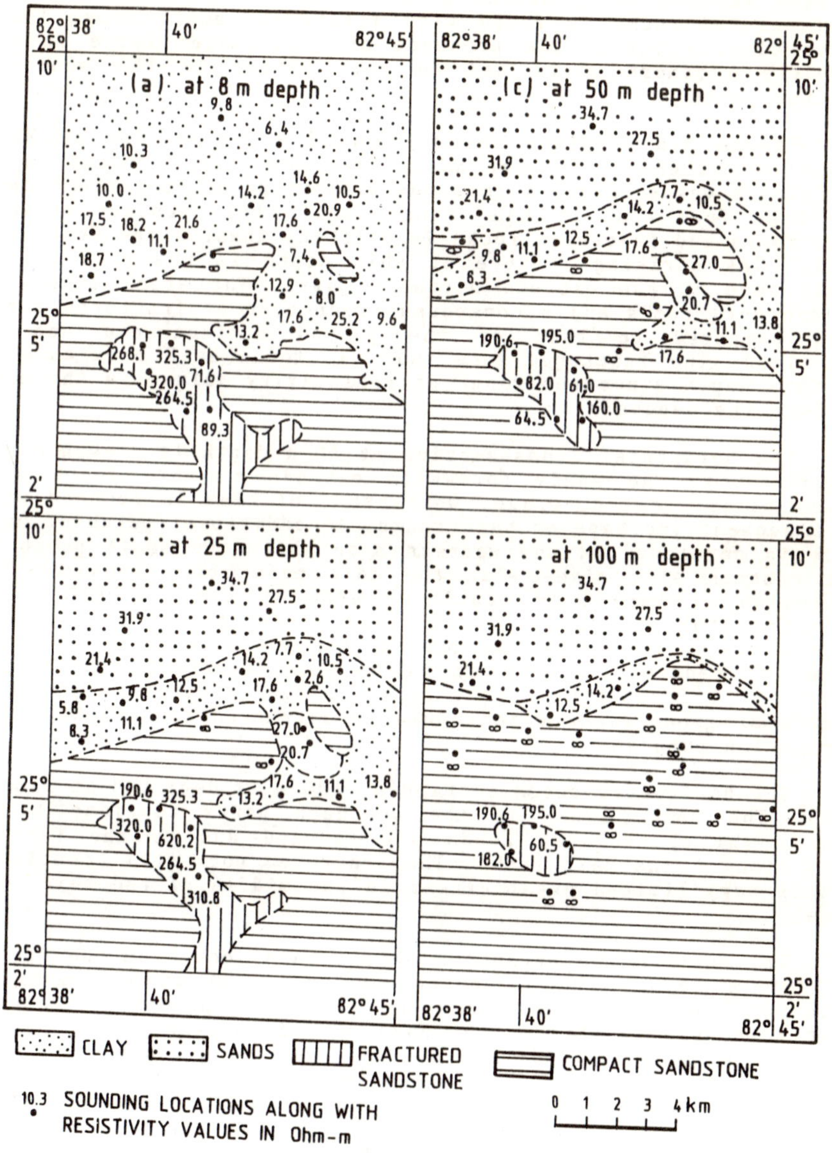

Fig. 3 - Horizontal Geoelectrical cross-sections at selected depth level of (a) 8m,(b) 25m,(c) 50m and (d) 100m.

alluvial cover observed in the southern central parts at the surface is completely absent at this depth and instead, a zone of fractured rocks is found with relatively small area. These facts clearly reveal the presence of groundwater table in the complete northern parts and to some extent in the southern central parts.

Figure 3(b) depicts the horizontal cross-section at 25 m depth. The boundary between hard rocks and clay is found to be shifting towards clayey formations without an appreciable change of shape from that of the ground, while the boundary between hard rocks and fractured rocks is shifting towards the fractured rocks. The striking feature at this depth is the presence of sandy zone in the extreme northern parts. The areal expansion of the hard rocks and a consequent reduction of fractured rocks when compared to that of the ground surface can not go unnoticed. The fractured zone that was found at 8 m extends up to this depth since the range of resistivity values is the same at both the depths.

The horizontal cross-section at 50 m depth is presented in Figure 3(c). The sandy formation extends to this depth with almost same areal coverage, while that of clay is reduced with a consequent increase of hard rocks. An additional patch of hard rock is developed on the western side while the small patch on the eastern side mixes with the main hard rock exposure. The intrusion of hard rock continues even in the southern parts reducing the fractured rock zone further. Moreover, the fracturing of the rocks has gone up considerably as can be seen from the reduced resistivity values. An important consequence of the more fracturing of the rocks is the presence of prolific source of groundwater which could be extracted for drinking and irrigation purposes.

The horizontal cross-section at 100 m is depicted in Figure 3(d). The sandy zones in the northern parts continue to be there, while theclayey zones are reduced due to the further expansion of the hard rocks. The fractured rock still continues with a very limited areal extent in the middle of the hard rock zone.

In general, the expansion of the hard rock from the ground to 100 m depth with a consequent reduction of the area of alluvium and fractured rocks is the most prominent feature mainly because the top of the hard rock is uneven and is having slopes in all directions. It may be noted here that the exploitable water depth is shallow in the northern parts compared to that in the southern central parts which is mainly because, the northern parts are not far away from river Ganga from where the recharge of the water makes these parts relatively more saturated. The southern parts where the scarcity of water is felt, are covered with hilly terrains and are obviously dry. Even in these parts the groundwater could be easily exploited from depths of about 35 to 50 m. Although sufficient information is obtained about the groundwater

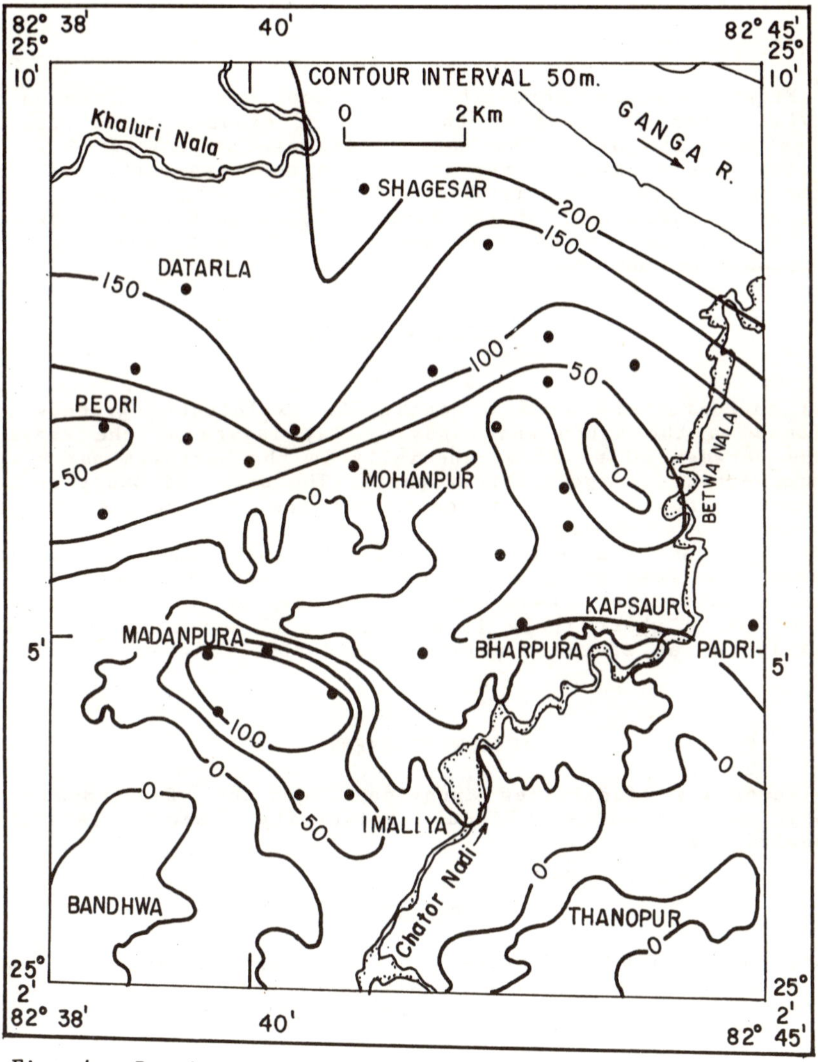

Fig. 4 - Depth contour map showing alluvium thickness above the resistant substratum.

resources from the cross-section analysis, up to 100 m, the topography of the hard rock below 100 m is not obtainable and hence there is a need for a depth contour map which gives, at a glance, the variation of the resistant substratum with depth in addition to the variation of thickness of alluvial cover and fractured rock on the top of resistant substratum.

Depth Contour Map :

Figure 4 shows the depth contour map. Maximum thickness of alluvial cover is found to the north of exposed rocks present in the central parts. The total thickness of fractured rocks including the alluvial cover is less in southern parts compared to the thickness of alluvial cover in the northern parts, the value being twice that of the former. It is this good thickness that is indicative of the presence of vast resources of exploitable groundwater.

CONCLUSIONS

The overall study reveals that the groundwater table is found approximately at 8 m depth. The exploitable groundwater is found in the sands and sandy - clay layers at the varying depths between 20 to 25 m especially in the northern parts and to some extent in the southern parts. The detailed study of the horizontal cross-sections reveals that the area covered by the hard rock is expanding with depth with the consequent decrease of the clay and fractured rock zones. Another important finding is that the fracturing is found to be more with depth which contains sufficient resources of groundwater in the southern central parts where there is a scarcity of water and it could be exploited at specific points from depths ranging from 35 to 50 m.

ACKNOWLEDGEMENTS

Necessary facilities provided by the Department of Geophysics, Banaras Hindu University are gratefully acknowledged.

REFERENCES

Ghosh, D.P., 1971. Inverse filter coefficients for the computation of apparent resistivity standard curves for a horizontally stratified earth. Geophys. Prospect. V. 19, p. 769-775.

Keller, G. V., and Frischknecht, F.C., 1966. Electrical methods in geophysical prospecting. Pergamon Press, New York, 517p.

Koefoed, O., 1979. Geosounding Principles, 1. Resistivity sounding measurements. Elsevier Scientific Publishing Company, Amsterdam, The Netherlands, 276 p.

Yadav, G.S., 1982. Development of theory of linear quadripole-dipole resistivity sounding techniques and Geoelectrical investigations for ground water in hard rock areas of Vindhyan terrain. Ph. D. Thesis, Banaras Hindu University, Varanasi, India.

Yadav, G.S., and Singh C.L., 1985. New sets of direct and inverse Schlumberger filters for resistivity sounding interpretation. Bolletino di Geofisica Teorica ed Applicata, v. 27, no. 108, p. 251-262.

Rijkswaterstaat, 1969. Standard graphs for resistivity prospecting. European Association of Exploration Geophysicists, The Hague, The Netherlands.

Zohdy, A.A.R., and Jackson, D.B., 1969. Application of deep electrical soundings for groundwater exploration in Hawaii. Geophysics, v. 34, p. 584-600.

A More Accurate Determination Of Aquifer Parameters Using The IP Sounding Method

Sayyadul Arafin
and
C.Y. Lee
UNIVERSITI SAINS MALAYSIA, MALAYSIA

ABSTRACT

The induced polarization (IP) method has been used in the present study to resolve the so-called saturated layer into the clay layer and the main water producing zones of the fractured limestone beds in northwest peninsular Malaysia. This information has been exploited in better estimating the hydrogeological parameters , such as the transmissivity, from the surface geoelectrical measurements.

INTRODUCTION

Determination of the fractured limestone zones by the DC resistivity method has not been very succesful in the north-west peninsular Malaysia. Identification and location of such zones are important as they are the major sources of groundwater in the region.

Unlike the DC resistivity method, the induced polarization method is seldom used in groundwater prospecting. A literature search shows that the IP method has been used intermettently over the last thirty years. Vacquier et al (1957) are the pioneers in this field. Sumi (1965), Bodmer et al (1968), Ogilvy and Kuzmina (1972), Olorunfemi (1984), Arafin and Lee (1987) and Seara and Granda (1987) have used the method by utilizing both the sounding and profiling techniques in various geological formations.

Many authors including Heigold et al (1979), Urish (1981), Schimschal (1984), and Ponzini et al (1984) have used surface geophysical methods to determine the hydrogeological parameters such as transmissivity and hydraulic conductivity. Their estimate of the hydrogeological parameters are based on the assumption that the so-called saturated layer contains clean sand only i.e. there is no clay in the aquifer. However, in practice, such a situation is not very often encountered while prospecting for groundwater.The present paper attempts to show that the IP method can be used to identify the fractured limestone zones at depth and separate the clay layer from the saturated layer, which is generally determined from the DC resistivity method. Thus the saturated thickness determined from the IP sounding curve is very likely to be more accurate and consequently, the estimate of the hydrogeological parameter is close to the true representative value.This information has been used here to estimate the transmissivity of the limestone aquifers in north-west Peninsular Malaysia from longitudinal conductance.

GEOHYDROLOGIC SETTING

The three major formations - the Setul Limestone, Kubang Pasu Formation and the Chuping Limestone constitute the surface and subsurface geology of Perlis, the north-

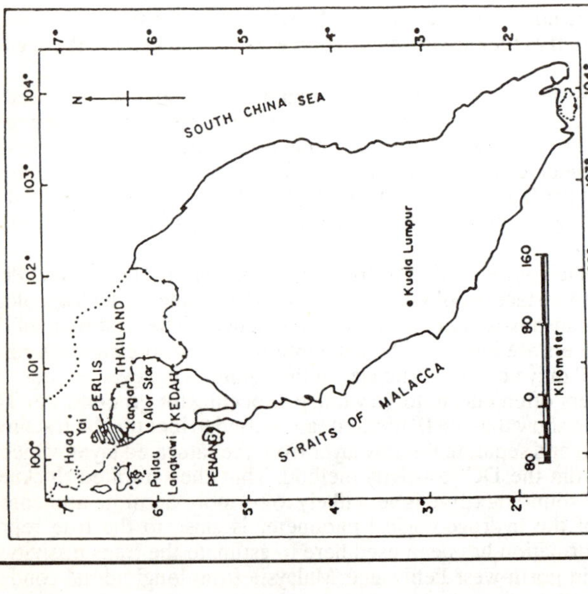

Figure 1: Location of the Study Area.(Shaded).

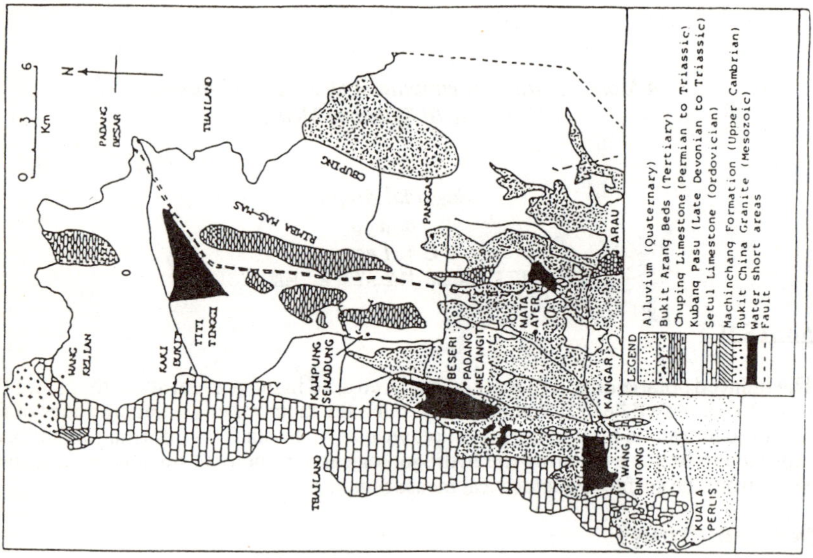

Figure 2: Geology of Perlis
(After Geological Survey of Malaysia)

ern-most state of Peninsular Malaysia (Figure 1). The Setul Formation limestone underlies the Kubang Pasu Formation, which underlies the Chuping Limestone. The Setul Limestone occurs in the west of Perlis (Figure 2) and is composed of hard, brittle, dark-coloured, thick-bedded, variably impure, crystalline limestone.The Chuping Limestone is composed of a pure light-coloured, massive or thickly-bedded limestone sequence (Jones, 1978). Evidence indicates that the formation gradually thins eastwards and is increasingly replaced in this direction by clastic sediments of the Kubang Pasu Formation. The Chuping Limestone hills develop a wide variety of caves, caverns and subterranean passages. The Kubang Pasu Formation consists of red and gray-coloured shale, mudstone and a variety of poorly sorted argillo-arenaceous deposits varying from muddy siltstone through wacke, subgreywacke and arkose to feldspathic quartzite.

Both the limestone formations lack primary permeability and porosity. Groundwater occurs in a semi-confined system below less permeable superficial deposits. The water -bearing zones are related to the intensity of secondary geological activity such as weathering, solution and fracturing. Two types of karst zones occur in the limestone formations of Perlis. The highly fractured zone with associated limestone gravel and rubble beds occur at shallow depths, and the zone of more isolated karst features of much more random occurence at great depth. The upper zone is believed to be part of an ancient karst landscape now buried by more recent alluvial and residual deposits. The Kubang Pasu Formation is generally unsuitable as a source for groundwater. However, yields ranging from 0.4 to 17.8 litres/sec have been reported from the relatively thin sandstone bands where they are factured (Ministry of Works and Public Utilities Report, 1983).

DATA ACQUISITION

The induced polarization sounding (IPS) data were obtained by using Scintrex IP equipment and employing the Wenner array. Care was take in the fields in laying the IP cables so that the electromagnetic coupling effect was minimized. The IP values were measured in the time domain. The vertical electrical sounding (VES) data were measured at the corresponding IP sounding points by using an ABEM (SAS 300) Terrameter. The Offset Wenner array was used in the case of the VES data. Most of the sounding points are located at or near the existing boreholes done by various government agencies including the Geological Survey of Malaysia and the Drainage and Irrigation Department.

INTERPRETATION OF VES AND IPS CURVES

We have presented here the integrated interpretation (both qualitative and quantitative) of two representative electrical sounding and induced polarization sounding curves. The data were interpreted on an IBM microcomputer.

ESP and IPS 139

Resistivity sounding (Figure 3) gives a three layer curve. The first two layers have almost the same resistivity (111.5 and 91.4 ohm-m). In modelling the IP curve the second layer from the resistivity model is split into two layers - the clay layer and a water-bearing layer. The chargeability of the water-bearing layer is 1 mV/V which is quite low

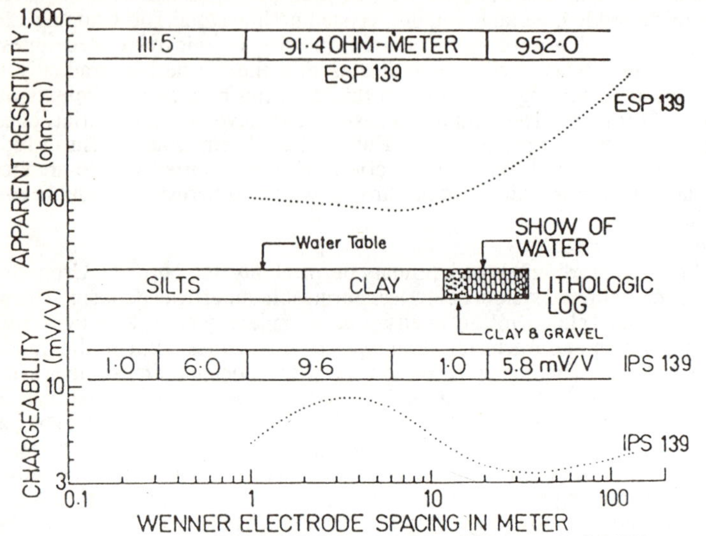

FIGURE 3 INTERPRETATION OF IPS & ESP 139 AT GS 798.

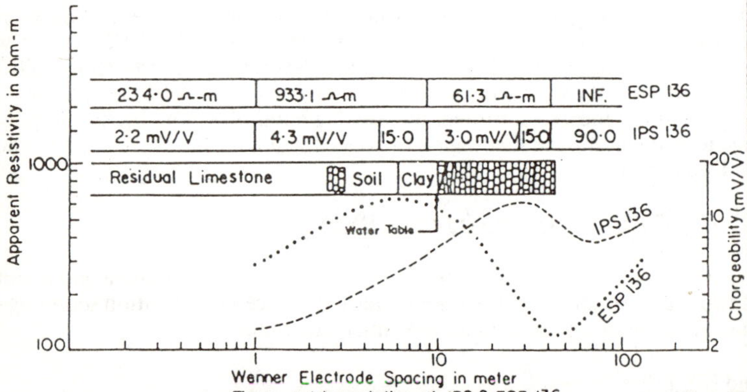

Figure 4 Interpretation of IPS & ESP 136 at borehole GS 790.

when compared with that of the water-bearing layer of the Setul Limestone Formation. Contrary to the majority of the resistivity sounding curves from the Chuping Limestone Formation which are of the KH-type, this is an H-type curve and yet shows the presence of water. However it is noteworthy here that the resistivity of the second layer is 91.4 ohm-m; it is unlikely that this is a clay layer because its resistivity lies at the upper extreme of the resistivity range for clays. This example clearly suggests that a combined interpretation of the resistivity and IP sounding data is very useful in groundwater prospecting in karstic areas. The true thickness of the aquifer is very likely to be less than what is determined from the VES model alone.

ESP and IPS 136

Both types of sounding curves at borehole GS 790 are four-layer curves. However, a six-layer model (Figure 4) is presented here for the IP sounding curve. The important feature of the curve is revealed unambiguously in the model. The IP value of the fourth layer is 3.0 mV/V, which is an indication of the presence of a water-bearing layer. The thickness of this layer is smaller than that shown in the corresponding resistivity sounding model.The thickness determined from the IP model is considered to be the true thickness of the aquifer.The model has clearly separated the clay layer from the water-bearing fractured limestone layer.

EMPIRICAL RELATIONSHIP BETWEEN TRANSMISSIVITY AND LOGITUDINAL CONDUCTANCE

In this study an empirical relationship has been established (Figure 5) between the longitudinal conductance (a parameter obtained from surface electrical measurements) and the transmissivity of the aquifers in Perlis. Table 1 shows the thicknesses and resistivities of the aquifers as determined from the combined interpretation (whenever appropriate) of the resistivity and IP sounding curves. Because of the better resolving power of the IP method the thicknesses of the aquifers determined by the integrated approach are more reliable than those done earlier (Arafin and Lee,1985). For bowl-shaped curves such as the ones in Perlis, longitudinal conductance can be determined uniquely from the VES curves. Figure 5 shows that the longitudinal conductance decreases log- linearly with increasing transmissivity. This means that with increasing transmissivity, either the bulk resistivity of the aquifer increases or the thickness of the aquifer decreases or both. If it is assumed that the thickness of the aquifer is constant, then an increase in the resistivity of the aquifer with an increase in transmissivity means that the aquifer material is of less clayey type and/or the electrolytic resistivity is higher indicating less salinity. The scatter in the values in Figure 5 is perhaps due to the complex hydrogeology of Perlis. The empirical relationship used here is of the type :

$$L = A.T^B \quad \text{.........................} \quad (1)$$

where L is the longitudinal conductance, T is the transmissivity and, A and B are constants. The values of A and B, determined from least squares analysis, are A = 0.032, B = - 0.371 and the value of the coefficient of correlation, r = - 0.89 .

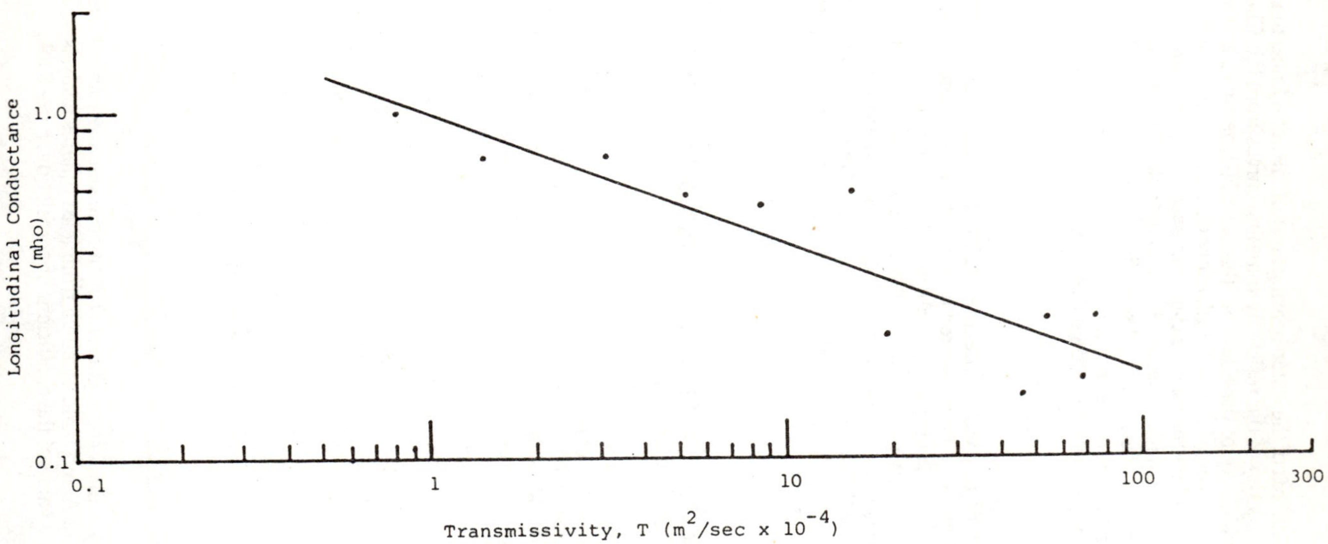

Figure 5: Empirical Relationship between Longitudinal Conductance and Transmissivity.

TABLE 1
Aquifer Parameters From VES And Pumping Tests Data

Borehole No.	ESP No.	Resistivity- (ohm-m)	Thickness (metre)	Longitudinal Conductance (mho)	Transmissivity $m^2/sec \times 10^{-4}$
787	96	26.7	26.07	0.98	0.8
786	201	24.0	17.60	0.73	1.4
795	29	51.2	38.50	0.75	3.2
790	136	61.3	21.00	0.57	5.2
778	197	28.7	15.50	0.54	8.7
779	147	22.8	13.48	0.59	15.4
780	137	20.0	4.56	0.23	19.4
863	193	210.0	32.00	0.15	46.3
865	194	110.0	29.00	0.26	55.6
798	139	91.4	21.10	0.16	69.9
771	195	53.7	14.52	0.27	75.2

CONCLUSION

The present study has shown that the induced polarization method is capable of resolving the saturated layer into a clay layer and the main water-producing zone which is the fractured limestone zone in this case. This particular aspect of the IP method has been exploited to better estimate hydrogeological parameters such as the transmissivity from surface geophysical measurements.

ACKNOWLEDGEMENT

We would like to thank Universiti Sains Malaysia for funding this project. Thanks are also due to the Geological Survey of Malaysia, Public Works Department and the Mines Research Institute for their cooperation in the field work.

REFERENCES

Arafin, M.S., and Lee, C.Y.,1985, A Resistivity Survey for Groundwater in Perlis Using Offset Wenner Array : IAHS Publication No. 161, p.221-232.

Arafin, M.S. and Lee, C.Y., 1987, Application of Resistivity and Induced Polarization Methods in Locating Karstic Aquifers in Perlis : Proc. Int'l. Conf. Groundwater and Environment, June,1987, Kuala Kuala Lumpur.

Bodmer, R., Ward, S.H., and Morrison, H.F., 1968, On Induced Electrical Polarization and Ground Water: Geophysics,V.33, p.805-821.

Heigold, P.C., Gielkeson, R.H., Cartwright, K., and Reed, P.C., 1979, Aquifer Transmissivity from Surficial Electrical Methods: Ground Water, V.17, p.339-345.

Jones, C.R., 1978, The Geology and Mineral Resources of Perlis, North Kedah and Langkawi Islands : Geol. Surv. Malaysia Memoir, V.17.

Ministry of Works and Public Utilities Report, 1983, Government of Malaysia.

Ogilvy, A.A., and Kuzmina, E.N., 1972, Hydrogeologic and Engineering Geologic Possibilities for Employing theMethod of Induced Potentials : Geophysics, V.37, p.839-861.

Olorunfemi, M.O. , 1985 , Computer Model Studies of IP and Resistivity Response of a Typically Saline Sandstone Aquifer (Wenner Electrode Array Depth Sounding) : Geoexploration, V.23, p.193-205.

Ponzini, G.S., Ostroman, A.,and Molinari, M., 1984, Empirical Relation between Electrical Transverse Resistance and Hydraulic Transmissivity : Geoexploration, V.22, p.1-15.

Schimschal, V., 1981,The Relationship of Geophysical Measurements to Hydraulic Conductivity at the Brentley Damsite, New Mexico : Geoexploration, V.19, p.115-125.

Seara, J.L., and Granda, A., 1987, Interpretation of IP Time Domain /Resistivity Sounding for Delineating Sea-Water Intrusions in Some Coastal Areas of North-East of Spain : Geoexploration, V.24, p.153-164.

Sumi, F., 1965, Prospecting for Non-metallic Minerals by Induced Polarization : Geophysical Prospecting, V.13, p.603-616.

Urish, D.W., 1981, Electrical Resistivity -Hydraulic Conductivity Relationships in Glacial Outwash Aquifer : Water Resources Research, V.13, p.1401-1408.

Vacquier, V., Holmes, C.R., Kintzinger, P.R., and Lavergne, M.,1957, Prospecting for Groundwater by Induced Electrical Polarization : Geophysics, V.22, p.660-687.

INTEGRATED GEOLOGICAL AND GEOPHYSICAL INVESTIGATIONS FOR DELINEATING KARSTIC ZONES IN CARBONATE FORMATIONS OF ANANTAPUR DISTRICT, A.P., INDIA

T. SHIBASAKI+, T. VENKATESWARA RAO*, B. VENKATANARAYANA*, C.P. GUPTA* and S. SANKARAN*

* National Geophysical Research Institute, Hyderabad-500 007
+ Quaternary Geology Laboratory, GRDC, Bandung, INDONESIA

ABSTRACT

Detailed geological and geoelectrical investigations were carried out to understand the process of karstification in lower Cuddapah carbonate rocks of Proterozoic age at Chillavaripalle, Anantapur district, Andhra Pradesh, India. The geological studies revealed presence of a few shear and fault zones. Geomorphological observations helped identify a few horizontal cavernous zones occurring at different levels above the present river bed. Regional and detailed geoelectrical sounding and profiling indicated the presence of cavernous zones at shallow depths with a good groundwater potential. A bore well drilled at one of the identified subsurface cavernous zones has given a high groundwater yield of 50,000 lph.

Another significant finding is the close correlation of the levels of cavernous zones and planation surfaces. This phenomenon appears to be a global one with similar observations in Japan, U.K., Africa and USA. The eustatic movements which occurred during Quaternary period in several parts of the world might have played a significant role in karstification of carbonate rocks.

INTRODUCTION

The term karst, indicating a specific morphologic feature developed in carbonate rocks, is synonymous with cavernous conditions in limestone formation. It implies carving of carbonate rocks through the leaching and dissolution of these rocks along shear zones, fault planes and other weak zones. Intense karstification gives rise to secondary void spaces which control the storage and movement of groundwater.

Integrated studies involving geomorphological, lithological and structural investigations followed by hydrogeological and geophysical (electrical resistivity) exploration led to the identification of karstic zones having good groundwater potential in Chillavaripalle, Anantapur district, Andhra Pradesh, India (Lat. 14° 38'45" N, Long. 77° 50' E). The location is shown in Fig. 1a.

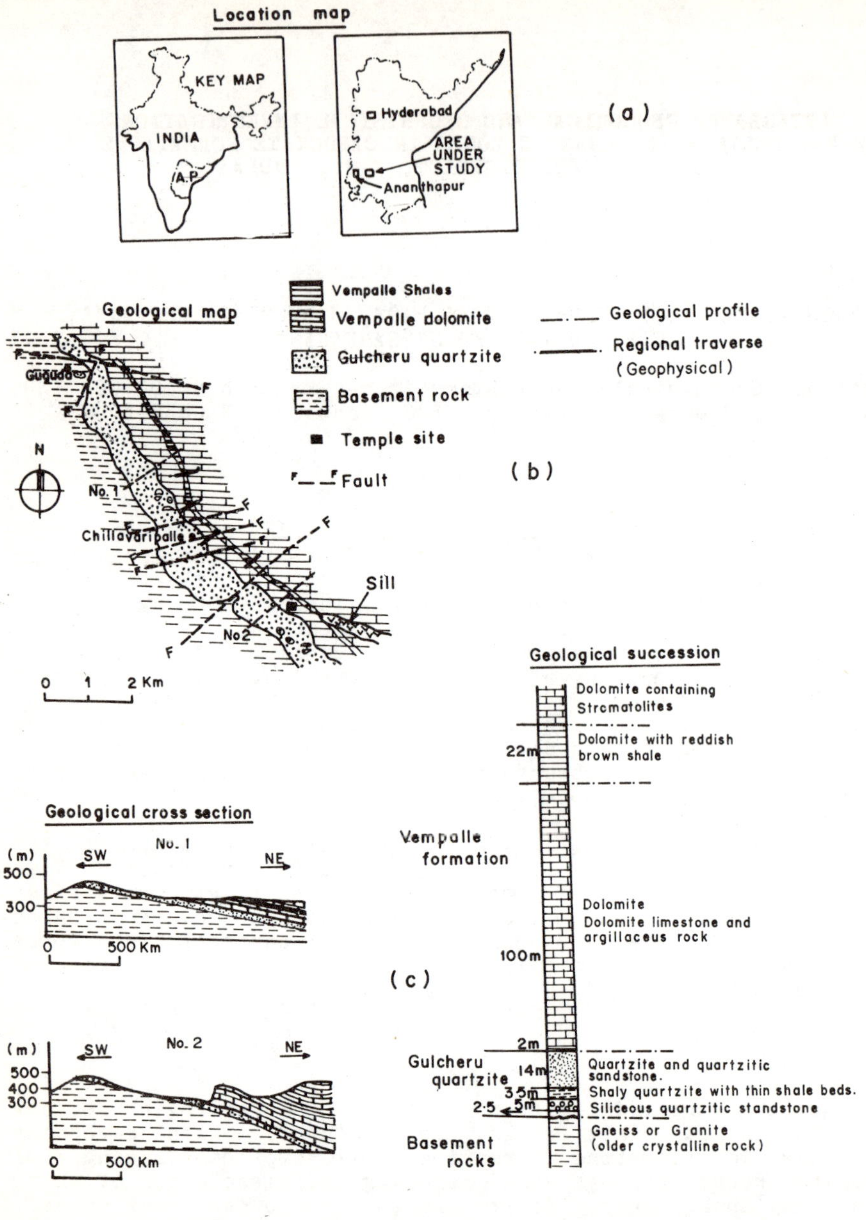

Fig. 1. LOCATION, GEOLOGY, AND STRATIGRAPHY OF THE AREA UNDER STUDY

GEOLOGIC SETTING OF THE AREA

Chillavaripalle situated about 20 km east of Anantapur town forms a part of Cuddapah basin. Huge calcareous and limestone deposits belonging to lower Cuddapahs of Proterozoic age are encountered. Balakrishna et al. (1979) presented a broad resume of the geological, structural and tectonic features of the area. One finds here rock formations belonging to Papagni series of lower Cuddapah group comprising older Gulcheru quartzites and younger Vempalle dolomites. They unconformably overlie the basement crystalline rocks (Fig. 1b). The Gulcheru series consist of basal conglomerates and quartzitic sandstones with intercalations of thin shale bands. They often exhibit well developed cross bedding and other sedimentary structures. They form cliffs and exhibit resistant topography due to their massive and hard nature.

The Vempalle series, comprising of dolomitic limestones (each bed being about 30 cm thick) with intercalations of quartzitic sandstone and shale beds conformably overlie Gulcheru series. Stromatolites and algal microbiota are also found. These formations show a gentle NE dip, but along the valley, near Katamiah temple steeply dipping beds are observed (Fig. 1c). Geological cross sections and stratigraphic succession are also shown in Fig. 1c.

A mega lineament (upto 300 km) and a parallel minor lineament trending WSW-ENE appear to occur in the northern and southern portions of the area, respectively. A major dip fault in which the NW block showing a downward displacement of 50 m is observed near the Katamiah temple. A few minor dip faults are also observed (Shibasaki et al., 1985).

KARST TOPOGRAPHY

Topography of the area is characterized by the presence of ridges and cuestas. The ridges are mostly parallel to the strike direction of Gulcheru quartzites and extend in NW-SE direction with an interval of several kilometers between them and the cuestas which decline gently towards NE.

The valley is formed by a gentle inclination in SW slopes and steep cliffs in NE slopes. Two or three topographic flat surfaces which appear to be the remanents of erosional river terraces are observed in the quartzitic slopes. On the other side, a cavernous system occurs in the dolomitic cliffs.

The topographic surfaces are more or less at the same horizontal levels as the cavernous zones (Fig. 2).

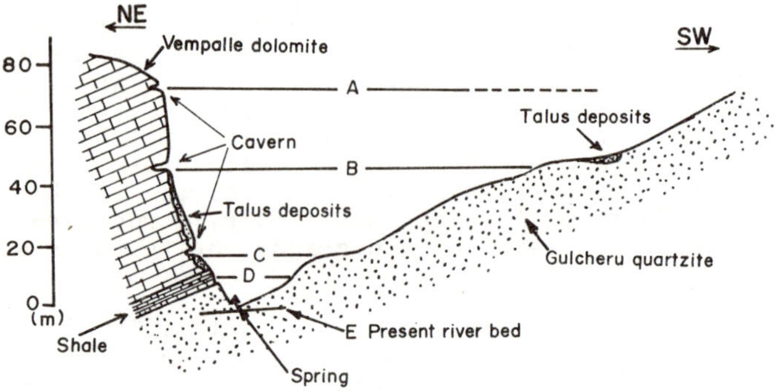

Fig. 2. SCHEMATIC GEOLOGICAL PROFILE (Temple site)

The topographic terraces A, B and C are formed at three levels viz., 75-80 m, 41-44 m and 17-18 m, respectively above the present river bed. Correspondingly, three cavernous levels A, B and C occur in the dolomitic cliffs. Though level A is a collapsed, level B is quite pronounced while level C is a relatively less developed. The cavernous feature at level B is controlled by three main vertical joint systems and cuts the dolomites obliquely.

GEOELECTRICAL SOUNDINGS

A number of vertical electrical soundings (VES) at an interval of 50 m, were carried out using Wenner electrode configuration along four regional geoelectric traverses, each separated by about a kilometer (Refer Fig. 1). These surveys have indicated the presence of low resistivity zones at shallow depths (5-10 m bgl) along almost all the traverses. Along some traverses, the low resistivity zone extends upto a depth of 30-40 m.

On the basis of the above inferences and available hydrogeological and geomorphological evidences such as the occurrence of springs and subsurface outflows in the Gulcheru quartzitic terrain, further detailed geoelectrical profiling along 52 traverses was carried out across the axis of the valley. Electrode separations ranged from 10 to 50 m, with a traverse interval of 20 m and station interval of 10 m. Apparent resistivity contour maps for each separation were prepared. One such map for a=20 m is shown in Fig. 3a.

A 100 ohm-m resistivity SE-NW contour line starting at station 8, divides the surveyed areas into two zones viz., a high resistivity zone-I (northeastrern portion) and a low resistivity zone-II (Fig. 3a). This indicates the existence of highly weathered and cavernous limestone formation in southwestern portion of the area which may be filled with low resistivity material. High resistivity zone-I indicates a barrier like feature which might have been caused by a dip fault. A few natural barrier like structures are inferred by hydrogeological observations also near the Katamiah temple. The high resistive zone-I (Fig. 3a) is traversed by a relatively low resistivity zone of about 200 ohm-m and divides it into two zones viz., A and B. The resistivity values in zone A range from 100-700 ohm-m and from 300-2,200 ohm-m for zone B, between traverses 20 to 40. This demarcation of A and B zones indicates the localized lithological variations in the dolomitic formations. The low resistivity zone in between zones A and B is probably caused due to a cavity formed by intense weathering and leaching.

Two low resistivity zones - one in the nothern corner extending in NE-SW (between traverses 47-52) and another in the southern portion between traverse 1-30 could be demarcated. The southern low resistivity zone with an areal extent of about 20,000 sq m might have been caused by intense weathering/karstification of dolomitic limestones. The nearest outcrops of quartzitic formation are found about 400 m away from this zone whose depth at this low resistivity zone is estimated to be around 50 m. The dip fault which approximately coincides with the traverse 12, may be a good aquifer. The rain water infiltered along the bedding planes might have helped formation of a cavernous zone.

The southern low resistivity zone observed along traverses 1, 3, 5 and 10 (Fig. 3b) extending upto a depth of 10 to 20 m could be correlated with the levels C and E (Fig. 2).

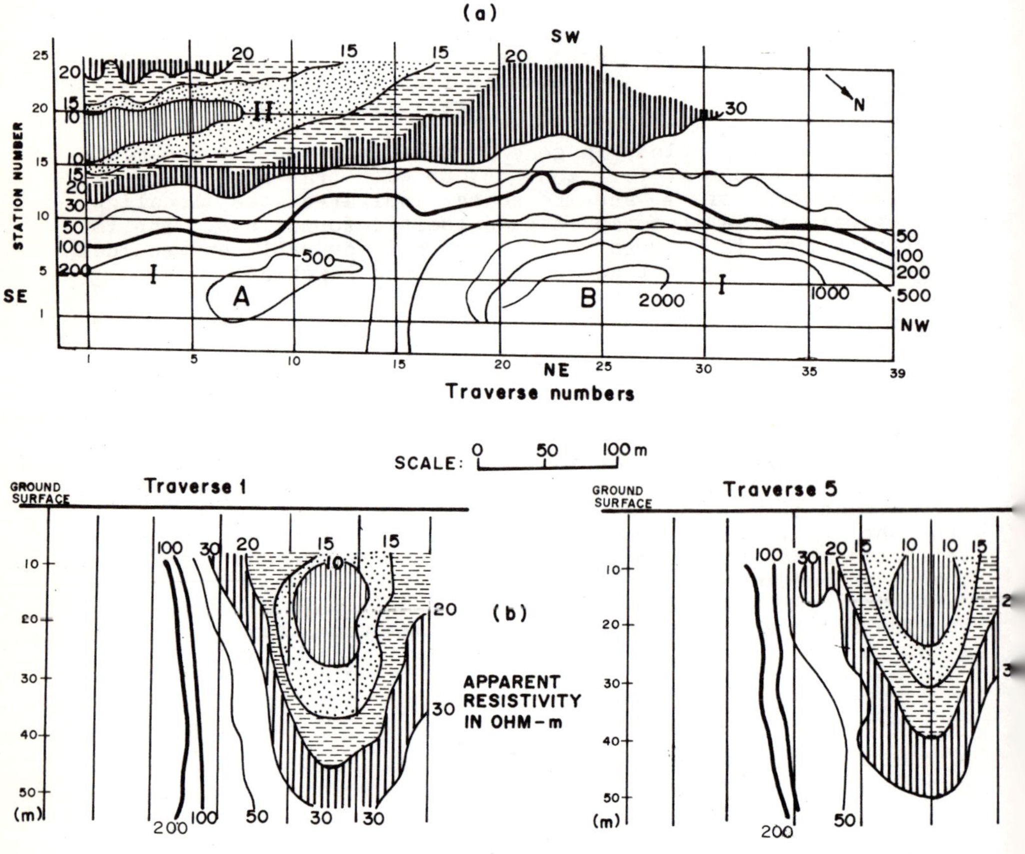

Fig. 3. (a) APPARENT RESISTIVITY CONTOUR MAP (for a = 20m)
(b) GEOELECTRIC SECTION ALONG SELECTED PROFILES

A schematic geological depth profile was drawn along NW-SE direction and is shown in Fig. 4 depicting a karstic zone at a depth of 18 m.

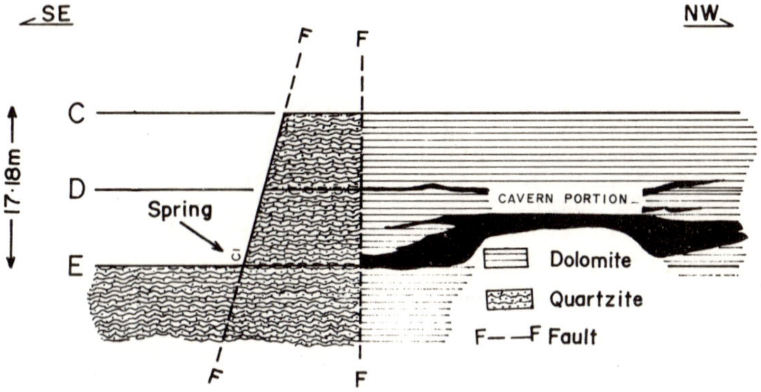

Fig. 4. SCHEMATIC GEOLOGICAL PROFILE ALONG THE VALLEY

DISCUSSION AND CONCLUSIONS

The karstification of carbonate rocks may be attributed to a number of factors such as tectonic movements, paleoclimatic conditions, influence of sea level fluctuations and eustatic movements of Pleistocene and Holocene times, vertical dynamic zoning and influence of carbon dioxide etc. The seat of karstification is usually lithological discontinuities, fractures or any such planes of weak zones. Loosing and gaining streams in the buried limestone terrains intensify the karstification.

At Chillavaripalle, river terraces, and planation surfaces can be correlated with the levels of cavernous zones. These are found to be inter-related and play a major role in karstification. Hall in Sierra Leone (Africa) has also identified such river terrace levels. Further, large scale sea level variations due to eustatic movements during geological past may also be responsible for karstification.

In conformity with the global picture, the cavernous zones of the Cuddapah basin could be classified into five levels. Level A is the highest cavern formation at 80 to 100 m above the present river bed. Level B is the middle cavern formation at 40 to 50 m. Levels C, D and E are the lower cavern formations occurring at heights between 0 to 20 m above the present river bed. Permanent springs are noticed at the same height of the lower cavern formation. This vertical arrangement of cavernous formations especially their relative heights above the present river bed is of vital significance as outlined below.

The heights of the cavern entrances of Taishaku karst plateau, West-Japan could also be divided into various levels similar to those of Cuddapah karst terrains (Kitabingo Plateau Research Group, 1969). The horizontal arrangements of caverns which have different heights were found along Tojo river and were confirmed not only in Japan but also in England (Sweeting, 1950) and in North America (Palmer, 1984).

The large number of caverns which are being formed at present are expected to be near the water table or close to the natural discharging points along the river. Conversely speaking, the cavernous formations distributed at the higher levels above the present water table indicates the position of older water tables of geologic past. The cavernous formation could be regarded as a kind of "fossil" of paleosystem of hydrological flow regime (Shibasaki, 1981).

These facts imply that the vertical distribution of the cavernous formation has resulted from regular geologic processes during the Quaternary period.

The reason for the coincidence of the phenomenon of coplanar arrangement of caverns and river terraces in different parts of the world is not yet very clear. From the field evidences, it may be inferred that these caverns were formed essentially by eustatic movements, and the global wide sea level changes since Middle Pleistocene. A detailed geochronological study could perhaps help our understanding of these complex processes.

ACKNOWLEDGEMENT

The authors express their deep sense of gratitude to Prof. V.K. Gaur, Director, National Geophysical Research Institute for providing all facilities and encouragement. The help received from Mr. P. Prabhakara Prasad, Dr. T. Yoshimura, Dr. H. Furukawa, Dr. N. Chuman, Dr. T. Sato and Mr. Y. Kato, Members, Indo-Japanese team is gratefully acknowledged. Sincere help received from Mr. G.R. Babu and Mr. K.N. Rao, is gratefully acknowledged.

REFERENCES

Balakrishna, S. Venkatanarayana, B. and Rao, M.N., 1979. A short review of geological studies in the Cuddapah basin, Second Workshop on the Status, Programmes and Problems in Cuddapah basin, IIPG, Hyderabad.

Hall, P.K. The diamond fields of Sierra Leone, Published by GS of Sierra Leone, pp. 1-133.

Kitabingo Plateau Research Group, 1969. On the age of the formation of limestone cave: An example, Jour. Geol. Soc., Japan, 75(5), pp 281-287 (in Japanese with English abstract)

Palmer, A.N., 1984. Geomorphic interpretation of karst features, 173-209, In Lafleur, R.G., ed., Groundwater as a Geomorphic Allen and Unwin, Boston.

Shibasaki, T., 1981. Groundwater basin and groundwater flow system: An approach to paleo-groundwater flow system, 109-135. In Sugiyama, R., Ikebe, Y., Hosino, M., and Shibasaki, T., eds., Fluid Dynamics in a Sedimentary Basin, Tokai Univ. Press (in Japanese with English abstract).

Shibasaki, T., Balakrishna, S., Yoshimura, T., Venkatanarayana, B., Furukawa, H., Venkateswara Rao, T., Chuman, N., Ramamohan Rao, Y. and Venkateswarlu, K., 1985. Karstification process of carbonate rocks in the Cuddapah sedimentary basin, Indian Peninsular Shield, J. Fac. Mar. Sci. Tech., Tokai Univ., no.21, pp. 31-45.

Sweeting, M.M., 1950. Erosion cycle and limestone caverns in Ingleborough district, Geogr. Jour., 115, 63-78.

HYDROGEOLOGICAL STUDY OF LATERITE AT A REPRESENTATIVE DAM SITE IN MALVAN VALLEY OF KONKAN REGION (MAHARASHTRA-INDIA) USING SEISMIC REFRACTION METHOD

Azher Ansari
Chief Research Officer

B.S.Shrotri
Assistant Research Officer

Central Water and Power Research Station, Pune-411 024

(Ministry of Water Resources, Government of India)

ABSTRACT

Konkan region in Maharashtra State comprises about 400 km long and 50 to 100 km wide belt on the West Coast, South of Bombay. This region is largely covered by laterite which is generally known for its high permeability, because of which most of the rainfall (about 200 cm anually) is drained to the Arabian Sea. Hence, the need to design appropriate water storage schemes. The peculiar cellular structure of the laterite cap here is a favourable factor for increasing the effective porosity of the formation and as such it acts as a conduit for vertical and/or horizontal movement of seepage water. From the seismic refraction survey conducted at Poip dam, it is concluded that providing a positive cut-off would be an impractical proposition (bedrock being 14-18 m deep). Hence, clay-blanketing is suggested in the reservoir area to prevent/reduce entry of water through the laterite.

1.0 Introduction

1.1 Konkan region comprises about 400 km long and 50 to 100 km wide belt on the west coast in the State of Maharashtra. This region is largely covered by laterite which is generally known for its high permeability. Moreover, due to the abrupt seaward slope of the Western Ghats (hills), most of the surface water is drained to the Arabian Sea, though the annual rainfall in this area is more than 200 cms. As a result there is scarcity of even drinking water for about six months in a year. Hence, the need to design appropriate water storage schemes as it has not been possible to store water satisfactorily by constructing dams in such areas without positive cut-off. Systematic studies with a view to ensuring designed storage capacity (retention) in these areas have not been carried out. Groutability of this formation is also poor. As such chemical grouting may not offer a solution eventhough it is an expensive proposition. There is a possibility that diaphragm walls could be constructed underground upto litho-marge against the direction of subsurface flow for creating underground storage. Such schemes are necessary in areas like Konkan in Maharashtra which are underdeveloped in this repsect though the land is very fertile for cash crops like cashew-nut.

1.2 The Poip Earthen Dam was completed in the year 1974. The maximum height of the dam is 12.36 m and length is 510 m with a waste weir of length 200 m on its right flank. The capacity of the reservoir is 1.16 M.cum. at Full Reservoir Level (FRL) 107 m. A 1.2 km long canal is provided to irrigate 175 hectares of land. Every year, after filling of the tank during monsoon, the reservoir level goes down below sill level within 3-4 months after monsoon (June-October) though as per design, it is expected to hold water till February end. As such, the reservoir does not serve its purpose of irrigation through the canal. But it is observed that in some areas, about 2 kms downstream of the dam, due to groundwater flow the wells are getting abundant water even in summer season which was not the case before the construction of this tank.

1.3 The geology of the area can be described in the following manner. A thick laterite cover exists here which is overlain by very thin cover made up of loose laterite soil. The upper hard portion of the laterite profile is concretionary which has hardened on exposure following erosion. This upper crust is rich in iron and kaolinitic nodules and appears to have a cellular structure. The above cap is underlain by a zone comprising litho-marge (kaolinitic clay material sometimes with small amounts of montmorillonite and micas) and highly decomposed bedrock (containing relicts of the parent rock material together with decomposing feldspars). The lithomargic clay here is yellow in colour which means that it is ferrugineous (i.e. hematitic). Underlying this zone, there is granulite (bedrock) which is the parent rock.

In order to study the hydrogeological behaviour of the laterite and to delineate the bedrock profile, the vicinity of Poip Dam can be considered as a representative of Malvan Valley and as such study was undertaken in this area.

2.0 Method Employed

2.1 A continuous refraction profiling system was employed with geophone interval of 5 metres. A 12-channel seismograph manufactured by SIE, Inc., Houston, Texas, USA was used. Shot points were placed at every 25 m so that each spread of 12 geophones was having 5 shots - 2 at both ends, one at the centre and 2 offsets. The shots were blasted by zero-delay electrical detonators using special gelatine.

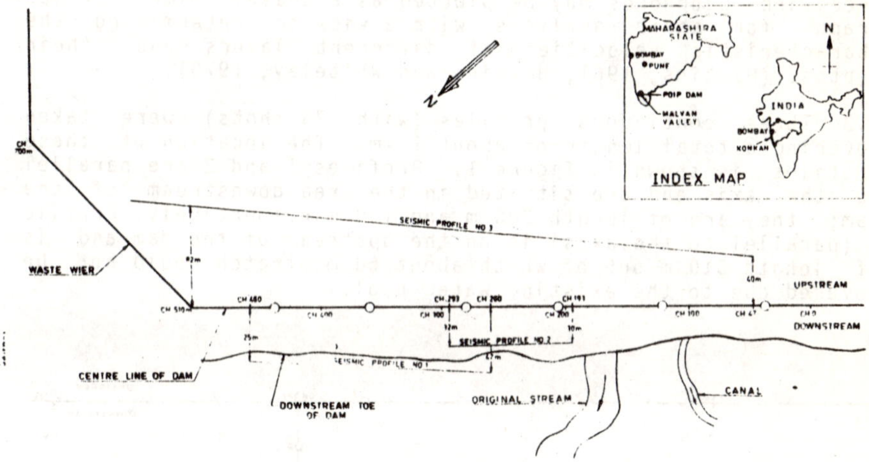

FIG.1. LOCATION OF SEISMIC PROFILES AT POIP DAM, KONKAN REGION (MAHARASHTRA)

2.2 Now-a-days most seismic refraction data are obtained with multi-channel equipment. These record the travel times of seismic waves between a seismic source (a shotpoint) and geophones. Usually there are 12 or 24 geophones which are generally deployed in a linear array, colinear with the source or shotpoint. For very shallow depth investigation, single channel seismograph also can be used. In the conventional seismographs, the multi-channel data may be recorded directly in analogue form with a dry,light-sensitive or photographic process, or on electro-sensitive conductive paper. The data may also be acquired in digital form either on magnetic tape or with digital memory registers. These are then available for subsequent analogue conversion and recording as above, or for some form of digital processing. (Dobrin, 1979).

The first arrival of seismic waves are indicated by the deflections of the galvanometer traces. These usually occur at successively later times with increasing distance from the source or shotpoint. It is very important that a good signal-to-noise ratio be obtained with sufficiently sharp deflections. This is to allow accurate picking of the time of arrival of the seismic waves arriving first at the geophone positions. It is on the interpretation of the travel times curves which in turn depend on the first arrivals that the accuracy of the results of the seismic refraction methods is based. Keeping this in view the travel times of the first-arriving seismic waves from the source point to the different receiving geophones may be plotted as a travel time-distance graph for their analysis with a view to determining the characteristic velocities of different layers and their depths. (Hawkins, 1961, Hawkins and Whiteley, 1979).

2.3 Three continuous profiles (with 75 shots) were taken covering a total length of about 1 km. The location of these profiles is shown in figure 1. Profiles 1 and 2 are parallel to the axis and are situated in the area downstream of the dam; they are of length 200 m and 100 m respectively. Profile 3 (parallel to the axis) is on the upstream of the dam and is of length 510 m out of which about 60 m stretch could not be covered due to the existing water pool.

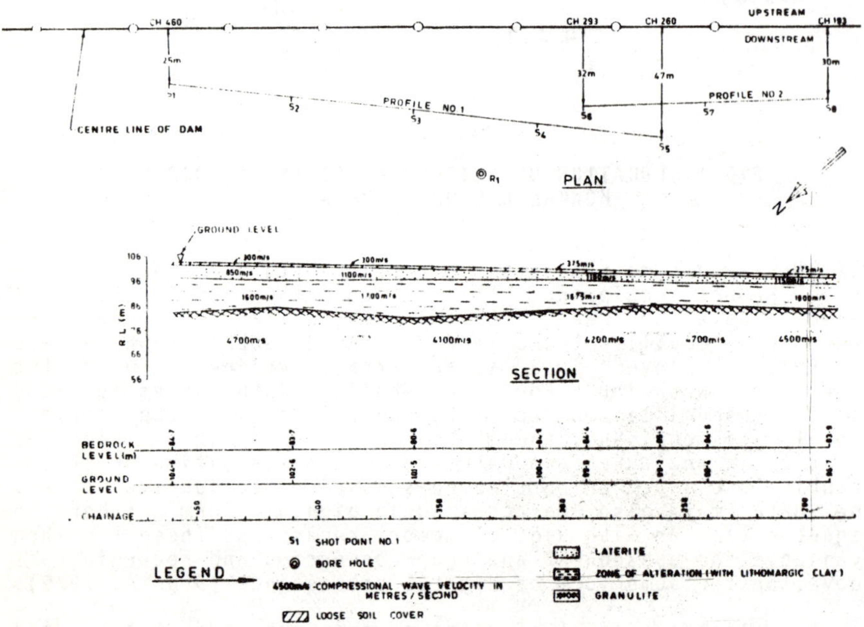

FIG.2. SEISMIC SECTION DOWNSTREAM OF POIP DAM, KONKAN REGION (MAHARASHTRA)

3.0 Presentation of Results

3.1 The thicknesses of different layers below the end shot-points of the 3 profiles are given in Table 1 alongwith description of layers and associated seismic velocities. The seismic section below each profile with chainages and reduced levels of ground and bedrock are shown in Figs.2 & 3.

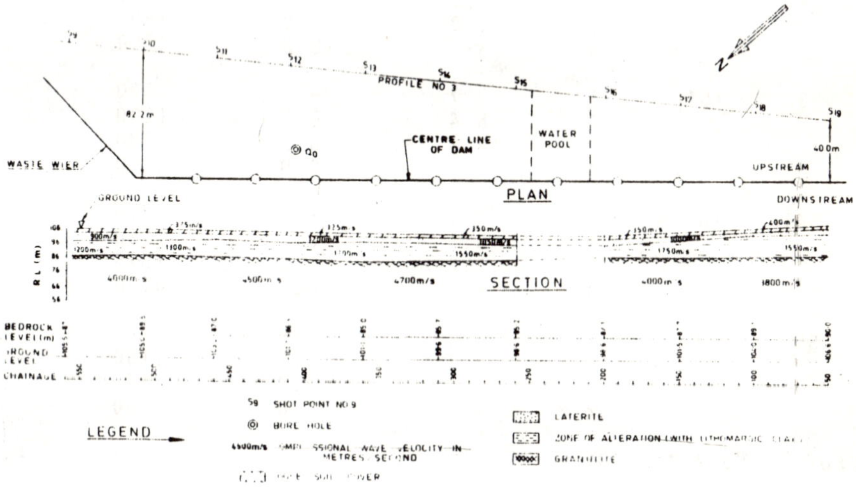

FIG.3. SEISMIC SECTION UPSTREAM OF POIP DAM, KONKAN REGION (MAHARASHTRA)

TABLE-1

Results of Seismic Survey Indicating Layer Thicknesses and Their Seismic Velocities

Shot No.	G.L. (m)	Layer* No.	Thickness of the layer (m)	Depth to the layer (m)	Velocity* of the layer (m/sec)	Remarks
1	2	3	4	5	6	7
Profile No:1**						*Footnote at the
S1	104.6	1	1.0	--	300	**Fig.2
		2	5.7	1.0	850	
		3	13.2	6.7	1600	
		4	--	19.9	4700	
S2	102.6	1	1.1	--	300	
		2	5.9	1.1	850	
		3	11.8	7.0	1600	
		4	--	18.8	4700	
S3	101.5	1	0.7	--	300	
		2	5.4	0.7	1100	
		3	14.8	6.1	1700	
		4	--	20.9	4700	
S4	100.4	1	0.7	--	275	
		2	3.1	0.7	1150	
		3	12.5	3.8	1675	
		4	--	16.3	4100	
S5	99.2	1	1.1	--	275	
		2	3.1	1.1	1150	
		3	9.8	4.2	1675	
		4	--	14.0	4200	
Profile No:2**						
S6	99.9	1	1.2	--	325	**Fig.2
		2	3.5	1.2	1200	
		3	10.8	4.7	1800	
		4	--	15.5	4700	
S7	98.4	1	0.9	--	325	
		2	2.6	0.9	1200	
		3	10.3	3.5	1800	
		4	--	13.8	4700	
S8	96.9	1	1.1	--	275	
		2	1.8	1.1	1150	
		3	10.1	2.9	1800	
		4	--	13.0	4500	

1	2	3	4	5	6	7
Profile No:3***						
S9	105.5	1	1.0	--	325	***Fig.3
		2	4.6	1.0	900	
		3	12.6	5.6	1300	
		4	--	18.2	4000	
S10	105.0	1	0.9	--	375	
		2	5.6	0.9	900	
		3	9.0	6.5	1300	
		4	--	15.5	4000	
S11	103.2	1	0.6	--	390	
		2	2.3	0.6	1000	
		3	13.3	2.9	1500	
		4	--	16.2	4500	
S12	101.7	1	0.8	--	325	
		2	5.1	0.8	1200	
		3	9.5	5.9	1700	
		4	--	15.4	4300	
S13	101.1	1	1.4	--	300	
		2	2.4	1.4	1050	
		3	12.5	3.8	1550	
		4	--	16.3	4700	
S14	99.6	1	0.8	--	350	
		2	2.7	0.8	1050	
		3	10.4	3.5	1550	
		4	--	13.9	4500	
S15	98.6	1	0.8	--	350	
		2	2.6	0.8	1050	
		3	10.0	3.4	1550	
		4	--	13.4	4500	
S16	98.6	1	0.6	--	400	
		2	4.2	0.6	1100	
		3	6.7	4.8	1750	
		4	--	11.5	4000	
S17	101.5	1	0.8	--	400	
		2	2.3	0.8	1000	
		3	10.7	3.1	1450	
		4	--	13.8	3700	
S18	104.0	1	0.6	--	400	
		2	1.3	0.6	1050	
		3	13.0	1.9	1550	
		4	--	14.9	1800	

1	2	3	4	5	6	7
S19	106.6	1	1.1	--	400	
		2	1.6	1.1	1050	
		3	14.0	2.7	1550	
		4	--	16.7	3800	

Foot Note

*Layer No	Seismic Velocity	Layer Identification
1	275 - 400 m/sec	Loose Lateritic Soil Cover
2	850 - 1200 m/sec	Cellular Laterite Cap
3	1300 - 1800 m/sec	Zone of alteration containing lithomargic clay and highly decomposed granulite(Bedrock)
4	3700 - 4700 m/sec	Granulite

4.0 Discussions of Results

4.1 From the results it can be seen that the first layer thickness along the three profiles varies from half a meter to one meter and its velocity is in the range of 275 to 400 m/sec. This first layer is attributed to the thin cover made up of loose lateritic soil. (Ghosh et.al, 1981).

4.2 The second layer below the thin soil cover varies in thickness from 1.5 to 6.0 m. This layer gives a seismic velocity in the range of 850-1200 m/sec. This is attributed to the concretionary upper crust of the laterite (refer 1.3). It has peculiar cellular structure which is confirmed by such a low seismic velocity.

4.3 The thickness of third layer is in the range of 9-14 metres except below shot-point S16 (refer fig.3) where it is less than 7 m. (This may be a local feature on left side of water pool on upstream). This third layer shows a seismic velocity ranging from 1300-1800 m/sec which can be attributed to saturated zone of alteration (refer 1.3) containing lithomargic clay and highly decomposed bedrock (granulite). The variation in seismic velocities of this layer is due to different degrees of leaching of the parent rock (granulite).

4.4 The seismic velocity of the bedrock below all the profiles is in the range of 4000 to 4700 m/sec except at two shot-points S17 and S18 (refer fig.3) where it is 3700 and 3800 m/sec which may be due to some local feature. This is the expected velocity range for granulite which constitutes the parent rock for the formation of laterite in the area.

4.5 The depth to the bedrock on the upstream of the axis (below profile 3) ranges from 11.5 to 18 m while on the downstream (below profile 1 & 2) it varies from 13 to 21 m. Hence, the bedrock appears to be dipping from upstream to downstream of the dam, the maximum dip being around Ch.360 m where the depth to bedrock is about 16 m on the upstream and about 21 m on the downstream (refer fig.2 & 3). On the left flank from Ch.150 m to 50 m, the reduced level of bedrock above mean sea level varies from 87 to 90 m on the upstream side (refer fig.3) indicating presence of a valley near the water pool. On the downstream side, the area between these chainages could not be covered by seismic profile due to thick jungle which suggests possible extension of the same valley facilitating seepage underground which is corroborated by a corresponding increase in the third layer velocity, on upstream and downstream both.

5.0 Conclusion

5.1 The peculiar cellular structure of the laterite cap here is a favourable factor for increasing the effective porosity of the formation. As such it may act as a conduit for vertical and/or horizontal movement of the water. The impounded water of the reservoir, therefore, may percolate vertically downwards in the upstream laterite layer and when it reaches the bedrock it travels horizontally to the downstream as the bedrock is dipping in that direction. After its arrival at the downstream, the seepage water seems to go towards the stream because of the topography as well as slope of the bedrock in the same direction.

5.2 As regards remedial measures, providing a positive cut-off would be a costly proposition (bedrock being 14 to 18 m deep). Hence, clay blanketing (which can prevent vertical entry of water through laterite in reservoir area) is suggested in the present situation. The extent of clay blanketing can be decided by the project authorities to suit the site conditions. The efficacy of it can be determined by carrying out pre-blanketing and post-blanketing tracer studies.

5.3 Keeping in view the behaviour of laterite cap due to its peculiar cellular structure and zone of alteration (made up of lithomargic clay, relicts of parent rock and highly decomposed upper portion of bedrock) which can facilitate movement of seepage water, the sites for future minor irrigation tanks in lateritic area in Malvan Valley may be restricted to areas where bedrock is available at shallow depth with a view to providing positive cut-off. The bedrock profile can be delineated before-hand accurately without drilling by employing seismic refraction method indicating also the nature of the formations from hydrogeological point of view.

5.4 Future Work : Similar work is proposed to be carried out in Devgadh Valley of the Konkan Region at a representative site as the parent rock there is altogether a different geological unit (mainly basalt) while in the above area it is granulite (a meta-sediment which is metamorphosed).

6.0 Acknowledgement

The authors are grateful to Dr.Z.S.Tarapore, Director and Mr.S.L. Mokhashi, Additional Director, Central Water and Power Research Station, Pune for taking keen interest in the work and for giving permission to publish this paper.

Thanks are due to Mr.K.S.Das, Assistant Research Officer, Central Water and Power Research Station, for assisting in the preparation of this paper.

7.0 References

1. Dobrin, M.B., 1978. Introduction to Geophysical Prospecting, McGraw Hill, New York, U.S.A.

2. Ghosh, N., Ansari, Azher, Wadhwa, R.S., 1981. Seismic Velocities from Seismic Refraction Measurements for Dredging and Marine Foundations, 1st Indian Conference in Ocean Engineering, I.I.T., Madras.

3. Hawkins, L.V., 1961. The Reciprocal Method of Routine Shallow Seismic Refraction Investigations. Geophysics 26, P.806-819.

4. Hawkins, L.V. and Whiteley, R.J., 1979. Shallow Seismic Refraction Methods in Exploration and Engineering, Lecture Notes, UNDP, CW&PRS, Pune.

BOREHOLE GEOPHYSICAL LOGGING TECHNIQUES FOR DELINEATING FRACTURED ZONES AND CAVITIES IN HARD ROCK AREAS FOR HYDROFRACTURING : A CASE STUDY

BY

H.N.PARIKH,	DIRECTOR	GWRDC LTD. GANDHINAGAR.
V.V.AYACHIT,	GEOHYDROLOGIST	AHMEDABAD .
N.C.V.N.CHAKRAVARTHY	GEOPHYSICIST,	GODHRA.
R-K-GARG,	JR.GEOPHYSICIST	AHMEDABAD.

A B S T R A C T

During severe scarcity conditions in the Saurashtra region of Gujarat State, deep drilling was taken up to explore the deeper aquifers in the Deccan trap formation. But in some bores the yields were meagre and hence it was decided to attempt hydrofracturing in the weak fractured and jointed zones and cavities by delineating them with the help of borehole geophysical logging techniques. GWRDC Ltd. was assigned the task of bore hole geophysical logging.Resistivity logging and caliper logging was undertaken in the bores selected for hydrofracturing. Correlation of these geophysical logs with oine another and with the lithologs of the bores has helped in delineating prominent fractured and jointed zones and cavities in the bores. The paper presents the analytical studies caried out with the borehole logging data and inferences arrived at from the study.

1.0 Introduction

Gujarat faced the worst draught of the century and the problem of drinking water shortage assumed such a large and alarming proportions that exploitation of water resources from the subsurface through deep drilling became inevitable in 1988. The sites for drilling were selected by careful and intensive geoelectrical resistivity surveyes in the drought hit Saurashtra region. Based on the observed geoelectrical resistivity variations of subsurface layers, a few sites were recommended and drilled upto depths ranging between 102.42 and 181.70m in the hard rock area. However, the yields from some of these bores were meagre and ranged between 8.2 to 129 litres per minute and were not in coherence with lithological layers demarcated from the drill cuttings. In the light of the geoelectrical resistivity variations and the drill cuttings collected at the time of drilling, which indicated weathered, vesicular and fractured trap formations, it was proposed to explore the possibility of hydrofracturing the relevant water bearing zones to enhance the well yields for rightful utilisation. In all, Eleven (11) sites in Bhavnagar,Junagadh and Amreli districts were proposed for exploitation of water through hydrofracturing. In order to delineate the zones suitable for hydrofracturing, it was decided to demarcate the water bearing fractured, jointed zones and cavities by undertaking borehole geophysical logging techniques.

GWRDC Ltd. a frontline organisation in the field of exploration and exploitation of ground water resources in Gujarat State, was assigned the task of attempting borehole geophysical logging techniques by GWSSB, an organisation that looks after the drinking water supply in the state. Though geophysical logging techniques are normally used in alluvial areas to delineate grannular zones of good quality, the task was taken up in view of the seriousness of the drinking water problem. Borehole geophysical logging was undertaken in eleven boreholes to delineate fractured zones and cavities.Resistivity and caliper logs were collected from the bores. Typical geophysical logs of five selected bores, their analysis and interpretation are discussed in this paper. Locations of the selected sites in their respective districts and in Gujarat State are presented in Fig. 1.

2.0 Borehole geophysical logging

Borehole geophysical logging techniques are deployed to determine the resistivity of formation waters and to locate aquifers in the subsurface. They help in supplementing either directly or indirectly the subsurface geological information. It also serves as an important complement to subsurface geophysical data. It also helps in arriving at 'insitu'physical properties of the rocks in the subsurface.Though the well logging methods do not directly measure the physical properties such as porosity, permeability and fractures in rocks, the analysis of various types of measurements can help in delineating and arriving at accurate analytical data. In addition to the porosity and permeability of aquifers,secondary porosity features such as joints, cracks and fractures can be demarcated by using complex geophysical logging methods. In the present study, resistivity logs were obtained by using normal and gradient methods. Emphasis was laid on caliper logging as the variations in the well diameter eventually reflect the nature and locations of weak fractured zones and cavities in the bores. By deploying complex geophysical logging methods and correlation of data of different logs help in arriving at specific inferences in relation to the nature and accurate location of weak fractured zones.

3.0 Analysis and interpretation of well logging data

Analysis and interpretation of geophysical logging data was attempted qualitatively in order to demarcate water bearing / low resistivity zones and their boundaries. Enough care was taken in delineating boundaries of various zones, keeping in view the theoritical curves against homogeneous resistive and conductive beds for different methods of logging.

3.1 Resistivity logs

Resistivity vriations observed from normal logs (N 64" and N 16") and lateral logs have been compiled depth wise to demarcate the probable low resistivity zones indicative of the presence of water or secondary porosity. The resistivity variations have been marked by observing (a) Sharp resistivity contrast (b) Low resistivity ranges and (c) Minimum resistivity values.

3.2 Caliper logs.

Fractured and jointed zones and cavities in trap rocks are loose and fragmented in nature. Drilling and development of the bore by high capacity compressor create cavities and lead to partial collapse of loose and weak parts of the water bearing zones of the bore and increase the diameter of the well. Caliper logging helps in locating such zones.

Further correlation of the caliper log data with the resistivity variations from lateral and normal logs leads to proper demarcation of weak fractured zones. Minor variations in the diameter of the well need not necessarily indicate weak zones, but moderate to larger variations are indicative of weak fractured zones and cavities.Variation in the well diameter are demarcated as minor, moderate and large and have been compiled for further necessary correlation with the rsistivity data.The compiled and analysed data of these five sites are discussed herewith.

3.3. Site : Dihor

Depth of the bore drilled is 102.42 Mts. Litholog of this bore indicates fractured trap at different depths.The yield obtained from this bore was 129 LPM. Resistivity logs obtained from this bore indicate low resistivity values ranging betwen 13.5 and 85.0 ohm.meters. Caliper log also indicated moderate to large variations in the diameter of the well. In all, 9 zones were demarcated from the normal log (N 16") and 10 zones from Normal (N 64") log, while the caliper log indicates 9 zones with moderate to large variations in the well diameter. Fig. 2 presents the litholog and geophysical logs of this bore. Table I-A presents the compiled variation in the resistivity and well diameter.

3.4 Site : Jilana

Depth of the bore is 134.11 Mts.Litholog of this bore indicates fractured and vesicular trap filled with secondary minerals at shallow and deeper horizones. A cavity was also struck at a depth of 45.0 Mts. at the time of drilling. The yield from this well was 48 LPM.....Resistivity log of this bore indicate a variation

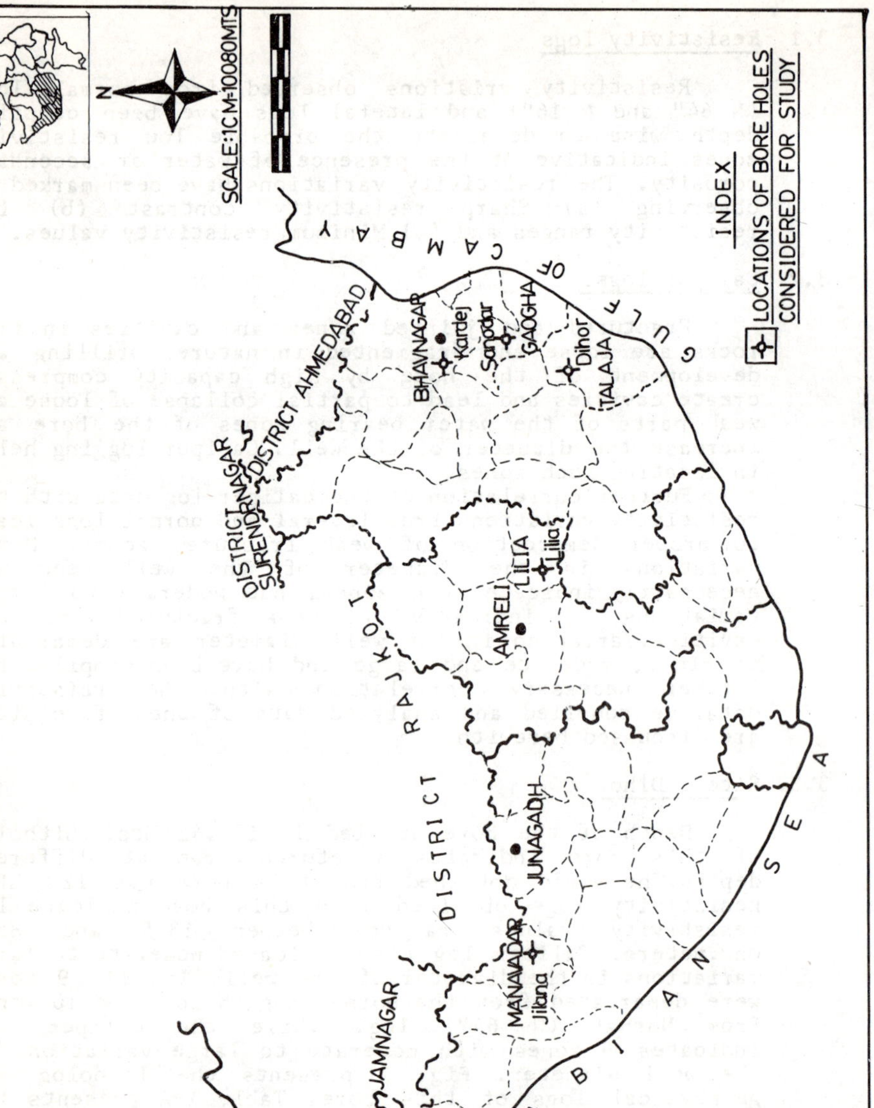

Fig.1 Map showing the locations of the sites selected for study.

Fig.2 Litholog and Geophysical logs of site : Dihor

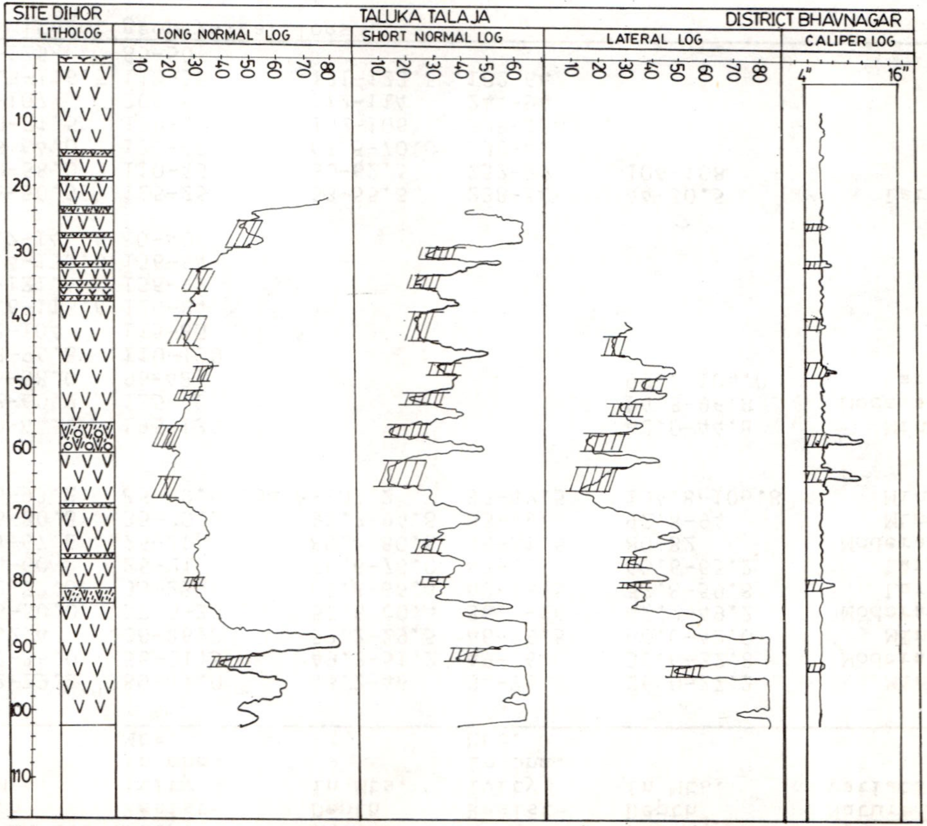

NORMAL LOG (N64")		NORMAL LOG (N16")		LATERAL LOG		CALIPER LOG	
Depth in Mts.	Resistivity in ohm. Mts.	Depth in Mts.	Resistivity in ohm. Mts.	Depth in Mts.	Resistivity in ohm. Mts.	Depth in Mts.	Nature of variation
Site :Dihor							
29.6-31.2	63-27.5	25.2-29.2	80-47.0	43.2-46	34-26	26.0-27.2	Minor
34.0-36.0	55-24.5	32.8-36.0	56-31.5	49.2-51.2	49-36	31.6-32.8	Moderate
39.6-43.6	40-24	40-42.4	30-26.0	53.2-29.5	46-29.5	40.6-42.0	Minor
47.2-49.2	50-31.0	47.6-50.0	37.5-33.0	57.6-60.4	36.5-16	47.2-49.2	Moderate
51.6-53.2	38-21	51.2-52.8	33-28	62.8-66.4	43-13.5	57.8-59.8	Large
56.4-58.4	41-15	56.4-60	25-21	76.6-78.0	47-31	63.6-65.2	Large
62.0-66.0	50-14.5	64.4-67.6	25-21	80.4-80.8	46-31.5	80-82	Moderate
74.0-76.0	48-26.0	79.6-80.8	35-30.5	93.2-94.8	85-49	92.8-94	Minor
79.6-80.8	41-30	91.6-93.2	85-40.5	104.8-107.2	85-42.5	104.8-106.8	Minor
90.4-92.2	65-40						
Site: Sanodar							
42.0-44.4	220-190	42.0-44.4	195-135			42.0-44.8	Minor
57.2-60.2	160-50	57.2-60.2	125-40			94.8-96.8	Moderate
66.0-69.2	110-85	65.6-68.0	96-48			99.2-104.0	Large
94.4-96.4	190-140	94.8-96.8	110-100				
99.2-103.6	160-112	99.2-104	115-85				
109.4-113.4	160-120	109.6-113.6	120-84				
118.0-121.6	175-125	118-121.4	136-92				
126.0-129.8	175-80	126.4-130.0	136-64				
133.8-145.6	90-54	134.2-144.4	80-40				
Site: Jilana							
45.5-50.5	50-25	48.5-50.5	125-25	53-55.5	238-50	44-50.5	Large
54.0-59.0	100-65	55.0-58.0	110-45	60-62.5	252-72	104-108	
61.0-64.5	140-80	60.5-64.0	120-60	67.5-70.0	330-42		
69.5-72.5	225-90	69.0-71.5	150-40	102-106	328-136		
102-106.5	345-125	103-107	200-80	112-114	244-94		
113.5-117.0	175-110	113.5-116.5	110-50	121-123.5	182-64		
121.5-123.5	125-110	122-125	85-50				

Table-I-A Analytical results of geophysical logs .

Fig.3 Litholog and geophysical logs of site : Jilana.

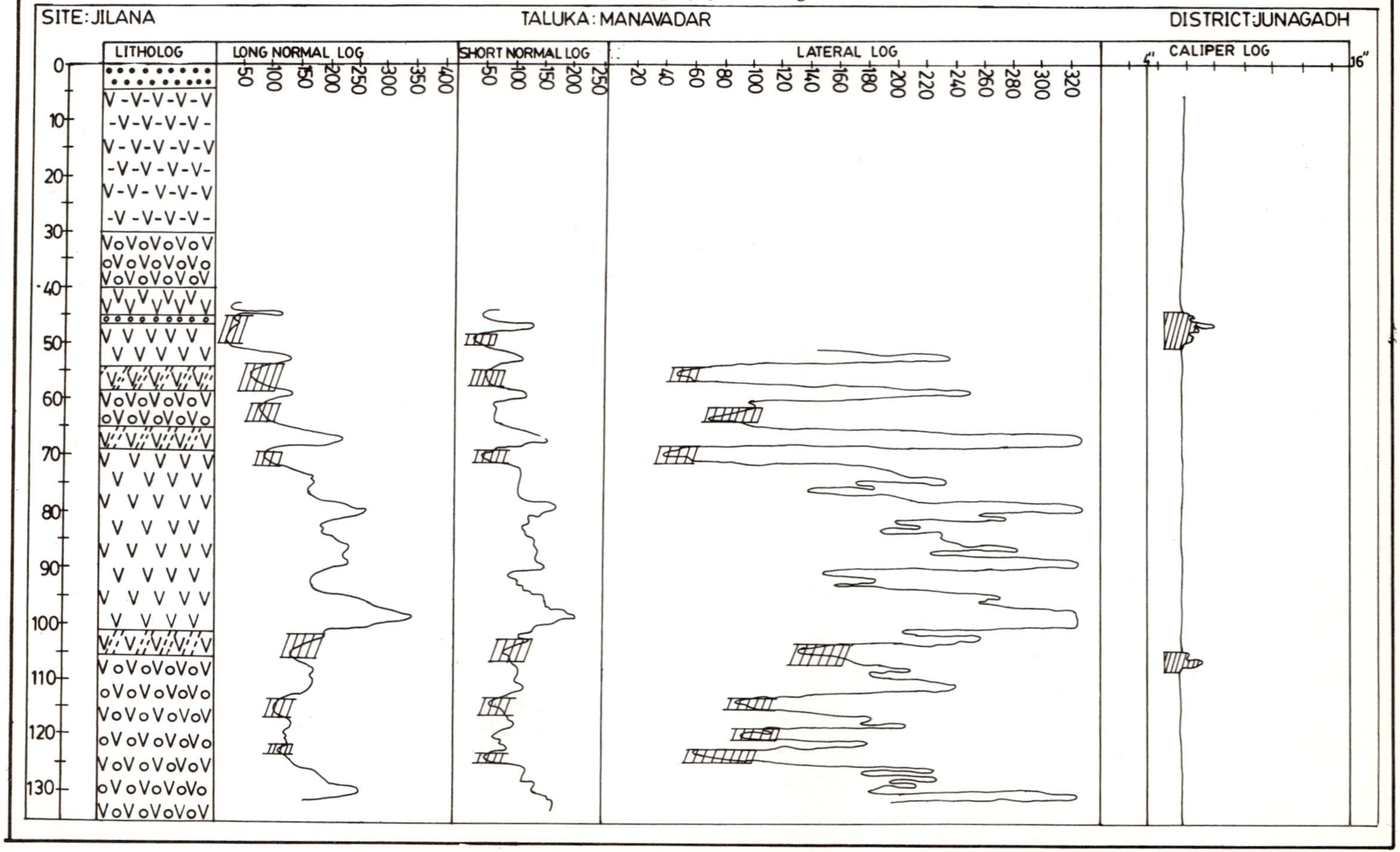

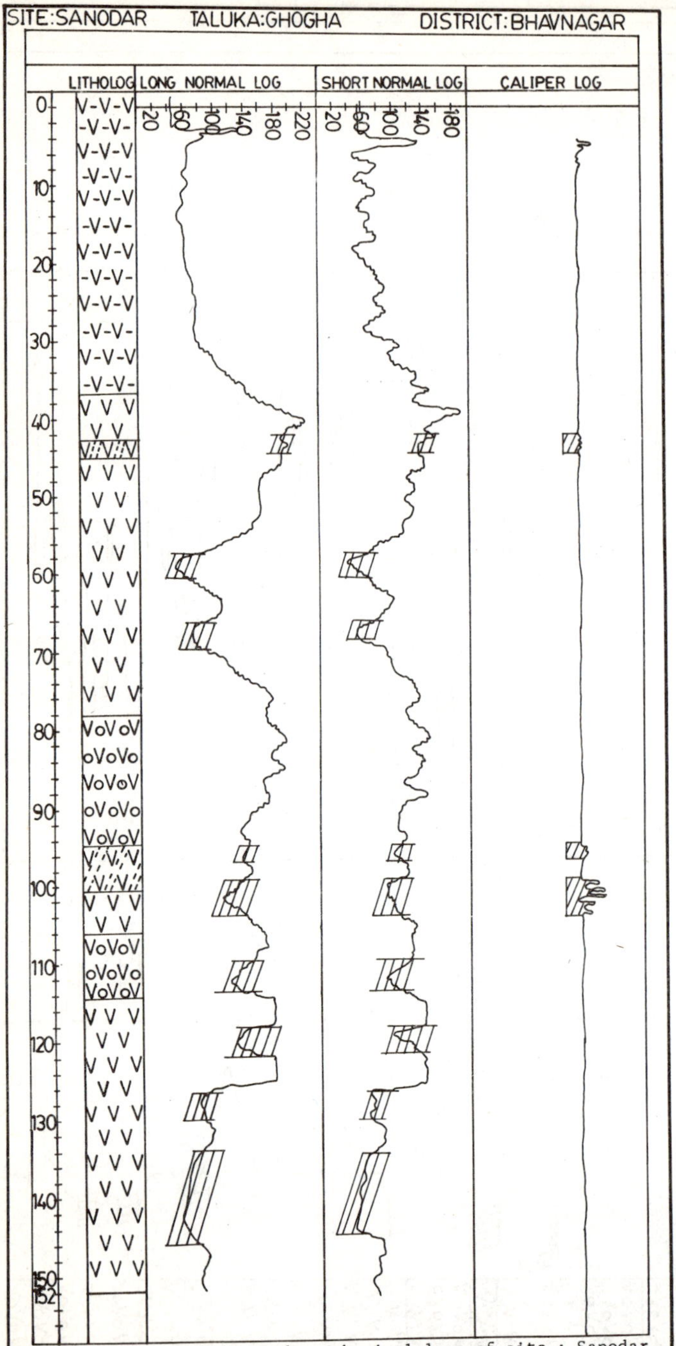

Fig.4 Litholog and geophysical logs of site : Sanodar.

Fig.5 Litholog and geophysical logs of site : Kadrej.

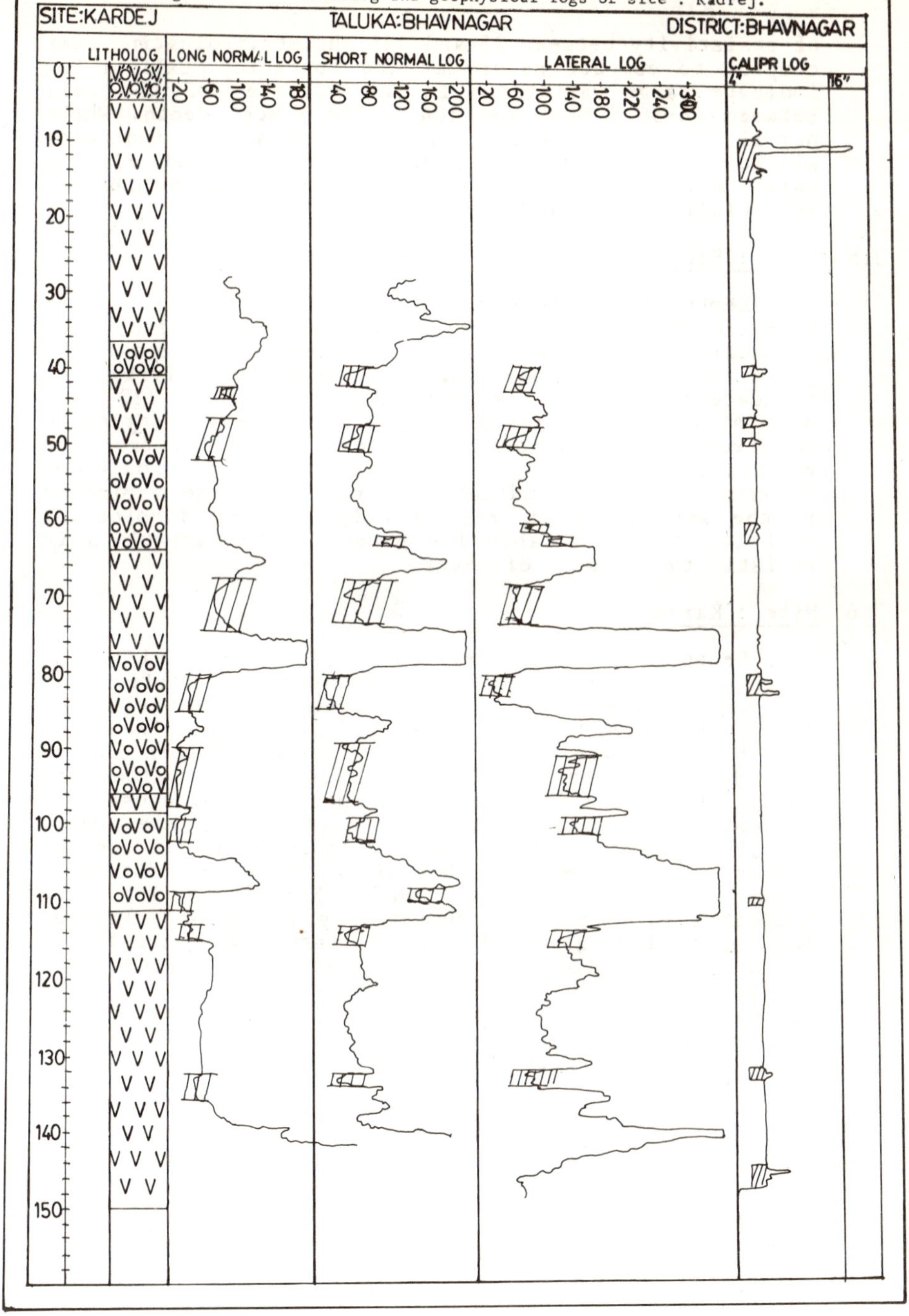

of resistivity between 25 and 345 ohm.Mts.indicating semi compact to compact nature of the formation. Caliper log indicate large variations in the diameter of the well between 44-50.5 Mts. and 104 - 108.0 Mts. depth. While normal log indicates seven zones, lateral log indicates six zones only. Fig. 3 presents the litholog and well log data and Table I-A indicates the variations in resistivity and well diameter.

3.5 Site : Sanodar

Depth of the bore is 169.59 Mts. Litholog of the bore indicates fractured and vesicular trap with secondary fillings at shallow and deeper horizons. The yield from this bore was 65 LPM. Resistivity log indicate variation in resistivity from 40 to 220 ohm. Mts. At depth, the resistivity drop has been observed to be abrupt. While 9 zones have been demarcated from the resistivity log. Caliper log indicates only 3 zones out of which two zones indicate moderate to large variation in the well diameter. Fig. 4 indicates the litholog and well log data and Table I-A presents the variations in resistivity and well diameter.

3.6 Site : Kardej

Depth of the bore drilled is 150.00meters. Litholog of this bore is indicative of fractured and jointed trap and zones filled with secondary minerals. The yield from this bore was 23 LPM only. Resistivity logs obtained from this bore indicate resistivity variations between 10 and 330 ohm. Mts. The variations in resistivity values are abrupt at certain depths. The caliper log indicates minor to moderate variations in the diameter of the well. This,in otherwords, indicates the semi-compact to compact nature of the formation. Normal (N 64") log indicated 9 zones of low resistivity, while Normal (N 16") log indicated 10 zones with resistivity variations. Lateral log also indicated 10 zones inferring resistivity variations whereas caliper log indicates 9 zones with variations in well diameter. Out of these only 4 zones indicate moderate to large variation. Fig. 5 presents the litholog and geophysical logs of this bore and Table I-B indicate the variations in the resistivity and well diameter.

3.7 Site : Lilia Mota

Depth of the bore is 160.97 Mts. Litholog of the bore indicates weathered and vesicular trap with secondary filling of minerals at shallow and deeper horizons. The yield from this well is 21 LPM. The resistivity logs indicate a resistivity variation between 50 to 670.0 ohm. Mts. while eleven zones have been indicated by normal (N64") log, 12 zones have been

NORMAL LOG (N64")		NORMAL LOG (N16")		LATERAL LOG		CALIPER LOG	
Depth in Mts.	Resist-ivity in ohm. Mts.	Depth in Mts.	Resist-ivity in ohm. Mts.	Depth in Mts.	Resist-ivity in ohm. Mts.	Depth in Mts.	Nature of variation
Site : Kardej							
42.4-44	140-80	39.6-42.2	220-44	39.6-42.8	100-60	9.6-14.8	Large
46.6-52.0	100-60	47.2-50.8	80-40	46.6-50.0	00-44	39.2-40.4	Moderate
67.6-74.4	142-80	62.0-63.2	120-92	60.4-61.2	90-74	46.0-47.2	Moderate
80.0-85.0	192-40	67.6-73.2	180-46	62.0-63.2	116-100	48.6-49.6	Minor
89.6-97.4	60-25	80-84.4	220-20	68.0-73.2	160-52	60-62.6	Minor
99.2-102.4	35-18	89.2-97.2	108-40	80.0-82.8	330-20	79.6-82.6	Large
108.4-111.2	120-10	99.2-102.4	84-48	91.2-96.0	212-112	109.2-110.0	Minor
112.8-114.8	30-20	108-110	200-140	99.2-101.6	208-132	131.2-138.	Minor
132.0-135.6	44-38	113.2-115.2	192-40	113.6-115.8	330-112	144-145.6	Moderate
		132.4-134	80-30	132-134	152-60		
Site:Liliamota							
48-50.8	110-90	48.8-51.6	190-80	50-52.4	240-100	10-20	Minor
54.6-58.2	90-75	54.0-58.0	125-60	54.2-58.4	200-80	78.8-80.4	Minor
61.6-65.2	80-65	61.6-65.2	110-50	63.2-66.0	140-50	82.8-85.2	Moderate
83.2-85.2	450-200	82.0-84	420-110	82.2-84.8	340-190	118.4-121.6	Minor
86.6-89.6	235-205	86.4-88	190-130	86.6-89.4	340-230		
96.0-97.6	350-255	95.2-97.2	220-160	96-97.2	340-270		
100-102.6	270-240	100-102	190-140	100-102.4	340-240		
116.4-120.8	420-150	115.6-119.2	240-90	105.4-107.6	310-260		
127.2-132	250-180	126.2-127.8	155-120	117.6-120.8	340-130		
140-142.8	370-240	129.2-134	140-120	127.2-128.6	330-210		
150-153.8	520-240	139.2-142.4	200-130	133.2-134.8	330-200		
		146.8-153.0	252-130	140.0-142.8	480-190		
				147.2-154.2	670-210		

Table I-B Analytical results of geophysical logs.

demarcated from normal (N16") log. Lateral log indicates 13 zones with resistivity variations, while caliper log indicated only one zone with moderate variation in well diameter. Fig. 6 presents the litholog and geophysical logs of the site and Table I-B indicates the variation in the resistivity values and well diameter.

Above discussion has revealed that the zones demarcated by caliper log have almost tallied with the resistivity logs and the zones demarcated with them.Though caliper log does not indicate any variation in the well diameter in some bores, resistivity logs have indicated marked lowering in resistivity values indicative of water bearing low resistivity zones.

4.0 Correlation of Resistivity and Caliper logs

Careful analysis and correlation of the data collected and compiled from the complex geophysical logs has indicated many a suitable zone for exploitation by hydrofracturing. After carefully studying and eliminating the zones which are likely to be non productive, feasible zones for hydrofracturing have been demarcated and have been presented in Table II. These zones have been demarcated either only on the basis of resistivity logs or caliper logs or both.

5.0 Conclusions

Based on the above study, following conclusions have been made.

(a) Correlation of lithological data with resistivity logs has indicated predominantly low resistive layers amidst the high resistivity layers, which are likely to be promising for exploitation through hydrofracturing.

(b) Caliper logs of these bores, though normally indicate the well diameter, clearly demarcated weak zones in the bore for further exploitation.

(c) Resistivity and caliper logs have shown a lot of coherance in delineating layers suitable for exploitation by hydrofracturing for better yields.

ACKNOWLEDGEMENTS:

The authors express their sincere thanks to Shri S.C.Sharma, Chief Geohydrologist, GWSSB for assigning this study to GWRDC Ltd. and valuable discussions during the compilation of the data.

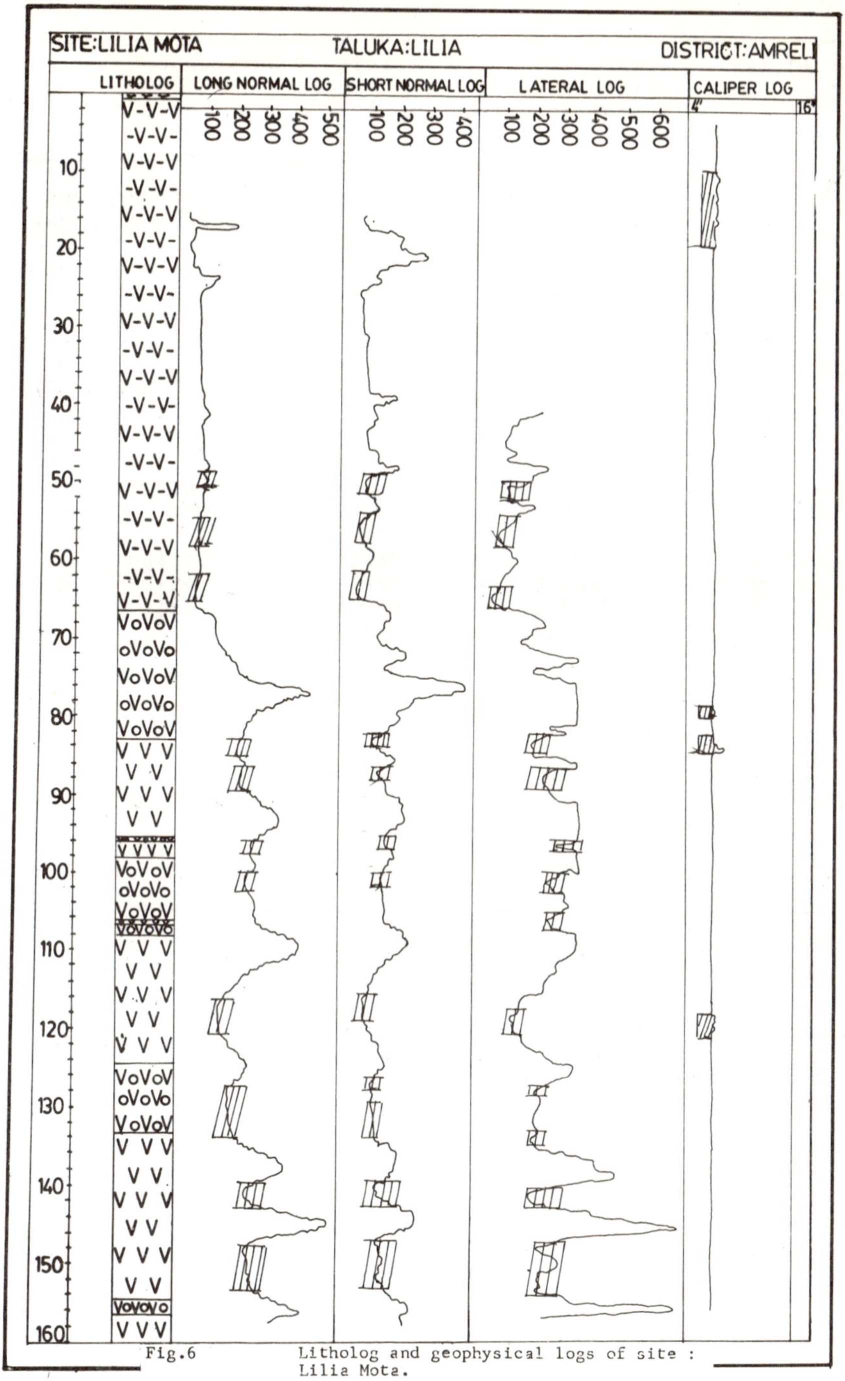

Fig.6 Litholog and geophysical logs of site : Lilia Mota.

Name of the site	Demarcated zones suitable for hydrofracturing From To	Basis of Demarcation
Dihor	47.2-50.0	Resistivity & Caliper logs.
	56.4-60.0	Resistivity & Caliper logs.
	62.0-66.0	Resistivity & Caliper logs.
	79.6-82.0	Resistivity & Caliper logs.
	91.0-94.0	Resistivity & Caliper logs.
	104.8-107.2	Lateral & Caliper logs.
Kardej	39.2-42.2	Normal logs.
	46.0-50.8	Resistivity & Caliper logs.
	80.0-83.0	Resistivity & Caliper logs.
	99.2-102.5	Resistivity logs.
	113.0-115.0	Resistivity logs.
	132.0-135.6	Resistivity & Caliper logs.
	144.0-145.6	Lateral & Caliper logs.
Sanodar	57.2-60.2	Resistivity & Caliper logs.
	99.2-104.0	Resistivity & Caliper logs.
	126.0-130.0	Resistivity & Caliper logs.
	134.0-144.0	Resistivity & Caliper logs.
Jilana	45.5-50.5	Resistivity logs.
	69.0-71.5	Resistivity & Caliper logs.
	103.0-107.0	Resistivity & Caliper logs.
	113.5-117.5	Resistivity log.
Lilia Mota	61.6-65.2	Resistivity & Caliper logs.
	83.2-85.2	Resistivity logs.
	116.4-120.8	Resistivity & Caliper logs.
	150.0-153.8	Resistivity & Caliper logs.

Table-II: Details of Demarcated zones for Hydrofracturing from Geophysical logs.

REFERENCES:

1. Bryan, F.L. 1950, ' Application of electrical logging to water well problems " - Water Well Journal Vol. No. 4.

2. Keys, W.S., and Mac. L.M., 1971, ' Application of bore - hole Geophysics to Water resources investigation' - USGS.

3. Jones, P.H., and Bufford, T.D., 1951, Electrical Logging applied to ground water exploration ' Geophysics Vol. 16.No.1.

HYDROGEOPHYSICAL STUDIES FOR CHARACTERIZING THE CRYSTALLINE AQUIFERS OF TAMBRAPARNI BASIN, TAMIL NADU, INDIA

A.BALASUBRAMANIAN[1], K.K.SHARMA[2] AND J.C.V.SASTRI[3]

[1]Department of Geology,
V.O.Chidambaram College, Tuticorin, India.

[2]Department of Applied Geology,
University of Madras, Madras, India.

[3]Department of Geology,
University of Mysore, Mysore, India.

ABSTRACT

Hydrogeophysical approach, deriving the empirical relationships between electrical resistivity measurements and aquifer parameters, can also provide the possible means of solving many groundwater problems of hardrock terrain. The present work is to elucidate the usefulness of Vertical Electrical Sounding (VES) for evaluating the hydrologic characteristics of crystalline aquifers of southern Tamilandu. Sequential, on the spot, measurements of VES, drilling details, pumping test analysis and water quality data have been utilized to correlate aquifer resistivity with transmissivity, hydraulic conductivity and specific capacity. Though these measurements are site-specific, it has been expressed in this work that reasonable conclusions could be derived using this approach.

INTRODUCTION

Vertical Electrical Soundings (VES) are widely used in groundwater investigations, which sometimes can replace test drilling. The measurements of aquifer resistivity are intuitively attractive for also estimating aquifer hydraulic conductivity because of the fundamental relation between hydraulic conductivity and electrical conductivity through their common dependence on tortuosity and porosity. Most of the applications of resistivity sounding are on a qualitative basis. Quantitative work has been limited to the application of these measurements for finding out the empirical relationships between aquifer resistivity, formation factor and hydraulic conductivity in a specific aquifer. A combination of field geophysical measurements, drilled hole details, aquifer tests and water quality studies can provide the most valuable solutions for predicting aquifer properties on a regional scale.

BACKGROUND

Limited attempts have been paid to correlate the resistivities and hydrogeologica properties of glacial outwash (Kelly, 1977; Heigold, et al., 1979; and Urish, 1981), Tertiary and Quarternary sediments (Ponzini, 1984) in various parts of the world. Verma et al.,(1980) tried to find the relationship between the longitudinal conductance of the weathered Precambrian metamorphics and the daily yield of wells. Most of these studies have been done either relating the aquifer parameters of adjacent locations or laboratory measurements of electrical resistivities of rocks nearby the VES points.

An attempt has been made in this study to correlate the geoelectric and hydrogeological parameters of a hardrock terrain in which the data of sequential investigations like VES, drilling, pumping tests and water quality were done at all individual sites.

GEOHYDROLOGY OF STUDY AREA

The Tambraparni river basin is located in the Tirunelveli and Chidambaranar districts of Tamilnadu. The river originates in the Westernghats and confluences with the Gulf of Manaar, 15 km south of Tuticorin. Except a narrow belt of Tertiary sediments near the coast, the basin exposes mainly charnockites and Peninsular gneisses of Archaean age. Groundwater occurs in the weathered, semi-weathered or fractured layers of these hard rocks, under water table conditions. The extent of weathering may vary from place to place and the thickness of the weathered zone ranges from 10 to 25 m. The depth to the water table ranges from 1m near the coastline to 8m bgl in the western parts of the basin. The valley bed is filled with present day sands and alluvium.

MATERIALS AND METHODS

VES data obtained in the field has been quantitatively interpreted for the determination of the geoelectric parameters like ρ_1, ρ_2, ρ_3, h_1 and h_2. The details of drilling made immediately after the soundings were compared with the results of geoelectrical surveys and the thickness of saturated aquifers have been evaluated. Using the pocket computer techniques of Balasubramanian et. al. (1989) the aquifer parameters of these stations have been computed. Water samples collected from these boreholes were analysed to derive the quality parameters.

Archie's (1942) formation factor (FF) has been computed using the expression

$$FF = Ro/Rw$$

where Ro is the resistivity of the saturated formation (ohmm) and Rw is the porewater resistivity (ohm m) which could be obtained using the relation

$$Ec = 10000 / Rw$$

where Ec is the electrical conductivity in micromhos. This formation factor, in turn, is related to the porosity of geological formations due to the fact that high porosity and fluid conductivities yield lower resistivities.

Aquifer hydraulic conductivities (K) have been determined using the saturated thickness (H) and transmissivity (T) by the relation K = TH. Relating aquifer transmissivity to Normalized Aquifer Transverse Resistance (NATR) is a widely used approach and the NATR values have been determined as a product of FF, Rw and H.

Kelly (op.cit) plotted the aquifer resistivities with the specific

TABLE 1. Geoelectrical and Aquifer Parameters used in Hydrogeophysical Studies

Sl. No.	AQUIFER RESISTIVITY A (Ohm mts)	FORMATION FACTOR F_f	AQUIFER TRANSMISSIVITY (m^2/d)	FIELD PERMEABILITY (m/d)	CALCULATED PERMEABILITY (m/d)	SPECIFIC CAPACITY ($m^3/d/mdd$)	NORMALIZED AQUIFER TRANSVERSE RESISTANCE (Ohm mts)
				GNEISSES			
1	125.00	13.13	11.05	0.37	0.90	3.42	2112.41
2	280.00	18.91	80.73	3.46	10.62	8.87	1879.81
3	57.00	2.20	2.16	0.08	0.05	2.72	1222.98
4	170.00	10.63	0.08	0.00	0.00	6.60	8690.03
5	285.00	19.90	55.22	10.60	1.38	86.17	10411.08
6	81.82	21.03	4.16	1.63	0.11	11.66	10315.22
7	69.95	3.02	0.932	0.08	0.02	3.52	1540.56
8	29.85	2.60	18.34	0.84	1.83	21.46	340.08
				CHARNOCKITES			
9	150.00	6.68	7.14	0.33	0.18	6.82	3879.74
10	137.50	24.34	0.11	0.01	0.00	38.38	10602.50
11	180.00	13.87	0.36	0.03	0.02	1.52	4229.24
12	110.00	6.83	3.86	0.16	0.09	3.14	3867.69
13	126.00	4.26	30.99	1.51	1.41	5.61	1360.81
14	65.00	2.26	425.92	125.71	53.24	197.12	262.52
15	35.00	1.28	0.99	0.16	0.04	40.67	464.64
16	48.00	2.23	0.57	0.07	0.06	38.13	323.79
17	9.00	1.22	74.52	0.00	8.77	9.19	150.57
18	60.00	7.06	19.87	1.02	1.24	13.90	1640.18
19	76.00	15.57	2.03	0.12	0.11	6.87	4295.45
20	62.20	4.26	0.87	0.03	0.10	0.77	525.77
21	249.67	21.84	176.82	22.37	4.07	80.87	13800.90
22	322.00	16.39	2.76	0.45	0.20	1.50	3331.76
23	153.80	16.16	1.07	0.07	0.03	3.36	7508.58

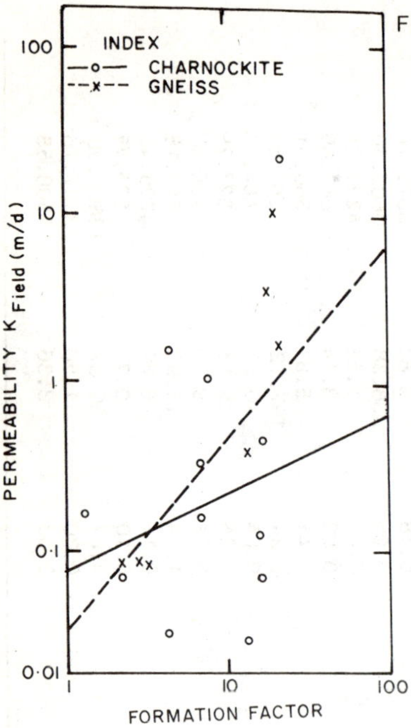

Fig.1 Relationship of formation factor to permeability.

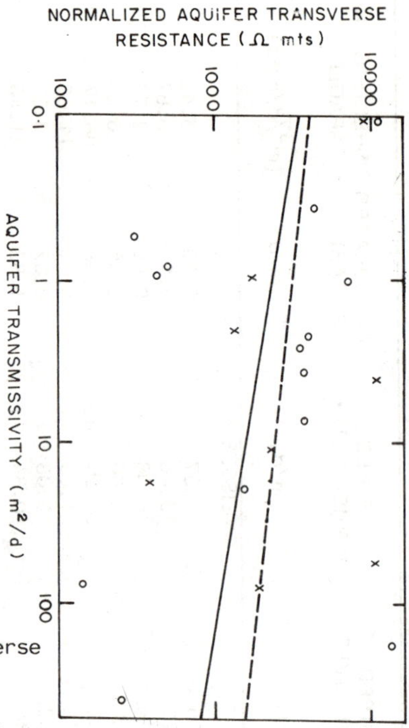

Fig.2 Crossplots of transverse resistance to aquifer transmissivity.

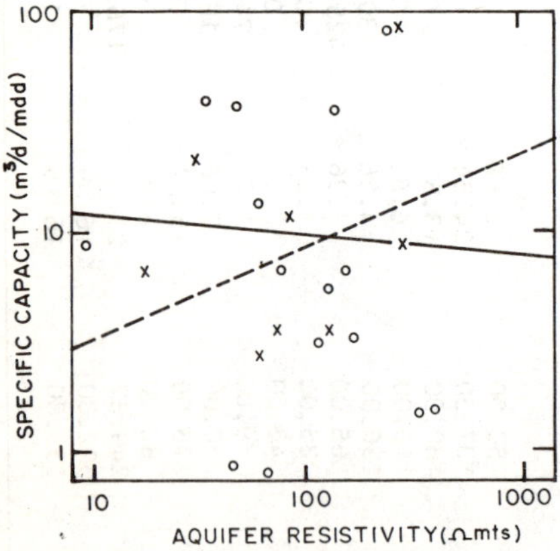

Fig.3 Relationship of aquifer resistivity to specific capacity.

capacity values of aquifers and has reported a positive correlation between them. Hence, specific capacities of these borewells have been computed and used in this study.

Table 1 shows the hydrogeologic and geoelectric parameters determined during this study in 23 locations of the basin, with reference to the lithology of aquifers classified as gneisses and charnockites.

RESULTS AND DISCUSSION

Fig.1 shows the resultant relationships of aquifer FF and K. Earlier studies on other terrain have also shown that the correlation between these two properties is positive denoting the fact that an increase in FF correspondingly increases the aquifer K (Jones and Buford, 1951; Henriet, 1976; Kelly, 1977; Worthington, 1977; Kosinski and Kelly, 1981; and Urish, 1981). It could be seen that (i) the behaviour of charnockites differs slightly from that of the gneisses, eventhough the relationships are positive, (ii) the intensity of weathering, density and influence of fractures and the water yielding capacity of gneisses are greater than that of charnockites and (iii) the scatter of plotted points show the site-specific, anisotropic behaviour of the aquifers.

The relationship derived between NATR and T of aquifers has been shown in Fig.2. This confirms the site-specific conclusions derived by Kelly (op.cit.) and Urish (op.cit). It could be seen that the behaviour of charnockites slightly varies from that of gneisses, eventhough they both show a negative correlation.

Fig.3 shows the relationships between specific capacity of aquifers with that of their resistivities. It could be seen that (i) positive correlation exists between these two parameters for gneisses and the behaviour of charnockite is reverse, i.e., negative and (ii) the influence of structural characteristics in these rocks is varying in holding and yielding groundwater, which is also evident from the point scatter.

CONCLUSION

This study elucidates the possibilities of correlating the aquifer constants with the geoelectric parameters obtained through surface geophysical soundings. Standard empirical relationships could be established through this approach. Although the relationships obtained in this study are site specific, the methodology could be useful for other geologic terrain.

ACKNOWLEDGEMENT

The authors are thankful to their parent organisations for providing the necessary facilities to carryout this work.

References:

Archie, G.E., 1942. The electrical resistivity log as an aid in determining some reservoir characteristics, Amer.Inst.Min.Met.Eng., Techn. Pub., 1422, Petroleum Tech., 8p.

Balasubramanian, A., and Sastri, J.C.V., 1981. Techniques of aquifer parameter evaluation using pocket computers, Proc. Volume. Int. G.W. Workshop 1989, NGRI, Hyderabad.

Heigold, P.C., Gilkeson, R.H., Cartwright, K., and Reed, P.C., 1979. Aquifer transmissivity from surficial electrical methods. Groundwater. V 17(4), p.338-345.

Henriet, J.P., 1975. Direct interpretation of the Dar Zarrouk parameters in groundwater surveys. Geophys. Prosp. V.24, p.344-353.

Jones, P.H., and Buford, T.B., 1951. Electrical logging applied to groundwater exploration. Geophysics. V.16(1), p.115-139.

Kelly, W.E., 1977. Geoelectric sounding for estimating aquifer hydraulic conductivity. Groundwaer. V.15(6), p. 420-425.

Kosinski, W.K., and Kelly, W.E., 1981. Geoelectric soundings for predicting aquifer properties. Groundwater. V.19(2), p.163-171.

Ponzini, G., Ostroman, A., and Molinari, M., 1984. Empirical relation between electrical transverse resistance and hydraulic transmissivity. Geoexploration, V.22, p. 1-15.

Urish, D.W., 1981. Electrical resistivity/hydraulic conductivity relations in glacial outwash aquifers. Wat. Res. Res., V.17(5), p.1401-1408.

Verma, R.K., Rao, M.K., and Rao, C.V., 1980. Resistivity investigations for groundwater in metamorphic areas near Dhanbad, India. Groundwater. V. 18(1), p. 46-55.

Worthington, P.F., 1977. Influence of matrix conduction upon hydrogeophysical relationships in arenaceous aquifers. Wat. Res. Res., V. 13(1), p. 87-92.

-o-

DECIPHERING GROUNDWATER USING REMOTESENSING TECHNIQUES

Om Prakash Dubey[1], Sri Niwas[2], Pravez Ahmed[2]

1. Civil Engineering Department, University of Roorkee, Roorkee-247 667
2. Earth Sciences Department, University of Roorkee, Roorkee - 247 667

ABSTRACT

For optimal management of resources accurate and timely information system is required. The conventional data collection systems are uneconomical difficult and provide information with time lag. Remotesening techniques can be organised in near real time for evaluation of water resources. In the present paper an analysis has been presented for evaluation of groundwater resources in part of Indogangetic plain. A linear mathematical model in terms of landcover system operator matrix has been developed for evaluation of groundwater recharge. The operator has been estimated from the landsat CCT, band 5 and band 7 imagery blow up at 1:250,000. The operator can be singular, nonsingular and square or rectangular. The model was calibrated using conventional, GLI and MPI techniques. It has been observed that MPI gives better results. The model is then used to predict the ground water level fluctuations for year 1977, which is comparable with observed data.

INTRODUCTION

India is an agricultural country. For agricultural development water is essential. In many parts surface water is not sufficient under such circumstances ground water is only alternative. For project planning available data is generally not sufficient and representative. Conventional system of data collection are costly and provide information with time log. As such most of the project planning work is carried out on questionable data. Remotelysensed data can be used to generate data base which can be used for ground water resources evaluation.

A study has been carried out in parts of Indogengetic plain bounded by latitude 29° 50' to 30° 30'N and longitudes 77° 30' to 78° 20'E. Remotelysensed data in terms of landsat CCT, B and W photoproduct at 1:250,000 scale, water table, Rainfall data etc. were integrated to develop a linear model to predict the fluctuations in the water table.

THE MODEL

For the development of model following assumptions have been made.

(1) The land cover is the most influencial parameter for recharge.

(2) Groundwater system can be represented by a linear system.

(3) Rainfall is the only recharging factor

(4) In an influencing zone, water table is recharged vertically.

under these assumptions landcover operator model can be represented as (Dubey, Sri Niwas and Awasthi, 1984) :

$$A\,R = G \tag{1}$$

where

A = Land cover operator system matrix

R = Vector representing recharge coefficient

G = Vector representing the recharge in groundwater.

ANALYSIS OF DATA

Supervised classification of CCT was carried out based on maximum likehood classifer on MIDAS using following steps.

(a) From raw plot and ground truth training samples were feeded using TEST Program.

(b) statistics of training set were tested using TSTAT Program.

(c) Landuse classification using maximum likehood classifer.

In this way landcover map was prepared for years 1973, 1974, 1975, 1976 and 1977.

CALIBRATION OF MODEL

The rainfall data, landcover system operator matrix and recharge volume for years 1973, 1974, 1975 and 1976 were calculated and vector R is evaluated using Inversion of landcover system operator matrix A.

The operator may be singular or nonsingular and square or rectangular Conventional Inversion, GLI (Generalised Linear Inversion) and MPI (MoorePen Rose Inverse, Moore 1920) was evaluated and vector R is evaluated as :

using conventional inverse,

$$R = A^{-1} G \tag{2}$$

using GLI,

$$A^{-} = V_p \hat{A}^{-1} U_p^T \tag{3}$$

where

A^{-} = GLI of A

V_p = p eigenvestors associated with rows

U_p = p eigen vector associated with columns

$\hat{A}^{-1}$ = Diagonal matrix p non zero eigen values of A.

using MPI,

$$A^{+} = A^{*} (A^{*} A A^{*})^{-1} A^{*} \tag{4}$$

where

A^{+} = MPI of A

A^{*} = Conjugate transpose of A

PREDICTION OF FLUCTUATIONS

Once the model is calibrated the fluctuation in groundwater table can be predicted using equation (1).

RESULTS AND CONCLUSION

At the time of calibration of model it was seen that conventional Inversion and GLI are not yielding satisfactory results. Only MPI was giving satisfactory results, and hence MPI has been used in predictive model. For year 1977 the observed and predicted values of fluctuations at some location are tabulated as under, from the table it is seen that the predicted as observed values are comparable.

ACKNOWLEDGEMENTS

The authors are thankful to IIRS, IRI, GWIO for their help in carrying out the study, their friends for live discussions, Shri Rajeev Grover for neet and timely typing the manuscripts.

Groundwater level fluctuations

Location	Fluctuation	
	Observed (m)	Predicted (m)
Roorkee	1.9	2.1
Belri	2.8	2.6
Imlikhera	1.4	1.2
Bhagwanpur	1.7	1.9
Bahbalpur	1.7	1.9
Manubas	1.2	1.8
Daulatpur	1.3	1.4
Aurangabad	1.4	1.8

REFERENCES

Dubey, O.P., Sri Niwas, Awasthi, A.K., 1984, Analysis of Remotelysensed Data for Groundwater studies of Piedmont Zone, Proc. 5th ACRS, Kathmandu, pp. 6-1-E-6-4.

Moore, E.H., 1920, On Reciprocal of the General Algebraic Matrix, Bull. Amer. Math. Soc., 26, pp. 394-395.

PROSPECTING GROUNDWATER IN PUSHKAR LAKE REGION THROUGH REMOTE SENSING - A CASE STUDY

Rajiv Sinha
Centre for Remote Sensing,
Civil Engg. Deptt.
University of Roorkee, Roorkee - 247 667

R.S. Tiwari
Department of Civil Engg.
University of Roorkee, Roorkee - 247 667

ABSTRACT

A major portion of earth's fresh water resources comesfrom groundwater acquifers. The importance of groundwater prospecting is of special importance at the present time because of the fact that surface water resources are depleting fast. This paper explores the effectiveness of remote sensing techniques over conventional ground-based methods in Pushkar lake valley region at Ajmer, Rajasthan for groundwater targetting. Besides visual interpretation methods digital image processing of data has been emphasized and has been found to be extremely useful for qualitative and to some extent quantitative analysis. The geological and geomorphological parameters of groundwater occurrence have been studied in detail to bring out the 'Clues' for groundwater prospecting.

INTRODUCTION

A proper understanding of groundwater resources of an area is necessiated not only by ever-increasing demand for usable water but also by adverse conditions of nature resulting in scarce availability of surface water. For judicious exploitation of groundwater resources, an accurate early assessment is important before commissioning an in-depth study. In conventional methods of groundwater prospecting, the success rate is often very low mainly because of the 'hit & trial' approach. The present paper evaluates the effectiveness of remote sensing techniques for groundwater investigations employing visual interpretation methods and digital processing approach. The digital analysis has been found to be much versatile. The different combination of enhancement and filtering techniques have been worked out to be instrumental in bringing out the parameters which serve as 'clues' for groundwater. The investigations presented in this paper form only a part of the larger project "Eco-regeneration of Pushkar Lake Valley Region" running under the Centre for Remote Sensing, Civil Engg. Deptt., University of Roorkee sponsored by Ministry of Human Resources, Govt. of India.

PARAMETERS AFFECTING GROUNDWATER OCCURRENCE

In general, the parameters that affect the groundwater potential of a region may be categorized as follows:

1.	Geology	:	Lithology, stratigraphy and structure, fracture joints and other partings etc.
2.	Geomorphology	:	Weathered pediment zones, landuse/landcover, alluvial features, burried channels etc.
3.	Drainage Network	:	Drainage pattern, drainage texture, drainage anamolies etc.
4.	Vegetation	:	Type, vigour, morphologic, taxonomic and spectral responses etc.
5.	Soil	:	Texture, moisture content etc.

In the present work, only two of the above parameters have been studied namely geological and geomorphological parameters. As to be explained later, these are the two most important parameters in the area under study. The other parameters have also been planned to be investigated in the later phases of the work to target precisely the sites for exploration.

CONVENTIONAL METHODS VIS-A-VIS REMOTE SENSING TECHNIQUES

The conventional methods for groundwater targetting

comprise of two categories of investigations - surface and subsurface. The surface investigations include field oriented geological mapping, geophysical and geochemical surveys, test hole drilling etc. whereas subsurface investigations include bore hole logging. The biggest hindrance encountered while using conventional methods are that they are too time-consuming and are very costly. Their applicability is further limited by lack of field outerops, surface or subsurface inhomogenities, problems in drilling etc. In Pushkar lake valley region at Ajmer, these problems are particularly more predominate as most of the area is covered by blown sand and good outcrops are difficult to ascertain. Moreover, the geology of the area, being a pre-cambrian formation, is relatively complex having undergone several phases of deformation, making the structural and spatial relations difficult to interpret in field.

On the other hand, remote sensing data-both aerial and satellite - are relatively simple and less expensive and provide other advantages such as synoptic view of the region, sequential data acquisition, multi-spectral coverage and computer compatibility. In the present study emphasis has been made to use satellite-based data to draw out qualitative and to some extent quantitative informations of the region. The informations are qualitative because of the fact that satellite data generally have a low resolution. Of late, high resolution data from TM, SPOT and IRS sensors are available which have made possible to have quantitative investigations with encouraging results. Another aspect which the user must bear in mind while ordering remote sensing data is cloud cover which often obscures useful informations. A proper planning can obviate this difficulty.

AREA OF STUDY

The area undertaken for the present study is the Pushkar Lake Valley region which is located between longitude $74^{O}15'$E to 75°E and latitude $26^{O}15'$N to $26^{O}45'$N. (Fig. 1). The area falls under the semi and zones of Rajasthan.

Scarcity of water is a major problem in this region and hence investigation, assessment and development of water supply is of special importance at the present time. The water supply of the region is characterized by its seasonal and highly fluctuating nature, both in space and time, and by its localised occurrence. Moreover, its usefulness is further limited due to high salinity of water found in them. The only river system in this region is the Luni river and its small tributary Saraswati both of which are only rain-fed and are ephemeral. The rainfall is scanty, variable and undependable.

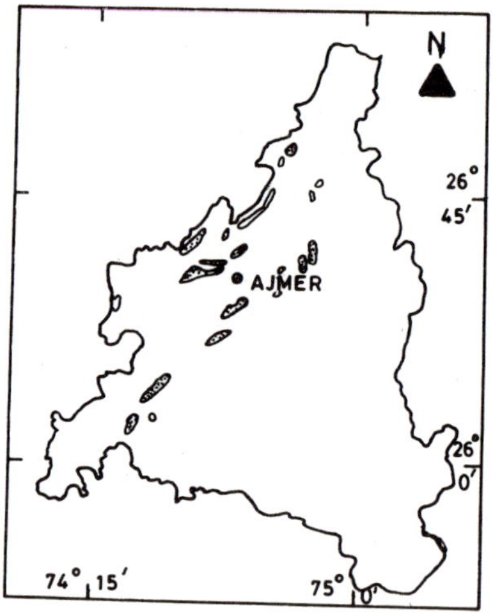

Fig. 1 Study area

Stratigraphically, two major geological sequences of Pre-cambrian age are exposed in the area - Banded Gneissic Complex (BGC) and Delhi Group of rocks. The younger Delhi Group of metamorphics are directly overlying the older BGC with a profound unconformity. The major rock types of Delhi Group are Quartzites, Calc-silicates, Schists, Granitic Gneisses etc. while the BGC is composed mainly of mafic Gneisses.

DATA COLLECTION AND ANALYSIS

The remote sensing data products which have been presently used in this study include multidate landsat imageries (MSS and TM) and one MSS CCT dated March 30, 1986. The processing of tape was done at Regional Remote Sensing Service centre (RRSSC), Dehradun. The funding for procusement of data products as well as processing was provided by the Centre for Remote Sensing, Civil Engg. Deptt., University of Roorkee, Roorkee.

A landuse map of the area was prepared by visual interpretation of Landsat imageries (Dikshit, 1987) in the early phases of the project "Eco-regeneration of Pushkar Lake Valley Region". From this map, nine landuse categories viz. cultivation, scrubs, vegetation, waterbody etc. were identified which were supplemented by toposheets and field visits. The groundwater investigations were taken up as follow-up activity of the project. The investigations carried out have been limited to two aspects only namely geology and geomorphology, as mentioned earlier.

GEOLOGICAL INVESTIGATIONS

The geological parameters that seem to have pronounced effect on groundwater occurrence in the region are lithology and linear structures and these have been analysed in detail.

Lithological Studies

An attempt was made to apply geological interpretation techniques for reconnaissance and lithologic mapping of the study area. The visual interpretation of B&W imageries of both MSS and TM sensors even at enlarged scales was not found helpful in picking up the lithological variations. The digital processing of the CCT, however, enhanced the display of information to help in delineating the major lithological units. The ground truth for identification of these units were obtained from published maps (Heron, 1953, GSI, 1980). Different combinations involving raw bands 1,2,4 (earlier 4,5,7), brightness indices, principal components etc. were used for lithological discrimination. The False Color composite made out of bands 1,2,4 was particularly useful for this purpose as this improved the separability of lithological units to a great extent. A remarkable distinction between two groups of rocks exposed in the area namely Delhi Group and BGC was noticed resulting due to different spectral signatures of these two litho units and also their structure and tectonic framework. The contact between these two units marked by Heron (1953) could easily be picked up on the enhanced images. The main lithologic types which could be distinguished include Quartzites, Granites, Granitic Gneisses, Calc. Schists, dry exposed sand etc. (Fig. 2). A large part of the area is covered with wind blown sand obscuring the identification of rock types.

Tectonic Studies - Lineament Analysis

The lineament analysis carried out through B&W TM imagery band 3 showed that most of the lineaments were aligned in NE-SW direction, parallel to the alignment of Aravalli ranges. Based on this observation, various filtering kernals of size 3x3 were decided to enhance the edges of the lineaments while carrying out the digital analysis of MSS CCT. The results obtained by applying following three kernals of raw band 4 of MSS were quite encouraging:

$$\text{(i)}\begin{matrix}1 & 0 & -1\\ 0 & 0 & 0\\ -1 & 0 & 1\end{matrix}\qquad \text{(ii)}\begin{matrix}-1 & 0 & 1\\ 0 & 0 & 0\\ 1 & 0 & -1\end{matrix}\qquad \text{(iii)}\begin{matrix}-1 & -2 & -1\\ 0 & 0 & 0\\ 1 & 2 & 1\end{matrix}$$

The first two are directional filtering kernals whereas the third one is the non-linear adge enhancement (Sobel). The first kernal enhanced the major and minor linear features on NE-SW direction which coincided with the regional trened of Aravalli ranges. The second kernal enhanced the features orinted mainly in NN-SE direction including some regional faults also which were not clearly visible on B&W imageries. The sobel filter very sharply enhanced the boundaries of hill ranges, river beds, lakes etc.

The tectonic details obtained from the processed outputs (Fig. 3) are much superior compared to those obtained from visual interpretation map prepared from B&W imageries. Three major lineaments have been picked up on this scale. The lineament I is running E-W alongwhich the river Luni is aligned. The lineament II is running approximately NW-SE

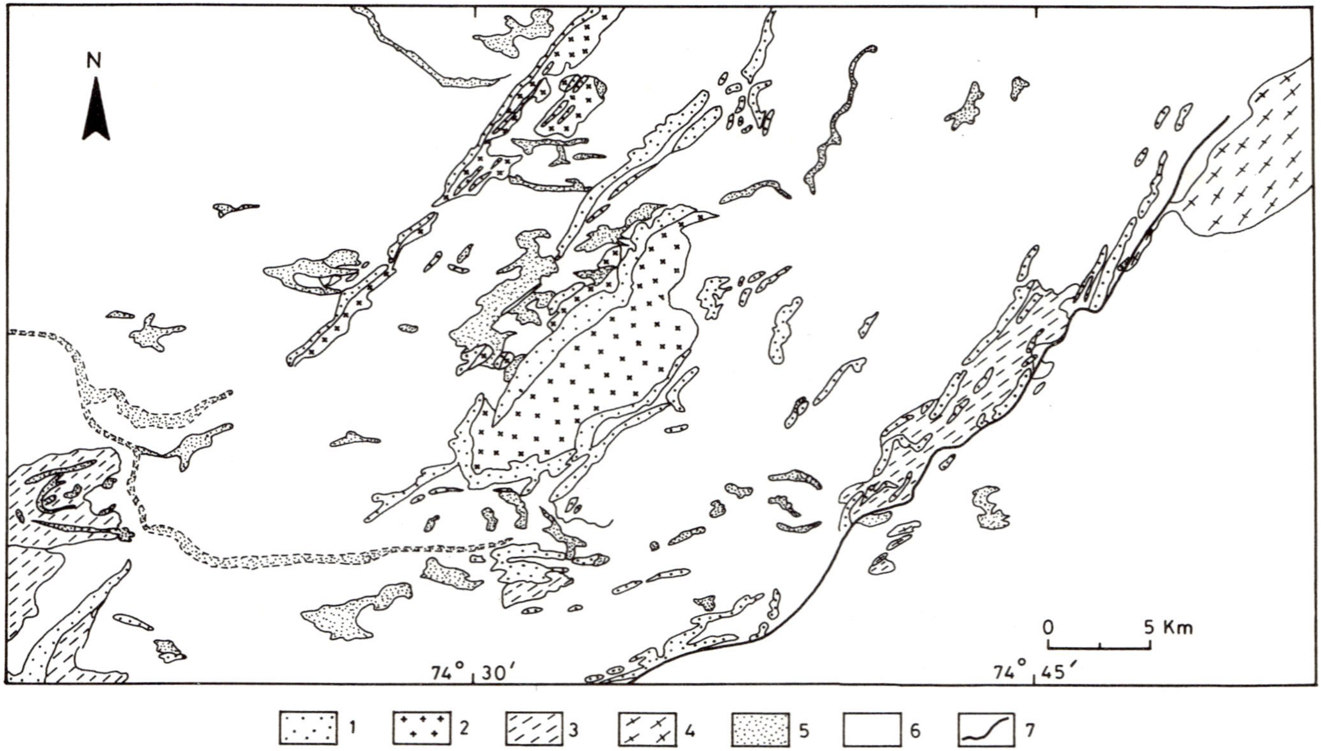

Fig. 2 Lithological map prepared from digitally processed images (1-Quartzites, 2-Granites of Delhi group, 3-Calc-Schists, 4-Pre-Aravalli granites, 5-Dry, barren sand, 6-Alluvium and wind blown sand, 8-contact between Delhi group and BGC)

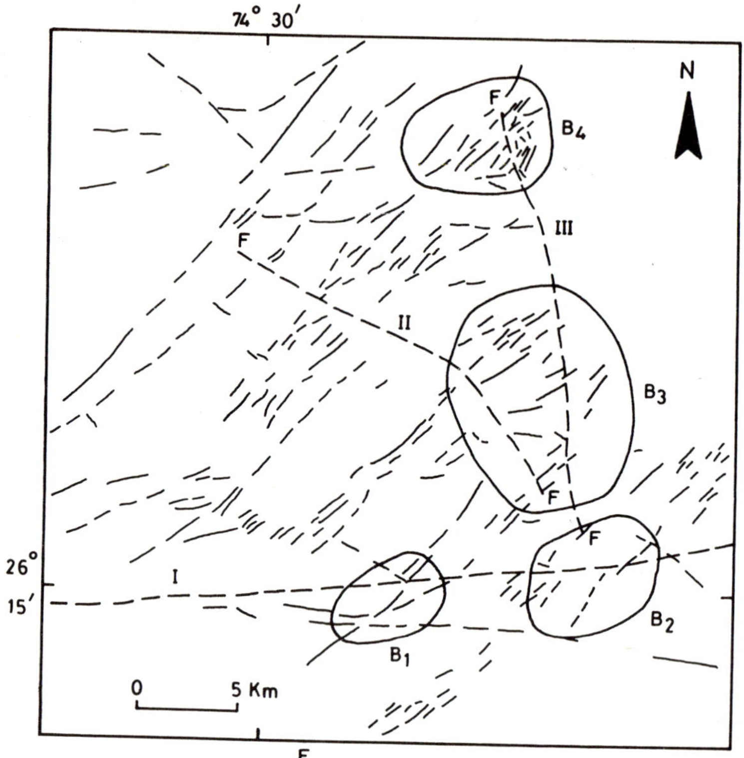

--- LINEAMENTS, F---F FAULTS, B1...B4 POTENTIAL BLOCKS

Fig. 3 Lineament map prepared from filtered images

and is in fact curvilinear showing the effect of deformational activities. Lineament III is running approximately N-S and this also is a bit curvilinear. A close observation revealed that lineament II and III are the traces of major faults which could be confirmed from some conclusive evidences such as shifting, truncation of hill ranges etc.

However, the filtered images posed some difficulties also during analysis mainly because of noise incorporated in the processed data. Many a times, it was difficult to assign any physical significance to the available linearities. At the same time, man-made features like roads etc. could not be distinguished on filtered images. These difficulties were overcome to a great extent by supplementing the details from the toposheets of the area and False color composites.

GEOMORPHOLOGICAL INVESTIGATIONS

The digitally processed data in conjunction with topographic sheets were used to map major geomorphological units in the study area with a view to highlight the geomorphic controls of groundwater in the region. Although the scales of satellite imageries are not ideal for geomorphological mapping, sufficiently good results were obtained with False color composites and hybrid color composites of combinations of brightness indices, principal components and filtered images. A geomorphological map (Fig. 4) was prepared utilizing the processed outputs.

The most striking geomorphic features in the area are the Aravalli ranges of varying heights and plateau. The area exhibits an inselberg topography typical of semi-arid regions with high, steep hills arranged in irregular groups of isolated peaks separated by vast exposures of sand. In fact, a major part of the area is covered by wind-blown sand distinguished by high reflectance in all bands resulting in high brightness indices also. The extent of blown sand implies that much of the precipitation penetrates into the ground and fails to be removed as runoff. A much smaller proportion of the area is capped by alluvium mainly along river Luni and its tributaries, delineated by the cultivations and small vegetations supported by them. A few patches of barren sand could also be demarcated, which are possibly having salt encrustations upon them. The identification of typical alluvial and eolian features was not possible on the scale of processed outputs. The selected areas are planned to be investigated further in detail in the later phases of the work.

The river Luni is the only river system of the region which rises from Aravalli ranges. It's small tributary Saraswati rises from Pushkar lake and joins it between Lesawa

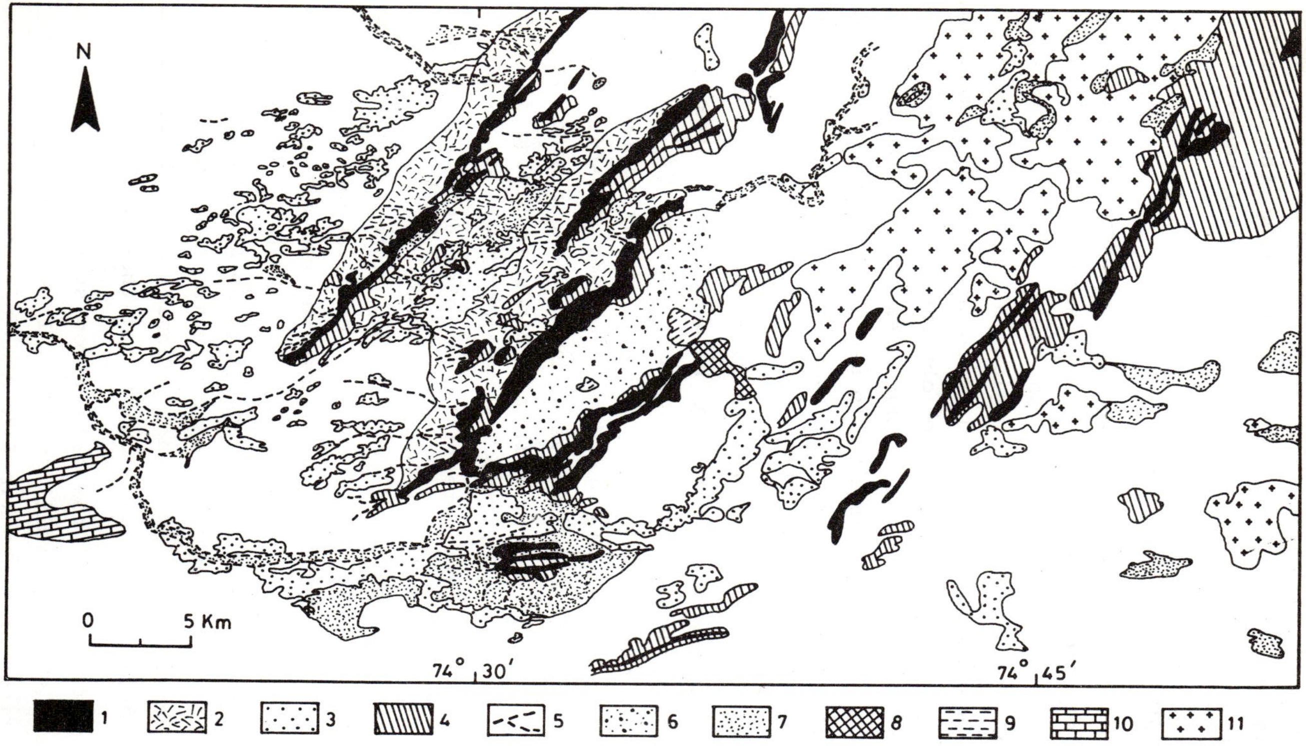

Fig. 4 Geomorphological map prepared from digitally processed images (1-High hills with dense vegetation, 2-Dissected slopes covered with barren sand, 3-Alluvium covered with small vegetation, 4-Sparsely vegetated area, 5-Dry/paleochannels, 6-Probable pediment zone, 7-Dry, barren sand, 8-Urban area, 9-Water body, 10-Weathered zone, 11-Cultivation)

and Ladpura. Observations on tectonic map (Fig. 3) shows that the river Luni and its main tributaries are structurally controlled trending along major lineaments. Apart from this, a large number of traces of paleochannels could be picked up in the processed outputs. All the present and paleochannels are completely dry and barren sands are exposed all along their course. A continuous tract of dissected slopes could be separated all along the Aravalli ranges. The toposheets also show the bifurcation of channels in this zone some of which could be traced from the processed outputs. This could well form a piedmont zone which is indeed a common feature in this type of terrain. The detailed investigations alongwith field checks, which have been planned in future shall confirm this opinion. The intermontane plains between Nagpahar and the hills south-west of Ajmer city have all the features common to a mountain pediment which also needs to be verified by the study of aerial photographs and field observations.

RESULTS AND CONCLUSIONS

The purpose of this paper was to draw some broad conclusions on the basis of analysis carried out so far on satellite data viz. B&W satellite imageries, False color composites and computer compatible tapes (CCT's). The results of digital processing which was widely used for analysis are best seen in color and hence could not be produced in this paper. However, the thematic maps prepared out of the processed outputs have been fairly good for depicting these results.

On lithologic ground, three zones namely Granitic terrain, sandy area and alluvium seem to be promising for groundwater. At first sight on the tectonic map (Fig. 3), four major blocks can be demarcated on the basis of frequency, spacing and intersecting of lineaments which will have a high groundwater potential. The potential of the lineaments for groundwater tapping can also be understood by observing on the toposheet a large number of wells of potable groundwater aligned along these linears.

The results of geomorphological investigations have been limited only to demarcate major geomorphic units. However, some broad comments can be made regarding further investigations to be carried out in later phases of the project. Identification of sand dune areas in eolian tract and meander deposits in alluvium could help in demarcating potential zones of groundwater because of their high recharge rates and moderatelly high permeabilities. The numerous paleochannels indicate near surface high moisture zones. Another potential area could be the zone of dissected slopes all along the hill ranges onto which the streams discharge most of their load.

It may be concluded that useful information for delineating the potential zones for groundwater can be obtained by using remote sensing technique. The MSS data is good for extracting qualitative and to some extent quantitaive informations. The digital proceeding techniques have a great potential for mapping the surface manifestations of groundwater occurrence and act as a wonderful tool for faster analysis. The meaningful interpretation afterwards aids in inferring subsurface features. A detailed investigation of major zones identified through satellite data analysis is desirable before pin-pointing sites for drilling wells. The analysis through aerial photographs and digital processing of high resolution data like TM & SPOT CCT's are strongly recommended in general as well as in particular for this area to have some concrete results.

ACKNOWLEDGEMENTS

The authors are grateful to MHRD, GOI for sponsoring the Centre for Remote Sensing under which the project is running. The authors express their sincere gratitude to Prof. R.P.Gupta and Prof. A.K.Jain both from Deptt. of Earth Sciences, U.O.R., Roorkee for their constructive suggestions during the preparation of this paper.

REFERENCES

Agrawal, C.S., 1988. Hydrological investigations in Bundelkhand region using remote sensing techniques. M.Sc. Dissertation, Deptt. of Civil Engg., University of Roorkee, Roorkee.

Bakliwal, P.C., Ramasamy, S.M. and Grover, A.K. 1985. Use of remote sensing in identification of possible target areas for groundwater, hydrocarbons and minerals in the Thar Desert, western India. Proc. Vol. prospecting in areas of desert terrain, April 14-17, 1985, Rabat, Morrocco (I.M.M. London).

Bulletin of Geological Survey of India, 1965, No.20, Series B. 'Geology and groundwater conditions of parts of Rajasthan.

Chandra, K. and Bhatnagar, S.P., 1987. Studies on the Ajmer eco-system : a comparative analysis. Eco-system series, Vol.1, Regeneration of Pushkar (Ajmer) Lake Valley Eco-system Project.

Dikshit, O., 1987. Eco-regeneration of Pushkar lake valley system. M.E. Thesis, Civil Engg. Deptt., University of Roorkee, Roorkee.

Heron, 1953. Memoirs of geological survey of India. Vol.79.

Issues in water resources development and management and role of remote sensing. ISRO-NNRMS-TP-67-86.

Jones, A.R., 1986. The use of thematic mapper imegery for geomorphological mapping in arid and semi-arid environments. Symposium on Remote Sensing for Resource Development and Management/Enschede/Netherlands/August 1988/pp.

Manual of Remote Sensing, 1975. Vol.II, First edition.

Naha, K., Mukhopadhyay, D.K., Mohanty, R., Mitra, S.K. and Biswal, T.K., 1984. Significance of contrast in the early stages of the structural history of the Delhi and the Pre-Delhi rock groups in the proterozoic of Rajasthan, western India.

Principles of Groundwater Hydrology, 1972. HEC IHD, Vol.10.

Proceedings of Symposium on Groundwater Studies in Arid and Semi-arid Regions, 1966. Deptt. of Geology and Geophysics, University of Roorkee, Roorkee.

Roy, A.K., 1984. Remote sensing in groundwater targetting. International Workshop on Rural Hydrogeology and Hydraulics in Fissured Basement Zones. Deptt. of Earth Sciences, University of Roorkee, Roorkee.

KRIGED ESTIMATES OF WATER-LEVELS FROM THE SPARSE MEASUREMENTS IN A HARD ROCK AQUIFER

M. THANGARAJAN AND SHAKEEL AHMED

National Geophysical Research Institute,
Hyderabad - 5000007, INDIA

ABSTRACT

The water-levels in Vaippar basin, a hard rock aquifer in southern India have been estimated at each node of the aquifer model from a limited number of observations. Universal Kriging was used firstly to estimate the water-levels and to calculate the estimation variance. A few more hypothetical locations were selected to reduce the estimation variance near the boundaries. A bivariate interpolation technique using a polynomial of higher degree was also applied to compare the results. Finally the estimations were made using theory of Intrinsic Random Functions of higher order (IRF-k). In this case first we tried to evolve a single generalised covariance for pre and post monsoon periods and then for all the periods. The generalised covariance model was tested by cross-validations for a large number of period before adapting it for a faster estmation of water-levels.

INTRODUCTION

Aquifer modeling utilizes a number of input/output variables as well as system parameters. Among them water-level is the most important one as it is the only magnitude which can be directly measured with a greater accuracy and also the calibration of an aquifer model mostly carried out against it. Although the water-levels are comparatively available at many more places, it is not possible to measure at all node points or discretized volumes. Thus an estimation at the unmeasured locations is imperative.

Kriging is a Geostatistical estimation technique based on the theory of regionalized variables introduced by Matheron (1971). The main idea is to consider a spatial variable as a realization of random function and thereby the theory of probability can be applied for the natural spatial data. A number of interpoltion techniques could have been used to estimate the water levels of Vaippar basin, however, kriging was chosen for the following reasons:

i. Kriging is best linear estimator wherein the estimates are made by using the adjacent known values,

ii. It takes into account the spatial structure of the variable and also provides the confidence interval of estimation.

Kriging technique was developed at the Institute of Geostatistics, Fontainblaue, France by Matheron and his group. The theory is well documented by David (1977), Journel and Huijbergts (1978) and Clark (1978). Recent works on the application of kriging in hydroscience are Delfiner and Delhomme (1973), Delhomme (1978) and Aboufirassi and Marino (1983).

WATER-LEVEL ESTIMATION IN VAIPPAR RIVER BASIN

The phreatic water levels of Vaippar basin were estimated using kriging methods. The study area is a semi-arid region and covers about 1350 sq km located in Ramnad and Tirunelveli districts of Tamilnadu, South India (Fig. 1). The hydrogeological set up of the basin is described by Thangarajan and Trippler (1984). The water levels were monitored monthly in 22 observation wells fairly distributed in the area. The water levels of January 1974 which varies from 50 to 173 m (AMSL) were chosen for kriging estimation.

The first step in kriging is the computation of semi-variogram. A variogram or semi-variogram shows the structure of the spatial correltion and the nature of spatial dependence of a variable. The semi-variogram of an intrinsic function is defined by:

$$g(h) = (1/2) \text{ Var } [Z(x+h) - Z(x)] \qquad \dots \quad (1)$$

Where h is the lag distance between observation points of the variable Z(x).

The plot of computed semi-variogram in the East-West direction is shown in Fig. 2(a). The parabolic behaviour of the variogram indicates that the surface is continuous with no abrupt discontinuities or change in the slope. The continuous rise of g(h) with distance h reflects the pronounced drift along West-East direction. A theoretical model with a finite range cannot be fitted with this semi-variogram.

The computed semi-variogram for the Noth-South direction is shown in Fig. 2(b). This variogram exhibits a hole effect due to small localized difference in water level. But it is difficult to conclude this as number of pairs in comuting the variogram are very less. The semi-variogram is parabolic at the origin indicating a moderate continuity of the phenomenon. Since the variogram in East-West shows the trend and the Noth-South direction only moderate continuity, semi-variogram of residual was computed. Gaussian Model with Sill 500 sq m and a range 8 km was fitted to the residual semi-variogram.

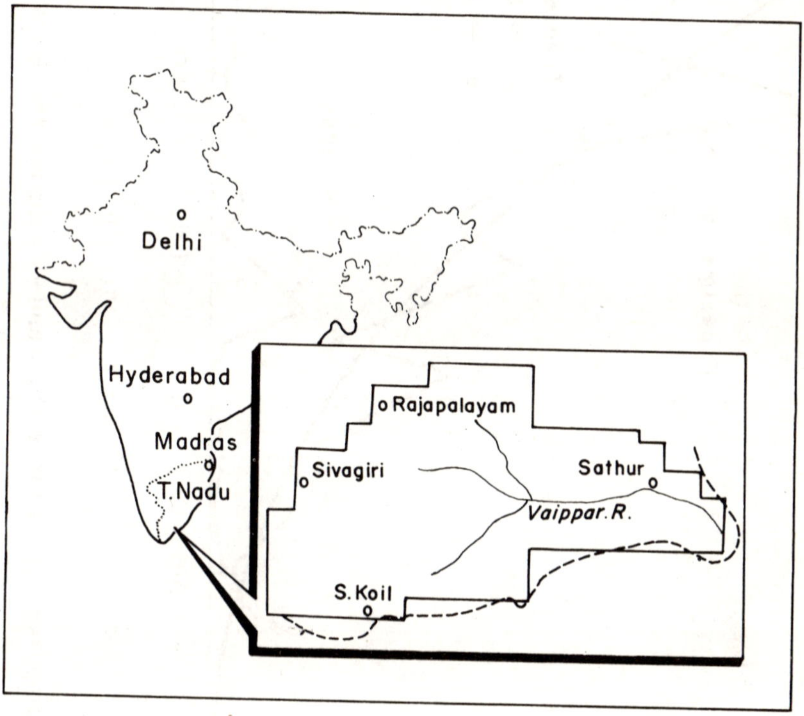

Fig. I Location Map Of Vaippar River Basin, India

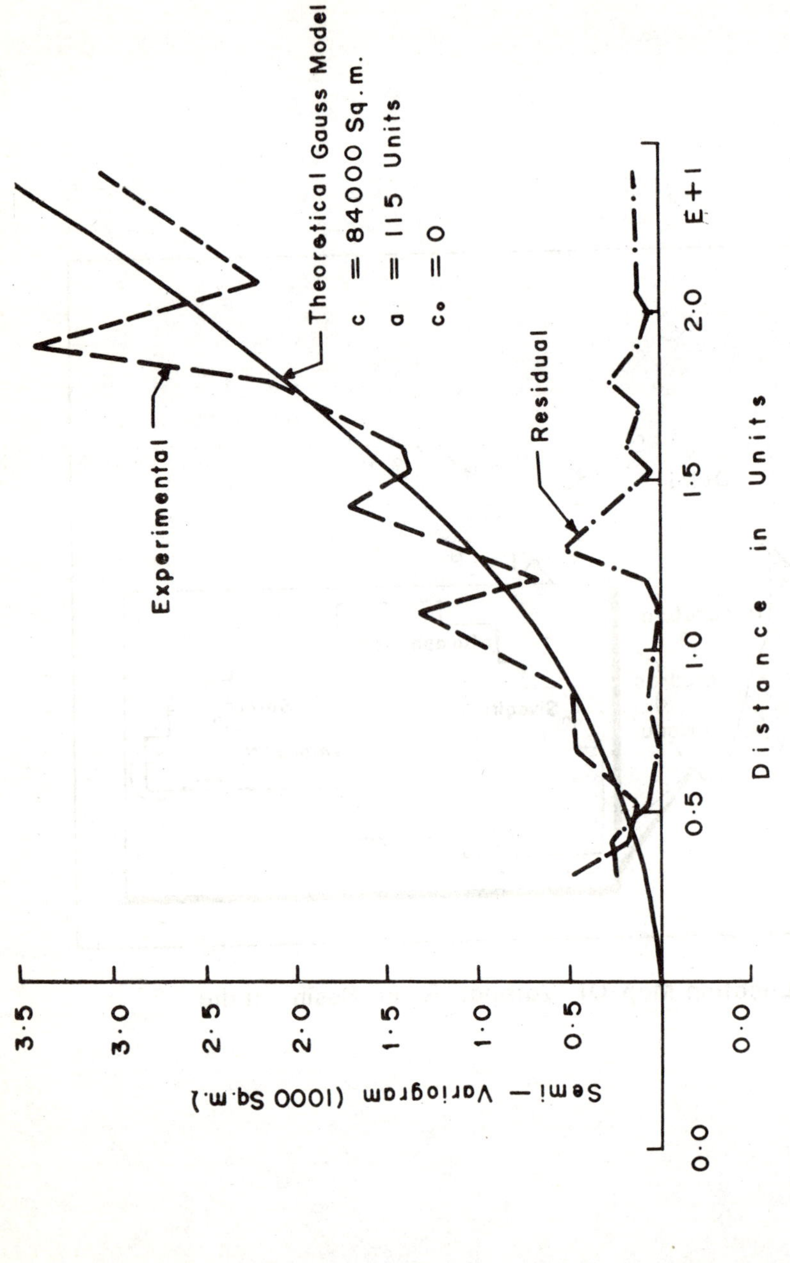

Fig. 2a Computed and Theoretical Semi–Variogram for E–W Direction

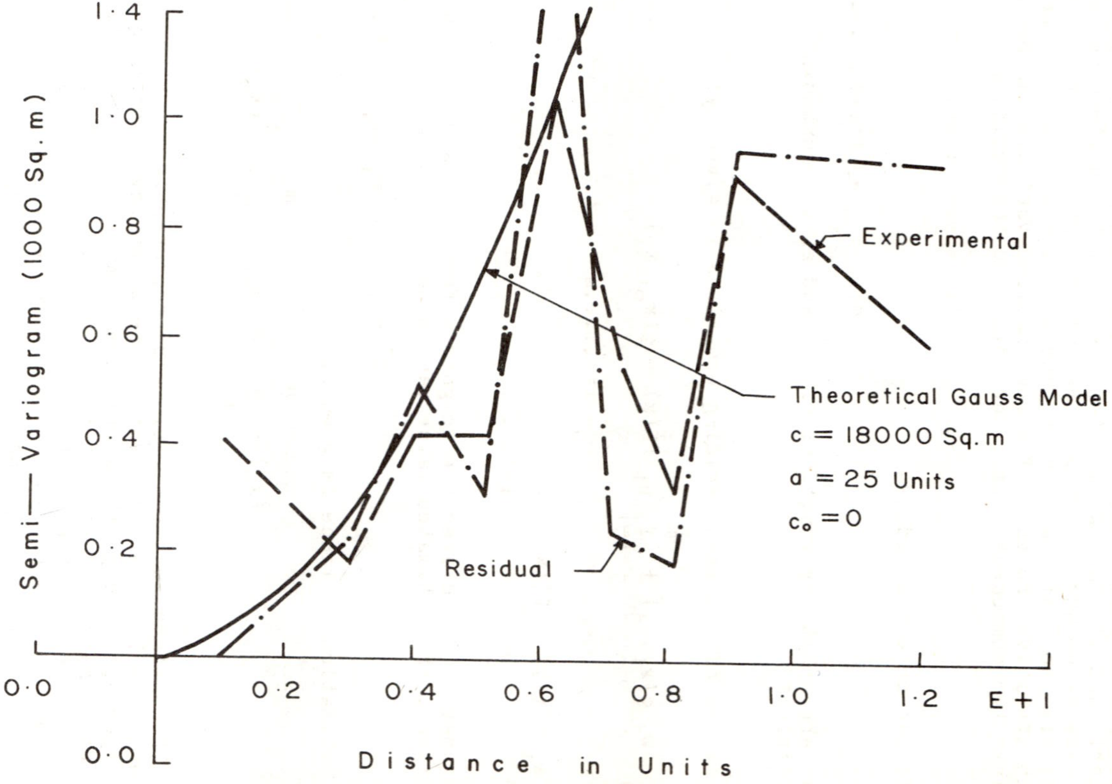

Fig.2b Computed and Theoretical Semi—Variogram for N—S Direction

Universal kriging has been applied to estimate the water levels of Vaippar basin as the phenomenon is non-stationary due to the presence of drift. The technique uses the residual of the variable and its variogram determined from the difference between the data value and the drift. Because a semivariogram is needed initially to calculate drift, the solution of universal kriging is an iterative process. The regionalized variable can be split up into two components viz., the drift and the residual

$$Z(x) = m(x) + R(x) \quad \ldots (2)$$

Drift can be expressed as

$$m(x) = \sum_{l=1}^{m} a_l \; f^l(x) \quad \ldots (3)$$

$f^l(x)$ are arbitrary functions of x and a_l are unknown coefficient

The system of universal kriging equations is given by:

$$\sum_{j=1}^{n} w_j \; g(x_i, x_j) + \sum_{l=1}^{m} \mu_l \; f^l(x) = g(x_o, x_i) \quad \ldots (4)$$

$$i=1,\ldots,n$$

$$\sum_{i=1}^{n} w_i \; f^l(x_i) = f^l(x_o) \quad \ldots (5)$$

$$l=1,\ldots,n$$

After having calculated the weights w from the above system of equations, the estimated value can be calculated as:

$$Z^*(x_o) = \sum_{i=1}^{n} w_i Z(x_i) \quad \ldots (6)$$

The estimation variance is gien by:

$$\sigma^2 (x_o) = \sum_{i=1}^{n} w_i \; g_{io} - \sum_{l=1}^{m} \mu_l f^l(x_o) \quad \ldots (7)$$

The kriged water level and topographic contours are presented in Fig. 3. In general the water level contours follow the topography except the water level contour of 80 m. This part of the region is characterized by steep topographic gradient which could not be taken into account without sufficient number of water level observations. The water level contours of 120 and 140 m deviate more from that of topographic contours in the South-Western region. This is clearly exhibited in the kriged standard deviation contour map shown in Fig. 4. The standard deviation is, obviously, high wherever observation points are not available. The estimated error map is a very powerful tool to evaluate the reliability of estimates produced by Universal kriging. The error map and semi-variogram can also be used for optimal data network design.

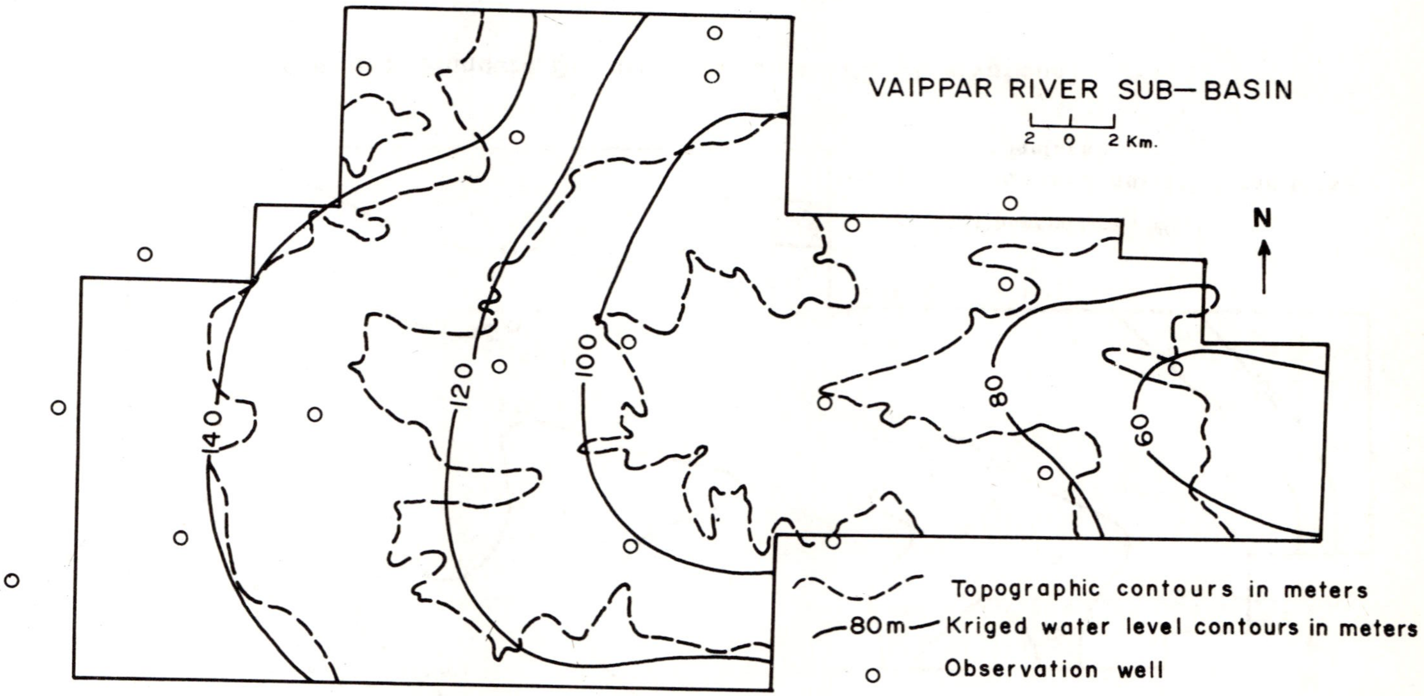

Fig.3 Kriged Water — Level Contours

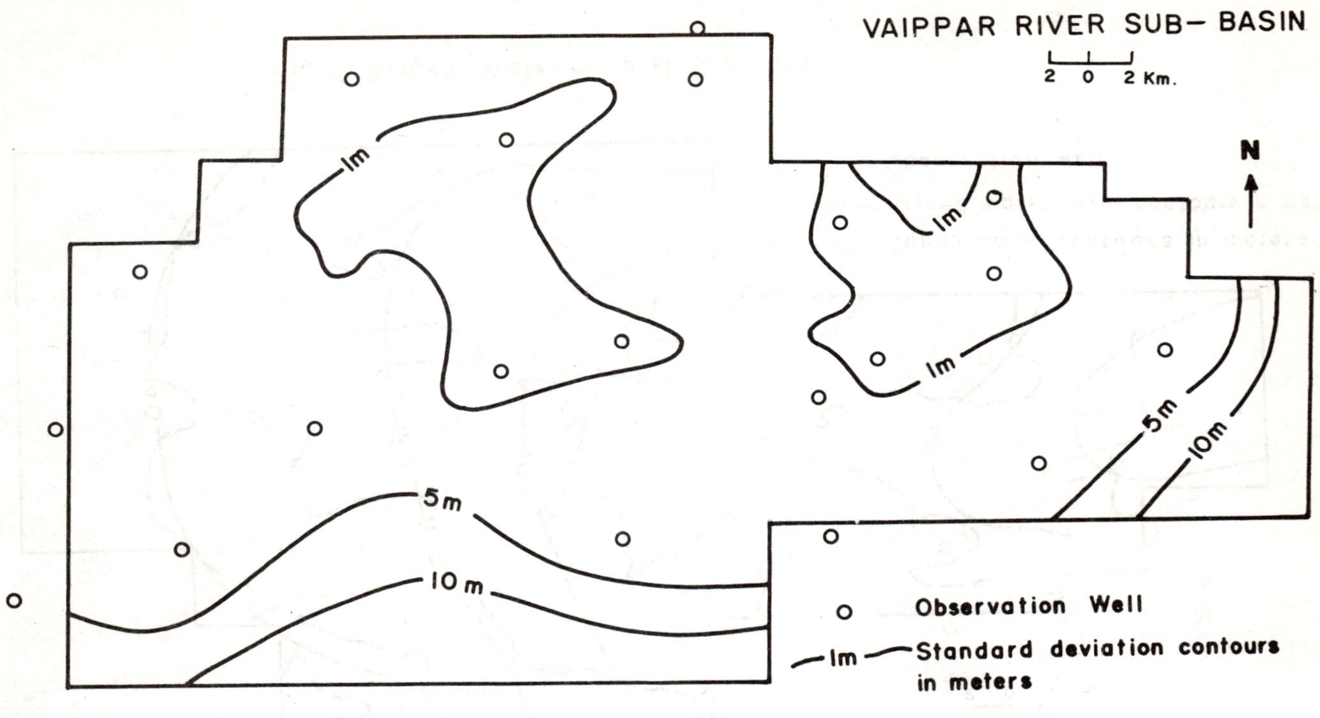

Fig. 4 Kriged Standard Deviation Contours (22 Observations)

In order to study the effect of additional data points on the reduction of kriging error, ten additional observations based on geometrical considerations were introduced more or less along the boundary. The kriged error map (Fig. 5) shows that error has been considerably reduced all along te boundaries. It may be concluded that kriged error estimates mainly depend on the correlation structure and geometrical disposition of observations. The estimation error can be reduced to a prescribed limit by taking more measurement in critical areas.

USE OF BIVARIATE INTERPOLATION TECHNIQUE

The same data were used in bivariate interpoltion technique to compare the results with that of kriging. Bivariate interpoltion and smooth surface fitting was developed by Akima (1978) for Z values given at points irregularly distributed in X-Y plane. The interpolating function is a fifth degree polynomial in x and y defined in each triangular cell which has projections of three data points in the x-y plane as its verteces. The polynomial is determined by the estimated values of partial derivations at each data point and by the given value of Z.

The estimated bivariate function is a combination of polynomials of fifth degree in x and y as shown below:

$$Z(x,y) = \sum_{i=1}^{5} \sum_{j=1}^{5-i} q_{ij}\, x^i\, y^j \quad ...(8)$$

The interpolated water level contours using bivariate function and Universal kriging technique are presented along with topography in Fig. 6. The contours clearly exhibit that at data-void places, kriged water levels closely follow the topography whereasbivariate estimates diverge. This bringsout clearly that when data is sparse, kriging technique performs beteer than the other interpolation technique. However, the interpolted water lvels from both the techniques are comparable in places of higher density of data. The major advantage of kriging is, in fact, that it provides error estimates while other techniques do not. Also kriging is a statistically optimal method of estimation, providing better estimates in areas of poor control provided a structrual model is precisely known which can only be achieved by further sampling in other techniques.

The phreatic water lvel contours obtained through Universal kriging in Vaippar river basin generally follow the topography which appears to be intuitively correct. However, the kriging has produced smooth water level surface as against undulated topography because it acts as apresampling low-pass filer which cannot take into account the undulating nature of water level surface in hard rock region. It has been found in the present case that a general reduction in interpolation error can be achieved by introducting at least 8-10 data points wihtin 8 km radius of estimation points. The kriged error map can also be effectively used for optimal data network design.

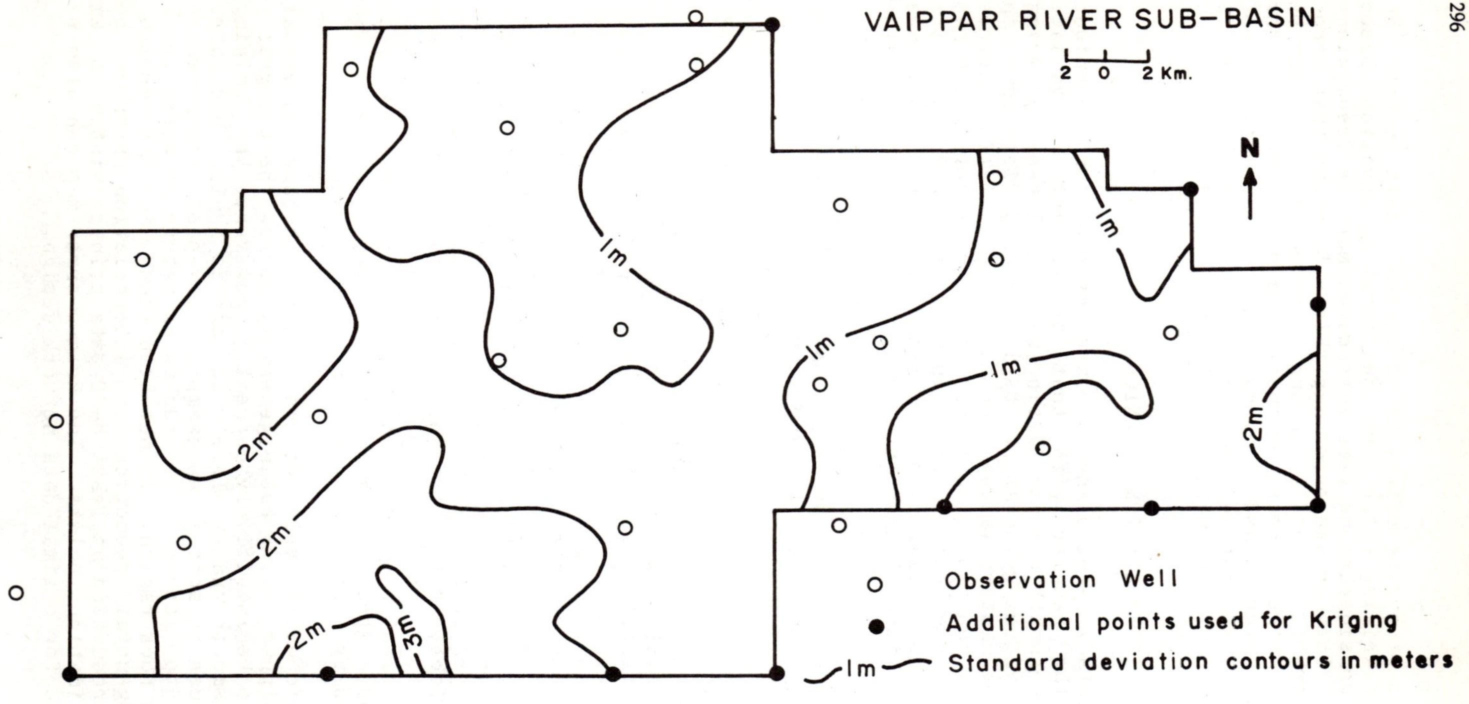

Fig. 5 Kriged Standard Deviation Contours with Additonal Data Point

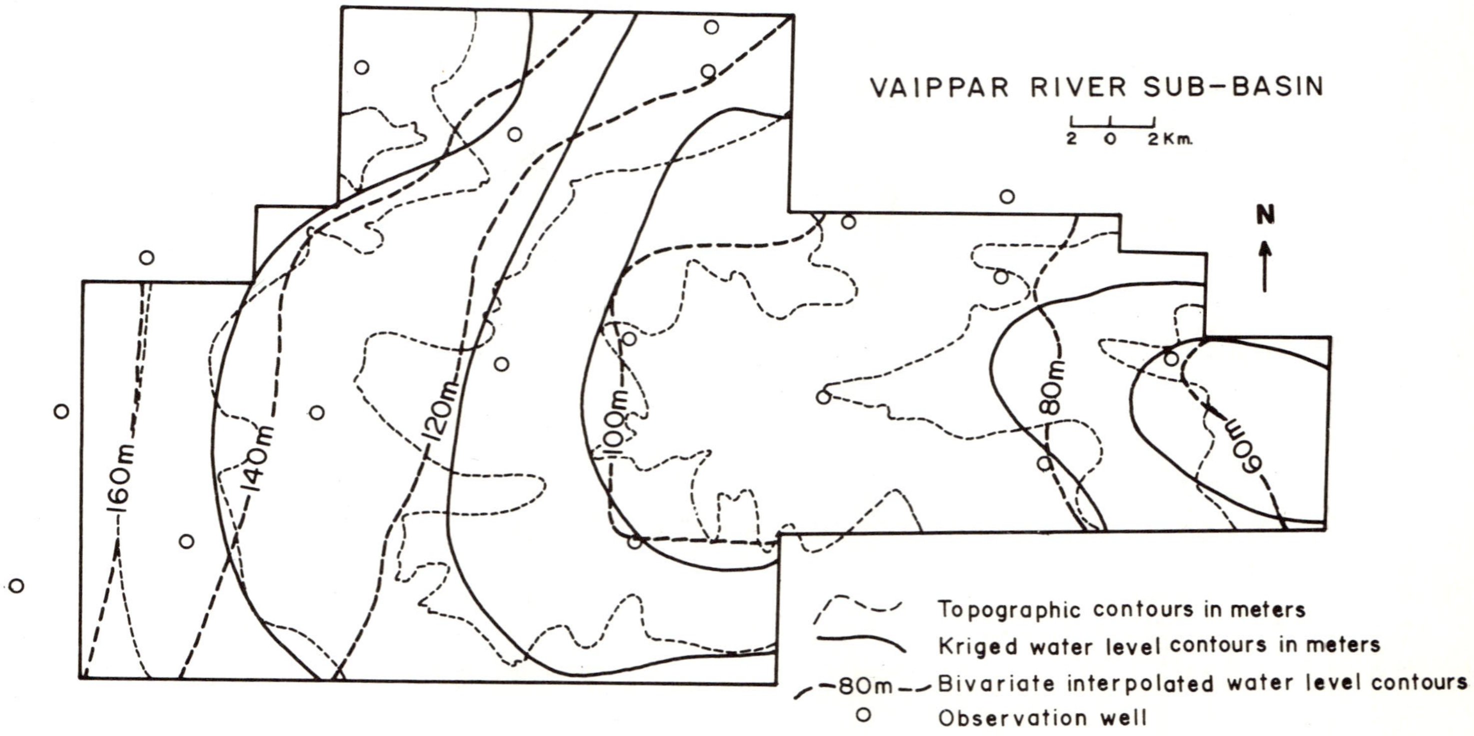

Fig. 6 Comparison of Kriged and Bivariate Interpolated Water – Level Contours

APPLICATION OF THEORY OF INTRINSIC RANDOM FUNCTION OF HIGHER ORDER (IRF-k)

As seen earlier in the use of Universal kriging, the calculation of drift coefficient and the residual variogram are dependent processes and even after a number of iterations we cannot arrive at correct drift coefficients. Another theory developed by Matheron (1973) uses a function known as Generalised Covariance in place of the variogram or the residual variogram. The non-stationary variable is transformed to an stationary one by successive differences or by a weighted combination of the known values, called Authorised Linear Combination (ALC) as:

$$Z(w) = \sum_{i=o}^{n} w_i Z(x_i) \qquad \ldots(9)$$

with the condition that,

$$\sum_{i=o}^{n} w_i = 0 \qquad \ldots(10)$$

$$\sum_{i=o}^{n} w_i f^l(x_i) = 0 \qquad \ldots(11)$$

where x is the spatial coordinate in one, two or three dimensions and $f^l(x_i)$ are polynomials.

This newly transformed variable also known as generalised increment becomes stationary and we can calculate the structural model in the form of generalised covariance, K(h). Thus we simply repalce g(h) by -K(h) in the equations of Universal kriging. The genarlised covariance is given in the form of :

$$K(h) = A_o \, \delta(h) + A_1 h + A_2 h^2 \ln h + A_3 h^3 + A_4 h^5 \qquad \ldots(12)$$

where $h = |x_i - x_j|$ and the coefficient A's are calculated by minimizing

$$\min \left[Z(w)^2 - \sum_{i=o}^{n} \sum_{j=0}^{n} w_i w_j (A_o \, \delta(h) + A_1 h + A_2 h^2 \ln h + A_3 h^3 + A_4 h^5)\right] \qquad \ldots(13)$$

This, of course, is a lengthy procedure but sophisticated computer programs are available to calculate the generalised covariance using authorised linear combinations. As one can see from equation 11 that this condition is quite easy to check when the measurements are at regular intervals but this seldom happens in parctice and hence a large number of checks are required.

The generalised covariance was calculated for the post and pre-monsoon water-level data of 9 years (i.e. from 1973 to 1981) and it was found that these covariances have a nugget effect term and a spline term having different coefficients (Ahmed, 1985) as:

$$K(h) = A_o \,\delta(h) + A\, h^2 \ln h \quad \text{... (14)}$$

It is obvious that the calculation of generalised covariance for each time step would be a cumbersome procedure and hence before proceeding to estimation it was thought to evolve a common structural model by simply averaging the coefficients of the corresponding terms.

First of all average generalised covariances for premonsoon and postmonsoon periods were seperately calculated and then a single one for all the periods was calculated. The three models are given below:

$$K(h)_{pre\text{-}monsoon} = 34.46\, \delta(h) + 3.45\, h^2 \ln h \quad \text{... (15)}$$

$$K(h)_{post\text{-}monsoon} = 51.15\, \delta(h) + 3.16\, h^2 \ln h \quad \text{... (16)}$$

$$K(h)_{average} = 42.81\, \delta(h) + 3.31\, h^2 \ln h \quad \text{... (17)}$$

A cross validation was used to verify the deviations in using the average models and the corresponding computed models. It was found that the results of the two models are quite comparable. Even a single model for all periods does not deviate the results much and can be easily utilized.

This study shows two important aspects viz., use of the theory of IRF-k to estimate the water-levels at the nodes of an aquifer model and that a single generalised covariance model can be used to estimate the water-levels for different time periods. This also helps in estimating water-levels in future without computing a generalised covariance. Fig. 7 shows the estimated water-level using two different models (case of January 1973 is shown for an example). Fig. 8 shows the corresponding map of standard deviation of estimation error. It can be seen that the values of estimation error is more when a model taken for post-monsoon is used. This is due to the fact that the generalised covariance calculated for the water-levels in post-monsoon period has a higher nugget effect. The higher value of nugget effect is the result of the large fluctuations in phreatic water-levels due to the rain fall recharge.

(A) With the generalised covariance of equation 16

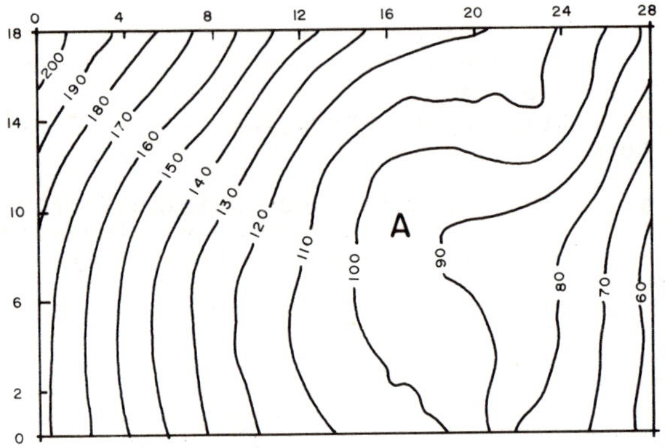

(B) With the generalised covariance of equation 17

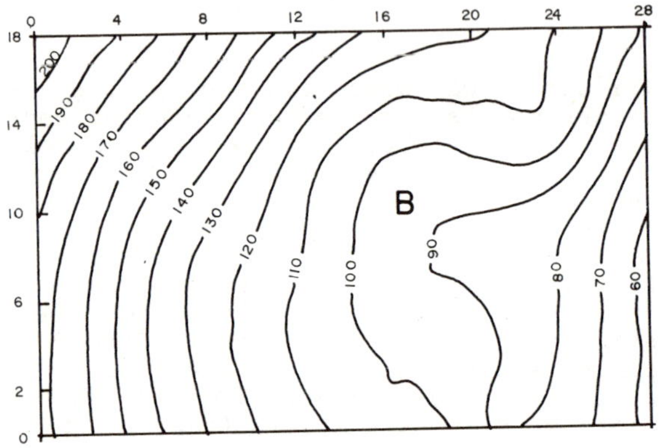

FIG. 7 ESTIMATED WATER LEVELS (JAN, 1973) USING IRF-k

(A) With the generalised covariance of equation 16

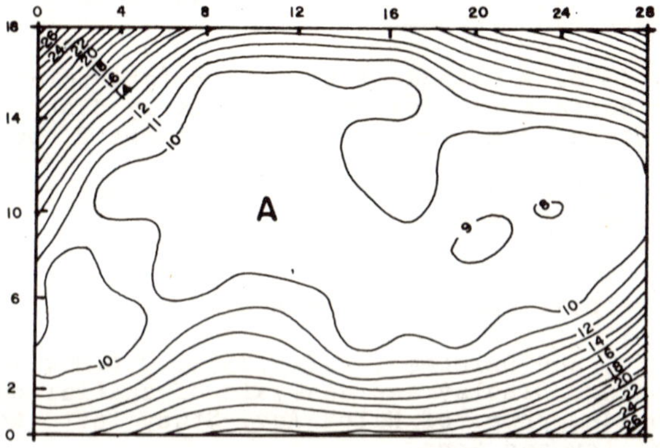

(B) With the generalised covariance of equation 17

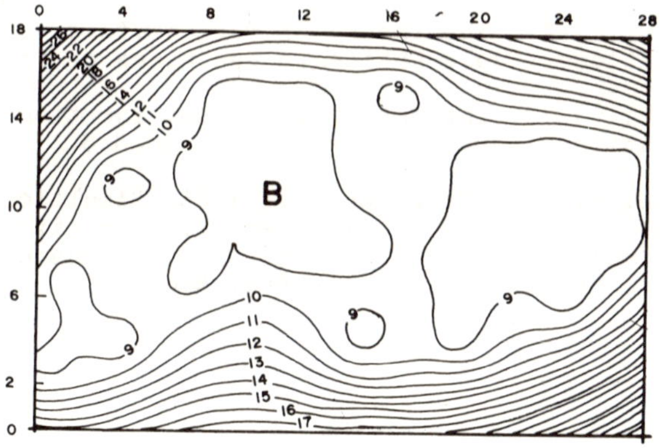

FIG. 8 STANDARD DEVIATION OF ESTIMATION ERROR (JAN, 1973)

REFERENCE

Aboufirassi, M., and Marino, M.A., 1983. Kriging of water levels in the Souss aquifer, Morocco, Math. Geol., v. 15, no.4, pp. 537.551.

Ahmed, S., 1985. Analyse Geostatistique de quelque Variables Hydrogeologiques, D.E.A. thesis, No. S-184, Juin 1985, Fontainebleau, France.

Akima, H., 1978. A method of bivariate interpolation and smooth surface fitting for irregularly distributed data points. Acm. Trans. Math. Software 2, pp. 160-164.

Clark, I., 1978. Practical geostatistics, Applied Sci. Publishers, London.

David, M., 1977. Geostatistical ore reserve estimation, Elsevier Sci. Pub. Co., Amsterdam.

Delfiner, P., and Delhomme, J.P., 1973. Optimum interpolation by kriging. Display and analysis of spatial data, pp. 96-114, Proc. of Nato. Adv. Study Inst. Nothingham, John Willey and Sons.

Delhomme, J.P., 1978. Kriging in the hydroscience, Adv. Water Res., 1(5), pp. 251-266.

Journel, A.G., and Huijbregts, Ch.J., 1978. Mining geostatistics, Academic Press, London.

Matheron, G., 1971. The theory of regionalised variables and its applications, Ecole Nationale des Mines de Paris.

Matheron, G., 1973. The intrinsic random functions and their applications, Adv. Appl. Prob. 5, 438-468.

Thangarajan, M., and Trippler, K., 1984. Simultion model studies of the phreatic aquifer, Vaippar river basin, South India, Tech. Rep. no.10174/84, Bundesanstalt fur Geowissenschaften und Rohstoffe, Hannover.

RADIAL VERTICAL ELECTRICAL SOUNDINGS : A Case Study in Groundwater Exploration.

V.R.Hegde, K.Gowdreddy and S.B.Chetti*

Dept. of Studies in Geology, Karnatak University, Dharwad-3

*H.K.E.S.Engineering College, Raichur - 584135

Groundwater occurrence in Hard rocks is mainly confined to weathered and fractured zone. The experience in borewell drilling in such terrains is that a borewell situated at a few meters away from a high yielding well fails to yield good water or sometimes even becomes practically dry. The failure or low yielding may either be due to isolation from the saturated fracture systems or due to low permeability of the minor fracture systems at depth. Hence, the determination of thickness of weathered zone and detection of the presence of fractures at depth play a major role in groundwater exploration in hard rock areas.

The crystalline rocks are generally inhomogeneous and anisotropic to the flow of electric current because of the presence of fractures/joints and minor fissures. As the anisotropy or inhomogeneity is irregular and confined to small zones, those can be detected and studied by the electrical resistivity methods. Because of the paradox of electrical anisotropy the maximum apparent resistivity (ρ_{al}^{-}) is obtained along the strike direction of the fractures/joints and the minimum apparent resistivity (ρ_{at}^{-}) is obtained at right angles to it. The paradox is explained by the fact that the current density along the fracture or plain of stratification is greater than that at right angles to this plain. The square root of the ratio of ρ_t and ρ_l is called the Co-efficient of electrical anisotropy which is given as

$$\lambda = (\rho_t / \rho_l)^{\frac{1}{2}}.$$

The change in the values of λ for the different electrode separation indicates the vertical inhomogeneity/anisotropy at different depths. The λ values at different depths can be determined by the Radial Vertical Electrical Soundings (RVES) which facilitate the study of degree of fracturing at depth and the direction of fractures.

Radial soundings along four major azimuths have been conducted at 11 evenly distributed localities in the north eastern part of the Sasive Halla Basin of Ron Taluk, Dharwad district. The area consists of granitic gneisses. The Schlumberger electrode configuration was adopted for conducting RVES and the maximum AB/2 separation was 90 mts. The polar diagrams are prepared by plotting the apparent resistivity values along

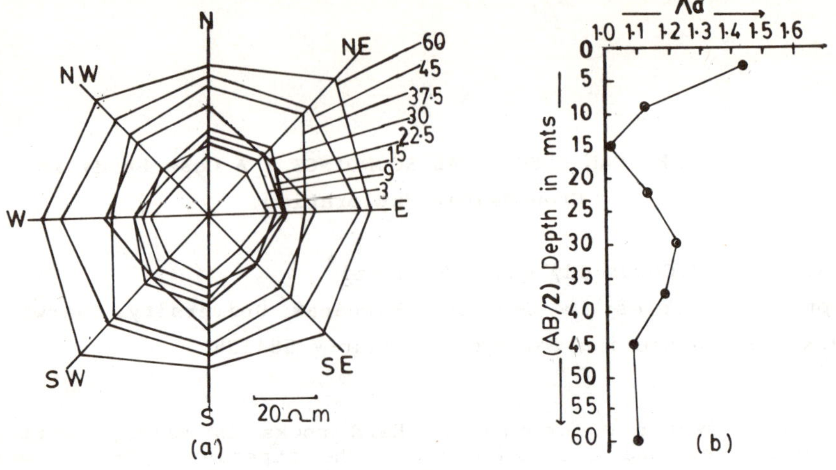

Fig I

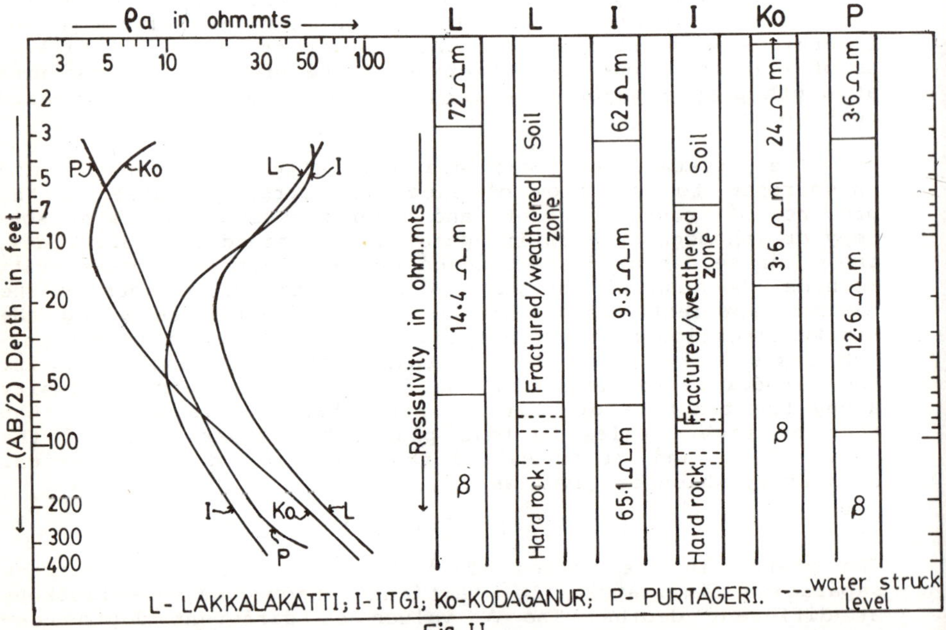

Fig II

Fig I (a) Anisotropy polygons constructed from the plotting of apparent resistivities along different azimuths. AB/2 values in meters.

(b) Variation of degree of fracturing (anisotropy, a) with depth.

Fig II Electrical sounding curves with electrical logs and lithologic logs.

different azimuths for given AB/2 separation and the co-efficient of apparent anisotropy (λa) is determined by taking the azimuthal ratios of maximum and minimum ρa values. The change of λa with depth (Fig.I) indicates that the degree of fracturing varies at depth and the fractures occur at different horizons. The orientation or the direction of elongation of the anisotropy polygons from circular shape indicates the major strike direction of the fractures. In the area the fracture systems striking N-S, E-W are dominant. However, at places fractures with NNE-SSW, NW-SE strike are also indicated.

The sounding curves at the same points interpreted by the partial curve matching techniques using Orellana and Mooney master curves and the auxillary graphs indicate that the top layer (soil) has the resistivity varying from 3.6 ohm. mts to 88 ohm.mts. The resistivity of the second and third layer attributed to the weathered and fractured granitic gneisses ranges from 2.76 ohm.mts to 20 ohm.mts. The depth to the hard and compact basement indicated by the asymptotic nature of VES curves varies from very near the surface to 60 mts(Fig.II).

The borewells drilled at Lakkalakatti and Itgi recommended on the basis of RVES have yielded 9.72 liters/sec (approx. 7000 gph) and 12.5 liters/sec (approx.10000 gph). The results of Radial vertical electrical soundings have clearly indicated not only the presence of fracture patterns, the intensity and degree of fracturing at different depths also. This has been further confirmed by the drill hole data and the striking of water at anticipated levels. The locii of high anisotropy have also been the zones of saturated fractures. The application of RVES has proved to be more dependable technique to detect the presence of saturated fracture systems in the groundwater exploration especially in Hard rock terrain.

DEEP GEOELECTRICAL SOUNDINGS AND SHAPES OF CURVES USED FOR RECOMMENDATIONS OF BORE SITES IN BASALTIC TERRAIN

H.N. PARIKH *A.J. MERH* *S.B. SINGH*

Gujarat Water Resources Development Corporation Ltd.
Ground Water Division-2, Opp. Jam Tower
Rajkot 360 001 - Gujarat - India

INTRODUCTION :

Geophysical methods provide indirect picture of subsurface formations by measuring the various physical parameters of the earths crust. These parameters are density, magnetic susceptibility, electrical conductivity and elasticity. Electrical Resistivity Survey by for is most suitable method for ground water investigation. The geoelectrical resistivity method when applied to collect resistivity values of surface to depth in a rock mass at a point is vertical electrical sounding (VES). The true resistivity values and thickness of layers are obtained by interpretation of the sounding curves. The Schlumberger configuration was used in the VES conducted in Jamnagar district for selection of bore hole sites in basaltic terrain.

WHY DEEP SOUNDING ?

The survey was required because, the geological formation in Saurashtra are mostly Deccan Trap Basalt hard rocks. These rocks are impervious and impermeable. Joints, fracture planes, veins, weathered flows between hard rock masses, intrusions in parent rocks (dykes) are channelways and receptacles of ground water. In 1984 to 1987 the rainfall in Rajkot reduced from 541 mm to 172 mm with shortfall every year. The water levels in openwells declined and availability of ground water depleted to the extent of 90%. Old bore holes of even 60 m depth dried up. Such drying of shallow and middle aquifer zones lead to deep scanning of rocks for ground water exploitation. Therefore remote sensing of villages was taken up first for searching of lineaments and potential zones, before deep soundings.

INSTRUMENTATION AND TECHNIQUE :

Deep electrical sounding was done by the DC digital resistivity meter, manufactured by Jaks Electronics of Hyderabad. The instrument is small and compact meter. In this meter the potential and current measurement ranges are 1 to 2000 MV and 1 to 2000 MA respectively. Accuracy of measurement is ± 2%. This instrument suited the speedy survey needed in scarcity conditions. During our survey one sounding normally took 2 to 2.1/2 hours when the spread was 500 M (AB/2) or more. The interpretation was done by curve matching technique using 62.5 bilog papers and master curves of Orellena and Mooney. Correlation of geological structure, existing well or bore well earlier drilling and sounding results were considered in bore hole recommendation.

DESCRIPTION OF SHAPES :

Figure No. 1 is prepared for showing different shapes of curves obtained in four talukas of Jamnagar district. These talukas are Bhanwad, Jamjodhpur, Jamnagar and Lalpur. The results of drilling on these locations are also shown. As such many soundings and bores are done in these talukas but few cases are described here.

FIG. 1

(I) DEEP SOUNDING IS NECESSARY FOR DEEP SCANNING IN BASALT.
(II) FOR DEEP SCANNING AB SHOULD BE 1000m. OR SO.
(III) THE CABLE AND BATTERY PACK SHOULD BE STRONG AND LEAK-PROOF.
(IV) THE RESISTIVITY METER SHOULD BE FAST AND ACCURATE.
(V) A CLEAR CORRELATION OF DIFFERENT DATA, EXISTING CONDITIONS AND PAST EXPERIENCE MUST IMPROVE DRILLING RESULTS.
(VI) WATCH OUT FOR SALINE WATER AND FURTHER RECOMMENDATION.

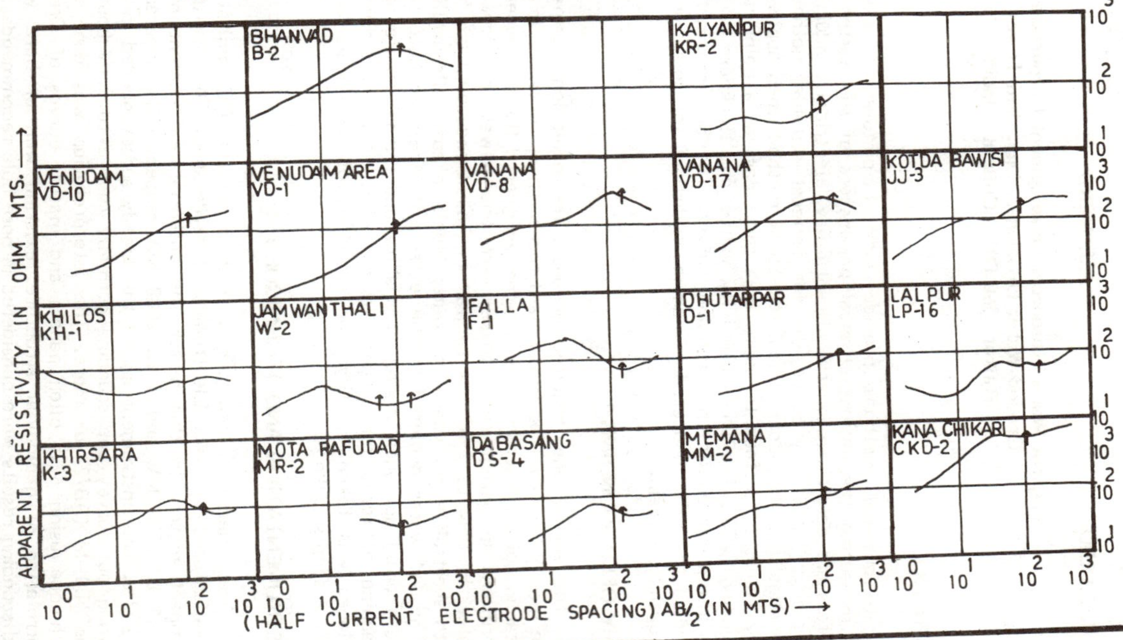

BHANVAD TALUKA :

The sounding at B2 VES is located at the foot hills of Barda ranges. A highly fractured zone can be seen at this location. The shape of curve is AK type and lowering in resistivity is seen after 150 meters. 210 Ohm. m. is the valve of the layer where water is struck, underlying a high resistivity layer of 490 Ohm.m. A 150mm bore drilled on this sounding to a depth of 153.35m gave a good quality of water at a discharge of 12.43 lps.

The curve shape at Kalyanpur is different from B2. It is AKH type curve. Lowering in resistivity is seen from 10 to 50 meters, but beyond this the curve is ascending. When the bore was drilled at this location it gave 1.41 lps. and 12.43 lps. discharge at 103.65 and 141.76 m respectively. A deep fracture must have encountered at this point. 175 Ohm. m. is the resistivity of water bearing zone.

JAMJODHPUR TALUKA :

The shapes of curves at Venudam and Vanana villages are shown in Fig. 1. Lowering in resistivity is seen after 130 meters. The discharge ranges vary from 0.8 lps to 8.35 lps in these bores.

At Kotda-Bawisi the point of sounding was selected on a lineament and drainage intersection. Discharges of two bore holes are 2.68 lps and 6.05 lps.

JAMNAGAR TALUKA :

In Jamnagar Taluka the examples can be divided into three classes - (i) Shape of curve giving good fresh water, (ii) Shape of curve giving non-potable water, (iii) Shape of curve giving inadequate water.

In (i) above category fall the villages Falla and Jam Vanthali. Here clear lowering in resistivity is seen. Both the curves are AKH type. The discharges obtained in bore holes are 1.11 lps and 2.18 lps.

In (ii) above, non-potable water was found in the bore hole at Khilos of Jamnagar. Here the lowering is not smooth. Abrupt breaks are seen at 70m and 300m.

In (iii) above, less discharge was obtained at Dhutarpur village. Lowering at 200 meter is seen which led to recommending the site.

LALPUR TALUKA :

(i) Shapes of curves which have given good discharge in bore holes while drilling are of Memana, Moti Rafudad and Khirasara. Clear lowering in resistivity are seen on the curves. In Khirasara a good correlation between a bioindicator, a lineament and sounding results, gave good discharge in the bore hole.

(ii) A typical case of Lalpur is worth describing here. In this case fresh and HOT ground water is obtained from a bore hole of 170 meters. At this site earlier a bore of 58 meters gave hot water. When this bore dried up it was deepened to 100 meters and hot water was found. After 1.1/2 year this 100 m bore also dried up and lastly the 170 m bore was drilled. During summer season the temperature of this water almost reaches 90°C. Curve shape at this site show depression between 130m to 220m. This may be a highly fractured zone and connected with a still deeper fracture. The resistivity value of this zone is 33 Ohm. m.

(iii) In two cases of Kana Chikari and Dabasang villages non-potable water was obtained in bore holes. This may be due to dykes entering into sea and leading to intake of saline water.

OBSERVATIONS :

(1) Deep soundings are necessary for deep scanning of basalt rocks.

(2) For deep scanning the electrode seperation should be about 1000m or so.

(3) For large electrode seperation care has to be taken in the larger spreads.

(4) The cable wires and battery packs should be strong and leak proof.

(5) Shapes of curves depend on various factors and do not guarantee large quantity of fresh water.

(6) Once a bore is failure or saline, great care should be taken in further recommendation.

(7) The resistivity meter should be accurate and fast in working.

(8) A clear correlation of different data, keen observation of existing conditions and past experience must improve the drilling results.

DESCRIPTION OF LINEAMENTS FOUND IN KALAVAD TALUKA OF JAMNAGAR DISTRICT AND DRILLING DRILLED FOR WATER SUPPLY

A.J. MERH *S.A. LAGHATE* *R.K. PATEL*

Gujarat Water Resources Development Corporation Ltd.
Ground Water Division-2, Opp. Jam Tower
Rajkot 360 001 - Gujarat - India

INTRODUCTION :

Kalavad taluka of Jamnagar district is covered by Survey of India toposheets No. 41 J/3, 41 J/4, 41 J/7, 41 J/8, 41 J/11 and 41 J/12. This taluka was investigated for water supply under the technology mission for drinking water supply and scarcity programme 1987-88. Villages were surveyed, involving remote sensing, geohydrological investigation and geophysical soundings. Purpose of this ground water investigation was to locate suitable sites for drilling or digging of bore well and establish a dependable source of water supply. Kalavad taluka comprise of deccan trap basalt hard rocks. These hard rocks are no good for a dependable aquifer. Ground water just do not occur every where in this rock type. A hydrogeologist has to do extensive walking in the nala and river cuttings to find out a suitable structure. The structure may be a geological structure or a man made open well or bore well which indicate potential site, for geophysical confirmation. Extensive walking without a pre-seen data consumes lot of time and energy in the crisis of water shortage. Here remote sensing comes as an aid and a guide. Remote sensing in laboratory is a skilled job. A satellite imagery is inspected on a High Magnification Enlarger. The imagery can be enlarged to the scale of a toposheet and the copying of interesting structures is done on toposheets simultaneously. High Magnification Enlarger instrument is available at Indian Space Research Organisation at Space Application Centre, Ahmedabad. Though the toposheets themselves give fair idea of some ground water potential areas, the imagery add some regional features which are not ordinarilly visible. Aerial photographs of Kalavad were not available. Otherwise aerial photographs come more handy, give minor details and reduce the search for an interesting structure.

Mr. D.G.A. Whitten with J.R.V. Brooks describes lineament as "A large scale linear feature which expresses itself in terms of topography, which is in itself an expression of the underlying structural features. Such features may include: Valleys controlled by faulting or jointing; fronts of mountain ranges; narrow straight mountains or hill ranges and ridges; lines of isolated hills; straight coastlines. Modern and ancient structural features associated with lineaments include fault zones, fracture (joint) zones, fold axes, linear igneous intrusions, lines of voleanoes or volcanic fissure erraptions".

A lineament is an interesting structure in basalt rock, important for ground water movement, stoppage and storage. Basalts are genetically formed by cooling of lava flows and the rock itself can be indentified by its dark tone. During cooling of lava and formation of a rock, joints develop. Joints have strike, dip and spacing. Some joints are structurally disturbed and they develop fractures. Such fractures and closely spaced joints has a linear extension. So they can be seen on an imagery as a line. A dyke is a later intrusion in the country rock giving a distinct linear effect on the surface. Dykes are important in ground water studies as they obstruct the normal ground water and surface water movement.

Larger dykes are very conspicuous on imagery and aerial photographs, as they appear as lineaments on these maps. A photogeologist sees these lineaments on the maps and transfer them to a scale suited for field checks. Technically these field checks are called Ground Truth Checks.

LINEAMENTS OF KALAVAD :

A map has been prepared on 1 inch = 2 miles with markings of all lineaments (Figure - 1). Here lineaments are shown as lines. The lineaments which are dykes are marked with a clear symbol. Fracture zones are also marked. The lineaments which are not identified clearly like hill ranges, break in slopes, tonal variations etc. are denoted as hatched lines. A brief description of the main lineaments on which bores are drilled is given as Statement No. 1. The details of lineaments and the bores drilled, are stated in this statement.

OBSERVATIONS :

(1) Most of the successful sites were falling on intersection of lineaments with the drainage courses.

(2) E-W trending lineaments are found to be more productive.

(3) The bores drilled on a lineament crossing a reservoir (dam or talav) are more productive.

(4) Weathered and fractured dykes are more productive.

(5) Fracture zones forming the courses of existing drainage are more productive.

(6) Pump tests are recommended on bore holes near lineaments which have given less discharge during drilling.

(7) Water supply bores sites are required near the village for easy approach. A lineament far away from the village is considered suitable for drilling, only if a site is not available near the village.

(8) If the site is selected accurately, and discharge is not noted during drilling, it is advised to conduct pump test after the monsoon.

ACKNOWLEDGEMENT :

The authors are thankful to Shri S.C. Sharma, the Chief Geohydrologist for encouragements. The authors are equally thankful to Shri G.D. Patel, Managing Director of Water Resources Development Corporation Ltd., Gujarat. The authors thank Dr. Baldev Sahay and Dr. R.K. Sood of ISRO, Ahmedabad for their co-operation in these studies.

FIG.-1

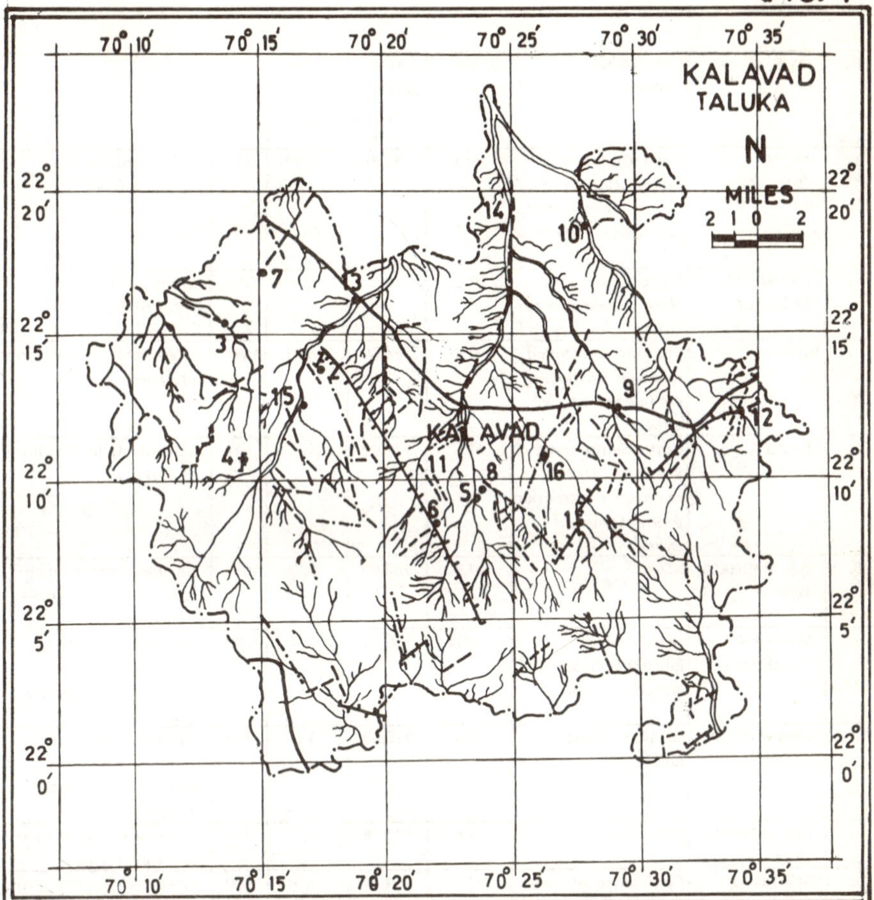

LEGEND

CONFIRM LINEAMENTS

—— DYKE

-·-·- CONTACT OF TWO LITHOUNITS

- - - - FRACTURE

-·-·- UNCONFIRMED LINEAMENTS

• DRILLING SITE

DRAINAGE

—— ROAD

PLEASE NOTE :-

LOCATION NUMBERS GIVEN ARE AS PER STATEMENT NO. 1

STATEMENT No. 1

SR. NO.	NAME OF VILLAGE	LINEAMENT CONTROL	GEOPHY-SICALLY CONFIRM	TOTAL DEPTH DRILLED m	DISCHA-RGE IN L P S	TDS IN ppm	DESCRIPTION
1.	MOTA BHADUKIA	Dyke Drainage Intersection, formation of three sided obstructive reservoir	NO	97.56	6.59	418	Lineament, in the form of Dyke
2.	LAXMIPUR DUDHALA	North East South West trending minor Dyke	YES	120.75	6.59	1490	Lineament, in the form of Dyke
3.	BANGA	Fracture controlled drainage on the confluence of two streams	YES	86.89	6.59	1496	Lineament, in the form of Dyke
4.	SAVLI	North South trending Dyke. Dyke drainage intersection down stream of Dyke	YES	137.00	4.07	444	Lineament, in the form of Dyke
5.	BALAMBHADI DAM-II	NO	YES	182.80	3.76	650	No lineament control but red tone vegetation cluster
6.	MAKRANI SANOSARA	Dyke drainage intersection down stream of Dyke	NO	150.91	3.76	700	The lineament is in the form of Dyke fractured and yellow coloured
7.	KHANKOTDA	Break in slope of hill	YES	120.00	3.76	300	Lineament is in the form of straight courses of small hill ranges
8.	BALAMBHADI DAM-I	NO	YES	173.78	2.71	745	Cluster of vegetation red toned area
9.	SISANG	Confluence of two	YES	219.51	1.51	900	Lineament, in the form of fractured course of the stream
10.	VIBHANIA	NO	YES	99.00	1.15	468	Minor Dyke could not be picked up
11.	SARVANIYA	Dyke NW - SE	YES	96.00	0.66	1108	Lineament as Dyke
12.	ANANDPUR	Dyke drainage	NO	178.00	0.32	-	Lineament as Dyke
13.	KHANDHERA	NO	YES	132.00	0.32	-	Flood Plain
14.	MOTI NAGJAL	NO	YES	100.00	0.22	468	Not on imagery
15.	NANI BHAGEDI	NO	YES	150.00	1.83	582	Confluence of Nalas
16.	KHIMANI SANOSARA	Trending NNE-SSW	YES	100.00	0.117	650	Lineament is in the form of lithocontact away from site

A BRISK GROUND WATER INVESTIGATION FOR LOCATING BORE HOLE SITE IS BASALTIC TERRAIN

A.J. MERH : ***S.A. LAGHATE :***

Gujarat Water Resources Development Corporation Ltd.
Ground Water Division, Opp. Jam Tower
Rajkot 360 001 - Gujarat - India

INTRODUCTION :

The drought years of 1984 to 1987 in Gujarat created acute shortage of drinking and irrigation water, in Saurashtra, the western portion of Gujarat STate. The villages of Rajkot, Jamnagar, Junagadh, Amreli, Bhavnagar and Surendranagar districts comprise of deccan trap basalts. Geohydrologically basalt rocks are poor aquifers. They do not possess primary porosity or permeability. Generally the water levels in open wells in basaltic terrain decline with approach of summer. In the year 1987 there was an acute decline in water level and ground water availability in open wells reduced to 80 to 90% from the normal yield. This resulted in drying up of wells and bore holes in the villages and a herculan task of providing drinking water to villages was put before the hydrogeologists of Gujarat Water Resources Development Corporation. List of villages were given to us by the respective Collectors of the district. They eagerly awaited our findings.

PROCEDURE ADOPTED :

The Remote Sensing of villages was done by the use of imagery and aerial photographs and important features were marked on toposheets by visual manual transfer or with the help of High Magnification Enlarger. Wherever aerial photographs were available, a tracing of the portion around the village was done for field checks. This information was collected by the geologists and the party rushed to field in the early morning.

After reaching the village the local Sarpanch (Village Head) of the village was contacted and the purpose of visit was explained to him. The questions regarding water supply were asked to him in his mother tongue. The questionairre is as under :

(1) What is the population of the village ?

(2) What is the existing facility of water supply ?

(3) How many bore holes are drilled in the past and what is their present condition ?

(4) Are there any private bore wells/open wells which are running good ?

(5) In the river or rivulet near by is there any geological structure which is different than other rock ?

(6) Where is the exposure of dyke, fracture, lineament around the village as seen on toposheet, image, aerial photosketch ?

The above querry gave the first hand information on existing condition. If some mechanical repairing or replacements on existing facility were needed, the point was noted and communicated. The next step was to walk in streams, nala cuttings, along lineaments or in the potential side of village as noted earlier from remote sensing or local inquiry. Particularly the side of dyke, calcite veins cutting the country rock, closely spaced joints/fracture planes, were looked for.

The next step was to check the extent of such anomalous structures. If it is a dyke then the geologist walked along it. A good well or a bore if present on the sides of the dykes, were seen and their correlation tried. If it was a lineament or fracture the geologist tried to locate the existing exploitation by farmers. Existing open well inventory on these interesting structures were done to see the importance of the particular structure. If the lineament marked by remote sensing could not be traced, the description of exposure there was noted for correlative study. The next step was to select a point for subsurface strata checking. This selection of point for sounding is the most important part of the game. Some clues on this point are indicated as under :

(i) Dyke drainage intersection point.

(ii) Point in the middle or side of a calcite vein.

(iii) Point on a fractured zone.

(iv) Point away from an existing good bore but falling on same condition.

Once the points for geophysical soundings were fixed, the soundings were conducted from the best point to good point. In our surveys we used a digital resistivity meter manufactured by Jaks Electronics of Hyderabad. Sounding by this meter using Schlumberger array was found to be suitable for this type of job. Interpretation of curves were done, a sketch plan of all observations was made and brief discussion amongst geologist, geophysicist and geohydrologist were arranged. Decision on site was taken based on the correlation between geological and geophysical data.

In drought condition where the villagers are waiting for your clearance, a quick decision on part of geohydrologist was expected. Thus as per the confidence of fieldworker, the shape and values of resistivity soundings the point was released or refused for drilling.

Though the survey was done by best possible method and spirit, the exact quantum of water from the point of drilling could not be predicted. The geophysical soundings indicate subsurface variation in resistivities. The true resistivity values or shape of curve cannot assure the solution of problem.

Regarding time frame of working, the experience during these surveys showed that five average sized villages could be covered in eight days of working. In one day maximum four geoelectrical soundings could be done. If it were deep soundings, three soundings are expected from the person doing this. We feel that for such fast survey there should be at least two days available for one field party to cover one village with present procedure. The field party can consist of one good geologist and one strong geophysicist or one good geologist knowing the handling of resistivity meter and interpretation.

The short form of the brisk survey is given on Statement No. 1.

STATEMENT No. 1 BRISK GROUND WATER INVESTIGATION FOR LOCATING BORE HOLE SITES [IN BASALTIC TERRAIN]

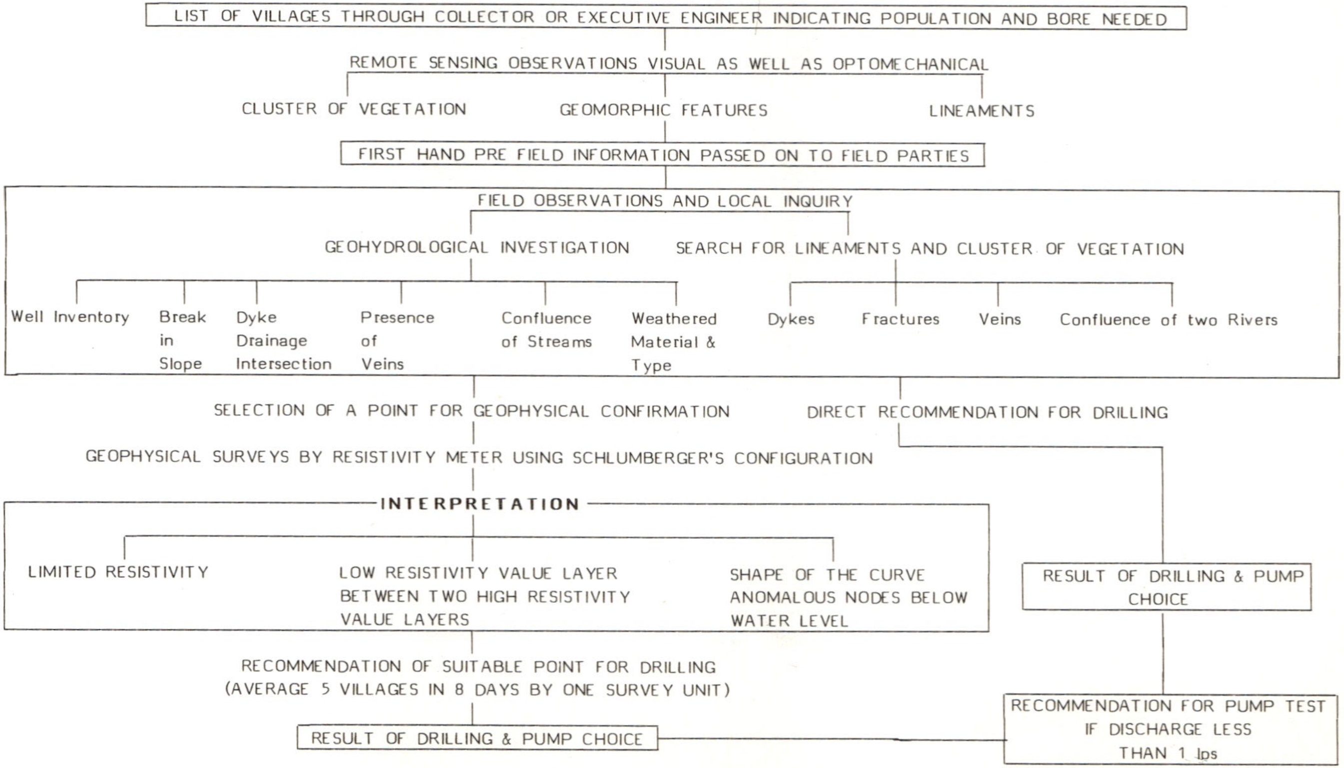

FRACTURE PATTERN MODELLING FOR GROUNDWATER TARGETTING IN HARD ROCK TERRAIN - A STUDY AIDED BY REMOTE SENSING TECHNIQUE

Ms.K. Usha I.I.T., Madras - 600 036

Dr.S M. Ramasamy, Anna University, Madras-600 025

Dr.SP.Subramanian, IIT., Madras-600 036.

I. INTRODUCTION

1.1 GENERAL

The ground water conditions in hard rock terrain is multivariate because of the heterogenous nature of the aquifer, owing to the varying composition, compaction, density of fracturing and degree of weathering. As a result, targetting of ground water in such terrain has proven to be a complex phenomena. At the sametime, the vast hardrock terrain of the global surface cannot be rejected as an unfavourable zone and instead some deeper level modern technologies will have to be invented for the same. Taking all the avaliable trends in ground water exploration in consideration, a new experimental work has been carried out in an area of about 800sq.km., in North-Arcot District of Tamil Nadu, which exposes a complexly folded and fractured crystalline terrain, using the modern remotely sensed data.

1.2 THE PREVAILING TECHNIQUES

Since olden days, until 19th century, the man has relied mostly on Dowsing technique and proximating groundwater using floral indicators. Subsequently, the water level fluctuation measurements, drilling and reconstruction of subsurface lithological formations, pump testing and analysis of well hydraulics and also Geophysical techniques were resorted to. Subsequent to the advent of the modern remote sensing techniques, the scientists have mainly concentrated on hydrogeomorphic mapping and the fracture pattern analysis. Even after these drives, the precisional mapping of groundwater pools in such hard rock terrain has remained a matter of great conjecture and speculation.

1.3 THE PRESENT MODELLING TECHNIQUE IN BRIEF

The above techniques of pre and post remote sensing days, could not yield much information on the ground water targetting because the lithology and deformational fabrics, which only holds and conducts ground water in hard rock regimes have not been given full attention, it deserves. Hence in the present model, the lithologic and the deformational fabric have been brought out in detail using the digitally enhanced and the enlarged formats of the Thematic Mapper Data. From this surface signatures, subsurface configuration of the folded structure, the nature and the morphology of the fractured system have been analysed. The various fracture density diagrams were drawn and these informations were integrated with water level and water table maps and the exact controls of ground water mobility and stability has been identified.

2. GEOLOGIC AND TECTONIC FRAME WORK

The pambar area, the arena of the present model, exposes the precambrian sequence of rocks viz, charnockite, various Gneisses and occasionally syenites. These rock types are folded to long, linear and open upright anticlines and synclines, with their axial trace disposed in NNE - SSW direction. The beds show moderate to marginally steeper slip at places. This folded structures are forming hill ranges in the Eastern fringe and also a vast plain track in the central and western part of the area.

3. FRACTURE PATTERN MODELLING

3.1 GROUNDWATER FLUCTUATION DATA

Over hundred open dug wells punctured to a depth of 10 to 20m have been examined and the water level fluctuations have been analysed, by making contour diagrams, which show that the waterlevel fluctuations vary from 6 to 18m. These wells are connected to the MSL and water table has been evaluated and contoured, which once again shows that, it varys from 300 to 450m above MSL.

3.2 MAPPING OF FOLDS AND FRACTURES

The structural trends on the basis of the soil tone characteistics, micro relief linearity, vegetation alignment and moisture rhythmicity have been interpreted from the remotely sensed data. Such structural trends were linked up with exposed structures and extrapolated to the partially soil covered areas, thereby the full limit of the exposed and buried structures were brought out (fig.1). Such reconstruction also shows that the rock types were folded to series of upright synclines and anticlines with their axial trace trending in NNE-SSW direction. All the linear fractures which are once again characterized with drainage, soil tone, vegetation, moisture and rock trend linearities in space imagery, have been marked as lineaments. These lineaments which regionally show polymodal orientation were superposed over the folded structures and grouped into different azimuthal sets depending upon their geometrical relationship and their morphology with folds (Anderson, 1951) as follows (fig.2)

1. WNW-ESE sets (Extension fracture) - orthogonal to the fold axis, show wide open morphology with prolific vegetation fills and with no lateral shifts of lithology.

2. NE-SW and NW-SE sets (Shear fractures) - conjugate to the fold axis, very tight, thin and linear, with NE-SW sets showing dextral geometry, NW-SE sets showing sinistral geometry.

3. NNE-SSW sets (Release fractures) - marginally opened up fractures mainly observed in the axial portions and closures of the folds.

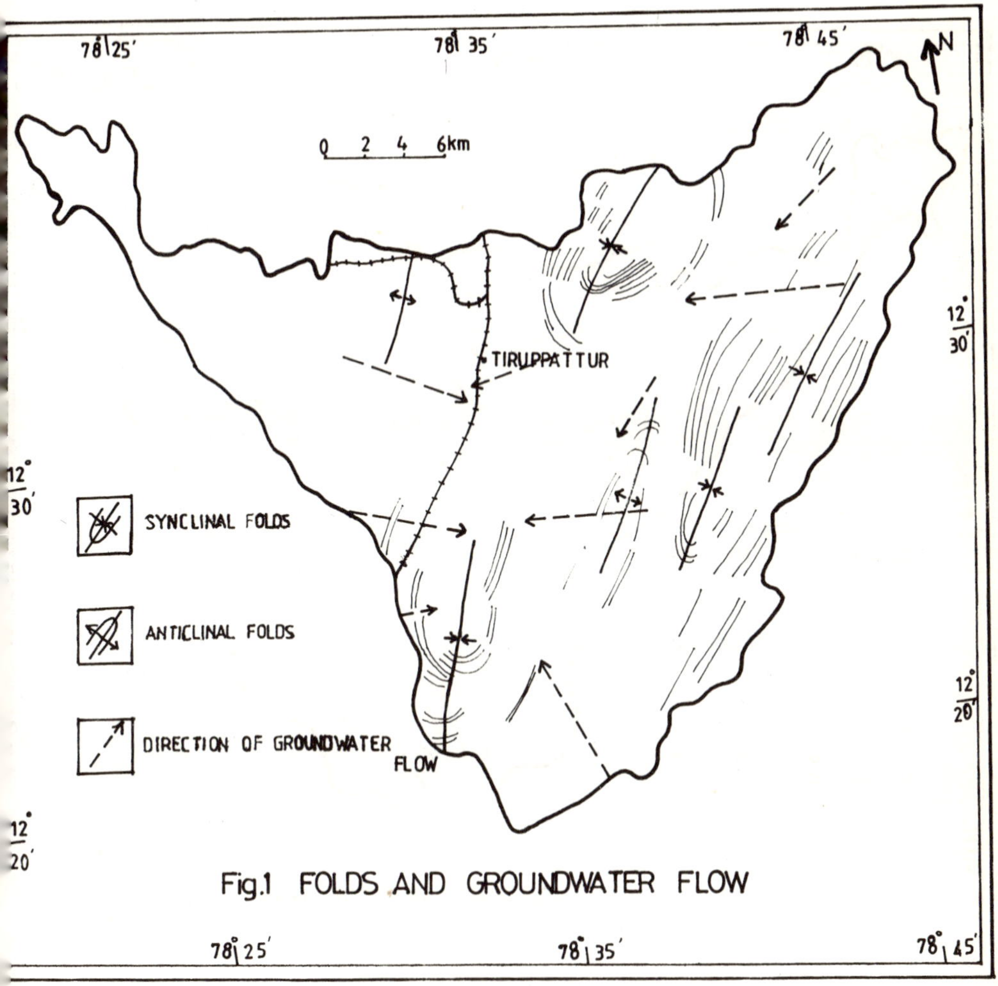

Fig.1 FOLDS AND GROUNDWATER FLOW

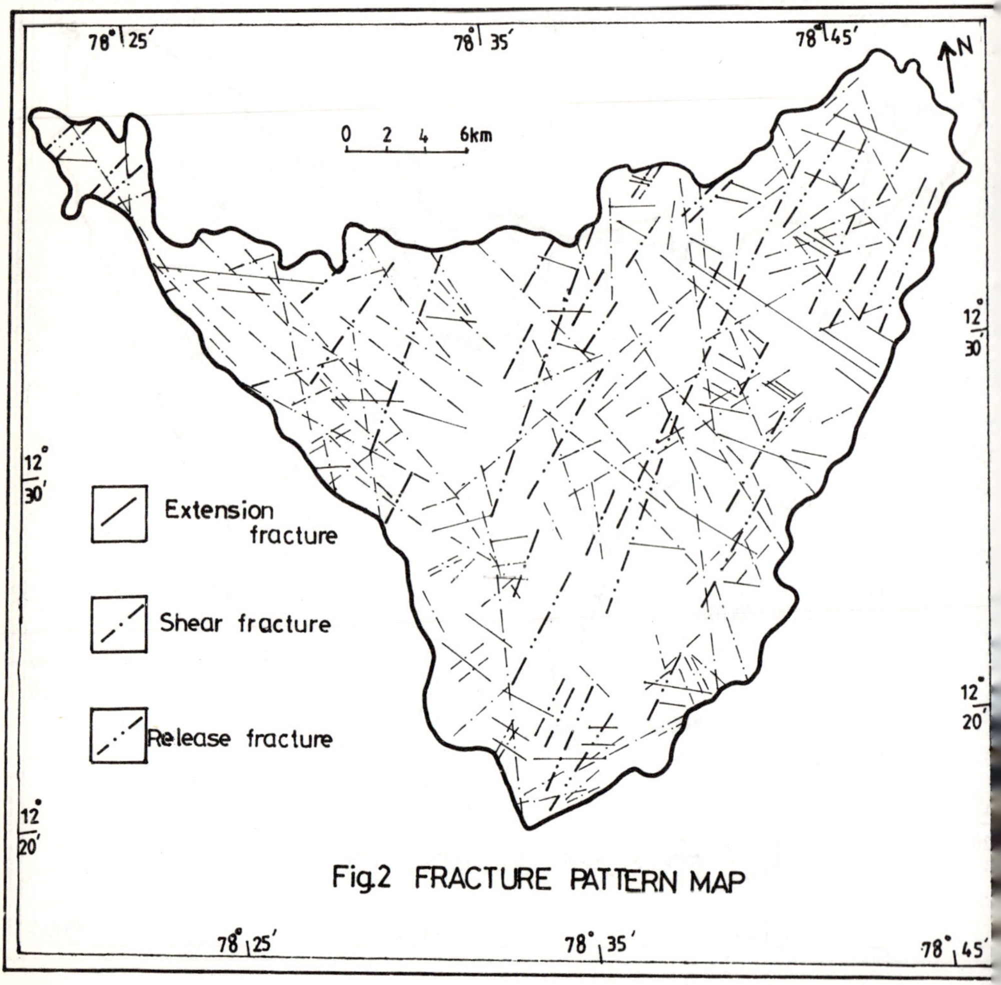

Fig.2 FRACTURE PATTERN MAP

3.3 ANALYTICAL TREATMENT OF FRACTURES

This different fracture systems were subjected to further analytical treatments by preparing fracture incidence maps of Extension, Shear and Release fractures. These were generated by griding the whole area into 1 sq.km cells and counting the length of fractures in each cells and contouring these values (fig.3). The orientation of these contours have been taken as anomaly axis (Haman, 1961; Ramasamy et al, 1983).

The fracture intersection of extension and shear, extension and release, shear alone, and shear and release were plotted on these grids and total number of such intersections were independently counted for each set and contoured.

3.4 IDENTIFICATION OF GROUNDWATER FLOWAGE ZONES

The water table map (fig.4) and the pattern of water table contours show that the groundwater flowage is in WNW-ESE direction i.e. from the peripheral part to the central part of the area. The integration of this flow direction with folded structure show that, the groundwater flowage is from anticlinal structures to synclinal domains (Fig.1). Once again this groundwater flow data was superposed over the anomaly axes derived for the extenstion fractures which are oriented in WNW-ESE direction, shear fractures which are oriented in NE-SW and NW-SE directions and release fractures which are disposed in NNE-SSW direction (fig.5). This analysis shows that, the extension fractures are acting as conduits for the groundwater flow in the study area.

3.5 IDENTIFICATION OF GROUNDWATER TARGETS

The analysis of structural trends and water level contour diagram suggests that the synclinal areas have got more potential for groundwater target in hardrock terrain. Then, various density contour diagrams were integrated with water level fluctuation data. Sequentially, first the fracture incidence map of extension failures were correlated with water level contours and wherever the fracture density maxima is coinciding with water level minima, were demarceted as groundwater potential zones in extension regimes. In otherwords, in these regimes, the extension factures are acting as zones for groundwater stability. In the sameway, the zones of stability in shear fracture domain, release fracture domain were also demarcated. In order to further probe it in detail, the water level data was correlated with intersection intensity contours prepared for various combinations and groundwater potential zones were demarcated. All these zones identified for various fractures combinations were ultimately welded together on a single map (fig.6), and the final map, showing the various groundwater zones along-extension fractures, shear fractures, release fractures, intersection of extension and shear fractures, intersection of extension and release fractures, intersection of shear fractures, intersection of shear and release fractures and intersection of extension,shear and release fractures, was prepared.

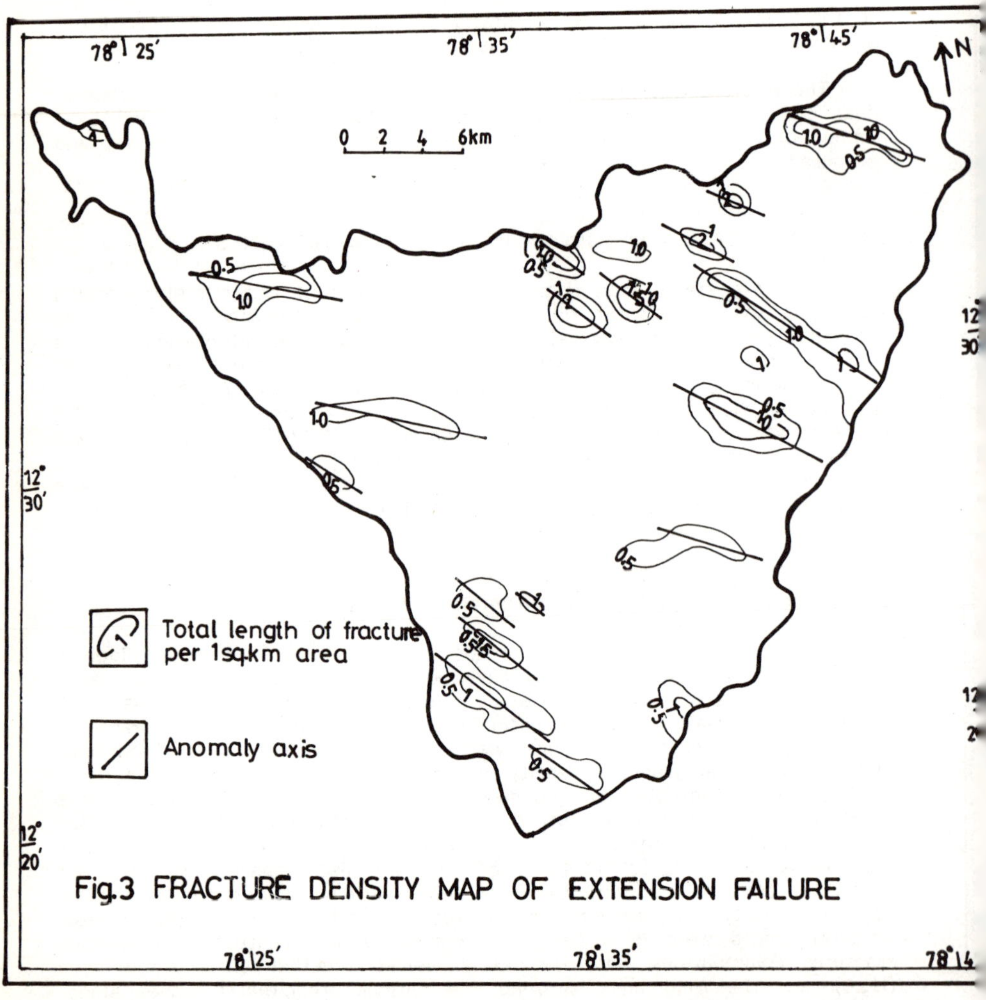

Fig.3 FRACTURE DENSITY MAP OF EXTENSION FAILURE

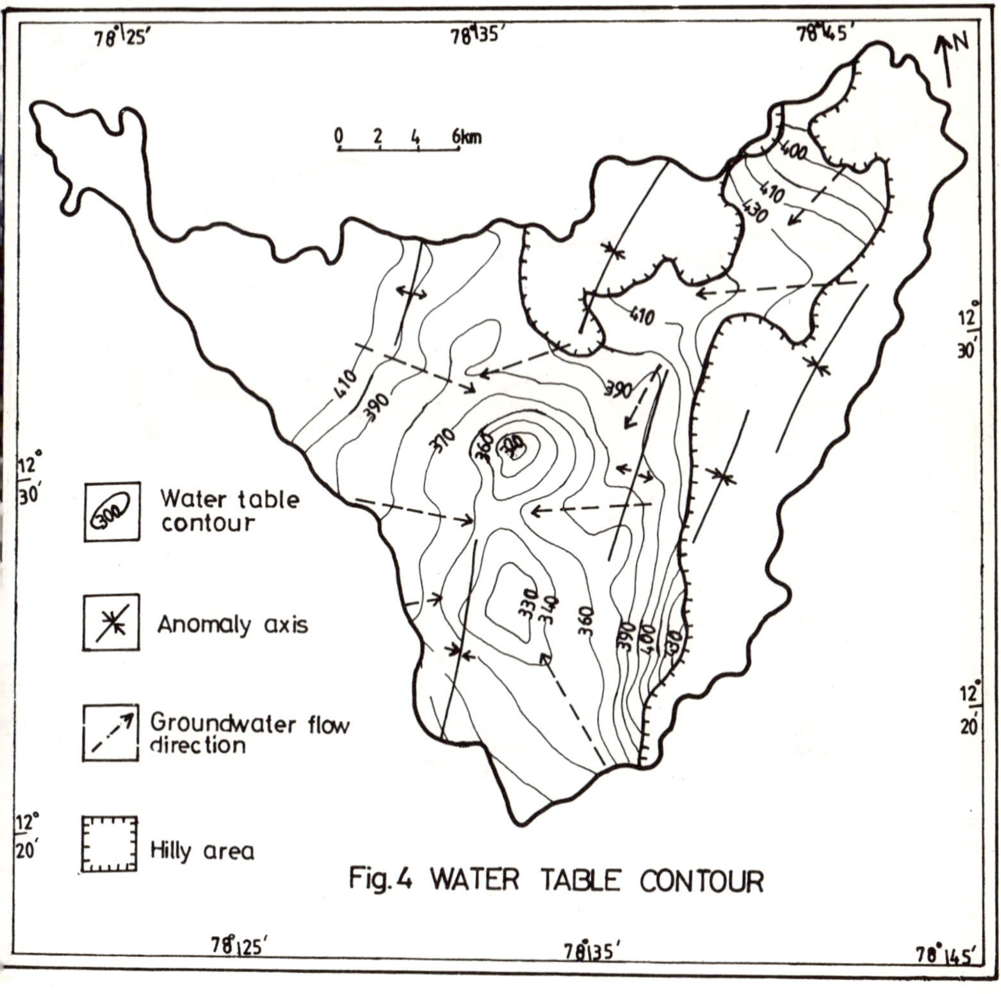

Fig. 4 WATER TABLE CONTOUR

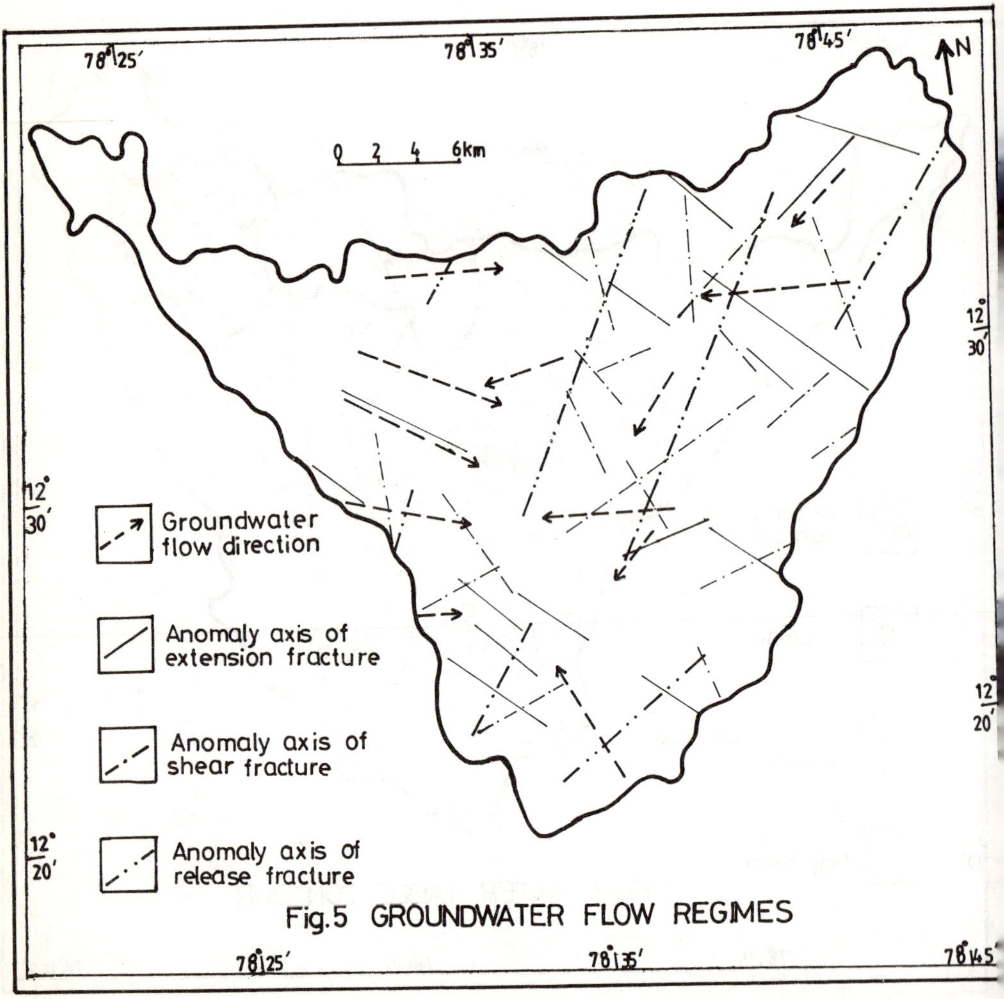

Fig.5 GROUNDWATER FLOW REGIMES

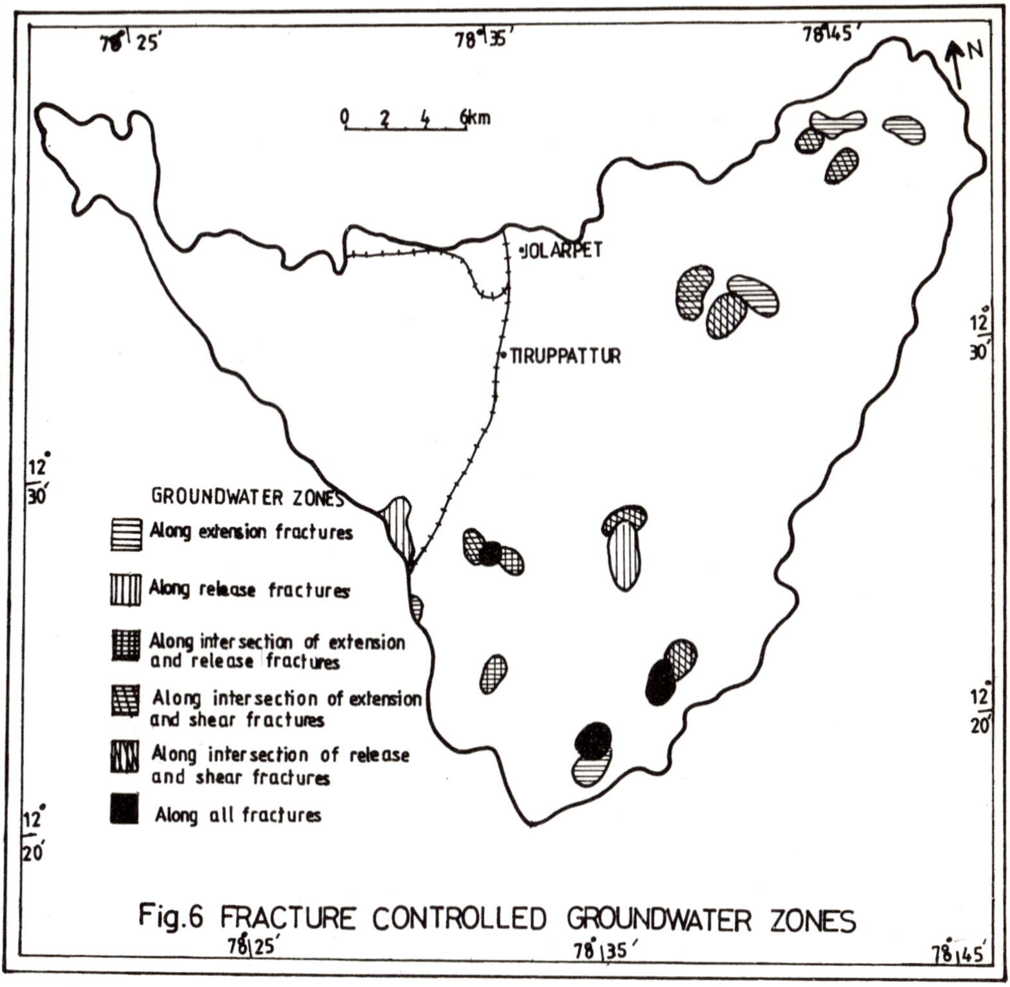

Fig.6 FRACTURE CONTROLLED GROUNDWATER ZONES

Now with this type of analysis, it can be accurately told to the common man, about the ground water storage in a particular zone. This will greatly help in precisely targetting the groundwater, so that during the groundwater exploration program, d igging of wells in negative areas will be avoided. This study further clearly demonstrates, that not all the fractures are water bearing and some of the fractures like shear are purely water barren fractures.

CONCLUSION

This sort of experimental models can be operationalised in a vast hard-rock track confidently, after selecting a critical area and studying it for the validation of the model.

REFERENCES

1. Haman, P.J., 1961, Lineament analysis on aerial photographs exemplified in the North Surgeon lake area, calgary Alberta. West Canadian Research Publication in Geology and Related Sciences, series 2 (1): 1-20.

2. Ramasamy, SM, and Balkiwal P.J, 1983, Lineament Fabric of Rajasthan and Gujarat, India, Geol. Surv. India, Records, Vol.113 pt.7.

ELECTRICAL RESISTIVITY SURVEY FOR GROUND WATER IN VARAHA RIVER BASIN, A.P., INDIA

T.V.S.R. Kshira Sagar & B. Nagamalleswara Rao
Department of Geology
Andhra University
Visakhapatnam - 530 003

INTRODUCTION

Taking its birth in Eastern ghat terrain, the Varaha river flows for a over a length of 15 Kms starting from Darlapudi village until its joins in the Bay of Bengal (fig.1). The river splits up into two minor streams at Penugollu and the two divides again merge South East of Pittalapalem. The drainage pattern of the river is in general dendritic, comprising of mostly second and third order streams.

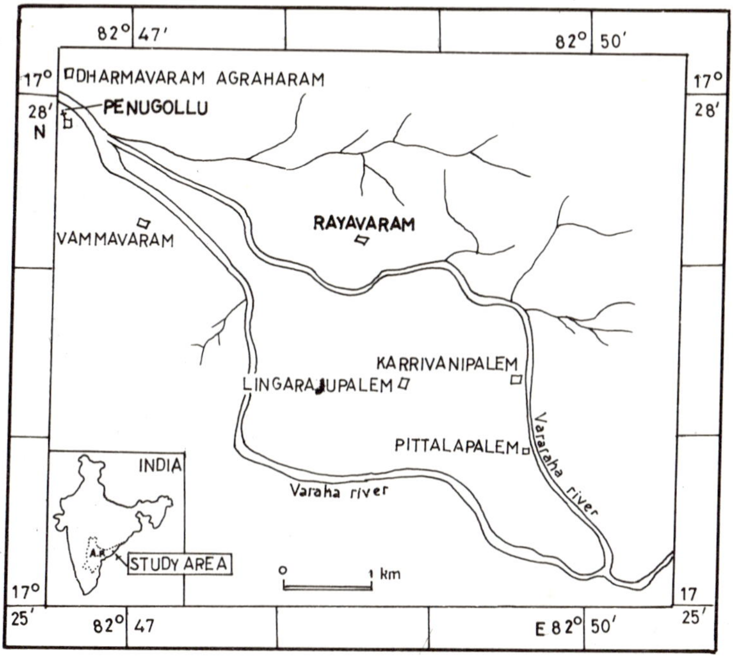

FIG. 1 LOCATION MAP

The river basin is under-lain by khondalitic suit of rocks comprising garnetiferous sillimanite gneisses and quartzites in abundance with charnokites in places. Granite intrusions are also noticed in the area. In order to understand the ground water potential of this area, as also its quality for domestic and irrigation purposes, an electrical resistivity survey has been carried out and compared with the available lithologs of the drillholes in the area to deliniate the different water resource zones.

EXPERIMENTAL

Sixteen Vertical Electrical Soundings (VES) were carried out in a four square kilometres study area. A direct current resistivity meter with constant current facility was used for the soundings. Schlumberger electrode configuration was employed with a maximum current electrode separation of upto AB/2 = 100 metres.

The VES curves of the sedimentaries are generally multi-layered with a few containing three layered H-Type sections. Some of the typical curves obtained are shown in figure 2. The apparent resistivity curves were interpreted

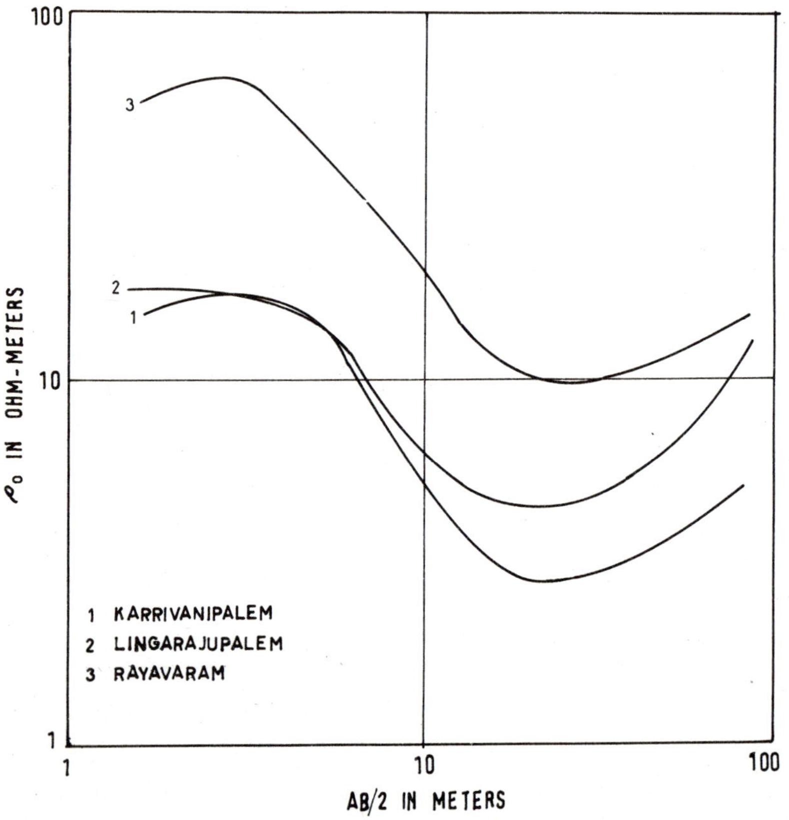

FIG. 2 TYPE CURVES IN THE AREA

using the theoretical curves for three layered medium prepared by Rijkswaterstat and the Auxiliary point charts (Bhattacharya and Patra, 1968). The interpreted sections were correlated with lithological formations exposed in the area in the vicinity of the sounding stations. The units obtained by the curve matching technique are as follows :

Top soil	:	10.5 to 48 Ohm-metres.
Sanday clay	:	2.4 to 9.6 Ohm-metres.
Semihard to Hard formations	:	24 to 240 Ohm-metres.

The higher order resistivity of the top layer is due to dryness of the soils. The geoelectrical sections of three sounding stations were correlated with the bore well sections as detailed in Table 1.

The resistivity survey has brought to light the existance of saline zones with calcarious shells south of Lingarajupalem and Karrivanipalem areas at shallow depths of 16-20 mts. below ground level. The low resistivity (2.5 ohm-metres) observed in this area is indicative of a marine environment, possisbly resulting due to palaeo-sealevel changes (Niyogi, 1970). The yielding capacity of tube wells vary from 165-1665 gpd depending on the hydrogeological conditions of the area. In the Varaha river basin the depth to static water varies from 5 to 10 metres.

CONCLUSION

The electrical resistivity survey has revealed the presence of a multilayered system characterised by alternating clay and sand zones along with their mixed proportions, besides weathered products of country rocks of khondalite group. The wide variation observed in the resistivity data reflects the nature of intermixing of sand and clay in zones. The shallow marine environment encountered south of Lingarajupalem and Karrivanipalem throws light on the possible paleo-sea level changes, which constitutes a vital aspect of academic interest. But for this saline zone the quality of the ground water in the study area in general is useful both for domestic and irrigation purposes.

REFERENCES

Bhattacharya, P.K. and Patra, H.P.., 1968, Direct current geoelectric sounding, Elsevier, Amsterdam.

Niyogi, D. 1970, Morphology and evolution of the Balasore shore line Orissa. Studies in Earth Sciences, West Commemoration volum, Today and tomorrow publishers, New Delhi, 289-304.

Table : 1

Lithology	Karrivanipalem		Lingarajupalem		Rayavaram	
	Depth in mts	Resistivity Ohm-meters	Depth in mts	Resistivity Ohm-meters	Depth in mts	Resistivity Ohm-meters
Top soil	0-22	14	0-6	20	0-13	48
Sandy clay	22-40	2.4	6-22	4	13-37	9.6
Semi hard and hard formations	40- ∞	24	22-37	∞	37- ∞	240

DIGITAL ENHANCEMENT OF SURFACE GEOLOGIC FEATURES IN A PART OF BANAS RIVER CATCHMENT FOR GROUND WATER POTENTIAL CHARACTERISATION.

O.N.Tiwari
Central Ground Water Board, C-13, Bani Park, Jaipur
302 016 India.

INTRODUCTION:

Since about two decades now, machine processing techniques for remotely sensed data have greatly advanced. Pattern recognition and enhancement techniques are application oriented operations. Digital enhancement of images and their interpretation has established its significance for lithological, structural and geomorphological components which are used as basic criteria in defining a terrain for its ground water potential. In light of above, an attempt has been made in the catchment area under the semi arid climate surrounding Mt. Abu, falls in parts of Gujarat and Rajasthan States, bounded by Latitudes $72^{0}10'$ to $73^{0}00$ North and Longitudes $24^{0}15'$ to $25^{0}30'$ East.

GEOLOGICAL SET-UP:

The area comprises aeolian and fluvial deposits of quaternary age, overlying Delhi supergroup of precambrians. Calc-gneisses, altered talcose schist, limestone, marble, quartzite and paragneisses with extensive intrusions of granite and lenses of basic and ultrabasic rock, constituted the precambrian assemblages. The area has been subjected to intense folding primarily on NNE-SSW axis with a subsequent folding phase having NW-SE axial trend. Dislocations trending NNE-SSW, sub parallel to regional folding, occures in the area.

MATERIAL & METHODS:

(a) Landsat MSS data in the form of CCT, black and white print and and FCC print pertaining to path row number 159-043 dated March 1975 have been interpreted alongwith Indian Remote Sensing Satellite (IRS) simulated images (LISS - I and LISS - II).

(b) A sub image of Landsat MSS (Bands 4,5,6 and 7) has been extracted from the computer compatible tape. A window of 512 x 512 pixel size has been displayed and digitally exercised on display device. Images generated are interpreted and supplemented with field observations.

(c) To achieve the ultimate objective i.e ground water potential characterisation, separate maps on lithology, lineament, drainage, geomorphology have been prepared. Geologic features favourable to groundwater have been orderly numbered. Finally number of coincidental occurrences of favourable features has been used in ground water potential characterisation.

DIGITAL ENHANCEMENT AND RESULTS

Although several Techniques exist that have been developed and proved in image analysis area, colour coding of resultant pixel value from arithmatic operations, statistical operations, neighbourhood point (matrix) operations and point operations remains the basic and important exercise. These operations have got a range of variations in their performance. In geologic remote sensing some of them have got established significance. These are mainly contrast enhancement, edge enhancement, principal component analysis and ratioing, which are used in the present study.

CONTRAST ENHANCEMENT :-

Contrast enhancement is most widely used point operation in image pixels for lithologic mapping. It is the process of re-distributing the brightness levels in pixels to utilise the entire dynamic range of the display device. Linear and non-linear stretch (algorithms) and process of histogram equalisation can be used for this technique to identify rocktype successfully. Histogram equalisation is particularly important in sudy of dark tonned areas. Contrast stretch provides better information on trend lines and lieaments as compared to original image. (Rakshit, et.al., 1985 and Ziong, 1984). In the present study contrast stretch helped in delineation of granitic and schistose rock with quartzite bands. Contrast demarcation of geomorphic features has been quick and accurate on linear contrast stretch. Histogram equalisation helped in the study of forested granite hills and schistose rocks.

EDGE ENHANCEMENT:

Edges are place at which there is an abrupt change in pixel intensity. It is normally done by comparing intensities or averages of intensities of pixels computed over its neighbourhoods. (Rosenfeld , 1984). Of interest here is the accurate detection of lithologic contacts, fault, and fractures. The image generated of Banas catchment helped in identification and better delineation of structural lineaments.

PRINCIPAL COMPONENT ANALYSIS:

Statistical, transformation of four bands data into principal component (PP) images is widely carried out in geologic remote sensing activities. It has been found that PP image II of them is significant in particular for drainage system and enhancement of negative relief. PP image I is exceptionally good for lineament and structural trends (Tiwari, 1987). In this study it has been used to extract additional lineaments..

RATIOING

Digital ratioing is division of two spatially registered pixels (Swain & Davis, 1978). It can be done in various combinations of bands in various combinations of arithmatic operations. The resultant data are rescaled to use the full dynamic range of the display medium using contrast enhancement techniques. Advantages other than enhancement of features is that it could minimize the first order effects of brightness variations due to topographic slope. Ratioing operation yielded minor differances in tonal reflections of rock units.

GROUND WATER:

Banas catchment is classified into four classes with varying

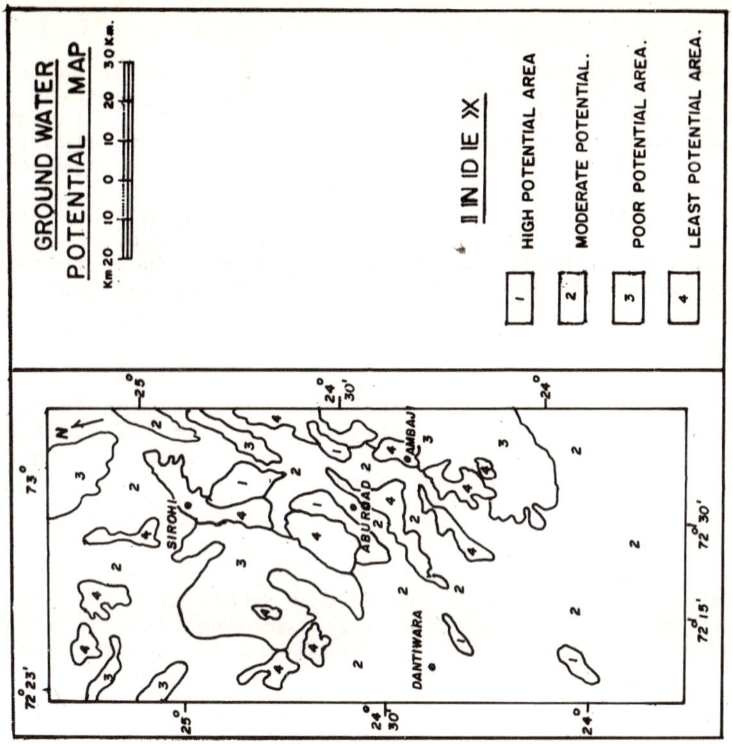
GROUND WATER POTENTIAL MAP
Km 20 10 0 10 20 30 Km.
INDEX
1 HIGH POTENTIAL AREA
2 MODERATE POTENTIAL.
3 POOR POTENTIAL AREA.
4 LEAST POTENTIAL AREA.
N
SIROHI
ABU ROAD
AMBAJI
DANTIWARA
73°
72° 30'
72° 15'
72° 23'
25°
24° 30'
24°

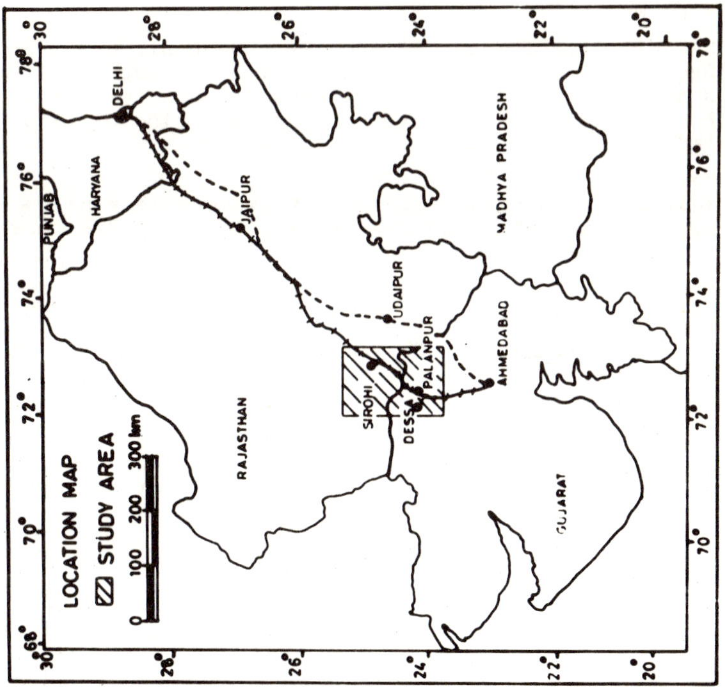
LOCATION MAP
STUDY AREA
0 100 200 300 km
DELHI
HARYANA
PUNJAB
JAIPUR
UDAIPUR
RAJASTHAN
SIROHI
DESSA
PALANPUR
AHMEDABAD
MADHYA PRADESH
GUJARAT

probability of ground water occurrences i.e. high, moderate, poor and least potential areas. High potential areas near Mt. Abu are having weathered schistose rock and overlying fluvial recharge zone, alongwith prominent lineaments. The zone is also identified in pockets of deep sandy alluvium alongwith lineaments, of southern plain region. Entire adjoining peneplained ocuntry is moderately potential. In hard rock areas, mainly fractured schistose, moderately potential zone occurs along the Banas river. Poor potential areas in the northern and central part are having combination of any two favourable features. Rest is defined as least potential area.

CONCLUSION:

The study has reflected that in general, linearly contrast stretched image contains better geological details which can be integrated to define area in terms of potentiality of ground water. Potential areas demarcated could be taken as sites for further geophysical and ground water drilling exploration.

ACKNOWLEDGEMENT:

Author is grateful to Dr. Baldev Sahai, Group Director, Space Application Centre, Ahmedabad, for providing basic facilities, and thankful to Shri R.K. Tiware, Scientist D 'Incharge', CGWB Jaipur.

REFERENCES:

1. PRATT, W.K., 1978. Digital Image Processing, John Wiley and Sons, New York, p.750.

2. RAKSHIT, A.M., Bose, P. and Krishnanunni, K., 1985. Geological evolution of multilevel multispectral remote sensing data - A case study of Amba Mata mineralised zone. Proc. Seminar on remote sensing in geological and mineral targeting. P. 6 (1-10).

3. Rosenfeld , A., 1984. Image processing in remote sensing data analysis ; the State-of-the-Art. Remote Sensing (editing) B.L. Deekshatulu and Y.R. Rajan (Eds.) Indian Academy of Sciences, P. 51-58.

4. Swain, P.H., and Davis, S.M., 1978: Remote Sensing - The quantitative approach. McGraw - Hill, New York P. 396..

5. Tiwari, O.N., 1987 . Multilevel geological Remote Sensing of Aravalli region, Unpublished Ph.D. thesis Delhi University, India. p.104.

6. ZIONG, Z., 1984 . Image processing for identifying lithology in Ying Xiongling area, Qaidam basin, proc. The seminar on remote sensing for geological application. P.263-267.

ROLE OF HYDROGEOLOGICAL SET UP IN THE YIELD OF WELLS IN GRANITE TERRAIN-CASE HISTORIES OF NIZAMABAD, WARANGAL AND HINDUPUR AREAS

Dr.M.Pandu Ranga Rao
Professor of Engineering Geology
Regional Engineering College, Warangal
India 506 004

Sri K.Radhakrishnam Raju
Research Scholar,Engineering Geology Lab
Department of Civil Engineering,
Regional Engineering College, Warangal
India 506 004

The need for exploration and evaluation of groundwater resources is of considerable importance in these days of serious water shortage owing to the vagaries of monsoon in different parts of Andhra Pradesh. As a result of which intensive groundwater surveys were conducted in the districts of Warangal, Karimnagar, Adilabad, Nalgonda, Hyderabad, Khammam, Nizamabad, Cuddapah, Nellore and Anantapur. A number of sites were located for tube wells, open wells, dug cum bore wells and infiltration galleries based on integrated surveys involving geological, hydrogeological and geophysical investigations in these areas. Remote Sensing Techniques have been utilised to evaluate the structural features of the terrain. All these areas investigated fall in the hard rock terrain. The investigations in these areas have been carriedout during the years 1968 to 1988.

In view of the extensive investigations carried out in hard rock areas in different parts of Andhra Pradesh, it is felt befitting to highlight the variations of hydrogeological set up in the same granite terrain. Nizamabad, Warangal and Anantapur districts have been chosen to study the occurrence of groundwater in granites and gneisses belonging to the unclassified crystallines of the Archaean age. It is to be noted that there is a substantial variation in the quantity and quality of groundwater occurrence in these Districts. Nizam Sugar Factory Farms have been taken as the type area for representing the hydrogeological set up in Nizamabad District, whereas Warangal City has been considered in the Warangal District while Hindupur area (under Nizam Sugar Factory Zone) has been chosen in Anantapur District for a comprehensive study of the hydrogeological set up.

NIZAM SUGAR FACTORY FARMS AREA

Detailed hydrogeological investigations were carried out in the Nizam Sugar Factory Farms, Nizamabad district for the location of tube wells for irrigation of Sugarcane. This area is under the ayacut of Nizam Sagar Reservoir for the past 40 years. Regular irrigation facilities are available for all the farms, but due to vagaries of monsoon, and due to insufficient inflows into the N.S.Reservoir, the tail end farms do not get enough water. In addition, due to sequential droughts the insufficient water in the Reservoir cannot supply assured water to the sugarcane crop through out the year. Hence it is felt essential to tap groundwater for assured irrigation as well as to supplement the surface water sources. Conjunctive use of surface and subsurface waters is felt ideal for assured irrigation of the Sugarcane crop.

Hydrogeologically the area is very favourable for tapping groundwater. The groundwater is under water table conditions and some times under semiconfined conditions in the granites and gneisses. These rocks are highly jointed and fractured. There are a number of mega and minor lineaments in the area. The sites selected along these lineaments and at the intersections of these lineaments have yielded very good and exciting results. The yields of the tube wells varied from 400 lpm to 2000 lpm. As many as 20% of the tube wells have registered high yields with reasonable drawdown. In fact there is no suitable submersible pump to bail out the water as all these tube wells are only 15 cm dia. The highest pump available for use in 15 cm dia tube well is only 20 HP, which could not do the job effectively. The very high rate of recuperation is attributed to the well defined fracture system with high hydraulic conductivity. The surface water in the near by fields and nalas available due to applied irrigation has helped in the recharge of the aquifer. The stabilised ayacut for the past 40 years might be responsible for leaching out the precipitated and deposited material along the jointed and fractured granites. No other explanation can account for such high discharge. Most of the tube wells are of 15 cm dia drilled to a depth of about 40 m to 55 m.

Contrary to this, the open wells in these areas are not highly productive. The top weathered morrum layer is admixed with higher percentage of clay derived by the disintegration and decomposition of feldspars present in the parent rock namely granites. So most of the open wells have relatively low yields and as such they are unecominical. A total of about 150 tube wells have been drilled in the areas. The percentage of success is fairly high, about 90%. Quite a few of them have yeilds ranging from 400 lpm to 1200 lpm.

WARANGAL CITY AREA

With the help of integrated surveys involving all the methods of exploration, a total number of about 80 nos. of 15 cm dia tube wells were drilled in the area for augmenting drinking water to the Warangal city during the drought periods. The depth of the tube wells range from 45 to 65 m. The yields range from 80 lpm to 400 lpm. In addition a total of 620 nos. of 11 cm dia tube wells have been drilled and are fitted with hand pumps for pumping water whenever necessary without depending on power. The yields from these tube wells range from 8 lpm to 80 lpm. The analysis of yields from all these tube wells indicate that they all fall under low yields.

This type of hydrogeological set up can be attributed to the physiographic set up of the city and also due to lack of proper recharge sources. The city of Warangal is on a drainage divide, separating Krishna and Godavari River basins. The depth of weathering and the development of fractured and jointed systems is rather low. Added to this, the sequential drought for the past one and half decade have contributed insufficient recharge to the groundwater basin. All these factors collectively contributed for the low groundwater potential in the area compared to the Nizam Sugar Factory Farms area which is under well developed Nizam Sagar ayacut. Considering similar climatalogical conditions in both the areas, the high groundwater potential is mainly attributable to the accelerated improvement in the secondary porosity of the rocks along the major and minor lineaments.

HINDUPUR AREA

This area falls in the Hindupur taluk of Anantapur district which is notorious for its drought proneness with very low precipitation. This area is drained by the major rivers namely Jayamangali and Pennar. Both of them are ephemeral with considerable thickness of sand deposits extending to a depth of about 10 to 12 m below bed level. Very rarely water flows through these rivers at this place. It is reported that these rivers flow few days in an year. The average rain fall in this area is very low. Hydrogeologically this area can be considered under very low groundwater potential. Groundwater can be tapped from the alluvial deposits of the paleo channels of the major river systems provided the water table is at a fairly reasonable depth. During drought periods, even these wells in this alluvium becomes totally dry due to the fall of water table below this zone. The availability of grounwater in these zones is highly fluctuating, depending upon the recharge conditions from the normal Monsoon. The other structure favourable for tapping groundwater in this area is through 15 cm dia tube wells. These wells have limited yields ranging from 40 lpm to 240 lpm. In some exceptional places the yields have gone up very high and touched 720 lpm. These tube wells are located along the major lineament with very good recharge sources. Here again, the major sources are, the river systems and the major tanks. Some major lineaments are identified in the area along with the highly developed joint system. Though the area is well drained by the river systems, the non-availability of adequate precipitation made the whole area to be classified under very low groundwater potential compared to the Nizam Sugar Factory area of Nizamabad district.

Though all these three areas fall under the same granite terrain, the hydrogeological set up varied considerably due to a large number of factors beyond human control. A number of Histograms can be proposed in each area to depict the % yields of tube wells which give a fairly good idea about its productivity.

REMOTE SENSING BASED APPROPRIATE METHODOLOGY FOR THE GROUND WATER EXPLORATION - A CASE STUDY FROM PRE-CAMBRIANS OF SOUTH INDIA

SM. Ramasamy, Scientist, Institute of Remote Sensing, Anna University, Madras-600 025.

S. Thillai Govindarajan, Deputy Director, Remote Sensing Laboratory, Public Works Department, Madras-600 005.

T. Balasubramanian, Assistant Director, Remote Sensing Laboratory, Public Works Department, Madras-600 005.

1.0. INTRODUCTION

Ground water exploration in hard rock areas still remains a matter of great gambling owing to the frequently changing lithology, multivariate deformative architecture and rhythmically structured lineaments. And at the same time the entire hard rock terrain can not be rejected as unfavourable zones as more 70% of the global surface is occupied by hard rock, more over the existing surface water resources are too in adequate to cope with phenomenal growth in population. As a result man started what all the possible technologies available right from the good old traditional to the most recent modern ones. Having the traditional water divining and floral indicators failed in ground water targetting the man jumped into modern geoscientific methods of water level maintaining and ground water prediction, drilling and reconstruction of subsurface architectures and pump tests and aquifer characteristic models. Subsequently, the modern remote sensing technology came into picture through which various geomorphic zones and lineaments were indexed as ground water path finders. But even after all these the precision limit in ground water targetting could not reach even 50 to 60%. The present paper hence deals with an appropriate remote sensing based technique of ground water exploration citing a case study conducted over the typical hard rock precambrians of South India.

2.0. REMOTE SENSING BASED APPROPRIATE METHODOLOGY

2.1. The present concept/methodology

In the present study a unique method of integrating ground water resource information system obtained from remotely sensed and proximally sensed/ground collected data. From the raw and digitally enhanced thematic mapper data lithological spectrum, deformational display of rocks and lineament fabric were extracted and these data were integrated by a multivariate analysis with water level, yield and transmissivity data and zones of ground water mobility and stability were identified.

The lithological discrimination has been done using the tonal, textural, vegetation and drainage grades and densities. Such lithologies form two characteristics domains viz. Sargurs and Charnockitic and Charnockitic and Khondolitic domains on the north and south respectively forming well defined hill ranges with the central valley/plain along Coimbatore - Tanjore alignment (Fig.1).

The structural trends of such lithologies were interpreted on the basis of relief, drainage, soil tone and vegetation linearities/curvilinearities from the satellite imagery and total frame work of such folds were brought out. Such frame work indicates that the South Indian part of the precambrians are folded into series of anticlines and synclines with NNW-SSE trending axial traces with synclines forming hill domains and anticlines forming valley domains. These regional anticlinal and synclinal folds do have poly modally oriented series of minor anticlines and synclines suggesting that minor folds are first generation folds of N-S compression and regional folds are the members of second major orogeny due to ENE-WSW compression (Ramasamy, etal, 1987).

The fracture controlled lineaments were interpreted on the basis of the same photo tone characters from the satellite data and such lineaments show a regional polymodel orientation. The study area has been digitised into number of grids of 100 Sq.Km. each, total number of lineaments per each grid were counted, plotted in the respective grid centre, contoured and thus isofracture map was prepared. The alignment of such contours were identified as lineament maxima (Heman, 1961) or fracture linearity (Fig.1).

Water level fluctuations for a period of three years for over 2500 observation wells were standardised, values were plotted in their respective geographical positions, contoured and from their contour linearity, ground water level linearity was structured out.

In the same way the yield and transmissivity data of over 2500 bore holes drilled in hard rock area were put into analytical rigor, and isoyield lines and isotransmissivity lines were drawn. Once again on the basis of their shape the yield linearity/maxima and transmissivity linearity were identified.

2.2. DATA INTEGRATION AND INTERPRETATION

2.2.1. Lithology and Ground water

The correlation of lithological domains with water level, yield and transmissivity maxima shows that except in Ootacamund massif no where the harder lithologies show positive correlation with ground water. It follows formit that only the softer lithologies of the valley domains form favourable ground water locales.

2.2.2. Folds and Ground water

The integration of folded structures with ground water level and yield shows that the water level linearity maxima faithfully follows the axis of the folds in Tirunelveli, Madurai, Pudukkottai and Ootacamund massif. The circular pattern of water level maxima northeast of Coimbatore shows the existance of a basin structure, a complimentary to Sankaridurg dome which holds the ground water.

Similarly the yield maxima exhibits striking linearity and curvilinearity with axis of the first generation folds too which got refolded to NNW-SSE trending axial trace, whereas the transmissivity value maxima does not show any relation with folded structures.

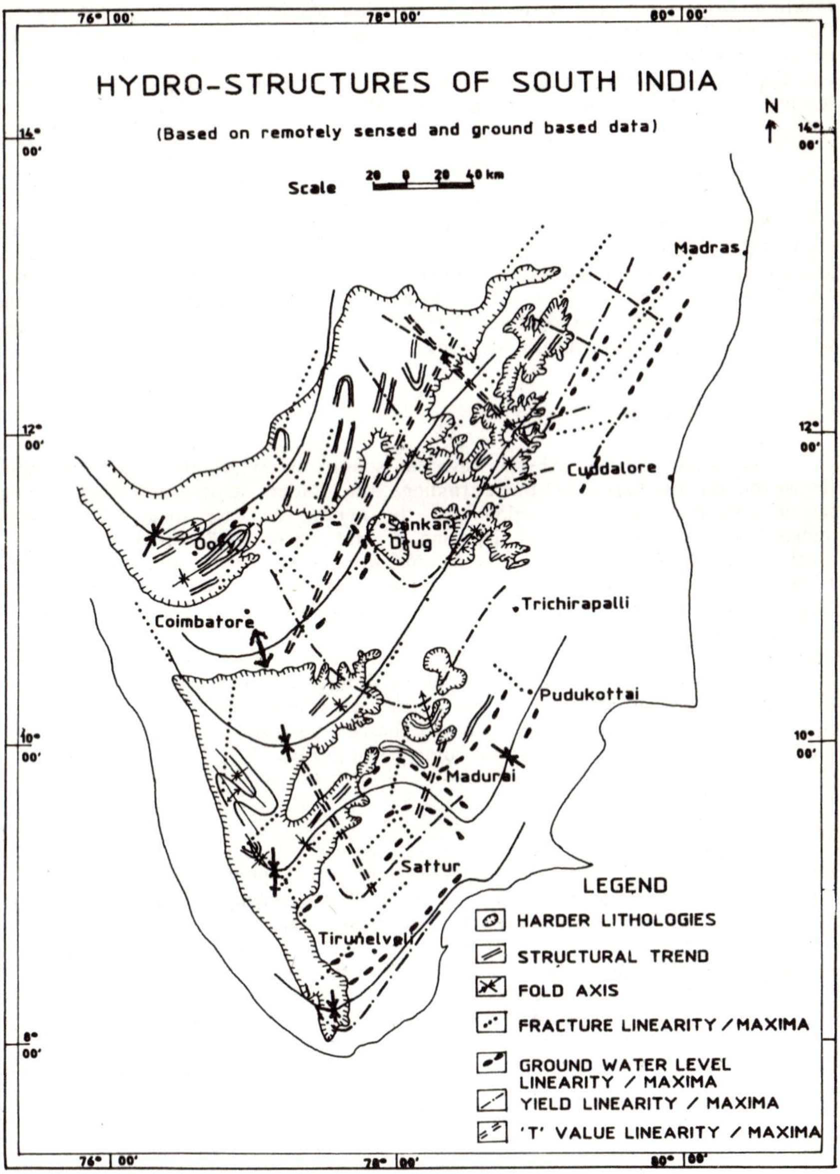

HYDRO-STRUCTURES OF SOUTH INDIA
(Based on remotely sensed and ground based data)
Scale
20 0 20 40 km
N
76° 00'
78° 00'
80° 00'
14° 00'
12° 00'
10° 00'
8° 00'
Madras
Cuddalore
Ooty
Sankari Drug
Coimbatore
Trichirapalli
Pudukottai
Madurai
Sattur
Tirunelveli
LEGEND
HARDER LITHOLOGIES
STRUCTURAL TREND
FOLD AXIS
FRACTURE LINEARITY / MAXIMA
GROUND WATER LEVEL LINEARITY / MAXIMA
YIELD LINEARITY / MAXIMA
'T' VALUE LINEARITY / MAXIMA

2.2.3. Fractures/Lineaments and Ground water

In the area between Madurai and Tirunelveli the concidence of fracture maxima with yield and water level maxima shows that the NE-SW trending lineaments have control over ground water; whereas in Dharmapuri - Kanchipuram sector lineament maxima shows one to one correlation with water level, yield and transmissivity suggesting that the groud water mobility is completely controlled by the fracture systems with area (Fig.1).

2.3. SYNTHESIS

Such multivariate analysis of remotely sensed data and ground based data shows that in southern part of the area south of Ootacamund - Cuddalore alignment the ground water is completely controlled by the folded structures except in Tirunelveli - Sattur area where fractures also contribute to a minimum whereas in the northern sector (Dharmapuri - Kanchipuram sector) the ground water is controlled by fracture systems.

3.0. CONCLUSION

The study flashes off the regional spectrum on ground water targetting. However on the basis of these findings the folded structures of the southern sector and the fractured regimes of the northern sector must be studied at micro-level for detailed planning and ground water exploration. Such micro-level study in the southern sector (Coimbatore - Tirunelveli sector) will include mapping of all lithological/structural trends and construction of fold styles including their fold morphology, dip amount linearity extent of fold axes and aerial extent of each folds and demarcate synclinal structures as potential ground water locales using large scale aerial photographs. Whereas on northern Dharmapuri - Kanchipuram sector all the fracture systems must be interpreted from aerial photographs and all open fractures which will normally be perpendicular to the folds, show wide open morphology possess prolific vegetation along them must be interpreted as water holding structures.

REFERENCES

Heman, P.J., 1961, "Lineament analysis on aerial photographs examplified in the north Surgeon Lake area, Alberta, West Canadian, Research publication of geology and related sciences, Series No.1, pp 1-20.

Ramasamy, SM., Panchanathan, S.P., and Palanivelu, R., 1987, Pleistocene earth movements in Peninsular India - Evidences from Landsat MSS and Thematic Mapper Data. Proc. IGARSS'87, Symp. AnmArbor, pp 1157-1161.

SESSION S-2

GROUNDWATER FLOW IN UNSATURATED/SATURATED ZONES : SOIL MOISTURE STUDIES ; RECHARGE ESTIMATION ; INTERACTION BETWEEN GROUNDWATER AND SURFACE WATER

MOISTURE VARIATION STUDY BY NEUTRON-NEUTRON LOGGING

I.R. Bavaji
Central Ground Water Board
West Central Region
Ahmedabad - 380 022

Abstract

The Ground water Scientists till date are using the same borehole logging instruments, initially designed for oil exploration purposes, directly or with a very little modification for ground water studies. Though these instruments are fabricated and manufactured keeping in mind their use in searching the hydrocarbons, the present paper attempts to solve one of the problems related to ground water - measusrement of soil moisture content in percentage using neutron-neutron logging tool.

The neutron-neutron logging tool, one of the most useful tools, available for porosity measurements of the oil bearing formations, is used by the author for moisture variation study in the zones above the water table. A scalar method of interpretation technique is used for understanding the log responses.

The study of neutron-neutron logs recorded in 4" diameter wells in alluvium at four different locations in Ghaghar River Basin on different dates reveals an interesting and valuable use of this technique. The log which records a continuous information about the moisture content in different zones has an edege over the other conventional methods.

INTRODUCTION

The soil moisture is commonly regarded as the sub-surface water in the zone of aeration. The zone of aeration is, infact, a zone of transition in which water is absorbed, held, or transmitted either downwards towards the water table, or upwards towards the soil surface and the atmosphere.

Some part of the precipitation water or irrigation water which reaches the ground surface may flow over the surface as overland flow, while the larger part will normally be absorbed by the surface layer as ground water recharge. This latter portion may flow laterally, close to the surface, as interflow, or it may move downwards through the soil, subsoil, and rock layers, eventually becoming part of one or more of the moisture zones. These zones are soil zone, capillary fringe, intermediate zone and water table. The soil zone forms the top of the ground surface and extends to few feet below ground level. The water which first reaches to soil zone moves out of it either by means of evaportranspiration to the atmosphere, or by means of downard percolation towards the water table. Immediately above the water table lies the capillary fringe, where substantial number of pores are full of water as a result of suction forces operating against the force of gravity. Between the soil zone and capillary fringe lies intermediate zone where the movement of water is gradually downwards.

If gravity were the only force to which water in the soil were subjected to, the soil moisture would drain completely after rainfall and the only water to be found would be that below the water table. But the fact is that, in natural condition, the soil always contains some moisture, even at the end of long dry periods.

The classical technique for measuring the soil moisture is by collecting the soil samples in the field and determining their moisture content in the laboratory. This method has limitations of depth and distructive in-situ conditions. However a quick, precise and non-destructive technique ie. moisture content measurement with the help of Neutron-Neutron (n-n) logging which overcomes these difficulties is slowly gaining importance in the hydrological studies.

In the present paper, the moisture contents in unsaturated zones derived from (n-n) logging carried out at Chhota Korwa, Ambala district, Haryana (in catchment area of Ghaghar River), in experimental plots at Ganganagar and in a sandy track at Rangmahal, in Ganganagar district of Rajasthan (in a flood plain of Ghaghar River) are discussed. The n-n logs recorded are interpreted using Scalar method of interpretation and the moisture content at different depths in percentages are arrived. This moisture content measurement is further used to study the movement of water front downward in the unsaturated zones. This study establishes the applicabiliyty of n-n logging for better understanding of hydrological conditions.

SOIL MOISTURE MEASUREMENT BY DIFFERENT METHODS

The water content of the soil may be measured either by a direct or indirect measurement.

Direct measurement:

This method includes moisture measurement by oven method and Lysimeter.

(a) Oven Method : One of the most accurate and most frequently used ways of measuring the water content of the soil is the gravimetric method which involves determining of the weight loss from a number of oven-dried field samples obtained by coring or augering.

(b) Lysimeter : The use of lysimeter represents an extension of the gravimetric method, where changes in the moisture content of a block of soil are determined by recording changes in the weight of the soil block.

Indirect Method:

Using Neutron soil moisture meter and n-n logging tool soil moisture can be measured indirectly.

(a) Neutron Soil Moisture Meter: The neutron scattering soil moisture meter with a neutron probe is extensively used now-a-days to study the soil moisture content. In this method a single calibration serves for a wide range of mineral soils and also the meter gives the moisture percentage on a volume basis, which is often the most useful result for hydrological work.

(b) Neutron-Neutron Logging: The Neutron-Neutron(n-n) logs are chiefly used for the measurement of the total porosity below the water table, however, it can also be used for estimating the moisture content against the zones above the water table. In n-n logging system, a neutron source and a detector are arranged in a probe so that the output is primarily a function of the hydrogen content of the borehole environment. The advantage of this method over others is that in addition to the volumetric information it also gives a continuous record.

INSTRUMENT USED IN PRESENT STUDY AND ITS APPLICATION

In present study, the n-n logging tool attached to a multichannel logging system of a logger manufactured by Gear Hart Owen Industries Inc. USA is used. This system provides a means for determining the formation hydrogen content. The tool contains a source of fast neutrons - Americium Beryllium with a thermal neutron detector at a fixed distance.

This tool is designed to count only neutrons, thereby eliminating the effects of gamma rays that may be emitted from the source and natural gamma radiation from the formations.

In n-n logging, neutrons are artificially introduced in to the rock-fluid system, and the effect of the environment on the neutrons is measured.

The emitted fast neutrons collide elastically with the surrounding elements they may strike. Because of this collisions, they are slowed down and thermalized. The element that thermalizes neutron most effectively is hydrogen.

Moisture Content Determination by Neutron-Neutron Log:

It is a common practice to use the response charts in calculating the formation porosity. However, a scalar method for estimating liquid filled porosity is also in extensive use. The scalar system of interpretation depends on locating two points of known porosity on the log and applying a logarithmic scale to properly fit these two points, figure 1. The intermediate porosity values for any depth can be read directly from the logarithmic scale. Core information can be used to exactly position the scale to the log. Where core data are not available, average porosity values can be assigned opposite both a clean shale and a dense limestone or sand. Normal apparent neutron porosity opposite a clean shale is from 30% to 40%, depending on the area. Values from 1% to 5% porosity are usually used for the low porosity zones.

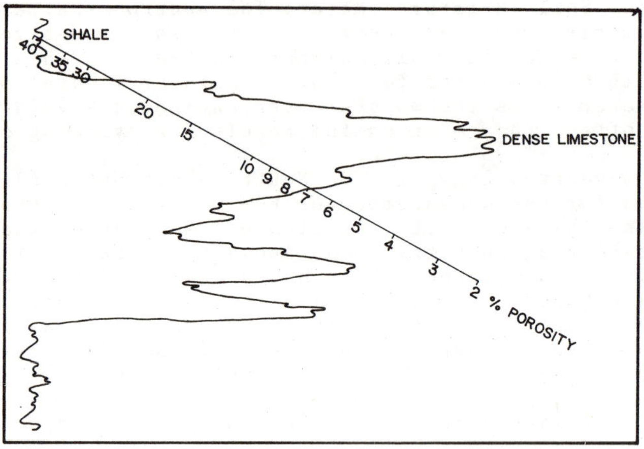

Figure.1. Neutron Porosity Determination using Scalar method

In the present study a hand drilling machine was used in constructing the wells and thus it was possible to collect the litho samples most accurately which permitted to assign them the appropriate values of porosity.

CASE STUDY

During the course of ground water studies in Ghaghar River Basin in northern India, under UNDP, figure 2, an indepth study using n-n logging was carried out for deciphering soil moisture content in different hydrological settings. For this purpose 4" dia wells were constructed at four areas. The sites for conducting this experiment were selected keeping in mind the following factors ie., depth to water level, soil texture, irrigated and non-irrigated areas, etc. The logging was conducted five times periodically between pre-monsoon and post-monsoon periods of the water year 1977.

Chota Korwa: In order to understand the soil moisture variation a well was constructed down to 18 m. In this area the rainfall is considerably good and the water levels are generally shallow(3 m). The litholog based on the drill cuttings is as follows, upto 3m silt; 3 to 4 m sand; 4 to 5 m clay; 5 to 6 m sand; and 6 to 18 m silty sand.

Figure.2. Location Map of Study Area.

To understand the soil profile and study the variation in moisture content above the water table the first n-n logging in this well was conduted on 22-7-1977. The critical examination of this log, figure 3, reveals the presence of the following zones. The soil zone OA is a shallow saturated layer existing below ground surface. Immediately below this, lies the transition zone AB in which the moisture content decreases rapidly from top to bottom. Below the transition zone, exits the transmission zone BC through which the water from the upper zones is transmitted with little or no change in moisture content to the underlying wetting zone CD. At the base of the wetting zone, exists wetting front DE which is charecterised by very steep moisture gradient and it marks the limit between wet soil and dry soil. The capillary fringe FWT is located just above the water table WT.

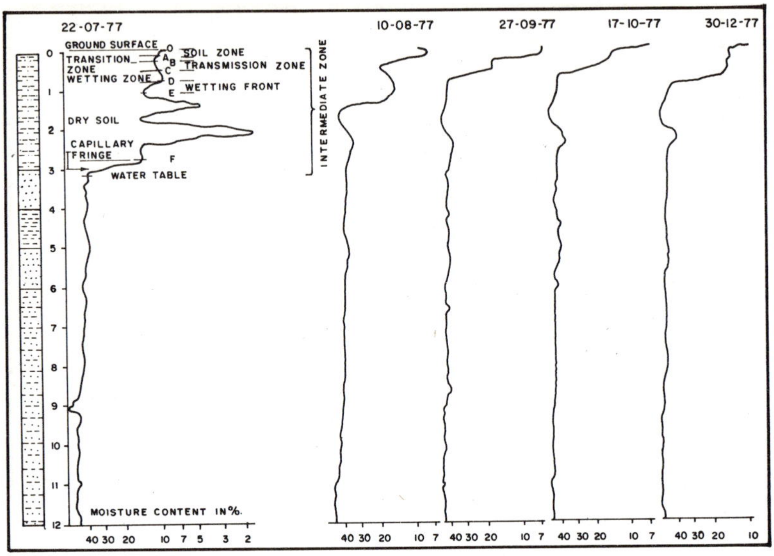

Figure.3. Neutron-Neutron Logs at Chhota-Korwa.

Taking one of the formation encountered and water table as reference the moisture content values (sand at 2 m depth: 1.5%; water table: 40%) are assigned to the log.

The month of July and August are the peak period of rainy season in the study area and by the time the second logging was attempted on 10-8-77 there was already a downfall of heavy rain.

In comparision to the moisture level recorded on 22-7-77, ~~the~~ log response on 10-8-77 showed a different picture. A remarkable increase in moisture content at all depths is observed on this log. The layers located just above and below two metres of depth and were characterised by very less amount of moisture content(less than 4%) on 22-7-77 are now below water table. The value of moisture content at 1 metre depth is increased from 13% to 15% and that at 0.3 m depth from 10% to 13%.

Further increase in moisture content was observed during the logging on 27-9-77. The log recorded on 17.10.77 shows that all the zones between ground level and water table maintained almost the same moisture content as that was on 27-9-77.

The logging carried out on 30-12-77 shows loosing tendency of the moisture content at all the depth. The moisture percentage depthwise recorded on different dates is presented in table no.1, below.

Table no. 1.

Depth m. bgl	Moisture in percentage volumetric 22-7-77	10-8-77	27-9-77	17-10-77	30-12-77
0.3	10.0	13.0	15.0	13.0	9.5
1.0	13.0	15.0	40.0	40.0	40.0
2.0	4.5	40.0	40.0	40.0	40.0
3.0	30.0	40.0	40.0	40.0	40.0
4.0	40.0	40.0	40.0	40.0	40.0
5.0	40.0	40.0	40.0	40.0	40.0
6.0	40.0	40.0	40.0	40.0	40.0

Ganganagar I: A well down to the depth of 15 m was constructed in an experimental plot in the campus of an agricultural farm in a cotton field at Ganganagar of Ganganagar district of Rajasthan State. The litholog shows the presence of clay from ground level to 1 m, 1 to 6 m sand with clay band, 6 to 8 m clay and 8 m downwards sand. The depth to water level recorded was 11 m on 25-7-77.

The first logging of this well was conducted on 25-7-77. The response of n-n log in the formations above the water level is attributed to seven groups, figure 4. These are AB, BC, CD, DE, EF, FG & GWT. The average moisture content values against these zones recorded on 25.7.77 are as follows: AB = 1.5%; BC = 5%; CD = 2.5%; DE = 3%; EF = 9%; FG = 2.5% and GWT = 10%. The second logging of this well carried out 18-8-77 indicates no

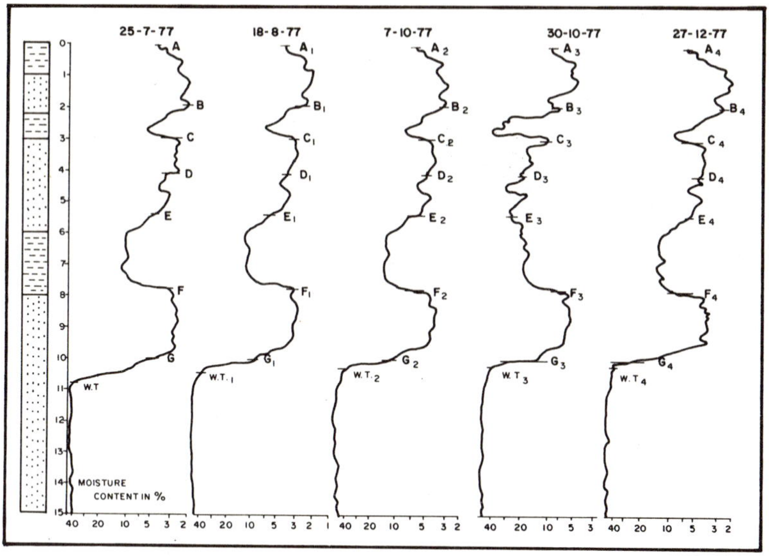

Figure.4. Neutron-Neutron logs at Gaganagar I.

change in moisture contents at any depth. In comparision to the log recorded on 25-7-77 and 18-8-77 the log recorded on 7-10-77 shows higher amount of moisture content at all the depths. The rise in water table WT1 to WT2 also indicates the contribution of surface water to ground water body. The surface water in the form of irrigation and rain between 18-8-77 and 7-10-77 after percolation and infiltration processes, increased the moisture content of all the zones before reaching to water table.

A remarkable increase in moisture content at all the depths was observed on the log recorded on 30-10-77. The moisture distribution in the soil profile and its higher percentages show the effect of irrigation. Due to the additional surface water, the zones at B3 C3 gained the maximum moisture and reached to the fully saturated level ie., 40%. The zone just above it and below the ground level ie., A3 B3 attained 5% moisture content. The moisture content in the zones further below and at C3 D3, D3 E3, E3 F3 has reached to 15% and that at F3 G3 to 5%. Though the increase in the moisture content was observed at all the depths no change in water level was noticed and the water table remained to the same level as that was recorded on 7.10.77. The movement of water front between surface layer and water table is clearly visible on this log.

The comparison of the log recorded on 27.12.77 to the log recorded on 30.10.77 permits to conclude that all the zones above water table have started loosing the moisture content and the water infiltrated reached the water table. The maximum decrease in moisture content was observed at the zone B4 C4 ie., from 40% to 7%. A comparative statement of moisture content with respect to depth is shown in table no.2, below.

Table no.2.

Depth m. bgl	Moisture in percentage volumetric 27-7-77	18-8-77	7-10-77	30-10-77	27-12-77
0.3	3.0	2.8	4.2	5.0	4.3
1.0	1.8	1.6	2.7	5.0	2.7
2.0	2.0	2.3	3.0	7.0	2.3
3.0	2.2	2.4	3.5	40.0	5.0
4.0	2.8	3.0	3.7	14.0	3.7
5.0	3.0	3.2	4.1	17.0	4.0
6.0	8.6	9.5	11.0	15.0	10.0
7.0	9.0	9.5	12.0	15.0	11.0
8.0	2.4	2.7	3.5	5.0	3.3
9.0	2.4	2.6	3.4	5.0	3.2
10.0	4.8	7.0	11.5	20.0	20.0
11.0	37.0	39.0	40.0	40.0	40.0
12.0	40.0	40.0	40.0	40.0	40.0
13.0	40.0	40.0	40.0	40.0	40.0
14.0	40.0	40.0	40.0	40.0	40.0

Ganganagar II: The Ganganagar II well was constructed in a sugarcane plot. The litholog here is quite different than that at Ganganagar I. Based on the interpretation of litho samples collected at the time of drilling it is found that the formation from ground to 1.5 m is of silty nature, 1.5 to 2.5 m is sandy, 2.5 to 4.0 m silt, 4.0 to 4.5 m clay and further below upto the water table mostly sandy. The correlation of the logs recorded in this wells on different dates, figure 5, and the comparision of moisture levels, table no.3, reveals the following.

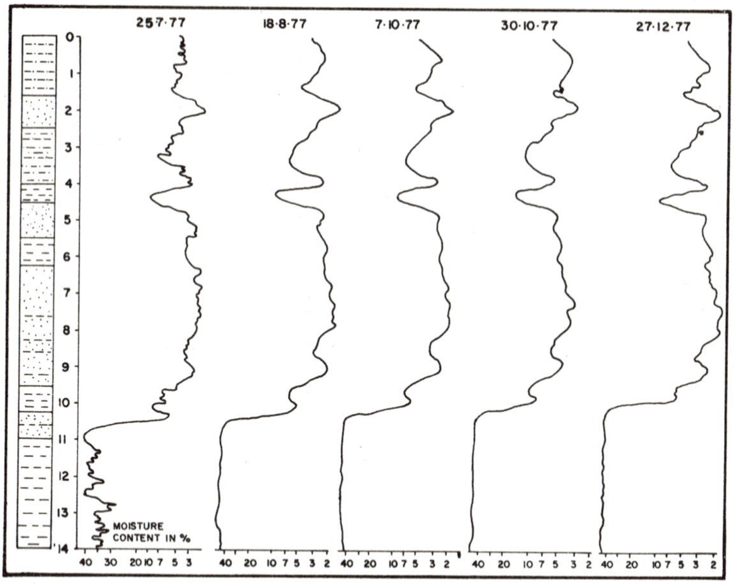

Figure.5. Neutron-Neutron Logs at Ganganagar II

A comparitive study of the logs recorded on 25-7-77 and 18-8-77 shows that the moisture content level, on 18-8-77 is reduced for the first 2 m of depth, while it remains at the same level for the depth range 3 to 4 m, whereas further below at the depth 5 to 6 m again it shows decreasing trend. At the depth of 7 m, no change is observed, however, the zone between 7 m and water table shows a decreasing tendency. The water table remains almost at the same level. The logging carried out on 7-10-77 and 30-10-77 shows a continuous rise in moisture content at all the depths.

A comparitively high amount of moisture (10%) against the zone located at 3 & 4 m, recorded on 30-10-77 is due to the presence of clayey layer just below 4 m. It is obvious that the moisture content of the sand layer will increase if a impermeable clay layer exists at its base.

Table no. 3.

Depth m. bgl	Moisture in percentage volumetric 25-7-77	18-8-77	7-10-77	30-10-77	27-12-77
0.3	3.5	2.4	3.0	3.5	3.3
1.0	4.0	2.7	3.3	4.2	3.0
2.0	1.8	1.3	1.7	3.3	1.8
3.0	5.0	5.2	5.5	10.0	5.0
4.0	2.6	2.7	3.5	10.0	3.1
5.0	2.8	2.2	2.5	4.7	2.3
6.0	2.9	2.0	2.3	4.2	2.1
7.0	1.9	1.8	2.0	3.3	1.7
8.0	2.0	1.6	2.2	4.2	1.5
9.0	2.4	1.7	2.2	4.0	2.2
10.0	5.3	4.4	6.4	11.0	40.0
11.0	40.0	40.0	40.0	40.0	40.0
12.0	40.0	40.0	40.0	40.0	40.0
13.0	40.0	40.0	40.0	40.0	40.0

Rangmahal: The well constructed at Rangmahal, district Ganganagar, Rajasthan is a case study of fine textured sandy formation having comparitively high permeability. The area is devoid of irrigation and the only possibility of increasing the soil moisture content is the rain water. The litholog of this well shows that the formation encountered above the water table is mostly sandy except a thin band of clay at 11.5 to 12.0 m The first logging in this well carried out on 26.7.77, figure 6 shows that the zones above the water table are characterised

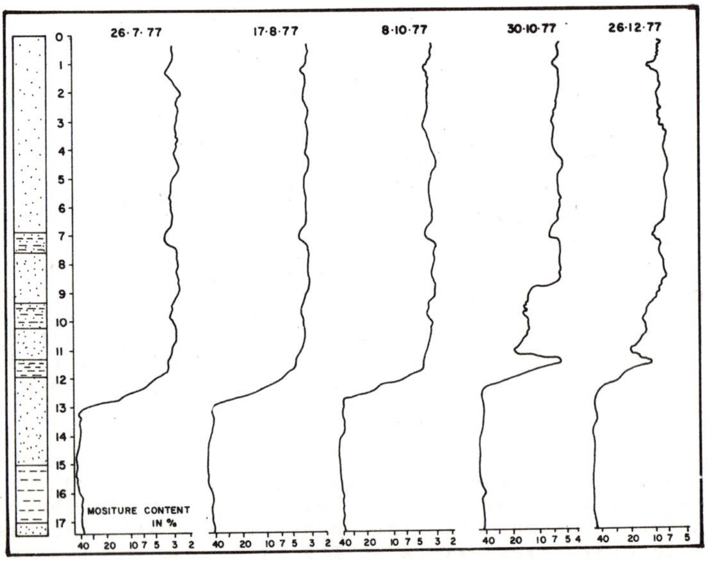

Figure.6. Neutron-Neutron Logs at Rangmahal.

with low moisture content (2% to 4%) and also there is not much variation in the moisture content at different depths. The logging carried out on 17-8-77 and 8-10-77 also reveals almost the same picture except a small increase in the moisture content due to occasional shower. But appreciable change in moisture content was observed on the logs recorded on 30-10-77 and 26-12-77. The moisture content in the depth range 9.0 to 11.5 m bgl was increased to a considerable extent. This is due to the rapid percolation of rain water through the sandy formation which has good vertical permeability and also due to the presence of a impervious clayey layer as a base at 11.5 to 12.0 m. The table no.4 shows the variation of moisture contents recorded on different dates.

Table no. 4.

Depth	Moisture in percentage volumetric				
m. bgl	26-7-77	17-8-77	8-10-77	30-10-77	26-12-77
0.3	2.6	2.8	3.3	5.3	9.0
1.0	2.7	2.8	3.4	5.3	7.0
2.0	2.3	2.8	3.7	5.3	6.2
3.0	2.5	2.7	3.9	6.0	6.8
4.0	2.6	2.8	3.5	6.0	5.8
5.0	2.7	2.9	3.2	5.0	5.2
6.0	3.0	2.9	3.3	5.0	5.2
7.0	3.4	3.6	3.8	6.2	7.2
8.0	2.5	2.9	3.0	5.0	6.2
9.0	2.4	2.8	3.1	7.5	6.2
10.0	3.0	3.4	3.7	14.0	9.8
11.0	3.0	3.4	3.6	15.0	11.0
12.0	4.4	5.5	5.7	11.0	17.0
13.0	40.0	40.0	40.0	40.0	40.0
14.0	40.0	40.0	40.0	40.0	40.0
15.0	40.0	40.0	40.0	40.0	40.0

CONCLUSIONS

On the basis of the case studies it is concluded that a continuous volumetric record of moisture content is possible by neutron-neutron logging. This method also permits to have a repeated record without disturbing the formation and the natural conditions. It is also concluded that the depthwise periodic variation in moisture content can be obtained using neutron-neutron logging.

The study also proves that even if the n-n logging tool is not properly calibrated, relative changes in moisture content can be obtained by periodic logging.

ACKNOWLEDGEMENTS

The author express his sincere gratitude to Shri B.P.C. Sinha, the Chief Hydrogeologist & Member, Central Ground Water

Board and Shri P.N. Phadtare, Director, Central Ground Water Board, West Central Region, Ahmedabad for kind permission to present the paper.

REFERENCES

Gardner, W.R., and Kirkham, D., 1952. Determination of soil moisture by neutron scattering. Soil Sci. 73.

Lynch, E.J., 1962. Formation evaluation. New York, Harper and Row, Publishers.

Meyer, W.R., 1963. Use of neutron moisture probe to determine the Storage Coefficient of an unconfined aquifer, in short papers in Geology, Hydrogeology and Topography. U.S.Geol. Survey. Prof. paper 450-E.

Scott Keys, W., and Mac Cary, L.M., 1971. Application of Borehole Geophysics to water resources investigations, Chapter E1, book 2, USGS publication.

STUDY OF SOIL MOISTURE MOVEMENT IN THE VADOSE ZONE USING NEUTRON PROBE

RAMESH CHAND
NATIONAL GEOPHYSICAL RESEARCH INSTITUTE
HYDERABAD - 500 007

ABSTRACT

Neutron probing of soils, a modern technique, is found to be useful in assessing the soil moisture variation in the vadose zone. Results of field studies in alluvial soil probed up to 250 cms during pre monsoon, monsoon and post monsoon periods under both natural and artificially ponded conditions, are discussed.

Soil moisture data was plotted against the depth probed. Infiltration rates and evaporation losses under different conditions were calculated. Field capacity values ranging from 30 to 45 percent of moisture by volume were inferred from these observations. Redistribution curves of soil moisture movement exhibited superposition of

a) fast process for short duration in shallow zones (60 cms),

b) slow process for long duration in depths varying from 60 to 120 cms and

c) negligible change in depths beyond 120 cms.

Moisture depletion and enhancement during dry spells and during rainstorms including ponding condition are clearly reflected on the profiles. Such studies of moisture movement in the vadose zone would enable planning the dry crop pattern and irrigation schedules of the area.

INTRODUCTION

Soil moisture movement studies provide potential information in the field of agriculture and hydrology. In the field of agriculture, information is needed for many diverse applications including irrigation scheduling while in the field of hydrology, the soil moisture movement study is important for understanding the mechanism of recharge through the soil and to provide soil moisture storage data for the water balance study. In this paper, investigations on the movement of soil moisture in the unsaturated zone during pre monsoon, monsoon and post monsoon periods under both natural and artificially ponded conditions, are presented.

Studies carried out clearly show the capabilities of neutron moisture probe, for accurate measurement of soil moisture in the vadose zone. The probe, one of the most sophisticated equipment, is found particularly suitable under different conditions,since they can measure repeatedly the same soil sample with little soil disturbance. It is easy to follow the moisture changes at different depths, locations and times.

Five access tubes for soil moisture observations were installed in the study area. Small infiltrometer of size 200 cm x 200 cm x 100 cm was also installed for accurate measurement of soil moisture changes under ponding conditions. Observation were made daily down to a depth of 250 cm using Troxler depth moisture gauge (Model 3321). The gauge usually referred to simply as the neutron probe, consists essentially a probe containing a fast neutron source, a slow neutron counter adjacent to the neutron source, and a probe carrier. Fast neutrons are emitted by an Americium 241: Beryllium radioactive source. The emitted neutrons collide with the nuclei of soil atoms of the surrounding soil, predominantly those of hydrogen in the soil water and cause the neutron to lose their kinetic energy. These slowed neutrons are detected by a helium (He-3) detector. Counts thus obtained are directly proportional to the moisture content. Basic principle, theory of neutronic method, design features, calibration procedure, merits and limitations of the probe were summarized in the earlier papers (Ramesh Chand, 1986).

INVESTIGATIONS CARRIED OUT

Five access tubes 300 cm long and 4 cm in diameter were installed. The Aluminium tubes were preferred as it absorbs very few thermal neutrons. The tubes were sealed at the bottom to prevent water seepage in the tube. One of the access tube was placed at the centre of the rectangular infiltro meter. The tank walls penetrated in to the soil to a depth of 60 cm and about 40 cm was left above the ground surface for use in studying soil moisture movement under controlled conditions. Experimental set up of field measurement under controlled conditions is shown in fig 1. Neutron probe is kept on the tube and observations of soil moisture content were taken by lowering the probe into the access tube. The thermalised neutron were counted at various depths by lowering the probe in steps. The probe directly compute the ratio of measurement counts to the standard counts and the moisture content is read in the units of pounds/cubic foot, inch/foot,

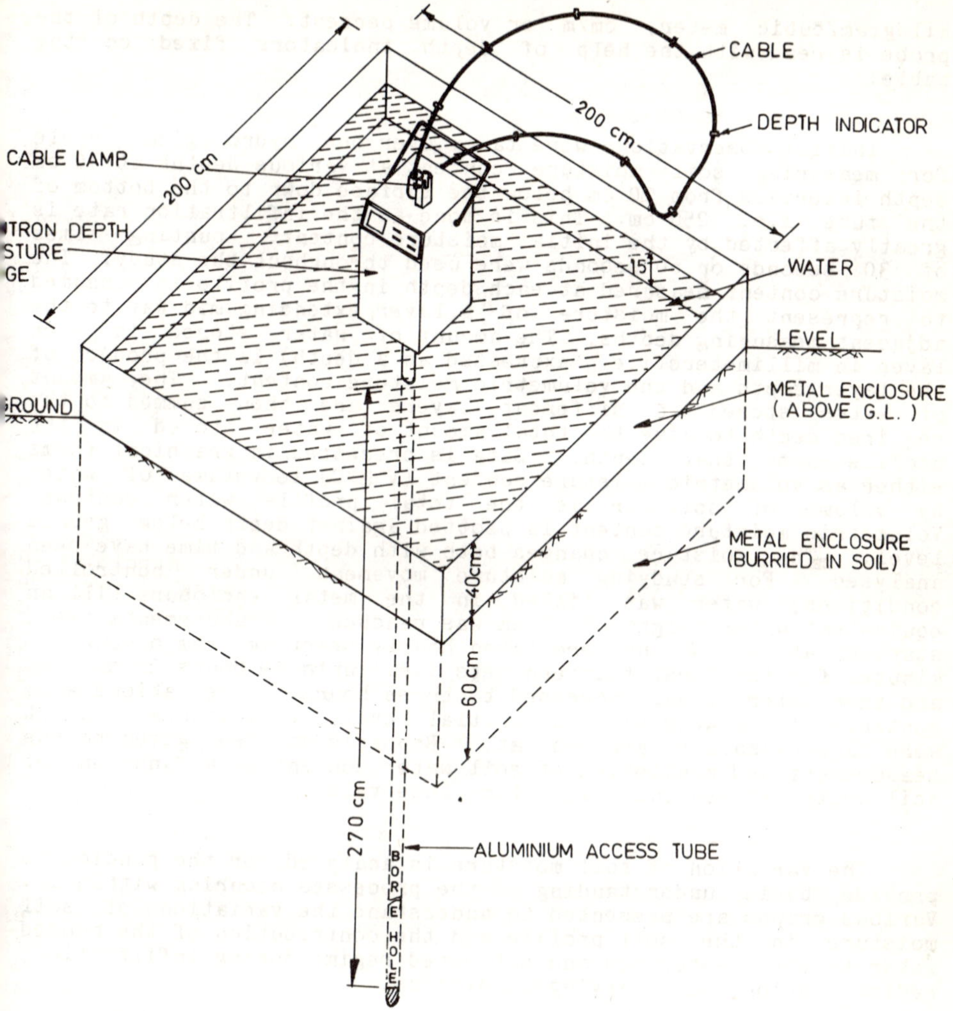

- EXPERIMENTAL SETUP OF FIELD MEASUREMENT FOR SOIL MOISTURE MOVEMENT STUDIES kilogram/cubic meter, cm/m or volume percent. The depth of the probe is read with the help of depth indicators fixed on the cable.

Initial observations were taken over one hydrological cycle for measuring soil moisture content at various depths at 10 cm depth intervals from 20 cm below the surface down to the bottom of the tube i.e. 250 cm. This is done as the infiltration rate is greatly affected by the initial moisture content. Counting times of 30 seconds or 60 seconds were used throughout the study. The moisture content measured at each depth in the profile is assumed to represent the moisture of a layer extending mid way to the adjacent measuring depths. The amount of water stored in each layer in millimeters(i.e. expressed as a depth) is the product of the layer depth and the volumetric moisture content. The amount of water stored in different layers are then summed to the required depth to give the total amount of water stored in the profile upto that depth. Data is presented in graphical forms either as volumetric moisture content (V M C) ie volume of water by volume of soil or as the total profile water content. Volumetric moisture content is plotted against depth below ground level. The moisture changes both with depth and time have been analysed. For studying moisture movement under controlled conditions, water was filled in the metal enclosure till an equivalent water height of 15 cm was reached. Measurements were started at 4.0 AM and were taken hourly starting from 6 hours 25 minutes for the first fourteen sets i.e. upto 19 hours 25 minutes and then interval was increased to three hours. Observations were continued for a week till the initial level of moisture content were approximately arrived at. Emphasis has been given to the measurement and monitoring of soil water content as a function of soil depth between successive time intervals.

The variation of soil moisture is analysed for the ponding to provide basic understanding of the processes occuring within it. Various graphs are presented to understand the variation of soil moisture in the soil profile and the contribution of the ponded water to the unsaturated and saturated regime during infiltration, redistribution under varying conditions.

RESULTS AND DISCUSSION

Some of the profiles are shown in Fig 2, 3 and 4. Several curves are getting superimposed on the same graph corresponding to different dates. The area under each curve represents the total moisture content. Actual evaporation was computed for the period 22nd April to 22nd May and 22nd May to 21st June as 5 cm and 1.9 cm respectively. Fig 2 shows the variation in moisture i.e. moisture depletion and enhancement during dry spells and during

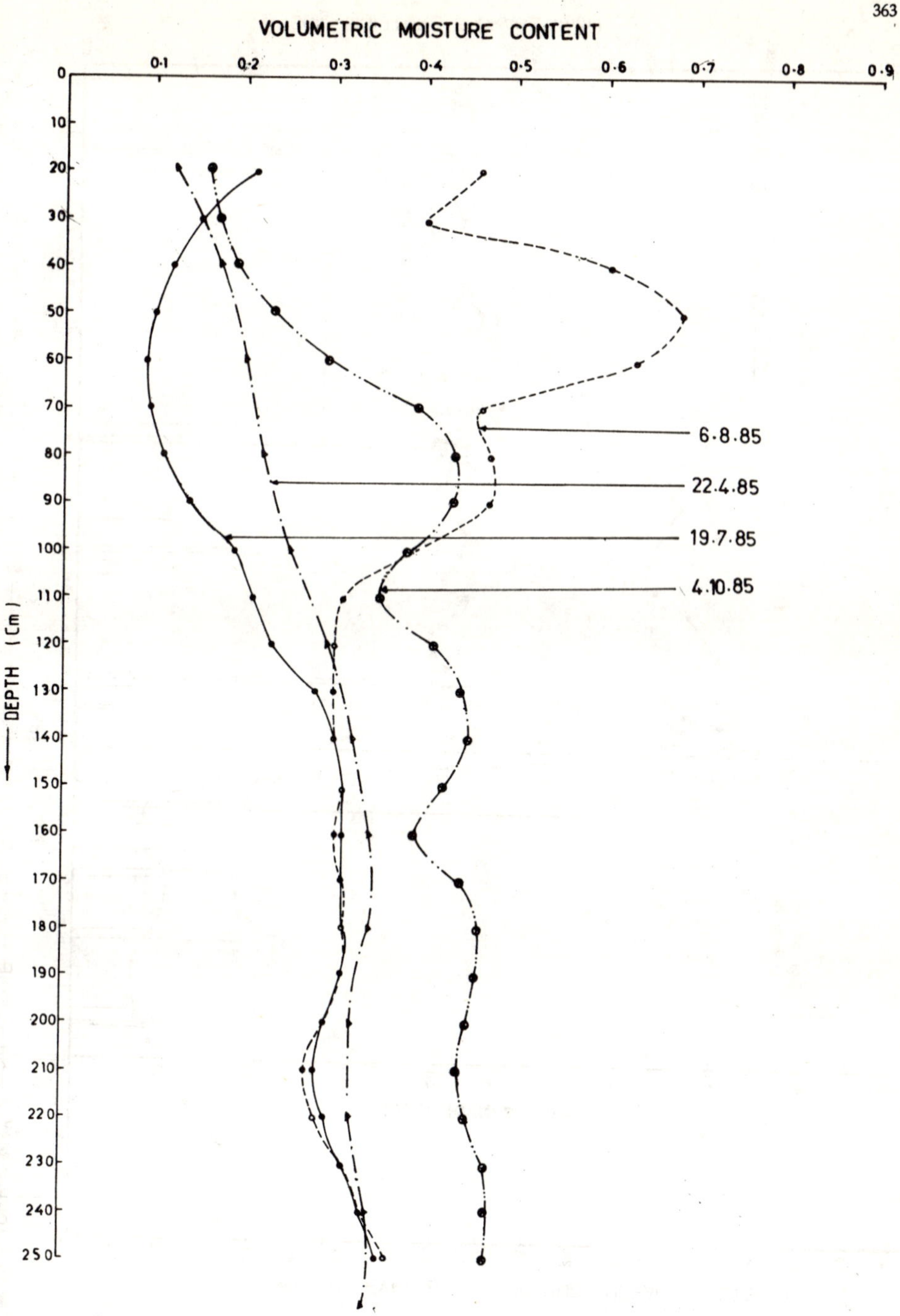

FIG.2 - MOISTURE CONTENT PROFILE FOR THE SITE 'A' FOR 22 APR., 19 JULY, 6th AUG & 4th OCT 1985

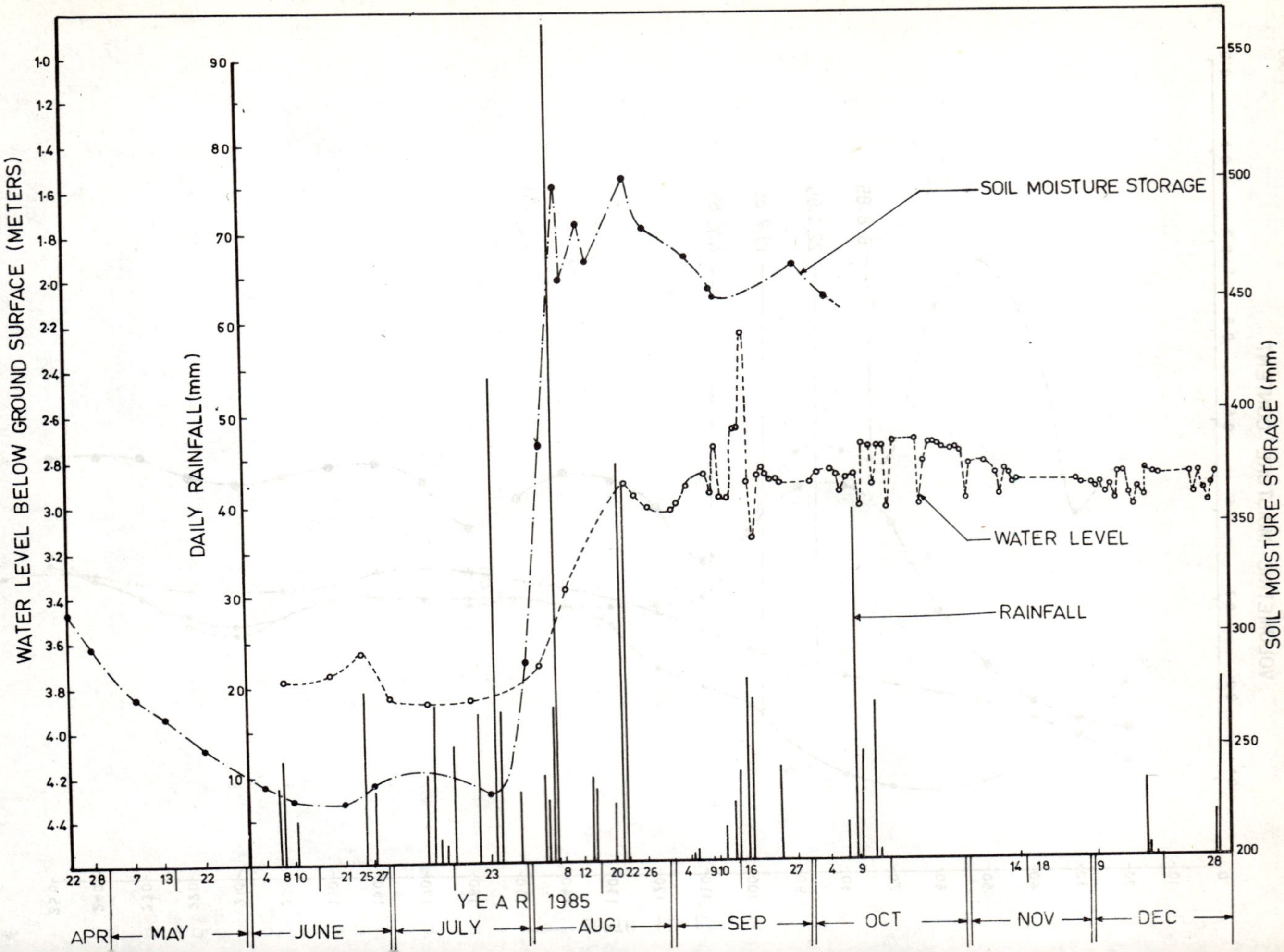

WATER LEVEL BELOW GROUND SURFACE (METERS)
DAILY RAINFALL (mm)
SOIL MOISTURE STORAGE (mm)
SOIL MOISTURE STORAGE
WATER LEVEL
RAINFALL
YEAR 1985
APR
MAY
JUNE
JULY
AUG
SEP
OCT
NOV
DEC

rainstorms. It is seen from the pre monsoon period profiles from this figure that there is not much variation in moisture content below 120 cm depth. Profile of 19th July shows depletion in moisture content above 120 cm depth, while profile of 6th August represents the amount of water gained. Soil returns to an almost identical moisture content after few days and retains its level. Rise in water table and increase in soil moisture storage against the precipitation of 1985 is shown in fig 3. Data is plotted from May 85 to Dec 1985. Rainfall storms are also shown in the figure. From the figure it is seen that between 22nd August and 30th Dec, the water table in the study area was less than 300 cm below the ground level. The areal variability at any one depth in the upper 120 cm shows a 2 - 3 fold increase with the change from the dry to wet stage.

Soil moisture content were also measured at one site where infiltrometer was installed after creating ponded conditions artificially. The moisture contents in the profile at various times before and after ponding are shown in fig 4. The variation is moisture content at various depth with time is shown in fig 5. Philip's two - parameter equation was applied to the data collected in the present study. As per this equation, the cumulative infiltration I at time 't' is approximated by

$$I = St^{1/2} + At$$

$$\text{where } S = \int_{\theta_0}^{\theta_s} \phi \, d\theta \text{ and } A = \int_{\theta_0}^{\theta_s} \psi \, d\theta + Ko$$

Ko is the hydraulic conductivity at θo. Theoretical and experimental analysis indicated that the value of A lies between (2/3) ks and (1/3) ks, but usually closer to (1/3) ks where ks is the saturated hydraulic conducti vity, ϕ and ψ are functions of K(θ) and ψ(θ) , Ko, usually makes a negligible contribution to A. Values of ϕ and ψ and therefore S and A can be calculated if the basic hydraulic properties, K(θ) and ψ(θ) are known for the soil. Experimental evaluation of these functional relationship for soils is time consuming and expensive. Attempts have therefore been made to evaluate S and A directly from infiltration data.

During the very early stages of infiltration when the contribution of gravity is small, the second term (At) of the infiltration equation is negligible relative to the first term and thus S was calculated as the slope of an I vs $t^{1/2}$ plot. It has been observed that the initial infiltration rate increases with the increase in depth of water above the profile, and this effect decreases as time tends to infinity.

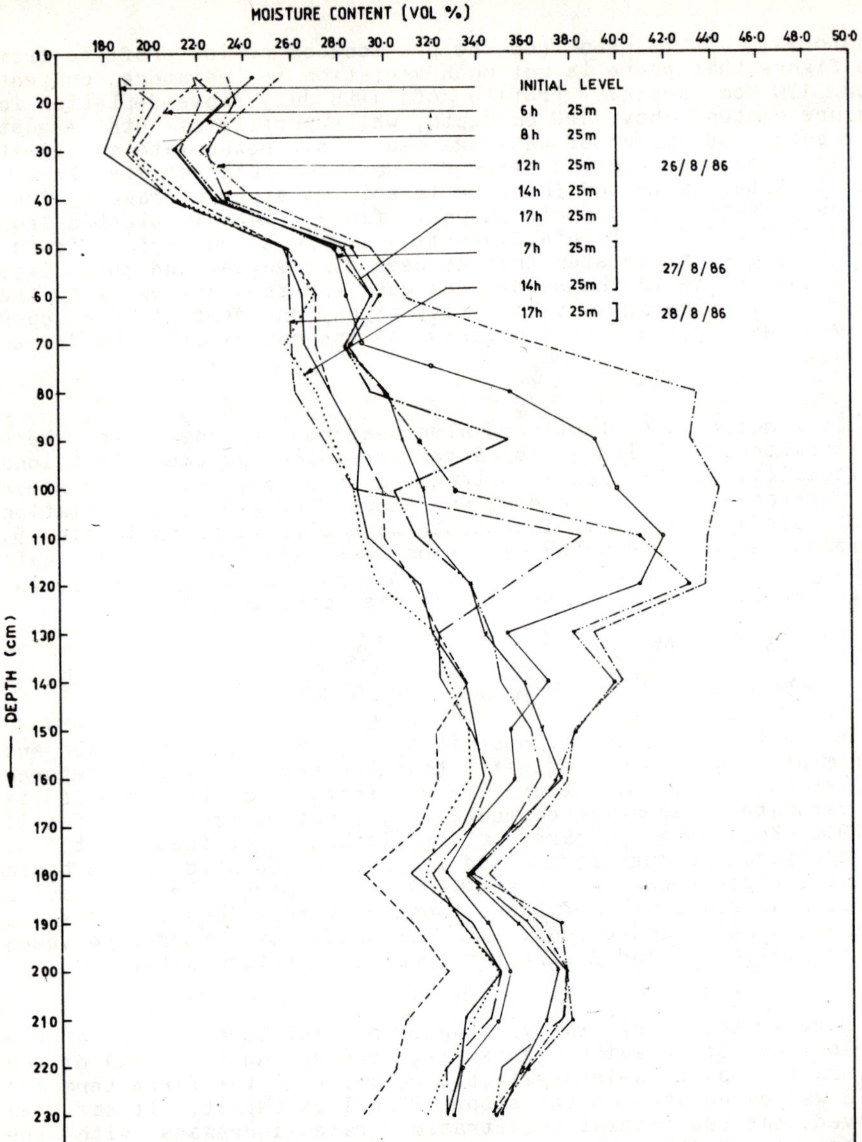

Fig. 4

MOISTURE PROFILES BEFORE AND AFTER PONDING

Calculation of S from I vs $t^{1/2}$ was justified by the linearity. Infiltration data was examined to determine whether the infiltration behaviour over the full time range (0 - 490 minutes) could be adequately described by using the parameters S and A, determined at the extremes of the range. Overall agreement was observed between the measured and calculated infiltration.

Between the start of ponding (4 hours 53 minutes) to the end (6 hours 25 minutes) the drainage loss was estimated at 15.1 mm. The drainage rate was about 9.8 mm/hour immediately after the ponding which increased to 23.3 mm/hour within first 3 to 4 hours and remained about 18.7 mm/hour later till the ponding height reduced to zero. Replenishment of the soil moisture is rapid after the ponding attained some height. In the first four hours after ponding the soil moisture storage increased by 138.4 mm and obsorbed 80.37% of ponded water. After the initial rapid replenishment, increase in soil moisture storage is small and slow in the succeeding hours (the storage increased by only 2.3 mm approximately equal to 1.5% of the ponding)

Experiments were also conducted keeping the water height constant in the infiltrometer. The head was kept constant till the column got saturated and started draining freely at its lower end with water column of 15 cm in the tank. Later it was allowed to lose its head. The drainage resulted into a decreasing height of ponding and with in nine hours the water in the infiltrometer infiltrated completely into the soil column. After 66 hours, the soil profile attained the initial moisture conditions.

Measurements have been carried out at air temperatures from 25.8° C to 32° C. Pan evaporation data was also collected to calculate evaporation from ponded surface. Which was found to be 2.1 mm and has been accounted.

CONCLUSIONS

The investigations carried out clearly bring out the potential of neutron depth moisture gauge for studying movement of soil moisture in the vadose zone. The probe allows measurements at different times and at different depths at number of locations and thus provides a versatile method for study of soil moisture distribution in space and time. The instrument though somewhat expensive, is very suitable for studying variations of moisture in the root zone.

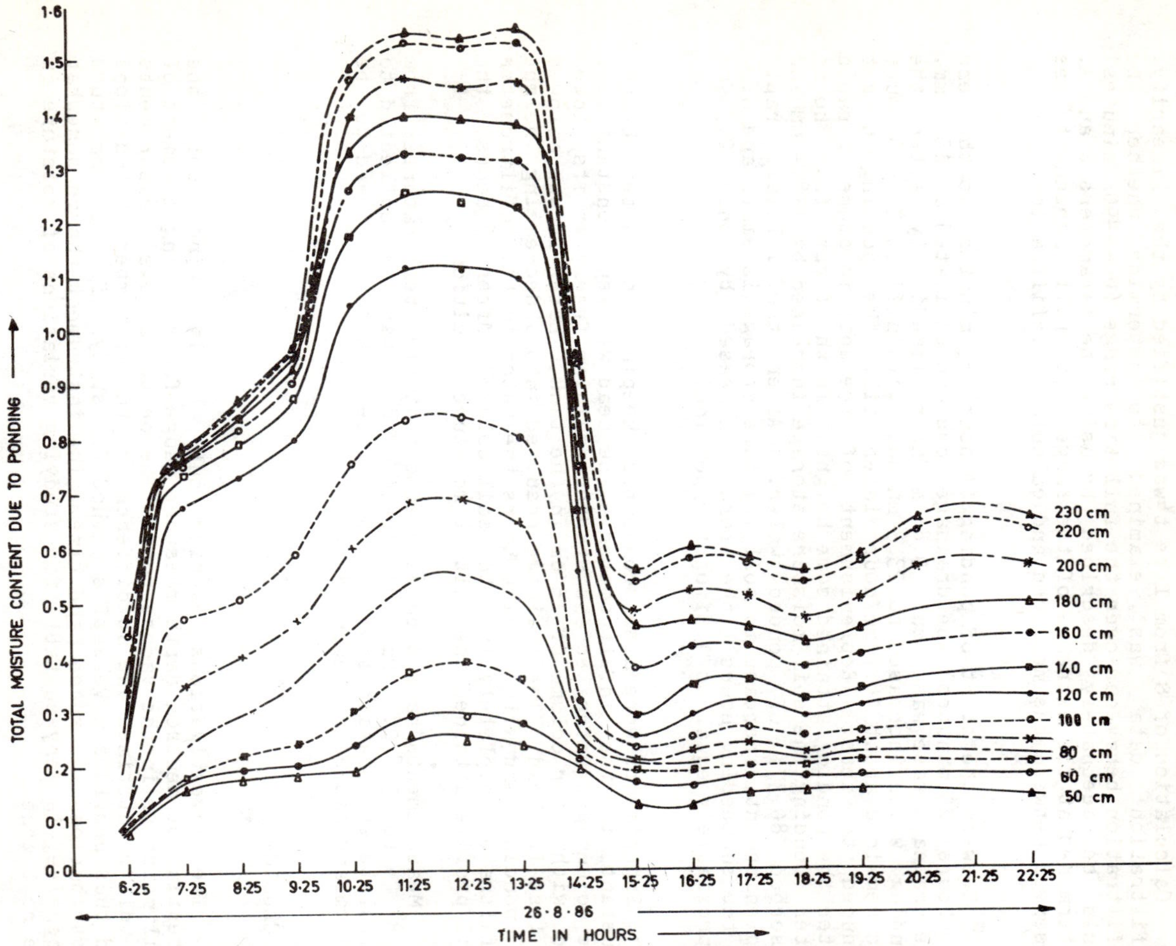

FIG.5 — VARIATION IN TOTAL SOIL MOISTURE CONTENT WITH TIME DUE TO PONDING IN THE PROFILE OVER DIFFERENT DEPTHS.

During the dry periods, the neutron probe provided data to compute actual evaporation and the studies also show tht drainage losses for the soil studied are relatively small in the zones below 120 cm depth and there is not much variation in the moisture content below this depth.

The velocity of infiltration and the velocity of drainage discharge in the profile are found varying with time. But the velocity of vertical flow over the distance between the centres of consecutive layers is found constant during one time interval. The infiltration rate was about 9.8 mm/hr immediately after ponding and increased to 23.3 mm/hr within 3 to 4 hours and remained about 18.7 mm/hr later till the ponding height reduced to zero. During this period of four hours after ponding, the soil moisture storage increased by 15.35 cm and absorbed 98% of ponded water. It is difficult to draw firm conclusions about the recharge mechanism from the soil moisture data alone. However, this study demonstrates the key role which soil moisture process studies can play in the investigation of the hydrology of the area. The understanding gained can be used to model the response of a basin to rainfall/irrigation.

Evaporation for the period 22nd April to 22nd May and 22nd May to 21st June 1985 was estimated to be 5 cm and 1.9 cm respectively. The infiltration data was fitted in the philip's two parameter equation and infiltration parameter S and A were evaluated as 0.3077 cm $min^{-0.5}$ and 0.0473 mm min^{-1} respectively.

REFERENCES

Bavel, C.H.M., Underwood, N. and Swanson, R.W. (1956), Soil moisture measurement by neutron moderation, Soil Sciences, V.82, pp.29-41.

Bell, J.P. and McCulloch, J.S.G. (1969), Soil moisture estimation by the neutron method in Britain, Journal of Hydrology, V.7, pp.415-443.

Ramesh Chand (1986), Soil moisture studies using neutron probe, Proc. C.B.I.P., 53rd Annual R and D Session, V.3, pp.287-296.

Ramesh Chand, Seth, S.M. and Singh, P. (1986), Study of soil moisture using neutron probe, Technical Report TN-28, NIH, Roorkee.

Ramesh Chand (1987), Field measurement of soil moisture movement in ponding condition, Technical Report, TR-12, NIH, Roorkee.

ENVIRONMENTAL ISOTOPE-AIDED STUDY ON RIVER WATER AND GROUNDWATER INTERACTION IN THE REGION OF SEOUL AND TAEGU.

J.S.AHN and I.S.SUH

Korea Advanced Energy Research Institute, Daejeon, Korea

J.S.HAHN

HANS Engineering Co.,Ltd and Seoul National University, Seoul, Korea

B. PAYNE and P.AIREY

International Atomic Energy Agency, Vienna, Austria.

ABSTRACT

Environmental isotope-aided studies on river water and groundwater interaction in the Han River Basin in areas especially occupied by crystalline rocks and limestone have been undertaken. The results indicates that the groundwater in the Seoul area is recharged by the Han River whereas in non-urban area it is replenished by the infilteration of precipitation, the crystalline rock aquifers are recharged by downward percolation of shallow groundwater stored in overlying alluvium. Older groundwater having a very low concentration of tritium, ranging from 0 to 2 TU, was observed at the lower Han River Basin near Bupyeong. It may indicate that the water sampled aws recharged at a much earlier time than the other groundwater sampled, maybe, as early as,the pre-thermonuclear period. The same interaction study conducted in the Karst area has shown that there is clearly no systematic differences of tritium level between surface water and groundwater and that the residence time of groundwater in limestone is very short, probably not longer than a few months. The area recharged by water from the overlying alluvium into the sedimentary rock aquifer were restricted only along the Kumho River channel and its tributaries in a clastic sedimentary rock area. Bedrock groundwater whose tritium level and stable isotope composition was less than 14 TU and - 6.8 respectively was observed in the central part of Taegu.

TOPOGRAPHY

The Korean peninsula projects southwestward about 1,200 km from the mainland of northeast Asia. It is separated from mainland China on the west by the Yellow sea, and from island of Japan on the east by the Eastern sea(see Fig.-1). The peninsula as such came into being relatively recently in terms of geologic time. The peninsula owes its present shape to the regional faulting and uplift that occured during Quaternary period and possible intermittent changes in sea level, uplift, and erosion since that time.[1] The rocks of the peninsula are related to those of mainland China and Japan.

The eastern part of South Korea is characterized by high mountains whose peaks stand 1,000 m or more above mean sea level. Most of valleys in the upper streams are narrow and "V" shaped. The western part of South Korea contains broader valleys and small mountains whose height does not exceed 1,000 m. The Taebaek mountain range, from 1,000 to 1,700 m high, runs from the demilitarized zone to Busan along the east coast in a north-westly direction. The mountain ridge is about 10 to 15 km, from the Eastern sea coast. The mountain slopes dip steeply to a distance of 5 to 7 km and then steeply from there to the sea a distance of 2 to 7 km. The land appears to have been eroded by a sea whose level was slowly and steadily dropping, and probably is related to the emergence of the east coast as the peninsula tilted to the west. The shallow dipping coastal slope has a beveled bedrock surface into which narrow stream channels have been cut. The streams that drain to the Eastern sea have relatively short, straight north-northeast courses in the mountains turning more easily as they cross the beveled surface and meet the sea. Generally, Gwangju, Charyeong, Sobaek, Noryeong, and Dukyou mountains strech south-westly from the Taebaek mountain range. Those mountains are composed of mainly Pre-Cambrian and Mesozoic granitic rocks which intruded into the older rocks. They are intensively metemorphosed and are deeply weathered.

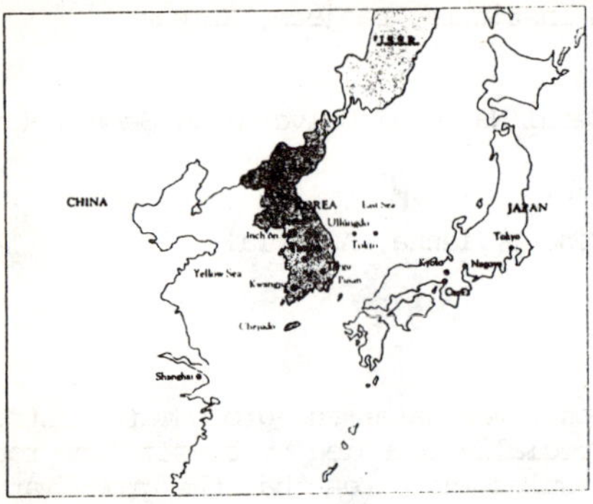

Figure 1 Location map of Korea.

The western and southern coast, on the other hand, have extremly irregular shorelines. The western shore is characterized by broad tidal mud flats. The southern shore is a typically submerged shoreline with downward, almost fiord like valleys and many islets and islands. Throughout the South Korean peninsula, the only large areas of low rolling topography occur where large bodies of granitic rock have been eroded.

CLIMATE AND HYDROLOGIC CHARACTERISTICS

The climate of southern Korea is characterized by a moderately cold,dry winter with long period of clear skies. Summer is humid and hot. Autumn and spring are very pleasant with mild temperatures and many days of sunshine. Summer temperatures reach 35°C and above;winter temperatures drop as low as minus 30°C in the high mountains.

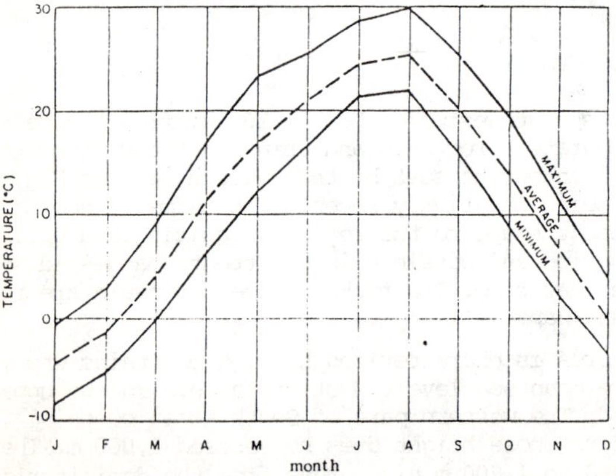

Figure 2. Average monthly temperature in Seoul area.

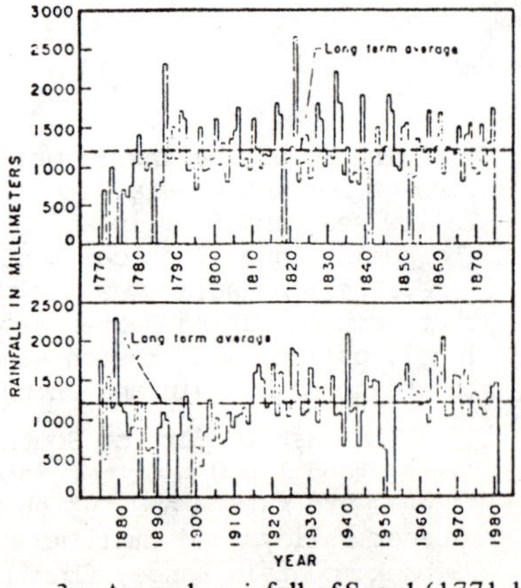

Figure 3. Annual rainfall of Seoul (1771-1980

The warmest and colest temperaures are experienced inland from the sea. Precepitation ranges from 1,000 m/m to more than 1,300 m/m in a year. About two thirds of the annual precipitation occurs during the summer monsoon seasons from mid-June to mid-September. The other months are characterized by a fairly uniform distribution of precipitation. Snow occurs from November to March in the winter. Although the Korean climate may be classified as humid and temperate, actually it is a combination of the two climates. The hot rainy summer season is subtropical, the rest of year is semi arid(see Fig.-3). This combination of climate has a profound effect on the occurrence and development of water resources, both surface and ground water.

The republic of Korea which covers an area of about 98,490 km^2 has a four main watersheds of which Han, Nakdong, Geom, and Yeongsan river basins account for about 64 % and the rest are comprised of several smaller basins which are only minorly significant to water management. Most of the rivers have relatively small drainage areas, are short in length and steeply sloped. The river level rises very abruptly after rainfall for those reasons, causing very frequent floods. Moreover the amount of water flowing varies greatly owing to the seasonal variation of the rainfall.

ISOTOPE HYDROLOGY IN SOUTHERN KOREA

The isotope hydrologic studies of the several areas in southern Korea have been conducted to interprete the hydrogeologic interaction between surface water and groundwater occured in crystalline, carbonate, clastic sedimentary, and volcanic basaltic rocks, by a team consisting KAERI's staff,Peter Airley of ACE of Australia, and Rryne Payne of IAEA, Vienna from 1980 to 1984. The water sample were analysed for Deuterium by Australian AAEC, for oxygen-18 by IAEA, Vienna, and for tritium and other chemical components by KAERI, Seoul, as part of IAEA's regional cooperative programme on isotope hydrology study.

1. Crystalline rock area (Seoul)

The aim of the study is to assess the possible threat to the groundwater of Han river basin from urban and industrial effects. The first study has been restricted to the shallow alluvial aquifer underlying the area shown in Figure-4.

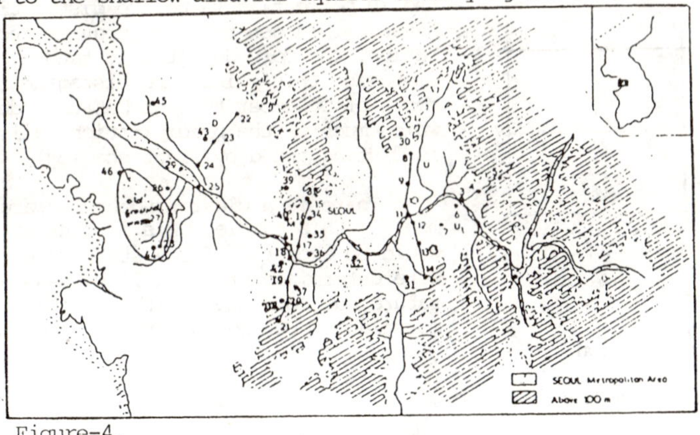

Figure-4.
Location of sampling points, Han River Valley. Elevation above 100m are distinguished by line shading. The approximate bounaries of the Seoul metropolitan area are also shown.

Three transects approximately normal to the river were established. One was located in the heavily urbanized region of Seoul(transect M) and the upper stream(U) and down-stream (D) of the capital city of Seoul. Environmental isotopes were used;

1. To determine the relative importance of the river and precipitation to the recharge at various locations within the basin and

2. To study the interaction of river and groundwater, and the dynamics of the the important transport processes.

1-1. Isotope composition of the recharge water

1-1-1. Tritium (precipitation)

Since none of the IAEA's environmental isotope monitoring stations[7,12] are within the study area, the input function had to be assessed from data collected at the closest station Pohang which had operated between 1961 and 1976. The

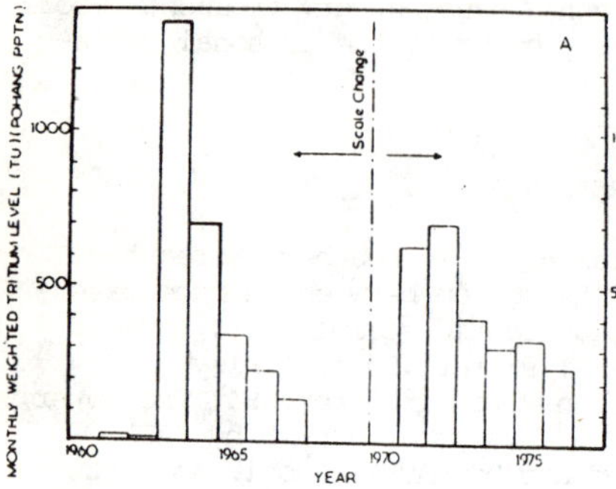

Figure-5. The variation with time of the monthly weighted tritium levels in Pohang rainfall.

The variation with time of the weighted annual means of Pohang rainfall is shown in Figure-5. The tritium levels of Pohang rain fall was measured maximum of about 1,400 TU in 1963 due to the effect of the detonation of thermonuclear devices. Since that time the concentration has decreased to less than 40 TU in 1976 and tritium levels of groundwater in alluvium and stream water of Han-river measured at every transects mentioned before were depicted in Figure-6.

1-1-2. Stable isotope ratios(precipitation)

Figure-7 illustrates the frequency distribution for the available D/H ratios from Pohang precipitation(1960-1967,1973-1975). The distribution for those months in which the precipitation exceeded the 90 percentile value and those for which it was less than the 10 percentile value are distinguished by hatching. In general, the heavier the precipitation, the more depleted it tends to be. To emphasise this point, the histogram for the month of December through February during which much of the precipitation falls as snow is distinguished by light shading.

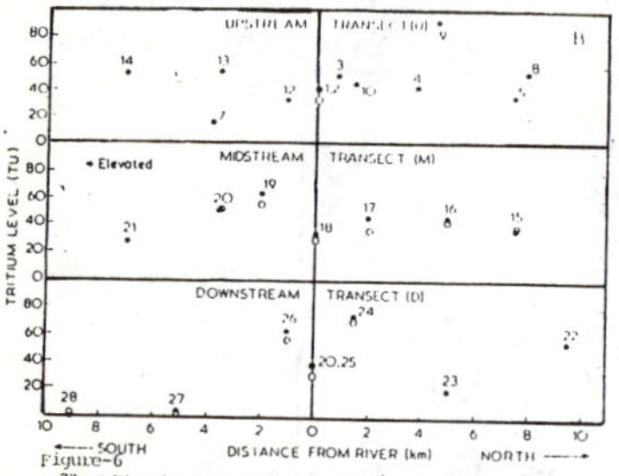

Figure-6
The tritium levels of groundwater (•, 0) and river water (•, 0) of the Han River Valley. Samples were collected during late March, early April (•) and in June (0.) The elevation of all sampling locations (except 21) was less than 100m.

There is a clear tendency towards enrichment in this sample. This is contrary to what would be expected if the stable isotope ratios were determined principally by seasonal temperature but consistent with the observed inverse correlation between D/H ratios and the amount of monthly precipitation.

1-1-3. Surface water of Han River

The variations of the tritium and D/H ratios with distance from 29 sampling points are shown in Fiure-8. The results have shown that the deuterium levels are appreciably more depleted thaneither the average Pohang precipitation (weighted mean 51.3 per mille) or the groundwater of the lower Han river basin (delta D= 53.6 mille) where direct rainfall recharge dominates. The effect is almost certainly due to altitude. The head waters are about 1,700 m above sea level; about 14 % of the catchment is above 800 m and 28 % more than 200 m above sea level. The D/H ratios in samples collected from (1) and (2) in March/April were -64.0 and - 62.3 per mille respectively. The sample collected at position (6) which is river water had Delta D value of -50.9 per mille.

Since the sample was collected from the bank during a period of low flow, and since the D/H ratio is close to the groundwater taken from a nearby well(No-3, delta D = -48.3 per mille), it is temting to suggest that there is significant groundwater input at the point. There is no statistical evidence for any variation in the D/H ratios of the river water during the June sampling period. The average value was -59.4 ± 0.8 per mil, where the standard deviation is of the population. The individual results are shown in Figure-8. Any groundwater input at point (6) would be diminished or eliminated by the higher level of the river during late spring. The available tritium and conductivity data are consistent with groundwater input near sampling point (6) during the April sampling period. The observed tritium levels at the upstream locations (1) and (2) were 40.6 and 40.8 TU separately. The increase of 43.5 TU at location (6) may reflect the higher tritium level in the groundwater,e.g. 56.3 TU at the position (3). The generally lower levels at further down-stream are due to surface water inflow. Although post-1976 data for Pohang are not available, evidence from Tokyo rainfall would suggest that modern tritium levels would be less than 25 TU.

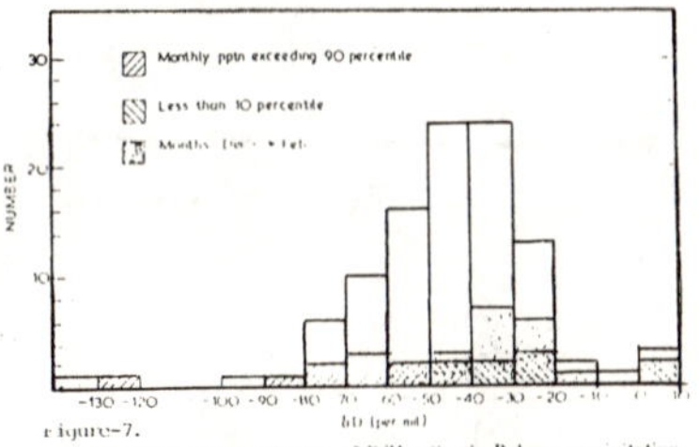

Figure-7.
The frequency distribution of D/H ratios in Pohang precipitation. Hatching indicates where monthly precipitation is greater than the 90 percentile value and less than the 16 percentile level. The months December to February duing when there is substantial snow are shown by light shading.

1-2. Isotopic composition of the groundwater

1-2-1. Stable isotope ratio

As shown in Figure-9, δD and $\delta^{18}O$ values satisfy the following equation; $\delta D = (6.2 \pm 0.5)\ \delta^{18}O + (3 \pm 0.5)$ per mil. More detailed study should be given to interprete the relation between delta D and delta ^{18}O. Figure-10 is a frequency distribution in which the wells sampled in the urbanized area are distinguished by hatching. Interestingly these data are grouped about a mean

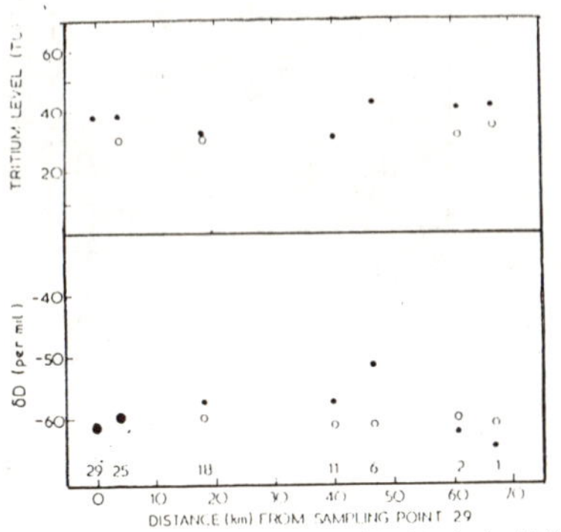

Figure 8. The variation in river water samples of the tritium levels and D/H ratios with distance from the sampling point 29 (Figure-?). ● April sampling, ○ June sampling

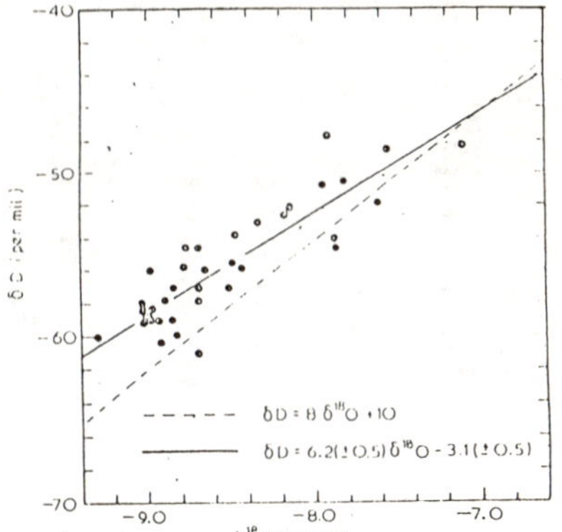

Figure-9
The variation of the δD values of groundwater samples with $\delta^{18}O$. The line of best fit and the meteoric line ($\delta D = 8\,\delta^{18}O + 10$) is shown.

which is more depleted than the average of all groundwater collected, but is close to the river value at the time of sampling. It would appear that the combined effect of increased pumping and reduced infilteration in the Metropolitan area has been to lower the groundwater table and to enhance the river recharge.

On the other hand, virtually all the wells sampled down-stream(transect D) were more enriched than river awter and within the range of values expected for precipitation. Recharge in the lower catchment appears to be dominated by rainfall. The contrast is the most clearly seen by comparing the urbanized mid-stream (M) and down-stream (D) transects (see Fig.-11). The up-stream transect (U) show

Table 1 Associated data (Upstream) Han R. Valley

Sample No	Season	Dist. from River (Km)	Depth (M)	Temp. (°C)	pH	Conductivity (U mho Cm^{-1})	Tritium (tu)	δD per mil	$\delta^{18}O$ per mil
01	autumn	River		19	9.4	67	35.2	-68.2	-9.4
	winter			-1	6.5	100	37.1	-65.8	
02	autumn	river		19	8.6	111	34.0	-66.2	-9.22
	winter								
03	autumn	0.8	0.8	15	6.8	588	36.1	-54.9	-7.64
	winter			7	5.9	606	43.1	-51.6	
04	autumn	3.8	3.6	18	6.4	200	34.5	-57.4	-7.88
	winter			5	5.8	132	36.4	-53.7	-7.27
05	autumn	7.5	16	17	6.2	156	31.7	-57.4	-8.14
	winter			9	5.7	122	37.2	-55.1	-7.94
06	autumn	river		19	8.9	100	49.7	-66.1	-9.18
	winter			-3	6.2	127	38.1	-63.9	-8.75
07	autumn	3.6	5	19	7.1	222	21.2	-58.3	-8.21
	winter			12	5.5	244	25.2	-56.7	-7.94
08	autumn	8.0	18.2	18	6.2	192	42.7	-55.8	-8.03
	winter			12	5.7	167	50.8	-54.2	-7.78
09	autumn	4.5	7.0	14	6.2	1,020	74.9	-51.9	-7.72
	winter			12	5.9	1,111		-50.3	-7.30
10	autumn	1.5	9.1	14	6.4	111	37.2	-55.9	-8.27
	winter			11	5.7	91	38.9	-55.6	-8.19
11	autumn	river		19	6.8	105	32.7	-65.4	-9.09
	winter			-2	6.2	147	36.7	-64.7	-9.02
12	autumn	1.0	5.5	13	6.1	909	46.1	-52.7	-7.60
	winter			12	5.9	625	54.6	-51.4	-7.44
13	autumn	3.5		13	6.5	227	57.1	-56.5	-7.93
	winter			12	5.8	167	63.5	-55.4	-7.78
14	autumn	7.0	18.2	14	6.4	200	52.2	-55.1	-7.97
	winter			12	6.2	167	52.5	-55.0	-7.90
15	autumn	7.5	7.5	15	6.5	926	32.3	-57.8	-8.21
	winter								
16	autumn	5.0	5.8	17	7.1	833	38.8	-58.3	-8.38
	Winter			8	5.5	769	37.2	-58.9	-8.44
17	autumn	2.0	6.2	15	7.0	556	36.9	-61.2	-8.49
	Winter			8	5.5	526	40.2	-59.2	-8.55
18	autumn	river		19	7.1	125	31.3	-66.7	-9.21
	winter			-2	6 1	196	37.2	64.3	-9.12
19.	autumn	2.0	6.0	16	6.2	588	49.2	-59.3	-8.18
	winter			14	5.5	526	60.5	-58.8	-8.49
20.	autumn	3.0	12.0	16	6.2	769	49.1	-58.0	-
	winter			13	5.3	476	37.2	-60.3	-8.42
21.	autumn	7.0	4.5	15	5.2	400	22.4	-65.0	-
	winter			8	5.5	909	19.9	-60.9	-8.61
22.	autumn	9.5	5.0	16	6.3	161	52.5	-53.8	
	winter			4	5.7	91	59.4	-51.6	-7.41
23.	autumn	5.0	21.2	14	6.0	435	23.0	-62.9	
	winter			8	5.6	100	23.7	-58.3	-8.04
24.	autumn	1.5	5.0	15	6.3	556	60.1	-52.9	
	winter			8	5.7	333	74.9	-52.6	-7.63
25	autumn	river		20	6.0	200	32.2	-63.9	-
	winter			-2	6.1	250	38.9	-64.0	-8.75
26.	autumn	1.0	15	13	6.5	185	50.1	-55.0	-
	winter			7	5.9	87	59.9	-54.5	-7.77
27.	antumn	5.0	12	13	6.8	233	0.33	-53.5	
	winter			11	6.1	213	0.45	-54.6	-7.88
28.	autumn	9.0	24.2	13	7.1	286	0.70	-53.4	-
	winter			11	6.9	250	0.10	-55.5	-7.89
29.	autumn	river		20	7.0	135	28.4	-63.7	-
	winter			-2	6.1	238	37.7	-64.1	-8.09

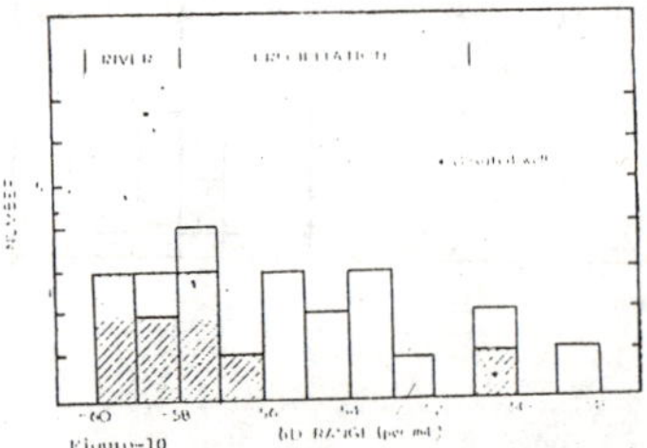

Figure-10
Frequency distribution of D/H ratio s in groundwater samples Those wells located in the metropolitan area are distinguished by shading. The well 21 (*) has an elevation of ~ 200m.

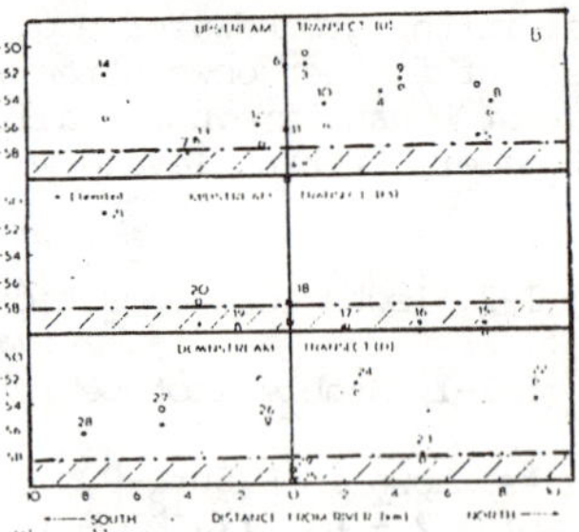

Figure 11.
The D/H ratios of the river water (▮ |) and ground water (●, 0) of the Han River Valley. Samples were collected during April (●) and in June (0). The sample points are numbered (Figure 4). The range of river δD values is shown by shading.

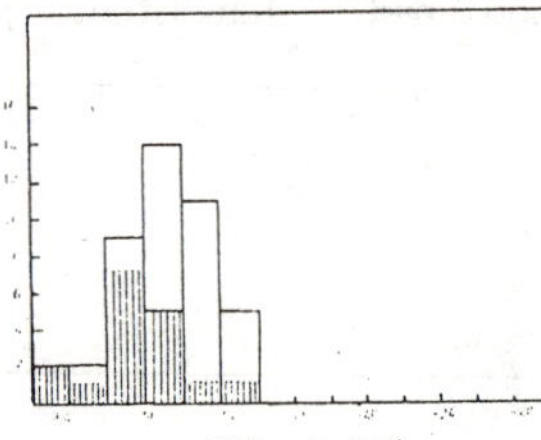

Figure-12
Frequency distribution of D/H ratios in ground water samples from the shallow alluvial rock aquifer collected during spring and summer (15/4/81 and 15/6/81). In both cases wells located in the metropolitan area are distinguished by shading.

mixed behaviour. That portion of the transect to the north of the river is in the Green Velt and exhibits D/H ratios more depleted than those from the southern section which includes wells in outlying urban area. It is tentatively proposed that, while recharge to the north is dominated by precipitation, the groundwater to the south is a mixture of precipitation and river infilteration.[3] The data from samples collected during the Autumn and Winter sampling period are listed in Table-1(1980). A distribution of stable isotope ratios is shown in Figure-12. Those wells in the Metropolitan area of Seoul are also shaded. The trend observed is confirmed; samples from urban areas tend to be depleted relative to those from open, agricultural areas of the basin.

In the former case the recharge is principally from the river, in the later, by direct infiltration of precipitation. A comparison of Figure 12 with Figure 10, indicates more "overlap" of the two classes samples in the autumn and winter than in the late spring and summer. This may indicate that the groundwater is dominated by recharge of summer rainfall as would be expected since 70 percent of the precipitation occurs between June and September.

If stable isotopic composition of groundwater collected in autumn and winter reflected that of the precipitation falling at the period of collection, substantial enrichment would have been expected. The conclusion that the isotope composition of the groundwater reflects that of the summer precipitation is important in assessing the tritium data.

1-2-2. Tritium

The groundwater tritium levels are shown in Figure 7. As discussed above, those wells tapping transect(M) are dominated by recharge from the river. The tritium values are found to increase regularly with distance from the river to a steady value on a broad maximum of about 60 TU. If the groundwater tritium can be correkated with the mean in Pohang rainfall levels, it follows from simple piston flow model that less than 8 years are required for the groundwater to traverse the distance from the river to sampling points(14) The tritium levels in those parts of the aquifer recharged directly by precipitation are much more variable. Insufficient data are yet available to delineate trends. The two very low values(27 and 28) are clearly significant 'Although these sampling points lie close to the coast, there is no evidence for sea water intrusion into the aquifer. The δD values (-55.8 and -54.5 mil)are typical of precipitation and to not exhibit the enrichment that would result from appreciable mixing with oceanic water (δD = 0). Moreover, the electrical conductivities of the samples are typical of the area. Obviously there must be another explanation for the low tritium values observed. It is tentatively suggested that the water is pre-nuclear.

1-2-3. Shallow alluvial aquifer

The results of the levels of environmental tritium in river the groundwater samples have been surveyed 4 times between March 1980 and February 1981. Data for the Han River samples are plotted as in Figure 13. They tend to exhibit a broad minimum in early autumn and a maximum in late winter or early spring. A similar trend is observed with the Pohang precipitation. The data presented in the Figure were found in available data. Interestingly the tritium variations in the shallow alluvial aquifers in the non-urban areas of the basin show analogous seasonal periodicy (see Figure 14). Possibly insufficient time has elapsed since recharge for the variability to have dispersed. Data from the Metropolitan groundwater support this idea (see Figure 15). The absence of an observed periodicy is consistent with the expected of dispersion associated with transport from the river.

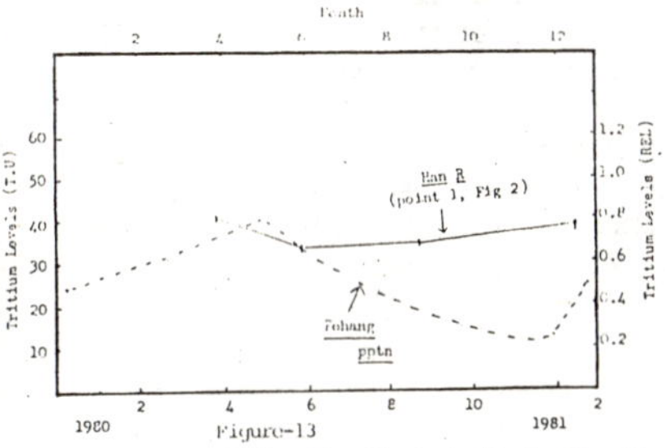

Figure-13

. The variation with time of the tritium levels of the Han River. For comparison the variation of monthly tritium levels in the Pohang precipitation are also shown;

There are, however, two serious problems with the simple hypothesis that the seasonal variation in the shallow non-urban

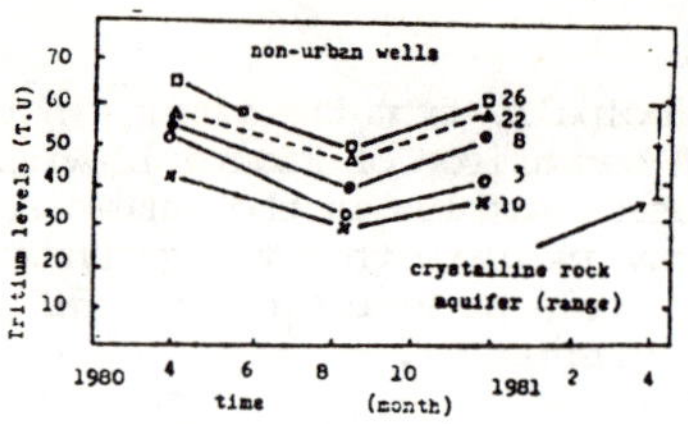

Fig. 14, The variation with time of the tritium levels on the shallow alluvial ground water in the non-urban region.

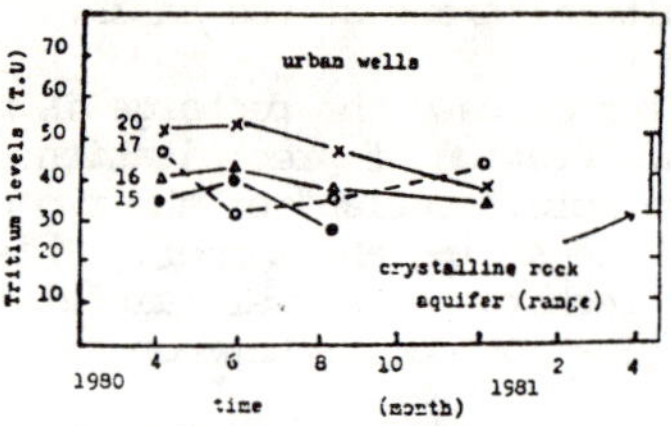

Fig. 15 The variation with time of the tritium levels in the shallow alluvial groundwater in the meteropolitan area, in the underlying crystallino rock aquifer.

groundwater is a simple consequence of the fact that the water is very recent. The first relates to the fact that the stable isotope ratio data indicates that the groundwater is derived predominently from summer precipitation. An annual cyclical variation would therefore not be expected. The second problem is potentially more serious. The tritium level in samples collected in autumn and winter are substantially higher than the river values. The most likely explanation is based on two facts. The tritium levels in Pohang rainfalll have been decreasing regularly from a maximum of about 1,400 TU in 1963 to less than 40 TU between 1973 and 1976. The distribution of precipitation is far from uniform; about 70 percent generally falls in the four months from June to September. During the rest of the year, the potential evaporation exceeds mean monthly rainfall. During the drier months the effect or irrigation may be to pump a proportion of older groundwater with higher tritium levels. It is not simply a question of stratification because much of the groundwater in the underlying crystalline rock aquifers is high in tritium.

1-2-4. Deep seated groundwater in crystalline rock aquifer

Groundwater from seventeen wells drilled in the crystalline rock aquifer was collected during April and June 1981. The location of the sampling points are already shown in Figure 4. A total of seventeen wells drilled in the underlying crystalline rock ranging in depth from 58 to 150 meters were sampled. The D/H distribution is shown in Figure 16 and Tabel 2. Again samples in the Metropolitan area are shaded. The distributions in both the shallow and deep wells are similar; in both cases, samples from the urban wells tend to be more depleted. The data support a hypothesis that groundwater in the crystalline rock aquifer is recharged by downward, direct percolation from the overlying alluvium and unconsolidated deposits. The tritium levels in samples from the deep wells tapping the crystalline rock aquifer are listed in Table 2. Those from non-urban area are plotted in Figure 14; those from urban area in Figure 15. Despite the depth of the wells, the tritium levels are entirely comparable to samples from the shallow alluvium in the same geographical location.

The low tritium levels recorded in the shallow wells, (27) and (28), throughout the entire sampling period are consistent with the observations in the deep well of (44) and (46) which is located downstream on the Han River. The region of older groundwater is shaded in Figure 4. In the other areas the results provided powerful evidence in general for the fact that groundwater in the two aquifer systems has the same origin. The areas showing low tritium levels can be explained as follows: The area, in general, overlain by thick impermeable clay deposits on the crystalline rock aquifer is an industrial site from which large amounts of deep seated groundwater is taken. Due to the heavy pumping of the groundwater stored in the crystalline rock aquifer that is not highly permeable, the water level declines very rapidly. Many industries in

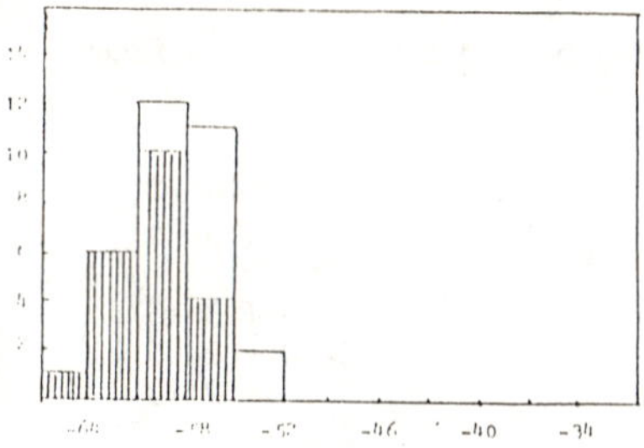

Figure 16.

Frequency distribution of D/H values in groundwater samples from the crystalline rock aquifer collected during autumn and winter (1/9/80, 15/1/81). In both cases wells located in the metropolitan area are distinguished by shading.

Table 2 Environmental isotope and associated data of deep wells

Sample No.	Description*	Date (1981)	Dist. from River (km)	Depth (M)	Temp. (°C)	Conductivity (umho Cm⁻¹)	Tritium** (TU)	δD per mil	δ18O per mil
30	gw (n)	Spring	11.3	58	14	133	44.6	-57.4	8.00
31	gw (s)	"	3.8	115	12	40	55.1	56.7	[illegible]
32	gw (s)	"	1.6	100	18	313	54.8	57.9	8.28
33	gw (n)	"	6.3	120	12	714	51.0	60.8	8.35
34	gw (n)	"	5.5	150	14	400	32.6	61.0	8.56
35	gw (n)	"	5.0	120	16	833	32.2	60.2	8.28
36	gw (n)	"	1.1	100	17	500	34.2	-61.3	8.61
37	gw (s)	"	3.2	121	11	45	49.9	-57.0	-7.94
38	gw (s)	"	6.3	100	8	142	35.5	64.5	8.83
39	gw (n)	"	6.4	60	14.5	588	47.5	58.3	8.12
40	gw (n)	"	4.8	150	10	333	33.0	63.3	8.70
41	River	"	River	-	7	142	34.8	63.1	8.93
42	gw (s)	"	1.4	150	15	833	46.9	-59.7	8.35
43	gw (n)	"	5.5	70	12	50	67.5	-54.2	7.81
44	gw (s)	"	9.5	120	15	125	0.4	-58.0	8.11
45	gw (n)	"	6.0	70	13	133	41.3	58.8	7.66
46	gw (s)	"	2.5	100	13	118	4.9	-57.1	8.17
30	gw (n)	Summer	11.3	58	14	76	63.3	-58.5	8.26
31	gw (s)	"	3.8	115	14	455	35.0	-55.2	-7.50
32	gw (s)	"	1.6	100	15	370	45.2	-58.7	8.21
33	gw (n)	"	6.3	120	14	1,851	24.6	-59.0	8.31
34	gw (n)	"	5.5	150	15	417	19.9	-62.0	8.81
35	gw (n)	"	5.0	120	16	1,111	36.1	-58.6	8.36
36	gw (n)	"	1.1	100	19	556	40.2	-60.9	-8.51
37	gw (s)	"	3.2	121	18	588	51.0	-56.4	8.13
38	gw (s)	"	6.3	100	26	222	42.2	-62.2	-8.64
39	gw (n)	"	6.4	60	17	588	53.4	-57.4	8.27
40	gw (n)	"	4.8	150	14	476	38.8	-62.4	8.77
41	River	"			25	222	41.1	-	8.69
42	gw (s)	"	1.4	120	20	1,041	44.2	-59.8	.
43	gw (n)	"	5.5	70	-		-	-	-
44	gw (s)	"	9.5	120	15	156	0.2	-52.1	8.22
45	gw (n)	"	6.0	70	20	222	47.6	-57.1	7.89
46	gw (s)	"	2.5	100	13	131	9.2	-58.0	7.89
50	River	"	-	-	-	50	38.2	-	-8.87
51	Sea Water	"	-	-	-	5,000	17.5	-	-0.52

the area try to deepen the existing wells or drill new wells. Because of an impermeable and thick clay bed ranging in thickness from 5 to 20 meters and poor storage of deep seated groundwater in the crystalline rock, precipiation and other surface water can not recharge the rock and alluvial aquifer.

As a Summary of the isotope hydrology study in the crystalline rock area of Seoul city; these are the result;

Analyses of the stable isotope data has confirmed the hypothesis that groundwater from the Metropolitan area is recharged from the river whereas that from the non-urban area of the basin is replenished by infiltrating precipitation. And groundwater of the Han River basin is dominated by recharge from summer precipitation. The tritium levels in groundwater samples from the non-urban area show a periodic annual variation similar to that observed in the river samples. The water must be current. In the dry months between October and May the tritium levels frequently are appreciably greater than the river values. It is tentatively concluded that during the dry part of the year, changes of the flow patterns due to extensive pumpage leads to the tapping of a proportion of older groundwater with higher tritium levels resulting from thermoneuclear testing in the mid-1960's. Both the tritium levels and the D/H distribution in samples from the underlying crystalline rock aquifer are similar to those of the shallow groundwater. It is concluded that water from shallow alluvial aquifer is the principal source of recharge, and that the rate of transport of transport of water through the rocks and geologic structures in the granitic, gneissic and schistic rocks is very rapid. There was no evidence of exchange between the groundwater and the river in the vicinity of transect D. Two samples collected near the coast exhibited very low tritium. The effect was not due to sea water intrusion but to the tapping of an older pre-nuclear source which is not very rapidly recharged by the impermeable alluvial aquifer which lies above.

2. Limestone area (Karst topography)

A total of 9 water samples including 3 from well water, 3 from river and 3 others from springs of cavernous limestone were sampled to determine the interaction between surface water and groundwater by measuring of level of tritium and stable isotope of the water sources. Water samples were collected once each season in 1982. The results had shown that there was clearly no systematic difference between the tirtium level of the river and the groundwater. Moreover, some evidence exists for a seasonal variation in the tritium levels. The two pieces of evidence strongly suggested that the residence time of the groundwater in the karst area was very small, probably not longer than a few months. The data are summarized as in Table 3.

Table-3. Environmental tritium levels in groundwater and river samples from Karst area of the Han River basin.

Period(1982)	Groundwater (TU)	River (TU)
Spring	34.4 ± 4* (5)#	33.0 ± 3 (3)#
Summer	34.3 ± 4 (6)	32.0 ± 2 (3)
Autumn	30.4 ± 3 (8)	27.4 ± 2 (3)
Winter	30.8 ± 6 (8)	29.9 ± 0.4(3)

Note; * Standard deviations
Sample numbers

3. Clastic sedimentary rock area(Taegu area)

Water samples from 20 sources located in the eastern part of Taegu City were collected seasonally in 1983 including 2 from river water, 3 from a shallow well drilled in alluvium, and 5 others from a deep well sunk in the Gyeongsang sedimentary formation. In addition to those samples, 16 other sampling sites were established in the western part of the City. (see Figure 17) Water samples from 16 sources comprising I from river water, 1 from the reservoir, 4 from a shallow alluvial well, and 9 from a deep well drilled in the sedimentary rock were collected in winter, spring, and summer in 1984.

Water samples collected at each period were analyzed for tritium, deuterium, and oxygen-18 for the following purposes of;

1) To characterise deep seated groundwater in clastic sedimentary rock from different sources and to relate them to the place of recharge.
2) To determine any possible interaction between surface water and groundwater stored in alluvial deposit and in clastic sedimentary rock aquifer.

The tritium levels of monthly rainfall measured from March of 1983 to June of 1984 in the investigation area ranged from 8 to 31.5 TU with an average of 15.7 TU. The tritium level of shallow groundwater excluding sample No.14 was always much higher than that of Kumho River water whose average level was about 19.3 plus a minus 2.9 (see Figure 18). Source(No.14) is located in the northern part of an industrial area from which large volumes of groundwater are pumped from the alluvial and clastic sedimentary rock and is the water from the shallow well sunk in the alluvium (see Figure 17). The fact that No.14 water has the same range of that of river water can be interpreted that waster stored in permeable alluvial deposits is hydraulically interconnected with river water(15).

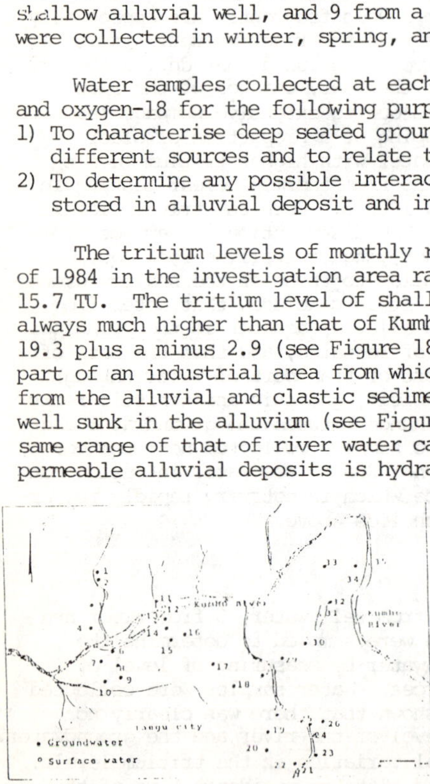

Figure 17. Map showing the water sampling points in Taegu area (Clastic sedimentary rock area).

The range of stable isotope(deuterium) and level of tritium found in shallow groundwater was about same range as that of river water in winter of 1983. And the tritium level in groundwater even in the rainy season was also the same range with the river water as in winter, whereas the stable isotope contents in shallow groundwater became much higher than that of surface water. It meant that the interaction between river water and shallow

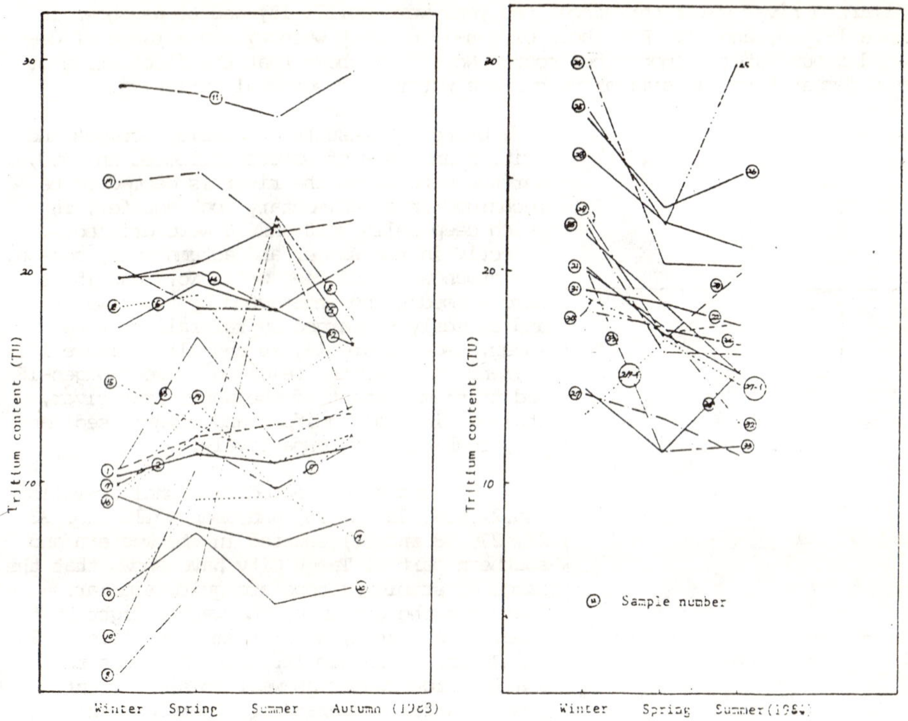

Figure 18. Seasonal variation of tritium content of surface water, shallow and deep seated groundwater in clastic sedimentary rock area/Taegu, from 1983 to 1984.

groundwater at site No.14 was very rapid. Also rainfall or surface water may be directly recharged into shallow alluvial aquifer specially in monsoon season. The tritium levels of sources No.11 and No.3 are higher than 25 TU throughout the year. It is considered that the alluvial aquifer still contains post-thermonuclear water due to the very slow discharge of shallow groundwater from the alluvial aquifer because well No.11 is installed in impermeable colluvial deposits and well No.3 is also located in a marginal zone of fluvial deposit containing some impermeable materials(see Figure 18). Water samples No.8 and No.10 were collected from the deep well located in a newly established industrial site from which large volumes of deep seated groundwater were abstracted and the depth of the wells ranges from 120 to 150 meters. The tritium level of those two samples in the winter of 1983 contained less than 3 TU but that of autumn increased abruptly to more than 22 TU which was similar to that of river water. In addition to the significant change of tritium level, the temperature of the groundwater also changed in total 7°C during spring and summer, which is rare in normal groundwater found in southern Korea(see Figure 18). It indicates that a great amount of stream water during the monsoon season caused to percolate and infiltrate the surface water into the openings of sedimentary rock aquifer and the recharged stream water mixed with the existing deep groundwater and finally formed a part of deep groundwater at the sedimentary rock aquifer specially in monsoon season; that is, surface water directly recharged in sedimentary rock aquifer through the geologic opening's in the area.

The groundwaters from the deep wells whose tritium level is higher or similar to river water throughout the year were No.17, 18, and 6. as shown in Figure 19, 20, and 21. But their oxygen-18 content were within a range of deep seated groundwater. Among them source No.17 had shown that the fluctuation of water temperature was similar to surface water with seasonal variation.

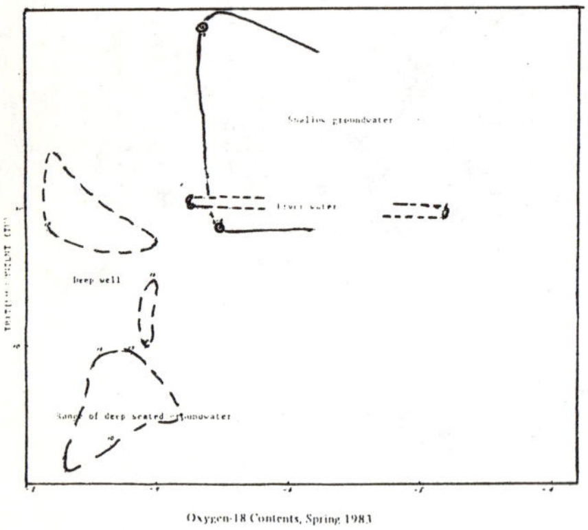

Figure 19. The relation between Oxygen-18 and Tritium content in Clastic sedimentary rock area.

Using the results of stable isotopes and tritium analyses of waters collected in 1983, surface water from the river is deemed to be recharged into sedimentary rock aquifer, in which deep wells No.8 and 10 were drilled, directly in the summer and autumn rainy season. The groundwater of deep wells No.7 and 15 is also seemed to be interacted by river water and directly recharged by rainfall or river water. (see Figure 19, 20 and 21). Figure 19, 20, and 21 shows the relation between oxgen-18 and tritium content of the water from river, shallow alluvial aquifer, and clastic sedimentary rock aquifer respectively.

The samples collected from shallow wells (No.22, 24, 25 and 34) and deep wells (No. 23, 26, 27, 28 and 33) located in the western and southern part of Taegu City have shown that the range of stable isotope are quite similar. Therefore the origin of the waters might be the same. Among them, water from deep wells No.23 and 26 has shown similar tritium levels to shallow groundwater as well as similar variation of water temperature. Therefore the shallow groundwater may infiltrated into the clastic sedimentary rock aquifer which lies underneath.

The water from deep well No.32 having sightly lower tritium levels than that of the nearby reservoir shows that the seasonal fluctuation of both water sources tends to be very similar. The deep well(No.32) drilled near the reservoir is covered with very thin top-soil. It can be interpreted that the reservoir water infiltrates directly into the sedimentary rock aquifer along the bedding planes and fractures. Also the stable isotope content and tritium levels of No.30(deep well) and No.31 (surface water) are very similar as well as similar in the seasonal fluctuctuation of their temperatures and conductivities. It can be predicted that the surface water may percolate directly into the bedrock aquifer.

The rest of the deep seated groundwater in the clastic sedimentary rock always shows lower tritium levels, about 15 TU, and fairly constant stable isotope(^{18}O) content throughout the entire measuring period as well as nearly constant variation of water temperatures from 16°C to 19°C like the other deep seated groundwater. Therefore no direct interaction can be observed between surface water and groundwater specially in central part of Taegu City. The phenomena of infiltration and recharge from river water into alluvial aquifer, from river water into underlain sedimentary rock aquifers, and from shallow groundwater into bedrock aquifer are restricted to the stream channel sides. The rest of the area has shown very low tritium levels and stable isotope content. These rock aquifers do not, therefore, interact much with river and shallow groundwater.

As a summary of the isotope hydrology study in the clastic sedimentary rock area of Taegu City; these are the results;

1) The area recharged directly from the water from overlying alluvial aquifer into bedrock aquifer are limited to the Kumho river and its tributary sides.
2) The shallow groundwater aquifer is directly rechaged by rainfall at sources No.24 and No.25.
3) Water Sources No.8 and No.10 of the sedimentary rock aquifer interacted closely with river water in rainy season.
4) Five kinds a deep seated groundwater are observed in Taegu area; they are,
   a. The bedrock groundwater occuring at the central parts of Taegu City whose tritium levels and stable isotope composition(oxgen-18) are less than 14 TU and-6.8 respectively.
   b. The bedrock groundwater which is directly recharged by surface water is observed at sites No.8, 10, 30 and 32.
   c. The groundwater whose tritium level is between 10 and 15 TU and oxygen-18 more than-7 is observed at sources No.5 and No.7.
   d. The bedrock groundwater reserved in the clastic sedimentary rock distributed along the river channel sides shows slightly higher tritium levels.
   e. The bedrock groundwater showing fairly constant isotope contents and low level of tritium in the central parts of Taegu City indicates very slow recharge of other sources of water into the sedimentary rocks.

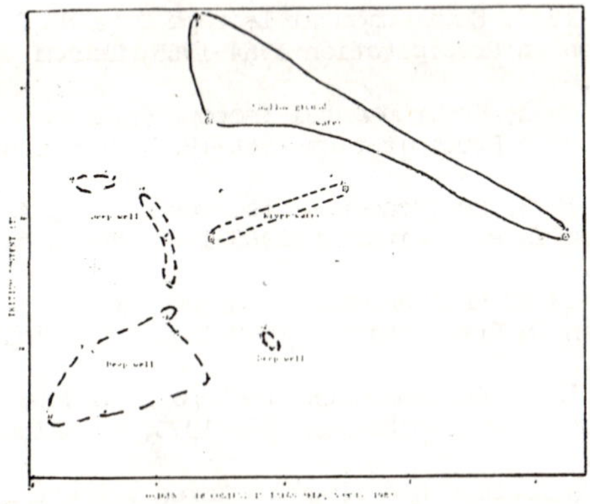

Figure 20. The relation between Oxygen-18 and tritium content in Clastic sedimentary rock area.

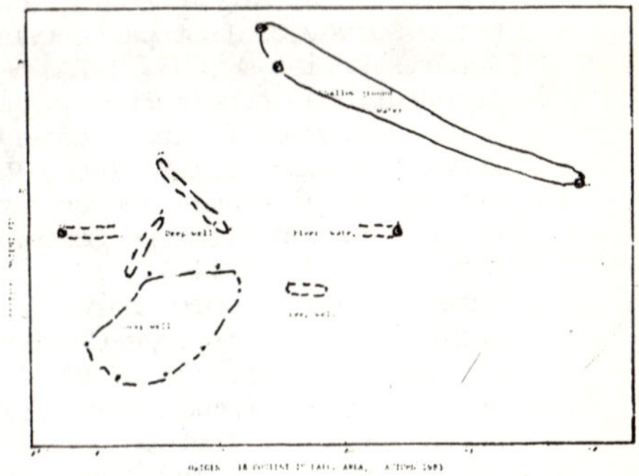

Figure 21. The relation between Oxygen-18 and Tritium content in Clastic sedimentary rock area.

REFERENCES

1. Callahan, J.T., Hahn J.S., et al., 1971, Groundwater of Han River Basin, Ministry of Construction/USBR and USGS, V.3 p.K25-K206.
2. Craig, H., Gordon, L.J., 1965, Deuterium and Oxygen-18 variations in the ocean and the marine atmosphere, Stable isotope in oceanographic studies and paleotemperatures, Proc. Conf. Nuclear geology Spoleto.
3. Eriksson, E., 1965, Deuterium and Oxygen-18 in precipitation and other natural waters, some theoretical considerations, Tellus 17, 498.
4. Dansgaard, W., 1965, The isotope composition of natural water with special reference to the Greenland ice cap; Meddel. Gronland, Kobenhavn, Bd. 165. No.2.
5. Deak, J., 1976, Study of the recharge of deep groundwaters and their connection with shallow groundwaters using environmental isotopes in the Nagykunsag region, Hungary : IAH, Memories V.XI p. 733-744.
6. Ahn, J. S., et al., 1983, Environmental isotope aided studies on river water and groundwater interaction in the Han River Basin : Jour. of Korea Assoc. of Hyro. Scie. V.16, No.4, p. 245-252.
7. International Atomic Energy Agency, 1969, Environmental Isotope Data No.1 : World Survey of Isotope Concentration in Precipitation 1953-1963, Technical Reports Series No.96, IAEA, Vienna.
8. International Atomic Energy Agency, 1970, Environmental Isotope Data No.2 : World Survey of Isotope Concentration in Precipitation 1964-1965, Technical Reports Series No.177, IAEA, Vienna.
9. International Atomic Energy Agency, 1971, Environmental Isotope Data No.3 : World Survey of Isotope Concentration in Precipitation 1966-1967, Technical Reports Series No.129, IAEA, Vienna.
10. International Atomic Energy Agency, 1973, Environmental Isotope Data No.4 : World Survey of Isotope Concentration in Precipitation 1968-1969, Technical Reports Series No.147, IAEA, Vienna.
11. International Atomic Energy Agency, 1975, Environmental Isotope Data No.5 : World Survey of Isotope Concentration in Precipitation 1970-1971, Technical Reports Series No.165, IAEA, Vienna.
12. International Atomic Energy Agency, 1979, Environmental Isotope Data No.6 : World Survey of Isotope Concentration in Precipitation 1972-1975, Technical Reports Series No.192, IAEA, Vienna.
13. Ahan, J. S., et al., 1984, Studies on isotope hydrology of Korea : Fourth Pacific Basin, Nuclear Conf. p. 340-349.
14. Pospisil p., 1974, A contribution of tritium analyses to the study of groundwater flow in the sediments of Zitny Ostrov : IAH memories Vol. XI. p. 601-610.
15. Payne, B.R., 1983, Interaction of surface water with Groundwater, Guide book on Nuclear techniques in Hydrology, No.91, IAEA, p. 319-326.

SOIL MOISTURE STUDIES IN SOILS OF DELHI TERRITORY USING THE RESISTIVITY METER

H. Chandrasekharan, B.R. Yadav, S.K. Tyagi and P.B.S. Sarma
Nuclear Research Laboratory and Water Technology Centre,
Indian Agricultural Research Institute, New Delhi-110 012

ABSTRACT

The utility of resistivity meter for the study of soil moisture in 21 villages of Nangaloi and Najafgarh blocks and IARI farm of Delhi region are presented and discussed in the light of other related soil water properties. Investigations reveal that for a given soil type and for a given range of electrical conductivity of soil saturation extract (EC_e) the bulk soil electrical conductivities (EC_a), measured by the resistivity meter, and the soil moisture contents have linear relationship. It is envisaged that more such investigations in different soil types may yield better understanding on the potentialities of resistivity meter in determining soil moisture contents.

INTRODUCTION

Soil moisture is one of the basic but important parameters used to estimate groundwater recharge, to schedule irrigation practices in agricultural fields and to solve some of the engineering problems. Present literature survey indicates a number of methods such as gravity, microwave (Srivastava et al, 1978), infra red (Athavale et al, 1978), neutron moderation (IAEA, 1968), gamma scattering (Danfors, 1975), thermal methods (Danfors, 1975), etc., to study the soil moisture qualitatively and quantitatively. The four electrode probe using the resistivity meter for such studies have been attempted by very few workers like Kirkham and Taylor (1950), Gupta and Hanks (1972) and recently by Goyal et al (1986); but still at the research level. This paper forms the initial part of the investigation to estimate/study the soil moisture using a D.C. Earth Resistivity Meter in sandy loam soils of Delhi region.

MATERIALS AND METHODS

Field investigations in 21 sites of Nangaloi and Najafgarh blocks of Delhi region (Fig. 1) were carriedout to study the soil moisture using the resistivity meter. The resistivity of each depth interval was measured using four stainless stell electrodes placed in Wenner configuration with the interelectrode spacings of 50, 100 and 150 cm. These measurements provide the conductivities (EC_a) of the soil matrix in 0-50, 0-100 and 0-150 cm depth ranges. The EC_a of of discrete layers, namely, 50-100 cm and 100-150 cm depth ranges were calculated using the equation (Halvorson and Rhoades, 1974)

$$EC_{a_i - a_{i-1}} = \frac{[\{EC_{(0 - a_i)} \cdot a_i\} - \{EC_{(0 - a_{i-1})} \cdot a_{i-1}\}]}{a_i - a_{i-1}} \qquad (1)$$

where a_i represents the depth of sampling and a_{i-1} represents the earlier depth of sampling. This equation is based on the assumption that the depth to which conductivities is measured is equal to the inter-electrode spacing and the soil resistances of a sequence of soil layers is assumed to behave like resistors in

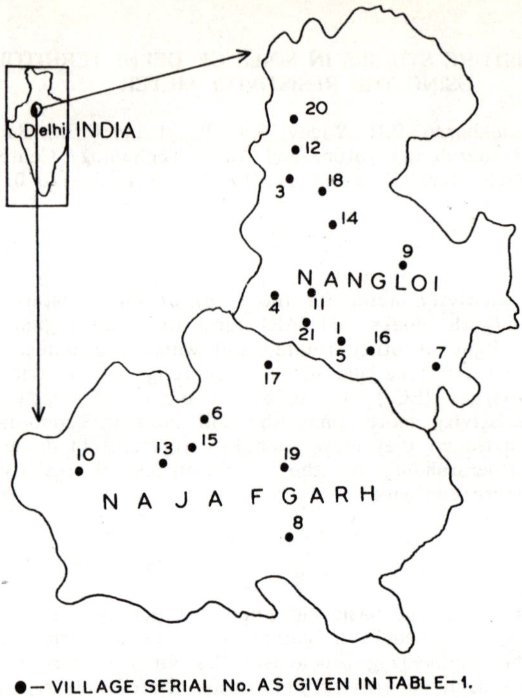

Fig. 1. Location of soil resistivity investigation.

parallel (Fig. 2) (Barnes, 1954). At a location at IARI farm, the soil resistivity investigations were carriedout at 30 cm depth interval upto 300 cm from ground zero. The EC_a of discrete layers were determined using equation 1.

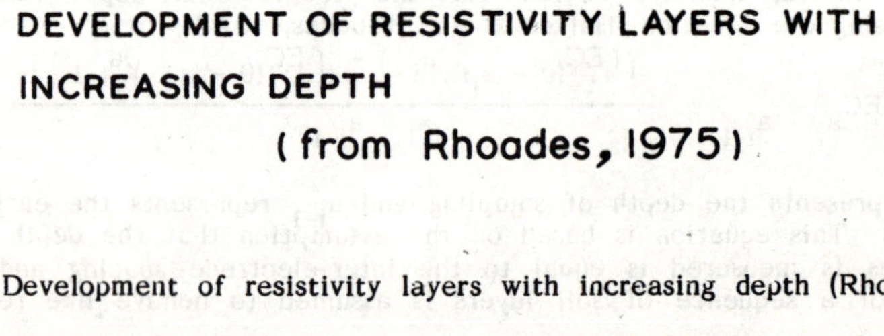

Fig. 2. Development of resistivity layers with increasing depth (Rhoades, 1975).

In addition to the above investigation, soil samples at corresponding intervals were collected from all locations and soil moisture contents (θ, in volume percentage), mechanical compositions of soils and electrical conductivity of saturation extract (EC_a) were determined by conventional methods. All data are presented in Tables 1 and 2.

RESULTS AND DISCUSSIONS

Simple plot of EC_a versus θ for all field values do not indicate any type of relationship between the two. This is expected because the EC_a values depend not only on the soil moisture content but also on the soil type, temperature, EC_e, etc. (Rhoades, 1975). On this basis, present investigations were carriedout on sandy loam soils during the summer months of 1987. EC_e values of soils were found to fall in three groups (Table 1); namely, 0-1 mmhos/cm, 1-2 mmhos/cm and > 2 mmhos/cm. The last group has only three samples and hence they are not considered in the present discussions. The plot of EC_a versus θ of the two ranges of EC_e yield the relationships (Fig. 3)

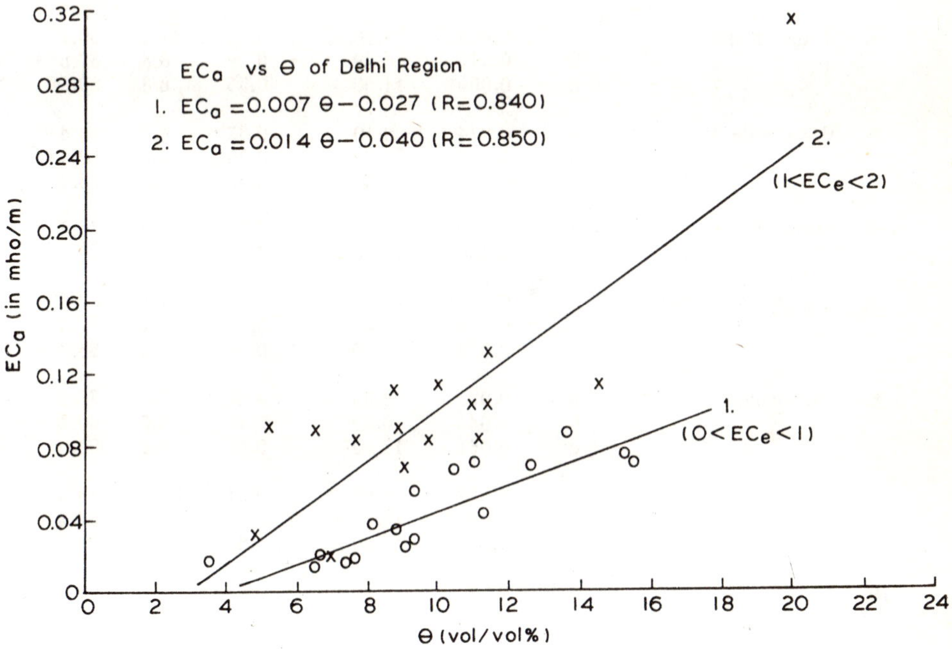

Fig. 3. Relationship between 'EC_a' and 'θ' for the locations in Nangloi and Najafgarh blocks, Delhi.

Table 1. Soil Moisture Studies Using Earth Resistivity Meter.

S.No.	Name of the village	Soil depth	EC_a (mho/m)	Moisture content (vol/vol %)	EC_e (mmho/cm)	Clay %	Sand %
1.	Bakarwala I	*a	0.019	6.90	1.74	7.3	65.4
		b	0.017	7.59	0.24	7.8	63.6
		c	0.087	2.60	1.80	10.8	63.6
2.	Ramkhera and Madanpurdabas	a	0.015	7.40	0.27	10.8	59.6
		c	0.072	9.20	0.87	12.0	60.4
3.	Jointi	a	0.016	3.50	0.29	25.6	72.6
		b	0.058	7.29	0.28	27.6	65.6
		c	0.514	11.00	0.30	25.8	68.2
4.	Tikari Kalan	a	0.024	9.10	0.35	11.8	60.1
		b	0.012	6.49	0.29	8.8	57.6
		c	0.069	11.00	0.55	8.8	59.6
5.	Bakarwala II	a	0.024	9.10	0.35	8.4	63.4
		b	0.012	6.49	0.29	6.6	62.4
		c	0.069	11.00	0.55	6.6	68.4
6.	Mitraon	a	0.020	6.60	0.39	3.3	75.2
		c	0.086	13.60	0.40		
7.	Nangloi Sahrawati	a	0.074	15.20	0.66	7.1	67.6
		b	0.054	23.09	0.53	5.4	67.6
		c	0.034	9.80	0.44	5.4	65.6
8.	Chhawala	a	0.028	9.30	0.46	5.3	70.9
		b	0.044	6.80	0.43	8.6	60.4
		c	0.051	17.60	0.96	8.2	60.8
9.	Kotla	a	0.027	5.80	0.51	4.3	73.4
		b	0.067	12.62	0.67	6.0	64.4
		c	0.182	14.50	0.81		
10.	Kajipur	a	0.55	9.30	0.65	7.8	66.6
		b	0.93	14.06	1.01	6.0	63.4
11.	Geora	a	0.056	16.50	0.76	28.8	63.2
		b	0.042	11.30	0.73	31.8	60.2
12.	Tatesher	a	0.082	7.60	1.40	29.6	63.4
		b	0.066	10.36	0.77	25.8	68.2

...contd.

Table 1 contd.

S.No.	Name of the village	Soil depth	EC_a (mho/m)	Moisture content (vol/vol %)	EC_e (mmho/cm)	Clay %	Sand %
13.	Kheredabar	c	0.037	8.10	0.86	4.6	64.4
14.	Kanjhawala (C)	a	0.130	11.40	1.30	34.3	54.4
		b	0.148	15.34	2.93	29.6	55.6
		c	0.517	15.40	3.20	27.6	50.6
15.	Sureda	a	0.112	10.00	1.65	8.1	66.5
		c	0.067	9.00	1.40	6.6	62.4
16.	Rauhela	a	0.030	4.80	1.48	5.6	65.6
		b	0.070	15.50	0.82	10.2	58.8
		c	0.311	20.00	1.50	8.8	59.6
17.	Dichaunkalan	a	0.088	6.50	1.70	6.8	73.6
		b	0.088	8.87	1.60	5.2	72.4
		c	0.082	11.10	1.20	5.2	70.4
18.	Ladpur	a	0.086	5.60	2.30	36.3	52.0
		b	0.110	8.75	1.70	41.6	45.6
		c	0.395	12.00	1.10	47.6	39.1
19.	Dinpur	a	0.102	11.40	1.10	7.5	62.9
		b	0.110	22.42	1.01	12.8	55.6
		c	0.103	11.00	1.40	14.8	55.6
20.	Punjabkhore	a	0.090	5.20	1.07	16.0	54.6
		b	0.082	9.74	1.07	10.8	61.6
		c	0.272	11.10	1.20	12.8	53.6
21.	Hirankudhana	a	0.087	16.00	1.64	14.8	58.6
		b	0.113	14.52	1.06	11.8	49.6
		c	0.400	14.80	0.20	16.8	51.0

* a : 0 - 50 cm
b : 50 - 100 cm
c : 100 - 150 cm

Table 2. Soil Moisture Studies of IARI Farm Using Earth Resistivity Meter.

S.No.	Soil depth (cm)	EC_a (mho/m)	Moisture content (vol/vol %)	EC_e (mmho/cm)	Clay %	Sand %
1.	0 - 30	0.009	1.70	0.34	28.32	64.68
2.	30 - 60	0.013	3.56	0.14	34.32	59.68
3.	60 - 90	0.023	3.51	0.14	29.32	63.68
4.	90 - 120	0.011	3.52	0.14	26.32	62.24
5.	120 - 150	0.029	7.12	0.14	27.32	58.74
6.	150 - 180	0.029	10.57	0.14	30.32	57.68
7.	180 - 210	0.019	10.70	0.17	28.32	59.68
8.	210 - 240	0.027	11.98	0.19	28.32	63.68
9.	240 - 270	0.038	14.02	0.19	26.32	63.68
10.	270 - 300	0.022	10.50	0.14	28.32	63.68

$$EC_a = 0.0070 - 0.027 \text{ for } 0 < EC_e < 1 \quad (R = 0.84) \tag{2}$$

and

$$EC_a = 0.014\,\theta - 0.014 \text{ for } 1 < EC_e < 2 \quad (R = 0.85). \tag{3}$$

Some of the EC_a values lying away from the regression lines are omitted. Their deviations may be attributed to reasons other than EC_e and moisture contents these are being studied. One of the reasons is that high EC_a values without high salinity is indicated in the high exchangeable sodium levels, that is, sodic soils (Rhoades, 1975). The studies conducted by Rhoades et al (1976) on the effects of water content on EC_a values yield the relation

$$EC_a = (EC_W).\theta.T + EC_S \tag{4}$$

where EC_W is the electrical conductivity of the soil water; EC_S is the surface conductance. EC_W and EC_S can be represented by EC_e and soil compositions. According to Gupta and Hanks (1972)

$$\left(\frac{EC_a}{EC_e}\right) = a\theta + b \tag{5}$$

where a and b are the empirical constants. In our study none of the above relations (equations 4 and 5) hold good for the field data. Obviously more such studies involving a large number of samplings need to be conducted for the possible relations between EC_a, θ, EC_e, etc., which may be used for any soil type in the unsaturated zone. The analyses of the EC_a and soil moisture data of depthwise samples for a location at IARI farm gives a fairly good correlation. The EC_a values for all samples lie in the range of 0-1 mmho/cm. The equation is

$$EC_a = 0.002\,\theta + 0.009 \quad (R = 0.79) \tag{6}$$

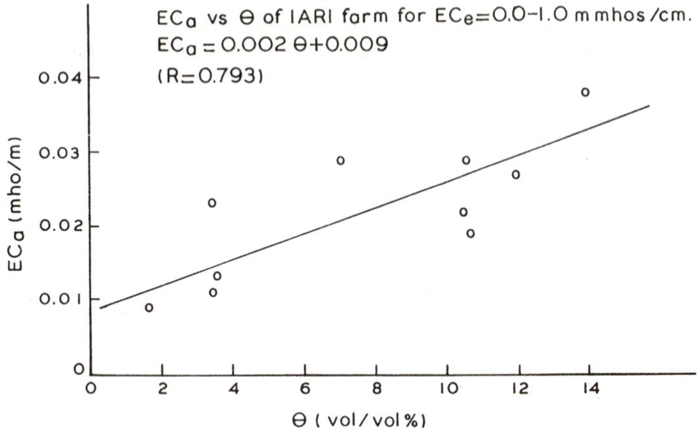

Fig. 4. Relationship between 'EC_a' and 'θ' for the location at IARI farm.

The moisture contents for this area determined by equation 3 do not match with the field data. This is mainly due to the difference in the soil compositions (Table 1). The soil compositions normally vary from place to place. It is not advisable to extrapolate the equation developed in one place to the other although the range of EC_e values may be the same. However, the present investigation indicates the possibilities of using the resistivity methods for determining soil moisture contents and has yielded positive results. Further studies on these lines will be useful for practical adoption of this technique on a large scale.

CONCLUSIONS

The present study gives fairly good correlation between the bulk soil electrical conductivity (EC_a) determined by the resistivity meter and the soil moisture content (vol/vol %) for EC_e ranges 0-1 and 1-2 mmho/cm under field conditions. The relationships are linear in the three cases studied.

ACKNOWLEDGEMENTS

Authors are very thankful to Dr. P.N. Tiwari, Project Director, NRL (IARI) for encouragements. The help rendered by our colleagues Shri P. Mookerjee, Dr. S.P.S. Tomar, Shri Badan Singh, Shri Om Pal Singh are duly acknowledged.

REFERENCES

Athavale, R.N., Murti, C.S., and Chand, R., 1978. Exploration and management studies of groundwater resources. Technical Report No. GH-9-NH.1, p. 29.

Barnes, H.E., 1954. Electrical subsurface exploration simplified. Roads and Streets, 140, p. 81-84.

Danfors, E. (Ed.), 1975. Soil water distribution - A state of art report. Nordic IHD Report No. 9, p. 167.

Goyal, V.C. Sinvhal, H. and Sinvhal, A., 1986. Use of resistivity technique in soil moisture studies - A preliminary evaluation. Hydrology (Journal of IAH Roorkee), V, IX No. 4, p. 48-55.

Gupta, V.C., and Hanks, R.J., 1972. Influence of water content on electrical conductivity of the soil. Soil Sci. Soc. Amer. Proc. V 36, No. 6, p. 855-857.

Halvorson, A.D., and Rhoades, 1974. Assessing soil salinity and identifying potential saline seep areas with field soil resistance measurements. Soil Sci. Soc. Amer. Proc. vol. 38, p. 576-581.

Kirkham, D., and Taylor, G.S., 1950. Some tests on a four electrode probe for soil moisture measurements. Soil Sci. Soc. Amer. Proc. V 14, p. 42-46.

Rhoades, J.D., 1975. Measuring, mapping and monitoring field salinity and water table depths with soil resistance measurements. Proc. 'Expert consultation on progonis of salinity and alkalinity', Rome 3-6, June, pp. 35.

Rhoades, J.D., Raats, P.A.C., and Prathar, R.J., 1976. Effects of liquid phase electrical conductivity on bulk soil electrical conductivity. Soil Sci. Soc. Amer. J. vol. 40, p. 651-655.

Srivastava, K.P., Jaswant Sing, Sanyal, G.S., and Usha K. Loomba, 1978. Microwave technique for soil moisture determination. Journal of Agriculture Engineering, vol. XV, No. 4, p. 181-188.

ESTIMATION OF RECHARGE AND POLLUTANT TRANSFER TO GROUNDWATER BY MODELLING OF UNSATURATED ZONE

Deepak Kashyap*, K.M.Mohan Rao** and Sulekha Gupta***

ABSTRACT

An attempt has been made to present basic equations governing flow and mass transport through the unsaturated zone and various possible strategies for their solutions. A case study is included. Further, the necessary enhancement of generally prevalent groundwater data collection programs, for implementing such models, is highlighted.

INTRODUCTION

The geological formation extending from ground to watertable is generally known as unsaturated zone since only a part of the pore space in this zone is filled with water. The partial saturation is associated with a subatmospheric pore pressure, the magnitude of which decreases as the soil moisture increases reaching a zero value at the saturation. The hydraulic conductivity of an unsaturated medium (generally termed as capillary conductivity) is not an intrinsic property of the medium and fluid, but increases with the increasing soil moisture.

The unsaturated zone acts as a conduct for vertical transfer of water and pollutants to watertable. It also accommodates the root zone of agricultural crops. Thus, the processes of infiltration, evapotranspiration, recharge, pollutant transfer to watertable from ground or below (e.g., from landfills) are essentially governed by the unsaturated zone. However, most of the research efforts in groundwater hydrology and hydraulics have been traditionally directed towards the saturated zone i.e., the zone below the watertable. The inherent snag in such priorities becomes evident when it is realized that an important input

---

* Reader
** Ex-Research Scholar
*** Research Scholar

Department of Hydrology
University of Roorkee,
Roorkee, U.P. 247 667, India

requirement for saturated flow models is the recharge from rainfall and applied irrigation at the watertable; whereas the data actually available are the rainfall and applied irrigation at the ground. An empirical and irrational transformation of the inputs at the ground into inputs at the watertable by means of some coefficients or mere watertable fluctuations, may render all the sophistication of saturated flow models futile.

MODELLING OF FLOW

The partial differential equation governing unsteady state flow of water in the unsaturated zone can be written as follows (Philip 1969)

$$\frac{\delta\theta}{\delta t} = \frac{\delta}{\delta x}[K\frac{\delta}{\delta x}(-h)] + \frac{\delta}{\delta y}[K\frac{\delta}{\delta y}(-h)] + \frac{\delta}{\delta z}[K\frac{\delta}{\delta z}(-h+z)] - S \quad ....(1)$$

where, x and y are the coordinates along principle permeability directions in horizontal plane; z is the coordinate along vertical direction (+ve upwards); t is time; θ, a function of space and time, is the volumetric moisture content, h is the associated subatmospheric (capillary pressure); K, a function saturated hydraulic conductivity (K_s), porosity (ϕ) and θ,is capillary conductivity; and S is a sink term representing space and time dependent volume of water abstracted per unit volume of soil per unit time.

The first and second term on the right hand side of eqn.(1) represent the horizontal flows and the third term represents the vertical flow. Absence of gravity in the horizontal plane generally renders the flow grdients in the horizontal plane to be small enough to be neglected in comparison to the flow gradients in the vertical direction. Further assuming the sink term to consist only of evapotranspiration, eqn.(1) can be simplified as,

$$\frac{\delta\theta}{\delta t} = \frac{\delta}{\delta z}[K\frac{\delta}{\delta z}(-h+z)] - E \quad .....(2)$$

Equation (2) can be rewritten in the following two univariate forms,

(i) $$C_u \frac{\delta h}{\delta t} = \frac{\delta}{\delta z}[K\frac{\delta}{\delta z}(-h+z)] - E \quad .....(3a)$$

where C_u known as specific moisture capacity equals d /dh.

(ii) $$\frac{\delta\theta}{\delta t} = \frac{\delta}{\delta z}[-D\frac{\delta\theta}{\delta z} +K] - E \quad .....(3b)$$

where, $D(L^2\ T^{-1})$, known as moisture diffusivity, equals (K dh/dθ)

Equation (3a) and (3b) are respectively known as the head form and the moisture content form of Richard's equation. These are non-linear second order partial differential equations in h and θ respectively. The non-linearity arises due to dependence of C_u,D and K on θ (and h). Further, the evapotranspiration may also be a non-linear function of h, though it is usually considered to be a linear function of θ. A solution of the above equations with appropriate boundary conditions can give a distribution of both capillary head (h) and moisture content (θ) in space and time.

Characteristics

The basic requirement of such a solution is to know the soil characteristics i.e.h vs θ and K vs h or K vs θ relationships. The characteristics show the phenomenon of hysterisis. The most reliable estimates of these characteristics can be obtained by generating adequate discrete point data through actual experimentation. However, such facilities may not be always available. Under such conditions these can be estimated by employing empirical or semi-empirical relations proposed by different investigators e.g., Gupta and Larson (1979), Ghosh (1980).

For K(θ) characteristic, the following equation is commonly used (Brooks and Corey, 1964).

$$K(\theta) = Ks \left(\frac{\theta-\theta r}{\phi-\theta r}\right)^4 \quad \theta \geq \theta r$$

$$\theta = 0 \qquad \theta < \theta r \qquad \ldots\ldots(4)$$

where, K_s, ϕ and θr are saturated hydraulic conductivity, porosity and residual moisture content respectively.

Rao(1986) has presented the following relation for h( ) characteristic

$$\theta = \phi - \frac{\theta r}{h_b} h \qquad 0 \leq h \leq h_b$$

$$\ldots\ldots(5)$$

$$= \exp[\ell n \frac{\phi-2\theta r}{h_b} h] + \theta r \quad h > h_b$$

Method of Solution

The solution techniques for Richard's Equation can be grouped under two heads i.e., analytical and numerical. As yet, no analytical solution is available; however, a quasi-analytical solution has been proposed by Philip (1969).

In case of numerical solutions, both finite differece and finite element techniques can be used for solving Richard´s equation.

Boundary Conditions: For arriving at a unique solution of the differential equation, it is necessary to assign boundary conditions at the ground and watertable. The boundary condition at the ground may be either Neuman type (when entire input at the ground infilters) or Dirichlet type(when ponding occurs over the ground). At the watertable the boundary condition is always Dirichlet type($\theta = \phi$ or h =O).

Finite Difference Solution: A finite difference form of eqn. (3), written for each node in the domain under consideration, leads to a system of non-linear simultaneous equations.To carry out the solution of this system of equations by an iterative scheme, a hint solution is required. Further, the system of non-linear simultaneous equations will not have a unique solution. Therefore, any inappropriate specification of hint solution may lead the iterative scheme to converge to an unrealistic solution. Thus, the task of specifying an appropriate hint solution,which is again dependent on the particular problem being solved, may be rather difficult. To overcome this problem the system can be linearised in the following two ways.

Explicit Linearization: In this method, the terms causing non-linearity, namely K, C and E are evaluated as per the known values of h (or) at the begining of a time step. The resulting system of equations will be linear in nature, having a tridiagonal coefficient matrix. To evaluate the unknowns, the tridiagonal system of equations can be solved by Thomas algorithm (Remson etal 1971). The algorithm is extremely stable with respect to round off errors.

Picard Iteration Method: In this method (Remson etal,1971) a hint solution is specified.Let the hint solution be $h^{(o)}$ and the values of K, C and E evaluated using $h^{(o)}$ be $K^{(o)}$, $C^{(o)}$ and $E^{(o)}$. These values substituted in eqn. (3) and written in a finite difference form results in a system of linear simultaneous equations. The coefficient matrix of the system will be tridiagonal and can be solved using Thomas algorithm. Solution of the system yields the value of $h^{(1)}$. Using $h^{(1)}$; $K^{(1)}$, $C^{(1)}$ and $E^{(1)}$ can be evaluated. Subsequently, $h^{(2)}$ is obtained on the same lines. As the iteration index approaches infinity, h is expected to converge to the true solution of the system of non-linear simultaneous equations. To avoid too large a number of iterations, which means more C.P.U. time, the iterations are stopped, when the following check is satisfied,

$$|h^{(\ell+1)} - h^{(\ell)}| < \varepsilon \qquad .....(6)$$

where, ε_o : prescribed tolerable error (small positive value).

Development of a Finite-differences Model

A numerical model was developed by the second author (Rao, 1986), as a part of his doctoral research, for solving head form of Richard´s equation by finite differences. The equation is initially linearized by explicit linearization. The solution so obtained is used as hint solution for implicit linearization. The model permits automatic identification and assignment of ground surface boundary condition, under no overland flow condition. Thus, the model identifies the ponded condition and computes the ponded depth. The model accounts for soil moisture dependent evapotranspiration from a time variant root zone.

The model has been employed for estimating return flows (recharge) from applied irrigation for rice-wheat combination on loamy sand (Kashyap, 1987). The results are as follows.

Month	Rainfall	Applied Irrigation	Infiltra- tion	PET_c^+	ET	Change storage	Outflow
June	43.20	200.00	243.20	112.65	63.27	178.39	5.64
July	374.20	150.00	524.20	163.00	122.25	203.17	204.46
Aug.	142.20	150.00	292.20	142.10	101.96	-140.16	334.84
Sept.	130.10	150.00	280.10	142.20	102.38	- 20.35	202.64
Oct.	9.60	0.0	9.60	55.65	24.26	- 80.43	66.70
Total	699.30	650.00	1349.30	615.60	414.12	140.62	814.28

+ Taken as PET ref.
Note: All the components are in mm.

Table 1 - ESTIMATION OF RETURNFLOWS (RICE CROP)

Month	Rainfall	Applied Irrigat- ion	Infiltra- tion	PETc	Evapot- anspi- ration	Change in storage	Return flow
Nov.	1.0	175.0	176.0	19.47	19.42	130.50	26.08
Dec.	0.0	150.0	150.0	32.55	32.55	92.35	25.10
Jan.	43.50	150.0	193.50	59.83	59.83	114.22	19.45
Feb.	13.0	75.0	88.0	52.64	52.64	- 86.35	121.71
Mar.	0.0	0.0	0.0	13.80	13.80	- 90.13	76.33
Total	57.50	550.0	607.50	178.29	178.24	160.59	268.67

Note: All the components are in mm.

Table 2 - ESTIMATION OF RETURNFLOWS (WHEAT CROP)

The model has also been employed for generating irrigation schedules for various predecided soil-moisture depletion levels for sugarcane crop. This leads to theoretical generation of crop-response function (Kashyap, 1987).

MODELLING OF MASS TRANSPORT

The partial differential equation governing convective-diffusive transport of conservative dissolved pollutants through an isotropic porous medium can be written as follows (Bear, 1979),

$$\frac{\delta}{\delta t}(C\theta) = \frac{\delta}{\delta x}(\theta D \frac{\delta C}{\delta x}) + \frac{\delta}{\delta y}(\theta D \frac{\delta C}{\delta y}) + \frac{\delta}{\delta z}(\theta D \frac{\delta C}{\delta z}) - \frac{\delta}{\delta x}(q_x C) - \frac{\delta}{\delta y}(q_y C) - \frac{\delta}{\delta z}(q_z C) \quad ......(7)$$

where, C a function of space and time, is solute concentration of the soil-solution; q, a function of space and time, is volumetric flux of solution per unit area or flow velocity and D, a function of soil characteristics, soil moisture and velocity, is combined diffusion dispersion coefficient.

Considering the flow in vertical direction to be dominant, as is in the case of unsaturated flow, eqn.(7) can be simplified as,

$$\frac{\delta}{\delta t}(C\theta) = \frac{\delta}{\delta z}(\theta D \frac{\delta C}{\delta z}) - \frac{\delta}{\delta z}(q_z C) \quad ......(8)$$

In eqn. (8), the two terms on the right hand side, represent the components of diffusive transport and convective transport respectively. Again, the diffusion-dispersion coefficient D can be written as, (Bear,1979)

$$D(V,\theta) = D_h(V) + D_p(\theta) \quad (9)$$

where, $D_h(V)$ is the mechanical dispersive coefficient and $D_p(\theta)$ is the molecular diffusion coefficient.

Further, the functional relationship between D_p and θ can be written as follows (Kemper and Van Schalk, 1966; Bresler,1973; David Russo, 1988)

$$D_p(\theta) = D_o\, a\, e^{b\theta} \quad (10)$$

where, D_o is molecular diffusion coefficient in a free water system; a and b are empirical constants, characterizing the soil.

The mechanical dispersion coefficient D_h in a given porous material depends on the average flow velocity and can be taken to be proportional to the first power of average flow velocity (Ogata,1970; Bresler,1973a; David Russo,1988)

$$D_h(V) = \lambda |V| \qquad (II)$$

where, V is average interstitial flow velocity and λ is dispersivity of the porous medium.

Coupled Solution

To solve eqn.(8) for concentration distribution in depth and time, an additional equation is required since the equation comprises of two unknowns C and q. Recognising that q_x, q_y and q_z can be described in terms of h ($q_x = -K\,\delta h/\delta x$, $q_y = -K\,\delta h/\delta y$ and $q_z = -K\,\delta h/\delta z$) equation 8 coupled with Richard´s equation can be solved for obtaining the desired concentration distribution and estimating the pollutant transfer to water table. Dirichlet´s boundary condition at the watertable and ground can be assigned, knowing the concentration of pollutant in the groundwater and at the ground respectively.

Method of Solution

Equation (8) is a second order non-linear partial differential equation, an analytical solution of which would be rather far-reaching. Use of finite difference methods for solving eqn.(8), introduces an artificial numerical dispersion in the solution, which may lead to erroneous results, depending on the scale of the problem being solved. Khaleel etal(1985) have overcome a similar problem while solving for mass transport in saturated media, by resorting to method of characteristics. The third author, as a part of her ongoing doctoral research, has developed a numerical model, based upon the same method, for solving equations (3a) and 8. The output includes, among others, the concentration profiles and the pollutant transfer to watertable at differenmt times.

DATA REQUIREMENT

Our prevalent data collection programs are oriented towards saturated flow modelling. Thus, for implementing unsaturated zone models, the programs have to be enhanced as follows.

The foremost data required for such a modelling are h(θ) and K(θ) characteristics. Experimental determination of these characteristics at large enough number of sites may not be possible due to elaborate nature of the experimental procedures. However, these can be estimated theoretically, knowing a few hydrauilic/mechanical properties of soil i.e., porosity, bubbling pressure, residual moisture content, saturated vertical conductivity, grain size distribution, bulk density, organic matter content etc. Thus, these soil properties should be determined by standard tests.Subsequently, the evolved characteristics should be cross checked by direct experimentation at a few representative sites.

Soil moisture profiles should be monitored periodically (on the same lines as the measurement of watertable/piezometric heads). Such data are necessary for calibration of the models as well as for assigning initial conditions.

The variation of evapotranspiration with soil moisture should be determined for different crop and soil conditions.

REFERENCES

Bear, J., 1979. Hydraulics of groundwater. McGraw Hill Inc.,p.233 and p.244.

Bresler, E., 1973. Simultaneous transport of solute and water under transient unsaturated flow conditions.Water Resour. Res., v.9, No.4,p.975-986.

Brooks, R.H., and Corey, A.T., 1964. Hydraulic properties of porous media. Hydrology papers- Colorado State University, Fort Collins,Colorado, p.27.

Russo, D., 1988. Numerical analysis of the nonsteady transport of interacting solutes through unsaturated soil 1 Homogeneous systems.Water Resour. Res., v.24, No.2, p. 271-284.

Ghosh, R.K., 1980. Estimation of soil moisture characteristics from mechanical properties of soil. Soil Sci. 130(2), p60-63.

Gupta, S.C. and Larson, W.E., 1979. Estimating soil water retention characteristics from particle size distribution, organic matter present and bulk density. Water Res. Res. 15(6), p.1633-1635.

Kashyap, D., 1987. Conjunctive use modelling for ground and surface water in Eastern Yamuna Canal Command area. Completion report submited to the CBIP (Govt. of India).

Khaleel, R., and Redell, D.L., March 1985. Miscible displacement in porous media : MOC solution, Journ. Irrigation and drainage Engineering. V.III, No.1.

Kemper, W.D., and van schaik, J.C., 1966. Diffusion of salts in clay-water systems. Soil Sci. Soc. Amer. Proc., V.30, p.534-540.

Ogata, A., 1970. Theory of dispersion in a granular medium. U.S. Goel Surv. Prof. Pap. 411-I.

Philip, J.R., 1969. Theory of infilteration. In : Advances in Hydrosciences. Ed. Ven Te Chow, Academic press, New York, p.215-296.

Pinder, G.F. and Cooper Jr. H.H., 1970. A numerical technique for calculating the transient position of the salt water front. Water Resour. Res., V.6, No.3, p.875-882.

Remson, I., Hornberger, G.M., and Molz., F.J., 1971. Numerical methods in subsurface hydrology. Wiley Interscience., p.389.

Rao, K.M.Mohan, 1976. Hydrological response of unsatunated zone up to water table. Ph.D. Thesis, University of Roorkee.

Hoover, G.J. and Grant, W.J., 1983. Numercial fitting of the Gardner equation to hydraulic conductivity and water retention data. Trans. ASAE 26(5), P.1401-1408.

A FUNCTIONAL RELATION FOR CAPILLARY SUCTION.VS.MOISTURE CONTENT

K.M.MOHAN RAO   DEEPAK KASHYAP   SATISH CHANDRA

Dept. of Hydrology
University of Roorkee
Roorkee (India)

National institute of
Hydrology
Roorkee (India)

ABSTRACT

A functional relation for the capillary suction.vs.moisture content, that can be adopted in the Richards' equation , is presented.The relation requires the soil properties-- Residual soil moisture content,Porosity and Air entry value.The relation was satisfactory, when adopted , in a field evidence simulation exercise conducted by the authors,earlier.

INTRODUCTION

The Richards equation (Philip,1969) is a powerful governing differential equation for mathematical modelling of flow in the unsaturated zone,with proven ability to simulate the field evidences as well(Mohan Rao et al,1986).However,one of the important input requirements for a unique solution of the equation is the specification of capillary suction.vs.moisture content relation (h-θ relation or h characteristic).A finite number of discrete point data on the characteristic,obtained experimentally,may be used for interpolation at non-tabulated points and for differentiation to determine the specific moisture capacity and moisture diffusivity.However,it is well known that the numerical differentiation may involve serious errors , especially when the discrete point data are not error free.Thus,the data generation requires , on one hand , elaborate experimentation and on the other,efficient numerical algorithms for interpolation and differentiation.It would certainly be desirable to adopt this way,if the said facilities are available.However,such facilities may not always be available. The present proposal is an attempt to mitigate the problem to a reasonable extent.

LITERATURE REVIEW

Brooks et al(1964),Taylor et al(1969),Clapp(1978) and few others proposed different kinds of empirical relations for the characteristic.Gupta et al(1979),Ghosh et al(1980) proposed methods for the evaluation of the characteristic , using parameters of the soil.Few investigators (eg:Hoover et al,1983) used least squares calibration on the available relations.A well sustained criticism is not possible as the characteristic forms a sub-model in the model describing the flow through the unsaturated zone and the success of the main model implicitly means the acceptability of the sub-model(the functional relation).

THE PROPOSED RELATION

Figure (1) shows the plot of a h characteristic, using the experimentally determined discrete point data , by Brooks et al (1964). It can be seen from the figure that with h (capillary suction head) increasing from zero to h_b (air entry value), θ (moisture content) reduces marginally from Ø(porosity) to (Ø-δ), where δ is a small value. For values of h greater than h_b , the variation in θ resembles the classical exponential decay. From the inspection of plots of h characteristics , using the experimentally determined discrete h and θ values, the δ may be found to be nearly equal to θ_r (residual moisture content), especially in case of coarse soils like sand. So, the general form of the characteristic may be approximated by a piecewise functional relation. The variation of θ in h=0 to h=h_b may be approximated by a relation, adopting θ_r for δ .

$$\theta = Ø - \theta_r . h/h_b \qquad 0 \leqslant h \leqslant h_b \qquad (1a)$$

The variation of θ , in values of h greater than h_b may be approximated by the classical exponential decay relation.

$$\theta = \exp(-\alpha h) + \theta_r \qquad \alpha > 0 \ \& \ h > h_b \qquad (1b)$$

The exponential decay relation ensures that at higher values of h, θ tends to θ_r . The constant α can be determined from the condition that θ is unique (since, the flow is assumed to be non-hysteritic) at h_b

$$\alpha = -\ln(Ø - 2\theta_r)/h_b \qquad (1bb)$$

Thus, the piecewise approximation will be:

$$\theta = Ø - \theta_r . h/h_b \qquad 0 \leqslant h \leqslant h_b \qquad (1c)$$

$$= \exp[\ln(Ø - 2\theta_r).h/h_b] + \theta_r \qquad h \geqslant h_b \qquad (1cc)$$

The piecewise functional approximation given by equation requires the well defined soil properties Ø , θ_r and h_b . The relation is amenable for differentiation and interpolation.

Figure (1) shows the plot of a reported experimental h and θ discrete point data by small circles. The firm-line curve in the figure has been plotted using the proposed approximation and with the reported experimentally determined Ø , θ_r and h_b . Comparison of the plots is satisfactory.

Figure (2) shows the simulated and observed moisture profiles of the field evidence simulation exercise conducted by the authors Mohan Rao et al, 1986) earlier, with Richards equation as the governing equation and the h characteristic adopted as proposed presently. The simulation is satisfactory.

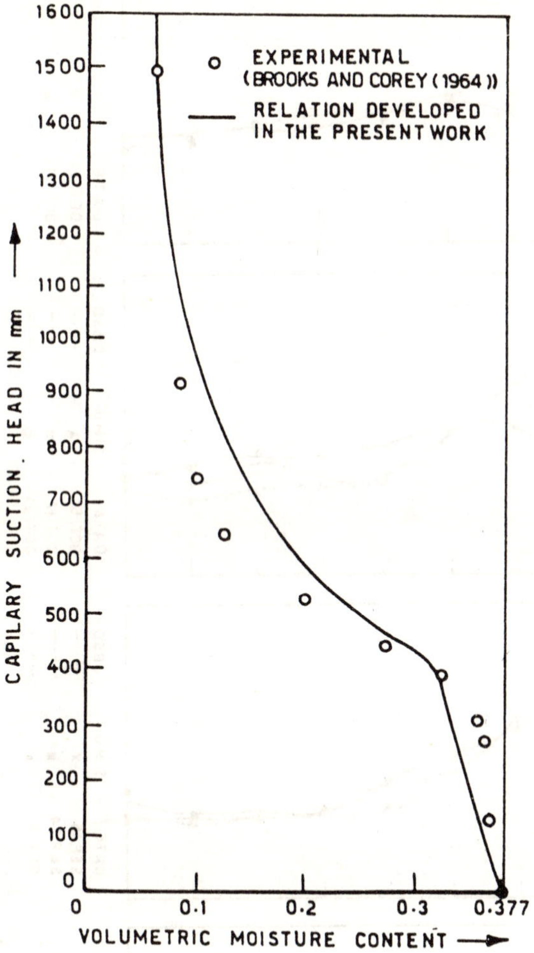

FIG. 1 – COMPARISON OF PLOTS OF PROPOSED 'h CHARACTERISTIC' RELATION WITH THE PLOT USING EXPERIMENTALLY DETERMINED DISCRETE POINT DATA

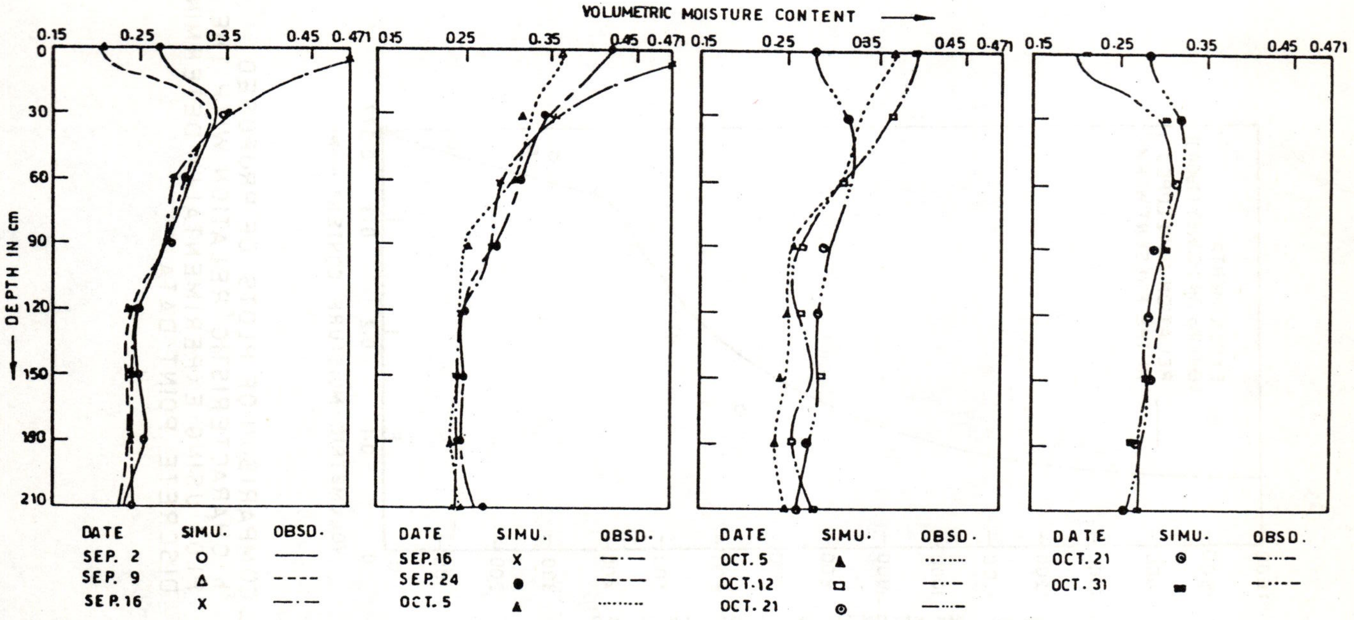

FIG. 2 _MODEL VERIFICATION WITH FIELD EVIDENCE: SIMULATED AND OBSERVED MOISTURE PROFILES

CONCLUSION

In case of the h characteristic , it is recommended to adopt refined experimental data with an efficient numerical algorithm for differentiation and interpolation. However, to cater the needs in the absence of such elaborate experimental facilities, the proposed relation may be adopted as it requires only few soil parameters. The exercise conducted , using the relation in the Richards' equation, for a field evidence simulation problem, proved satisfactory.

REFERENCES

Brooks,R.H. and Corey,A.T. 1964. Hydraulic properties of porous media ; Hydrology papers : Colorado State University. Fort Collins,Colorado.27 pp.

Clapp,R.B. and Horenberger,G.M.1978.Empirical equations for some soil hydraulic properties. Water Res. Research.14(4):6Ø1-6Ø4.

Ghosh,R.K. 198Ø.Estimation of soil moisture characteristics from mechanical properties of soils. Soil Science.13Ø(2):6Ø-63.

Gupta,S.C.and Larson,W.E.1979.Estimating soil water retention characteristics from particle size distribution,organic matter percent and bulk density. Water Res. Research.15(6):1633-1635.

Hoover,J.R.and Grant,W.J. 1983.Numerical fitting of the Gardener equation to hydraulic conductivity and water retention data. Trans.A.S.A.E.26(5):14Ø1-14Ø8.

Mohan Rao,K.M. Kashyap,D. and Chandra.S.1986. Hydrological response of unsaturated zone upto water table. Ph.D.,thesis. University of Roorkee,Roorkee(India).266pp.

Philip,J.R. 1969. Theory of infiltration ; In:Advances in Hydrosciences.Ed:Ven Te Chow.Academic Press,New York.215-296.

Taylor,G.S.and Luthin,J.N. 1969. Computer methods for transient analysis of watertable aquifers. Water Res. Res.5(1):144-152.

CONJUNCTIVE USE FOR IRRIGATION DEVELOPMENT IN ALLUVIAL REGIONS - PROBLEMS, PROSPECTS AND RESEARCH ISSUES -

by

Dr. T. Prasad

Professor of Civil Engineering & Director
Centre for Water Resources Studies
(PATNA UNIVERSITY)
Bihar College of Engineering, Patna.

ABSTRACT

The alluvial regions, particularly the lower Ganga basin such as North Bihar, have high agropotential on account of congenial agroclimate, fertile lands and ample water resources. This potential, however, is far from realized and the agricultural productivity is very low. In view of the prevailing hydrometeorology, irrigation is a sine qua non for agricultural development of this region to its maximum feasible limit. Irrigation from surface water sources, however, is not optimally effective due to hydrological constraints and is even counter-productive on account of its adverse effect on water table equilibrium and drainage. Conjunctive use of surface and ground waters for irrigation is the most suitable strategy in this situation. In view of the availability of adequate groundwater of good quality at low lift, this strategy is highly prospective in North Bihar and will make for sustainable irrigation on long term basis, leading to a high level of agricultural performance and production. This strategy has not been, and cannot be, adopted on account of certain unresolved physical and socio-economic issues inhibiting conjunctive irrigation. Resolution of these issues having policy implications is essential to facilitate implementation of conjunctive irrigation and consequent agricultural development of this region.

INTRODUCTION

The alluvial regions, particularly the lower Gangetic plains, typically have abundance of waters, both in the surface streams as well as groundwater aquifers. This abundance, however, is only in the annual aggregate sense and is hydrologically characterised by temporal non-uniformity of river flows and active interaction between the surface water and ground water phases. Under this prevailing situation, the topographical hydrological, geomorphological and hydrogeological features characteristic of these regions give rise to complex problems, such as recurrent floods, frequent droughts and wide spread waterlogging. In view of these problems, the agricultural potential of this region on account of very congenial agro-climate, most fertile tracts of land and ample water resources remains far from realized. Agriculture which is the mainstay of the economy suffers from low productivity, resulting in poor economic performance and poverty in this region.

North Bihar, constituting 52442 sq. km. area of Bihar,

bounded by Nepal on its north, the state of U.P. to the west, the state of West Bengal to the east and the river Ganga on the south (Fig. 1) is typically characteristic of this region. It is presently the poorest region in India, having a per capita annual income at 1985-86 prices of approximately Rs. 1200 which is 50% of the national per capita income. This is paradoxical in view of the fact that 80% of the geographical area of the land is culturable with a high degree of fertility, agroclimate is congenial to sustain at least 250% cropping intensity and surface water as well as ground water are plentiful. The poor economic performance of the region is explained by a low agricultural productivity sustaining a high density of population (604 persons/sq.km), a major portion of which (about 80%) depends on agriculture and agriculture-related activities.

A number of physical factors have combined to diminish the agricultural productivity in North Bihar. An average annual rainfall of 130 cm occurs in the region with substantial annual variability and marked seasonal non-uniformity. On an average, 85% of the annual rainfall takes place in less than 4 monsoon months. Thus, on the one hand, the water requirements of the crops in various cropping seasons are not only not fully met by the natural rainfall occurrences inspite of ample aggregate annual rainfall amount but also agricultural drought of various magnitudes are frequently caused in the region. On the other hand, coupled with the extant topographical features of the region, this gives rise to recurrence of devastating floods and problematic waterlogging conditions. The annual flood damages of North Bihar, largely on account of loss of crops and agricultural production, constitute about 40% of the national figure. Waterlogging turns the depressed areas into what are known as 'chaurs' which almost totally go out of agricultural production, and diminishes the agricultural yields from other areas.

Projects have been implemented and are envisaged for this region primarily for flood control and irrigation. The former objective is sought to be achieved almost exclusively from embankments and the latter largely from surface sources. Generally, waterlogging and groundwater utilization are not in the purview and hence are not integral parts of the projects. If at all, drainage schemes are prepared separately and groundwater is assigned only a supplemental role to increase the effective command. Conjunctive use of surface and ground waters has not, as a rule, been planned and resorted to.

This paper investigates the problems, prospects and the unresolved issues in adopting conjunctive use as a strategy for irrigation development in such a region. As a case study, it has taken up an area which is proposed to be served by an irrigation-cum-flood control scheme, Bagmati Project, which is in the initial stage of implementation.

FEATURES OF THE STUDY AREA AND THE PROJECT

Bagmati Project has been formulated to provide irrigation

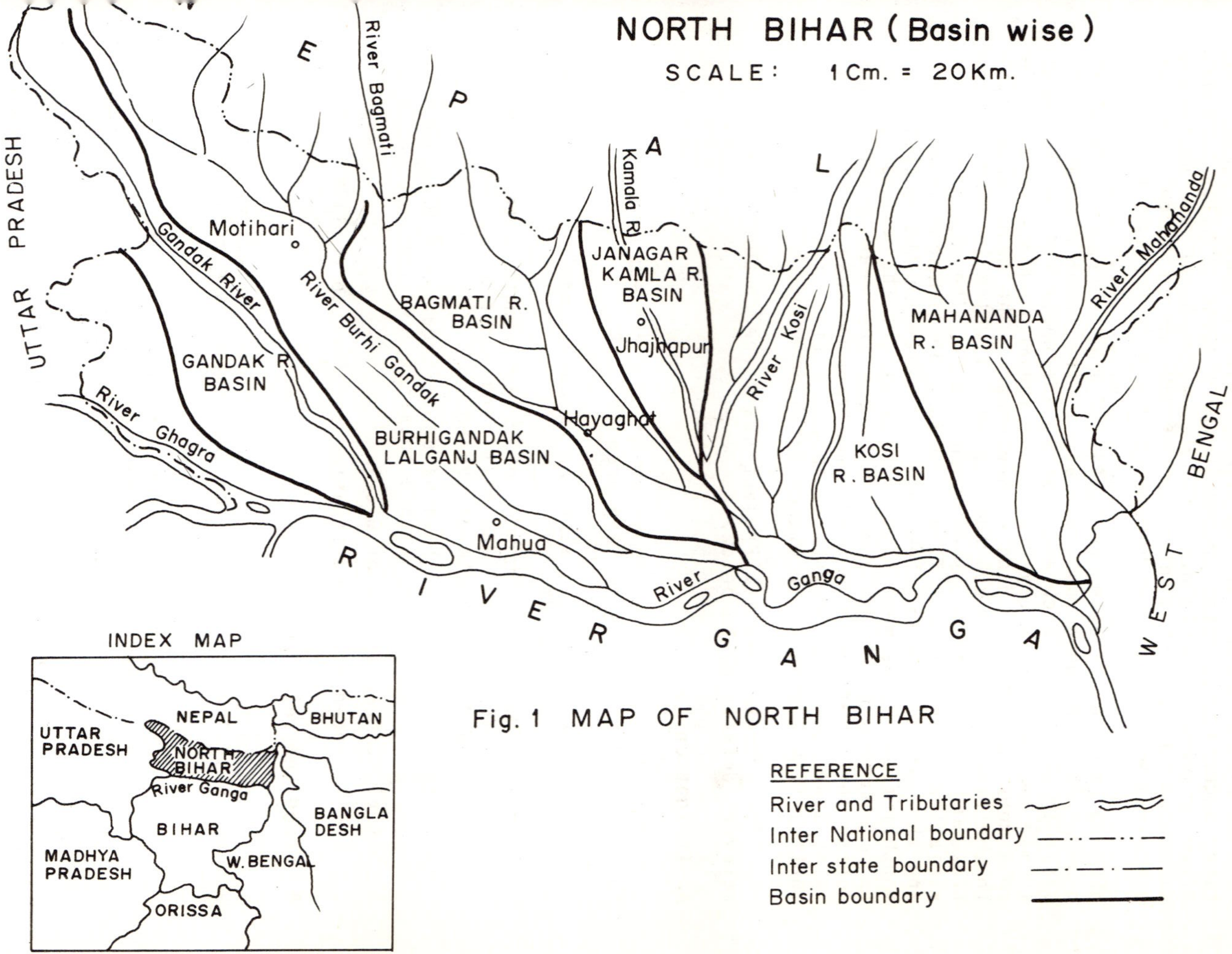

Fig. 1 MAP OF NORTH BIHAR

and flood control benefits (1) to an area lying in the lower reaches of the Bagmati and adjoining river basins (Fig. 2). The river Bagmati originates in the Shivpuri range of hills in Nepal. After traversing a length of 195 km in Nepal in southward direction, it covers a distance of 388 km in North Bihar in south-easterly direction before it joins the river Kosi. It drains a total area of 13690 sq. km. out of which approximately 6000 sq. km., constituting about 44% of the total, lies in North Bihar. The average annual rainfall in the lower reaches of the basin in Bihar is 120 cm and that in the upper reaches in Nepal is 145 cm. Approximately 85% of the annual rainfall takes place in 4 monsoon months of June to September. The mean annual flow in the river is around 200 cumecs, 80% of which is mean monsoon flow and 20% is mean non-monsoon flow. The coefficient of variation of mean annual flow is 0.30. The monsoon meteorology and the variability of flows are reflected in the analysis of monthly flows. The mean monthly flow during the monsoon period is 375 cumecs which is about 10 times the mean monthly flow during the non-monsoon period. The maximum mean monthly flow, which takes place in August is about 35 times the minimum mean monthly flow occurring in either March or April.

The proposed Bagmati Project has three components with the relevant details given below:

A. Irrigation

This is the major component of the Project for which the following primary measures are proposed,

(i) A barrage at a site 1.5 km downstream of Dheng Railway Bridge No. 89

(ii) A left bank canal of 35 km length with a design discharge of 29.7 m^3/s and branch canals of an aggregate length of 195 km.

(iii) G.C.A. is 1.01 lakh ha on left bank and 0.47 lakh ha on right bank, with a total of 1.48 lakh ha.

(iv) C.C.A. is 70% of G.C.A.

(v) Intensity of cultivation is taken to be 117%

B. Flood Control

Flood control is sought to be achieved exclusively through embankments. The following major embankments are proposed in the Project.

(i) Left Bagmati embankment of length 79.3 km.
(ii) Right Bagmati embankment of length 77.5 km.
(iii) Embankment in doab of rivers Bagmati and Lalbakeya of length 21.0 km.
(iv) Belwa Dhar right embankment of length 52 km.
(v) Belwa Dhar left embankment of length 36 km.
(vi) Back embankment along Lakhandeyi on both banks of length 24 km.

INDEX MAP – BAGMATI PROJECT COMMAND AREA

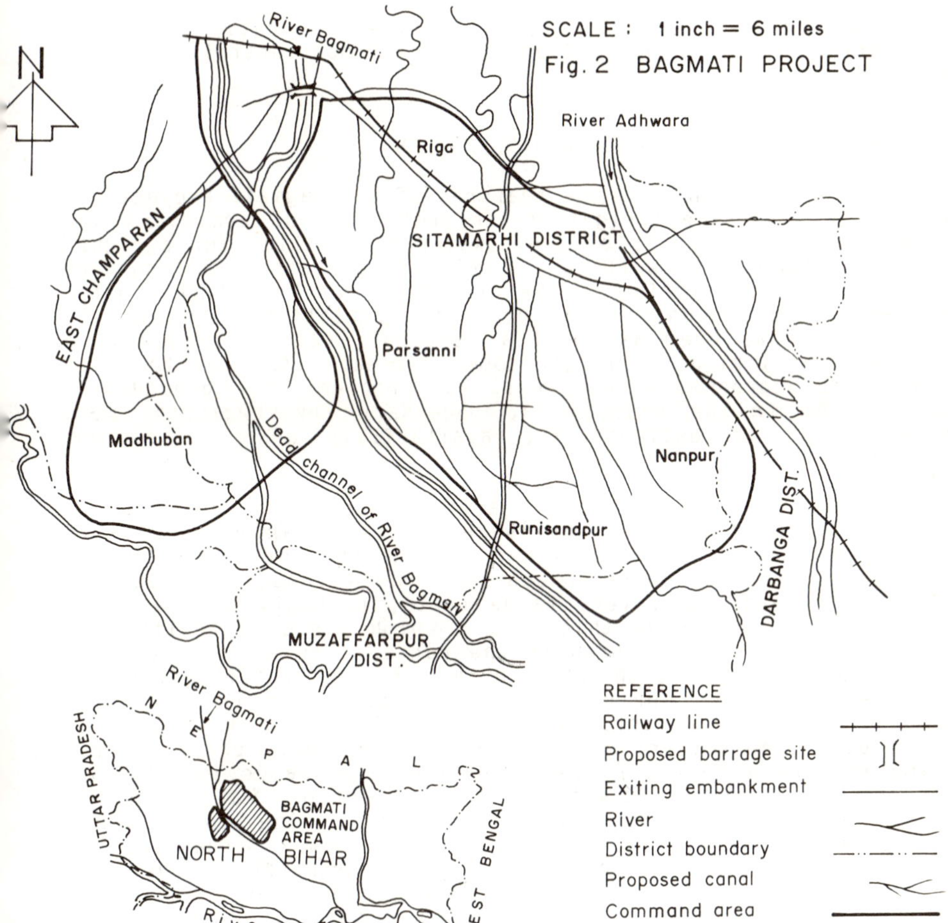

C. Drainage

Drainage of only surface water inundating for prolonged periods low lying areas and 'chaurs' is provided for in the Project. For this, trunk channels of a total length of 311 km, draining 6 chaurs with an aggregate area of 777 ha are proposed.

The Project envisages a benefit-cost ratio for the irrigation sector to be 1.94 and for the flood control sector to be 1.28.

GROUNDWATER AVAILABILITY AND UTILIZATION

As can be seen from the features of the Bagmati Project, it is mainly a surface irrigation-cum-flood control project. The potential and prospects of groundwater utilization for agricultural development as well as its role in maintaining a long-term hydrological equilibrium in the area have not been investigated or taken into account in the project formulation. If at all, groundwater has been assigned a marginal role in providing a supplemental irrigation in the command area. In order to fully assess the prospects as well as problems in groundwater utilization in the study area, necessary investigations were carried out to estimate the availability and its current utilization.

The aquifer in the study area exists in unconfined and semi-confined states with the water table at an average depth of 3 m below the ground surface, fluctuating through 1 to 2 m from pre-monsoon to post-monsoon seasons. Although the saturated alluvium extends to a very large depth, probably in the range of 1000 m, the average thickness of high yielding aquifers in several layers may be taken in the range of 50 m. The groundwater is recharged both vertically, through infiltration and percolation of incident rainfall and seepage from inundation, as well as horizontally through subsurface flow from the north to the south. The latter, originating from a highly pervious sub-Himalayan high rainfall zone constitutes a major part of the recharge. A year round water table gradient of ½ m per km from north to south corroborrates this fact. The total annual recharge is estimated of the order of 60 cm. Investigations carried out regarding the quality of groundwater in the study area (2) establish that it is excellent to good for agricultural purposes in all seasons and hence will not be a constraining consideration in extensive as well as intensive development of groundwater utilization in the study area.

In the background of the groundwater availability as indicated above, the current utilization of groundwater is very low inspite of the fact that canal irrigation has not yet been introduced in the area. It is estimated that, at the present, less than 25% of safe yield is being used for agriculture by way of private and public tubewells, covering less than 20% of cultivated area.

PROBLEMS IN DEVELOPMENT OF EXTENSIVE GROUNDWATER UTILIZATION

In spite of attractive prospects and potentials in groundwater utilization to boost agricultural production in this region, there are several reasons for the existing situation of low utilization of groundwater. A prime factor is the preponderance of marginal and small farmers, having holdings less than 2 ha, in the area. More than 90% of the holdings, covering about 56% of the total cultivated area, are owned by such farmers who simply cannot afford to provide themselves with this facility. Added to this is the lack of incentive or positive disincentive on account of environmental and hydrological factors. Fertile lands and normally adequate rainfall in the monsoon season ensure a subsistence level to these farmers from the Kharif crops. Residual soil moisture and the winter rains are also generally suitable for the Rabi crops. Thus, this normal congeniality of agro-climate and agro-meteorology do not adequately motivate the farmers to be enterprising enough to go in for tubewells. This is corroborrated by the fact that over 70% of the existing tubewells were installed in this area in the wake of a severe drought experienced in 1979-80. Even this adversity and a general evidence that tubewell irrigation leads to a definite and substantial improvement in returns from agriculture have not stirred the small and marginal farmers into endeavour to take recourse to tubewells. There is a circular cause and effect relationship between economic status and private initiative in groundwater use for irrigation. A positive disincentive for farmers to invest in tubewell in this area is the recurrent occurrence and persistent threat of occurrence of floods and consequent damage to crops.

Another important factor which has inhibited development of tubewell irrigation in this area is the paucity of adequate and reliable source of energy supply required for operating the tubewells. Electrical energy is either non-existent, inadequate or erratic and hence is not relied upon for the purpose. This has been a major cause for the non-performance or poor performance of public tubewells in the region. Diesel operated tubewells are more common. The inconvenience, the unreliability in diesel availability and the cost involved, however, have been deterrent factors in more wide spread adoption of tubewell irrigation by means of diesel pumps.

PROSPECTS OF CONJUNCTIVE USE

In this context, the desirability, the prospects and the imperativeness of conjunctive use and management of surface and ground waters for optimum irrigation development become rather obvious (3). The irrigation intensity of 117%, as envisaged in the Bagmati Project, is rather low in view of the congeniality of the agro-climate, the fertility of land and the aggregate availability of water. It can optimally be increased to an average of 250%, i.e. it can be more than doubled. This can, however, be achieved only if groundwater is also utilized in conjunction with surface water, particularly for the rabi crops when the river has deficient flow. Irrigation from surface water sources alone cannot ensure a higher irrigation intensity through the proposed Project in which no storage is provided for. The barrage, an important component of the Project, will only serve to divert the river flow into the network of canals by creating necessary head at the headworks. Consequently, the Project is not likely to meet the needs of Rabi irrigation on account of the deficiency of non-monsoon flow in the river. Also, on account of substantial variation of annual flows, co-eff. of variation being 0.30, the effective irrigation intensity will fall below the envisaged 117% in deficient rainfall years. Thus, groundwater utilization must be made an integral part of irrigation strategy in the study area in order to achieve the feasible optimum irrigation intensity which is more than double the Project irrigation intensity and also to achieve stability in irrigation intensity over the years. Keeping in view the estimated annual recharge of the groundwater and the increase in return flow from irrigation, groundwater in the study area is capable of fulfilling this conjunctive role.

Conjunctive use of groundwater alongwith surface water for irrigation is imperative from the consideration of waterlogging and subsurface drainage. Water table exists at shallow depths throughout the study area and the low-lying pockets have been turned into what are called 'chaurs' - area of permanent surface inundation. This situation will be exacerbated by irrigation exclusively from surface waters and in the long run, such irrigation may even be counterproductive. Groundwater utilization lowers the watertable, helps subsurface drainage and is thus an antidote for this impending adversity. Thus, it is necessary to resort to ground water utilization to maintain the subsurface hydrologic equilibrium for optimum agricultural production, and making irrigation sustainable on a long-term basis.

If groundwater is utilized for irrigation in a major way, seepage of canal water becomes a source of replenishment of groundwater and is thus not to be considered as a loss of valuable water having the potentiality to irrigate fields. In fact, seepage from canals is incorporated as an essential and desirable process in the conjunctive use strategy, obviating the necessity to line the canals which is a costly proposition.

Tubewell irrigation also enhances the farmers' access to and control of water for irrigating their fields, enabling them to adopt high yielding varieties of crops and more remunerative cropping patterns. The contentious issue of distribution of canal water among the users and the conflicting interests of users in the head reaches and those in the tail reaches is effectively resolved. Certain high patches of land in the command area, which cannot get the benefit of canal irrigation due to adverse gravity, can also be provided with irrigation by the conjunctive use strategy. Thus, it enhances net effective command of irrigation and promotes better cropping pattern uniformly through the command area.

Apart from these prospects, it has been experienced, such as in Gandak and Kodi Projects (4,5), that irrigation exclusively by surface waters in the alluvial plains invariably leads to the rise of water table, resulting in encroachment of the root zones by water. Often, it causes salinity or alkalinity of the land and contributes to waterlogging conditions, particularly in low lying areas of the command. The agricultural productivity has been adversely affected, rendering exclusive surface irrigation progressively counterproductive. Clearly, conjunctive irrigation is the effective antidote for the situation.

UNRESOLVED ISSUES IN CONJUNCTIVE USE

Although there has been increasing realization of the suitability of conjunctive use as a strategy for irrigation development in the alluvial plains, it has hardly been implemented so far. This is on account of the fact that there are unresolved issues related to conjunctive use - both in the physical as well as socio-economic domains (6). Conjunctive use signifies an integrated use of the two interacting sources - surface water and ground water - for irrigation as distinct from supplemental use or alternative use of groundwater in time and space. The following issues related to conjunctive use are identified as unresolved,

A. Physical

(i) What is the mechanics, the magnitudes and the rates of seepage and infiltration from various sources under shallow water-table conditions? What are the various practically feasible ways to accelerate or decelerate these rates?

(ii) How does the ground water table respond to various processes active in the alluvial plains, such as infiltration, seepage, evaporation, evapotranspiration and groundwater discharge?

(iii) How does the use of groundwater for irrigation from various aquifers affect the groundwater table, ground water regime and other related processes?

(iv) While the methodology of designing an irrigation system based only on surface water or exclusively on ground water is well established, that of designing an irrigation system based on conjunctive use is not quite known or established. In designing the system, the various physical factors and the objectives, as indicated earlier, are to be taken into account.

(v) What should be the criteria of operating the irrigation system designed for conjunctive use in order to achieve the objectives optimally?

B. Socio-economic

(i) What should be the concept of ownership of surface water sources and ground water sources? What should be the role of public versus private initiative and enterprise in the development, utilization and management of the two sources for irrigation?

(ii) How should the economics of use of the surface water and ground water sources for irrigation be determined and regulated? How should the question of subsidy, incentive and disincentive in relation to the use of the two sources for irrigation be resolved in order to promote the desired conjunctive use?

(iii) What should be the administrative and management structure in conformity with the conjunctive irrigation?

(iv) There are various externalities relating to agricultural economics, land laws, other inputs for irrigated agriculture, and, most dominantly, availability of power for groundwater use, having an impact on conjunctive use of surface and ground waters for irrigation? How are these externalities, primarily availability of power, to be managed?

These interconnected issues need to be investigated, researched and resolved in order to make conjunctive irrigation professionally operational for the alluvial plains. In particular, a methodology for designing a conjunctive irrigation system in the alluvial plains with shallow water tables, based on the physical factors mentioned above, has to be developed. Similarly, a procedure for on line operation of a conjunctive irrigation system, so designed, has to be evolved in order to maximise irrigation intensity, to achieve long-term hydrologic equilibrium in the subsurface regime and to ensure efficient drainage.

CONCLUSION

In view of the hydrometeorological features prevailing in the alluvial region (such as in North Bihar), irrigation constitutes an essential requirement for optimal agricultural development. The natural endowments of this region, such as the agronomic properties of the soil, the agroclimate and the aggregate availability of water can be fully harnessed into sustained maximum agricultural production only through controlled and assured supply of water to the crops. Irrigation, thus, is imperative for this region in order to realize its immense agricultural potential.

Irrigation exclusively by surface water is not wholly effective or optimally productive in this region on account of the seasonality of flows in the rivers and other ecological factors. Moreover, it is likely to be counterproductive due to gradual rise of water table almost invariably caused by it, resulting in salinity and waterlogging conditions. As there is ample groundwater of good quality available at relatively low lifts, and as there is substantial hydrologic interaction constantly going on between the surface water and groundwater regimes, conjunctive use of these two sources is an obvious strategy of irrigation for optimum agricultural development in this region. In particular, this strategy will ensure maximum irrigation and cropping intensities commensurate with the agroclimate, agronomic properties of the soil and total water resource availability in the region and will make for sustainable irrigation on a long-term basis by maintaining surface water - ground water equilibrium and ensuring efficient drainage

There are, however, several physical and socio-economic issues related to conjunctive irrigation still unresolved, which inhibit adoption and implementation of this strategy in professional practice. Several physical processes related to the interaction of surface water - ground water regimes in the alluvial plains and their response to conjunctive uses are not quantitatively understood due to which design methodology and operating procedure of a conjunctive irrigation system are not yet developed. Also, suitable policies regarding concepts of ownership of surface and ground water sources, relative economics of the use of water for agriculture from these two sources, administrative and management structure, and energy issues have to be evolved and prescribed for promoting conjunctive irrigation in the alluvial plains.

Widespread adoption of conjunctive irrigation, alongwith effective water resource management in the flood prone regions in eastern India, would lead to phenomenal increase in agricultural production and much higher levels of income from the farm sector, leading to economic emancipation and rightful prosperity of this region (7).

ACKNOWLEDGEMENT

This paper is based on research carried out by the Centre for Water Resources Studies, Bihar College of Engineering, Patna University on the topic, "Problems and Prospects in Groundwater Utilization for Agricultural Development - A Case Study in a North Bihar Region", with the support of a grant from the Ford Foundation.

REFERENCES

1. Bagmati Project Report, 1981. Water Resources Dept., Govt. of Bihar

2. Prasad, T. and Prasad, V.S., 1988. Groundwater Quality Investigations in Bagmati Command Area. Proc. of the National Seminar on Management of Water and Wastewater Systems, held at Bihar College of Engineering, Patna, 27-28 February, 1988.

3. Prasad, T. and Prasad, V.S., 1988. Conjunctive Use as a Strategy for Irrigation Development in North Bihar, Proc. of the National Seminar on Water Resources Development and Management (With Special Reference to Groundwater), Regional Engg. College, Rourkela, Feb. 1988.

4. Agnihotri, A.K., and Joshi, P.K., 1985. Regional Disparities and Soil Degradation Vis-a-Vis Canal Irrigation : Problems and Appraisal, Proc. of the All India Symposium on Formulation and Appraisal of Irrigation Projects, Water Resources Studies Programme, Patna Univ., March 1985.

5. International Workshop on Post-Fact Evaluation of a Water Resources Project - A Symposium Devoted to the Study of the Performance of the Kosi Project, 1979. Recommendations of the Workshop held at Bihar College of Engineering, Patna.

6. Prasad, T., 1988. Conjunctive Use of Surface and Ground Waters for Agricultural Development in Eastern India - Research Issues, A paper presented at the Groundwater Workshop held at N.D. Univ. of Agriculture and Technology, Faizabad, U.P., 5-6 April 1988.

7. Rogers, P. and Prasad, T., 1988. Flood Control and Rural Development in Eastern India. A paper presented at the National Workshop on Technology Delivery Systems for Rural Development, Parliament House Annexe, New Delhi, October 13-15, 1988.

COMPUTER SIMULATION OF AN INJECTED TRACER MOVEMENT

N.M.MAHAJAN
Research Officer
Central Water and Power Research Station,
Pune-411 024, India

ABSTRACT

Problems of seepage hydraulics have been solved using mathematical models written in finite difference form and digital computer programs. These models are the solutions to the two dimensional flow through porous media which give distribution of groundwater potential at each nodal point. An attempt is made in this paper to utilize the model in a modified form to trace the path of an injected tracer by computing true velocities (porous velocities) from the knowledge of groundwater head distribution and permeabilities and thus Darcy velocities. The streamlines are computed by approximating polygonal succession of straight lines. In each square or a rectangular element velocity vector of streamlines is assumed parallel to the resultant of the porous velocities. The transit time through each segment is computed and totalled to obtain travel time of tracer particle from an injection borehole to the observation borehole. Two cases have been studied, in the first case movement of tracer is observed in an isotropic aquifer, in the second case, tracer movement is observed in a stratified aquifer. The studies indicated that the rate of travel of a tracer is proportional to the permeability of the aquifer and in case of a stratified aquifer the inputs of the tracer at the observation points are in proportion to the respective permeabilities of the strata through which tracer travels. These studies are useful in predicting contaminant movement.

1.0 INTRODUCTION

Tracers such as strong electrolytes, organic dyes and radioisotopes have been used for the estimation of seepage losses through hydraulic structures such as dams and canals (1). Point Dilution and Multiwell Methods are employed for this purpose. In the former method a tracer is uniformaly distributed throughout the volume of water in the borehole. Then if only steady state horizontal flow exists the dilution of the tracer in a borehole is related to the filtration or Darcy velocity in a water bearing formation. In the latter method a predetermined quantity of tracer is introduced in the borehole called an injection borehole and the arrival of the tracer is monitored in the boreholes situated in the probable direction of the subsurface flow. The velocity of the subsurface flow can be determined by knowing the time of arrival of the peak activity and the distance betweeen injection and observation holes. In a uniform isotropic aquifer multiwell method gives only one peak on tracer concentration versus time (break-through) curve. However, in case of a stratified aquifer number of peaks are observed on the breakthrough curve. Possibly the rate of travel of a tracer is proportional to the permeability of the strata through which it travels and the inputs of the tracer at the observation holes are in proportion to the respective permeabilities of the strata.

Recently, a mathematical model has been developed for the assessment of seepage losses from unlined canals of trapezoidal shape and having impermeable layer below its bottom (2). This model is a solution to the two dimensional steady state flow through a porous medium of the canal. Finite difference Successive Over Relaxation method and digital computer program have been used to solve the problem (3). This model gives groundwater potentials (heads) in the problem domain of the canal.

Suitable modifications were made in the program written for the assessment of seepage losses through canals. With the help of this program it has become possible to trace the path of an injected tracer in an aquifer by computation of streamlines. It has also been possible to calculate travel times of the tracer particles from an injection borehole to the observation boreholes.

2.0 THE PROBLEM

For a simple case of canal for which there is an impermeable layer below its bottom, the problem can be formulated in a following way. It is assumed that the water table slope at a sufficiently long distance is constant and flow is uniform. The lateral boundary can be represented by a vertical equipotential. For practical purpose this distance can be taken as 10d, where d is the width of a canal at its bottom. Within the canal the head remains constant (4) Fig.1.

FIG.1 FORMULATION OF CANAL SEEPAGE PROBLEM

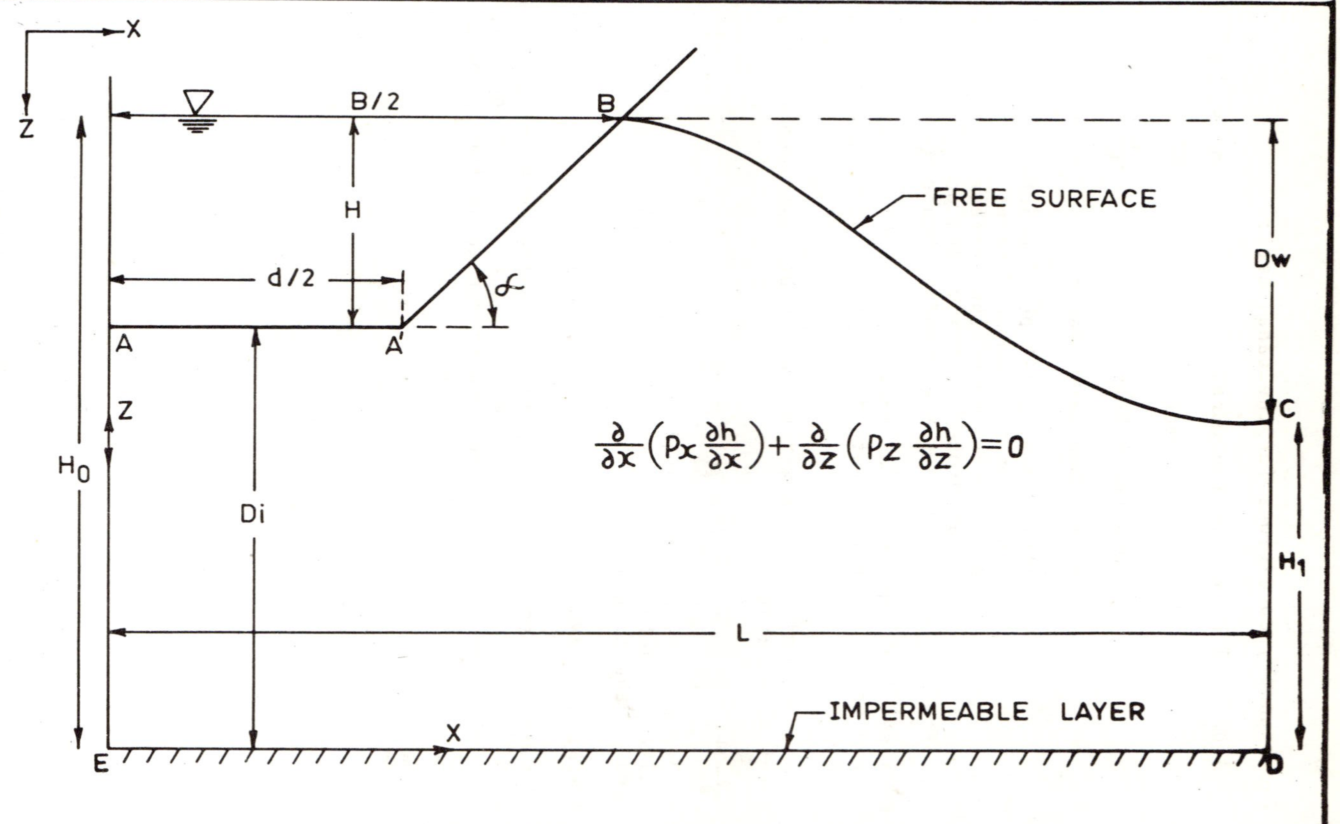

FIG. 1 FORMULATION OF CANAL SEEPAGE PROBLEM

The boundary conditions for the canal are :

Along	AA'B	within the canal	$h = Ho$
	AE	axis of symmetry	$\frac{\partial h}{\partial x} = 0$
	DE	impermeable base	$\frac{\partial h}{\partial z} = 0$
	CO	undisturbed water table	$h = H1$
	BC	water table, position of which is unknown but two conditions apply, pressure on the surface is atmospheric $h = z$ and no flow across the boundary hence	$\frac{\partial h}{\partial n} = 0$

Because the position of free surface BC is unknown a trial and error (iterative method) has to be used to find its position.

Flows predominantly in x-z plane can be described by the differential equation

$$\frac{\partial}{\partial x}\left(Px\ \frac{\partial h}{\partial x}\right) + \frac{\partial}{\partial z}\left(Pz\ \frac{\partial h}{\partial z}\right) = 0 \qquad \ldots(1)$$

where Px and Pz are permeabilities in x and z directions respectively Equation (1) allows variable permeabilities to be included in the solution.

3.0 NUMERICAL APPROACH

A numerical solution using finite difference method can be obtained by creating a nodal mesh in the problem domain. The node arrangement and the mesh superimposed in the problem domain of the canal is shown in Fig.2.

FIG.2 FINITE DIFFERENCE MESH IN PROBLEM DOMAIN OF THE CANAL

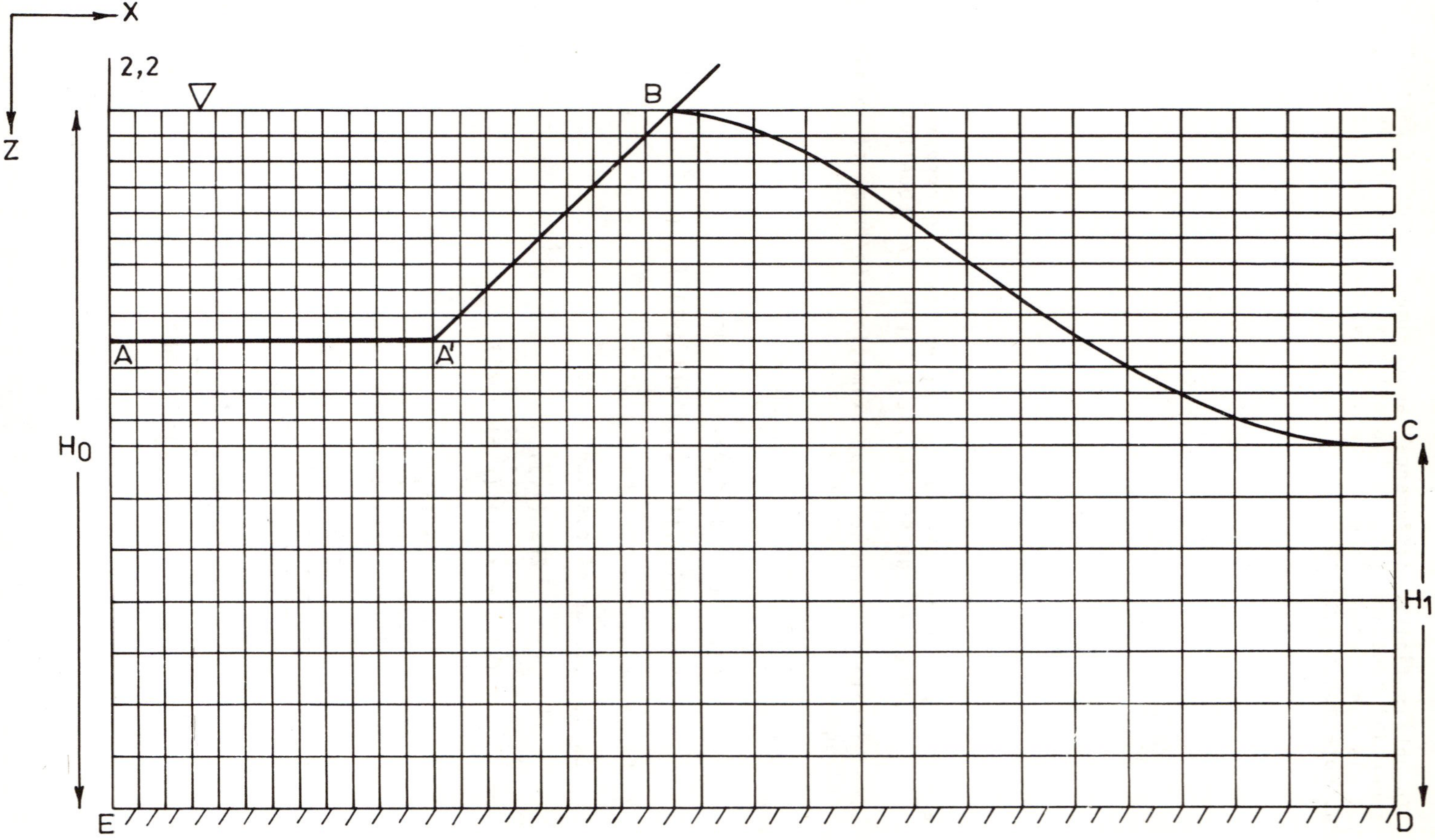

FIG. 2 FINITE DIFFERENCE MESH IN PROBLEM DOMAIN OF THE CANAL

To obtain finite difference approximation of a partial differential equation (1), a typical five point star shown in Fig.3 is considered. To describe the value of head at any point notation h(i,j) is used.

FIG.3 A FIVE POINT STAR REPRESENTING FINITE DIFFERENCE APPROXIMATION OF PARTIAL DIFFERENTIAL EQUATION

The finite difference approximation in terms of heads for the equation (1) is given by

$$h(i,j) = AB/CX \qquad \ldots(2)$$

where

$$AB = A_{i,j}\, h(i+1,j) + B_{i,j}\, h(i,j-1) + C_{i,j}\, h(i-1,j) + D_{i,j}\, h(i,j+1) \qquad \ldots(3)$$

$$CX = A_{i,j} + B_{i,j} + C_{i,j} + D_{i,j} \qquad \ldots(4)$$

CX is made up of constants and

$$A_{i,j} = \frac{2P_x\,(i,j)}{(x_{i+1} - x_{i-1})\,(x_{i+1} - x_i)} \qquad \ldots(5)$$

$$B_{i,j} = \frac{2P_z\,(i,j-1)}{(z_{i+1} - z_{j-1})\,(z_j - z_{j-1})} \qquad \ldots(6)$$

$$C_{i,j} = \frac{2P_x\,(i-1,j)}{(x_{i+1} - x_{i-1})\,(x_i - x_{i-1})} \qquad \ldots(7)$$

$$D_{i,j} = \frac{2P_z\,(i,j)}{(z_{j+1} - z_{j-1})\,(z_{j+1} - z_j)} \qquad \ldots(8)$$

FIG. 3 A FIVE POINT STAR REPRESENTING FINITE DIFFERENCE APPROXIMATION OF PARTIAL DIFFERENTIAL EQUATION

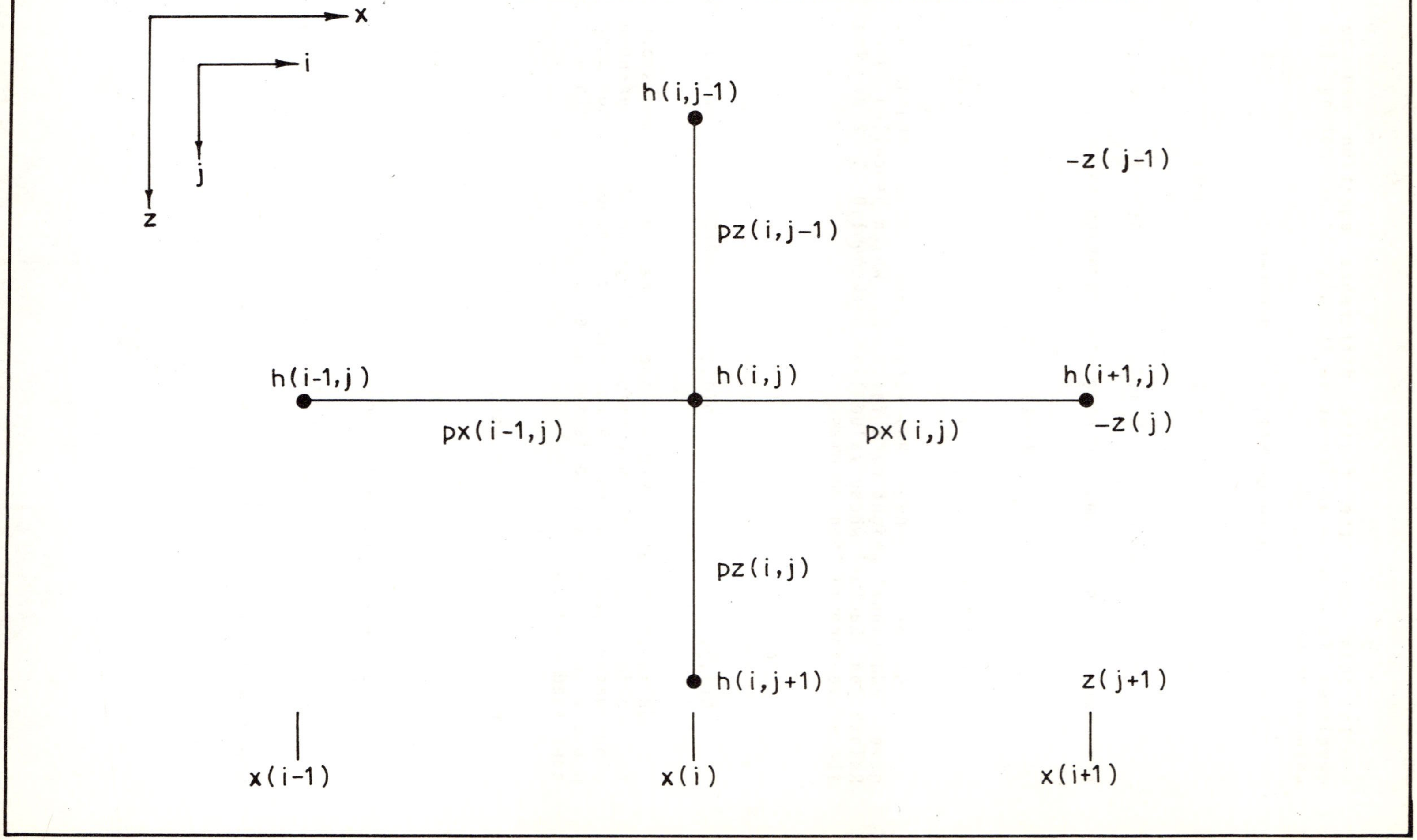

In this form the finite difference equation can be developed to provide an iterative formula for estimating the unknown head at node i,j.

If the nth estimate of h(i,j) is denoted by

$$h^{(n)}(i,j) = AB/CX - h^{(n-1)}(i,j) + h^{(n-1)}(i,j) \qquad \ldots(9)$$

the change in the estimate during nth iteration is

$$AB/CX - h^{(n-1)}(i,j) \qquad \ldots(10)$$

In Successive Over Relaxation (SOR) this change is given additional weight by a factor w, which is called 'Over Relaxation Factor' whose value lies between 1.0 & 2.0. The iterative formula then becomes

$$h^{(n)}(i,j) = (1 - w)\ h^{(n-1)}(i,j) + wAB/CX \qquad \ldots(11)$$

SOR is an iterative process and, therefore it is necessary to provide a test for successful convergence to the unknown groundwater head values. A conventional method is to stop process when the balance is satisfied at any node within a specified limit.

This balance can be expressed as

$$h(i,j) - AB/CX = 0 \qquad \ldots(12)$$

and therefore a suitable test can be

$$|CX\ h(i,j) - AB| \leq \text{Error} \qquad \ldots(13)$$

where error can be 0.00001 or even less.

The equations developed above have been written in digital computer program, which calculates groundwater heads at each nodal point in the problem domain of the canal and the quantity of flow. It also determines free surface to a reasonal accuracy. A typical computer output is shown in Table-1.

4.0 INJECTED TRACER MOVEMENT

The aim of finding the tracer movement is to compute the path of the streamlines and the travel time of tracer particles from an injection borehole to the observation borehole. While doing this it is assumed that the dispersion is negligible, molecular diffusion does not take place, the tracer is not adsorbed by the porous medium and it does not modify hydrodynamic characteristics of the water. The change in concentration of a tracer has also not been taken into account.

From the knowledge of groundwater heads and the permeabilities Darcy velocities at each nodal point have been calculated. The relation between Darcy velocity and groundwater heads has the following form in the computer notation.

$$VEX\ (I,J) = \frac{PX\ (I,J) * (H\ (I,J) - H\ (I+1,J))}{X(I+1) - X(I)}$$

$$VEZ\ (I,J) = \frac{PZ\ (I,J) * (H\ (I,J) - H\ (I,J+1))}{Z(J) - Z(J+1)}$$

where VEX and VEZ are Darcy velocities in X and Z directions respectively while other terms have usual meanings, true velocities (porous velocities) of the tracer have been computed by the relation

$$VEPX\ (I,J) = VEX\ (I,J)/P \quad \text{and} \quad VEPZ\ (I,J) = VEZ\ (I,J)/P$$

where VEPX and VEPZ are porous velocities in X and Z directions respectively, and P is fractional porosity. Resultant porous velocities were then calculated.

TABLE-1

TRIAL NO = 4
GROUNDWATER POTENTIALS

	2	3	4	5	6	7	8	9	10	11	12
2	10.00	10.00	10.00	10.00	10.00	9.81	9.68	9.59	0.00	0.00	0.00
3	10.00	10.00	10.00	10.00	9.90	9.77	9.66	9.55	9.42	9.30	9.20
4	10.00	10.00	10.00	9.92	9.84	9.73	9.63	9.52	9.41	9.30	9.19
5	9.94	9.94	9.91	9.86	9.78	9.70	9.60	9.50	9.39	9.29	9.18
6	9.90	9.89	9.86	9.81	9.75	9.66	9.57	9.48	9.38	9.28	9.18
7	9.82	9.81	9.78	9.74	9.68	9.61	9.53	9.44	9.35	9.26	9.16
8	9.75	9.75	9.72	9.69	9.63	9.57	9.50	9.42	9.33	9.24	9.14
9	9.71	9.70	9.68	9.64	9.60	9.54	9.47	9.39	9.31	9.22	9.13
10	9.67	9.67	9.65	9.61	9.57	9.51	9.45	9.38	9.30	9.21	9.12
11	9.66	9.65	9.63	9.60	9.55	9.50	9.44	9.37	9.29	9.20	9.12
12	9.65	9.64	9.62	9.59	9.55	9.50	9.43	9.36	9.28	9.20	9.11

	13	14	15	16	17	18	19	20	21	22
2	0.00	0.00	0.00	0.00	0.00	0.00	0.00	0.00	0.00	0.00
3	9.09	8.99	8.80	8.64	0.00	0.00	0.00	0.00	0.00	0.00
4	9.09	8.99	8.79	8.60	8.34	8.10	7.88	7.69	0.00	0.00
5	9.08	8.98	8.78	8.58	8.34	8.10	7.87	7.64	7.31	7.00
6	9.08	8.97	8.77	8.57	8.33	8.09	7.86	7.62	7.31	7.00
7	9.06	8.96	8.76	8.55	8.32	8.08	7.84	7.59	7.30	7.00
8	9.05	8.95	8.75	8.53	8.31	8.07	7.83	7.57	7.29	7.00
9	9.04	8.94	8.74	8.52	8.30	8.06	7.82	7.56	7.28	7.00
10	9.03	8.93	8.73	8.52	8.29	8.05	7.81	7.55	7.28	7.00
11	9.02	8.93	8.72	8.51	8.29	8.05	7.80	7.55	7.28	7.00
12	9.02	8.92	8.72	8.51	8.28	8.05	7.80	7.54	7.28	7.00

FLOW TO RIGHT ACROSS SECTION I = 15 EQUALS .6221 CUBIC METERS / DAY

SUCCESSFUL CONVERGENCE TO FREE SURFACE POSITION

Streamlines were constructed from various preset points of tracer particles in the injection borehole by approximating a polygonal succession of straight lines (5). In each either a square or rectangular element, velocity vector was assumed to be constant and the streamlines were then assumed to proceed by straight line segments parallel to the resultant velocity vector. A special test was incorporated in the program for finding the direction of streamlines.

The travel times through each segment were computed and totalled to obtain the time required for a tracer particle to travel from injection borehole to observation borehole.

The streamlines were stopped either after reaching a node with zero velocity or reaching an observation borehole.

5.0 TEST EXAMPLE

The canal modelled for tracing the movement of an injected tracer has the following geometrical dimensions.

The width of the canal at the bottom	d =	4.0 m
The width of the canal at the water surface	B =	8.0 m
The depth of the water in the canal	H =	2.0 m
The side of the canal makes an angle of 45° with the horizontal	α =	45°
The distance of constant groundwater table from the centreline of the canal	L =	30.0 m
Groundwater level at a distance of L from the canal	Dw =	3.0 m
The depth of an impermeable layer below the canal bottom	Di =	8.0 m

Two cases have been studied for this canal, in the first case the problem domain was taken as homogeneous and isotropic and in the second case a two layer situation has been introduced.

In the first case both horizontal and vertical permeabilities were kept as 1.0 m/day. The results obtained are given in Table-2.

In the second case upto 4.0 m from the surface of water both the permeabilities were kept as 1.0 m/day from 4.0 m onwards the values of permeabilities were reduced to 0.5 m/day. The results are given in Table-3. Streamlines for both isotoropic condition and for layered model are shown in Fig.4.

FIG.4 MOVEMENT OF AN INJECTED TRACER

6.0 DISCUSSIONS OF THE RESULTS

The results given in Tables 2 and 3 and also shown in Fig.4 indicate that the travel times of the tracer particles are influenced by the distribution of groundwater heads and also by the permeability values.

The real velocity of the subsurface flow through porous medium is given by, the distance travelled divided by the time taken. Due to non uniform pore distribution, there is a local variation in the velocity as tracer takes tortuous path. The average of all these velocities give transit velocity. Estimation of real velocities is possible by determining the transit time of a tracer from an injection borehole to an observation borehole.

From the Fig.4 it can be seen that in case of an isotropic medium the streamlines are more or less parallel to each other, and the transit times of tracer particles from an injection borehole to the observation borehole are more or less same.

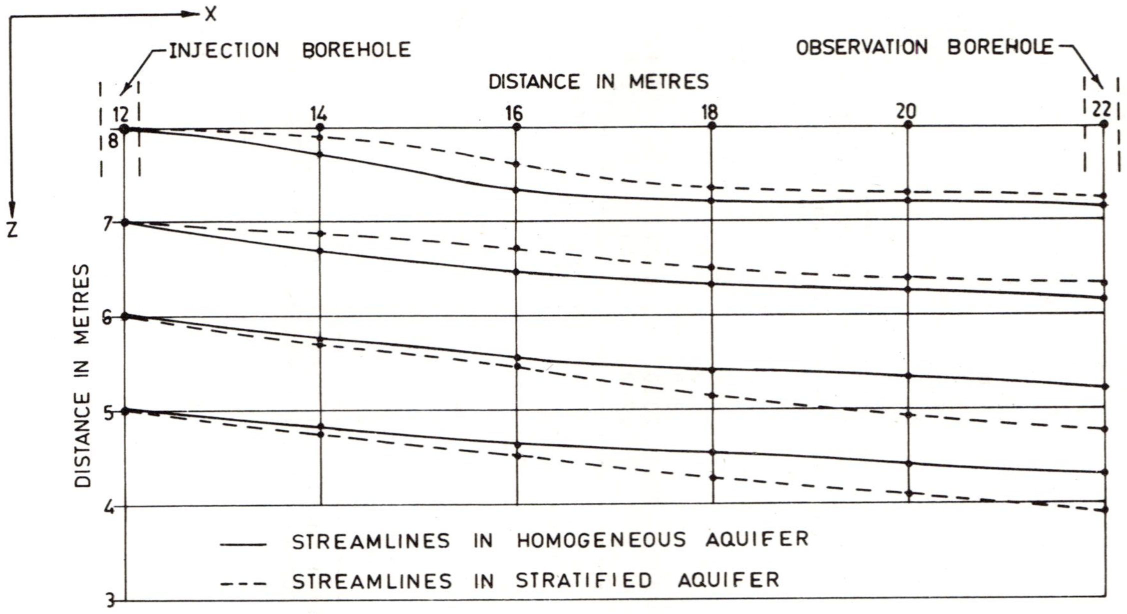

FIG. 4 MOVEMENT OF AN INJECTED TRACER

TABLE - 2

Tracer Particle No.	Coordinates Meters X	Z	Time in Days
1	12.0	8.0	0.0
1	14.0	7.72	5.28
1	16.0	7.33	3.99
1	18.0	7.23	4.38
1	20.0	7.19	4.58
1	22.0	7.13	4.85
Time required for Tracer Particle No.1 to travel from injection borehole to observation borehole, a distance of 10.0 m is 23.00 days.			
2	12.0	7.0	0.0
2	14.0	6.74	4.98
2	16.0	6.44	4.32
2	18.0	6.31	4.44
2	20.0	6.24	4.57
2	22.0	6.15	4.72
Time required for Tracer Particle No.2 to travel from injection borehole to observation borehole, a distance of 10.0 m is 23.03 days.			
3	12.0	6.0	0.00
3	14.0	5.76	4.88
3	16.0	5.53	4.51
3	18.0	5.40	4.50
3	20.0	5.31	4.56
3	22.0	5.22	4.60
Time required for Tracer Particle No.3 to travel from injection borehole to observation borehole, a distance of 10.0 m is 23.05 days.			
4	12.0	5.0	0.0
4	14.0	4.79	4.88
4	16.0	4.61	4.62
4	18.0	4.49	4.56
4	20.0	4.41	4.55
4	22.0	4.32	4.53
Time required for Tracer Particle No.4 to travel from injection borehole to observation borehole, a distance of 10.0 m is 23.14 days.			

TABLE - 3

Tracer Particle No.	Coordinates Meters X	Z	Time in Days
1	12.0	8.0	0.0
1	14.0	7.88	5.10
1	16.0	7.69	5.24
1	18.0	7.34	3.85
1	20.0	7.26	4.18
1	22.0	7.19	4.44
Time required for Tracer Particle No.1 to travel from injection borehole to observation borehole, a distance of 10.0 m is 22.81 days.			
2	12.0	7.0	0.0
2	14.0	6.87	5.01
2	16.0	6.70	4.91
2	18.0	6.49	4.12
2	20.0	6.38	4.20
2	22.0	6.29	4.34
Time required for Tracer Particle No.2 to travel from injection borehole to observation borehole, a distance of 10.0 m is 22.58 days.			
3	12.0	6.0	0.0
3	14.0	5.73	9.94
3	16.0	5.45	9.60
3	18.0	5.14	8.46
3	19.30	5.00	5.49
3	20.0	4.93	2.95
3	22.0	4.76	8.34
Time required for Tracer Particle No.3 to travel from injection borehole to observation borehole, a distance of 10.0 m is 44.78 days.			
4	12.0	5.0	0.0
4	14.0	4.76	9.86
4	16.0	4.52	9.44
4	18.0	4.28	8.68
4	20.0	4.09	8.47
4	21.05	4.0	4.40
4	22.0	3.93	3.88
Time required for Tracer Particle No.4 to travel from injection borehole to observation borehole, a distance of 10.0 m is 44.73 days.			

In case of a stratified medium of differing permeabilities, the tracer seems to circulate with different velocities. Therefore the measurement of tracer concentration in the observation borehole has multi-peaks depending upon the number of layers. The tracer concentration at the peaks will be different which depend upon flow rates in the layers and the volume of the water in the observation borehole, which means inputs of tracer in the observation borehole are in proportion to the permeabilities of the layers through which tracer travels.

7.0 CONTAMINANT MOVEMENT

In a homogeneous isotropic porous medium all fluid particles move at identical velocities along parallel streamlines. The driving force of the contaminant movement is due to hydraulic gradients which depend upon the distribution of groundwater heads. The model presented above can be used to predict the contaminant movement only approximately by the application of Darcy's law in macroscopic sense. Because 'Darcy'flow makes no allowance for variations in the subsurface flow velocity within the pore, due to drag effects of pore wall on the fluid. However, the transport of contaminant in the medium depends upon two more factors, which are molecular diffusion and hydrodynamic dispersion.

In order to predict the contaminant movement in the porous medium it will be necessary to modify the model by taking into account dispersion phenomena. The model can be modified by neglecting molecular diffusion and tranverse dispersion and taking into account only longitudinal dispersion. The model will allow the direct calculation of the concentration of contaminant at any time and any position along a streamline (6) by solving the equation :

$$\frac{\partial C}{\partial t} + U \frac{\partial C}{\partial S} = \alpha L \cdot U \cdot \frac{\partial^2 C}{\partial S^2} \qquad \ldots(14)$$

where

C is the concentration of the contaminant

t is the time

U is average pore water velocity

S is the curvilinear distance along the streamline

αL is longitudinal dispersivity (dimension L)

8.0 CONCLUSIONS

The mathematical model developed for the estimation of seepage losses through unlined canal having impermeable layer at its bottom is successfully modified to compute the streamlines for tracing the path of an injected tracer. It has been observed that the arival of the tracer in an observation borehole has a direct relation with the distribution of groundwater heads and the permeability of the medium through which it passes. The influence of dispersion will have to be included in the model to study the movement of contaminant in the porous medium.

9.0 ACKNOWLEDGEMENT

Author is grateful to Dr. Z.S.Tarapore, Director and Shri S.L.Mokhashi, Additional Director for their encouragement and permission to publish this paper.

Thanks are due to Mrs.V.M.Bendre, Chief Research Officer and Shri R.K.Prasad, Senior Research Officer for their help during running the programs on CYBER system of Central Water and Power Research Station, Khadakwasla, Pune.

Sincere thanks are due to Shri Azher Ansari,Chief Research Officer, Mrs.S.R. Pawar and Shri J.S. Borje for their help in preparing this paper.

10.0 REFERENCES

1) **Mahajan, N.M.,** 1984. Application of Radioisotope Techniques and Borehole Geophysics in Water Resources Research Ph.D. Thesis, University of Poona, India.

2) **Mahajan, N.M.,** 1985. Canal Leakage Models. M.Sc. Thesis University of Birmingham, United Kingdom.

3) **Rushton, K.R.,** and **Redshaw, S.C.,** 1978. Seepage and Groundwater Flow, John Wiley and Sons.

4) **Rushton, K.R.,** 1985. Personal Communications.

5) **Sauty, J.P.,** 1972. Computer Simulation of Pollution Front Movement. Groundwater Pollution in Europe. Port Washington, New York, U.S.A. Water Information Centre.

6) International Hydrological Program, 1980. Aquifer Contamination and Protection. UNESCO Report, France.

SESSION S-3

ACQUISITION OF GEOHYDROLOGIC DATA

AN INTERACTIVE GROUNDWATER DATABASE SYSTEM

A.K. DIKSHIT and S. RAMASESHAN

Department of Civil Engineering, Indian Institute of Technology, Kanpur-208016, INDIA

ABSTRACT

Site specific water resources data are needed for resource management including design, development and operation of water resources systems. To illustrate the development of a database and the use of Database Management System (DBMS), a groundwater database management system has been developed for a part of Uttar Pradesh using dBASE III PLUS language and data from Central Ground Water Board for a year and U.P. Ground Water Investigation Organisation for a decade on IBM-PC/XT. A multiple file structure with a master file and other subsidiary files (viz., water data file etc.), each including a common field (viz. Station Code) to establish a relationship among various databases, has been used. A menu based interactive program using dBASE III PLUS has been developed to edit, update, sort and manipulate multiple databases and retrieve data as required. The capabilities of the system has been illusrated with examples. The transportation of data to other programs for analyses and graphics is possible. However, two graphic programs, one for X-Y plot and other for contouring have been developed for post processing of retrieved data.

INTRODUCTION

Water resources systems probably encompass the broadest spectrum of resources from ocean to lakes to rivers to reservoirs to canals. The activities in this area has been directed towards water quantity and quality through prediction, planning, design, operation and management. During last three decades, the introduction of digital computers and its application to water resources tecnology has led to dramatic changes in methodologies and practices in this field. Computer systems and programs have become an indispensable tool for simulation and optimisation, precipitation and flood forecasting, data collection and management activities, operation of river related facilities, treatment plants etc. Thus computers have come to exert a major influence on all of these aspects.

A problem common with most of the water resources systems is the extensive formulation necessary to describe the system as well as the large amount of data required to describe inputs and outputs of the system. In general it is difficult to describe the system and its inputs and outputs in terms of the complex relationships between system and its subsystems. Before the introduction of digital computer, the data and the information describing the system were to be stored only as manuals, reports and drawings. Digital computers have made possible the storage and manipulation of very large amount of information, but the amount of data required to completely describe a comprehensive water resources system may be so large as to defy the capabilities of even very large computer systems. Hence trade-offs become necessary between the accuracy of discription and cost, or between the amount of data stored and cost.

WATER RESOURCES DATABASE MANAGEMENT SYSTEMS

A water resources data bank provides the linkage among those who acquire and those who use water resources data. In other words, its primary objective is to make water data available to those who need it for the planning, design and development of water resources and for operational and management studies. A data bank in the field of hydrology and water resources is more complex as it has to store - (i) hydrologic data both for surface and subsurface hydrology, (ii) meteorologic data, (iii) ecologic data, (iv) geologic data, (v) geographic data, (vi) landuse data, (vii) engineering data, (viii) socio- economic data etc. The basic organisation of a typical water resources data bank can be represented by Figure 1 which is self explanatory.

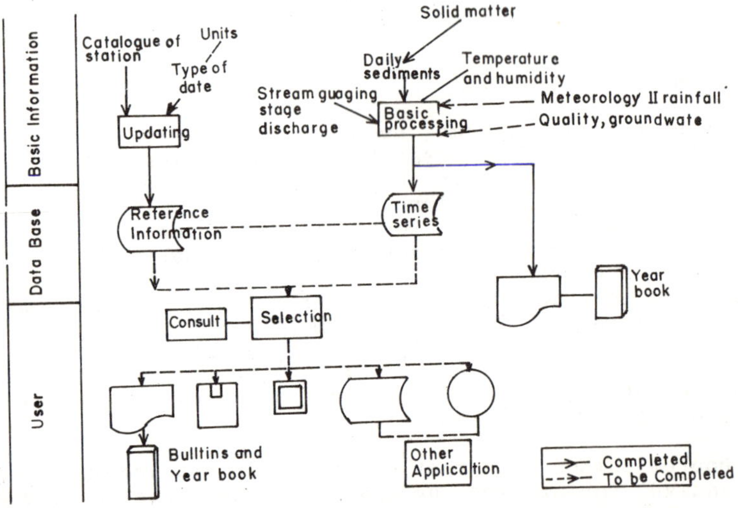

Figure 1. Basic organisation of a typical water data library (WMO, 1981)

Figures 2 and 3 show systematic functions of a data library. These are:
(i) data acquisition and management including collection of field data, standardisation of data collection formats, methods of collection, measurement and observation as well as establishment of a systematic data acquisition network. (ii) processing of raw data through quality control, editing, and analysis; check for consistency and missing data, time-series, regression and correlation analyses etc., and (iii) storage and retrieval of processed data using a database management system (DBMS) and dissemination of information to users in the forms of computer listings, microfiche, floppies, paper tapes, magnetic tapes etc. Data may also be presented in analogue form such as graph, map and drawing for better understanding.

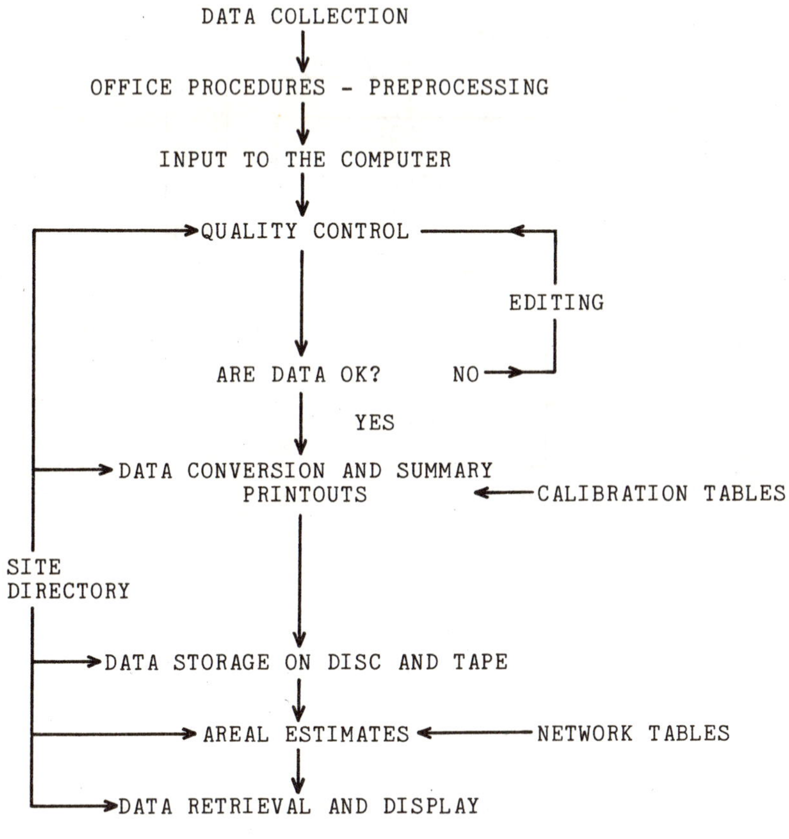

Figure 2. Flow chart showing the generalf water resources data processing frame work (I.H., 1981)

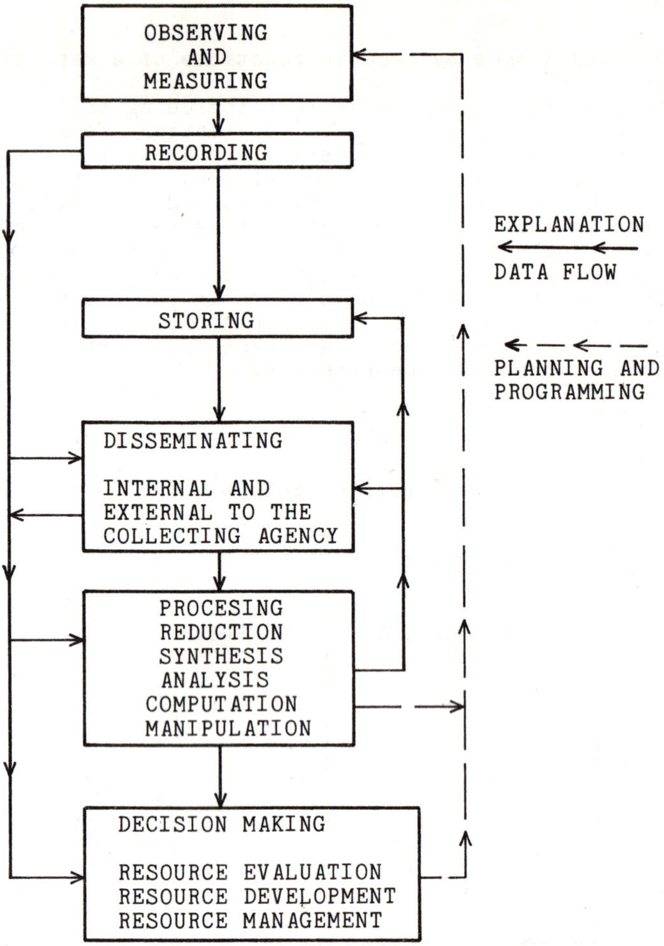

METHODS OF DATA FLOW		FORMS OF DISSEMINATION	
RADIO	TELEGRAPH	PUBLICATION	MANUAL OUTPUT
MICROWAVE	MAIL	MICROFORM	MACHINE COMPATIBLE
TELEPHONE	MESSENGER	PRINTOUT	MEDIA

Figure 3. Systematic functions of a typical data bank (WMO, 1981)

Two types of data banks are possible - (i) centralised data bank and (ii) data bank network. A development in growing needs and importance in data acquisition, management and dissemination is the establishment of centralised data bank. The basic concept is to maintain data at a central location rather than at each user location. Another type of data bank consists of a network of small data. Thus a large number of manageable data banks are coordinated to a central data system. This central system does not hold data but acts as the focal point of entire network.

GROUNDWATER DATABASE MANAGEMENT SYSTEM

Groundwater plays an important role in the development of a region in the country. The states with sufficient groundwater potential have a strong economic potential compared to the states having limited groundwater. Any design, development or planning involves decision making which calls for the need of huge amount of data immediately. In India, the Central Groundwater Board and the State Ground Water Departments are collecting hydrogeological, hydrological and hydrochemical data for the development and management of groundwater resources. Generally, it is difficult to locate and retrieve a particular information in desired format resulting in unnecessary delay in the process of design or in developing or specifying operational policy for a particular project. Present study is concerned with the systematic creation and management of a database whereby the data can be manipulated interactively and retrieved easily. It is proposed to use generally easily available software for PC's in the development of the Groundwater Database (GWDB) system.

In particular a versatile database management package dBASE III PLUS was available for the study and GWDB has been implemented in dBASE III PLUS in a PC/XT. The dBASE III PLUS is primarily capable of performing following tasks:

1. Add new data to the database
2. Sort the database in some meaningful order
3. Search the database(s) onto formatted reports
4. Print data from database(s) onto formatted reports
5. Edit and update data on the database
6. Delete data from the database.

It is a powerful and flexible system on micro/mini computers. dBASE III PLUS has two major advantages over its predecessor:

1. Much improved interface betwen data user and database(s)
2. More powerful as a programming language. New commands, functions and debugging utilities added to programming language make dBASE III PLUS a very powerful database management system.

Structure of the Database GWDB

To computerise all the information available with reference to groundwater, the structure of the database files has to be decided before its actual implementation. In other words, comprehensive data formats are to be designed first as they are intimately tied to the development and management of the database system. These data formats are such that a minimum computer storage is required and they are dependent upon the data acquisition system too.

Keeping in view the way of data storage at various stations of Central Groundwater Board (CGWB) and at those of Groundwater Investigation Organisation (GWIO, U.P., data formats are to be finalised. CGWB collects depth to water table and water quality data at all hydrograph stations for the months of January, April, June, August and November. Whereas GWIO collects maximum and minimum depth to water table, thus only two times a year. The data collected by each of two agencies are site location and identification, geohydrological characteristcs, well construction history, periodic water levels etc. Accordingly, a system with three database files has been developed. These are:

1. Master file
2. Water data file
3. Water quality file

Master File

This file provides space for site specific and time independent data including regional and location code, the district name, block name, station name, site description including latitude, longitude, ground level at measuring stations, datum etc. Station code is used as a key for relating master file to rest of files. The designed structure of this file has been shown in Table 1.

Water Data File

Starting with station code, this database file stores the time dependent data of depth to water table below ground level including the month and year of measurement and remarks, if any. Table 2 displays the structure of this file.

Table 1. Structure of master file

Field	Field Name	Type	Width	Dec
1	CODE	Character	12	
2	DISTRICT	Character	12	
3	BLOCK	Character	10	
4	STATION	Character	15	
5	LATITUDE	Character	9	
6	LONGITUDE	Character	9	
7	GL	Numeric	8	3
8	DATUM	Numeric	5	2
9	LOCATION	Character	80	
10	TYPE	Character	5	
11	DIAMETER	Numeric	5	2
12	DEPTH	Numeric	6	2
13	LITHOLOGY	Character	8	
** Total **			165	

Table 2. Structure of water data file

Field	Field Name	Type	Width	Dec
1	CODE	Character	12	
2	YEAR	Character	5	
3	MONTH	Character	3	
4	OBSERVATION	Numeric	6	2
5	REMARK	Character	4	
** Total **			31	

Implementation of GWDB for a Part of U.P.

In case of water resources database systems, it is essential that information be divided into separate databases for better management and maintenance of each of these database files. For GWDB system all groundwater data is stored in a set of three database files. All the database files have a field called CODE to relate them with master file. The common field that relates the three databases is an important one. A code is ideal common field because it does not take up much disk space. It is also very easy to asign each station a unique code number. This common but unique field is referred to as a KEY field.

It must be kept in mind that for general file maintenance such as additing new reords or editing existing data, the three databases will be treated as separate. However, there will be times when it is needed to retrieve information simultaneously from all databases. Tables 3 and 4 show sample dataset from master file and water data file respectively.

Data Retrieval from GWDB

Data retrieval from GWDB means to search or query database(s) to view records that match some criterion. The specific search criterion may be - to search all water table records for some specified district or block or station or some combination, to list all stations where water table is shallow (say less than or equal to 2 meter below ground level) or to list all stations where water depth is more than 10 Meter or so, to list all water records for specified year (for a specified district too) etc. dBASE III PLUS imposes no limit to ways in which one can access data stored in databases. Sample outputs are shfown in Table 5 and Table 6.

Table 3. Sample data set from master file

DISTRICT	BLOCK	STATION	LATITUDE	LONGITUDE	GL	DATUM
DEHRADUN	DOIWALA	DOIWALA	30 10 09	78 06 57	475.735	1.30
DEHRADUN	DOIWALA	DATO	30 14 31	77 57 43	749.673	1.30
DEHRADUN	DOIWALA	HARRAWALA	30 15 42	78 05 35	558.653	1.00
DEHRADUN	DOIWALA	KANSROO	30 05 20	78 07 57	414.301	1.20
DEHRADUN	DOIWALA	RISIKESH	30 06 37	78 17 24	338.994	1.20
DEHRADUN	DOIWALA	RAIWALA	30 01 59	78 12 35	351.679	1.50
DEHRADUN	SAHASPUR	HARBRATPUR	30 26 19	77 44 09	441.866	1.50
DEHRADUN	SAHASPUR	JHAJHARA	30 20 41	77 55 40	546.750	1.80
DEHRADUN	SAHASPUR	MAJARA	30 17 34	78 00 12	607.551	1.50
DEHRADUN	SAHASPUR	SAHASPUR	30 23 28	77 48 32	476.626	2.50

Table 4. Sample data set from water data file

CODE	YEAR	MONTH	OBSERVATIO	REMARK
DDDOIWALA	1972	7	28.10	MAX
DDDOIWALA	1973	5	34.18	MAX
DDDOIWALA	1974	--	32.25	MAX
DDDOIWALA	1975	--	0.00	DRY
DDDOIWALA	1976	--	33.00	MAX
DDDOIWALA	1977	--	0.00	DRY
DDDOIWALA	1972	--	19.54	MIN
DDDOIWALA	1973	--	16.21	MIN
DDDOIWALA	1974	8	18.80	MIN
DDDOIWALA	1975	--	18.85	MIN

Table 5. Sample output

. LIST CODE,BLOCK,LATITUDE,LONGITUDE,GL FOR DISTRICT=SAHARANPUR

RECORD	CODE	BLOCK	LATITUDE	LONGITUDE	GL
11	SRAURAGABAD	BAHADRABAD	29 59 15	78 00 30	298.003
12	SRBAHADRABAD	BAHADRABAD	29 55 10	78 02 36	273.759
13	SRDOLTAPUR	BAHADRABAD	29 56 10	77 58 17	272.702
14	SRHARIDWAR	BAHADRABAD	29 59 43	78 09 22	290.400
15	SRJWALAPUR	BAHADRABAD	29 55 03	78 06 13	283.093
16	SRPATHRI	BAHADRABAD	29 51 02	78 04 49	256.555
17	SRSULTANPUR	BAHADRABAD	29 45 35	78 06 27	242.098
18	SRBALIAKHERI	BALIAKHERI	29 55 15	77 39 03	276.762
19	SRSAHARANPUR	BALIAKHERI	29 57 55	77 34 02	275.353
20	SRBAHBALPUR	BHAGWANPUR	29 59 15	77 48 37	278.010

Table 6. Sample output

.list for year='1977'.and.observatio =2.00

RECORD	CODE	YEAR	MONTH	OBSERVATIO	REMARK
6	DDDOIWALA	1977	--	0.00	DRY
48	DDKANSROO	1977	--	1.09	MIN
95	DDJHAJHARA	1977	--	1.30	MIN
107	DDMAJARA	1977	--	0.51	MIN
177	SRPATRI	1977	--	1.21	MIN
185	SRSULTANPUR	1977	--	1.21	MIN
217	SRBAHBALPUR	1977	--	1.57	MIN
246	SRDHOLKHAND	1977	--	0.95	MIN
291	SRDEOBAND	1977	--	1.65	MIN
304	SRRAJUPUR	1977	--	1.48	MIN
361	SRKANDLANA	1977	--	1.22	MIN

Water table data has been stored directly into databases. It is, however, possible to derive water table elevation or water table fluctuation data from water table, ground level and datum information by making coordinated use of various databases simultaneously. This is one of strong features of a database system. Table 7 lists a report of code number, latitude, longitude, ground level, depth to water table below datum and deriveed water table elevation from master file and water data file. If one is interested in knowing pattern of depth to water table variation or changes in water table elevation at a particular site and to perform a time-series analysis, data retrieval is very easry and straight forward. A typical set of outputs is given in Table 8.

In order to make GWDB system more effective and user oriented, a computer code has been written using dBASE III PLUS commands. This allows user to interact with GWDB system for specifying the data retrieval and other needs through menu selections rather than through programming like commands illusrated in Table 9. These menus free the user from memorising commands and rules of syntax.

Table 7. Sample output

```
.LIST CODE,LATITUDE, LONGITUDE, GL,(B- OBSERVATIO-DATUM),
(GL+DATUM-B- OBSERVATIO) FOR (B- YEAR)=197
```

RECORD	CODE	LATITUDE	LONGITUDE	GL	(B- OBSER-VATIO-DATUM)	(GL+DATUM-B-OBSERVATIO)
1	DDDOIWALA	30 10 09	78 06 57	475.735	26.80	448.935
2	DDDATQ	30 14 31	77 57 43	749.673	13.50	736.173
3	DDHARRAWALA	30 15 42	78 05 35	558.653	15.00	543.653
4	DDKANSROO	30 05 20	78 07 57	414.301	2.52	411.781
5	DDRISIKESH	30 06 37	78 17 24	338.994	5.25	333.744
6	DDRAIWALA	30 01 59	78 12 35	351.679	17.95	333.729
7	DDHARBRATPUR	30 26 19	77 44 09	441.866	6.86	435.006
8	DDJHAJHARA	30 20 41	77 55 40	546.750	11.61	535.140
9	DDMAJARA	30 17 34	78 00 12	607.551	9.61	597.941
10	DDSAHASPUR	30 23 28	77 48 32	476.626	6.18	470.446
11	SRAURANGABAD	29 59 15	78 00 30	298.003	5.85	292.153
18	SRBALIAKHERI	29 55 15	77 39 03	276.762	3.69	273.072
19	SRSAHARANPUR	29 57 55	77 34 02	275.353	5.60	269.753
21	SRBUDWASAHID	30 05 14	77 52 53	329.557	4.41	325.147
22	SRBHAGWANPUR	29 56 14	77 48 55	276.140	3.32	272.820
24	SRMANUBAD	29 59 52	77 54 39	295.455	12.60	282.855
25	SRBHAILA	29 42 09	77 36 54	256.271	3.25	253.021
31	SRGANGOH	29 47 15	77 15 55	257.119	7.68	249.439

Table 8. Sample output

```
.SELECT A
.LIST CODE,DISTRICT, BLOCK, STATION,B- YEAR, (B- OBSERVATIO-DATUM) FOR
 DISTRICT=SAHARANPUR
```

Record	CODE	DISTRICT	BLOCK	STATION	B- YEAR	(B- OBSERVA-TIO-DATUM)
11	SRAURANGABAD	SAHARANPUR	BAHADRABAD	AURANGABAD	1972	5.85
12	SRBAHADRABAD	SAHARANPUR	BAHADRABAD	BAHADRABAD	1975	4.50
13	SRDOLTAPUR	SAHARANPUR	BAHADRABAD	DOLTAPUR		- 1.40
14	SRHARIDWAR	SAHARANPUR	BAHADRABAD	HARIDWAR	1974	7.18
15	SRJWALAPUR	SAHARANPUR	BAHADRABAD	JWALAPUR	1971	12.47
16	SRPATHRI	SAHARANPUR	BAHADRABAD	PATHARI	1974	2.60
17	SRSULTANPUR	SAHARANPUR	BAHADRABAD	SULTANPUR	1974	2.64
18	SRBALIAKHERI	SAHARANPUR	BALIAKHERI	BALIAKHERI	1972	3.69
19	SRSAHARANPUR	SAHARANPUR	BALIAKHERI	SAHARANPUR	1972	5.60
20	SRBAHBALPUR	SAHARANPUR	BHAGWANPUR	BAHBALPUR	1974	2.65
21	SRBUDWASAHID	SAHARANPUR	BHAGWANPUR	BUDWASAHID	1972	4.41
22	SRBHAGWANPUR	SAHARANPUR	BHAGWANPUR	BHAGWANPUR	1972	3.32
23	SRDHOLKHAND	SAHARANPUR	BHAGWANPUR	DHOLKHAND	1974	3.30
24	SRMANUBAD	SAHARANPUR	BHAGWANPUR	MANUBAD	1972	12.60
25	SRBHAILA	SAHARANPUR	DEOBAND	BHAILA	1972	3.25
26	SRBHANERAKHA	SAHARANPUR	DEOBAND	BHANERAKHAS	1974	2.05
27	SRDEOBAND	SAHARANPUR	DEOBAND	DEOBAND	1971	3.00
28	SRRAJUPUR	SAHARANPUR	DEOBAND	RAJUPUR	1971	1.00
29	SRSAKHANKHUR	SAHARANPUR	DEOBAND	SAKHANKHUR	1971	2.80

Table 9. Typical menus used in GWDB
Table 9(A). Main menu

```
----------------------------------------------------

                 GWDB MANAGEMENT MENU
                 ====================

                 What do you want do-

                   1. Add new records
                   2. Edit data
                   3. Print report
                   4. Retrieve data
                   5. Exit the GWDB system

    Enter your choice (1 to 5) from above:
-------------------------------------------------------
```

Table 9(B). Sub menu 1

```
-------------------------------------------------------

                  ADD NEW RECORDS
                  ===============

             Which database file you want to
             append -

             I).  GWIO files :
                      1. GWIOF1 - Master file
                      2. GWIOF2 - Water data file
                      3. GWIOF3 - Water quality file
             II). CGWB file :
                      4. CGWBF1 - Master file
                      5. CGWBF2 - Water data file
                      6. CGWBf3 - Water quality file

   Enter the serial no. of data base in use :
-------------------------------------------------------------
```

POST PROCESSING OF RETRIEVED DATA

Data held in database files could be used to export data to Pascal, Fortran programs etc. to perform some specified job. The post processing of data needs the data to be exported from GWDB system to some other analysis softwares which essentially yield the information in most useful forms such as :

1. High and low valuesof a particular data field(s) e.g. maximum/ minimum water table for particular district
2. Preparation of annual statistics and publishing tables of actual or processed data
3. Preparation of graphs and maps from actual or analysed data

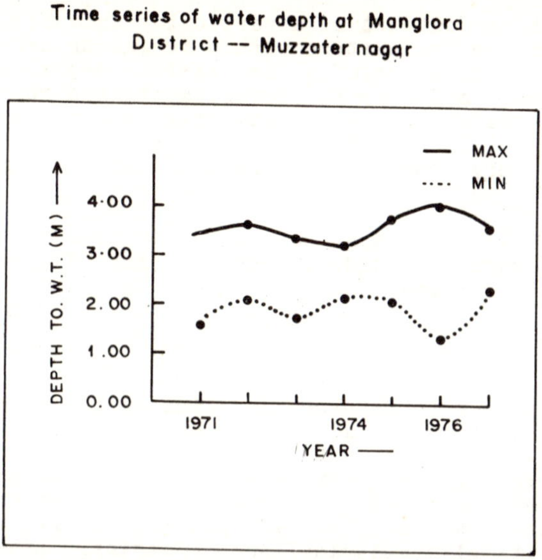

Fig. 5. Time – series of depth to water depth at Manglora

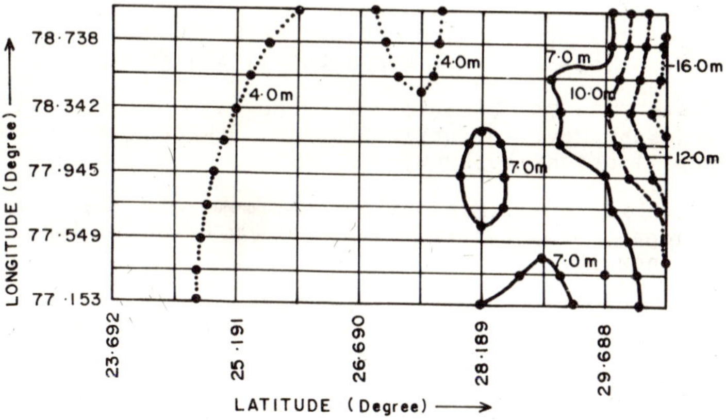

Fig. 6. Contours of depth to water table below G.L.

4. Plotting contours of a particular data field to clearly visualise the pattern within region or among regions
5. The hydrologic studies to assist decision makers for optimal management, development and monitoring of the groundwater resources as well as to estimate water balance for each administrative unit or drainage basin at a very short notice and low cost.

Some utilities for post processing of data viz., getting a X-Y plot (Figure 5) and a contour map (Figure 6) have been developed in this study.

CONCLUSIONS

It is possible to develop and implement a multiple file groundwater database in a personal computer. dBASE III PLUS or equivalent laguage may be used for database management including storage, updating and retrieval of data. Since the DBMS languages provide for report generation, graphics and export of data to other systems, it is possible to generate formatted reports, results of statistical and numerical analyses, charts and maps from DBMS and finally a menu based interactive program will facilitate easy and efficient database management.

sReferences:

CGWB, 1984. Hydrogeological and Hydrochemical Characteristics of Permanent Hydrograph Stations in Uttar Pradesh (1982), Central Ground Water Board, Northern region, Govt. of India, Ministry of Irrigation.

Dikshit, A.K., 1987. A Groundwater Database Management System, M.Tech Thesis, Civil Engg. Dept., IIT Kanpur.

GWIO, Part I, 1979. Location of Hydrograph Stations in Uttar Pradesh, Govt. of Uttar Pradesh, Groundwater Investigation Organisation, Lucknow, U.P.

GWIO, Part II, 1979. Minimum and Maximum Water Level in Uttar Pradesh, Groundwater Investigation Organisation, Lucknow, U.P.

Mason, F.D. 1975. Support in the Implementation of a National Water Data Exchage (NAWDEX), Final Report (June 1974-May (1975)) PRC System Sciences Company.

Roy, Gautam, 1983-84. Data Systems and Library, Status Report SR-4, National Institute of Hydrology, Roorkee.

Simpson, A, 1986. Understanding dBASE III PLUS, B.P.B. Publication, New Delhi.

USGS, 1971. Design Characteristics for a National System to Store, Retrieve and Disseminate Water Data, U.S. Geological Survey, Washington, DC.

BOREHOLE RESISTIVITY LOGGING IN PVC SCREENED WELL

S.V.V. PRASADA RAO, G. VARADARAJAN, P. PRABHAKARA PRASAD, K. SUBRAHMANYAM and T. SESHUNARAYANA

NATIONAL GEOPHYSICAL RESEARCH INSTITUTE, HYDERABAD

ABSTRACT

A method of obtaining resistivity logs in oreholes with PVC well screens is described. Normal resistivity logs run in a PVC screened well in unconsolidated formations exhibits almost the same curve characteristics as that of the open mud filled hole. The long normal curve is almost identical in both the screened and uncased well and the short normal curve shows a slight increase in resistivity in the screened well as compared to the open well. This method is particularly useful for logging in collapsing formations viz., coastal sediments. Periodic logging in PVC screened wells will enable the measurement of changes in porosity and water salinity during the life span of the well.

INTRODUCTION

Resistivity logs are generally recorded in the open holes immediately after drilling. Open hole' indicates that the formation in the wall is exposed as opposed to cased hole in which a tube of metal casing lines the well. A successful attempt was made to get valid logs in a well cased with perforated PVC tubes. Perforated casing provided ohmic contact for current injection and voltage measurement while at the same time it prevented the collapse of the formations. If one used a steel casing, it would have screened the formation from the sonde, making measurement impossible. In order to confirm the workability of the method, model tank experiments were conducted. The results obtained were later tested in the borehole with PVC casing in a sedimentary area near Agra.

The model tank experiments were conducted by using the tank set up available at NGRI, consisting of a constant current source and an accurate digital voltmeter.

The borehole logging resistivity work was conducted using model No 3200 PLS manufactured by Gearhart-Owen industries Inc, Texas, USA.

EXPERIMENTAL STUDIES IN THE LABORATORY AND THE FIELD

Experimental studies were carriedout in a model tank filled with water of resistivity, 23.6 ohm-m. Firstly, the resistivity of water was measured with perforated PVC tubes. The dimensions of the tube are:- length 55 cm, inner diameter 1.6 cm, perforation diameter 0.5 cm, distance between two successive perforations along the length is 3.5 cm and along the circumference 0.8 cm. Constant current of magnitude 86.4 micro amps. and of frequency 25 Hz was passed between two insulated pointed electrodes completely immersed in water and placed in the two perforated tubes and the resulting potential drop was measured through two insulated point electrodes placed in the other two perforated PVC tubes. Wenner configuration was used and the distance between the electrodes was 12 cm. The calculated resistivity was 24.2 ohm-m. The same experiment was repeated without PVC tubes and the calculated resistivity turned out to be of 23.6 ohm-m. From this experiment it was concluded that under experimental conditions the resistivity is almost same with perforated PVC tubes or without them.

Encouraged by the results of model tank experiments, the authors attempted to obtain resistivity logs in a borehole with PVC well screens at Nagalapatta village, Agra Dist., U.P. A borehole of dia 0.203 m drilled upto a depth of 50 m. with bentonite mud of resistivity of 4.5 ohm-m was used. The resistivity of the water used for drilling is 8.75 ohm-m. The formations encountered in the borehole are clays, fine sands, and sandstones. Five PVC pipes of dia 0.15 m and of length 6.00m.and with slots each of dia 0.025 m, spaced 0.045 m along the diagonal and 0.04 m along the circumference were lowered in the borehole one after another after joining them with couplers of 0.20 m length without slots. Thus the PVC screen has 0.15 m length of blank pipe both above and below the junction of two tubes. Short normal and long normal curves were recorded with the sonde in the screened part of the well and also in the open hole from 28 m to 50 m in a 2 ohm-m per division calibration.The same experiment was repeated in the uncased hole without disturbing the instrumental set up. The recorded logs are presented in Fig. 1 and Fig. 2. The entire logging operations were completed in less than an hour after drilling. Hence it is assumed that the invasion effects in the borehole is same for both types of logging operations.

DISCUSSION OF THE RESULTS

Comparing the curve obtained in open well with that of the screened well in short normal mode, it is observed that the apparent resistivity values in the later are higher. Comparison between the open well and screened situations in the long normal mode, we have almost the same readings in both. This means that the screen which is at a close distance from the borehole axis has negligible influence on the electric log in the long normal mode. This is as it should be vide current thinking on the subject of radius of investigation (Roy and Dhar, 1971).

STATE : UTTAR PRADESH, INDIA
LOCATION : NAGAL PATTA
TOTAL DEPTH LOGGED : 50 m
TYPE OF MUD : BENTONITE
DEPTH OF PVC CASING : 28 m
RESISTIVITY OF DRILLING WATER : 8.75 Ω m
RESISTIVITY OF MUD : 4.5 Ω m
DIAMETER OF BOREHOLE : 0.203 m (8")

FIG. 1 SHORT NORMAL — FIG. 2 LONG NORMAL

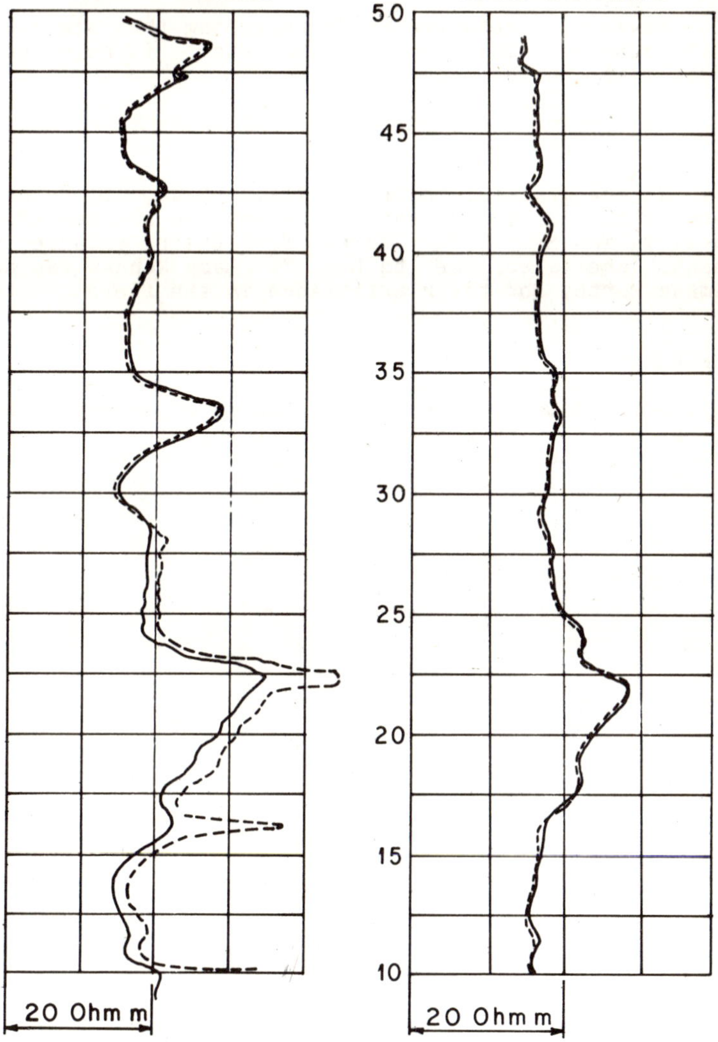

- - - - CASED DOWN LOG
——— UNCASED DOWN LOG

SHORT NORMAL AND LONG NORMAL LOGS IN A CASED AND UNCASED WELL

CONCLUSIONS

It is possible to record resistivity logs in a cased hole with suitable well screens.

The Long normal curve presents identical curve characterstics both in cased and uncased well.

The Short normal curve exhibits small deviations in the apparent resistivity values.

However, a detailed study regarding the dimensions of the well screens, invasion effects etc. is required for drawing additional conclusions.

ACKNOWLEDGEMENTS

We are grateful to Prof. V.K. Gaur, Director, NGRI for permitting us to present the paper. We are thankful to Ms. M. Kousalya and Mr. M.P. Sarma for their valuble suggestions. Thanks are also due to Mr. Y. Rama Mohana Rao and Mr. V.V. Ramana Murthy for their assistance in the field.

REFERENCE

Roy, A., and Dhar, R. L., 1971, Radius of investigation in DC Resistivity well logging. Geophysics, v. 36, no. 4, p. 754-756.

TECHNIQUES OF AQUIFER PARAMETER EVALUATION USING POCKET COMPUTERS

A.BALASUBRAMANIAN[1] AND J.C.V.SASTRI[2]

[1]Department of Geology, V.O.C.College, Tuticorin, India.
[2]Department of Geology, University of Mysore, Mysore, India.

ABSTRACT

Pocket computer techniques of aquifer parameter evaluation involves both simple and iterative computations. Simple methods enable in approximating the properties like transmissivity and storage coefficients, using least squares approach. Parameter estimation with a limited data could be attempted through iterative techniques. Automated fitting of pumping test data can be done through sensitivity analysis. 'On the spot' analysis and interpretation of aquifer tests could be done using these pocket computer techniques.

INTRODUCTION

Aquifer parameters are evaluated by conducting pumping tests at selected locations of various lithologies. Though most of the constants are site specific, specially in hardrock terrain reasonable conclusions could be derived when they are used for a regional study. Hydrogeologists continually seek and test simple, quick and inexpensive methods for the determination of aquifer properties using pumping test data. Most of the available methods are time-consuming curve-matching types and in these procedures there is room for personal error in judgement and hence in computations.

Numerous computational methods have been proposed during the last two decades (Walton, 1970; Saleem, 1970; Ryzner, 1980; Mc Elwee, 1980a and Mukhopadhyay, 1985). Adequate software for this analysis is available as packages which could be normally used in mainframe, mini or personal computers. Such capability is costly and is not always readily available to the hydrogeologists either in the office or in the field. Listings, detailed descriptions of such computer techniques could be seen in Holzschuh (1976), Mc Elwee (1980b), Bradbury and Rothschild (1985) and be used only in a computer having enough memory. An attempt has been made to simplify the computational algorithm so as to evaluate the aquifer constants, even in the field, using the pocket computers like HPC, CASIO-PB-100, FX-790, etc.

POCKET COMPUTER TECHNIQUES

Pocket computers are normally encountered with 'Step' memories ranging from 554 to 8000. Each command, operator or variable name in BASIC language uses one step memory and the variables can only be one-dimensional, if needed. All the automated techniques specially used for the evaluation of aquifer constants could be done using these pocket computers through the development of suitable simple algorithms. Of these, two direct and two iterative techniques are discussed below.

A. Direct Methods :

1. Jacob's straight line method :

Aquifer transmissivity (T in gpd/ft) and storage coefficient (S) are determined in this approach using the equations

$$T = 114.59 * Q / b \quad \ldots 1$$

and

$$S = 0.0125 * T / [r_w^2 * EXP (a/b)] \quad \ldots 2$$

where Q is the pumping rate (gpm); r_w is the radius of the well (ft); a and b are the intercept and slope of the straight line fitted between drawdown (ft) and pumping time (hrs) in a semi-log graph (keeping time in log-axis). These could be evaluated by plotting the data or through least squares analysis. As the plots of initial data points may not coincide with the straight line fixed manually, a few of the initial values could also be deleted in the numerical analysis. The reliabilities of the estimates, amount of data points to be considered and the simple ways of approximating T and S have been studied and shown in Table 1 using the pumping phase data of Walton (1970). It is of interest to note that T and S are found to be closer to that of curve-matching results even when less number of data points are considered.

2. Theis Recovery Method :

The same procedure of fixing the straight line between s"(residual drawdown m) and t/t" (where t is the time in days since pumping started; and t" is the time in days since pumping stopped) on a semi log graph could be adopted numerically. As the slope of the straight line is equal to $2.303 * Q/(4 * \pi * T)$ in metric units, T could be evaluated very easily.

B. Iterative Methods :

1. Specific Capacity Method :

Theis et.al (1963) described a method for determining the T (m^2/d) of an aquifer from the specific capacity of wells using the relation

$$T = 2.303\, Q/(4\pi s).\ \log [(2.25\, Tt/(r_w^2.S)] \quad \ldots 3.$$

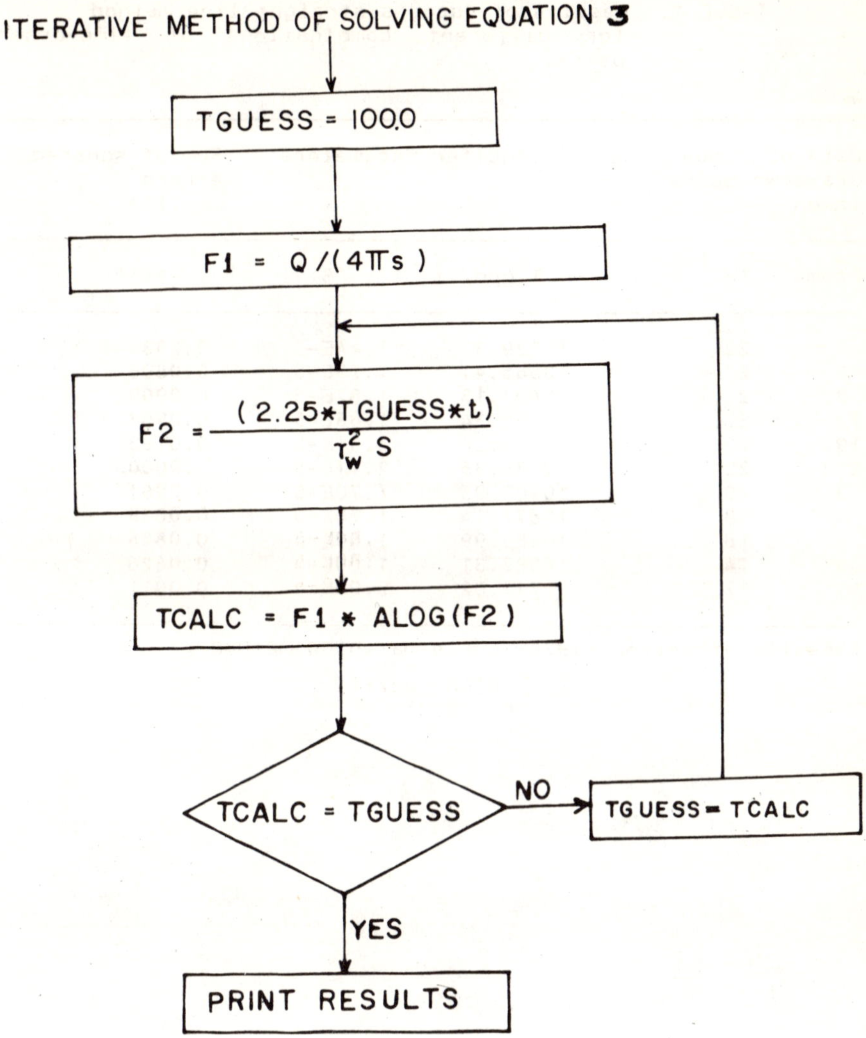

FIG. 1. FLOW CHART OF SPECIFIC CAPACITY METHOD

TABLE 1. Results of Jacob's straight line method for different combinations of time-drawn pairs.

Data of Time-drawdown pairs taken		Aquifer Parameters		Sum of squared errors (ft)
From	To	T gpd/ft	S	(ft)
1	22	11529.32	1.44E-5	0.2935
5	22	10565.41	1.75E-5	0.0806
9	22	10407.13	1.82E-5	0.0909
13	22	10350.34	1.85E-5	0.0963
19	22	9618.24	2.31E-5	0.0193
21	22	9885.35	2.11E-5	0.00003
3	20	10907.95	1.70E-5	0.0961
5	18	10677.14	1.70E-5	0.0839
7	16	10485.99	1.80E-5	0.0825
9	14	10682.31	1.80E-5	0.0426
11	12	9717.02	2.00E-5	0.0011

(Results of Walton (1970) curve-matching method

T = 10100 gpd/ft.
S = 0.00002)

where Q is the pumping rate (m^3/d); r_w is the radius of the well (m); t is the time duration in days of pumping; s is the drawdown in m and S is the storage coefficient.

When all other parameters are known except T and S, reasonable approximations could be made by assigning appropriate storage coefficients to the aquifer. As T appears on both sides, this equation could only be solved through an iterative technique, by initially guessing one T for the computation of T. The computation should be repeated by replacing computed T to the guessed T till both of them become equal as shown in the flow chart (Fig.1). In pocket computers convergence is achieved within 3 seconds and 10 iterations irrespective of the magnitude of guessed T values. For the data of Walton (op.cit) a T value of 9057.81 gpd/ft is obtained for the storage coefficient of 0.000021.

2. Sensitivity analysis :

Mc Elwee (1980a) proposed a technique and an algorithm to automatically fit experimental pumping test data to the Theis equation by obtaining the 'best' transmissivity and storage coefficient in the least squares sense through the use of sensitivity analysis. Detailed mathematical descriptions of the analysis and programs could also be seen in Mc Elwee (1980b), Balasubramanian (1986) and Subba Rao and Gokhale (1987).

In this analysis, more than 5 one-dimensional arrays are to be used apart from the lengthy computer listing. Though the method is very suitable, the processing needs a personal computer. By taking all these limitations and the capacity of pocket computers, a simple computer code in BASIC has been structured and shown in Appendix-A. Fig.2 shows the 'best fit' drawdown obtained for the data of Walton (op.cit) through this scheme.

CONCLUSION

The table given above gives all the results obtained through different techniques cited above.

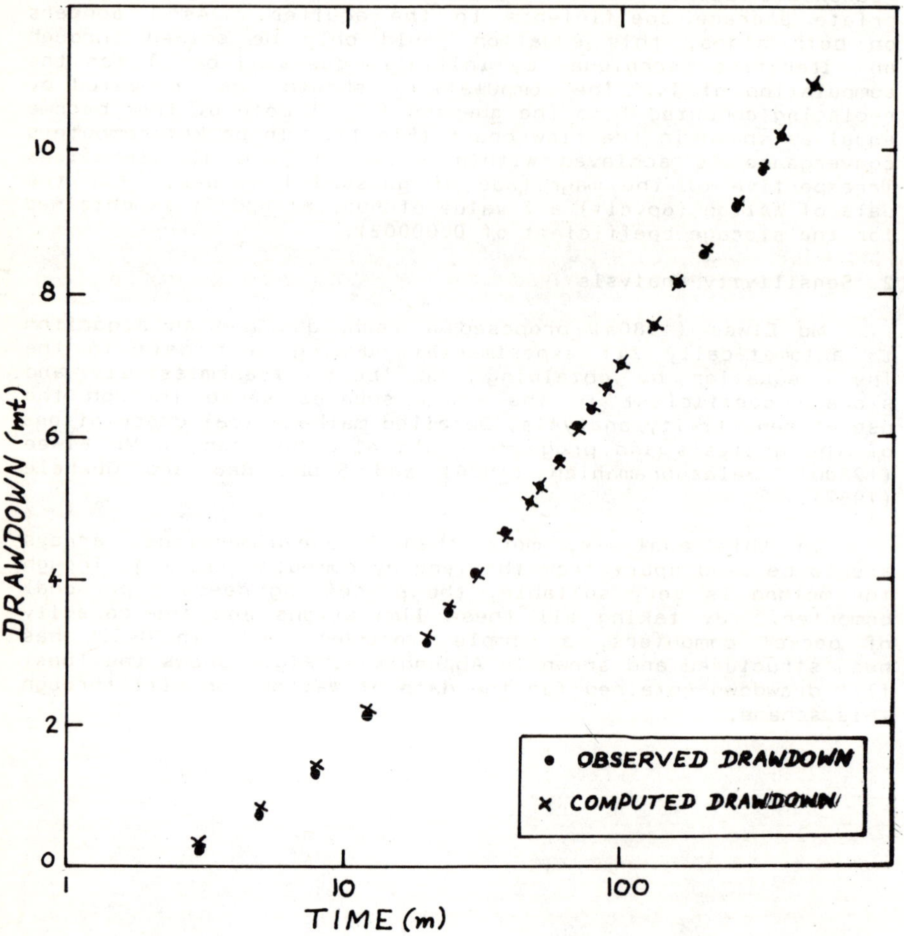

FIG. 2 AUTOMATED FIT OF PUMPING TEST DATA

Method	Duration of Analysis per Well (sec)	Aquifer parameters T (gpd/ft)	S (dimensionless)
1. Graphical Results of Walton	3600 (Manual)	10,100.00	2×10^{-5}
2. Jacob's (error 0.00003)	120	9,885.35	2.1125×10^{-5}
3. Iterative-specific capacity method.	45	9,057.81	2.1×10^{-5} (assigned)
4. Sensitivity analysis	300	9,910.00	2.1×10^{-5}

It is of interest to note that the results obtained through the method 2 is not only close to the results of methods 1 and 4 but also uses (i) a less computing time, (ii) simple computational steps and (iii) less memory constants.

Programmable pocket computers cost twenty times less than those of personal computers and have the limitations of (i) using BASIC language, (ii) less memory cells and (iii) displaying the results. The algorithms and techniques detailed in this work could be utilized to evaluate the aquifer constants not only in an office where a limited facility is available, but also in the field immediately after conducting the pumping tests.

ACKNOWLEDGEMENTS

The authors thank their Heads of Departments and Institutions for providing the necessary facilities to carryout this work.

References :

Balasubramanian, A., 1986. Hydrogeological investigations in the Tambraparni river basin, Tamilnadu, Unpublished Ph.D. Thesis, University of Mysore, Mysore, 349p.

Bradbury, K.R., and Rothschild, E.R., 1985. A computerized technique for estimating the hydraulic conductivity of aquifers from specific capacity data. Groundwater, V.23(2), p.240-246.

Holzschuh, J.C., 1976. A simple computer program for the determination of aquifer characteristics from pump test data. Groundwater. V.14(5), p.283-285.

Mc Elwee, C.D., 1980a. Theis parameter evaluation from pumping tests by sensitivity analysis. Groundwater. V.18(1), p. 56-60.

Mc Elwee, C.D., 1980b. The Theis equation; evaluation sensitivity to storage and transmissivity, and automated fit of pump test data. Kasas Geol. Sur. GW Series-3, 39p.

Mukhopadhyay, A., 1985. Automated derivation of parameters in a non-leaky confined aquifer with transient flow. Groundwater. V.23(6), p.806-811.

Rayner, F.A., 1980. Pumping test analysis with a handheld calculator. Groundwater. V.18(6), p.562-568.

Saleem, Z.A., 1970. A computer method for pumping test analysis. Groundwater. V. 8(5), p.21-24.

Subba Rao, T.V., and Gokhale, K.V.G.K., 1987. Estimation of aquifer parameters from Theis equation using sensitivity analysis. Geophy. Res. Bull., V. 25(1), p.21-26.

Theis, C.V., Brown, R.H., and Meyer, R.R., 1963. Estimating transmissibility of a watertable aquifer from specific capacity of wells (in methods of determining permeability, transmissibility and drawdown, compiled by Ray Bentall). USGS - Wat. Sup. Pap. 1536-I, p.332-336.

Walton, W.C., 1970. Groundwater resource evaluation, McGraw Hill Book Co., New York, 664p.

APPENDIX-A

SENSITIVITY ANALYSIS

Definition of Variables : N - Number of time-drawdown pairs

Array Z (I) - I = 1 to N Pumping times (T)

I = N + 1 TO 2N Pumping drawdowns (L).

T - Transmissivity (L^2/T); Q - Pumping Rate (L^3/T)

R - Radius of the well (L); F-Sum of squared errors.

K - Guessing T; H - Guessing Storage Coefficient

(The Program)

```
10   INPUT N, Q, R, K, H:  L = Ø
15   FOR I = 1 TO N : J = N+I:  INPUT Z(I), Z(J): NEXT I
20   G = K: V=H
25   A = Ø : B = Ø : C = Ø:  D = Ø : E = Ø
30   FOR I = 1 TO N : J = N+I : X = Z(I) : Y = Z(J):
     F = 2 * N+I
35   U = R * R * H/(4 * K * X): IF U > 1.0 THEN 55
40   W = -LOG(U) - .57721566 + .99999193 * U - .24991055*U*U
45   W = W + .05519968 * U * U * U - .00976004 *U↑4
50   W = W + .00107857 * U↑5 : GOTO  65
55   W = [EXP (-U)/U) * (U * U + 2.334733 * U + .250621)
60   W = W / (U * U + 3.330657 * U + 1.681534)
65   M = [Q/(4 * 3.14159 * K)] * W : Z (F) = M
70   O = [Q/(4 * 3.14159 * K↑2)] * [-W + EXP (-U)]
75   P = [Q/(4 * 3.14159 * K * H)] * EXP (-U)
80   A = A + P↑2 : B = B+O↑2: C=C+O*P : D=D+P* (M-Y)
85   E = E + O * (M-Y) : NEXT I
90   M = (B * D - C * E)/(A * B - C * C) : IF M >(2 * H)
     THEN 100
95   IF M < (-.95 * H) THEN 105
100  M = 2 * H : GOTO 110
105  M = -.95 * H
110  H = H + M : P=(E-M * C)/B:IF P > (2 * K) THEN 120
115  IF P< (-.95 * K) THEN 125
120  P = 2 * K : GOTO 130
125  P = -.95 * K : F = Ø
130  K = K+P : PINT " ITER,  TRANS, STORCO = "; L,K,H
135  L = L+1: IF  L > 100  THEN STOP
140  IF [ABS (G-K)/K] >.001  THEN 20
145  IF [ABS (V-H)/H] > .001  THEN 20
150  FOR I = 1  TO  N: J=N+I : L=2 * N+I : X=Z(J): Y=Z(L)
155  F = F + (X-Y)↑2:  PRINT Z (I), X, Y: NEXT I
160  F = SQR (F/N) : PRINT F : STOP
```

ACQUISITION OF HYDROGEOLOGIC DATA FOR THE MAPPING OF SHALLOW UNCONFINED DECCAN BASALTIC AQUIFERS FROM MAHARASHTRA, INDIA

by

Himanshu Kulkarni and S.B. Deolankar
Department of Geology, University of Poona,
Pune - 411 007, India

Abstract

The low-Permeability Deccan basalts of Maharashtra, India form shallow unconfined aquifers tapped by means of large-diameter dugwells. The heterogeneity of these aquifers necessitates their detailed hydrogeological mapping (on the scale of 1:5000). The basic hydrogeological map projects information relevant to the geological framework of the area,such as the lithological contacts,the delineation of fracture zones and dykes and last but not least, the location of dugwells which form the fundamental and in-depth data acquisition unit. The data acquired through the unit, such as the subsurface lithological sections, the seasonal water levels and the distribution of the areas through which water enters each well, can be used to prepare the ground water maps. The hydrogeological maps and the ground water maps can be effectively used as a base to the detailed hydrogeological investigations of an area.

INTRODUCTION

A hydrogeological map represents the mapping of the physical state of ground waters within their geological framework (Ambroggi and Margat, 1960; Day, 1983). In context to the Deccan basaltic aquifers, this information includes the lithological boundaries between basaltic flows, the dykes and fracture zones and the dispositions of the fundamental data-acquisition units (wells) through which the detailed acquisition of data is being done. The more specific information like the specific yields of wells, the aquifer thickness and the water table behaviour can be supplied through the supplementary maps or ground water maps which include the well-inventory map, the water table maps and the fence diagrams.

The shallow unconfined Deccan basaltic aquifers form a very important source of water supply, especially in the rural sector of Maharashtra. The heterogeneity of the shallow unconfined Deccan basaltic aquifers is mainly related to the fact that the conditions influencing the development of ground water resources within these basalts change locally due to the uneven distribution of Permeability.

In order to identify and describe the shallow basaltic aquifers, it is necessary to study the hydrogeological significance of the structures and features associated with the different types of basaltic flows. These include the sheet jointing, the geometry of different joints, the thicknesses of the individual basaltic flows and the nature of the contact between the flows. Therefore, the mapping of these shallow unconfined aquifers should be undertaken with a view to study the exact relationship between the lithology of different types of basalt, the characteristics that identify them as aquifers and finally, their hydrogeological behaviour as reflected through the examination of wells and several other evaluation techniques.

The foremost factor in the mapping of the shallow unconfined Deccan basaltic aquifers is that of scale. Conditions of heterogeneity may not be fully represented on a regional scale. Therefore, a more appropriate approach involves identifying a small region or setting with some distinctive characteristics with a recurrence of hydrogeological conditions, which could help in better generalisations. The hydrogeological mapping of a smaller unit (third or fourth order drainage basin), on a scale of 1 : 5000 can incorporate a great degree of detail.

The present paper is an attempt to explain a systematic approach towards data-acquisition for mapping the shallow unconfined Deccan basaltic aquifers from Maharashtra, India. The methodology of data acquisition is documented in the present paper through the different maps depicting the hydrogeological behaviour of the unconfined Deccan basaltic aquifers from the Pabal area of Maharashtra State, India.

METHODOLOGY OF DATA-ACQUISITION

The methodology for data acquisition must essentially include a continuous updating of the data. This updating, steadily but surely, brings about a refinement in the study. The hydrogeological map and the ground water maps, together, help in understanding the qualitative and quantitative interrelationship between the aquifers over an area. The following points have to be considered for the data acquisition leading to the preparation of the hydrogeologic maps.

HYDROGEOLOGICAL MAP

The first step in the hydrogeological mapping of an area involves the preparation of a base map which is an enlargement of toposheets on an appropriate scale (1:5000). This base map must contain details like the topographic contours, the drainage, the roads and any other prominent landmarks which can be used for reference. Further, all the dugwells are precisely numbered and located on this map with the help of a compass. This is followed by recording the geological information onto this map.

During the well location programme, the contacts between different basaltic flows can be clearly marked on the map by observing carefully their dispositions on the surface and in different dugwells. The disposition of the contacts between basaltic flows can be continuously modified on the basis of repeated observations in the field. Another salient feature that controls the movement of ground water in the Deccan basaltic terrain is the presence of certain linear features in the form of fracture zones and basic dykes. The Compact basalts exhibit close spaced linear trends in the form of lineaments. The distribution of these lineaments, in the field, is in the form of zones. The intensity, frequency and the close-spaced nature of jointing observed along these zones imply that these may be referred to as "fracture zones".

Gale (1982) has pointed out that when the lengths of joints in different sets of a fracture zones are much greater than the spacing between the members of each joint set, they form continuous networks for ground water flow. The strike-frequency data of the joints in these fracture zones and the details of the linear continuity and joint spacing of each set can be recorded at some sample localities along these fracture zones. The basic dykes, which also form linear features, can form either barriers or favourable channels for ground water movement. The data regarding the jointing pattern of these dykes can similarly be acquired as in the case of fracture zones.

All these data regarding the delineation of these dykes and fracture zones from the Deccan basaltic terrain can be recorded from surface exposures and from the well sections. The details regarding the contacts between basaltic flows and the disposition of dykes and fracture zones can be further corroborated and updated from the study of high-scale (1:20,000) Aerial photographs.

An alternating sequence of the Amygdaloidal and the Compact types of basalts is widely observed over the Deccan basaltic terrain. Within this sequence, a typical lithological setup includes an Amygdaloidal basalt, which is exposed over a contiguous area and is overlain & underlain by Compact basalts. The contact of the Amygdaloidal basalt with the underlying Compact basalt is sharp, whereas its contact with the overlying Compact basalt is marked by a red tuffaceous layer. This red tuffaceous layer extends over large distances and can be used as a reliable marker horizon in mapping the contact between these two flows over an area.

Such a typical lithological setup was observed at and around Pabal village from the Shirur taluka of Pune district, Maharashtra State (Fig. 1). The hydrogeological mapping of the shallow unconfined Deccan basaltic aquifers was carried out in the Pabal area on a scale of 1 : 5000. Fig. 1 shows a portion of the entire hydrogeological map of the area to maintain the original scale of mapping at 1 : 5000.

GROUND WATER MAPS

Ground water maps aim at giving a clear picture regarding the ground water conditions in an area. For such maps, the well inventory forms the main phase of data acquisition. During this inventory, information regarding the diameter and depth of each well, the lithological section exposed therein and the seasonal water level in each well is collected. The excavated rock samples collected near each well are very useful in deciding the exact nature of the lithology along the well-face. The well-head elevation for each well is carefully recorded during the course of this well inventory. Other details like the mode of draught and the average pumpage, the cropping pattern and the agricultural support of each well were also noted. The major stress, however, in the well inventory, was on the observations relating to the areas or zones through which the inflow into each well occurs. The well inventory data was constantly updated on the basis of the repeated examination of the dugwells.

In addition to the vertical distribution of inflow zones represented in the hydrogeological map, the directions of inflows into each well can be recorded during the well inventory. The inflow into wells takes place through arcuate segments of the circumference of the well. The directions of inflows into each well can be well documented by dividing a circle (representing a well) into eight radial sectors of 45° each, corresponding to the directions North, Northeast, East, Southeast, South, Southwest, West and Northwest. These inflow directions, radialy plotted can be effectively used to contemplate the ground water flow directions and as a corroboration to the water table contour and depth-to-water maps. The details regarding the subsurface data can be carefully observed and documented in the form of the well inventory map.

Well Inventory Data Map

Fig. 2. represents a portion of the well inventory map of the Pabal area on the scale of 1 : 5000. As already mentioned, 210 wells in the area were serially numbered and plotted on the map. The map depicts the actually measured well sections that show the disposition of the basaltic flows encountered in the wells. Such well sections were

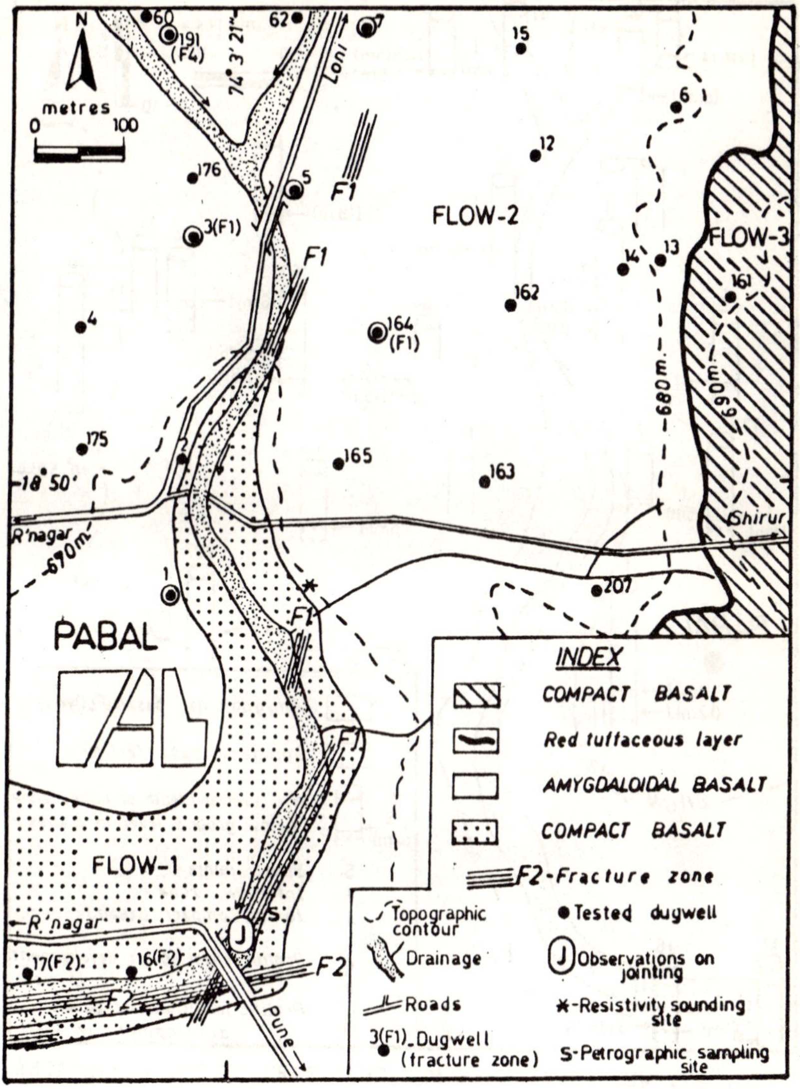

FIG. 1 : HYDROGEOLOGICAL MAP OF A PART OF THE PABAL AREA.

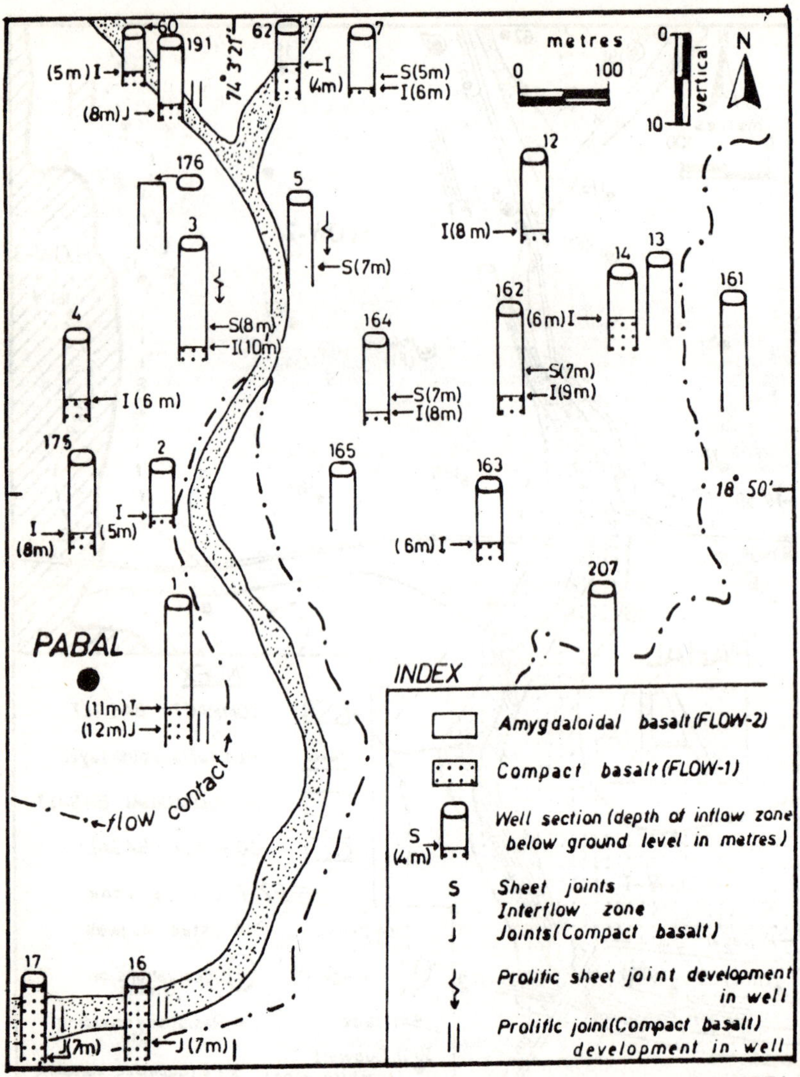

FIG. 2 : WELL-INVENTORY MAP OF A PART OF THE PABAL AREA.

depicted at the respective wells locations. It is generally believed that water enters a well through the entire surface area below the water table, as tapped by the well. The well inventory from the Pabal area, however, revealed that the inflow into a well occurs only through certain areas or zones of the well section. These zones can be referred to as "inflow zones". These inflow zones can be regarded as zones of high Permeability when compared to the remaining portion tapped by the well (Kulkarni, 1987).

Water Table Contour Map

The water table contour map is prepared with a view to find out the general directions of ground water flow and to locate the probable recharge-discharge areas. The assumption of a fair degree of isotropy and homogeneity is a prime necessity for any water level data analysis (Cedergren, 1967; and Fetter, 1980). It is extremely difficult to find the isotropic and homogeneous conditions in the Deccan basaltic terrain on a regional scale. However, it is quite possible to find some degree of isotropy and homogeneity over a contiguous area, on a local scale. For instance, in the Pabal area, the ground water circulation within the Armygdaloidal basalt is probably controlled by the network of sheet joints and by the interflow contact zone mentioned earlier. A fairly uniform and prolific development of these inflow zones over a contiguous area is evident in areas where the Amygdaloidal basalt shows a greater thickness, e.g. the Khatal area (Fig.3). Conceivably, a fair degree of homogeneity, as regards the circulation of ground water, could exist in areas like the Khatal.

Another important factor in the analysis of water level data is that due consideration must be given to the ground water potential in the wells. Rushton and Redshaw (1979) indicate that the abstraction of ground waters from unconfined aquifers causes significant changes in the ground water potential in the vicinity of the abstraction well. In the Deccan basaltic terrain of Maharashtra, due to the varied, continuous and greater abstraction from the wells during the summer period, the ground water potential is drastically reduced, and hence, it is pointless to use the summer water level data for the preparation of the water table contour map. On the contrary, the winter water levels represent a more realistic ground water potential which is significantly lesser affected by the draught for irrigation. Therefore, the water level data should preferably show a minimum of water level fluctuations and consequently represent the maximum ground water potential in different dugwells. The fluctuations in the ground water potentials for the wells from the Pabal area were found to be minimum during the Winter season when all the inflow zones into the wells were working. Thus, the variation in ground water potentials in different dugwells, under such conditions

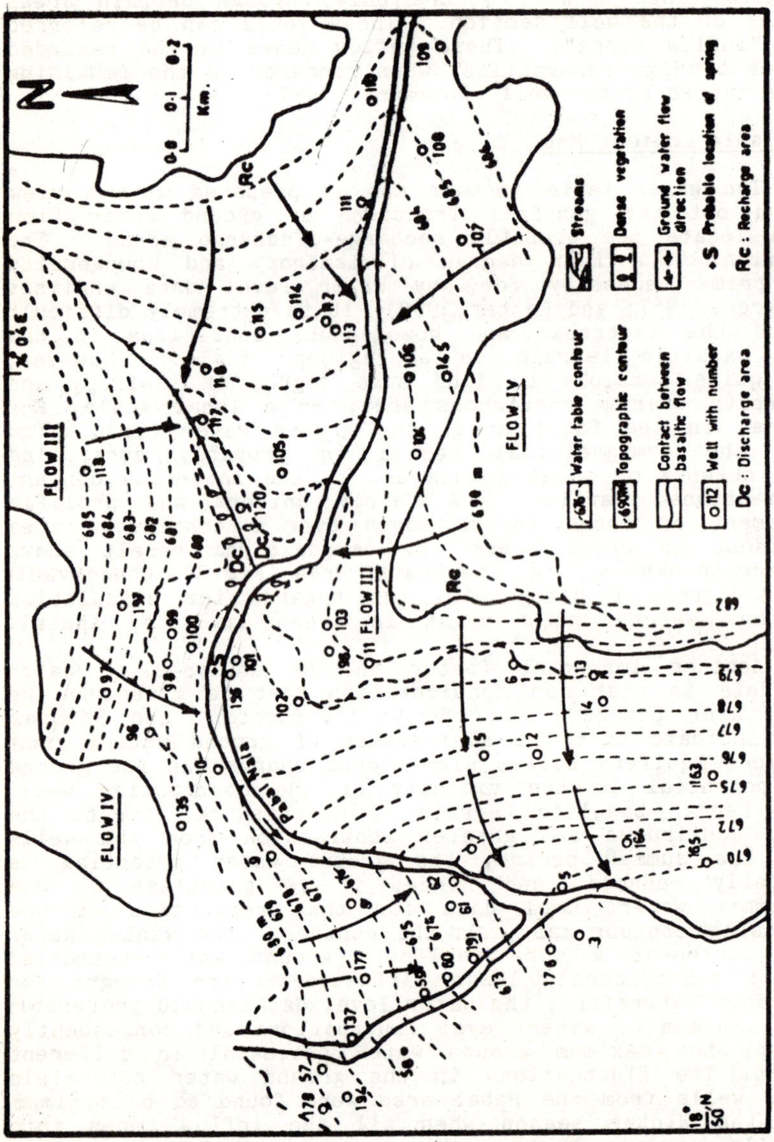

FIG. 3 : WATER TABLE CONTOUR MAP OF KHATAL AREA.

can highlight the probable recharge and discharge areas and also the probable directions of ground water movement.

Fig. 3 shows the water table contour map (of 1m. contour interval) for a part of the Pabal area, constructed by plotting the reduced levels using the well head elevations w.r.t. mean sea level. For this purpose, the water levels for the winter (October - November) of 1984 were used. The Pabal area receives an average annual rainfall of 550mm. During the monsoon of 1984, the Pabal area, however, received an above normal rainfall of 717mm. Consequently, the water levels during the following winter (Oct.-Nov.) of 1984 were found to be at a maximum. The fluctuations in these levels were also minimum during this period. It was contemplated that the variation in the ground water potential in different dugwells, under these conditions, could throw light upon the recharge and discharge conditions and the general ground water flow directions.

For depicting the disposition of the water table contours, consideration was given to factors like the topography, the drainage, the lithology and to the physical manifestation of recharge and discharge characters. The general directions of ground water flow have been deduced from the water table map by drawing them at right angles to the water table contours.

Depth-to-water Maps

The depth-to-water maps represent the actual depth of the water table from the ground surface, without any reference to the topographic elevation. The depth-to-water maps give an idea as to how deep the water table is from the surface. These maps can be constructed on lines similar to the water table contour maps and can be used to corroborate the recharge-discharge conditions deduced from the water table contour maps.

Fence Diagram

The Deccan basaltic lava flows are generally considered to be horizontal in their disposition, with respect to a particular contour. This implies a constant thickness to any basaltic flow over a contiguous area. Surface investigation and well-section data, however, show that the actual thicknesses of the basaltic flows over an area varies from place to place. This is as a result of the undulations in the boundaries representing the contacts between the basaltic flows.

Fig. 4 shows the actual flow contacts superimposed on the assumed horizontal contacts. The contact between the Amygdaloidal basalt (flow 2) and the underlying Compact basalt (flow 1) is exposed on the surface at an elevation

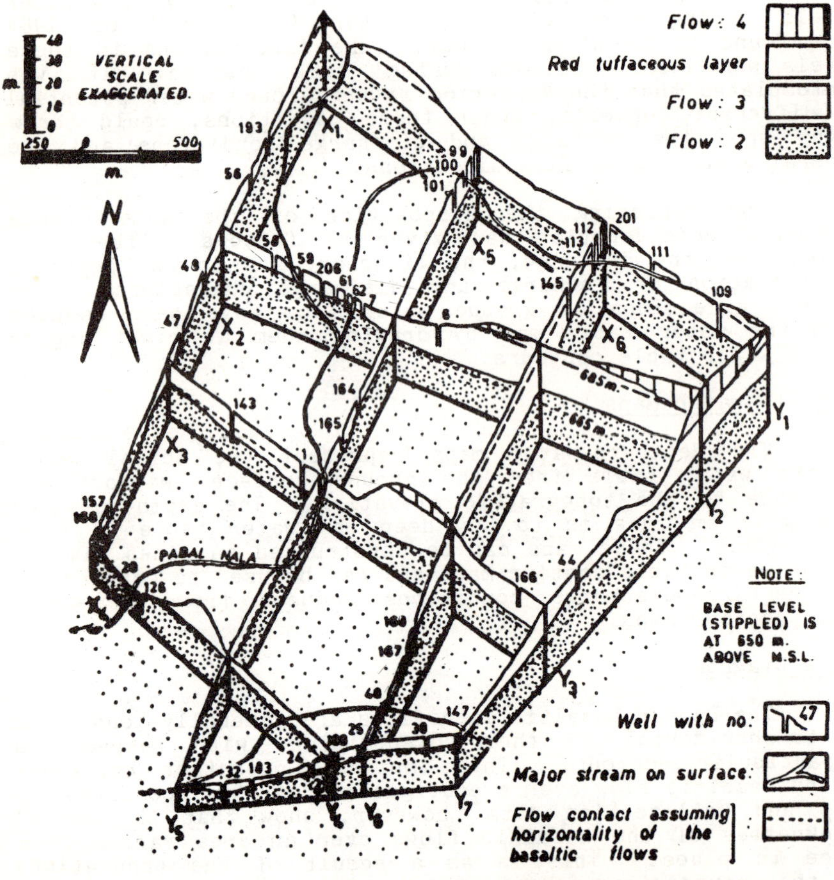

FIG. 4 : FENCE DIAGRAM FOR THE PABAL AREA.

of 665m. In reality, as shown in the figure, this contact is encountered at elevations higher as well as lower than 665m (above m.s.l). Similarly, the contact between the Amygdaloidal basalt (flow 2) and the overlying Compact basalt (flow 2) is exposed at elevations higher or lower than 685m. (above m.s.l) at many places. The fence diagram thus depicts the undulations in the boundaries representing the basaltic flow contacts and brings out the variation in the thickness of the Amygdaloidal basalt (flow 2) over the area.

DISCUSSION

In case of the low-Permeability Deccan basalts, the data acquisition has to be on a local scale (like 1:5000) in order to exemplify their heterogeneity. The data acquired can be documented as systematic maps, the fundamental ones in the present case being the hydrogeological map (Fig.1) and the well inventory map (Fig.2). In the present case, these maps (Fig. 1 and Fig.2) depict that most of the wells have encountered either the Amygdaloidal basalt (flow 2) or the underlying Compact basalt (flow 1). However, some of the wells have encountered both, the Amygdaloidal and the Compact basalts together. The hydraulic continuity between these two basaltic units (flows) is indicated by the distribution, disposition and frequency of the inflow zones (sheet joints, the interflow zone and the joints of the Compact basalt) in wells. This distribution of inflow zones in wells can be effectively used to study the well-yields and aquifer characteristics. For instance, the variation in the Specific capacities of wells numbered 1 and 3, which are situated close to each other (Fig.2), can be related to the type of inflow zones tapped by these wells. Well numbered 1 taps the interflow zone and the joints of the Compact basalt as inflow zones and has a Specific capacity of 12.67 litres per minute per metre of drawdown (LPM/m.). On the other hand, well numbered 3, which taps the well developed sheet joints of the Amygdaloidal basalt as inflow zones shows a higher Specific capacity of 114 LPM/m. Such variations can be conveniently studied from the fence diagrams, which clearly depict the actual disposition of the basaltic flows over an area. The analysis of the well yields and aquifer characteristics, alongwith the type and disposition of the inflow zones can give a concrete idea about which of the basalt flows forms the dominant unit within the lithological setup.

The water level data acquired can be presented as ground water maps, which can be used to draw broad inferences regarding the hydrogeological conditions in an area. In a water table contour map, the divergence of flow lines indicates a recharge area, whereas the convergence of the flow lines depicts a discharge area (Fetter, 1980). Similarly in a depth-to-water map, the recharge areas are characterised by a deeper water table below the land surface and, a shallower water table below the land surface indicate a discharge area.

In summing up, hydrogeological mapping forms a key to the final hydrogeological inferences in a ground water resource evaluation. The acquisition of basic data, that goes into the preparation of the hydrogeological maps, has to be achieved through a systematic and unbiased approach. Such an approach has to involve a continuous data acquisition through repeated field visits, leading to the updation and refinement in the observations, so as to bring out more reliable and realistic hydrogeological inferences in the study.

Acknowledgement

The first author gratefully acknowledges the Council of Scientific and Industrial Research, New Delhi for financial assistance in the form of Fellowships and Associateship.

REFERENCES

Ambroggi, R. and Margat, J.,1960. Legende generale des cartes Hydrogeologiques du Maroc (General legend for hydrogeological maps of Morocco). No.50 IASH,32pp.

Cedergren, H.R.,1967. Seepage, Drainage and Flow Nets. John Wiley and Sons, New York.

Day, J.B.W.,1983. The role of hydrogeological maps in ground water resource evaluation and development, Proc. International Conference on Ground water and Man. Sydney. p. 51-67.

Fetter (Jr.) C.W.,1980. Applied Hydrogeology. Charles E. Merril Publ. Co., Ohio. 488pp.

Gale, J.E.,1982. Fundamental hydraulic characteristics of fractures from field and laboratory investigations, Proc. AWRC Conf. on Ground water in Fractured Rocks, Canberra. p. 79-94.

Kulkarni, H.C.,1987. A study of the Deccan basaltic unconfined ground water system in the Pabal area of the Shirur taluka, Pune district, Maharashtra State. Unpubl. Ph.D. thesis, University of Poona. 285p.

Rushton, K.R. and Redshaw, S.C.1979. Seepage and Ground -water flow. John-Wiley and Sons, New York. 339p.

AUTHOR INDEX